The Encyclopedia of **Microscopy and Microtechnique**

The Encyclopedia of
Microscopy and
Microtechnique

EDITED BY

Peter Gray

ANDREY AVINOFF PROFESSOR OF BIOLOGY
UNIVERSITY OF PITTSBURGH

 VAN NOSTRAND REINHOLD COMPANY
New York   Cincinnati   Toronto   London   Melbourne

Van Nostrand Reinhold Company Regional Offices:
New York Cincinnati Chicago Millbrae Dallas

Van Nostrand Reinhold Company International Offices:
London   Toronto   Melbourne

Library of Congress Catalog Card Number: 73-164
ISBN: 0-442-22812-0

Manufactured in the United States of America

Published by Van Nostrand Reinhold Company
450 West 33rd Street, New York, N.Y. 10001

Published simultaneously in Canada by Van Nostrand Reinhold Ltd.

15   14   13   12   11   10 9   8   7   6   5   4   3   2   1

Library of Congress Cataloging in Publication Data

Gray, Peter, 1908–
   The encyclopedia of microscopy and microtechnique.
   Includes bibliographies.
   1. Microscope and microscopy—Technique—Dictionaries.
I. Title. [DNLM: 1. Microscopy—Encyclopedia.
QH 203 G781e 1973]
QH203.G8   502'.8   73-164
ISBN 0-442-22812-0

Preface

Few instruments have so diverse a set of users as does the microscope. There are not only medical and biological scientists in all the numerous disciplines into which they are subdivided but also, amongst others, atmospherologists, bakers, chemists, metallurgists, textile workers, the manufacturers of paints, and those who examine paintings. All these make use of a profusion of instruments for manipulating both photon and electron beams and record the results by methods varying from photography to stereology. The objects that they examine may be temporarily or permanently prepared by dispersion, grinding, sectioning, incineration and a multitude of other methods. A worker in any field may require information not only about instruments and techniques but also guidance as to which of them can profitably be applied to specific objects and specific problems. This diversity presents difficulties for the encyclopedist.

I have arranged the articles in this encyclopedia in alphabetical order whether they are about instruments in theory or practice (e.g., resolution, scanning electron microscope), techniques (e.g., microspectrophotometry, rock section grinding, photomicrography), disciplines (e.g., atmospheric microscopy, insect histology), individual subjects or materials intended for examination (e.g., clay minerals, pineal, gastrotricha) or reagents (e.g., fixatives, oxazine dyes). It is to be hoped that the index will be adequate to assist the reader in tracking his way through this maze of information.

It should perhaps be explained that this volume is not a second edition of Clark's Encyclopedia of Microscopy which appeared more than a decade ago in the same series. It has however, Dr. Clark's blessing and one or two contributors are common to both volumes. Neither is this volume in any sense a rehash of my own "Microtomist's Formulary and Guide" (Blakiston, 1954) though a few tables and illustrations that are still relevant are reproduced through the courtesy of McGraw Hill Book Company.

The greatest debt that I have to acknowledge is, of course, to the several hundred contributors from more than a score of countries who have prepared articles in the fields of their specialization. These have of necessity been drawn not only from universities and other scholarly institutions but also from the research laboratories of numerous great industries. All have proved helpful, courteous and understanding. It also gives me great pleasure to express my thanks to my secretaries Miss Audrey Taylor, who assisted in the organization of the volume as a whole, and Mrs. Ida Finifter, who was involved in the laborious process of preparing the index.

PETER GRAY

Contributors

E. R. ACKERMAN, National Center for Atmospheric Research, Boulder, Colorado. ATMO-
  SPHERIC MICROSCOPY
W. G. ALDRIDGE, The University of Rochester, Rochester, New York. INDIUM STAINING
E. E. ANDERSON, Duke University, Durham, North Carolina. BRYOPHYTA
H. M. APPLEYARD, Halifax, Yorkshire, England. TEXTILES
B. BACCETTI, Istituto di Zoologia, Siena, Italy. SPERMATOZA
R. BARER, The University of Sheffield, Sheffield, England. PHASE CONTRAST MICROSCOPES
J. W. BARTHOLOMEW, University of Southern California, Los Angeles, California. BACTERIA
H. W. BEAMS, University of Iowa, Iowa City, Iowa. GOLGI APPARATUS
Y. BEN-SHAUL, Tel-Aviv University, Tel-Aviv, Israel. LICHEN (part)
G. G. BERG, The University of Rochester, Rochester, New York. PHOSPHOTASES
L. R. BERKOWITZ, University of California, Los Angeles, California. NERVOUS SYSTEM
Q. BONE, The Marine Laboratory, Plymouth, England. AMPHIOXUS
B. BOOTHROYD, The University of Liverpool, Liverpool, England. MICROINCINERATION
R. BOUYER, Laboratoire de Physique, Poitiers, France. RESOLUTION
E. B. BRAIN, Royal College of Surgeons, London, England. DECALCIFICATION
V. BRAITENBERG, Max-Planck-Institut für Biologische Kybernetik, Tübingen, Germany.
  GOLGI METHODS
R. B. BRUNSON, University of Montana, Missoula, Montana. GASTROTRICHA
R. BULGER, University of Maryland, Baltimore, Maryland. KIDNEY
A. R. BURDI, University of Michigan, Ann Arbor, Michigan. SKELETON
J. CACHON, Laboratoire Maritime, Villefranche-sur-Mer (A.M.), France. RADIOLARIA (part)
M. CACHON, Laboratoire Maritime, Villefranche-sur-Mer (A.M.), France. RADIOLARIA (part)
E. CAPANNA, Università di Roma, Rome, Italy. LYSOSOMES
A. R. CAVALIERE, Gettysburg College, Gettysburg, Pennsylvania. ASCOMYCETES
M. DE CECCATTY, Université Claude-Bernard, Villebranne, France. PORIFERA
J. K. CHOI, Chon-nam University Medical School, Kwangju, Republic of Korea. BLADDER
S. CHROMÝ, Forschungsinstitut für Baustoffe, Brno, Czechoslovakia. REFRACTIVE INDEX
  DETERMINATION
R. A. CLONEY, University of Washington, Seattle, Washington. UROCHORDATA
M. COLE, Metals Research Limited, Cambridge, England. QUANTIMET
O. R. COLLINS, University of California, Berkeley, California. MYXOMYCETES
J. J. COMER, Air Force Cambridge Research Laboratories, (AFSC), Bedford, Massachusetts.
  CLAY MINERALS
J. O. CORLISS, University of Maryland, College Park, Maryland. PROTOZOA
D. CORNELIUS, Cincinnati Art Museum, Cincinnati, Ohio. PAINTINGS AND OTHER WORKS
  OF ART

D. P. Costello, University of North Carolina, Chapel Hill, North Carolina. NEGATIVE STAINING TECHNIQUES (part)

P. Davison, Boston Biomedical Research Institute, Boston, Massachusetts. COLLAGEN

J. G. Delly, McCrone Research Institute, Chicago, Illinois. MOUNTANTS

H. W. Denyer, 45 Grove Road, Chertsey, Surrey, England. ROTIFERS

H. M. Dott, Animal Research Station, Cambridge, England. REPRODUCTIVE ORGANS, MALE

M. E. Downey, Smithsonian Institute, Washington, D.C. ECHINODERMATA

M. K. Dutt, University of Delhi, Delhi, India. FEULGEN STAINS

V. B. Eichler, Wichita State University, Wichita, Kansas. EMBRYOLOGICAL TECHNIQUES

H. Elias, 463 Marietta Drive, San Francisco, California. STEREOLOGY

A. I. Farbman, Northwestern University, Chicago, Illinois. TONGUE

R. L. Fernald, University of Washington Laboratories, Friday Harbor, Washington. INVERTEBRATE LARVAE

P. M. Fogden, Nuffield Unit of Tropical Animal Ecology, Uganda, East Africa. RECONSTRUCTION

R. H. Fox, Pittsburgh and Allegheny Crime Laboratory, Pittsburgh, Pennsylvania. FORENSIC MICROSCOPY (part)

E. A. C. Follett, Institute of Virology, Glasgow, Scotland. VIRUS

L. L. Franchi, University of Birmingham, England. OVARY

B. Francis, The Energen Foods Co., Ltd., Ashford, Kent, England. BAKERY PRODUCTS

R. Friedenberg, Sinai Hospital of Baltimore, Baltimore, Maryland. OSMIUM STAINING (part)

T. Fujita, Niigata University, Niigata, Japan. PANCREAS

H. M. Fullmer, University of Alabama, Birmingham, Alabama. ELASTIC TISSUE

C. C. Fulton, Route #1, Box 981C, Venice, Florida 33595. CHEMICAL MICROCRYSTAL IDENTIFICATION

P. Galle, Université de Paris, Paris, France. CYTOCHEMICAL ANALYSIS BY X-RAY SPECTROGRAPHY (part)

M. Galun, Tel-Aviv University, Tel-Aviv, Israel. LICHEN (part)

R. Gander, Wild Heerbrugg, Ltd., Heerbrugg, Switzerland. PHOTOMICROGRAPHY

W. A. Gibson, National Institute of Dental Research, National Institutes of Health, Bethesda, Maryland. TRACHEA

D. J. Goldstein, The University of Sheffield, Sheffield, England. SCANNING MICROSPECTROPHOTOMETRY

W. R. Graham, Coordinated Science Laboratory, University of Illinois, Urbana, Illinois. FIELD ION MICROSCOPY (part)

P. Gray, University of Pittsburgh, Pittsburgh, Pennsylvania. CARMINE, FIXATIVE FORMULAS, HEMATOXYLIN STAINING FORMULAS, HISTORY OF MICROTOMY AND MICROTECHNIQUE (part), MAGENTA, MOUNTANT FORMULAE, OXAZINE DYES, PARAFFIN SECTIONS, POLYCHROME STAINS, FORMULAE AND METHOD, SAFRANIN, THIAZINE DYES, WHOLEMOUNTS ZOOLOGICAL MATERIALS

F. Gray, 5131 Ellsworth Avenue, Pittsburgh, Pennsylvania. HISTORY OF MICROTOMY AND MICROTECHNIQUE (part)

J. Grehn, 633 Wetzler, Germany. MICROPROJECTION

W. Gumpertz, Wild-Heerbrugg Instruments, Farmingdale, New York. BINOCULAR CAMERA LUCIDA

E. Gurr, Michrome Laboratories, London, England. THEORY OF STAINING

J. D. Harkin, Tulane University, New Orleans, Louisiana. MYELIN

R. Hasson, Centre d'Études Nucléaires de Saclay, Gif-sur-Yvette, France. MOLYBDENUM

W. Hay, University of Illinois. FORAMINIFERA

C. Henley, University of North Carolina, Chapel Hill, North Carolina. NEGATIVE STAINING TECHNIQUES (part)

B. HONIGBERG, University of Massachusetts, Amherst, Massachusetts. SILVER PROTEINATE STAINS

G. HUMASON, Oak Ridge Associated Universities, Oak Ridge, Tennessee. HEMOPOETIC TISSUES, LEUCOCYTES

W. HUNN, E. Leitz, Inc., Rockleigh, New Jersey. METALLOGRAPHIC MICROSCOPES

F. HUTCHINSON, Yale University, New Haven, Connecticut. FIELD ION MICROSCOPY (part)

W. F. INGRAM, Ingram Laboratories, Inc., Griffin, Georgia. ROCK SECTION GRINDING AND POLISHING

J. ISINGS, Central Laboratorium, t.n.o., Delft, The Netherlands. COTTON

H. A. JAMES, University of Bridgeport, Bridgeport, Connecticut. ACANTHOCEPHALA

D. B. JONES, University of New York, Syracuse, New York. SILVER STAINS

J. KAMAN, Vysoká škola veterinárni, Brno, Czechoslovakia. INJECTION

F. H. KASTEN, Louisiana State University Medical Center, New Orleans, Louisiana. ACRIDINE DYES

D. P. KNIGHT, Physiological Laboratory, Cambridge, England. TRACERS

H. KOMNICK, Universität Bonn, Bonn, Germany. CALCIUM, SODIUM AND CHLORIDE LOCALIZATION

M. J. KOPAC, New York University, Washington Square College, New York, New York. MICROMANIPULATION

B. M. KOPRIWA, McGill University, Montreal, Canada. RADIOAUTOGRAPHY

E. KOZLOFF, University of Washington Laboratories, Friday Harbor, Washington. MESOZOA

D. KRINSLEY, University of Cambridge, Cambridge, England. SAND

N. KRISHNAN, Case Western Reserve University, Cleveland, Ohio. CHITIN

F. KRÜGER, Biologische Anstalt Helgoland, Hamburg-Altona, Germany. DISSECTING MICROSCOPES

M. KUHNERT-BRANDSTÄTTER, Institut für Pharmakognosie der Universität Innsbruck, Innsbruck, Austria. MICROMELTING POINT DETERMINATION

J. LAI-FOOK, University of Toronto, Toronto, Canada. INSECT HISTOLOGY

N. R. LERSTEN, Iowa State University, Ames, Iowa. WHOLEMOUNT, BOTANICAL

J. H. LIPPS, University of California, Davis, California. MICROFOSSILS

J. P. LODGE, JR., National Center for Atmospheric Research, Boulder, Colorado. ATMOSPHERIC MICROSCOPY (part)

Z. LODIN, Czechoslovak Academy of Sciences, Bŭdejovická, Czechoslovakia. NERVE AND GLIA CELLS IN SECTIONS

F. N. LOW, University of North Dakota, Grand Forks, North Dakota. BLOOD

A. LUPULESCU, Detroit, Michigan. THYROID

C. MACKIE, University of Victoria, Victoria, Canada. COELENTERATA

P. G. MAHLBERG, Indiana University, Bloomington, Indiana. LATICIFERS

R. MARTOJA, Université de Paris, Paris, France. CYTOCHEMICAL ANALYSIS BY X-RAY SPECTROGRAPHY (part)

M. D. MASER, Northeastern University, Weston, Massachusetts. PLASTIC EMBEDDING, ULTRA-THIN SECTIONING

D. S. MAXWELL, University of California, Los Angeles, California. FIXATION

H. V. MAYERSBACH, Medizinische Hochschule Hannover, Hannover, Germany. GLYCOGEN

D. W. MAYFIELD, Technicon Instruments Corporation, Tarrytown, New York. AUTOMATED HISTOLOGICAL EQUIPMENT AND TECHNIQUE (part)

G. A. MEEK, The University of Sheffield, Sheffield, England. TRANSMISSION MICROSCOPY

R. H. MEHLEN, Wake Forest College, Winston-Salem, North Carolina. TARDIGRADA

D. MENTON, Washington University, St. Louis, Missouri. HAIR (part)

E. MIDDLETON, Imperial Cancer Research Fund, London, England. FREEZE-DRYING, FIXING AND EMBEDDING

M. P. Mohn, University of Kansas Medical Center, Kansas City, Kansas. SKIN

G. Moment, Goucher College, Towson, Maryland. ANNELIDA

C. W. Monroe, Tufts University, Boston, Massachusetts. HYPOPHYSIS

R. T. Moore, New University of Ulster, Coleraine, N. Ireland. FUNGI

M. L. Moss, College of Physicians and Surgeons, Columbia University, New York, New York. ENAMEL, TOOTH

T. Mulvey, The University of Aston in Birmingham, Birmingham, England. HISTORY OF ELECTRON MICROSCOPY

M. Nahmacher, E. Leitz, Inc., Rockleigh, New Jersey. CAMERA ATTACHMENTS

C. W. Oatley, Cambridge University, Cambridge, England. SCANNING ELECTRON MICROSCOPE

M. T. O'Hegarty, University College, Dublin, Ireland. MITOCHONDRIA

D. Ollerich, University of North Dakota, Grand Forks, North Dakota. PLACENTA

C. F. Oster, Eastman Kodak Company, Research Laboratories, Rochester, New York. PHOTOGRAPHIC MATERIALS (part)

O. Pflugfelder, University Hohenheim, Stuggart, West Germany. ONYCHOPHORA

P. Pierce, Pittsburgh Plate Glass Industries, Springdale, Pennsylvania. PAINT MICROSCOPY

E. Pihl, Monash University, Victoria, Australia. HEAVY METAL LOCALIZATION

J. J. Poluhowich, University of Bridgeport, Bridgeport, Connecticut. RHYNCOCOELA

W. W. Prichard, Jr., Pittsburgh and Allegheny Crime Laboratory, Pittsburgh, Pennsylvania. FORENSIC MICROSCOPY (part)

G. Prenna, Instituto di Anatomia Comparata della Università, Pavia, Italy. FLUORESCENT SCHIFF REAGENTS

H. R. Purtle, Medical Museum, Armed Forces Institute of Pathology, Washington, D.C. HISTORY OF MICROSCOPES

M. Reeve, University of Miami, Miami, Florida. CHAETOGNATHA

S. Reinius, University of Uppsala, Uppsala, Sweden. OVIDUCT

C. C. Remsen III, Woods Hole Oceanographic Institution, Woods Hole, Massachusetts. FREEZE-ETCH TECHNIQUES

R. V. Rice, Mellon Institute, Pittsburgh, Pennsylvania. SMOOTH MUSCLE

E. M. Rodriguez, National University of Cuyo, Mendoza, Argentina. NEUROSECRETORY STRUCTURES

C. Rohde, Northern Illinois University, Dekalb, Illinois. ARTHROPOD SECTIONS

A. H. Rose, Bath University of Technology, Bath, England. YEAST

H. E. Rosenberger, Bausch & Lomb, Inc., Rochester, New York. ZOOM MICROSCOPES

L. Sakovich, University of California, San Francisco, California. NITROCELLULOSE EMBEDDING AND SECTIONING

P. Satir, Zoological Institute, University of Tokyo, Tokyo, Japan. CILIA

R. L. De C. H. Saunders, Dalhousie University, Halifax, Canada. X-RAY MICROSCOPY

A. Schaefer, Wild Heerbrugg, Ltd., Heerbrugg, Switzerland. FLUORESCENCE MICROSCOPY, TECHNIQUES

H. Schiechl, Universität Graz, Graz, Austria. CHROMIUM SALTS

L. Schmid, University of Minnesota, Minneapolis, Minnesota. MOLLUSCA

A. B. Segelman, The State University of New Jersey, New Brunswick, New Jersey. BOTANICAL DRUGS

A. M. Seligman, Chief Research On Ecology and Cell Biology, Johns Hopkins University School of Medicine, Sinai Hospital, Baltimore, Maryland. OSMIUM STAINING (part)

M. Sensenbrenner, Centre de Neurochimie du C.N.R.S., Strasbourg, France. NEURON AND GLIA CELLS IN CULTURE

R. Seymour, Ohio State University, Columbus, Ohio. SAPROLEGNIALES

A. Sharma, University of Calcutta, Calcutta, India. CHROMOSOMES (part)

A. K. SHARMA, University of Calcutta, Calcutta, India. CHROMOSOMES (part)

M. N. SHERIDAN, The University of Rochester, Rochester, New York. ELECTRIC ORGANS

S. SHOSTAK, University of Pittsburgh, Pittsburgh, Pennsylvania. MESOGLOEA

M. T. SILVA, Universidade do Porto, Porto, Portugal. URANIUM SALTS

J. J. SKVARLA, University of Oklahoma, Norman, Oklahoma. POLLEN

B. SLAVIN, University of Southern California, Los Angeles, California. FAT

E. M. SLAYTER, Children's Cancer Research Foundation, Inc., Boston, Massachusetts. OPTICAL MICROSCOPE

C. SMITH, South Hadley, Massachusetts. THYMUS

H. M. SMITH, Eastman Kodak Company, Research Laboratories, Rochester, New York. HOLOGRAPHIC MICROSCOPY

K. G. V. SMITH, British Museum (Natural History), London, England. INSECT WHOLEMOUNTS

E. SOLCIA, University of Pavia, Pavia, Italy. HEMATOXYLIN

D. F. SOULE, University of Southern California, Los Angeles, California. ECTOPROCTA (part)

J. D. SOULE, University of Southern California, Los Angeles, California. ECTOPROCTA (part)

B. M. SPINELL, Eastman Kodak Company, Research Laboratories, Rochester, New York. PHOTOGRAPHIC MATERIALS (part)

G. W. W. STEVENS, Kodak Limited, Research Laboratory, Middlesex, England. MICRO-PHOTOGRAPHY

N. STRAUSFELD, Max-Planck Institute, Tübingen, Germany. GOLGI METHOD INVERTEBRATE APPLICATION

J. STÜBNER, Ernst Leitz, Wetzlar, West Germany. MICROPROJECTION

W. C. STUVER, Pittsburgh and Allegheny County Crime Lab, Pittsburgh, Pennsylvania. FORENSIC MICROSCOPY (part)

G. SVIHLA, Argonne National Laboratory, Argonne, Illinois. ULTRAVIOLET MICROSCOPY

C. E. TOBIN, University of Colorado, Denver, Colorado. LUNG

M. TROTTER, Washington University, St. Louis, Missouri. HAIR (part)

M. UHLIG, Wild Heerbrugg, Ltd., Heerbrugg, Switzerland. DRAWING WITH THE AID OF THE MICROSCOPE

A. VARSCHAVSKY, Universidad de Chile, Santiago, Chile. NON-FERROUS METALS

V. I. VEINBLAT, All-Union Research Institute "Microbes", Saratov, USSR. AGAR-AGAR

I. VERES, Agarununiversität, Gödöllö, Hungary. ULTRAMICROBIOPHYSICS

K. VICKERMAN, University of Glasgow, Glasgow, Scotland. BLOOD PARASITES

G. VRENSEN, Katholieke Universiteit, Nijmegen, The Netherlands. ENDOPLASMIC RETICULUM

E. E. WAHLSTROM, University of Colorado, Boulder, Colorado. POLARIZATION MICROSCOPES

H. WARTENBERG, Institut der Universität, Bonn, West Germany. PINEAL

M. WEBB, Natal, South Africa. POGONOPHORA

J. A. WESTFALL, Kansas State University, Manhattan, Kansas. NEMATOCYST

H. E. WILLIAMS, Technicon Instruments Corporation, Tarrytown, New York. AUTOMATED HISTOLOGICAL EQUIPMENT AND TECHNIQUE (part)

H. WINCHELL, Yale University, New Haven, Connecticut. MICROPROBE

J. J. WOLKEN, Carnegie-Mellon University, Pittsburgh, Pennsylvania. MICROSPECTROPHOTO-METRY

C. L. F. WOODCOCK, University of Massachusetts, Amherst, Massachusetts. PLASTID

F. B. P. WOODING, ARC Institute of Animal Physiology, Cambridge, England. MILK

R. M. WYNN, University of Illinois, Chicago, Illinois. AMNION

S. YAMAGUTI, Kyoto, Japan. PLATYHELMINTHES

W. ZEIDLER, Commonwealth Science and Industry Research Organization, North Ryde, Australia. COAL AND COKE PREPARATION FOR MICROSCOPIC EXAMINATION

D. J. ZINN, University of Rhode Island, Kingston, Rhode Island. PSAMMON

a

ACANTHOCEPHALA

All members of the phylum Acanthocephala (spiny-headed or thorny-headed worms) are pseudocoelomate, obligatory parasites whose common names denote their most obvious characteristic—a spiny, eversible proboscis (Fig. 5). The spines or hooks of the proboscis are recurved, and the entire structure can be withdrawn into a muscular sheath. Sexes are separate and easily distinguished by their well developed, distinctive reproductive organs; other internal organs are either reduced or absent, as is the entire digestive system. Such extremes of organ specialization and degeneration have made it difficult to relate acanthocephalans to other animal groups.

Adult acanthocephalans (Figs. 3, 4) infect the intestine of all classes of vertebrates. They are found rarely in man, more frequently in domesticated animals, and most commonly in wild animals, especially fish. Their typical life cycle has been succinctly summarized by Nicholas (1967). Reproduction is entirely sexual; following copulation and internal fertilization, the female releases shelled acanthors (Fig. 6) into the host's intestine. The acanthors, passed in the host's feces, infect arthropods, in which they develop through a larval form, acanthella, to the cystacanth stage (Figs. 1, 2). The cystacanth may infect the definitive host directly when ingested, or pass through paratenic hosts in which the cystacanth penetrates the tissue but does not develop into a reproductively mature adult.

Although the significance of serology, ecology, biochemistry,[1] life-cycle studies or any other aspect of an organism cannot be overlooked, it remains a fact that the identification of acanthocephalans, and presumably most other animals, still relies most heavily on anatomical features (Van Cleave, 1948). "One of the difficulties of working with Acanthocephala . . . is the problem of technique" (Bullock, 1969); and the quality of any study is generally proportional to the care and experience of the investigator in applying the best available techniques. Historically, the finest available study on collection, preservation, and mounting of Acanthocephala is H. J. Van Cleave's (1953) monumental treatise, "Acanthocephala of North American Mammals." Since this publication, there have been numerous modifications and sophistications of his techniques, but except for those special treatments employed in preparation of specimens for electron microscopy, there have been few significant changes. More recent, invaluable, technique-oriented literature, devoted

---

[1]See Florkin and Sheer (1969) for a comprehensive review on metabolism and histochemistry of acanthocephalans.

to Acanthocephala, is that of Nicholas (1967) and Bullock (1969).

Collection of specimens. Ordinarily, acanthocephalans are recovered as a matter of course at necropsy. Unlike trematodes or cestodes, some may be pigmented (orange to yellow) although most are whitish and variously compressed to round in cross section. Lack of recognition of an acanthocephalan at this point in procedure has contributed to many poor or taxonomically useless preparations. Proboscis hooks are a major identifying characteristic and must be kept intact on a proboscis free of host tissue. There is no simple method of removing stubbornly embedded worms, and many have narrow and delicate proximal constrictions of the proboscis which tend to break. Teasing with fine dissecting needles and forceps is tedious but has proved to be a successful way to remove the parasite from its host. A gentle brushing with a fine paintbrush usually completes the process. Intestinal tissue may be digested from the worm in a 1 % pepsin in 1 % hydrochloric acid solution at temperatures of 50–56°C. This digestion technique, so successful for removal of trematode and cestode egg shells and cysts, or for recovery of *Trichinella* larvae, is disproportionately time consuming, and although it produces a clean proboscis, internal organs are often distorted or destroyed. Acanthellae or cystacanths are easily removed from their cysts or from the hemocoel of the arthropod. Notes should be taken on color, shape, and other characteristics of the living worms as well as their location in the host. If eggs are to be examined, it is best to collect them prior to fixation. West (1964) cleaned debris from freshly obtained females in Ringer's solution, blotted them dry, punctured the pseudocoel, and smeared its content on slides. Some slides were examined as fresh preparations; others were fixed. Preparations for ultrastructural studies require techniques beyond the limits of this work, and reference to Wright and Lumsden (1970) will be helpful.

Relaxation and prefixation. Specimens removed from hosts are transferred to shallow dishes with a volume of Ringer's solution (isosmotic to the host) sufficient to cover them and in which further examination and the final cleaning process can be accomplished. Placing worms with a withdrawn proboscis into distilled water usually causes eversion, but there is no predictable time. Sometimes this technique will not work, and the added pressure of a coverslip must be used. Specimens in distilled water should be placed in a refrigerator to reduce the hazards of organ degeneration or cuticle sloughing, and the more quickly a specimen is processed from host to finished slide, the less likely will be its loss.

Fixation. Almost any standard fixative—Bouin's,

1

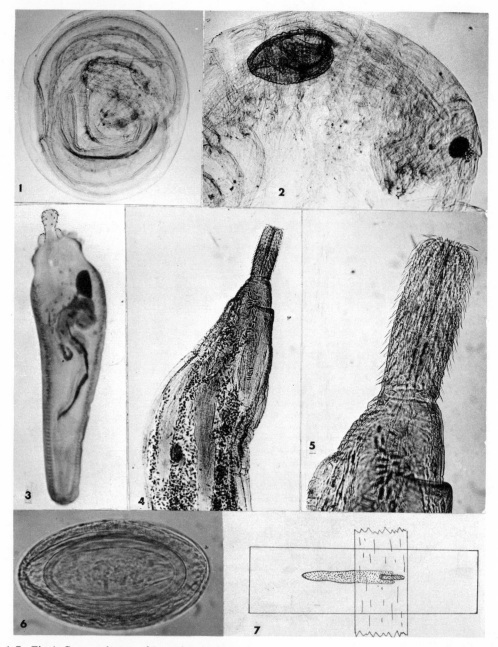

Figs. 1–7   Fig. 1: Cystacanth stage of *Leptorhynchoides* sp. from *Lepomis* (sunfish). Fig. 2: Cystacanth of *Polymorphus* sp. in hemocoel of the amphipod intermediate host, *Hyalella*. Fig. 3: Male *Oncicola* sp. taken from cat intestine.(Photo retouched.) Figs. 4 and 5: Anterior aspects of female *Tanaorhamphus* sp. taken from the intestine of the American eel, *Anguilla rostrata*. Fig. 6: Shelled acanthor (egg) from *Macrancanthorhynchus hirudinaceus*: unstained preparation mounted in glycerin. Fig. 7: Method of applying pressure to proboscis using Saran Wrap strip (vertical to long axis of worm on slide).

Zenker's, 10% buffered formalin, FAAb (Demke's fluid modified by the addition of 5% glycerol)—seems to work well. Hot fixatives often cause the cuticle to peel. Because a worm is apt to contract or withdraw its proboscis at any time prior to being killed, place it on a slide and spread a narrow strip of Saran Wrap across the slide (Fig. 7). If the specimen tends to withdraw its proboscis during fixation, a gentle but firm pressure on each end of the Saran Wrap strip helps to keep the proboscis extended until fixation of the musculature has occurred. Usually this will occur in less than one minute. Complete fixation is needed, but overfixing, like excess heat, may damage the specimen. Prior to staining, it is beneficial to rinse the specimen in a clean wash that approximates the water or alcoholic content of the fixative. Washing may be omitted, but it seems to result in clearer differentiation and keeps some stains from precipitating.

Staining. Whole mounts, most commonly used for

Table 1[a]

1. Fix specimens of Acanthocephala in a mixture of 95% AFA and 5% glycerin (AFAG) for 4–12 hr.
2. If necessary, preserve (store) in a 5% solution of glycerin in 70% alcohol.
3. To stain, cover specimens well with Mayer's paracarmine in a suitable container preferably with a ground-glass cover and allow to stand overnight (approximately 12 hr). Length of time required will depend on the size and staining capacity of the worm.
4. Carefully pipette off or decant stain. Transfer fluids (not specimens) in subsequent steps. Keep dishes covered and avoid handling as much as possible.
5. Wash off uncombined stain with two or more changes of 70% alcohol. Allow at least 1 hr between each change.
6. Differentiate with 1% acid alcohol (HCl—70% ethanol). Observe this process under the microscope at intervals until parenchyma and muscles are nearly free from stain and reproductive organs and nuclei are well stained.
7. Stop destaining by pouring off the acid alcohol and washing with two or three changes of 1% basic alcohol (NH₄OH—85% ethanol). Allow at least 1 hr in basic alcohol.
8. Wash with 95% alcohol and counterstain in alcoholic fast green in 95% alcohol. The more dilute the stain, the better will be control over staining. Caution: Overstaining is easy at this point.
9. Dehydrate with 95% and absolute alcohols, 1 hr or longer in each grade; longer for larger specimens.
10. Clear in oil of wintergreen (methyl salicylate) or xylol.
11. Remove specimens from clearing oil to a dilute mixture of mountant (mountant to solvent 1 : 10). Allow to evaporate slowly under cover until residue with specimens is viscous or tacky. This process may take up to several weeks but may be hastened in a vacuum.
12. Place specimen on a clean slide after infiltration (step 11). Center and orient it on the slide. Add mountant of normal consistency (permount or other synthetic resin) and cover with a clean coverslip. If the specimen is thick, support the coverslip with glass rods or other suitable supports. Fill any remaining space under coverslip with mountant using a fine pipette.
13. Label. Indicate name of specimen, fixative, stain, and mountant solvent employed. Keep slide flat in tray until mountant is thoroughly hardened which may take a week or more.
14. Store in *horizontal* position.

[a]For general formula and technique see: Peter Gray, "The Microtomist's Formulary and Guide," New York, Blakiston Co., Inc., 1954.

demonstration and taxonomic purposes, should be counterstained. However, choice of stains and staining techniques is a matter of personal preference. Bullock (1969) prefers Lynch's precipitated borax carmine. A Mayer's paracarmine with a fast green counterstain is excellent, and is a modification of that used for tapeworms by Corrington (1935) (Table 1).

Dehydration, Clearing, Mounting. In the procedure outlined in Table 1, dehydration parallels staining. From the fast green counterstain in 95% alcohol, a minimum of three absolute alcohol washes should be used. Some investigators stab their specimens with minuten pins or slash the body wall. By using the techniques outlined in Table 1, few specimens have been lost. Once in a while a worm may become opaque but can usually be saved by reprocessing. Clearing with terpineol, methyl salicylate, or xylol is satisfactory. Clearing seldom causes opaque chalking or whitening of the specimen. However, CAUTION: When specimens of Acanthocephala are placed directly in a mountant from the clearing agent, they will, if not punctured, become opaque-white and useless for most purposes. However, if a cleared specimen is transferred to a *dilute* solution of the mountant and its solvent allowed to evaporate slowly in a covered dish, the infiltration of mountant into the worm gradually replaces the clearing agent and the specimens remain clear and usable. Infiltration of acanthocephalans in vacuum, whether for whole mounts or for sectioning, has proven to be time saving and has produced good results. When the mounting medium containing the specimen in process has thickened to a consistency of heavy syrup through evaporation of its solvent, transfer the specimen to a glass slide with fresh mountant of thinner consistency and cover. The slides should be permanently marked with a diamond stylus and set aside to dry. Synthetic resins are preferred to balsam, but in all instances the freshly prepared slides should be frequently examined and fresh mountant added as the solvent evaporates. A labeled and properly accessioned preparation (see Van Cleave, 1953) should complete the processing and provide the basis for subsequent interpretation of those features needed to recognize even the most well established genera of Acanthocephala.

HUGO A. JAMES

References

Bullock, W. L., "Morphological Features as Tools and as Pitfalls in Acanthocephalan Systematics," in "Problems of Systematics of Parasites," G. D. Schmidt, ed., Baltimore, Univ. Park Press, 1969.
Corrington, J. D., "Explorations of the interior II— tapeworms," *Practical Microscopy*, **1**(4):4–9 (1935).
Florkin, M. and B. T. Sheer, ed., "Chemical Zoology, Vol. III, Echinodermata, Nematoda and Acanthocephala," New York, Academic Press, 1969.
Nicholas, W. L., "The Biology of the Acanthocephala," in "Advances in Parasitology," Vol. 5., Ben Dawes, ed., New York, Academic Press, 1967.
Van Cleave, H. J., "Expanding horizons in the recognition of a phylum," *J. Praasitol.*, **34**(1):1–20 (1948).
Van Cleave, H. J., "Acanthocephala of North American mammals," *Ill. Biol. Monogr.*, **23**:1–179 (1953).
West, A. J., "The acanthor membranes of two species of Acanthocephala," *J. Parasitol.*, **50**(6):731–734 (1964).
Wright, R. D., and R. D. Lumsden, "The acanthor tegument of *Moniliformis dubuis*," *J. Parasitol.*, **56**(4): 727–735 (1970).

ACRIDINE DYES[1]

Introduction. The use of fluorochromes in microscopic staining adds higher orders of sensitivity and selectivity to the staining-properties which have been recognized by many workers. The acridine dyes undoubtedly constitute the most important group of fluorescent dyes for the modern microscopist and histochemist. These are basic dyes of yellow, orange, or brown colors which are derived from the parent compound *acridine*, shown in its salt form.

summary of staining and fluorescent studies with Acridine orange. The volumes by Harms (1965) and Lillie (1969) are useful references. The early contributions to fluorescence microscopy are summarized by Haitinger (1938) in various monographs. For more details about using the acridines as Schiff-type reagents in the detection of DNA (fluorescent-Feulgen) and polysaccharides (fluorescent-PAS), the reviews by Kasten (1960; 1964) should be checked.

The acridines are one of the dye groups exhibiting the phenomenon of metachromasy, a term referring to a

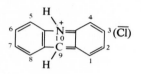

ACRIDINE

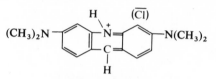

ACRIDINE ORANGE

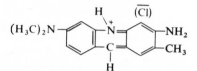

CORIPHOSPHINE O

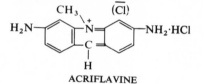

ACRIFLAVINE

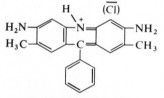

FLAVOPHOSPHINE N

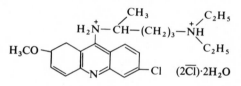

QUINACRINE

The acridines are planar molecules which are structurally related to the xanthene dyes and are readily synthesized from m-diamine precursors. The derivatives all contain amino groups in the 3,6 positions and are subdivided into diphenylmethane and triphenylmethane groups. Many of these dyes prove to be active antiseptics (Acriflavine, Proflavine), and useful in control of the malaria trypanosome (Atabrine or Quinacrine) and amoebic dysentery (Rivanol). Other acridines are valuable in the dyeing of silks (Acridine orange) and leathers (Phosphine).

For biological staining and general applications in histochemistry and cytochemistry, the most important dye in this group is Acridine orange (C. I. 46005), whose chemical structure is illustrated. Other dyes of interest include Acridine yellow (C. I. 46025), Acriflavine or Proflavine (C. I. 46000), Coriphosphine O (C. I. 46020), Flavophosphine N (C. I. 46065), Phosphine GN (C. I. 46045), and Quinacrine. A lack of space prevents an exhaustive account of all the reported uses of these dyes. Workers who wish to gain further information are advised to consult the paper by Kasten (1967) which is a recent

pronounced shift in absorption peak to lower wavelengths after binding to a variety of polyanions like chondroitin sulfate and nucleic acids, either in the test tube or in fixed cells. The same dyes exhibit this phenomenon in free solutions according to the dye concentration. In dilute dye solutions, acridines exhibit a single absorption band (α band) due to monomer formation. At higher dye concentrations, β and γ bands appear, apparently from the presence of dye dimers and polymers; it is the presence of these additional bands which produces the metachromatic color shift. The most interesting fluorescence metachromatic results are seen after dye binding to nucleic acids, which results in brilliant green, yellow, or orange emissions. This application will be discussed in reference to fluorochroming with Acridine orange (AO).

Acridine Orange (C. I. 46005). This important dye has an absorption maximum at approximately 497 n. Its structure is shown above. Acridine orange (AO) has a molecular weight of 302 and is approximately 5 % soluble in water. It has been employed as a *vital dye* to distinguish live from dead cells (Strugger, 1948). According to Strugger's (1940) early studies, AO fluoresces green in living cells and red in dead cells. The binding is far more complicated, as shown by the elegant work of Robbins and associates (1963; 1964). According to the conditions employed, treatment of living cells with AO may cause a binding to multivesicular bodies containing acid phosphatase (lysosomes), producing orange-fluorescing par-

[1]Work supported by Public Health Service Research Grants No. NS-09524 from the National Institute of Neurological Diseases and Stroke and No. CA-12067 from the National Cancer Institute. It has also benefited from USPHS Training Grant 5-T01-DE-0024-04 from the National Institute of Dental Research.

ticles. This binding is compatible with cell viability. Alternatively, AO may produce a general reddening of the cytoplasm, apparently due to prior release of lysosomal enzymes. This kind of reaction is indicative of cell death. Details of this application *for visualization of intact lysosomes and dispersed lysosomal enzymes* are given in the Appendix under Technique 1. An exhaustive account of *vital fluorochroming* appears in the monograph by Stockinger (1964).

Probably the most popular and important use of AO is as *a fluorochrome for nucleic acids* (see Appendix, Technique 2). Under appropriate conditions, AO binds to DNA or polynucleotide chains by intercalating between adjacent base pairs. When a sufficient number of dye sites become saturated in this way, the DNA-AO dye complex is seen as a green to green-yellow fluorescence. Following this initial type of binding, interaction occurs with phosphate groups as a stacking phenomenon on the DNA surface to produce a red fluorescence. This second kind of interaction may take place immediately with RNA, polynucleotides, and polyanions, inducing a red fluorescence. In fixed and stained preparations, cell nuclei, with their DNA, fluoresce green. Chromosomal DNA observed in mitotic figures often fluoresces yellow to yellow-green. Color variations likewise occur in RNA sites; nucleolar RNA may appear orange and contrasts markedly with the brick-red fluorescence of the cytoplasm (Armstrong, 1956). All of these color variations reflect the fact that dye binding and induced fluorescence are sensitive indicators of molecular size and shape (Schümmelfeder et al., 1957) and not qualitative markers of the two nucleic acids. *It is worth emphasizing again that AO does not distinguish between RNA and DNA.* There have been misunderstandings about this point, although in practice using commonly studied cells and tissues, the net result is the same, i.e., high molecular weight nucleic acid (DNA) fluoresces green and the lower molecular weight nucleic acid (RNA), fluoresces red. Treatments which cleave away pieces of DNA (depolymerization) or separate the double helix of DNA into single strands (denaturation) cause the induced fluorescence to change from green to red. Likewise, when unusually high molecular weight RNA is treated with AO, the induced fluorescence is green rather than red, which one might have expected if the fluorescence color was due simply to qualitative chemical differences between RNA and DNA.

An unusual application of AO fluorochroming has been to analyze the orientation of DNA in chromosomes using polarized fluorescence microscopy (MacInnes and Uretz, 1966). Fluorescence phenomena resulting from AO binding to purified viruses have been investigated using a number of DNA and RNA viruses. It has been shown with the aid of suitable controls that induced fluorescence of AO-treated viruses reflects molecular complexity in the same way as indicated above in higher cells, i.e., single-stranded DNA and most double-stranded RNA viruses fluoresce red whereas double-stranded DNA and unusually high molecular weight double-stranded RNA viruses fluoresce green (Mayor, 1963). *Intracellular alterations in nucleic acids induced by viruses and the subsequent formation of viral inclusions are nicely followed by AO fluorochroming* (Armstrong and Niven, 1957).

The above discussion is not meant to discourage the use of AO in studying nucleic acid patterns in fixed cells. Except for unusual instances, the staining technique under controlled conditions can be expected to yield valuable information about the qualitative intracellular and histo-

logic distribution of RNA and DNA. New and unusual cellular material should also be investigated with appropriate nucleases and other selective extraction procedures in combination with AO fluorochroming to avoid erroneous interpretations (see Kasten, 1965; Kasten and Churchill, 1966). It should be mentioned that unexpected color reactions may reflect the simultaneous presence of acid polysaccharides (viz., heparin, hyaluronic acid), unusual proteins, or internal protein-nucleic acid interactions which complicate the picture. In spite of the problems alluded to above, the AO–nucleic acid fluorochroming technique has proved to be of value in the detection and screening of cancer cells in human exfoliative cytology (Bertalanffy et al., 1956). This practical application is made possible because cancer cells frequently contain nuclei with large amounts of DNA and cytoplasm with increased quantities of RNA. In some hands, the method has proved useful as a rapid screening test where large numbers of slides must be examined.

AO has been employed in the histochemical identification of acid mucopolysaccharides. Saunders (1964) demonstrated that AO binds to acid mucopolysaccharides, which could be selectively extracted in graded concentrations of sodium chloride. The method differentiates between the presence of hyaluronic acid, chondroitin sulfuric acid, and heparin. Earlier, Hichs and Mattaei (1958) found that AO could be combined with hematoxylin to induce specific fluorescence of mucin and likewise be visible for routine transmitted light. Similar results have been obtained with Coriphosphine O (see below).

An interesting application of AO fluorochroming was reported by Schümmelfeder and Stock (1956). They found that by adjusting the pH conditions, they could *employ AO to determine the isoelectric point of proteins in tissue sections.* This use of AO does not seem to have been followed up by others could but prove to be a useful tool.

Some Applications of Other Acridines. Some of the acridines, like Proflavine (C. I. 46000), have been shown to be viral *mutagenic agents* (see Dulbecco and Vogt, 1958). This early study suggested that the dye acts on nucleic acid, a fact which is in harmony with our present understanding of the dye binding process.

It has been pointed out by Keeble and Jay (1962) that prolonged exposure of AO-fluorochromed cells to the activating ultraviolet or blue-violet light affects the complex with RNA. As a result, sites rich in RNA have their fluorescence changed gradually from red to green, which makes it difficult to separate from the green fluorescence of DNA. They recommend the use of *Coriphosphine O* (C. I. 46020), which has the same staining properties as AO but is resistant to the radiation.

Coriphosphine O was shown by Kuyper (1957) to bind to purified mucopolysaccharides at pH 4 so that a sharp differentiation could be made between mucoitin sulfuric acid, chondroitin sulfuric acid, and heparin without interference from nucleic acids.

A procedure utilizing purified Neutral Acriflavine (C.I. 46000) has been reported to *localize RNA and DNA at both the light and electron microscopic levels* (Chan-Curtis et al., 1970). Neutral Acriflavine, which contains a quaternary amine and two primary amines, behaves as a ligand that binds metals to form coordination complexes. Using this principle, the authors synthesized an *Acriflavine-phosphotungstate complex* and showed that it binds and induces fluorescence to nucleic acids and sulfated cerebroside esters. At the electron microscope level, the electron-dense complex binds selectively to the two nucleic acids.

Quinacrine mustard and other quinacrine derivatives have been shown to bind to chromosomes to form distinctive chromosome bands. These fluorescing regions apparently contain a type of heterochromatin whose nature is still under investigation (Caspersson et al., 1968). Acriflavine has been recommended by Harcroft (1953) as *a fluorochrome to detect specific granules of the islets of Langerhans.*

Acridines as Schiff-type Reagents in Cytochemistry. Beginning in 1958, a series of publications appeared which pointed out the availability of several dozen basic dyes which could be substituted for basic fuchsin in the preparation of Schiff's reagent (see the review by Kasten, 1960). A lack of space here prevents an extensive discussion of the chemical bases for the specificity of these Schiff-type reagents for tissue polyaldehydes. From over 400 dyes studied, *more than 40-odd dyes could be employed in the Feulgen reaction for DNA or the PAS reaction for polysaccharides* (Kasten, 1959). Fortunately, some of the reagents are fluorescent (Kasten et al., 1959) and impart brilliant colors, according to the reaction and tissue employed. They could also be employed in multiple staining reactions using complementary colors. In addition to enhanced sensitivity and selectivity, the slide preparations are permanent. The following acridines were found to form Schiff-type reagents: Acridine yellow (C. I. 46025), Acriflavine (C. I. 46000), Chrysophosphine 2G (C. I. 46040), Coriphosphine O (C. I. 46020), Flavophosphine N (C. I. 46065), Phosphine 5G, (C. I. 46035), Phosphine GN (C. I. 46045), Proflavine (C. I. 46000), and Rheonin AL (C. I. 46075).

The best acridine to be used in the Feulgen reaction is Acriflavine, which generally emits a yellow to yellow-green fluorescence; for the PAS reaction, Flavophosphine N is highly recommended because of its brilliant yellow-orange fluorescence. The Appendix contains a resume of techniques for both the fluorescent-Feulgen and fluorescent-PAS reactions. Another practical summary of the uses of Schiff-type reagents appears in Kasten, 1964 (p. 390).

APPENDIX

Technique 1. *Acridine Orange Vital Stain for Lysosomes* (after Robbins and Marcus, 1963; Robbins et al., 1964).

Prepare stock solution of AO by adding 1 mg dye to 2 ml. of Gey's balanced salt solution (BSS); adjust to pH 7.3 with minimum volume of $1N$ HCl or $1N$ NaOH; working solution is made by adding 1 ml stock AO to 99 ml BSS; adjust pH to 7.3 if necessary; coverslip cultures of living cells are drained of medium, washed with BSS at pH 7.3, and immersed in working AO solution for 5 min; rinse a few seconds in BSS and invert on glass slide in BSS solution; seal edges of coverslip with fingernail polish; examine immediately by fluorescence microscopy; lysosomal particles or dispersed lysosomal enzymes fluoresce bright red and nuclei fluoresce green; slide preparation is temporary.

Technique 2. *Acridine Orange Binding to Nucleic Acids* (see Kasten, 1967).

Prepare 0.1 % stock solution of AO in McIlvaine's buffer pH 4.0; working AO solution is made by adding 5 ml stock AO to 45 ml of buffer at pH 4.0; cell or tissue section preparation (previously fixed in Carnoy's acetic-alcohol 1:3 or methanol) is brought to water and then immersed in buffer; stain in working AO solution 10 min; wash in two changes of buffer for 2 min each; mount in buffer and seal coverslip with fingernail polish; examine immediately in fluorescence microscope; nuclei and

chromosomes fluoresce green to yellow-green (DNA), nucleoli and cytoplasm fluoresce in shades of red to orange (RNA); unusual results should be reexamined using appropriate cytochemical controls.

Technique 3. *Fluorescent-Feulgen Reaction for DNA* (after Kasten, 1964, pp. 391–392).

Fix tissues in Carnoy's acetic-alcohol (1:3), 10 % neutral formalin (wash well), or formol-acetic-alcohol; using sections, cellular imprints, or coverslips of cultured cells, slides are placed in $1N$ HCl at 60°C for 10 min (Carnoy's), 12 min (FAA), or 15 min (formalin); tap water 1 min; stain 45 mins in 0.25 % Acriflavine-SO$_2$ (dye solution saturated with SO$_2$ gas for 1–2 min or add a few drops of thionyl chloride to 40 ml of dye solution); wash in running water until no further color is extracted; 5 min in SO$_2$-water; dehydrate in ascending alcohols; xylene; mount in nonfluorescent mounting medium, as Fluormount of G. Gurr. In fluorescence microscope, DNA fluoresces yellow to orange. Slide preparation is permanent.

Technique 4. *Fluorescent—PAS* (induced fluorescence specific for polysaccharides and glycoproteins; after Kasten, 1964).

Bring fixed sections or cell preparations to 70 % ethanol; treat in periodic acid solution 5 min (0.4 g periodic acid, 35 ml absolute ethanol, 5 ml of 0.2M sodium acetate, 10 ml distilled water); rinse in 70 % ethanol; treat in reducing bath 1 min (1 g potassium iodide, 1 g sodium thiosulfate, 50 ml 70 % ethanol, 0.5 ml $2N$ HCl, ignore deposit of sulfur); dip in 70 % ethanol; 15 min in 0.1 % Flavophosphine N-SO$_2$ (dye solution saturated with SO$_2$ gas for 1–2 min or add a few drops of thionyl chloride to 40 ml of dye solution); wash in running tap water; dehydrate in ascending alcohols; xylene; mount in nonfluorescent mounting medium, as Fluormount of G. Gurr. Sites of polysaccharides and glycoproteins fluoresce a bright yellow, orange, or brown according to the component stained. Slide preparation is permanent.

FREDERICK H. KASTEN

References

Armstrong, J. A., *Exp. Cell Res.*, **11**:640 (1956).
Armstrong, J. A., and Niven J. S. F., *Nature*, **180**:1335 (1957).
Bertalanffy, F. D., et al., *Science*, **124**:1024 (1956).
Caspersson, T., et al., *Exp. Cell Res.*, **49**:219 (1968).
Chan-Curtis, V., et al., *J. Histochem. Cytochem.*, **18**:609 (1970).
Dulbecco, R., and Vogt., M., *Virology*, **5**:236 (1958).
Haitinger, M., "Fluoreszenzmikroskopie. Ihre Anwendung in der Histologie und Chemie," Leipzig, Akadem. Verlagages, 1938.
Harcroft, W. S., *Nature*, **168**:1000 (1951).
Harms, H., "Handbuche der Farstoffe für Mikroskopie," Staufen Verlag, 1965.
Hichs, J. D., and Matthaei, E., *J. Pathol. Bacteriol.*, **75**:473 (1958).
Kasten, F. H., *Histochemie,* **1**:466 (1959).
Kasten, F. H., *Int. Rev. Cytol.*, **10**:1 (1960).
Kasten, F. H., in E. V. Cowdry, ed., "Laboratory Technique in Biology and Medicine," 4th ed., Baltimore, Williams & Wilkins, 1964.
Kasten, F. H., *Stain Technol.*, **40**:127 (1965).
Kasten, F. H., and Churchill, *J. Histochem. Cytochem.*, **14**:187 (1966).
Kasten, F. H., et al., *Nature*, **184**:1797 (1959).
Kasten, F. H., *Int. Rev. Cytol.*, **21**:141 (1967).
Keeble, S. A., and Jay, R. F., *Nature,* **193**:695 (1962).
Kuyper, C. M. A., *Exp. Cell Res.*, **13**:198 (1957).

Lillie, R. D., ed., "Conn's Biological Stains," 8th ed., Baltimore, Williams & Wilkins, 1969.

MacInnes, J. W., and Uretz, R. B., *Science*, **151**:689 (1966).

Mayor, H. D., *Int. Rev. Exp. Pathol.*, **2**:1 (1963).

Robbins, E., et al., *J. Cell Biol.*, **21**:49 (1964).

Robbins, E., and Marcus, P. I., *J. Cell Biol.*, **18**:237 (1963).

Saunders, A. M., *J. Histochem. Cytochem.*, **12**:164 (1964).

Schümmelfeder, N., and Stock, K. F., *Z. Zellforsch. Mikr. Anat.*, **44**:327 (1956).

Schümmelfeder, N., et al., *Naturw.*, **44**:467 (1957).

Stockinger, L., *Protoplasmatologia*, Vol. 2, part 1.

Strugger, S., *Jena Z. Naturw.*, **73**:97 (1940).

Strugger, S., *Can. J. Res.*, Sec. E, **26**:229 (1948).

See also: CHROMOSOME TECHNIQUES, FEULGEN STAINS, FLUORESCENCE MICROSCOPY.

AGAR-AGAR

AGAR-AGAR is a composite organic compound containing 0.4–0.9% of total nitrogen, 0.03–0.009% of amino nitrogen, 70–75% of polysaccharide, 11–22% of moisture, and 2–4% of ash. The monosaccharides, galactose, arabinose, xylose, and a small amount of free amino acids form a part of agar-agar. The polysaccharide of agar-agar is a mixture of agarose (mol. wt. $\simeq$ 120,000) and agaropectin (mol. wt. $\simeq$ 12,000), which differ in the content of sulfate groups, diffusivity, viscosity, capability to keep water, energy of activation of the process of gelatinization, electrophoretic motility, presence of hydrogen links, and chemical composition. The basic and most valuable component of agar-agar is agarose. Its molecule is built from D-galactose and 3,6-anhydro-L-galactose. On mild acid hydrolysis agarose turns into disaccharide-agarobiose II with a high yield, and on fermentative hydrolysis, into neogarobiose III evidencing by this a high regularity of its structure. It is suggested that the agarose molecule is built from the residues of galactose linked alternatively by $\alpha-1 \rightarrow 4$ and $\beta-1 \rightarrow 3$ bonds; all the monosaccharide residues are 3,6 anhydro-derivatives of L-isomer, while the residues substituted in position 3 are a nonmodified D-isomer. The content of SO_4 groups in agarose is not great and their exact arrangement is not disclosed. Much more residues of sulfuric acid in the form of combination with calcium are found in agaropectin, the detailed structure of which is not known exactly.

On commercial scale agar-agar is produced from red algae of various genera (*Ahnfeltia, Gracilaria, Gelidium* etc.). There are two basic methods for extracting agar-agar from the algae.

Freezing-thawing technique: Dried algae are steeped in lime white, then washed in water and boiled two to three times in weak solutions of lime. The resultant solution of agar-agar is filtered and cooled until gel is formed. The gel is cut into thin strips, placed on wooden lattices, and frozen at -7 to $10°C$. The frozen gel is thawed, washed, and dried in the flow of warm air.

Diffusion-vacuum technique: Steeped and washed algae are boiled two to three times under pressure. The filtrate is cooled, and the gel is dialyzed against running water until it becomes colorless and translucent.

The gel is melted and condensed under vacuum, then dried and ground into powder.

To produce highly purified agar-agar there are several different methods, the basis of which is the removal of contaminating substances. With this end in view the following methods are applied: (a) lasting electrodialysis; (b) gel washing with a weak solution of acetic acid and distilled water; (c) fractionation of hot solution of agar-agar (after preliminary treatment with adsorbed carbon) with ethanol and acetone; (d) repeated freezing and thawing of agar-gel, and so on. As a rule the best purification results are obtained with a combination of several methods. In this case 90% of nitrous substances are removed from agar-agar. A high degree of purity of agar-agar is reached as a result of removal of agaropectin from it or isolation of agarose in free condition. To this end pure agar-agar is ground with dimethyl sulfoxide at $60–80°C$; and agarose is precipitated with three volumes of acetone. In this case almost all agaropectin is left in residuum. In thus extracted agarose is contained 1% of ash and 0.97–1.32% of SO_4 groups, and in agaropectin, 7–33% of ash and 4–6% of SO_4 groups (depending on the material used).

For the removal of agaropectin from purified and EDTA-treated agar-agar it is subjected to adsorption on $Al(OH)_3$-gel. After this, agar-agar, consisting mainly of agarose, is washed with water, frozen and thawed, and dried with acetone.

Agar-agar is used for the stabilization of sera, emulsions, the sizing of expensive yarn, the tightening of paper and leather, the production of photo-emulsions, and cellophane, and as anticorrosive coatings and boiler compounds. Agar-agar is used in cookery, therapeutic nutrition, and pharmaceutics. In bacteriology agar-agar is used for the preparation of semisolid, solid, and dry culture media. In biochemistry and allied fields of science it is used for the preparation of holding media, in which analytical and preparative separation by way of electrophoresis or immunofiltration of composite mixtures of biopolymers are performed. Recently agar-agar and agarose have been finding wider application in the production of molecular sieves, which make it possible to separate quite easily the polymers of different molecular weights. In immunochemistry agar-agar is used for immunochemical determinations of antigens and antibodies and for other purposes.

V. I. VEINBLAT

References

Barteling, S. J., "A simple method for the preparation of agarose," *Clin. Chem.*, **1969**:(10)1002–1005.

Hickson, T. G. L., and A. Polson, "Some physical characteristics of the agarose molecule," *Biochim. Biophys. Acta,* **165**(1):43–58 (1968).

Kochetkov, N. K., A. F. Bochkov, B. A. Dmitriev, A. I. Usov, A. S. Chizhov and V. N. Shibaev, "Chemistry of carbohydrates," in "Chemistry," Moscow, 1967.

Tagawa, S., "Chemical studies of the production of agar-agar," *J. Shimonoseki Univ. Fish.*, **17**(2):35–86 (1968) (Japan).

Veinblat, V. I., and N. P. Borisova, "Methods for the preparation of agar-agar," *Laboratornoe Delo*, **1969**(6): 357–358 (USSR).

AMNION

Amnion, in common with somatic embryonic tissues, contains a high proportion of water. Furthermore, its single-layered epithelium and its delicate thin connective tissue render it susceptible to damage during fixation and dehydration. During preparation for histologic or ultra-structural examination, therefore, amnion must be handled carefully with fine instruments, dehydrated by serial passage through numerous changes of alcohol of gradually increasing concentrations, and maintained within narrow limits of pH and osmolality. Since the epithelium varies from placental to reflected portions and at various times during gestation, anatomic assessment of its physiologic state requires examination of numerous specimens to avoid sampling errors. Unlike the placenta, however, the amnion exhibits similar histologic and ultrastructural features throughout all mammalian orders.

In working with the amnions of man and large experimental animals, best results are obtained by removing the tissues through a hysterotomy incision, opening the uterus as quickly as possible, and gently separating amnion from chorion in a large volume of fixative before transferring smaller pieces to fresh solutions of the same fixative. For study of large areas of amnion in which it is desirable to preserve histologic relations, the tissues should be collected in sterile Hank's solution at 4°C before they are cut into smaller pieces. With smaller animals, best results are obtained by dripping fixative over the membranes through the opened uterus. After the membrane has hardened slightly, it is excised, cut into smaller pieces, and fixed in the same solution for an appropriate length of time. For study of cells desquamated into the amniotic cavity, amniotic fluid is removed at laparotomy or amniocentesis. Pellets are prepared by centrifugation with glutaraldehyde in cacodylate buffer for 45 min, washing in cacodylate-buffered sucrose, and postfixation in buffered 1 % osmic acid. For ultrastructural study of tissue cultures, monolayer spreads are best fixed in 4 % glutaraldehyde in isotonic Tyrode's solution.

In general, the best fixation of amnion for electron microscopy is obtained with glutaraldehyde followed by osmic acid. Satisfactory results have been reported with concentrations of glutaraldehyde varying from 2 % to 5 %. Optimal results are usually obtained with 3 % glutaraldehyde. The best buffers have proved to be $0.05-0.1M$ cacodylate at pH 7.3–7.5 for 60–90 min at 0°–4°C. Almost equally good results are obtained with 0.075–$0.1M$ phosphate buffer. Tissues are best rinsed for 6–12 hr in several changes of buffered $0.25M$ sucrose, or left overnight in that solution, and secondarily fixed in 1 % osmic acid in $0.05M$ cacodylate or $0.1M$ phosphate buffer for 60 min.

Amnion should be dehydrated in an ethanolic series utilizing 5 % increases in concentration followed by three changes of absolute alcohol and two of propylene oxide. For standard electron microscopy, Araldite has been used most extensively and most successfully. Ultrastructural details are best demonstrated after double staining with lead citrate and uranyl acetate.

For localization of enzymes optimal initial fixation consistent with preservation of ultrastructural integrity is obtained with 4 % formaldehyde prepared from powdered paraformaldehyde at 4°C in $0.05M$ cacodylate buffer at pH 7.4 for 90 min. The specimens are then washed in four changes of cold $0.25M$ sucrose in the same buffer or left overnight in this solution. Before incubation the tissues are transferred to Tris-maleate buffer at pH 7.2 and brought to room temperature. The specimens are then incubated with occasional agitation at 37°C for 15–25 min in the appropriate medium. For demonstration of nucleoside phosphatases a medium containing the following reactants has proved most successful: Tris-

maleate buffer $(0.08M)$; lead nitrate (2.5×10^{-3}); and magnesium chloride $(0.05M)$. Reactions for adenosine triphosphatase are carried out by the addition of adenosine triphosphate disodium salt (8.3×10^{-4}) to the incubation medium. Other enzymes can be localized by substituting equimolar concentrations of such substrates as adenosine diphosphate, adenosine monophosphate, and sodium beta-glycerophosphate for the ATP. Substrate-free media are used as controls. Rat liver, which is known to be rich in these enzymes, serves as a control for the method.

After incubation the tissues are washed in distilled water and postfixed in 1 % osmic acid in $0.05M$ cacodylate or $0.075M$ phosphate buffer at pH 7.4–7.5. Some specimens are briefly washed in 2 % ammonium sulfide before final fixation to precipitate the reaction products for conventional histologic identification. Because the amnion is very thin, tissue blocks have proved as good as frozen sections for localization of the enzymes and much better for preservation of ultrastructural details.

RALPH M. WYNN

References

French, G. L., A. H. MacLennan, and R. M. Wynn, "Nucleoside phosphatases in human amnion: ultrastructural localization," *Obstet Gynecol.* **37**:173 (1971).

Novikoff, A. B., E. Essner, S. Goldfischer, and M. Heus, "Nucleosidephosphatase Activities of Cytomembranes," in "The Interpretation of Ultrastructure," R. J. C. Harris, ed., New York, Academic Press, 1962.

Wynn, R. M., and G. L. French, "Comparative ultrastructure of the mammalian amnion," *Obstet. Gynecol.,* **31**:759 (1968).

AMPHIOXUS

Amphioxus (*Branchiostoma lanceolatum* (Pallas)) has long been a favorite study of microscopists and many fixation, embedding, and staining procedures have been employed in light microscope investigations.

More recently, electron microscopists have investigated various features of the animal, and in this article both light and electron microscopical techniques will be considered.

Routine light microscopical investigations upon animals anesthetized with 7.5 % magnesium chloride or MS 222 (Sandoz) have generally used seawater Bouin's fluid or seawater formalin as the fixative, though for particular purposes (e.g. silver impregnation of nervous tissue) other fixatives such as 5 % glutaraldehyde in seawater, or methanol–acetic acid–formalin (80 : 5 : 5) are useful.

The lamellar myotomal muscles are difficult to section well after embedment in paraffin wax following any fixation, and celloidin embedding, double embedding with celloidin-paraffin, or polyesterwax embedding (Steedman, 1957) are generally more successful than routine paraffin wax treatment. Polyesterwax sections are liable to become detached in certain solutions, and it is advantageous to smear well-cleaned slides with albumen before flooding with amylopecten. After draining, leave overnight in a coplin jar with a few drops of 40 % formalin before dewaxing. Good results for light microscopy of the lamellar muscles are obtained from thick sections of araldite-embedded material, stained with toluidine blue.

Unlike tunicates, amphioxus is easy to stain routinely, and good general preparations may be obtained by any

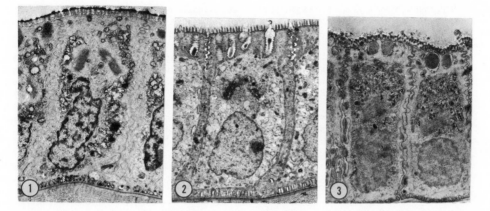

Fig. 1 Rostral epithelial cells of amphioxus fixed in 5% cacodylate-buffered glutaraldehyde in seawater, stained with uranyl acetate and lead citrate. This shows the effects of different embedding media upon preservation and staining: 1, Durcupan; 2, Araldite; 3, methacrylate/styrene. (All ×50,000 approximately.)

of the common procedures such as Azan, Mallory, or H and E. In sections, nephridia are particularly well stained with Linder's hematoxylin; in whole mounts with Delafield's hematoxylin. Whole mounts are often useful and informative, particularly of skin, pharyngeal apparatus, atrial musculature, and buccal cirrhi. These may be stained simply with Delafield's hematoxylin, or by various reduced silver techniques, such as those of Bodian and Holmes. Holmes' whole mount preparations of the entire gill apparatus, or the pterygial muscles, are successful after Bouin fixation; in my hands the best results have been obtained for peripheral nerves by doubling the concentration of silver in the impregnating bath, and washing out the 20% silver nitrate prior to impregnation for 48 hr. The peripheral nervous system (apart from the neurons of the gut wall) is also well stained in whole mounts with methylene blue, by immersing preparations pinned out on wax plates, in a dilute solution of the dye in seawater. Such preparations may be fixed in the usual manner in 10% ammonium molybdate, but better results are obtained by an initial fixation for 5–30 min in a mixture of equal parts of 5% glutaraldehyde and 5% formaldehyde (made up in seawater), before rinsing and placing overnight in 10% ammonium molybdate.

For observation and photography, methylene blue whole mounts may be made more transparent by transferring them after dehydration to mixtures of xylene and methyl salicylate, methyl benzoate, or benzyl benzoate (it is necessary to experiment with the proportions of these to obtain the best results; 3 parts of xylene; 7 of methyl salicylate may be tried), but if left in such solutions, the dye fades; therefore it is wise to return to xylene, in which the dye is permanent (if not exposed to sunlight), as soon as possible.

For central and peripheral nerves in sections (preferably 25–30 μ thick), excellent results are given by Palmgren's earlier method (Palmgren, 1948), which also gives a good general picture. For peripheral nerves, and for the "ventral root" processes of the myotomal muscles, it is advantageous to reduce the concentration of pyrogallol in the reducer by 50% or more from that in the original method.

Little histochemical work has been done upon amphioxus, but the standard methods for cholinesterase reveal

its presence at motor end plates both at the junction of the spinal cord and the "ventral root" myotomal processes, and at the notochordal synapses. The composition of Reissner's fiber and the neurosecretory cells of the brain have been investigated with Gomori's chrome-alum hematoxylin.

Investigations upon amphioxus larvae have been either upon material specially fixed (usually in Flemming) and subsequently stained with iron hematoxylin, or upon material from plankton collections fixed in seawater formalin. It is possible that glutaraldehyde-fixed material embedded in araldite would give better results for light microscope investigations. For taxonomic purposes, myotome counts are important, and it has been found useful to clear larvae in benzyl alcohol and stain lightly with light green and chlorazol black, or alternatively, to employ polarized light.

Initial electron microscope investigations of amphioxus were upon osmium-fixed material embedded in methacrylate; more recently both osmium fixatives and glutaraldehyde have been employed, and araldite, epon, and water soluble resins have been used for embedding.

Some fixatives which have been found to give good results with amphioxus material are:

1. Eakin and Westfall (1962)
   Equal parts 4% OsO_4 and 8% $K_2Cr_2O_7$ made up in seawater. Diluted to make final concentration of salts isotonic with seawater. pH 7.6.
2. Flood (1966) (Note: poor preservation of muscle mitochondria)
   12.5 ml buffer (1 part 2.72% KH_2PO_4: 4 parts 3.56% Na_2HPO_4 $2H_2O$)
   12.5 ml salt solution (0.15% KCl, 11.7% NaCl)
   25 ml distilled water
   1 g OsO_4
3. Flood (1968)
   50 ml cacodylate buffer (pH 7.5) (20 parts 0.2M sodium cacodylate: 1–2 parts 0.1M cacodylic acid)
   1 g NaCl
   250 mg $CaCl_2$
   0.5 g OsO_4
4. 5% glutaraldehyde in seawater, buffered to pH 7.5 with cacodylate buffer (in 3). If the stock 25%

glutaraldehyde solution is shaken with metallic zinc before dilution, and kept over zinc for some hours, the pH will be around 7.0, and little buffer is required to make the final fixative solution pH 7.5.

Unless high concentrations of calcium chloride (0.5 %) are used in osmium fixatives, poor preservation of mitochondria may occur in muscle. The osmotic pressure of the fixative is more important with glutaraldehyde fixation (Bone and Denton, 1971). In most investigations, sections have been stained with uranyl acetate followed by lead citrate, in the usual manner, but good results are also obtained by staining with uranyl acetate in the block, provided this is not too large.

Finally, both for electron microscopy and light microscopy, amphioxus material has been found to resemble in its staining properties the tissues and cells of vertebrates, rather than invertebrates, so that routines which were initially devised for higher vertebrates, such as silver impregnation methods, or cholinesterase techniques, may generally be used for amphioxus material with little variation; special methods which work well with teleost material will be found in general to work equally well with amphioxus.

Q. BONE

References

Bone, Q., and E. J. Denton, *J. Cell Biol.* **49**:571–581 (1971).

Eakin, R. M., and J. A. Westfall, *J. ultrastr. Res.,* 6:531–539 (1962).

Flood, P. R., *J. Comp. Neurol.,* **126**:181–217 (1966).

Flood, P. R., *Z. Zellforsch.,* **84**:389–416 (1968).

Palmgren, A., *Acta Zool.,* **29**:377–392 (1948).

Steedman, H. F., *Nature* (London), **179**:1344 (1957).

ANNELIDA

Annelid worms constitute a relatively small phylum of approximately 8000 species divided into several very small and problematical classes and three major ones: the Polychaeta, which are almost exclusively marine; the Oligochaeta, which are freshwater or terrestrial and include the earthworms; and the Hirudinea or leeches. The importance of annelids is chiefly ecological, notably in soil formation. As pointed out by Darwin long ago, depending on soil type and moisture, earthworms bring 2–18 tons of earth to the surface per acre per year. Polychaetes have been much used in research on fertilization and the early development of the egg, and oligochaetes in the study of regeneration.

All annelids are segmented, as is obvious in earthworms, and the body consists of a tube within a tube with the gut a straight passageway from mouth to anus. The coelomic cavity lying between gut and body wall is lined with a peritoneal epithelium and the cavity of each segment separated from those adjacent by a septum which, in many species, is incomplete so that body fluids pass freely from one compartment into the next. When the body wall muscles contract, this fluid acts as a skeleton giving rigidity to the worm. Like all protostomes, the branch of the animal kingdom to which annelids belong, there is a ventral ganglionated nerve cord extending the length of the body and connected to the brain, or suprapharyngeal ganglion, by a pair of nerves. The mouth develops from the embryonic blastopore (hence the name proto-stome), the cleavage of the egg is spiral, and the aquatic larva, when present, is a minute, top-like trochophore.

The polychaetes are characterized by paired, leg-like lateral appendages, one pair to a segment, called parapodia, which bear chitinous setae. Many polychaetes possess large, brilliantly colored gills. The sexes are separate. Members of the first order, Errantiformes, swim or crawl freely or live in mud burrows on marine bottoms, in coral reefs, or wharf pilings. There is little differentiation of the segments except for the head, which commonly bears chitinous jaws, palps, and in some species, well developed eyes. Familiar examples include *Nereis* (*Neanthes*), the clam worm, *Eunice*, the palolo worm of the South Seas, and *Manayunkia*, a small, freshwater worm reported from a few rivers and the Great Lakes. Members of the other order, the Sedentariformes, live in tubes from which their gills can usually be extended. Jaws and palps are missing but minute eyes may be present on the gills. Familiar examples include *Sabella*, the fan worm, and *Chaetopterus*, a highly luminous form with elaborate parapodia, which lives in a U-shaped parchment tube buried in the mud except for the tips of the tube.

Oligochaetes lack parapodia, are hermaphroditic, and in the breeding condition, bear a ring-shaped swollen glandular zone called the clitellum which secretes a proteinous capsule over the fertilized eggs. The Opisthoform order includes the earthworms of which the two most easily identified and commonly available are *Lumbricus terrestris*, the nightwalker or dew-worm, and *Eisenia foetida*, called by Izaak Walton the brandling but also known as the tiger earthworm because of its transverse cinnamon brown stripes, one to a segment. It is the only earthworm with this trait. *Lumbricus* can be identified by the way the posterior half inch becomes flattened when the worm moves backwards, and by the creases of the prostomium (a kind of proboscis) which extend back to touch the second segment. *Eisenia* is common around barnyards, but rare in gardens, where the common angleworms are species of the genera *Allelobophora*, *Helodrilus*, and *Octolaseum*.

The Plesioporoform order includes many small, transparent, freshwater species notable for reproducing primarily asexually by fission. In the slow type as seen in *Dero*, a single segment undergoes an explosive series of cell divisions, the posterior portion of the segment develops into the six or so head segments characteristic of the species while the anterior portion forms many small tail segments. The worm then separates into two, the anterior with an old head and a new tail, the posterior vice versa. The rapid type is the same except that additional fission planes usually form before the original separates. In the slow type and most of the rapid the number of segments anterior to the fission plane, termed *n*, remains constant but in some, e.g. *Stylaria*, *n* declines by one after each separation. Thus the fission plane moves anteriorly until *n* reaches 14–17, when the anterior worm grows many new segments before the process recommences.

The leeches form two groups, those without teeth, which include many small species, and those having three chitinous teeth, which include *Hirudo*, the medicinal leech still occasionally used as the source of the anticoagulant hirudin, and *Hemopsis*, somewhat smaller and found attacking, men, cattle, frogs, fishes, etc. The internal anatomy of leeches is difficult to dissect and there are

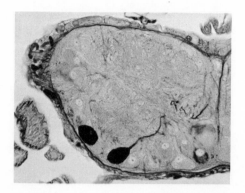

Fig. 1 Neurosecretory cell in the hypopharyngeal ganglion. *Eisenia foetida*. Gomori aldehyde fuchsin after Bouin's fixation.

many more external circular wrinkles than actual segments judged by the number of ventral ganglia.

The small classes of annelids include the Archiannelida, which some zoologists have regarded as primitive and others as degenerate. The Echiuroidea, with *Echiurus* used as bait in the codfish industry, and *Bonellia*, famous for probably the most extreme form of sexual dimorphism in the animal kingdom, and the Sipunculoidea (Gephyrea) are usually regarded as separate phyla with close annelid affinities.

The histological structure of annelids can be found in Andrew, 1959. Most of the body mass of earthworms consists of smooth muscle running longitudinally. Its cellular organization has long remained a subject of controversy. The nervous system is notable for the handsome neurosecretory cells (Fig. 1) and for the three giant nerve fibers which extend the entire length of the animal in the ventral nerve cord. They conduct with extraordinary rapidity and enable the entire body of the worm to contract simultaneously when in danger. Each fiber is divided by a synaptic septum whenever it passes from one segment into the next. Such giant flat synapses are remarkable for their speed of transmittal and for conducting in both directions. Excretion is affected by a pair

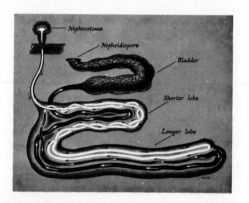

Fig. 2 Earthworm nephridium. Arrows indicate direction of flow. (From G. B. Moment, "General Zoology," Boston, Houghton Mifflin Co., 1967.)

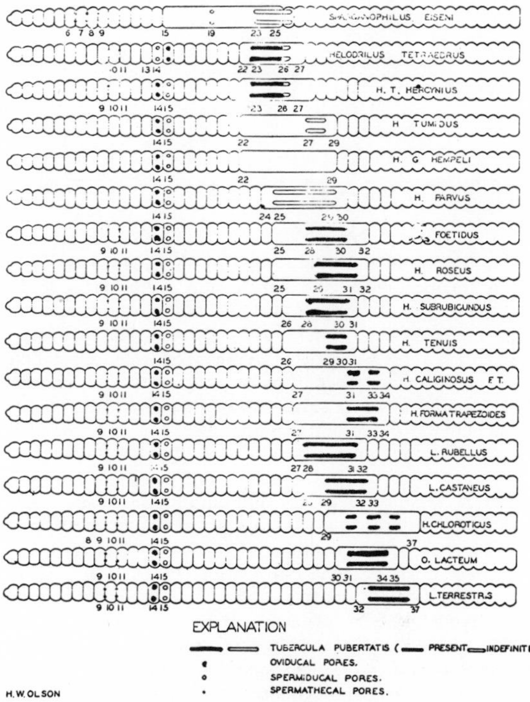

Fig. 3 Diagrammatic key to the earthworms of New York. It will be found serviceable throughout most of North America. (From Olson, 1940.)

of nephridia (many in some tropical species) located in each segment except the most terminal ones. Each nephridium bears a striking resemblance to the vertebrate nephron having a ciliated funnel, the nephrostome, removing coelomic fluid from one segment and passing it through a long, triply coiled tubule in the postjacent segment to a bladder and thence to the exterior. (Fig. 2.) The vascular system is closed and the hemoglobin is dissolved in the blood except in a few marine polychaetes with erythrocytes, e.g. *Glycera*.

The best anesthetic for earthworms is 0.2 % chlorotone. Small freshwater species like *Stylaria* will withstand 0.1 % chlorotone for several hours without noticeable injury even to growing zones. Earthworms can be cultured indefinitely in generous amounts of the soil or rubble in which they are found and should be maintained at 25°C for regeneration. *Stylaria*, *Aeolosoma*, and other freshwater species flourish in boiled lettuce. *Enchytraeus*, the small white oligochaete much used for tropical fish food, does well in earth with weekly admixtures of oatmeal or mashed potatoes.

GAIRDNER B. MOMENT

References

Andrew, W., "Textbook of Comparative Histology," New York, Oxford Univ. Press, 1959.

Dales, R. P., "Annelids," New York, Hillary House, 1963.

Laverack, M. S., "The Physiology of Earthworms," New York, Macmillan, 1953.

Moment, G. B.. "The relation of body level, temperature, and nutrition to regenerative growth," *Physiol. Zool.* **26**(2):108–117 (1953).

Olson, H. W., "Earthworms of New York State," New York, Amer. Museum Novitates no. 1090 1940, pp. 1–9.

Stephenson, J., "The Oligochaeta," Oxford, Clarendon Press, 1930.

ARTHROPODA (SECTIONS)

The problems in preparing sections of small arthropods and their developmental stages for study under the light microscope involve the cuticle and its penetration by reagents, orientation in the embedding medium, bonding properties of the medium, and the preparation of intact sections. Many workers have either perforated the cuticle to permit the entrance of reagents or they have removed the organs for processing, with neither procedure readily adaptable to microarthropods. Paraffin and paraffin-celloidin combinations have been used in the preparation of serial sections. Both media lead to particular difficulties. Mixtures of plastic and paraffin have been made in order to utilize the bonding properties of the former and the ribboning qualities of the latter (DeGiusti and Ezman, 1955; Woodring and Cook, 1962).

Difficulties with the plastic-paraffin mixtures in a study involving the life history stages of mites led to the elimination of the paraffin and the substitution of an adhesive compound to the plastic block in order to obtain a ribbon (Rohde, 1965). The technique particularly involved the processing of the eggs, larvae, nymphs, and adults of *Fuscuropoda marginata* (Acarina: Uropodidae), a mite with a hard cuticle, and adult *Caloglyphus mycophagus* (Acarina: Acaridae) (Rohde and Oemick, 1967), a mite with a soft cuticle. The technique was further tested on the eggs of *Drosophila melanogaster*, the proventriculus of *Periplaneta*, and the larvae of Amphioxus with excellent results.

Eggs. The chorion of arthropod eggs interferes with penetration by the usual reagents, and most investigators have had the choice of either making an opening through this covering or of dissolving it away. The latter, following Slifer (1945), was accomplished with a dechorionating fluid consisting of 0.4 % aqueous solution of sodium hypochlorite using commercial Clorox bleach which contains 5.25 % by weight of the active ingredient. Action is improved by the addition of a wetting agent such as Tergitol (3 drops per 100 ml of fluid).

Submergence of the eggs is followed by release of the chorion in 3–5 min. This begins with a longitudinal split which progresses toward one end and results in release of the egg. Gentle agitation will assist the separation. Two water rinses precede fixation.

Standard arthropod fixatives were tested and rejected either because of the alteration in form or lack of penetration; only fixatives containing dioxane produced obvious penetration. Most satisfactory for the preparation of both whole mounts and sections was the fluid used fresh as follows:

Picric acid in dioxane (28 g dry crystals/100 ml)	80 ml
Formalin	10 ml
Formic acid[a]	7 ml
Chloroform	3 ml

[a]Acetic acid may be substituted

The time for fixation may vary up to 1 hr and should be followed by several rinses with storage in ethyl cellosolve or the eggs may be processed immediately for embedding, preceded with several rinses in tetrahydrofuran.

Larvae, Nymphs, and Adults. These stages are treated alike and immersed whole in tetra-Bouin fixative, without vacuum, for 3–4 hr. The use of tetrahydrofuran in histology is described by Salthouse (1958), and the substitution of this solvent for alcohol in the alcoholic Bouin formula was done by Woodring and Cook. It was the fixative of choice for all stages except the egg. The fixative is replaced with pure tetrahydrofuran with subsequent changes as necessary to remove picric acid. This step may vary in time as a matter of convenience but 12 hr is preferred. The specimens may be stored in cellosolve or processed immediately.

Preparation of the Monomers. Preparation of the monomers is given by Rohde (1964) or may be obtained from the manufacturer (Rohm and Haas Co., Philadelphia, Pa.). The inhibitor-free monomers of ethyl and butyl methacrylates should be washed, filtered, and stored in the refrigerator over Drierite. Consistent results during sectioning can be obtained using a 1:9 mixture of ethyl and butyl methacrylates.

A 2 % accelerated mixture of the monomers (benzoyl peroxide, volume to weight) (Lucidol Division, Buffalo, N.Y.) is placed to the one-half level in No. 1 gelatin capsules held upright in wood trays, the caps replaced and the trays placed in a 60°C oven for overnight polymerization. They are stored for later use.

Embedding. The best penetration was achieved by beginning with 2 % methacrylate mixture in tetrahydrofuran and then successively through 5, 10, 20, 40, 50, 60, 80, and 90 % mixtures, at 5-min intervals, to the pure methacrylate mixture.

Immature stages and adults are passed through two changes of 1:1 solution of tetrahydrofuran and methacrylate mixture over a period of 1–2 hr to the pure methacrylate mixture for a similar period of time.

The remaining procedure is common to all specimens passed through tetrahydrofuran to the methacrylate mixture. It is replaced with a 2 % accelerated methacrylate mixture for 1 hr prior to the transfer to the previously prepared gelatin capsules. Capsules are filled with the accelerated mixture and the specimens transferred singly, oriented, the caps replaced, and the trays placed in a 60°C oven for overnight polymerization (or about 12 hr). A minimum time of 48 hr is allowed for post-oven polymerization, after which the capsules are removed by soaking in warm water and stripping them away from the plastic cylinder.

Sectioning. The singly embedded specimen is prepared for sectioning by trimming away excess matrix (leaving a broad base) and cemented to a suitable holder for clamping in the microtome (20 × 10 mm lucite sticks have served admirably). Final trimming under the stereomicroscope is best done the following day to allow thorough hardening of the household cement.

A layer of Tackiwax (a multipurpose adhesive, primarily beeswax, in general laboratory use), equal to the vertical face of the block, is applied with digital pressure along the underside of the mount and trimmed to conform to its sides.

Sections within the range of one or more microns can be cut with a rotary microtome (Spencer Model 820 was used) at room temperatures varying from 20 to 30°C. Glass knives are recommended and can be cut from plate glass scraps following the directions of Pease (1960). A special microtome clamp for the glass knife is necessary to hold it in a vertical cutting position (a suitable clamp has been developed by Frank Fryer Co., Carpentersville, Illinois), a simple device which can be made by a good machinist.

When ribboning begins, the leading edge of the ribbon should be grasped with a fine-tipped forceps and gently stretched with a vertical pull as the ribbon leaves the knife. The holding capacity of the adhesive is usually not more than 40–50 sections. The problem of static electricity may be serious at times and can be controlled to a large extent by employing some air-ionizing device like the Neutra Stat (Eberhard Co., Ann Arbor, Michigan).

Mounting. Ribbons are floated on water over a microslide previously coated with albumen, flattened in a xylene-saturated atmosphere, and dried overnight on a warming table. The Tackiwax and plastic are removed in passing the mount through xylene to the staining procedure of choice.

Comments. This kind of embedding medium provides the bonding necessary, especially with yolky eggs, to maintain the integrity of fragile or refractory specimens during sectioning. Specimens with a dense, hard cuticle will occasionally reveal improper bonding, a condition which can be eliminated with a brief post-fixation exposure to a 2 % solution of sodium hypochlorite in water, inserted in the schedule.

There is also a distinct advantage during orientation in being able to see the embedded specimen in the clear field provided by the plastic matrix.

The use of dioxane and tetrahydrofuran eliminates alcohol in the schedule because of its shrinkage effects, a matter of particular significance relating to small specimens.

The use of this technique need not be limited to arthropods but should be restricted to specimens not much larger than 2 mm in diameter or at least within the length of the cutting edge provided by the glass knife. Specimens preserved in alcohol may be processed through 100 % alcohol to the methacrylate mixture leading to the steps described.

CHARLES J. ROHDE, JR.

References

DeGiusti, D. L., and L. Ezman, "Two methods for serial sectioning of arthropods and insects," *Trans. Amer. Microscop. Soc.,* 74(2):197–201 (1955).
Pease, D. C., "Histological Techniques for Electron Microscopy," New York, Academic Press, 1960 (274 pp).
Rohde, C. J., "Some techniques in the preparation of stained whole mounts and serial sections of mite embryos and adults," *Proc. 1st Int. Congr. Acarology, Acarologia,* 6:208–214 (1964).
Rohde, C. J., "Serial sections from plastic-embedded specimens: arthropods in methacrylate," *Stain Technol.,* 40(1):43–44 (1965).
Rohde, C. J., and D. A. Oemick, "Anatomy of the digestive and reproductive systems in an acarid mite (Sarcoptiformes)," *Acarologia,* 9(3):608–616 (1967).
Salthouse, T. N., "Tetrahydrofuran and its use in insect histology," *Can. Entomol.* 90(9):555–557 (1958).
Slifer, E. H., "Removing the shell from living grasshopper eggs," *Science,* 102:282 (1945).
Woodring, J. P., and E. F. Cook, "The internal anatomy, reproductive physiology and molting process of *Ceratozetes cisalpinus* (Acarina:Oribatei)" *Ann. Entomol. Soc. Amer.,* 55 (2):164–181 (1962).
See also: CHITIN, INSECT HISTOLOGY.

ASCOMYCETES

The Ascomycetes are a rather large group of fungi with a range of morphological diversity from yeast-like forms to those producing a myriad of multicellular ascocarps. They vary in nutrition and habitat as well, being symbiotic, saprophytic, or parasitic, terrestrial, aquatic, or marine. Each type of ascomycete is collected and cultured according to its particular habitat, mode of growth, and nutrient requirement.

This article will consider the aspects of collection and some methods of investigating two groups of these fungi having relatively wide mycological interest, the freshwater and marine Pyrenomycetes and other ascocarpic forms. Many phases of investigation relating to these fungi are applicable to terrestrial Ascomycetes as well. Methods depart only in aspects of collection and culturing techniques.

Collection and Culture. Several Ascomycetes occur as saprophytes or weak parasites infesting moribund species of freshwater or marine phanerogams. Methods of harvesting and examining these fungi are relatively simple. Plants such as bulrushes, cattails, or eelgrass afloat at the edge of a body of water are gathered either from the shore or by boat. Those rooted to the bottom of the water are collected by severing the culms with a probe and collecting them as they surface.

Marine species are also collected on driftwood along the shores or by the use of "trapping" techniques (Johnson and Sparrow, 1961). Trapping methods employ small panels of various types of wood attached in a linear fashion to a polyethylene or nylon line and submerged at or below the low tide limit for a period of 2–4 months. Once panels are harvested, they may be examined for fungal growth or incubated for an additional period of time according to methods described by Johnson et al. (1959).

With a dissecting scope and high-intensity light source, clusters of ascocarps are located on the substrate and subsequently removed with a spearhead dissecting needle. Occasionally, culms heavily infested with algae, or panels infested with fouling organisms must be scraped before the underlying ascocarps are exposed.

Fresh mounts may be made by placing specimens in a drop of fresh water (or seawater when marine species are being examined). Semipermanent mounts are made by placing ascocarps into a drop of lactophenol (phenol crystals 20 g; lactic acid 20 g; glycerol 40 g; dist. water 20 g; cotton blue or acid fuchsin .01 g) or Hoyer's medium. Hoyer's medium is made by soaking 30 g of flake gum arabic in 50 ml of distilled water for 24 hr, dissolving 200 g of chloral hydrate into the mixture, and then stirring in 20 ml of glycerol. The mixture is ready to use after settling. Semipermanent mounts can be made by soaking ascocarps in a drop of 2 % KOH on a slide, blotting up excess liquid, and adding a drop of Hoyer's medium. By lightly tapping the coverslip with the handle end of a dissecting needle or the erasure end of a pencil, the ascocarps are crushed and the centrum exposed. Spores of several of the marine species have gelatinous appendages which are best observed in a seawater mount under reduced light intensity. These processes are deliquescent, in most cases, short-lived, and unfortunately not retained satisfactorily in any known mounting medium.

Pure cultures are initiated by seeding spores or centrum cells from several ascocarps onto low nutrient level, buffered seawater agar media and incubating at room temperature. White birch applicator sticks, balsa strips, filter paper, or toweling paper may serve as the substrate. The most useful account of the isolation and culture of lignicolous marine fungi appeared in a paper by Kirk (1969).

Embedding and Sectioning. To prepare specimens for embedding, ascocarps are killed and fixed for a minimum of 24 hr in 4% formalin (seawater formalin with marine forms). The fructifications are then transferred to a depression plate and washed in fixative. Foreign matter such as wood fragments, algae, and calcareous remains, in the case of marine collections, are teased free from the venters with a micropipette. Because of their minute size, the fruiting bodies are best left in the depression plates during the dehydration series. The various concentrations of alcohol are then siphoned from, or added to the plates containing the fructifications. The subsequent dehydration process, including the preliminary infiltration of the specimens with paraffin, may follow the standard tertiary butyl alcohol series outlined by Johansen (1940).

The embedding process follows the methods outlined by Cavaliere (1966). Depression plates are rubbed with glycerol (to prevent paraffin from adhering to them), a 3–4 mm square of colored tin foil is placed in the bottom center of each depression, and a few drops of melted paraffin are poured into the plate. With the aid of a flamed micropipette and dissecting needle, 15–25 ascocarps are transferred from melted paraffin (terminal step in the dehydrating series) to the depression. The venters are arranged on their sides above and within the visible borders of the tin foil. The fruiting structures are then covered with additional paraffin and the plates allowed to cool. When the paraffin form is removed from the depression plate, the foil indicates the exact area containing the ascocarps and serves as a guide while trimming and mounting the paraffin on a wooden block.

The ascocarps of several genera are hyaline and virtually impossible to detect once placed into the fixative. In such cases, the venters may be stained in one of several alcoholic stains. A 0.25% solution of erythrosin in 25% alcohol renders the ascocarps quite visible throughout the additional dehydrating process.

Sections are best cut at 10 μ and stained. Delafield's hematoxylin for three minutes serves well as a cytological and nuclear stain. Sections are then dehydrated and mounted in either balsam or one of the several commercially prepared mounting media.

Characteristics of the venter and neck, the centrum structure, and spores are best studied in sections cut longitudinally through the neck and venter. This prevents possible misinterpretations of structures caused by oblique sections.

A. R. CAVALIERE

References

Cavaliere, A. R., "Marine Ascomycetes: ascocarp morphology and its application to taxonomy. I," *Nova Hedwigia*, **10**:387–398 (1966).

Johnson, T. W. Jr., H. A. Ferchau, and H. S. Gold, "Isolation, culture, growth and nutrition of some lignicolous marine fungi," *Phyton, Int. J. Exp. Bot.,* **12**:65–80 (1959).

Johnson, T. W., Jr., and F. K. Sparrow, Jr., in "Fungi in Oceans and Estuaries," J. Cramer, ed., Weinheim, 1961 (668 pp).

Kirk, Paul W., Jr., "Isolation and culture of lignicolous marine fungi," *Mycologia*, **61**:174–177 (1969).

See also: FUNGI, MYXOMYCETES, SAPROLEGNIALES, YEAST.

ATMOSPHERIC MICROSCOPY

The microscopic examination of dusts began with the invention of the compound microscope in 1590. Nearly all of the early use of the microscope was biological, but the early microscopists were forced to learn to identify and reject the dust particles that inevitably invaded their samples. However, there appears to have been little interest in systematic dust microscopy until the nineteenth century.

Darwin, during the voyage of the *Beagle*, noted that even far out at sea the ship occasionally became dusty. He swept up dust samples from the deck and collected dust from the air using strips of gauze attached to the ship's wind vane. Darwin sent these samples to Professor Ehrenberg in Vienna (Darwin, 1909). Ehrenberg reported that the dust appeared to have come from the Sahara (Ehrenberg, 1844; Meldau, 1962). Ehrenberg's identification of the particles suggests that a body of knowledge already existed concerning particle morphology; however, no written documentation of this can be found.

Since 1832, when Darwin made his collections, the microscopy of atmospheric particles has undergone considerable development and improvement. Great effort has been expended on the development of nonmicroscopic techniques as well, but the microscope still provides information unattainable by any other means.

Sampling. In some gross analogy to Heisenberg's uncertainty principle, the act of sampling invariably modifies the sample. The problem may be treated in two ways. Either the modification must be minimized and allowed for, or the modification can be used and accentuated for analytical purposes.

The essence of successful microscopy, in this field as in others, is in sample preparation. Lodge (1962) provided a useful classification of sampling techniques for atmospheric particles. He noted, first, that sampling techniques could be regarded as pure or impure. A pure technique involves only one basic physical principle (e.g., gravitational sedimentation), whereas an impure technique involves many. Filtration, for example, may involve as many as a half-dozen separate phenomena. Furthermore, the mechanisms can be classified into slightly overlapping categories of gravitational-inertial methods, diffusional methods, gradient methods, and sieving methods. Filtration, which is extremely impure, involves all of these principles and requires a separate category.

Gravitational and inertial methods are lumped together because of the well known principle of relativity that within a system gravity and acceleration are indistinguishable. Gravity is a significant natural mechanism for removing particles from the air, and Darwin, as already noted, made early use of it. Subsequent reports span the entire period from Darwin to the present. Miquel (1879) used a clean glass plate as the collecting surface, as did Paxton (1951). An oil coating to retain the particles was added by Leifman (1907) and by Mitchell (1914). Tissandier (1880) spread sheets of paper; Burke (1953) and Kalmus (1954) substituted membrane filter; and paper with an adhesive surface has been widely used (see e.g., Gruber and Jutze, 1957). Use of a standardized glass jar as a receptacle seems to have been initiated by the Miner's Phthisis Prevention Committee of South Africa (1919), and such jars, with modifications, are still in use (see e.g., Degroot et al., 1966). The need for some liquid in the jar is still debatable (Nader, 1958); this probably depends on jar geometry and local wind velocity. Dustfall devices provide qualitative, but not quantitative, information on nearby dust sources.

A much more sophisticated gravitational sampler was developed by Lapple (cited by Robinson et al., 1959). Air was drawn into and flowed uniformly through a horizontal channel with a very large ratio of width to height. Large particles settled out almost at once, while progressively smaller particles traveled greater distances down the channel. In this way, sampling was combined with separation of particles according to settling velocity and thus, approximately, to size. (In view of the shape of the apparatus, it is not surprising that it was nicknamed the "angleworm kennel.")

This device was limited to the sampling of particles larger than a few micrometers by the slow sedimentation of smaller particles. This limitation can be overcome if gravity is augmented by wrapping the channel around a cylinder or cone which can then be rapidly rotated. Several geometries have been used: the Conifuge of Sawyer and Walton (1950); the Aerosol Spectrometer of Goetz and Preining (1960); and unnamed devices developed recently by Flachsbart and Stöber (1969) and by Hochrainer and Brown (1969). For some types of studies, it is useful to have particles separated according to size. For many types of microscopy, however, it is far easier to collect all particles in a restricted area so that, to the greatest extent possible, every field of view contains representatives of all particle sizes.

The centrifugal devices clearly depend on the inertia of the particles as do numerous other devices in which the principal mechanism of collection is the ability of air molecules to change direction more rapidly than larger particles. Brown (1965) states that priority in this discovery should go to Miquel (1879). Actually, inertia was the principal collection mechanism of the gauze used by Darwin, so the point is in doubt. The first widely used device involving change of airflow direction was the Kotzé konimeter, developed for the Miner's Phthisis Prevention Committee of South Africa (1919). A jet of air was drawn through an orifice and impinged on an adhesive-coated glass slide. The impinging stream changed direction by at least 90°, and particles down to a few tenths of a micrometer, unable to follow the abrupt direction change, were caught in the adhesive. The original device was subsequently used by Brown and Schrenk (1942), and improved (or at least modified) versions have been described by Hill (1917), Lehman and Lowe (1934), Hasenclever (1954), and Rowley and Jordan (1943). Studies of the efficiency of the device have been made by Greenburg (1925), Katz et al. (1926), Davies et al. (1951), Beadle (1951), Ranz and Wong (1952), and numerous subsequent investigators.

The basic impactor was also modified to permit the sampling of smaller particles by the addition of a layer of moist blotting paper in the inlet tube. Because of the added humidity, water condensed on the particles as they left the jet and increased their size. The penalty, of course, was a very wet sample and possible changes in chemistry and size distribution. Impactors employing this principle were developed by Owens (1922, 1923), Gurney et al. (1938), and Hatch and Thompson (1934). Studies of the efficiency of this device were made by Davies et al. (1951) and Drinker and Hatch (1954).

Any impactor has a threshold particle size below which it does not collect particles and a region above the threshold in which collection efficiencies are low. May (1945) conceived the idea of putting several impactors of increasing efficiency in series. The first, with a low jet velocity, collects only the largest particles. Those particles that escape are passed on to a second stage with a higher jet velocity, which takes a smaller size fraction, etc. This was the first cascade impactor. Numerous modifications have been made to this impactor, including those of Sonkin (1946), Laskin (1950a), Wilcox (1953), May (1956), Anderson (1958), Mitchell and Pilcher (1959), and Lippman (1961). A single-stage impactor which nevertheless accomplishes significant size classification was described by Dessens (1961).

Impactors have the advantage of collecting a sample directly on a microscope slide or other substrates suitable for immediate examination. However, many impactors require an adhesive surface that may inhibit subsequent microscopic manipulation, and all can give misleading results if a large sample is collected. If the collected sample becomes thick enough, additional particles strike previously collected particles rather than the sampling surface and have a high tendency to bounce off and be carried on to the next stage. Finally, the extremely rapid velocity changes, including the final abrupt landing on the collection surface, tend to deaggregate or shatter fragile particles (Beadle, 1951).

The many devices which collect particles by bringing them into contact with liquid can probably be classified as inertial, although next to filtration, this type of sampler is probably the least pure. Early forms simply bubbled the sampled air through the liquid, with varying degrees of success. Early users of the technique was Tissandier (1874) Hill (1917), Meyer (1921), and Naeslund (1932). A scrubber in which the liquid was sprayed, presumably into better contact with the particles, was developed by Palmer (1916); this method was used for a time by the American Public Health Association as a standard sampling technique.

Winslow (1909) suggested that the collection efficiency of liquid devices could be improved if the incoming air made forceful contact with a submerged surface. Modifi- (1913) and the Committee for the Investigation of Atmospheric Pollution (1916). Finally, Greenburg and Smith (1922) developed a version in which a glass plate was suspended a short distance below the intake nozzle; they called the device an impinger. The Greenburg-Smith impinger became very popular and was subsequently used by many workers including Greenburg (1932), Greenburg and Bloomfield (1932), DallaValle (1937), and Brown and Schrenk (1938). Katz et al. (1926) and Hatch et al. (1932) published studies of the efficiency of the impinger. The possible shattering of aggregates and fragile particles was investigated by Ficklen and Goolden (1937), Watson (1939), Beadle (1939), Silverman and Franklin (1942), and Anderson (1939). Littlefield et al. (1937) developed a much smaller apparatus in which the sample was impinged onto the flat bottom of the liquid container rather than onto an integral target which was part of the jet. This apparatus, called the midget impinger, is in wide use today.

The impinger has been shown by Schadt and Cadle (1957) to have for small particles a rather low and unpredictable collection efficiency which depends on the wetability of the particles. Nevertheless, it has been so widely used that numerous standards are based on results obtained with it, and its continued use is highly likely. For microscopy it is often unsatisfactory, since collected particles may well dissolve in the collecting medium.

Difficulty has been experienced, because of the impact

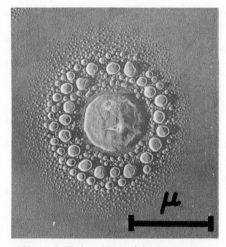

Fig. 1 Sulfuric acid special appearance.

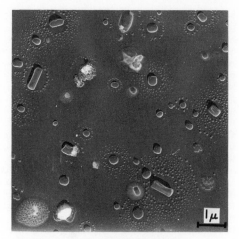

Fig. 4 Sulfuric acid (Panama).

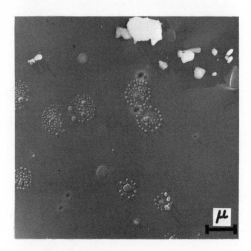

Fig. 2 Sulfuric acid (Antarctic).

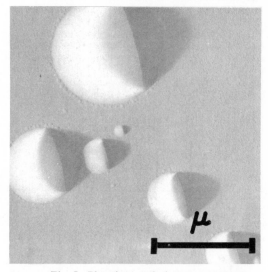

Fig. 5 Phosphate typical appearance.

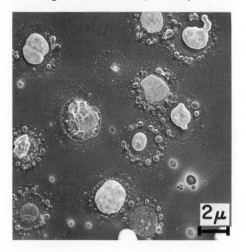

Fig. 3 Sulfuric acid (Halemauman Volcano fume, Hawaii).

of the particles on the target, in recovering viable micro-organisms from impinger collections. On the other hand, the early versions which lacked a target had even poorer collection efficiencies. An apparently satisfactory compromise was developed by Shipe et al. (1959), Tyler and Shipe (1959), and Tyler et al. (1959) in which the air sample is brought into the sampling liquid tangentially, causing rotation of the sample, which greatly prolongs the contact time between sample and liquid. The device does not seem to be in present use.

Diffusional sampling means, simply, the collection of small particles by virtue of their brownian motion. This is an important mechanism in the operation of membrane filters, but no sampler based explicitly on diffusion appears to be in use for purposes of microscopy.

Gradient sampling methods use the displacement of particles by a gradient in various properties. Photophoresis, the movement of particles in an intense light beam (gradient of illumination), and diffusiophoresis, motion caused by a gradient in the partial pressure of a vapor, are little used, although the latter was probably on operating mechanism of the "ball condenser" of Quitman and Cauer (1939) cited below.

Thermophoresis is much more widely used. When particles pass through a strong thermal gradient, they tend to move away from the hotter and towards the colder region. The phenomenon was first noted by Aitken (1883–84), and his experiments were later repeated by Bancroft (1920). The first actual sampling by this means appears to have been done by Whytlaw-Gray and Lomax (cited by Green and Watson, 1935); their apparatus was modified by Green and Watson (1935), and a further modification was patented by Whytlaw-Gray et al. (1936). Numerous modifications have been made to the original instrument, including provisions for moving the collecting surface to obtain a time record, for oscillating the collecting surface to get a more even deposition, and for increasing the air-flow. Among the modifications are those of Donoghue (1953), Walkenhorst (1955), Burdekin and Dawes (1956), Beadle and Kitto (1952), Hasenclever (1962), Cember et al. (1953), Laskin (1950b), Lauterbach et al. (1952), Orr and Martin (1958), Balashov et al. (1961), Watson (1936–37), Hamilton (1956), Bredl and Grieve (1951), Kethley et al. (1952), and Wright (1953). Quitman and Cauer (1939) and Lodge and Tufts (1955) created thermal gradients by chilling the collecting surface rather than by electrically heating the repelling element of the sampler.

Collection efficiency studies have been made by Walton et al. (1947), Schadt and Cadle (1957), and Watson (1958). For reasons that are not completely understood, the thermal precipitator is highly efficient at all particle sizes. Its chief drawback is its low sampling rate. Cartwright (1954) and Billings et al. (1961) reported that deposition on electron microscope grids was uneven, presumably because of temperature gradients between the grid wires and the center of the grid holes. The present authors have failed to observe this effect on silicon monoxide coated grids, probably because of the conductivity of the sub-strate. Questions of errors introduced by a particle over-lap have been discussed by Hasenclever (1955), Ashford (1960), and Dawes and Maguire (1960).

If particles carry, or can be given, an electric charge they will move in a potential gradient, and electrostatic forces are generally stronger than thermal ones. The typical electrostatic precipitator consists of a source of electrons, usually a wire or sharpened point maintained at a high negative potential, and two parallel conducting plates of opposed electrical charge. Alternatively, the negative or repelling electrode may be a central cylinder surrounded by a tube which constitutes the collecting surface. This geometry is possible since the corona tends to charge all particles negatively.

Probably the simplest electrostatic precipitator for purely qualitative purposes is a plastic rod charged by

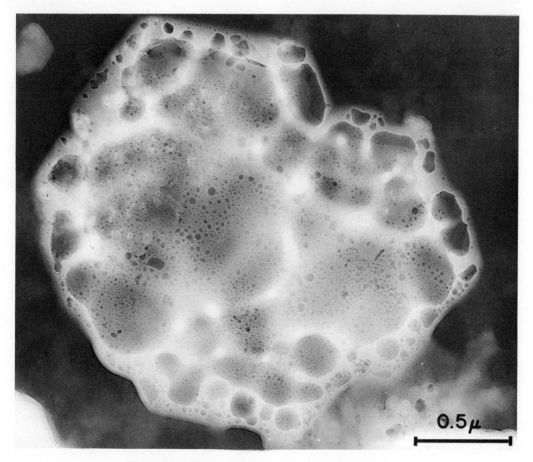

Fig. 6 Phosphate particle (Carribean coast of Panama).

rubbing with a suitable cloth. This technique has actually been used to sample bacteria in operating rooms by O'Connell et al. (1960). William Gilbert's *De Magnete* provides evidence of some knowledge of the phenomenon in the seventeenth century, but Hohlfeld (1824) seems to have priority in actual use of the effect (White, 1963). Sir Oliver Lodge (1886) discussed an early electrostatic precipitator for smoke control, but the technique was not in regular use until Cottrell (1911) built the first commercial electrostatic precipitator as an air cleaner. Precipitators were subsequently used for sampling particles by Tolman et al. (1919), Bill (1919), Lamb, et al. (1919), Shibusawa and Niwa (1920), Drinker et al. (1923), Drinker and Thomson (1925), Drinker and Hatch (1936), Barnes and Penney (1938), Lea (1943), Clayton (1947), and Brown et al. (1951). Modifications to increase the convenience of precipitators were introduced by Schadt et al. (1950), Beadle et al. (1954), Hosey and Jones (1953), and Binek et al. (1963).

The precipitator is not convenient for direct collection on either microscope slides or electron microscope grids, since the collecting surface must conduct electric current. Techniques for getting samples onto electron microscope grids were outlined by Riedel (1943), Lipscomb et al. (1947) Fraser (1956), Thürmer (1960), Billings et al. (1961), Walton et al. (1947), and Walkenhorst (1955). Instru-

ments specifically designed to permit collection on surfaces suitable for microscopy have been designed by Woolrich (1961), Robinson (1961), Billings and Silverman (1962a), Mercer et al. (1963), Binek et al. (1963), and Liu et al. (1967). The efficiency of the precipitator was the subject of investigation by Lapple (1951) Friedlander et al. (1952), White and Baxter (1960), Underwood (1962), Mercer (1957), and Penney (1962).

The electrostatic precipitator is nearly ideal for microscopy if techniques are used to permit collection on suitable substrates and if allowances are made for the fact that most precipitators collect small particle with higher efficiencies than larger ones. Sampling rates are high, and particles are handled quite gently in comparison with inertial devices. The principal disadvantage is the considerable quantity of ozone generated by the corona, which may modify the chemistry of the collected particles, particularly organic substances.

Sieving occurs when the particles are collected on a porous surface because the pores are smaller than the particles. It is an important collection mechanism for many types of membrane filters.

As noted above, filtration is an extremely impure, albeit useful, method of sampling. The collection mechanism depends upon the particular type of filter. Large particles may be collected by sieving, smaller ones by

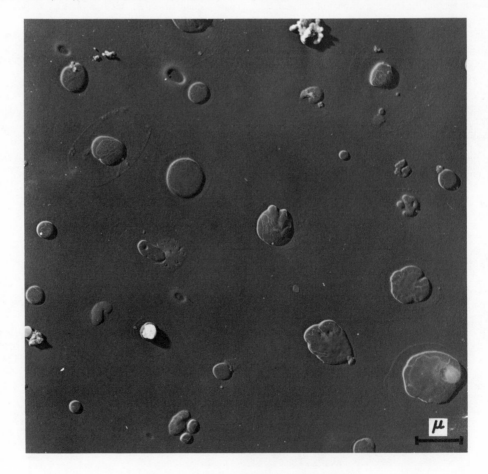

Fig. 7   Ammonium sulfate (Ft. Sherman).

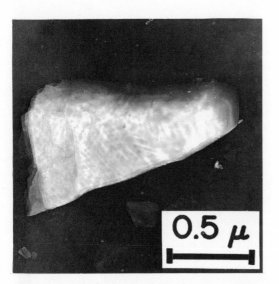

Fig. 8   Ammonium sulfate (Halemauman Volcano).

impaction or other inertial phenomena, and the smallest by diffusion or electrostatic precipitation—some types of filters accumulate very strong static charges.

It was discovered very early (Coulier, 1875; Ditman, 1912) that a tube stuffed with cotton wool was an efficient filter for airborne particles and the collected sample could be analyzed gravimetrically. This system was poorly suited to microscopy, and attempts were made to use paper filters instead of cotton wool (Möller, 1894; McKay 1922). However, paper surface is a poor substrate for microscopy, and the recovery of particles from a paper surface is difficult. If the particles under study are insoluble, a soluble filter material can be used and subsequently removed. Frankland (1886) appears to have been the first to draw an air sample through a deep bed of sugar. Nitrocellulose (Fritzsche, 1898), collodion (Hahn, 1908), and salicylic acid (Briscoe et al., 1936) were subsequently used. Of these, sugar was the most widely used, possibly because the particles of interest were insoluble in water and because sugar can be contained in remarkably high purity. Beginning in 1911 with the Miner's Phthisis Prevention Committee of South Africa (Miner's Phthisis Prevention Committee of South Africa, 1916 and 1919) a tube packed with sugar became a standard filtration

Fig. 9   Calcium sulfate (Allrook Forest).

method for industrial hygiene applications. In the United States, the technique was used by Lanza and Higgins (1915) in their study in Joplin, Missouri, and subsequently by Fieldner et al. (1921).

All of these techniques left the particles in a solution, which might or might not be desirable. Matthews and Briscoe (1934) suggested the use of volatile substances such as anthracene, benzoic acid, or naphthalene, which would evaporate completely after serving as the filter. Holt (1951a, 1951b, 1951c) subsequently used discs of sintered naphthalene which were coherent and consequently could be handled more easily than a long glass tube full of loose crystals. In recent years, all of these techniques seem to have lost favor.

Still other investigators have attempted to use soxhlet thimbles for particle collection (The Chicago Association of Commerce Committee of Investigation on Smoke Abatement and Electrification of Railway Terminals, 1915; Trostel and Frevert, 1923; Bloomfield and Dalla-Valle, 1935). Thimbles of paper and of more refractory materials are still used for gravimetric collection of particles, particularly in sampling from concentrated sources, but the thimble geometry is totally unsuited to good microscopy.

The needs of World War II led to the development of high-efficiency fibrous filters of a variety of materials (Green, 1936; Green and Thomas, 1955; Stafford and Smith, 1951; Billings and Silverman, 1962b; Blasewitz and Carlisle, 1951; Sehl and Havens, 1951; Billington and Saunders, 1947; Thuman, 1959; Cadle and Thuman 1960; Thomas, 1952; and Van Orman and Endres, 1952). The materials of these filters were increasingly exotic, ranging from mixtures of natural organic fibers and asbestos through stainless steel wool to highly purified fibrous polystyrene. The war also led to the perfection of the cellulose ester membrane filter (Goetz, 1953; First and Silverman, 1953), which was finally a filter material truly suited to microscopy. The membrane filter was an aerogel of tortuous pores with a very high porosity and with surface openings small enough to capture by sieving most particles of light microscopic size. The filter was relatively flat and could be made optically transparent by filling its pores with a liquid of the same refractive index. Fortunately, this was very close to the refractive index of microscope immersion oil. Finally, a filter, which for many purposes was still better, was recently discovered and given the trade name "Nuclepore" (Price and Walker, 1962; Fleischer et al., 1965). This material has the disadvantage of a certain amount of stress birefringence, but is exceedingly flat and is transparent without further treatment (Spurny et al., 1969; Frank et al., 1970).

For special purposes, use has been made of deliberately inefficient filter materials such as nylon mesh (Delany et al., 1967; Prospero and Bonatti, 1969) and spider silk (Dessens, 1949).

Microscopy. For the experienced microscopist, the

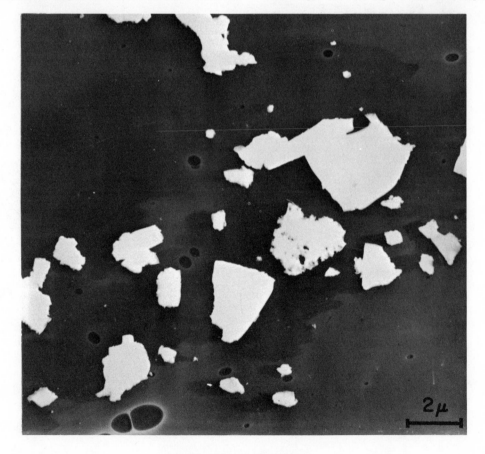

Fig. 10   Lava ash (Philippines).

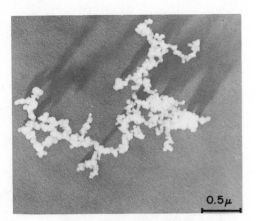

Fig. 11 Soot (Chiva Chiva).

Obviously the simplest thing to do with a collection of particles and a microscope is to use the latter to look at the former. Given adequate experience at examining known samples, a remarkably high percentage of larger atmospheric particles can be recognized. Particles in the size range of some tens of micrometers, such as are found in samples collected by sedimentation (dustfall), are often best examined with a stereoscopic microscope. Smaller particles require the larger magnifications available with single-objective instruments. McCrone et al. (1967) have compiled a sizable atlas of photomicrographs of particle species likely to be found in atmospheric dust samples.

McCrone also describes many of the techniques available to resolve close decisions. In general, these are based on the refractive index or indices, birefringence, and occasionally on interference figures, and therefore require the use of a polarizing microscope.

Cadle et al. (1953) developed a technique for sectioning particles collected on a membrane filter.

Phase contrast microscopy has been exploited extensively by Schmidt (1955, 1960), Schmidt and Heidermanns (1958, 1959), and by Beyer (1965). Dispersion staining (Crossmon, 1957, 1964, 1966; Grabar, 1962; McCrone et al., 1967) is particularly useful because it allows recognition of all particles of a given species in a mixture.

However, there are many situations in which purely

collection of a valid sample is more than half of the problem. Unfortunately, there are more people collecting samples than there are microscopists, and therefore there is a tendency to avoid microscopy and rely on bulk chemical analysis, thereby losing a great deal of information.

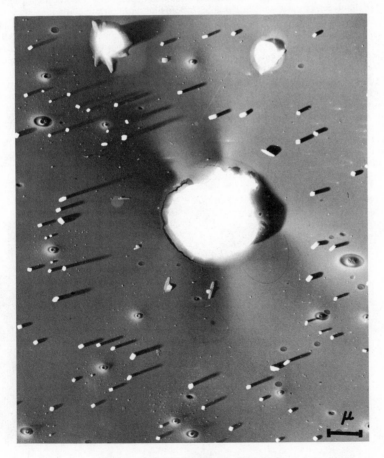

Fig. 12 Sodium chloride (Ft. Sherman).

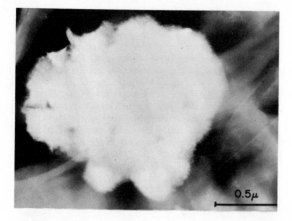

Fig. 13   Replica ammonium sulfate (polystyrene filter).

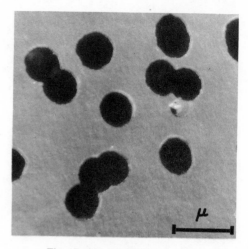

Fig. 15   Nuclepore filter.

observational methods may be inadequate or too costly. It may therefore be desirable to investigate the chemical and physical properties of the particle. The use of melting-point determinations is well known but not especially useful in the examination of inorganic salts. Woodcock among others (Woodcock and Gifford, 1949; Woodcock, 1950, 1952, 1953) used the characteristic humidity at which soluble salts deliquesce as a basis for identifying sea salt.

Characteristic chemical reactions may also be used, although they have the disadvantage of destroying the particle. The pioneers in this work were Chamot and Mason (1958), who systematized the chemical reactions that could be carried out under the microscope by adding microdrops of selected reagents to the specimen. Another early contributor was Feigl (1954), who devised or collected and systematized extremely sensitive methods of chemical analysis carried out on glass slides and on filter paper. Still another contributor was Winckelmann (1931, 1933, 1935), who detected very small amounts of iron in small drops of aqueous solution by placing them on a gelatin surface previously impregnated with a specific reagent.

However, the techniques of Chamot and Mason are more tedious than morphological identification, and those of Feigl and Winckelmann do not involve microscopy. The first workers to combine these concepts were Crozier and Seely (1950). They showed that halide-containing particles could be identified by impacting them onto a gelatin surface containing glycerol, as a moistening agent, and the reagent mercurous fluosilicate. They demonstrated that particles as small as 0.2 μm in diameter could be reorganized by the characteristic bluish-white spot of mercurous chloride. Larger particles of sodium halides also gave a central nucleus of sodium fluosilicate. Seely (1955) subsequently used impregnated gelatin layers as collecting surfaces for micrurgic tests similar to those of Chamot and Mason. Other techniques with preimpregnated gels were described by Fedele and Vittori (1953), Vittori (1954, 1955), Lodge and Fanzoi (1954), Farlow (1954, 1957a, 1957b), Gerhard and Johnstone (1955), Lodge et al. (1960), Lodge and Frank (1962), and Anyz (1966).

The gelatin method has the advantages that generally little further manipulation is necessary once the sample is

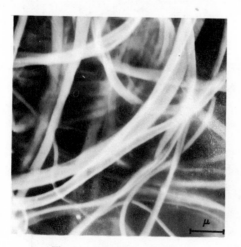

Fig. 14   Polystyrene filter.

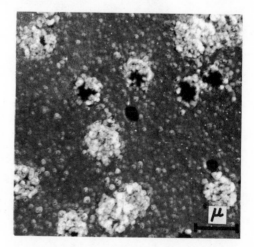

Fig. 16   Filtration mechanism on nuclepore filter.

taken and that a gel sensitized for a given species gives a unique reaction which is easy to recognize. The disadvantages are the extreme ease of contamination, the sensitivity of the gel to heat, and in many cases, the impermancy of the samples. For example, Pidgeon (1954) showed that Seely's chloride reaction spots rapidly peptized and vanished; it is almost always mandatory to examine collections within an hour after sampling if the smallest particles are to be recognized.

Pidgeon also followed up on the early work of Seely (1952) which had suggested that original particle size (more correctly, the mass of reacting ion) could be inferred from the size of the reaction spot which resulted.

Lodge (1954) demonstrated that similar tests could be carried out on particles previously collected on membrane filters. If the filter were placed on a suitable reagent solution, the reagent would diffuse up through the pores to cause the characteristic reaction on the surface of the filter. The pore structure, like the gelatin, inhibited lateral diffusion to localize the reaction in the vicinity of the reacting particle. The consequences were then examined microscopically after making the filter transparent with immersion oil. The work was extended by Tufts (1959a, 1959b, 1960). Frank et al. (1970) show that at least some of the same tests could be run on Nuclepore filters. A method for preimpregnating a filter for the specific detection of sulfate was reported by Lodge and Parbhakar (1963) and improved by Lodge and Frank (1966). These methods have had substantial analytical use in the laboratory (Lodge and Baer, 1954; Lodge et al., 1954) and in the field (Lodge, 1955; Lodge and Pate, 1966; Byers et al., 1957). Lodge et al. (1956) demonstrated that the relationship on the gelatin between particle size and reaction spot size also existed on the membrane filter. Hence the technique can be used for both identification and determination of a measure of size distribution.

The early use of the electron microscope recapitulated the history of light microscopy; it too was dominated by biologists. However, carbon black was examined at a very early date (Columbian Carbon Company Research Laboratories, 1942; Anderson and Emmett, 1948). On the basis of this work McCabe et al. (1949) recognized soot as an ingredient of industrial haze. Sodium chloride was recognized by its cubic shape and its volatility under the electron beam (Burton et al., 1947; Watson and Preuss, 1950; Lodge and Tufts, 1955; Balk and Colvin, 1961). Burton et al. (1947) investigated the behavior of several salts in the electron beam. Bush (1955–56) and Cartwright et al. (1956) attempted to characterize atmospheric particles in terms of their destructibility in the electron microscope.

Possibilities of morphological identification seem limited under the electron microscope because of the absence of the clues of color, birefringence, etc. which are so prominent in light microscopy. However, the number of species present in the atmosphere in high concentration is rarely large, and some species may be eliminated from consideration by knowledge of the nature of the sampling location. Sulfuric acid is particularly easy to recognize (Fig. 1). It was first noted by Waller et al. (1963); Frank and Lodge (1967) repeated the work and extended it to a number of other substances which can be identified at sight. Sulfuric acid is particularly ubiquitous and has been found in the Antarctic (Fig. 2), Hawaii (Fig. 3), and Panama (Fig. 4). Phosphoric acid is also readily recognized; Fig. 5 shows the foamy structure characteristic of phosphates. Figure 6 displays a phosphate particle collected near the Caribbean coast of Panama. Other recognizable species are ammonium sulfate (Figs. 7 and 8) calcium sulfate (Fig. 9), lava (Fig. 10), soot (Fig. 11), and sodium chloride (Fig. 12).

Within limitations, physical and chemical changes may also be used for identification in electron microscopy. Kumai (1951), Kuroiwa (1953), and Nakaya (1957) examined particles, exposed them to air saturated with water vapor, and reexamined them to distinguish between hydroscopic and nonhydroscopic particles. Monkman (1955) treated particle collections with various gaseous reagents such as hydrogen chloride and observed the resulting changes. Tufts and Lodge (1958) used chemical reactions, analogous to those previously used on membrane filters, to identify chloride and sulfate particles. Koenig (1959) and Steele and Sciacca (1966) identified silver iodide by treating it with photographic developer, which causes the formation of a characteristic minute ribbon of silver. Löffler and König (1955) noted that vacuum-evaporated zinc or cadmium would not deposit on a freshly prepared silicone monoxide substrate. How-

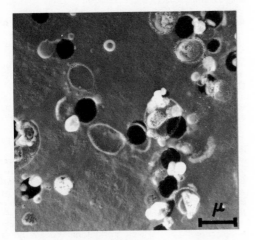

Fig. 17   Particles on nuclepore filter.

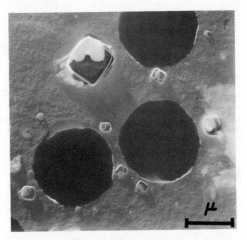

Fig. 18   Particles (Hawaii).

ever, if the substrate was brought into contact with liquid water, the metals would deposit on the areas which had been moistened. This was made the basis for distinguishing between particles collected dry and those which were collected as aqueous droplets.

Electron microscopy, no matter what the final technique is to be, is simplest if collections are made directly on the substrate mounted on a grid. However, sometimes samples are presented which are not so conveniently arranged. Fraser (1953), Kalmus (1954), and Schlipköter et al. (1959) have described techniques for transferring particles from membrane filters to electron microscope grids. However, none has been completely satisfactory. Frank and Lodge (1966) managed to replicate and extract a few large particles collected on fibrous polystyrene filters (Fig. 13). The same technique also made visible the filter structure (Fig. 14). Particles captured on Nuclepore filter surfaces are easily retrieved by a replication technique (Spurny et al., 1969; Frank et al., 1970). The technique has been used to study filter structure (Fig. 15), in laboratory investigations of filtration mechanisms (Fig. 16), and in field studies involving collections of particles (Figs. 17 and 18).

EVELYN R. (FRANK) ACKERMAN
JAMES P. LODGE JR.

References

Aitken, J., "On the formation of small clear spaces in dusty air," *Trans. Roy. Soc. Edinburgh,* 32(Part 2): 239–272 (1883–84).

Anderson, A. A., "New sampler for the collection, sizing, and enumeration for atmospheric particles," *J. Bacteriol.,* 76:471–84 (1958).

Anderson, E. L., "The determination of industrial gas dispersoids," *J. Ind. Hyg. Toxicol.,* 21:122 (1939).

Anderson, R. B., and P. H. Emmett, "Measurement of carbon black particles by the electron microscope and low temperature nitrogen absorption isotherms," *J. Appl. Phys.,* 19:367–373 (1948).

Anyz, F., "A contribution to the detection of the aerosols containing sulfate particles," *Tellus,* XVIII(2):216–220 (1966).

Ashford, J. R., "Some statistical aspects of dust counting," *Brit. J. Appl. Phys.,* 11:13–21 (1960).

Balashov V., J. G. Brading, and R. E. G. Rendall, "Dust sampling instruments for metalliferous mines," *J. Mining Vent. Soc. South Afr.,* 14:98–100 (1961).

Balk, P. and J. Colvin, "Note on an indirect measurement of object temperatures in electron microscopy," *Kolloid Z.,* 176:141 (1961).

Bancroft, W. D., "Thermal filters," *J. Phys. Chem.,* 24:421–436 (1920).

Barnes, E. C., and G. W. Penney, "An electrostatic dust weight sampler," *J. Ind. Hyg. Toxicol.,* 20:259 (1938).

Beadle, D. G., "The shattering of dust particles by the inpinger," *J. Ind. Hyg. Toxicol.,* 21:109–120 (1939).

Beadle, D. G., "An investigation of the performance and limitations of the konimeter," *J. Chem. Met. Mining Soc. South Afr.,* 51:265–283 (1951).

Beadle, D. G., and P. H. Kitto, "A modified form of thermal precipitator," *J. Chem. Met. and Mining Soc. South Afr.,* 52:284–311 (1952).

Beadle, D. G., P. H. Kitto, and P. J. Bliqnaut, "Portable electrostatic dust sampler with electronic air flow," *A.M.A. Arch. Ind. Hyg. and Occup. Med.,* 10(5): 381–389 (1954).

Beyer, H., "Theorie und Praxis des Phasenkontrastverfahrens," *Akad. Verlaggsgesellschaft Geest und Portig K. G., Leipzig* (1965).

Bill, J. P., "The electrostatic method of dust collection as applied to the sanitary analysis of air," *J. Ind. Hyg.,* 7:223–342 (1919).

Billings, C. E., W. J. Megaw, and R. D. Wiffen, "Sampling of submicron particles for electron microscopy," *Nature,* 189:366 (1961).

Billings, C. E., and L. Silverman, "Aerosol sampling for electron microscopy," *J. Air Pollution Control Ass.,* 12 (12):586–590 (1962a).

Billings, C. E., and L. Silverman, "High temperature filter media performance with shock wave cleaning," *Int. J. Air Water Pollution,* 6:455–466 (1962b).

Billington, N. S., and D. W. Saunders, "Air filtration," *Institution of Heating and Ventilating Engineers,* 15: 46–95 (1947).

Binek, B., K. Spurny, and J. Pixova, "Elektrostaticke impaktory k zachycovani Vzorku aerodisperznich skodlivin," *Pracovni Lekarstvi,* 15:415–419 (1963).

Blasewitz, A. G., R. V. Carlisle, and others, "Filtration of radioactive aerosols by glass fibers," U.S. Atomic Energy Commission, Hanford Works, HW-20847, April 1951.

Bloomfield, J. J., and J. M. DallaValle, "The determination and control of industrial dust," U.S. Pub. Health Service Bull. 217, 1935.

Bredl, J., and T. W. Grieve, "Thermal precipitator for bulk samples," *J. Sci. Instr.,* 29:21–23 (1951).

Briscoe, H. V. A., J. W. Matthews, P. F. Holt, and P. M. Sanderson, "Porous solid filters for sampling industrial dusts," *Trans. Inst. Mining Met.,* 46:145–153 (1936).

Brown, C. E., and H. H. Schrenk, "A technique for use of the impinger method," U.S. Bureau of Mines Inf. Circ. 7026, 1938 (p. 20).

Brown, C. E., and H. H. Schrenk, 1942. "Standard methods for measuring the extent of air pollution," U.S. Bureau of Mines Inf. Circ. 7210, May 1942.

Brown, J. K., A. D. Hosey, and H. H. Jones, "A lightweight power supply for an electrostatic precipitator," *A.M.A. Arch. Ind. Hyg. Occup. Med.,* 3:198–203 (1951).

Brown, K. M., "Evaluation of dust in mine air with particular reference to the use of the electron microscope," Doctorate Thesis, University of Wales, 1965.

Burdekin, J. T., and J. G. Dawes, "The use of a sizeselector for dust-sampling with the thermal precipitator," *Brit. J. Ind. Med.,* 13:196–201 (1956).

Burke, W. C., "Size determination of silica particles collected on membrane filters," *Amer. Ind. Hyg. Ass. Quart.,* 14:299–302 (1953).

Burton, E. F., R. S. Sennet, and S. G. Ellis., "Specimen changes due to electron bombardment in the electron microscope," *Nature,* 160:565–567 (1947).

Bush, A. F., "Strange air-borne particles," *A.M.A. Arch. Ind. Health,* No. 1, 1–2 (1955–56).

Byers, H. R., J. R. Sievers, and B. J. Tufts, "Distribution in the atmosphere of certain particles capable of serving as condensation nuclei," in "Artificial Stimulation of Rain," H. Weickmann and W. E. Smith, ed., New York, Pergamon Press, 1957.

Cadle, R. D., and W. Thuman, "Filters from submicrondiameter organic fibers," *Ind. Chem.,* 52:315–316 (1960).

Cadle, R. D., A. G. Wilder, and C. F. Schadt, "A technique for collecting, mounting, and sectioning airborne particulate material," *Science,* 118:490–491 (1953).

Cartwright, J., "The electron microscopy of airborne dust," *Brit. J. Appl. Phys.,* 5:109–120 (1954).

Cartwright, J., G. Nagelschmidt, and J. W. Skidmore, "The study of air pollution with the electron microscope," *Quart. J. Roy. Meterol. Soc.,* 82:82–86 (1956).

Cember, H., T. F. Hatch, and J. A. Watson, "Dust sampling with a rotating thermal precipitator," *Amer. Ind. Hyg. Ass. Quart.,* 14:191–194 (1953).

Chamot, E., and C. Mason, "Handbook of Chemical Microscopy," Vol. II, New York, John Wiley & Sons, Inc., 1958, 1960

The Chicago Association of Commerce Committee of Investigation on Smoke Abatement and Electrification of Railway Terminals, "Smoke Abatement and Electrification of Railway Terminals in Chicago." 227–236 (1915).

Clayton, G. D., "Improvement of the MSA electrostatic precipitator," *J. Ind. Hyg. Toxicol.*, 29:400–402 (1947).

Columbian Carbon Company Research Laboratories, *Columbian Collodial Carbon,* 2, 3 (1940, 1942).

Committee for the Investigation of Atmospheric Pollution, "First Report," suppl. to *Lancet,* 1916, 40 pp.

Cottrell, F. G., "The electrical precipitation of suspended particles," *J. Ind. Eng. Chem.,* 3, 542–550 (1911).

Coulier, M., "Sur une nouvelle propriété de l'air," *J. Pharm. Chimie,* 22:165–175 (1875).

Crossmon, G. C., "New developments in dispersion staining microscopy as applied to industrial hygiene," *Amer. Ind. Hyg. Ass. Quart.,* 18:341–344 (1957).

Crossmon, G. C., "Microscopic identification of toxic dusts," *Occup. Health Rev.,* 16(3):3–7 (1964).

Crossmon, G., "Theory and practice of dispersion staining," *Microchem. J.,* 10:273 (1966).

Crozier, W. D., and B. K. Seely, "Some techniques for sampling and identifying particulate matter in the air," *Proc. First Nat. Air Pollution Symp.,* Stranford, Calif., Stranford Press, 1950 (pp. 45–49).

DallaValle, J. M., "Note on comparative tests made with the Hatch and the Greenburg-Smith impingers," U.S. Public Health Service, Reprint #1848, 1937 (p. 61).

Darwin, C., "The Voyage of the Beagle," 2nd ed., Charles W. Cliot, L. L., ed., New York, P. F. Collier and Son, 1909.

Davies, C. N., M. Aylward, and D. Leacey, "Impingement of dust from air jets," *A.M.A. Arch. Ind. Hyg. Occup. Med.,* 4:354–397 (1951).

Dawes, J. G., and B. A. Maguire, "The thermal precipitator and the P.R.U. hand pump; a critical study," Safety in Mines Research Establishment Report No. 187, 1960.

Degroot, I., W. Loring, A. Rehm, S. Samuels, and W. Winkelstein, "People and air pollution: A study of attitudes in Buffalo, New York," *J. Air Pollution Control Ass.,* 16:245–247 (1966).

Delany, A. C., A. C. Delany, D. W. Parkin, J. J. Griffin, E. D. Goldberg, and B. E. F. Reimann, "Airborne dust collected at Barbados," *Geochim. Cosmochim. Acta,* 31:885–909 (1967).

Dessens, H., "The use of spiders' threads in the study of condensation nuclei," *Quart. J. Roy. Meteorol. Soc.,* 75:23–26 (1949).

Dessens, J., "Un capteur classeur de particules a lame unique," *J. Bull. Obs. Puy de Dome,* 1–13 (1961).

Ditman, N. E., "Quality of dust in some dusty trades," *J. Ind. Safety,* 2(3):48–53 (1912).

Donoghue, J. K., "A modification in the hot-wire thermal precipitator," *J. Sci. Instr.,* 30:59(1953).

Drinker, P., and T. Hatch, "Industrial Dust," New York, McGraw-Hill Book Company Inc., 1936; 2nd Ed. 1954.

Drinker, P., and R. M. Thomson, "Determination of suspensoids by alternating current precipitators," *J. Ind. Hyg.,* 7:261–272 (1925).

Drinker, P., R. M. Thomson, and S. M. Fitchet, "Atmospheric particulate matter: II the use of electric precipitation for quantitative determinations and microscopy," *J. Ind. Hyg.,* 5:162–185 (1923).

Ehrenberg, Ch.-G. "Über einen die ganze Luft längere Zeit trübenden Staubregen im hohen atlantischen Ocean in 70° 43′ N.B. 26° W.L. und dessen Mischung aus zahbreichen Kieselthieren." *Verh. d. Königl Preuss. Akad. der Wissenschaften zu Berlin,* 194–196 (1844).

Farlow, N. H., "An improved halide ion-sensitive sampling surface for water aerosols," *Rev. Sci. Instr.,* 25:1109–1111 (1954).

Farlow, N. H., "Quantitative determination of chloride ion in 10^{-6} to $^{-12}$ gram particles," *Anal. Chem.,* 29:883–885 (1957a).

Farlow, N. H., "Chromatographic detection of mixed halide ions in 10^{-10} gram particles," *Anal. Chem.,* 29:881–883 (1957b).

Fedele, D., and O. Vittori, "Determination delle particelle di cloruro nell' atmosfera e metodi di misura," *Riv. Metereol. Aero.,* 13(4):9–13 (1953).

Feigl, F., "Spot Tests," Vol. I, 4th English Ed., New York, Elsevier, 1954 (pp. 58–59).

Ficklen, J. B., and L. L. Goolden, *Science,* 85:587–588 (1937).

Fieldner, A. C., S. H. Katz, and E. S. Longfellow, "The sugar tube method of determining rock dust in air," U.S. Bureau of Mines. Tech. Paper 278, 1921.

First, M. W., and L. Silverman, "Air sampling with membrane filters," *A.M.A. Arch. Ind. Hyg. Occup. Med.,* 7:1–11 (1953).

Flachsbart, H., and W. Stöber, "Preparation of radioactively labeled monodisperse silica spheres of collodial size," *J. Colloid and Interface Sci.,* 30:568–573 (1969).

Fleischer, R. L., P. B. Price, and R. M. Walker, "Tracks of charged particles in solids," *Science,* 149:383–393 (1965).

Frank, E. R., and J. P. Lodge, "Characterization of polystyrene fibers by electron microscopy," *J. Microscop.,* 5(1):95–96 (1966).

Frank, E. R., and J. P. Lodge, "Morphological identification of airborne particles with the electron microscope," *J. Microscop.,* 6 (4):449–456 (1967).

Frank, E. R., K. R. Spurny, D. C. Sheesley, and J. P. Lodge, "The use of nuclepore filters in light and electron microscopy of aerosols," *J. Microscop.,* 9(6): 735–740 (1970).

Frankland, L. F., "A new method for the quantitative estimation of the microorganisms present in the atmosphere," *Phil. Trans. Roy. Soc. London,* 178: 113–152 (1886).

Fraser, D. A., "Absolute method of sampling and measurement of solid air-borne particulates," *A.M.A. Arch. Ind. Hyg. Occup. Med.,* 8:412–419 (1953).

Fraser, D. A., "The collection of submicron particles by electrostatic precipitation," *Amer. Ind. Hyg. Ass. Quart.,* 17(1):73–79 (1956).

Friedlander, S. K., L. Silverman, P. Drinker, and M. W. First, "Handbook on Air Cleaning," Washington, D.C., U.S.A.E.C., 1952.

Fritzsche, P., "Colorimetric determination of smoke density," *Z. Anal. Chem.,* 37:92–94 (1898).

Gerhard, E. R., and H. F. Johnstone, "Microdetermination of sulfuric acid aerosol, *Anal. Chem.,* 27:702–703 (1955).

Goetz, A., "Application of molecular filter membranes to the analysis of aerosols," *Amer. J. Public Health,* 43:150–159 (1953).

Goetz, A., and O. Preining, "The aerosol spectrometer and its application to nuclear condensation studies," in "Physics of Precipitation," H. Weickmann, ed., Geophysical Monograph No. 5, Publ. No. 746 of American Geophysical Union, Washington, D.C., 1960 (pp. 164–184).

Grabar, D. G., "Application of dispersion staining to microscopic identification of settled dust," *J. Air Pollution Control Ass.,* 12 (12):560–566, 592 (1962).

Green, H. L., British Patent 482,137, 1936.

Green, H. L., and D. J. Thomas, British Patent 727,975, 1955.

Green, H. L., and D. J. Thomas, British Patent 727, 975, 1955.

Green, H. L., and H. H. Watson, "Physical methods for the estimation of the dust hazard in industry," Med. Res. Coun. Sp. Rept. Ser. No. 199, London, His Majesty's Stationery Office, 1935.

Greenburg, L., "A review of the methods used for

sampling aerial dust," *U. S. Public Health Repts.* **40**(16):765–786 (1925).

Greenburg, L., "The impinger dust sampling apparatus as used by the United States Public Health Service," *U. S. Public Health Reps.* **47**(12):654–675 (1932).

Greenburg, L., and J. J. Bloomfield, "The impinger dust sampling apparatus as used by the U. S. Public Health Service," *U. S. Public Health Repts.*, **47**(12):654–675 (1932).

Greenburg, L., and G. Smith, "A new instrument for sampling aerial dust, U. S. Bur. Mines R. I., 2392, 1922 (3 pp.).

Gruber, C. W., and G. A. Jutze, "The use of sticky paper in an air pollution monitoring program," *J. Air Pollution Control Ass.*, **7**:115–117 (1957).

Gurney, S. W., C. R. Williams, and R. R. Megis, "Investigation of the characteristics of the Bausch and Lomb counter," *J. Ind. Hyg. Toxicol.*, **20**:24 (1938).

Hahn, M., "Process for the rapid determination of the dust content of technical gases," *Abstr. J. Soc. Chem. Ind.*, **27**:1144–1145 (1908).

Hamilton, R. J., "A portable instrument for respirable dust sampling," *J. Sci. Instr.*, **33**: 395–399 (1956).

Hasenclever, D., "Bestimmung des Feinstaubgehalts der Luft," *Chem.-Ingr.-Tech.*, **26**:180–187 (1954).

Hasenclever, D., "Untersuchungen über die Eignung verschiedener Staubmessgeräte zur betrieblichen Messung von mineralischen Stäuben," *Staub*, **41**:388–435 (1955).

Hasenclever, D., "Ein neuer Thermalpräzipitator mit Heizdraht und seine Liestung," *Staub*, **22**:99–102 (1962).

Hatch, T., and E. W. Thompson, "A rapid method of dust sampling and approximate quantitation for routine plant operation," *J. Ind. Hyg.*, **16**:93–99 (1934).

Hatch, T., H. Warren, and P. Drinker, "A modification form of the Greenburg-Smith impinger for field use, with a study of its operating characteristics," *J. Ind. Hyg.*, **14**:301–311 (1932).

Hill, E. V., "Quantitative determination of air dust," *The Heating and Ventilating Magazine*, **14**(6):23–33 (1917).

Hochrainer, D., and P. M. Brown, "Sizing of aerosol particles by centrifugation," *Environ. Sci. Technol.*, **3**:830–835 (1969).

Hohlfeld, M., "Das Niederschlagen des Rauches durch Electricität," *Kastner Arch. Natur.*, **21**:205–206 (1824).

Holt, P. F., "The determination of the mass concentration of air-borne dusts; electrical sampling pump for use with volatile filters," *Instr. Mining Met. Bull.*, **539**:15–20 (1951a).

Holt, P. F., "Dust in industrial atmospheres V. Determination of mass concentration by the volatile filter method," *Metallurgie*, **44**:52–54 (1951b).

Holt, P. F., "Study of dust in industrial atmospheres. IV. Determination of mass concentration," *Metallurgie*, **43**:309–310 (1951c).

Hosey, A. D., and H. H. Jones, "Portable electrostatic precipitator operating from 110 volts A-C or 6 volts D-C, *A. M. A. Arch. Ind. Hyg. Occup. Med.*, **6**:49–57 (1953).

Kalmus, E. H., "Preparation of aerosols for electron microscopy," *J. Appl. Phys.* **25**:87–89 (1954).

Katz, S. H., G. W. Smith, and W. M. Myers, "Determinations of air dustiness with the sugar tube, Palmer apparatus, and impinger, compared with determinations with the konimeter," *J. Ind. Hyg.*, **8**:300 (1926).

Kethley, T. W., M. R. Gordon, and C. Orr, "A thermal precipitator for aerobacteriology," *Science*, **116**:368–369 (1952).

Koenig, L. R., "Submicron determination of silver iodide," *Anal. Chem.*, **31**(10):1732–35 (1959).

Kumai, M., "Electron-microscope study of snow crystal nuclei," *J. Metereol.*, **8**:151–156 (1951).

Kuroiwa, S., "Electron-microscope study of atmospheric

condensation nuclei," in "Studies of Fogs," T. Hori, ed., Japan, Tanne Trading Co. Ltd., 1953 (pp, 349–382).

Lamb, A. B., G. L. Wendt, and R. E. Wilson, "A portable electric filter for smokes and bacteria," *Trans. Amer. Electrochem. Soc.*, **35**:357–369 (1919).

Lanza, A. J., and E. Higgins, "Pulmonary disease among miners in the Joplin, Mo. district and its relation to rock dust to mine," U. S. Bureau of Mines Technical Paper 105, 1915.

Lapple, C. E., "Processes use many collector types," *Chem. Eng.*, **58**:144–151 (1951).

Laskin, S., "The modified cascade impactor," Report U. R.-#129, Univ. of Rochester Atomic Energy Project, Rochester, New York, 1950a.

Laskin, S., "Oscillating thermal precipitator," Report U. R.-#126, Univ. of Rochester Atomic Energy Project, Rochester, New York, 1950b.

Lauterbach, K. E., R. H. Wilson, S. Laskin, and D. W. Meier, "Design of an oscillating thermal precipitator," Report U. R.-#199, Univ. of Rochester Atomic Energy Project, Rochester, New York, 1952.

Lea, W. L., "A portable electrostatic precipitator," *J. Ind. Hyg. Toxicol.*, **25**(4):152–156 (1943).

Lehman, H., and F. Lowe, "Selective dust sampling: modified Ziess konimeter," *Iron and Coal Trades Rev.*, **128**:840 (1934).

Leifman, K., "Über den Nachweis von Russ in Luft," Habilitationschrift, University of Halle, 1907 (pp, 3–31).

Lippmann, M., "A compact cascade impactor for field survey sampling," *Amer. Ind. Hyg. Ass. J.*, **22**:348–353 (1961).

Lipscomb, W. N., T. R. Rubin, and J. H. Sturdivant, "An investigation of a method for the analysis of smokes according to particle size," *J. Appl. Phys.*, **18**:72–79 (1947).

Littlefield, J. B., F. L. Feicht, and H. H. Schrenk, "Bureau of Mines midget impinger for dust sampling," U. S. Bureau of Mines Report Invest. 3360, 1937.

Liu, B. Y. H., K. T. Whitby, and H. S. Yu, Henry, "Electrostatic aerosol sampler for light and electron microscopy," *Rev. Sci. Instr.*, **38**(1):100–102 (1967).

Lodge, J. P., "Analysis of micron-sized particles," *Anal. Chem.*, **26**:1829–1831 (1954).

Lodge, J. P., "A study of sea-salt particles over Puerto Rico," *J. Metereol.*, **12**(5):493–499 (1955).

Lodge, J. P., Jr., "Identification of aerosols," in "Advances in Geophysics," vol. IX, H. E. Lansberg and J. Van Mieghem, ed., New York, Academic Press, 1962 (pp. 97–130).

Lodge, J. P., and F. Baer, "An experimental investigation of the shatter of salt particles on crystallization," *Anal. Chem.*, **11**:420–421 (1954).

Lodge, J. P., and H. Fanzoi, "Extension of the gelatine method for the detection of micron-sized particles," *Anal. Chem.*, **26**:1829 (1954).

Lodge, J. P., J. Ferguson, and B. Havlik, "Analysis of micron-sized particles, determination of sulfuric acid aerosol," *Anal. Chem.*, **32**:1206–1207 (1960).

Lodge, J. P., and E. R. Frank, "The detection and estimation of particulate formaldehyde," *Proc. 1st Czech. Conf. Aerosols*, 169–171 (1962).

Lodge, J. P. and E. R. Frank, "An improved method for the detection and estimation of micron-sized sulfate particles: Correction," *Anal. Chim. Acta*, **35**:270–271 (1966).

Lodge, J. P., J. E. McDonald, and F. Baer, "An investigation of the Melander effect," *J. Meteorol.*, **11**:318–322 (1954).

Lodge, J. P., and K. J. Parbhakar, "An improved method for the detection and estimation of micron-sized particles," *Anal. Chim. Acta*, **29**:372–374 (1953).

Lodge, J. P., and J. B. Pate, "Atmospheric gases and particulate in Panama," *Science*, **153**(3734):408–410 (1966).

Lodge, J. P., H. F. Ross, N. K. Sumida, and B. Tufts, "Analysis of micron-sized particles. Determination of particle size," *Anal. Chem.* 28:423–424 (1956).

Lodge, J. P., and B. Tufts, "An electron microscope study of sodium chloride particles as used in aerosol generation," *Colloid Sci.,* 10(3):256–261 (1955).

Lodge, O., "The electrical deposition of dust and smoke with special reference to the collection of metallic fume and a possible purification of the atmosphere," *J. Soc. Chem. Ind.,* 5:572–576 (1886).

Löffler, H. J., and H. König, "Zur electronenmikroskopischen Abbildung von Tropfen aus Wasser and Wässrigen Lösungen," *Kolloid-Z.* 142:65–73 (1955).

McCabe, L. C., P. P. Mader, H. E. McMahon, W. J. Hamming, and A. L. Chaney, "Industrial dusts and fumes in the Los Angeles area," *Ind. Eng. Chem.,* 41:2486 (1949).

McCrone, W., R. Draftz, and J. Delly, "The Particle Atlas," Ann Arbor, Michigan, Ann Arbor Science Publishers Inc., 1967 (406 pp.).

McKay, R. J., "Recent progress in smoke abatement and fuel technology in England," Mellon Inst. Indust. Res. *Smoke Invest. Bull.,* 10 (1922).

Matthews, J. W., and H. V. A. Briscoe, "Volatile solid filters for dust-sampling," *Trans. Instr. Mining Met.,* 44:111–112 (1934).

May, K. R., "The cascade impactor, an instrument for sampling coarse aerosols," *J. Sci. Instr.,* 22:187–195 (1945).

May, K. R., "A cascade impactor with moving slides," *A. M. A. Arch. Ind. Health,* 13:481–488 (1956).

Meldau, R., "Eine physikalisch-geometrische Ordnung der Staubmorphologie," *Staub,* 22(8):296–300 (1962).

Mercer, T. T., "Charging and precipitation characteristics of submicron particles in the Rohmann electrostatic particle separator," Report U.R.-#475, Univ. of Rochester Atomic Energy Project, Rochester, New-York, 1957.

Mercer, T. T., M. I. Tillery, and M. A. Flores, "An electrostatic precipitator for the collection of aerosol samples for particle size analysis." LR-7, Lovelace Foundation for Med. Res. and Ed., Albuquerque, New Mexico, 1963.

Meyer, A. L., "A method for determining the finer dust particles in air," *J. Ind. Hyg.,* 3(2):51–56 (1921).

Miners' Phthisis Prevention Committee (of South Africa). General Report of the Miners' Phthisis Prevention Committee published in 1916 by Pretoria (The Government Printing and Stationery Office), 199 pp.

Miners' Phthisis Prevention Committee (of South Africa). Final Report of Miners' Phthisis Prevention Committee published in 1919 by Pretoria (The Government Printing and Stationery Office), 111 pp.

Miquel, P., "Étude dur les poussières organisées de l'atmosphère," *Ann. d'hyg.,* S.3, 2:226–333 (1879).

Mitchell, J. P., "Determination of dust fall in the neighborhood of cement plants," *Ind. Eng. Chem.,* 6:454–459 (1914).

Mitchell, R. I., and J. M. Pilcher, "Improved cascade impactor for measuring aerosol particles size in air pollutants, commercial aerosols, and cigarette smoke," *Ind. Eng. Chem.,* 51:1039–1042 (1959).

Möller, K., "Quantitative determination of the dust content of the air," *Gesundh. Ing.,* 17:373–376 (1894).

Monkman, J. L., "Gas chamber microapparatus in identification of airborne pollutants," *Anal. Chem.,* 27:704–708 (1955).

Nader, J. S., "Dust retention efficiencies of dustfall collectors," *J. Air Pollution Control Ass.,* 8:35–38 (1958).

Naesland, C., "Contribution to the methods used for rapid determination of dust in air," *J. Ind. Hyg.,* 14:113–116 (1932).

Nakaya, U., "Electron-microscope studies on the nuclei of sea fog and snow crystals," in "Artificial Stimulation of Rain," H. Weickmann and W. E. Smith, ed., New York, Pergamon Press, 1957.

O'Connell, D. C., N. J. G. Wiggin, and G. F. Pike, "New technique for the collection and isolation of airborne microorganisms," *Science* 131:359–360 (1960).

Orr, C., and R. A. Martin, "Thermal precipitator for continuous aerosol sampling," *Rev. Sci. Instr.,* 29:129–130 (1958).

Owens, J. S., "Dust in expired air," *Proc. Roy. Soc.,* A101:18–37 (1922).

Owens, J. S., "Jet dust counting apparatus," *J. Ind. Hyg.,* 4:522–534 (1923).

Palmer, G. T., "Apparatus for determination of aerial dust," *Amer. J. Public Health,* 6:54–55 (1916).

Paxton, R. R., "Measuring rate of dustfall," *Rock Products,* 54(2):114, 116, 118 (1951).

Penney, G. W., "Role of adhesion in electrostatic precipitation," *A.M.A. Arch. Ind. Health,* 4:301–305 (1962).

Pidgeon, F. D., "Controlling factors in identification of microscopic chloride particles with sensitized gelatine films," *Anal. Chem.,* 26:1832–1835 (1954).

Price, P. B. and R. M. Walker, "Electron microscope observation of a radiation-nucleated phase transformation in mica," *J. Appl. Phys.,* 33:2625–2628 (1962).

Prospero, J. M., and E. Bonatti, "Continental dust in the atmosphere of the eastern equatorial Pacific," *J. Geophys. Res.,* 74(13):3362–3371 (1969).

Quitman, E., and H. Cauer, "Verfahren zur chemischen; Analyze der Nebelkerne der Luft," *Z. Anal. Chem.,* 116:81–91 (1939).

Ranz, W. E., and J. B. Wong, "Jet impactor for determining the particle-size distribution of aerosols," *A.M.A. Arch. Ind. Hyg. Occup. Med.,* 5:464–477 (1952).

Riedel, G., "Ein elektrischer Kernfäller zur Gewinnung übermikroskopischer Präparate," *Kolloid Z.,* 103:228–232 (1943).

Robinson, E., J. A. MacLeod, and C. E. Lapple, "A meteorological tracer technique using uranine dye," *J. Metereol.,* 16:63–67 (1959).

Robinson, M., "A miniature electrostatic precipitator for sampling aerosols," *Anal. Chem.,* 33:109–113 (1961).

Rogers, G., "Progress report to the Committee on Standard Methods for the Examination of the Air," *Amer. J. Public Health,* 3(1):78–86 (1913).

Rowley, F. B., and R. C. Jordan, "A new type adhesive impingement dust counter," *J. Ind. Hyg. Toxicol.,* 25:293–302 (1943).

Sawyer, K. F., and W. H. Walton, "The conifuge, a size-separating sampling device for airborne particles," *J. Sci. Instr.,* 27:272–276 (1950).

Schadt, C. L., and R. D. Cadle, "Critical comparison of collection efficiencies of commonly used aerosol sampling devices," *Anal. Chem.,* 29:864–868 (1957).

Schadt, C. L., P. L. Magill, R. D. Cadle, and L. Ney, "An electrostatic precipitator for the continuous sampling of sulfuric acid aerosols and other air-borne particulate electrolytes," *Arch. Ind. Hyg. Occup. Med.,* 1(5):556–564 (1950).

Schlipköter, H. W., H. Steiger, H. F. Esser, and E. G. Beck, "Elektronenoptische Untersuchungen von Grubenstäuben unter Verwendung von Membranfiltern," *Staub,* 19:320–322 (1959).

Schmidt, K. G., "Die Phasenkontrastmikroskopie in der Staubtechnik," *Staub,* 41:436–467 (1955).

Schmidt, K. G., "Asbestsorten, ihre Untersuchung mit optischen Mitteln und ihre krankmachende Wirkung," *Staub,* 20:173–180 (1960).

Schmidt, K. G., and G. Heidermanns, "Zur Technik der Staubmikroskopie mit Phasenkontrast und Grenzdunkelfeld," *Staub,* 18:236–246 (1958).

Schmidt, K. G., and G. Heidermanns, "Untersuchung von Staubproben mit dem Phasenkontrastmikroskop, insbesondere bei Verwendung von Membranfiltern," *Staub,* **19**:413–416 (1959).

Seely, B. K., "Detection of micron and submicron chloride particles," *Anal. Chem.,* **24**:576–579 (1952).

Seely, B. K., "Detection of certain ions in 10^{-10} to 10^{-15} gram particles," *Anal. Chem.,* **27**:93–95 (1955).

Sehl, F. W., and B. J. Havens, "A modified air sampler employing fiberglass," *A.M.A. Arch. Ind. Hyg. Occup. Med.,* **3**:98–100 (1951).

Shibusawa, M., and Y. Niwa, "A new electrical precipitation treater," *J. Amer. Inst. Elec. Eng.,* **39**:890–903 (1920).

Shipe, E. L., M. E. Tyler, and D. H. Chapman, "Bacterial aerosol samplers II. Development and evaluation of the Shipe sampler," *Appl. Microbiol.,* **7**: 349–354 (1959).

Silverman, L., and W. Franklin, "Shattering of particles by the impinger," *J. Ind. Hyg.,* **24**: 80–82 (1942).

Sonkin, L. S., "A modified cascade impactor: A device for sampling and sizing aerosols of particles below one micron in diameter," *J. Ind. Hyg. Toxicol.,* **29**: 269–272 (1946).

Spurny, K., J. Lodge, E. Frank, and D. Sheesley, "Aerosol filtration by means of Nuclepore filters," *Environ. Sci. Tech.,* **3**(5):453–464 (1969).

Stafford, E., and W. S. Smith, "Dry fibrous air filter media: performance characteristics," *Ind. Eng. Chem.,* **43**:1346 (1951).

Steele, R. L., and F. W. Sciacca, "Characteristics of silver iodide ice nuclei originating from anhydrous ammonia–silver iodide complexes. Part II, Thermal Systems," Mechanical Engr. Tech. Paper No. 66–3 and Atmospheric Science Tech. Paper No. 78, Colorado State Univ., 1966 (13 pp.).

Thomas, D. J., "Fibrous filters for fine particle filtration," *J. Inst. Heating Ventilating Engrs.* (London), **20**:35–55; disc. 55–70 (1952).

Thuman, W. C., "Organic fiber filters," Final Report Project No. SU-2261, Stanford Research Inst. Contract No. A.F. 19(604)-2644, 1959.

Thürmer, H., "Der Fraktionierungseffekt im Thermalpräzipitator und die Folgerungen für ein elektronenmikroskopisches Kornanalysenverfahren," *Staub,* **20**: 6–8 (1960).

Tissandier, G., "Les poussières atmosphériques," *Comp Rend.,* **78**:821–824 (1874).

Tissandier, G., "Les poussières de l'atmosphère," *Rev. Sci.,* S. **2**(17;814 (1880).

Tolman, R. C., L. H. Reyerson, A. P. Brooks, and H. D. Smyth, "An electrical precipitator for analyzing smokes," *J. Amer. Chem. Soc.,* **41**:587–589 (1919).

Trostel, L. J., and H. W. Frevert, "Collection and examination of explosive dusts in air," *Ind. Eng. Chem.,* **15**:232–236 (1923).

Tufts, B. J., "Determination of micron-sized particles. Detection of potassium ion," *Anal. Chem.,* **31**:242–243 (1959a).

Tufts, B. J., "Determination of particulate lead content in air. Results of tests in city traffic," *Anal. Chem.,* **31**: 238–241 (1959b).

Tufts, B., "A method for identifying particulate fluoride compounds," *Anal. Chim. Acta,* **23**:209–214 (1960).

Tufts, B., and J. P. Lodge, "Chemical identification of halide and sulfate in submicron particles," *Anal. Chem.,* **30**:300–303 (1958).

Tyler, M. E., and E. L. Shipe, "Bacterial aerosol samplers, I. Development and evaluation of the all-glass impinger," *Appl. Microbiol.,* **7**:337–349 (1959).

Tyler, M. E., E. L. Shipe, and R. B. Painter, "Bacterial aerosol samplers, III. Comparison of biological and physical effects in liquid impinger samplers," *Appl. Microbiol.,* **7**: 355–362 (1959).

Underwood, G., "Removal of sub-micron particles from industrial gases, particularly in the steel and electricity industries," *Int. J. Air Water Pollution,* **6**:229–263 (1962).

Van Orman, W. T., and H. A. Endres, "Self-charging electrostatic air filters," *Heating, Piping, Air Conditioning,* **24**:157–163 (1952).

Vittori, O., "La rivelazione di particelle constitute da particolari sostanze chimiche nell' indagine sui nuclei di condensazione e sui nuclei di sublimazione artificale," *Riv. Meteorol. Aero.,* **14**(2);17–22 (1954).

Vittori, O. "Sur la détermination de la distribution des diamètres des particules atmosphériques de chlorure avec la réaction de Liesegang," *Geofis. Pura Appl.,* **31**:90–96 (1955).

Walkenhorst, W., "Staubprobenahme mit dem Thermalpräzipitator unter besonderer Berücksichtigung elektronenmikroskopischer Auswertung," *Staub,* **40**:241–252 (1955).

Waller, R. E., A. G. F. Brooks, and J. Cartwright, "An electron microscope study of particles in town air," *Int. J. Air Water Polution,* **7**:779–786 (1963).

Walton, W. H., R. C. Faust, and W. J. Harris, "A modified thermal precipitator for the quantitative sampling of aerosols for electron microscopy. Porton Tech. Paper No. 1, Ser. 83, 1947.

Watson, H. H., "A system for obtaining from mine air, dust samples for physical, chemical, and petrological examination," *Trans. Instr. mining Met.,* **46**:155–175 (1936–37).

Watson, H. H., "A note on the shattering of dust particles in the impinger," *J. Ind. Hyg. Toxicol.,* **21**:121–123 (1939).

Watson, H. H., "The sampling efficiency of the thermal precipitator," *Brit. J. Appl. Phys.,* **9**:78–79 (1958).

Watson, J. H. L., and L. Preuss, "Motion picture studies of electron bombardment of colloidal crystals," *J. Appl. Phys.,* **21**:904–907 (1950).

White, H. J., "Industrial Electrostatic Precipitation," Reading, Mass., Addison-Wesley Publication Co., 1963.

White, H. J., and W. A. Baxter, Jr., "Electrostatic precipitators," *Mech. Eng.,* **82**:54–56 (1960).

Whytlaw-Gray, R., H. L. Green, R. Lomax, and H. H. Watson, British Patent #44551, 1936.

Wilcox, J. D., "Design of a new five-stage cascade impactor," *A.M.A. Arch. Ind. Hyg. Occup. Med.,* **7**:376–382 (1953).

Winckelmann, J., "Über einige neue Methoden der präparativen Mikrochemie," *Microchemie,* **10**:437–439 (1931).

Winckelmann, J., "Über einige neue Methoden der qualitative Mikroanalyze," *Mikrochemie,* **12**:119–128 (1933).

Winckelmann, J., "Aus der Praxis der Tüpfelanalyze," *Mikrochemie,* **16**:203–210 (1935).

Winslow, C. E. A., "A method of determining the number of dust particles in air," *Amer. J. Public Hyg.,* **19** (N.S. **5**):87–88 (1909).

Woodcock, A. H., "Sea salt in a tropical storm," *J. Meteorol.,* **7**:397–401 (1950).

Woodcock, A. H., "Atmospheric salt particles and raindrops," *J. Meteorol.,* **9**:200–212 (1952).

Woodcock, A. H., "Salt nuclei in marine air as a function of altitude and wind force," *J. Meteorol.,* **10**:362–371 (1953).

Woodcock, A. H., and M. M. Gifford, "Sampling atmospheric sea salt nuclei over the ocean," *J. Marine Res.,* **8**:177–197 (1949).

Woolrich, P. F., "A new electrostatic sample collector permitting direct microscopic examination," *Amer. Ind. Hyg. Ass. Quart.,* **15**:267–268 (1961).

Wright, B. M., "Gravimetric thermal precipitator," *Science,* **118**:195 (1953).

AUTOMATED HISTOLOGICAL EQUIPMENT AND TECHNIQUE

Processing of Specimens for Diagnosis. For optimum results in the processing of any type of surgical or autopsy tissue, it is important that the specimens be kept moist and refrigerated until they are trimmed, described, sectioned, and fixed.

After sections are cut, they must be immediately submerged in the correct fixative. Some stains require a special fixative. It is essential, therefore, that the pathologist adivse his histologist of any special work that will be required before processing has begun. A Zenker type of fixative is considered one of the very best for most types of tissue since it is excellent for preserving nuclear detail and compatible with a wide variety of stains (Technicon Corporation, Tarrytown, New York). Another frequently used fixative is 10 % formalin which is especially useful in preserving all kinds of fat tissues including the myelin in nerve fiber sheaths. Alcohol is preferred for preservation of mucus, glycogen, and some pigments.

It is not necessary to wash tissues fixed in formalin and alcohol. When tissues are fixed in Zenker's, however, they should be washed in water and placed in 80 % alcohol for storage unless the sections are going to be processed immediately.

Processing, if possible, should be done automatically in an Autotechnicon. The tissues are dehydrated in graded alcohol or a commercial dehydrant that does not require dilution (S-29 dehydrant, Technicon Corporation, Tarrytown, New York). After dehydration, tissues go into a clearing agent and then into paraffin.

Automated Histological Equipment and Techniques. In addition to the standard histology laboratory equipment, the automated histology laboratory should have an automatic tissue processor, a good microtome for cutting paraffin sections, as stated previously, a slide dryer, and a unit for staining slides.

The tissue-processing units should consist of a processor equipped with heat and vacuum to allow for fast or prolonged processing schedules. Several timing discs to be specifically notched for programming the various schedules are necessary. Convenient ranges from 1 to 16 hours are essential for all types of processing schedules.

An additional safety device should be incorporated on tissue processors such as the "Tissue Safe-Guard" device on the Autotechnicon Ultra (Fig. 1) which is described as follows:

A. When there is a power failure and the vacuum head of the Ultra which is holding the receptacle basket is in the raised or partially raised position, the system will:

1. index to the next beaker, if the head was rising at the time, and the unit will shut down after the head has seated itself over the beakers;
2. complete the descent of the head if the head was in the descending portion of the cycle. The Ultra will again shut down after the head seals itself on the beakers.

Table 1   Most Common Fixatives

Fixatives	Advantages	Disadvantages
1. Technicon FU-48 (Zenker type)	a. Rapid penetration b. Casts no precipitate c. No wash required after fixation d. Indefinite shelf life e. No excessive hardening f. Excellent for all tissues	None
2. Formalin calcium (formalin) 10%	a. Fairly rapid b. Best for phospholipids c. Preserves mitochondria	a. Cannot be used to fix calcium
3. Formol saline	a. No excessive hardening b. Tissue can be left for long periods of time	a. Fixes nuclear chromatin b. Penetrates slower than Zenker's
4. Neutral formalin buffered	a. No excessive hardening b. Tissue can be left for long periods of time	a. Penetrates slower than Zenker's
5. Zenker fluid	a. Excellent for lymph nodes	a. Requires wash after use b. Hardens tissues
6. Bouin's	a. Excellent for connective tissue	a. Should not be used routinely b. Requires wash after use
7. Orth's fluid	a. Excellent for demonstration of chromatin cells and glycogen	a. Requires long wash after use b. Must be made fresh c. Not good for routine surgicals
8. Helley's fluid	a. Good for bone marrow	a. Requires wash b. Hardens
9. Formal alcohol	a. Excellent for skin biopsies	a. Hardens tissue excessively
10. Acetone	a. Excellent for enzyme studies	a. Not good for fixing surgical tissues
11. Muller's fluid	a. Excellent for bone specimen	a. Not good for other tissues
12. Carnoy's fluid	a. Penetrates rapidly b. Excellent nuclear fixation	a. Destroys cytoplasmic elements b. Causes shrinkage c. Not good for routine work d. Hemolyzes RBCs

Fig. 1   Autotechnicon Ultra tissue processor.

B. The time delay between the loss of power and the time the battery power comes into the circuit can be adjusted from 0 to 30 min. The use of a time delay allows a reasonable period of time to pass in the hope that power would be restored prior to the safeguard circuit being energized. If power is restored in the time delay period, then the unit can continue to run through its normal cycle rather than being shut down.

Tissue processors should possess an added feature such as the "Tissue Safe-Guard" device for the following reasons:

1. To prevent tissue from drying and being completely destroyed by being inadvertently arrested in the atmosphere during a tissue-processing schedule.
2. To delete from an additional surgical operation of the patient whose tissue was lost due to arresting in the atmosphere (hung up in mid air on the Autotechnicon).
3. To delete from morphological alterations of the tissue cells that could be produced by arresting in the atmosphere. This alteration could definitely affect the pathological diagnosis of the specimen.
4. To prevent all of the aforementioned if there was a power failure and the specimen was changing to the next beaker.

Most processors are equipped with time-delay devices to make it possible to set a schedule for weekends and holidays (Technicon Corporation, Tarrytown, New York). The machine is then set for the night or weekend (if for the weekend, the time-delay mechanism will be set), and upon completion of the processing, which consists of fixation, dehydration, clearing, and infiltration with paraffin, the technician removes the processing basket from the Autotechnicon, and the tissues are embedded in paraffin. Care must be taken at this point to check identification numbers carefully.

It is very important when embedding, in order to insure that the pathologist is given every opportunity to see the true picture of the cells and structures of the tissues, that the area in question be placed with the correct side down for cutting. To a "practiced" eye this can usually be determined by an area of discoloration, a lesion, or merely an area that is different from normal. Some pathologists will notch the tissue or put a dot of dye on the specimen to indicate the correct side for embedding.

When the paraffin blocks are cooled and lined up for cutting, they should be placed in numerical order to insure correct marking of the slides.

Microtome knives must be sharp. Improperly sharpened knives will result in compression of tissues, thick and thin areas, ragged, torn sections, and lines throughout the finished section which may make diagnosis more difficult than necessary and photography not up to standard. Follow the directions on your knife sharpener and check the edge under a microscope to assure that sharpening is complete.

When cutting tissues, a constant-temperature water bath is essential. Distilled water should be used, and the surface should be cleared of any dust or tissue fragments. Superimposing cut tissue fragments onto another section is not only extremely poor technique, but it could cause disastrous results in diagnoses. The water surface should be cleaned thoroughly and frequently during cutting.

The sections are carefully lifted from the knife edge and placed in the water bath with a gentle pulling motion to straighten the sections and get them to spread flat as they are eased and positioned onto the slides.

An adhesive is usually used to affix the tissues to the slide. Egg albumin and glycerine or gelatin are the most popular types. Some commercial preparations may be obtained and are good. Care must be taken that the adhesive used is fresh and free from contaminants. It should be kept under refrigeration to prevent deterioration and bacterial growth. A small crystal of thymol will prolong the life of the adhesive.

The water bath should be rinsed and cleaned thoroughly each day.

The microtome must be thoroughly cleaned, greased, and oiled after use. Correct care of this instrument will ensure its use for a long time, and it will remain an accurate and dependable machine.

As the tissues are being cut, they may be placed in a staining rack or slide holder which fits in the slide drier and the Autotechnicon staining machine. The slide holder will eliminate the transfer of slides from a box to a slide holder.

A good method for drying slides is to use a commercially available slide dryer (Technicon Corporation, Tarrytown, New York). The dryer is operated at the melting point of the paraffin and after 7–10 min slides may be taken directly from the dryer and placed in the

Table 2   Most Common Dehydrants

Dehydrants	Advantages	Disadvantages
1. Technicon S-29	a. Rapid dehydration b. No shrinkage c. Penetrates rapidly d. Does not harden tissue e. Nonhydroscopic f. Nontoxic g. Dehydrates fatty tissue well h. Does not remove stains i. Completely soluble with water j. Very little evaporation	None
2. Isopropynol alcohol		
3. Ethyl alcohol	a. Dehydrates rapidly	a. Hardens tissue
4. Dioxane	a. Does not remove stains b. Completely soluble with water	a. Causes slight shrinkage b. Hardens tissue c. *Very toxic* d. Excessive evaporation e. Causes kidney and liver damage
5. Acetone	a. Rapid penetration b. Completely soluble with water	a. Slow dehydration b. Causes shrinkage c. Removes stains d. Excessive evaporation e. Toxic
6. Butyl alcohol	a. Dehydrates rapidly b. Does not remove stains	a. Very toxic b. Not completely soluble with water c. Causes slight shrinkage
7. Amyl acetate and butyl acetate	a. Good dehydration	a. Slow penetration
8. Tetrahydroferan (THF)	a. Fairly rapid b. Soluble in water c. Can be used for dehydration and clearing	a. Very *toxic* b. NOTE: (comment) c. Ether-like odor d. Some shrinkage
9. Methanol	a. Does not remove stains b. Completely soluble with water	a. Shrinkage b. Slow penetration

Autotechnicon for staining (Figs. 2 and 3). Reagents will be explained later in this article.

After the slides are stained, they are removed and placed in a beaker of clearing agent. A commercially available clearant called Paraway[1] or xylene are the most commonly used.

Sections are now ready for cover glasses. This simple procedure must be carefully carried out. It is important not to have too much mounting medium under the cover glass and just as important to have enough. Practice will enable the technician to judge how much to use. Care should be taken to eliminate any air bubbles under the cover glass. As the cover is placed onto the tissue, gentle pressure will eliminate the bubbles and also affix the coverslip to the slide properly.

Label the slide carefully and neatly.

The finished product should result in:

1. well processed sections;
2. beautifully cut and stained sections;
3. neat, clean, clearly marked slides.

Reagents. All reagents for processing and staining specimens may be purchased from the Technicon Corporation in Tarrytown, New York.

The hematoxylin used in the Autotechnicon staining system must be a progressive-type stain not requiring decolorizing (Technicon Corporation, Tarrytown, New York).

If commercial processing reagents are not used, standard fixatives, dehydrants, and clearants should be sought possessing flash points above 80°F.

Freezing and Cryostat Methods. An entirely different approach to preparing sections is accomplished by the freezing or cryostat method. A block of tissue (fixed or fresh, but normally fresh) is quickly frozen, usually employing liquid carbon dioxide. A cryostat, in which a microtome is placed in a chamber or cabinet and temperatures are maintained at all times below freezing, is most univerally used today.

This method is equivalent to infiltrating the tissue with ice. In the frozen state, sections can be cut as thin as 5–16 μ. These sections are not as perfect or as thin as those which can be cut after paraffin embedding, but there are many advantages to the frozen-section method. For one thing, it is a very rapid method, and the surgical pathologist may be able to return a definitive, positive diagnosis to the surgeon during the course of the operation, often within 2–5 min after he receives the tissue.

An additional advantage relates to the fact that frozen sections, not requiring treatment by fat-solvent dehydrating substances, still contain all the liquid normally present in them and can be stained for this liquid.

---

[1]Technicon Corporation, Tarrytown, New York.

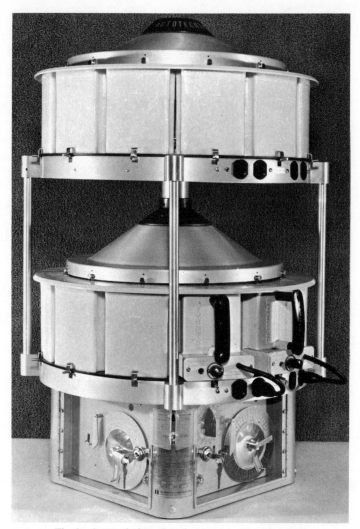

Fig. 2 Autotechnicon duo tissue processor and stainer.

Automatic Histological Equipment. In the preparation of tissue specimens for an automatic tissue processor, care must be taken to cut the sections the correct thickness according to the time schedule which will be selected for processing.

Usually, one hour of processing for each millimeter in thickness is adequate; however, one must keep in mind that as thickness of tissue increases, the rate of penetration slows down.

This is especially important for sections over $3\frac{1}{2}$ mm in thickness which are classed as difficult.

If tissue sections are going to be processed overnight, it is best to use the 16 hr clock disc even if the tissue processor is equipped for short runs of less time. The reason for this is that using a short timing schedule overnight would require using the time-delay device which is intended for use only on 16–17 hr schedules on weekends and holidays due to solidifying of paraffin, which usually takes approximately 6–8 hr to melt.

As specimens are cut and labeled, they are placed in receptacles with their identifying number tag, or the number may be written on the outside of many types of receptacles. Whatever type is used, the identifying number never leaves its specimen. The tissues are immediately placed in a tissue carrier basket which is immersed in a fixative. The tissues must not be left unprotected, and they must be placed in the fixative immediately after they are trimmed.

When the basket is full, it is attached to the basket holder in the automatic tissue processor or to the vacuum head in the Autotechnicon Ultra. The basket is allowed to descend into the first beaker, which contains fixative. Set the machine and make sure the temperatures are accurate and beakers are filled to their correct volumes.

Paraffin temperatures should be 59–60°C and not higher than one degree over the melting point of the paraffin, as this may make the tissue brittle and difficult to cut. Check all the reagents if you have not already done this,

Table 3   Most Common Clearing Agents

Clearing Agents	Advantages	Disadvantages
1. Technicon UC-670 clearant	a. No shrinkage of tissue b. Ho hardening c. Soluble in Technicon dehydration fluids, S-29, and ethyl alcohol and paraffin d. Penetrates rapidly e. Excellent for fatty tissues f. Nonflammable g. Nonhygroscopic h. Nonvolatile i. Excellent for routine or special work	None
2. Technicon C-650 and C-550 (for fatty tissues)	a. No shrinkage of tissue b. No hardening c. Soluble in Technicon dehydration fluids, S-29, and ethyl alcohol and paraffin d. Penetrates rapidly e. Excellent for fatty tissues f. Nonflammable g. Nonhygroscopic h. Nonvolatile i. Excellent for routine or special work	None
3. Technicon Paraway	a. Slow evaporation rate gives technician time to cover slip neatly, lasts longer b. No shrinkage or distortion c. No fire hazard d. Mixes readily with most other solvents and dehydrants e. Refractive index 1.499 in comparison with xylene's 1.5 f. Paraffin solvency 8.5% compared to xylene's 8% g. Noncombustible, has high flash point at 160°F compared to xylene's 75°F	None
4. Xylene (xylol)	a. Penetrates rapidly b. Good for routine work	a. Hardens tissue excessively b. Slight shrinkage c. Slightly toxic, kidney and liver damage
5. Chloroform	a. Good clearing agent for animal tissue b. Can be used for routine or special work	a. Very toxic b. Volatile c. Has strong ethereal smell
6. Toluene	a. Can be used for routine or special work	a. Clears slowly b. Toxic c. Causes liver and kidney damage
7. Benzene	a. Can be used routinely	a. Very toxic b. Slight shrinkage
8. Acetone	a. Soluble in water b. Penetrates rapidly	a. Toxic b. Causes shrinkage c. Evaporates excessively d. Fire hazard e. Removes stains
9. White gasoline	a. Can be used for routine work b. Penetrates rapidly	a. Fire hazard b. Some shrinkage c. Odor d. Evaporates fast

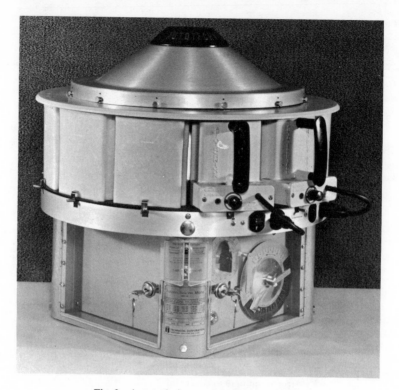

Fig. 3   Autotechnicon mono tissue processor.

Tested Schedules for the Autotechnicon Ultra (17 hr)

Solutions	Time	Beaker no.
Technicon Fixative	4 1/2 hr	1
Distilled Water	1/2 hr	2
Technicon Dehydrant	1 hr	3
Technicon Dehydrant	1 hr	4
Technicon Dehydrant	1 hr	5
Technicon Dehydrant	1 hr	6
Technicon Dehydrant	1 hr	7
Technicon Dehydrant	1 hr	8
Technicon Clearant UC-670	1 hr	9
Technicon Clearant UC-670	1 hr	10
Paraffin	1 hr	11
Paraffin	2 hr	12

Tested Schedules for the Autotechnicon Ultra (4 hr)

Solutions	Time	Beaker No.
Technicon FU-48	40 min	1
Technicon S-29	10 min	2
Technicon S-29	10 min	3
Technicon S-29	10 min	4
Technicon S-29	20 min	5
Technicon S-29	20 min	6
Technicon S-29	20 min	7
Technicon Clearant UC-670	10 min	7
Technicon Clearant UC-670	15 min	9
Technicon Clearant UC-670	25 min	10
Paraffin (56–58°C M.P.)	20 min	11
Paraffin (56–58°C M.P.)	40 min	12

Tested Schedules for the Autotechnicon Ultra (8 hr)

Solutions	Time	Beaker No.
Technicon FU-48	2 hr	1
Technicon S-29	10 min	2
Technicon S-29	20 min	3
Technicon S-29	20 min	4
Technicon S-29	20 min	5
Technicon S-29	30 min	6
Technicon S-29	45 min	7
Technicon Clearant UC-670	15 min	8
Technicon Clearant UC-670	30 min	9
Technicon Clearant UC-670	50 min	10
Paraffin (56–58°C M.P.)	1 hr	11
Paraffin (56–58°C M.P.)	1 hr	12

Tested Schedules for the Autotechnicon Ultra (3 hr)

Solutions	Time	Beaker No.
Technicon FU-48	30 min	1
Technicon S-29	5 min	2
Technicon S-29	10 min	3
Technicon S-29	10 min	4
Technicon S-29	10 min	5
Technicon S-29	10 min	6
Technicon S-29	25 min	7
Technicon Clearant UC-670	5 min	8
Technicon Clearant UC-670	15 min	9
Technicon Clearant UC-670	20 min	10
Paraffin (56–58°C M.P.)	15 min	11
Paraffin (56–58°C M.P.)	25 min	12

Tested Schedules for the Autotechnicon Ultra (2 hr)

Solutions	Time	Beaker No.
Technicon FU-48	15 min	1
Technicon S-29	5 min	2
Technicon S-29	5 min	3
Technicon S-29	5 min	4
Technicon S-29	10 min	5
Technicon S-29	10 min	6
Technicon S-29	15 min	7
Ultra Clearant UC-670	5 min	8
Ultra Clearant UC-670	10 min	9
Ultra Clearant UC-670	10 min	10
Paraffin (56–58°C M.P.)	10 min	11
Paraffin (56–58°C M.P.)	20 min	12

Automatic Staining With the Autotechnicon Mono or Duo[a]

1. "Paraway"—2 min
2. 0.25% iodine in Technicon dehydrant—2 min
3. Technicon dehydrant—2 min
4. Distilled water—2 min
5. Technicon hematoxylin solution—2 min
6. Distilled water—2 min
7. Technicon lithium carbonate 0.01N[b]—2 min
8. Distilled water—2 min
9. Technicon eosin solution—2 min
10. Distilled water—2 min
11. Technicon dehydrant—2 min
12. Technicon dehydrant[c]—2 min

[a]This represents the Autotechnicon's 12-stage setup for staining. The reagent used is shown for each stage together with the immersion time required for each solution.

The Autotechnicon completely automates the staining cycle. Slides are advanced from stage to stage precisely. When the cycle is completed, the Autotechnicon stops. Sections are now ready for fresh clearing in "Paraway" and application of the cover glasses. Time-consuming manual procedures are done away with, and technicians are free to handle other duties.

Intensity of coloration may be increased or decreased at will, by increasing and decreasing the length of immersion time. With Technicon Hematoxylin and Eosin the immersion time in either stain merely affects the intensity of the color reaction, not its quality of differentiation.

[b]An approximately 0.01N solution is prepared by dissolving one Technicon lithium carbonate vial in a liter of distilled water.

[c]After the twelfth stage, slides are automatically transferred to the No. 1 beaker of "Paraway." After 3 min immersion, they may put in another change of "Paraway," from which the cover glasses are applied. This is accomplished by having 12 notches on the disc instead of the usual 11.

and make sure the levels are where they should be to cover all the receptacles in the carrier basket.

Now, set the timing clock. Be sure the disc is in place and the automatic shutoff cam down. The timing pin should be against the timing disc; it should not be in a notch. Press the on switch and check the vacuum; the automatic Autotechnicon will now take over.

When the processing is completed and the carrier

basket is in the last paraffin and the time is up on infiltration of the paraffin, press the raise button, which will lift the basket drive. Relase the button at the top position, remove the carrier basket, place it in a fresh paraffin.

Embed the tissues carefully and number the molds so that all tissues will be identified correctly. This is an extremely important step. A tissue mix-up is inexcusable, and should never happen.

The tissue blocks are cooled and are then ready for cutting on the microtome.

Arrange the cutting materials, number the slides, set up the tissue flotation bath, and have at hand the necessary accessories such as camel's-hair brushes, mounting adhesive, or if you prefer, use gelatin in the water bath. If the combination of egg albumin and glycerine is used, it should be put on the slide very sparingly or a blue cast will be stained on the finished slide.

As the tissues are being cut, the slides may be placed in the staining racks. When a slide rack is full, it should then be placed in the slide drier (tissue sections must be dried

Fig. 4 Technicon lab aid filing system.

Table 4 Most Common Paraffins

Paraffins	Advantages	Disadvantages
1. Paraplast (Sherwood Labs.)	a. Several plastic polymers b. Elasticity c. Excellent for routine work d. Excellent for eyes	None
2. Most refined paraffins	a. Good for routine work	

Table 5 Most Common Stains

Stains	Advantages	Disadvantages
1. Technicon basophilic dye (hematoxylin)	a. Highly selective b. Stains only basophilic components c. Never has to be filtered, free of precipitate d. Very stable e. Stain penetration is progressive f. Standardized so that every batch of dye will possess exactly the same characteristics as every other batch	None
2. Technicon acidophilic dye (eosin)	a. Highly sensitive b. Stains on acidophilic components c. Never has to be filtered, free of precipitate d. Very stable e. Stain penetration is progressive f. Standardized so that every batch of dye will possess exactly the same characteristics as every other batch	None
3. Harris hematoxylin	a. Satisfactory for routine work	a. Costly to make up b. Not standardized c. Not stable d. Requires filtering each day
4. Delafield's hematoxylin	a. Satisfactory for routine work	a. Costly to make up b. Not standardized c. Not very stable d. Requires filtering often
5. Harris-Lillie stain	a. Satisfactory for routine work	a. Requires filtering each day b. Stable for about two weeks
6. Mayer's hematoxylin	a. Satisfactory for routine work b. Progressive stain	a. Costly to make up b. Must be filtered frequently
7. Aqueous eosin	a. Satisfactory for routine work	a. Not stable b. Requires constant filtering c. Not standardized d. Costly to make up and time consuming
8. Alcoholic eosin	a. Satisfactory for routine work	a. Not stable b. Requires constant filtering c. Not standardized d. Costly to make up and time consuming
9. Picro-eosin	a. Satisfactory for routine work	a. Not stable b. Requires constant filtering c. Not standardized d. Costly to make up and time consuming

Table 6 Most Common Mountants

Mountants	Advantages	Disadvantages
1. Technicon mounting medium	a. Hardens quickly b. Will not crack with age c. Dries fast d. Will not darken or turn yellow with age e. Will not crystallize f. Will not acidify with age g. High refractive index h. Highly developed synthetic resin	None
2. Other synthetic resins	a. Dry fast b. Normally satisfactory for routine work	a. Low refractive index b. Yellow with age c. Some cracking with age
3. Canada balsam and gum damar	a. Harden quickly	a. Yellow with age

before staining, or they will fall off the slides). It usually takes 8–10 min in a slide drier. If the slides are placed in an oven, one hour is usually enough time, overnight if the oven is 40°C. When the slides are removed from the slide drier, they may be placed in the staining machine.

The Mono or Duo Autotechnicon may be equipped with a one hour clock for staining. The positions for staining require only two minutes in each reagent. This is an excellent setup because one may process many racks at a time. A technician must be on hand to remove the slide racks as they come out of the number 1 beaker, which is also the beginning position. This is only necessary when more than one rack is stained at a time. When one rack only is stained, the cutoff cam should be in operation, which will disengage the clock, and the slides will remain in the number 1 position until the lid is raised and the slides are removed. From here, it is a good practice to clear the slides with additional Paraway, then mount with coverslips.

Clean off any excess dye around the tissues and clean off any excess mountant.

Tag and identify the slides and submit them to the pathologist with the necessary record sheets.

Filing System. An often neglected subject in the histology area is the filing system. This storage of information is very important due to the necessity of many times having to recut old blocks for verification of diagnosis and follow-up programs.

After the paraffin sections have been cut, the blocks should be placed in cardboard boxes, identified, and carefully filed in the drawer of a metal slide file in numerical order according to case numbers. (See Fig. 4.)

Microscopic slides should be filed according to case and case number after they have been read by the pathologist. When this system of filing is used, it is a very easy and simple procedure to find old blocks for recutting for special stains or deeper cuts and slides for review. The files are also equipped with storage space for projection slides and Kodachromes. Any combination of drawers may be purchased as needed.

HAZEL E. WILLIAMS H.T. (ASCP)
DONALD W. MAYFIELD B.S., H.T. (ASCP)

b

BACTERIA

The term "bacteria" includes all microorganisms which are prokaryotes with the exception of the blue-green algae. Prokaryotic cells are characterized by the absence of a nuclear membrane and the mitotic process, and they have no structured chromosomes, mitochondria, or chloroplasts. The genetic material is a long (600–1000 μ) molecule of double-stranded DNA with its ends joined to form a circle, and appears on electron microscopy as a fibrous structure with no separation from the surrounding cytoplasm. Two major subdivisions of bacteria are recognized: (1) the true bacteria which can be cultured on artificial growth media, and (2) the rickettsiae and chlamydiae, all of which are strict cellular parasites.

Culture Media. Most nonpathogenic species of bacteria can be grown on solid-nutrient agar or in liquid-nutrient broth. Bacteria of human or animal origin, especially pathogenic species, usually require complex media such as blood agar, brain-heart-infusion agar, or comparable broths. All such media can be obtained from biological supply houses either ready to use or in a dehydrated form with instructions for rehydration and sterilization. Available, also, are prepackaged and sterilized plastic Petri dishes costing only a few cents each. The culturing of anaerobic bacteria requires some method of excluding air. The simplest is a disposable anaerobic system using a plastic bag and prepackaged chemicals which remove O_2 within the system as well as generating required CO_2. The rickettsiae and chlamydiae can only be cultured in living biological systems such as the yolk sac of a fertilized chicken egg. Saprophytic bacteria such as found in soil and water are usually grown at room temperature, while organisms of animal or human origin are grown at 37°C.

Microscopy. For most routine and clinical observations of bacteria the optical microscope is used with a 1.3 numerical aperture condenser, combined with a 1.25 NA oil immersion objective, a system which can achieve a resolving power of about 0.2 μ. Such resolution is adequate where only the determination of size, shape, or staining reaction of the bacterial cells is required. The average size of bacteria (0.5–10 μ), however, dictates that studies of the details of cytological structure must take advantage of the 0.0002–0.0008 μ resolving power of the electron microscope.

Preparation of Slides. If the bacteria are in liquid suspension a loop (0.01 ml or less) is placed on the surface of a clean 1×3 in. glass slide, spread over a 1–2 cm^2 area, air-dried, and heat-fixed by passing three or four times through the flame of a Bunsen burner, smear side

up. If the organisms are taken from a solid culture medium, then a loop of water should first be placed on the slide, and enough cells from the culture added to produce the first traces of turbidity. After spreading in a thin smear, air drying, and heat fixing, the smear should be sufficiently thin that the printed page can easily be read through it. For most purposes bacteriological smears do not require coverslips and the immersion oil is placed directly on the smear and the objective lowered into it.

The Gram Stain. The most useful of the bacteriological stains is the Gram procedure (see Bartholomew, 1962), which separates all bacteria into two groups, gram-positive and gram-negative, and it is an important step in the identification of unknown organisms. The procedure requires four reagents: Hucker's crystal violet, Burke's iodine, safranin, and 95% ethanol. The stock solution of Hucker's crystal violet is made by adding 20 ml of 95% ethanol to 2 g of certified crystal violet chloride. For use, dilute 1 part of stock solution to 4 parts of 1% ammonium oxalate in distilled water. Burke's iodine is prepared by placing 2 g of KI into a mortar with 1 g of iodine and mixed by grinding with a pestle. While grinding add 1 ml, then 5 ml, of distilled water. After the KI-I are dissolved, add an additional 14 ml of water and mix. Pour into a 100 ml reagent bottle and rinse the mortar and pestle with the water needed to bring the volume to 100 ml. The safranin stock solution is prepared by adding 100 ml of 95% ethanol to 2.5 g of certified safranin chloride. For use, 10 ml of the stock solution is diluted with 90 ml of distilled water. The procedure for a heat-fixed smear is as follows. Flood the smear with Hucker's crystal violet for 1 min then wash gently (a 5 sec wash in a beaker with tap water running into it at a rate of 30 ml/sec). Over-washing which might occur when the smear is washed in the tap stream could cause gram-positive organisms to appear gram-negative. Apply Burke's iodine for 1 min and wash. The wet smear should be decolorized with 95% ethanol by passing the slide through three Coplin dishes (75×25 mm) decolorizing about 10 sec in each. An alternate method is to use a dropper and run the 95% ethanol over the smear for a period of 20–30 sec. Do not let the smear dry between the iodine and decolorizer as this will increase the required decolorization time. Over-decolorization will cause gram-positive organisms to appear gram-negative. Wash, and flood with safranin for about 1 min, then wash briefly, blot dry, and examine. Gram-positive organisms will appear blue to black and gram-negative organisms red.

Most eukaryotic cells (except yeast) are gram-negative and therefore the taxonomic value of differentiation by the Gram procedure is greatest for prokaryotic cells, and especially for the true bacteria since all of the rickettsiae and chlamydiae are gram-negative. Gram differentiation of prokaryotic cells is more fundamental than just a staining reaction since it correlates also with physiological properties as well as with difference in cell wall chemistry and structure. The majority of cells of living things lose the primary dye on decolorization and are gram-negative. It is the dye retention ability of gram-positive cells which is unique and which needs explanation. Almost every researcher studying this retention has concluded that the cell wall is the most likely responsible agent, and that differences in cell wall chemistry and structure are involved. However, the exact role of the cell wall in differentiation is unknown (for a discussion see Bartholomew et al., 1965).

The Acid-fast Stain. Another important differential stain is the acid-fast stain which is used mostly to identify bacteria belonging to the genus *Mycobacterium*. Although some of the mycobacteria are saprophytic, this group also includes the organisms causing tuberculosis and leprosy. Mycobacteria are difficult to stain, but once stained they resist decolorization with acid alcohol while other bacteria and all other cells do not. The most used acid-fast stain is the Ziehl-Neelsen procedure, which employs three reagents, Ziehl's carbol fuchsin, acid alcohol, and Loeffler's methylene blue. Liquid phenol (90%) is helpful and can be prepared by heating phenol crystals to 45°C and adding warm distilled water to 10%. On cooling this 90% phenol will remain liquid and can be used for preparing other phenol solutions. Begin by disolving 0.3 g of certified basic fuchsin chloride with 10 ml of 95% ethanol, and after the dye is dissolved add 100 ml of 5% phenol. Ziehl's carbol fuchsin must be aged several days before use, and will keep for about two months. The acid alcohol consists of 3 ml of hydrochloric acid (conc.) in 97 ml of 95% ethanol. Loeffler's methylene blue is prepared by dissolving 0.3 g of methylene blue chloride with 30 ml of 95% ethanol, which is then added to 100 ml of 0.01% sodium hydroxide. This stain may deteriorate in a few weeks' time. The procedure for heat-fixed smears begins by flooding the slide with Ziehl's carbol fuchsin and heating gently (do not boil) for 5 min. Do not let the volume of the stain become reduced during heating or a red precipitate will form which may confuse interpretation of the slide. Cool the slide and wash in water for as short a time as possible. Decolorize the wet smear by using a dropper to run acid alcohol over it for from 10 to 30 sec depending on the smear thickness. Wash gently and counter stain 1 min with Loeffler's methylene blue. Wash, blot dry, and examine. The acid-fast organisms should be bright red, and the rest of the material blue.

Bacterial Spore Stains. The presence of bacterial spores should be determined by using a 24–48 hr culture grown on a solid culture medium. A successful prodecure is that of Schaeffer and Fulton (1933). Prepare a heat-fixed slide and flood with 5% aqueous malachite green chloride. Heat two or three times to steaming for a total time of 1–2 min, cool, and wash gently. Flood with 0.5% safranin for 1 min, wash lightly, blot dry, and examine. The spores should be green, the vegetative cells red. An alternate procedure which does not require heating is that of Bartholomew and Mittwer (1950). The slide is passed through the flame of a Bunsen burner, smear side up, for 20 passages, and cooled. Stain 10 min at room temperature with a saturated (10%) solution of aqueous malachite green chloride. Wash briefly, and stain for 15 sec with 0.25% safranin, wash 1 sec, blot dry, and examine. The spores stain green, the vegetative cells red.

Stains for Capsules. The demonstration of well developed capsules on bacterial cells is not difficult. A negative stain using India ink (Butt et al., 1936) is simple and quick. Place a drop of 6% dextrose on one end of a slide and add organisms to a light turbidity. Add a drop of India ink (Higgins waterproof ink is acceptable) and mix. Spread in a thin smear with the thin edge of another glass slide as for thin blood smears, and air dry. Fix 1 min with absolute methanol, drain dry, and stain 1 min with any aqueous basic dye such as Loeffler's methylene blue. Wash, blot dry, and examine. The background should be dark, the cells blue, and the capsules colorless. In this procedure shrinkage of vegetative cells may cause a thin clear area to appear around the cell even though no capsules are present. As an alternate procedure (Anthony, 1931) suspend the organisms in a drop of serum on a slide, spread into a thin smear, and air dry (do not heat fix). Stain 1 min with Hucker's crystal violet (see Gram stain) and wash with 20% copper sulfate, blot dry, and examine. The copper sulfate decolorizes the capsular material faster than the protein background and cells. The cells and background should be deeply stained, the capsules should be light blue to colorless.

Staining of Rickettsiae and Chlamydiae. Although the rickettsiae and chlamydiae are bacteria, because of their small size (0.3–1.0 μ) they are difficult to demonstrate with the optical microscope, and because slides are usually prepared from yolk sac material this also tends to mask the organisms since both organisms and background material are gram-negative. As yet no spectacular differential staining procedure is available; however, the Giménez (1964) modification of the Macchiavello stain is probably the best. The reagents needed are a buffered carbol fuchsin, and an 0.8% aqueous solution of malachite green oxalate. The stock carbol fuchsin solution is prepared by dissolving 10 g of basic fuchsin chloride with 100 ml of 95% ethanol, adding 250 ml of 4% phenol, and diluting with distilled water to make 1 liter, and storing at 37°C for 2 days. A 0.1M sodium phosphate buffer, pH 7.45, is prepared by mixing 3.5 ml of 0.2M NaH_2PO_4, 15.5 ml of 0.2M Na_2HPO_4, and 19 ml of distilled water. The working dye solution is prepared by adding 4 ml of the stock carbol fuchsin to 10 ml of the buffer, and filtering. This working stain must be filtered before every use, and will keep for about 40 hr. A thin smear is prepared using yolk sac tissue from which the yolk has been drained as much as possible, then placed on a glass slide, air-dried, and heat-fixed (or heat fixation may be omitted). The staining procedure (except for *R. tsutsugamushi*) begins with covering the tissue with the working carbol fuchsin stain for from 1 to 2 min. Wash thoroughly with tap water and cover with malachite green oxalate for 6–9 sec, wash, restain with the malachite green for a similar time, wash, blot dry, and examine. Rickettsiae and chlamydiae will be stained deep red, other cells greenish blue, and the background material green. For *R. tsutsugamushi* the procedure is varied in that following the carbol fuchsin stain the slide is washed, then 4–6 drops of 4% aqueous ferric nitrate ($Fe(NO_3)_3 \cdot 9H_2O$) are placed on the smear and immediately washed away, followed by 0.5% aqueous fast green FCF for 15–30 sec, washed, blotted dry, and examined. The rickettsiae are reddish black while the background is green.

Rickettsiae and chlamydiae can be stained using a prolonged Giemsa stain as described below for mycoplasmas. Specific immunofluorescent procedures also are available for various rickettsiae and chlamydiae species (Cherry et al., 1960).

Stains for Mycoplasmas and L-phase Bacterial Variants. Mycoplasmas are true bacteria which do not have cell walls, and which are not known to have originated from, or to be able to revert to, forms with cell walls. The L-phase growth variant of bacteria also does not have cell walls, but is known to have originated from, or to be able to revert to, normal bacterial cells. Both are delicate and are destroyed if the usual heat-fixed smears are attempted. The following applies specifically to myco-plasmas, but in general the same staining methods can be used for L-phase variants. The simplest method for observing mycoplasmas with the optical microscope is the stained agar block method of Dienes (see Madoff, 1960). Mycoplasmas are grown on a meat-infusion, 10% serum, 1.5% agar medium. This medium is inoculated by cutting an agar block from an older culture and placing it with the organisms down onto the surface of fresh agar (2 mm thick or less) in a Petri dish and streaking the block over the surface. After 2–4 days' incubation small "fried egg" colonies should be visible. The Dienes stain is prepared by dissolving in 100 ml of distilled water 2.5 g of methyl-ene blue chloride, 1.25 g of azure II, 10 g of maltose, 0.25 g of Na_2CO_3, and 0.2 g of benzoic acid. Using a cotton applicator, apply a thin film of stain to one side of a coverslip, and air dry. A block of agar should be cut from the freshly grown culture and placed culture side down (and next to the stain) in the center of the coverslip. The agar block should be about 1 mm smaller than the coverslip. Place both, coverslip up, onto a glass slide and seal with paraffin and 10% vaseline. After a few minutes observe with the oil immersion objective. Mycoplasmas will stain bright blue, and as time passes will retain the dye better than any other bacteria which might be present. Autolyzed bacterial cells, other cellular material, and debris will stain with a pink hue.

A more difficult procedure, but one which yields better results, is the agar-fixation method of Klieneberger-Nobel (1962). The reagents needed are Bouin's solution (75 ml of saturated aqueous picric acid, 25 ml of formalin, 5 ml of glacial acetic acid), Giemsa stain, and a phosphate buffer pH 6.8. The stock Giemsa solution is prepared by heating 33 ml of glycerol to 55°C and slowly adding 0.5 g of Giemsa stain (powder form, certified) with stirring. Hold at 55°C for 3 hr, cool, and add 33 ml of absolute methanol with mixing. Store overnight in a desiccator to allow sediment to settle, then decant into small (30 ml) bottles to capacity and tightly stopper. Store two weeks and filter before use. Coplin dishes should be used for staining and these should be filled with 1 part of the stock Giemsa diluted with 50 parts of the phosphate buffer pH 6.8. Best results are obtained when the organisms are allowed to grow at the agar-coverslip interface. Organisms are inoculated onto the surface of a sterile coverslip and covered with meat-infusion, 10% serum, agar; or a block of freshly inoculated agar medium is cut out and placed, organisms down, on the surface of a sterile coverslip. The agar of such preparations should not be over 2 mm thick and they should be incubated in a moist chamber for 2–3 days. After growth, the coverslip with the agar is fixed in Bouin's solution for 8–18 hr, then the agar is peeled away. Wash the coverslip until the picric acid color disappears, then stain for 8–24 hr with the 1 : 50 Giemsa. Cautiously remove the stain with filter paper, but before the coverslip can dry pass through 19 : 1 acetone-xylene,

then 14 : 6, then 6 : 14, and then three times through xylene alone. Mount in Canada balsam.

If the organisms are in frozen sections a stain which will differentiate them from other cells and background material should be used such as the differential stain, or the immunofluorescent stain used by Goodburn and Marmion (1962) for chick embryo lung tissue. Details for other specific immunofluorescent stains can be found in Cherry et al. (1960), and in Purcell et al. (1969).

Electron Microscopy. Although the optical microscope is an important instrument for studying the size, shape, and staining reactions of bacteria, the electron microscope must be used if details of structure are to be resolved. The methods used for the transmission electron microscope include the direct disposition of cells onto coated grids, shadowing, and surface replication. However, most fine detail is observable only through more sophisticated techniques such as ultra-thin sectioning. The fixation and embedding of bacteria for ultra-thin sectioning should generally follow the standard technique of Kellenberger and Ryter (see Kellenberger, et al., 1958), or the more recent technique which uses glutaraldehyde as well as osmium tetraoxide for fixation (Ryter and Jacob, 1966). The details of these procedures are too involved to be given here, but can be found in the above references. Most of the techniques of electron microscopy are of sufficient technical difficulty that they should not be attempted except under the direction of a qualified electron micro-scopist. An appreciation of these techniques can be obtained by reference to the books of Pease (1964), or Sjöstrand (1967).

JAMES W. BARTHOLOMEW

References

Antony, E. E., "A note on capsule staining," *Science,* **73**:319–320 (1931).
Bartholomew, J. W., "Variables influencing results, and the precise definition of steps in Gram staining, as a means of standardizing the results obtained," *Stain Technol.,* **37**:139–155 (1962).
Bartholomew, J. W., T. Cromwell, and R. Gan, "Analy-sis of the mechanism of Gram differentiation by use of a filter-paper chromatographic technique," *J. Bacteriol.,* **90**:766–777 (1965).
Bartholomew, J. W., and R. Mittwer, "A simplified bacterial spore stain," *Stain Technol.,* **25**:153–156 (1950).
Butt, E. M., C. W. Bonynge, and R. L. Joyce, "The demonstration of capsules about hemolytic strepto-cocci with India ink or azo blue," *J. Infect. Dis.,* **58**: 5–9 (1936).
Cherry, W. B., M. Goldman, T. R. Carski, and M. D. Moody, "Fluorescent antibody techniques," Public Health Service Publication No. 729, Washington, D.C., U. S. Govt. Printing Office, 1960.
Giménez, D. F., "Staining rickettsiae in yolk sac cul-tures," *Stain Technol.,* **39**:135–140 (1964).
Goodburn, G. M., and B. P. Marmion, "A study of the properties of Eaton's primary atypical pneumonia organism," *J. Gen. Microbiol.,* **29**:271–290 (1962).
Kellenberger, E., A. Ryter, and H. Séchaud, "Electron microscope study of DNA-containing plasma. II. Vegetative and mature phage DNA as compared with normal bacterial nucleoids in different physiological states," *J. Biophys. Biochem. Cytol.,* **4**:671–678 (1958).
Klieneberger-Nobel, E., "Pleuropneumonia-like Organ-isms (PPLO) Mycoplasmataceae," London–New York, Academic Press, 1962.
Madoff, S., "Isolation and identification of PPLO," *Ann. N. Y. Acad. Sci.,* **79**:383–392 (1960).

Pease, D. C., "Histological Techniques for Electron Microscopy," 2nd ed., London–New York, Academic Press, 1964.

Purcell, R., R. M. Chanock, and D. Taylor-Robinson, "Serology of the mycoplasmas of man," in L. Hayflick (ed.), "The Mycoplasmatales and the L-phase of Bacteria," New York, Appleton-Century-Crofts, 1969.

Ryter, A., and F. Jacob, "Étude morphologique de la liaison du noyau a la membrane chez *E. coli* et chez les protoplastes de *B. subtilis*," *Ann. Inst. Pasteur (Paris)*, **110**:801–812 (1966).

Schaeffer, A. B., and M. D. Fulton, "A simplified method of staining endospores," *Science*, **77**:194–195 (1933).

Sjöstrand, F. S., "Electron Microscopy of Cells and Tissues. Vol. I., Instrumentation and Techniques," London–New York, Academic Press, 1967.

BAKERY PRODUCTS

Introduction. Prior to describing techniques that can be applied to baked goods, the general technology of these products is best discussed, together with some indication of the problems of structure they pose.

Types of Baked Goods. In general, baked goods are derived from mixtures of wheat flour, salt, water, sugar, and fats, in varying proportions, the principal ingredient being wheat flour. Flours and starches derived from other cereals may be used in small amounts, e.g. rye meal, oat meal, and maize starch. Further, legume-based raw materials, such as soya flour, may be employed.

According to the mixing procedures used, the proportions of the basic ingredients, the use of yeast or raising agents to aerate the dough either before or during baking, and the methods of forming the dough piece, dough sheet, or final liquid or semiliquid mix, a baked food of a certain type is produced.

Thus wheat flour, water, yeast, and salt in the correct proportions provide a dough that can be fermented, moulded, proved, and baked to provide bread. Again, wheat, flour, water, sugar, fats, and eggs can be mixed to produce a batter that can be baked to produce cake. Further, wheat flour, water, sugar, and fats, together with aerating agents, can be mixed in a great variety of proportions, especially with regard to the amount of fat and sugar, and can be baked to produce the considerable range of baked products known as biscuits.

The general nature of the problem of structure in baked goods is to ascertain the distribution within the product of the principal components supplied in the original mix and to determine any modifications these components have undergone and any structures to which they have given rise.

Considering, for example, bread production, the unique property of undenatured wheat protein to form a cohesive, visco-elastic protein complex, generally termed gluten, on hydration with the right proportion of water, forms the basis of the bread-making industry. Starch is embedded in this gluten complex and following baking becomes swollen and gelatinized.

Therefore the final shape and size of a loaf of bread depends, first, on the expansion, during fermentation and proofing by the production of carbon dioxide from yeast, of a hydrated gluten network in which starch is embedded; and second, on the baking process further expanding the whole structure, gelatinizing the starch which absorbs water, swells, and forms a firm gel, and at the same time expanding and denaturing the gluten network. Thus during baking, starch granules present assume an important role and provide a major part of the structure of bread crumb, so bringing to completion the potential loaf structure present in the dough by virtue of a three-dimensional gluten network containing starch granules.

The description of the nature of bread raises a number of points on which microscopical techniques can be brought to bear. For example, is there a continuous protein network surrounding all starch granules in bread? Does this protein network completely line the gas cells formed in bread dough during fermentation? How is fat or lipid distributed in bread crumb and in bread dough?

In the case of biscuits, while some contain gluten networks, in many others the presence of considerable proportions of sugar and fat render these networks unlikely. In these cases there is often a low ratio of water to wheat flour, and this not only limits the development of gluten networks but also has consequences for starch gelatinization at the baking stage, for without the appropriate mass ratio of water to starch the degree of starch swelling and gelatinization will be considerably modified. This has consequences for the structure of the biscuit. It raises questions concerning for example, the extent of the damage to the starch and the technique that could be applied to determine it. Thus, in the case of biscuits microscopical techniques are frequently being applied to heterogeneous mixtures of components supplied in the mix.

Histological Techniques and Baked Products. Botanical and zoological specimens selected for histological examination are subject normally to a standard series of treatments, well delineated elsewhere. These specimens are well able to withstand this treatment due to the presence of strong cell walls, and the aggregation of cells into tissues. Even so, distortion and the production of artifacts is an ever-present possibility, and the interpretation of results requires skill and caution: The history of cytology is littered with the bones of dead and gone controversies spawned from the over-eager interpretation of artifacts.

Baked products by virtue of their desirable attributes, such as lightness, softness, and crispness, are often very fragile. The possibility of distortion and the potentiality for artifact production are therefore very much greater in baked goods during their examination by histological techniques. The crucial problem of the application of histological technique to the study of the structure of baked goods is the avoidance of distortion, destruction, and artifact. This is well shown in the following review of work using the light microscope.

Techniques Using the Light Microscope. Many workers have been concerned with evaluating artifacts produced. Butterworth and Colbeck (1938) concluded that for doughs ordinary methods of staining and preparation were unsuitable because such methods interfered greatly with the true structure of the material examined. They found that when dough films were teased out with a mounting needle in the presence of liquid (stain, water, or glycerine) starch cells in considerable numbers were readily detached from the gluten structure and dispersed in the liquid. Burhans and Clapp (1942), working on dough and bread, also enumerated the difficulties and considerations that prevent the use of many standard procedures and make the problem of true structural representation a difficult one. Sandstedt et al. (1954) avoided the distortions introduced by the use of aqueous solutions by resorting to plastic-embedded specimens where the embedded structure was demonstrated by grinding the surface of the block to expose the structure on the surface, then staining and examining by reflected light. In the case of determining fat distribution in doughs and baked goods, Platt and Fleming (1923) outlined precautions needed to prevent the fat from forming globules, while Hanssen and Dodt (1952) claimed that these precautions were not strictly necessary.

Hanssen (1954) has given a useful summary of microscopy in the field of biscuit making, and in 1957 published details of a fluorescent method for observation of changes in structure of dough from cereal flour.

During the nineteen fifties work also began at the British Baking Industries Research Association, Chorleywood, England, on aspects of biscuit dough and biscuit microstructure. Bradshaw (1951) described the application of the ultraviolet microscope to the examination of the structure of biscuits, and Ottaway and Groves (1955) published a technique for the examination of the microstructure of biscuit doughs.

While work on mainland European biscuits by Hanssen and on bread by American workers, was available, the application of techniques to the study of the structure of the three principal United Kingdom biscuit types, namely "short sweet," "semisweet" or "hard sweet," and "cream cracker," had not fully taken place.

In 1962 Francis and Groves published a technique using gelatin embedding. The specimen was stained prior to embedding, using 2% aqueous Fast Green FCF for differentiating protein, and Sudan III or Oil Red O in gelatin, for staining fat. The fat stains work from a colloidal suspension and the disposition of fat is not so affected as when fat stains are employed made up in fat solvents; the method of making this type of fat stain is given in Gurr (1958).

The prestained specimen was then brought through a suitable embedding series culminating in 20% gelatin; after removal from the mold, the block was stored in 5% formalin. The block was well rinsed with distilled water prior to sectioning using a Sartorius freezing microtome. After floating off with distilled water, the sections were fixed on with 1% gelatin solution, dried overnight at 25°C, then immersed in 10% formalin for 10 min, rinsed with distilled water, and permanently mounted in Farrant's medium.

This technique was applied to typical biscuit types made in the United Kingdom, known as "hard sweet," "short sweet," and "cream cracker," and also applied to wafers, crispbread, cake, and bread.

In this work (Francis and Groves, 1962) the authors discuss the problem of artifacts, distortion, and destruction. They concluded that the general picture of bread crumb structure found was similar to that obtained by previous workers, and therefore the overall picture obtained of the structure of "hard sweet," "short sweet," "cracker," and wafer biscuits, and cakes, was substantially correct. As "hard sweet" biscuits, for example, contain a continuous protein network akin to that in bread, this was a reasonable assumption. Nevertheless, the authors emphasized three points to be borne in mind when interpreting sections of baked goods prepared by the gelatin embedding technique.

First, the use of aqueous media results in swelling of the specimen, especially the starch. Secondly, shrinkage of the specimen occurs after transfer and fixing to the slide, with consequent rupture of attenuated protein films, or with gluten strands detaching themselves from starch granules.

Thirdly, the role of sugar in baked-goods structure is at present indeterminable; the use of aqueous media means that any sugar present dissolves in the reagents used and cannot be detected in the final sections. Where a biscuit contains a large proportion of sugar, therefore, the interpreter of the structure must bear in mind that much of the structure may be missing. The location of oligosaccharides such as sugar is not a problem confined to baked goods, but has also proved difficult in the development of histochemical methods for the detection of these substances in plant and animal tissues, and no satisfactory method exists.

While this work applied the gelatin embedding/freezing microtome technique to the study of the structure of certain hitherto uninvestigated biscuit types, it was not applied to the dough mixing and machining stages of biscuit manufacture.

This gap was filled in the late nineteen sixties by further work at the Flour Milling and Baking Research Association, Chorleywood, in collaboration with the University of Leeds, by Flint et al. (1970), who published techniques for the study of the microstructure of biscuits and their doughs. They were concerned with studying the changes occurring in the structure of doughs and biscuits at various stages of manufacture. "Cream cracker," "hard sweet," ("semisweet"), and "short sweet" biscuits and their doughs were again chosen for study.

Specimens of plant or animal tissues are normally fixed prior to embedding. The situation is somewhat different with baked goods. In their case the process of baking acts in a way analogous to a fixative in that the final structure is set, and components such as protein are denatured. For baked goods it is not strictly necessary to employ a fixative, but in the case of doughs, and certainly in the case of wheat kernels, some workers consider it desirable to use a fixative in the normal way.

Flint et al. put their dough pieces, approximately 3 cm by 2 cm in labeled cassettes which were then placed in cool (5°C) 4% aqueous glutaraldehyde solution buffered to pH 6.8. This fixative made the dough samples more cohesive, and therefore easier to section, by acting as a cross-linking agent between the protein molecules. Fixation was complete after 10–14 days. Prior to sectioning, the dough samples were taken from the fixative and cut into cubes of edge measurement less than 1.0 cm.

The authors employed a refrigerated compartment the temperature of which could be accurately controlled, termed a cryostat, in which a retracting microtome was mounted and which could be operated by controls passing through the walls.

The 1 cm cubes of dough were frozen in position on the microtome specimen holder by means of a jet of liquid carbon dioxide, and the specimen holder was then fitted to the microtome. Sections of dough 8–10 μ thick were cut from the specimen and carefully transferred to slides, previously coated with adhesive, while still in the cryostat. In the case of the dough specimens the sectioning technique relies on the presence of water in the specimen acting as the support media. In order, therefore, to apply this technique to biscuits it was necessary to condition them.

Conditioning was achieved by placing the biscuits over a humectant solution for two days to allow the moisture content to increase slowly. Next, small squares, size 1.0 cm, were cut from the biscuits and placed in cooled (5°C) fixative solution for two days. Finally, traces of air remaining in the biscuit samples were removed by gently drawing a vacuum over the fixative solution. Biscuit samples were then sectioned in a similar way to the dough specimens, the sections being 12–15 μ thick; they were also transferred to prepared slides while still in the cryostat.

Both dough and biscuit sections were allowed to thaw and dry. Those intended for fat distribution studies were thawed and dried at 5°C and stored at this temperature to prevent possible migration of the fat. Remaining sections were dried at 37°C and stored at ambient temperature.

Flint et al. employed the following staining techniques: A solution of Ponceau 2R was used for protein, and for demonstrating damaged starch, chlorazel violet R was satisfactory, being taken up by the damaged starch but not by the undamaged granules. (See Figs. 1 and 2.) Fat was stained with a 70% alcoholic solution saturated with Oil Red O and Sudan IV; the fat staining was carried out at 5°C at which temperature the fats are quite unable to flow. However, the rate of diffusion of the dyestuffs in the fat, and therefore the rate of staining, was greatly decreased at this temperature so the staining time was increased from 10 min to 1 hr. Yeast cells, present in "cream cracker" biscuits, were stained a blue-green color, using Fink and Kühle's methylene blue.

The technique described by Flint et al. involving the use of water, carefully introduced into the biscuit sample, as the sole supporting medium, while sectioning with a microtome in a cold environment, is essentially an extension of the freezing microtome technique. It has the considerable advantage of avoiding embedding in gelatin by maintaining the integrity of the section by virtue of keeping it at a low temperature until it has been transferred and fixed to the slide. Thus the technique reduces the possibilities of distortion of the structure occurring during embedding.

A certain amount of destruction and distortion is unavoidable with some baked products in the actual process of sectioning itself, and it is here that the skill and care exercised by the operator of the microtome in sectioning and handling sections becomes paramount.

Many of the observations made by Flint et al. on the general structure of doughs and biscuits, "hard sweet," "short sweet," etc., confirm the results of Francis and Groves; for example, the presence of a continuous protein network in "hard sweet" biscuits and "cream crackers," and the absence of such a network in "short sweet" biscuits.

A similar technique involving the use of low-temperature sectioning has been used by Wehrli and Pomeranz (1970). They were concerned with studying the interactions in dough and bread between the following wheat flour components: starch, glycolipid, and gluten. They carried this out with a tritium-labeled glycolipid, using autoradiographical techniques.

Micro-mixing and micro-baking techniques were employed, the labeled glycolipid being mixed in with all the other dough ingredients.

Small pieces of dough (immediately after mixing and after fermentation) and of bread were placed on the freezing table of a microtome that had been precooled with dry ice (AO Spencer automatic clinical microtome with freezing attachment). The pieces were attached to the table with a few drops of water. Sectioning was carried out in a cold room at −15°C. Knife and dough temperatures were kept below −30°C by placing chunks of dry ice under the freezing table. Above that temperatures no satisfactory sections could be made. The sections, 5 μ thick, were transferred by a brush with about 10 camel hairs, from the knife to a microscope slide covered with a thin layer of glycerol. Thus, no fixation or embedding was employed, and as bread contains about 40% water, no conditioning was necessary. The authors considered that artifacts from fixation and embedding were eliminated and that the sectioning procedure minimized diffusion, redistribution, and leaching of the labeled glycolipid (see also Stumpf, 1969).

After standing for a day in a box at room temperature, the slides were preheated to about 60°C, and were dipped into a Kodak N-2 nuclear track emulsion at about 40°C. The emulsion was allowed to dry for 2 hr in a box at room temperature, and the box with slides was stored

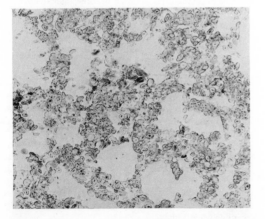

Fig. 1 Cross section of a commercial "short sweet" biscuit, stained with chlorazol violet R to show the distribution of damaged starch. This shows the center of the biscuit. The starch damage has gradually increased but is less than 50%. Courtesy Flour, Milling and Baking Research Association, Chorleywood, England. (×56.)

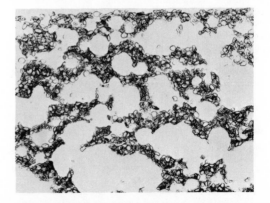

Fig. 2 Cross section of pilot bakery "hard sweet" biscuit, stained with chlorazol violet R to show the distribution of damaged starch; protein matrix is thin and delicate. At the top there is a thin crust where damage is low, but this increases until at the bottom of the picture damage is 50%. Courtesy Flour, Milling and Baking Research Association, Chorleywood, England. (×56.)

in the refrigerator at 40°C for 6 days. To develop the photographic emulsion, the slides were dipped consecutively for 5 min into Eastman Developer D-19, twice for 15 sec into distilled water, for 10 min into Kodak Rapid Fixer with Hardener, and washed for 25 min in running water. All operations beginning with dipping into the emulsion were made in complete darkness.

The sections were stained with iodine by dipping the slides for 10 sec in 10 % Lugol iodine solution. Sections were then examined under a Leitz-Wetzlar microscope, and pictures taken with the Orthomat microscope attachment camera at a magnification of 1000×. For comparison, nonradioactive dough and bread were processed and compared.

As far as the general microscopical results were concerned, Wehrli and Pomeranz concluded that these were basically similar to those published by other authors. Starch granules stained blue were embedded in an unstained protein matrix. In the dough, black points produced by the labeled glycolipid were distributed both in the gluten matrix and in the starch granule. In the bread, most of the black points appeared in the gelatinized starch.

Summary and Conclusions. It is clear that over the decade 1960–1970 work on the microstructure of baked goods has ceased to employ the classical fixation and embedding techniques and evolved towards the use of water, either present in or infiltrated into the product, as the sole support medium during microtomy carried out on a freezing microtome situated in a low temperature area.

Subsequent affixing, drying, staining, and examination with the light microscope have employed typical techniques and the stains appropriate to the identification of protein, starch, and fat.

It is apparent from the published work reviewed here, concerned with the microstructure of baked goods as revealed by histological techniques employing protein, fat and starch stains, and examined by means of the light microscope, that there is a limit to the information so to be obtained. The overall microstructure of the several types of baked product and their doughs is now generally established. Refinements of technique will most probably occur by the use of more specific stains and reagents. These could be derived from histochemical and cyto-chemical techniques and designed to locate precisely components and sites of interaction in the protein, starch, and fat complexes. The ingenious combination of microtomy and autoradiography employed by Wehrli and Pomeranz foreshadows these future developments.

BRIAN FRANCIS

References

Bradshaw, R. C. A., B.B.I.R.A., Chorleywood, England, Research Report No. 16, 1951.
Burhans, M. E., and J. Clapp, *Cereal Chem.,* **19**:196 (1942).
Butterworth, S. W., and W. J. Colbeck, *Cereal Chem.,* **15**:475 (1938).
Fink, H., and R. Kühles, *Wschr. Brau.,* **50** (via *Stain Technol.,* **36**:329 (1961).
Flint, O., R. Moss, and P. Wade, *Fd. Trade Rev.,* **40**:32 (1970).
Francis, B., and C. H. Groves, *J. Roy. Microscop. Soc.,* **81**:53 (1962).
Gurr, E., "Methods of Analytical Histology and Histochemistry," London, Leonard Hill Books Ltd., 1958.
Hanssen, E., Sucker- und Süsswaren Wirtschaft, **10**:1 (1954).
Hanssen, E., *Z. Lebensmitt. Untersuch.,* **106**:196 (1957).
Hanssen, E., and E. Dodt, *Mikroscopie,* **7**:2 (1952).
Ottaway, F. J. H., and C. H. Groves, B.B.I.R.A., Chorleywood, England, Research Report No. 28, 1955.
Platt, W., and R. S. Fleming, *Ind. Eng. Chem.,* **15**:390 (1923).
Sandstedt, R. M., L. Schaumburg, and J. Fleming, *Cereal Chem.* **31**:43 (1954).
Stumpf, W. E., *Science,* **163**:958 (1969).
Wehrli, H. P., and Y. Pomeranz, *Cereal Chem.,* **47**:221 (1970).

BINOCULAR CAMERA LUCIDA

The drawing attachments discussed here are of the "Treffenberg" type, developed by Wild Heerbrugg in the early 1960s. They are commercially available from several other microscope manufacturers as accessories for their standard instruments. They differ basically from the earlier "camera lucidas" in that they permit simultaneous binocular vision and are therefore referred to as "binocular camera lucida attachments."

As applied to the microscope, according to Gage (1947), the *camera lucida* is a device that causes the magnified virtual image of the object under the microscope to appear as if projected on the table or drawing board, where it is visible with the drawing paper, pencil, dividers, etc. by the same eye, and in the same field of vision. The microscope image appears like a picture on the drawing paper.

The function of the drawing attachment is, to a certain extent, analogous to that of a photomicrographic camera. It serves as an optical recording accessory, whereby the graphic action is done by the observer's hand. It is the tool of choice where outline drawings, rather than accurate photomicrographs, are preferred. Consequently, it has found widespread use among biological illustrators.

Some obvious advantages of the drawing attachments are:

1. Irrelevant "clutter," which is sometimes disturbing in photomicrographs, is eliminated in the drawing and points of interest can be emphasized.
2. Related structures that appear at slightly different focus levels in the specimen can be outlined in one plane so that their connection is evident. Such multiple-focus drawings are of great value in the study of relatively thick transparent specimens, such as chromosome plaques. They also serve for elucidating spatial relationships in superimposed images obtained from a series of consecutive sections (Zimmermann, 1966).
3. Use of the drawing attachment facilitates comparison and measurements, for instance, comparison of growth rates of living specimens, recorded at various stages of development. Measuring with a ruler placed on the drawing pad is a straightforward procedure, once the ruler is calibrated against an object micrometer.

As we have seen, there is a functional similarity between the drawing attachment and a photomicrographic camera. However, the underlying design concepts of the two devices are quite different.

In the camera, the microscope image is simply projected

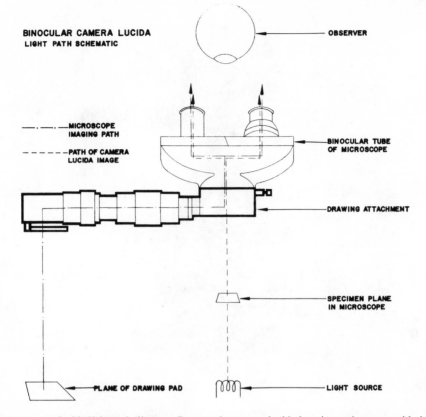

BINOCULAR CAMERA LUCIDA
LIGHT PATH SCHEMATIC

OBSERVER

MICROSCOPE
IMAGING PATH

PATH OF CAMERA
LUCIDA IMAGE

BINOCULAR TUBE
OF MICROSCOPE

DRAWING ATTACHMENT

SPECIMEN PLANE
IN MICROSCOPE

PLANE OF DRAWING PAD

LIGHT SOURCE

Fig. 1   Binocular camera lucida light path diagram. Because the camera lucida imaging path merges with the imaging path of the microscope *prior* to their entering the binocular, both images appear to the observer in a common plane in true inocular vision.

into the film plane. The drawing apparatus does not project an image originating *from* the microscope, it rather adds an external image, that of the drawing, *to* the one normally observed in the microscope eyepiece. This added image appears sharp and superimposed with the microscope image. The observer views both images in his binocular, quite comfortably and without fatigue. He can work in a well-lit room, as the ambient light will not diminish the visibility of his image.

Because the microscope image and the drawing are superimposed, perfectly matched structures, there is no confusion caused by side-inversion or by upside-down images, as encountered by conventional microprojection.

The drawing attachment described here fits easily between the binocular tube and the tube support of a Wild compound microscope. It is a tubular piece, protruding horizontally to the left or the right hand of the observer. The portion inserted into the observation path contains a beamsplitter, and the opposite end of the tube is equipped with a first surface mirror, inclined at a 45° angle. A sliding lens system inside the tube produces a variable enlargement effect.

It is the function of the mirror to pick up the pencil drawing while the drawing's enlargement may be changed to a blowup ratio of 1 : 2, so that the drawing may be enlarged to twice the size of the observed image. Auxiliary devices extend the blowup capability to the proportion of 1 : 5.

Initial focusing of a pencil tip on the pad is carried out while the microscope lamp is turned off. Conversely, the initial focusing operation in the microscope is done while the image emanating from the pad is blacked out. Finally, both images are viewed simultaneously and the optimum blending conditions for both the microscope lamp and the desk lamp (an adjustable tensor lamp is preferred) are established by trial and error. (The optimum light balance has to be reestablished whenever the objective power is changed.) The operator can now proceed with his tracing job, making sure, of course, that the drawing pad is maintained in a fixed position, preferably taped down to prevent it from moving.

Similar "drawing tubes" for camera lucida viewing with *stereomicroscopes* are also available. Because each ocular in the stereomicroscope produces an image of the specimen from a different angle, both images together give the impression of 3D-vision, but obviously, the drawing picked up from the drawing tube can be superimposed with only one of the two members of the stereopair. In other words, we cannot expect to obtain a stereo-outline on the drawing pad. Nevertheless, the drawing tube for stereomicroscopes has proven to be of great value, since it enables the observer to trace the specimen directly at variable scale ratios, much like the true binocular camera lucida.

W. GUMPERTZ

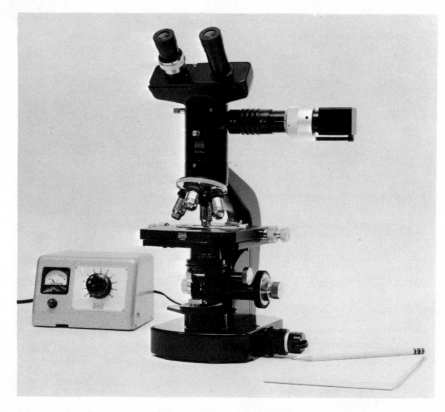

Fig. 2 Drawing attachment "binocular camera lucida" with additional 2.5× enlargement piece, on Wild M20 Research Microscope. The entire tube assembly is rotable, so that the operator can face the microscope stage. Moving the sleeve horizontally on the tube in direction V+ increases the enlargement ratio; in direction V— the enlargement of the drawing is decreased. With every change of the sleeve position, sharp focus on the drawing is reestablished with the knurled ring. The drawing pad may be taped to the table to prevent it from moving while tracing is in process. A desk lamp (not shown in picture) is required to illuminate the drawing pad.

References

Gage, S. H., "The Microscope," 17th ed., Ithaca, N. Y., Comstock Publishing Co., 1947 (p. 318).

Zimmermann, M. H., "Analysis of complex vascular systems in plants," *Science,* **72**:152 (1966).

See also: DRAWING WITH THE AID OF THE MICROSCOPE.

BLADDER

The capability of active sodium and passive water transport of the toad and turtle bladders together with their structural simplicity has led to the wide use of these organs in the study of electrolyte transport and morphological study in relation to their functions. Amphibian bladders are transparently thin, yet their histology is not as simple as it is assumed to be. The bladder wall is composed of mucosa, submucosa, and serosa. The mucosal epithelium consists of one or two layers of several cell types. Lamina propria and submucosa contain varying amounts of collagen, vessels, and smooth muscles. Serosal cells cover the coelomic surface of the tissue with their attenuated but uninterrupted cytoplasm. Following are the techniques used for studies on the bladders of toads and turtles, both in vivo and in vitro.

Electron Microscopic Techniques. (Fig. 2) The abdomen of pithed amphibia is cut open, avoiding bleeding, to minimize deterioration of osmium solution during fixation. Transparently thin bladder is an excellent organ for *in situ* fixation. It can be performed with an injection syringe by applying a few drops of 2% osmium in s-collidine on the serosal surface and by simultaneously squirting fixative on the subjacent mucosal surface from within the bladder. In a few minutes after *in situ* fixation the hardened fixed tissue is cut out with sharp scissors and transferred onto the fixative dropped on a dental wax plate, on which the tissue is cut into pieces with a sharp razor blade. These are transferred to fresh cold fixative for 1–2 hr of further fixation. (Formaldehyde or glutaraldehyde can also successfully be used as primary fixative or for histochemical purposes by introducing them into the bladder lumen through the cloaca.)

Fixed tissues are rinsed briefly with 30% ethanol and are dehydrated in 60%, 70%, 80%, 90%, and two changes of 100% ethanol, each for 5 min, and two changes of propylene oxide for 20 min each. Subsequent infiltration and embedding of the tissue, and polymerization processes are performed with good result by the method of Luft (1961). But prolonged infiltration time is desirable (for 4 hr with Epon or 6 hr with Araldite) and harder blocks are recommendable (Epon mixture A(6) : B(4) + 1.5% v/v

DMP-30), since submucosa of the bladder contains considerable amounts of tough collagen.

Orientation of the tissue within the polymerized resin in a gelatin capsule is required so as to cut the whole thickness of the tissue. Tissue with some amount of its surrounding resin is cut out with a dental disc and is mounted on an aluminum rod with synthetic glue. The aluminum rods remain tight in the chuck of the microtome during sectioning, whereas plastic rods often become loose.

Before sectioning, the surfaces of the block are trimmed with a clean razor blade under a binocular microscope but the resin surrounding the tissue should not be removed completely. In the trimming process the quality of the blocks can be judged; good blocks should have smooth and shiny cut surfaces. Bad blocks may be too soft or brittle, or may show frosted cut surfaces.

Submucosa with its quantities of tough collagen is particulary difficult to cut smoothly. Good blocks and well selected sharp knives overcome this difficulty. Orientation of the block is also important. The short axis of the tissue sheet in the block face should be vertical to the knife edge of the microtome, permitting the first reach of mucosal epithelium. Thus chatter effects and tears are at least confined to the connective tissue zone, often saving the epithelium, the major topic of interest in the study. For the study of membrane structures, sections with silver interference colors are suited but often show damage. Gold sections are too thick but are good for low-magnification work. Straw-colored sections are a good compromise for moderate-magnification and high-resolution work.

Stained sections may show some differences with each stain. For example, the fine structure of tight junctional complexes of the bladder mucosa may be demonstrated more readily with the uranyl solution than with lead solution.

Light Microscopic Techniques (Fig. 1). Light microscopic preparations of osmium-fixed, epoxy- or methacrylate-embedded tissues provide excellent cytological detail and bridge the gap in magnification and resolution between the light and electron microscopes. Methacrylate-embedded sections have an advantage over epoxy-embedded sections in that most of the histochemical staining methods used in paraffin sections can be applied with little or no modification. For example, methacrylate-embedded sections of 1.5–2 μ thickness stained with the PAS method followed by staining in slightly alkaline 1% toluidine blue in 50% ethanol solution reveal intensely colored PAS-positive substances of the bladder tissue.

Radioautography is applicable on epoxy or methacrylate-embedded tissue with excellent tissue preservation and precise localization of reduced silver grains.

Electrophysiological Techniques and Correlation Studies of Structure and Function. Isolated bladders of amphibia and turtles survive for many hours (more than 24 hr) in aerated Ringer's solution with well preserved function and fine structure. Correlation studies of structure and function can best be made on the bladders in vitro. Sheets of isolated bladder up to 10 cm^2 are mounted and bathed in aerated Ringer's solution in a lucite chamber. Active sodium transport from mucosal to serosal media is accurately measured by the short-circuiting technique of Ussing and Zerahn (1951), and water movement across the bladder wall can be measured with the volume chamber of Edelman et al., (1963) or Bentley's (1960) sac technique.

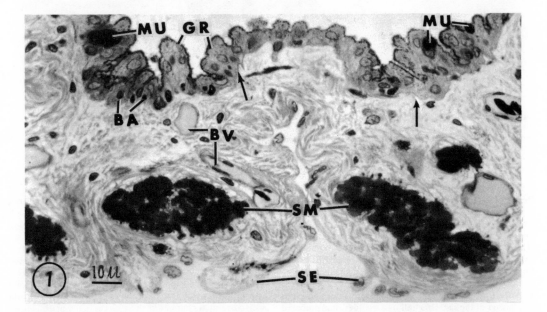

Fig. 1 Light microscopic picture of full thickness of toad bladder taken from a section of osmium-fixed methacrylate-embedded tissue, stained with PAS method and counterstained with alkaline toluidine blue. Mucosal epithelium consists of granular cells (GR) with many PAS granules (black), mitochondria-rich cells (not identified), mucous cells (MU) filled with PAS-stained globules and basal cells (BA). Submucosa (middle) contains blood vessels (BV), smooth muscles (SM), and various cellular elements and abundant collagen bundles. Serosal cells (SE) cover the coelomic surface of the bladder. Arrows indicate basement membrane.

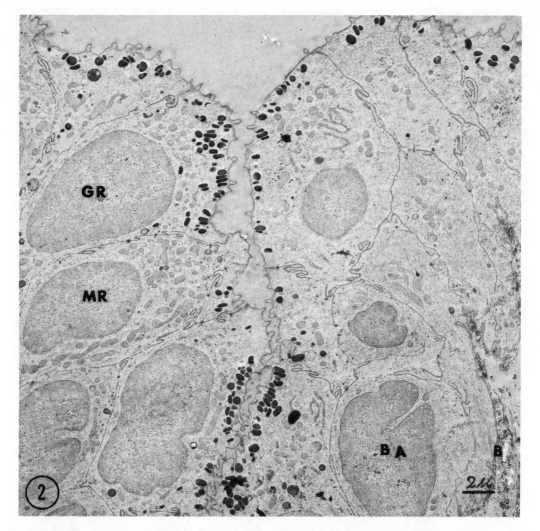

Fig. 2 Electron micrograph of mucosal epithelium at the folded portion of toad bladder. Specimen is prepared with the method described in the text. Araldite-embedded, uranyl acetate-stained. The lumen is at the top and down the center. Granular cells (GR), mitochondria-rich cells (MR), and basal cells (BA) are shown, and basal lamina (B) is noted at the right lower corner.

For the correlation study of structure and function, Bentley's sac technique seems to be well suited and widely used. One half of a bladder is tied over the tip of a hollow glass with serosal surface outward and filled with the appropriate Ringer's solution, and is immersed in aerated Ringer's solution. The rate of water net flow from mucosal to serosal is measured by the decrease in weight of the bladder and by reading tube tied over the sac. Short-circuit current across the bladder wall can be measured by using polyethylene agar bridges placed inside and outside media of the sac. One half of a bladder thus prepared is used for the experiment, while the other half serves as the control.

With these techniques, effects of hormones, chemicals, or ions added to the immersing media on short-circuit current and on water net flow can be monitored, and tissues under these functional states can thus be obtained

for morphological studies. For example, the vasopressin added to the serosal media causes an increase of sodium and water net transport across the bladder, during which the tissue fixed for morphological studies shows widening of the intercellular spaces and swelling of the mucosal cells, mainly of granular cells. Amphotericin B added to the mucosal media increases short-circuit current across the bladder at initial period of its addition, may affect mainly mitochondria-rich cells before other cells. Following withdrawal of calcium ions from the bathing media well correlated changes in short-circuit current and structures of the toad bladder are noted, which is another good example of such correlation works.

Oxygen Consumption Studies. For oxygen consumption studies of the bladder tissue Warburg apparatus has been used, and it has been found that a significant portion of the total oxygen consumption of the bladder tissue is accounted for by active sodium transport. With this technique, metabolic rate of the tissue or isolated mucosal cells of the bladder can be correlated with the active sodium transport or uptake under conditions of various substrates.

JAE KWON CHOI

References

Bartoszewicz, W., and R. J. Barrnett, *J. Ultrastruct. Res.,* **10**:599 (1964).

Bennett, H. S., and J. H. Luft, *J. Biophys. Biochem. Cytol.,* **6**:113 (1959).

Bentley, P. J., *J. Endocrinol.,* **21**:161 (1960).

Carasso, N., et al., *J. Microscopie,* **1**:143 (1962).

Choi, J. K., *J. Cell Biol.,* **16**:53 (1963).

Choi, J. K., *J. Cell Biol.,* **25**:175 (1965).

Edelman, I. S., et al., *Proc. Nat. Acad. Sci. U. S.* **50**:1169 (1963).

Edelman, I. S., et al., *J. Clin. Invest.,* **43**:2185 (1964).

Hays, R. M., et al., *J. Cell Biol.,* **25**:195 (1965).

Jard, S., et al., *J. Microscopie,* **5**:31 (1966).

Keller, A. R., *Anat. Rec.,* **147**:367 (1963).

Leaf, A., *Ergeb. Physiol.,* **56**:216 (1965).

LeFevre, M. E., et al., *Amer. J. Physiol.,* **219**:716 (1970).

Luft, J. H., *J. Biophys. Biochem. Cytol.,* **9**:409 (1961).

Matty, A. J., and F. E. Guinness, *J. Anat.* (London), **98**:271 (1964).

Pak Poy, R. K. F., and P. J. Bentley, *Exp. Cell Res.,* **20**:235 (1960).

Peachey, L. D., and H. Rasmussen, *J. Biophys. Biochem. Cytol.,* **10**:529 (1961).

Pease, D. C., ed., "Histochemical Techniques for Electron Microscopy," 2nd ed., New York, Academic Press, 1964.

Saladino, A. J., et al., *Amer. J. Pathol.,* **54**:421 (1969).

Sharp, G. W. G., and A. Leaf, *Physiol. Rev.,* **46**:593 (1966).

Ussing, H. H., and K. Zerahn, *Acta Physiol. Scand.,* **23**: 110 (1951).

See also PLASTIC EMBEDDING.

BLOOD

Study of the fine structure of blood cells and platelets dates back to the early days of electron microscopy. Beginning with the work of Wolpers in 1941 (1), the early work in this field was done before development of adequate techniques for thin sectioning. This phase of investigative effort gave rise to an intensive program in the laboratory of M. Bessis in Paris. By 1950 Bessis (2) was able to review 13 contributions by himself and his collaborators. At this time he quoted a comparable number of publications from other laboratories. These early "presection" studies utilized spreads (smears), metallic impregnations, and shadow casting in preparations that were fixed with osmium vapors or alcohol. A few fractionated preparations showed the contour characteristics of granules and mitochondria. Transmission micrographs of thin, impregnated cells that revealed internal structure were rare. Although this era of electron microscopy was an impressive and highly praiseworthy beginning in its day it soon became outmoded by newer methods.

Revolutionary advances in technique occurred during a five year period beginning in 1949. Plastic embedding (3), controlled fixation for fine structure (4), and ultramicrotomy (5,6) were developed to everyday practicability in this relatively short period of time. Under the stimulus of these technical developments the formed elements of blood were extensively studied in thin sections of plastic-embedded tissue. Early studies by these methods included those of De Robertis (7), Watanabe (8), Kautz and De Marsh (9), Sheldon and Zetterquist (10), Bernhard et al. (11) Miller (12), and others. The rapid technical progress that characterized the early fifties was illustrated by the electron microscopic blood atlas of Low and Freeman (13). This volume was devoted to the fine structure of normal and leukemic human blood cells and contained 150 pages of electron micrographs of sectioned blood cells. It made clear that technical problems no longer hindered the study of the fine structure of blood.

The technique developed by Freeman during the course of this work (13) made possible the collection and concentration of large numbers of leukocytes and platelets for electron microscopic study. Peripheral blood was collected in the usual fashion (14) (using instruments coated with nonwetting agents), ice-cooled, and centrifuged at 1500 rpm for 15 min at $0°C$ (R.C.F. = 265). The buffy coat, consisting of the formed elements of the blood other than erythrocytes, was aspirated and transferred directly to the fixative. This consisted of 5 cm^3 of 1 % veronal-buffered (pH = 7.4) OsO_4 at $5-10°C$. Fixation lasted one-half hour. Between each successive step of fixation, dehydration, and plastic infiltration the specimen was centrifuged at 1500 rpm (R.C.F. = 385). The supernatant fluid was decanted after each centrifugation. After addition of the subsequent fluid the tube was agitated to keep individual cells separate and to discourage pellet formation. N-butyl methacrylate and methyl methacrylate in proportions of 4 : 1 were used for embedding. The low viscosity of the pure methacrylates made it possible for the individual cells to settle, after gentle tapping, to the bottom of the 00 gelatin capsules used as embedding vessels. The remainder of the technique was routine for electron microscopy.

The perfection of fixation, embedding, and sectioning techniques for electron microscopy made possible the observation of blood cells and platelets in far greater detail than with light microscopy. Nevertheless it added surprisingly little that was new to our information about these structures. The well recognized specific granulations were readily confirmed as were the less well known organoids such as mitochondria, Golgi zone, and the recently rediscovered endoplasmic reticulum. In blood cells the endoplasmic reticulum tended to be dispersed in apparently isolated vesicles. Some of these contained smaller membrane-bound vesicles, named compound vacuoles by Low and Freeman (13).

Shortly after these early advances in the electron microscopy of blood, improvements were made in embedding media. Epoxy resins (Araldite [15], Epon [16], Maraglas [17],) and polyesters (Vestopal-W [18], Selectron [19]) were substituted for the earlier acrylic esters (n-butyl methacrylate, methyl methacrylate). Although improved embedding resulted from the use of these new plastics, their high viscosity made the processing of individual cells a difficult procedure. It was discovered with techniques comparable to those used for methacrylate that indivdual cells would not settle in the highly viscous monomers and that mechanical prodding by centrifugation resulted in frequent membrane ruptures. This discouraged the use of free cells. Older techniques using free cells were soon replaced largely by the use of hemopoietic tissues as sources of supply. Although Anderson's (20) notable study of normal and leukemic leukocytes was made on isolated cells in epoxy embedments (Epon-Araldite) the bone marrow, lymph nodes, thymus, spleen, etc. soon became convenient sources of blood leukocytes and platelets. The introduction of aldehyde fixation for fine structure (21), now used routinely prior to osmication, also had its effects on the interpretation of fine structure in blood cells (20). A vast number of studies of the formed elements of blood in electron microscopy were made in the last decade, of which only a few may be mentioned here. These were chosen as examples of different technical approaches with the purpose of aiding the reader to select the methods best suited to his interests.

Methacrylate-embedded tissue was used principally in the decade prior to 1960 and its possibilities were well illustrated in the Low-Freeman atlas (13). Although largely supplanted in electron microscopy by epoxy or polyester embedments, methacrylate is still the recommended embedding medium for autoradiographic studies (22). Epoxy embedments such as Epon 812 (16) or Epon-Araldite mixtures (23) were used with considerable success in whole blood studies (20). Earlier workers aspirated the centrifuged buffy coat(13) but an interesting variation was introduced by Watanabe et al. (24). These workers removed the plasma of centrifuged whole blood and replaced it with 6.25% glutaraldehyde in $0.1M$ phosphate buffer at pH 7.4. This procedure was carried out at 4°C and lasted 30 min. It solidified as well as fixed the buffy coat, which could then be removed as a disc and handled much the same way as any other soft tissue.

Blood cells obtained from any of the hemopoietic organs are best fixed by perfusion with aldehydes (25) followed by postosmication in the manner now standard for most electron microscopic studies (26). An important consideration for good aldehyde perfusion is the introduction of 0.1% procaine to the "washout" solution to prevent vasoconstriction as recommended by Forssman and co-workers (27). Staining and microtomy present no special problems for blood cells and platelets. It usually suffices to follow methods standard for electron microscopy (28).

Considerable attention has been paid in recent years to the specific granulations of blood cells. The high-resolution study of eosinophil granules by Miller et al. (29) revealed a crystalline lattice with a $\sim 30\text{Å}$ repeating period in rodents and ~ 40 Å in human cells. Bainton and Farquhar have published a number of studies inquiring into the origin, enzyme content, and primary lysosome nature of the granules of polymorphonuclear leukocytes (30,31). They have also investigated the granule enzymes of eosinophilic leukocytes (32). As many as four types of granule have been described in human neutrophils (33). The ultrastructure of thoracic duct lymphocytes has recently been reported by Wivel et al (34). Each of the studies cited above utilized techniques chosen as best suited to the particular subject to be investigated. The illustrations in these papers are representative.

An interesting recent development concerning the fine structure of blood cells is the use of the scanning electron microscope (SEM) by Bessis and co-workers (35,36). Preparations for SEM must be dried without change of surface contours (preferably by freeze-drying or by the critical point method) and the surface of the specimen must be electrically conductive (by coating with metal or carbon (37).) The SEM is then able to record the surface contours of the specimen at practical magnifications of from 20 to 20,000 (or even greater). At lower magnifica-

tions the great depth of field which at $100\times$ approaches 1 mm enables the SEM to outrank white light micrography in which the depth of field is of the order of only $10\text{--}30\ \mu$. At higher magnifications the depth of field is still great enough to provide a clear picture in depth at a resolution of about 250 Å. Secondary electron emission (those electrons emitted by the specimen when bombarded by a probe at from 1 to 30 kv) is the mode most commonly used in SEM of soft tissues. Blood cells may be fixed in 1% glutaraldehyde in $0.1M$ cacodylate or phosphate buffer for one hour(38). After two centrifugations at 1000 g and resuspensions in distilled water the cells are spread on a clean coverslip which is fragmented, glued to an aluminum stub, and shadowed with gold or gold-palladium. The process of phagocytosis of erythrocytes by macrophages as well as calciform erythrocytes and prickle cells (35) have revealed very clear details of their three-dimensional form and surface detail.

In summary, leukocytes and platelets are amenable to study by methods that are now standard for soft tissues in electron microscopy (28). Whenever hemopoietic or lymphoid organs offer an adequate supply of a particular cell type, whether developmental or mature, these tissues may be used without particular precautions. If peripheral blood is called for, then simple centrifugation of whole blood, followed by fixation of the buffy coat, is the method of choice.

Frank N. Low

References

1. Wolpers, C., "Zur Feinstruktur der Erythrocyten membran," *Naturwiss.*, 29:416 (1941).
2. Bessis, M., "Studies in electron microscopy of blood cells," *Blood*, 5:1083 (1950).
3. Newman, S. B., E. Borysko, and M. Swerdlow, "New sectioning techniques for light and electron microscopy," *Science*, 110:66 (1949).
4. Palade, G. E., "A study of fixation for electron microscopy," *J. Exp. Med.*, 95:285 (1952).
5. Hillier, J., and M. E. Gettner, "Sectioning of tissue for electron microscopy," *Science*, 112:520 (1950).
6. Porter, K. R., and J. Blum, "A study in microtomy for electron microscopy," *Anat. Rec.*, 117:685 (1953).
7. De Robertis, E., "Electron microscope studies of circulating blood cells," *Proc. 4th Int. Cong. Soc. Hematol.*, 67–85 (1952).
8. Watanabe, Y., "An electron microscopic study of the leukocytes in the bone marrow of the guinea pig," *J. Electron Micr.* (Japan), 2:34 (1954).
9. Kautz, J., and Q. De Marsh, "Electron microscopy of sectioned blood and bone marrow elements," *Rev. Hematol.*, 10:314 (1955).
10. Sheldon, H., and H. Zetterquist, "Internal ultrastructure in granules of white blood cells of the mouse," *Bull. Johns Hopkins Hosp.*, 96:135 (1955).
11. Bernhard, W., F. Haguenau, and R. Leplus, "Coupes ultrafines d'elements sanguins et de ganglions lymphatiques étudiées au microscope électronique," *Rev. Hematol.*, 10:267 (1955).
12. Miller, F., "Elektronenmikroscopische Untersuchungen an weissen Blutzellen," *Verh. deutsch. Ges. pathol.*, 40:208 (1956).
13. Low, F. N., and J. A. Freeman, "Electron Microscopic Atlas of Normal and Leukemic Human Blood," New York, The Blakiston Division, McGraw-Hill Book Co., 1958.

14. Low, F. N., "Mitochondrial structure," *J. Biophys. Biochem. Cytol.*, Suppl. **2**:337 (1956).
15. Glauert, A. M., and R. H. Glauert, "Araldite as an embedding medium for electron microscopy," *J. Biophys. Biochem. Cytol.*, **4**:191 (1958).
16. Luft, J. H., "Improvements in epoxy resin embedding methods," *J. Biophys. Biochem. Cytol.*, **9**:409 (1961).
17. Freeman, J. A., and B. O. Spurlock, "A new epoxy embedment for electron microscopy," *J. Cell Biol.*, **13**:437 (1962).
18. Ryter, A., and E. Kellenberger, "L'inclusion au polyester pour l'ultramicrotomie," *J. Ultrastruct. Res.*, **2**:200 (1958).
19. Low, F. N., and M. R. Clevenger, "Polyester-methacrylate embedments for electron microscopy," *J. Cell Biol.*, **12**:615 (1962).
20. Anderson, D. R., "Ultrastructure of normal and leukemic leukocytes in human peripheral blood," *J. Ultrastruct. Res.*, Suppl. 9 (Sept., 1966).
21. Sabatini, D. D., K. G. Bensch, and R. J. Barrnett, "Cytochemistry and electron microscopy: the preservation of cellular ultrastructure and enzymatic activity by aldehyde fixation," *J. Cell Biol.*, **17**:19 (1963).
22. Caro, L. G., "Progress in high-resolution autoradiography," *Prog. Biophys. Molec. Biol.* (Butler and Huxley) **16**:171–190 (1966).
23. Mollenhauer, H. H., "Plastic embedding mixtures for use in electron microscopy," *Stain Technol.*, **39**:111 (1964).
24. Watanabe, I., S. Donahue, and N. Hoggatt, "Method for electron microscopic studies of circulating human leucocytes and observations on their fine structure," *J. Ultrastruct. Res.*, **20**:366 (1967).
25. Rosen, W. C., C. R. Basom, and L. L. Gunderson, "A technique for the light microscopy of tissues fixed for fine structure," *Anat. Rec.*, **158**:223 (1967).
26. Karnovsky, M. J., "A formaldehyde-glutaraldehyde fixative of high osmolality for use in electron microscopy," *J. Cell Biol.*, **27**:137A (1965).
27. Forssman, W. G., G. Siegrist, L. Orci, L. Girardier, R. Pictet, and C. Rouiller, "Fixation par perfusion pour la microscopie electronique: essai de generalisation," *J. Microscopie*, **6**:279 (1967).
28. Pease, D. C., "Histological Techniques for Electron Microscopy," 2nd ed., New York, Academic Press, 1964.
29. Miller, F., E. DeHarven, and G. E. Palade, "The structure of eosinophil leucocyte granules in rodents and in man," *J. Cell Biol.*, **31**:349 (1966).
30. Bainton, D. F., and M. G. Farquhar, "Origin of granules in polymorphonuclear leucocytes. Two types derived from opposite faces of the Golgi complex in developing granulocytes," *J. Cell Biol.*, **28**:277 (1966).
31. Bainton, D. F., and M. G. Farquhar, "Differences in enzyme content of azurophil and specific granules of polymorphonuclear leukocytes. I. Histochemical staining of bone marrow smears," *J. Cell Biol.*, **39**:286 (1968). "II. Cytochemistry and electron microscopy of bone marrow cells," *J. Cell Biol.*, **39**:299 (1968).
32. Bainton, D. F., and M. G. Farquhar, "Segregation and packaging of granule enzymes in eosinophilic leukocytes," *J. Cell Biol.*, **45**:54 (1970).
33. Daems, W. Th., "On the fine structure of human neutrophilic leukocyte granules," *J. Ultrastruct. Res.*, **24**:343 (1968).
34. Wivel, N. A., M. A. Mandel, and R. M. Asofsky, "Ultrastructural study of thoracic duct lymphocytes of mice," *Amer. J. Anat.*, **128**:57 (1970).
35. Bessis, M., J. Dobler, and P. Mandon, "Discocytes et échinocytes dans l'anémie a cellules falciformes. Examen au microsope électronique à balayage," *Nouv. Rev. franc. Hematol.*, **10**:63 (1970).
36. Bessis, M., and A. De Boisfeury, "Étude des differentes étapes de l'érythrophagscytose par microcinématographie et microscope électronique à balayage. *Nouv Rev. franc. Hémat.*, **10**:223–242 (1970).
37. Boyde, A., and C. Wood, "Preparation of animal tissues for surface-scanning electron microscopy," *J. Microscopy*, **90**:221 (1969).
38. Luse, S. A., "Preparation of biologic specimens for scanning electron microscopy," *Proc. Cambridge Stereoscan Colloq.*, 149 (1970).
See also: HEMOPOITIC ORGANS, LEUCOCYTES.

BLOOD PARASITES

The blood of vertebrates may contain various animal parasites, namely extracellular protozoa: trypanosomes and trypanoplasms; intracellular protozoa: malaria parasites and related Haemosporidia, hemogregarines (coccidians), and piroplasms; helminths: blood flukes (schistosomes) and microfilariae (larval nematodes). Prokaryote blood parasites include the extracellular spirochaetes (*Borrelia* spp.) and various eperythrocytic (*Bartonella*, *Haemobartonella*, *Eperythrozoon*) and intraerythrocytic (*Anaplasma*, *Aegyptianella*) forms of dubious taxonomic status. Special microscopical techniques have been developed to assist diagnosis of infection with these parasites in smears of peripheral blood. Schistosomes are not to be found in circulating blood, and methods for handling these parasites are similar to those employed for other trematodes.

Withdrawal of peripheral blood for examination should be repeated at different times of day, for the blood cycle of certain parasites shows marked periodicity. Thus microfilariae of *Wuchereria* and *Brugia* are most abundant in blood of the human host between 22.00 and 2.00 hr, those of *Loa* between 10.00 and 14.00 hr. Also the schizogony of malaria parasites tends to be completed at a particular time of day.

Light microscope examination of fresh blood easily reveals motile parasites (trypanosomes, spirochaetes, exflagellating microgametocytes of haemosporidians, microfilariae). Dark ground illumination is useful for detecting movement of such parasites in low numbers where several fields have to be scanned. For diagnosis of scanty infections, concentration techniques are profitably employed. The most rapid of these involves centrifugation of citrated blood in a microhaematocrit tube and then examination of the tube at a total magnification of 80–160 × for movement of microfilariae or trypanosomes at the junction of cells and plasma (Goldsmid, 1970). Usually, however, microscopic detection and identification of blood parasites relies upon the stained blood smear. Romanowsky staining is most frequently used in survey work and has the advantage that it can be performed on air-dried smears.

In 1891 Romanowsky found that a mixture of 1% aqueous eosin and saturated aqueous methylene blue could be used to demonstrate the hitherto unseen nucleus of the malaria parasite, *Plasmodium*, staining the nucleus carmine violet and the parasite's cytoplasm blue. Subsequent improvements on Romanowsky's original mixture came from an understanding of the conversion of methylene blue to other thiazine dyes ("polychromed methylene blue to other thiazine dyes ("polychromed methylene blue") under the action of various "ripening" agents (Nocht, 1898), and of the role of these derivative dyes in staining. A further advance was the development of a stable stock solution of the dye mixture (Leishman, 1901). Giemsa (1901) showed that the Romanowsky effect was obtained with the nucleus of *Plasmodium* using a methylene azure (azure B)/eosin mixture lacking methylene blue, and he attributed the red nuclear staining to the metachromatic properties of azure B. It is interesting to note that while the chromatin of leukocytes is stained purple by this mixture the chromatin of protozoa is stained red, and whereas staining of the former occurs in the absence of eosin, eosin is essential for staining of the latter. The suggestion (Unna and Tielemann, 1918) that in protozoan nuclei it is the basic protein rather than the nucleic acid which binds the eosin and the eosin then binds the red imino base of the azure dye, seems unlikely; the kinetoplast of trypanosomes contains DNA but no basic protein (Steinert, 1965), yet like the nucleus it stains red. To obtain blue, rather then gray, cytoplasmic staining, Giemsa found that methylene blue had to be added to his azure B-eosinate. Azure A was present as a contaminant of azure B in Giemsa's experiments and such traces are thought to ensure optimum cytoplasmic staining (Saal, 1964); indeed the presence of azure A eosinate is essential if the stippling of malarious host red cell cytoplasm is to be revealed. In alkaline solution methylene blue (tetramethyl thionine) undergoes demethylation to produce azure A (dimethyl thionine). Such demethylation occurs on a large scale in Leishman's mixture.

Romanowsky staining is conducted on air-dried films of blood on grease-free slides. The films may be thin or thick. Thin films are made by placing a drop of blood at one end of a slide and allowing the blood to run along the narrow edge of a second slide held at an acute angle (30°) to the first, then spreading the blood evenly along the first slide. The resulting film is dried rapidly by waving in the air. Such films give the best morphological preservation of the parasites. For simply detecting parasites in mammalian blood, thick films are used. Three to four drops of blood are spread with a needle into a circular area 1–1.5 cm across on the slide and allowing to dry at 37°. The film should be thin enough to allow newsprint to be read through it. Thick films can also be made with buffy coats from centrifuged blood and other material in which parasites have been concentrated. Thin films may be stored in a desiccator over silica gel for up to one year and thick films similarly for up to three years.

The variations of Romanowsky staining technique are legion but the stains most commonly used for diagnostic and descriptive work are those of Giemsa and Leishman. The former is preferable for routine usage because it is more stable and keeps indefinitely provided that the stock bottle is tightly stoppered to prevent absorption of moisture. Leishman's stain requires more technical skill than Giemsa's, and the mixture is more sensitive to atmospheric CO_2. Practical details have been summarized by Shute (1966) and discussed at length by Shute and Maryon (1960).

Giemsa's stain may be purchased in solution or made up as follows. Giemsa powder 3.8 g (*or* azure II 0.8 g, azure eosinate 3.0 g), glycerol 250 ml, methanol 250 ml, are placed in a methanol-washed 500 ml flask with 50 methanol-washed 5 mm glass beads. The flask is tightly sealed and shaken vigorously for a few minutes every 15 min, six shakings in all. The stain can be used in 24 hr.

It is diluted 1 : 20 with distilled water buffered with phosphate 0.7 g KH_2PO_4, 1.0 g Na_2HPO_4 in 1 liter at pH 7.2 for use. Unless the distilled water is neutral or slightly alkaline, details of the stained parasite do not stand out; e.g. host cell stippling, helpful in diagnosing the species of young *Plasmodium* trophozoites, may be absent.

To stain thin films, 20 sec fixation in 1–2 drops of methanol is adequate. After evaporation of the methanol, the diluted stain is poured over the film and left for 20–60 min, then washed off with a stream of buffered distilled water from an aspirator (1 sec only). After drying in an upright position the film may be mounted in a neutral mounting medium. Thick films are not fixed but submerged in Giemsa diluted 1 : 20 with buffered (pH 7.2) normal saline and left for 20 min. The stain is then slowly replaced by distilled water, and the slides drained dry. Stained parasites and leukocytes are seen against a clear background composed of red cell ghosts. The use of buffered saline (rather than buffer alone) as a stain diluent prevents staining of the red cells and distortion of leukocytes so that if needs be a differential leukocyte count can be made. Blood with nucleated erythrocytes does not make satisfactory thick smears.

Leishman's stain is preferably made up from powder in the laboratory by shaking 0.3 g crystals with 200 ml good-quality methanol in the presence of glass beads and observing the same precautions as in the preparation of Giemsa. Powdered Giemsa substituted for Leishman crystals gives a more stable stain. The alcoholic stain serves first as fixative for the films (Leishman's technique is not suitable for thick films). Seven to eight drops are left on the blood smear for not more than 20 sec. then 12–15 drops of buffer (pH 7.2) are added and mixed thoroughly with the stain using a pipette. After 20 min the film is washed and dried as for Giemsa preparations The 2–3 min recommended for fixation in most textbooks is too long: Evaporation over this period results in precipitation of the stain, especially in hot climates; moreover host cell stippling is more easily revealed in slightly underfixed malarious blood.

After Romanowsky staining procedures, the dried films are best mounted in neutral mounting medium and not Canada balsam. If kept unmounted and examined by oil immersion the oil may be removed by drawing lens tissue soaked in neutral xylene across the slide.

The information available from Romanowsky-stained smears can be increased by recourse to reflex microscopy, where incident light is reflected from the surface of various structures. The specimens must then be mounted in a medium with the same refractive index as glass (in order to obtain optical homogeneity of media) and viewed under oil immersion. Thus the physiologically different classes of cytoplasmic granules of trypanosomes can be differentiated (Michel, 1964) and contaminating infections of eperythrozoa are readily demonstrated using this technique (Westphal, 1963).

Most morphological descriptions of blood parasites are based upon Romanowsky preparations, often without proper allowance being made for the artifacts introduced by the drying out of cells. Thus keys for the identification of mammalian trypanosomes on the size, shape, and position of the kinetoplast pertain to Romanowsky-stained organisms: in wet-fixed flagellates the kinetoplast may present quite a different picture. Giemsa staining may be applied to malaria parasites fixed in Bouin's fluid, then washing out the picric acid in the usual way. The smears are left in the diluted stain for 2–3 hr, rinsed briefly, and then transferred through graded acetone/ xylene mixtures to pure xylene followed by mounting in a neutral medium. Wet fixation gives a better preservation of the amoeboid form of malaria trophozoites, and merozoites stand out more clearly in mature schizonts, though corpuscle stippling may be lost.

Polarized light is useful in the detection of pigment in malaria parasites, especially in differentiating pigment from dense deposits resulting from cytochemical tests, e.g., lead sulfide in the Gomori reaction for acid phosphatase. The free pigment or that of the erythrocytic parasites is doubly refractile, but pigment taken up by phagocytes is not (Kósa, 1925).

Culture and vector forms of trypanosomes are not easy to stain satisfactorily using the technique for thin smears: phosphate-buffered formalin (10%) should be substituted for methanol in fixation (2 min) of dried smears (Lehmann, 1964). Trypanosome smears may also be wet-fixed in buffered formalin and then stained by the Giemsa technique given above for wet-fixed malaria parasites.

Microfilariae, on account of their size, call for special treatment. Basically the technique of diagnosis is the thick blood smear stained with either Giemsa or haemalum. This method suffers from two major drawbacks. Firstly, for such large parasites the amount of blood examined is small (less than 60 mm³), and secondly the nuclear stains employed tend to obscure morphological details which are of use identifying the worm. Recent interest has centered upon means of concentrating the worms for microscopial study and upon special techniques to reveal details of anatomical interest.

If the microhaematocrit technique is used for concentrating microfilariae for diagnosis, examination of the tube must be made immediately after removal from the centrifuge, as the larvae tend to disperse quickly from the interface. If larger volumes of blood are available these may be hemolyzed with an equal volume of 2% saponin in physiological saline, and after centrifuging at 2000 rpm. for 10 min, the sediment examined for microfilariae (Ho Thi Sang and Petithory, 1963). If hemolysis is performed in a syringe, the needle may then be replaced by a circular holder containing a 25 mm 5 μm-porosity membrane filter and the hemolyzed blood forced through the filter. After washing with physiological saline, then formol saline to fix the microfilariae present, the filter is removed, stained in Giemsa's stain for 1 hr, rinsed, and dried thoroughly. Mounted in immersion oil or in a clearing-mounting medium (e.g. "Permount," Fisher Scientific Co.), the microfilariae stand out stained against the transparent filter background (Chularerk and Desowitz, 1971).

In order to identify microfilariae, staining techniques other than Giemsa and hemalum are useful. The pattern of structures on the cephalic space of microfilariae appears to be characteristic for the genus, and these structures are most easily revealed in the larvae by treatment with basic dyes at low pH, after oxidation or sulfation of the blood film (Laurence and Simpson, 1969). Thick blood films are dehemoglobinized in tap water and fixed overnight in 1% trichloracetic acid in 80% ethanol. They are oxidized by immersion in 10% peracetic acid for 30 min. After washing out the acid in running tapwater for 5 min the smear is rinsed in distilled water. Sulfation is carried out after dehydrating the fixed smears through graded ethanols and immersing in glacial acetic acid for 1 min. A mixture of equal parts of concentrated

sulfuric acid and glacial acetic acid is allowed to react with the smear for 10–20 min and then it is rinsed in glacial acetic (2–3 min), running tapwater, and distilled water. Brilliant staining of the hooks, spines, and oral ring on the surface of the cephalic space can be obtained after oxidation by staining with 1% basic fuchsin in 0.05N HCl for 30 min. The pharyngeal thread, nuclei, excretory vesicle and pore, and anal vesicle may also be stained.

Acridine orange (0.05% w/v in veronal acetate buffer at pH 4.0 for 30 min) may be used to stain the trypanosome kinetoplast and demonstrate the absence of kinetoplast DNA in dyskinetoplastic trypanosomes by fluorescence microscopy (Baker, 1961). Walker (1964) recommends euchrysine as a vital fluorochrome stain at pH 7.2 giving similar results.

Optical whiteners (as used in washing powders) have been employed as fluorochromes for detecting trypanosomes in blood. Solutions of 0.5% aqueous Tinopal AN and of saturated aqueous CH 3558 (both manufactured by J. R. Giegy, Basle, Switzerland) will render trypanosomes in dried smears brilliantly fluorescent by UV microscopy (Herbert et al., 1967).

The fluorescent antibody technique (FAT) has been used for the detection and identification of blood protozoa or antibodies elicited by them. This technique has also been used for the determination of antigenic similarities and differences between closely related parasite species and the measurement of antibody levels during an infection. Thus the FAT will distinguish the different groups (subgenera) of mammalian trypanosomes, e.g. fluorescein-conjugated antiserum prepared from a rabbit infected with *Trypanosoma (Trypanozoon) brucei* will react with *T.(T) rhodesiense* but not with *T. (Duttonella) vivax* or *T.(Nannomonas) congolense*. An indirect FAT can be used to diagnose infection with a given trypanosome species (Sadun, et al., 1963; Bailey, et al., 1967). The indirect technique is used on serum samples to supplement classical blood smear examinations for malaria (Voller, 1971). Fluorescent antibody studies show cross reactions within the primate malarias, so smears of the simian malaria *Plasmodium cynomolgi* can be used as the basis of the test. Among rodent blood parasites there is some cross reaction not only within the rodent malarias but between rodent malarias and piroplasms, though homologous titers are always highest (Cox and Turner, 1970).

For electron microscopy of thin sections heavily infected blood may be allowed to clot and the clot fixed as a piece of tissue in the case of erythrocytic parasites. More usually, however, blood is collected in a heparinated syringe and then squirted into fixative. Centrifugation then produces a pellet which can be handled as a piece of tissue in subsequent processing. Alternatively the blood may be centrifuged first and the pellet fixed subsequently, or the pellet produced by filtering off the plasma through membrane filter. Concentration of parasites of infected corpuscles by filtration gives random sedimentation of different stages of the parasite, while centrifugation results in sedimentation according to mass and therefore non-random distribution of stages in the resulting pellet. Differential centrifugation may be used to separate non-infected from infected corpuscles and extracellular parasites from blood cells, the trypanosomes and microfilariae sedimenting in the buffy coat of the blood. For very scanty infections (e.g. bird trypanosomes) centrifugation in microhaematocrit tubes is an advantage. Pathogenic trypanosomes can be separated from infected blood by passing this over anion exchange resins which retain the erythrocytes (Lanham, 1968); the trypanosome suspension is then concentrated by centrifugation or filtration before fixation. Apart from methods involved in concentrating the organisms, blood parasites can be treated like other microorganisms or tissues in processing for electron microscopy.

KEITH VICKERMAN

References

Bailey, Cunningham, and Kimber, *Trans. Roy. Soc. Trop. Med. Hyg.,* **61**:696 (1967).
Baker, *Trans. Roy. Soc. Trop. Med. Hyg.,* **55**:518 (1961).
Chularerk and Desowitz, *J. Parasitol.,* **56**:623 (1971).
Cox and Turner, *Bull. World Health Org.,* **43**:337 (1970).
Giemsa, *Centralbl. f. Bakt.,* 1 Abt., **31**:429 (1901).
Goldsmid, *J. Clin. Pathol.,* **23**:632 (1970).
Herbert, Lumsden, French, and Paton, *Vet. Rec.,* **81**:638 (1967).
Ho Thi Sang and Petithory, *Bull. Soc. Exot.,* **56**:197 (1963).
Kósa, *Virchow's Arch. Pathol. Anat.,* **258**:186 (1925).
Lanham, *Nature,* **218**:1273 (1968).
Laurence and Simpson, *Trans. Roy. Soc. Trop. Med. Hyg,.* **63**:801 (1969).
Lehmann, *Trans. Roy. Soc. Trop. Med. Hyg.,* **58**:366 (1964).
Leishman, *Brit. Med. J.,* **2**:757 (1901).
Michel, *Z. Tropenmed. Parasitol.,* **15**:400 (1964).
Nocht, *Centralbl. f. Bakt.,* 1 Abt., **24**:839 (1898).
Romanowsky, *St. Petersburger Med. Woch.,* **16**:297 (1891).
Saal, *J. Protozool.,* **11**:573–85 (1964).
Sadun, Duxbury, Williams, and Anderson, *J. Parasitol.,* **49**:385 (1963).
Shute, *Trans. Roy. Soc. Trop. Med. Hyg.,* **60**:412 (1966).
Shute and Maryon, "Laboratory Technique for the Study of Malaria," London, Churchill, 1960.
Steinert, *Exp. Cell Res.,* **39**:69 (1965).
Unna and Tielemann, *Centralbl. f. Bakt.,* 1 Abt., **80**:66 (1918).
Voller, *Trans. Roy. Soc. Trop. Med. Hyg.,* **65**:111 (1971).
Walker, *Int. Rev. Cytol.,* **17**:51 (1964).
Westphal, "Einführung in die Reflexmikroskopie," Stuttgart, Thieme-Verlag, 1963.

BOTANICAL DRUGS

Since very ancient times, the plant kingdom has been a traditional as well as an invaluable source of useful medicinal agents (drugs). A drug may be defined as being any substance which affects the structure or any function of the body of man or other animals. One outstanding feature of the higher plants especially is their ability to synthesize and subsequently accumulate within certain plant organs a vast array of organic substances (phytochemicals) representing nearly all the recognized structural classes of chemical compounds. Many of these chemicals are drugs. Briefly stated, a *botanical drug* represents plant material containing substances with drug activity. Since most botanical drugs (botanicals) normally consist of only the certain separated and dried plant parts from a particular plant species, they may be conveniently classified as leaf drugs, root and rhizome drugs, bark drugs, seed drugs, flower drugs, or fruit drugs, and these are available in commerce either in their more or less unbroken forms or ground to powders of varying coarseness. Examples include American hellebore (dried rhizome and roots of *Veratrum viride* Ait.), Peruvian bark (dried

bark of the stem or of the root of *Cinchona succiruba* Pavon et Klotzsch), and Jimson weed (dried leaf and flowering or fruiting tops with branches of *Datura stramonium* L.). Of equal importance are a number of botanicals which do not represent well-defined plant tissues but which are instead derived from plant secretions. Opium (air-dried milky exudate obtained by incising the unripe capsules of *Papaver somniferum* L.) and benzoin (dried balsamic resin from *Styrax benzoin* Dryander) are well-known examples. Today, the majority of botanicals are not employed *per se*, but are rather extracted by various means in order to obtain their drug-active constituents which in turn are incorporated into dosage forms (tablets, injections, ointments) intended for medicinal use.

Regardless of how a botanical drug is to be ultimately utilized, its authenticity (botanical origin) must first be established. Furthermore, it must meet certain standards of purity such as freedom from rodent filth, insect contamination, manure, and other extraneous matter. These standards and pertinent information, including detailed microscopic descriptions of entire as well as powdered botanical drugs, are published in the official compendia (pharmacopoeias and formularies) of the nations of the world and in standard pharmacognosy textbooks.

Initial Handling. (1) Gross contamination (insect parts, rodent droppings, manure) is detected by means of a hand lens or stereoscopic microscope (10–60× magnification). (2) If the sample is present in the entire (unground) state, adequately thin sections may be cut free-hand, but it is usually necessary to soften the plant material by soaking it in water or water-glycerin (1 : 1) mixtures prior to sectioning. However, when the sample is extremely woody or if it consists of entire seeds or flowers, sectioning becomes difficult. Therefore botanicals are most conveniently prepared for study by first reducing the well-mixed sample to a fine powder (using a mortar or grinding mill) and then passing the ground material through a 60 mesh (or finer) sieve to obtain a homogeneous powder.

Preliminary Treatment. (1) Oily powders should be defatted by mixing with ether or chloroform in a watch crystal and then pouring off the solvent. The small amount of solvent remaining on the powder quickly evaporates. (2) A small amount of powder is mixed with a few drops of water. Powders which form a mucilaginous mass in this manner must subsequently be mounted in 95% alcohol rather than in water or in chloral hydrate solution as described below for the majority of powders.

Microscopic Examination. (Figs. 1–6) The use of a compound microscope (100–400× magnification) fitted with a standardized ocular micrometer for making accurate measurements is desirable. Mounts are prepared by mixing less than one milligram of powder with 1–2 drops of water or chloral solution (chloral hydrate 50 g, water 20 ml, glycerin 5 ml) on a microscope slide and then covering with a cover glass. The treatment of prepared mounts with specific reagents is best effected by placing 1–2 drops of the reagent along one side of the cover glass and then drawing the reagent into the mount by means of a filter paper placed at the opposite side of the cover glass.

1. *Water Mounts* are particularly useful for noting the type, size, and appearance of *starch* grains. The presence of starch is confirmed by treating the mount with iodine reagent (iodine 0.3 g, potassium iodide 1.5 g, water 100 ml.) Starch grains are colored blue to black in this manner. *Pollen* grains, sometimes

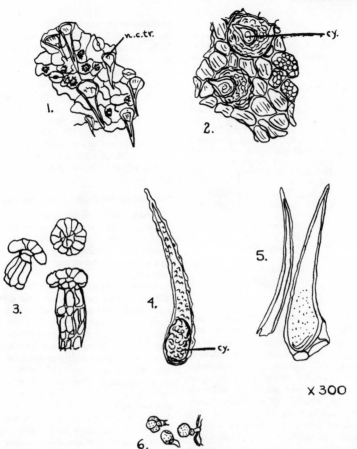

Figs. 1–6 Selected characteristic cellular elements from a chloral hydrate-cleared mount of *Cannabis sativa* L. (marijuana) (× 300). 1: Lower epidermis of a bracteole in surface view showing anomocytic stomata, covering noncystolithic trichomes (n.c.tr.) and calcium oxalate crystals in the underlying mesophyll. 2: Upper epidermis of a bract in surface view, showing covering trichomes containing cystoliths (cy.) and part of the underlying palisade. 3: Fragments of multicellular, multiseriate glandular trichomes. 4: Warty-walled covering trichome containing a cystolith (cy.). 5: Covering noncystolithic trichomes. 6: Small glandular trichomes. (Drawn after Jackson and Snowdon, by permission of the publishers.)

confused with starch grains, may be detected in additional water mounts by the use of Calberla's reagent (glycerin 5 ml, 95% alcohol 10 ml, water 15 ml, saturated aqueous solution of basic fuchsin 2 drops). Pollen grains are stained pink to red while starch is not affected.

2. *Chloral Mounts* Chloral hydrate solution serves to expand shrunken cells, dissolves starch and chlorophyll, and generally clears the tissues. Gently warming (not boiling) the mount over a microflame accelerates the clearing action. Chloral mounts are especially well suited for noting *plant hairs* as well as the numerous types of *calcium oxalate crystals* which occur in many botanical powders. The presence of calcium oxalate is confirmed and distinguished from *calcium carbonate* by treating the mount with one drop of glacial acetic acid. Calcium oxalate crystals are unaffected by this reagent but calcium carbonate masses (cystoliths) dissolve with effervescence when contacted by the acid. *Lignified elements* are detected by treating additional mounts with phloroglucinol-HCl reagent (saturated aqueous solution of phloroglucinol in 20% HCl). Lignin-containing cells are colored red to violet-red and nonlignified cells remain unstained.

Additional reagents which detect specific chemical-type compounds present in tissues are helpful in confirming tentative identifications of botanicals. These reagents as well as less commonly employed techniques such as microsublimation and vein islet, stomatal index, and palisade ratio determinations are dealt with in standard reference works.

In summary, the microscopic examination of botanical drugs is carried out to note those histological features and histochemical reactions which are quite constant for each botanical drug. By comparing the data from these studies with published "keys," descriptive monographs, and authentic material, an investigator can verify an unknown botanical drug with respect to its botanical origin and purity.

ALVIN B. SEGELMAN

References

Harris, K. L., ed., "Microscopic-Analytical Methods in Food and Drug Control," Food and Drug Technical Bulletin No. 1, Washington, D.C., U.S. Department of Health, Education and Welfare, Food and Drug Administration, 1960.

Jackson, B. P., and D. W. Snowdon, "Powdered Vegetable Drugs: An Atlas of Microscopy for Use in the Identification and Authentication of Some Plant Materials Employed as Medicinal Agents," New York, American Elsevier Publishing Company Inc., 1968.

Segelman, A. B. and F. H. Pettler, "A highly reliable method for the identification of marijuana, using combined chromatography–microscopy techniques" *in* Abstracts of Papers, 119th Annual Meeting, Houston, Texas. Amer. Pharm. Assoc. **2**(1):95, 1972.

Youngken, H. W., "A College Textbook of Pharmaceutical Botany," 6th ed., Philadelphia, The Blakiston Co., 1938.

Youngken, H. W., "Textbook of Pharmacognosy," 6th ed., Philadelphia, The Blakiston Co., 1950.

See also CHEMICAL MICROCRYSTAL IDENTIFICATION.

BRYOPHYTA

The division Bryophyta (bryophytes) includes three classes, Musci (mosses), Hepaticae (liverworts), and Anthocerotae (hornworts). There is an alternation of generations between an independent, green, haploid gametophyte, and a diploid sporophyte which remains attached to the gametophyte for its entire life and is at least partially dependent upon the gametophyte for its nutrition. The sporophyte consists of a capsule, borne on an elongated seta, and a foot, which attaches the sporophyte to the gametophyte. A columella may be present or absent. Spores develop from sporocytes, which undergo meiosis, each sporocyte producing four haploid spores. When mature, capsules dehisce, usually by means of longitudinal valves, a differentiated lid or operculum, or rarely, by breaking irregularly. In liverworts and hornworts the spore, upon germination, develops into a rudimentary germling, which usually produces only a single leafy or thallose gametophyte. In mosses the spore develops into a much-branched, filamentous protonema, which usually produces numerous gametophytic plants. Rarely, the protonema in mosses is thalloid or prothallial. Gametophytes produce male and female sex organs, antheridia and archegonia, respectively bearing sperm and egg. Upon wetting, mature antheridia discharge sperm, which may swim directly to the archegonium or may be dispersed short distances by splashing rain drops. After fertilization, the zygote develops into the diploid sporophyte, completing the cycle.

Collection and Preservation

Bryophytes are probably the easiest plants to collect, handle, and preserve. Lacking true roots, the gametophytes are only lightly anchored to the substrate by thin rhizoids. They occur on bark of both living and dead trees, dead or decaying wood, humus, soil, rocks, and rarely on dung or decaying animal flesh. They may be gathered by hand or by scraping with a small knife. Dirt and debris should be removed and plants of some soil-growing species may require washing to rid the material of dirt and sand.

Ordinary paper bags can be used for collecting and are preferred to plastic bags if the plants are to be dried later. Collection data can be written on the side of the bag and the paper will absorb some moisture. Some collectors prefer to collect in specially folded packets that can be prenumbered or bear specially printed habitat data that can be checked or circled.

Plants can be kept alive for a number of days or even weeks by placing them in polyethylene bags in a damp but not wet condition, tying the ends loosely, and refrigerating, preferably under continuous light, between 0° and 5°C. It may be necessary to open up the bag periodically to allow air circulation and to moisten plants slightly with an atomizer. This is an exceedingly useful procedure for storing plants for mitotic and meiotic studies.

Methods for Studying Unfixed Material

Living plants of bryophytes can be studied microscopically with a minimum of preparation, and with very little dissection. Dried plants can be quickly restored to almost their natural state merely by soaking in water. Hot water, just below the boiling point, will hasten the recovery.

Whole plants of most leafy hepatics and a few of the smaller mosses can be mounted directly in a drop of water on a slide and topped with a cover glass. Large plants may require removal of leaves from the stem, which can be done with fine forceps, or if the leaves are tiny they can be scraped off with a scalpel, holding the stem near the tip with forceps and scraping down the stem. The coarse stems can then be removed, leaving only the desired leaves. The leaves of most bryophytes are only one cell thick and thus provide excellent microscopic contrast in a water mount. Cell walls, which are often surface-ornamented, pores, chloroplasts, oil bodies, and other organelles show up brilliantly in nearly all species without staining or at most by using a soft blue filter.

If a stain is necessary, e.g. to bring out pores in the cell walls, a weak aqueous solution (usually no stronger than 0.1%) of methylene blue, bismarck brown Y, or a comparable basic dye may be employed. These dyes are especially useful in studying the difficult moss genus *Sphagnum.*

Thallose liverworts too opaque for light to pass through or other thick parts, such as stems, midribs, and setae, can be sectioned free-hand. Freezing microtome techniques are also useful for obtaining thin sections.

Often it is desirable to make permanent slides of whole mounts, dissected material, or free-hand sections. Mounting in glycerin jelly is the oldest method and perhaps the most permanent, but it is tedious, messy, and difficult. It is prepared by dissolving 10 g of gelatin into 60 cm² of distilled water, which requires 3 or 4 hr at room temperature. Do not heat the water. When the gelatin is thoroughly dissolved, add 70 cm² of glycerin and 1.5 g of phenol crystals. Warm gently for 15 to 20 min, stirring constantly, until the phenol is completely dissolved. While still warm, filter through two or three layers of cheesecloth into an open container that can be covered. The mixture will solidify upon cooling and should be stored in a cool place.

For use, cut out a small square of the solidified glycerin jelly and place on a clean slide. Warm gently until it melts. Do not overheat. It should remain semiviscous. Place the plant material which has been dissected and thoroughly soaked in water (boiling the material in water will help eliminate troublesome air bubbles), into the melted glycerin jelly and apply a cover glass. Avoid an excess of the melted glycerin jelly around the margin of the cover glass. This prevents a proper set. The mount should be set aside for a week or so, allowing the jelly to set, and then ring the slide with a sealer such as gold size, asphaltum, or King's cement. Slides should be stored flat.

A second and much simpler permanent mounting medium is Hoyer's fluid (Anderson, 1954). It may be somewhat less permanent than glycerin jelly, especially with some species, but its simplicity and its time-saving advantages are very appealing. It is prepared by dissolving 30 g of gum arabic (U.S.P. flakes) into 50 cm^2 of distilled water at room temperature. Powdered gum arabic should not be used, as it produces excessive air bubbles, which are difficult to get rid of. Gum arabic dissolves slowly, requiring several hours or even overnight. Stirring should be done slowly and intermittently to avoid excessive air bubbles. When the gum arabic has completely dissolved add 200 g of chloral hydrate in small amounts while stirring slowly. When all of the chloral hydrate is in solution, add 20 cm^2 of glycerin. Set the mixture aside for 24–48 hr, allowing the air bubbles to dissipate. The final mixture should be clear, only faintly yellow, and free of sediment.

Dissections or whole mounts can be transferred from water directly to Hoyer's fluid on a clean slide. Material should be thoroughly wetted before transferring to Hoyer's. Dry plants should be soaked in water for a period of time, or preferably, boiled for a few seconds before mounting. After the cover glass is applied, the slide should be heated, but not boiled, and placed on a warming plate for a few days until the outer margin of the medium has hardened. Ringing with gold size or asphaltum is preferred, but not required. An alternative to ringing is as follows. An excess of Hoyer's is used so that after adding the cover glass and pressing it down somewhat, a raised ring of the fluid forms around the margin of the cover glass. This ring of Hoyer's will harden and act as a barrier against checking.

Use round cover glasses in making permanent whole mounts. Square or rectangular cover glasses check at the corners and tend to raise as the medium dries.

Methods for Studying Fixed Material

There are no special techniques for killing and fixing, dehydrating, embedding, microtoming, mounting, and staining bryophytes. Thoroughness and diligence in cleaning plants, which often grow prostrate on soil or sand, before fixing cannot be overemphasized. Soil grains, especially sand, can wreck material that is being microtomed, not to speak of the knife itself!

The most acceptable fixatives are formalin-acetic-alcohol, Kraft's formalin–chromic acid mixture, and for chromosome work, Carnoy's fluid. Dehydration is accomplished through a tertiary butyl alcohol series, embedding in tissuemat or one of the acrylic plastics. Staining with iron hematoxylin or Harris' hematoxylin (especially useful for sections of sex organs), counterstained with fast green or orange G, or quadruple stain (safranin, fast green, orange G, and crystal violet) is customary. These methods are described elsewhere in this volume.

Embedding and sectioning is generally unsatisfactory for mitotic and meiotic studies of bryophytes because the chromosomes have a strong tendency to clump and stick together. Nearly all chromosome studies of bryophytes have utilized squash methods (Anderson and Crum, 1958).

Meiotic squashes are made by first isolating the dividing sporocytes in a drop of fixative. Selection of capsules at the proper stage must be done by trial and error. In hepatics, meiosis occurs usually shortly after the capsule assumes its mature shape, nearly always before there is any elongation of the seta, if any, and before there is any coloration of the capsule wall. In mosses, meiosis occurs after the capsule attains its mature shape, but long after the seta elongates. If present, the degree of coloration of the annulus in mosses can be used as a guide to the approximate stage of meiosis, which occurs usually just before or about the time faint color begins to show in the annulus.

A clean capsule is placed in a drop of fixative (Carnoy's: 3 parts absolute ethyl alcohol to 1 part of glacial acetic acid) on a chemically clean slide. The tip of the capsule is cut off and discarded. In hepatics, which lack a columella, the entire content consists of sporocytes which can be squeezed into the fixative, discarding the capsule wall. In mosses, there is a central columella and the sporocytes are in a cylindrical sheath surrounding it. Dissect out the columella, discard the capsule wall, free the sporocytes from the columella, which is also discarded, leaving only a mass of sporocytes in the fixative. Add additional fixative as needed. After the sporocytes have been isolated and all debris carefully removed, allow the fixative to evaporate until nearly all of the liquid is gone. Then apply a drop of dye.

The most suitable stain for bryophyte chromosomes is aceto-orcein, prepared by saturating 40% acetic acid with a good grade of synthetic orcein, which is much superior to the natural dye. About 2 g of dye will more than saturate 100 cm^2 of 40% acetic acid at room temperature. The mixture should be shaken periodically over at least 24 hr and then filtered. The mixture should not be heated. In time, a scum tends to form over the surface of the solution, necessitating periodic filtering.

After the dye is applied, add a cover glass, remove excess dye, and tap the cover glass gently to spread the sporocytes into a single layer. Examine to determine if they are in the desired stage. If so, they can now be squashed, which (1) separates and spreads the chromosomes, (2) flattens the configuration so the chromosomes lie in one plane, and (3) rids the cytoplasm of oil bodies and other materials that tend to obscure the chromosomes. Squashing is accomplished by alternately tapping on the cover glass and applying firm pressure directly above the mass of sporocytes, which can be done with the slide either in the upright position or inverted on a flat surface on blotting paper. Examine the sporocytes at intervals while squashing to determine progress. When completed, the cover glass can be sealed with petroleum jelly, paraffin, or other suitable sealer.

It is best to study the slide immediately; in time, the cytoplasm gradually takes up stain, which destroys contrast. Most preparations will last two or three days or even longer if refrigerated. If desired, permanent preparations can be made by following any one of the methods described in this volume for making permanent squash or smear preparations.

Somatic chromosome preparations can be made by following essentially the same procedures outlined above. Gametophytes of bryophytes lack meristems. They grow from an apical cell, cutting off leaf primordia at fixed intervals. Mitosis is usually studied from developing leaf primordia. The extreme tips of actively growing stems and branches are excised and placed in a drop of Carnoy's on a slide. In thallose liverworts a segment of tissue from the tip of a thallus branch can be used. The remaining steps are the same as outlined for sporocytes, except that mitotic material requires much longer and much firmer tapping and pressure in order to spread the chromosomes properly. For some reason, bryophyte tissue does not respond to the macerating and hydrolyzing techniques that are so successful in higher plant squashes of root and stem tips. Similarly, prefixation procedures seem to have little or no effect on bryophyte chromosomes.

Culture Techniques

Bryophytes are normally cultured from spores, from specially differentiated gametophytic propagules, or from gametophytic or sporophytic fragments. Regeneration from sporophytic parts, a process called apospory, is more difficult, but results in a diploid gametophyte and is a convenient way to produce polyploids.

Natural Substrates With this method, plants are grown on soil, stones, or peat in ordinary porous pans, flower pots, or plastic boxes. A useful substrate mixture consists of equal parts of peat, loamy soil, and quartz sand (Szweykowski and Krzakowa, 1969). After planting, the container is covered with a glass plate and placed in an inch or so of water. Plants should be sprayed once or twice daily with distilled water by means of an atomizer. Light intensity and temperature range vary with different species. Attempt to match as nearly as possible the natural habitat of the species. Periodic weeding is necessary to eliminate higher plants and weedy moss species that tend to overrun cultures. Autoclaving the substrate before planting will reduce contaminants.

Components of the natural substrate may be mixed with 1 % agar. Humus, soil, ground bark or wood, peat, and even small pebbles can be boiled with enough agar to form a firm substrate upon hardening. Within limits imposed by the coarseness of the medium, aseptic techniques can be used. A further modification involves straining the substrate mixture after thorough boiling and adding the resulting liquid to an agar mixture. The mixture can then be autoclaved and used for pure cultures.

Nutrient Solutions Nutrient agar (1–1.5 % or stronger, if desired) is the usual substrate, either in Petri dishes or in flasks. Some commonly employed nutrient solutions are as follows:

BENECKE—NH_4NO_3, 0.2 g; $CaCl_2$, 0.1 g; KH_2PO_4 0.1 g; $MgSO_4$, 0.1 g; $FeCl_3 + H_2O$, 0.005 g.

IVERSON (1957)—Benecke's solution, but ferric chloride is replaced by $MnSO_4$, 0.0002 g; $ZnCl_2$, 0.0002 g; $Na_2B_4O_7$, 0.002 g; ferric citrate, 0.00002 g/1000 cm^2; 1 % sucrose.

BOLDS BASAL MEDIUM (Bischoff and Bold, 1963)—Six stock solutions are prepared by dissolving the indicated weight of the following salts into 400 ml distilled or deionized water: $NaNO_3$ 10.0 g; KH_2PO_4, 7.0 g; K_2HPO_4, 3.0 g; $MgSO_4$ $7H_2O$, 3.0 g; $CaCl_2 \cdot 2H_2O$, 1.0 g; NaCl, 1.0 g. Ten milliliters of each stock are used or each liter of final solution.

Minor (trace) elements are supplied by the following four stocks: EDTA STOCK SOLUTION—50 g EDTA (Ethylenediaminetetraacetic acid) and 31 g KOH, diluted to 1 liter with deionized or glass-distilled water. H-FE STOCK SOLUTION—4.98 g $FeSO_4 \cdot 7H_2O$, diluted to 1 liter with acidified water. Acidified water is prepared by adding 1 ml concentrated H_2SO_4 to 999 ml deionized or glass-

distilled water. H-BORON STOCK SOLUTION—11.42 g H_3BO_3, diluted to 1 liter with deionized glass-distilled water. H-H_5 STOCK SOLUTION—8.82 g $ZnSO_4 \cdot 7H_3O$; 1.44 g $MnCl_2 \cdot 4H_2O$; 0.71 g MoO_3; 1.57 g $CuSO_4 \cdot 5H_2O$; 0.49 g $Co(NO^3)_2 \cdot 6H_2O$, all diluted to 1 liter with acidified water (as above).

One milliliter of each of the trace element stock solutions is added to a liter of the final solution.

VOTH (1941)—Use the following quantities of a 0.5M solution of each of the following salts: KNO_3, 1.6 cm^2; $Ca(NO_3)_2$, 1.4 cm^2; $Mg(NO_3)_2$, 1.2 cm^2; KH_2PO_4, 0.8 cm^3; $MgSO_4$, 1.6 cm^2; all added to 1 liter of water.

The nutrient agar solution should be sterilized and poured into sterile Petri dishes or flasks. Upon hardening it can be inoculated with spores, propagulae, tissue fragments, etc. For much culture work, spores are preferred. Capsules should be first "sterilized" by rinsing in a 5 % "Chlorox" (about 5 % sodium hypochlorite) solution in distilled water for about 5 min, then rinse in several changes of distilled sterile water (Kelley and Postlethwait, 1962). The capsule is broken open in the last rinse and plates or flasks are inoculated by using a platinum loop. Standard aseptic procedures are followed. It is usually necessary to transfer growing plants at least once a month. Inoculations with stem and leaf fragments, pieces of setae, etc., are more susceptible to contaminations of bacteria and fungi, because a weaker Chlorox solution is usually required. Use as strong a solution as the material can withstand or decrease the rinse time in the 5 % solution. Increase the number of subsequent rinses in distilled water. It is almost impossible, however, to prevent some contamination when plant pieces are used as the inoculum.

Instead of agar, sterilized glass fiber, vermiculite, quartz sand (Schelpe, 1953), and similar substrates may be used. Water and additional nutrient solution can be added to these types of substrates. Unlike agar, they do not dry out, and less maintenance of cultures is required.

LEWIS E. ANDERSON

References

Anderson, L. E., "Hoyer's solution as a rapid permanent mounting medium for bryophytes," *Bryologist*, **57**: 242–244 (1954).

Anderson, L. E., and H. Crum, "Cytotaxonomic studies on mosses of the Canadian Rocky Mountains," National Museum of Canada, Bulletin 160, Contributions to Botany, 1958 (89 pp).

Bischoff, H. W., and H. C. Bold, "Phycological studies. IV. Some soil algae from Enchanted Rock and related algal species," The Univ. Texas Publication No. 6318, 1963.

Iverson, G. G., "Pure culture of *Frullania*," *Bryologist*, **60**:348–358 (1957).

Kelley, A. G., and S. N. Postlethwait, "Effect of 2-chloroethyltrimethylammonium chloride on fern gametophytes," *Amer. J. Bot.* **49**:778–786 (1962).

Schelpe, E. A. C. L. E., "Techniques for the experimental culture of bryophytes," *Trans. Brit. Bryol. Soc.*, **2**: 216–219 (1953).

Szweykowski, J., and M. Krzakowa, "The variability of *Plagiochila asplenioides* (L.) Dum. as grown in parallel cultures under identical conditions," *Bull. Soc. d. Amis Sci. et d. Lettres de Poznan*, Ser. D., **9**:85–103 (1969).

Voth, P. D., "Gemmae-cup production in *Marchantia polymorpha* and its response to calcium deficiency and supply of other nutrients," *Bot. Gaz.*, **103**:310–325 (1941).

See also: HEMATOXYLIN STAINS, PARAFFIN SECTIONS, POLYCHROME STAINS, SAFRANIN, WHOLEMOUNTS, BOTANICAL.

C

CALCIUM, SODIUM, AND CHLORIDE LOCALIZATION

Inorganic ions are involved in numerous biological processes, e.g. in osmoregulation, in nerve and muscle function, as activators of enzymatic reactions, and as buffer systems of body fluids. The study of their concentration and localization in tissues and cells is therefore of great interest in cell physiology, and attempts to localize electrolytes histochemically go back to the beginning of this century. Since then, the methods have not changed in principle. Normally they consist in fixing fresh tissue with a precipitating agent, alone or combined with another fixative (for light microscopical methods see Barka and Anderson, 1963). Up to this day, the histochemical precipitating methods are still limited by three main factors: the sensitivity and specificity of the precipitating reactions, and the possibility of diffusion artifacts. The latter point holds true not only for free ions but also for bound ones, which may be released during the procedure. In spite of these limitations, the methods have been demonstrated to yield reliable and useful results. If the procedures are carefully performed and the outcome is cautiously evaluated, the distribution pattern of the precipitates may be taken as an equivalent of the electrolyte distribution.

The current methods of histochemical localization of calcium, sodium, and chloride with the electron microscope are based on a combination of precipitating reactions used in analytical chemistry and techniques of tissue preparation for electron microscopy, so that the electrolytes in question are precipitated before or during tissue fixation as more or less insoluble compounds which are readily detectable in ultrathin sections by their contrast.

The success of these methods depends on several preconditions:

1. The concentration of the electrolytes in question must be—at least locally—sufficient for the sensitivity of the precipitating reaction.
2. To avoid diffusion artifacts one must take care for rapid precipitation. Therefore quite fresh tissue, tissue blocks as small and thin as possible, and fairly fast penetrating agents should be used. Application of cold reagent and shaking at least at the beginning of fixation may favor penetration and precipitation.
3. The precipitates must be practically insoluble in the fluids used during the embedding procedure. As it is advisable to remove any excess of precipitating reagent from the tissue by washing after fixation, the sensitivity of the histochemical method normally does not depend on the sensitivity of the precipitating reaction but on the solubility of the precipitates. Therefore washing and dehydration must be carried out as fast as possible in the cold with a minimum of fluid. The relative loss of reaction product by dissolution during the embedding procedure is reasonably reduced when processing an excess of tissue pieces (Komnick and Bierther, 1969).
4. The precipitating reaction must be of adequate specificity. If possible, additional specifying and control reactions should be performed. A simple and specific negative control reaction for free ions in general (not for a special kind) is a thorough rinse of prefixed tissue to remove the electrolytes in question prior to the histochemical procedure. At sites with precipitates of suitable density positive control can be obtained by selected area electron diffraction or by electron microprobe analysis.
5. For detection in the electron microscope, the reaction product should contain a heavy metal or have otherwise a contrast-producing character.
6. Last but not least, a minimum degree of ultrastructural preservation has to be assured so that precipitate location can be associated with fine structure.

  I. Methods for calcium
    a. Oxalate method (after Carasso and Favard, 1966)
      1. Immerse thin tissue slices in $1-3 mM$ ammonium oxalate for 1 hr (neutralized sodium or potassium oxalate may be used instead).
      2. Fix with 2% glutaraldehyde in $0.05M$ cacodylate buffer, pH 7.2, for 45 min.
      3. Wash with cacodylate buffer for 18 hr (solubility of calcium oxalate: 3.1×10^{-5} moles per liter H_2O at room temperature).

      Steps 2 and 3 may be omitted.

      4. Postfix with 1% osmium tetroxide in veronal buffer, pH 7.2, for 30 min.
      5. Dehydrate in graded alcohols.
      6. Embed in suitable plastics for ultramicrotomy.
Result: Granular or splinter—like fine precipitates of calcium oxalate giving fairly high contrast. Precipitates in the terminal sacs of sarcoplasmic reticulum have been identified by microprobe analysis (Podolsky et al., 1970).
Remarks: Other authors have combined the fixing and precipitating solutions, and added oxalate to the solutions used for washing and postfixation in order to depress the solubility of the reaction product (Böck, 1970; Constantin et al., 1965; Komnick, 1969).

Special modifications have been employed for the demonstration of the calcium pump in muscle cells (Heumann, 1969; Heumann and Zebe, 1967).

b. Phosphate method (Carasso and Favard, 1966).
1. Fix thin slices with 2% glutaraldehyde in Sörensen's phosphate buffer, pH 7.8–8.2, for 45 min.
2. Wash in three changes of distilled water adjusted to pH 8.0 with diluted NaOH (10 min each).
3. Substitute calcium with lead by incubation in 5% lead acetate at $+37°C$ (10–15 min).
4. Postfix with 1% osmium tetroxide in veronal buffer, pH 7.2 (30 min).
5. Dehydrate in graded alcohols, embedding in plastics.

Result: Fine precipitates of high contrast. As judged from experiments on skeletal muscle the phosphate method seems to be less sensitive than the oxalate method.

Remarks: Calcium phosphate of the intramitochondrial granules can be demonstrated either directly or after experimental accumulation by treatment with phosphate-buffered fixative and by staining the ultrathin section with current lead methods (e.g. Greenawalt et al., 1964; Peachy, 1964). A similar staining effect probably due to calcium has been observed after fixation with the hexahydroxoantimonate-osmium tetroxide solution (see the following section and e.g., (Böck, 1970; Bulger, 1969; Zadunaisky et al., 1968).

II. Methods for sodium (Komnick 1962)
1. Fix thin tissue slices in osmium tetroxide/potassium hexahydroxo-antimonate solution for 2 hr. This solution is prepared as follows: Dissolve 2% $K[Sb(OH)_6]$ in distilled water by heating to 60–80°C with constant stirring. Cool to room temperature and add 1–2% OsO_4. Adjust pH to 7.4–8.5 by adding drops of either 0.01N KOH or 0.01N CH_3COOH depending on the delivery or age of the antimonate. Avoid low pH because of possible formation of antimony acid.
2. Rinse briefly in ice-cold distilled water or in a suitable buffer free of sodium, calcium, and magnesium. (Solubility of the precipitated $Na[Sb(OH)_6]$: 0.0564 g per 100 ml H_2O at 18°C.)
3. Dehydrate quickly in graded alcohols, embed in plastics.

According to the modification of Zadunaisky (1966) tissue slices are fixed with 3% glutaraldehyde in 0.1M potassium phosphate buffer, pH 7.4, containing 2% $K[Sb(OH)_6]$, subsequently rinsed with 10% sucrose in potassium phosphate buffer, and postfixed with 1% osmium tetroxide in the same buffer.

Result: Fine granular precipitates of $Na[Sb(OH)_6]$ with high contrast. The reagent is not specific in the presence of calcium and magnesium.

Remarks: According to Bulger (1969) the reactive concentration of Ca^{++} is approximately 0.001M, and that of Mg^{++} 0.1M. Correspondingly Tandler et al. (1970) have shown by microprobe analysis that precipitates within cell nuclei contain sodium, calcium, and magnesium, whereas in mouse vas deferens the precipitates were found to contain sodium only (Lane and Martin,

1969). Hartmann 1966 was able to identify the precipitates in astrocytes as sodium salt by selected area electron diffraction. According to Amakawa et al. (1968) the silver grains in autoradiographs of ^{22}Na-treated tissue fixed with the histochemical sodium reagent coincide with the precipitate location. Legato and Langer (1969) studied the distribution of sodium and calcium in the myocardium fixed by vascular perfusion with the osmium/antimonate solution. Discrimination between the different precipitates was achieved by removal of calcium by chelation with EGTA and EDTA before applying the fixative.

III. Methods for chloride (Komnick, 1962; 1963; Komnick and Bierther, 1969)
1. Fix thin tissue slices for 1–2 hr in freshly prepared 1–2% osmium tetroxide in 0.05–0.1M cacodylate–acetic acid buffer, pH 6.4–6.6, containing 0.5–1.5% silver lactate.
Preparation of the fixative and buffer:
Stock solution A : 0.4M sodium cacodylate
Stock solution B : 0.4M acetic acid
0.1M buffer solution
pH 6.4 : 25 ml A + 7.0 ml B + dist. water to 100 ml;
pH 6.6 : 25 ml A + 6.0 ml B + dist. water to 100 ml.
For 0.05M buffer dilute with dist. water in the proportion of 1 : 1.
Preparation of the fixative:
Dissolve 1–2% osmium tetroxide in the buffer solution first and then 0.5–1.5% silver lactate.
2. Rinse briefly in ice-cold buffer (solubility of the precipitated AgCl: 0.15 mg per 100 ml dist. water at $+ 20°C$)
3. Dehydration and specification:
10% acetone, 5 min
30% acetone, 5 min
50% acetone, 5 min
  containing 0.1N nitric acid 2 × 5 min
70% acetone, 2 × 5 min
Perform steps 1–3 under red save light!
4. Further dehydration and embedding (under normal light):
90% acetone 2 × 5 min
100% acetone 3 × 5 min
(or change to 100% alcohol, depending on the kind of plastics used).
Embed in plastics.

Result: Fine granular precipitates of AgCl, which have been identified by selected area electron diffraction and control experiments. The precipitated AgCl is reduced quantitatively to colloidal silver by heavy irridation with the electron beam (Komnick and Bierther, 1969).

Remarks: The method is also suitable for the demonstration of iodide in thyroid gland. Treatment of the fixed tissue with 5% ammonium carbonate dissolves the silver chloride precipitates and leaves that of silver iodide (Van Lennep et al.).

H. KOMNICK

References

Amakawa, T., V. Mizuhira, K. Uchida, S. Shiina, and K. Tsuzi, "Evaluation of the sodium-ion detection method ^{22}Na-electron microscopic autoradiography," *J. Electron Microscopy,* **17**:267 (1968).

Barka, T., and P. J. Anderson, "Histochemistry: Theory, Practice, and Bibliography," New York, Harper and Row, 1963.

Böck, P., "Elektronenmikroskopischer Nachweis von Na⁺, Ca⁺⁺ und Cl⁻ in der lactierenden Milchdrüse des Meerschweinchens," *Cytobiologie*, **2**:68–82 (1970).

Bulger, R. E., "Use of potassium pyroantimonate in the localization of sodium ions in rat kidney tissue," *J. Cell. Biol.*, **40**:70–94 (1969).

Carasso, N., and P. Favard, "Mise en évidence du calcium dans les myonèmes pedonculaires de ciliés péritriches," *J. Microscopie*, **5**:759–770 (1966).

Costantin, L. L., C. Franzini-Armstrong, and R. J. Podolsky, "Localization of calcium-accumulating structures in striated muscle fibers," *Science*, **147**:158–159 (1965).

Greenawalt, J. W., C. S. Rossi, and A. L. Lehninger, "Effect of active accumulation of calcium and phosphate ions on the structure of rat liver mitochondria," *J. Cell. Biol.*, **23**:21–38 (1964).

Hartmann, J. F., "High sodium content of cortical astrocytes," *Arch. Neurol.* (Chic.), **15**:633–642 (1966).

Heumann, H.-G., "Calciumakkumulierende Strukturen in einem glatten Wirbellosenmuskel," *Protoplasma*, **67**:111–115 (1969).

Heumann, H.-G., and E. Zebe, "Uber Feinbau und Funktionsweise der Fasern aus dem Hautmuskelschlauch des Regenwurmes, *Lumbricus terrestris* L.," *Z. Zellforsch.*, **78**:131–150 (1967).

Komnick, H., "Elektronenmikroskopische Lokalisation von Na⁺ und Cl⁻ in Zellen und Geweben," *Protoplasma*, **55**:414–418 (1962).

Komnick, H., "Zur funktionellen Morphologie der Salzsäure-Produktion in der Magenschleimhaut. Histochemischer Chloridnachweis mit Hilfe der Elektronenmikroskopie," *Histochemie*, **3**:354–378, (1963).

Komnick, H., "Histochemische Calcium-Lokalisation in der Skelettmuskulatur des Frosches," *Histochemie*, **18**:24–29 (1969).

Komnick, H., and M. Bierther, "Zur histochemischen Ionenlokalisation mit Hilfe der Elektronenmikroskopie unter besonderer Berücksichtigung der Chloridreaktion," *Histochemie*, **18**:337–362 (1969).

Lane, B. P., and E. Martin, "Electron probe analysis of cationic species in pyroantimonate precipitates in Epon-embedded tissue," *J. Histochem. Cytochem.*, **17**:102–106 (1969).

Legato, M. J., and G. A. Langer, "The subcellular localization of calcium ion in mammalian myocardium," *J. Cell Biol.*, **41**:401–423 (1969).

Peachy, L. D., "Electron microscopic observation on the accumulation of divalent cations in the intramitochondrial granules," *J. Cell Biol.*, **20**: 95–111 (1964).

Podolsky, R. J., T. Hall, and S. L. Hatchett, "Identification of oxalate precipitates in striated muscle fibers," *J. Cell Biol.*, **44**:699–702 (1970).

Tandler, C. J., C. M. Libanati, and C. A. Sanchis, "The intracellular localization of inorganic cations with potassium pyroantimonate. Electron microscope and microprobe analysis," *J. Cell. Biol.*, **45**:355–366 (1970).

Van Lennep, E. W., G. Young, and H. Komnick, "Electron microscopic demonstration of iodide in the rat thyroid gland (in prep.).

Zadunaisky, J. A., "The localisation of sodium in the transverse tubules of skeletal muscle," *J. Cell Biol.*, **31**:C11–C16 (1966).

Zadunaisky, J. A., J. F. Genarro, N. Bashirelahi, and M. Hilton, "Intracellular redistribution of sodium and calcium during stimulation of sodium transport in epithelial cells," *J. Gen. Physiol.*, **51** (Suppl.):290–302 (1968).

CAMERA ATTACHMENTS

Introduction. The history of the photomicrographic camera attachment began with the invention of the daguerreotypy. The first photomicrographs were probably shown by A. Donna in 1840 in Paris at the Academy of Science. He had taken these photographs with a "camera obscura" which could be called the first photomicrographic apparatus. Since then the use of photography in combination with the microscope has become commonplace.

In the early days of photomicrography, it was customary to develop special units for photomicrography which were used for this purpose exclusively. The advent of the 35 mm film format and the design of the modern microscope stand, equipped with a focusable stage and a fixed microscope tube, have advanced the use of camera attachments. Even more sophisticated and heavier camera attachments can safely be added to a fixed microscope tube, and today each microscope in a laboratory can easily be converted into a camera microscope at relatively low cost.

Principle. In a compound microscope, the intermediary image is located at the front focal plane of the eyepiece. The eyepiece projects this image to infinity. The human eye, being accommodated for infinity, focuses the parallel pencil of rays onto the retina (Fig. 1A). In photomicrography, the retina is replaced by the photographic emulsion. In order to obtain a sharp image at the film plane, the lens of the eye must also be replaced by a photographic objective which is set for infinity (Fig. 1B). Without such objective it would be necessary to refocus the specimen in order to obtain a sharp image at the film plane (Fig. 1C). This, however, is not permissible because the microscope objective would be used with a free working distance and consequently with a tube length for which it is not corrected. As a result, the image quality would be impaired mainly by spherical aberration.

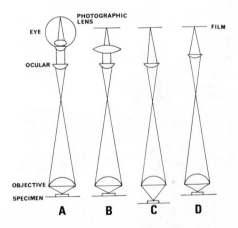

Fig. 1 Path of rays in a microscope for visual observation (A); equipped with a camera attachment containing a photographic objective (B); with a camera attachment without a photographic objective (C); and with a camera attachment containing a photographic eyepiece with adjustable eyelens (D).

Instead of a photographic objective, one could also use a special photographic eyepiece with an adjustable eyelens to focus the image onto the film plane (Fig. 1D).

Manual 35mm attachments. For the majority of applications, the 35mm film format should be given preference. It is inexpensive, and a wide variety of black and white and color film emulsions is available to suit almost all applications. The exposure times are considerably shorter than for the larger formats, and the overall time required to take a photomicrograph, develop the negative, and produce a print is much shorter by comparison.

The decision whether the 35mm camera attachment should have automatic features rests on two questions: the quantity of photographs to be taken, and the variety of different microscopic techniques which are routinely employed. If only an occasional photomicrograph is required or if many photographs of the same type of specimen using the same technique are needed, a manual 35mm attachment with an attachable exposure meter will serve the purpose. Of course, the microscope must be equipped with a trinocular tube or an interchangeable straight photographic tube. It is also advantageous if the microscope has a focusable stage and a fixed tube. With such a microscope, the relatively heavy camera attachment does not rest on the coarse and fine adjustment, and stability of focus is assured.

When selecting a 35mm camera attachment (Fig. 2), the following thoughts should be considered:

a. Camera Body. In photomicrography, the camera body usually serves only as a film transport housing, and most features of the camera are not used, frequently not even the camera shutter. However, the advantage of a

sophisticated 35mm camera body for photomicrography lies in the fact that it can also be used for general photography, gross photography, and close-up photography by employing the various lenses and close-up attachments offered by the manufacturer.

Some camera bodies for photomicrography are equipped with a slot into which a plastic strip with a descriptive text can be inserted. This text is photographed on the same negative as the object.

b. Photographic Objective. This objective guarantees that the microscope objective is used with a tube length for which it is corrected. Another important advantage of this lens is that it provides an exact magnification factor which can be used to calculate the magnification on the negative. This factor (Mc) is given as:

$$Mc = \frac{fc \,(\text{mm})}{250 \,(\text{mm})}$$

in which fc stands for the focal length of the camera objective. For most purposes, it is practical to use a lens with a focal length of approximately 80 mm. The camera attachment then has a magnification factor times 1/3. The negative is now enlarged three times, and the specimen appears on the print under the same magnification as through the binocular tube of the microscope if the print is viewed under normal viewing distance of 250 mm. If a photomicrographic eyepiece with an adjustable eyelens is used instead of a camera lens, the magnification cannot be given accurately and one has to expect an error of up to 15%.

c. Focusing Telescope. Conventional binocular tubes cannot be used conveniently for focusing the specimen for photomicrography because the focus changes in such tubes as one changes the interpupillary distance. In other words, a photograph is not necessarily in focus when the image in the binocular tube is in focus. A photomicrographic attachment should therefore be equipped with a focusing telescope. A reticle in the image plane of the telescope is used to establish parfocality between the film plane and the visual image. A fine double line or double circle just within the resolving power of the human eye serves this purpose best. The reticle should also delineate the area which is photographed.

Recently, binocular tubes have become available in which the focus does not change with a change of interpupillary distance. The image in such tubes is always parfocal with the film plane, eliminating the need for a separate focusing telescope in the attachment. In this case, the reticle is placed into one of the eyepieces in the binocular tube. This eyepiece is equipped with an adjustable eyelens in order to focus the reticle. The development of these tubes with focus compensation for interpupillary distance adjustment has made photomicrography much more convenient.

d. Beam Splitter. A beam splitter, in most cases a prism, directs the light into the focusing telescope. For most applications, it is best to use a prism which directs approximately 20% into the telescope and 80% up to the camera. Such a prism can remain in the path of rays and its effect on the image quality is negligible especially if it is located in a parallel beam. This is the case as long as the camera attachment contains a photographic objective. A prism offers the additional advantage that one can observe the specimen while it is being photographed; a feature which is important with mobile specimens or, for example, in hot stage microscopy. With low light levels as, for

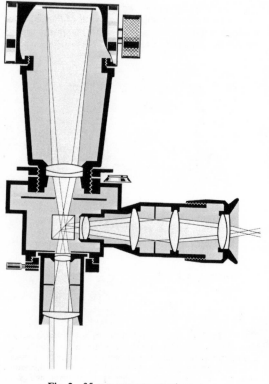

Fig. 2   35 mm camera attachment.

example, in fluorescence microscopy, a 100% deflecting prism is preferred because in these cases, 20% of the light in the telescope would not provide a bright enough image for focusing.

e. Shutter. The focal plane shutter in the camera body should be used only if the entire camera and the photographic objective are shock-mounted. Otherwise, it is preferable to use a shutter which is separately mounted in the camera attachment in order to avoid vibrations. This shutter should also be shock-mounted if one wants to use short exposure times. Rigidly mounted shutters usually cannot be used for exposure times of 1/25 sec or shorter, since the shutter vibrations will be apparent on the photograph. Shutter speeds from 1 sec to 1/125 sec are sufficient for most applications. Since exposure times longer than one second are used quite frequently, it is important that time (T) and bulb (B) settings are provided on the shutter. Photomicrography of mobile specimens requires the use of flash equipment, and therefore, shutters should be synchronized for flash.

f. Exposure Meters. For most applications an exposure meter is essential. A window with a receptacle for the photosensor is a very convenient feature since the photosensor can remain attached to the unit. The accuracy of exposure time determination is greatly improved if only the light from the center of the image reaches the sensor. The specimen detail which is of most interest is usually positioned in the center of the image; it alone should determine the exposure time. Such a detailed measurement is especially important in photographing high-contrast specimens, for darkfield and polarized light microscopy.

Camera attachments for larger film formats. Black and white and color sheet film in large formats especially $3\frac{1}{4} \times 4\frac{1}{4}$ in. and 4×5 in. are still occasionally used today if very high definition is required as, for example, in chromosome studies or for the production of large color prints. Most camera attachments for larger film sizes (Fig. 3) are, however, used with Polaroid film. The same considerations as for 35mm attachments also apply here. The only difference is that such units should be equipped with a camera lens which provides a $1\times$ magnification.

Automatic 35mm camera attachments. In spite of all the refinements in the design of manual camera attachments, it takes time and concentration to take a photomicrograph. This becomes very apparent when one has to take a large number of photographs employing a variety of different microscopical techniques which all require special attention with respect to exposure time determination and film material. In such cases, a fully automatic camera attachment is invaluable (Fig. 4). Everyday practice proves that an automatic camera attachment should be automated to the highest degree and it should be applicable for the widest variety of uses.

Analog computer methods are a practical approach to automatic exposure time determination. In such systems, a certain small portion of the light coming from the microscope is deflected to a photosensor where an exactly proportional current is produced (Fig. 4). This current charges a capacitor which is matched to the sensitivity (ASA rating) of the film emulsion. This means that the capacitor requires the same time to get charged as the film emulsion needs for correct exposure. As soon as the capacitor is fully charged, the shutter is closed and the film advance is triggered. This means that the exposure time determination takes place during the actual exposure,

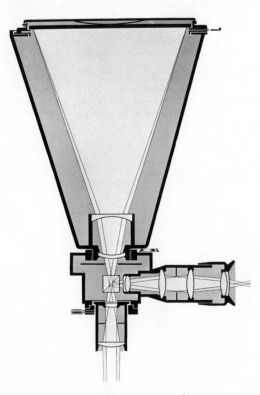

Fig. 3 4×5 in. camera attachment.

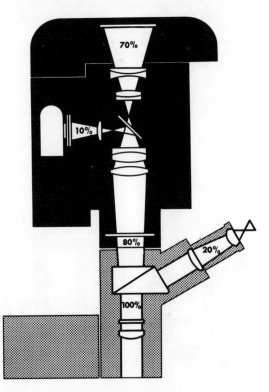

Fig. 4 Automatic 35 mm camera attachment.

not before. The capacitor is variable to allow proper exposure for different ASA ratings. The microscopist, using an automatic camera attachment, can devote his whole attention to the microscopy of the specimen, and only has to trigger the camera to obtain a photomicrograph. Such additional features as outlined below are important because they determine the usefulness of the automatic camera for special applications.

a. Photosensor. Photo resistors are inexpensive but slow, and can only be used if the light levels are relatively high. Automatic attachments which are equipped with highly sensitive photomultipliers function at much lower light levels and can be used for techniques such as polarized light and fluorescence microscopy.

b. Shutters. Small electromagnetic shutters are superior to the conventional mechanical shutters because they work essentially vibration free.

c. Exposure Time Range. Frequently automatic camera attachments do not permit the use of very short exposure times. This is a serious disadvantage, since one must use neutral density filters to reduce the light to levels manageable by the camera. There are, however, automatic cameras on the market which function with exposure times as short as 1/200 sec. Very long exposure times usually do not prove to be any problem. Many automatics function with exposure times up to half an hour.

d. Exposure Time Determination. As long as a specimen has low or medium contrast, which is true for most stained biological specimens and specimens in phase contrast, an exposure meter which integrates over the total photographic frame will give adequate results. An exposure time is found which does not underexpose the dark or overexpose the light areas in the specimen. In this case, an automatic camera would be of limited value because once the exposure time for a given type of specimen is established, it could be used again as long as the transformer setting for the light source is not changed. If specimens of high contrast are to be photographed (e.g., darkfield, fluorescence, polarized light microscopy, high magnifications on heavily stained specimens, high contrast metallurgical specimens) an integrating exposure meter will not provide good results. The integrated exposure time will be too long for the light specimen detail and too short for the dark specimen detail. This problem is aggravated if the specimen detail of interest does not cover the whole frame, which is frequently the case in darkfield and fluorescence microscopy. The coverage of the photographic frame will affect the exposure time. In these cases, the best exposure meters and even a semi-automatic or fully automatic camera will not solve the problem. What is needed is a camera which is equipped for detail measurements or in other words equipped with an exposure measuring device which permits determination of the proper exposure for that specimen detail which is of most interest. Usually a very small area (1/100 of the total frame) is used for the detail measurement.

Camera stands. If utmost flexibility with respect to differing photographic tasks is required of one and the same photomicrographic instrument, a separate camera stand (Fig. 5) should be used instead of a true camera attachment. A typical camera stand usually has many accessories permitting one to use different film formats (35mm and 4 × 5 in.). Movie cameras and 4 × 5 in. bellows cameras, which are really too heavy to be attached directly to the microscope, can be attached to such a stand. In low-power photomicrography, it is easier to use a groundglass screen for focusing, and the camera attachments for photographic stands are usually equipped with these screens.

Camera stands can be used for both photomicrography and photomacrography in transmitted and incident light. For photomicrography, the microscope is placed directly on the stand under the camera. An additional advantage is that microscopes of different manufacture can be used. For photomacrography, the microscope is removed and accessories such as an apparatus for transmitted light can be added.

The availability of accessories such as an automatic 4 × 5 in. camera, equipment for transmitted light photomacrography, and the number of macro objectives available should help one decide which unit offers the most capabilities for the laboratory.

Conclusion. It seems to be more practical to have separate microscopes set up for different applications such as brightfield, fluorescence, interference, polarized light

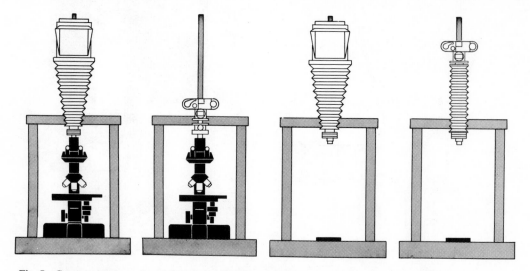

Fig. 5 Camera stands equipped for photomicrography and photomacrography using different film formats.

microscopy. By comparison an integrated camera microscope would have to be equipped with a multitude of accessories for these techniques. Each time the technique is different it would be necessary to set up the instrument for the special task at hand. This requires time and effort. It is therefore much more practical to use manual or automatic camera attachments to convert any microscope in the laboratory into a camera microscope.

As a general rule one should try to use the camera attachments which are offered by the manufacturer of the microscope. The adaptation of attachments of different manufacture is often difficult, not only mechanically but also optically.

MANFRED NAHMMACHER

Bibliography

Claussen, H. C., "Mikroskope," in "Encyclopedia of Physics," Vol. XXIX, Optical Instruments, Berlin, Springer-Verlag, 1967.

Kingslake, R., "Applied Optics and Optical Engineering," Vol. IV, Optical Instruments, Part I, New York, Academic Press, 1967.

Michel, K., "Die Wissenschaftliche und Angewandte Photographie," Vol. X, Berlin, Springer-Verlag, 1957.

CARMINE

There is considerable confusion in the literature between the terms "cochineal," "carminic acid," and "carmine." A number of coccid homopteran insects produce a red pigment, of which that from the European oak parasite *Kermes ilicis* has been used from classical antiquity. The insect is scarce and a poor pigment producer so that red, as distinct from the muddy brown-red derived from madder, cloth was usually the prerogative of royalty. The discovery by the early explorers of Mexico of *Coccus cacti*, enormously productive of dye and massively infecting cacti over large areas, made it possible for European dyers to produce purple, red, and scarlet clothes as easily as those of other colors.

The dried bodies of *Coccus cacti*, either entire or powdered, are properly known as "cochineal." Aqueous "extract of cochineal" was used by dyers to produce a crimson-red with alum mordants and black with iron mordants. The addition of tin to an alum mordant, a discovery said to have been accidentally made through the use of a pewter dyeing vat, produced a true scarlet.

An alcoholic extract of the defatted dried cochineal insects is properly "tincture of cochineal" and the residue obtained from its evaporation was originally called "carminic acid." This is the product intended in all early formulas. Some contemporary "carminic acid" is obtained from extract of cochineal, but from the microscopist's point of view appears no different.

"Carmine" is a product of variable, and ill defined, composition. It is primarily an aluminum lake of carminic acid. However, bright scarlet samples (e.g., the "carmin opt. rubr." of many European manufacturers) usually also contain tin, while the calcium-tin lake is the "carmine lake" of artists.

Staining formulas and methods

The six divisions of the carmine formulas here employed are based entirely on the ingredients, the most widely known being the first two classes of "alcohol carmines" and "alum carmines." Considerable confusion has been occasioned by the fact that Grenacher, in 1879, published formulas for each of these two divisions and early workers almost invariably employed the alum carmine, while modern workers seem to prefer the "alcoholic borax carmine."

The early employment of carmine for staining materials before sectioning was necessitated by the fact that no method had been worked out for attaching sections to slides, so that the fewer manipulations which were undertaken in the sections, the more chance there was for preserving the whole. The straight alum carmines are best employed for direct staining from exceedingly dilute solutions, a 1% solution of ammonia alum being the customary diluent. Alcoholic carmines, particularly that of Grenacher 1879, are most employed for the preparation of wholemounts of small invertebrates. The formulas of Mayer and Mayer 1829a, are the best devised for small marine invertebrates. Though the borax-carmine of Grenacher is commonly made today by the method here given for the "working formula direct," the original method of Grenacher was to prepare the dry stock and to make up working formulas from it in various strengths of alcohol. This gives far better control of the process, since the solubility of the dry stock is a direct function of the concentration of alcohol employed. When any of the alcoholic carmines are used, they are differentiated with a 0.1% solution of hydrochloric acid in 70% alcohol.

The aceto-carmines, which form the next class, are more widely employed in botanical than in zoological techniques, and their most valuable application is the staining of unfixed nuclei. They should be confined if possible to this use, for their preservation as permanent objects is difficult. Their only other use, besides the counting of chromosomes, is in the diagnostic staining of parasitic platyhelminthes.

Picro-carmines are warmly recommended to the beginner, for it is almost impossible to overstain in them. The picric acid, moreover, acts as a fixative. It is possible to take a small living invertebrate, throw it into the stain for ten minutes or so, and remove it fixed and stained. The original formula of Ranvier 1889 called for a preparation of a dry stock and the preparation of a working solution from this. Until quite recent times the dry stock could be purchased. It undoubtedly makes a better solution and keeps better than do the formulas prepared as solutions directly.

Iron carmines had a brief vogue in the first decade of the present century and then again fell into disfavor. They are nuclear stains strongly resembling the reactions of the iron hematoxylins. The formula of De Groot 1903 is the one most usually recommended. Ammonia carmines and hydrochloric carmines are no longer very well known, though the formula of Hollande 1916 gives excellent nuclear staining. Carmines of the final class, which contains those formulas that cannot reasonably be fitted into the previous classes, are rarely used today. The only formula finding any great acceptance is the "lithium carmine" of Orth (1892), which was rediscovered in Germany in the 1920s. It is most warmly recommended by Spielmeyer, 1924, for counterstaining sections of the nervous system.

Alum Carmines

Anderson 1926 (*J. Path. Bact.*, **29**: 117)

FORMULA: water 95, abs. alc. 10, acetic acid 5, carmine 1, ammonium alum 3.5, calcium hypochlorite 0.1

PREPARATION: To the carmine suspended in the alc., add the hypochlorite suspended in 5 water. Dissolve the alum in 90 water, bring to boil, add carmine mixture, boil 1 min, cool, filter. Add acid.

Arcangeli 1885a (*Proc. verb. Soc. tosc. Sci. nat.*, **4** : 233)
FORMULA: water 100, ammonium alum 15, boric acid 2, carmine 0.25
PREPARATION: Boil 10 min. Filter.

Arcangeli 1885b (*Proc. verb. Soc. tosc. Sci. nat.*, **4** : 233)
FORMULA: water 100, ammonium alum 15, salicylic acid 0.25, carmine 0.25
PREPARATION: As Arcangeli 1885a.

Czokor 1880 (*Arch. mikr. Anat.*, **18** : 413)
FORMULA: water 100, cochineal 2, potassium alum 2, phenol 0.25
PREPARATION: Suspend the cochineal and alum in 200 water. Boil until volume is reduced to 100. Leave 2 days, filter, and add phenol to filtrate.

Gower 1939 (*Stain Tech.*, **14** : 31)
FORMULA: water 100, alum 5, residue from preparation of Schneider 1880 (*q.v.*) 0.5
NOTE: This was originally recommended for trematodes, but is an excellent general-purpose, wholemount stain.

Grenacher 1879 (*Arch. mikr. Anat.*, **16** : 465)
FORMULA: water 100, carmine 1, ammonium alum 10
PREPARATION: Add the carmine to the boiling alum solution. Cool. Filter.
METHOD: [water] → suitable dilution of stain, until sufficiently stained → balsam, via usual reagents.
NOTE: The larger the animal, the greater the dilution. For protozoans use full strength overnight; for a large leech use 1 : 5000 for 2 months. This stain is frequently confused with Grenacher 1879. Mahrenthal's carmine (Rawitz, 1895, p. 60) is 4 parts of this solution with 1 of 95% alc.

Guyer 1930 (Guyer, 1930, p. 9)
FORMULA: water 100, potassium alum 6, cochineal 6
PREPARATION: Boil 30 min. Dilute to 150. Boil until reduced to 100. Cool. Filter.

Kirkpatrick cited from 1938 Carleton and Leach cit. Cappell (Carleton and Leach, 1938, p. 105)
FORMULA: water 100, acetic acid 2.5, ammonium alum 2.5, cochineal 2.5, salicylic acid 0.1
PREPARATION: Soak cochineal 20 min in 10 water with 2.5 acetic acid. Add 40 water and boil 1 hr. Dissolve alum in 50 water, bring to boil, and add to boiling cochineal. Boil 1 hr. Cool, make up to 100, filter, add salicylic acid.

Mahrenthal (see Grenacher 1879 [*NOTE*])
Mayer 1892 *Carmalum-auct.* (*Mitt. zool. Stat. Neapel.*, **10** : 482)
FORMULA: water 100, potassium alum 5, carminic acid 0.5
USE: Full strength, usually for 1–10 days, on embryos prior to embedding and sectioning.

Mayer 1897 *Carmalum-auct.* (*Z. wiss. Mikr.*, **14** : 29)
FORMULA: water 100, potassium alum 5, carmine 2
PREPARATION: Boil 1 hr. Cool. Filter.

Partsch 1877 (*Arch. mikr. Anat.*, **14** : 180)
FORMULA: water 100, potassium alum 6, cochineal 3, salicylic acid 0.25

PREPARATION: Boil alum and cochineal 2 hr. Cool. Filter. Add salicylic acid.

Rabl 1894 (*Z. wiss. Mikr.*, **11** : 168)
FORMULA: water 100, potassium alum 4, cochineal 4
PREPARATION: Suspend cochineal and alum in 130 water. Boil until reduced to 100. Cool. Filter.
NOTE: The formula of Böhm and Oppel, 1907, p. 98, is in every way identical.

Rawitz 1895 (Rawitz, 1895, p. 61)
FORMULA: water 100, potassium alum 5, carminic acid 0.5

Rawitz 1899 (*Anat. Anz.*, **15** : 438)
FORMULA: water 50, glycerol 50, ammonium alum 6.5, carminic acid 0.7
PREPARATION: Dissolve the dye with heat in the alum solution. Filter. Add glycerol.

Alcoholic Carmines

Grenacher 1879 *Alcoholic borax carmine—compl. script.* (*Arch. mikr. Anat.*, **16** : 448)
PREPARATION OF DRY STOCK: Boil 250 water, 8 carmine, 10 sodium borate for 30 min. Cool overnight. Filter. Evaporate filtrate to dryness.
WORKING FORMULA FROM DRY STOCK: 30%, 50%, or 70% alc. 100, dry stock to sat.
PREPARATION WORKING FORMULA DIRECT: Boil 50 water, 1.5 carmine, 2 sodium borate for 30 min. Cool. Add 50 70% alc. Leave 2–3 days. Filter.
REAGENTS REQUIRED: A, any selected working formula; B, 0.1% hydrochloric acid in 70% alc.
METHOD: [Alcohol of lower concentration than selected working formula] → A, not less than 12 hr → B, until pink and translucent → balsam, via usual reagents.
NOTE: Though it is customary to prepare the working solution direct, much better results may be obtained from dry stock in various strengths of alcohol. The solubility decreases very rapidly with increasing alcoholic content so that power of the stain may be accurately controlled.

van Mahrenthal (see Grenacher 1879 [*NOTE*])
Mayer 1881 Alcoholic cochineal—compl. script. (*Mitt. zool. Stat. Neapel.*, **2** : 14)
FORMULA: water 30, 95% alc. 70, cochineal 10
PREPARATION: Digest 1 week. Filter.
METHOD: [marine invertebrate larvae, formaldehyde or alcohol preserved] → stain, 1–10 min → 70% alc., until differentiated → balsam, via usual reagents.
NOTE: For the material indicated this stain gives better results than other carmine formulas.

Mayer 1892a Paracarmine—compl. script. (*Mitt. zool. Stat. Neapel*, **10** : 491)
REAGENTS REQUIRED: A, 70% alc. 100, aluminum chloride 0.5, strontium chloride 4, carminic acid 1; B, 0.1% strontium chloride in 70% alc.
METHOD: [small invertebrates] → 50% alc. → A, until required structures are stained, 5 min to 1 week → B, if differentiation necessary → balsam, via cedar oil.
Mayer 1892b (*Mitt. zool. Stat. Neapel.*, **10** : 498)
FORMULA: water 50, 95% alc. 50 nitric acid 0.3, cochineal 5, calcium chloride 5, aluminum chloride 0.5
PREPARATION: Grind the dry ingredients to a paste with the acid. Mix solvents with paste, bring to boil, leave 5 days, filter.

Schwarz 1933 (*Z. wiss. Mikr.*, **50** : 305)
FORMULA: water 50, methanol 50, carmine 2, sodium borate 2
PREPARATION: Boil the carmine and sodium borate 1 hr in 100 water. Evaporate to 25. Cool. Dilute to 50 and add 50 methanol. Leave 1 day. Filter.

Seiler 1881 (Seiler, 1881, p. 62)
FORMULA: water 100, 95% alc. 50, sodium borate 1, carmine 0.6

Spuler cited from 1907 Böhm and Oppel (Böhm and Oppel, 1907, p. 99)
FORMULA: water 100, cochineal 10, 95% alc. *q.s.*
PREPARATION: Boil the cochineal in 100 water until reduced to 50. Add 95% alc. until ppt. appears. Filter. Evaporate to 100.

Aceto Carmines

Belling 1921 (*Amer. Nat.*, **54** : 573)
PREPARATION: To 50 Schneider 1880 add Belling 1921 until ppt. appears. Then add 50 Schneider 1880.
METHOD: [smears for chromosome counts, or fresh cestodes for diagnosis] → stain, on slide → examine.

Henneguy 1887 (Lee and Henneguy, 1887, p. 88)
STOCK FORMULA: water 100, potassium alum, 6, carmine 2, acetic acid 25
PREPARATION OF STOCK: Boil the dye and alum in the water 1 hr. Cool. Add acid; leave 10 days, filter.
WORKING SOLUTION: A, water 99, stock 1
METHOD: [95% alc.] → A, until stained → distilled water → balsam, via usual reagents
NOTE: This is the only aceto-carmine well adopted for general staining of wholemounts.

Nickiforow cited from 1900 Pollack (Pollack, 1900, p. 76)
FORMULA: water 100, carmine 3, sodium borate 10, ammonia *q.s.*, acetic acid 0.5
PREPARATION: Boil dye and sodium borate 1 min in 200 water. Add ammonia drop by drop until all carmine is dissolved. Evaporate to 100. Add acid.

Schneider 1880 (*Zool. Anz.*, **3** : 254)
FORMULA: water 55, acetic acid 45, carmine 5
PREPARATION: Boil 15 min under reflux. Cool. Filter.
METHOD: as Belling 1921
NOTE: This formula is frequently (cf. Gatenby and Painter 1937, p. 685) attributed to Belling 1921 (*q.v.*).

Sémichon 1924 (*Rev. path. veg.*, **11** : 193)
REAGENTS REQUIRED: A, water 50, acetic acid 50, carmine 5; B, 70% alcohol
PREPARATION OF A: Digest ingredients 1 hr at 90°C. Cool. Filter.
METHOD: [living material] → A, till stained → B, till differentiated → balsam, via usual reagents.

Zacharias 1894 (*Zool. Anz.*, **11** : 62)
REAGENTS REQUIRED: A, water 70, acetic acid 30, carmine 0.5; B, 1% acetic acid; C, 1% ferric ammonium citrate
PREPARATION OF A: As Sémichon 1924.
METHOD: [whole objects to be sectioned, or sections] → water → A, 2–5 hr → B, rinse → C, 2–3 hr → distilled water, thorough wash → balsam, or paraffin, via usual reagents

Picro Carmines

Arcangeli cited from circ. 1890 Francotte (Francotte, p. 217)
FORMULA: water 100, picric acid 1, carmine 0.5
PREPARATION: Boil 10 min. Cool. Filter.

Bizzozero cited from 1889 Friedländer (Friedländer, 1889, p. 89)
FORMULA: water 100, 95% alc. 20, picric acid 1, carmine 1, ammonia 6
PREPARATION: Dissolve carmine in ammonia. Add 100 water. Dissolve picric in 100 water. Add to carmine and evaporate to 100. Cool; filter. Add alc. to filtrate.

Francotte circ. 1890 (Francotte, p. 216)
FORMULA: water 100, carmine 1, ammonia 5, picric acid 1, chloral hydrate 1
PREPARATION: Dissolve carmine in ammonia. Dissolve picric acid in 50 water and add to carmine. Dilute to 100. Add chloral hydrate.

Friedlander 1889 (Friedlander, 1889, p. 88)
FORMULA: water 50, carmine 1, ammonia 1, sat. sol. picric acid 100
PREPARATION: Dissolve carmine in ammonia. Add water and picric solution.

Gage 1880 cited from circ. 1890 Francotte (Francotte, p. 213)
FORMULA: water 100, carmine 1, ammonia 50, picric acid 1
PREPARATION: Dissolve carmine in ammonia. Add picric dissolved in water. Leave 1 day. Filter. Evaporate to dryness. Dissolve residue in 100 water.

Guyer 1906 cited from 1930 ips. (Guyer, 1930, 239)
FORMULA: water 50, sat. aq. sol. picric acid 50, ammonia 5, carmine 1
PREPARATION: Dissolve carmine in ammonia. Add water and picric solution. Leave 2 days. Filter.

Jensen 1937 (*Zbl. Bakt.*, **139** : 333)
STOCK SOLUTIONS: I, water 100, 0.1, carmine 0.5; II, water 100, picric acid 0.5, magnesium oxide 2
WORKING SOLUTION: stock I, 80; stock II, 20

Legal 1884 (*Morph. Jahrb.*, **8** : 353)
FORMULA: Grenacher 1897 90, sat. aq. sol. picric acid 10

Löwenthal cited from 1900 Pollack (Pollack 1900, 79)
FORMULA: water 100, sodium hydroxide 0.5, carmine 2, picric acid *q.s.*
PREPARATION: Dissolve carmine and alkali in 50 boiling water. Boil 15 min. Dilute to 100 and cool. Add picric acid in excess of saturation.

Malassez cited from 1877 Frey (Frey, 1877, p. 96)
FORMULA FOR DRY STOCK: water 100, ammonia 2, carmine 0.5, picric acid 2.5
PREPARATION FOR DRY STOCK: Dissolve carmine in ammonia and water. Add picric. Shake; allow to settle; and decant. Evaporate supernatant to dryness.
FORMULA FOR WORKING SOLUTION: water 100, dry stock 2
PREPARATION OF WORKING SOLUTION: Mix. Leave 1 week. Filter.

Mayer 1897 (*Z. wiss. Mikr.*, **14** : 23)
FORMULA: Mayer 1897 10, 0.6% magnesium picrate 90

Neuman cited from 1928 Schmorl (Schmorl 1928, p. 120)
REAGENTS REQUIRED: A, Grenacher 1879 100, picric acid 1.25% B, 2% hydrochloric acid in glycerol; C, anhydrous glycerol
METHOD: [sections] → 5–10 min → B, 10 min → C
NOTE: To prepare balsam mounts, substitute a sat. sol. picric acid in abs. alc. for C above and clear in clove oil.

Oppier cited from 1928 Schmorl (Schmorl 1928, p. 341)
FORMULA: carmine 0.5, ammonia 0.5, picric acid 0.005, water 100

Orth cited from 1904 Besson (Besson 1904, p. 255)
FORMULA: Orth (1892) 65, sat. aq. sol. picric acid 35
NOTE: This same formula is given by Squire, 1892 but without acknowledgement to Orth.

Ranvier 1875 (Ranvier 1875, p. 100)
PREPARATION OF DRY STOCK: Dissolve 5 carmine in 50 ammonia. Add 500 sat. aq. sol. picric acid. Evaporate to 100. Cool 24 hr. Decant supernatant liquid which is evaporated to dryness.
WORKING SOLUTION: water 100, dry stock 5
PREPARATION OF WORKING SOLUTION: Boil 10 min. Cool. Filter.
METHOD: [anything, living or dead] → stain, until done → 70% alc. until color clouds cease → balsam, via usual reagents
NOTE: The original formula called for a prolonged period of putrefaction in the course of manufacture. This stain, now practically unknown, should be placed in the hands of every beginner. A live Cyclops may be dropped into the solution and removed, fixed, and stained 10 min later; so may any other object.

Rutherford cited from 1878 Marsh (Marsh 1878, p. 79)
FORMULA: sat. sol. picric acid 100, carmine 1, ammonia 2, water
PREPARATION: Dissolve carmine in ammonia and 5 water. Raise picric solution to boiling and add carmine solution. Evaporate to dryness. Dissolve residue in 100.

Squire 1892a (Squire, 1892, p. 35)
FORMULA: ammonia 1.5, carmine 0.5 water 2.5 sat. aq. sol. picric acid 100
PREPARATION: Dissolve dye in ammonia. Add picric solution.

Squire 1892b (Squire, 1892, p. 35)
FORMULA: water 100, sodium hydroxide 0.05, carmine 0.5, picric acid *q.s.*
PREPARATION: Dissolve the carmine in the boiling alkali sol. Add just enough sat. aq. sol. picric acid to redissolve the ppt. first formed.

Squire 1892c (Squire, 1892, 35)
FORMULA: Orth (1892) 25, sat. aq. sol. picric acid 75
NOTE: See also Orth (1904) note.

Vignal cited from 1907 Böhm and Oppel cit. Henneguy (Böhm and Oppel, 1907, p. 118)
FORMULA: water 100, picric acid 2, carmine 1, ammonium hydroxide 5
PREPARATION OF DRY STOCK: Mix all ingredients. Leave 2–3 months in closed bottle. Evaporate at room temperatures to 80. Decant and evaporate supernatant to dryness.
WORKING SOLUTION: water 100, dry stock 1

Weigert 1881 (*Virchows Arch.*, **84** : 275)
FORMULA: sat. aq. sol. picric acid 100, carmine 1, ammonia 2, acetic acid *q.s.*
PREPARATION: Dissolve carmine in ammonia. Add picric solution and leave 24 hr. Then add just enough acetic acid to produce permanent ppt. Filter.

Iron Carmines

de Groot 1903 (*Z. wiss. Mikr.*, **20** : 21)
FORMULA: water 100, hydrochloric acid 0.05, potassium alum 2.5, ferric alum 0.05, carminic acid 0.5
PREPARATION: Dissolve ferric alum in 10 water. Add dye and dilute to 100. Heat to 60°C and add potassium alum. Cool. Filter. Add acid to cold filtrate.

Hansen 1905 (*Z. wiss. Mikr.*, **22** : 85)
FORMULA: water 100, sulfuric acid 1, ferric alum 3, cochineal 3
PREPARATION: Boil all ingredients 10 min. Cool. Filter.

Lee 1902 cited from 1905 *ips.* (Lee 1905, p. 170)
REAGENTS REQUIRED: A, Benda 1893; B, 0.5% carminic acid in 50% alc.
METHOD: [sections] → water → A, some hours → rinse, 50% alc. → B, 1–2 hr → wash, 50% alc. → balsam, via usual reagents

Peter 1904 (*Z. wiss. Mikr.*, **21** : 314)
REAGENTS REQUIRED: A, water 100, cochineal 6, hydrochloric acid 0.3; B, 2.5% ferric alum
PREPARATION OF A: Boil cochineal in 150 water until reduced to 30. Dilute with hot water to 100. Cool. Filter. Add acid.
METHOD: [whole objects to be sectioned, or sections] → A, 48 hr at 37°C → rinse → B, some hours → balsam, or paraffin, via reagents
RESULT: Said by Lee (Lee 1905, p. 171) to stain yolk granules red on a gray background.

Spuler 1901 cited from 1910 *ips.* (Ehrlich, et al., 1910, **1** : 240)
PREPARATION OF STOCK SOLUTION: Boil 20 cochineal 1 hr in 100 water. Filter, saving both residue and filtrate. Boil residue in 100 water 1 hr. Filter. Reject residue. Mix filtrates and reduce to 100. Add 95% alc. until ppt. appears. Filter. Reduce filtrate to 100. Filter.
REAGENTS REQUIRED: A, stock 35, 50% alcohol 65; B, 0.75% ferric alum
METHOD: 70% alcohol → A, 48 hr at 37°C → wash → B, 24 hr

Wellheim 1898 (*Z. wiss. Mikr.*, **15** : 123)
REAGENTS REQUIRED: A, 0.01% ferric chloride in 50% alc.; B, 0.5% carminic acid in 50% alc.; C, 0.1% hydrochloric acid in 70% alc.
METHOD: [sections] → 50% alc. → A, overnight → rinse, 50% alc. → B, 1–2 hr → wash, 50% alc. → C, until differentiated → balsam, via usual reagents

Ammonia Carmines

Beale 1857 cited from 1880 Beale (Beale, 1880, p. 125)
FORMULA: carmine 0.5, ammonia 1.5, water 44, 95% alc. 12, glycerol 44
PREPARATION: Dissolve carmine in ammonia. Add other ingredients.

Cole 1903 (Cross and Cole 1903, p. 169)
FORMULA: water 100, carmine 1, ammonium hydroxide 2, sodium borate 6
PREPARATION: As Beale 1857.

Frey 1877 (Frey 1877, p. 94)
FORMULA: water 40, glycerol 40, 95% alc. 20, carmine 4, ammonia *q.s.* to dissolve
PREPARATION: As Beale 1857.

Gerlach 1858 cited from 1892 Squire (Squire 1892, p. 31)
FORMULA: water 100, ammonia 1, carmine 1
NOTE: Squire (*loc. cit.*) and Lee (Lee, 1905, p. 172) recommended that the solution be allowed to grow mold, evaporated to dryness, and then redissolved in distilled water.

Hoyer cited from 1900 Pollack (Pollack 1900, p. 78)
PREPARATION OF DRY STAIN: Dissolve 2 carmine in 4 ammonia and 16 water. Boil until smell of ammonia not apparent. Cool. Add 80 95% alc. Filter.
WORKING SOLUTION: water 100, powder from above 0.5

Meriwether 1935 (*Bull. Int. Ass. Med. Mus.*, **14** : 64)
FORMULA: water 75, ammonia 25, potassium chloride 6.25, potassium carbonate 1.9. carmine 2.5
PREPARATION: Boil the salts and the dye in the water 2 min. Cool and add ammonia.

Smith cited from 1903 Cross and Cole (Cross and Cole, 1903, p. 171)
FORMULA: water 50, glycerol 25, 95% alc. 25, carmine 1, ammonia 1.5, sodium borate 1 ,
PREPARATION: As Beale 1857.

Squire 1892 (Squire, 1892, p. 31)
FORMULA: water 40, 95% alc. 20, glycerol 40, carmine 2.5, ammonia 5
PREPARATION: As Beale 1857.

Hydrochloric Carmines

Hollande 1916 Chorcarmin—compl. script. (*C. R. Soc. Biol. Paris*, **79** : 662)
REAGENTS REQUIRED: A, water 90, 95% alc. 10, carmine 7, hydrochloric acid 2.5 water *q.s.* to bring filtrate to 90, 70% alc. 10; B, 3% ferric alum; C, 1% pyridine
PREPARATION OF A: Grind the carmine with the acid. Wash out mortar while grinding with 12 doses each of 10 water. Boil washings until reduced to 70. Cool. Filter, Dilute to 90. Add alc.
METHOD: [sections] → water → A, 2 hr → water, rinse → B, until black, few minutes → B, fresh solution, until differentiated, 1/2–2 hr → C, thorough wash → running water, 15 min → balsam, via usual reagents

Kingsbury and Johannsen 1927 (Kingsbury and Johannsen, 1927, p. 44)
FORMULA: water 30, 95% alc. 70, carmine 2, hydrochloric acid 3
PREPARATION: Boil all ingredients under reflux. Cool. Filter.

Langeron 1942 (Langeron, 1942, p. 517)
REAGENTS REQUIRED: A, water 2.5, 90% alc. 100, carmine 2.5, hydrochloric acid 2.5; B, 0.5% hydrochloric acid in 80% alc.

PREPARATION OF A: Grind carmine, acid, and water to a paste. Leave 1 hr. Transfer to reflux flask with alc. Reflux on water bath until solution complete.
METHOD: [whole objects] → 70% alc. → A, until stained → B, until differentiated → balsam, via usual reagents

Mayer 1881 (*Mitt. zool. Stat. Neapel*, **2** : 1)
FORMULA: 95% alc. 90, water 15, carmine 4, hydrochloric acid 1, ammonia *q.s.* to give pH 7
PREPARATION: Heat water, carmine, and acid to 80°C. Add alc. and reflux 10 min. Adjust to pH 7 with ammonia.
NOTE: In 1883 (*Mitt. zool. Stat. Neapel*, **4** : 521) Mayer directed that the carmine, acid, and water be boiled to complete solution, cooled, and the alc. added before neutralization.

Mayer 1883 (see Mayer 1881 [*NOTE*])
Meyer 1885 (*Ber. deuts. bot. Ges.*, **10** : 363)
FORMULA: 90% alc. 50, carmine 1.25, hydrochloric acid 5, chloral hydrate 60
PREPARATION: Reflux the carmine with the acid and alc. for 30 min Cool. Filter. Add chloral hydrate.

Schwartz 1934 (*Z. wiss. Mikr.*, **50** : 305)
PREPARATION OF DRY STOCK: Boil 1 carmine and 4 ammonium alum in 150 water until reduced to 75. Filter. Evaporate filtrate to dryness.
WORKING SOLUTION: water 60, methanol 40, hydrochloric acid 0.2, dry stock 8

Snow 1963 (*Stain Tech.*, **38** : 9)
FORMULA: Carmine 4, hydrochloric acid, water 15, 85% alc. 85
PREPARATION: Boil carmine, acid and water for 10 min with frequent stirring. Cool, add alcohol, and filter.
METHOD: [fixed plant material] → 70% alc. until all fixative removed → stain 24 hr at 60°C → 70% alc. rinse → [macerate, if necessary, in 45% acetic acid] → prepare squash in usual manner

Other Carmine Formulas

Arcangeli 1885, boric carmine (*Proc. verb. Soc. tosc. Sci. nat.*, **4** : 233)
FORMULA: water 100, boric acid 4, carmine 0.5
PREPARATION: Boil 10 min. Cool. Filter.
Best 1906 potash carmine see Best 1906
Cuccati 1886, soda-carmine (*Z. wiss. Mikr.*, **3** : 50)
FORMULA: water 90, 95% alc. 10, acetic acid 0.5, sodium carbonate 6.5, carmine 1.7, chloral hydrate 2
PREPARATION: Boil the carmine with the carbonate in 35 water. Cool. Add the alc, to cooled solution. Leave overnight; filter, and bring filtrate to 100. Add the acid and chloral hydrate.

Cuccati cited from 1889 Friedländer, soda-carmine (Friedländer, 1889, p. 90)
FORMULA: water 100, carmine 4, sodium carbonate 13, abs. alc. 7, acetic acid 1.5, chloral hydrate 2.5
PREPARATION OF DRY STOCK: As Cuccati 1886, save that the final solution is evaporated to dryness.
PREPARATION OF WORKING SOLUTION: 80% alc. 100, powder *q.s.*

Francotte circ. 1890, boric-carmine (Francotte, p. 209)
FORMULA: water 25, 95% alc. 75, boric acid 5, carmine 0.4
PREPARATION: Boil under reflux 15 min.

Frey 1877, borax-carmine (Frey, 1877, p. 95)
FORMULA: water 3, sodium borate 3.5, carmine 0.6, 95% alc. 60
PREPARATION: Boil dye with water and borax. Cool. Filter. Add alc. to filtrate; leave 24 hr. Filter.

Fyg 1928a, chrome-carmine (*Z. wiss. Mikr.*, **45** : 242)
FORMULA: water 100, chrome alum 6, carmine 1
PREPARATION: Add carmine to boiling alum solution. Boil 15 min. Cool. Filter.
NOTE: Copper alum may be substituted for chrome alum. Copper alum (Merck Index 1940, p. 165) is prepared by fusing together 34% potassium alum 32% cupric sulfate, 32% potassium nitrate, and 2% camphor.

Fyg 1928b, soda-carmine (*Z. wiss. Mikr.*, **45** : 242)
FORMULA: water 100, sodium bicarbonate 5, carmine 0.5
PREPARATION: Boil 5 min. Cool. Filter.

Haug cited from 1900 Pollack, ammonia-carmine (Pollack 1900, p. 80)
FORMULA: carmine 1, ammonium chlorate 2, water 100, ammonia 0.25, lithium carbonate 0.5
PREPARATION: Boil carmine in chlorate solution 15 min. Cool. Add other ingredients. Filter.

Kahlden and Laurent 1896, lithium-carmine (Kahlden and Laurent, 1896, p. 56)
FORMULA: sat. aq. sol. lithium carbonate 100, carmine 3
PREPARATION: Boil 15 min. Cool. Filter.

Mayer 1897, magnesia-carmine (*Z. wiss. Mikr.*, **14** : 23)
FORMULA: magnesium oxide 0.2, carmine 2, water 100
PREPARATION: Boil 5 min. Cool. Filter.
NOTE: Magnesium oxide very readily turns to the carbonate which cannot be used for this preparation. Hence the insistence on fresh magnesia (*magnesia usta* —"burnt magnesia" in pharmacopoeial Latin) in the original formula.

Mayer 1902, aluminum-carmine (*Mitt. zool. Stat. Neapel*, **10** : 490)
FORMULA: water 100, aluminum chloride 1.5, carminic acid 0.5

Orth cited from 1892 Squire, lithium-carmine (Squire 1892, p. 33)
FORMULA: water 100, lithium carbonate 1.5, carmine 2.5
PREPARATION: Boil. Cool. Filter.

Rawitz 1899, aluminum-carmine (*Mitt. zool. Stat. Neapel*, **10** : 489)
FORMULA: water 50, glycerol 50, aluminum nitrate 2, cochineal 2
PREPARATION: Boil the cochineal with the water and nitrate 1 hr. Cool. Filter. Add glycerol.

Schmaus cited from 1896 Kahlden and Laurent, uranium-carmine (Kahlden and Laurent, 1896, p. 158)
FORMULA: water 100, sodium carminate 1, uranium nitrate 0.5
PREPARATION: Boil 30 min. Cool. Filter.

Thiersch cited from 1871 Robin, ammonia-carmine (Robin, 1871, p. 318)
FORMULA: water 65, ammonium acetate 2, carmine 3, oxalic acid 3, 95% alc. 35
PREPARATION: Dissolve acetate in 15 hot water. Add carmine, heat to solution. Add oxalic acid dissolved in 50 water slowly and with constant stirring. Cool. Filter. Add alc. to filtrate.

Thiersch cited from 1877 Frey, ammonia-carmine (Frey, 1877, p. 94)
FORMULA: water 40, 95% alc. 60, carmine 1.25, ammonia 1.25, oxalic acid 2
PREPARATION: Dissolve carmine in ammonia with 5 water. Filter. Add acid dissolved in 40 water. Add alc.; leave 1 day; filter.

References

To save space the undernoted books are referred to in the above article by author and date only. Literature references in the article are cited in full.

Beale, S., "How to Work with the Microscope," 5th ed., London, Harrison, 1880.
Besson, A. "Technique Microbiologique et Sérothérapique," Paris, Ballière, 1904.
Böhm, A., and A. Oppel, "Manuel de Technique Microscopique, Traduit de l'Allemand par Étienne de Rouville," 4th ed., Paris, Vigot, 1907.
Carleton, H. M., and E. H. Leach, "Histological Technique," 2nd ed., London and New York, Oxford University Press, 1938.
Cross, M. I.,* and M. J. Cole. "Modern Microscopy," 3rd ed., London, Bailliere, Tindall and Cox, 1903.
Ehrlich, P., R. Krause, M. Mosse, H. Rosin, and K. Weigert, "Enzyklopadie der Mikroskopischen Technik," 2nd ed., 2 vol., Berlin, Urban und Schwarzenberg, 1910.
Francotte, P., "Manuel de Technique Microscopique," Paris, Lebègue, circ. 1890.
Frey, H., "Das Mikroskop und die Mikroskopische Technik," 6th ed., Leipzig, Wilhelm Engelmann, 1877.
Friedländer, C., "Mikroskopische Technik zum Gebrauch bei Medicinischen und Pathologisch-anatomischen Untersuchungen," 4th ed., by C. J. Eberth, Berlin, Fischer, 1889.
Gray, P., "The Microtomist's Formulary and Guide," New York and Toronto, The Blakiston Company, Inc., 1954.
Guyer, M. F., "Animal Micrology," 3rd ed., Chicago, University of Chicago Press, 1930.
Kahlden, C. von, and O. Laurent, "Technique Microscopique," Paris, Carré, 1896.
Kingsbury, B. F., and O. A. Johannsen, "Histological Technique," New York, Wiley, 1927.
Langeron, M., "Précis de Microscopie," 6th ed., Paris, Masson, 1942.
Lee, A. B., "The microtomist's Vade-mecum," 8th ed., London, Churchill, 1905.
Lee, A. B., and L. F. Henneguy, "Traité des Methodes Techniques de l'Anatomie Microscopique: Histologie, Embryologie et Zoologie," Paris, Dorn, 1887.
Marsh, S., "Section Cutting," London, Churchill, 1878.
Pollack, B., "Les Méthodes de Préparation et de Coloration du Système Nerveux, Traduit de l'Allemand par Jean Nicolaide," Paris, Carré et Naud, 1900.
Ranvier, L., "Traité Technique d'Histologie," Paris, Savy, 1875.
Rawitz, B., "Leitfaden für Histologische Untersuchungen," Jena, Fischer, 1895.

---

*Spence (*in litt.* 1955) states that "M. I. Cross is now known to be F. W. Watson Boker the elder."

Robin, C., "Traité de Microscope; son Mode d'Emploi; ses Application à l'Etude des Injections; à l'Anatomic Humaine et Comparée; à la Pathologie Médico-chirurgicale; à l'Histoire Naturelle Animale et Végétale; et à l'Économic Agricole. Paris, Baillière, 1871.

Schmorl, G., "Die Pathologisch-histologischen Unter. suchungs-methoden," 15th ed., Leipzig, Vogel, 1928-

Seiler, C., "Compendium of microscopical technology," Philadelphia, Brinton, 1881.

Squire, P. W., "Methods and Formulae Used in the Preparation of Animal and Vegetable Tissues for Microscopical Examination," London, Churchill, 1892.

CHAETOGNATHA

There is virtually no information available concerning techniques in the handling of live specimens of Chaetognatha because they have proved notoriously resistant to laboratory culture. The few exceptions are noted in Reeve (1970) where a technique for the culture of *Sagitta hispida* was described. Essential elements in the technique were exercise of greatest care at every stage (capture, transport, aquarium design) to avoid abrasion damage, and the provision of live microzooplankton for food. Development of such techniques are essential to a better understanding of chaetognath form and function since microscopic examination has hitherto relied almost entirely on preserved material.

Examination and photography of both surface and internal structural detail in live animals is greatly facilitated by phase contrast microscopy, a technique limited by the relatively large size (often over 10 mm) of some species and lack of very low power phase objectives. Gross examination in this range must depend on the poorer brightfield image of the stereoscopic (dissecting) microscope. Photography in the macro range, always difficult with low-contrast highly transparent unstained living animals, has yielded good quality results using modern macro systems such as the Nikon "Multiphot" with "Macro-Nikor" lenses and diascopic illumination assembly. Immobilization or killing of live animals without morphological distortion is difficult. Most standard narcotizing agents, as well as fixatives, produce varying degrees of bodily opacity, muscular contraction and shrinkage, loss of body turgor and hence characteristic form, and other effects. This may be advantageous in the case where contraction of the head musculature reveals the normally obscured grasping spines and teeth (of taxonomic importance) but usually, of course, is not. The series of delicate sensory hairs, for instance, which project from the chaetognath body surface, are mostly completely obliterated by the routine formalin preservation of plankton samples. In the latter case, where a cheap, long-lasting fixative/preservative is mandatory, recommendations are awaited from the SCOR committee which is currently testing a wide range of possible compounds using plankton samples from various localities. Short-term immobilization of live animals for purposes of observation or photography can be accomplished by bubbling carbon dioxide through the water containing them. Experimentation under the particular conditions employed will be necessary to adjust too rapid revival, or death and contraction at the other extreme. Immobilization or death by application of cold is effective, at least for warm-water species, (though limited where no provision for avoidance of rapid warm-up is possible),

but heat death is immediately followed by opacity and loss of body turgor.

Standard techniques for fixation and staining of invertebrate tissue have long been employed in cytological studies of chaetognaths with acceptable results using light microscopy. Several specialized studies using the transmission electron microscope have been made, and are cited by Cosper and Reeve (1970) who employed the scanning electron microscope. Osmium/mercuric fixation, followed by lyophilization and metallic coating in vacuo caused obvious and serious artifacts in the appearance of the general body surface and sensory hairs, although giving a good impression of the hard parts (spines and teeth). Subsequent experiments in fixation using an isotonic sodium cacodylate buffered 4% glutaraldehyde solution (after Manton and Parke, 1965) achieved good results, but preparation for scanning electron microscopy still produced artifacts. It is to be expected that much more experience will be accumulated within the next few years, as maintenance of living material in the laboratory becomes routine.

MICHAEL REEVE

Contribution no. 1569 of the Rosenstiel School of Marine and Atmospheric Science, University of Miami.

References

Cosper, T. C., and M. R. Reeve, *Bull. Marine Sci.*, **20**: 441–445 (1970).

Manton, I., and M. Parke, *J. Marine Biol. Assoc. U,K.*, **45**:743–754 (1965).

Reeve, M. R., *Nature* (London), **227**:381 (1970).

CHEMICAL MICROCRYSTAL IDENTIFICATIONS

Aspects, Origins, and Modern Authorities. The microscope should be used throughout chemistry; it is surely an obvious instrument (and not a new one) simply for taking a closer and better look at small things. Chemists should turn to it as naturally as to beakers or the Bunsen burner, whenever it will help. It has applications requiring knowledge and study, too, particularly those with the polarizing microscope, which is by far the most valuable kind of microscope to a chemist. Its special use, discussed here, is in a branch of chemistry that is absolutely basic— the making of identifications. Even with the marvelous new machines that draw their curves based on physical properties of substances, there is still frequent need for *chemical* identifications of chemical substances.

The recognition of chemical substances by observing their characteristic modes of crystallization dates back at least to Robert Boyle (1), in the latter half of the seventeenth century, at the very beginning of scientific chemistry. The microscope, invented between 1590 and 1609, antedated chemistry, and so was available to the "natural philosophers" of Boyle's time; but in those days, it did no more than magnify. Even so, it helped the beginning of chemistry by recognitions based on microcrystal forms. Now, we can also observe properties of birefringence, dichroism, signs of elongation, and often others, made visible to us by polarized light, besides having available to us *reagents* for chemical attacks on the identity of chemical substances. The earliest observations of Boyle and of Baker (2), connecting chemical identity and crystal form, were but little based upon reagents, and consisted mostly of observing the crystallization of a substance from a saturated solution or the crystalline deposit left on drying. Of course, in those days, any understanding of optical crystallography was still in the future, and only the forms as seen in ordinary light, and the manner of crystallization, could be noted.

It was not until 1808 that there was a definitive discovery of polarized light, generally credited to Malus in that year. In 1828, Nicol invented his remarkable prism. The first real polarizing microscope was apparently one put together by Talbot, the photography inventor, in 1834, although various microscopic observations of polarization properties had already been made by others. Thereafter, the science of optical crystallography was developed, and became a recognized means of identifying chemical substances *already* crystalline, particularly minerals. Modern books of special assistance to chemists are those by Hartshorne and Stuart (3), A. N. and H. Winchell (4), and Chamot and Mason (5).

However, although some knowledge of optical crystallography is essential to good use of modern microcrystal tests, it is not their basis. This is, rather, the use of definite chemical reagents to *produce* crystals that can then be distinguished from the crystals of related substances by their forms and optical properties. It seems often not to be realized—even by chemists—that most of the microcrystal tests are strictly *chemical* tests, entirely in the usual pattern of adding a reagent and observing its effects. However, they far excel other chemical tests in their value for identifications. The reactions of chemical *precipitation* given by a reagent are general for a certain *class* of substances, but the crystals given by a particular reagent differ greatly from substance to substance, when examined using the polarizing microscope, and can be distinguished by *individual* characteristics.

This use of crystal-producing reagents for chemical identifications with the microscope began in the 1830s, but became a definite science only in the 1860s, with books by Helwig (6) and Wormley (7), in the field of toxicology.

At least as early as 1838 the identification of the plant alkaloids in poisoning cases had been recognized as a special and difficult problem. Drastic preliminary treatment as for metallic poisons was obviously out of the question; and how were alkaloids to be distinguished from "ptomaines" and harmless natural organic bases, and individually identified? Both Helwig and Wormley gave excellent microcrystal tests for the individual identification of strychnine, morphine, nicotine, and some others—tests still of use and value; and though for two or three of the alkaloids studied, their results were not very satisfactory, they knew they were on the right track and had solved the problem for the most-used alkaloidal poisons. Today there are hundreds of drugs in use that are essentially similar to alkaloids in properties (though often entirely synthetic), which can be identified in the same way. Moreover, the microcrystal methods have now been greatly extended to cover substances of other types, as will be noted here.

Both Helwig's and Wormley's books contained also tests for the most important inorganic poisons. Later in the century, others extended the coverage of microcrystal tests over the whole field of the usual inorganic cations and anions. The culminating microcrystal work of the century was Behrens'. His inorganic work (8) was published in 1893–1895, and the German book went through four editions. The fourth, edited by Kley, appeared in 1921. Early in the new century, Schoorl applied microcrystal tests to the usual systematic inorganic qualitative analysis (9). This is certainly valuable on occasion, but perhaps the greatest value of such tests is in providing immediate identifications without first requiring eliminations of other reactive ions, when one already has a reasonably good idea of the salt present: for example, making it unnecessary to go through all the preceding separations of systematic qualitative analysis to identify a salt of sodium or potassium. Two very important modern books are Volume II of the Chamot and Mason "Handbook" (10), and Korenman's book, in Russian (11). (For some other references, see the chapter on "Inorganic Tests" in Reference 25).

Any amount of research in inorganic microcrystals can still be done. The field of *organic* identification by microcrystals uses mostly *inorganic* reagents, and much more needs to be known about them. When they are used for inorganic tests, the whole procedure is inorganic. More knowledge along the line of quick, simple, and certain identification of common powders or particles, both organic and inorganic, will be of great value to forensic chemistry. Thus far, little has been done in studying the inorganic applications of the newer reagents that use other media than predominantly aqueous solutions.

In the latter part of the last century, and in this one, the crystal tests for alkaloids were also studied further, again and again. Important works were published by Stephenson (12) and Amelink (13). They selected reagents and then applied them systematically to alkaloids available commercially. Others started with such alkaloids and gave the best crystal tests they knew for each one. Small books by Ducloux (14) and Bastos (15) may be mentioned. A recent Russian work is by Pozdnyakova (16).

These procedures are by no means out of date, and of course the tests have more and more come to take in not merely the natural plant alkaloids, but also all modern drugs having properties similar to those of the alkaloids (organic bases, precipitable from aqueous solution by the crystal-forming "alkaloidal reagents"). This work has culminated in that of Clarke (17, 18). He has tried nearly all modern basic drugs with at least 27 or 28 selected reagents (occasionally one or two more). In the two books cited, he gives the two best microcrystal tests found for each drug in his survey. In earlier articles on certain groups, he sometimes tabulated all the results, not merely the two best tests, but much of his data remains unpublished.

Of course, microcrystal tests should be and have been used for other substances besides alkaloids—even beyond the most extended meaning of "alkaloid." Behrens went on from inorganic to organic microcrystal identification tests, and did not confine himself to alkaloids, but to "the most important organic compounds" (19).

In botanical microchemistry crystal tests have been used for alkaloids, but also for other substances, such as volatile amines, decomposition bases, and lichen acids. A large book, concerned with microscopic botanical examinations (not solely with microcrystal tests), originally by Tunmann, revised and added to by Rosenthaler, was published in 1931 (20). (For some additional references in this field, see the chapter "Botanical Microchemistry" in reference 25.)

Rosenthaler's Toxicologic Microanalysis (21), treated various classes of substances, not just alkaloids. In recent years, Dutch chemists have contributed various articles on microcrystal tests for acidic drugs. Amelink's small book (22) covers four classes of acidic drugs, and van der Wegen's booklet (23) includes a number of such drugs along with alkaloids and similar bases. The present writer compiled a list (27)—in part from Clarke's work, but mainly from his own, with re-agents of a different kind—of two suggested tests for each of 286 substances, by no means all alkaloids or even bases. However, such selected tests for alphabetized substances of numerous kinds are chiefly reminders: A chemist should know the comparisons of allied substances by chemical groups.

The work of the present writer has been to extend high-quality crystal tests over as many different kinds of organic substances as possible, particularly those in use as drugs. This work has been incorporated in book form (25). Tests more or less of the alkaloidal type have been extended from alkaloids and similar bases, not only to other bases that are simpler, more water-soluble, and more difficult to precipitate, but even to all other organic substances that have *any* basic qualities, although the whole molecules, because of other groups, may be neutral or even acidic. Oxygen-containing compounds can often react by the formation of oxonium salts. Crystals of the free substance may be obtained with acids as well as with bases. Iodine-iodide reagents give crystals, not only with bases, but with many neutral and acidic substances. There are excellent crystal tests for barbiturates and other acid drugs, especially of iodine-reaction and free-acid nature. There are even excellent crystal tests for steroids. The principles of these new developments will be mentioned in the following paragraphs.

Principles of Modern Microcrystal Tests. The *best* microcrystal tests are not drastic reactions involving intramolecular change, but simple addition reactions—the addition compounds form easily, and under reversed conditions, just as easily decompose again. The organic bases in general form salts, like ammonia, by addition of the whole acid. With a simple acid, the salt may be soluble in water; with a heavy, complex anion of large volume, the salt will commonly be rather insoluble. When the base is already in an acidified solution, the addition of a reagent with a heavy anion may cause precipitation simply by anion exchange, which comes essentially to the same thing as addition of the heavy acid.

The alkaloids themselves, for the most part, have rather heavy and complex molecules; they are soluble readily enough in dilute, simple acids, and are readily precipitated, often in crystalline form or with the precipitate quickly becoming crystalline, by quite a variety of complex acids and anions, especially complexes of heavy metals with negatively charged halogens or pseudo-halogens. This is the simple secret of the special success of the aqueous tests with the alkaloids and other organic bases of a similar degree of complexity. Simpler, more water-soluble, and partly acidic bases were not even precipitated from aqueous solutions by the best crystallization reagents for the alkaloids, while exceptionally complex and heavy bases might give noncrystallizing precipitates with the same reagents.

It should have been obvious long ago that the range of greatest effectiveness could be shifted simply by altering the *medium* in which the precipitation takes place; but in fact, aside from the use of alcohol in some cases to *increase* solubility, only the aqueous medium was used for nearly 100 years from the time such tests began. Needed were media still dissolving the substances and the precipitating agents, but *decreasing* the solubility of the addition products, as compared to the aqueous medium.

A little over 30 years ago, the use of *laboratory acids* as media finally started. At first it was simply a matter of modifying aqueous tests with HCl; then full-strength concentrated hydrochloric acid was tried as the medium. The present writer soon extended the study to syrupy phosphoric acid, glacial or high-concentration acetic acid, and sulfuric acid in high concentration (in this last case, the fully concentrated acid cannot be used because it causes decomposition of most of the precipitating anions and substances tested).

Aside from its acidity, and specific effects of the chloride anion, concentrated HCl has only a minor effect on the solubility of the addition products, and even makes those of strong bases more soluble. But it does have a great effect in enhancing basicity of very weak bases and suppressing any acidic qualities, thereby causing amphoteric substances to unite more readily with precipitating anions. The two other mineral acids mentioned (H_3PO_4 and H_2SO_4) also show this effect. Even more importantly, syrupy H_3PO_4 as the medium makes the addition products a great deal more insoluble than in water. Thus the best crystallizing reagents with alkaloids (bromaurates, chloraurates, bromoplatinates, etc.; in fact all those compatible with this medium) immediately became available for crystal tests of simpler, more water-soluble, and partly acidic substances with some basic qualities. The laboratory acids are used, not only singly, but also in combination. Acetic acid makes the products much more soluble than in water, and is often used with the other acids for a certain solubilizing effect.

The technique of not *first* putting the substance into solution, but rather adding the reagent directly to the dry substance, with a cover glass, has been perfected. This greatly facilitates the trial of numerous different combinations of acids as test media for any particular substance.

There is not space here to describe all the different crystal-forming reagents and their variations in media and otherwise, but four other types of addition reactions will be mentioned, three of them involving iodine.

Iodine-iodide reactions of *bases* are of great value, but may be examples of the usual anion exchange or addition (the anion here being $I_3{}^-$, or perhaps one of greater complexity). However, iodine, generally with iodide, also forms crystalline additive compounds with many *non-bases*, frequently substances definitely of acidic or anionic nature. Phosphoric acid may be used to enhance this kind of precipitation, also.

Some iodine compounds of nonbases are a kind called the "blue-iodo" type, but those obviously of this kind are only a minority of the iodine "acidic" compounds. Sometimes the same substance gives blue-iodo crystals and also others very different. The whole subject is obscure, but it seems probable that the blue-iodo crystals should be considered a separate type, distinct from most of the iodine-reaction crystals of the barbiturates, for example.

Iodine also builds crystal complexes with both an acidic or amphoteric substance and (at the same time) a cation or simple organic base of iodine-combining properties, such as K^+, $NH_4{}^+$, Rb^+, or ethylamine. The test of caffeine with iodine-KI and ethylamine, for example, results in some of the most extraordinary polarizing crystals ever seen for transmitting light in only one direction through the crystal. So far, these are only aqueous reactions (with NaH_2PO_4 buffer recommended).

A different type of crystal-forming additive reaction is given by nitro-organics (e.g., picric acid) with hydrocarbons (as the substances to be identified, purified, etc., by means of the crystals). Some relatives of the hydrocarbons, e.g., indole and α-naphthol, give similar reactions. The nitro-organics are, of course, also reagents for bases, as weighty acids or anions. Unfortunately they are not compatible with a mineral acid medium and cannot be used in syrupy phosphoric acid. Recently the writer has been experimenting with certain mixtures of organic solvents as media. Some beautiful new crystals have been obtained, but a medium for drastically lowered solubility of addition products of nitro-organics remains to be discovered.

Value and Uses of Microcrystal Tests. The possibility of using microcrystal tests should always be considered when any chemical identifications have to be made. Microcrystal tests have been used considerably in botanical microchemistry but surprisingly little in most other biochemistry. Probably the past fixation upon alkaloids has prevented the realization that perfectly good identification tests are available for relatively simple bases, e.g., adenine and creatinine.

Where the microcrystal tests have been especially successful has been in drug analysis, narcotics examination, law-enforcement cases involving drugs, and toxicology. These are forensic sciences. Many legal cases involve the question of chemical identity much more than quantity, as regards narcotics, poisons, potent drugs, adulterants, contaminants, substitutes, etc.; and they require exacting proof of the identity. That microcrystal tests have been and still are important in such fields, in spite of neglect otherwise, speaks strongly for their reliability for exact identification.

In fact, very small differences show up plainly in the microcrystal tests. For example, it has happened several times at the Office of the Chief Medical Examiner in New York City that a toxicological residue thought to be quinine from fluorescence, chromatography, and UV data, was then shown to be quinidine by crystal tests, which easily distinguish the two. With the crystal tests, there is no difficulty in distinguishing amphetamine from the closely related base of decomposition, phenethylamine. In fact, the crystals easily distinguish dl-amphetamine from the separate isomers, d- and l-amphetamine, and can even distinguish the d- from the l- in micro quantities, by mixing the unknown with the known isomer to form the racemate (Clarke's method).

The microcrystal tests are as simple and direct as tests can be, and this is important forensically, not only because the analyst is much less likely to make a mistake, but also because it is much easier to explain the findings to others, a judge and jury perhaps, who may have little or no chemical knowledge.

The crystals of the actual test may be photographed, and are quite objective. Sensitivity and selectivity allow tests on microscopic particles, often even in mixtures, without first putting everything into solution and then trying to make separations.

The advantages of microcrystal tests may be listed as follows:

1. The very high assurance of correct identification.
2. The *direct* character of the tests and conclusions.
3. Their simplicity, convenience, and speed.
4. High sensitivity, generally of the order of a few hundredths of a microgram.

5. Selectivity; noninterference or no great interference of most impurities.
6. The tests are objective; they may be photographed, and in general anyone can be shown the correspondence of unknown and known.

It is hoped that the use of the modern tests will be developed more and more.

CHARLES C. FULTON

References[1]

1. R. Boyle (1627–1691). "The Works of the Honourable Robert Boyle," in 6 vols.; 3rd reproduction of 2nd ed. (London 1772), published by George Olms, Hildesheim, 1915. References to crystal forms are scattered in many places.
2. H. Baker, "The Microscope Made Easy," London, 1742; "Employment for the Microscope," London, 1753.
3. N. H. Hartshorne and A. Stuart, "Crystals and the Polarizing Microscope, A Handbook for Chemists and Others," 3rd ed., London, Arnold, 1960.
4. A. N. and H. Winchell, "Microscopical Characters of Artificial Inorganic Solid Substances: Optical Properties of Artificial Minerals," 3rd ed., New York, Academic Press, 1964; A. N. Winchell, "Optical Properties of Organic Compounds," 2nd ed., New York, Academic Press, 1954.
5. É. M. Chamot and C. W. Mason, "Handbook of Chemical Microscopy," Vol. I, 3rd ed., New York, Wiley, 1958 (2nd printing, 1966). (Not only optical crystallography, this introduces microcrystal tests, but they are mainly deferred to Volume II [reference 10].)
6. A. Helwig, "Das Mikroskop in der Toxikologie," Mainz, Zabern, 1865.
7. T. G. Wormley, "Micro-Chemistry of Poisons, Including Their Physiological, Pathological, and Legal Relations," New York, William Wood, 1867 (some copies dated 1869); 2nd ed., Philadelphia, Lippincott, 1885.
8. H. Behrens, "A Manual of Microchemical Analysis," New York, Macmillan, 1894. (This was first published in French in "Encyclopedie Chimique," M. Fremy, ed., Paris, 1893.) "Anleitung zur Mikrochemischen Analyse," 1st ed., 1895. P. D. C. Kley, ed., "Behrens-Kley Mikrochemische Analyse," counted as the 4th ed. of Behrens' inorganic work in German; Leipzig and Hamburg, Voss, 1921.
9. N. Schoorl, "Beiträge zur mikrochemischen Analyse," Weisbaden, Kriedel, 1909. (The parts were also published as a series of papers in Z. Anal. Chem., 1907–1909.)
10. É. M. Chamot and C. W. Mason, "Handbook of Chemical Microscopy, Volume II, Chemical Methods and Inorganic Qualitative Analysis," 2nd ed., New York, Wiley, 1940 (8th printing, 1963).
11 I. M. Korenman, "Microcrystalloscopy," (in Russian), Moscow, 1955.
12. C. H. Stephenson, "Some Microchemical Tests for Alkaloids," including "Chemical tests of the alkaloids used," by C. E. Parker. Philadelphia, Lippincott, 1921.
13. F. Amelink, "Schema zur mikrochemischen Identifikation von Alkaloiden," *translated by* (übersetzt von) Marga Laur, (in German), Amsterdam, N.V.D.B. Centen's Uitgevers Maatschappi, 1934.
14. E. H. Ducloux, "Notas Microquímicas Sobre 'Doping'," Buenos Aires, Peuser Lda, 1943 (paperback, 58 pp. text, 302 photographs of crystals).

---

[1]Half or more of the cited works are out of print, but can often be found in libraries.

15. M. L. Bastos, "Microquímica De Alguns Alcalöides," Boletin do Instituto de Química Agrícola, No. 51, Rio de Janeiro, 1957 (paperback, 134 pp., 63 photographs of crystal tests).

16. V. T. Pozdnyakova, "Mikrokristalloskopicheskie Reaktsii na Alkaloidy" (in Russian: transliterated title), Kiev, 1960 (163 pp.).

17. E. G. C. Clarke, "The isolation and identification of alkaloids," in "Methods of Forensic Science," Vol. I, F. Lundquist, ed., New York, Interscience (Wiley), 1962. (Here "alkaloids" includes synthetic drugs of similar properties.')

18. E. G. C. Clarke, ed., assisted by J. Berle, "Isolation and Identificaiton of Drugs," London, Pharmaceutical Press, 1969.

19. H. Behrens, "Anleitung zur mikrochemischen Analyse der wichtigsten organischen Verbindungen," Heft 1, 3, 4 (Heft 2 concerned fibers), Hamburg and Leipzig, 1895–1897. P. D. C. Kley, ed., "Behrens-Kley Organische Mikrochemische Analyse," Leipzig, Voss, 1922 (counted as the 2nd ed. of Behrens' organic microcrystal work).

20. O. Tunmann and L. Rosenthaler, "Pflanzenmikrochemie," Berlin, Gebrüder Borntraeger, 1931 (1047 pp.).

21. L. Rosenthaler, "Toxikologische Mikroanalyse," Berlin, Gebrüder Borntraeger, 1935. (Reissued by Edwards Brothers, Ann Arbor, Mich, 1946.)

22. F. Amelink, "Rapid Microchemical Identification Methods in Pharmacy and Toxicology: Sulfonanimides, Sulfones, Barbiturates, Hydantoins," B. Kolthoff, trans., Netherlands University Press and Interscience (Wiley), New York, 1962.

23. Th. P. A. van der Wegen, "Kristalreacties op een aantal alkaloïden, barbitalen, sulfonamiden en enige andere stoffen," Amsterdam, D. B. Centen's Uitgevers Maatschappij, 1960 (paperback, 86 pp.).

24. C. C. Fulton, "Microcrystal tests,' (Table 13, pp. 461–496 in "Handbook of Analytical Toxicology," Irving Sunshine, ed., Cleveland, Ohio, Chemical Rubber Company, 1969.

25. C. C. Fulton, "Modern Microcrystal Tests for Drugs," New York, Wiley, Interscience, 1969.

CHITIN

The name "chitin" was given by Odier (1823) to a cuticular component insoluble in hot aqueous alkali. It is now known to be a neutral mucopolysaccharide consisting predominantly of unbranched chains of β-(1 → 4)-2-acetamido-2-deoxy-D-glucose residues. Chitin from a variety of sources has this structure of repeating units of N-acetyl glucosamine (Fig. 1). It is a colorless amorphous solid, insoluble in water, dilute acids, dilute and concentrated alkalies, alcohol, and other organic solvents.

Chitin occurs in the exoskeleton and peritrophic membrane of all arthropods, in the molluscan radula, chaeta of annelids, and the cell walls of fungi and some algae. Wherever chitin occurs as part of the skeletal framework, it is always associated with a protein as glycoprotein. In the exoskeleton of arthropods and the pen of *Loligo* (Cephalopoda) chitin is covalently linked with protein through aspartyl and histidyl residues (Hackman, 1960). The drastic treatment required to separate the protein for purification (repeated digestion with N NaOH at 100°C or treatment with 5% HCl at 100°C for several days) invariably results in some degradation of the chitin.

Histochemical tests such as chitosan, NaOH–sulfuric acid, and Diaphanol–iodine–zinc chloride have been employed to demonstrate chitin in tissues. The chitosan test described by Campbell (1929) involves treatment with saturated aqueous alkali at 160°C for 20 min. The material surviving the treatment is chitosan, a deacetylated product of chitin. It gives a characteristic violet color reaction with dilute iodine and 1% sulfuric acid and dissolves in 3% acetic acid. This test is specific for chitin. Unfortunately, it completely destroys all cells and is also difficult to apply on trace amounts or dispersed materials. The other tests lack specificity. Chitin does not react with the periodic acid-schiff reagent (PAS) but becomes reactive after deacetylation (Brunet, 1952). Nonacetylated glucosamine residues when present in chitin give a positive PAS reaction. The PAS can sometimes be used to detect newly secreted chitin which contains a large number of nonacetylated residues.

Chitin protein fibers are resolved with difficulty by electron microscopy after heavy-metal staining. Using the conventional double staining with uranyl acetate and lead citrate, Locke (1961) observed an apparently parabolic fibrous pattern of chitin (Fig. 2) caused by overlapping images of fibers arranged in planes or laminae parallel to the surface (Bouligand, 1965). Staining with 2% aqueous potassium permanganate followed by lead citrate also gives good resolution of the fibers (Neville and Luke, 1969). The diameter of microfibrils measured from micrographs of insect chitin is in the range of 25–45 Å.

The great stability of chitin permits the use of a number

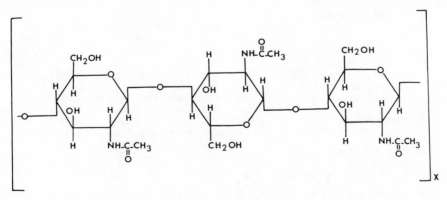

Fig. 1 Structure of chitin.

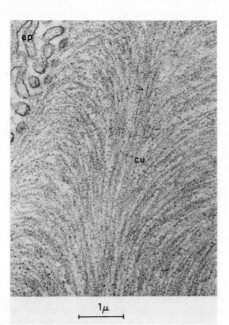

Fig. 2 Electron micrograph of the larval cuticle of *Calpodes* cut obliquely to the surface and stained with uranyl acetate and lead citrate. The microfibrils are stained by the heavy metals. cu, cuticle; ep, epidermis. (×31,000.)

of fixatives commonly employed in histological preparations. The choice of an embedding medium for chitinous structures depends on the hardness of the material. Soft chitinous structures can be sectioned after embedding in paraffin or ester wax while hard materials require embedding in synthetic resin or double embedding in nitrocellulose and paraffin. Backel's (1959) technique is useful to eliminate compression and rolling of sections when cutting whole insects or large chitinous structures. An adhesive tape is applied to the block face before each section is cut and the sections are attached to the slide with 2% celloidin in ethanol. The tape is stripped off after immersion in chloroform and the adhesive and paraffin are dissolved with xylene. Thin sections for electron microscopy and light microscopy are easily obtained from chitin embedded in epoxy resins.

N. KRISHNAN

References

Beckel, W. E., *Nature,* **184**:1584 (1959).
Bouligand, Y., *Compt. Rend. Acad. Sci. Paris,* **261**:3665 (1965).
Brunet, P., *Science,* **116**:126 (1952).
Campbell, F. L., *Ann. Entomol. Soc. Amer.,* **22**:401 (1929).
Hackman, R. H., *Aust. J. Biol. Sci.,* **13**:568 (1960).
Locke, M., *J. biophys. Biochem. Cytol.,* **10**:589 (1961).
Neville, A. E., and B. M. Luke, *Tissue and Cell,* **1**:689 (1969).
Odier, A., *Mem. Soc. Hist. Natur.,* **1**:29 (1823).
See also: INSECT HISTOLOGY.

CHROMIUM SALTS

Chromium Salts in Light Microscopy. Chromium salts are applicable to light microscopy both for tissue fixation and for the histochemical demonstration of many biological substances. The most widely used salts are chromic oxide (CrO_3) and potassium dichromate ($K_2Cr_2O_7$); the latter compound in pH ranges below 3.0 has the same effect as CrO_3. Both compounds have a slow tissue penetration.

Chromium as Fixing Agents. Chromium salts alone (either as aqueous solutions of from 0.3% to 0.5% CrO_3 or 0.5% to 5% $K_2Cr_2O_7$ or as Muller's fluid (H. Muller, 1857), are today seldom in use, though in the case of the Marchi method it is best to fix tissues in 3% $K_2Cr_2O_7$ for about 14 days (1).

More frequently the Cr salts are mixed with other reagents, especially with acetic acid (K.v. Tellyesniczky, 1898) (5 ml acetic acid + 100 ml 3% aqueous $K_2Cr_2O_7$ solution, mixed shortly before using. Fixation time is 1–2 days with subsequent 1 day washing in water), or with formol and acetic acid (Wittmaack, 1906) (5 ml acetic acid + 85 ml 3% aqueous $K_2Cr_2O_7$ solution + 10 ml formol). Fixation time is 6–12 hours. The tissue must be immersed in 5% Na_2SO_4 before the subsequent 24 hr washing in water. Chromium salts are also used with formol alone (80 ml 3% aqueous $K_2Cr_2O_7$ solution + 20 ml formol) (Kopsch-Regaud's fluid, 1896). The fixation in this fluid results in a good preservation of cytoplasm and mitochondria.

The best results are obtained with Cr-Os mixtures. Typical mixtures consist of 15 ml 1% CrO_3 + 4 ml 2% OsO_4 together with either 1 ml acetic acid (Flemming, 1882) or with 1–3 drops of acetic acid (Benda, 1901). Champy's mixture (1901) (7 ml 1% CrO_3 + 7 ml 3% $K_2Cr_2O_7$ plus 4 ml 2% OsO_4) gives a good fixation of mitochondria and Golgi apparatus.

Chromium Salts as Histochemical Reagents. The technique of reacting tissue with $K_2Cr_2O_7$ and subsequently staining with hematein (Smith-Dietrich, 1908; Baker, 1961) is useful for the specific demonstration of phosphatides (especially compounds with unsaturated fatty acids). The very complicated procedure, which is given in detail in (2, 3), must be accurately followed to assure specific results. Frozen sections must be used.

The histochemical demonstration of nucleic acids is based on staining with a complex of gallocyanin with Cr and Al by the method of Einarson (4, 5). To render this easy method specific to DNA, the RNA must be removed with specific enzymes or vice versa. The specificity of the reaction also depends on the use of very pure chemicals and the fixation of the tissue in ethanol.

Treatment of the tissue with CrO_3 and subsequent staining with Zn^+ Dithizon (6) gives specific staining of nucleoli.

Polysaccharides, especially glycogen, are demonstrable with Bauer's reaction (7), which uses a freshly prepared 4% aqueous CrO_3 solution as oxidizing agent. In this case the result depends on the length of oxidation. The time is not critical if CrO_2Cl_2 is used as oxidizing agent (8). The dehydrated tissue is treated from 1 min to 20 hr with a 0.5–2% solution of CrO_2Cl_2 in CCl_4 (desiccated by addition of P_2O_5 to the solution), and subsequently washed in CCl_4, hydrated, and stained with Schiff's reagent.

Cells of the suprarenal medulla show a yellow color after treatment with $K_2Cr_2O_7$; this is caused by the oxidation products of epinephrine and norepinephrine. The exact technique for this so-called chromaffin reaction is as follows (9): The fresh tissue is immersed 1–2 days in a mixture of 10 ml 5% aqueous $K_2Cr_2O_7$ solution + 1 ml 5% aqueous $KCrO_4$ solution (pH 5–6). Frozen sections are prepared after washing for 1 hr in distilled water. $K_2Cr_2O_7$ is used in the following manner (10) for the specific demonstration of the norepinephrine: Paraffin sections of suprarenal medulla, fixed in a solution of 5% glutaraldehyde (buffered with $0.1N$ cacodylate to a pH of 7.2–7.4), are immersed in an aqueous solution of $3·5\%$ $K_2Cr_2O_7$ for 6–12 hr. The norepinephrine-containing cells are blackened. In young animals, the stain must be intensified by after-treatment with $AgNO_3$ in NH_3. Epinephrine is not demonstrable by this technique. This method was worked out for use in electron microscopy (11).

Chromium Salts in Electron Microscopy. Chromium salts serve either as fixing agents or as a contrast stain. By use of the chromium compounds under certain conditions it is possible to contrast many cell structures specifically. Of the various Cr salts only $K_2Cr_2O_7$ or CrO_3 are used often, rarely CrO_2Cl_2.

Chromium Salts as Fixing Agents. 1. 1.2–2.4% $K_2Cr_2O_7$ or 0.5–3% CrO_3 in aqueous solution or Müller's fluid (see above) is seldom in use today in electron microscopy. The tissue, when fixed with these fixing fluids, shows disruption of the plasma and destruction of the membrane structure and swelling of the nucleus and mitochondria.

2. $K_2Cr_2O_7$ mixed with other fixing agents is more often used in electron microscopy. In the case of Regaud's fluid (see above) the fixation time is several days. Another fixing fluid consists of 3% CrO_3 + 4% formol + 0.85% NaCl (adjusted to a pH of 3.5) (12). If the fixation time in this mixture is 15–30 min, followed by a 30 min washing in distilled water, the tissue is far better preserved than with Cr salts alone.

3. Dalton's Cr-Os fixative is widely used (13). This is a mixture of 1% OsO_4 + 1% $K_2Cr_2O_7$ + 0.85% NaCl (pH is adjusted with KOH to 7.2). This solution is stable against reducing influences. Fixation time is 2–24 hr, and over this period there is no disruption of cytoplasma. The fixed tissue is stable against embedding artifacts. The disadvantages are the low electron optic contrast, and it is very hard to secure ultrathin sections. A slight variation of this mixture has been proposed for the fixation of nerve-tissue (14) to avoid swelling, which is observed after fixation with Dalton's mixture.

In general one can say that the fixation with Cr salts does not yield as good results in comparison to OsO_4 or other fixing agents. The contrast is low, the membrane structures of cells and mitochondria are often destroyed. Moreover the tissue is difficult to section.

Chromium Salts as a Contrast Stain. CrO_2Cl_2 is very often used; its production and application are described in (15). It must be handled carefully because it is very toxic. It produces a high contrast, especially of the membranes of the endoplasmic reticulum and mitochondria, as well as of cell membranes. On the other hand, many substances are extracted from the cells (as the core of viruses and mucous granula).

HANS SCHIECHL

References

1. Harman, P. J., and P. W. Bernstein, *Stain Technol.,* **36**:49 (1961).
2. Sinapius, D., *Histochemie,* **2**:217 (1961).
3. Baker, J. R., *Quart. J. Microscop. Sci.,* **87**:441 (1946).
4. Einarson, L., *Am. J. Pathol.,* **8**:295 (1932).
5. Pakkenberg, A., *J. Histochem. Cytochem.,* **10**:367 (1962).
6. Studzinsky, G. P., H. E. Reidbord, and R. Love, *Stain Technol.,* **42**:301 (1967).
7. Bauer, H., *Z. mikr.-anat. Forschg.* **33**:143 (1933).
8. Mende, T. J., and E. L. Chambers, *J. Histochem. Cytochem.,* **5**:606 (1957).
9. Hillarp, N. A., and B. Hökfelt, *Endocrinology,* **55**: 255 (1954).
10. Jones, P. A., *Stain Technol.,* **42**:1 (1967).
11. Wood, J. G., and R. J. Barrnett, *Anat. Rec.,* **145**: 301 (1963).
12. Low, F. N., and J. A. Freeman, *J. Biophys. Biochem. Cytol.* **2**:629 (1956).
13. Dalton, A. J., *Anat. Rec.,* **121**:281 (1955).
14. Forssman, W. G., *Experientia,* **21**:358 (1965).
15. Bullivant, S., and J. Hotchin, *Exp. Cell Res.,* **21**:211 (1960).

CHROMOSOME TECHNIQUES

Various methods of analysis have been devised to study the *chromosome*, carrier of the material basis of heredity, the *gene*. In the lower organisms or prokaryotes with no distinct nuclear membrane, the chromosome is represented by the DNA molecule alone. This structure is referred to as the genophore as compared to the complex organization known as the chromosome in higher forms of life. The structure of the genophore, being at the level of molecular dimensions, is beyond the resolution of the light microscope and has been studied mainly through electron microscopy.

The chromosome of the eukaryotes, on the other hand, has RNA, basic and non-basic proteins, in addition to a fundamental composition of DNA and exhibits a fibrillar constitution. The uninemic and multipepliconic nature of its constitution is mostly advocated. Under the light microscope, without any treatment, chromosomes in the living state are difficult to analyze due to the approximately similar phase maintained by the nuclear and cytoplasmic components. Only under phase contrast and interference lenses, through magnifications of this phase difference, can chromosomes be studied in the living state.

For observation under the light microscope, chromosomes have to be processed prior to staining. The procedure involves principally fixation, and in certain cases, a preliminary pretreatment in certain specific compounds. Fixation is performed to secure an instantaneous killing and precipitation of protein matter, together with simultaneous prevention of bacterial autolysis and protein denaturation of the tissue. Inclusion of heavy metallic salts, as those of osmium and chromium, intensifies the stain in addition to precipitation. The recommended fixatives, like Flemming's, Champy's, La Cour's, or Nawaschin's fluids, contain one of these compounds in addition to ethyl alcohol, acetic acid, or formalin. Non-metallic fixing fluids were used principally by Carnoy including chloroform as well.

For somatic chromosomes, where the chromosome morphology can be distinguished clearly, it is preferable to follow a pretreatment schedule, ensuring metaphase arrest and differential contraction of chromosome segments, which result in the clarification of chromosome details. Pretreatment agents affect the cytoplasmic viscosity, thereby disrupting the spindle. The more efficient pretreatment agents are aqueous solutions of colchicine (0.05%), 8-hydroxyquinoline (0.002M), p-dichlorobenzene, aesculin, isopsoralene, coumarin, umbelliferone, and α-bromonaphthalene, the latter compounds being applied in saturated solution at 12–14°C, the periods of treatment ranging from 30 min to 4 hr prior to fixation for 1 hr to overnight in acetic alcohol. To soften tough tissues in temporary preparations, they are to be heated in N HCl and 2% acetic-orcein mixture (9 : 1) for a few seconds before staining in 1% acetic-orcein solution. In temporary preparations of meiotic chromosomes, smearing of the anthers or testicular tissue in 1% solution of carmine in 45% acetic acid, with or without prior fixation, is recommended. Acetic-carmine can often be substituted by acetic-lacmoid, lactic-orcein, or propionic-carmine solutions, all these compounds serving the dual purpose of fixation and staining. For the study of chromosomes in development, function, and action, salivary gland chromosomes of *Drosophila* and lampbrush chromosomes of vertebrate oocytes are studied where the simple squash schedule and isolation method respectively are applied.

In order to obtain permanent preparations after fixation, the materials require washing in water, followed by progressive dehydration through different grades of ethyl alcohol, with final treatment in absolute alcohol for overnight. Alcohol-chloroform grades, with increasing concentrations of chloroform, are used for gradual embedding with paraffin. Embedding in paraffin involves three to four changes in pure, liquefied paraffin at intervals of 1–2 hrs. each, followed by the preparation of paraffin blocks. Celloidin may be employed as an alternative embedding agent in a slightly modified schedule. Sections of required thickness, cut by microtomy, are mounted on slides and brought down to water through descending grades of xylol and ethyl alcohol. The next step is usually washing in water and staining in aqueous solution of crystal violet (a triphenyl methane dye, which can be substituted by other dyes of the same series). The slides are rinsed in water to remove the superficial stain, mordanted for specific staining in potassium iodide and iodine (1 : 1 in 100 cm³ of 80% ethanol), dehydrated very rapidly in two quick changes of absolute ethanol, differentiated in clove oil, cleared in xylol, and mounted in balsam. Several variants of this method are in practice.

Of all the stains applied thus far in chromosome study, Feulgen reaction is the most important one, where the exact chemical basis is known to be an aldehyde reaction for chromosomal DNA. In Feulgen staining, the method involves the hydrolysis of the tissue in normal hydrochloric acid for 10–12 min, followed by staining in fuchsin–sulfurous acid, in which the chromosomes stain magenta. For temporary preparations, the tissue can be mounted in 45% acetic acid immediately after Feulgen staining. The principle of hydrolysis is to liberate aldehyde groups of deoxy sugar, which react later with fuchsin sulfurous acid to give the color reaction of DNA. The specificity of the test and its stoichiometric correlation have led to its application in quantitative study. The preparation of leucofuchsin solution requires a critical procedure since the basic fuchsin solution, itself magenta-colored, is decolorized by the addition of SO_2 water or a mixture of potassium metabisulfite and N HCl. In addition to Fuelgen, pyronin–methyl green staining is applied for differential localization of RNA and DNA in the cell.

In spite of the general similarity of methods, for the study of plant, animal, and human chromosomes, particularly in the fixing chemicals, there are differences in the procedures applied to human and mammalian materials. The most prevalent method for mammalian chromosome study is the leucocyte culture technique, using a specific synthetic medium for culturing a few drops of peripheral blood. The most common practice is to use AB serum and a few drops of peripheral blood to which phytohemagglutinin—a mucoprotein isolated from the red kidney bean, *Phaseolus vulgaris*—has been added. This compound combines the property of agglutinating red blood corpuscles with that of inducing division in leucocytes. After a specific period (60 hr or so) of culturing, colcemid is added to secure the maximum number of metaphase plates and the cells are harvested after 2 hr or so of further incubation. The final preparation involves hypotonic treatment, fixation in acetic-methanol (1 : 3), and staining in Giemsa or a suitable stain with mild centrifugation between the steps to obtain the maximum number of cells. Air-dried smears can be prepared after fixation by spreading a drop or two of the fluid on a grease-free slide.

Absorption measurements for quantitative estimation of DNA in chromosomes are carried out through ultraviolet and ordinary light microspectrophotometers. In both, special built-in microscopes with phototubes and galvanometric attachments over the eyepiece are used, the entire microscope serving as the microspectrophotometer. Glass lenses, opaque to short UV wavelengths, are replaced by quartz lenses in UV microscopes and the monochromatic beam is split into two beams, one falling directly on the photoelectric cell and the other indirectly through the material, where at 265 mμ purine and pyrimidine components show strong absorption as worked out by Caspersson (1947). Bier-Lambert's law is followed for absorption measurement $I_x = I_o - 10^{kcd}$ (where $I_x =$ changed intensity of I_o, the incident intensity; $d =$ the thickness of c, the concentration of absorbing materials in grams per 100 cm³, and $K =$ the extinction coefficient). Fluorometric methods, based on the secondary fluorescence through fluorochrome (acridine dyes), are also adopted. Chilled or freeze-dried preparations of unfixed materials are generally employed for such quantitative studies. Living tissues can also be studied through phase and interference microscopy, in which, by suitable adjustment of lenses, minute phase changes of cellular constituents are converted to visible intensity changes. In addition, the molecular orientation of chemical constituents of chromosomes as shown by Astbury (1945) may be worked out through an analysis of X-ray diffraction patterns in X-ray microscopy (vide Oster, 1955).

Electron microscopy is an effective tool in the ultrastructural analysis of chromosomes. The limit of resolution of an electron microscope is very high and biological macromolecules cut even at a thickness of 5–10 Å can be conveniently resolved. In common with other cellular components, for electron microscopy, chromosomes need special fixation, embedding, and polymerization before section cutting on an ultramicrotome. For fixation, buffered osmium tetroxide (2%) with added calcium in containers is deemed to be the best and the period may extend up to 4 hr. Optimum chromosome fixation can be obtained if 2% lanthanum nitrate and 3% potassium dichromate are added to the osmium fixative. A prefixation in formaldehyde or glutaraldehyde avoids the undesirable effects of osmium fixation. Formaldehyde fixation, as it is, is not recommended for chromosome study. Potassium permanganate has also been used as a fixative but manganese dioxide granules formed in the background present a serious limitation of its use. As embedding medium, n-butyl and methyl methacrylate (8 : 2) was considered adequate but the polymerization damage had to be avoided. With methacrylate monomer, dehydration in ethanol is preferred. Final embedding is carried out in small gelatin capsules using a mixture of pure methacrylate monomer and a catalyst (1 : 1), such as 2% 2,4 or 1,2-dichlorobenzylperoxide. Polymerization can be carried out against an ultraviolet source for 48 hr or at 47°C for 48 hr followed by keeping at 22°C for 24 hr. Araldite, an aromatic epoxy resin, may also be used after dilution with a short chain mono-epoxide like propylene oxide, prior to and during embedding. In araldite embedding, a hardener, dodecemyl succinic anhydride (DDSA); a plasticizer, n- or dibutyl phthalate (DBP); and an accelerator, trimethanine methyl phenol (964C) are added to the final mixture in the proportions araldite—10, DDSA—10, DBP—1, and 964C—0.5 (Glauert and Glauert, 1958). The former three are mixed at 60°C 15 min before use, and after stirring thoroughly, the last ingredient is added. In Luft's (1961) modification, the proportions used are araldite—27, DDSA—23, and DMP—1.5–2% V/V added just before embedding. The procedure involves dehydration of small bits of tissue (0.1–0.2 mm) in absolute ethanol and propylene oxide successively, keeping 10–15 min in each. Treatment in fresh propylene oxide for one hr is followed by addition of an equal quantity of resin mixture. After 3–6 hr, the material with a little mixture is transferred to gelatin capsules, which are filled with the resin mixture and kept at 35°C, 45°C, and 60°C successively for overnight at each for complete polymerization. Embedding of cultures, smears, and suspensions can be done in vinyl cups immediately after fixation (Anderson and Drane, 1967). In the widely followed Epon 812 embedding procedure (Luft, 1961), after osmium fixation and dehydration in ethanol, two changes (15–30 min each) are given in propylene oxide, followed by treatment for 1 hr in resin-accelerator mixture (A-Epon 812, 62 cm³; DDSA, 100cm³ and B-Epon 812, 100 cm³ and MNA, [methyl endomethylene tetrahydrophthalic anhydride], 89 cm³ mixed in the proportion of 2 : 1 and added to 1.5% V/V of accelerator DMP-30) and propylene oxide (1 : 1). The tissue is transferred to pure resin-accelerator mixture (20–30 min) in gelatin capsules and set directly at 60°C for overnight or after the 3-stage curing programe. Modifications include vacuum incubation at 60°C for 24–36 hr (Porter, 1964) and use of water-soluble embedding media for cytochemical work, like Aquon (Gibbons, 1960), glycol methacrylate (Rosenberg et al., 1960), and Durcupan (Straubil, 1960).

Sections (0.1–0.01 μ) are cut in ultramicrotome (Porter Blum or LKD) with a glass or diamond knife; spread quickly in water, where the color change from gold to silver interference color in reflected light through fluorescent lamp is noted; floated in hydrolyzing solution (for GMA) without supporting grids; and after incubation, further floated in distilled water and lifted up on formvar (polyvinyl formal plastic)-coated grids. Epon sections can also be mounted on 300 mesh copper grids but for araldite-embedded materials, even a carbon film may provide sufficient support. During staining, the tissue is generally exposed to heavy metal salts to secure ion complexes within chromosomes. For contrast chromosome staining, treatment of the moistened material in 7.5% aq. uranyl acetate solution at 20°C for 45 min is followed by immersion in 0.2% aq. lead citrate solution for 10–60 sec (Reynolds, 1963). Other modifications are also available (Salpeter, 1966). Schedules for other epoxy embedding media, such as Maraglas, involves certain changes in procedure as well. A variation of Sparvoli et al.'s method (1965) by Sorsa and Sorsa (1967) for Drosophila salivary gland smears involves initial dehydration of materials on the slide, after removal of coverslip in ethanol series (50, 60, 90, and 99.5%), all saturated with uranyl acetate. Plastic capsules, filled with araldite final embedding mixture (CIBA-Cy 212) are inverted over the material on the slide and kept at 80°C for 1½ hr for penetration and the final polymerization is carried out at 50°C for 2–3 days.

Lately, electron microscopy at the level of the chromosome has been combined with autoradiography since high resolution autoradiography permits a correlative study of the chromosome structure with function and also with its chemical constitution. The principle of autoradiography, in general, involves localization of radioactive material in the tissue with the aid of photographic development. The emulsion is brought into contact with the radioactive material, so that the ionizing radiation activates the corresponding silver halide crystals in the emulsion, forming latent images which appear as spots at specific loci after development. In light microscopy, the different stages include incorporation of a tracer in the tissue, fixation, paraffin embedding or smearing, application of emulsion either as a liquid or as a thin film, drying, exposure, and photographic processing. Staining may be performed before or after coating with emulsion.

For incorporation, ^{32}P, ^{35}S, etc., obtainable in their respective acid forms or tagged isotopes as tritium-labeled (^3H) thymidine or ^{14}C-labeled adenine, thymine, etc. can be utilized in the medium or injected into the body. Amino acids, labeled with radioactive compounds such as ^{35}S, can be used for the protein constituent of chromosomes and can be incorporated through the culture medium or injected directly. The concentration varies with the material and the object but in general, the effective doses are: ^{32}P—2 μc as orthophosphate per cm³ for 2 days to one month; ^{35}S—1 μc as sodium sulfate per cm³ for 20 days; ^{14}C adenine—0.3 μc/g for 24 hr or 20 μc per cm³ for 24 hr, or 0.5 μc/cm³ in medium for human bone marrow culture. Tritiated thymidine (1 μc/cm³) can be added 5–6 hr before the termination of the culture. In human material, colcemid may be added 2 hr before harvesting followed by usual hypotonic treatment, fixation, air drying, and staining in 2% acetic-orcein solution. Smears or paraffin sections (5 μ) may be prepared in the usual manner in grease-free dry slides coated with a mixture of photographic gelatin (5 g), chrome alum (0.5 g) and water (1 liter). However, with Kodak emulsion such slides are not required.

All the stages of emulsion coating have to be carried out in a cool, dark chamber. (1) In the liquid emulsion method, the emulsion is melted at 40–42°C. Slides are dipped into it for 4–5 sec, then removed and dried against a stream of air, kept in tapesealed slide boxes in a cool dark chamber for 2–3 weeks, developed in Kodak (D-11 or D-19) developer for 2 min, fixed in Kodak acid fixer for 2–5 min, rinsed in running water, and dried. For Feulgen reaction, staining precedes the application of emulsion whereas for toluidine blue, methyl green–pyronin, and Giemsa, poststaining is preferable. (2) In the stripping film method, 40 mm × 40 mm squares are cut in AR 10 films mounted on glass plates, kept for 3 min in 75% ethanol, and then transferred to absolute ethanol. A single square is lifted out with forceps and floated with emulsion side down in distilled water, where the film spreads out automatically. The slide with the material is placed just below the film under water and lifted out so that the film adheres to the material. Air drying and the subsequent steps are the same as the liquid emulsion method.

To have a comparative picture and to work out the correct period of exposure and demarcation of area in high resolution studies, the observation of thick sections under light microscope before that of ultrathin ones is preferable. Tritiated compounds having a short range of radiation (0.018 MeV–1μ) do not exceed the thickness of emulsion. A uniform thickness with fine-grained emulsion (Ilford K5, Gavaert 307, Kodak NTE L 4) is quite convenient. Ultrathin sections require 3–4 min of exposure as against two weeks necessary for thicker ones. The methods of fixation and embedding are similar to those described previously. Of the several available methods, the one adopted by Stevens (1966) is adequate. Blocks are trimmed to a suitable size for obtaining ultrathin sections and transferred by means of a damp fine-haired nylon brush to a drop of water placed at the edge of a subbed slide. Several ultrathin sections are cut (600–1000 Å) and shifted in the boat. Another thick section is cut and transferred in the same way on the previous slide. Both slide sections are dried at 45°C as well as at 60–80°C and the position of the desired area is marked under a microscope. The intervening ultrathin sections are then mounted on electroplated Athene-type copper grids coated with colloidion with a thin carbon layer (Caro, 1962) or covered with 0.25–0.5% parlodion-carbon and dried for additional strength (Stevens, 1966). Selection of the proper emulsion for coating is a very important factor in high resolution autoradiography. A 50% solution of Ilford L4 in water can be prepared at 45°C followed by cooling at 20–24°C. A thin platinum, silver, or copper wire is dipped in the solution and a thin layer is allowed to form around it, which, on application to a slide, spreads out in a fine layer of emulsion. Agar surface can be substituted for a glass slide and several other modifications are also prevalent (Stevens, 1966).

After suitable exposure, the grids are kept in absolute ethanol to prevent swelling and resultant loss of grains, before developing. Several developers are used, such as Microdol X, and the dried grids are floated in an inverted position on the convex surface of the developer for 5–6 min at 22–24°C, followed by immediate transfer to distilled water for a few seconds. The grids are then placed in the fixer with the sections lying at the top to remove the exposed silver halide crystals and later the gelatin. After 10 min, the grids are rinsed in water (5 min), washed thoroughly by flushing, dried, and stored in a dust-free chamber. For staining, the grids are floated for 30 min on distilled water at 37°C, transferred to 0.5N acetic acid at 37°C for 15 min, rinsed in water and floated again in distilled water for 10 min at 22–25°C, moistened and stained in an inverted position in 7.5% aq. uranyl acetate for 20 min at 45°C, dried on filter paper, moistened again, and stained with 0.2% lead citrate for 10–60 sec to a few minutes (Stevens, 1966). Staining may be performed prior to the application of emulsion, but in that case, a fine layer of carbon is deposited through evaporation on the

stained section for screening it from the emulsion and for forming a base for the emulsion layer.

Chromosomes can also be studied following isolation procedures, either directly through micrurgical methods or indirectly through chemical treatment (Claude and Potter, 1943; Kroeger, 1966). Fixed chromosome preparations may be isolated by fixation in 45% acetic acid and removal from the cell with a needle. For unfixed preparations, chromosomes may be squeezed out from the punctured salivary gland in sugar medium. Alternatively, the glands may be treated with pronase, followed by homogenization and differential centrifugation or dipped in 0.25% dried egg white for 2–3 hr and the chromosomes isolated with a needle. In the chemical method of isolation, the chilled tissue is ground in a mortar with dry sand for a few minutes; 0.9% NaCl (pH 7.4) is added slowly (6 times); the tissue is centrifuged for a minute; the residue is discarded; the supernatant is again centrifuged (10 min); the sediment is suspended in saline solution and centrifuged at 1500 rpm. The chromatin threads are isolated through differential and repeated centrifugation. Modifications of this method have been developed by Mirsky and others.

Lately cell hybridization, through somatic fusion of malignant and nonmalignant cell lines, has been achieved (Harris and Watkins, 1965), with the objective of converting the malignant lines to nonmalignant ones; as, for example, between the B.82 mutants of Earle's L line and A 9 line of mouse. In a schedule adopted by Engel et al. (1969), the medium used is 4×10^{-7} aminopterin, $3 \times 10^{-6}M$ glycine, $1.6 \times 10^{-5}M$ thymidine, and $1 \times 10^{-4}M$ hypoxanthine, in which the hybrid lines could grow but not the parent ones. Cell fusion was obtained by exposure to Sendai virus (6000 HAU and 4000 HAU), previously inactivated through UV treatment. Methods for molecular hybridization at the chromosome level, such as of mouse chromosome DNA with complementary RNA at the site of occurrence on chromosomes, permitting localization of segments meant for specific gene action, have also been devised (Gall and Pardue as cited in Sharma and Sharma, 1972). This method has opened up new possibilities in cancer therapeutic research.

A. K. Sharma
Archana Sharma

References

Astbury, W. T., *Nature,* **155**:201 (1945).
Caspersson, T., *Symp. Soc. Exp. Biol.,* **1**:127 (1947).
Claude, A., and J. S. Potter, *J. Exp. Med.,* **77**:345 (1943).
Darlington, C. D., and L. F. LaCour, "The Handling of Chromosomes," London, Allen and Unwin Ltd., 1966.
Engel, E., B. J. McGee, and H. Harris, *Nature,* **223**:152 (1969).
Harris, H., and J. F. Watkins, *Nature,* **205**:640 (1965).
Kroeger, H., p. 61 in "Methods in Cell Physiology," Vol. 2, New York, Academic Press, 1966.
Oster, G., in "Physical Techniques in Biological Research," Vol. 1, New York, Academic Press, 1955.
Prescott, D. M., and M. A. Bender, p. 381 in "Methods in Cell Physiology," Vol. 1, New York, Academic Press, 1964.
Schultz, B., in "Physical Techniques in Biological Research," Vol. 3B, New York, Academic Press, 1969.
Sharma, A. K., and A. Sharma, "Chromosome Techniques—Theory and Practice," London, Butterworths, 1972.
Sjöstrand, F. S., in "Physical Techniques in Biological Research," Vol. 3C, New York, Academic Press, 1969.

Stevens, A. R., in "Methods in Cell Physiology," Vol. 2, New York, Academic Press, 1966.

Yunis, J. J., "Human Chromosome Methodology," New York, Academic Press, 1965.

See also: FEULGEN STAINS, FLUORESCENT SCHIFF-TYPE REAGENTS.

CILIA

Cilia are long and slender centriolar-based, motile processes of eukaryotic cells. Historically, it has been customary to distinguish between cilia and eukaryotic flagella on the basis of length, number, and mode of beat. Cilia are usually shorter than flagella. There are often many cilia along one cell border, like the hairs of an eyelash, from which the term "cilium" is derived. Cilia beat with flexural or oar-like motion, while flagella beat with a sinusoidal-like undulation. The beat may be planar or in three dimensions. Cilia and flagella are homologous organelles that possess the same main structural arrangement. The mechanism of motion for both organelles is almost surely the same, on the basis of several criteria, so that, in present usage, the distinction between them has become blurred. Motile cilia are found on many animal cells including cells of human trachea, oviduct, male reproductive tract, and brain. Motile flagella are found on sperm of most animals and many plants. Protozoan taxonomy is in part based on these locomotory appendages: ciliates, such as *Paramecium*, and flagellates, such as *Euglena* are well-known groupings. In these groups, and in certain other invertebrates, such as ctenophores (comb jellies) the cilia form the main motile appendages and ciliary responses often determine organism behavior. In ctenophores, cilia reach lengths of millimeters. In nematodes, insects, and flowering plants, cilia are extremely limited in distribution or entirely absent. Light microscopy of cell organelles may be said to have begun with the study of cilia by Antony van Leeuwenhoek in 1675. Leeuwenhoek used a single simple lens, with which he obtained excellent images of both protozoan cilia and sperm tails. Since about 1830, cilia have been favored organelles for certain microscopic and microsurgical techniques because of their accessibility and their motility.

Over extended distances, individual cilia may beat in synchrony or they may beat out of phase one with the next, to give rise to metachronal waves. Metachronal waves, which move at speeds of up to 0.5 mm/sec, can run either parallel or perpendicular to the plane of beat. In 1928, von Gelei devised a technique for quick fixation of the metachronal wave. This technique has since been modified and has been successfully adapted to electron microscopy. Since cilia change position so rapidly, it is often desirable to stop them instantaneously so that morphological or physiological changes are minimized, and quick fixation techniques are generally useful. One simple method of quick fixation of cilia requires rapid application of osmium tetroxide to the specimen being studied. After this treatment the cilia are preserved in their metachronal wave positions, where one wavelength includes all ciliary beat phases. Such material is suitable for study of the stroke form of the cilium by light microscopy of whole-mounted preparations, by conventional thin section electron microscopy, or, after critical point drying, by scanning electron microscopy.

Stroke form has also been analyzed by dark field microscopy, by phase contrast high-speed cinematography, and by ordinary bright field light microscopy with stroboscopic illumination. In dark field, the cilium appears as a white, bright line against a black background. Multiple exposure of a single photographic plate can capture swimming cells for up to several seconds, and rates of progression can be measured or single bending waves can be traced as they move along the ciliary shaft. In this way, it has been possible to demonstrate that bending waves consist of circular arcs separated by straight regions. Alternatively, in dark field the camera shutter can be left open and the path taken by a ciliated cell can be determined. In positive phase contrast microscopy, a single cilium appears as a gray line on a white background. Internal extensions of the organelle, such as the basal body, can sometimes be seen.

Cilia beat about 20–30 times a second, and bending waves move up the shaft at speeds of several hundred microns per second. High-speed cameras, capable of exposing over 1000 frames per second, are easily attached to standard microscopes and are used to capture the beat. A frame-by-frame analysis is often undertaken. Alternatively stroboscopes may be substituted for normal continuous microscope illumination. If the stroboscope is operated at the frequency of ciliary beat or at simple multiples or fractions of the frequency, the motion of the cilia stops. At multiples of the frequency, multiple images of each cilium will be obtained; at the true frequency, a single clear image is seen. Small changes in flash rate produce a slow motion either in forward or backward direction so that stroboscopic illumination is useful in following stroke form and progression, as well as measuring beat frequency.

Microsurgical methods have been applied to single cilia to sever the organelle from the underlying cell. Glass or fine tungsten needles are generally used in amputation experiments, but recently elegant laser microbeam methods have been developed. Ciliated cells have been suspended in a nontoxic blue dye and irradiated with a short pulse of red laser light. The laser emission is directed down a microscopic tube and is focused by the microscope to a diameter of about 2 μm. The damaged point is quite localized. This technique has made it possible to demonstrate that certain cilia swim for a number of strokes after they are detached from the cell body, and therefore to settle the question of whether the cilium alone is an active organelle.

Amputation of cilia is often a prelude to studies of regrowth and morphopoiesis, since cilia regenerate and grow back to their original length after surgery, if the cell body is intact. These studies may be carried out on single cells, or on standarized cell cultures, and may involve bulk preparation of cilia. Cilia can sometimes be removed from cells, such as *Chlamydomonas*, by mechanical shearing in a blender. *Tetrahymena* cilia are shed in a buffered ethanol-EDTA solution after the addition of appropriate amounts of calcium chloride. Such cilia are often fractionated further and may be used to prepare tubulins, which are similar globular proteins of about 60,000 daltons that comprise the ciliary microtubules, or dynein, the main ciliary ATPase, or other constituents. Electron microscope assays have been used to demonstrate that dynein is localized as rows of projections, "arms," along the peripheral microtubules. In isolated *Tetrahymena* cilia, it is possible to remove and reconstitute these rows by changing the ionic composition of the surrounding medium.

One additional important preparative technique, derived from similar work on muscle, is glycerination.

Either isolated cilia or ciliated cells can be treated with solutions containing high concentration of glycerin and then washed and reactivated with ATP. Glycerination apparently degrades and then removes the ciliary membrane. More recently, the detergent triton has been used in place of glycerin in the preparation of reactivatable membrane-free axonemes.

Microelectrode studies of ciliated cells such as *Paramecium* have been used to reveal mechanisms that control ciliary behavior. Intracellular recordings can be made with conventional KCl capillary electrodes of about 20 megohm tip resistance. Mechanical stimuli are applied to the cell by means of a glass stylus driven by a piezoelectric crystal. When applied at the anterior end of *Paramecium*, stimuli of sufficient strength elicit a transient depolarization of the cell membrane and reversal of ciliary beat direction. Attempts are now being made to record intracellularly from single sperm cells.

Much of our current knowledge of ciliary structure and function comes from a combination of physiological or histochemical micromethods or chemical analysis and transmission electron microscopy. Electron microscope preparative techniques that have been useful in such studies include sectioning of plastic-embedded material, negative staining, and freeze-etching. After fixation with glutaraldehyde or quick fixation with osmium fixatives, cilia are usually embedded, sectioned, and stained for morphological study. The small diameter of the organelle (about 0.25 μ) effectively prevents useful light microscope examination of internal structure and the resolution of the electron microscope is necessary, although, over 50 years ago, using conventional light microscopes, careful observers of fixed or squashed preparations noted that cilia are composed of a number of fibers. A typical thin-sectioned preparation of cross-sectioned cilia for transmission electron microscopy is seen in Fig. 1. In such a preparation, all motile cilia, from whatever source, reveal for the greater part of their length, a unique internal arrangement of "fibers" called the 9 + 2 pattern because there are nine doublet microtubules arranged around a central singlet pair. Certain variations on this pattern occur in the nonmotile cilia-like derivatives and in tails of motile sperm cells. In one variation, known as the 9 + 0 pattern, the central microtublues are missing.

The microtubule is a proteinaceous cytoplasmic organelle common to such structures as cilia, centrioles, and the mitotic spindle. In 0.1 μm or thinner stained sections, ciliary microtubules have an electron-opaque wall surrounding an electron-transparent core. The central pair of microtubules is usually somewhat less stable than the peripheral doublets. The central microtubules have the ability to bind colchicine and so resemble cytoplasmic microtubules more closely than do the peripheral doublets. Serial section reconstructions have been used to follow changes in microtubule position and pattern along the axoneme. This has led to development of a sliding microtubule theory of ciliary motility. According to this theory, microtubules do not change length as the cilium bends.

Microtubules are seen to good advantage in negative stain preparations of cilia. Cell fractions containing cilia are generally mixed with an electron-dense material, such as phosphotungstic acid, and allowed to dry on a carbon-coated electron microscope grid. Often the ciliary membrane that surrounds the microtubules is broken and the microtubules fray out. The microtubules exclude the stain to reveal an internal arrangement of rows of microfilaments of the globular subunits of tubulin.

PETER SATIR

References

Holwill, M., *Physiol. Rev.,* **46**:696, 1966.
Satir, P., *Protoplasmatologia,* **IIIE** (1965).
Sleigh, M. A., "The Biology of Cilia and Flagella," New York, Pergamon Press, 1962.

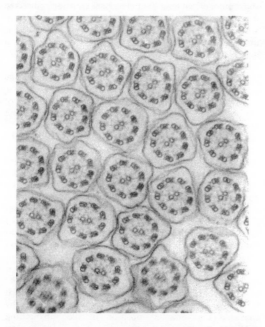

Fig. 1 Transmission electron micrograph of cilia from mussel gill. Cross sections of epon-embedded material showing standard morphology. ($\times$42,700.)

CLAY MINERALS

Interest in studying the clay minerals was fostered by the introduction of the electron microscope. By the mid 1940s instruments were available commercially which were capable of providing useful magnifications up to 33,000 diameters. For the first time one could observe the size and shape, and eventually surface features, of the small crystalline particles of which clay aggregates are composed. Because of their widespread occurrence and importance these minerals have been the subject of intensive study by electron microscopy, X-ray diffraction, and other techniques for the past 30 years. One can readily understand the importance of the electron microscope to the mineralogist concerned with the origin, growth, and alteration of these minerals, the soil chemist concerned with the role of clay minerals in holding and releasing nutrients to plants, the ceramist who literally builds with clay particles, the paper manufacturer who coats his paper with clay, the engineer interested in soil mechanics, and many others.

Many examples of electron microscope studies of clay minerals are found in the scientific literature. Some of these are listed in an article on the electron microscopy of minerals which appeared in an earlier edition of this encyclopedia (1). Other articles, special publications, and books are available which present typical electron micrographs and, sometimes, diffraction patterns of the various clays (2–17). Some of these are particularly valuable as guides because of the discussion of the techniques used in obtaining the micrographs or diffraction patterns.

While many of the earliest techniques for preparing clay minerals for electron microscopy are still in use, there has been within the last decade a remarkable increase in image quality and information available because of certain refinements over some of the earlier methods, the introduction of new techniques, and improvements in the design and performance of the electron microscope. To emphasize the importance of these changes examples are

Fig. 2b Extraction replica showing highly oriented mullite needles developed upon heating kaolinite to 1200°C for 20 hr.

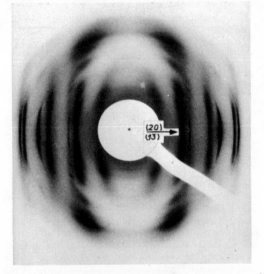

Fig. 1 Oblique texture electron diffraction pattern of beidellite. Specimen tilted ∼60° about axis normal to the electron beam. (Courtesy of A. Oberlin, Laboratoire d'Etude Des Microcristaux, Faculté des Sciences, Orléans, France.)

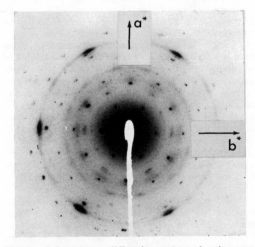

Fig. 2a Selected area diffraction pattern showing meta-kaolinite and highly oriented spinel phase (arcs) formed upon heating kaolinite to 850°C for 20 hr.

given to show how selected area diffraction, decoration techniques, greatly increased resolving power, and scanning electron microscopy contribute to a better understanding of the clay minerals.

Electron Diffraction. A widely used technique for preparing clay minerals for examination in the microscope is to disperse and mount the particles on thin supporting membranes covering 200-mesh specimen grids. Originally this method was used to obtain information on size and shape. As the electron microscope evolved into the highly versatile instrument it is today, it became a simple matter to obtain electron diffraction patterns of individual clay particles, including those only a few hundred angstroms in diameter. This led to new uses of the electron microscope, the identification of clays by their diffraction patterns, the detection of phase changes as in thermal transformations, or the attainment of crystallographic data necessary in determining the structure of clay minerals (18–25).

Mering and Oberlin applied the oblique texture pattern method of Pinsker to the study of the stacking of random layer lattices in some clays of the montmorillonite group (26). The diffraction pattern of beidellite in Fig. 1 was obtained by tilting the dispersed particles ∼60° about an axis normal to the electron beam. Instead of the pattern normally obtained from these platy particles, which show only (hk0) reflections, the resulting pattern shows the reflections (hk1) from which an approximate value for the c parameter and the angle β can be obtained.

Figures 2a and 2b illustrate the use of electron microscopy and selected-area electron diffraction in studying phase changes in clays. The diffraction pattern illustrates a stage in the thermal transformation of kaolinite where the orientation relationship between the newly formed spinel phase and the residual metakaolinite phase is seen (27). The preferred orientation of mullite particles obtained at higher temperatures is seen in the micrograph,

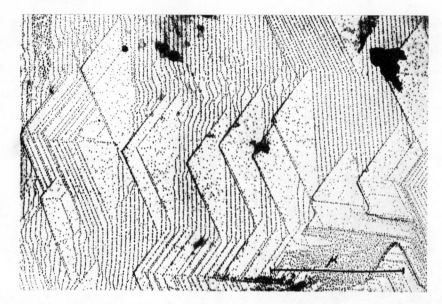

Fig. 3 Gold-decorated surface of a single crystal of dickite. (Courtesy of G. Gritsaenko, Institute of Ore Deposits, Geology, Petrology, Mineralogy and Geochemistry, USSR Academy of Sciences, Moscow.)

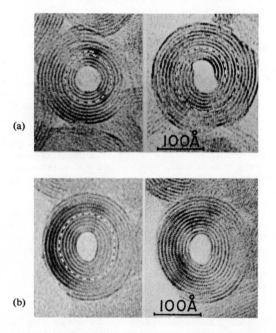

Fig. 4 Cross-sections of chrysotile from Transvaal, South Africa. The circumferential fringes correspond to the 7.3 Å-spacing. Radial (020) planes are also visible. (a) Spirally-wound layers. (b) Cylindrical tubes. (Courtesy of K. Yada, Research Institute for Scientific Measurements, Tohoku University, Sendai, Japan.)

Fig. 2b, of an extraction replica of a thermally transformed kaolinite crystal.

Gold Decoration Technique. The technique of gold decoration developed for ionic crystals by Bassett (28) has been used very successfully by Gritsaenko (29) to study growth steps on minerals with a layer structure. The distribution of the gold particles along growth steps shows clearly the outcropping of elementary layers in kaolinite, dickite, halloysite, and all crystals which exhibit polytype modifications. In Fig. 3 is seen the gold-decorated surface of a single crystal of dickite in which the transition from steps 14 Å in height to pairs of 7 Å steps can be seen. In principle, it is possible to determine the growth rate ratios for a crystal along different crystallographic directions by observing the height and frequency of steps decorated by the gold.

Direct Imaging of Lattice Planes. Remarkable improvements in instrumentation have resulted in an extension of the practical limits of resolution to about 2 Å. When conditions are favorable, the atomic structure within a clay particle can be resolved, so we may expect a reexamination of many clay minerals for features which could not be seen before. Yada has obtained micrographs of chrysotile in which the lattice fringes corresponding to the 7.3 Å c parameter are observed parallel to the fiber axis (30). In some places dislocations in the (001) planes were seen. Fringe systems corresponding to the (020) and (110) planes of 4.5 and 4.6 Å were observed sometimes in the middle of the fibers. Taking advantage of improved techniques for sectioning materials he cut undistorted cross-sections of the fibers, revealing the hollow form of chrysotile. Particularly interesting is the fact that although he did find spirally wound layers, Fig. 4(a), many of the fibers were simple cylindrical tubes as shown in Fig. 4(b). Another interesting application of high resolution electron microscopy was made by Brown and Rich, who cut sections of

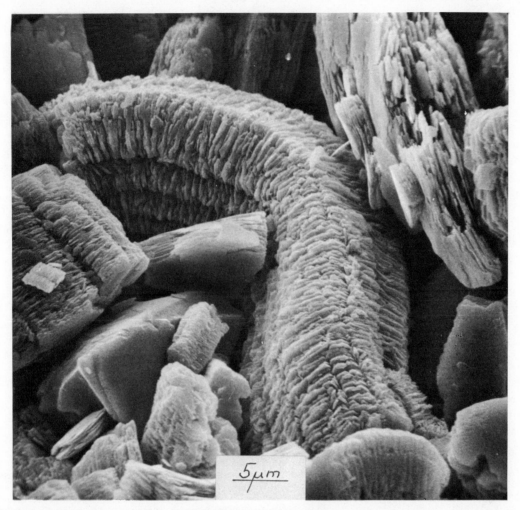

Fig. 5 Scanning electron micrograph of kaolinite "stacks." (Courtesy of J. Brown, Georgia Institute of Technology.)

muscovite normal to the (001) planes and observed differences in the intensity of the 10 Å spacing fringes between untreated and barium-treated specimens (31).

Scanning Electron Microscopy. The preshadowed carbon replica is used to observe surface details on clays or products containing clay particles (32). The resolution is limited to 20 Å by this method because of structure inherent in the replicating and shadowing materials. Unfortunately the preparation of these replicas is time-consuming and destructive to the sample. The most exciting recent achievement in instrumentation in electron microscopy is the scanning electron microscope, in which bulk specimens can be examined directly after the application of a thin conductive coating applied by vacuum evaporation (33). The image formed by backscattered or secondary electrons reveals details as small as 150 Å. This instrument is ideally suited for the examination of specimens which have surface details too small to be resolved by the light microscope but which are too rough for satisfactory handling by the replica technique. A good example is shown in Fig. 5, a scanning electron micrograph of a long "stack" of kaolinite platelets. The large depth of field renders all parts of the specimen equally sharp.

It is evident that electron microscopy will continue to play an important part in clay mineral studies. With the improved resolution and flexibility of present electron microscopes there are many problems to be studied which could not be dealt with adequately using earlier instruments and techniques. A few examples are the efforts of weathering and cation substitution in clays, and the interaction of clays with organic materials. The current interest in oceanography and environmental studies will undoubtedly provide a stimulus for such studies.

JOSEPH J. COMER

References

1. Bates, T. F., "Electron microscopy of minerals," pp. 187–200 in "The Encyclopedia of Microscopy," G. L. Clark, ed., New York, Van Nostrand, 1961.
2. Pinsker, Z. G., "Electrographic and electronomicroscopic studies of clay minerals," *Trudy Biogeokhim. Lab., Akad. Nauk SSSR,* **10**:116–141 (1954).
3. Bates, T. F., "Electron microscopy as a method of identifying clays," *Proc. 1st Nat. Conf. Clays and Clay Minerals, Calif. Dep. Natur. Resources Div. Mines Bull.,* **169**:130–150 (1955).

4. Taggart, M. S., W. O. Mulligan. and H. P. Studer, "Electron micrographic studies of clays," *Nat. Acad. Sci.–Nat. Res. Council Publ.,* **395**:31–64 (1955).

5. Dwornik, E., and M. Ross, "Application of electron microscope to mineralogic studies," *Amer. Mineral.,* **40**:261–274 (1955).

6. Bates, T. F., and J. J. Comer, "Electron microscopy of clay surfaces," *Nat. Acad. Sci.–Nat. Res. Council Publ.,* **395**:1–25 (1955).

7. Neuwirth, E., "The determination of clay minerals with the electron microscope," *Tschermaks mineralog. u. petrog. mitt.,* **5**:347–361 (1956).

8. Bates, T. F., "Selected electron micrographs of clays and other fine-grained minerals," Circular No. 51, Mineral Industries Expt. Station, College of Mineral Industries, The Pennsylvania State University, 1958.

9. Comer, J. J., "The electron microscope in the study of minerals and ceramics," *Amer. Soc. Testing Materials, Spec. Tech. Publ.,* **257**:94–120 (1959).

10. Suito, F., "Electron microscopy of clay minerals," *Nendokagaku no Shimpo,* **1**:166–180 (1959).

11. Eitel, W., "Electron microscopy of dispersed systems," pp. 390–406 in "Silicate Structures," Vol. I, New York, Academic Press, 1964.

12. Brown, J. L., "Laboratory techniques in the electron microscopy of clay minerals," pp. 148–169 in "Soil Clay Mineralogy," C. I. Rich and G. W. Kunze, ed., Chapel Hill, Univ. North Carolina Press, 1964.

13. Bates, T. F., "The application of electron microscopy in soil clay mineralogy," pp. 125–147 in "Soil Clay Mineralogy," C. I. Rich and G. W. Kunze, ed., Chapel Hill, Univ. North Carolina Press, 1964.

14. Beutelspacker, H. and H. W. Van der Marel, "Atlas of Clay Minerals and Their Admixtures," New York, Elsevier, 1968.

15. Grim, R. E., "Clay Mineralogy," 2nd ed., New York, McGraw-Hill, 1968 (pp. 165–184).

16. Gritsaenko, G. S., B. B. Zyvagin, R. V. Boyarskaya, et al., "Methods of the Electron Microscopy of Minerals," Moscow, Nauka, 1969.

17. Gard, J. A., "The electron-optical Investigation of Clays," London, Mineralogical Society (Clay Minerals Group), 1971.

18. Ross, M. and C. C. Christ, "Mineralogical applications of electron diffraction," *Amer. Mineral.,* **43**:1157–1178 (1958).

19. Pinsker, Z. G., "Electron Diffraction," London, Butterworth's Scientific Publications, 1953 (pp. 281–287).

20. Zussman, J., G. W. Brindley, and J. J. Comer, "Electron diffraction studies of serpentine minerals," *Amer. Mineral.,* **42**:133–153 (1957).

21. Chapman, J., and J. Zussman, "Further electron optical observations on crystals of antigorite," *Acta Cryst.,* **12**:550–552 (1959).

22. Gard, J. A., and J. M. Bennett, "A goniometric specimen stage and its use in crystallography," pp. 593–594 in Proceedings of the Sixth International Congress for Electron Microscopy," Tokyo, Maruzen Co., Ltd., 1966.

23. Zyvagin, B. B., "Electron Diffraction Analysis of Clay Mineral Structures," New York, Plenum Press, 1967.

24. Radczewski, O. E., and H. J. Balder, "X-ray and electron optical investigations of soil from Schlettaer. High resolution diffraction patterns of individual crystals," *Fortschr. Miner.,* **37**:74–78 (1959).

25. Nakahiro, M., and Kato, "Thermal transformations of pyrophyllite and talc as revealed by x-ray and electron diffraction studies," pp. 21–27 in Proceedings of the Twelfth National Conference on Clays and Clay Minerals," New York, Pergamon Press, 1964.

26. Mering, J., and A. Oberlin, "Electron-optical studies of smectites," pp. 3–25 in "Proceedings of the Fifteenth Conference on Clays and Clay Minerals," New York, Pergamon Press, 1967.

27. Comer, J. J., "New electron-optical data on the kaolinite-mullite transformation," *J. Amer. Ceram. Soc.,* **44**:561–563 (1961).

28. Bassett, G. A., "A new technique for decoration of cleavage and slip steps on ionic crystal surfaces," *Phil. Mag.,* **3**:1042–1045 (1958).

29. Gritsaenko, G., and N. Samotoyin, "Decoration method applied to the study of the relationship between microcrystals surface microtopography and the crystal structure of kaolinite and dickite," pp. 595–596 in "Proceedings of the Sixth International Congress for Electron Microscopy," Tokyo, Maruzen Co., Ltd., 1966.

30. Yada, K., "Study of chrysotile asbestos by a high resolution electron microscope," *Acta Cryst.,* **23**:704–707 (1967).

31. Brown, J. L., and C. I. Rich, "High-resolution electron microscopy of muscovite," *Science,* **161**:1135–1137 (1968).

32. Bates, T. F., and J. J. Comer, "Electron microscopy of clay surfaces," *Nat. Acad. Sci.–Natl. Res. Council Publ.,* **395**:1–25 (1955).

33. Kimoto, S., and J. C. Russ, "The characteristics and applications of the scanning electron microscope," *Amer. Scientist,* **57**:112–133 (1969).

COAL AND COKE PREPARATION FOR MICROSCOPIC EXAMINATION

Introduction. The microscopic examination of a coal or coke sample provides information regarding composition, structure, and technological properties. Polished blocks are prepared for examining in vertically incident reflected light, and translucent thin sections (5–25 μm thick) for use in transmitted light. Polished thin-sections are used to correlate observations made in reflected and transmitted light. For certain special investigations, or where samples are required for demonstration purposes, it is possible to prepare thin-sections (several inches across) of large areas. These sections are either embedded in a block of transparent colorless plastic or prepared as flexible thin-

Fig. 1 Trimming block of bituminous coal with diamond saw.

sections from which regions of interest can be removed for chemical analysis.

Coke cannot be rendered translucent in thin-section, but thin-sections, especially large-area sections, are still useful for revealing details of the pore structure, a factor upon which many of the technological properties depend. Pore structure can also be revealed by impregnation with plaster or by replication.

Sampling. Coal samples may be in the form either of lumps or of crushed particles. Lump samples, trimmed to about 6–8 cm, are used mostly to study variation within coal seams. If the sample is soft (e.g., brown coals) it can be trimmed with a hand saw, but bituminous coals, anthracites, and cokes are hard, and must be trimmed with a diamond saw (Fig. 1). If a large lump is to be subdivided into a number of smaller blocks, the pattern of cuts should be staggered (see Fig. 2) to ensure that the minimum amount of information is lost as a result of the saw cuts. If a sample is fragile it must be strengthened by impregnation with plastic (see later) before being trimmed.

A representative sample of bulk coal is made by crushing the coal to less than 750 μm, keeping the proportion of particles smaller than 100 μm, to a minimum. The crushed coal is then poured through a sample splitter (Fig. 3) and recycled until a sample of 5–10 g is obtained.

Embedding. The sample, whether crushed or in lump form, must first be embedded in plastic. This strengthens it, provides a suitable block for handling, and also, in the case of particulate material, serves as the medium within which the particles are dispersed.

The choice of embedding medium depends on the sample (see Table 1). "Astic" is most satisfactory for general use because it has approximately the same grinding and polishing characteristics as bituminous coal.

Resins which must be heated before they will harden, or which set exothermically, cannot be used with temperature-sensitive materials such as the brown coals. Other resins, because they expand or contract during hardening, cannot be used where preservation of original structure is important, as with many coke samples. Porous and fragile samples such as cokes and chars must be thoroughly impregnated with resin before they can be handled and the resin used must be sufficiently fluid to penetrate the structure. The epoxy resins, while satisfactory for embedding bituminous coals and anthracites, cannot be used with brown coals; these contain adsorbed moisture which impairs the setting characteristics of the resins.

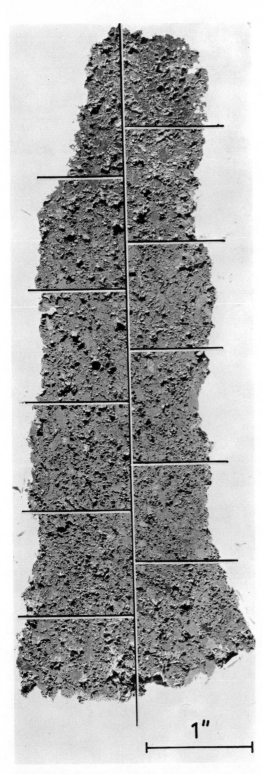

Fig. 2 Subdividing large sample using staggered pattern of cutting.

Fig. 3 Sample splitter.

Table 1   Materials used for Embedding and Replicating Coal and Coke Samples

Name	Type	Preparation	Properties
Astic[a]	Polyester	Three liquids. Mix in proportions: 97% resin; 2% catalyst; 1% accelerator.	Cold-setting (max. temp. ~rise 40°C depending on amount of accelerator). Forms tough, colorless, glass-clear solid. Setting time: 1–2 hr. Good general-purpose resin.
Araldite[b]	Epoxy	Two liquids. Various mixtures depending on type of Araldite. Suitable resin prepared from 50% resin No. 121 and 50% hardener No. 9.	Cold-setting. Forms hard solid, light amber in color. Setting time: 1–2 hr. Cannot be used with brown coals.
Zinc cement[c]		Powder + liquid. Mix to a paste (approx. 1 : 1).	Cold-setting. Forms very hard, opaque, white, heat-resistant solid. Setting time: 30 min.
Plaster of Paris (dental plaster)[d]		Powder. Mix to a paste with water.	Cold-setting. Forms soft, opaque, white solid. Useful only for embedding and grinding. Cannot be polished.
Araldite[b]	Epoxy	Two liquids. Various mixtures depending on type of Araldite. Suitable resin prepared from 50% resin B and 50% hardener No. 901. Heat resin in crucible until liquid, then add hardener, pour into mould, and allow to set in oven at 30–40°C.	Hot-setting. Forms very hard, heat-resistant solid, light amber in color. Setting time: overnight.
Mowilith No. 20[e]	Polyvinyl acetate	Lump or powder. Low melting point. (Overheating causes blisters and darkening of color.)	Hot-setting. Forms very tough, glass-clear solid. Setting time: 5–10 min.
Caedax[f]	Synthetic resin	Honey-colored liquid. Thickens when heated. (Overheating causes darkening and brittleness.)	Hot-setting. Forms glass-clear solid. Refractive index 1.54. Very fluid initially. Used for cementing coverslips.
Canada balsam	Natural resin	Honey-colored liquid. Thickens when heated. (Overheating causes darkening and brittleness.)	Hot-setting. Forms glass-clear solid. Refractive index 1.54. Very fluid initially. Used for cementing coverslips.
Lakeside –70C	Synthetic resin	Stick-form. M.p. = 80°C. (Overheating causes darkening and brittleness.)	Hot-setting. Forms light-brown solid. Refractive index 1.54. Used for cementing. Used for impregnating when dissolved in alcohol.
Cera Carnauba wax	Natural wax	Lump form crushed to powder. Melted by heating.	Hot-setting. Forms dark-gray solid. Can be stained blue or black with nigrosin. Used only for impregnation.
Vinamold[g]	Synthetic "rubber"	Cut into small pieces and melt in stainless steel, aluminum, or tin saucepan (copper or zinc vessels must not be used). Pour into mold quickly and continuously.	Hot-setting. Vinamold No. 1028 forms a flexible, medium-hard, red solid. M.p. = 120°C. Used for replication, and is reusable.

[a]®John Morris Pty. Ltd., 61 Victoria Ave., Chatswood, N.S.W., Australia.
[b]®C.I.B.A. Ltd., Basle, Switzerland.
[c]Dental cement—The S.S. White M/g & (G.B.) Ltd. London, W.1, England.
[d]Dental plaster—A fine-grade plaster of Paris.
[e]®Hoechst AG, Frankfurt, Germany.
[f]®Bayer, Leverkusen, Germany.
[g]®Selley's Chemical Manufacturing Co. Pty. Ltd., Melbourne, Vic., Australia.

Lump samples are embedded by placing them in a disposable mold (readily fabricated from aluminum foil) and pouring in the resin. Crushed samples are mixed thoroughly with the fluid resin and the mixture is allowed to solidify in a mold. A reference number, stamped onto one side of the mold before pouring, will be transferred permanently to the block as the resin sets. The resultant block, known as a "grain mount," is particularly useful in quantitative petrographic analysis of coals. It is therefore desirable to use a standard mold size (say 1 in. cube) so that approximately the same number of particles can be analyzed in different preparations. Fluid resin is poured into the mold until it is about one-quarter full, and a portion of the crushed coal is thoroughly mixed in. The remainder of the sample is then added slowly, more resin being poured in as required to prevent the mixture from becoming too dense. The aim is to obtain a uniform dispersion in which the particles are close-packed but not overlapping (Fig. 4). However, some stratification usually occurs because the particles settle as the resin hardens. For quantitative analysis on the sample surface, it is essential to cut the microscopic section parallel to one of the vertical cube faces in the solidified block to allow for this.

Fragile or brittle lump samples must be strengthened by impregnation with resin before trimming and embedding. This is almost always necessary when embedding porous samples, in order to prevent fragmentation of the material during grinding and polishing.

Vacuum Impregnation. Although in some cases adequate impregnation can be achieved simply by immersing the sample in the fluid resin and relying on surface tension and capillary action to draw it into the pores, best results are obtained by carrying out the immersion procedure under vacuum. Releasing the vacuum forces the fluid resin deeply into the pore structure.

Simple equipment for vacuum impregnation (see Fig. 5) consists of a glass desiccator, 10 in. in diameter, the lid of which has been modified to take a funnel and tap assembly. The desiccator chamber can be connected by means of the tap to a vacuum line, to the funnel, or to the atmosphere. The sample, contained in a mold, is placed in the desiccator and the chamber evacuated using a simple water pump which can reduce the pressure sufficiently in about 5 min. Semi-anthracites need a higher vacuum than other coals and must be evacuated for longer periods.

During evacuation the resin is prepared and poured into the funnel. When evacuation is complete, the resin is poured in gently until it fills the mold, the rate of flow being controlled by the tap setting. Air is then admitted slowly, the lid removed, and the funnel and tap assembly cleaned immediately with acetone. The resin is left to harden. A number of samples can be vacuum-impregnated in a single operation by arranging them in a circle in the desiccator and rotating the tap assembly to each one in turn.

Brown coals contain moisture, much of which must be removed before they are impregnated and embedded. Coal containing less than about 30% moisture need only be air-dried until "dry" to sight and touch, but one containing over 30% requires more elaborate treatment. The sample is wrapped in cotton gauze and immersed in ethyl alcohol for a period ranging from several hours for small samples (up to about 2 in.) to several days for large ones. The effect of thus immersing the brown-coal sample is to replace most of its moisture content by alcohol. During the operation the alcohol is agitated frequently, and

Fig. 4 Particles of bituminous coal in polished surface of a grain mount.

(a)

(b)

Fig. 5 Equipment for vacuum impregnation. (a) General view. (b) Detail of funnel and tap assembly.

because it becomes diluted with water, fresh alcohol must be added. The immersion procedure is now repeated using acetone, which in turn displaces the alcohol.

Polyvinyl alcohol is added to the final acetone bath until the solution becomes viscous. The acetone is then rapidly evaporated off under vacuum and a final coating of viscous polyvinyl acetate solution is applied to the sample surface. When the acetone solvent has evaporated the sample is coated with "Astic" resin to strengthen it and finally mounted in "Astic."

Grinding and Polishing. One face of the mounted sample, whether grain mount or lump, is chosen for grinding and polishing. To give an approximately plane surface this face is first cut on a diamond saw, the surface being cooled with flowing water in order to prevent thermal alteration. The sawn surface is ground on successively finer grades of carborundum paper (ranging from 220-grit, through 400-grit to 600-grit) which is laid on a flat surface such as a sheet of plate glass. Water flows continuously over this and grinding is carried out by drawing the block steadily across the paper with straight, not circular, movements. After every few strokes the block is rotated a few degrees, and is thoroughly cleaned after each change of carborundum paper.

Paper of 220-grit grade is used until a plane, evenly ground, surface is obtained, after which 400-grit paper is used until inspection with a hand lens ($10\times$) shows that the surface has an even-textured, matte finish, free of major defects. Fine grinding is carried out on the 600-grit paper until the surface shows a very fine matte texture which may even have a slight polish. At this stage it is possible to polish the surface, but best results are obtained if the surface is first reground with 400-grit carborundum and then 600-grit carborundum, embedded in wax laps.

Preparation of Wax Laps. These may be prepared by

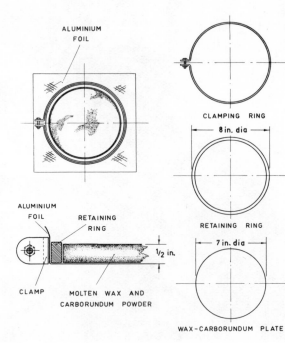

ALUMINIUM FOIL

CLAMPING RING

8 in. dia

RETAINING RING

7 in. dia

ALUMINIUM FOIL   RETAINING RING

$\frac{1}{2}$ in.

CLAMP   MOLTEN WAX AND CARBORUNDUM POWDER

WAX-CARBORUNDUM PLATE

Fig. 6 Former for preparing wax lap.

pouring a mixture of Mobil wax No. 2305* and the abrasive powder into a former (Fig. 6) consisting of a brass retaining ring placed on a sheet of aluminium foil, the edges of which are folded up around the ring and clamped to it by an external clamping ring. This circular dish is placed on a hot plate at about 80°C. A small amount of wax is melted in a copper ladle, abrasive is mixed into it, and the mixture is then poured into the warmed mold and stirred to promote settling of the powder. The process is repeated with more wax and abrasive until the mold is full, and the mixture is given a final stir to ensure that a dense layer of abrasive settles at the base. To prevent cracking, the wax should be allowed to cool slowly (preferably overnight). The mold is then dismantled, and the wax disc inverted to make use of the lower surface for grinding.

The amount of abrasive required is determined empirically. If there is too little, no grinding action is obtained; if too much, the powder will not bond tightly in the wax and will pluck out and damage the sample. Grinding on the wax laps is usually done by hand, but can, if desired, be done on rotating tables.

Polishing the Sample. After the fine grinding, initial polishing is carried out on a rotating metal lap (6–8 in. diameter) covered with wet linen cloth, using chrome oxide powder (e.g., Buehler AB chrome oxide) as the polishing agent. The consistency of the chrome oxide/water slurry and the speed of rotation of the lap must be adjusted to the type of sample. Soft materials such as brown coal require only a thin creamy slurry and speeds of about 100 rpm. Harder samples require a thicker slurry, and the rotation of the laps should be increased (about 500 rpm for bituminous coals; 800 rpm for anthracites and cokes).

As soon as inspection under a microscope shows that the ground matte texture has disappeared, final polishing is carried out, using a rotating metal lap covered with a water-saturated cotton cloth (e.g., Buehler AB Selvyt*). Magnesium oxide powder (e.g., Buehler AB Magomet Polishing Compound*) is used as the polishing agent. (This should be stored in an airtight container to prevent formation of magnesium carbonate, which impairs the polishing characteristics.) The final polishing stage should be as short as possible (not more than 2–3 min), otherwise excessive relief will develop on the surface.

Anthracites do not respond well to polishing with magnesium oxide, and it is preferable to use a diamond paste (e.g., Buehler AB Metadi Polishing Compound*) on a high-speed (1000 rpm) metal lap covered with a nylon cloth (e.g., Buehler AB Nylon*). The paste should contain 1 μm diamond powder and, during polishing, should be lubricated with kerosene or a commercially available lapping oil (e.g., Buehler AB Automet Lapping Oil*). This "wet" stage, which should be as short as possible, is followed by "dry" polishing on a rotating lap covered with nylon cloth using slightly moistened γ-alumina (e.g., Linde B† or Buehler AB Gamma Micropolish*) with a particle size of about 0.05 μm. Finally, the surface is buffed by hand until it is dry, using clean cotton cloth (Selvyt*) moistened slightly with alcohol. The polished surfaces are now ready for examination by reflected light microscopy.

*®Mobil Oil Company.
*®Buehler Ltd., Evanston Ill., U.S.A.
†®Alan H. Reid (Sales) Pty. Ltd. Clyde, N.S.W., Australia.

Sectioning. In the preparation of thin-sections, the same grinding and polishing procedures are adopted but the polished surface must be cemented to a clean glass slide for sectioning and thinning. Canada Balsam, Lakeside-70, or Araldite are commonly used cements. Polyvinyl acetate (Mowilith-No. 20) is also suitable, and because it is readily soluble (in acetone), it is the cement which *must* be used when preparing flexible thin-sections (see later). Astic is unsuitable because it will not stick to the glass.

Araldite and Canada Balsam are applied to the glass slide as liquids. Lakeside-70 and Mowilith No. 20 are placed as solid lumps on the slide and melted by warming on a hot plate. At the same time the polished surface of the block is warmed by suspending it a few inches above the hot plate, a necessary precaution against chilling (and hence cracking) of the cement when block and slide are brought together.

When the cement is liquid and the block warm, the slide is inverted (i.e., turned cement-side down) and placed gently on the polished surface to establish a continuous film of cement, free of bubbles (which can be removed by gentle pressure), between glass and block. If this operation is not immediately successful the glass and block must be rewarmed and the process repeated. The slide is then placed on a cold surface (a slab of marble is ideal) to cool the cement rapidly, pressure being applied evenly to obtain satisfactory adhesion between slide and sample.

When the cement has set, the slide is clamped into the vise of a diamond cutoff machine (Fig. 7) and the block is cut away to leave a slice about 60 μm thick on the slide. This slice is ground to uniform thickness on a rotating steel lap using 220-grit carborundum; it is then carefully ground with 400-grit followed by 600-grit carborundum, until it is about 5–25 μm thick. Grinding is normally carried out with water, but if the coal contains much clay, carbonates, or water-soluble constituents such as sodium chloride, then grinding is done with kerosene or on dry paper. Brown coals are of the correct thickness when the section appears translucent reddish-brown under the microscope; bituminous coals, when vitrinite appears

Fig. 7 Sectioning block on diamond saw cutoff machine.

translucent red and exinite translucent yellow under the microscope; anthracite and coke remain opaque in thin-section, but some semi-anthracites may be rendered translucent by very careful grinding.

Most thin-sections are completed at this stage by attaching a glass coverslip to protect the section and make it suitable for study with the corrected objectives normally supplied with transmission microscopes. The coverslip is smeared with Caedax which is very fluid (Table 1), and is then warmed on the reverse side to drive off some of the volatile constituents. (Care must be taken to avoid ignit-

ing the Caedax.) Meantime the glass slide is warmed slightly and the Caedax is poured from the coverslip onto the thin-section leaving only a thin film adhering to the slip itself, which is then immediately lowered cement-side down, onto the section and pressed gently to expel bubbles. After this it may be pressed more firmly onto the section. The Caedax sets hard on cooling and the slide is then cleaned with acetone and labeled ready for use.

Polished Thin-Sections. Doubly polished thin-sections are those which have both lower and upper faces polished. Thin-sections whose lower surfaces are ground but not

Fig. 8   Large-area thin-section of coke permanently mounted in plastic.

polished are easier to prepare, but the overall quality is not as good as that of doubly polished preparations.

Polished thin-sections are prepared as described above except that instead of attaching a coverslip when thinning is complete, the top surface is polished for viewing directly with the uncorrected objectives supplied with reflected light microscopes. To polish a thin-section, the slide is inserted into a suitable Perspex holder to facilitate handling and is first polished with chrome oxide, the consistency of the slurry and the speed of the lap having been adjusted to the hardness of the sample. Final polishing is carried out by hand using fine (0.05 μm) γ-alumina on cotton cloth stretched over plate glass. The section must be examined frequently under the microscope until a scratch-free surface is obtained, after which it is carefully cleaned and labeled ready for use.

Preparation of Very Small Samples. It is sometimes necessary to prepare thin-sections or polished surfaces of very small samples such as individual particles of coal. If the sample is large enough to handle, a plastic block (e.g., a 1 in. cube) should be prepared in a mold and a hole 2–3 mm deep and 3–5 mm in diameter drilled in one face. The sample is added to resin which is poured into the hole, and when this has set, the surface is ground, polished and sectioned as described above. If the sample is too small to handle then it should be transferred to a strip of adhesive tape (e.g., Sellotape or Scotch Tape), which is then placed at the bottom of a mold, sample-side uppermost, and resin is poured in to fill the mold. When the block has set, it is carefully ground, polished, and sectioned.

To embed loose grains of coal or other small samples with a particular orientation, a small amount of resin is first poured into a mold. When the resin becomes gelatinous the sample is placed in it and manipulated into the required orientation. If required, more resin is added to fill the mold. The resultant block is sawn, ground, polished, and sectioned as before.

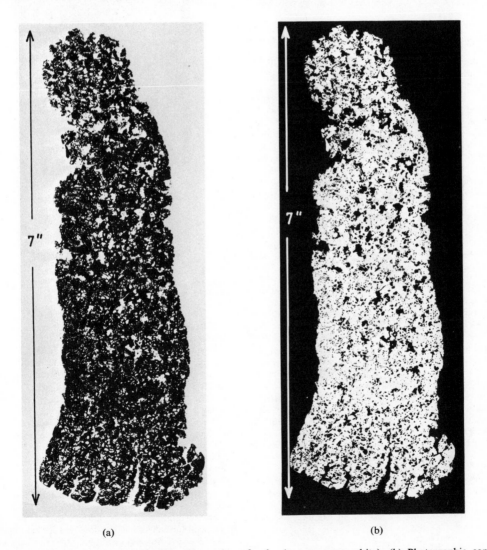

(a)

(b)

Fig. 9   Pore structure in coke. (a) Flexible thin-section of coke (pores appear white). (b) Photographic contact print from the flexible thin-section (pores appear black).

Flexible Thin-Sections and Large-Area Thin-Sections.
Flexible thin-sections, prepared for examination under the microscope, can be subsequently cut with scissors and areas of special interest removed for chemical or other types of analysis. The initial stages of preparation follow those given for a normal thin-section, except that when the polished surface is attached to a glass slide, a readily soluble cement such as Mowilith No. 20 (Table 1) must be used. The section is then thinned by grinding, but instead of using a coverslip, the surface is coated with a film of 5 % polyvinyl alcohol in water, applied from an atomizer. When the film is dry, the section is immersed in acetone for several hours until the Mowilith dissolves. The section should preferably be placed face down on a wire gauze to prevent it from curling as it is being freed from the slide. A layer of amyl acetate (dissolved in acetone) is then applied to the exposed surface forming a protective backing which keeps the section flat. Such sections, which may be of any size from standard up to many inches across, can conveniently be stored between sheets of transparent plastic stapled together.

Large or small thin-sections, permanently embedded in a transparent plastic block are particularly suitable, for demonstration purposes. The embedding block is first sectioned parallel to the polished surface to produce a slice about 5 mm thick. This is placed, polished surface uppermost, in a suitably sized mold which is then filled with a clear colorless resin such as Astic. When this has set, the mold is stripped away and the exposed face of the section is cut with a diamond saw to leave a slice as thin as possible on the block. The slice is then carefully ground to the correct thickness and protected by a layer of Astic. Embedded in this way, the section can be made clearly visible by polishing the surfaces of the Astic block (Fig. 8).

Replication and Impregnation of Cokes. The pore structure of coke is important in determining such properties as its suitability for use in blast furnaces, and can be studied in detail in large-area thin-sections prepared as previously described. The pores show up clearly in photographic contact prints made from flexible thin-sections, appearing black on a white background (Fig. 9). Alternatively, the pore structure can be made visible by impregnating the sample with plaster of Paris and preparing a polished surface. Such information, however, refers only to one plane within the sample.

Details of the structure of the bulk sample can be revealed by replicating the sample with such materials as Vinamold* and fine plaster of Paris (dental plaster) to provide a permanent record of the sample morphology.

The Vinamold replica can be readily stripped from the sample and will reproduce the finest details. It is flexible, tough, and unaffected by water and most laboratory chemicals. It is particularly useful in preparing a three-dimensional negative impression from a coke surface in which the depth of each pore is reproduced as a protuberance. These protuberances can be rendered more clearly visible by preparing a positive Vinamold replica from the negative one and using this to produce a new negative impression with plaster of Paris (Fig. 10).

Fine plaster of Paris is also a satisfactory material for making positive replicas. The powder must always be added to an excess of water; the water should not be added to the powder. A film of oil should be smeared

Fig. 10 Three-dimensional negative replica of coke, showing pore structure as protuberances.

over the surface of the mold to prevent sticking when the plaster hardens.

Where fine detail is required, an initial layer of very thin plaster should be poured into the mold. This will harden more rapidly if a solution containing approximately 1 % sodium chloride is used instead of fresh water to make the plaster. Once this layer has set, the mold can be filled with thicker plaster and left to harden.

If the morphology of a sample is complex and careful molding of the plaster is necessary to reproduce it, then the rate of hardening of the plaster can be slowed down by mixing with a dilute solution of borax. If a realistic reproduction of the sample is required (Fig. 11) the plaster replica can be painted, or preferably, a pigment of graphite powder can be added while the plaster is being mixed.

Acknowledgements. The author thanks Dr. A. R. Ramsden for valuable discussions and for help in preparing this article for publication, and Mr. E. G. Shrubb for photographic services.

W. ZEIDLER

---

*®Selley's Chemical Manufacturing Co. Pty. Ltd., Melbourne, Australia.

Fig. 11 Original and replica of coke sample.

Reference

"International Handbook of Coal Petrography," 2nd ed., 1963. Centre National de la Recherche Scientifique, 15 Quai Anatole France, Paris (7ᵉ). France.

COELENTERATA

Culture. Many marine species can be grown in their natural water or in commercially available artificial sea-waters and fed on *Artemia* larvae or other crustacean food. Colonial hydroids can be grown on glass slides held vertically in racks. The water should be renewed regularly and kept moving in the culture vessels. Hydras are mass-cultured for experimental work in distilled or deionized water containing 100 mg/liter $CaCl_2$, 100 mg/liter $NaHCO_3$, and 50 mg/liter disodium EDTA (Loomis et al., 1956). Addition of 0.5% glycerin to the culture water rids green hydras of their algal symbionts (Whitney, 1907).

Grafting Hydras. Set pieces of human hair upright in wax in a Petri dish, with 8–10 mm of hair above the wax. Thread the bits of hydra onto the hairs. Lower the water level until the tip of the hair is just above the surface, to prevent the bits from coming off. Fusion is complete within a few hours (Diehl et al., 1966).

Separation of Germ Layers: Tissue Culture Methods. The ectoderm can be pulled off the still viable endoderm of a hydra following immersion in a salt solution containing 0.1% NaCl, 0.2% $CaCl_2 \cdot 2H_2O$, 0.01% $KHCO_3$, and 0.03% $MgSO_4 \cdot 7H_2O$ (Haynes et al., 1963). In larger forms the layers can be separated by microdissection. Hydra mesogleas can be isolated by freezing to $-40°C$, thawing, and agitating to remove the dead cells (Hausman et al., 1969). Cell culture methods are available for *Hydra*,

Tubularia, and sea anemones (references cited by P. Tardent, 1965).

Anesthesia. For marine forms use equal parts of sea-water and isotonic magnesium chloride. For water of 35 °/oo salinity use 7.5% $MgCl_2 \cdot 6H_2O$ in distilled water. For coastal water of lower salinity, dilute the magnesium solution accordingly (e.g., 6.7% for salinity of 32°/oo). For hydra use 0.05% chloretone or Gray's menthol/chloral hydrate mixture (q.v.). Anesthetics should be added gradually *after* the specimen is fully extended and while the water is slowly circulating. Cooling to 5°C at the the same time frequently helps. Small specimens can be immobilized in the extended state by quick-freezing or by sucking them down rapidly onto a millipore filter in a vacuum filtration apparatus.

Perfusion and Injection. Direct injection through the mouth or body wall is feasible with larger coelenterates but difficult with hydra. Josephson's hydra-holder permits irrigation of the interior with artificial media (Fig. 1).

Microscopy of Living Specimens. Delicate forms should be examined alive where possible. Phase contrast can be used where the object is really thin and obviates the need for vital staining (see however methylene blue for nerves, below). Slightly thicker specimens such as small medusae benefit from staining with brilliant cresyl blue or neutral red. The microscope lamp should be fitted with a heat-absorbing filter.

Maceration. Osmic-acetic methods are still the best. O. and R. Hertwig soaked bits of sea anemone for 5–10 min in seawater containing 0.2% acetic acid and 0.04% OsO_4. (Small medusae require only 1 min.) K. C. Schneider used 4% acetic in the mixture. Transfer to 0.2% acetic acid (1.0% in Schneider's method) for 6–24 hr or until maceration has gone far enough. Wash in distilled water, tease if necessary, and examine in water or 50% glycerin. Tap the coverslip to loosen the cells. If

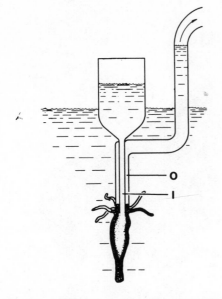

with Heidenhain's hematoxylin, dehydrate, and mount in balsam. With larger polyps and medusae, delicately strip off layers of cells with watchmaker's forceps and treat as above. Bouin-fixed tissues are sticky and should be hardened in 70% alcohol before stripping.

Sections for Optical Microscopy. Conventional methods apply. Severe shrinkage of the mesoglea accompanies wax-embedding and may distort the form, but the cells themselves suffer no more than do those of other animals so treated. Frozen sectioning and Peterfi double embedding (q.v.) are useful for delicate forms, including embryos. Sections of 0.5μ osmium-fixed, plastic-embedded material give a superior cytological picture and are increasingly used.

Electron Microscopy. A. E. Dorey's (1965) fixative is preferable to simple buffered osmium fixatives for routine work:

Solution A: 4 or 5% osmium tetroxide
Solution B: potassium chromate 2%, calcium chloride 5 millimolar, 0.1N HCl to bring the pH to 7.5

Mix A and B in equal parts before use, adding 75 mg powdered sucrose to 0.5 ml mixture and fix at 0°C for 30 min.

Aldehyde fixation followed by osmium has been used successfully by a number of workers on both freshwater and marine forms.

Conventional staining and embedding methods apply.

Special Methods for Nerves. Osmic-acetic maceration (above) and methylene blue vital staining are easiest. Unna's rongalit-reduced methylene blue (q.v.) should be used if the oxidized dye fails to work. Nerves in plexiform arrangements are easier to stain than those in bundles. Tissue pieces should be thin and well extended as the nerves are very small and have to be studied at high magnifications. Holmes' silver method works well on coelenterates fixed in a 3 : 1 mixture of saturated picric acid and 40% formaldehyde. For details and references see Mackie (1960). Another method using the same fixative is given by Titschack (1968).

Methods for Nematocysts. Small pieces of fresh tissue such as tentacle ends can be squashed under a coverslip in a little water. Many nematocysts come free and can be

Fig. 1 Josephson's hydra-holder. The hydra's lips are held between the two concentric glass tubes by suction applied through the outer tube (O). Perfusion fluid is admitted via the inner tube (I). Josephson and Macklin, 1969.)

phase contrast is not available the tissue can be stained with a water-fast dye such as carmalum or hemalum. The method requires ad hoc adjustment for each form. The cell layers separate easily, but epitheliomuscular cells are held together by desmosomes and are hard to isolate; by loosening them, it is possible to get a good view of the nerves and interstitial cells.

Whole Mounts. Small specimens or tissue strips, variously stained, can be mounted whole in aqueous or xylene-soluble media. Such preparations can be much more revealing than sections. For microanatomy of *Hydra* or *Cordylophora* first feed the animal to repletion to distend the body wall. Fix in F.W.A. diluted 1 : 3 with culture water. Cut out pieces of body wall, stain them

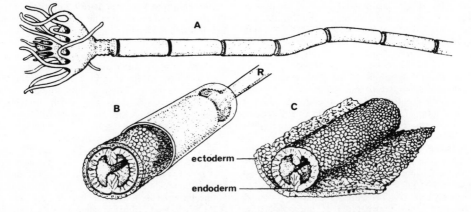

Fig. 2 Preparation of isolated ectoderm for study of interstitial cell distribution (*Tubularia*). A. The stem is cut into 3.5 mm long pieces. B. After immersion in the stain-fixative, the soft tissues are pushed out of the perisarc tube with a glass rod (R). C. The ectodermal sheet is separated from the endoderm. (Tardent, 1954.)

observed discharging. Undischarged nematocysts are obtainable in bulk from sea anemones such as *Diadumene*, *Aiptasia* by removing the acontia and placing them in 1.0M sodium citrate. The capsules extrude and can be concentrated by centrifugation (Yanagita, 1959). Smears of anemone nematocysts dried on a slide discharge rapidly on addition of distilled water, slowly if breathed upon or put into 1.0M sucrose (Robson, 1953). Nematocysts can be freed from alcohol-fixed tissue by digestion in 1% pepsin in 0.1M HCl for about 1 hr at 37°C.

For electron microscopy, explode nematocysts in a drop of water on a coated grid and dry in air. No staining is needed.

Methods for Interstitial Cells. Hydra I-cells can be selectively destroyed by exposing the animals to X-irradiation (5000–6000 r) or by placing the animal in 0.01% nitrogen mustard (Diehl et al. 1964). I-cells are basiphillic and can be seen in whole mounts and sections by staining with methylene blue, toluidine blue, etc. For *Tubularia* and *Hydra*, Tardent uses alcoholic toluidine blue mixed with water and acetic acid in the proportions 1 : 10 : 0.5 to demonstrate interstitial cells in the separated ectoderm layer. (Fig. 2).

G. O. MACKIE

References

Diehl, F. A., et al., *J. Exp. Zool.*, **155**:253 (1964).
Diehl, F. A., et al., *J. Exp. Zool.*, **163**:125 (1966).
Dorey, A. E., *Quart. J. Microscop. Sci.*, **106**:147 (1965).
Hausman, R. E., et al., *J. Exp. Zool.*, **171**:7 (1969).
Haynes, J. F., et al., *Science*, **142**:1481 (1963).
Josephson, R. K., and M. Macklin *J. Gen. Physiol.*, **53**:638 (1969).
Loomis, W. F., et al., *J. Exp. Zool.*, **132**:555 (1956).
Mackie, G. O., *Quart. J. Microscop. Sci.*, **101**:119 (1960).
Robson, E. A., *Quart. J. Microscop. Sci.*, **9**:229 (1953).
Tardent, P., *Roux' Archiv.*, **146**:593 (1954).
Tardent, P., in "Regeneration in Animals," V. Kiortsis and H. A. L. Tranpusch, ed., Amsterdam, North Holland, 1965.
Titschack, H., *Z. Zellforsch.*, **90**:347 (1968).
Whitney, D. D., *Biol. Bull.*, **13**:291 (1907).
Yanagita, T. M., *J. Exp. Biol.*, **36**:478 (1959).
See also: MESOGLOEA, NEMATOCYST.

COLLAGEN

Collagen is the major protein of the mammal and comprises more than 90% of dense connective tissue. In vertebrates it is usually recognized under the microscope as an extracellular, straight or wavy ensheathed fiber approximately 0.5 μ wide. The fiber is inelastic, white, and birefringent in contrast to the yellow *elastin* fibers. In ligaments and tendons collagen fibers are assembled into parallel bundles up to some hundred microns in diameter. Collagen fibers are prominent constituents of dense and loose connective tissue; they may occur in parallel orientation as in tendons, in variously oriented lamellae as in the corneal stroma, in an irregular layered feltwork as in the dermis, or in apparently random array as in loose connective tissue. The fibrils are stained by acid stains, e.g., by Mallory's procedure.

Mechanical teasing particularly of the acid-swollen fiber will reveal a substructure of parallel fibrils down to 50 mμ in width; these usually show a banded (64 mμ period) appearance in the electron microscope after metal shadowing or staining with heavy metals. In certain tissues where collagen fibrils are very thin or are associated with a high polysaccharide content this branding may not be detectable.

Native stretched collagen fibers show a characteristic X-ray diffraction fiber diagram with a prominent 64 mμ meridional periodicity in the low-angle pattern, and a 2.9 Å meridional spacing in the wide-angle diagram. From the analysis of the diagrams and from physicochemical studies on the soluble monomers extracted from collagen fibers the structure of the protein that comprises the collagen fibril, the tropocollagen molecule, has been deduced. Tropocollagen is a stiff rod 280 mμ by 1.5 mμ built from three helical polypeptide (α) chains that run colinearly the length of the molecule. Each α chain is wound in a three-residue per turn left-hand helix with a pitch of 8.6 Å; the three chains, associating through hydrogen bonds, then coil about each other like a rope, in a right-hand triple helix with an 86 Å period. Usually two distinguishable α chains are found occurring in the ratio 2 : 1, α1 : α2.

From the collagenous tissues of a young animal usually less than 1% of the collagen can be extracted as neutral salt-soluble tropocollagen, and this extract includes those molecules most recently synthesized. Treatment with acetic acid results in the cleavage of intermolecular aldol or aldimine linkages, with the solubilization of a few percent of "acid-soluble" tropocollagen. The native triple-helix structure of the molecule is preserved through these extraction procedures. These molecules have a molecular weight close to 280,000.

Tropocollagen molecules from mammalian skin or tendon normally contain 0.5% carbohydrate bound through O-glycosidic linkage to hydroxylysine residues; the remainder of the molecule is protein and comprises 33 moles percent glycine and 15–22% pyrrolidines (proline or hydroxyproline). The content of hyroxyproline and hydroxylysine, residues that are hydroxylated after polypeptide synthesis, distinguishes collagen from most other proteins. Tropocollagen from some other tissues (e.g., basement membrane) contains a much higher percentage of carbohydrate.

Breakage of hydrogen bonds by heat or reagents such as urea results in the separation of the polypeptide chains constituting the tropocollagen molecule in a cooperative denaturation process that results in a large change in viscosity and optical rotation. In the course of fiber maturation cross-links are formed between and within the tropocollagen molecules so that on denaturation dimeric (β) chains, trimeric (γ) chains, and higher aggregates are detectable as well as α chains. Intramolecular γ chains spontaneously renature under appropriate conditions, but although α chains refold on cooling, their reassociation to a triple helix can be effected only under special annealing conditions. The biosynthesis of cross-links is blocked by administration of osteolathyrogens such as semicarbazide or β-aminopropionitrile to an animal. Denaturation of the *intact* collagen fiber also may be effected by heat and it results in a contraction of the fibril. The association with other molecules appears to stabilize the triple-helix structure so that the shrinkage temperature, T_s, of a tendon, for example, may be 10–30° higher than the denaturation temperature of the tropocollagen molecules in solution. The latter denaturation temperature varies with the pyrrolidone content of the collagen and the solvent; many fish tropocollagens denature at about 25°C, while mammalian preparations melt out 10–15° higher.

The native collagen fibers may be dissolved by collagenases, or various proteases, most effectively by the action of pepsin in acetic acid. Pepsin solubilization results from the cleavage of "telopeptides", short terminal regions of the α chains on which are sited modified amino acid residues through which intermolecular bonds are formed. The ordered structure of the main body of the molecule is stable (below the denaturation temperature) to all proteases except for a few collagenases; some of these cleave the molecule transversely at one or two special sites and others cleave the molecule in many places releasing small peptide fragments.

The tertiary structure of the native molecule is such that under appropriate solvent conditions the tropocollagen, extracted by neutral or acidic solvents, or solubilized from mature fibers by pepsin, can be caused to reaggregate into fibrils morphologically identical to the native fibrils. In the native fibril the individual asymmetric molecules lie in a polarized, parallel manner, each staggered at integral multiples of 64 mμ from its neighbors (see Fig. 1). The earlier belief that this longitudinal displacement was one quarter of the molecule length gave rise to the "quarter-stagger" hypothesis that Schmitt and his co-workers first proposed to explain the 64 mμ fiber periodicity that is smaller than the monomer length. Later it was shown that the molecule length is not an integral multiple of the fiber period. Hence, there is a gap between the end of one tropocollagen molecule and the start of the next. It is not known if the native fiber in vivo is built up by the staggered accretion of molecules, or if protofilaments are built up from the overlapping lengthwise polymerization of individual molecules and these protofilaments then associate in a randomly staggered array.

By manipulation of the solvent conditions tropocollagen may be precipitated in novel forms; in the presence of certain acidic polymers the precipitates are revealed under the electron microscope as long symmetrical "FLS" filaments showing 220–280 mμ periodicity, where successive molecules with either polarity overlap by consistent lengths that vary with the experimental conditions; from dilute acetic acid solutions a crystallite of laterally associated polarized molecules (the Segment Long Spacing, SLS form) may be precipitated by 0.2% ATP. By positive staining of the segments with phosphotungstate or uranyl ions there may be visualized by electron microscopy a characteristic series of transverse bands that reflect the distribution of regions rich in arginine and lysine or dicarboxylic acids over the length of the tropocollagen molecule.

The banding patterns shown by SLS segments from collagens of different species show a remarkable consistency from man, mammals, fish, to coelenterates. These observations imply a consistent distribution of charged groups and therefore a remarkable evolutionary stability of the collagen molecule; this stability is borne out by recent investigations of peptide sequences within the α chains of collagen from various species.

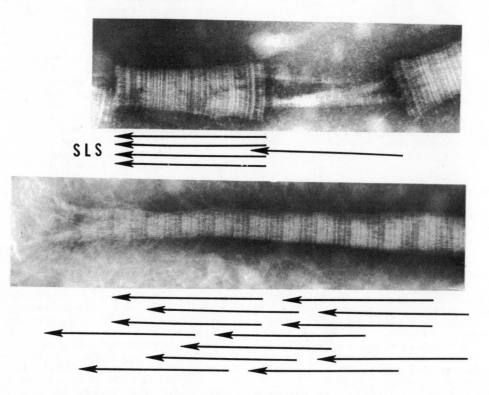

Fig. 1 Electron micrograph of negatively stained rat tendon tropocollagen precipitated in the SLS form (top) and the native fiber form in which the 64 mμ periodicity is evident (bottom). The arrows represent the tropocollagen molecules (280 mμ long) with the arrowhead at the amino termini of the peptide chains. The diagrams show how these aggregate structures arise by the ordered association of the molecules.

SLS segments are formed less readily from collagen molecules carrying a large carbohydrate content, but frequently the polysaccharide may be excised by protease treatment.

The short telopeptides removed by certain proteases from native tropocollagen may affect the solubility characteristics and change the aggregative properties: Under certain conditions nonpolarized SLS and 64 mμ period fibers may be precipitated. The amino-terminal telopeptides themselves are the vestiges of longer peptides cleaved extracellularly from the nascent "procollagen". In rare veterinary diseases this peptide scission does not occur and abnormal collagen fibers form.

Two classes of antibodies to collagen have been prepared in rabbits: One class is species-specific and reacts with the terminal enzyme-susceptible telopeptides, the other reacts with inner regions of the molecules and cross-reactions with collagens of different species are observed.

PETER F. DAVISON

References

Gallop, P. M., Blumenfeld, O. O., and Seifter, S., "Structure and Metabolism of Connective Tissue Proteins," *Annual Review of Biochem.*, **41**, 1972.

Kefalides, N. A., "Chemical Properties of Basement Membranes," *Intern. Rev. Exp. Path.*, **10**, 1, 1970.

Ramachandran, G. N., and B. Gould, "A Treatise on Collagen," New York, Academic Press, 1967.

Veis, A., "The Macromolecular Chemistry of Gelatin," New York, Academic Press, 1964.

COTTON

Structure of Natural Cotton Fiber. A cotton fiber is formed by elongation of a single epidermal cell of a cotton seed. During the elongation and the following period of cellulose deposition, the cells are almost cylindrical, but when the cotton boll opens, the cylinders collapse, forming twisted, ribbon-like tubes. Each cotton fiber is composed of 25–40 concentric layers. The outermost layer, or primary wall, consists of pectins and cellulose and is covered with a wax cuticle. The remaining layers together constitute the secondary wall. The first layer of the secondary wall (S1 layer), immediately adjacent to the primary wall, has a somewhat coarser structure than the inner layers.

During the growth of the cotton fiber, cellulose is deposited in the form of fibrils about 0.1 μm thick, which run in spirals in the growth layers. In a 1 cm length of fiber, the spirals reverse 20–25 times from clockwise to counterclockwise. This structural feature is responsible for the twisted appearance of cotton fiber, which is easily seen in the ordinary, or better, phase contrast microscope (Fig. 1). and by which it can readily be distinguished

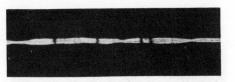

Fig. 2   Cotton fiber between crossed polarization.

from other technically important fibers. The reversals of spiral occur almost simultaneously in all growth layers. In the polarizing microscope (with crossed polars) the bands in which spiral reversal occurs appear dark. With a red first-order plate inserted into the microscope, the regions in which the spirals are counterclockwise appear blue, whereas those in which they are clockwise appear yellow. For all orientations of the polars the fibers remain bright (Fig. 2).

For the investigation of the surface structure of cotton fibers, the scanning electron microscope is still better suited. It clearly reveals the general shape of the fiber, important details such as surface folds and cracks, and structural irregularities resulting from damage (Fig. 3). For the investigation of the refractive index variations along the fiber axis the dispersion staining technique of McCrone can be used.

Structure of Mercerized Cotton Fiber. Mercerization is a process in which cotton fiber is treated with a fairly concentrated (18–25%) sodium hydroxide solution. This process is attended with several characteristic microstructural changes. Mature fibers lose their twisted

Fig. 3a   Cotton fiber, scanning electron micrograph. (×1150.)

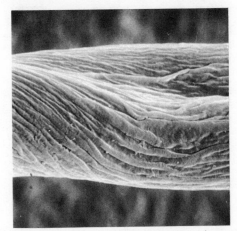

Fig. 3b   Cotton fiber, scanning electron micrograph. (×2900.)

Fig. 1   Cotton fiber, normal light microscopy.

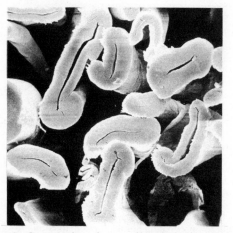

Fig. 4 Cross section of cotton fibers, scanning electron micrograph. (×1150.)

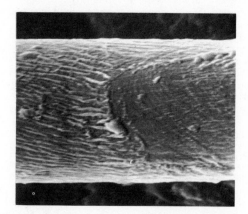

Fig. 6 Mercerized cotton, scanning electron micrograph. (×2900.)

appearance, whereas immature fibers, by contrast, become more twisted than before mercerization. The lumina of the fibers almost disappear, their cross-sectional appearance becoming more uniform (Figs. 4 and 5). The fiber surface loses much of its structural detail such as spiral and transverse folds (Fig. 6).

Microchemical Investigation of Cotton Fiber. The structure of the fiber wall is to some extent revealed by allowing fibers to swell in cuprammonium solution. (To prepare this solution, copper hydroxide is precipitated from 5% copper sulfate solution by adding sufficient 10% sodium hydroxide, the copper hydroxide is washed with distilled water and redissolved in 25% aqueous ammonia to give a solution containing 2 g of CuO per 100 ml.) In this solution cotton fibers swell very strongly, and the swelling is balloon-like, the constrictions being due to remnants of the primary wall. The spiral structure, in particular that of the S1-layer, can be seen clearly. (Given sufficient time, the fiber dissolves completely.) In a solution of cadmium ethylene diamine a similar but less pronounced and rapid swelling is observed. Mercerized

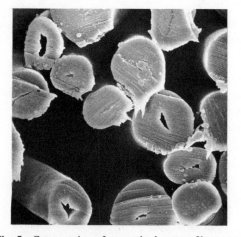

Fig. 5 Cross section of mercerized cotton fibers, scanning electron micrograph. (×1150.)

fibers swell more or less uniformly in both reagents, and they show no "ballooning."

Histochemical Detection of Cellulose. When cotton fibers are treated with a zinc-chloro-iodine solution, the cellulose is stained violet to purple. Mercerized fibers are stained more intensely.

Histochemical Detection of Wax Cuticle. The wax of the cuticle can be made visible in cross-sectional or longitudinal views by staining it with the dye Sudan III.

Tests for Damage to Fibers. Extrusion test. When short fiber parts (ca. 25 μm) are immersed in 18% sodium hydroxide, the primary wall remains intact, but water and hydroxide ion diffuse into the interior of the fiber, giving rise to an internal pressure by which some of the cellulose material is extruded at the ends of the fiber, and in places where the primary wall is slightly damaged. When the primary wall is severely damaged or entirely absent, it cannot resist the internal pressure, and no extrusion is observed.

Congo Red Test. Although Congo red has a great affinity for cellulose, it cannot penetrate the primary wall, except in places where the latter is damaged. The fibers to be examined are first soaked in 9% sodium hydroxide containing 0.5% of a wetting agent. After washing they are immersed in a concentrated Congo red solution, rinsed, and suspensed in 18% sodium hydroxide. By varying the concentration of the sodium hydroxide in which the fiber is soaked before staining, it is possible to differentiate between mechanical, heat, bleach and other chemical damage. For instance, soaking in 9% sodium hydroxide reveals mechanical damage, whereas chemically damaged parts are stained only after soaking in 11% sodium hydroxide.

Electron Microscopy of Cotton. *Preparation of Ultrathin Sections (Rollins et al, 1966).* A bundle of parallel cotton fibers is tied at each end with a nylon thread, laid on a glass microscope slide, and coated with a quick-drying nitrocellulose lacquer. When the lacquer is dry, the bundle is embedded in prepolymerized methacrylate resin, a coverslip is put on, and the slide is heated at 65° for 1 hr. From the embedded bundle ultrathin sections can now be cut. The methacrylate embedding resin consists of a 4:1 mixture of methyl and butyl methacrylate, containing 1%

of a catalyst 2,4-dichlorobezoyl peroxide in dibutylphtalate.

Expansion Technique (Rollins et al., 1966). Bundles of cotton fiber one-half the size normally used in sectioning are boiled in 50% (v/v) aqueous methanol containing 0.5% Deceresol wetting agent. After cooling and standing overnight the bundles are removed from the solution and excess liquid is shaken off. The moist bundles are then put on a microscope slide and immediately embedded in sufficiently (but not overly) viscous prepolymerized methacrylate. Next, a coverslip is put on, some more methacrylate is run around the edges of the coverslip, and the slide is heated at 70°C for at least 2 hr. The embedded fibers are sectioned and freed from polymethacrylate. After shadowing the fiber sections in the usual way they can be examined in the electron microscope. By this technique an exploded view of the growth layers is obtained.

Ion Etching (Spit). Ultrathin cross sections of fibers are put on electron microscope grids without carrier film. The grids are placed on the anode of a gas discharge chamber, the sections facing upwards. With a distance between the electrodes of 25 mm, a discharge current of 3 mA is applied for 3 min in oxygen. By this method it is possible to differentiate between untreated and cross-linked cotton fibers.

Preparation of Fibers for Scanning Electron Microscopy. Two pieces of double-sided adhesive tape are attached to both sides of an object stub coated with an electrically conductive carbon glue. The fibers to be examined are impregnated with a very dilute solution of the antistatic Duron in chloroform, rinsed with chloroform, and stretched onto the stub by means of the adhesive tape. Threads of a highly viscous polyvinyl alcohol solution are drawn across the fibers at spacings of about 2 mm. Some PVA solution is also applied around the edge of the stub. When the PVA solution has dried, the fibers are cut loose from the adhesive tape, the remnants of which are removed. Finally, the stub is coated with a gold layer of about 400 Å and the fibers are examined in the SEM at an acceleration voltage of 10 kV.

Auxiliary Techniques. *Dispersion Staining.* Cotton fibers are embedded in a cargille medium, the refractive index of which is chosen such that its wavelength dispersion curve and that of the object intersect at $\lambda_0 = 580$ nm. When viewed with a central stop in the microscope objective lens, the fiber then has a magenta color. For untreated cotton the refractive index of the medium has to be $nD = 1.59$, and for mercerized cotton 1.54.

Fluorescence Dichroism. Cotton fibers are stained with a 1% Euchrisine 2GNX solution at 60° for 1 hr. After washing with distilled water until the washings remain colorless, the fibers are examined in a fluorescence microscope equipped with a rotating analyzer, by means of which the direction of extinction can be determined. Since the dye penetrates only the outer layer to a depth of about 1 μm, the method can be used to differentiate between the outer layer and the inner secondary layers.

J. Isings

References

Clegg, G. J., *J. Text. Inst.,* **31**:63 (1946).
Koch, P. A., "Rezeptbuch für Faserstofflaboratorien," Berlin, Springer-Verlag, 1960.
Isings, J., *Text. Res. J.,* **34**:236 (1964).
Rollins, M. L., J. H. Carra, E. J. Gonzales, and R. J. Berni, *Text. Res. J.,* **36**:185 (1966).
Spit, B. J., *Proc. Eur. Reg. Conf. on Electr. Micr.,* **1**:564 (Delft, 1960).
Stoves, J. L., "Fibre Microscopy," London, National Trade Press, 1957.

CYTOCHEMICAL ANALYSIS BY X-RAY SPECTROGRAPHY

Principles of the Method and its Application

Microanalysis by electronic waves, introduced by R. Castaing in 1945, is a method permitting the determination of the elementary chemical composition of very small volumes of matter (less than a cubic micron) present in any sample.

Such an analysis can be carried out on classical histological sections or on ultrafine sections. In the first case, the observation of the sample is made with a light microscope. In the second case a normal transmission electron microscope is used.

All the elements of atomic number equal to or greater than 4 (beryllium) can be discovered and ultimately quantified; only the three first elements, hydrogen, helium, and lithium, cannot be analyzed. Figure 1 shows an apparatus in which an electronic wave microanalyzer is associated with an electron microscope.

General Principles of Analysis. It is known that when a beam of electrons accelerated by a sufficiently high difference in potential impinges on a specimen, x-rays are emitted. The wavelength of the emitted x-rays is not random but is arranged in accordance with a spectrum (Fig. 2) in which two parts can be distinguished: a continuous spectrum for which the interactions between incident electrons and bombarded nuclei are responsible, and a line spectrum, called characteristic, related to the interactions between incident electrons and peripheral electrons of bombarded atoms.

Only the line spectrum is of interest; it is known in fact, since Moseley, that the wavelength of each line is characteristic of the emitter element, that is, of the atom bombarded. This is readily explicable by the theory of Bohr. In this theory each atom consists of an extremely small nucleus around which spin a certain number of electrons in determinant orbits designated by the letters K, L, M (Fig. 3); on each of these orbits the electrons possess a determined energy which increases from the deep layers towards the superficial layers.

Let us suppose that in the course of an interaction between the incident electron and an electron of the layer K of an atom in the specimen, this electron should be ejected; the atom thus rendered unstable returns to stability after an extremely brief interval. The sestitation of the gap left in the planetary structure results from the passage of a more peripheral electron, of layer L for example, to layer K (Fig. 4); in the course of this transition the energy of the electron passes from the value W_L to a lower value W_K. Given the principle of the conservation of energy, the energy $W_L - W_K$ thus lost must reoccur in some form. There is a strong probability that it reappears in the form of an electromagnetic radiation, that is, in the form of a proton of energy $h\gamma$ (h = Planck's constant, γ = the frequency of the radiation), such that $h\gamma = W_L - W_K$. The value $W_L - W_K$ being of a magnitude characteristic of any given atom, the frequency γ of the radiation or again the wavelength $\gamma = c/\gamma$ (c = the

Fig. 1 Photograph of an electron microprobe associated with an electron microscope.

speed of light) will also be characteristic. The wavelength emitted in the course of electronic transitions in the atoms are such that they generally lie in the region of x-rays in the scale of electromagnetic radiation. This wavelength can be determined by x-ray spectrography, and this in turn allows the emitter element to be determined. One can deduce from this what is the elementary chemical even in the smallest volume of matter present under the electronic beam.

The extreme simplicity of the characteristic X spectrum, which in general consists of only a few lines for each element, in general eliminates all risk of ambiguity in the results obtained from this type of spectral analysis; the determination of a single wavelength usually allows the emitter element to be determined. If the impact of the

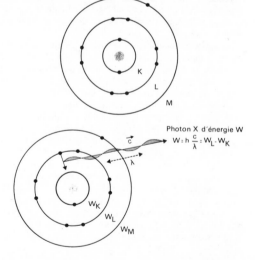

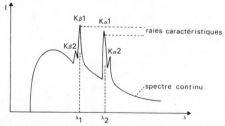

Fig. 2 Spectral disposition of X-rays emitted by an anticathode. The wavelengths λ_1 and λ_2 are characteristic of atoms present in the anticathode.

Figs. 3 (top) **and 4** (bottom) Fig. 3: Distribution, according to Bohr, of the sodium atom with its 11 peripheral electrons. Fig. 4: Emission of a characteristic spectra ray; the photon is emitted following the passage of an electron from layer L to layer K. In the course of this transition the electron loses the energy $W_L - W_K$. This electron energy is liberated in the form of an electromagnetic energy; that is, in the form of an X photon.

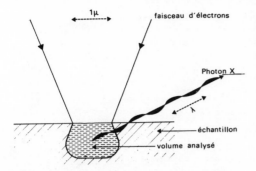

Fig. 5 Penetration of an electron beam in the anti-cathode. The volume analyzed here is of the order of magnitude of a cubic micron [*faisceau*: beam, *échantillon*: specimen].

electron beam on the preparation is very small, (1 μ^2), a microscopical volume of matter (1 μ^3) will be thus subjected to analysis (Fig. 5). the penetration of electrons of 30 kV energy in the usual metals being of the order of one micron.

The original apparatus constructed by Castaing for the study of metallic specimens consisted of an electronic microprobe and an x-ray spectrograph. In order that this apparatus could be adapted to the study of ultrathin biological sections it is apparent that it is necessary to add to the original apparatus a third element permitting the observation of the specimen on the level of the electronic microscope. In the special case of cellular studies, it is actually necessary to be able to establish a straight-line correlation between the ultrastructural image and the results of the analysis.

Contemporary apparatus consist of three principal elements:

1. An electronic microprobe the role of which is to produce an extremely fine beam of electrons of which the surface of impact on the specimen must also be as small as possible (less than a square micron).
2. An x-ray spectrograph designed to measure the wavelength of the characteristic spectral lines.
3. A supplementary electronic microscope that produces an image of the preparation as does a transmission electron microscope.

Description of the Apparatus. *The Electronic Microprobe.* A diagram of the microprobe is given in Fig. 6. It is seen that it consists of an incandescent electron-emitting filament, a Wehnelt cylinder, and an anode pierced by a hole. The acceleration potential between the filament and the anode can vary from 5 kV to 40 kV. These together form the electron gun. Two magnetic lenses act on the beam in such a manner that the surface of impact on the specimen is as small as possible.

The X-ray Spectrograph. It is known that a beam of x-rays falling on a crystal is capable of being "reflected" selectively if the angle of incidence agrees with Bragg's equation:

$$2d \sin \theta = n\lambda$$

d being the distance beteen two atoms of the crystal, λ the wavelength of the X photon, and n an integer (Fig. 7).

In the absence of these conditions there is no reflection. It is then necessary, by placing an x-ray detector (proportional gas flux counter) in the direction θ, to measure the angle corresponding to the selective reflection of Bragg. Generally, curved crystals are used to permit a better focus. The specific angle of reflection being thus measured, it is possible to deduce the wavelength of the incident photon by utilizing Bragg's equation.

The Set-up for Electron Microscope Observation. To permit observation, at the ultrastructural level, of an ultrafine histological section intended for analysis, supplementary electron optics are added to the preceding equipment.

It must first, however, be pointed out that the filaments of an electronic microprobe (electron gun, electron-magnetic lens) can be used as an electron microscope condenser; to do this it is only necessary to place the second lens slightly out of focus. Completing this assembly to produce a classical electronic microscope, it is necessary to add two supplementary lenses playing respectively the roles of the objective lens and the projector, and the fluorescent screen on which to project the image (Fig. 6). This apparatus can, if desired, be utilized as an electron microscope or as a microanalyzer. In the first case, the two upper magnetic lenses function as a double condenser; the passage of the electron beam thus permits the image to be obtained on the screen due to the two lenses (objective and projector) situated in the lower part of the apparatus. When an inclusion to be studied appears in the preparation, the electronic trajectory is modified by changing the current that passes in the second condenser lens. The impact surface of the beam on the preparation progressively retreats and becomes concentrated on the inclusion to be analyzed. The apparatus can then function as a microanalyzer; the second lens of the electron microscope condenser is now playing the role of the objective lens of the microprobe and the two lower lenses are functionless. Under these conditions the analysis is carried out in plain view on ultrafine sections.

Apparatus for Scanning Analysis. It is possible to attain images showing the distribution of atoms in a preparation. For this an x-ray spectrograph is tuned to the spectral ray of the element sought. A special apparatus allows the scanning of a given surface of a preparation with the microprobe. Under these conditions the intensity of the signal received by the x-ray spectrograph varies in proportion as the concentration at each point of the element sought. This signal is transmitted to an oscillograph screen on which the distribution of the element on the surface under examination is thus visible (Fig. 8).

Preparation of the Specimen. The specimen intended to be analyzed must be prepared in such a way that the fine structure is not destroyed in the course of the analysis. Thus it must be capable of resisting a vacuum of 10^{-5} mm Hg within the apparatus as well as the bombardment by the electron beam which may produce excessive heating. The specimen must, in addition, be a conductor, otherwise it would be capable of developing a static charge under the impact of the beam which would produce instability of the latter. Finally, it must be arranged on a support containing no traces of impurities.

The practical problems vary according to whether the analysis is made on thick sections (thickness greater than 1 μ) or ultrathin sections.

Working with Standard Histological Sections. Any histological section can be studied no matter what has been the preparation (fixation, embedding) to which it has been subjected.

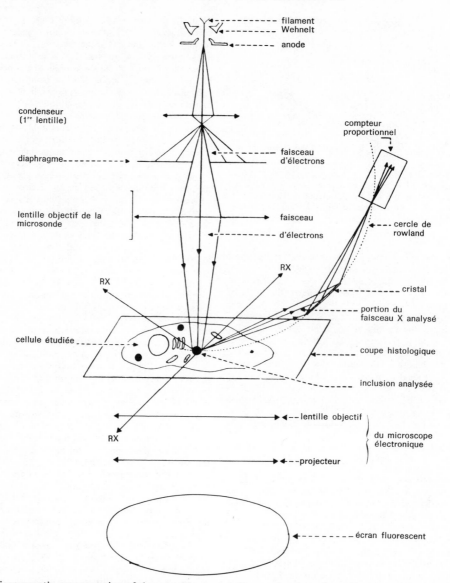

Fig. 6 Diagrammatic representation of the apparatus associating an electron microprobe, an electron microscope, and an X-ray spectrograph (crystal + proportional counter). The apparatus functions here as a microanalyzer. It is evident that the beam of electrons is focused on an intracellular vacuole [*lentille*: lens, *microsonde*: microprobe, *étudieé*: studied, *fasiceau*: beam, *écran*: screen].

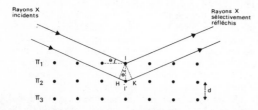

Fig. 7 Selective reflection of X-rays by a crystal. The difference in the direction of the two luminous rays is $HI + IK \simeq 2d \sin \theta$. When $2d \sin = n\lambda$ (condition of Bragg), the two are in phase and one observes a selective reflection. When λ does not satisfy Bragg's relation, there is no reflection.

The sections cannot be placed on glass slides because the glass, containing numerous mineral elements, is likely to produce extremely annoying parasitic spectra. It is best to use slides of heat-resistant transparent plastic; Mylar or Araldite are, in this regard, preferable to methacrylate. In order to render the specimen conductive, it is necessary to coat the slide with a very fine layer (100 Å) of a metal which is a good electrical conductor (gold, aluminum), or with carbon. This coat is produced by evaporation in high vacuum; when the section is laid on this support a second fine coat of metal or of carbon is applied on top.

Under these conditions, the section lies between two coats that at the same time conduct both heat and

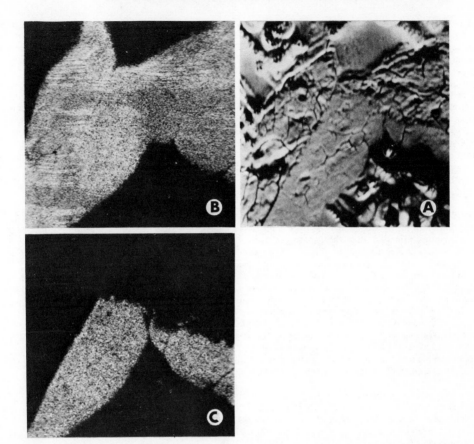

Fig. 8 Polished surface of a human renal calculus; scanning of surface of about 200 μ^2. A: morphological image obtained with the secondary electrons; B: image representing the distribution of calcium; C: image representing the distribution of phosphorus.

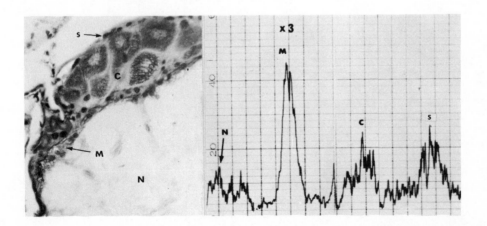

Fig. 9 Distribution of a diffusible ion, potassium, in four layers of an insect nerve ganglion. Fixed by freeze-drying. S: neutral sheath; C: cortex; M: acid mucopolysaccharides; N: neurophile. On the recorded graph the concentration of potassium in the acid mucopolysaccharide layer must be multiplied by 3 in consequence of the different adjustment in the sensitivity of the apparatus.

electricity, and the examination can be carried out without difficulty.

Working with Ultrafine Sections. The problems are more simple, since the temperature rise of the specimen under the impact of the electron beam is less. It is possible to analyze sections prepared by classical electron microscope techniques, no matter what embedding material is used. It may be desirable to coat the sections thinly with metal or carbon but this precaution is not indispensable since the contamination of the section under the electron beam is usually sufficient in itself to make the preparation resistant and to stabilize the beam.

Examples of Application

With Thick Sections. At the structural level, the demonstration of atoms by x-ray spectrography is possible in all structures provided that these atoms occur as organic compounds (for sulfur in proteins or acid mucopoly-saccharides; for phosphorous in phospholipides) or in mineral structures (for sulfur and phosphorus as sulfates and phosphates or cations in mineral concretions), provided only that there is enough of them. The decision as to whether the atoms in question are mineral or organic in nature, fortunately possible by comparisons with normal preparations and preparations subjected to demineralizing agents, is very frequently indispensable.

Spectrographic analysis can be used as an independent method or as a method complementary to cytochemical techniques.

The demonstration of *diffusable ions* by a cytochemical method is extremely difficult and limited in scope. The difficulty consists not in the fixation of the ions *in situ*, possible and even satisfactory by freeze-drying, but in their detection. Microprobe analysis is a palliative for this difficulty. Sections made of fixed materials must be thick enough to allow a suffecent concentration of the required ions and the examination is preferably made in an anhydrous environment. The demonstration of diffusable ions is thus possible in proportion as the localization is sought at the level of structures of which the surface

on the sections is greater than the beam of the microprobe ($1-2 \mu$) (Fig. 9).

Mineral concretions are often, at least in invertebrates, of complex composition. Very few cytochemical methods can be selectively used for ion determination. In addition, it is difficult to identify the various possible salts by their solubility since this can be modified either by the presence of an organic matrix or by the existence of another very soluble salt, the solution of which involves the destruction of the crystal. Only the electron probe thus permits the development of a complete inventory of cations. Sulfur and phosphorus anions are detectable by this method but carbon is more difficult. It must be remarked that paraffin

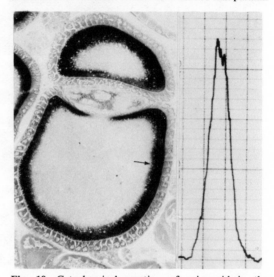

Fig. 10 Cytochemical reaction of uric acid in the oocytes of an insect (arrow) (Cantacuzene and Martoja, 1967). The microprobe analysis shows here a significant quantity of potassium; this ion disappears when the sections are treated by a solvent of urates. The microprobe thus permits the indication of the *molecular* nature of the compounds studied.

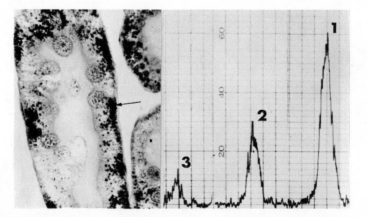

Fig. 11 Effects of a demineralizing agent and of a solvent or urates on the concentration of potassium in crystals of the excretory apparatus of insects (arrow). 1: concentration of crystals not having received any treatment; 2: concentration of crystal prior to the action of nitric acid; 3: concentration of the crystals after the action of piperazine. These results, supplementary to those obtained by cytochemistry, allow the conclusion of the presence of potassium linked in part to uric acid and in part to other ions (Ballan and Martoja, 1971).

sections are well adapted to the action of demineralization agents or elective solvents of organic composition (urates, guanates); it is possible in these cases to go beyond the mere inventory of ions and to demonstrate the probable molecular structure of the crystal (Figs. 10, 11).

The presence, in crystals, of a *water-soluble* salt of neutral or acid pH can cause the total disappearance of the crystalline structure under the effect either of the water hydration of the sections, or of chemical reactions. Microprobe analysis, insofar as it does not necessitate hydration, allows the conservation of crystals and their analysis.

Pigments (melanines, ommochromes) accumulate in numerous invertebrate organs. These pigments can exist in some species and be lacking in the homologous organs of other species; however, these organs contain mineral concretions. Even allowing for the existence of mineral salts in pigmented structures, the application of cytochemical techniques is practically impossible since the natural color masks the color produced by the reaction and few techniques permit the preliminary decoloration of the pigment. The electron microprobe is the only possible apparatus for both analysis of pigment grains and the demonstration that some among them, rich in mineral salts, can be considered as pigmented crystals (Fig. 12).

Some cations of primary biological interest (Ca, Fe, Mg, Cu) can be bound to complex organic molecules. Since they are impossible to demonstrate by simple cytochemical methods, these are said to be "masked." Chemical unmasking has been proposed only for iron and the success of the technique is a gamble. The identification of masked iron and calcium has been, up to the present, by microincineration. But if the iron is easy to identify on a spodogram the identification of calcium is less easy. In this case, as in the case of other masked metals, spectrographic analysis appears to be the ideal solution.

The microprobe permits the identification of biologically important elements that are *difficult or impossible to demonstrate by cytochemistry*. Thus, the production of crystals from diffusable potassium rarely gives positive results. Moreover this element plays a major role in the constitution of insect concretions, and the electron probe represents in this case the only method of identification.

The demonstration of metals such as bismuth, gold, mercury, copper, silver, aluminum, silicon, and lead, relies on methods which are for the most part suspect. The importance in pathology of the accumulation of cations normally present at oligoelement state forbids error,

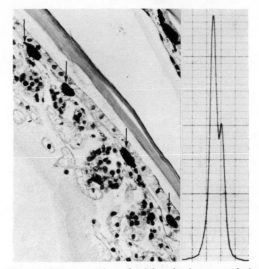

Fig. 12   Demonstration of calcium in the accumulation of pigment grains in an interstitial tissue of a crustacean (arrows).

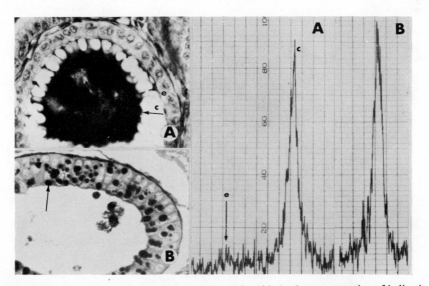

Fig. 13   Demonstration of iodine in two types of a vertebrate thyroid. A: the concentration of iodine in the colloid (c) and of the epithelium (e) of a Gnathostome thyroid. B: the concentration of iodine in the intra-epithelia granules (arrow) of an Agnathe thyroid (preparations M. Gabe).

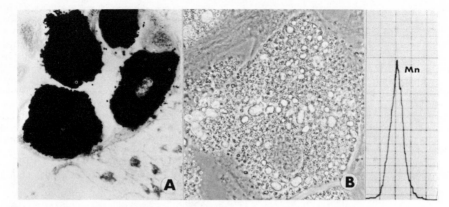

Fig. 14   Demonstration of manganese in the intestinal cells of an insect. A: cytochemical reaction with phosphates (after Gouranton, 1968); B: intestinal cell on phase contrast (note the numerous crystals).

and the utilization of the microprobe is indispensable.

In other cases, no method at all is available for cytochemical detection. Research on iodine is of undoubted importance in thyroid function, and what normally requires autoradiography or absorption spectrography becomes simple and rapid with the microprobe (Fig. 13). The same is true of manganese, which is dealt with in histochemical texts only in relation to the interference it gives with other cations (Cu, Zn, Al, Ag), but which nevertheless is found in important quantities in certain insect organs together with accumulations of other elements (Fig. 14).

In other cases, *cytochemical methods lack specificity* and the use of complementary techniques such as autoradiography or x-ray spectrographic analysis must always be considered. Such a problem arises, for example, for *calcium*, though widely distributed in animal tissues. In cytochemistry it can be detected either by lake methods or by substitution methods. The first show the cation but give positive results with other metals such as magnesium; the second also show the anion of the salt. Their application to calcified organs of vertebrates is valid in most cases where the mineral composition has been previously determined by a chemical method, but their utilization for purposes of analysis on a yet unknown biological material can give rise to errors. Numerous results obtained on organs of invertebrates must be accepted with even more reserve since in many cases magnesium and potassium, and even iron or manganese, form a large part of these mineral concretions. These remarks apply also to certain techniques transferred to the electron microscope that result from the classic substitution methods. The alleged localization of calcium by these methods rests on a postulate under which calcium would be the only cation associated with the anions PO_4 or CO_3.

The positive cytochemical result obtained at the level of *structures deprived of crystals or of concretions* always runs the risk of being considered an artifact. Spectrographic analysis alone can confirm the result. Such a case is found in the chorion of the eggs of *Orthoptera*.

Either used by itself, or associated with cytological techniques, x-ray spectrograph analysis can go beyond a simple inventory of ions. Actually, the *concentration of the same atom* in different structures can be compared by means of a microprobe if the biological specimens have been subjected to the same technical conditions. This possibility of comparison is interesting since cytochemical reactions give intensities of staining that are independent

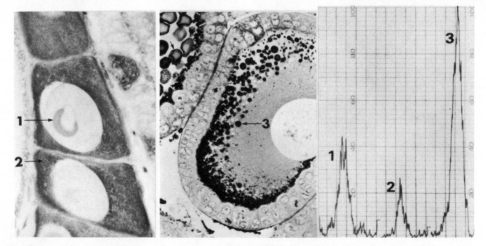

Fig. 15   Comparative concentrations of protein and sulfur of three structures of the female genital apparatus of an insect. 1: nucleolus; 2: cytoplasm of the oocyte; 3: yolk platelet.

of the quantity of the substance disclosed (detection of mineral substances) or when the reactions lead to a coloration proportional to the quantity of the substance sought, but are with difficulty distinguished by eye (proteins); microprobe analysis prevents, in the latter case, the necessity of using histophotometric techniques (Figs. 15, 16).

Finally, as a spinoff of the technique, the analysis of ions permits the consideration, on a solid base, of the problem of the *specificity of certain cytochemical methods.* Thus, the Titan Yellow method which is supposed to be specific for the demonstration of magnesium gives an intense stain with protein structures, whereas spectrography never discloses magnesium ions. Methods for the detection of anions by the substitution of cobalt or silver give different results for the same structure; knowledge of

ionic composition of the structure in question permits the determination of which of the cytochemical methods is in error. Ionic analysis similarly permits the explanation of certain anomalies; thus, decalcification allows to remain, in certain cases, a major reactivity with substitution methods which must, in reality, be attributed to mineral phosphorus.

On Ultrasections. The electron microprobe allows the analysis of extremely small inclusions (less than 1000 Å).

For example, after the injection of a mammal with water-soluble salts of gold it is possible to see with the aid of the electron microscope mitochondrial anomalies in proximal tubular cells of the kidney; these mitochondria contain extremely small dense inclusions. When the impact point of the microprobe is directed against one of these inclusions, it emits x-rays of which the spectral analysis demonstrates the characteristic line of gold.

The role of mitochondria in the mechanism for the intracellular concentration and elimination of gold by the kidney has indeed been studied by this method (Fig. 17).

The electron probe therefore appears very important in the analysis of mineral or organic compounds. This method, which demonstrates atoms, represents a complement to the cytochemical techniques which yet remain the only ones utilizable in the detection of certain compounds or of certain ions. The electron probe, in association with various methods of extraction, permits the elucidation of the chemical nature of the compounds studied, as well as their chemical composition. It is a palliative for the inadequacies of classical chemical analysis since this last is rendered difficult by the small size of animals or by the difficulties in the removal of indistinct organs. Associated with the simplest techniques of optical or electron cytology, x-ray spectrograph analysis thus permits the closing of important gaps in our knowledge. It is destined to become a fundamental method for the histologist, physiologist, and pathologist.

PIERRE GALLE
ROGER MARTOJA
(*trans. from French*)

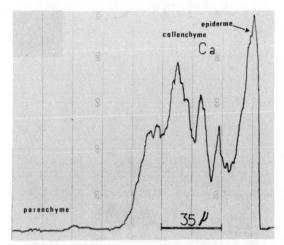

Fig. 16 Comparative concentrations of the calcium in three tissues of a plant organ (Roland and Bessoles, 1968).

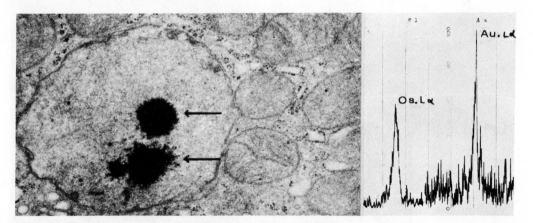

Fig. 17 Giant mitochondria in the cytoplasm of a proximal tubular cell of the kidney of a rat which had received injections of gold salts. The intramitochondrial inclusions emit, in addition to the characteristic lines of the osmium used for fixation, the characteristic line of gold (Galle, 1970).

d

DECALCIFICATION

Decalcification or demineralization in histological terminology refers to the solution of the mineral component of bone and dental tissues leaving the organic constituents intact. The process must be employed when thin sections are required to study the microscopic features of the hard and soft tissues simultaneously. The practical application of decalcification is influenced by the following factors:

1. The thickness of the specimen selected for treatment.
2. Choice and method of using the decalcifying agent.
3. Determination of the completion of the process.

Although speed is important it is also necessary to ensure that anatomical characteristics are accurately preserved and that the quality of the staining is fully maintained.

Thickness of the Specimen. Calcified tissues should be demineralized as quickly as possible, because prolonged treatment, especially with some reagents, may produce artefacts. Rapid solution of the mineralized constituents depends largely on the "minimum diffusing distance" (Brain, 1966), i.e., the last part of a specimen to be decalcified to the nearest point on the outer surface (Fig. 1). When the distance is reduced uniformly to less than 1.5 mm the maximum number of hydrogen ions operate at the same time, so demineralization occurs rapidly (Fig. 2a). When the minimum diffusing distance exceeds 1.5 mm the center of the specimen is gradually isolated (Fig. 2b) and the time required to complete the process is in-

creased. This is because a large proportion of the hydrogen ions and also molecules of undissociated acid diffuse into and out of the specimen in the later stages of the process, without coming into contact with the calcified residue.

Decalcifying Agents. Two types of reagent are employed: acids and complexing agents.

Inorganic acids, e.g., nitric and hydrochloric, are rapid in their action. However, when used in high concentrations or at elevated temperatures they tend to cause the tissues to swell and disintegrate.

Organic acids, such as formic and lactic, decalcify slowly even when employed in high concentrations. Un-

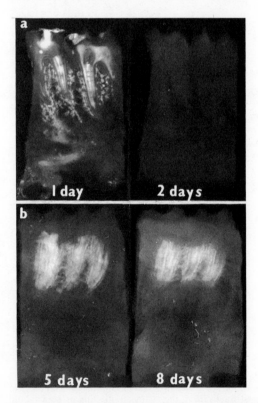

Fig. 2 The pattern of demineralization as shown by X-rays of similar specimens treated in 4N formic acid. a: minimum diffusing distance 1.5 mm; 1 and 2 days immersion. b: minimum diffusing distance 4 mm; 5 and 8 days immersion.

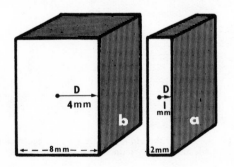

Fig. 1 Diagram illustrating the minimum diffusing distance (D) in specimens with dimensions of: a, 12 × 10 × 2 mm, and b, 12 × 10 × 8 mm.

like the inorganic acids, however, they do not alter or impair the staining properties of hematoxylin and eosin.

Many of the organic and inorganic acids have been used in combination or mixed with other reagents in attempts to produce more satisfactory decalcifying fluids.

The other method exploits the ability of certain complex organic substances to bind inorganic cations under neutral or even alkaline conditions. The most widely used of these substances, known as chelating agents, for decalcification purposes is ethylene-diamine-tetra-acetate (EDTA), which has a strong affinity for calcium ions. It demineralizes hard tissues slowly at temperatures below 20°C. The large number of reagents and special mixtures available for decalcification greatly exceeds the requirements for routine purposes. Many of them are similar in composition and performance, while the value of others is extremely doubtful. Routine demineralization of bone, dentine, and cementum can be carried out with either of two solutions, namely:

1. $4N$ formic acid, and
2. 10 % EDTA, pH7.4.

A formic acid solution is satisfactory when laboratory temperatures do not exceed 20°C, while EDTA may be used within a range of 5–60°C. Tissues treated at 60°C can be decalcified in approximately half the time required at 20°C, without impairing the histological features (Fig. 3).

Tooth enamel and other highly calcified tissues may be preserved for section cutting in paraffin wax after treatment in special solutions (Brain, 1967).

It is claimed that the rate of decalcification can be increased by applying mechanical agitation, treatment under vacuum, electrolysis, or ultrasonics. Experiments indicate, however, that the time saved when using these refinements is negligible compared with immersing the tissues in an adequate concentration and volume of the decalcifying fluid at atmospheric pressure. Shaking the container once or twice daily ensures that the products of demineralization are well distributed in the solution.

Determination of the Completion of Decalcification. It is essential to know that decalcification is complete before proceeding to the stage of dehydration. The presence of minute calcified residues renders section cutting difficult or even impossible. The end point of demineralization may be ascertained as follows:

1. By chemical tests.
2. With x-rays.

Chemical tests rely on the precipitation of calcium oxalate (Arnim, 1935), or the use of calcein (Brain 1966), which gives a green fluorescence in the presence of calcium. With both methods the decalcifying fluid is tested daily until the precipitate or the fluorescence is absent for several days, when decalcification is deemed to be complete.

In the absence of experience these tests are not as dependable as x-rays, where the decalcification process is followed visually (Fig. 2) and the end point ascertained within a short time of its occurrence.

E. B. BRAIN

References

Arnim, S. S., "A method for the preparation of serial sections of teeth and surrounding structures of the rat," *Anat. Rec.,* 62:321 (1935).

Brain, E. B., "Preparation of Decalcified Sections," Springfield, Thomas, 1966.

Brain, E. B., "Rapid demineralisation for microscopy of tooth enamel and associated structures," *Brit. Dent. J.,* 123:177 (1967).

Brain, E. B., and J. E. Eastoe, "Studies in the decalcification of dental tissues for histological purposes." *Brit. Dent. J.,* 122:277 (1962).

Hahn, F. L., and F. Regades, "Demineralisation of hard tissues," *Science,* 14:462 (1951).

Hallpike, C. S., "X-ray control of decalcification in the histology of bone," *J. Pathol. Bacteriol.,* 38:249 (1934).

Jaffé, "Methods for the histological study of normal and diseased bone," *Arch. Pathol.,* 8:817 (1929).

Lillie, R. D., "Studies in the decalcification of bone," *Amer. J. Pathol.,* 20:291 (1944).

Morse, A., "Formic acid–sodium citrate decalcification and butyl alcohol dehydration of teeth and bones for sectioning in paraffin," *J. Dent. Res.,* 29:143 (1945).

Nikiforuk, G., and Sreebny, "Demineralisation of hard tissues by organic chelating agents at neutral pH," *J. Dent. Res.,* 32:859 (1953).

Richman, L. M., M. Gelfand, and J. M. Hill, "A method of decalcifying bone for histologic section," *Arch. Pathol.,* 44:92 (1947).

Thorpe, E. J., B. B. Bellamy, and R. F. Sellars, "Ultrasonic decalcification of bone," *J. Bone Surg.,* 45:1257 (1963).

Wellings, A. W., "Practical Microscopy of the Teeth and Associated Parts," London, Bale, 1938.

Williams, A., "The preparation of combined hard and soft tissues for histologic study," *D. Cosmos,* 69:715 (1927).

Wilson, R. A. J., "Decalcification in vacuo," *Amer. J. Clin. Pathol. Tech. Sect.,* 12:74 (1942).

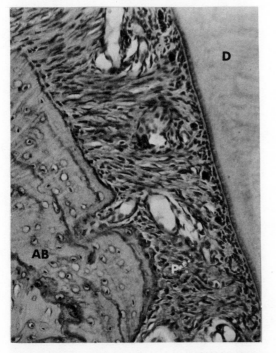

Fig. 3 Section from a tooth of a monkey showing alveolar bone (AB), periodontal membrane (PM), and dentine (D) ($\times$120). Stained hematoxylyn and eosin. Specimen decalcified in EDTA at 60°C.

DISSECTING MICROSCOPES

The normal microscope provides only a flat image. From early times the desire and necessity arose for instruments which would provide a three-dimensional impression of the observed object. In the beginning of this century a very suitable construction was made available in the Greenough stereomicroscope. This consists of two microscope-tubes inclined towards each other at an angle of 15°C and directed at the object. With three-dimensional structures, the two microscopes show two pictures which are not identical, and which present to the observer a stereoscopic effect. In the beam of the two microscopes are inserted prismatic reflections, which erect the inverted picture of the normal microscope, a fact which provides real relief for manipulation under the microscope. Besides this the prisms permit adjustment for the difference in space between the eyepieces and the difference in space between the eyes of the observer. In modern binoculars the construction of the image-reversing prisms additionally brings the eyepieces into a comfortable position for the observer. Figure 1 shows the passage of light through a microscope constructed on the Greenough principle.

Stereomicroscopes of this construction permit the production of an optimum beam of light and the creation of pictures of the highest quality, which may be magnified up to 200 times. A disadvantage of the Greenough binocular lies in the fact that in changing to another magnification it is necessary to change both objectives. This change can of course be simplified, as shown in Fig. 2.

A simpler magnification change for stationary objects

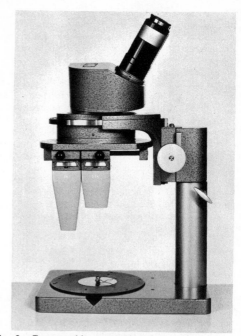

Fig. 2 Greenough's stereomicroscope with equipment for changing the objectives.

has been provided in the last ten years by the stereomicroscopes developed with a common primary objective for the light in both tubes. In them the beams of light do not project the object coaxially, but are diverged from the optical axis. In this way there are produced in the two tubes images which are not identical, and which give a truly three-dimensional appearance.

In this construction separated intermediate objectives are placed above the common primary objectives to produce the two beams, and these provide the further magnification and enlarged image for the inverted image prism and lead on the eyepieces. The intermediate objectives which lie above the chief objective in the tube can be easily manipulated, so that each change in magnification is easily obtained. It is significant and most advantageous that when the magnification is changed the distance from the object is not altered. The beam of light in a binocular with a single primary common objective is shown in Fig. 3.

A further development of this principle in construction is provided by the introduction of pancratic intermediate objectives (300 m systems) in the tube. The use of graduated changes of magnification provides a continuous progression of magnification, which can be useful for many purposes. Additional variations of magnification can also be produced by varying the power of the ocular and of the front of the primary objective.

Since the two beams of light do not enter the primary objective coaxially, they do not provide a completely optimal image. Because of this, in this type of construction there is a limit to the useful increase in magnification from above. But there is no real disadvantage in this, since with increasing magnification of the objectives their definition is so far reduced that the occurrence of a three-dimensional impression becomes impossible. One can assume that a magnification of about 8–100 $\times$ is the limit of

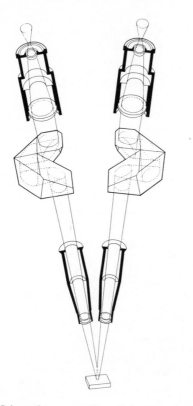

Fig. 1 Schematic presentation of the light beam in Greenough's stereomicroscope.

practical stereoscopic enlargement. Higher magnifications necessary for special problems require the use of more powerful oculars.

The decisive advantage of stereomicroscopes with a common primary objective is that at all magnifications the

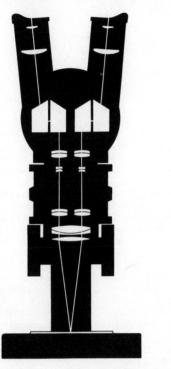

Fig. 3 Schematic presentation of the light beam in a stereomicroscope with a common principal objective.

distance between the objective and the object remains constant. In modern instruments the long distance of the primary lens to the object can therefore be used for all magnifications.

At first the construction of the stand of the stereomicroscope was confirmed to the model of the monocular microscope: that is, the object lay raised on a table which was illuminated from underneath by a mirror, over which —by means of the fine adjustment—the double tube was brought into position. This method of construction was sufficient for simple observations. If necessary, light reflected from the mirror could be diverted, so that the object appeared in reflected light.

The production of the upright enlarged plastic image by the Greenough tube soon led to its general use in work with small and very small objects, a development which today is continued in the field of technical preparations. In such manipulations under the stereomicroscope the condition arises that the object being worked on is raised above the work area on the table of the instrument. Manufacturers usually provide an arm rest for their instruments, which raises the hands to the height of the table of the binocular, and in this way make the work easier. In modern binoculars there is generally an arrangement for observations in transmitted light, but normally the foot of the apparatus serves equally as a work and an observation surface. An example of such a modern binocular is shown in Fig. 2. A column rises from the foot which serves to adjust the height of this optical system.

These simple binocular tripods require the object under investigation to be adjusted on to the table of the apparatus. Limiting of observations to the optical axis, which results from rigid joining of the binocular tube to the table, is in most instances clearly purposeless. For this reason, at the beginning of this century, P. Mayer constructed a table over which the column of the microscope could be moved in two directions perpendicular to each other. In this way a large number of objects dispersed over

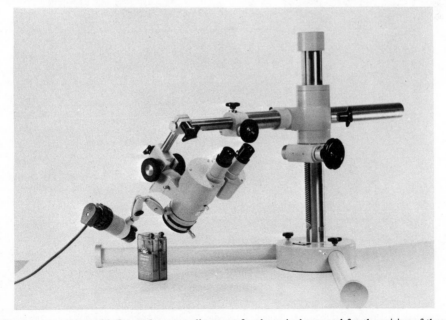

Fig. 4 Large post support with fine and coarse adjustment for the swivel arm and for the raising of the swivel arm.

the table surface can be examined without having to move them. In this construction the principle is employed of giving the optical system so much movement that it can be brought to the object, instead of being forced to bring into the optical axis small objects, which are difficult to maneuver.

A further development of this principle, in the hands of various manufacturers, resulted in stands which permit inspection of layer areas and examination of objects of any size desired. In these so-called "pillar-supports" the stereomicroscope is secured to a long swivel-mounted arm —up to 30 cm long—which is carried from a vertical supporting shaft in a gimbal. The supporting shaft permits

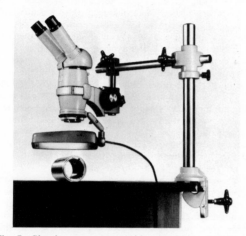

Fig. 5 Simple post support with table lamp. A light is fixed to the binocular tube.

the raising of the swivel arm. The swivel arm permits a variation of the distance of the binocular from the supporting shaft. Besides this, the gimbal can revolve around the supporting shaft. These devices permit control of a very wide field within which the binocular can be adjusted to any desirable point.

The vertical supporting pillar is furthermore firmly anchored to a very heavy foot, which even when the swivel arm is fully extended can support the weight through the binocular tube without tipping. A better solution is an angular foot, as the apparatus in Fig. 4 shows; this furnishes a firmer stand for the binocular in its enclosed quadrant. A further possibility is provided by strong table clamps; by this has the advantage that it does not significantly cut down the work area.

In simpler models (Fig. 5) the bearing and the swivel arm are made as simple round posts, to which is clamped the gimbal. In this apparatus the adjustment of the binocular to a desired point is somewhat involved. In better models (Fig. 4) a groove with coarse and fine adjustment prevents movements of the swivel arm around its axis (Langsachse) so that the distance of the binocular from the post can be varied without trouble.

For exact focus the coarse and fine adjustments which are attached to the tube are used; when these are insufficient, it is possible to vary the height of the swivel arm above the stage by a screw situated on the support post (Fig. 4).

Wobbling of the swivel arm was for many years a difficulty. Improvement in this is provided by the introduction of a very precise model, but the price for such a stage is high. The author has demonstrated that this can easily be prevented by using a rectangular, in place of a cylindrical arm between roller bearings as shown in Fig. 6. This form is not only technically easier to construct, but also prevents wear since the movement of the swivel arm

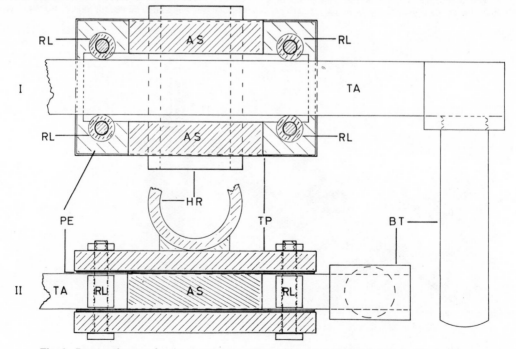

Fig. 6 Rectangular arm for the post support. The swivel arm is placed between roller bearings.

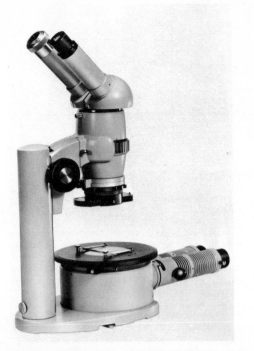

special lighting equipment, specially made for stereomicroscopic observation, is used. Low-voltage lamps have proved particularly valuable. They permit a very high light density, which can be adjusted at need. These small lamps are without exception so made that they can be fixed to the binocular and moved about. An example of this is shown in Fig. 4. For more diffused light, fluorescent lamps serve well, as in Fig. 5. The mounting of the source of light on the binocular must permit illumination from all angles in order to be able to find the most favorable lighting for the field of vision. In cases where transmitted illumination is necessary, there are available specially illuminated tubes with lamps fixed underneath. An arrangement of this kind is seen in Fig. 7.

FRIEDRICH KRÜGER

Fig. 7 Stereomicroscope with zoom objective and adjustment for polarization and with a lamp for transmitted light.

DRAWING WITH THE AID OF THE MICROSCOPE

From time immemorial living processes and natural phenomena have been graphically recorded for teaching, documentation, religious, and decorative purposes. Depending on requirements, either particular features of the object observed are represented, or a detailed reproduction is attempted. Whereas in the first instance the reproduction of the features may be unconsciously influenced by subjective interpretation of the object the second case involves absolutely accurate representation. For scientific study purposes only the latter alternative can be of interest. The more unfamiliar the object, the more critical becomes this need for faithful rendering.

can be made by simple pulling or pushing on the binocular. This arrangement has the advantage that the observer need not hold his hands far from his field of observation and, in case of necessity, can very quickly change the field of observation.

For observations under the stereomicroscope by reflected light the primary illumination is, in general, not similar to that used with a regular mciroscope but at high magnification requires strong reflected light. Usually,

Immediately after the invention of the microscope, which opened the door to the previously unexplored micro-world, people wanted to record and communicate what they saw, and above all to discuss it with others. Even Leeuwenhoek himself, generally recognized as the inventor of the microscope, made amazingly accurate drawings, of which his "little animals in water" are the most famous. Since, however, the observation and subsequent drawing of the microscope image requires great

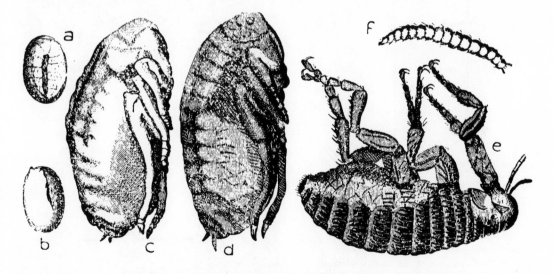

Fig. 1 "Development of the mosquito," Leeuwenhoek's drawings.

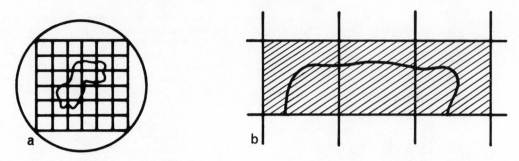

Fig. 2   a: Eyepiece graticule with grid; b: drawing surface with grid.

powers of observation and memory and is moreover inconvenient and time-consuming, a succession of more simplified and easy-to-use drawing accessories has been developed for the microscope. The present article is concerned with this development and the state it has reached today (Fig. 1).

Photography is today the best reproduction technique available. Drawing has consequently lost much of its popularity, but should not be completely given up, because it still has many advantages, such as the following:

When checking through specimens, one is often looking for specific features. These characteristics can be captured with a few brief strokes of the pencil and numerous details can be arranged on one page for comparison purposes. The advantages of drawing over photography are here clear—saving of time and money, simplicity, universal applicability.

The wealth of information contained in a photograph can lead to confusion in its evaluation. In teaching especially, drawing can be used to expound basic differences in shape and substance, and is immensely useful in circumventing the inadequate depth of field offered by photography.

Drawing devices have recently acquired considerable new significance by use in a reversed manner. Information such as lines, texts, numbers, letters, or other images and objects can thus be superimposed on the microscope image for purposes of measurement, comparison, and marking. The drawing devices are therefore being employed as a notation technique in photography, the newest form of drawing.

Drawing with Auxiliary Markings in the Field of View. If a graticule with appropriate guide marks (e.g., squares) is placed in the intermediate image plane of the microscope, and squared drawing paper is also used, a fairly accurate scale reproduction of the image can be obtained by looking into the microscope, memorizing the contents of one square, and filling in from memory the corresponding area on the drawing paper (Fig. 2).

Drawing without Supplementary Apparatus. If one looks with one eye into the microscope eyepiece and with the other eye at a drawing surface beside the microscope, then the brain registers the two different images as being superimposed. The microscope must, however, be defocused sufficiently for the microscope image to appear sharp in the drawing plane when both eyes are accommodating.

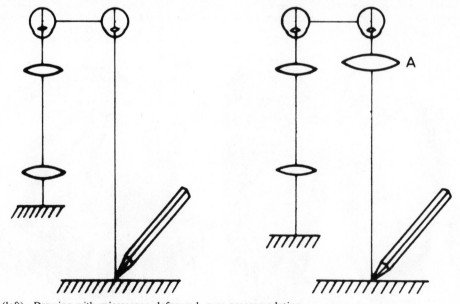

Fig. 3 (left)   Drawing with microscope defocused, eyes accommodating.
Fig. 4 (right)   Drawing with microscope focused normally, eyes not accommodating, supplementary lens A, focal length = drawing distance.

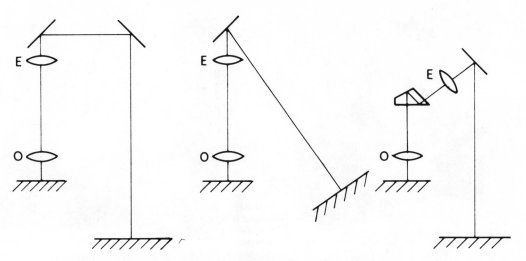

Fig. 5   Various arrangements with projection mirrors. O = objective; E = eyepiece.

The greater the drawing surface to eye distance, which necessarily depends on the size of the microscope, the smaller is the effort needed to accommodate.

The accommodation difficulties can be alleviated by wearing glasses which have a lens in only one side, the focal length of this lens being equal to the drawing distance (Figs. 3 and 4).

Drawing with a Projection Mirror. A quick, simple, and accurate method is to trace a projected image. The eyepiece is used as a projective and the ray bundle leaving it is mirrored onto a drawing surface. The image contours can then be quite comfortably traced in this plane. If two mirrors are used, it is possible to project the image onto the flat table surface beside the microscope, whereas if only one mirror is employed and the microscope functions vertically, an undistorted projected image can only be obtained in an inclined plane. The lateral inversion of the image and the use of an inclined tube are generally unimportant.

The disadvantage of this method is that a very strong light source is needed if, at high magnifications, a sufficiently bright image is to be obtained. Nevertheless with

Fig. 6   Equipment for drawing with conventional microscope and projection mirror.

Fig. 7 Wild microscope with projection mirror for drawing.

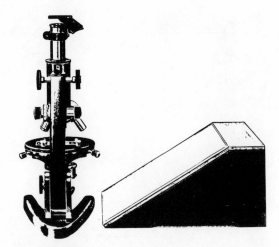

Fig. 9 Drawing prism and drawing board, as used formerly.

reflected light techniques or special microscopy methods the image intensity is sometimes still inadequate, even in a darkened room (Figs. 5, 6, and 7).

Drawing with a Drawing Prism. A special drawing prism manufactured by the firm of Zeiss enables the drawing surface to be reflected into the image while allowing simultaneous observation of the image through the eyepiece. This special prism is so positioned in the direct vicinity of the exit pupil of the eyepiece that one part of the pupil of the eye receives the ray bundle from the microscope and the other part receives the bundle from the drawing surface. Therefore one sees a superimposition of two images. The brightnesses of these two images must be matched, either by altering the illumination intensity or by using filters. The drawing plane is, however, inclined towards the microscope in order to avoid distortion. Special adjustable drawing tables were formerly manufactured for this purpose (Figs. 8 and 9).

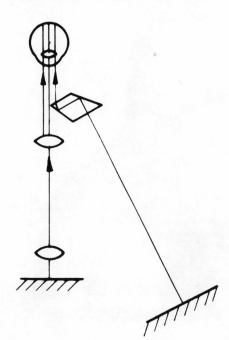

Fig. 8 Zeiss drawing prism.

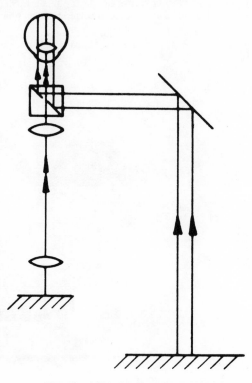

Fig. 10 Abbé drawing apparatus.

Fig. 11 Microscope and Abbé drawing apparatus, horizontal drawing surface.

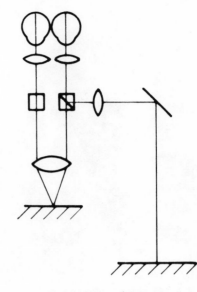

Fig. 13 Stereomicroscope.

Drawing with the Abbe Drawing Apparatus. This is an improved version of the drawing prism method. In this system also, a small beam-splitting cube is positioned in the region of the exit pupil of the eyepiece, but the beam-splitting plane of this cube is not coated in the central region of the exit pupil. The light from the drawing surface is directed via a mirror to the beam-splitting plane of the cube. The mirror again reflects the light parallel to the light path in the microscope and into the eye. The light from the microscope has, therefore, an unimpeded path into the eye, while light from the drawing plane is reflected symmetrically into the eye by the mirrored surface around the exit pupil. The brightnesses of the images formed are again matched either by adjusting the illumination or by using filters. The image of the drawing

area can be adjusted to eliminate distortion by tilting the first mirror. Here, as also in the previously mentioned instance of the drawing prism, the microscope must be defocused, by an amount equal to the drawing distance, from the position where the specimen is in sharpest focus (Figs. 10 and 11).

Drawing with the Treffenberg Drawing Apparatus. Wild Heerbrugg Ltd. brought a considerable improvement to the field of drawing with the microscope when they manufactured the Treffenberg drawing apparatus. With this equipment an upright, laterally correct image of the

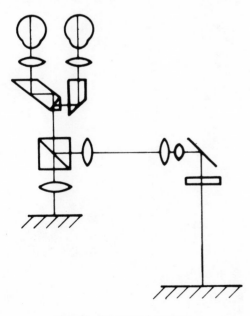

Fig. 12 Normal microscope.

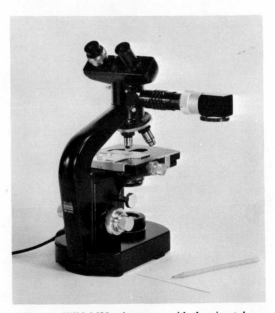

Fig. 14 Wild M20 microscope with drawing tube.

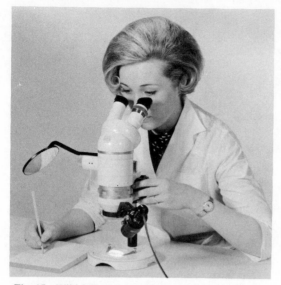

Fig. 15 Wild M5 stereomicroscope with drawing tube.

drawing plane is "mixed" with the intermediate image of the specimen. This is done by inserting between the objective and the eyepiece of the microscope a tube adapter having a semitransparent prism. The arm of the apparatus protrudes sideways and contains a 45° mirror, which directs into the beam-splitting cube the ray bundle received from the surface of the table. The arm also contains a second image-forming system, which is used for focusing the drawing plane and for the production of an image scale steplessly variable in the range 1 : 2 without altering the scale of the microscope image. In this way it is possible to carry out binocular observation of the microscope image as in normal microscopy and simultaneously to see an image of variable brightness of the drawing surface with drawing material. The variation is again achieved by altering the light intensity, using filters or a set of polarizing filters. This drawing accessory for microscopes is at present the best available and has, in a modified form, aroused particular interest in the field of stereo microscopy. More uses have been found for these drawing accessories than were originally envisaged; in particular, their application in the superimposition of scales and marks (Figs. 12–15).

M. UHLIG

e

ECHINODERMATA

The pentamerous Echinodermata (see Gray, 1970) have been used extensively as experimental animals in studies relating to various fields of biology, particularly cell physiology. It would not be possible to include here all of the techniques evolved in the course of these studies. The discussion is therefore limited to those basic techniques pertinent to the study of echinoderms as echinoderms.

All echinoderms should be preserved in 70% ethyl alcohol, not in formaldehyde or isopropyl alcohol. Delicate calcareous structures such as pedicellariae and small spines and plates will eventually be destroyed by formaldehyde, while echinoderms preserved in isopropyl alcohol deteriorate in time. Most echinoderms (except holothuroids) can be preserved well enough for most purposes by soaking overnight (or several days for large specimens) in ethyl alcohol and then drying. This effects a tremendous saving in storage space, alcohol, and expensive containers. If it is later necessary to examine soft tissues, the specimen may be soaked for about 24 hr in a saturated detergent solution, which will restore the soft tissue to nearly its original pliability and simplify dissection. Of course special preservatives may be required for the histological study of soft tissues, where decalcification is desirable; one of the best such preservatives is Bouin's fluid.

The arrangement and type of plates in echinoids, asteroids, and most ophiuroids can be examined by saturating the whole specimen or the part to be examined in full-strength liquid household bleach and rinsing in running water. Depending on the size of the specimen and the thickness of the tissue, the amount of time the specimen is left in the bleach will vary, but progress should be checked frequently by rinsing in tap water to prevent complete disarticulation of the plates. If the specimen is fairly robust, brushing with a stiff toothbrush will help speed the process. Preferably, the specimen should be completely dry before such cleaning is started.

Echinoderm gametes have been extensively used for experimental work because of the ready availability of ripe material and ease of obtaining fertilization in the laboratory. By far the best-known of echinoderms used for such studies is the common Western Atlantic sea urchin, *Arbacia punctulata*, abundant, easily obtained, and with a long breeding season. The best method of obtaining the eggs and sperm is by passing a 10 v electric current through the animal while it is completely immersed in seawater. Lead electrodes may be applied to any two points on the urchin's test, and shedding can thus be controlled. If the gonads are ripe, the males will shed a

thin white stream of sperm through the gonopores, and the females a thicker reddish stream of ova. The same method can be used to obtain gametes from sand dollars and other sea urchins. Another method, preferred by some investigators, is to inject about 0.5 ml. of isotonic potassium chloride solution through the peristomial membrane with a fine hypodermic needle, invert the injected animal over a beaker of seawater with gonopores immersed. Shedding induced by this method may be stopped by rinsing the animal in tap water.

The most commonly used starfish is *Asterias forbesi*; starfish eggs are sensitive, and care must be taken in their handling. Starfish will not generally shed gametes under laboratory stimuli, but a preparation of radial nerve extract from the starfish can induce shedding. As little as 0.15 ml/g of a 5 mg % seawater extract of lyophilized nerves injected into the perivisceral coelom of a starfish of moderate size will induce shedding into seawater if the gonads are ripe. Usually, it is necessary to dissect out the gonads, wash them free of body fluid in filtered seawater, and shake them gently in a dish of clean filtered seawater to start the flow of eggs through the gonoduct. Techniques for obtaining and culturing echinoderm eggs and sperm may be found in Harvey (1956) and Galtsoff, et al. (1959).

A recent paper by Kanatani and Shirai (1971) has introduced a new and more efficient method for induction of oocyte maturation and spawning in asteroids. This method uses a seawater solution of l-methyladenine. It has proved very effective on a number of starfish species, and will undoubtedly replace the radial nerve extract method eventually. Concentrations of 1-methyladenine at 3×10^{-7} m in buffered seawater were sufficient to induce spawning and oocyte maturation.

Many useful techniques and a great deal of information on the general biology and physiology of echinoderms can be found in Boolootian (1966).

Study of the pattern of ciliary currents on the body surface of echinoderms utilizing powdered carmine particles, fine graphite, or black sand grains has led to some interesting results regarding feeding habits and respiration. This simple technique of dropping colored particles on various parts of the body wall and observing their progress across the surface has been used for only a few species, again mostly echinoids and asteroids.

For studies of soft tissue, it is best to use live material. Specimens should be immersed in magnesium chloride solution (8% in tap water) for 15–30 min, or until flaccid. Dissection can then proceed according to the type of animal being used. For asteroids, the best method for examining internal organs is to cut off the tip of one arm, and with an iridectomy scissors, make an incision along the dorsal mid-arm to the disc. Gonads and pyloric caeca can be lifted out with forceps. Remove all arms, cut around the periphery of the disc, and carefully lift the disc dorsum, separating the tissues from it with a fine scalpel, to examine other internal organs. For echinoids, it is necessary to cut through the test, around the ambitus. The same method may be used with ophiuroids, first removing the arms. In dissecting holothuroids, the incision should be to the left or right of the mid-dorsal interradius, cutting longitudinally in both directions from the middle.

Echinoderm systematics are based mainly on hard-part morphology, and in the study of plates, spines, pedicellariae, mouthparts, and other structures, ordinary liquid household bleach (Clorox preferred) is an echinologist's best friend. Holothuroid platelets, embedded in the skin, are essential for the identification of the species. A small piece of skin from the body wall is placed on a glass slide and a drop of dilute bleach (10%) added; when most of the tissue has bubbled away, draw off the bleach solution and flood the sample with fresh water. Repeat as often as needed until only the platelets are left. The same procedure may be followed with echinoid and asteroid pedicellariae, which can be removed from the specimen with fine forceps under the microscope.

Growth lines in echinoderm plates have been studied extensively only in echinoids. A simple technique for examining annular growth lines in echinoid plates, by first scorching and then dipping in xylene, was published by Jensen (1969), but an even simpler method for relatively thin plates is to place a drop of light machine oil on the plate; it is then ready for immediate examination under the microscope. For thicker plates, the older method of grinding down the plate (Deutler, 1926; Moore, 1935) is still necessary, but when the plate has been ground sufficiently thin, the lines will show readily in oil or xylene.

Thin sections of spines should be prepared as one would prepare soft-rock thin sections. Cut the spine in the desired plane on a rock saw, mount the section on a glass slide with epoxy, allow to set firmly, and polish on a glass plate with number 800 grinding powder. Wash, and the slide should be ready for microscopic examination without further preparation. In some cases, staining with eosin will emphasize certain lines and structures, particularly for the purpose of making photographs.

Echinoderm larvae may be prepared as one would any unicellular organism. It is well to remember, however, that the pluteus larva of echinoids and ophiuroids has a tiny, delicate skeleton which may be crushed if a coverslip is dropped on it; the coverslip should be supported, either by ringing or some other method, about 0.2 mm above the surface of the slide. Specimens may be stained with borax carmine, cleared in xylene, and permanently mounted in balsam or Clearite. The pluteus skeleton shows up beautifully under polarized light.

The water vascular system, unique to the echinoderms, is an inviting field for the physiologist. Although many studies on the histology of the tubefeet have been made, little attention has been paid to function of the system as a whole. The preparation of tubefeet for histological study involves removal of the tubefoot from live material and fixing it for about 8 hr in Heidenhain's "Susa" fluid made up in seawater; this fluid will nicely decalcify the tissue as it is being fixed. The tubefeet may then be embedded by Peterfi's colloidin/paraffin method and subsequently in wax for serial sectioning. Mallory's triple stain or Mason's trichrome are suitable general stains. This treatment should also work well for other parts of the water vascular system.

A good stain for general histological study of soft tissues is Harris' hematoxylin. For more specialized stains, consult any good handbook on histological techniques (see references).

There has recently been a revival of interest in the structure of echinoderm calcite. Each individual unit of an echinoderm skeleton is a single crystal of calcium carbonate, with certain peculiarities of structure in some of the hard parts. For an introduction to techniques for X-ray diffraction studies and electron microscopy, see Donnay and Pawson (1969), Towe (1967), Nissen (1969), and others. Such complex techniques are too specialized to include here.

MAUREEN E. DOWNEY

References

Boolootian, R. A., "Physiology of Echinodermata," New York, Interscience Publishers, John Wiley & Sons, 1966.

Dales, R. P., ed., "Practical Invertebrate Zoology: A Laboratory Manual," Seattle, University of Washington Press, 1970.

Deutler, F., "Uber das Wachstum des Seeigelskeletts," *Zool. Jb. Anat. und Ontog. der Tiere*, **48**:199–200 (1926).

Donnay, G., and D. L. Pawson, "X-ray diffraction studies of echinoderm plates," *Science*, **166**:1147–1150 (1969).

Galtsoff, P. S., F. E. Lutz, P. S. Welch, and J. G. Needham, "Culture Methods for Invertebrate Animals," New York, Dover Publications, Inc., 1959.

Gray, P., "Handbook of Basic Microtechnique," New York, McGraw-Hill Book Co., 1958.

Gray, P., ed., "The Encyclopedia of the Biological Sciences," 2nd ed., New York, Van Nostrand Reinhold, 1970.

Harvey, E. B., "The American Arbacia and Other Sea Urchins, Princeton, N.J., Princeton University Press, 1956.

Hyman, L. H., "The Invertebrates: Echinodermata. The Coelomate Bilateria," Vol. IV, New York, McGraw-Hill Book Co., Inv., 1955.

Jensen, M. "Age determination of echinoids," *Sasria*, **37**:41–44 (1969).

Kanatani, Haruo, and Hiroko Shirai, "Chemical Structural Requirements for Induction of Oocyte Maturation and Spawning in Starfishes," *Development, Growth and Differentiation*, **13**(1):53–64 (1971).

Knudsen, J. W., "Biological Techniques,," New York Harper and Row, 1966.

Lee, A. B., "The Microtomist's Vade-Mecum," 10th ed., Philadelphia, P. Blakiston's Son & Co., Ltd., 1937.

Millott, N., ed., "Echinoderm Biology. Symposia of the Zoological Society of London Number 20," London, Academic Press, 1967.

Moore, H. B., "A comparison of the biology of *Echinus esculentus* in different habitats," Part II, *J. Marine Biol. Ass. U.K.*, **20**:109–128 (1935).

Nissen, H., "Crystal orientation and plate structure in echinoid skeletal units," *Science*, **166**:1150–1152 (1969).

Towe, K. M., "Echinoderm calcite: single crystal or polycrystalline aggregate," *Science*, **157**:1048–1050, (1967).

Weesner, F. M., "General Zoological Microtechniques," Baltimore, The Williams & Wilkins Co., 1960.

ECTOPROCTA

Living marine Ectoprocta (Bryozoa) collected in the field must be treated in different ways, depending upon the techniques to be used in their study. The calcareous cheilostome and cyclostome material may be initially fixed in 70% ethanol or isopropyl alcohol or in neutral formalin (formaldehyde-seawater 1 : 9). For taxonomic study, the specimens are then air dried, sorted, and mounted slides for examination of the exoskeletal structure with a dissecting microscope capable of graduated magnification to 50–100×. Often specimens will be covered with sand grains, silt, or other debris. This can usually be removed by an air jet or careful brushing with a camel's hair brush.

External organic material may be removed by soaking the specimen in a solution of potassium hypochlorite or commercial sodium hypochlorite commonly called "clorox." The technique called "calcining," incinerating the specimen with an alcohol lamp and a blowpipe, is useful for revealing structural details. This is a quick method, but it should be used only when there is an abundance of material available because specimens so treated disintegrate rapidly.

Dried specimens are mounted on glass slides with "white" casein glue (Wilhold, Elmer's, etc.). This type of adhesive holds specimens firmly to the slides, it is water soluble so that specimens can be soaked off if necessary, and it does not turn acid causing the ultimate breakdown of the calcareous exoskeleton. The slides are labeled with collecting data.

To reduce glare from calcareous specimens for photomicrography with the light microscope it is sometimes necessary to coat the material with fumes of ammonium chloride from a dual-chambered container of hydrochloric acid and ammonium hydroxide. After use, this sublimate is easily blown or rinsed from the specimens without harm (see Bassler, 1953).

Cheilostomes, cyclostomes, and the uncalcified ctenostomes to be sectioned for histological studies are routinely fixed in neutral formalin (formal-seawater) or alcohol as soon as possible in the field. Material to be used for wholemounts or sectioning is examined wet with the dissecting microscope and extraneous debris (sand, etc.) is removed with a camel's hair brush or fine needle. For wholemounts the wet specimens are stained with azocarmine (1% in 70% ethanol), dehydrated, cleared, and mounted under a coverslip with a synthetic resin. For sections, the fixed calcareous specimens are decalcified with EDTA or in trichloracetic acid (10% aqueous), washed overnight, dehydrated, and infiltrated in a vacuum oven with paraffin or epoxy embedding. The soft-bodied ctenostomes, obviously, do not require decalcification. Wet specimens, for routine storage, are kept in 70% ethanol.

For special purposes, transmission electron microscopy or histochemical studies, specific fixatives are used, such as glutaraldehyde or osmium tetroxide, the choices dependent upon the protocol to be followed. For scanning electron microscopy, dried calcareous material, both chloroxed or unchloroxed specimens are given a very thin coating of gold in a vacuum evaporator before being placed in the specimen holder.

JOHN D. SOULE
DOROTHY F. SOULE

Reference

Bassler, R. S., "Treatise on Invertebrate Paleontology," pt. G, Bryozoa, New York, Geological Society of America, 1953 (pp. 1–253).

ELASTIC TISSUES

Elastic tissues are part of the connective tissues of the body. They are generally fibrous in form and may branch, but they may be lamellar such as in the aorta. They are refractile, generally slightly yellow in color, and nonbirefringent in contrast to collagen. Elastic tissues derive their name on the basis of physical elasticity, which contrasts with collagen. Tissues containing elastic tissue include skin, ligaments, tendons, trachea, blood vessels, and fibroelastic cartilages. They generally comprise only a small percentage of the total volume of any tissue. Figure 1 illustrates the typical histological appearance of elastic

Table 1   Amino Acid Composition of the Elastic Fiber and Its Components[a]

	Elastic Fiber	Amorphous Component After Enzymatic Digestion[b]	Microfibrils After Enzymatic Digestion[c]
CM-cysteine	—		—
Hydroxyproline	16.4	11.3	1.7
Aspartic acid	16.1	5.4	92.5
Threonine	13.8	8.6	47.3
Serine	16.1	9.4	52.8
Glutamic acid	24.6	14.7	98.3
Proline	90.2	110	73.5
Glycine	305	324	142
Alanine	212	223	82.6
Cystine/2	10.2	5.0	56.3
Valine	130	140	69.7
Methionine	0.7	—	13.0
Isoleucine	29.6	26.2	43.8
Leucine	60.3	67.8	65.5
Tyrosine	9.4	8.2	27.6
Phenylalanine	34.4	30.9	32.8
Isodesmosine/4	3.9	4.1	—
Desmosine/4	5.5	6.5	—
Hydroxylysine	—	—	0.7
Amide nitrogen	n.c.	n.c.	n.c.
Lysine	11.9	7.6	36.7
Histidine	3.7	0.7	11.5
Tryptophan[d]	n.c.	n.c.	11.9
Arginine	11.1	5.2	42.3

[a]Values are given as residues per 1000 and are the averages of 2 or more determinations. Analyses of the elastic fiber and its amorphous component were corrected for hydrolytic losses. Analyses of microfibrils are uncorrected. Methionine includes methionine sulfoxide, and in the alkylated sample, the degradation products of the carboxymethyl sulfonium salt of methionine. Cystine includes cysteic acid. A dash indicates that the amino acid was either entirely absent or present at less than 0.1 residue 1000. n.c. = not calculated; CM = cysteine : carboxymethylcysteine.

[b]Obtained as a residue after digestion of microfibrils with chymotrypsin and trypsin.

[c]Soluble peptides obtained after chymotryptic and/or tryptic digestion of elastic fibers.

[d]Determined spectrophotometrically.

(Table condensed from Ross and Bornstein, *The Journal of Cell Biology*, **40** : 366–381 (1969). Used by permission.)

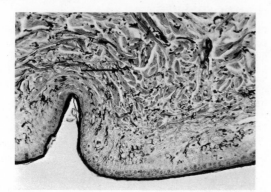

Fig. 1 Section of formalin fixed aged human skin stained with aldehyde fuchsin after peracetic acid oxidation. Elastic fibers such as those identified by the arrow and keratin stain very dark. Epithelial cells on the surface and collagen interspersed with elastic fibers stain less intensely with the fast green-orange G counterstain. (×210.)

fibers in normal human dermis as viewed by light microscopy. Elastic fibers are interspersed with collagen fibers, connective tissue cells, ground substance, and small amounts of carbohydrates, proteins, and minerals.

Figure 2 reveals the appearance of elastic fibers at the electron microscopic level. Note the amorphous appearance. A fibrillar sheath frequently may be observed at the periphery of elastic fibers viewed at either the light or electron microscopic levels. At the light microscopic level, sheaths at the periphery may sometimes be stained with

methods that identify acid mucopolysaccharides. Sometimes, in certain structures such as the aorta, collagen fibers may enter and become incorporated within elastic fibers.

Elastic Tissue Composition. Elastic fibers contain a characteristic protein called elastin. The amino acid content is characterized by relatively low percentages of the aromatic amino acids and high percentages of glycine, valine, and alanine (Table 1). Desmosine and isodesmosine are amino acids peculiar to elastic tissue. These are derived from lysine during the course of histogenesis and are concerned with cross linking of the elastin molecules. The precise amino acid content has not yet been established due to the extremely difficult task of obtaining a pure sample of elastic tissue from any source. The task is further complicated by the remarkable insolubility of elastin in many organic and aqueous acid, neutral, or alkaline solutions. Calcium binding to elastin has been related to β-turns of the elastin molecule favored by the high glycine content. Elastic fibers are digested by elastase presently derived from pancreas. The staining characteristics of elastic fibers vary from site to site, which promotes the assumption that the composition of elastic tissues may not be uniform.

Elastic Tissue Stains. Useful and reliable stains for the identification of elastic fibers at the light microscopic level include the Taenzer-Unna orcein, Weigert's resorcinfuchsin, Verhoeff's iron-hematoxylin, Gomori's aldehyde fuchsin and Fullmer-Lillie's orcinol–new fuchsin. The most selective of these is orcinol–new fuchsin. The only constituent in vertebrate tissue other than elastic tissue presently known to stain with orcinol–new fuchsin is the organic matrix of enamel which remains after demineralization. Although orcinol–new fuchsin is highly selective for elastic tissue, a disadvantage is that elastic tissue is

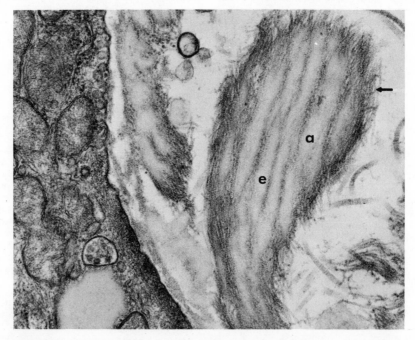

Fig. 2 Elastic fiber (e) from a developing artery. During histogenesis, elastic fibers develop from two morphologically distinguishable components. One (arrow) is fibrillar, and the other is amorphous (a). Mature elastic fibers appear homogeneous throughout (such as appears at a) except some fibrils may persist at the periphery. (×51,500.) (Courtesy Ross & Bornstein.)

generally less intensely stained than with other methods such as orcein or aldehyde fuchsin.

For ease of preparation, the use of orcein and aldehyde fuchsin is advocated for most laboratories. The disadvantage of orcein and aldehyde fuchsin stains is nonspecificity of staining. For example, aldehyde fuchsin stains certain mucins and certain other structures, and orcein stains nuclei and certain other structures. These attributes contribute to uncertain significances of staining reactions, and lead to confusion in the literature. Alterations in staining reactions are especially confronted during studies of pathological tissues.

Mechanism of Staining Reactions. Although the molecular structures of the various elastic tissue stains are unknown, on the basis of the constituents employed, it may be assumed that aromatic phenols or amines with ferric salts in an acid ethanolic solution appear to be conducive to staining.

Elastic tissue stains do not identify elastic tissues specifically. Rather, staining is the result of combination of the dyes with elastic tissues generally in an acid ethanolic solution, and the means by which elastic fibers react with the elastic tissue stains is unknown. Blockage of aldehyde, carboxyl, amino, and unsaturated functional groups has no effect on the subsequent orcein staining reaction with elastic fibers. Experiments have not been conducted to determine the possibility that a peculiar structure or molecular configuration is responsible for staining characteristics of elastic tissues. Not all elastic fibers stain in an identical fashion. For example, rodent elastic tissue has a readily detectable native aldehyde whereas human elastic tissues do not. Furthermore, elastic tissues of the dermis may stain with certain dyes with greater difficulty than those of the aorta in the same species or individual. It is interesting that acetylated or benzoylated collagen is intensely stained with one of the most reliable methods for staining elastic fibers; namely, Taenzer-Unna orcein. The

significance of this is unknown; however, it must not be interpreted that acetylated collagen is elastic tissue.

Aldehyde Fuchsin Stain. Among the elastic tissue stains, aldehyde fuchsin requires special mention. Aldehyde fuchsin is a dye of unknown structure formed at 25°C by addition of paraldehyde to an acidic ethanolic solution of basic fuchsin (C.I. 42510). Two to three days in a stoppered container are required to provide sufficient dye concentration to effect maximum reliable staining. Dependable stainability progressively diminishes after 7–10 days. On this basis, it is assumed that aldehyde fuchsin is a transient structure formed under the prescribed conditions. As a practical procedure in our laboratory, the dye is

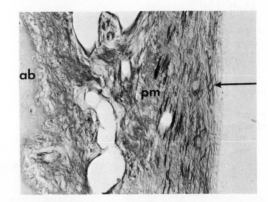

Fig. 3 Section from an upper human molar tooth stained with the Oxone-aldehyde fuchsin-Halmi method for the identification of oxytalan (dark-staining) fibers in the periodontal membrane (pm). Some oxytalan fibers (arrow) enter the cementum of the tooth, and many others are dispersed in an apico-occlusal direction. At the left is aleveolar bone (ab). (×160.)

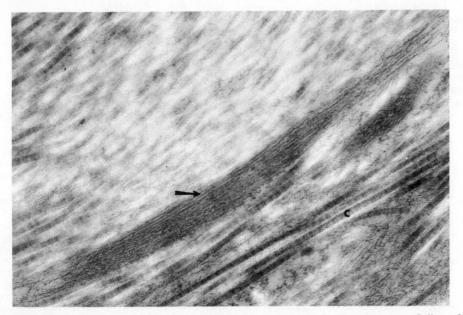

Fig. 4 Oxytalan fiber (arrow) from human periodontal membrane. Note the fibrillar appearance. Collagen fibers (c). (×48,250.) (Courtesy Dr. J. Goggins.)

made up on Friday, used for one week, and discarded. It should not be kept in a refrigerator in an attempt to prolong usefulness of the stain. Under these conditions, substantial red stain (basic fuchsin) is detectable, sometimes promoting confusion of stained sections. In addition to being an excellent stain for elastic fibers, aldehyde fuchsin stains certain mucins without prior oxidation of tissue sections. After oxidation of tissue sections with peracetic or performic acids, potassium persulfate (Oxone: E. I. DUPONT Co., Wilmington, Delaware), or less satisfactorily with permanganate or bromine, other ground substance, other mucins, basement membranes, oxytalan fibers, and several other tissue constituents are readily stained.

Elastic Fiber Histogenesis. The development of elastic tissues has been studied by correlative electron microscopic and biochemical methods. During the course of histogenesis, elastic fibers in bovine ligamentum nuchae appear to develop from two morphologically distinct constituents. Distinct microfibrillar and amorphous components are observed. The microfibrillar elements precede deposition of the amorphous component. Fusion of the central constituents follows, although fibrils may be seen at the periphery of fibers indefinitely. The amino acid content of the two constituents of elastic fibers is quite different (Table 1). Microfibrils have relatively lower valine, alanine, and glycine, and relatively higher aspartic acid, threonine, serine, glutamic acid, cystine, methionine, arginine, and lysine contents.

Oxytalan Fibers. Oxytalan fibers demand particular attention within the content of this article inasmuch as comparative anatomical, staining, and digestion experiments permit the assumption that oxytalan fibers represent a particular kind of modified elastic fiber required in certain peculiar anatomic locations such as periodontal membranes of man. Oxytalan fibers are selectively (but not specifically) stained by aldehyde fuchsin after an oxidation such as Oxone (Fig. 3). Elastic fibers are also stained with aldehyde fuchsin under identical conditions; however, an oxidative preparatory step is not required. A further distinguishing feature of oxytalan fibers is the digestability of the aldehyde fuchsin stainable component of oxytalan but not of elastic fibers by beta-glucuronidase. Elastase will digest oxytalan fibers provided they are preoxidized with either peracetic acid or potassium persulfate (Oxone). Oxytalan as well as developing elastic fibers appear to be comprised of fibrillar and nonfibrillar components as viewed electron microscopically (Fig. 4).

HAROLD M. FULLMER

References

Fullmer, H. M., "A comparative histochemical study of elastic, pre-elastic and oxytalan connective tissue fibers," *J. Histochem. Cytochem.*, **8**:290:295 (1960).

Fullmer, H. M., "The histochemistry of the connective tissues," pp. 1–76 in D. A. Hall, ed., "International Review of Connective Tissue Research," Vol. 3, New York, Academic Press, 1965.

Fullmer, H. M., and R. D. Lillie, "The mechanisms of orcein staining," *J. Histochem. Cytochem.*, **4**:64–68 (1956).

Fullmer, H. M., and R. D. Lillie, "The staining of collagen with elastic tissue stains," *J. Histochem. Cytochem.*, **5**:11–14 (1957).

Fullmer, H. M., and R. D. Lillie, "The oxytalan fiber: a previously undescribed connective tissue fiber," *J. Histochem. Cytochem.*, **6**:425–430 (1958).

Lillie, R. D., "Histopathologic Technic and Practical Histochemistry," 3rd ed., New York, McGraw-Hill, 1965.

Lillie, R. D., "Exploration of dye chemistry in Taenzer-Unna orcein type staining," *Histochemie*, **19**:1–12 (1969).

Mammi, M., L. Gotte, and G. Pezzin, "Evidence for order in the structure of α-elastin," *Nature*, **220**:371–373 (1968).

Miller, E. J., and H. M. Fullmer, "Elastin: Diminished reactivity with aldehyde reagents in copper deficiency and lathyrism." *J. Exp. Med.*, **123**:1097–1108 (1966).

Miller, E. J., G. P. Martin, C. E. Mecca, and K. A. Piez, "The biosynthesis of elastin cross links," *J. Biol. Chem.*, **240**:3623–3627 (1965).

Narayanan, A. S., and R. A. Anwar, "The specificity of purified porcine pancreatic elastase," *Biochem. J.*, **114**:11 (1969).

Partridge, S. M., "Isolation and characterization of elastin," pp. 593–616 in E. A. Balazs, ed., "Chemistry and Molecular Biology of the Intercellular Matrix," Vol. I, New York and London, Academic Press, 1970.

Ross, R., and P. Bornstein, "The elastic fiber. I. The separation and partial characterization of its macromolecular components," *J. Cell Biol.*, **40**:366–381 (1969).

Sandberg, L. B., N. Weissmann, and W. R. Gray, "Structural features of tropoelastin related to the sites of cross-links in aortic elastin," *Biochemistry*, **10**:52–56 (1971).

Sandberg, L. B., N. Weissmann, and D. W. Smith, "The purification and partial characterization of a soluble elastin-like protein from copper-deficient porcine aorta," *Biochemistry*, **8**:2940–2945 (1969).

Urry, D. W., "Neutral sites for calcium ion binding to elastin and collagen: A charge neutralization theory for calcification and its relationship to atherosclerosis," *Proc. Nat. Acad. Sci.*, **68**:810–814 (1971).

Visser, L., D. S. Sigman, and E. R. Blout, "Elastase I and II," *Biochemistry*, **10**:735–752 (1971).

See also: POLYCHROME STAINS.

ELECTRIC ORGANS

The electric organs which are present in elasmobranch and teleost fishes produce electric shocks of considerable intensity and are used to stun or kill prey or for self-defense.

In the elasmobranch, *Torpedo marmorata*, the electric organs lie on either side of the head between the gill region and the anterior portion of the pectoral fin. They consist of vertical stacks of electroplaques, each possessing a noninnervated dorsal surface and a richly innervated ventral surface. The plaques are derived from embryonic muscle in which contractile elements have vanished. Homologs of the sarcoplasmic reticulum and the motor end-plate of striated muscle are associated with the dorsal and ventral surfaces respectively. The remainder of the cytoplasm possesses the usual cytoplasmic organelles, although they are encountered infrequently due to the large cell diameter (3–4 mm) (Sheridan, 1965).

Torpedo which were obtained by air from Naples survived well in filtered seawater at 10°C. At the time of use, the fish was pinned to a dissecting board and immediately pithed with a blunted dissecting needle inserted into the cranial cavity. A steel rod approximately 2 mm in diameter was then inserted into the spinal canal to abolish all reflex activity. The skin was removed from the dorsal surface of the electric organ by sharp dissection, and columns of electroplaques were removed and placed in one of several fixatives. Fixaton was carried out for 3 hr at 2°C in (1) 2% OsO_4 containing $0.25M$ sodium chloride and $0.33M$ urea; (2) 2% OsO_4 in Veronal-acetate buffer (Palade, 1952); (3) 10% formalin in phosphate buffer (Pease, 1962) followed by postfixation in 2% OsO_4 in phosphate buffer (Millonig, 1961a); and (4) 1% potassium permanganate (Luft, 1956). After completion of fixation the tissues were rinsed in cold 25% acetone and placed in 50% acetone. Here, the columns were cut transversely into smaller stacks more conducive to subsequent dehydration and embedding. Dehydration was completed through 70% acetone in the cold, and the remaining dehydration through absolute acetone was done at room temperature. Tissue was then embedded in Araldite (Robertson, et al., 1963).

Morphological results from the above fixation procedures were uniformly good. Contrast was generally poor, presumably due to low tissue mass, and was enhanced with a lead hydroxide–tartrate complex (Millonig, 1961b) or in alcoholic phosphotungstic acid (Pease, 1964). Little or no tissue disruption occurred. Membrane structure was best revealed in the potassium permanganate fixed material.

Fixation techniques have improved recently and mixtures of glutaraldehyde and formaldehyde, introduced by Karnovsky (1965), are being used routinely with excellent results in many tissues. There is every reason to expect that they are capable of producing excellent results with electric organ.

The embedding agent in our work with electric organ was Araldite obtained from the British based Ciba firm. Unfortunately Araldite is no longer available directly from them. Araldite from other sources has not been used. Dow Epoxy Resin (DER) (Lockwood, 1964) is a routinely employed embedding agent with excellent penetrating and sectioning qualities.

Acetone was used in dehydration because its miscibility with Araldite permitted placing the tissue directly into a 50 : 50 mixture of Araldite and absolute acetone. Alcohol dehydration and propylene oxide treatment is recommended for DER embedment.

Michael N. Sheridan

References

Karnovsky, M. J., "A formaldehyde-glutaraldehyde fixative of high osmolality for use in electron microscopy," *J. Cell Biol.*, (abstr.) **27**:137a (1965).

Lockwood, W. R., "A reliable and easily sectioned epoxy embedding medium," *Anat. Rec.*, **150**:129 (1964).

Luft, J. H., "Permanganate: A new fixative for electron microscopy," *J. Biophys. Biochem. Cytol.*, **2**:799 (1956).

Millonig, G., "Advantages of a phosphate buffer for OsO_4 solutions in fixation," *J. Appl. Physics*, **32**:1637 (1961a).

Millonig, G., "A modified procedure for lead staining of thin sections," *J. Biophys. Biochem. Cytol.*, **11**:736 (1961b).

Palade, G. E., "A study of fixation for electron microscopy," *J. Exp. Med.*, **95**:285 (1952).

Pease, D. C., "Buffered formaldehyde as a killing agent and primary fixative for electron microscopy," *Anat. Rec.*, **142**:342 (1962).

Pease, D. C., "Histological Techniques for Electron Microscopy," 2nd ed., New York, Academic Press Inc., 1964.

Robertson, J. D., T. W. Bodenheimer, and D. E. Stage, "The ultrastructure of Mauthner cell synapses and nodes in goldfish brains," *J. Cell Biol.*, **19**:159 (1963).

Sheridan, M. N., "The fine structure of the electric organ of *Torpedo marmorata*," *J. Cell Biol.*, **24**:129 (1965).

EMBRYOLOGICAL TECHNIQUES

I. Introduction

This article includes selected techniques for laboratory investigations with vertebrate embryos. Although many manuals and handbooks of microscopial technique may include a section on general procedures for fixing or staining embryonic tissue, a variety of specific techniques for use especially with embryos is not generally available in a single source.

The limitation of space in the present volume must, of course, result in the selection of only a few procedures for inclusion in this article; therefore, the selection was made in consideration of the following criteria: (1) inclusion of those techniques which would be of most general interest and use to a large number of investigators and teachers, and (2) selection of some specific methods for use with embryos of each vertebrate class. Documentation of the techniques from the original literature has been included where reference to the literature might be useful. Many procedures which could have been included in this article, but were not, are ones which the author is either less familiar with personally, or considered less important according to the criteria stated above.

The collecting of healthy, intact embryos is paramount to the subsequent preparation of their tissues, and the first section which follows considers some of the methods employed in obtaining the more commonly used vertebrate embryos. The embryos of fish and reptiles have not been used as extensively in descriptive or experimental embryological studies as embryos of other vertebrate classes, and the relatively little coverage of techniques for obtaining or using these embryos in the present chapter reflects the paucity of reports dealing with these forms in the literature.

The reader interested in a discussion of morphological and cytochemical stains of particular relevance to embryonic tissues will find that these areas have not been covered here. The great variety of stains and staining techniques which would need to be discussed for the specific requirements of different body tissues and their components, and the modifications necessary for any one tissue in embryos of different classes, or of different ages in some cases, would be beyond the scope of the article. There are numerous handbooks of histologic staining methods as well as several professional journals devoted exclusively to the study of morphological and histochemical stains and staining techniques which can be consulted.

II. Methods of Obtaining Embryos

A. Induced Ovulation and Artificial Fertilization of Amphibians. No doubt, one of the greatest boons to experimental vertebrate embryology was the discovery that amphibian eggs could be obtained in nearly any month of the year by inducing ovulation of fertile adult females with hormones. No longer need experiments be confined to those months in which fertilized eggs could be collected from the field.

1. *Procedure for Anura.* Since normal fertilization is external in anura, occurring as the eggs leaving the cloaca of the female are contacted by the sperm from the male which is tightly grasping the female in amplexus, the method of artificial fertilization is quite simple. The female may be induced with hormones to ovulate eggs, which are then "stripped" from her body by gentle

abdominal pressure; the eggs are then fertilized with a sperm solution prepared from testes of mature males. Many hundreds or thousands of eggs, depending on the species, can thus be obtained from each female frog most of the year. The instructions given below are based on the method of Rugh (1934; 1937) for the common leopard frog, *Rana pipiens*.

a. Removal of the pituitary gland. The frog pituitary gland is the most effective source of hormone with the capacity to stimulate ovulation in other frogs, and Rugh has found that female glands are approximately twice as potent as male glands in their action. The number of glands required varies with the season, fewer being required as the season for natural egg-laying approaches (see Table 1). The donor frogs are decapitated behind the tympanic membrane and the lower jaw is removed. With a small pointed scissors inserted into the vertebral canal, a cut is made forward and lateral in the floor of the cranium to each side. The flap of bone is carefully deflected forward, exposing the ventral brain surface. The

Table 1

Months	Number of Female Glands Needed
Jan.–Feb.	4–5
Mar.–Apr.	2–3
May–Aug.	—[a]
Sept.–Dec.	5–6

[a]Induced ovulation is poor during this period, since ovulation normally occurs in April. However, by storing frogs in a cold (38–40°F) dark room in early March, ovulation may be inhibited during this period as necessary. Use 3–4 female frog pituitaries for inducing ovulation in frogs so treated.

pink-colored oval tissue lying either on the brain just posterior to the optic chiasm (Fig. 1), or in a small depression on the lifted flap of bone, is the anterior pituitary. This tissue must be carefully removed and placed in a small volume (about 3 ml) of aerated tap water or 10% Holtfreter's solution,[1] and all adhering tissue should be removed.

b. Injection of pituitary glands. Mature, healthy female frogs to be used as the source of eggs may be maintained in the refrigerator at 4°C to simulate the hibernating state until they are needed. However, the frogs must be allowed to warm to room temperature before the pituitary glands, which are the source of the ovulation hormone, are injected.

The glands are drawn into a hypodermic syringe with about 2–3 ml of the aqueous solution before attaching an 18 or 20 gauge needle to the syringe. Insert the needle superficially through the ventral abdominal skin and muscle, being careful not to injure the median abdominal vein on the belly or the deep viscera. As the syringe is emptied, the glands are macerated in the bore of the needle, and the small pieces will be absorbed through ciliated peritoneal funnels to the venous sinuses. The injected frogs are maintained at room temperature in a small amount of aerated tap or pond water, and will usually produce eggs capable of being fertilized in 48 hr.

c. Testing for ovulated eggs. After being ovulated, the eggs pass first into the body cavity, then into the upper end of the oviducts to be enveloped in several jelly layers. Holding the frog belly up, bend the legs forward, and gently press the abdomen behind the forelegs. If eggs emerge from the cloaca, place the frog back in the container and prepare the sperm solution. If no eggs appear, wait 12–24 hr. A second injection may be necessary if no eggs can be obtained 60–72 hr after injection of the glands.

d. Preparation of sperm solution. Viable sperm can be obtained from adult male *Rana pipiens* throughout the year. Remove the testes (oval yellow bodies located near the kidneys, one on either side of the spinal column) from one or two freshly killed frogs for each female injected. Each pair of testes should be dissected into 10 ml fresh pond water, 10% Holtfreter's solution, or aerated tap water, and completely macerated with scissors or scalpel to release the sperm. Allow the milky-colored suspension to stand for 5–10 min at room temperature before using.

---

[1]Stock Holtfreter's solution: 3.50 g NaCl; 0.05 g KCl; 0.01 g $CaCl_2$; 1000 ml water. After cooling and before use add 0.02 g $NaHCO_3$ per 100 ml solution.

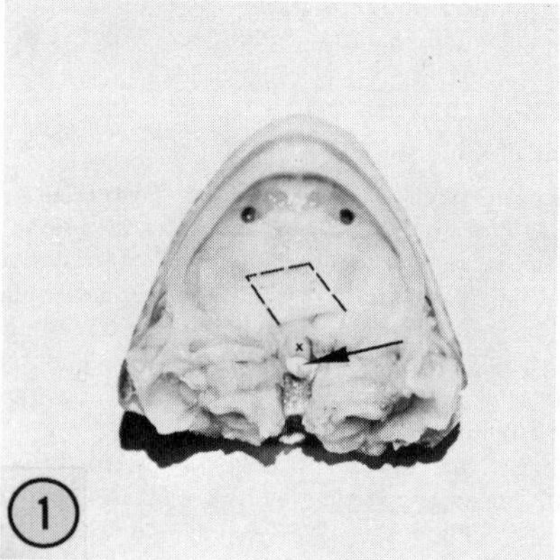

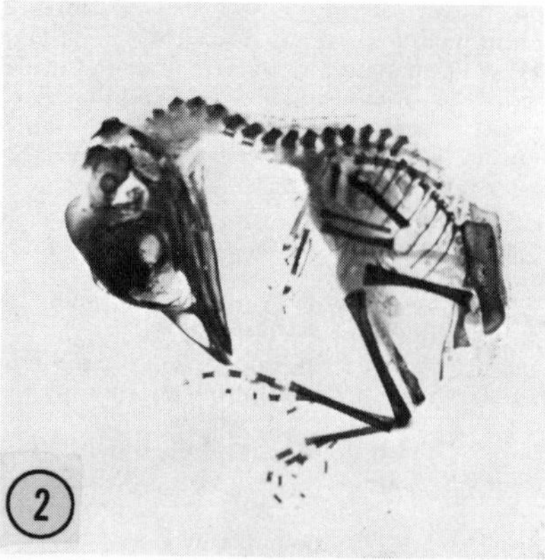

Figs. 1 and 2 Ventral view of upper jaw of adult *Rana pipiens*. A flap of skin and bone (outlined) has been lifted to expose ventral surface of the brain, including anterior lobe of pituitary (arrow) and optic chiasm (x). **Fig. 2** Ossified skeleton of a 14-day chick embryo stained with Alizarin Red S dye by the technique described in the text.

e. Laboratory artificial fertilization. A female which has tested positive for ovulated eggs is "stripped" of her eggs as described above in section c. As pressure is applied with the fingers around her abdomen, two "strings" of eggs will emerge from the cloaca—one from each oviduct. The frog should be moved so the eggs will lay in rows or spirals in several dry bowls or large petri dishes, avoiding dense groups of eggs. If necessary several dishes are used to avoid crowding the eggs. Each egg should have contact with the sperm solution which is then pipetted over them. During the next 5 min, the sperm solution in the dishes may be re-pipetted over the eggs to insure complete fertilization of the eggs. No other water is added during this time, for the jelly surrounding the eggs will rapidly imbibe water, and the sperm may be prevented from reaching the egg surface by the swelling of the jelly. However, 5–10 min after the sperm solution was first applied to the eggs, fresh pond water or aerated tap water to a depth of several inches must be added to prevent desiccation and to allow respiration by the eggs. All eggs which have successfully been inseminated will rotate within the jelly envelope so that the more dense white yolky pole is down, and the darkly pigmented hemisphere is above by the end of the first hour. At this time the eggs should be divided into small groups so there is approximately 500 ml of water per 25 eggs, and the water is subsequently changed daily. The first cleavage will occur about 90 min after rotation, at room temperature, and the embryos will hatch in about four days.

f. Care of the tadpoles. The animals should be reared at a temperature of 18–25°C. After hatching, the embryos should be separated from the gelatinous egg coverings, and this is best accomplished by carefully drawing the embryos into an eye dropper, and placing them into a rearing tank or aquarium (clean plastic dishpans are ideal) with approximately 3 in. of aerated water. As the animals get larger, more water can be added. Although the tadpoles do not eat for several days after hatching, a small amount of washed canned spinach or boiled lettuce can be added within a week after the eggs are fertilized. The animals may be maintained on these greens until metamorphosis, which occurs approximately 100 days after the eggs are fertilized, if the larvae are fed maximally in uncrowded aquaria at room temperature.

2. *Procedure for Urodela.* It is generally more difficult to obtain embryos of urodeles by artificial means in the laboratory than it is to obtain embryos of anurans, for the eggs of most salamanders are fertilized internally. Spermatophores, containing the mature sperm, are deposited in the water by the males of most urodele species. These packets of sperm are subsequently picked up by the female and retained in the cloaca until release of the sperm.

Induced egg-laying of urodeles in the laboratory requires either (1) stimulation of the males to form and release the spermatophores, and the subsequent stimulation of mature females to pick up the spermatophores and undergo ovulation; or (2) stimulation of females suspected of carrying activated sperm to deposit fertilized eggs. Both sexes may be stimulated by hormonal methods and/or by creating stimulatory environmental conditions. Anterior pituitaries of a variety of mature amphibians are effective, as are extracts of mammalian anterior pituitary glands. Since the eggs of urodeles are usually laid individually over a period of several days, large numbers of embryos of the same age cannot be obtained unless one has a quantity of breeding females. However,

eggs laid earliest may be held back in their development by refrigeration for several days. The water in the aquaria used for breeding should be shallow, and provisions should be made to allow the animals to leave the water and crawl onto sand or a floating piece of bark. Water plants should be provided for attachment of the eggs, although often eggs laid following injection of hormone will be dropped to the bottom of the aquarium. Water temperature of 18–20°C has been found to be satisfactory for species of *Triturus* and *Ambystoma*.

Patch (1933) has stimulated male *Triturus viridescens* to deposit spermatophores by three methods: (1) implanting amphibian anterior pituitary glands (one lobe on each of 2 or 3 days); (2) injecting gonadotrophic hormone extracted from human pregnant urine (4–6 daily injections, concentration not reported); and (3) pairing males with females and allowing normal courtship (preliminary courtship can be shortened with implantation of pituitary glands). He subsequently picked up the spermatophores or clusters of spermatozoa in a glass pipette and transferred them to the cloacas of females stimulated to ovulate by the implantation of pituitaries daily for 2–6 days. Fertile eggs should be laid within one or two days following the introduction of sperm into the body of the female, although additional hormonal stimulation may be required if egg-laying is delayed for several days.

Clearly, the procedures for obtaining fertile eggs of urodeles in the laboratory are not as well developed or refined as the procedures for inducing ovulation and effecting artificial fertilization of anuran eggs. An alternative in areas where the desired species are indigenous is to collect the fertilized eggs in the natural habitat.

Urodele larvae will nip the tails and other soft tissues of other larvae if not properly fed. They will normally eat living food, such as young *Daphnia*, *Enchytrea* or *Tubifex* worms, mealworms, or *Drosophila* larvae or adults. Water should be cleaned within several hours after feeding. If the diet is not adequate for the growing larvae, some cannibalism will be noted.

B. Collection of Embryos of Birds (and Reptiles). Although the chick embryo is the most commonly used avian species, the information which follows can generally be applied to any other species of birds, and with some modifications, to reptilian species, since both reptiles and birds develop from a cleidoic egg. The discussion in this section, however, focuses on the chick embryo in particular. Fertilized hen eggs can usually be procured from local hatcheries, and fertility can be verified by the technique of "candling" 48 hr after the eggs are laid. The egg is placed over a hole in an otherwise light-proof container containing a bright light source (such as a 100 W light bulb) in a darkened room. In fertilized eggs of two days incubation or more the blastoderm of the developing embryo can be clearly seen in the transmitted light through the shell. Fertile eggs should be maintained in incubators at 102–103°F with a relative humidity of at least 60 %. During the incubation period the eggs should be rotated twice daily to prevent adhesion of the egg membranes to the shell.

It is possible to remove the embryo by cracking the egg in the manner of a breakfast chef, but this is a rather haphazard procedure for obtaining clean, intact embryos. Rather, it is suggested that a hole be made in the broad end of the egg, which is the location of a large airspace. The egg is then submerged in a small fingerbowl of warm saline, and the air is allowed to escape from the puncture. In this manner the yolk will fall away from the upper surface of the shell, and the embryo, which lies uppermost on the yolk, will not be damaged as the top surface of the shell is chipped away. Without puncturing the yolk, enough of the shell should be removed to allow the contents to be gently emptied into the saline. Once in the saline, the embryo will again rotate to a position uppermost on the yolk where it can conveniently be removed.

C. Collection of Ova and Young Embryos in Mammals. The mammalian species most commonly used for classroom purposes have been those which are frequently found in laboratories and easily maintained for experiments (i.e., mouse, rat, and rabbit), or those which are raised for their commercial value (i.e., pig). In order to obtain fertilized ova or embryos of a particular stage, several features of the reproductive cycle must be known for the individual species, namely: (1) the length of the estrus cycle; (2) the time of ovulation relative to coitus; (3) the position of the ova in the reproductive tract at the time of fertilization, and at subsequent pre-implantation stages; and (4) the rate of development of the blastocyst.

1. *Procedure for Mouse.* The location of mouse ova at selected early stages while they are moving down the oviduct has been described by Lewis and Wright (1935). These investigators report that by 72 hr post-coitus all fertilized ova reside in the uterus. Most eggs recovered at this time are at the morula stage of development. By 82 hr after copulation the majority of ova are blastocysts. Using the data of these workers, tubal eggs or blastocysts of various stages can be obtained from the mouse by flushing the appropriate segment of the female mouse reproductive system containing the desired stages with warm saline.

To obtain cleavage stages or embryos of the mouse, the female is killed at the desired time after copulation, and the uterine horns are removed. The uterus should be placed in fixative if the embryos are less than one week old; if more than one week has elapsed post-coitus, the uterine wall should be slit longitudinally along the anti-mesometrial side, and opened to expose the decidual swellings before immersing in fixative. Fekete *et. al.* (1940) report that a modification of Zenker's fixative[2] is generally superior to several fixatives tested with early mouse embryos. After 3–6 hr in the fixative, the embryos are washed thoroughly in running water overnight, and then may be transferred to 70% alcohol where, under a dissecting microscope, the decidua may be opened and the embryos removed. Embryos of less than six or seven days should remain attached. The orientation of younger embryos can only be approximately determined—their anterior-posterior axis is usually oriented transverse to the long axis of the uterus, but this is subject to some variability.

2. *Procedure for Rabbit.* Unlike most mammals, ovulation in the rabbit is not regular and spontaneous, but rather is dependent on coital stimulation. As a rule, the mature female will ovulate approximately 10 hr after copulation. Because fertilization occurs within one to two hours after ovulation, the early stages of rabbit embryos are therefore quite uniform. Gregory (1930) has found that fertilized eggs enter the uterus by 80 hr after coitus, and are blastocysts at this time. He found Bouin's

fluid superior to other fixatives tested for tubal eggs and blastocysts.

Pincus (1940) obtained supranormal numbers of ova from rabbits (as many as 82 from a single female) by utilizing gonadotrophic hormone injections. The eggs obtained by this procedure are fertilizable.

3. *Procedure for Pig.* Prior to three days after coitus, the fertilized ova of the pig are usually found only in the oviducts, where they may be expelled by washing the tubes with warm, normal saline. A detailed method has been described by Heuser and Streeter (1929). Developing ova beyond the four-cell stage may be obtained from the uteri, following the procedure of these investigators, after the third day. Heuser and Streeter (1929) explain in detail their method of photographing eggs and embryos of the pig, and include an atlas of embryonic development from the one-cell tubal egg to the 23-somite stage, supplemented by a description of the principal stages of development of the early stages to the appearance of limb-buds.

III. Histological Preparation of Embryonic Tissues

A. Preparation of Embryos *In Toto*. 1. *Cleared Whole Mounts.* Embryos of vertebrates are mounted whole when it is desirable to preserve and study the gross relationships of the body parts. The embryos are usually stained with alum hematoxylin or a carmine stain, dehydrated, and cleared prior to being mounted in a medium which has a refractive index sufficiently high to allow a transparent preparation. The thickness of the specimen will, of course, determine the suitability of obtaining a mount satisfactory for observation under moderate magnification.

In preparations which are thicker than about 0.5 mm, some type of support is necessary for the cover glass to prevent this thin cover from breaking and to contain the mounting medium, unless slides with a recessed "well" are available. Chips of glass from a crushed microscope slide are suitable for thinner specimens; one chip is placed under each corner of a square coverslip, and the parallel faces of the broken glass piece maintain a uniform distance between the slide with the specimen and the overlying coverslip. To obtain a greater separation between slide and coverslip, plastic tubing of sufficient diameter to contain the specimen may be cut to the required thickness with a razor blade to form a ring in which the specimen and mounting medium are placed. Alternatively, plastic curtain rings have been found to be satisfactory for large specimens, or some investigators may desire to purchase ready-made mounting rings from a supply house. Regardless of the type of support used, its thickness should be at least 2–4 mm greater than the thickness of the specimen, as some shrinkage occurs while the mounting medium dries, and a coverslip placed upon the embryo with insufficient allowance for this shrinkage may slip or break, often resulting in an accumulation of bubbles within the mounting medium.

Although most vertebrate embryo specimens are large enough to handle without difficulty, it has been found to be more successful to pour the fluids from the glassware than to transfer the embryo itself.

The directions which follow describe the procedure for preparing whole mounts of the embryonic chick, but with minor modification they are applicable to embryos of any vertebrate classes. Chick embryos have traditionally been studied at the following "representative" stages of development: 18 hr (primitive streak stage), 24 hr, 33 hr, 48 hr, and 72 hr of incubation. It is during these stages of development that most of the organ rudiments are

---

[2]Zenker's fluid modification: Equal parts of the following solutions are mixed before use. Solution A: potassium dichromate, 4 g; water, 100 ml. Solution B: bichloride of mercury, 4 g; glacial acetic acid, 20 ml; water, 100 ml.

formed, although their complexity and differentiation continues during subsequent developmental stages.

The youngest stage is successfully fixed *in ovo*, after much of the overlying shell has been removed. The fixative (Bouin's, Tellyesniczky's or Goldsmith's fluid)[3] is pipetted onto the blastoderm to preserve structure as well as to harden the fragile tissue. After the initial application of fixative, the embryo is removed by cutting beyond the border of the area pellucida about 1 cm, and the specimen is transferred to fresh fixative in a petri dish or Syracuse glass for 8–12 hr.

The embryo of 24 hr or older should be separated from the yolk by cracking the egg and carefully pouring the contents into a dish of saline. Since the contents will rotate in the solution bringing the embryo to lie uppermost upon the yolk, it is easy to remove the specimen, cutting well beyond the sinus terminalis. The specimen is then brushed with a fine paintbrush onto a glass slide which is submerged below the fluid, and transferred to clean saline where much of the adhering yolk can be removed by squirts of saline from a pipette. The cleaned specimen should then be floated onto a clean slide. A ring of filter paper large enough to cover the area opaca, but with the center removed over the area pellucida, is placed on top of the preparation to keep it flat in the fixative. The fixatives mentioned above are suitable, but older embryos should remain submerged on the slide in the fixing solution for at least 24 hr. The filter paper and vitelline membrane may be removed, and the tissue trimmed after fixation.

Embryos for whole-mount preparations must be completely dehydrated in a series of alcohol from 35% to 100%. Since both recommended fixatives contain picric acid, bleaching of the yellow color is necessary, and this is accomplished in several changes of 70% alcohol to which either a few drops of saturated lithium carbonate (filtered) or 3% ammonium hydroxide has been added during dehydration. The embryo may be stained after removal of the yellow color by first hydrating, then using the desired aqueous stain, then dehydrating through two changes of absolute alcohol. When dehydration is completed the preparation is transferred to a clearing solution, preferably cedarwood oil or methyl salicylate overnight, and then transferred to a fresh change of the clearing agent for 24 hr more, or until the specimen is translucent. After a 30 min transfer to xylene, the specimen should be mounted on a slide prepared with the necessary coverslip support, as described above.

2. *Staining Skeletal Elements.* The preparation of whole mounts of embryos which have bone or cartilage stained, and the skin, muscles, and other soft tissues of the body rendered transparent by suitable hydrolysis and subsequent clearing, has been a technique of value to both descriptive and experimental embryologists. The procedures for staining bone and cartilage in amphibian, chick, and mammalian embryos are given below, with the references cited for the original description of each process. Although the author is unaware of any particular techniques of this type designed exclusively for embryos of fish or reptiles, the method of Hollister (1934) for staining adult fish skeletons with Alizarin dye would probably be acceptable for older fish embryos.

a. Staining cartilagenous skeleton of embryonic am-phibians (van Wijhe, 1902). Any size anuran or urodelean embryo which has developed a cartilagenous skeleton may be used. Embryos fixed in a mercuric fixative (i.e., Zenker's fluid) give superior results; any mixture containing picric acid should be avoided. After fixation, the mercuric chloride must be completely removed in several changes of 70% alcohol. Van Wijhe's stain consists of 0.1 ml Toluidine Blue and 0.1 ml hydrochloric acid in 100 ml 70% alcohol. The embryo should remain in the stain until completely saturated with the color—from one to several days depending on size. As there is no danger of overstaining, it is best to leave the specimen in the stain for at least 48 hr, even for a small embryo.

The stain is removed from the soft tissues with a solution of 0.1% hydrochloric acid in 70% alcohol. Several washings in this acid alcohol are necessary until no more stain is emitted from the specimen. At this time place the specimen in fresh acid alcohol to complete the differentiation of the stained and unstained tissues. The last trace of unbound stain will usually be removed after sitting in the final acid alcohol wash for about 7–10 days. The animals may be dehydrated in an increasing series of alcohols, to which 0.1 ml hydrochloric acid has been added per 100 ml of each alcohol used in the series. After clearing in xylene, the specimen may be mounted on a microscope slide as a permanent whole mount, provided that the mounting medium is acidic (i.e., salicyclic balsam) to insure preservation. For a modification of this method, see Hamburger (1960, p. 196).

b. Staining bone of embryonic amphibian skeleton (Gray, 1929). A freshly killed specimen should be bound with silk thread to a glass slide of appropriate size, which will facilitate transferring the specimen through subsequent steps without changing its body configuration. Gray suggests a 10% dilution of Lugol's iodine solution[4] in 95% alcohol for the initial treatment. Using approximately 500 ml of this iodine solution per animal, the submerged specimen is placed *in the dark* for 12 hr or more to promote hardening. The hardened specimen should be washed twice in 90% alcohol, approximately 12 hr per wash while still in the dark, to prevent it from fragmenting.

The washed specimen is placed in approximately 100 ml of 5% potassium hydroxide to which is added 1 ml saturated Alizarin Red S in absolute alcohol. The time required in the dye mixture depends on the size of the embryo, and as in the case of staining of cartilage, allowing more time than is necessary for staining will do no harm. At least one week is suggested for older larvae of anurans, in which thorough penetration of the dye requires extra time. At such time when the bones of the limb are seen as dark shadows within the pink flesh using transmitted light, the specimen may be transferred to a 5% potassium hydroxide solution for hydrolization of muscle and skin and also for removal of the pink color from noncalcified tissue. This decolorizing solution should be changed frequently until it no longer extracts color from the specimen. In some cases two or three months may be required for removal of all excess stain.

After all pink color has been removed, the specimen will retain a yellowish-brown color which is removed in equal parts of 5% potassium hydroxide and 1% ammonium hydroxide. The solution should be changed as it

---

[3]Formulas for these and other fixatives mentioned in the text will be found in Section III, B, 1.

[4]Lugol's iodine solution: 6.0 g iodine; 4.0 g potassium iodide; 100.0 ml water.

becomes discolored. The skin and muscles will be completely bleached in a few weeks to several months, again depending on the size of the specimen. Following complete bleaching, the ammonia is removed by several daily changes of 5% potassium hydroxide until the smell of ammonia has disappeared. At this time 5% glycerol is added to the potassium hydroxide, with several changes at daily intervals. The concentration of glycerol is increased at 10% intervals until the specimen is finally in pure glycerol, in which its skin and muscles are perfectly transparent and the regions of bone appear bright red. The specimens should be placed in several changes of glycerol until no diffusion currents with residual potassium hydroxide are apparent, and then they may be preserved permanently in a container of pure glycerol.

c. Staining cartilage of chick embryo skeletons (Lundvall, 1904; 1906). This procedure is ideal for chick embryos of 9–12 days of incubation. Fixation in Bouin's fixative for one or two days is suggested. The embryos are washed in 70% alcohol, to which 3% ammonium hydroxide or a few drops of saturated lithium carbonate (filtered) has been added. Several changes in this alcoholic decolorizing solution should be made until the yellow has disappeared from the specimen. The fixed and bleached specimen must now be cleared of all skin, feathers, and fatty tissue. A fine forceps should be used for the careful removal of these tissues.

The stain is prepared as 0.25% Methylene Blue (or Toluidine Blue) in 70% alcohol to which 3% (by volume) hydrochloric acid has been added. The embryo should be left in this stain three days to insure overstaining, as subsequent destaining will occur in 70% alcohol (several changes of alcohol, 2–3 days total). The soft tissues of the specimen will be made further transparent while dehydrating in 95% alcohol (two changes of two hours each) and absolute alcohol (three changes of two hours each). The embryo will become cleared in 3 parts methyl salicylate to which 1 part benzyl benzoate has been added. After one change of the clearing agent, the transparent specimen, through which the blue cartilagenous areas are clearly seen, may be stored in pure methyl salicylate.

d. Staining bone of embryonic chick skeleton. The following procedure, modified from Cumley, Crow, and Griffin (1939), has proven successful with chick embryos older than 10 days of incubation (see Fig. 2).

The embryos are fixed in 95% alcohol for 48 hr, followed by two changes of 70% alcohol of 2–3 hr each. If desired, the viscera can be removed through a ventral abdominal incision while in the 70% alcohol, and the skin and feathers may be carefully removed with a sharp forceps, although in young embryos this is not necessary or recommended. The specimens are then transferred to a 1% solution of potassium hydroxide until the bones clearly stand out as white, while the cartilage remains opaque. This will take less than one week, although the exact duration is variable and depends on the development of the specimen. Fresh potassium hydroxide may be added daily, but should be added at least every other day.

A saturated solution of Alizarin Red S made in 95% alcohol is added one drop at a time to the last change of potassium hydroxide until the solution appears dark pink. As the bone takes up the dye, add more as needed until the bone is intensely stained a deep red. If excessive stain is added at one time, soft tissues of the body may become stained.

The staining procedure will take several days. When the skeleton is intensely stained, the specimen should be transferred to an 80% solution of glycerine (made up in water which contains 1% potassium hydroxide) for clearing. The specimen should clear in this solution for 2–3 days, at which time it is progressively transferred to more concentrated changes of glycerine. The stained and cleared embryo may be permanently stored in 100% glycerine to which a crystal of thymol has been added.

e. Staining of cartilage and bone in mammalian embryos. Williams (1941) reports a technique which has been successfully used to stain embryonic bone and cartilage of several mammalian forms, among which are human, cat, pig, and rat. Following fixation in 10% formalin for at least one week, the specimen is washed for 24 hr in 70% alcohol to which a few drops of ammonium hydroxide are added. The first stain is made of 0.25 g Toluidine Blue in 100 ml 70% alcohol to which 2 ml 0.5% hydrochloric acid has been added. The embryo remains in the stain for approximately one week, and is then hardened and destained in 95% alcohol (4 changes) for 3 days.

The specimen is macerated in a 2% aqueous solution of potassium hydroxide for approximately one week following the destaining in the 95% alcohol, until the limb bones become clearly visible through the skin. At this time the embryo is placed in a fresh solution of 2% potassium hydroxide into which a saturated alcoholic solution of Alizarin Red S is added, drop by drop, until the hydroxide turns a deep wine red. The bones will be well stained after 24 hr in this solution. If the soft tissues have picked up any of the red color, they may easily be destained in 90% alcohol acidified by 1% sulfuric acid. Following the final destaining, the embryo is dehydrated in a series of alcohols, followed by three changes of benzene. The specimen is cleared in methyl salicylate, and may be stored permanently in a fresh and pure solution of this clearing agent.

The result of this procedure is an embryo in which the soft tissues are transparent, ossified tissues are stained a deep red, and cartilaginous tissues are stained a dark blue. In addition to the color differentiation of these two tissues, the degree of ossification or chondrogenesis at the given developmental stage is indicated by the intensity of the stain. By omitting either the Toluidine Blue or the Alizarin Red S stain, a preparation with only bone or cartilage stained may be obtained.

Staples and Schnell (1964) have modified the technique of Crary (1962) to provide rapid clearing following Alizarin Red S staining of fetal bone in mammals. These authors present data for rapid processing of fetal and newborn hamsters, mice, rats, and rabbits, and recommend their procedure when rapid processing of large numbers of specimens is necessary.

A procedure directed specifically toward the staining of bone of human fetuses younger than 18 weeks of age is described by Richmond and Bennett (1938). Although this procedure requires approximately 9–10 weeks of processing, the staining obtained for the resolving of the smallest bones is reportedly excellent.

B. Sectioned Material. 1. *Choice of Fixatives and Embedding Media.* The proper choice of fixative is critical to the successful preparation of tissue, as fixation results not only in a change in the appearance and texture of most material, but also in the physical properties and reactivity of the tissue. Before choosing a fixative one must first consider whether the shape and gross morphological characteristic of the body or the finer detail of the tissues is of greater importance, for most fixatives compromise one for the other. Moreover, the choice of fixing fluid must be carefully considered, for an inappropriate fixative may affect the successful embedding, sectioning, or staining of the embryonic material.

Since most embryo specimens are fixed by immersion in the fixing fluid, rather than by perfusion of the body, the rate of fixative penetration is of paramount concern. Chemicals of low molecular weight with few reacting radicals penetrate most rapidly, and the rapid penetration of the fixative will eliminate autolytic post-mortem changes which could ruin an otherwise good specimen. Therefore, if the embryo to be fixed is larger than about 2 mm, it will be necessary to provide an entrance through the body wall for the fixative to enter the body cavity in order to allow proper fixation of the internal organs. Formaldehyde, with a molecular weight of only 30, is a commonly used additive to many fixatives because it penetrates tissues rapidly and allows good retention of physical characteristics. As Gray (1954, Ch. 18) shows, there are nearly 800 solutions which are recommended as fixatives for various tissues or for special effects. A discussion of the advantages and disadvantages of the various classes of fixative agents, as well as the recipes of all specific fixing fluids mentioned in the present chapter, may be found in Gray (1954). Further information on all steps in the preparation of microscope slides for the novice as well as the serious technician is also presented in that reference.

The fixing solutions listed below were chosen due to their proven success in practice with vertebrate embryonic material. Some notes providing useful information accompany each fixative listed.

a. Tellyesniczky's fluid is a good general fixative for use with all vertebrate embryos.

Potassium dichromate	3.0 g
Glacial acetic acid	5.0 ml
Distilled water	100.0 ml

The solutions are mixed just before using. The specimens should be fixed for 24–48 hr, changing the solution once or twice during this period. The fixed material should then be washed in running water for at least 6 hr before dehydration in alcohols.

Smith's modification (see Davidson, 1932) is better for frog eggs and tadpoles less than 10 mm in length.

Potassium dichromate	0.5 g
Glacial acetic acid	2.5 ml
Formalin	10.0 ml
Distilled water	87.5 ml

Material is fixed for 24 hr, washed as above, and transferred to 3% formalin, which is changed several times until no more color is removed. The specimens may then be stored in 3% formalin.

b. Zenker's fluid preserves embryonic material up to about 3 cm well.

Potassium dichromate	2.5 g
Mercuric chloride	
(corrosive sublimate)	5.0 g
Sodium sulfate	1.0 g
Glacial acetic acid	5.0 ml
Distilled water	100.0 ml

The first two ingredients are dissolved in water which is heated. No metal objects should be used as these will be corroded by the sublimate. Acetic acid is not added until just before use. Embryos should be fixed 2–24 hr, until opaque throughout, then washed for 10–20 hr to remove the sublimate. The washed specimens are dehydrated in an increasing series of alcohols. All sublimate has been removed from the specimens when tincture of iodine (or Lugol's iodine solution) added to the 95% alcohol during dehydration does not lose its red color upon standing for several hours.

c. *Goldsmith's fluid* is recommended for avian embryos which are to be later stained in Heidenhain's iron-hematoxylin.

Chromic acid, 1% solution	75.0 ml
Potassium dichromate, 2% soln.	20.0 ml
Glacial acetic acid	5.0 ml

The fixation is diluted with an equal amount of water before use.

d. *Lavdowsky's fluid* (Mossman modification) is recommended for mammalian embryos, as it has good penetration properties and will decalcify any lime salts which might be present.

Formalin, commercial	10.0 ml
95% alcohol	30.0 ml
Glacial acetic acid	10.0 ml
Distilled water	50.0 ml

Tissue should be fixed 24–48 hr, replacing with 70% alcohol after a brief rinse in water.

e. *Worchester's fluid* is recommended as a general fixative for eggs of fish and amphibians, and for the embryos of these vertebrates.

Mercuric chloride	
(corrosive sublimate)	5.4 g
Formalin, 10% aqueous	90.0 ml
Glacial acetic acid	10.0 ml

Fixed eggs and embryos are washed to remove the sublimate as recommended for Zenker's fixative.

f. *Rabl's fixatives* are reported to be excellent for embryos of all vertebrate classes. Blastoderms and young embryos of fish through mammals may be fixed in the solution which follows.

Platinum chloride, 1% solution	10.0 ml
Picric acid, saturated aqueous	20.0 ml
Distilled water	70.0 ml

The stock solution of platinum chloride should be kept in the dark. Embryos are fixed for 12–24 hr, followed by washing in water for several hours. For larger embryos, such as avian embryos over 80 hr of age, Rabl recommends the following fixative:

Mercuric chloride, saturated	
aqueous	25.0 ml
Picric acid, saturated aqueous	25.0 ml
Distilled water	50.0 ml

Embryos should be fixed about 12 hr, and washed in water for 2 hr more before beginning dehydration in alcohol.

g. *Bouin's fixative* is a very popular fixing fluid for routine work, but is probably one of the less ideal fixatives for vertebrate embryos. Although there have been numerous modifications of this fixative, Bouin's original formula is given here:

Picric acid, saturated aqueous solution	75.0 ml
Formalin	25.0 ml
Glacial acetic acid	5.0 ml

Tissues are fixed for 4–12 hr, and then placed in several changes of 70% alcohol until all yellow is removed. This process may be hastened with the addition of a few drops of saturated lithium carbonate to the alcohol, but the mixture must be filtered before use. It is essential that all picric acid be removed for good results in staining. A combination of dioxane and Bouin's fixative has been found useful for frog eggs in advanced stages of division and this is described in a later section dealing with fixatives employed when yolk is present.

Following any fixation the specimens must be washed free of excess or unreacted chemicals in the solution, as these might react undesirably with substances used in subsequent steps. All water must then be removed by dehydrating the tissue, usually by transferring the specimen through an increasing series of alcohols as in preparation of whole mounts; and finally the specimen should, in most cases, be cleared before being embedded. Because the optimum solutions for washing, dehydration, and clearing and the times required for each process vary so much with procedure and material, no general instructions are given here. The washing procedure, for example, varies greatly depending on the type of fixative employed; dehydration time depends to a great extent on the size of the specimen and degree of cavitation of the body; and whether clearing is necessary or not depends on the type of embedding, as nitrocellulose embedding commonly does not require any clearing. It is suggested that the interested reader refer to a more specialized reference for his particular needs.

Two types of embedding are commonly used for sectioning animal tissue—paraffin and nitrocellulose. Embedding provides a supporting medium which is necessary for obtaining very thin sections of the specimen, and provides a matrix in which the sections can be handled. We shall consider here only the comparative advantages and disadvantages of paraffin vs. nitrocellulose embedding, and again leave the details of the method to more specialized references.

Paraffin embedding is the most popular method for routine slide-making, as it is relatively inexpensive, easy, and rapid, and requires little expensive or elaborate equipment beyond the standard oven and rotary microtome. Sections from 2 to 30 μ can be cut. A decided advantage for paraffin embedding is that the sections adhere to form a ribbon when cut, thus insuring that the serial order of the sections is maintained. On the other hand, there are some decided disadvantages to the paraffin method. One is that paraffin embedding is limited to small specimens, preferably with one dimension less than 2 mm in order to allow rapid penetration of the paraffin and exchange with existing fluids. Although paraffin can penetrate thicker tissue with longer time, the elevated temperature required to keep the paraffin melted tends to harden tissue with exposure, thereby producing an undesirable effect for sectioning. A vacuum oven will enhance the penetration of the paraffin while reducing the time required for the tissue to remain at the elevated temperature. A second advantage is that paraffin embedding often introduces artifacts, especially of shrinkage or separation of the tissue while the sections are mounted on the slides. To some degree the amount of this artifact depends on the suitability of the fixing fluid. In addition to the above, static electricity in the room may electrify the thin ribbon of sections as it forms on the microtome knife. This may be eliminated by increasing the humidity in the vicinity of the workspace, but may prove ineffectual if the static is too great, and one may simply have to wait for a better day.

Nitrocellulose, as it is used in histological procedures, is a partially nitrated form of cellulose which differs from the explosive (gun cotton) in that the latter is more completely nitrated. The most commonly used trade names of nitrocellulose for histological use are Celloidin and Parlodion. In general, because nitrocellulose embedding is more tedious and requires more expensive equipment than needed for the paraffin method, the use of this plastic will be limited to specimens where special requirements are necessary. With embryonic material, nitrocellulose embedding would be necessary where sections of intact large specimens are required. One such example is the use of large mammalian embryos, because very large specimens can be sectioned successfully by this method. However, the cutting of thin sections (less than about 10 μ) is difficult. Since infiltration and embedding of the specimen takes place at room temperature, one does not run the risk of hardening the tissue as in the previous method, and the consequent preservation of histological detail is excellent. It should be noted, however, that the procedure of embedding into the final nitrocellulose block may take days or weeks, as opposed to hours with paraffin. Since nitrocellulose sections are cut and removed from the knife individually, special efforts must be made to insure proper serial order. Finally, it is often difficult to stain the tissue without staining the nitrocellulose matrix of the section.

2. *Dealing with Yolk in Eggs and Young Embryos.* Yolk is more plentiful in the eggs and embryonic bodies of some vertebrates than in others, and in large quantities may present technical difficulties for many histological techniques which would otherwise be successful. The two main difficulties which must be overcome when dealing with yolky material are: (1) The selection of a fixative which will not harden the yolk platelets, thus causing severe fracturing during sectioning; and (2) the selection of a stain which does not bind to the yolk so heavily as to obscure other cellular components. Since the eggs of most reptiles and birds are heavily yolked, and the part of the blastoderm containing the embryo can be successfully separated from the bulk of the yolk-containing egg, the young embryos of these classes of vertebrates are most successfully preserved by first separating the embryo from the yolk rather than trying to work with the entire egg. In eggs and early embryos of fish and amphibians, however, the yolk is of much greater distribution, particularly in the young embryonic body, and must be given special consideration.

a. Fixatives. The first three fixatives listed in the preceding section (Tellyesniczky's fluid; Tellyesniczky, Smith modification; and Zenker's fluid) are each suitable for preserving yolky material in fish, amphibian eggs, and embryos. In addition, several investigators recommend Worcester's fluid (formula also given above) for excellent tissue preservation in these animals. It should be recalled that the mercuric chloride in tissue fixed with Zenker's or Worcester's fluid must be thoroughly removed by washing, and removal of the sublimate may be facilitated by iodine in the higher grades of alcohol. If the fixed yolky material is to be stored for several days or longer before dehydration, storage in 5% formalin is recommended over the usual 70% alcohol, since the formalin will not result in hardening of the yolk.

Many investigators recommend the Bouin-dioxan mixture of Puckett (1937) which avoids shrinkage and hardening often encountered with other fixatives. Puckett reports that his procedure is equally applicable for blastulae and gastrulae of the frog and large pieces of ovary which contain ripe ova. These tissues may be sectioned with little or no cracking, as the yolk does not become hard or brittle with this treatment. His fixative consists of Bouin's fixative and dioxan mixed in the proportions of 2 parts fixative to 1 part dioxan.

Gregg and Puckett (1943) have modified Worcester's fluid to make a fixative they regard as most satisfactory for yolk-laden amphibian eggs. Their modification is as follows:

Saturated aqueous mercuric chloride	90 ml
Formalin	8 ml
Glacial acetic acid	2 ml

Small egg masses with adherent jelly are fixed for two days in a large volume of this fixative. The fixed eggs may then be stored in 5% formalin; good results in sectioning are reported for eggs fixed even after a year's storage in the formalin. After washing for an hour, the eggs are de-jellied by shaking in a test tube partially filled with water, and the vitelline membrane is removed by shaking the jelly-free eggs in a partially filled test tube of 1% NaOCl solution for several minutes. Subsequent treatment of the eggs is similar to the routine procedures following fixation in corrosive sublimate.

b. Dehydration and clearing agents. Most techniques use a conventional alcohol series for dehydration of the fixed tissues, but they avoid the use of xylene or toluene for clearing. Some of the various methods which are applicable to eggs and early embryos of fish and amphibians will be considered here.

Amyl acetate is an excellent agent for clearing, as well as storage, of fixed yolky material. Barron (1934) recommends that the tissue be placed in equal parts of amyl acetate and absolute alcohol for 2 hr following complete dehydration through lower grades of alcohol. The specimens are then transferred to pure amyl acetate for 24 hr, followed by infiltration with amyl acetate and warm paraffin for 2 hr and final embedding in paraffin. Exposure to melted paraffin is limited to 1 hr, with several changes during this period. Barron (1934) has also worked out a procedure for use of amyl acetate in embedding in celloidin. Drury (1941) recommends clearing in pure amyl acetate for 24 hr following dehydration to 95% alcohol. The clearing agent is changed one time during this period. After a rinse in toluene, the cleared tissue is embedded in paraffin at a temperature which is slightly higher than the minimum melting point of the paraffin, again limited to 1 hr.

Aniline oil also prevents hardening of yolk. Following the 70% alcohol in the dehydration process, Bragg (1938) transfers the tissue to a mixture of 70% alcohol and aniline oil (1 : 2) for several hours, followed by a mixture of 95% alcohol and aniline oil (1 : 2) for a similar period. Final clearing in pure aniline oil for approximately one hour is followed by a mixture of aniline oil and toluene (1 : 1) and then pure toluene, each for one or more hours. To avoid long exposure to heat, paraffin (m.p. 53°C) is dissolved in toluene until the solution is saturated, and the

tissue is placed in this solution for 1–4 hr. It is then infiltrated with melted 53°C paraffin with several changes over 3–4 hr (a vacuum oven would be advantageous to reduce this time considerably). Final embedding is in the same paraffin with 1% bayberry wax added to allow thin sections to be more easily cut.

c. Embedding. Embedding yolky material in paraffin, as considered above, is acceptable provided that: (1) The *temperature* is kept to a minimum (slightly above the minimum melting point during infiltration, but at a temperature allowing the paraffin to become completely molten during the final change); and (2) the *time* is also kept to a minimum to prevent hardening (usually less than one hour for all changes). Use of a vacuum oven to further reduce time and to enhance infiltration is recommended. Puckett (1937), however, has developed a special method, employing dioxan not only for fixation (as described above), but also for washing, dehydration, and infiltration before embedding in hard paraffin. His embedding method utilizes consecutive changes in several paraffins of increasing melting points which apparently make embedding time less critical.

d. Stains. It has been reported by Faris (1924) that Janus green–neutral red avoids heavy staining of yolk in amphibian embryos while it selectively stains cell membranes and fibers in growing nerves and muscles. Faris adds that fibrils in growing neuroblasts and striations in embryonic muscle are clearly differentiated by the Janus green in embryonic amphibian tissue, and by modifying the staining time, fish embryonic tissues may also be stained successfully. His procedure should be consulted for use of these dyes.

In general, the counterstains commonly used with the hematoxylins do not produce good differentiation between ground cytoplasm and yolk granules in heavily yolked material. The two procedures given below are therefore a welcome addition to the small numbers of counterstains applicable for use with yolk-containing tissue. Slater and Dornfeld (1939) have combined Harris' hematoxylin, safranin O, and fast green to make a permanent triple stain. According to their procedure, paraffin-embedded sections are stained for 5 min in Harris' hematoxylin, rinsed in water, destained in 35% acid alcohol, and made blue in tap water. The slides are then put in a 1% solution of safranin O which is made in aniline water, for 5 min, then washed in tap water. They are further counterstained for 1–2 min in 0.5% fast green made in 95% alcohol. Timing for this step is critical—when ground cytoplasm is stained green while yolk granules remain bright red, the tissue should be washed and dehydrated with absolute alcohol. It should then be cleared in oil of cloves or xylol and mounted as usual. These authors report that following fixation in Puckett's Bouin-dioxan fixative all components are sharply defined; yolk granules appear red against a green cytoplasm, mitotic figures stand out (mitotic chromosomes = purple; spindle fibers = green), as do interphase nuclei (= blue) and nucleoli (= red).

Gregg and Puckett (1943) recommend a procedure for staining yolk-laden eggs of amphibians in Delafield's hematoxylin with eosin–orange G counterstain. The counterstain is made as follows:

1% eosin in 95% alcohol	10 ml
Saturated solution of orange G in 95% alcohol	5 ml
95% alcohol	45 ml

Iron hematoxylin, a favorite in many histological labs, is generally inferior as a stain for material containing much yolk since it has such a great affinity for the protein components of the yolk that other cytoplasmic structures are usually obscured.

IV. Special Staining Techniques for Embryos

A. Vital Staining of Vertebrate Embryos. Vital staining is the selective application of relatively nontoxic dyes or inert particles to the plasma membrane and cytoplasm of living cells. Thus, *vital staining* should clearly be discriminated from *intra vitam* staining, which involves the introduction, by injection or ingestion, of dyes or particulate material in suspension. The former is particularly useful in following one or more stained cells in a larger population during their migration or displacement in embryonic development, particularly the early morphogenetic movements; the latter allows the visualization of cell activity during such processes as ingestion, absorption, or excretion. The discussion here will be limited to vital staining.

1. *Technique for Applying Dyes.* The technique of localized vital staining in fertilized eggs of vertebrates was worked out by Vogt in 1925. His method involved the application of thin pieces of agar with 1 % solutions of Neutral Red, Nile Blue Sulfate, or Bismark Brown stain. A small piece of the stained agar, once placed in contact with the blastomeres of a developing amphibian embryo, will impart the color of the agar film to the cell, with no diffusion to adjacent cells. Thus, by using several differently colored dyes, or by spacing the marks in definite patterns on the surface of the blastula, the paths of many blastomeres can be followed as the cells move away from their positions on the surface where they were marked.

Lindahl (1932) has described a more precise, but also more involved method of applying very small areas of stain, i.e., 5–10 μ in diameter. His technique was devised for marking blastomeres of invertebrates of later cleavage stages. A capillary tube is drawn out in a very small flame to the diameter of the desired mark, and the large end of the tube is sealed in the flame. Again, a solution of 1 % dye in agar is used, and the microtip of the capillary tube is placed in the warm agar, allowing some to withdraw into the capillary through the small opening in the tip. Once cooled, the dyed agar in the capillary pipette may then be applied to the blastomeres with a steady hand, or may be inserted in a micromanipulator for more precise control.

Oppenheimer (1936) recommends that cellophane soaked in 1 % aqueous solution of vital dye (Nile Blue Sulfate, Neutral Red) is preferable to stained agar in marking fish embryos. Her procedure should be consulted before working with these animals.

The same vital dyes discussed above have been applied to the chick embryo as a means of following cell movements during morphogenesis. However, because these weak concentrations of dyes fade readily in the chick embryo, and this form is unable to tolerate dyes at greater concentration, the results are not as good by this method in chick as in amphibian embryo (see Appendix C in Spratt, 1947, for disadvantages of vital staining of chick blastoderm).

2. *Procedure for Preserving Stained Tissue.* Adams (1928) reports a simple method which provides good preservation of Nile Blue Sulfate vital stained tissue of urodelean amphibians. Pure acetone is used for dehydration instead of alcohols, since the latter leaches out the color of vital dyes. However, Stone (1932) reports that the procedure of Adams is not satisfactory for heavily yolked cells, and describes the following procedures. Tissue with vital stain is fixed in Zenker–acetic acid for two hours, followed by one hour in running water. The washed tissue is then transferred to a 1 % aqueous solution of phosphomolybdic acid for two hours. Dehydration in an alcoholic series is now possible provided that a small quantity (0.1 %) of phosphomolybdic acid is added to each change of alcohol. Following a gradual and complete dehydration (1/2 hr in each of the following alcohols with acid: 50 %, 70 %, 80 %, 95 %, 100 %), Stone recommends that the embryos be transferred for 1/2 hr to equal parts absolute alcohol (with the 0.1 % phosphomolybdic acid added) and cedarwood oil. The specimens are cleared in this mixture overnight, then infiltrated in paraffin and sectioned in routine fashion. After mounting, the paraffin is removed with xylol, and then coverslipped in the usual manner. This procedure preserves stained granules in the cytoplasm, as well as nuclear and cell boundaries.

Weissenberg (1937) has found that Bismark Brown vital stain can be preserved in eggs of the lamprey and amphibians if they are fixed in a mixture of 3 parts absolute alcohol and 1 part glacial acetic acid. The lamprey egg is fixed well in ten minutes; the eggs of other anamniotes may require an adjustment from this time. Following fixation, the eggs are washed in absolute alcohol for ten minutes, and are subsequently cleared in cedarwood oil. If sections are desired, the cleared eggs may be rinsed briefly in xylol, and embedded in paraffin. The stain remains preserved for several years in eggs and early embryos which remain in cedarwood oil or have been sectioned as directed above.

3. *Technique for Applying Carbon Particles.* Spratt pioneered the method of inserting particles of charcoal on the blastoderm of young chick embryos, and this method has proven to be superior to that of staining with vital dyes in certain species, notably the chick. The early chick embryo (less than 24 hr of incubation) must be explanted for the successful use of this technique, for the wounds created in the vitelline membrane when the carbon particles are applied *in vivo* often result in abnormal development or death. The procedure is described in Spratt (1946; 1947).

The application of the carbon-marking technique to the developing fish has been described by Brummett (1954), and his description should be consulted for this class of vertebrates.

Since all mammalian embryos develop *in utero*, with the exception of the monotremes, which are too scarce for laboratory use, the procedures described above are not applicable to mammals. They would be, however, equally useful for reptilian embryos, and it should be possible to use either the agar-dye or carbon particle method for these forms.

B. Chromosome Preparations from Embryonic Tissue.
1. *Amphibian Tail Tips.* The tailfin of amphibian embryos is a good source for obtaining mitotic figures of somatic cells. A technique for preparing and preserving mitotic figures in amphibian tailfins has been useful in descriptive (see Fankhauser and Humphrey, 1959) as well as experimental (see Costello et al., 1957) studies, and is a technique easily applicable for student laboratory experience.

a. Procedure for tailfins of anurans. Embryos of *Rana pipiens* at stage 23–25 (Shumway, 1940) that have a thin tail transparent at the posterior margin should be chosen. The embryos are fixed for about two hours in Bouin's fixative. The fixed embryo is then transferred to 70% alcohol made with saturated lithium carbonate (filter before use) until the yellow has been removed, and then hydrated through several intermediate alcohols. The tissue will become more flexible if allowed to remain in water for a day. The epidermis covering the tail is stripped off the embryo's body by first making a shallow cut around the animal's body from the point where the dorsal tailfin joins the head, ventral to the cloaca on each side, and second, removing the margin of the dorsal and ventral tailfins completely so that the epidermis on each side of the embryo's tail may be stripped separately from the underlying muscle. It may be necessary to separate the two sides of the tissue with a fine-pointed scalpel blade before they can be peeled from the body. If the epidermis is not easily removed, a glass needle may be used to separate it from underlying tissues.

The connective tissue and muscle are cleaned from the inner surface of the tailfin epithelium before the tissue is mordanted in 2.5% ferric alum (overnight) and then stained with Heidenhain's iron hematoxylin until the tissue appears completely black. Since this stain is a regressive one, the tissue requires destaining in a fresh solution of the mordant until good differentiation is obtained. The tissue should be placed on a glass slide for microscopic examination. The chromosomes will appear black (Fig. 3). as will any remaining yolk granules. The tissue can be made permanent by dehydrating through an increasing series of alcohols, clearing in xylol, and mounting between a slide and coverslip with a synthetic resin.

b. Procedure for tailfins of Urodeles. Several species in the genera *Ambystoma* (*Amblystoma*) and *Triturus* (*Taricha*) have become standard material for embryological research, and the tailfins of the common species in both these groups are suitable for chromosome demonstration. Larvae which are about to commence feeding (approximately Harrison stage 46 for *Ambystoma maculatum* [see Hamburger, 1960, p. 213, for staging] and

equivalent developmental stage for other species) are used. The procedure of Costello and Henley (1949) is described here.

The distal one-third of the tail should be removed and transferred in a wide-mouth pipette to a dish with fresh Bouin's fixative modified by the addition of 1 g urea and 5 ml extra acetic acid per 100 ml standard Bouin's fixative. The tail tips remain in the fixative for 2–6 hr, and are then washed in 70% alcohol overnight or until the yellow color is removed. The tissue is stained in Harris's acid hematoxylin which has been diluted 1 : 2 or 1 : 3 with distilled water before use. Usual staining time is 15–30 min, but this is variable. Since Harris's hematoxylin is a progressive stain, destaining is not necessary. The stained tissue should be transferred to an alkali solution (a few drops of 1% sodium bicarbonate in tap water) for bluing. If desired, the tail tips may be counterstained with erythrosin or acid fuchsin. The tissue is subsequently dehydrated in alcohols, cleared in xylol, and mounted on a slide with a synthetic resin and coverslipped. In addition to excellent preparation of chromosomes, this procedure also stains nucleoli well.

2. *Germinal Vesicles.* The "germinal vesicle" is the immature nucleus of the oocyte and, as such, contains the diploid ($2N$) number of chromosomes. Prior to the first meiotic division, when half the chromosomes are lost to the first polar body, many changes can be noted in the maturation of the chromosomes within the germinal vesicle (see Duryee, 1950) with the light microscope.

An excellent procedure for laboratory study of the amphibian germinal vesicle is given in Rugh (1962), and that reference should be consulted for specific formulas of solutions and general precautions to be considered when performing experiments with the germinal vesicle. The following method is generally applicable to urodeles as well as anuran species. Both references given above were consulted for the following account.

The ovary of a sexually mature frog is removed and

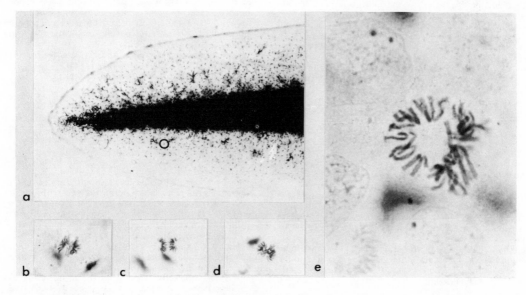

Fig. 3 Preparation of embryonic frog tailfin stained to reveal mitotic figures. a. Tail as mounted on slide with one mitotic figure indicated (circle), 10×. b.–d. Examples of mitotic figures from this tailfin, 80×. e. High-power view of paired chromosomes in one mitotic figure, in which the diploid number of chromosomes ($2N = 26$) may be discerned, 320×.

placed in amphibian Ringer's solution. A small piece of ovary, containing 20–30 eggs, is placed in several changes of Duryee's Nuclear Medium (N-medium),[5] each change for 3–5 min. A few eggs are then isolated in a Syracuse dish with fresh N-medium, and under low power of a dissecting microscope a tear is made with sharp-pointed forceps through the follicle cell layer and oocyte wall in the lower animal hemisphere near the equator of the egg. The large germinal vesicle will be extruded from the egg with other egg contents, which will flow freely through the opened wound. The germinal vesicle should be immediately cleaned of all adhering yolk by applying fresh N-medium to it in a gentle stream from a fine pipette. After all yolk has been removed, the isolated germinal vesicle is transferred to a clean Syracuse dish for examination of the unfixed germinal vesicle (for procedure, see Rugh, 1962, p. 172), or to any clean glassware suitable for fixing, if a permanent mount is desired.

Permanent mounts may be made using the LaCour (1941) acetic-orcein stain-fixative method for chromosomes. The germinal vesicle should be placed in a solution containing 0.5% orcein in 45% glacial acetic acid for 30–60 sec. Rapid dehydration by alcohols is followed by mounting in an alcohol-free medium. Acidified 1% Methyl Green is also a useful stain. Rugh further reports that Bouin's fixation followed by Mayer's fixative and Mayer's haemalum stain produces satisfactory permanent mounts of amphibian germinal vesicles.

Victor B. Eichler

References

Adams, A. E., "Paraffin sections of tissue supravitally stained," *Science*, **68**:303–304 (1928).

Barron, D. H., "Amyl acetate: a useful solvent for embedding masses," *Anat. Rec.*, **59**:1–3 (1934).

Bragg, A. N., "The organization of the early embryo of *Bufo cognatus* as revealed especially by the mitotic index," *Z. Zellforsch.*, **28**:154–178 (1938).

Brummett, A. R., "Carbon marking technique in *Fundulus*," *J. Exp. Zool.*, **125**:447–486 (1954).

Costello, D. P., and C. Henley, "Heteroploidy in *Triturus torosus*. I. The incidence of spontaneous variations in a 'natural population,'" *Proc. Amer. Phil. Soc.*, **93**:428–438 (1949).

Costello, D. P., C. Henley, and D. E. Kent, "The effects of P[32] on mitosis in tail-tips of larval salamanders," *J. Exp. Zool.*, **136**:143–170 (1957).

Crary, D. D., "Modified benzyl alcohol clearing of alizarin-stained specimens without loss of flexibility," *Stain Technol.*, **37**:124–125 (1962).

Cumley, R. W., J. F. Crow, and A. B. Griffin, "Clearing specimens for the demonstration of bone," *Stain Technol.*, **14**:7–11, (1939).

Dividson, M. H., "Practical suggestions for embryological technique. IV. Frog eggs and tadpoles," *Turtox News*, **10**:203–204 (1932).

Drury, H. F., "Amyl acetate as a clearing agent for embryonic material," *Stain Technol.*, **16**:21–22 (1941).

Duryee, W. R., "Chromosomal physiology in relation to nuclear structure," *Ann. N.Y. Acad. Sci.*, **50**:920–953 (1950).

Fankhauser, G. and R. R. Humphrey, "The origin of spontaneous heteroploids in the progeny of diploid, triploid, and tetraploid axolotl females " *J. Exp. Zool.*, **142**:379–422 (1959).

Faris, H. S., "Neutral Red and Janus Green as histological stains," *Anat. Rec.*, **27**:241–243 (1924).

Fekete, E., O. Bartholomew, and G. D. Snell, "A technique for the preparation of sections of early mouse embryos," *Anat. Rec.*, **76**:441–447 (1940).

Gray, P., "The preparation of alizarin transparencies," *Mus. J. [London]*, **28**:341–342 (1929).

Gray, P., "The Microtomist's Formulary and Guide," New York, The Blaskiston Co., Inc., 1954.

Gregg, V. R., and W. O. Puckett, "A corrosive sublimate fixing solution for yolk-laden eggs," *Stain Technol.*, **18**:179–180 (1943).

Gregory, P. W., "The early embryology of the rabbit," *Carnegie Contrib. Embryol.*, **21**:141–168 (1930).

Hamburger, V., "A Manual of Experimental Embryology," Chicago, Univ. of Chicago Press, 1960.

Heuser, C. H., and G. L. Streeter, "Early stages in the development of pig embryos, from the period of initial cleavage to the time of the appearance of limb-buds," *Carnegie Contrib. Embryol.*, **20**:1–30 (1929).

Hollister, G., "Clearing and dyeing fish for bone study," *Zoologica*, **12**:89–101 (1934).

LaCour, L., "Acetic-orcein: a new stain-fixative for chromosomes," *Stain Technol.*, **16**:169–174 (1941).

Lewis, W. H., and E. S. Wright, "On the early development of the mouse egg," *Carnegie Contrib. Embryol.*, **25**:115–143 (1935).

Lindahl, P. E., "Zur experimentallen Analyse der Determination der Dorsoventralachse beim Seeigelkeim. I. Versuch mit gestrecken Eiern," *Roux' Arch. f. Entw.-mech.*, **127**:300–322 (1932).

Lundvall, H., "Ueber Demonstration embryonaler Knorpelskelette," *Anat. Anz.*, **25**:219–222 (1904).

Lundvall, H., "Weiteres uber Demonstration embryonaler Skelette," *Anat. Anz.*, **27**:520–523 (1906).

Oppenheimer, J. M. "Processes of localization in developing *Fundulus*," *J. Exp. Zool.*, **73**:405–444 (1936).

Patch, E. M., "Fertility and development of newt eggs obtained after anterior lobe implants," *Proc. Soc. Exp. Biol. Med.*, **31**:370–371 (1933).

Pincus, G., "Superovulation in rabbits," *Anat. Rec.*, **77**:1–8 (1940).

Puckett, W. O., "The dioxan-paraffin technic for sectioning frog eggs," *Stain Technol.*, **12**:97–98 (1937).

Richmond, G. W., and L. Bennett, "Clearing and staining of embryos for demonstrating ossification," *Stain Technol.*, **13**:77–79 (1938).

Rugh, R., "Induced ovulation and artificial fertilization in the frog," *Biol. Bull.*, **66**:22–29 (1934).

Rugh, R., "Ovulation induced out of season," *Science*, **85**:588–589 (1937).

Rugh, R., "Experimental Embryology," 3rd ed., Minneapolis, Burgess Publishing Co., 1962.

Shumway, W., "Normal stages in the development of *Rana pipiens*. I. External form," *Anat. Rec.*, **78**:139–147 (1940).

Slater, D. W., and E. J. Dornfeld, "A triple stain for amphibian embryos," *Stain Technol.*, **14**:103–104 (1939).

Spratt, N. T., Jr., "Formation of the primitive streak in the explanted chick blastoderm marked with carbon particles," *J. Exp. Zool.*, **103**:259–304 (1946).

Spratt, N. T., Jr., "Regression and shortening of the primitive streak in the explanted chick blastoderm," *J. Exp. Zool.*, **104**:69–100 (1947).

Staples, R. E., and V. L. Schnell, "Refinements in rapid clearing technic in the KOH-Alizarin Red S method for fetal bone," *Stain Technol.*, **39**:61–63 (1964).

Stone, L. S., "Selective staining of the neural crest and its preservation for microscopic study," *Anat. Rec.*, **51**:267–273 (1932).

Vogt, W., "Gestaltungsanalyse am Amphibienkeim mit örtlicher Vitalfärbung. I. Methodik," *Roux' Arch. f. Entw.-mech.*, **106**:542–610 (1925).

---

[5]Duryee's N-medium: 6.6 g NaCl; 0.14 g KCl; 1000 ml glass distilled water.

Weissenberg, R., "Bismark-brown as a vital dye for localized staining on the egg of lamprey and the opportunity of preserving it for paraffin sections," *Collecting Net,* **12**:131 (1937).

van Wijhe, J. W., "A new method for demonstrating cartilaginous mikroskeletons," *Proc. Roy. Acad. Sci. Amsterdam,* **31**:47–50 (1902).

Williams, T. W., Jr., "Alizarin Red S and Toluidine Blue for differentiating adult or embryonic bone and cartilage," *Stain Technol.,* **16**:23–25 (1941).

ENAMEL (Tooth)

Enamel is a hard tissue found on the external coronal surfaces of teeth or toothlike structures. Mature enamel contains relatively large, well oriented hydroxyapatite crystallites, embedded in an extremely small amount of organic matrix. Crystallite orientation patterns vary topographically and with species. The crystallites are often arranged in crystallite domains or "prisms." The prisms may alternate with narrow, relatively unmineralized areas called "prism sheaths."

The amount of organic matrix as it was originally deposited by the matrix producing cells (ameloblasts) decreases during enamel maturation.

1. Visual Inspection. This is the simplest way to study the surface of enamel. Odontometric studies of tooth size and shape, genetic studies of cusp patterns, and pathologic studies of gross defects may be performed this way.

2. Low power stereomicroscopy is also restricted to studies of the surface of enamel. It permits the observation of perikymata, the external evidence of the terminal locations of ameloblast groups.

This method may also be used for surface wear studies, and studies of the distribution of enamel lamellae. The latter structures may be made more visible after superficial decalcification and immersion in a silver nitrate solution. They can be defined as areas with a high content of organic material and very little mineral, probably reflecting an area of tension in the developing enamel which failed to calcify. Enamel lamellae usually extend from the dentino-enamel junction to the external enamel surface.

3. Light Microscopy. Here two avenues of approach are open. *A. Routine Histologic Sections after Decalcification.* Routine decalcification techniques usually do not leave the very delicate enamel matrix intact. After the process of decalcification, topics of study concerning enamel are limited to the matrix of hypomineralized enamel, where a relatively large amount of organic matrix is present; the matrix of developing enamel, especially in marsupials; and amelogenesis. In the study of enamel formation routine cytology of the formative cells, the ameloblasts; specific histochemistry of these cells and of their products and autoradiographic studies of both organic and anorganic components are possible.

Histochemical studies revealing some characteristics of more mature enamel matrix may be performed on slightly decalcified sections. Some decalcification techniques have been developed in which it is possible to leave components of the enamel matrix intact. A useful application of this technique has been the differentiation between enamel cracks and "true enamel lamellae."

B. Undecalcified Sections. Most light microscopic studies of enamel are done on undecalcified material after the preparation of ground thin sections. It is important to mention here that the plane of section strongly influences

the observations, because of the often complicated courses and shapes of the prisms.

The following techniques are used to study specific problems.

a. Transmitted light, revealing directions and shapes of the enamel prisms; incremental Retzius lines, zones representing a specific period in the development of the tooth, during which enamel was not formed "normally." These zones may show different degrees of calcification.

Transmitted light is also used for the observation of enamel tufts, structures representing small areas of tension hypocalcified during the development of the enamel. Tufts in general do not extend beyond one-third of the enamel width starting from the dentino-enamel junction, where they originate.

Other structures observable in this way are enamel lamellae and spindles. Spindles represent small cytoplasmic projections of some odontoblasts (the formative cells of the adjacent dentin tissue). These cell processes were trapped in the enamel during its formation and may be seen as a continuation of dentinal tubules crossing the dentino-enamel junction.

b. Incident light will serve to demonstrate Hunter-Schreger bands. These are alternate zones of prisms, reflecting incident light in different ways. This optical difference is caused by the differential course of the groups of prisms in these zones, so that some groups may be cut longitudinally, some transversely and some obliquely.

c. Phase-microscopy, used for more detailed studies of prisms and the transverse striations of these prisms, observed in sections where they are longitudinally cut. Phase contrast is also used for the study of surface-replicas and of etched surfaces of ground sections.

d. Polarized light is a most useful method for detailed study of crystallite orientation, prism orientation, and Hunter-Schreger bands. It is critical to remember that enamel normally has a negative birefringence, due to the extremely high concentration of mineral.

Fine details of crystallite orientation in relatively small enamel areas are discernible, particularly with the use of a first-order red plate in the optical train. The same technique permits semiquantitative studies of pathological decalcification (caries) as well as of developmental hyper- and hypomineralization.

Immersion of the ground sections in fluids of different refractive index is an additional aid.

e. Micro-radiography, using "soft" X-rays, permits semiquantitative observations of degrees of calcification in normal, developmental and pathological studies.

It is also used to study Hunter-Schreger bands.

f. Fluorescence microscopy: Directly or after impregnation with fluorescent dyes the organic component of enamel may be studied, as well as the primary enamel cuticle, found on the external surface of enamel as the final product of the ameloblasts, and hypomineralized Retzius striae.

A tetracycline administration in vivo may result in an area of hypocalcification in enamel, while subsequent administrations of this drug may actually render such an area fluorescent.

g. Interference microscopy and differential interference microscopy permit surface studies of ground sections rendering even greater detail of enamel structures than phase contrast microscopy.

4. Scanning electron microscopy is a nondestructive technique. When used with intact or fractured enamel surfaces, it has proved valuable in the visualization of some aspects of amelogenesis and of enamel structures. Primarily it demonstrates structures otherwise only determinable after model reconstruction.

5. Electron Microscopy. Both undecalcified and carefully decalcified enamel may be studied. This method may provide and has already provided a detailed knowledge of many of the ultrastructural aspects of amelogenesis and enamel maturation. It is augmented by the increasing use of autoradiography. Studies of the variation of the crystallite orientation reach a fine degree of precision in these studies. Major new insights have been reached by the integration of E. M. evidence with previous light microscopic findings.

Letty Moss Salentijn
Melvin L. Moss

ENDOPLASMIC RETICULUM

One of the most prominent contributions of electron microscopy to cytology is the discovery of a coherent system of membranes present in the cytoplasm of virtually all plant and animal cells: *the endoplasmic reticulum* (ER) (Porter, 1953). These endomembranes enclose a more or less ordered and continuous "reticulum" of cavities, thus introducing two distinct cellular phases separated by membranes, In electron micrographs the ER is observed as membrane-limited cisternal or vesicular elements (Figs. 1–6), whose coherence can only be deduced from serial sections. As could be concluded from phase contrast studies of cultured cells, this electron microscopic picture reflects a real cellular structure present in the living cell.

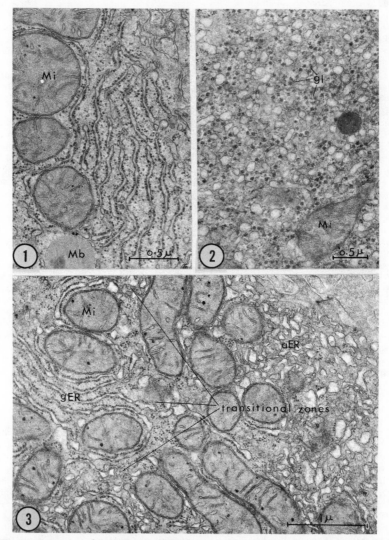

Figs. 1–3   Different appearance of the endoplasmic reticulum in liver parenchymal cells. Fig. 1: Groups of associated cisterns of the granular reticulum. (Mi: mitochondria; Mb: microbody.) Fig. 2: Vesicular form of the agranular reticulum with glycogen particles (gl) lying in between. (Mi: mitochondrion.) Fig. 3: Illustration of the transition of the different forms of the endoplasmic reticulum (transitional zones). (gER: granular reticulum; aER: agranular reticulum; Mi: mitochondria.)

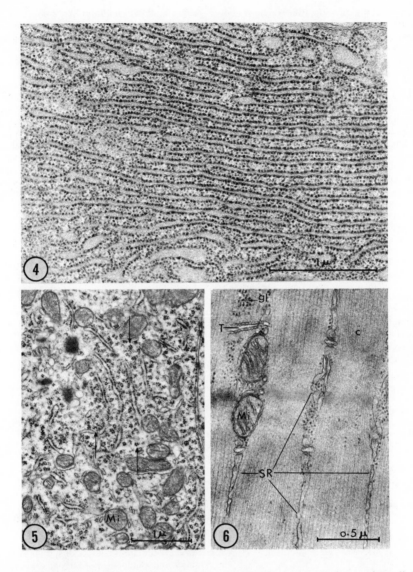

Figs. 4–6 Fig. 4: Parallel-oriented cisterns of the granular reticulum in the pancreatic exocrine cell. Fig. 5: Nissl-body (associated cisterns of granular reticulum) in the cerebellar Purkinje cell. Note the rosette-like polysomes (arrows). Mi: mitochondria.) Fig. 6: The sarcoplasmic reticulum (SR): a specialized form of the agranular reticulum in muscle cells. (T: transversal system; gl: glycogen, Mi: mitochondria, c: contractile apparatus.)

Description of the morphology of the ER are mainly based on electron microscopic observations of ultrathin sections prepared according to current methods of electron microscopy (cf. Sjöstrand, 1967; Wischnitzer, 1967; Reimer, 1967; Kay, 1965; Pease, 1964; and Hayat, 1970). The specific influence of pH, osmolarity, and ion content of the fixing medium on the ER-appearance has to be considered critically as is recently emphasized once more by Stockmann et al. (1970). Comparative studies with different fixatives are recommended. The contribution of the technique of freeze-etching is recently discussed (Moor, 1969).

The ultrastructural appearance and the extent of the ER greatly vary from cell type to cell type and also in correlation with changing physiological conditions (cf. Fawcett, 1966). Depending on the shape of the membrane-limited cavities *tubular* (Figs. 2, 3, 6) and *cisternal* (Figs. 1, 3, 4, 5) forms of the ER can be distinguished. Moreover, in many cell types the endomembranes—especially the cisternal form—are studded with ribosomes. For this reason a distinction is made between the *granular* (Figs. 1, 4, 5) (rough ER, ergastoplasm, chromophilic substance, Nissl substance) and the *agranular* (Figs. 2, 3, 6) (smooth ER) form of this cytoplasmic structure. These classifications are, however, arbitrary, and many transitional forms between the types mentioned are often encountered (Fig. 3). Continuity between the ER and the nuclear envelope has often been established and the latter is generally considered as an integral part of the ER. Moreover, continuity of the ER with the Golgi-membranes and the plasmamembrane has often been claimed. Nevertheless, with respect to the origin of the ER no definite statements can be given. It is considered either as a derivative of the nuclear envelope or as an outgrowth of the plasma membrane.

The ER membranes segregate the cytoplasm into distinct compartments; the *groundplasm s.s.* and the *reticuloplasm* (content of the ER cavities). This structural compartmentalization links up very well with the biochemical and physiological concepts of segregation of enzymes and their substrates as well as with the existence of intracellular selective permeability barriers (cf. Howland, 1968). Moreover, the extensive surface area of the endomembranes is considered to be a suitable support for the organized association of multienzyme assemblies.

The functions in cellular metabolism ascribed to the ER are numerous. The most important are briefly summarized in Table 1.

An important tool for the investigation of the functional significance of the ER is the method of cell fractionation. Differential and/or gradient centrifugation of tissue homogenates results in a separation into distinct subfractions, composed mainly of one type of cell organelle. The ER is regained as the *microsome fraction*, composed of smooth and rough vesicles (cf. Palade and Siekevitz, 1956). Further separation of this fraction into *smooth* and rough microsomes and separation of the latter into *membranes* and *ribosomes* can be considered as a routine technique (cf. Dallner et al., 1966, Glaumann and Dallner, 1970) (See Fig. 7). Biochemical estimations on these fractions have revealed and will continue to reveal a wealth of information on the biochemical and enzymic properties of the ER.

Quantitative changes in, for example the metabolism of proteins, glycogen, and steroids are often attended by changes in the ER which are essentially quantitative in nature (cf. Rohr et al., 1970 a, b). Investigation of such changes in structurally intact cells significantly contributes to our understanding of the functional aspects of the ER. Based on the principle (of Delesse) that volume composition can be directly estimated from areal profile composition, a number of *stereological procedures* have been worked out (cf. Weibel, 1969; DeHoff and Rhines, 1968). These procedures allow a statistically correct analysis of the volume density, surface density, and ribosomal density of the ER by means of simple and little time consuming point and intersection countings (cf. Weibel, 1969). Correlation of these quantitative morphological data of the ER with biochemical and physiological data obtained on microsome fractions will be a very useful tool in elucidating the structure-function relationships of the ER in a number of metabolic processes.

Besides the *morphometric approach*, the methods of *cytochemistry* and *autoradiography* at the electron microscopic level offer possibilities for the study of functional aspects of the ER which are relatively unexploited.

Table 1   Brief Summary of the Functions of the Endoplasmic Reticulum

Metabolism of Proteins	
(Blobel and Sabatini, 1970; Bresnick and Schwartz, 1968; Elson, 1967; Lipman, 1969; Sabatini and Blobel, 1970)	*ribosomes*: polymerization of amino acids
	gER membranes: segregation of proteins formed, especially in secretory cells
	aER membranes: contain proteolytic enzymes for degradation of proteins
Metabolism of Carbohydrates	
(Coimbra and Leblond, 1966; Maddaiah and Madsen, 1970; Stadhouders, 1965; Steiner, 1966; Vrensen and Kuyper, 1969; Vrensen, 1970a,b)	*gER and ribosomes*: involved in the repletion of glycogen, likely by *de novo* synthesis of glycogenetic enzymes, e.g., glycogen synthetase (E.C.2.4.1.11). Mainly in liver cells.
	aER: storage and degradation of glycogen. Membranes contain glucose-6-phosphatase (E.C.3.1.3.9.)
Metabolism of Lipids	
(Reid, 1967; Siekevitz, 1963; Stein and Stein, 1969)	*aER membranes*: contain a number of specific enzymes for the biosynthesis and metabolism of fatty acids, phospholipids, and steroids
Detoxification of Drugs	
(Jones and Fawcett, 1966; Orrenius et al., 1968; Siekevitz, 1963)	*aER membranes*: contain specific enzymes for the hydroxylation, oxidation, reduction, etc., of a variety of compounds of both biological and nonbiological origin
Transport	
(Cardell et al., 1967; Caro and Palade, 1964; Jamieson and Palade, 1968a,b)	*ER membranes*: intracellular transport of proteins and lipids.
Osmotic Properties	
(Bielka, 1969; De Robertis et al., 1970; Giese, 1968)	*ER membranes*: maintenance of the difference in osmotic and ionic properties of groundplasm and reticuloplasm
Conduction of Excitation	
(Bielka, 1969; De Robertis et al., 1970; Giese, 1968)	*aER membranes*: in muscle cells the sarcoplasmic reticulium (specialized form of aER) provides for a rapid conduction of the excitation from the plasmamembrane to the contractile apparatus.

ER: endoplasmic reticulum.
gER: granular form of the endoplasmic reticulum.
aER: agranular form of the endoplasmic reticulum.

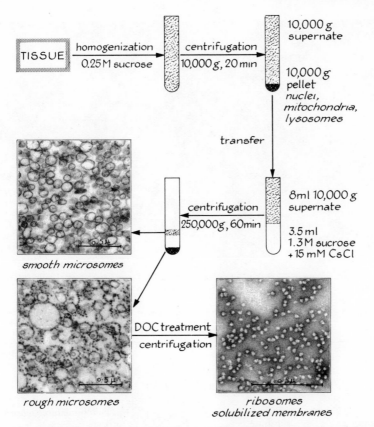

Fig. 7 Schematic representation of a method for the separation and subfractionation of the endoplasmic reticulum. (Micrographs: courtesy of Dr A. L. H. Stols. References: Dallner et al., 1966; Glaumann and Dallner, 1970.)

Although progress is made in developing methods for the specific staining of substances of biological interest (cf. Zobel and Beer, 1965), it is likely that the method of electron microscopic localization of enzymes will prove to be the more promising route of investigation. A generalized scheme for electron microscopic localization of enzymes and some of the basic prerequisites are summarized in Table 2. The opposite demands of adequate preservation of ultrastructure and of enzyme activity imply that a detailed study of optimal conditions is necessary for every new enzyme to be investigated. Until now, the method is of value only for enzymes bound to membranes, organelles, and particles.

The method of *submicroscopic autoradiography* too is a powerful tool for the study of the numerous functions of the ER in the intact cell. The appropriate administration of tritiated precursors to animals and plants often results in a metabolically directed incorporation of the radioactivity into one species of macromolecules (Leblond and Warren, 1965). The radioactivity, thus incorporated, can be localized submicroscopically by application of a suitable photographic emulsion over ultrathin sections, prepared from this material. Recent development of reliable and reproducible methods for emulsion application, photographic development (cf. Salpeter and Bachmann, 1964, Kopriwa, 1967; and Vrensen, 1970) and analysis of the final autoradiographs (cf. Salpeter et al., 1969; Williams, 1969) permits a rather strict localization of certain metabolic processes and even permits quantitative

evaluation. The technical problems of localizing soluble compounds (cf. Roth and Stumpf, 1969) restricts this method to macromolecules preserved *in situ* during the fixation procedure.

Future contributions to our understanding of the ER will be correlative in nature. Therefore, the methods of cell fractionation, morphometry, cytochemistry, and autoradiography must be used concurrently in the investigation of structure-function relationships of the endoplasmic reticulum.

G. F. J. M. VRENSEN
A. M. STADHOUDERS

References

Arnold, M., "Histochemie." Berlin–Heidelberg–New York Springer-Verlag, 1968.

Bielka, H., "Molekulare Biologie der Zelle," Jena, VEB Gustav Fischer Verlag, 1969.

Blobel, G., and D. D. Sabatini, "Controlled proteolysis of nascent polypeptides in rat liver fractions. I. Location of the polypeptides within ribosomes," *J. Cell Biol.*, **45**:146–157 (1970).

Bresnick, E., and A. Schwartz, "Functional Dynamics of the Cell," New York and London, Academic Press, 1968.

Cardell, R. R., S. Badenhausen, and K. R. Porter, "Intestinal triglyceride absorption in the rat. An electron microscopic study," *J. Cell Biol.*, **34**:123–155 (1967).

Caro, L., and G. E. Palade, "Protein synthesis, storage

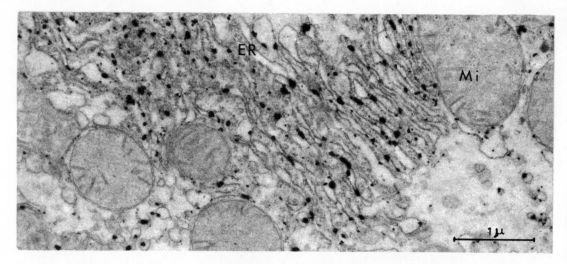

Localization of glucose 6-phosphatase in the endoplasmic reticulum (ER) of liver parenchymal cells. (Mi: mitochondria.)

Table 2   Generalized Scheme for Cytochemical Localization of Enzymes at the Electron Microscopic Level

Prerequisites		Procedure
1. Tissue preservation	Pre-fixation	Fixation in aldehyde solutions at appropriate pH and osmolarity
2. Enzyme localization	Incubation	Incubation mixture: *substrate*, cofactors, pH etc.
a. preservation of enzyme activity	substrate ↓ *enzymic reaction*	
b. immediate capture of stain precursor to minimize diffusion effects	stain precursor ↓ *capture reaction* stain deposit	*capture reagent*, heavy metal salts azo dyes, formazon salts
c. deposit of electron dense stain of minimal size		
3. Visualization of cellular ultrastructure	Post-fixation and embedding	Fixation in OsO₄, dehydration, and embedding

←——→ : opposite requirements.
References: Arnol, 1967; Daems et al., 1969; Holt, 1959; Reimer, 1967.

and discharge in the pancreatic exocrine cell. An autoradiographic study," *J. Cell Biol.,* **20**:473–495 (1964).

Coimbra, A. and C. P. Leblond, "Sites of glycogen synthesis in rat liver as shown by electron microscope radioautography after administration of glucose H³," *J. Cell Biol.,* **30**:151–175 (1966).

Daems, W.Th., E. Wisse, and P. Brederoo, "Electron microscopy of the vacuolar apparatus," pp. 64–112 in J. T. Dingle and H. B. Fell, ed., "Lysosomes in Biology and pathology," Vol. I, Amsterdam and London, North-Holland Publ. Co., 1969.

Dallner, G., "Studies on the structural and enzymic organization of the membranous elements of liver microsomes," *Acta Pathol. Microbiol. Scand. Suppl.,* **166** (1963).

DeHoff, R. T., and F. N. Rhines, ed., "Quantitative Microscopy," New York, McGraw-Hill Book Co., 1968.

De Robertis, E. D. P., W. W. Nowinsky, and F. A. Saez, "Cell Biology," Philadelphia, W. B. Saunders Company, 1970.

Elson, D., "Ribosomal enzymes," pp. 407–474 in D. B.

Roodyn, ed., "Enzyme Cytology," London and New York, Academic Press, 1967.

Fawcett, D. W., "An Atlas of Fine Structure. The Cell, Its Organelles and Inclusions," Philadelphia, W. B. Saunders Company, 1966.

Giese, A. C., "Cell Physiology," Philadelphia, W. B. Saunders Company, 1968.

Glaumann, H., and G. Dallner, "Subfractionation of smooth microsomes from rat liver," *J. Cell Biol.,* **47**:34–48 (1970).

Holt, S. J., "Principles and potentialities of cytochemical staining. Methods for the study of intracellular enzyme distributions," pp. 44–53 in E. M. Crook, ed., "The Structure and Function of Subcellular Components," Cambridge University Press, 1959.

Howland, J. L., "Introduction to Cell Physiology: Information and Control," New York, The Macmillan Company, 1968.

Jamieson, J. D., and G. E. Palade, "Intracellular transport of secretory proteins in the pancreatic exocrine cell. III. Dissociation of intracellular transport from protein synthesis," *J. Cell Biol.,* **39**:580–588 (1968a).

Jamieson, J. D., and G. E. Palade, "Intracellular transport of secretory proteins in the pancreatic exocrine cell. IV. Metabolic requirements," *J. Cell Biol.,* **39**:589–603 (1968b).

Jones, A. L., and D. W. Fawcett, "Hypertrophy of the agranular endoplasmic reticulum in hamster liver induced by phenobarbital" (with a review on the functions of this organelle in liver), *J. Histochem. Cytochem.,* **14**:215–232 (1966).

Kopriwa, B. M., "A semiautomatic instrument for the radioautographic coating technique," *J. Histochem. Cytochem.,* **14**:923–928 (1966).

Kopriwa, B. M., "The influence of development on the number and appearance of silver grains in electron microscopic radioautography," *J. Histochem. Cytochem.,* **15**:501–515 (1967).

Leblond, C. P., Warren, K.B., ed., "The use of Radioautography in Investigating Protein Synthesis," New York and London, Academic Press, 1965.

Lipmann, F., "Polypeptide chain elongation in protein biosynthesis," *Science,* **164**:1024–1031 (1969).

Maddaiah, V. T., and N. B. Madsen, "Studies on the biological control of glycogen metabolism in liver. III. Subcellular distribution of glycogen metabolizing enzymes," *Can. J. Biochem.,* **46**:521–525 (1970).

Moor, H., "Freeze-etching," *Int. Rev. Cytol.,* **25**:391–412 (1969).

Orrenius, S., Y. Gnosspelius, M. L. Das, and L. Ernster, "The hydroxylating enzyme system of liver endoplasmic reticulum," pp. 81–96 in P. N. Campbell and F. C. Gran, ed., "Structure and Function of the Endoplasmic Reticulum in Animal Cells," Oslo, Universitets Vorlaget, and London and New York, Academic Press, 1968.

Palade, G. E., and P. Siekevitz, "Liver microsomes. An integrated morphological and biochemical study," *J. Biophys. Biochem. Cytol.,* **2**:171–200 (1956).

Porter, K. R., "Observations on a submicroscopic basophilic component of the cytoplasm," *J. Exp. Med.,* **97**:727–750 (1953).

Porter, K. R., "The endoplasmic reticulum: some current interpretations of its forms and functions," pp. 127–154 in T. W. Goodwin and O. Lindberg, ed., "Biological Structure and Function," Vol. I, London and New York, Academic Press, 1961.

Reid, E., "Membrane systems," pp. 321–406 in D. B. Roodyn, ed., "Enzyme Cytology," London and New York, Academic Press, 1967.

Rohr, H. P., J. Strebel, and L. Bianchi, "Ultrastrukturell-morphometrische Untersuchungen an der Rattenleberparenchymzelle in der Frühphase der Regeneration nach partieller Hepatektomie," *Beitr. Pathol.,* **141**:52–74 (1970).

Rohr, H. P., A-C. Hundstad, L. Bianchi, and H. Eckert, "Morphometrisch-ultrastrukturelle Untersuchungen über die durch die Tageszeit induzierten Veränderungen der Rattenleberparenchymzelle," *Acta Anat.,* **76**:102–111 (1970).

Roth, L. J., and W. E. Stumpf, ed., "Autoradiography of Diffusable Substances," New York and London, Academic Press, 1969.

Sabatini, D. D., and G. Blobel, "Controlled proteolysis of nascent polypeptides in rat liver cell fractions. II. Location of the polypeptides in rough microsomes," *J. Cell Biol.,* **45**:158–172 (1970).

Salpeter, M. M., and L. Bachmann, "Autoradiography with the electron microscope. A procedure for improving resolution, sensitivity and contrast," *J. Cell Biol.,* **22**:469–477 (1964).

Salpeter, M. M., L. Bachmann, and E. E. Salpeter, "Resolution in electron microscope radioautography," *J. Cell Biol.,* **41**:1–20 (1969).

Siekevitz, P., "Protoplasm: endoplasmic reticulum and microsomes and their properties," *Ann. Rev. Physiol.,* **25**:15–40 (1963).

Stadhouders, A. M., "Particulate glycogen," Thesis, Nijmegen, 1965.

Stein, O., and Y. Stein, "Lecithim synthesis, intracellular transport and secretion in rat liver. IV. A radioautographic and biochemical study of choline-deficient rats injected with choline-3H," *J. Cell Biol.,* **40**:461–483 (1969).

Steiner, D. F., "Insulin and the regulation of heaptic biosynthetic activity," pp. 1–60 in R. S. Harris, I. G. Wool, and J. A. Loraine, ed., "Vitamins and Hormones," Vol. 24, New York and London, Academic Press, 1966.

Stockmann, F., L. Bianchi, H. P. Rohr, and H. Eckert, "Morphometrische Untersuchungen an der Rattenleberparenchymzelle nach Anwendung verschiedener Fixationspuffer," *Experientia,* **26**:174–176 (1970).

Vrensen, G. F. J. M., "Some new aspects of efficiency of electron microscopic autoradiography with tritium," *J. Histochem. Cytochem.,* **18**:278–290 (1970).

Vrensen, G. F. J. M., "Further observations concerning the involvement of rough endoplasmic reticulum and ribosomes in early stages of glycogen repletion in rat liver," *J. Microscopie,* **9**:517–534 (1970a).

Vrensen, G. F. J. M., "The influence of cycloheximide on the deposition and localization of glycogen in the liver of fasted rats," *J. Microscopie,* **9**:749–764 (1970b).

Vrensen, G. F. J. M., and Ch. M. A. Kuyper, "Involvement of rough endoplasmic reticulum and ribosomes in early stages of glycogen repletion in rat liver," *J. Microscopie,* **8**:599–614 (1969).

Weibel, E. R., "Stereological principles for morphometry in electron microscopic cytology," *Int. Rev. Cytol.,* **26**:235–302 (1969).

Williams, M. A., "The assessment of electron microscopic autoradiographs," pp. 219–272 in R. Barer and V. E. Cosslett, ed., "Advances in Optical and Electron Microscopy," Vol. 3, New York and London, Academic Press, 1969.

Zobel, C. R., and M. Beer, "The use of heavy metal salts as electron stains," *Int. Rev. Cytol.,* **18**:363–400 (1965).

f

FAT

Fat, known also as adipose tissue, presents to the microscopist a formidable tissue to study because of certain technical difficulties. This is due primarily to the large bulk of stored fat within fat cells which oft-times defies fixation, embedding, and sectioning procedures. Nevertheless, several biological microtechniques are presently in use to study this tissue to one's best advantage.

In mammals, fat is distributed in the subcutaneous region of the skin and in the immediate vicinity of many visceral organs, e.g., intestines, kidney, heart. It consists mainly of fat cells surrounded by connective tissue cells and fibers. In addition, it is permeated by nerve fibers, blood, and lymphatic vessels. The typical fat cell is polygonal or spherical in shape and consists of a large central lipid droplet surrounded by a thin rim of cytoplasm beneath the cell membrane. The diameter of such cells varies with the species and age of the animal, and can reach a size of 120 μ.

For light microscopic study, one can utilize techniques which allow for the visualization of fat cells with or without the presence of their large intracellular lipid droplets. Routine histologic procedures which use paraffin or celloidin as the embedding medium for tissue result in the loss of fat from fat cells due to the solubility of fat in solvents such as ethanol, xylene, toluene, ether, or benzene (Fig. 1). In order to visualize the intact fat droplet one must be able to preserve it throughout all phases of the microscopic technique. An aqueous fixative such as formalin (10%) is preferred by most microscopists who wish to preserve fat. This is followed by cutting frozen sections on a cryostat or CO_2-freezing microtome at the thickness desired, usually 10–15 μ. Alternatively, one may avoid cutting frozen sections by embedding the fixed tissue in a variety of media such as carbowax, polyvinyl alcohol, or ester waxes (Humanson, 1967). Sudan colorants, such as Sudan Black and Oil Red O, are used routinely to demonstrate neutral fat in adipose cells because of their solubility within the fat droplets (Humanson, 1967; Lillie, 1965; Thompson, 1966). Another technique to study fat, entails the use of osmium tetroxide as a stain for fat droplets in cells (Humanson, 1967; Lillie, 1965; Thompson, 1966). Osmium tetroxide reacts with fat by virtue of its oxidizing effect on the double bonds of unsaturated fatty acids, and causes a blackening of the lipid droplet due to osmium reduction. Subsequent embedding in paraffin or other media can then be accomplished. For example, Fig. 2 shows a micrograph of fat cells initially fixed in 3% buffered glutaraldehyde followed by exposure to 1% buffered osmium. In this case, the tissue was embedded in an epoxy resin and sections were cut at 1 μ on an ultramicrotome.

Adipose tissue can also be studied as a whole mount preparation which obviates any embedding or sectioning procedures. For this, small pieces of fat (preferably mesenteric fat) are dissected from the animal and placed

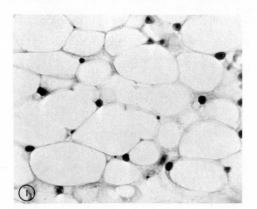

Fig. 1 Histologic section (7 μ thick) of mouse fat cells prepared in a manner which results in the loss of intracellular fat. ($\times 360$.)

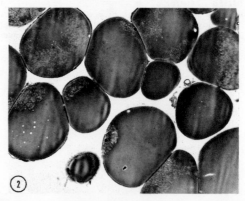

Fig. 2 One micron thick plastic section of fat cells fixed in glutaraldehyde followed by exposure to osmium tetroxide. In this preparation the fat is retained except for small areas of extraction. ($\times 480$.)

on a slide. A stain, either one which·stains nuclei (hematoxylin) or one which stains nuclei, blood vessels, and connective tissue (Masson's trichrome), is placed on the tissue. The tissue is rinsed, flooded with a mounting medium, and covered with a thin cover glass. In such a preparation, fat cells can be visualized in their entirety by careful focusing of the microscope.

One can use electron microscopy of adipose tissue although certain limitations make visualization difficult. Firstly, the large size of the fat cell (up to 120 μ) presents a restricted view of the non-lipid-droplet portion of the cytoplasm. Only in the area of the peripherally placed nucleus can one obtain a fair quantity of cytoplasm in which a variety of organelles can be viewed. Secondly, proper fixation is important for electron microscopic analysis. Cold buffered osmium tetroxide appears to be the preferred fixative for adipose tissue (Napolitano, 1965; Sheldon et al., 1962; Williamson, 1964). Primary fixation by gluteraldehyde followed by post-fixation in cold osmium has also been used with success perhaps overcoming some of the extraction properties of osmium (Stein et al., 1970). It is also important to limit the size

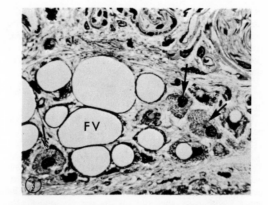

Fig. 3 One micron thick section of highly active fat cells from the mesentery of a mouse starved for 3 days. In this case, some cells contain fat vacuoles (FV) where fat has been extracted. Note several fat cells which are devoid of fat (arrows) due to extreme lipid mobilization. ($\times$600.)

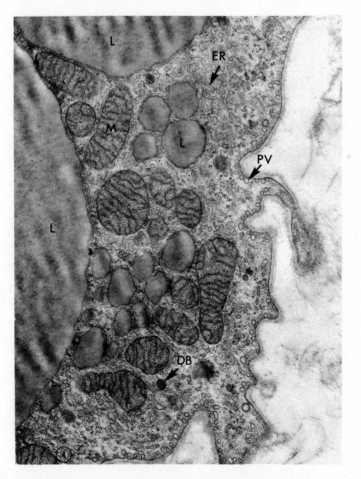

Fig. 4 Electron micrograph of a portion of an osmium-fixed fat cell undergoing lipid mobilization. With the reduction in volume of intracellular fat one can clearly visualize other subcellular structures, i.e., L = lipid droplets, M = mitochondria, ER = endoplasmic reticulum, DB = dense bodies, PV = pinocytotic vesicles. ($\times$19,200.)

of the tissue block to about 1 mm³ so that proper penetration of the fixative can be obtained. Tissue blocks can be dehydrated in graded ethanol and embedded in Epon, Araldite, Vestopol W, or Maraglas plastic mixtures (Pease, 1964). Lastly, it appears necessary to limit the size of the block face during sectioning to as small a size as possible (i.e., 0.5–1 mm²). In this manner, evenly cut sections can be obtained and chattering of the lipid droplets by the glass knife may be prevented.

One can also observe fat cells in a highly active metabolic state by fasting laboratory animals for 3–5 days or by chronic underfeeding whereupon large quantities of fat are mobilized from the cells (Napolitano and Gagne, 1963; Sheldon et al., 1962; Williamson, 1964). In this way, many more cytoplasmic constituents (e.g. mitochondria, endoplasmic reticulum, lysosomes) can be studied with ease since the fat droplet no longer predominates in the cytoplasm (Figs. 3 and 4). Conversely, fat cells can be studied histologically or cytologically as they are becoming differentiated during fetal and neonatal stages in laboratory animals (Napolitano, 1963). In this way, one may trace the cellular and subcellular changes occurring during fat synthesis in these cells.

Since normal fat cells are metabolically active in carbohydrate, protein, and lipid metabolism, one can study a variety of enzymes and products associated with these processes using a number of histochemical and cytochemical techniques. Again, the most preferable tissue is the fat of the mesenteric spread because of its relative thinness and ease of handling. Those enzymes which have been demonstrated include esterases, phosphatases, dehydrogenases, and oxidases, the methods for which are readily available (Barrnett, 1963; Thompson, 1966). In addition, several other chemical components of these cells can be selectively demonstrated using a variety of histo- and cytochemical techniques. Methods are also available for the demonstration of glycogen, a variety of lipids, and sulfhydryl groups of proteins (Barrnett, 1963; Thompson, 1966).

BERNARD G. SLAVIN

References

Barrnett, R. J., p. 3 in L. W. Kinsell, ed., "Adipose Tissue as an Organ," Springfield, Illinois, Charles C. Thomas, 1962.
Humanson, G. L., "Animal Tissue Techniques," San Francisco and London, W. H. Freeman and Company, 1967.
Lillie, R. D., "Histopathologic Technique and Practical Histochemistry," New York, McGraw-Hill Book Company, 1965.
Napolitano, L. M., J. Cell Biol., 18:663 (1963).
Napolitano, L. M., p. 109 in A. E. Renold and G. F. Cahill, eds., "Handbook of Physiology, Section 5: Adipose Tissue," Washington, D.C., American Physiological Society, 1965.
Napolitano, L. M., and H. T. Gagne, Anat. Rec., 147:273 (1963).
Pease, D. C., "Histological Techniques for Electron Microscopy," New York and London, Academic Press, 1964.
Sheldon, H., C. H. Hollenberg, and A. I. Winegrad, Diabetes, 11:378 (1962).
Stein, O., R. O. Scow, and Y. Stein, Amer. J. Physiol., 219:510, (1970).
Thompson, S. W., "Selected Histochemical and Histopathological Methods," Springfield, Illinois, Charles C. Thomas, 1966.
Williamson, J. R., J. Cell Biol., 20:57 (1964).

FEULGEN STAINS

Feulgen's "nuclear reaction" became known after Feulgen and Rossenbeck (1924) found that when hydrolyzed tissue sections are stained by leucobasic fuchsin (Schiff reagent), a magenta color develops at sites where aldehyde molecules from DNA are released. The essential basis of this reaction is that following hydrolysis in mild hydrochloric acid, or any other inorganic or organic acid, preferential removal of purines results, releasing aldehyde molecules from DNA that react with the dye molecules of the Schiff reagent, yielding a purple or a magenta color. Thus this reaction is a specific method for the in situ localization of DNA of the animal and plant cell nuclei.

The most important step in the cytochemical detection of DNA after Feulgen procedure is, therefore, preparation of the Schiff reagent. One of the current methods for its manufacture is according to de Tomasi (1936). There are various other methods such as the use of thionyl chloride (Barger and DeLamater, 1948; Dutt, 1967a) or dithionite (sodium hydrosulfite) (Alexander et al., 1950) which are required for the release of SO_2 which turns basic fuchsin solution into a colorless product. The latter two methods are more handy. The author has, however, modified the method of de Tomasi in such a manner that Schiff reagent prepared by his modified method is also equally convenient and is as potent a reagent as that obtained after the methods employing thionyl chloride or dithionite.

Preparation of Feulgen Stain. Dissolve 0.25 g of basic fuchsin by pouring over it 200 ml of boiling distilled water. Filter rapidly and then add 5 ml of N HCl and 1.0 g of potassium metabisulfite to the filtrate and shake well. This solution should be left in the refrigerator for ½ hr—the time required to bleach basic fuchsin solution. Afterwards, shake the dye solution well with 0.25 g of activated charcoal (Coleman, 1943) for a minute and then filter. The filtrate is now colorless. The Schiff reagent prepared in this way remains perfectly potent for more than a year if kept constantly in the refrigerator. It has been found that basic fuchsin of British Drug Houses Ltd., London, is particularly suitable. The dye product of E. Merck is also suitable but it requires shaking twice with activated charcoal in order to obtain a colorless reagent. The successive treatments with charcoal, however, do not hamper the staining potentiality of the reagent (Dutt, 1967a).

Conn (1961) stated that the dye basic fuchsin contains a mixture of three different dyes of the triphenylmethane series, viz., pararosaniline, rosaniline, and new fuchsin, which differ from one another with respect to the absence or presence of one or three substituent methyl groups. Thus in the routine preparation of Schiff reagent, any of the four dyes, including basic fuchsin, can be used.

Fixing and Staining Procedure. For the detection of DNA, tissues or smears may be fixed in either acetic acid–alcohol (1:3) or any other cytological fixative except those that contain picric acid. Fixation in formalin yields better results than fixation in acetic acid–alcohol. For quantitative study, fixation in 10% neutral formalin (Pearse, 1961) has been recommended by various authors, although fixation in 40% formaldehyde solution (w/v) yields far better results than fixation in 10% formalin. Salts of heavy metals, such as chromic acid, uranium acetate, or nitrate and platinum chloride, in combination with 10% formalin or acetic acid–alcohol yields somewhat better Feulgen stainability than after fixation in 10% formalin or after acetic acid–alcohol alone (Dutt, 1968b; 1969a).

Hydrolysis and Staining. The fixed tissue sections or smears are hydrolyzed in N HCl at 60°C for 7 min and then stained by Schiff reagent for 20 min at room temperature and under laboratory light. It has been seen that a longer staining period is not necessary since optimum staining is possible within the time prescribed above. Itikawa and Ogura (1954) suggested hydrolysis in $5N$ HCl at room temperature. Hydrolysis for 15 min in $5N$ HCl results in perfect Feulgen staining. Staining in darkness is not necessary, although some authors have stated that staining should be done in a cool and dark place. However, it has been seen that temperatures varying between 40° and 60°C are useful in showing an enhanced Feulgen staining of animal material (Dutt, 1968c).

Influence of pH on Feulgen Stain. The usual pH of the Feulgen stain is 2.0–2.3. At this pH stainability is not optimum, although for routine staining this pH is all right. However, to obtain optimum stainability, the hydrogen ion concentration of the Schiff reagent can be made less acidic by the addition of weak solutions of various alkaline substances (Itikawa and Ogura, 1954; Swift, 1955; Walker and Richards, 1959; Dutt, 1963, 1967b; 1968a).

Schiff-type Dyes. A number of potentially suitable Schiff-type dyes have now been found after Ostergren (1948) discovered that dyes such as brilliant cresyl blue, indoin blue, neutral red, safranin, and thionin can be used in place of basic fuchsin for the preparation of Feulgen stain (Kasten, 1959, 1960). The author (Dutt, unpublished) made studies to find out if the pH of some of the already known dye solutions such as, azure A, azure 11, toluidine blue O, thionin, brilliant cresyl blue, cresyl violet, safranin, and phenosafranin has anything to do in increasing the staining potentiality of these reagents. It has been found that all these dye reagents are susceptible to change of pH. Some become more reactive at slightly higher pH while others are more reactive at lower pH. The high and low pH is based on the initial pH when it is prepared.

Influence of Light during Feulgen Staining. To enable the cytochemist for obtaining better Feulgen staining, a new technique has now been devised (Dutt, 1969b). The technique is as follows: Following routine hydrolysis, the sections are stained by the conventional Schiff reagent and staining is carried out under light obtained from a Hanovia mercury arc lamp. The photochemical effect of this teatment is to change the color of the reagent to deep purple. The treatment is administered for 5–10 min during staining; afterwards staining under the laboratory conditions is carried out for 5–10 min, the slides are then treated with the usual SO_2 water for 15 min, dehydrated through different grades of alcohol to xylol and then mounted in D. P. X., made by B. D. H. Staining carried out by this technique shows a far better staining intensity as compared with that obtained after the conventional procedure. The wavelengths of light particularly suitable for obtaining better results are 360 nm and 390 nm.

In conclusion it can be said that for routine studies in detecting DNA cytochemically, the method of preparation of the Feulgen stain as suggested at the beginning may be quite adequate. But for the cytochemist, employing this dye reagent for quantitative purposes, much serious consideration of the variables would be needed before optimum staining is achieved.

MIHIR K. DUTT

References

Alexander, J., K. S. McCarty, and E. Alexander-Jackson, "Rapid method of preparing Schiff's reagent for the Feulgen test," *Science,* 111:13 (1950).

Barger, J. D., and E. D. DeLamater, "The use of thionyl chloride in the preparation of Schiff's reagent," *Science,* 108:121–122 (1948).

Coleman, L. C., "Preparation of leucobasic fuchsin for use in the Feulgen reaction," *Stain Technol.,* 13:123–124 (1943).

Conn, H. J., "Biological Stains," 7th ed., Baltimore, The Williams & Wilkins Co., 1961.

de Tomasi, J. A., "Improving the technic of the Feulgen stain," *Stain Technol.,* 11:137–144 (1936).

Dutt, M. K., "Changes in pH of the Feulgen stain and their effect on the staining of biological material," *J. Histochem. Cytochem.,* 11:390–394 (1963).

Dutt, M. K., "Modifications of Feulgen stain prepared by the use of thionyl chloride," *The Nucleus,* 10:28–32 1967a).

Dutt, M. K., "Microspectrophotometric determination of *in situ* Feulgen reaction by Schiff reagent with enhanced pH caused by different chemicals," *The Nucleus,* 10:168–173 (1967b).

Dutt, M. K., "Increase of the amount of DNA-Feulgen in mammalian tissue by Schiff reagent at less acid pH," *Experientia,* 24:615–616 (1968a).

Dutt, M. K., "Use of a combination of metallic compounds and acetic acid–alcohol as fixatives for mammalian tissue in microspectrophotometry," *Experientia,* 24:934 (1968b).

Dutt, M. K., "*In situ* Feulgen reaction with Schiff reagent at different temperatures," *Experientia,* 25:783 (1968c).

Dutt, M. K., "*In situ* Feulgen reaction on mammalian tissue fixed in chromium and platinum containing fixatives," *Sci. Cult.,* 35:338–339 (1969a).

Dutt, M. K., "Optimum Feulgen staining of DNA through the influence of light," *The Nucleus,* 12:154–158 (1969b).

Feulgen, R., and H. Rossenbeck, "Mikroscopisch-chemischer Nachweis einer Nucleinsaure von Typus der Thymonucleinsaure und die darauf beruhende elektive Farbung von Zellkerzen in mikroskopischen Praparatan," *Z. Pyhsiol. Chem.,* 135:203–248 (1924).

Itikawa, O., and Y. Ogura, "Simplified manufacture and histochemical use of the Schiff reagent," *Stain Technol.,* 29:9–11, (1954).

Kasten, F. H., "Schiff-type reagents in cytochemistry. I. Theoretical and practical considerations," *Histocheimie,* 1:466–509 (1959).

Kasten, F. H., "The chemistry of Schiff's reagent," *Int. Rev. Cytol.,* 10:1–100 (1960).

Ostergren, G., "Chromatin stains of Feulgen type involving other dyes than fuchsin," *Hereditas,* 34:510–511 (1948).

Pearse, A. G. E., "Histochemistry: Theoretical and Applied," 2nd ed., London, J. A. Churchill & Co., 1961.

Swift, H., "Cytochemical techniques for nucleic acids," pp. 51–92 in E. Chargaff, and J. N. Davidson, ed., "The Nucleic Acid," Vol. 2, New York, Academic Press, 1955.

Walker, P. M. B., and B. M. Richards, "Quantitative microscopical techniques for single cells," pp. 91–138 in J. Brachet and A. E. Mirsky, ed., "The Cell," Vol. 1, New York, Academic Press, 1959.

See also: CHROMOSOME TECHNIQUES, FLUORESCENT SCHIFF-TYPE REAGENTS.

FIELD ION MICROSCOPE

Introduction. The field ion microscope (FIM), invented by Prof. E. W. Müller in the 1950s, is the most powerful existing microscope, being capable of magnifications in excess of a million and a resolution of individual surface atoms for a wide range of metal specimens. The image is obtained by the radial projection of ionized gas atoms onto a fluorescent screen. The ions are produced by the application of a high electric field at the hemispherical surface of the sharply pointed metal specimen.

Principles. The field ion specimen, generally formed by electro-polishing a fine wire of the metal to a tip with radius in the range 200–2000 Å (1 Å = 10^{-8}cm), is placed in the microscope chamber opposite a fluorescent screen and the system evacuated (see Fig. 1.) A suitable imaging gas such as helium is introduced into the microscope to a pressure of around 1 mtorr. A voltage of 5–30 Kv between the tip (positive) and fluorescent screen produces a high electric field at the specimen because of the small radius of curvature of the tip. Image gas molecules polarized by the high field will be attracted to the specimen surface, ionized preferentially in the regions of highest ionization probability, and projected radially onto the fluorescent screen to form an image of the tip surface.

Field Ionization. Imagine an atom of ionization potential I at a distance x from the tip surface and with an electron in an energy level as shown in Fig. 2. The presence of an electric field raises the energy of this electron to a point where it is separated from the conduction band in the metal only by the potential barrier which is shaded in the figure. Now, by a purely quantum mechanical process known as "tunneling" there is a finite probability that the electron initially in the atom will materialize on the left side of the barrier and within the metal. There is no classical analogy for this process—it is a purely quantum mechanical effect.

The probability of tunneling increases very rapidly as either the height or the thickness of the barrier decreases. Surface detail is provided in the image by variations in electric field (variation in barrier height) or in extension of the electron cloud (variation in barrier thickness). For a gas atom of high ionization potential, such as helium, a very large electrical field is required to raise the energy of an electron in a nearby atom high enough so that the electron can tunnel directly into the conduction band; tunneling into an energy level below the conduction band is forbidden. With an atom of lower ionization potential, an equivalent tunneling probability will occur with a lower electric field so that an image may be formed at a lower electric field.

If the electric field is increased sufficiently, ionization will occur at greater and greater distances from the surface, and details of surface structure will become blurred. The electric field giving the sharpest image is known as the Best Image Field (BIF); values of the BIF for commonly used gases are: He, 4.4 V/Å; Ne, 3.5 V/Å; H_2, 2.2 V/Å; A, 1.9 V/Å, and Xe, 1.1 V/Å.

The Image. The ions thus formed travel out radially from the specimen to the screen, producing a projection image (closely related to a stereographic projection) as shown in Fig. 3. The close relation between the image and a direct view of the surface is easily seen by looking at a ball model of a hemispherical tip (Fig. 4). In general the bright spots correspond to the atoms which protrude furthest from the surface.

Magnification. This is given approximately by the ratio of the tip to screen distance, say 5 cm, to the tip radius, say 500 Å, which in this case would be about a million.

Resolution. A number of factors control this, but in practical cases the dominant consideration is the mean thermal energy of the image gas molecule at the time of ionization. The component of thermal velocity parallel to the tip surface will cause a point source of field ionization to show on the screen as a disc of diameter $\delta \simeq (6.4 \times 10^{-4} Tr/F)^{1/2}$Å where T is the effective temperature of the gas molecules at the surface of the tip, r the tip radius in Angstroms, and F the electric field at the tip in volts/Å. To minimize thermal velocities the specimen is cryogenically cooled using liquid nitrogen (77°K), hydrogen (20°K), or helium (4°K). The gas molecules accommodate to nearly the tip temperature quite rapidly, and a resolution of 2–3 Å is readily obtainable with helium as the image gas and practical tip radii.

Image Recording. At a gas pressure of the order of 1 mtorr, total ion currents are the order of 10^{-9} ampere, which gives images with helium or hydrogen which can be seen only with the dark-adapted eye. They may be photographed with large aperture lenses and fast films with exposure times in the range 10 sec to 10 min. Increasing the gas pressure increases the image brightness, but pressures above 1–10 mtorr are not practical because of scattering of the accelerated ions on their passage to the screen, with consequent deterioration in image quality.

When heavier image gases (such as neon) are used, the image becomes very dim because of the low efficiency of such ions in producing light in a fluorescent screen. Three methods of intensifying the image have been used. (1) So-called "converter grids" may be suspended above the screen to convert the incident ions into electrons, which have high luminescent efficiency. (2) Commercial image intensifiers akin to television cameras may be used to view the screen. (3) The most effective device is the "channel plate"; this converts the incident ions into electrons and then multiplies the electrons by a cascade process to give overall light gains up to 10^6. The recent commercial availability of these relatively inexpensive devices has given great impetus to the use of image gases other than helium or hydrogen.

Field Evaporation. The application of a sufficiently high voltage to the specimen will remove the most protruding surface atoms, which experience the highest field, by ionizing them and pulling out the ions. The successive removal of the most protruding surface atoms in this way produces an end form which is atomically smooth. Continued field evaporation gives layer-by-layer removal of the surface atoms, which allows inspection of a volume of material. If the evaporation field and imaging field of the specimen are close enough, then the surface may be continuously observed during removal of successive atomic layers. When the evaporation field exceeds the imaging field, application of a pulsed desorption field in addition to the constant imaging field allows continuous observation of the surface during controlled field evaporation.

Vacuum Considerations. The resolution of the FIM is such that the interaction of many gases, even common air, with the surface is a serious complication. Fortunately, when an image gas of very high ionization potential (i.e., helium) is used, the various contaminant gases are ionized before they can strike the specimen surface. Contaminant atoms can reach the tip by migrating down the shank, but reasonable microscopy can be done at

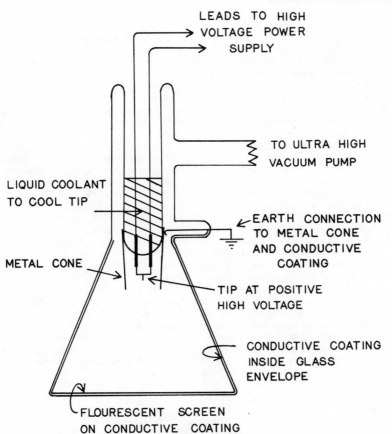

LEADS TO HIGH
VOLTAGE POWER
SUPPLY

TO ULTRA HIGH
VACUUM PUMP

LIQUID COOLANT
TO COOL TIP

EARTH CONNECTION
TO METAL CONE
AND CONDUCTIVE
COATING

METAL CONE

TIP AT POSITIVE
HIGH VOLTAGE

CONDUCTIVE COATING
INSIDE GLASS
ENVELOPE

FLOURESCENT SCREEN
ON CONDUCTIVE COATING

Fig. 1 A schematic drawing of a simple field ion microscope.

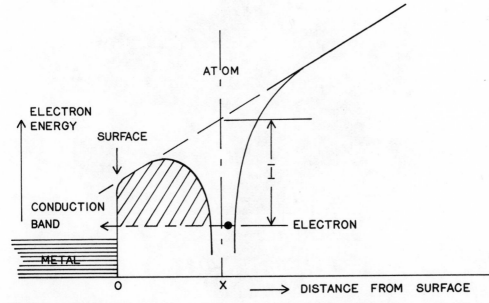

ELECTRON
ENERGY

ATOM

SURFACE

CONDUCTION
BAND

ELECTRON

METAL

O   X   DISTANCE FROM SURFACE

Fig. 2 An energy level diagram for the process of field ionization of an atom near a metal surface. Electron energy is plotted vertically, and distance from the metal surface, horizontally. The slope of the potential outside the metal is equal to the applied electric field—the higher the field, the steeper the slope.

background pressures (before the image gas is added) of 10^{-7}–10^{-8} torr, obtained with diffusion or getter ion pumps. If the field is removed during microscopy, ultrahigh vacuum (10^{-10} torr) is needed to keep the surface clean, which usually requires either baking the microscope, or chilling to temperatures of around 20°K. The image gases must, of course, be quite pure. Helium is normally admitted by allowing it to diffuse through the walls of a heated Vycor tube. Hydrogen can be admitted through a palladium tube. Other gases such as neon or argon must be carefully purified by storage over chemical getters.

Applications. The basic capability of the field ion microscope is that of examining, with atomic resolution, the surface of a conductor. Insulators cannot be used as specimens, because the electrons tunneling into the tip must be carried away.

The widest application of the FIM to date has been in metallurgical studies of the surface and bulk properties of metal lattice imperfections. Point defects, line dislocations, and grain boundaries can be studied using controlled field evaporation to reveal the three-dimensional nature of the defect. Such studies are without doubt complicated by the high field stresses present during the imaging and evaporation process; however, much detailed information has been obtained by the technique. Radiation damage studies of metals and studies of the properties of alloys are other subjects of interest for which the FIM is finding increased applicability.

Surface studies such as measurement of binding energies and surface diffusion energies on single crystal planes are yielding information on the little understood and most important problem of surface interactions. Several studies have been conducted for single atoms adsorbed on a surface, and are now being extended to studies of thin film structure. Chemisorption studies of gases on metal surfaces are of much interest to surface scientists, and the FIM offers an interesting method of approaching this problem.

The possibility of using the high resolution of the FIM to determine the structure of molecules, particularly macromolecules of biological importance, has been the object of some work. The basic difficulty is that the high electric field strips the molecule from the tip before the ion image can be formed. In principle there are several ways around this problem. Unmistakable evidence for the detection of macromolecules on an FIM tip has been obtained, but as yet no new structural information has been gained by this method.

A most ingenious modification of the FIM, again by Prof. E. W. Müller, is the "atom probe." In the screen of an FIM is placed a small hole, of area equivalent to a few atoms, leading to a time-of-flight mass spectrometer. The tip is adjusted so that a bright atom in a FIM pattern is superposed on the hole, and a pulse of voltage is placed on the FIM. The mass spectrometer signal which coincides with the disappearance of the bright image gives direct information on its charge-to-mass ratio, and

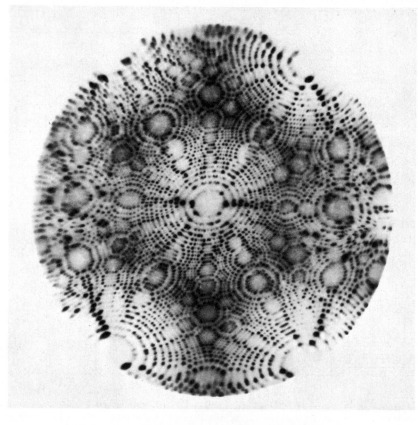

Fig. 3   A field ion micrograph of a tungsten tip.

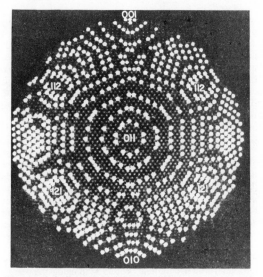

Fig. 4 A ball model of a crystalline tungsten tip. The more protruding "atoms" have been marked with fluorescent paint, and the picture taken with ultraviolet light. (From E. W. Mueller, in "Advances in Electronics and Electron Physics," 13 p. 145, Academic Press Inc., New York, 1960.)

therefore its chemical identity. This device will certainly have numerous applications in metallurgy and in surface physics and chemistry.

W. R. GRAHAM
F. HUTCHINSON

References

Gomer, R., "Field Emission and Field Ionization," London, Oxford University Press, 1961.

Hren, J. J., and S. Ranganathan, "Field Ion Microscopy," New York, Plenum Press, 1968.

Müller, E. W., *Adv. Electron. and Electron Phys.*, **XIII**:83 (1960).

Müller, E. W., and T. T. Tsong, "Field Ion Microscopy," New York and Amsterdam, Elsevier Publishing Co., 1969.

FIXATION

Introduction

The demands of high resolution light microscopy and the need for permanent preparations in the last century led to the development of a constellation of procedures generally subsumed under the heading of "fixation." The requirements of a fixative came to be recognized as multiple:

1. "Killing" (inactivation) of enzyme systems, especially those which cause autolytic changes in tissue structure.
2. Rendering insoluble the components of tissue structure to the greatest extent possible, thus retaining them in position through subsequent dehydration, embedding, and staining procedures.
3. Osmotic desensitization, or alteration of membrane permeability characteristics to prevent swelling or shrinkage during subsequent treatment.

4. Hardening of normally soft tissues to permit section cutting with or without embedment in a supporting medium.
5. Prevention of tissue or section changes due to bacterial or fungal activity.
6. Preparation of structural components for staining (mordanting), changes in refractive index of components, and other alterations to enhance subsequent visualization of structure.

These considerations are all applicable to the rigorous demands of electron microscopy, although as technology vastly increased the microscopist's resolving power, a diminishing number of fixation procedures proved adequate. Many chemical mixtures meet some or all of these requirements but only a very few avoid unacceptable side effects in terms of severe distortion of the structural elements of tissue.

Formulations of the multitude of general and special purpose fixatives developed for light microscopy, together with critiques and rationales for their use are readily available in the standard works on microscopic technique (see references 5, 19, 49, 69, 78, 99, 139, and others) and need not be reviewed here. This chapter will concern itself principally with the fixatives of electron microscopy. Since its demands are more rigorous than those of general light microscopy, the relevant considerations may be taken as generally applicable to light microscopic technique. The reader is referred to general discussions of electron microscopic fixation (references 2, 6, 13, 35, 65, 84, 85, 94, 110, 117, 126, 145, 149), and to such authors as Glauert (40), Pease (100), and Sjostrand (130) for the formulation of standard fixatives.

The contemporary microscopist is often concerned less with the simple descriptive exercise than with the compatibility of several methods of study. His choice of procedure may be limited by the simultaneous requirements of specific light and electron microscopic, and biochemical techniques. Fixation methods combining electron microscopy with biochemical studies (see, e.g., reference 71), histochemistry (47, 116), special light microscopy (9, 118, 128), including metallic impregnation (10, 52–54, 147) have been developed and exploited, to mention only a few. Such problems of compatibility have been determinants in establishing the emphasis of this chapter.

A large number of the references (by no means intended to be exhaustive on the subject of fixation) are drawn from the literature on fixation of the vertebrate nervous system. This tissue is notoriously sensitive to post-mortem change, preservation of its structural complexity is at best a challenge, the problems of extracellular space shrinkage and swelling are serious, and many studies of fixation generally have employed it as a model.

Most of the commonly used fixative reagents meet the outlined requirements to a degree. Those which alter protein molecule configurations to render them insoluble or which stabilize water-lipid interfaces, may be expected to alter enzyme activity, and thus to serve as tissue killing agents, and to affect bacterial and fungal lytic enzymes as well. Protein and lipid stabilization similarly "hardens" tissue, and alters the refractive index of components. Alteration of protein conformation may also expose reactive groups for subsequent staining.

Fixation in the sense of inactivation of enzyme systems, and stabilization and preservation of the structural components of tissue may be accomplished by either *physical* or *chemical* means. The physical methods include heating (thermocoagulation of proteins) or desiccation (usually freeze-drying). In the former case enzymes are destroyed through rearrangement of the protein molecules. In the latter they are rendered ineffectual by limiting substrate mobility through first freezing then removing water. These methods have important but limited application.

This article deals only with the chemical methods of fixing tissue—the result of chemical reactions between fixing reagents and tissue components. A variety of chemicals may be employed, not one of which is ideally suited for all three classes of tissue constituents: proteins, lipids, and polysaccharides. Usually, the reagents of choice are simply assumed to act as killing as well as stabilizing agents and no further attention is paid to this aspect of fixation. Such assumptions warrant reconsideration, however. A distinction should be drawn between the desirability (or lack of it, for the histochemist) for efficient killing of tissue, and maximum structural stabilization against extraction or reorganization into artifact during subsequent procedures. These two considerations may work at cross purposes, and intelligent fixative selection and design should take both aspects of fixation into account.

The protein fixatives are imperfectly classified as either *coagulating* or *denaturing*. Coagulating fixatives cause a coarse polymerization of protein molecules resulting in the formation of a precipitate similar to the effects of thermocoagulation. The denaturing of protein by chemical fixatives is more suited to fine cytological, histochemical, and electron microscopic work since, although the protein components of tissue are rendered insoluble by internal rearrangement, cross-linking and fixative binding, they do not form precipitates in the usual sense of the word.

Commonly used coagulating reagents in histology include methanol, ethanol, nitric acid, trichloracetic acid, mercuric chloride, picric acid, and chromic acid. These are important constituents of such general purpose fixatives as Gilson's, Heidenhain's, Zenker's, Carnoy's, Fleming's, and Bouin's solutions.

The noncoagulating (denaturing) reagents most widely employed have been osmium tetroxide, potassium dichromate, acetic acid, and formaldehyde (to which must be added other aldehydes recently introduced in electron microscopic histochemistry). The success of osmium tetroxide, the aldehydes, and dichromate in electron microscopy is due to their limited reactivity with protein, serving to kill and stabilize without extensive polymerization, precipitation, and consequent structural disorganization of the cell. It is thus no accident that "delicacy" of fixation was long recognized as a feature of fixative mixtures using principally these reagents: Helly's, Zenker-formal, Fleming's, Champey's, Hermann's, and Bouin's.

The early fixative combinations are largely supplanted in contemporary work by denaturing agents in mixtures designed to specific tonicity (osmolality) and pH requirements through the use of buffers and other dissociated salts and/or colloids. The present chapter will deal further with only some of these denaturing chemical fixatives and additives.

Criteria for the evaluation of the quality of fixation are specified by the investigator's needs, but in general morphological terms, some may be set forth. Light microscopically the standard has been the degree to which fixed tissue retains the features of the living cell as seen with phase microscopy, or resembles other preparations of "unfixed" cells (e.g., frozen cells). To this the electron microscopist has added the specifications that "well-fixed" material should exhibit membranes free of discontinuities, absence of swelling on the part of organelles, etc.—all of which suggest the retention of a high degree of "order." The reader is referred to the discussions of the subject by Palade (94), Palay et al. (95), Pease (100), Peters (104), and Sjostrand (130) for more detailed treatment.

Aldehydes

Introduction. The use of formaldehyde was introduced in electron microscopy by Holt and Hicks (57), Pease (see 100), and Richardson (109). Pease pointed out the need to use freshly prepared formaldehyde solution, free of methanol preservative. In appropriate concentration and buffered to near physiological pH, it has proven a rapidly penetrating and effective killing agent with excellent properties of protein stabilization. The very slow penetration of the previous fixative of choice (osmium tetroxide) and its limited protein stabilizing properties could be circumvented, and this latter reagent has generally come to be used as a secondary treatment for lipid stabilization and "electron staining."

The Yale group under the guidance of Professor Barrnett explored other possible aldehydes in the quest for structural fixatives with *limited* killing capacity. This combination of features is desirable from the standpoint of the histochemist who wishes to retain enzyme activity in material otherwise suitable for ultrastructural study. The results of the survey by Sabatini et al. (116, 117) revealed that the most satisfactory compromise between retention of enzyme activity and adequate structural stabilization was obtained by using glutaraldehyde, a small dialdehyde whose superior protein-fixing properties apparently stem from its extraordinary ability to form cross-links between protein molecules, but whose reactivity involves only the most exposed and sensitive amino (and some other) groups, thus presumably leaving some enzyme-reactive sites available for substrate attachment.

It is worth noting the cross-linking capacities of common aldehydes at this point. Bowes and Cater (16) state that the moles of cross-links per 10^5 g for formaldehyde = 5, glutaraldehyde = 10, acrolein = 11, and adipicaldehyde = 2. Further, the cross-links of glutaraldehyde and acrolein are very stable, moreso than those of formaldehyde (see also 99).

Details of the chemistry of aldehyde fixation of protein are not entirely known (see 6, 16, 99). Aldehydes are incorporated with protein into a larger, relatively stable molecule (aldehyde "fixation by protein," to the tanner). Most of the study of this phenomenon has been related to the tanning industry and involves concentrations, temperatures, and pH conditions outside the range of interest to the microscopist. It does seem clear that under conditions of tissue fixation, the principal reaction of aldehydes with protein is with NH groups (see 16 and 99). In the case of dialdehyde one or both aldehyde groups may react, in the latter case producing more and firmer cross-links than is the case with formaldehyde (16, 116).

In addition to reducing protein solubility this fixing process causes a net loss of positive charges associated with the protein molecule, effectively increasing its basophilia (see 16), and "acidifying" the environment. The change in shape in the protein molecule is related to the protein species, hence the effect on a given enzyme cannot be predicted.

While aldehydes were initially regarded as superior to osmium tetroxide only from the standpoint of penetration and protein stabilization, they may prove as well superior in the fixation of some lipids (122 and 133), and formaldehyde would appear to be more toxic to some enzyme systems than osmium (18) although clearly not all.

Formaldehyde. The rapid penetration of formaldehyde and its excellent killing and stabilizing properties are generally utilized to rapidly kill tissue for subsequent osmium tetroxide treatment, or for other study (histochemical, biochemical, metallic impregnation, etc.) concommitant to osmium post-fixation. Pease (100) emphasizes the desirability of avoiding undue delay in this post-fixing osmium treatment wherever possible, and appears generally to regard formaldehyde as of most use as a killing agent. Short exposure time to formaldehyde does not make much use of its protein-fixing properties (including the sluggish and weak cross-linking potential) according Pearse (99). Most of the formaldehyde-protein links formed are easily broken in subsequent treatment of the tissue (including washing). Two to four hours appear necessary for major amino binding to occur (106).

Formaldehyde solutions are easily perfused through the vascular tree, and it rapidly hardens tissue to a consistency at which small blocks may be removed with minimal mechanical damage. It can be used with all conventional biological buffer systems which are free of amino groups (with the possible exception of acetate-veronal, 58), but must be used in a methanol-free form for electron microscopy. The gas is volatile and it can be used as a vapor-phase fixative quite satisfactorily.

The formaldehyde molecule in aqueous solution probably exists in multiple forms, principally monomeric and polymeric hydrates (99). However, there is no evidence that fixation is the consequence of other than single molecule attachment to reactive sites, principally NH groups of proteins. It is effective in cross-linking through methylene bridge formation between protein end groups, and apparently fixes lipids and polysaccharides principally through their attachment to stabilized protein moieties (see Pearse, 99).

Ideally formaldehyde is used for killing and fixing in buffered mixtures adjusted for tonicity and in dilute concentrations (2%–4%). However, otherwise unobtainable material, such as autopsy specimens in conventional 4% commercial formaldehyde can be post-fixed (glutaraldehyde then osmium or directly in osmium), and surprisingly useful electron micrographs can be obtained (137 and 63). Even material embedded in paraffin can be retrieved for electron microscopy if necessary (25). The concentration of formaldehyde in fixatives is not critical, and it contributes little to the "effective" tonicity of fixative solutions. Presumably adequate fixation can be obtained at concentrations under 2% and perfusion at full strength (37%, 12M) fixes brain well although with some shrinkage (45).

The widest use of formaldehyde appears to be related to its use in combined study with the electron microscope and methods for light microscopy developed for use with formaldehyde (such as metallic impregnation methods), and in conjunction with other reagents, especially glutaraldehyde, with inferior killing properties with respect to some enzymes but superior protein stabilizing characteristics. Formaldehyde was regarded as inferior to glutaraldehyde by Sabatini et al. (116) in respect to the preservation of some enzyme activities, although in other enzyme systems (or in systems studied nonhistochemically) formaldehyde is a less efficient enzyme inhibitor than glutaraldehyde (see 60, 99).

Glutaraldehyde. Sabatini et al. (116, 117) introduced this small dialdehyde as a primary fixative for electron microscopic histochemistry. Glutaraldehyde in about 6% concentration in dimethylsodium arsenate–HCl buffer ("cacodylate") proved a satisfactory compromise between the two requirements of protein stabilization and the preservation of enzymatic activity. It forms strong cross-links almost to a quantitative maximum, and the reactions of its aldehyde groups are presumed to be identical to that of formaldehyde (principally NH groups). This denaturation through polymerization is apparently less disruptive than that of formaldehyde, for, histochemically, greater activity of some enzyme systems is retained. Glutaraldehyde has a much slower penetration rate than formaldehyde (see 60) although this is not a serious drawback, as it is easily delivered by vascular perfusion. It is preferable to formaldehyde for protein fixation, and to both formaldehyde and osmium tetroxide for phospholipids (89, 114). Some organelles (e.g., cytoplasmic microtubules) are better preserved by this fixative than any other.

Glutaraldehyde-fixed tissue has a crisp texture which makes it possible to trim and block small tissue samples without mechanical damage, and to produce thin slices suitable for histochemical incubation. The tissue accepts osmium post-fixation well, and after glutaraldehyde fixation tissue may be stored (refrigerated) in buffer for months with little further loss of enzyme activity.

Glutaraldehyde probably exists in aqueous mixtures for the most part as small polymers. The monomer tends to become degraded to glutaric acid or to form these polymers with time, and pH and temperature elevation; problems related to storage and purification have therefore been given some investigative attention (16, 36, 59, 99, 111, 144). The proclivity to polymerize, indeed, has led Peterson and Pease (105) to employ this aldehyde with urea as an embedding mixture.

Protein-monomer reactions may be of paramount importance in histochemical procedures, but cross-linking and other reactions related to protein stabilization may rely at least in part on the presence of dimers or oligomers. Robertson and Schultz (111), for example, maintain that the "quality" of protein preservation depends on short polymer reactions with protein, and these authors urge that highly purified glutaraldehyde for general fixation be avoided. Wrench (150) also supports the notion of glutaraldehyde polymers being of primary importance in good structural preservation. These suggestions may be an example, however, of the need to distinguish between the killing and the stabilizing actions of fixative. Superior morphological preservation may indeed attend the use of "contaminated" glutaraldehyde due to a superior killing action—that is, attachment of larger molecules to protein may facilitate cross-linking and may more effectively block lytic enzyme activity (through stearic hindrance of substrate access to enzyme-reactive sites, for example) than the highly purified monomer. Such an effect would clearly be contrary to the needs of the histochemist, however, who desires to retain maximum enzymatic activity.

More careful work is clearly needed to understand the nature of the fixing process employing glutaraldehyde. Methods of study are at hand and for the interested reader references on glutaraldehyde purification (36, 59, 111, 144), assay (28, 60), and protein binding (16, 99) are included.

The success of glutaraldehyde intelligently applied has led to its rapidly becoming the most widely employed general purpose fixative. It is presently most commonly used in concentrations from 2% to 6%, usually in "phosphate" buffer solutions, with or without additives. It is compatible with other biological buffers (including in our experience, acetate-veronal, but see below under this heading), except those containing amino groups (such as "tris").

No fixative should be elected without consideration of the experimental needs, and indiscriminate use of glutaraldehyde as the sole primary fixative of choice is subject to this reservation. Frequently it will emerge as the ideal fixative but should do so on its merits and with appreciation of its limitations. Being a dialdehyde it may be bound to tissue through one or both of its reactive groups. In the former instance, however, a free aldehyde is introduced which is not reflective of tissue structure. Application of periodic acid-Schiff methods, for example, must take this into account. Glutaraldehyde appears to interfere with some silver impregnation procedures in the nervous system which were developed for use with formaldehyde-fixed material. The slow penetration of glutaraldehyde may prove a handicap in some circumstances. This reagent cannot be used in the vapor phase as a fixative due to its low volatility. A final consideration is the recognition that the introduction of glutaraldehyde by Sabatini et al. (116) was the result of a compromise between maximum morphological stabilization, and preservation of some specific enzyme activities. Where one requirement can be sacrificed other fixatives or additives to glutaraldehyde may be considered (see Fixative Selection and Design, below). Where the demands of protein stabilization are paramount, and enzyme activity can be neglected or it is desired to arrest it as completely and quickly as possible, glutaraldehyde alone is not necessarily the fixative of choice, although it may well be, given appropriate additives.

A recent report suggests a modification of glutaraldehyde with peroxides to produce a hydroxyalkylperoxide fixative (103). This application of glutaraldehyde to tissue preservation remains to be more widely tested.

Acrolein (Acrylic Aldehyde). This compound was introduced by Luft (75) as an electron microscopic fixative in an early effort to employ a more effective protein cross-linking reagent than osmium tetroxide. It has never achieved widespread use, probably in part due to its extremely noxious odor and lacrimatory effect. In a fume hood for immersion procedures it can be tolerated, however. It is as efficient a stabilizer of protein as glutaraldehyde but is much more toxic to enzyme systems, possibly due to its reactivity with sulfhydryl groups. It does not preserve cytoplasmic microtubules (124) although it is as satisfactory in most regards as formaldehyde. Acrolein has the benefit of being a more potent enzyme-killing agent than either formaldehyde or glutaraldehyde and warrants consideration as an additive to mixtures of either (119, 120, 124, 129).

Hydroxyadipaldehyde (Adipic Dialdehyde, HAA). The aldehyde tested by Sabatini et al. which preserved the greatest enzymatic activity was HAA. It is probably not coincidental that it is simultaneously a poor cross-linking reagent and generally quite inferior as a preservative of cellular structure. Its applicability thus seems limited to certain histochemical and related problems. Torack and associates (140–142) have employed this aldehyde to assess the distortion of water compartments in the nervous system during fixation, but there is nothing to commend it yet as a general purpose fixative or fixative additive.

Osmium Tetroxide

Osmium tetroxide (incorrectly "osmic acid") has been long known to the light microscopist as the most delicate and life-like preservative of cellular structure. Its use was restricted, however, by the intense black reaction product which limits the utility of subsequent staining procedures. It was classified as principally a lipid fixative (see 148), and for the light microscopist is unsurpassed in this respect. Its superior qualities of cytological preservation happily combine with its mass and consequent electron scattering power, and Palade (94) in his classical paper on fixation firmly established its preeminence in electron microscopy as fixative and stain. It is compatible with most buffer systems familiar to the biologist although it is most commonly used in acetate-veronal, or one of the phosphate buffers. Its effectiveness is unimpaired by inorganic additives and even sucrose may be added (just before use) for osmotic reasons (23) with little loss of the osmium-tissue reaction. Dalton (26) combined this interesting reagent with another lipid fixative, dichromate, to enhance this aspect of osmium fixation and to capitalize on the slight buffering capacity of dichromate-hydroxide mixtures. This latter fixative combination is of particular benefit to the investigator who must work with myelin.

Osmium tetroxide is a noxious agent to work with (note the warnings offered by Pease, 100). It renders tissue so black and brittle that block trimming and orientation are difficult. It penetrates very slowly (see, e.g., 29, 84), and is difficult to deliver satisfactorily by vascular perfusion (95). It is inferior to aldehydes for protein stabilization, and some structural artifacts may occur with its use (143). For these reasons it is most commonly used presently following aldehyde killing and prefixation, to retain lipids, further stabilize proteins, and stain the tissue. Other procedures (histochemical incubation, metallic impregnation, etc.) can be interposed between these two stages. The simultaneous fixation by glutaraldehyde and osmium mixtures has not been widely explored but this method has been reported (56). The hazards of sequential treatment are not great. Some lipid moieties may be extracted in the aldehyde stage, and the aldehydes may not be as toxic to some enzyme systems as osmium tetroxide (but see 18 for example). However, when required, "post-mortem" continuation of enzyme activity in aldehyde fixatives may be reduced by appropriate aldehyde mixtures or additives, and by tissue cooling. Lengthy stays in aldehydes are to be avoided if osmium post-fixation is intended, as lipid extraction may continue with time. This is evident in the case of central nervous system myelin. Stored overnight or longer in fixative and/or buffer, the myelin suffers extensive lamellar separation which is not evident in tissue post-fixed in osmium after only about an hour's exposure to the aldehydes. Apparently some structural lipid is lost on prolonged aldehyde soaking. Bodian (12) also reported changes in the structure of synaptic vesicles due to delays in osmium postfixation.

There is some lipid loss during osmium tetroxide fixation (30, 89, 121), but with appropriately modified aldehyde prefixation treatment, quantitative or nearly quantitative preservation of some lipids can be obtained (see 122 and 133 for example). The association of osmium and cholesterol can be stabilized and capitalized upon as in the ingenious tracer studies of Higgins et al. (55).

There have been many efforts to elucidate the chemistry of osmium fixation, although with anything but unanimous agreement on the conclusions to be drawn.

One possible explanation of disputes and uncertainties may be related to failure to distinguish between the reactions of which osmium tetroxide is capable in the test tube, and those which may be relevant to its lipid, lipoprotein, phospholipid, and protein stabilizing properties in tissue, and further to distinguish these properties from those which relate to its contribution to the electron microscopic image (properties as an "electron stain"). Each of these may prove quite a different class of reactions. Osmium tetroxide was thought to be active as a fixative at one time, principally through its reactivity with double bonds of unsaturated fatty acids (see 1, 2, 110). Such osmium ester formation may not be the only reaction of significance to fixation, however. It clearly stabilizes protein to a degree and certainly "stains" it: Sulfur linkages and amino groups react with osmium tetroxide.

Structural stabilization of cellular constituents may occur with little or no "electron staining"; the latter may be a secondary effect of prolonged exposure to reasonably high concentrations of osmium. The chemistry of osmium-lipid complexing is reviewed by Riemersma (110) and some of the problems related to interpretation of osmium-fixed lipids are provocatively discussed by Korn (67). These papers and the comments of Pearse (99), Pease (100), and Sjostrand (130) are also recommended. Recent work (72) suggests that particular attention should be paid to the hydrogen bonding of the "osmium VIII" form of osmium tetroxide to polar lipid groups and protein in explaining the *staining* action of this reagent. It remains possible on the other hand that the *stabilization* of lipoprotein may not be limited to this specific reaction.

Osmium tetroxide has been employed in vehicles other than water for particularly difficult material. Afzelius (2) and Dunn (33) have satisfactorily used a carbon tetrachloride vehicle for sea urchin eggs and reptilian retina respectively, and osmium tetroxide in acetone is a convenient method of freeze-substitution fixation (146).

Special purpose applications of osmium tetroxide have been developed. Prolonged exposure selectively "stains" the Golgi apparatus (37) very intensely, recapitulating the experience of the light microscopist's special methods for this organelle. In conjunction with zinc-iodide a special fixing-staining method for synaptic terminals has been developed (3,4). Hanker et al. (51) and Seligman et al. (127) report an intriguing use of tissue-bound osmium to add osmium "staining" through the use of ligands, eliminating the need for subsequent addition of lead or uranium atoms for purposes of contrast.

Similar in many respects to osmium tetroxide but not widely explored or evaluated as a fixative is ruthenium tetroxide. It is said to produce a greater degree of polar staining of lipids and to emphasize the laminar character of membranes (39).

Other Fixatives for Electron Microscopy

Potassium Permanganate. This compound (.5%–3%) in buffer solutions was introduced by Luft (74) as a special purpose fixative and "stain." Its similarities to osmium tetroxide are discussed by Riermersma (110). Its general applicability is limited (see 100), although if its artifacts are noted it can be useful for special purposes. It has been employed as a control for osmium artifact (115) and it has proven of use in the study of special synaptic structures (135). It may prove useful, however, to further distinguish between the beneficial effects of permanganate fixation and the presence of manganese ions in fixative solutions. It has been denied that this potent oxidizing agent is a "fixing" agent at all (e.g., ribosomes are not seen after fixation with permanganate) and it is not supposed to attach to lipids. As long as the tissue is killed and stained and structural components are rendered insoluble, however, the point seems inconsequential. Alcohol dehydration may serve to coagulate otherwise "unfixed" protein components in this method.

Robertson has extensively studied the structure of biological membranes with this fixative and describes a modification of the osmium perfusion method of Palay et al. suitable for use with permanganate (112).

Phosphotungstic Acid. This complex molecule is too poorly understood to warrant thorough treatment here. It has been employed as a polysaccharide fixative, but its effects on tissue components are pH dependent although there is not general agreement on the interpretation of its action. The papers of Pease (see 101, 102) and the relevant comments and references (41) should be examined for the uses and problems associated with this "fixing" and "staining" reagent.

Potassium Dichromate. Chromic acid and potassium dichromate have been used as lipid fixatives in light microscopy but little for electron microscopy, except for the combination of potassium dichromate with osmium tetroxide (26). Gonatas et al. (44) have successfully employed it with formaldehyde as an initial fixative followed by Dalton's chromium-osmium mixture.

Uranium. The nuclear stain of choice for electron microscopy is usually a uranium salt, generally aqueous or ethanolic (or methanolic) uranyl acetate. Superior results are obtained with *en bloc* staining prior to embedment, and this has been attributed to a combined "fixing" and "staining" of nuclear polypeptides and proteins (138).

Picric Acid. This reagent is classed as a coagulating fixative, and is further reputed in the light microscopic literature to be a slowly penetrating one. However, it has been very successfully employed as a fixative additive (28, 61, 152, and in our hands) in conjunction with glutaraldehyde and/or formaldehyde and deserves further examination. It may add to the protein stabilizing effect of aldehydes in principally bridging intramolecular amino groups, and thus enhance as well the killing action of the fixative. Its penetration rate is said to be intermediate between formaldehyde and osmium tetroxide (29) but seems to us sufficiently rapid to permit excellent immersion fixation (small blocks) and perfusion delivery with aldehyde-buffer mixtures. It is only slightly soluble and may be used in final concentrations, saturated or half saturated.

Buffers[1]

Palade (94) firmly established the desirability of pH control during fixation for electron microscopy. While some of the reported effects of fixative pH far from neutrality may now be attributable to early embedment procedures (100) all subsequent major fixative innovations have incorporated provision for pH control. The rationale for selection of a specific buffer species or concentration, however, is often absent although some considerations are warranted.

The selection of buffer mixtures for fixation should be conditioned by the needs of the investigator with principal concern for the primary purpose of pH regulation, so the pK and effective range are of paramount importance. The effects of the ion species may be placed secondary, and the concentration employed should relate to the needed buffering capacity. Often it seems, buffer species are selected for the "physiological" properties of the ions, and the concentrations are determined by the osmotic requirements. Such incongruities are unnecessary and may limit the flexibility inherent in rational fixative design.

The more closely the buffer pH approximates the desired pH value, the greater the protection against shifts of pH in either direction, and the lower the requisite concentrations of buffer in the fixative for a given fixative-tissue volume ratio. Since fixation is accompanied principally by shifts to the acid (94), any disparity in buffer pK and desired pH value of the fixative should be such that the pK is somewhat lower than the desired pH. That is, the buffer ideally should have a pK at or slightly lower than the desired pH. However, the pH desired must fall within the flatter part of the titration curve, which is generally a narrower range than that given in textbooks as "effective range (E.R.)." This latter term includes all parts of the titration curve which depart essentially from the curve of strong bases titrated with strong acids, and at the extremes of "effective range" buffers are very weak protective additives.

If an appropriate buffer for the desired pH range is selected, and reasonably large fixative-to-tissue volume

ratios can be maintained (normally 20 : 1 or more), low concentrations of buffer are probably adequate. Osmolality may be adjusted by the addition of other substances (see Fixative Selection and Design, below). Although most investigators prefer to fix tissue at pH 7.0–7.4, it is clear that some shift in pH is tolerable. Although fixation at pH below 6.0 or above 8.5 is certainly attended by artifact production, the acidification of tissue and fixative solution during fixation is not enormous and modest protection seems all that is required.[2]

Phosphate Buffers. The selection of phosphate buffers seems commonly predicated on the notion that it is preferable to use fixatives containing physiologically compatible ions. This may have evolved from consideration of the slow penetration of osmium tetroxide as a fixative and the thought that tissue may be exposed to the buffer ions for some time prior to fixative exposure and fixation. The validity of assuming this sort of behaviour on the part of diffusing ion mixtures in the fixation process remains to be established and in any case may not be as relevant to aldehyde fixation as to that with osmium tetroxide.

Palade rejected phosphate buffering as a first choice in part because of the possibility of insoluble phosphate salt formations. Others appear to be concerned about the commonly known effect of phosphate ions on mitochondria (the "Lehninger effect" of mitochondrial swelling). Indeed, mitochondrial swelling appears to be a greater problem with fixatives containing phosphate buffer mixtures than with most others. The use of phosphate buffers is precluded if histochemical procedures relying on phosphate precipitates (such as acid phosphatase) are contemplated.

The use of the term "phosphate buffer" alone does not suffice to describe adequately the vehicle employed by the investigator. Several varieties of buffer employing phosphate salts have been used in contemporary light and electron microscopy, and collectively they probably comprise the most popular class. The descriptive term "Sorensen's buffer" too is inadequate, inasmuch as it may refer to any one of several buffers bearing this pioneer's name. "Sorensen's phosphate" buffer may be taken to refer to the popular buffer mixture of the monobasic and dibasic sodium phosphates (E.R. = 5.7–8.0). This buffer is commonly stated to be advantageous because it has three pK values, one associated with each stage of dissociation of the trivalent phosphate ion. It should be noted, however, that two of these pK values lie far outside the range of normal biological interest (pH 2.0–12.4). It

---

[1]pH is the notation introduced by Sorensen to describe the hydrogen ion concentration (ions per liter), and is equal to the negative $\log_{10}$ of that concentration. A unit decrease in pH, then, indicates a tenfold increase in hydrogen ion concentration. An excellent brief discussion of pH and its measurement is given in reference 7, and textbooks of physical chemistry may be consulted for details of the buffer phenomenon. The Henderson-Hassalback equation is derived from the law of mass action and describes the relationship between the dissociation constant pK' of the buffer base (or acid), pH and concentrations (actually activities) of a weak acid (HA) and an ionized salt of the weak acid (BA):

$$\mathrm{pH} = pK' + \log \frac{[BA]}{[HA]}$$

The half titration of a buffer salt with its acid is at a pH at the midpoint of its effective range, i.e., where it possesses the greatest reserve alkalinity and acidity against pH shifts. At this point the pH and pK' have equal numerical values. This pH may be determined either empirically by titration or theoretically from concentrations, dissociation constants, activities, and the law of mass action. Knowledge of the relevant pK' values and effective pH range of buffers is essential in selecting buffer mixtures in fixative design.

---

[2]"Fixation" of egg albumin in an equal volume of fixative without buffer (4% formaldehyde and 0.5% glutaraldehyde in distilled water adjusted to pH 7.3 with NaOH) is accompanied by a shift in pH to only 6.7 after 15 min. Other investigators fixing in the absence of buffers (85, 12) have noted effects attributable to hypotonicity but have not reported effects that might be related to excessive acidification. The fixative: "tissue" ratio in this observation is far less than ideal, and normal use of mildly buffered mixtures should afford ample protection. However, what cannot be inferred is the extent of possible highly local and transient pH shifts in or near cells exposed momentarily to small volumes of fixative during normal procedures. These may be more severe and require greater buffer concentration than one might guess from the crude measurement reported here.

should also be noted that "Sorensen's phosphate buffer" may be interpreted as referring to monopotassium-disodium phosphate buffer (E.R. = 5.0–8.2), and equally useful to the microscopist. Other "phosphate" buffers which may be employed, or referred to as "phosphate buffer" include McIlvaine's (citric acid–dibasic sodium phosphate, E.R. = 2.6–7.0), Kalthoff's (borax–monopotassium phosphate, E.R. — 5.8–9.2), Teorell and Stenhagen's (citric acid-phosphoric acid—orthoboric acid—sodium hydroxide, E.R. = 2.0–12.0), or Lewis' acid phosphate (mono- and dibasic potassium phosphate, E.R. = 1.5–3.5). McEwen's buffer-salt mixture has been used as a vehicle and is called a phosphate buffer although it is a phosphate–carbonate-saturated carbon dioxide system. For electron microscopy a popular variant was introduced by Millonig (86) employing monosodium phosphate and sodium hydroxide. The formulations of these various phosphate buffers are readily available in standard works (7, 19, 43, 99, 100, 130).

Acetate-Veronal Buffer.[3] The Michaelis sodium acetate–sodium veronal (sodium barbital)–hydrochloric acid buffer[4] (E.R. = 2.6–9.4) was employed by Palade (94) and has remained many electron microscopists' standard. This buffer's popularity probably stems from its early widespread use after Palade's study of electron microscopic fixation and from the "physiological" character of its constituent ions. It is the basis for several modifications through addition of sucrose for osmotic purposes (Caulfield's, 23) and of other salts (Zetterquist's, 153).

Reference to this buffer should be explicit, however, as "Michaelis' barbital," or "veronal" may imply sodium barbital–hydrochloric acid (E.R. = 6.8–9.6).

Bicarbonate Buffer. This term may refer to any one of several buffers employing bicarbonate ions, not all of which are of use to the general microscopist, and it is to be

---

[3]Holt and Hicks (58) stated that this buffer cannot be used with (commercial) formaldehyde due to veronal-formaldehyde complex formation which alters the buffering characteristics to that of a sodium acetate–HCl mixture. In our laboratory, titration curves plotted with this acetate-veronal-HCl buffer (0.1M) with and without freshly prepared formaldehyde (4%, after Pease, 100) fail to duplicate this loss of buffering capacity by the addition of formaldehyde, and we can neither explain the discrepancy of these observations nor at present agree that formaldehyde must not be used in this buffer. There is a slight shift in the empirically determined pK under these circumstances (about +.4), but the effective range and buffering capacity is essentially unchanged and entirely suitable for biological fixation. Methanol (1%) is not accountable for the Holt and Hicks phenomenon. Glutaraldehyde (6%) does not alter the buffer titration curve in the slightest and apparently may be used in this buffer quite satisfactorily. The investigator wishing to employ aldehyde and acetate-veronal-HCl mixtures is advised to check the buffering capacity of his prepared fixative, until the cause of these discrepant observations can be elucidated, however.

[4]"Michaelis' buffer" is not sufficiently descriptive, and while it is commonly taken in electron microscopic literature to refer to this buffer it is not safe always to do so. Further confusion may arise for the novice by the perpetuation of an apparent typographical error, origin unknown, which has led to the occasional reference to "Michael's buffer," which presumably but not certainly refers to this acetate–veronal–hydrochloric acid mixture.

hoped that reports on methodology will come to be more specific. The Delory and King buffer mixture of sodium carbonate–sodium bicarbonate for example has an E.R. of 9.2–10.7, and is probably useless as a buffer for general fixation purposes. It is not always clear, either, if "bicarbonate buffer" may not in fact refer to McEwen's phosphate-carbonate-CO$_2$ systems (79, 112), or a variant thereof: Gonzales-Aguilar reports adjusting pH with NaHCO$_3$ (46), or NaHCO$_3$ and CO$_2$ (45), and Wood and Luft used bicarbonate-HCl as a buffer (149).

Schultz and Case (124) state that cytoplasmic microtubules are not preserved with bicarbonate buffer vehicles, although they do not specify what buffer was employed. The investigator should also beware of the possible formation of insoluble carbonates in designing a fixative with other additives.

Collidine Buffer. This refers to s-Collidine (a methylated pyridine) used as a buffer with hydrochloric acid (E.R. = 6.4–8.3), and introduced to electron microscopy by Bennett and Luft (8). It does not complex with osmium tetroxide, and buffering capacity is unaffected by formaldehyde or glutaraldehyde. The warnings of Pease (100) regarding purity should be noted, and old ("straw-colored") s-collidine will retain its buffering capacity (which therefore cannot be taken as a test of purity) despite the contamination due to aging. Luft and Wood (76) believe there is more extraction of protein with its use than with some other buffers.

Arsenic Buffers. A useful buffer is obtained with Plumel's mixture of dimethyl sodium arsenate (sodium cacodylate) and hydrochloric acid ("cacodylate buffer," E.R. = 5.0–7.4). This was popularized in electron microscopy by the study of Sabatini et al. (116). It is unaffected by the addition of aldehydes or osmium tetroxide. It has the disadvantages of cost and toxicity (and must be handled with the care accorded any arsenical).

The use of the term "cacodylate buffer" should be avoided, since there is also a well known cacodylic acid–sodium hydroxide buffer (E.R. = 5.2-7.2), although this author knows of no experience in its use in electron microscopy. Wood and Luft (149) examined the use of an "arsenical acid–arsenate" buffer with good results. Although details of its characteristics were not given, they were thought to be similar to "phosphate" buffers.

The excellent results obtained with the use of sodium cacodylate–HCl buffers in general fixation suggest that the presence of "physiological" ions is not an important consideration in fixative buffer selection. The toxic nature of the cacodylate ion suggests that it, like other arsenicals, may act to inhibit enzyme systems, probably also by attaching to SH groups. In this regard it might add to the fixative composition an additional potentially useful killing action not available in the phosphate, veronal, carbonate, or other "physiological" buffers. However, cacodylate ions are not totally effective against lytic enzymes, since tissue is quite satisfactory for use with the acid phosphatase procedure even after tissue has been stored for months in the sodium cacodylate buffer. In this context the arsenical acid–arsenic buffer of Wood and Luft probably deserves wider experimental evaluation.

Dichromate Buffer. This potassium dichromate-potassium hydroxide buffer (E.R. = 5.6–7.6) is another cytotoxic buffer, combining buffering in the physiological range with some lipid-binding capacity. It has been most widely used in Dalton's osmium (1%–2%)–dichromate mixtures. The buffer concentration is usually 1%. It has been used as a buffer fixative with formaldehyde as well (44) with good effect, and chromate fixation and buffering remains a common feature of myelin studies.

"Tris" Buffer and Other Amine-containing Buffers. "Tris" buffer refers to the use of tris (hydroxy methyl) amino methane, with hydrochloric acid, or with maleic acid and HCl (tris-maleate buffer, 43). These have an E.R. of 5.0–7.2 and 5.2–8.6 respectively. They are popular for use in histochemical procedures, but are mentioned here only to discourage their use as fixative vehicles. The amino group apparently is susceptible to osmium and to aldehyde fixatives, and all buffering capacity is lost in such fixative mixtures. Similarly, aldehydes especially should not be used with any other amine-containing buffer salt, and osmium fixation should similarly avoid untried organic buffers until tested for buffer inactivation or osmium reduction.

Fixative Selection and Design

Several excellent studies of the effects of varying the parameters of fixation have been published and should be consulted by the interested investigator (35, 65, 84, 85, 126, 145, 149). The data of Maser et al. (83) is very useful in solving fixative design problems, and the general comments in the chapters by Glauert (40), Pease (100), and Sjostrand (130), along with listing of fixative formulations is recommended. The study of Dempster (29) among others is useful in comparing fixative penetration rates (see also 125).

Fixatives. If protein preservation is important an aldehyde fixative is indicated, preferably one capable of rapid and durable cross-linking. Such preservation with retention of enzyme activity suggests glutaraldehyde, but if enzyme activity is not desired, addition to glutaraldehyde of formaldehyde, acrolein, picric acid, or other additional killing agent may be considered.

Many investigators find the stabilizing effect of glutaraldehyde sufficient in concentrations as low as 0.5%, and employ it with formaldehyde (0.5% and 4% or 2% and 2%). Pearse (99) comments on the possible advantages of dilute glutaraldehyde fixatives. High concentrations (6% or more) of glutaraldehyde are sometimes used, although some workers claim 0.25% glutaraldehyde with formaldehyde is sufficient to obtain the beneficial effects of its superior protein-stabilizing properties. Maunsbach (85) was of the view that about 1% was the lowest satisfactory concentration to be employed alone. It is likely that purity (or lack of it) of the glutaraldehyde, and tissue-fixative volume ratio are relevant at low concentrations, since the "fixing" process binds the aldehyde, effectively removing it from the fixing solution. Formaldehyde is usually employed in the conventional 4% concentration although it can be used full strength if shrinkage can be tolerated (45). It is quite satisfactory with glutaraldehyde in 2% concentration, although there seems to be no particular benefit to using lower concentrations. Reese and Karnovsky (108) used a dilute aldehyde fixative for perfusion fixation, presumably as a killing agent, followed by a soaking period in concentrated fixative mixture, probably thereby obtaining maximum aldehyde-protein binding and cross linking. Schultz and Case (125), however, reverse this sequence.

Not sufficiently explored as yet are the possible beneficial effects of toxic ions as "killing" agents in the form of arsenical buffers, a salt of mercury, chromium, copper, etc. More dilute buffer vehicles than commonly employed would permit "osmotic room" for such additives, or for other enzyme poisons.[5]

For the preservation of lipids, osmium tetroxide with or without dichromate is the fixative of choice. Its benefits may be enjoyed by post-fixation after aldehyde killing and initial stabilization, but most structural lipids are not sufficiently fixed by aldehydes alone to withstand the dehydrating and embedding procedures which must follow. The additional stabilization of osmium treatment as well as its staining properties are highly desirable therefore, if the osmium is not deleterious to other concurrent methods of tissue study, such as special staining methods for light microscopy, including metallic impregnation, histochemistry, etc. Many of these procedures can be carried out prior to the osmium post-fixation, and delicate emulsions can be protected from tissue-bound osmium (and aldehydes) by an intervening carbon layer (118).

The satisfactory preservation of myelin for electron microscopy seems closely related to the problem of lipid preservation. Thus Dalton's fixative alone or after aldehyde prefixation seems the method of choice, and Luse (77) introduced an interesting modification employing Saponin added to the fixative and a toluene clearing step, which seems to emphasize further the importance of structural lipid preservation for the maintenance of the morphological integrity of myelin.

Buffers, Osmolality, and Additives. The selection of buffer species may be conditioned by the considerations included in the section on buffers in this chapter, and by the impressions of investigators comparing the effects of different buffers (35, 145, 149). The buffer should offer ample protection against pH shifts particularly in the acid direction, in the range desired. However, it probably need not be used in as high concentrations as is commonly the case (up to $0.1M$). Osmotic considerations may be dealt with quite apart from buffer concentrations.

It is commonly accepted that the tonicity of a fixative should be isotonic or preferably somewhat hypertonic to plasma for optimal fixation (21, 66, 85, 126, and others). It is furthermore noted but less often clearly stated that it is the buffer concentration and salts added that are of significance in this respect, and that the aldehyde components of fixatives are essentially irrelevant to the consideration of "effective tonicity." In fact, the vehicle for aldehyde fixation should be adjusted to the desired tonicity (by calculation, 83, or by freezing point depression measurements) quite apart from the aldehyde. Freezing point depression measurements are quite meaningless, if made on a complete aldehyde fixative (buffer, salts and aldehyde) insofar as the osmotic behavior of the cell is concerned. The cell membrane once exposed to fixative will no longer behave as a classical semipermeable membrane, as its structure is fundamentally altered. Membrane transport systems are partially or wholly destroyed introducing a further departure from the normal. Probably more important to consider in the use of freezing point depression as an estimate of

---

[5] Repeated reference has been made to the possible desirability of considering the enzyme-killing aspect of fixation. No more specific definition of this problem is as yet possible. Which enzyme systems might be critical is conjectural, and certain dangers are inherent in interruption of a chain of enzymic reactions which, for example, might lead to local accumulations of osmotically active substrates.

"osmolality" of fixatives are the factors discussed by Dormandy (32).

Freezing point measurements are reported by the machine in terms of equivalent concentrations of sodium chloride in water, and are of use only in respect to the salts and water employed in the fixative. Osmolality is a function of the vapor pressure of the water, which is in turn a function of the particles held in solution. The addition of aldehyde "antifreezes" renders freezing point a doubtful indicator of the "osmolality" of the solution.

With appropriately adjusted vehicle tonicity the aldehyde concentrations can be varied within wide limits (see "Fixatives" above).

The adjustment in tonicity of fixing vehicle is commonly accomplished by the addition of monovalent ions, usually sodium chloride (85). However, other ions may be useful in the fixative, and the tonicity "gap" between buffer concentration and final vehicle osmolality may be made up by some of these. The beneficial effects of potassium ions should be considered, among the monovalent ions available (14, 15, 17, 112). Probably more important as ionic additives are the divalent cations.

Calcium is probably the first choice for a cationic additive to fixatives (see 14, 15, 17, 27, 95, 97, 112, 114, 140). It is not always clear if its stated virtues are to be attributed to the ion species or to a nonspecific effect of divalent cations generally. However, a number of structural features are said to be preserved by its presence including cytoplasmic microtubules and chromosomal microfilaments. Millonig (87) claims that divalent cations speed protein fixation (while imparting some granularity). It is usually stated that calcium helps stabilize membrane structure. Palay et al. (95) reported this to be true in brain, but required a concentration of 0.5% $CaCl_2$ for this effect. Gobel (42) reported significant improvement in membrane preservation at a concentration of 0.135%.

Other divalent cations are seldom added deliberately to fixatives although their use might be considered. Manganese ion may well prove specifically useful for the catechol amine-related synaptic vesicles of the brain, and the protective effect of magnesium against phosphate-induced mitochondrial swelling should be recalled (68). Some divalent cations may prove useful as enzyme inhibitors, for tissue killing.

Sucrose was an early choice (23) for osmotic adjustment of osmium tetroxide and may be employed with other fixatives as well, although in much higher concentrations for a given osmotic effect than is required of ionizing salts. Millonig (87) suggests it may interfere with protein stabilization, however. Colloid osmotic pressure has long been recognized as of importance in perfusion methods. Presently, inert plasma substitutes appear favored over the older gum acacia additives. Polyvinyl pyrolidone (PVP) or Dextran are widely used (see 13, 132). Little benefit may accrue from such use in immersion fixation but its use seems indicated in perfusion fixation. It is probably of use at least initially to help limit organ swelling during high-pressure perfusion, and to facilitate exchange and flow of fixative through extravascular spaces and its return to the venous channels, possibly according to the Starling extracapillary circulation model. Such replenishment of tissue fixative is clearly desirable in that it assures a steady concentration of fixative and buffer during the initial phases of fixation and stabilization.

The addition of Saponin or Digitonin for lipids has been referred to. Dermer (30) and Elbers et al (34) should be consulted for methods of phospholipid preservation as well.

Special Tissue Modifications. Most of the uses of fixatives referred to here have been developed for general, mature mammalian tissues. Special tissues have been handled in large part by only slightly modifying their basic formulations (e.g., see Glauert, 40, on plant and microbial preservation).

Embryonic and newborn tissues seem to benefit from somewhat less hypertonic fixatives (21, 134, 136) than those satisfactory for adult tissues. Buffers of about 0.067–0.075M seem adequate for newborn brain and fetal tissues. It was popular for a time to elevate the pH of embryonic fixatives (pH 8.0 or higher) but this does not seem critical if appropriate tonicity is achieved.

Marine animals pose special problems. The tonicity and pH of body fluids of these forms vary enormously with species, and the problems are compounded by the different osmotic environments of various organs (e.g., gills and skin v. brain). Available data on the osmotic environment of tissues and species of interest must enter fixative design consideration. The usual aldehyde-buffer combinations are often employed, with osmolality adjustments made with sodium chloride (90, 91) or salt and urea in shark (73), for example, with reasonably good effect. Marine as well as other poikilotherm animals can be cooled, and immersion fixation works well with their tissues; perfusion fixation may thus not always be preferred. In fish, the interposition of the gill vasculature between the heart and brain, and the sparse vascularity of the brain in these forms work against effective perfusion fixation. However, with careful attention to detail, strikingly good perfusion fixation of fish brain is possible (64, 112, 113). Small swimming animals, marine and freshwater, are often fixed by addition of fixative to their normal medium (e.g., 123).

Insect tissues are commonly fixed with little or no modification of the conventional mixtures (glutaraldehyde-phosphate buffer being generally the most popular class). In the case of invertebrates, and possibly non-mammalian vertebrate forms, further experimentation may well reveal that additional modification of "standard" fixatives will provide superior preservation. One is frequently assailed with vague dissatisfaction in comparing invertebrate with ideally preserved mammalian tissues, both prepared with "mammalian" fixatives, although the illustrations of Chalazonitis (24) are an example of superb fixation in snail, with a near "mammalian" mixture.

The preservation of nuclear material in cells has not received the attention its importance warrants. Fixative development has largely proceeded on criteria related to cytoplasmic organelle and membrane structure. Chromate and aldehydes generally present nuclear material in an aggregated or "clumped" form, while osmium preservation displays it in a finely granular, scattered form. There are clearly differences in the disposition of nuclear contents relative to the functional or differentiated status of the cell, but the interpretation of the responses to fixatives is still not as clear as might be desired. The interested reader is referred to some pertinent literature (48, 81, 82, 98).

Delivery of Fixatives

Vapor Fixation. When it is desirable to minimize diffusion problems either with or without freezing, fixation by vapor offers some advantages. This serves to maximize the fixative concentration relative to the water in the system. Of the fixatives discussed, only osmium tetroxide and formaldehyde are used in this form. Acrolein is too unpleasant for the purpose, and the volatility of the others is too low to be of use. Osmium tetroxide in this form has been applied to myelin to demonstrate the structure-altering properties of this fixative in water: Only after vapor fixation are the X-ray diffraction characteristics retained (88). Some of the disordering of protein due to aldehyde fixation (70) might be avoided by this method of handling as well.

Immersion Fixation. For general purposes immersion is employed where perfusion is impractical, although this is to be regarded as the second choice of methods. The procedure must emphasize minimal delay between interruption of normal circulation and tissue metabolism, and contact with the fixative. Dripping of fixative on the surface of exposed but otherwise normal tissue was a popular method for initial osmium tetroxide fixation. The penetration of fixative by this method is very slow indeed (18, 29, 84). It may be slightly improved by pretreatment with hyaluronidase (96), and possibly with the use of DMSO, presently employed in histochemistry to facilitate substrate entrance (51, 80), but at best little is to be expected of drip-fixing in an organ (such as cerebral cortex) whose blood supply enters from the surface, even if rapidly penetrating aldehydes are used.

Some improvement is to be noted in immersion fixation methods, if the tissue will tolerate "mincing" in fixative (see 100), so that fixative may penetrate from multiple surfaces. Mincing should be done in relatively large volumes of fixative, preferably cold. With small drops of mincing fluid, the percent water loss through evaporation is unacceptably large and the consequent elevation of fixative tonicity severely damages the tissue. For the same reason tissue must be kept submerged in fixative at all times. The slightest degree of surface drying of small blocks is catastrophic.

Mechanically distortion must be held to an absolute minimum in this procedure, and some tissues do not tolerate mincing well at all. It is then preferable to fix large blocks in cold, rapidly penetrating fixatives, then after hardening to trim off mechanically damaged surfaces and divide the remaining tissue into blocks of useful size. The element of mechanical damage must be particularly impressed on the surgeon securing a biopsy specimen for the investigator (see the comments of Pease, 100). With proper attention to this matter surgical biopsies can yield exquisite light and electron microscopic preparations. Punch biopsies probably are not as satisfactory principally for the reason of mechanical damage.

Perfusion. Perfusion fixation is the method of choice for general purposes where practical, serving to minimize disruption of metabolic processes prior to fixative exposure. Certain organs (e.g., liver, 130, kidney, 84) may be individually perfused, and greater control over the hydrodynamics of the procedure may be exerted with respect to the organ of interest. For other tissues, including the nervous system, perfusion of the whole animal or major regions is elected. The head, cervical region, and upper extremities may be selectively perfused through the heart, ascending or descending aorta by clamping the descending aorta inferior to the cannula. During whole-body perfusion the resistance of the visceral vascular shunt is easily elevated by application of a clamp to the free margin of the lesser omentum occluding the portal vein,

prior to the initiation of perfusion. This is particularly useful in perfusions in which spinal cord fixation is desired.

Many reports detailing perfusion procedures have been published, and the beginning investigator is urged to consult some of these (11, 22, 46, 62, 64, 65, 92, 95, 107, 112, 113, 125, 126, 147, 151). The present description of methodology includes factors considered by these authors in its design, as well as the results of experiments and observations in our laboratory.

The careful study of Palay et al. (95) popularized the notion of the induction of vasodilation prior to perfusion, by the use of sodium nitrite. All efforts to improve the reliability of the notoriously capricious osmium tetroxide perfusion method may be warranted, but such treatment has not been generally found necessary with aldehyde perfusion, and indeed may be deleterious to central nervous system perfusion (see Cammermeyer, 22, for alternative treatment). Central nervous system vasculature can be more selectively opened by carbon dioxide inhalation, also employed by Palay et al. In any case, a vasodilator can be of benefit only in insuring that the vascular bed of interest is open at the moment of initiation of the perfusion. Fixation and agonal constriction of vascular smooth muscle surely become of paramount importance in the determination of vascular patency over any preperfusion myogenic or neurogenic vascular wall status.

The vessels must be fixed while dilated, after which pressure may be allowed to drop to minimize tissue distortion due to fixative extravasation. It is also essential to initiate perfusion while the animal's arterial pressure is at a physiological level. A drop in pressure at the initiation of perfusion will lead to vascular collapse and the attendant problems of reopening arterial pathways with the perfusate. This introduces not only the problem of vessels fixed in a constricted or collapsed state but the vascular reopening problem analyzed by Burton (20). This reference is recommended for appreciation of the relationship of transmural pressure and wall tension to vascular patency. These factors are essential considerations in the design of perfusion technique. Since critical closing pressures cannot be predicted in a given vascular bed at the moment of perfusion, the problem is best circumvented by initiating perfusion with the animal in no danger of vascular collapse, that is, in a state of good cardiovascular function.

With sufficient flow rates it is not necessary to wash out the vascular tree with saline or other rinse solution, and for the reasons above related to maintenance of vascular patency it is probably undesirable to do so. Bodian (11) recognized the need to minimize this phase of perfusion fixation, and it has been reemphasized by Karlsson and Schultz (65). Such a procedure delays the introduction of fixative, and the changeover from saline to fixative may be accompanied by vascular collapse prior to the introduction of the fixative.

Perfusion of the upper half of the animal may be accomplished through the abdominal aorta, making it unnecessary to enter the thoracic cavity. We prefer not to use this method, however. There is shunting of fixative through the low-resistance pulmonary circulation, and it is difficult to obtain a sufficiently wide opening in the venous system. Especially in larger animals, it is a simple matter to introduce a tracheal cannula to provide positive pressure ventilation in an open chest preparation. The gas mixture can be controlled to elevate blood carbon dioxide partial pressures, and this can be monitored if desired. We ventilate animals with a prepared gas mixture of 20% oxygen, 10% carbon dioxide, and 70% nitrogen. With the thorax widely opened and the animal adequately ventilated, the operator may proceed at leisure.

The pericardium is widely incised, the heart delivered, and the animal is respired until good cardiovascular function returns after the surgical shock (1–2 min). A regular strong heartbeat and (if recorded) physiological arterial pressures are taken as indications of this state. The lungs may be allowed to deflate momentarily and a clamp applied to the descending aorta if desired.

With cardiac recovery, it is necessary to accomplish the next steps in rapid sequence. The right atrial wall is grasped with toothed forceps and widely opened. The apex of the heart is immediately seized and elevated, and a scissor incision made in the left ventricular wall near the apex. The fixative line has previously been filled with fixative at the desired initial pressure and clamped for quick release (hemostatic forceps). The cannula is inserted through the ventricular incision towards the base of the aorta. The myocardium is immediately clamped around the cannula (sponge forceps) sealing against backflow leakage at the incision and providing a firm grip for retaining the heart and cannula in a position in the chest which precludes tension or kinks in the line and aorta. As soon as the cannula is in place the line clamp is released and fixative in large volume fills the vascular tree at full perfusion pressure. With a little practice and an assistant the procedure can be carried out in seconds with a negligible fall in arterial pressure. The animal will stiffen almost immediately and the atrial effluent will be clear in a few moments.

Metal cannulae (stainless steel or aluminum) are used because of the large bore diameter compared to wall thickness and for fear of breakage if a glass cannula were to be vigorously clamped. The choice of cannula size is determined empirically according to whether upper-half or whole-body perfusion is intended, and animal size. In general a bore diameter no less than one-half the diameter of the aorta is used for upper-half perfusion, and between this size and aortic diameter for whole-body perfusion. No tubing or other stricture in the fluid pathway should be smaller than the cannula bore, if the perfusion pressure gauge readings are to be meaningful. For this reason as well, a saline flush is avoided, as the addition of a three-way stopcock of conventional size introduces a segment of about 16 gauge, and the pressure drop through such a stricture at high flow rates is very substantial indeed. The amount of pressure drop across this or other strictures is a function of flow rate, and proximal pressure readings will yield little information on the pressure status of the animal's vascular tree.

Initial perfusion pressures should be set at whatever values are necessary in order to keep measured arterial pressure (femoral or brachial artery) *at least* at normal mean arterial pressure, and preferably initially near systolic pressure, *during steady fixative flow*. Monitored venous pressure (superior or inferior vena cava) should not exceed 10 mm Hg or so; if it does, the aperture for atrial outflow is not adequate, and small-vein transmural pressure will probably be too high. For intracranial fixation the pressures in the subarachnoid space appear to be critical. The pressure may be monitored via a polyethylene tube sealed in place through a trephine hole

with dental cement or dental wax. As the vasculature becomes fixed open, the resistance to flow is localized in the small vessels; extravasation of fixative increases rapidly and the pressure rises in the subarachnoid space and will, unchecked, approach arterial perfusion pressure. This can induce a tamponade effect occluding the venous pathways and fixative stasis results, obviating any benefit of further perfusion. Under repeatable circumstances, a few measurements during perfusion will indicate when the perfusion pressure should be reduced to eliminate this effect, and subarachnoid space pressure measurements need not always be made subsequently. Routine perfusions can be performed, other things being equal, with adjustments based on measured perfusion pressures (aneroid manometer in the fixative line) and a floating ball flowmeter (in the air line driving the fixative from a reservoir bottle).

Perfusion pressure of sufficient magnitude and subject to rapid control are best provided by a laboratory compressed air line or a source of bottled gas. Pressure is regulated with a variable bleed valve in the air line, and driving pressure is led (in our routine perfusions) through a flowmeter to the fixative reservoir bottle. The perfusing pressure is read on the fixative outflow side of the system at a point distal to which the narrowest part of the system is the perfusion cannula. The overall arrangement is rather similar to that described by Wright and Sanderson (151).

The duration of the perfusion must be determined by the investigator's experience. Seven to fifteen minutes has proven sufficient in our experience with the nervous system, after which tissue is blocked and treated with osmium tetroxide.

DAVID S. MAXWELL

References

1. Adams, C. M. W., Y. H. Abdulla, and D. B. Bayliss, *Histochemie,* **9**:68–77 (1967).
2. Afzelius, B. F., pp. 1–19 in R. J. C. Harris, ed., "The Interpretation of Ultrastructure," New York, Academic Press, 1962.
3. Akert, K., and C. Sandri, *Brain Res.,* **7**:286–295 (1968).
4. Akert, K., and K. Pfenninger, pp. 245–260 in S. H. Barondes, ed., "Cellular Dynamics of the Neuron," New York, Academic Press, 1967.
5. Baker, J. R., "Principles of Biological Microtechnique," London, Methuen, 1958 (pp. 19–151).
6. Baker, J. R., and J. M. McCrae, *J. Roy. Microscop. Soc.,* **85**:391–399 (1966).
7. Bates, R. G., pp. J-190–J-194 in H. A. Sober, ed., "Handbook of Biochemistry," Cleveland, Chemical Rubber Company, 1968.
8. Bennett, H. S., and J. H. Luft, *J. Biophys. Biochem. Cytol.,* **6**:113–114 (1959).
9. Berkowitz, L. R., O. Fiorello, L. Kruger, and D. S. Maxwell, *J. Histochem. Cytochem.,* **16**:808–814 (1968).
10. Blackstad, T. W., pp. 186–216 in W. J. H. Nauta and S. O. E. Ebbesson, ed., "Contemporary Research Methods in Neuroanatomy," New York, Springer-Verlag, 1970.
11. Bodian, D., *Anat. Rec.,* **65**:89–97 (1936).
12. Bodian, D., *J. Cell Biol.,* **44**:115–124 (1970).
13. Bohman, S.-O., and A. B. Maunsbach, *J. Ultrastruct. Res.,* **30**:195–208 (1970).
14. Bourke, R. S., and D. B. Tower, *J. Neurochem.,* **13**:1071–1097 (1966).
15. Bourke, R. S., and D. B. Tower, *J. Neurochem.,* **13**:1099–1117 (1966).

16. Bowes, J. A., and C. W. Cater, *J. Roy. Microscop. Soc.*, **85**:193–200 (1966).
17. Bulger, R. E., and B .F. Trump, *J. Ultrastruct. Res.*, **28**:301–319 (1969).
18. Burkel, A., and H. Schiechel, *J. Histochem. Cytochem.*, **16**:157–161 (1968).
19. Burstone, M. S., "Enzyme Histochemistry and Its Application in the Study of Neoplasms," New York, Academic Press, 1962.
20. Burton, A. C., *Physiol. Rev.*, **34**:619–642 (1954).
21. Caley, D. W., and D. S. Maxwell, *J. Comp. Neurol.*, **138**:31–48 (1970).
22. Cammermeyer, J., *Acta Neuropathol.*, **11**:368–371 (1968).
23. Caulfield, J. B., *J. Biophys. Biochem. Cytol.*, **3**:827–829 (1957).
24. Chalazonitis, N., pp. 229–243 in S. H. Barondes, ed., "Cellular Dynamics of the Neuron," New York, Academic Press, 1969.
25. Clark, T. S., and S. P. Rochlani, pp. 232–233, Abstr. in C. J. Arceneaux, ed., "Proc. 28th Ann. EMSA Meeting," Baton Rouge, Claitor's Publ. Div., 1970.
26. Dalton, A. J., *Anat. Rec.*, **121**:281 (Abstr.) (1955).
27. Davies, H. G., and M. Spenser, *J. Cell Biol.*, **14**:445–458 (1962).
28. del Cerro, M. P., and R. S. Snider, p. 555, Abstr. in *Proc. 4th European Regional Conference on Electron Microscopy* (1968).
29. Dempster, W. T., *Amer. J. Anat.*, **107**:59–72 (1960).
30. Dermer, G. R., *J. Ultrastruct. Res.*, **22**:312–325 (1968).
31. Dermer, G. B., *J. Ultrastruct. Res.*, **27**:88–104 (1969).
32. Dormandy, T. L., *Lancet*, **Feb. 1967**:267–270.
33. Dunn, R. F., *J. Ultrastruct. Res.*, **16**:651–671 (1966).
34. Elbers, P. F., J. T. Ververgaert, and R. Demel, *J. Cell Biol.*, **24**:23–30 (1965).
35. Ericsson, J. L. E., A. J. Saladino, and B. F. Trump, *Z. f. Zellforsch.*, **66**:161–181 (1965).
36. Fahimi, H. D., and P. Drochmans, *J. Histochem. Cytochem.*, **16**:199–204 (1968).
37. Friend, D. S., and J. Murray, *Amer. J. Anat.*, **117**:135–150 (1965).
38. Frigerio, N. A., and M. J. Shaw, *J. Histochem. Cytochem.*, **17**:176–181 (1969).
39. Gaylarde, P., and I. Sarkony, *Science*, **161**:1157–1158 (1968).
40. Glauert, A. M., pp. 166–212 in D. H. Kay, ed., "Techniques for Electron Microscopy," 2nd ed., Philadelphia, F. A. Davis and Co., 1965.
41. Glick, D., and D. C. Pease, Letters to the editors, *J. Histochem. Cytochem.*, **18**:455–458 (1970).
42. Gobel, S., *J. Ultrastruct. Res.*, **15**:310–325 (1966).
43. Gomori, G., pp. 138–146 in S. P. Colowick and N. O. Kaplan, ed., "Methods in Enzymology," Vol. I, New York, Academic Press, 1955.
44. Gonatas, N. K., S. Levine, and R. Schoulson, *Amer. J. Pathol.*, **44**:565–583 (1964).
45. Gonzales-Aguilar, F., *J. Ultrastruct. Res.*, **29**:76–85 (1969).
46. Gonzales-Aguilar, F., and E. de Robertis, *Neurol.*, **13**:758–771 (1963).
47. Graham, R. C., Jr., and M. J. Karnovsky, *J. Histochem. Cytochem.*, **14**:291–302 (1966).
48. Granboulan, N., and P. Granboulan, *Exp. Cell Res.*, **34**:71–87 (1964).
49. Gray, P., "The Microtomists' Formulary and Guide," New York, Blakiston Co., 1954.
50. Hanker, J. S., C. Deb, H. L. Wasserkrug, and A. Seligman, *Science*, **152**:1631–1634 (1966).
51. Hanker, J. S., C. J. Kusyk, D. H. Clapp, and P. E. Yates, *J. Histochem. Cytochem.*, **18**:673 (Abstr.) (1970).
52. Heimer, L., *Brain Res.*, **5**:86–108 (1967).
53. Heimer, L., pp. 162–172 in W. J. H. Nauta and S. O. E. Ebbesson, ed., "Contemporary Research Methods in Neuroanatomy," New York, Springer-Verlag, 1970.
54. Heimer, L., and A. Peters, *Brain Res.*, **8**:337–346 (1968).
55. Higgins, J. A., N. T. Florendo, and R. J. Barrnett, *J. Cell Biol.*, **47**:87a–88a (Abstr.) (1970).
56. Hirsch, J. G., and M. E. Fedorko, *J. Cell Biol.*, **38**:615–627 (1968).
57. Holt, S. J., and R. M. Hicks, *J. Biophys. Biochem. Cytol.*, **11**:31–45 (1961).
58. Holt, S. J., and R. M. Hicks, *Nature*, **191**:832 (1961).
59. Hopwood, D., *Histochemie*, **11**:289–295 (1967).
60. Hopwood, D., *J. Anat.*, **101**:83–92 (1967).
61. Ito, S., and M. J. Karnovsky, *J. Cell Biol.*, **39**:168a–169a (Abstr.) (1968).
62. Johnston, P. V., and B. I. Roots, *J. Cell Sci.*, **1**:377–386 (1967).
63. Kadin, M. E., L. J. Rubenstein, and J. S. Nelson, *J. Neuropathol. Exp. Neurol.*, **29**:583–600 (1970).
64. Kaiserman-Abramoff, I. R., and S. L. Palay, pp. 171–204 in R. Llinás, ed., "Neurobiology of Cerebellar Evolution and Development," Chicago, Inst. Biomed. Res. AMA/ERF, 1967.
65. Karlsson, U., and R. L. Schultz, *J. Ultrastruct. Res.*, **12**:160–186 (1965).
66. Karnovsky, M. J., *J. Cell Biol.*, **27**:137a–138a (Abstr.) (1965).
67. Korn, E. D., *Science*, **153**:1491–1498 (1966).
68. Kroll, A. J., and T. Kuwabara, *J. Cell Biol.*, **15**:29–35 (1962).
69. Lee, B., (Bolles Lee's) "The Microtomists' Vade-Mecum," 11th ed., by J. B. Gatenby and H. W. Beams, Philadelphia, Blakiston, 1950.
70. Lenard, J., and S. J. Singer, *J. Cell Biol.*, **37**:117–121 (1968).
71. Levy, W. A., I. Herzog, K. Suzuki, R. Katzman, and L. Scheinberg, *J. Cell Biol.*, **27**:119–132 (1965).
72. Litman, R. B., and R. J. Barrnett, *Anat. Rec.*, **163**:313–314 (Abstr.) (1969).
73. Long, D. M., T. S. Bodenheimer, J. F. Hartmann, and I. Klatzo, *Amer. J. Anat.*, **122**:209–236 (1968).
74. Luft, J. H., *J. Biophys. Biochem. Cytol.*, **2**:799–801 (1956).
75. Luft, J. H., *Anat. Rec.*, **133**:305 (Abstr.) (1959).
76. Luft, J. H., and R. L. Wood, *J. Cell Biol.*, **19**:46a (Abstr.) (1963).
77. Luse, S. A., *J. Ultrastruct. Res.*, **4**:108–112 (1960).
78. McClung, C. E. "McClung's Handbook of Microscopical Technique," 3rd ed., ed. R. M. Jones, New York, Hoeber, 1950.
79. McEwen, L. M., *J. Physiol.*, **131**:678–689 (1956).
80. Makita, T., and E. B. Sandborn *J. Histochem. Cytochem.*, **18**:686 (Abstr.) (1970).
81. Marinozzi, V., *J. Roy. Microscop. Soc.*, **81**:141–154 (1963).
82. Marinozzi, V., and A. Gautier, *J. Ultrastruct. Res.*, **7**:436–451 (1962).
83. Maser, M. D., T. E. Powell, and C. W. Philpott, *Stain Technol.*, **42**:175–182 (1967).
84. Maunsbach, A. B., *J. Ultrastruct. Res.*, **15**:242–282 (1966).
85. Maunsbach, A. B., *J. Ultrastruct. Res.*, **15**:283–309 (1966).
86. Millonig, G., *J. Appl. Phys.*, **32**:637 (Abstr.) (1961).
87. Millonig, G., pp. 21–22, Abstr. in R. Uyeda, ed., "Proceedings of the Sixth International Congress on Electron Microscopy," Vol. 2, Tokyo, Maruzen Co., Ltd., 1966.
88. Moretz, R. C., *J. Cell Biol.*, **39**:95a (Abstr.) (1968).
89. Morgan, T. E., and G. L. Huber, *J. Cell Biol.*, **32**:757–760 (1967).
90. Mugnaini, E., *Sarsia* (Bergen), **29**:221–232 (1967).

91. Mugnaini, E., and F. Walberg, *Z. f. Zellforsch.,* **66**:333–351 (1965).

92. Nevis, A. H., and G. H. Collins, *Brain Res.,* **5**:57–85 (1967).

93. North, R. J., *J. Ultrastruct. Res.,* **16**:83–95 (1966).

94. Palade, G. E., *J. Exp. Med.,* **95**:285–298 (1962).

95. Palay, S. L., S. M. McGee-Russell, S. Gordon, and M. A. Grillo, *J. Cell Biol.,* **12**:385–410 (1962).

96. Pallie, W., and D. C. Pease, *J. Ultrastruct. Res.,* **2**:1–17 (1958).

97. Pappius, H. M., and K. A. C. Elliott, *Can. J. Biochem. Physiol.,* **34**:1007–1022 (1956).

98. Patrizi, G., and M. Poger, *J. Ultrastruct. Res.,* **17**:127–136 (1967).

99. Pearse, A. G. E., "Histochemistry, Theoretical and Applied," 3rd ed., Vol. I, Boston, Little Brown and Co., 1968 (pp. 70–105).

100. Pease, D. C., "Histological Techniques for Electron Microscopy," 2nd ed., New York, Academic Press, 1964 (pp. 14–81).

101. Pease, D. C., *J. Ultrastruct. Res.,* **15**:555–588 (1966).

102. Pease, D. C., pp. 36–37 in C. J. Arceneaux, ed., "Proc. 26th EMSA Meeting," Baton Rouge, Claitor's Publ. Div., 1968.

103. Peracchia, C., B. S. Mittler, and S. Frenk, *J. Cell Biol.,* **47**:156a (Abstr.) (1970).

104. Peters, A., pp. 56–76 in W. J. H. Nauta and S. O. E. Ebbesson, ed., "Contemporary Research Methods in Neuroanatomy," New York, Springer-Verlag, 1970.

105. Peterson, R. G., and D. C. Pease, pp. 334–335, Abstr. in C. J. Arceneaux, ed., "Proc. 28th Ann. EMSA Meeting," Baton Rouge, Claitor's Publ. Div., 1970.

106. Pizzolato, P., R. D. Lillie, R. Henderson, and P. Donaldson, *J. Histochem. Cytochem.,* **18**:674–675 (Abstr.) (1970).

107. Raimondi, A. J., G. Vailati, and S. Mullan, *Riv. Pat. Nerv. Ment.,* **85**:116–132 (1964).

108. Reese, T. S., and M. J. Karnovsky, *J. Cell Biol.,* **34**:207–217 (1967).

109. Richardson, K. C., *Anat. Rec.,* **139**:333 (Abstr.) (1961).

110. Riemersma, J. C., pp. 69–99 in D. F. Parsons, ed., "Some Biological Techniques in Electron Microscopy," New York, Academic Press, 1970.

111. Robertson, E. A., and R. L. Schultz, *J. Ultrastruct. Res.,* **30**:275–287 (1970).

112. Robertson, J. D., *J. Cell Biol.,* **19**:201–221 (1963).

113. Robertson, J. D., T. S. Bodenheimer, and D. E. Stage, *J. Cell Biol.,* **19**:159–199 (1963).

114. Roozemond, R. C., *J. Histochem. Cytochem.,* **17**:482–486 (1969).

115. Rosenbluth, J., *J. Cell Biol.,* **16**:143–157 (1963).

116. Sabatini, D. D., K. Bensch, and R. J. Barrnett, *J. Cell Biol.,* **17**:19–66 (1963).

117. Sabatini, D. D., F. Miller, and R. J. Barrnett, *J. Histochem. Cytochem.,* **12**:57–71 (1964).

118. Salpeter, M. M., and L. Bachmann, *J. Cell Biol.,* **22**:469–477 (1964).

119. Sandborn, E. B., "Cells and Tissues by Light and Electron Microscopy," Vol. I, New York, Academic Press, 1970.

120. Sandborn, E. B., P. F. Koen, J. D. McNabb, and G. Moore, *J. Ultrastruct. Res.,* **11**:123–138 (1964).

121. Saunders, D. R., J. Wilson, and C. E. Rubin, *J. Cell Biol.,* **37**:183–187 (1968).

122. Scallen, T. J., and S. E. Dietert, *J. Cell Biol.,* **40**:802–813 (1969).

123. Scharrer, E., *Z. f. Zellforsch.,* **61**:803–812 (1964).

124. Schultz, R. L., and N. M. Case, *J. Cell Biol.,* **38**:633–637 (1968).

125. Schultz, R. L., and N. M. Case, *J. Microscopie* **92**(part 2):69–84 (1970).

126. Schultz, R. L., and U. Karlsson, *J. Ultrastruct. Res.,* **12**:187–206 (1965).

127. Seligman, A. M., H. L. Wasserkrug, C. Deb, and J. S. Hanker, *J. Histochem. Cytochem.,* **16**:87–101 (1968).

128. Sevier, A. C., and B. L. Munger, *Anat. Rec.,* **162**:43–52 (1968).

129. Siegesmund, K. A., *Anat. Rec.,* **162**:189–196 (1968).

130. Sjostrand, F. S., "Electron Microscopy of Cells and Tissues," Vol. I, New York, Academic Press, 1967 (pp. 138–176).

131. Sober, H. A., ed., "Handbook of Biochemistry," Cleveland, Chemical Rubber Company, 1968 (pp. J-195–J-199).

132. Sotello, C., and S. L. Palay, *J. Cell Biol.,* **36**:151–179 (1968).

133. Sterzing, P., J. Scaletti and L. Napolitano, *J. Cell Biol.,* **47**:203a (Abstr.) (1970).

134. Sumi, S. M., *J. Ultrastruct. Res.,* **29**:398–425 (1969).

135. Taxi, J., and B. Droz, pp. 175–190 in S. H. Barondes, ed., "Cellular Dynamics of the Neuron," New York, Academic Press, 1969.

136. Tennyson, V. M., pp. 47–116 in W. A. Himwich, ed., "Developmental Neurobiology," Springfield, C. C. Thomas, 1970.

137. Terry, R. D., pp. 335–347 in O. T. Bailey and D. E. Smith, ed., "The Central Nervous System. Some Experimental Models of Neurological Diseases," Int'l. Acad. Pathol. Monograph No. 9, Baltimore, Williams and Wilkins Co., 1968.

138. Terzakis, J., *J. Ultrastruct. Res.,* **22**:168–184 (1968).

139. Thompson, S. W., "Selected Histochemical and Histopathological Methods," Springfield, C. C. Thomas, 1966 (pp. 3–23).

140. Torack, R. M., *Z. f. Zellforsch.,* **66**:352–364 (1965).

141. Torack, R. M., *J. Ultrastruct. Res.,* **14**:590–601 (1966).

142. Torack, R. M., M. L. Dufty, and J. M. Haynes, *Z. f. Zellforsch.,* **66**:690–700 (1965).

143. Tormey, J. M., *J. Cell Biol.,* **23**:658–664 (1964).

144. Trelstad, R. L., *J. Histochem. Cytochem.,* **17**:756–757 (1969).

145. Trump, B. F., and J. L. E. Ericsson, *Lab. Invest.,* **14**:507/1245–585/1323 (1965). Also published as chapter in G. F. Bahr and E. H. Zeitler, ed., "Quantitative Electron Microscopy," Baltimore, Williams and Wilkins.

146. Van Harreveld, A., J. Crowell, and S. K. Malhotra, *J. Cell Biol.,* **25**(No. 1, part 1):117–137 (1965).

147. Westrum, L. E., and R. D. Lund, *J. Cell Sci.,* **1**:229–238 (1966).

148. Wigglesworth, V. B., *Proc. Roy. Soc.,* (B)**147**:185–199 (1957).

149. Wood, R. L., and J. H. Luft, *J. Ultrastruct. Res.,* **12**:22–45 (1965).

150. Wrench, C. P., pp. 300–301, Abstr. in C. J. Arceneaux, ed., "Proc. 28th EMSA Meeting," Baton Rouge, Claitor's Publ. Div., 1970.

151. Wright, G., and J. M. Sanderson, *J. Pathol.,* **100**:295–305 (1970).

152. Zamboni, L., and C. deMartino, *J. Cell Biol.,* **35**:148a (Abstr.) (1967).

153. Zetterquist, H., "The ultrastructure organization of the columnar absorbing cells of the mouse jejunum." Privately printed for the Department of Anatomy, Karolinska Institutet, Stockholm, by Aktiebolaget Godvil, Stockholm, 1956 (83 pp.).

FIXATIVE FORMULAS

From the standpoint of the working technician fixative formulas may be divided into (1) fixatives for materials to be examined in the electron microscope, (2) histological fixatives containing unusual ingredients,

(3) histological fixatives in non aqueous solvents, and (4) histological fixatives in aqueous solvents.

1. Fixatives for Ultramicroscopy. These have been adequately dealt with in the preceding article and will not be further recorded here.

2. Fixatives Containing Unusual Ingredients. It will be pointed out under 4 below that the great majority of fixatives can be prepared from few ingredients. A few fixatives, however, made from nonstandard ingredients are worth recording:

Becher and Demoll 1913 (Becher and Demoll, 1913, 48)
FORMULA: water 230, copper sulfate 5, mercuric chloride to sat. 40%, formaldehyde 20, acetic acid 1.25.

Benoit 1922a (*C.R. Soc. Biol., Paris*, **86**:1101)
FORMULA: water 250, osmic acid 1.25, mercuric chloride 3.1, potassium dichromate 3.75, uranium nitrate 2.0, sodium chloride 0.5

Benoit 1922b (*C.R. Soc. Biol., Paris*, **86**:1101)
FORMULA: A. water 250, mercuric chloride 3.1, potassium dichromate 3.75, uranium nitrate 2.0, sodium chloride 0.5; B. 2% formaldehyde

Carpenter and Nebel 1931 (*Science*, **74**:154)
FORMULA: water 250, ruthenium tetroxide 0.1, formic acid 1

Cohen 1935 (*Stain Technol.*, **10**:25)
FORMULA: water 225, chromium sulfate 11.25, 40% formaldehyde 25, picric acid to sat.

Davenport, Windle, and Rhines 1947 (Conn and Darrow 1947, 1C₂, 24)
FORMULA: paranitrophenol 12.5, water 112, 95% alc. 112, formamide 25

Friedenthal 1908 cited from *circ.* 1938 Wellings (Wellings *circ.* 1938, 25)
FORMULA: trichloroacetic acid 156, osmic acid 4, platinic chloride 4, chromic acid 8, uranium acetate 78

Gilson 1890 (*Cellule.*, **6**:122)
FORMULA: water 80, 95% alc. 20, zinc chloride 5, acetic acid 1.25, nitric acid 1

Juel cited from 1915 Meyer (Meyer 1915, 198)
FORMULA: water 125, 95% alc. 125, zinc chloride 5, acetic acid 5

Kenyon 1896 (*J. Comp. Neurol.*, **13**:296)
FORMULA: water 200, copper sulfate 5, potassium dichromate 9.5, 40% formaldehyde 50

Kingsbury 1912 (*Anat. Rec.* **6**:48)
FORMULA: water 225, mercuric chloride 7.5, copper sulfate 2.5, copper dichromate 6.25, 40% formaldehyde 25

Lison 1931 (*Arch. Biol. Paris*, **41**:343)
FORMULA: water 250, lead acetate 10, 40% formaldehyde 25

Lo Blanco 1890 (*Mitt. zool. Stat. Neapel* **9**:443)
FORMULA: water 250, mercuric chloride 1.75, copper sulfate 25

MacFarland and Davenport 1941 (*Stain Technol.*, **16**:53)
FORMULA: water 112, 95% alc. 112, chloral hydrate 12.5, formamide 25

Nelis 1900 (*Bull. Acad. Belg. Cl. Sci.*, **72**:6)
FORMULA: water 230, mercuric chloride 17.5, 40% formaldehyde 35, acetic acid 1.25, copper sulfate 1.25

Petrunkewitsch 1933a (*Science*, **77**:117)
FORMULA: water 100, ether 4.5, phenol 3, copper nitrate 3, nitric acid 2

Petrunkewitsch 1933b (*Science* **77**:117)
FORMULA: water 100, 96% alc. 150, ether 12.5, copper nitrate 5, nitric acid 7.5, paranitrophenol 12.5

Rawitz 1909 (*Z. wiss. Mikr.*, **25**:385)
FORMULA: water 100, 95% alc. 125, phosphotungstic acid 10, acetic acid 25.

Schiller 1930 solution 131—auct. (*Z. Zellforsch.*, **11**:63)
FORMULA: water 250, mercuric chloride 0.15, potassium dichromate 5, uranium acetate 0.2, magnesium acetate 2.5

Stappers 1909a (*Cellule.*, **6**:48)
FORMULA: water 225, mercuric chloride 7.5, copper sulfate 2.5, copper dichromate 6.25, 40% formaldehyde 25

Stappers 1909b (*Cellule.*, **25**:356)
FORMULA: water 220, copper nitrate 5, 40% formaldehyde 30

Zirkle 1928 (*Protoplasma*, **4**:201)
FORMULA: water 225, chromium sulfate 12.5, 40% formaldehyde 25, copper hydroxide to excess

3. Fixatives in Nonaqueous Solvents. These are for the most part used for the fixation either of nematodes or of various arthropods. They cannot be recommended for other purposes. They are, however, of some importance and the more usual are here recorded:

Bauer 1933 (*Z. Zellforsch.*, **33**:143)
FORMULA: dioxane 212.5, picric acid 75, 40% formaldehyde 25, acetic acid 12.5

Bensley 1910 (*Anat. Rec.*, **4**:379)
FORMULA: water 125, 95% alc. 125, mercuric chloride 32, potassium dichromate 2.5

Brasil 1904 (often called "Duboscq-Brasil" or "Alcoholic Bouin") (*Arch. Zool. exp. gén.*, **4**:74)
FORMULA: 80% alc. 150, picric acid 1, 40% formaldehyde 60, acetic acid 15

Claverdon 1943 (*Science*, **97**:168)
FORMULA: isopropyl alc. 137.5, acetone 75, picric acid 12.5, 40% formaldehyde 12.5, acetic acid 12.5

Davenport and Kline 1938a (*Stain Technol.*, **13**:160)
FORMULA: n-butyl alc. 150, n-propyl alc. 50, acetic acid 25, trichloroacetic acid 25

Davenport and Kline 1938b (*Stain Technol.*, **13**:160)
FORMULA: n-butanol 150, n-propyl alc. 50, trichloroacetic acid 25, formic acid 12.5

Davenport, McArthur, and Bruesch 1939 (*Stain Technol.*, **14**:22)
FORMULA: n-butanol 160, n-propyl alc. 65, trichloroacetic acid 12.5, formic acid 12.5

Eltringham 1930a fixative B—auct. (Eltringham 1930, 44)
FORMULA: water 145, 95% alc. 100, mercuric chloride 10, acetic acid 9

Eltringham 1930b fixative D—auct. (Eltringham 1930, 46)
FORMULA: 95% alc. 200, chloroform 37.5, picric acid 2, acetic acid 12.5

Emig 1941 (Emig 1941), 71
FORMULA: methanol 75, water 67, copper acetate 0.75, 40% formaldehyde 75, acetic acid 7.5

Grapnuer and Weissberger 1933 (*Zool. Anz.*, **102**:39)
FORMULA: methanol 50, dioxane 200, paraldehyde 5, acetic acid 12.5

Hetherington 1922 (*J. Parasitol.*, **9**:102)
FORMULA: abs. alc. 100, chloroform 75, acetic acid 25, phenol *q.s.* to make 250

Hosokawa 1934 cited from 1942 Langeron (Langeron 1942, 841)
FORMULA: methanol 250, 40% formaldehyde 12.5 glacial acetic acid 2.5

Huber cited from 1943 Cowdry *cit*. Addison (Cowdry 1943, 95)
FORMULA: 95% alc. 250, mercuric chloride 7.5, tri-chloroacetic acid 3.75

Jackson 1922 (*Bull. Int. Ass. Med. Mus.*, **8**:125)
FORMULA: 90% alc. 200, 40% formaldehyde 25, acetic acid 25

Kingsbury and Johannsen 1927 (Kingsbury and Johannsen 1927, 9)
FORMULA: abs. alc. 200, chloroform 35, picric acid 2, acetic acid 15

Lendrum 1935 (*J. Path. Bact.*, **40**:416)
FORMULA: water 10, abs. alc. 75, chloroform 10, penol 4, acetic acid 3

Lepine and Sautter 1936 (*Bull. Histo. Fdk.*, **12**:287)
FORMULA: abs. alc. 80, mercuric chloride 20, acetone 80, acetic acid 80

Ohlmacher 1897 (*J. Exp. Med.*, **3**:671)
FORMULA: abs. alc. 200, chloroform 37.5, mercuric chloride 50, acetic acid 12.5

Petrunkewitsch 1901 (*Zool. Jahrb.*, **14**:576)
FORMULA: water 150, 95% alc. 100, mercuric chloride to sat., acetic acid 45, nitric acid 5

Potenza 1939 cited from 1942 Langeron (Langeron 1942, 419)
FORMULA: methanol 50, dioxane 200, paraldehyde 5, acetic acid 12.5

vom Rath 1895 (*Anat. Anz.*, **11**:286)
FORMULA: abs. alc. 250, mercuric chloride 1.25, acetic acid 5

Saling 1906 cited from 1937 Gatenby and Painter (Gatenby and Painter 1937, 389)
FORMULA: water 125, 95% alc. 100, mercuric chloride 5, nitric acid 10

Sannomiya 1926 (*Folia. Anat. Jap.*, **4**:363)
FORMULA: abs. alc. 250, acetic acid 12.5, sulfosalicylic acid 7.5

Schaudinn 1893 (*Miami Univ. Bull.*, **57**:19)
FORMULA: 60% alc. 250, mercuric chloride 5.5

Spuler 1910 (Ehrlich, Krause, et al. 1910, **2**:521)
FORMULA: abs. alc. 250, mercuric chloride 10, acetic acid 7.5

Sz.-Györgyi 1914 (*Z. wiss. Mikr.*, **31**:23)
FORMULA: acetone 185, mercuric chloride 6, 40% formaldehyde 60, acetic acid 7.5

4. Fixatives in Aqueous Solvents. The writer long ago pointed out (Gray, 1933 *Jr. Roy. Micros. Soc.* **53**:13) that the great majority of fixatives could be prepared as required from standard aqueous solutions that were stable. This list was expanded and inserted as an appendix to the tenth and eleventh editions of Lee's "Microtomist Vade-mecum" (Philadelphia, Blakiston, 1937 and 1950). This list was very greatly expanded and used by Gray 1954 in the "Microtomist's Formulary and Guide" (Philadelphia, Blakiston, 1954). It is here considerably contracted and reproduced since many letters to the author have testified as to its utility. Two solvents, water and alcohol (not exceeding 50% of the whole) are listed together with the following "Basal Fixative Solutions."

It is to be hoped that persons using an author's name in relation to a fixative will also include the date. It may be pointed out that the brothers Bouin were responsible for five fixative mixtures of which only the one given in the table below is now commonly employed. The formulas for these, and several hundred other fixatives are to be found in Gray 1954 (*op. cit.*). The conditions of space do not make it possible to include literature references for the more than 500 formulae that follow. These literature references are to be found in full in Gray (*op. cit.*).

In view of the danger to health of using osmic acid, those who are preparing fixatives for purely optical histological examination should read the paper by Olney 1953 (*Turtox News*, **31**:29) drawing attention to the fact that with a great majority of cases nitric acid may be substituted for this dangerous reagent.

List of Solutions

The following solutions are of sufficient strength to require dilution for the preparation of any of the fixatives shown. The first two columns in this tabular material show the two most commonly employed diluents. The last column shows such other ingredients as must be added to secure the required fixative. It is understood that when the ingredient is liquid, a volume is indicated and when the ingredient is solid, a weight is indicated.

Solution 1   2% osmic acid
NOTE: This should be prepared in chemically clean glassware in filtered, triple-distilled water, to which has been added enough potassium permanganate (approximately 0.01%) to give a faint pink color. This pink color should be maintained by the addition of a few drops of potassium permanganate solution whenever necessary.

Solution 2   1% platinic chloride
Solution 3   7% mercuric chloride
Solution 4   a saturated aqueous solution of picric acid
Solution 5   2% chromic acid
Solution 6   7.5% potassium dichromate
Solution 7   7.5% potassium dichromate and 3% sodium sulfate
NOTE: This is a triple-strength Müller's 1859 fixative. The previous solution (7) may be substituted for it by those who do not believe in the efficacy of the sulfate content of Müller.

Solution 8   40% formaldehyde
Solution 9   glacial acetic acid

PETER GRAY

Alphabetical Table of Fixatives that can be Prepared from Stock Solutions

Reference	H₂O	95% alc.	1 — 2% OsO₄	2 — 1% H₂PtCl₆	3 — 7% HgCl₂	4 — Sat. sol. picric acid	5 — 2% CrO₃	6 — 7.5% K₂Cr₂O₇	7 — sol. 6 +3% Na₂SO₄	8 — 40% formaldehyde	9 — acetic acid	Other Additions
Aigner 1900	43							33			1	
Allen 1918						75	1.5			25	5	2 urea
Allen 1929						75	1			15	10	1 urea
Allen 1929	25					50	1			20	5	1 urea
Allen 1929						75				15	10	1 urea
Allen 1929a						75				10	10	
Allen 1929b										5	5	
Allen and McClung 1950						50					10	40 dioxane
Altmann 1890	60				20						20	
Altmann 1890	60				20							20 formic acid
Altmann 1890	15		50					35				
Altmann 1894	16.5		50					33.5				
Ammermann 1950	87									12		1.2 chrome alum
Andriezen 1894	80		2.5					17			0.8	
Aoyama 1930	85									15		1 CdCl₂
Apáthy 1893	31	40	50		50							
Apáthy 1896	25				29						5	
Apáthy 1897		50	25		50							
Apáthy (1920)					50							
Apáthy and Boeke (1910)	10				85							0.25 NaCl
Apáthy and Boeke (1910)	80		25		14							0.5 NaCl
Armitage 1939						40					10	4 HNO₃
Arnold 1888	85						15				0.5	5 HNO₃
Bachuber 1916						75				15	10	50 dioxane
Baker 1944	90		20									
Baker 1945	35		50				40					
Baker and Thomas 1933	25							25			5.2	0.4 CaCl₂
Baley 1937	52				20			18		10		
Barret 1886a	30	50	10				10					
Barret 1886b	92		10				8					
Barret 1886c	92		5				13					
Bartelmez 1915a		90									10	
Bartelmez 1915b		95									5	
Bartelmez 1915c		60									10	
Bataillon 1904	45						45				10	chloroform 30

Basal Solutions (columns 1–9)

	1	2	3	4	5	6	7	8	9	10	11	Additive
Becher and Demoll 1913a	87.5			12.5							0.1	
Becher and Demoll 1913b	35	30								85	1.6	
Becher and Demoll 1913c	50		10			18	26				2	
Becher and Demoll 1913d	50.5				28	21.5					2	
Becher and Demoll 1913e	52		20			21.5	24.5				10	
Béguin (1910)				40		90					0.6	
Belar 1929a	40		20							60	15	
Belar 1929b	30	120	5								5	$5\ ZnCl_2$
Belar 1929c	50	50								10	16	
Belling 1928	24	90		50							1	
Benario 1894	9					90					1	
Benda 1901	41			38							2.5	
Benda 1903a	65		20					10				
Benda 1903b	25		20	25								50 pyrolignious acid
van Beneden and Heyt 1887		50	25					10			50	
Benoit 1922a	37					18		20				$0.8\ UO(NO_3)_2\ 6\ H_2O\ 0.2\ NaCl$
Benoit 1922b	72					18		20				$0.8\ UO(NO_3)_2\ 6\ H_2O\ 0.2\ NaCl$
Bensley 1911	52		20					22			0.3	$0.5\ FeCl_3$
Bensley and Bensley 1929			100									
Bensley and Bensley 1938	7					70		30				
Bensley and Bensley 1938						63		30				
Bensley and Bensley 1938	53					68		32			5	
Bensley and Bensley 1938	50		20					27			0.5	
Berkely 1897	75		12.5					35				
Besta 1905	80									25		$4\ Sn_4Cl(NH_4)Cl_2$
Besta 1910a	100									20		2 acetaldehyde
Besta 1910b							95			5		$4\ (NH_4)Mo_7O_{27},\ 4\ H_2O$
Bigelow 1902	87											
Bignami 1896						14	95					
Bing and Ellerman 1901	87.5									10		0.8 NaCl
Bizzozero 1885	96		12.5								0.25	90 acetone
Blanc 1895	20						4					0.75 NaCl
Bles 1905	69	70								7		$0.12\ H_2SO_4$
Bock 1924	35								27	4	3	
Boeke 1910	63	55								10	5	
Böhm and Davidoff 1905	15								27	10	5	
Böhm and Oppel 1907	78					70				10	5	
Böhm and Oppel 1907									12	5	5	
Böhm and Oppel (1910)a						95						
Böhm and Oppel (1910)b							100					
Bonn (1937)	71			20							3	
Bonner (1915)	69			22							3	
Borgert 1900						75					25	
Bouin 1897							75			25	5	

Alphabetical Table of Fixatives that can be Prepared from Stock Solutions—continued

			Basal Solutions									
			1	2	3	4	5	6	7	8	9	
Reference	H_2O	95% alc.	2% OsO_4	1% H_2PtCl_6	7% $HgCl_2$	Sat. sol. picric acid	2% CrO_3	7.5% $K_2Cr_2O_7$	sol. 6 +3% Na_2SO_4	40% formaldehyde	acetic acid	Other Additions
Boule 1908	63	..								22	5	..
Bowie 1925		38			70			30			2	..
Bradley 1948											12	50 chloroform
Bühler 1898					100							..
Burckhardt 1892	50		2.5				2.5				8.2	..
Burckhardt 1897	40		3.5				30	20			5	..
Burke 1933	68									27		5 pyridine
Busch 1898	85		15									1 $NaIO_3$
Cajal 1890a	53							47				..
Cajal 1890b	60		8					32				..
Cajal 1891	60		5					35				..
Cajal 1893	56.8			3.2						40		..
Cajal 1914	75	25								15		..
Cajal (1933)	33		10					52		15		..
Cajal 1933a	58		8					32				5 $FeCl_3$
Cajal 1933b	66		8					26				0.5 $K_3Fe(CN)_6$
Calvet 1900	90						8.5					1.25 HCl
Carazzi (1910)	75	7			29						0.5	1 NaCl
Carazzi (1920)	40	30			30						3.3	..
Carleton and Leach 1938					90							..
Carnoy 1887a		75								10	25	..
Carnoy 1887b		60									10	30 chloroform
Carnoy and Lebrun 1887		30			25						30	30 chloroform
Carter 1919	7	50						40			3	..
de Castro 1916	100									15		1.5 urea nitrate
Catcheside 1934			8				90			1		1.2 maltose
Caullery and Mesnil 1905			26	26			40				5.5	..
Chamberlain 1906	40		2.5				50	15			10	..
Champy 1911	43		22				20	16				..
Champy 1913a	33		22				20					..
Champy 1913b	20						50					..
Champy 1913c	60							40				35 pyrolignious acid
Champy 1913d	50		50									1.5 NaI
Cholodkowsky (1910)					100						0.5	..

Author										Added reagents
Chura 1925	54				5.5	25			20	0.4 each $CrF_3.4\,H_2O$ and $NH_4Cr(SO_4)_2. 12\,H_2O$
Ciaccio 1910	30			25		53		16	1	0.4 formic acid
Claussen 1908	75			25						
Cohen 1934	90								1	3.6 $2H_3PO_4.H_2O$
Coker 1902	94	25	5.2				20	10		
Cole 1884	55								0.2	
Colombo 1904	40	20	40						0.1	
Cori 1890	85	0.5	87.5	12.5					0.0125	
Cori (1910)	10	0.06	60	1.6					10	
Cox 1891										0.8 K_2CrO_4
Cox (1895)	72	20								
Cox 1896	20		14							
Crétin 1938			40							
Cummings 1925	45		40	12.5		45			5	
Czermak 1893	87.5		45	12.5			15	4	1	
Dahlgrens 1897	35		45				15		5	0.5 trichloroacetic acid
Davenport, Windle, and Beach 1934	90									
Davidoff 1889a			75						25	
Davidoff 1889b					75				25	
Dawson and Friedgood 1938	38	45	52		50			10		0.7 NaCl
Deegener 1909									5	
Demarbaix 1889	71			26				6	3	
Destin (1929)	50			50				6	2	
Destin (1943)	46			46				10	2	
Dietrich (1946)	60	30							2	
Dobell 1919		32			68			12	10	iodine to wine color
Dominici 1905	8		80							
Drüner 1894	20	2.5	72					10	5	1.6 HNO_3
Duggar 1909a	41		49	25					0.7	
Duggar 1909b	65	10						10	0.2	
Durig 1895	50			40						
Ehlers (1910)	65							15	0.2	
Eisath 1911	48			37	25		6.25			
Eisig 1878	70		85	25				10		
Eltringham 1930a	5				100				5	10 HNO_3
Eltringham 1930b					100					2 H_2SO_4
Erlanger 1892		5							1	5 tannic acid
van Ermengen 1894	70	30						1	1	0.5 NaCl
Ewald 1897	100	2.5								
Faber-Domergue 1889	30	50							20	
da Fano 1920	100							15		1 $Co(NO_3)_2\,6\,H_2O$
Farkas 1914		50	25						10	0.002 NaI

Alphabetical Table of Fixatives that can be Prepared from Stock Solutions—continued

Columns 1–9 below are the **Basal Solutions**.

Reference	H_2O	95% alc.	1 2% OsO_4	2 1% H_2PtCl_6	3 7% $HgCl_2$	4 Sat. sol. picric acid	5 2% CrO_3	6 7.5% $K_2Cr_2O_7$	7 sol. 6 +3% Na_2SO_4	8 40% formaldehyde	9 acetic acid	Other Additions
Farmer and Shove 1905a		85									15	
Farmer and Shove 1905b		71									29	
Faussek 1900	50										0.25	
Favorsky 1930		95					50				5	
Fick 1893	75		5.5				25				0.1	
Fischler 1906	72		0.5				20			0.2	2.5	
Fish 1895a	65								35			
Fish 1895b	55							40		5		
Fish 1896a	80				7					20	1	6 $ZnCl_2$, 35 NaCl
Fish 1896b	85					8					0.1	
Flemming 1882a	82.5		5				12.5				0.1	
Flemming 1882b	75		5			20					0.1	
Flemming 1882c	95						5				0.1	
Flemming 1884*	45		20				37.5				5	
Flemming (1910)	55		10				37.5				5	
Flesch 1879	85		5				12.5					
Foa 1891	35			30					35			
Foa 1895	27			50				33		0.1		0.6 NaCl
Fol 1884	85		1			10	12.5					
Fol (1885)	73					10	12.5				1	
Fol 1896a	95		5									
Fol 1896b	73		0.25			10	12.5				1	
Fol (1910)a	17					65	17					1.3 H_2SO_4
Fol (1910)b	72.5						7.5				1	
Fontana 1912	84									16	0.8	
Francotte 1890	63		20			17						2 H_2SO_4
Frenkel 1893	75		25								0.5	0.75 $PdCl_2.2 H_2O$
Friedländer 1889	91						8				1	
Friedmann (1900)	62		4				50				4	
Gage 1890	33	50				17						
Galesescu 1908	65		0.8		25		10				0.8	
Gatenby 1937	20						50	24				6 HNO_3
Gates 1907	65		5				35				0.5	

* FWA, a fixative designation still sometimes occurring in literature is "Flemming without acetic" and refers to this formula with the acetic acid left out.

Table (rotated on page). Reference list with associated formula values and reagent notes.

Reference												Reagents
Gates 1908	65								35		0.5	
Gates and Latter 1927	50							32	50		1	
van Gehuchten (1927)	58	10					95			5		
von Gelei and Horvath 1931	35						24		14	25	2.5	
Gerhardt 1901	100											
Gerlach 1872		34				95	66				4	2 (NH₄)₂ Cr₂O₇
Giemsa (1909)		6									0.4	
Giesbrecht (1910)	65	2					30		38		5	1.2 HNO₃
Gilson 1898	46			11								
Goldsmith	70	8		22								
Golgi 1880	58	16.5		26								
Golgi 1900	73			27								
Golgi 1903	75						24				1.5	
Goto 1898	55				90				15	10		4.5 glycerol
Graf 1897	20	30								8	2	
Graf 1898	63				34		70				3	3.2 oxalic acid
Gregg and Puckett 1943											10	
Gulick 1911	37.5								37.5			
Gulland 1900	34	90					70			34	4	5 formic acid
Guthrie 1928a	75								28		0.5	5 formic acid
Guthrie 1928b	50	25										
Guyer 1930	10			10			70	25		20		0.2 trichloroacetic acid
Hamann 1885	40	25		22			25			25		
Hamilton 1878	27						17			34	5	
Hartz 1950	60							30		5		0.5 NaCl
Harvey 1907a	25						100					
Harvey 1907b	23						50			20	1	2 trichloroacetic acid
Haver 1927							100			50		1.25 NaCl
Heidenhain 1888	80						57			15	5	0.5 NaCl
Heidenhain 1896	15			10			50	27		10	5	5 trichloroacetic acid
Heidenhain 1909	30			22			63			20	4	2 trichloroacetic acid, 0.5 NaCl
Heidenhain 1916a	45						65			20	0.5	
Heidenhain 1916b	60			13			50			3.5	1.6	40 acetone
Heidenhain 1916c	19						15	35		0.5	3	
Heidenhain 1916d							43				3	
Heidenhain 1916e	37	20	37.5				97	30		5	5	
Heinz 1901	99	1.2					70				0.1	
Held 1897												
Held 1909a												
Held 1909b												
Held (1933)												
Helly 1903												
Hermann 1889												
Hertwig 1879												

173

Alphabetical Table of Fixatives that can be Prepared from Stock Solutions—continued

Reference	H₂O	95% alc.	2% OsO₄ (1)	1% H₂PtCl₆ (2)	7% HgCl₂ (3)	Sat. sol. picric acid (4)	2% CrO₃ (5)	7.5% K₂Cr₂O₇ (6)	sol. 6 +3% Na₂SO₄ (7)	40% formaldehyde (8)	acetic acid (9)	Other Additions
						Basal Solutions						
Hertwig 1885	85		15								1	
Hertwig 1892	50										0.2	
Hertwig 1905	42				30					10	3	
Hill 1910	77		1		17							5 HNO₃
Hirschler 1918			50		50							
Hoehl 1896	60		10				50				2	
Hoffmann 1908	25	25		25				30			15	
Hofker 1921		80								10	10	10 trichloroacetic acid
Hornell 1900	10	100										
Hoyer 1899	50				21			20				
Huber (1943)		57			43							1.5 trichloroacetic acid
Hultgren and Anderson (1910)	65					75			35	4		
Ingelby 1925			5					67		25	5	2 KBr
dell'Isola 1895	25				35					15		
"J.A." 1937	35				10					10		
Jäger (1928)	56	34									0.1	
Johnson 1895	55		10	7.5				23			5	
Jolly 1907	45		15				30				1	
Jones (1915)	30	68								2		
Joseph 1918	72							28		20	10	
Kahlden and Laurent 1896	84					8.5	7.5					
Kahle 1908	60	26								12	2	
Kaiser 1891	50				47					3	3	
Karpenchenko 1924	74		40				20			3		
Kaufman 1929	17		10				38			5		
Kerschner 1908	80											20 formic acid
King (1910)	77.5						12.5				10	
Klein 1878	60	35					5.5					
Kleinenberg 1879						100						2 H₂SO₄
Kohn 1907	51				18			31			5	
Kolmer 1912	17				17			52		4	10	
Kolmer (1938)	59							24		3.6	9.2	0.75 UO(NO₃)₂·6 H₂O, 5 trichloroacetic acid
Kolossow 1897	35		12.5					53				
Kolossow 1898	75		25									2.5 UO₂(NO₃)₂·6 H₂O

Reference											Amount	Reagent
Kolster (1910)	40					99					1	0.5 NaCl
Kopsch 1896a	53						40		40	20		
Kopsch 1896b	50						47					2.8 HNO₃
Kostanecki and Siedlecki 1896						50			25	10		
Krueger 1911	18		50			63	27				4.4	
Kulschitzsky 1897	35					5	40	50		25	1	
Lachi 1895	52	10			30		40				0.3	
LaCour 1929	63	12			20		8				0.6	0.6 urea
LaCour 1931	34	60			16		5				0.25	0.04 saponin
Laguesse 1901	36	32			32						0.4	
Laguesse (1933)	38										5	
Laidlaw (1936)	78									5		
Landsteiner 1903						57			17	5		
Lane (1937)	35	10				70	30				7	
Lang 1878	20				25	100					30	8 NaCl
Langendorf (1910)			50			77					3	
Langeron 1942	30		27			50	67			10	2	
Lapp (1910)	60				5	2					1	
Lavdowsky 1893	100		10		10						2	
Lavdowsky 1894a	80			2.5							5	
Lavdowsky 1894b	80	2		18		75					0.5	
Lavdowsky 1894c											0.2	
Lehrmitte and Guccione 1909			25			50		50			5	
Lenhossék 1897a	40			25		75						
Lenhossék 1897b	50					50					5	4.5 NaCl
Lenhossék 1898	50					36						1 KI
Lenhossék 1899	50			50		50	15			1		
Lenoir 1926	82	10			8	35				10	0.83	
Levi 1918	61	19			20		18			10		
Levitsky 1910								85		9	4.5	5 formic acid
Lillie (1929)	69				17	65					17	
Lillie 1944	17					66				5		
Lillie 1948	17		35		25		25					45 sea water
Lo Bianco 1890a	25				25		25					
Lo Bianco 1890b	40	2					75	50				
Lo Bianco 1890c	25	1			50			50			5	
Lo Bianco 1890d	25				50							
Lo Bianco 1890e	50				5						100	2 H₂SO₄
Lo Bianco 1890f	50					28	27				10	
Lo Bianco 1890g	5						27					
Lo Bianco 1890h	35	10										
Lo Bianco 1890i	63											
Long and Mark 1912												
Löwenthal 1893												

Alphabetical Table of Fixatives that can be Prepared from Stock Solutions—continued

			Basal Solutions									
			1	2	3	4	5	6	7	8	9	
Reference	H_2O	95% alc.	2% OsO_4	1% H_2PtCl_6	7% $HgCl_2$	Sat. sol. picric acid	2% CrO_3	7.5% $K_2Cr_2O_7$	sol. 6 +3% Na_2SO_4	40% formaldehyde	acetic acid	Other Additions
Löwit 1887	99				1.2							2.4 Na_2SO_4. 10 H_2O, 1 NaCl
Lukö 1910	20	80								10	5	
Mann 1894a	50		50									0.75 NaCl
Mann 1894b	25		25		50							0.9 NaCl
Marchi 1886	63		17.5						22			
Marchoux and Simond 1906a			28	28			43				6	
Marchoux and Simond 1906b			14	14	50		22				3	
Marchoux 1910	29						16.5			48	3	
Marina 1897		95								5		
Marrassini (1910)	50				14	75						
Masson (1947)	45		10				5	37		25		3 chrome alum
Maximow 1909	90									10		
McClung and Allen 1929	6									7.5		2 HNO_3
Meeker and Cook 1928	80				57		10	27		10	10	2 Na_2SO_4.10 H_2O
Merkel 1870	50		10	10								
Mettler 1932a	60							40		50		
Mettler 1932b	58		10					32				
Mettler 1932c	70						20					
Meves and Duesberg 1908a	90		5								2.5	
Meves and Duesberg 1908b	65		3	11							1.5	
Meves 1910	59		20				18				5	
Michaelis (1948)	61				21	20				8	1.2	
Milovidov 1928		25	2.5					6.5				
Mingazzini 1893	46				50						25	
Mislawsky 1913	48		15				25			20		
Möller 1899	40						40	32		20		
Mottier 1897	50						50	32			5	
Müche 1908	65								35		2.5	
Müller 1859	12							8		80		
Müller 1899	60								30	10		
Murray 1919a	65								35			
Murray 1919b	60			40								
Murray 1919c			100									
Nakamura 1928	13		20					27				
Nassanow 1923	20		30				15	25				

Author	(1)	(2)	(3)	(4)	(5)	(6)	(7)	(8)	(9)	(10)	(11)	Reagent note
Navashin 1912	35						40			20	5	
Nemeč (1937)	47						46			7		
Newton and Darlington 1919	47	20					30				3	6.4 HNO₃
Nicholas 1891	70	25		2.5	50							
Niessing 1895	25	20			30		15			10	5	
Novak 1901	42				23		5				3	
Nowack 1902	45										23	
Orr 1900	20	80							30		0.2	
von Orth 1896	60									10		5 formic acid
Oxner 1905a	47	10						23			5	
Oxner 1905b	47	10						23				
Oxner (1910)											3	
von Pacaut 1906				4.5	97	50	0.15					
Pappenheim 1896	80				50					2		
Parker and Floyd 1895	55		98		50					20		6 KI, 4 urea
Penfield 1930	37											4 HNO₃
Perényi 1882	45		30				10					6.4 HNO₃
Perényi 1888	48		40				15					0.05 AgNO₃
Perriraz 1905			30			20						
Pfeiffer and Jarisch 1919	45	4.8			34			42		10		
Pfuhl 1932	80									20	5	0.2 formic acid
Pianese 1899a	85	20		30			4					0.8 CoCl₂.6 H₂O, 0.2 formic acid
Pianese 1899b	30	20										
Pietschmann 1905	25		95		5						5	
Podhradszky 1934	85	20								10	2	5 HgCl₂
Podwyssozki 1886	50						50					
Pritchard 1873	62.5		25				15					
Rabl 1884												0.1 formic acid
Rabl 1894a					25							
Rabl 1894b				12.5	25							
vom Rath 1895a	20	10		40		50					1.6	
vom Rath 1895b	95	10				45					1	
vom Rath 1895c		6				100					1	
Rawitz 1895a	50					20						0.8 HNO₃
Rawitz 1895b	60	0.5									0.04	(sea water may be used)
Rawitz (1905)	90	5				90	40					5.6 HNO₃
Régaud 1910a	10							30		20		
Régaud 1910b	66							40				
del Rio-Hortega 1925										10		1.5 UO₂(NO₃)₂.6 H₂O
del Rio-Hortega 1928								80		10		6 chloral hydrate
del Rio-Hortega 1929								24		10		
Rohl (1910)	85						15					0.1 formic acid
Romeis 1918a	20				70					10		2 trichloroacetic acid, 0.6 NaCl
Romeis 1918b	61				34					24		1.6 trichloroacetic acid

Alphabetical Table of Fixatives that can be Prepared from Stock Solutions—continued

			Basal Solutions									
			1	2	3	4	5	6	7	8	9	
Reference	H₂O	95% alc.	2% OsO₄	1% H₂PtCl₆	7% HgCl₂	Sat. sol. picric acid	2% CrO₃	7.5% K₂Cr₂O₇	sol. 6 +3% Na₂SO₄	40% formal-dehyde	acetic acid	Other Additions

Reference	H₂O	95% alc.	2% OsO₄	1% H₂PtCl₆	7% HgCl₂	Sat. sol. picric acid	2% CrO₃	7.5% K₂Cr₂O₇	sol. 6 +3% Na₂SO₄	40% formaldehyde	acetic acid	Other Additions
Romeis 1948a	43	45	..	..	..	..	..	..	..	10	2	..
Romeis 1948b	45	..	..	..	..	..	..	..	..	10	5	..
Roskin 1946a	6	..	..	..	..	..	..	40	..	24	5	..
Roskin 1946b	11	..	..	..	..	65	..	..	..	24	..	0.24 trichloroacetic acid
Roskin 1946c	14	..	..	..	80	65	..	..	..	..	6	..
Roskin 1946d	22	..	66	..	..	..	..	12	..	..	..	..
Röthig 1900	..	10	..	..	90	..	..	..	..	..	..	..
Röthig 1904	41	..	..	..	31	..	15	..	..	3	10	..
Ruffini 1905	60	..	..	..	35	..	..	..	..	..	5	..
Ruffini 1927	36	..	..	..	17	..	36	6.5	..	..	5	6.5 formic acid
Ruge (1942)	97	..	..	..	..	..	..	..	..	2	1	..
Russel 1931	63	..	..	..	..	..	..	33	..	4	..	..
Rutherford (1878)	45	..	..	..	..	..	50	5	..	..	..	..
Saling 1906	28	40	..	..	28	..	..	..	..	..	..	4 HNO₃
Salkind 1917	15	..	..	..	57	..	..	27	..	..	..	14 chloral hydrate
Sánchez (1933)	63	..	..	..	..	..	..	27	..	..	..	
Sansom (1928)	..	65	..	..	..	..	..	..	..	10	5	30 chloroform
Schaffer 1908	17	51	..	..	..	..	..	..	..	32	..	..
Schaffer 1918	15	50	..	..	..	..	..	..	..	35	..	..
Schaffner 1906	85	..	..	..	..	..	15	..	..	..	0.7	..
Schaudinn 1893	..	65	..	..	35	..	..	..	..	..	..	..
Schaudinn 1900	..	35	..	..	65	..	..	..	..	..	..	..
Scheuring 1913	..	48	..	..	..	..	..	..	..	48	4	..
Schmorl 1928a	37	..	..	..	63	..	..	..	..	..	..	0.5 NaCl
Schmorl 1928b	57	..	..	..	43	..	..	..	..	..	0.6	..
Schmorl 1928c	16	..	..	..	64	..	..	..	..	20	..	0.5 NaCl
Schrieber 1898a	90	..	..	..	..	..	..	4.5	..	6.5	..	..
Schreiber 1898b	75	..	..	..	..	..	..	24	..	1.5	..	..
Schridder (1928)a	60	..	..	..	..	..	..	..	30	10	..	..
Schridder (1928)b	65	..	..	..	..	..	..	..	35	..	..	..
Schridder (1928)c	..	..	100	..	..	..	..	..	..	..	..	..
Schuberg 1903	10	..	7.5	..	..	85	..	..	..	..	..	..
Schuberg (1920)	5	..	5	..	..	87	..	..	..	..	3	..
Schultze 1904a	60	..	..	..	..	..	..	40	..	..	..	2 H₂SO₄
Schultze 1904b	45	..	25	..	..	4	..	30	..	..	..	..

Reference												Notes
Schwartz 1888	90		10								1	
Severinghaus 1932	44		24					20		20	1.2	
Showalter 1926	88		1.2				20				0.1	
Simons 1906	87.5						12.5					
Smirnow 1895	53		7.5					50		5	2.5	
Smith 1912	88							14		10		
Smith 1930a	90											
Smith 1930b	47		16					53			0.6	0.6 saponin
Smith 1935a			16				25	3.2			0.4	0.04 saponin
Smith 1935b			2.5				22	8.5			0.6	
Spuler 1892						100						
Spuler 1910a	45				45					10	1	
Spuler 1910b					35	35				35	1.5	
Spuler 1910c	47			17	30				23		4	
Stieve (1946)	59				17					20	4	
Stieve (1948)a	10				66	85				20	4	0.3 trichloracetic acid
Stieve (1948)b	9									6		
Stilling 1895	58						2.5	32		10		
Stowell 1884	28	60										
Strauss 1909		97								3	3	
van der Stricht 1895	40		20	40								
Strong 1895a	50							47		3.5		
Strong 1895b	25							40		35		
Swank and Davenport 1934a	90									10	1	0.25 KClO₃
Swank and Davenport 1934b	73		16.5									
Swank and Davenport 1934c	90		17							10	2	
Swank and Davenport 1934d	67		6.25						14		1	0.6 KClO₃
Swank and Davenport 1935	93		20							12		
Sypkens 1904	40										4	4 formic acid
Szepsenwol 1935	88						36			8		
Szmonowicz 1896			14		84							
Szüts 1913	20		25	12.5	43					25		
Takahashi 1908	72		12				3.75				1.6	0.02 HCl
Taylor 1924	66			20							1	1.2 maltose
Tellyesniczky 1898a					50	50				1		
Tellyesniczky 1898b	60		16					40		5		
Timofecheff (1933)	42		25		75				42			
Tschassownikow (1910)										2.5		
Vassale and Donaggi 1895	53							47			2.5	5 acetaldehyde
Veratti (1900)	60		12.5	2.5				24				
Viallane 1883a	50		50									
Viallane 1883b	65							6.5				35 formic acid
Vialleton 1887	80					14					10	1 H₂SO₄
Virchow (1897)	80						10					

Alphabetical Table of Fixatives that can be Prepared from Stock Solutions—continued

Reference	H₂O	95% alc.	1 2% OsO₄	2 1% H₂PtCl₆	3 7% HgCl₂	4 Sat. sol. picric acid	5 2% CrO₃	6 7.5% K₂Cr₂O₇	7 sol. 6 +3% Na₂SO₄	8 40% formaldehyde	9 acetic acid	Other Additions
Vlakovic (1926)	··	85	··	··	··	··	15	··	··	··	··	··
Völker (1910)	··	··	··	··	45	45	··	··	··	··	5	··
Wenrich and Geiman 1933	32	20	··	··	46	··	··	··	··	··	2	··
Whitehead 1932	56	··	··	··	··	··	··	32	··	12	··	··
Wilhelmi 1909	··	77.5	··	··	··	··	··	··	··	7.5	··	2.4 HNO₃
Windle 1926	52	··	8	··	··	··	··	··	··	··	··	··
Winge 1930	45	··	··	··	··	··	38	40	··	10	7.5	··
Winiwarter (1930)	36	··	20	40	··	25	··	··	··	··	··	4 trichloroacetic acid
Wistinghausen 1891	75	··	··	··	··	25	··	··	··	··	··	0.5 H₂SO₄
Wittmaack (1910)	20	··	··	··	··	··	··	67	··	10	3	··
Wlassow 1894	65	··	2	··	2.5	··	··	33	··	··	··	0.8 NaCl
Wlassow (1910)	100	··	··	··	50	··	··	1	··	··	··	3 NaCl
Woltereck (1910)	10	40	··	··	··	··	··	··	··	··	10	··
Worcester (1929)	100	··	··	··	··	··	19	··	··	9	10	8.1 HgCl₂
Yamanouchi 1908	78	··	3.8	··	··	··	··	··	··	··	0.1	0.1 chloroform
Zacharias 1887	··	80	0.5	··	··	··	··	··	··	··	20	··
Zacharias 1888	··	80	0.75	··	··	··	··	··	··	··	20	··
Zenker 1894	··	··	··	··	70	··	··	··	30	··	5	··
Zieglwallner 1911	3	50	20	··	17	··	··	··	··	··	10	··
Zietschmann 1903	62	··	16	··	··	··	··	··	22	··	··	··
Zweibaum 1933	60	··	6	··	··	··	20	14	··	··	··	··

FLUORESCENCE MICROSCOPY

1. The Nature of Fluorescence. In many chemical compounds irradiated with short-wave light (fairly high-energy photons), some electrons are transposed into higher-energy states from which they eventually return in several steps to their original energy levels, simultaneously emitting light of a longer wavelength (fairly low-energy protons) (Fig. 1). The emission of light of longer wavelength than that absorbed is called *fluorescence.* The light absorbed is known as *exciting light,* and that emitted as *fluorescence light.* With the right combination of material, fluorescence is obtained at any wavelength of light, but the fluorescence light is always of longer wavelength than the exciting light and is not monochromatic. Thus certain substances (e.g., chlorophyll) fluoresce in red light but the fluorescence light lies in the invisible (infrared) region; blue, visible fluorescence light can be obtained by irradiation with short-wave (ultraviolet) rays.

2. Primary and Secondary Fluorescence. Many naturally occurring substances (especially polymeric carbohydrates, e.g., cellulose and starch, but also many fats, and albumen) show marked fluorescence; this is *primary fluorescence.* Others do not fluoresce naturally but can be observed in the fluorescence microscope after staining with fluorescing dyes, termed *fluorochromes* (e.g., acridin orange, fluorescein, berberine sulfate, auramine, morine, and thioflavine). This induced effect is *secondary fluorescence.*

3. Blue-Light and UV Fluorescence. Blue-light and UV fluorescence are defined by the wavelength of the exciting light.

The exciting light which produces *blue-light fluorescence* is short-wave visible light, usually around 400 nm. Blue-green light is also used, e.g., in FITC fluorescence (see later). To permit observation of the blue-light fluorescence, always weak in comparison with the exciting light, the latter must be completely filtered out using a yellow barrier filter. Unfortunately this yellow filter cuts out all short-wave light, including the blue and blue-green colors particularly common in primary fluorescence, so these cannot be observed (Fig. 2). It also causes alteration of the other fluorescence colors, e.g., from white to yellow and from pale red to orange.

In *UV fluorescence* a UV spectral region around 366 nm (occasionally 335 or 312 nm) is used as the exciting light. The barrier filter must again absorb the exciting light as completely as possible, but, unlike the one in blue-light fluorescence, can be almost colorless (Fig. 3). The true fluorescence colors (even blue and blue-green) can thus

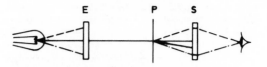

Fig. 2 Filter arrangement for blue-light fluorescence. E = BG 12 exciting filter; S = OG 530 barrier filter; P = specimen; ——— = blue; – – – = yellow; –·– = remaining colors.

be observed. Unfortunately fluorochromes are quickly destroyed by the intense UV radiation used, and the fluorescence rapidly fades, so photography (which requires long exposure times) is difficult or impossible. It is also often difficult in UV fluorescence to obtain fluorescence images on a completely black background, since short-wave exciting light causes unwanted fluorescence in many substances (lens cements, mounting media, immersion oils).

The rather vague term *mixed-light fluorescence* occurs in the literature. The exciting light for this covers the whole UV and blue wavelength ranges. Since, however, a yellow barrier filter is needed to absorb the visible part of the exciting light, mixed-light fluorescence has the same disadvantages as blue-light fluorescence although, particularly if mixed dyes are used, brighter images result due to the wider spectral range of the exciting light.

4. Fluorescence Microscopy in Transmitted Light. Fluorescence investigations can be made in bright or dark field. True *phase-contrast observation* is impossible, because the fluorescing objects behave like self-luminous bodies and a phase diaphragm therefore cannot be imaged on a phase plate in or above the objective. This does not, of course, exclude the use of phase-contrast equipment with strongly reduced white light to produce an additional image, in order, for example, to locate the fluorescing particles relative to a biological section. The fluorescence observations are then made in bright field, and the simultaneous phase-contrast observations in transmitted light.

Observations in polarized light are rare except for some natural fibers where the fluorescence light is itself polarized (*difluorescence*) so that conclusions can be drawn about the structure of the fluorescent or fluorochromed material.

Bright-field fluorescence remains the commonest form of fluorescence observation in microscopy. Apart from the use of extra filters, the procedure is the same as for normal bright field. It is important to establish Köhler illumination using UV or blue light (with an evenly fluorescing auxiliary specimen, e.g., Scotch tape) and not with normal white light, since with most condensers the

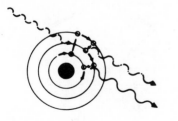

Fig. 1 The nature of fluorescence (shown on the Bohr atomic model).

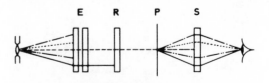

Fig. 3 Filter arrangement for UV fluorescence. E = UG 1 exciting filter; S = GG 13 comb. barrier filter; R = BG 38 red-absorbing filter; P = specimen; – – – = UV; ——— = red; – · – = blue; ...... = yellow; –··–··– = remaining colors.

focal length for UV differs from that for visible green light, for example. The condenser diaphragm is not set as for Köhler illumination, since the fluorescent parts of the specimen must behave as self-luminous bodies emitting more or less spherical wavefronts, irrespective of the illumination aperture. The aim is not to match the illumination aperture to the objective aperture but to concentrate as much exciting light as possible in that part of the specimen under observation. Therefore a condenser with a high numerical aperture is used to fully illuminate this part; even with medium-power dry systems, condenser immersion is quite common. Either special fluorescence-free immersion oil or purest paraffin oil, having similar qualities, is employed.

Bright-field fluorescence offers minimum light loss and is therefore generally preferable to other methods, especially for weak fluorescence (for exceptions, see section on FITC fluorescence). A very good barrier filter is, however, essential, since here most of the UV passes directly into the objective. If this UV is not filtered out, there is a risk of damage to the eye, and photomicrographs will show a fogged, instead of a black, background (with color film this is mostly blue).

In bright-field fluorescence it is also highly important that all optical components up to and including the barrier filter itself (an often-neglected component) have no autofluorescence.

Dark-field fluorescence is also used frequently today (see Fig. 4). It is best set up using an auxiliary specimen, as in normal dark-field. Use of a dark-field (usually immersion) condenser prevents exciting light from entering the objective directly; the "dark aperture" must always therefore be larger than the objective aperture. The barrier filter requirements are less stringent here than in bright-field. Should the barrier filter be completely forgotten, there is less danger of eye damage; any autofluorescence of the objective is hardly noticeable. Thus dark-field fluorescence is frequently employed in conjunction with fluorite objectives (semi-apochromats and apochromats), since these have a strong bluish autofluorescence. With simultaneous illumination by UV (through appropriate filters) and reduced white light (second light source), dark-field fluorescence has the advantage that the background remains black; with suitable adjustment of the lamps, both fluorescing and nonfluorescing struc-

tures and particles can be observed together. The same effect is obtainable by using (in the exciting filter) glasses which transmit some visible light. However, dark-field fluorescence involves an appreciable light loss and is only recommended for strongly fluorescing specimens. A special method, FITC fluorescence, only possible in dark-field, is described later.

5. Reflected-Light Fluorescence Microscopy. *Fluorescence in reflected light* has recently been used for the investigation of opaque specimens, e.g., rocks and minerals. It is also sometimes preferable to transmitted-light fluorescence for studying transparent or translucent specimens, especially when a thickish specimen absorbs much of the exciting light (particularly with UV fluorescence). If transmitted light is employed under these circumstances, the only particles which fluoresce are situated deep in the specimen (Fig. 5). If reflected light is used, particles near the surface fluoresce and can be readily observed. A disadvantage of reflected-light fluorescence in bright-field is the presence of the beam splitter, which reduces both exciting and fluorescence light, but equipment is now being built to incorporate interference splitting layers, which simultaneously function partly as combined exciter and barrier filters (see Fig. 6).

6. Light Sources. Halogen lamps (e.g., the quartz-iodine lamp) are generally used for blue-light fluorescence because of their good cost/performance ratio. High-pressure mercury vapor and xenon burners can be used, though the former emits relatively little blue light.

The strong double peak at about 366 nm in the output of mercury vapor burners is most frequently used for pure UV fluorescence (see Fig. 7). The 405 and 435 nm lines are on the limit of the visible range and the fluorescence produced by them is no longer true UV-fluorescence. The shorter wavelengths 334 nm (low intensity) and 313 nm (insufficient transmission of normal exciting filters) are of little significance for fluorescence microscopy.

7. Filters. Each fluorescence outfit must contain at least one exciting filter and one barrier filter. The former transmits only those wavelengths needed to excite the specimen's fluorescence. Its transmission curve should be as steep as possible on the longer wavelength side so as to absorb as little fluorescence light as possible.

Many fluorescence outfits contain a wide range of

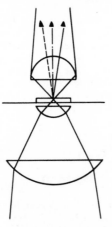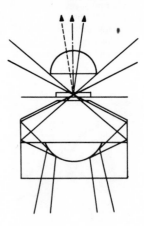

Fig. 4 Ray paths for bright-field and dark-field fluorescence.

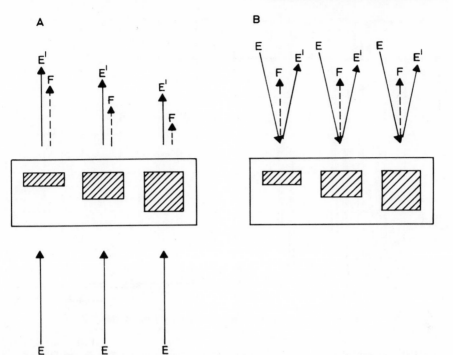

Fig. 5 Dependence of fluorescence intensity on specimen thickness. (A) for transmitted-light fluorescence, and (B) for incident-light fluorescence. E = exciting light; E′ = exciting light transmitted or reflected; F = fluorescence light.

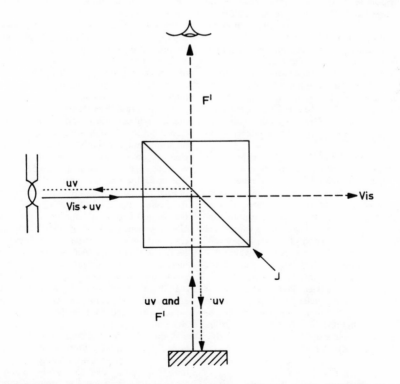

Fig. 6 Partition of exciting and fluorescence light by an interference layer in incident-light fluorescence. uv = UV light; Vis = visible primary light; F′ = fluorescence light.

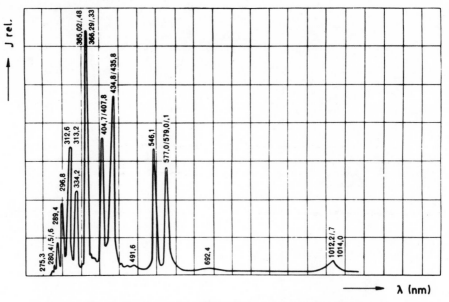

Fig. 7 Spectrum of the HBO-200 mercury vapor burner.

exciting filters. With blue-light fluorescence, if the exact wavelengths of exciting and fluorescence light are known (which is the case for certain fluorochromes), the best exciting filter can be selected. However, a choice of exciting filters for UV fluorescence with the mercury vapor burner is usually pointless, since the spectral line at 366 nm is generally employed and a filter can only transmit more or less UV of this wavelength. Thus the filter with maximum transmission at 366 nm is best suited.

The barrier filter should absorb as much exciting light and transmit as much fluorescence light as possible. The combined transmission of exciting and barrier filters should be less than $10^{-3}\%$ of the amount of light received when the transmittance for fluorescent light exceeds 50%. Where these conditions are fulfilled there is only one ideal barrier filter for each exciting filter; any other combination is a compromise between fluorescence intensity and background darkness.

The commonest combination for *blue-light fluorescence* is exciting filter BG 12,[1] 4 mm with barrier filter OG 530[1] (formerly OG 1), 2 mm (see Fig. 8). Maximum transmission of the exciting filter is at approximately 400 nm i.e., in the very short-wave blue range. If a mercury vapor lamp is used, the lines at 366, 406, and 435 nm are employed; with quartz-iodine or low-voltage lamps the high transmission between 400 and 450 nm is important. The exciting and barrier filter curves cross at about 500 nm, the combined transmission here being $0.1\% \times 0.1\% = 10^{-4}\%$; the background will thus be completely black. At 510 nm, however, the barrier filter transmission is already nearly 1%; shorter-wavelength fluorescence can no longer be distinguished, and the short-wave components of mixed-light fluorescence colors are missing, which factors contribute to the production of false coloration by the deep orange barrier filter.

Another disadvantage of barrier filter OG 530, strong red autofluorescence, is often countered by placing an additional filter (e.g., GG 4[1]) before the barrier filter. The GG 4 has no autofluorescence, and filters out that short-wave light which would excite the barrier filter. This is especially important for work involving a relatively high proportion of UV.

For *UV fluorescence*, type UG 1 glasses (see Fig. 9) are usually employed as exciting filters. Their transmission maximum corresponds with the strong 366 nm line of the mercury vapor burner, but they almost totally absorb the emission at 406 nm. Filter thicknesses of 4 mm are necessary for bright-field fluorescence, but two 2 mm filters give better cooling and are normally used, since high-absorption filters get very hot. For the same reason one or two heat-absorbing filters (having very high UV transmission) must always be placed between exciting filter and lamp. A single 2 mm UGI filter is usually sufficient as an excitation filter for dark-field fluorescence because no direct UV enters the optics.

Barrier filters rarely consist of pure glass, since glasses of high UV absorption are either tinted or are very thick. Recourse is made to a filter (e.g., GG 13 comb. of Wild Heerbrugg) with a UV-absorbent foil cemented between UV-absorbent glasses (e.g., GG 385[1]). High UV absorption with minimum filter autofluorescence is thus obtained, so that the barrier filter can remain in position for increased safety in bright-field work. The curve in Fig. 9 shows that filter UG 1 has, besides its UV-transmission maximum, considerable long-wave red transmission at around 750 nm. This colors the background red and must be reduced by a bluish red-absorbing filter (e.g., BG 38[1]).

A special method often applied in immunofluorescence is *FITC fluorescence*. Fluorescein isothiocyanate is used as the fluorochrome; its exciting and fluorescing wavelengths are close together (495 and 520 nm respectively). Normal glasses cannot be used to obtain a combined transmission below $10^{-3}\%$, therefore an interference

---

[1]Reference numbers of glasses manufactured by Schott & Gen., Mainz.

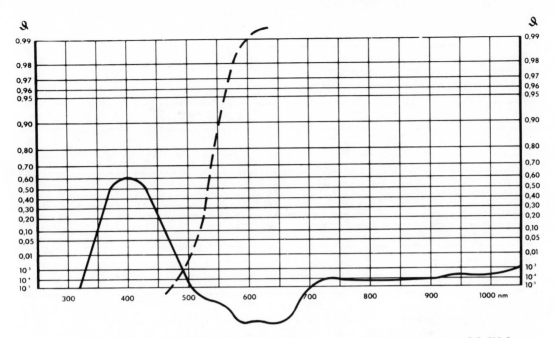

Fig. 8 Transmission curves for the FITC-fluorescence filters. ——— = BG 12.4 mm; ——— = OG 530 2 mm.

filter with a sharp cutoff is employed as the exciting filter (Fig. 10). This itself has a high transmission in deep red which cannot be suppressed by the red-absorbing filter alone, so dark-field must be used to prevent much exciting light (and particularly red light) from entering the optics; the remaining red light which does penetrate can be filtered out normally. If the red-absorbing filter is omitted, the fluorescing particles appear green, any others deep red (dark-field effect), and the whole on a black background.

8. Optical Equipment. The optical components of a fluorescence microscope must meet specific requirements.

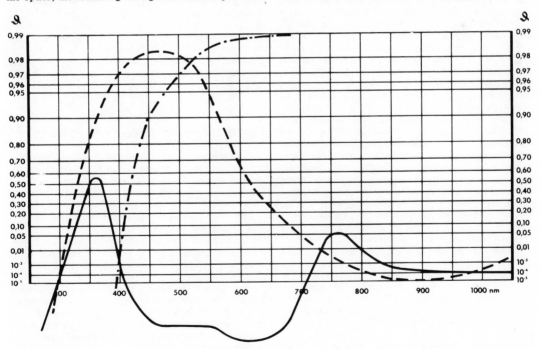

Fig. 9 Transmission curves for the UV fluorescence filters. ——— = UG1 4 mm; —·—· = GG13 comb.; ——— = BG 38 2.5 mm.

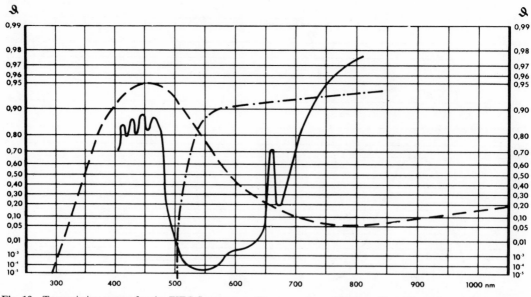

Fig. 10 Transmission curves for the FITC-fluorescence filters. ——— = FITC exciting filter; –·–· = FITC barrier filter; – – – = BG 23 red-absorbing filter.

The illumination optics (collector and condenser) must transmit a very high proportion of the exciting light, a condition easily attainable with blue-light, but not with UV fluorescence. UV-lamp collectors are of normal design, but are made of quartz, more because of its high heat resistance (smaller coefficient of expansion) at the high working temperatures of gas discharge lamps than because of its high UV transmission (a property of many of today's optical glasses). A multiple filter holder for easy working is just as necessary as a good UV barrier filter for protection, particularly of the specimen, between observations. Beam-directing mirrors are usually surface-aluminized, the coating having a high UV reflectance, thus avoiding the excessive UV-absorption by normal mirror glass. The condenser may be of quartz; high apertures are preferred.

Various condensers show a markedly different suitability, particularly for UV fluorescence. This is due not only to the UV-transmission characteristics of the optical glasses, but also to those of the cement layers which may be very UV-absorbent or have bright autofluorescence. Thus aspherical condensers, having no cement layer, are especially suitable for UV fluorescence. These have a transmission of over 85 % at 366 nm, while that of aplanatic/achromatic condensers for the same wavelength is usually below 26 %. Even an expensive quartz condenser can offer little more than a simple swing-out condenser, until shorter wavelengths are employed (e.g., with UG5[1] filters, having a transmission maximum of 321 nm), when the transmission of optical glasses decreases considerably. The very expensive quartz or quartz-glass slides are advantageous for short-wave UV observations; the UV transmission of normal slides varies greatly and a few trials with slides from different manufacturers are recommended. These are carried out by normally focusing a good fluorescence specimen, then holding several slides between lamp and condenser. The decrease in fluorescence image brightness indicates the degree of UV absorp-

tion. The UV transmission of the coverslip is unimportant since the UV is only required as far as the specimen (unlike UV microscopy).

The low intensity of most fluorescence effects suggests the use of objectives of maximum aperture. Thus, for dark-field fluorescence, fluorite systems (apochromats or semi-apochromats) are used almost exclusively. In bright-field fluorescence, however, the autofluorescence of the fluorite becomes noticeable. Since manufacturers started using only synthetic fluorite this has, however, been greatly reduced. It disappears completely in blue-light fluorescence if the UV part is filtered out of the exciting light before reaching the specimen.

A. SCHAEFER

FLUORESCENT SCHIFF-TYPE REAGENTS

1. Generalities. The fluorescent Schiff-type reagents, included in the larger chapter of Schiff-type reagents, were introduced in histochemistry by Ornstein et al. (15), Kasten et al. (7, 8, 9, 10), Prenna et al. (19, 20, 21, 23, 24, 25, 26, 27, 29, 32), Ruch (35), and Stoward (37).

Any basic fluorescent stain without acid groups and with at least one primary amine group, after addition of SO_2, can theoretically be used as a fluorescent Schiff-type reagent. These reagents, similarly to standard Schiff reagent, are employed in histochemistry to demonstrate aldehydic groups.

Since many histochemical techniques are based on the demonstration of aldehydes, already present or produced in tissues, the importance of specific fluorescence reactions is evident, given the high sensitivity of fluorescent techniques and the possible quantitative cytofluorometric applications.

The mechanism of reaction obtained with fluorescent Schiff-type reagents is still not known with certainty.

Some observations made by Prenna et al. (24, 32) and Stoward (37) seem to give support to the alkyl sulfonic acid theory (6, 14) rather than to the sulfinic acid theory (40).

A list of fluorescent Schiff-type reagents at our disposal includes acridine, thiazole, diphenylmethane, azine derivatives etc., with various fluorescence color (Table 1). Due to their high fluorescence intensity, thiazole derivatives are particularly recommended (Figs. 1 and 2). This list, most probably, has to be extended, since other Schiff-type reagents, excited at suitable wavelengths, may be discovered as fluorescent. For example, it has recently been demonstrated that structures, stained by conventional Schiff reagent, show an intense red fluorescence (18) when excited with green light.

The use of pure products in the preparation of the

Table 1   Fluorescent-Schiff Type Reagents

Dye	Class Dye	Primary Amine Groups	Fluorescence Color[a]	Author(s)[b]	Cytofluorometric Applications[b]
Acridine Yellow C.I. 46025	Acridine	2	Yellow	Kasten et al., 1959 (7, 8, 10)	Böhm et al., 1968 (1, 2)
Acriflavine C.I. 46000	Acridine	2	Yellow-Orange	Ornstein et al., 1957 (15)	Prenna et al., 1963 (19, 22, 23); Böhm et al., 1968 1, 2)
2'-(p-aminophenyl)-6-methyl-benzothiazole or dehydrothio-p-toluidine	Thiazole	1	Bluish-White	Prenna et al., 1964 (25, 26)	Prenna et al., 1966 (30)
2'-(p-aminophenyl)-6-methyl-2,6'-bibenzothiazole	Thiazole	1	Bluish-White	Prenna, 1964 (20)	—
2,5-bis-[4'-amino-phenyl-(1')]-1,3,4-oxidiazole or BAO	Oxidiazole	2	White	Ruch, 1966 (35)	Ruch, 1966 (35). Yataghanas et al., 1969 (41)
Auramine O C.I. 41000	Diphenyl-methane	1	Green-Yellow	Kasten, 1959 (7, 8)	Ruch et al., 1963 (36); Bosshard, 1964 (3); Rozanov et al., 1967 (34); Van Dilla et al., 1969 (39); Khavkin et al., 1970 (11); Kudryavtseva et al., 1970 (12)
Chrysophosphine 2G C.I. 46040	Acridine	2	Yellow-Orange	Kasten, 1959 (7, 8)	—
Coriphosphine O C.I. 46020	Acridine	1	Yellow-Orange	Kasten et al., 1959 (7, 8, 10)	Böhm et al., 1968 (1, 2)
Flavophosphine N C.I. 46065	Acridine	2	Yellow-Orange	Kasten et al., 1959 (7, 8, 10)	—
Neutral Red C.I. 50040	Azine	1	Reddish	Kasten et al., 1959 (7, 8 10)	—
Phenosafranin C.I. 50200	Azine	2	Red	Kasten et al., 1959 (7, 8, 10)	—
Phosphine 5G C.I. 46035	Acridine	2	Yellow-Orange	Kasten, 1959 (7, 8)	—
Phosphine GN C.I. 46045	Acridine	2	Yellow-Orange	Kasten et al. 1959 (7, 8, 10)	—
Rheonin AL C.I. 46075	Acridine	1	Yellow-Orange	Kasten, 1959 (7, 8)	—
Rhodamine 3GO C.I. 45210	Xanthene	1	Yellow-Orange	Kasten et al., 1959 (7, 8, 10)	—
Rivanol	Acridine	2	Yellow	Prenna et al., 1964 (24)	Prenna et al., 1968 (27)
Safranin O C.I. 50240	Azine	2	Reddish	Kasten et al., 1959 (7, 8 10)	—

[a]The listed fluorescence color is purely indicative, since it may vary according to the excitation and barrier filters adopted, and to the concentrations used.

[b]For conciseness, mention is made only of the first author; in parentheses, the bibliographic references.

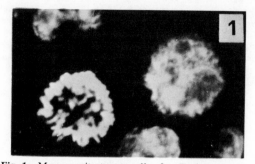

Fig. 1   Mouse ascite tumor cells after Feulgen reaction with dehydrothio-p-toluidine-SO$_2$ at 0.001% concentration.

reagent is important. Commercial products are generally satisfactory, but some acridine derivatives may require purification. In general, the usual procedure of preparation of Schiff's reagent can be followed. Better results are obtained, however, saturating the solution by bubbling SO$_2$. The concentrations of reagents vary between 0.05% and 0.01% and below, according to purpose required: lower concentrations are used for quantitative methods. The reactions are carried out in the usual manner. Accurate rinsing in water and sulfurous water is recommended. Further details can be obtained by consulting original works published by the above-mentioned authors.

The specificity of the reactions, verified extensively by many authors, is held to be good, but Stoward (37) (who prefers to call these reagents pseudo-Schiff) considers that reactions are specific only if they are executed in a particular way and after methylation.

2. Applications in Qualitative Histochemistry. The most common application of fluorescent Schiff-type reagents is in the Feulgen and PAS reaction for the demonstration of DNA and polysaccharides respectively. These reagents can also substitute standard Schiff in other histochemical reactions, such as ninhydrin-Schiff, performic acid Schiff, peracetic acid Schiff reaction, etc.

The reaction product shows a bright fluorescence, varying in color according to the reagent used. In some cases (e.g., using thiazole derivatives) the preparations appear colorless when examined with an ordinary microscope.

An interesting application is multiple staining reactions (7, 8, 9): double staining for DNA and polysaccharides or for different polysaccharides (in the sequence Feulgen-

PAS or Alcian-PAS) or triple staining for DNA, polysaccharides, and proteins, as conceived by Himes and Moriber (5). In these multiple staining reactions the fluorescent Schiff-type reagents can be variously combined among themselves, with standard Schiff reagent, nonfluorescent Schiff-type reagents, or with other histochemical stains. It is thus possible to obtain reaction products which are either all fluorescent, but showing a variety of fluorescence colors, or products differently combined with nonfluorescent stains. Obviously in the latter case the preparations have to be examined with a microscope equipped with a double illuminating system: visible light and fluorescence. A further possibility is in combined phase and fluorescence microscopy.

It is also advantageous to couple the sensitivity of these fluorescent techniques with the better optical resolution of thin resin sections, according to the experience of McNary et al. (13): extremely fine cytological details can thus be observed (Fig. 3). The fluorescent Schiff-type reagents, however, have their main application in detecting substances at low concentration. Explicative examples are papers published by Prenna and Gerzeli (28) and Danilina et al. (4). These authors demonstrated the presence of DNA in material which gave negative results when treated with conventional Feulgen reaction.

3. Applications in Quantitative Histochemistry. As conceived and prospected by Ornstein et al. (15), cytofluorometry can today be considered one of the most important methods of quantitative histochemistry with undoubtable promising development. Cytofluorometry was previously handicapped by the lack of satisfactory stain techniques; the employment of fluorescent Schiff-type reagents gave a decisive impulse to this relatively new quantitative method.

The theoretical basis of fluorescence measurements by means of the fluorescence microscope have been discussed by Rigler (33); and instrumentation and measuring procedures have been described by Prenna and Bianchi (22), Thaer (38), Prenna (21), and Böhm and Sprenger (2). The advantages offered by cytofluorometry can be thus summarized: (a) higher sensitivity with respect to absorption cytophotometry; (b) absence of distributional

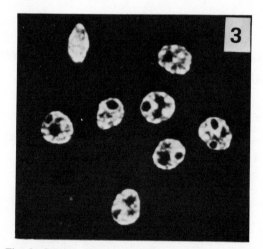

Fig. 3   Section of rat pancreas embedded in Araldite after Feulgen reaction with Acriflavine-SO$_2$ at 0.01% concentration.

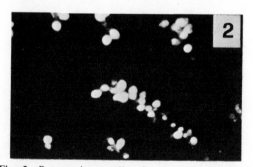

Fig. 2   Rat mucipars cells after PAS reaction with dehydrothio-p-toluidine-SO$_2$ at 0.001% concentration.

and cytometric error; (c) simple instrumentation and rapid measuring rate. Possible sources of error such as photodecomposition (21, 22), autoabsorption (42), partial nonlinearity between fluorescence intensity and concentration of fluorescent substances (17), and metachromatic phenomena when using certain fluorochromes (22), can be avoided under determined conditions.

Numerous data in literature (1, 2, 3, 12, 19, 21, 23, 27, 30, 31, 34, 35, 36, 41) definitely demonstrate that cytofluorometric measurements are correct and equivalent to those obtained with the scanning absorption method. Table 1 includes a list of authors who employed the fluorescent Schiff-type reagents in cytofluorometry.

Mention should be made of some recent cytofluorometric applications of the fluorescent Schiff-type reagents such as the quantitative study of DNA distribution along single chromosomes (11), and the employment of Feulgen fluorescent reactions in the new techniques of pulse cytophotometry in cytology automation (39). Moreover, it is probable that a technique could be developed for measuring different substances in the same cell, adopting multiple stain reactions with various fluorescence colors.

Finally, keeping in mind the peculiar characteristics of fluorescence photometry, it should be pointed out that the use of the fluorescent Schiff-type reagents, in connection with a high-sensitivity system such as photon counting, should permit the measuring of such extremely limited quantities of DNA and polysaccharides, otherwise impossible to evaluate (16).

G. PRENNA

References

1. Böhm, N., W. Sandritter, and E. Sprenger, pp. 48–54 in "Cytology Automation," D. M. D. Evans, ed., Edinburgh and London, E. & S. Livingstone Ltd., 1970.
2. Böhm, N., and E. Sprenger, *Histochemie,* **16**:100–118 (1968).
3. Bosshard, U., *Z. Wiss. Mikroskop.,* **65**:391–408 (1964).
4. Danilina, A. N., M. A. Panov, and R. B. Chesin, pp. 137–140 in "Electron and Fluorescence Microscopy of the Cell," Akad. Nauk USSR, ed., Moscow, 1964 (in Russian).
5. Himes, M., and L. Moriber, *Stain Technol.,* **31**:67–70 (1956).
6. Hörmann, H., W. Grassman, and G. Fries, *Justus Liebigs, Ann. Chem.,* **616**:125–147 (1958).
7. Kasten, F. H., *Histochemie,* **1**:466–509 (1959).
8. Kasten, F. H., *Int. Rev. Cytol.,* **10**:1–100 (1960).
9. Kasten, F. H., *Acta Histochem.,* Suppl. **3**:240–247 (1963).
10. Kasten, F. H., V. Burton, and P. Glover, *Nature,* **184**:1797–1799 (1959).
11. Khavkin, T. N., B. N. Kudryavtsev, L. B. Berlin, and M. V. Kudryavtseva, *Tsitologiya,* **12**:1209–1212 (1970) (in Russian).
12. Kudryavtseva, M. V., B. N. Kudryavtsev, and Yu. M. Rosanov, *Tsitologiya,* **12**:1060–1068 (1970) (in Russian).
13. McNary, W. F., Jr., R. C. Rosan, and J. A. Kerrigan, *J. Histochem. Cytochem.,* **12**:216–217 (1964).
14. Naumann, R. V., P. W. West, F. Tron, and G. G. Gaeke, *Anal. Chem.,* **32**:1307–1311 (1960).
15. Ornstein, L., W. Mautner, B. J. Davis, and R. Tamura, *J. Mt. Sinai Hosp.,* **24**:1066–1078 (1957).
16. Pearse, A. G. E., and F. W. D. Rost, *J. Microscop.,* **89**:321–328 (1969).
17. Perrin, M. F., *C. R. Acad. Sci.* (Paris), **178**:1978–1980 (1924).
18. Ploem, J. S., *Z. Wiss. Mikroskop.,* **68**:129–142 (1967).
19. Prenna, G., *Riv. Istoch. norm. pat.,* **9**:211–213 (1963).
20. Prenna, G., *Riv. Istoch. norm. pat.,* **10**:469–474 (1964).
21. Prenna, G., *Mikroskopie,* **23**:150–154 (1968).
22. Prenna, G., and U. A. Bianchi, *Riv. Istoch. norm. pat.,* **10**:645–666 (1964).
23. Prenna, G., and U. A. Bianchi, *Riv. Istoch. norm. pat.,* **10**:667–676 (1964).
24. Prenna, G., U. A. Bianchi, and L. Zanotti, *Riv. Istoch. norm. pat.,* **10**:389–404 (1964).
25. Prenna, G., and A. M. De Paoli, *Riv. Istoch. norm. pat.,* **10**:185–186 (1964).
26. Prenna, G., and A. M. De Paoli, *Istituto Lombardo* (*Rend. Sc.*) *B,* **98**:267–273 (1964).
27. Prenna, G., and A. M. De Paoli, *Riv. Istoch. norm. pat.,* **14**:169–170 (1968).
28. Prenna, G., and G. Gerzeli, *Riv. Istoch. norm. pat.,* **10**:607–612 (1964).
29. Prenna, G., N. Piva, and L. Zanotti, *Riv. Istoch. norm. pat.,* **8**:427–446, (1962).
30. Prenna, G., and A. Thaer, *Riv. Istoch. norm. pat.,* **35**:127 (1966).
31. Prenna, G., and L. Zanotti, p. 231 in "Abstracts of the Second International Congress of Histo- and Cytochemistry. Frankfurt/Main," Berlin, Springer-Verlag, 1964.
32. Prenna, G., L. Zanotti, and U. A. Bianchi, *Riv. Istoch. norm. pat.,* **10**:539–552 (1964).
33. Rigler, R., Jr., *Acta physiol. Scand.,* **67**(Suppl. 267): 1–122 (1966).
34. Rozanov, J. M., and B. N. Kudryavtsev, *Tsitologiya,* **9**:361–367 (1967) (in Russian).
35. Ruch, F., pp. 281–294 in "Introduction to Quantitative Cytochemistry," G. L. Wied, ed., New York and London, Academic Press, 1966.
36. Ruch, F., and U. Bosshardt, *Z. Wiss. Mikroskop.,* **65**:335–341 (1963).
37. Stoward, P. J., *J. Roy. Microscop. Soc.,* **87**:237–246 (1967).
38. Thaer, A., pp. 409–426 in "Introduction to Quantitative Cytochemistry," G. L. Wied, ed., New York and London, Academic Press, 1966.
39. Van Dilla, M. A., T. T. Trujillo, P. F. Mullaney, and J. R. Coulter, *Science,* **163**:1213–1214 (1969).
40. Wieland, H., and G. Scheuing, *Ber. dtsch. chem. Ges.,* **54**:2527–2555 (1921).
41. Yataghanas, X., G. Gahrton, and B. Thorell, *Exp. Cell Res.,* **56**:59–68 (1969).
42. Zanotti, L., *Acta Histochem.,* **17**:353–370 (1964).

FORAMINIFERA

The foraminifera are testaceous rhizopods belonging to the Order Foraminiferida. They are characterized by grano-reticulate pseudopods and the presence of a test which may be agglutinated with a variety of cements or formed of calcium carbonate. Because the tests are commonly preserved as fossils in rocks of marine origin, and because they are too small to be ground up in the drilling of wells, foraminiferal tests are widely used by micropaleontologists to determine the age and environment of deposition of geologic deposits.

Biologists, interested in the living protoplasm and only incidentally concerned with the test, use microscopic techniques which differ markedly from those employed by micropaleontologists, who are concerned only with the tests. Biological studies typically employ phase contrast light microscopy, transmission electron microscopy of ultramicrotome sections of fixed and embedded protoplasm, and scanning electron microscopy of fixed and freeze-dried specimens. Paleontologists routinely

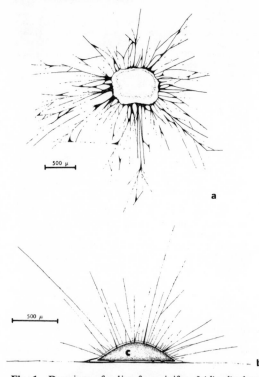

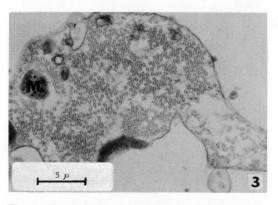

Fig. 3 Transmission electron micrograph of a thin section of a pseudopod of *Iridia diaphana* showing a high density of microtubules in the groundplasm. M is a mitochondrion (from Marszalek, 1969).

Fig. 1 Drawings of a live foraminifer, *Iridia diaphana*. a: top view showing grano-reticulate pseudopods extending outward in all directions from the test. b: Side view showing rigid pseudopods extending upward into the water; the main cytoplasmic mass (c) is indicated inside the test (from Marszalek et al., 1969).

study the tests of foraminifera with stereoscopic binocular microscopy, by transmission electron microscopy of carbon replicas of natural or polished and etched surfaces and by scanning electron microscopy.

Problems inherent in the microscopy of foraminifera

for biological purposes are evident from Fig. 1a and b, diagrammatic representations of *Iridia diaphana*, a common foraminifer of the shallow water areas of south Florida. The test of this species is a simple dome of chitinous material with agglutinated diatom frustules, sponge spicules, and other foreign objects. It extends grano-reticulate pseudopods in all directions along the substrate, forming a pseudopodial net which may have an area of several square centimeters. As seen in the side view (Fig. 1b), the pseudopods are semirigid and some are directed into the water to varying distances. The relatively large size of the entire animal contrasts with the extreme thinness and delicacy of the pseudopods so that it is not possible to see both the whole test and the entire pseudopodial net at one magnification. Figure 2 shows a portion of the pseudopodial net seen in phase contrast light microscopy. Observation of the protoplasm inside the test is even more difficult because the relatively thick test wall obscures all finer internal detail. The mineral part of the test can be removed by dissolution with weak acid without lethal effect on the animal, but in many forms the chitinous test liner is so thick that the internal features are still obscured to the light microscopist. Best

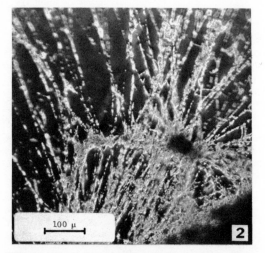

Fig. 2 Phase contrast photomicrograph showing part of the pseudopodial net of *Iridia diaphana* (from Marszalek, 1969).

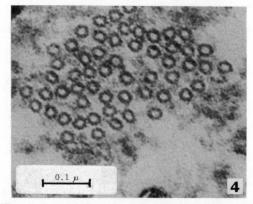

Fig. 4 Transmission electron micrograph of a thin section of a pseudopod of *Iridia diaphana*, showing microtubule substructure of 9–13 microfilaments (from Marszalek, 1969).

Fig. 5 Scanning electron micrograph of part of the fixed and freeze-dried pseudopodial net of *Iridia diaphana*. The granules seen in the thinnest pseudopods are about the same size as mitochondria. Note the complex branching and anastomising, and the thick nodes at junctions.

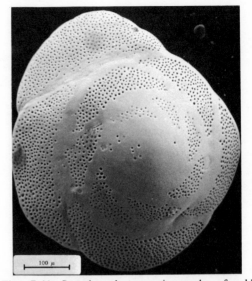

Figs. 7–11 Scanning electron micrographs of gold-palladium coated foraminiferal tests. Fig. 7: *Eponides turgidus*; Fig. 8: *Uvigerina flintii*; Fig. 9: *Peneroplis bradyi*; Fig. 10: *Homotrema rubrum*; Fig. 11: *Articulina antillarum*. (Specimens provided by W. D. Bock.)

results for phase microscopy of living specimens are obtained using a water immersion objective lowered directly into the salt water medium in which the foraminifera live.

Preparation of specimens for transmission electron microscopy is as follows (after Marszalek, 1969): Specimens are fixed in 3 % glutaraldehyde in sea water, followed by 1 % osmium tetroxide in sea water. The fixatives are adjusted to pH 7.2 with sodium cacodylate as buffer minimizing solution of calcium carbonate tests. After fixation the specimens are imbedded in a 1 % agar solution (following Haller et al., 1961). The agar block is dehydrated in graded ethanol and then embedded in a hard resin which is cured for ultramicrotomy. Cut sections are stained as usual with uranyl acetate and lead citrate. Examples of the results of this technique are shown in Figs. 3 and 4.

For scanning electron microscopy of the protoplasm, the technique of Small and Marszalek (1969) is followed. After fixation with glutaraldehyde and osium tetroxide as described above, or alternatively with Parducz' fixative,

the specimens are rinsed in distilled water, quick-frozen over liquid nitrogen, and freeze-dried. After sublimation is complete, a thin film of gold-palladium is deposited over the surface of the specimens and they are ready for immediate observation. With this method even the pseudopodia which originally extended through the water can be observed, as seen in Fig. 5.

Fig. 6 X-ray photomicrograph of *Globoquadrina dutertrei* (d'Orbigny) obtained with a projection X-ray microscope (Bé et al., 1969). Maximum length of test is 525 μ.

Fig. 8

Fig. 9

Fig. 11

Paleontologists use stereoscopic binocular microscopy with reflected light for routine examination of foraminiferal tests. Adhering foreign material is removed from fossil tests by ultrasonic treatment. The tests may then be mounted on opaque cardboard or plastic microscope slides providing a black background in a depression. The cardboard slides may have large black background areas marked off into squares or other subdivisions which may

be numbered, and the depressions are protected by glass or plastic covers. Foraminiferal tests are attached to the slides by gum tragacanth or other water-soluble adhesive; later removal of this adhesive for scanning electron microscope investigation may be accomplished with an oxidizing agent such as Clorox, but this treatment may destroy the chitin of agglutinated tests and cause them to disintegrate. Ordinarily only the surface features of foraminiferal tests are visible in reflected light.

Nondestructive examination of the internal features is achieved by immersing the test in glycerine or castor oil. The test becomes transparent and with transmitted light internal features can readily be observed in small specimens. Larger specimens may become only translucent or may remain opaque, but can be examined nondestructively by X-ray microscopy (see Fig. 6).

Destructive methods of examining the internal features of foraminiferal tests include thin sectioning by mounting the specimen on a glass slide and grinding a section using abrasive powder, following the usual technique for preparing a petrographic thin section. Serial sections are made by taking peels as the specimen is ground away. Careful crushing of a specimen between glass slides can expose internal features directly to view. Better results can be obtained by dissolution of the calcium carbonate of parts of the test, using minute amounts of acid on a brush with only one or a few bristles. Some micropaleontologists have designed minute grinding wheels for cutting away parts of the test.

Illustration of the tests of foraminifera was, until recently, best accomplished by drawings. The problems

Fig. 10

Fig. 12 Transmission electron micrograph of a carbon replica of the surface of *Quinqueloculina seminulum*, showing the "tile roof" pattern characteristic of miliolid foraminifera (from Hay et al., 1963).

of depth of field inherent in light microscopy of such thick objects caused photography to be a relatively poor technique except for unusually flat specimens. The surface features of many species require a large numerical aperture for resolution, but the thickness of the specimens requires a small diaphram opening. Consequently it was impossible to adequately portray many species by photomicrography.

Since 1967 the scanning electron microscope has become the standard tool for illustration of foraminiferal tests. The tests are cleaned, mounted on a specimen stub using double sticky tape, and coated first with carbon, then with gold-palladium. The carbon is more mobile, and penetrates to regions not reached by gold-palladium to provide a conductive surface over the entire area which may be irradiated by the electron beam. Typical results are shown in Figs. 7–11.

Details of test wall ultrastructure may be observed using either direct examination by scanning electron microscopy or one- or two-stage carbon replicas in transmission electron microscopy. Combination of both techniques is advisable because the ultrastructural details lie near the limit of resolution of the scanning electron microscope and the carbon replicas commonly represent such small areas as to be difficult to interpret. An example of a carbon replica of the surface of the test of a miliolid foraminifer is shown in Fig. 12.

W. W. HAY

References

Bé, A. W. H., W. L. Jongebloed, and A. McIntyre, "X-ray microscopy of recent planktonic foraminifera," *J. Paleontol., 46*:1384–1396, pls. 161–167 (1969).

Haller, G. de., C. F. Ehret, and R. Naef, "Technique d'inclusion et d'ultramicrotomie, destinée a l'étude du developpement des organelles dans une cellule isolé," *Experientia, 17*:524–529 (1961).

Hay, W. W., K. M. Towe, and R. C. Wright, "Ultra-microstructure of some selected foraminiferal tests," *Micropaleontology, 9*:171–195, 16 pls. (1963).

Marszalek, D. S., "Observations on *Iridia diaphana*, a marine foraminifer," *J. Protozool., 16*:599–611, 21 figs. (1969).

Marszalek, D. S., R. C. Wright, and W. W. Hay, "Function of the test in foraminifera," *Trans. Gulf Coast Ass. Geol. Socs., 19*:341–352, 22 figs. (1969).

Small, E. B., and D. S. Marszalek, "Scanning electron microscopy of fixed, frozen, and dried protozoa," *Science, 163*:1064–1065, 8 figs. (1969).

FORENSIC MICROSCOPY

In an era when the crime rate continues to soar, and more sophisticated instrumentation is being developed, the backbone of any scientific criminal investigation remains as it has been throughout history, the use of the microscope. The development of microscopic instrumentation has progressed from the simple hand lens to sophisticated instruments such as the scanning electron microscope. The experienced criminalist utilizes many magnifying instruments in this broad range of the simple to complex. Crime laboratories certainly have other instrumentation, such as the emission spectrograph, spectrophotometers for determining absorption spectra with visible, ultraviolet, and infrared light, electrophoresis and thin layer chromatography equipment, as well as the gas chromatograph. These augment the microscope, but at least 70% of the criminalist's work can be done with the use of microscopes alone.

Data and information developed by the criminalist are used daily to determine how a crime was perpetrated and what physical evidence, if any, connects a suspect with a crime. Of equal importance, the physical evidence can also exonerate an innocent individual.

It is the opinion of most investigators that while much of the gross evidence of a crime can be concealed or destroyed by the perpetrator, certain types of evidence by the nature of their seeming unimportance or minutia are too small for the criminal to be aware of at the time he is committing a crime. Therefore, most criminals unknowingly leave behind at a crime scene certain microscopic traces, as well as taking away others. This subtle evidence, when properly examined and interpreted, forms the foundation of most trace evidence analyses. The results of the large volume of work done through microscopic investigative procedures are entered into the courts and given substantial weight in the proper adjudication of criminal offenses, where a qualified examiner can often turn the smallest traces from a crime situation into a monumental piece of evidence at the time of trial. Although the bulk of forensic microscopy is directly related to criminal offenses, the criminalist/microscopist can also bring his knowledge and expertise into the courts in civil ligitations.

In a forensic laboratory the microscope is used to enlarge materials for one or both of two major purposes: identification of a material, and/or comparison of one material to another, to determine if they are of a common origin. In accomplishing these objectives it is necessary to employ a variety of microscopes, including the standard bright field microscope, the stereoscopic microscope, dual comparison microscopes (using transmitted light and reflected light), the polarizing microscope, and the scanning electron microscope.

The applications of bright field microscopy in scientific

criminal investigation are many and varied. One such application is the examination and identification of animal, vegetable, and synthetic fiber materials. In cases of pedestrian "hit and run" accidents, fibers are often collected from an automobile suspected of being involved in an accident. When mounted on a glass slide and observed at 100 and 450 magnifications, the fibers are examined for their class, color, and morphology. In many cases, the type of fiber can be identified without further testing. The microscopist then observes fibers from the clothing of the victim. If the microscopic appearance of the fibers indicates that they are of the same class, color, and morphology as those from the suspect vehicle, a comparative study utilizing the comparison microscope is made.

The comparison microscope is constructed from two standard light microscopes connected by a common optical bridge allowing the viewer to examine two separate samples simultaneously. Fiber samples on two microscope slides appear adjacent to one another, allowing accurate comparison of their color, diameter, and texture (see Fig. 1). The use of the comparison microscope is imperative in comparing subtle similarities or differences in two such samples. Human hair also can be examined by comparison microscopy to determine its origin.

By the addition of a micro-hot stage (such as the Kofler) to the bright field microscope, materials can be identified or categorized by their melting points. In some instances, the microscopic examination of synthetic fibers enables the microscopist to categorize fibers as having the same microscopic appearance, but does not afford identification of a fiber. For example, Creslan and Dacron sometimes have a highly similar microscopic appearance. They can be differentiated, however, by placing small specimens of each fiber on a hot stage, and observing them during heating. Different types of synthetic fibers are characterized by their physical behavior such as swelling, change of color, and melting which occurs as the temperature is increased.

The identification and grouping of blood in violent crimes is often of prime importance in criminal investigations as connective evidence to a suspect. Blood evidence can constitute a major part of the criminalist's work.

Methods of analysis of grouping dried blood differ from those used diagnostically in the medical sciences, although both are based on the same biochemical property of blood: the group specific antibody-antigen reaction. The microscope is essential in both the identification and grouping of dried bloodstains.

In a sensitive and specific method of the identification of dried blood, a crust is placed upon a microscope slide and warmed when chemical reagents are added. A chemical complex is formed with the heme moiety of the blood, and the microscopic, feathery orange crystals which result can be observed at $100\times$.

The immunological techniques used for the grouping of blood into the ABO classification are carried out by taking dry blood samples and reacting them with either specific antisera or with whole blood cells. In the Lattes (or crust) method, dried blood samples are placed on three areas of a microscopic slide and covered with coverslips, and suspensions of washed erythrocytes of groups A, B, and O are added to samples one, two, and three, respectively. After incubation for a short period of time, microscopic examination will reveal agglutination (or clumping) of those erythrocytes which have come into contact with antisera specific for its group. The Lattes method therefore tests for the presence of group specific antibodies from the sera of the dried blood. (See Figs. 2 and 3).

In the ABO and other classifications the presence of group specific antigenic factors—present on the cell wall of the red cells—can be demonstrated by the immunological techniques known as absorption-inhibition and absorption-elution. These techniques are more involved than the Lattes test, but have the added advantages that very small samples can be examined, and that a grouping of dried blood is more likely to be accomplished, since the antigens are less susceptible to destruction (e.g., by aging or oxidation) than are antibodies.

Pieces of bloodstained cotton thread as small as one-tenth centimeter are tested in the absorption-elution method, which utilizes the temperature dependence of the antibody-antigen reaction. The bloodstained threads are placed in the dimpled wells of microscope slides. Anti-a and anti-b sera are added to the wells respectively, and the slide is incubated at 4°C for several hours. If the sera and antigens are complementary, a reaction takes place and the sera are absorbed onto the thread. The threads are then carefully washed to remove superfluous (nonabsorbed) sera, suspensions of washed erythrocytes of groups A and B are added respectively, and the slides are warmed and examined microscopically. Agglutination of the erythrocytes indicates the presence of group specific sera, and the group of the dried blood can be determined.

The immunological techniques for grouping dried blood can also be applied to other biological fluids and secretions, such as sweat, saliva, and seminal fluid, and to human hair.

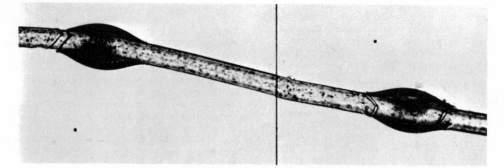

Fig. 1 Known standard Dacron fiber (left) and suspect fiber (right) as viewed through the comparison microscope at $140\times$.

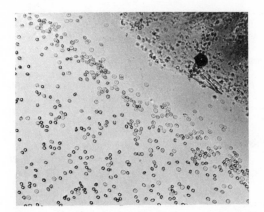

Fig. 2 Normal (nonagglutinated) human red blood cells at $52\times$.

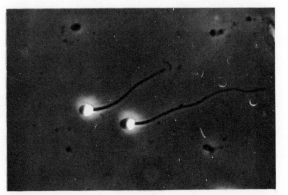

Fig. 4 Human spermatozoa at $300\times$, photographed using phase contrast illumination.

Forensic microscopy is used for the medicolegal determination of the cause of death through the histological study of tissues removed at autopsy. The phase microscope is a variety of bright field microscope used in the examination of unstained histological materials. An important forensic application of phase microscopy is its use in the identification of spermatozoa (see Fig. 4). The increased contrast which the phase optics render allows the microscopist to identify spermatozoa directly from a vaginal smear, without going through the procedures of staining.

The bright field microscope is a valuable tool for the identification of drugs such as narcotics, sympathomimetics, stimulants, barbiturates, sedatives, hallucinogens, and tranquilizers as their normal dosage forms or in toxicological specimens. Drugs are identified by the morphology of microcrystals formed upon the addition of reagents such as iodides or salts of various heavy metals. These microcrystals are usually observed at 100 magnifications. Microcrystal tests for the identification of drugs are sensitive and require very little of the sometimes minute amount of available material; thus, that which remains can be presented as evidence in court. Usually other tests supplement microcrystal tests; however, many

microcrystal tests are so highly characteristic that two or three of them made on the same substance will often identify it with certainty. The specificity of the microcrystal identification can be further enhanced with the polarizing microscope.

The polarizing microscope is a bright field microscope which uses polarized light for sample illumination. The characteristic physical parameters of refractive index, anistrophy (or its lack), degree of birefringence, pleochromism, angles of extinction, sign of elongation, and interference figures of crystalline substances can be determined with the polarizing microscope.

It is of important evidential and investigative value to determine if the physical properties of microscopic glass fragments collected from a suspect's clothing can be related to those of broken glass found at a crime scene. The crystalline properties of different glasses permits their characterization by measurement of refractive indices.

The Becke-line method for refractive index commonly utilizes the polarizing microscope. To determine the refractive index of a chip of glass, it is crushed, and a small fragment is placed onto a microscope slide. A fluid of known refractive index is added, and the microscope is focused. The Becke-line is a white halo which appears around the perimeter of the glass fragment. If, when focusing up on the fragment, the Becke-line appears to move into the piece of glass, the glass has a higher refractive index than the surrounding fluid. The fluid is then removed, and ones of higher refractive index are added successively until a point is reached where the Becke-line can no longer be seen. At this point the glass has the same refractive index as the surrounding fluid. This technique is sensitive, and small differences in refractive indices of samples can readily be distinguished.

Refractive indices are usually measured at 5890 Å (sodium D line). Additional measurements at two other wavelengths (e.g., 4860 Å and 6560 Å) and a mathematical combination of the three values yields a parameter known as the dispersive power which further characterizes a glass.

If a suspect chip of glass and a standard sample, collected at a crime scene, are viewed on the same microscope slide, a very accurate comparison of their refractive indexes and dispersive powers can be made. When glass is manufactured, the homogeneity of the melt assures quite a constant composition. Thus the physical properties, including refractive index and dispersive power, of

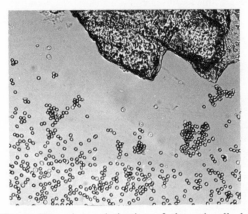

Fig. 3 Here the agglutination of the red cells has resulted from the reaction of a group specific antibody from a suspect bloodstain.

glass samples from a common origin (e.g., a window pane) are the same. (This does not imply, however, that identical physical properties of two glass samples establishes proof of common origin.)

Particulate materials such as soil particles and paint pigments can also be identified and compared by the properties which they exhibit under polarized light. Refractive indices and dispersion properties play important roles in identification and comparison to determine a common source.

In instances where the melting point and microscopic examinations of man-made fibers are not sufficient for identification, the polarizing microscope may be utilized to determine the refractive indices of the ordinary and extraordinary rays which are characteristic of many types of fibers.

The stereomicroscope is a binocular instrument with a normal range of magnification from $10\times$ to $40\times$. A three-dimensional view, allowing full depth perception, results from paired objectives. Illumination by reflected light, low magnification, and a wide field of view permit the criminalist to rapidly examine large, opaque objects such as clothing and tools. Thus, the stereoscopic microscope serves a very important function during the initial search for trace (microscopic) evidence.

Before a hair comparison, fiber comparison, or blood grouping can be made, suspect hairs, fibers, and bloodstains must be located and collected. For example, clothing of a victim and/or a suspect, and objects suspected of having been used as weapons are carefully examined with the stereomicroscope for such trace evidence.

After this preliminary examination, clothing is vacuum swept with a special type of filtering device which collects dust and minute particulate material on a filter paper. Observation of the filter paper permits the separation of items of interest which can then be examined at higher magnification or mounted for bright field observation.

The excellent resolution at high magnification of the stereomicroscope permits detailed examination of microscopic particles of glass, safe insulation, soil, paint, animal (e.g., insects), and vegetable materials. For example, soil is examined for its color and texture, and its mineral, animal, and vegetable inclusions are noted.

Paint has considerable evidential value. Microscopic chips of paint from a jimmied door or window are often found in a burglar's clothing. Paint chips and smears are also found on a pedestrian struck by an automobile,

and cross-transfer of paint from one vehicle to another commonly occurs when two automobiles collide. Paint particles collected from the clothing of a burglary suspect (or a victim of a hit and run accident) are microscopically compared with known samples from the crime scene (or the suspect vehicle). In such a paint examination, not only can gross color and texture be examined, but with multiple layers of paint, chips can be placed on edge next to each other and compared with respect to sequence of layering and thickness, texture, and color of each layer (Fig. 5). If ten layers or a paint chip found on a suspect are indistinguishable from ten paint layers from a burglarized establishment, this is overwhelming evidence of their common origin.

Forensic laboratories responsible for the identification of counterfeit drugs use the wide field stereoscopic microscope to examine the external characteristics of tablets and capsules. Tablets are examined to detect microscopic marks made by imperfections in the dies with which they are manufactured. These microscopic marks are compared to marks on authentic samples and the origin of a product can often be determined. Capsules are similarly examined to detect minute marks on their surfaces. The size, shape, and spacing of the letters of any monograms and the color and texture of the contents are noted. Counterfeit drug products can be detected by their deviation from the genuine product in external appearances and by the composition of the formulated ingredients.

Marijuana can be identified by the morphology of fragments of its leaves, seeds, and flowers, as well as the characteristic effervescence of CO_2 from its cystolith hairs upon addition of dilute acid. (See Fig. 6.)

A further use of the stereomicroscope is in examination of questioned documents. The questioned-document examiner determines the authenticity, origin, and age of writings, signatures, and printed or typewritten documents. The direction, change in direction, and pressure points of written strokes are easily distinguished with the aid of the stereomicroscope.

An important and well known tool in forensic microscopy is the dual comparison microscope with reflected light illumination. Similar in construction to dual comparison microscopes which use transmitted light, reflected light comparison microscopes are used at lower magnifications (normally between 10 and 20 diameters), and are fitted with carefully matched objectives. This

Fig. 5 A cross section of a paint chip of known origin (left) and of a paint chip collected from a suspect's clothing, at $40\times$, photographed through the stereoscopic microscope.

Fig. 6 A fragment of a marijuana leaf, showing the "cystolith hairs." 18×.

insures equal magnification and uniformity of resolution about the field. The reflected light dual comparison microscope is used primarily to relate tool marks left at a crime scene with a suspect tool and to relate bullets and cartridge casings to a firearm.

Toolmark comparisons are based on the property that each tool possesses certain imperfections on its surfaces. These marks or imperfections can result from manufacturing, use and wear, and neglect; thus on the working surface of a tool are a combination of machine marks, casting flaws, scratches, compression areas, and rust and dirt. Regardless of its source, the negative impression of an imperfection can be transferred to any material softer than the tool itself. Therefore, on the surface of a jimmied window, a drilled safe door, or a cut wire, a tool leaves what can be regarded as a fingerprint of its working edge.

Using the comparison microscope, the criminalist can compare the microscopic striations left in toolmarks at a crime scene (or associated with a crime) with test tool marks, which he makes with a suspect tool in wax or lead. A sufficient correspondence in both the gross dimensions and fine striations of the suspect and test toolmarks enables the examiner to determine whether or not a tool made the suspect mark.

Fig. 7 Two bullets, as viewed by the comparison microscope at 16×. The congruence of the fine striations (or individual characteristics) in the rifling impressions indicate that both bullets were fired by the same gun.

In firearms examination it can be determined whether or not a bullet or cartridge case was fired from a particular firearm. On the surface of the barrel, chamber, and firing mechanism of a firearm are minute imperfections resulting from manufacturing, use, wear, and neglect. As a "soft metal" bullet passes through the barrel of a firearm, it is scraped by the rifling lands (which impart a twist to the bullet) and by any imperfections in the barrel. The latter produce a series of striations on the bullet which are known as "individual characteristics" and which enable an identity to be established between a bullet and the barrel of the firearm.

When a bullet is recovered at a crime scene or from a victim, and a suspect weapon is found, the following procedure is employed: The suspect weapon is test fired under controlled conditions, and the undamaged bullets are collected. An examination is made to determine the general characteristics of caliber, and the number, direction of twist, pitch, and width of the rifling impressions. If all the general characteristics of the test bullets and evidence bullet are consistent, then the examination is continued using the comparison microscope. The examiner first determines the reproducibility of the striations or individual characteristics by comparing several test bullets. If it is found that a "match" is easily determinable between two test bullets, then the evidence bullet is substituted for one of these. If in comparing the individual characteristics of the test and evidence bullets through a rotation of 360°, the firearms examiner finds that they have sufficient markings in common, he can state that the evidence bullet was fired by the suspect firearm. (See Fig. 7.)

Also of importance in firearms examinations are spent cartridge casings, which retain firing pin impressions and breech-face markings. The pressure produced when a cartridge is fired not only propels the bullet through the barrel, but also slams the cartridge casing back against a metal retainer called the breech-block. Consequently, machine marks or imperfections due to wear located on the breech-block will be imprinted as negative impressions on the primer cap and head of the cartridge case. An automatic weapon imparts additional identification features in the form of marks made by the extractor and ejector as the spent casing is mechanically extracted and ejected. Firing pin impressions, and breech-face, ejector, and extractor markings have specific "individual characteristics," and can be examined and compared by comparison microscopy to relate cartridge cases to the weapon which fired them.

A nonoptical microscope, with electronic and photographic readout, just beginning to be used in forensic investigation is the scanning electron microscope. Until recently, the expense and complexity of this tool restricted it to research applications, but today it is used in routine examinations.

The scanning electron microscope has features no optical microscope can approach—a much greater depth of field and a range of magnification from about 20 to 20,000× with excellent resolution. The high resolution photographs of opaque surfaces which can be obtained lend the scanning electron microscope especially to the identification of marijuana, firearms examination, and paint chip, document, and toolmark examinations. The photomicrographs from the scanning electron microscope and from optical microscopes (most of which are equipped with trinocular heads, permitting the convenient

mounting of cameras) serve as records, and can be presented as evidence in court.

Because of a decade of revolutionary Supreme Court rulings, great importance is being placed upon the scientific facets of criminal investigation.

Physical evidence can stand on its own in criminal and civil courts, and can help to corroborate or contradict the testimony of witnesses, police, and other investigative agencies. When properly examined and with correct emphasis placed upon its meaning, this evidence can be the unbiased witness in judicial and other legal proceedings.

W. C. STUVER
W. W. PRICHARD, JR.
R. H. FOX

References

Kirk, P. L., "Crime Investigation," New York, Interscience Publishers, Inc., 1953.
Kirk, P. L., "Density and Refractive Index," Springfield, Ill., C. C. Thomas, 1951.
Nicol, J. D., p. 338 in "Encyclopedia of Microscopy," edited by George L. Clark, ed., New York, Van Nostrand, Reinhold, 1961.
O'Hara, C. E., and J. W. Osterburg, "An Introduction to Criminalistics," New York, MacMillan Company, 1949.
O'Hara, C. E., "Fundamentals of Criminal Investigation," 2nd ed., Springfield, Ill., C. C. Thomas, 1970.
Osterburg, J. W., "The Crime Laboratory," Bloomington, Ind., Indiana University Press, 1968.
"Proceedings of the Workshop on Forensic Applications of the Scanning Electron Microscope 1971," Chicago, IIT Research Institute, 1971.
Soderman, H., and J. J. O'Connell, "Modern Criminal Investigation," 5th ed., rev. by C. E. O'Hara, New York, Funk and Wagnalls, 1962.
Svensson, A., and O. Wendel, in "Techniques of Crime Scene Investigation," 2nd rev., American ed., J. D. Nicol, ed., New York, American Elsevier, 1965.
Walls, H. J., "Forensic Science," London, Sweet and Maxwell, 1968.
See also: BOTANICAL DRUGS, CHEMICAL MICROCRYSTAL IDENTIFICATION.

FREEZE-DRYING, EMBEDDING AND FIXATION

Pearse (1968) has fully reviewed the history and rationale of freeze-drying. The aspects of freeze-drying to be dealt with in this article concern comparisons of the drying forces at play in efficient and inefficient freeze-drying systems, and a routine method is described based upon these considerations.

The process consists of the following:

1. Quenching the tissue at low temperature.
2. Drying the tissue at low temperature under high vacuum.
3. Vapor fixation where necessary.
4. Embedding the tissue.

Quenching. Tissue is quenched at low temperature to minimize disorganization of structure and diffusion of elements. Unfortunately the process results in the formation of disruptive ice crystals from the unbound water. To minimize this factor the size of the tissue should be

Fig. 1 Top to bottom, left to right: stirrer and thermocouple, tissue platform, Isopentane container, liquid nitrogen container; ice reference junction, temperature meter; instruments, including sharpened cork borer for obtaining plugs of tissue 4 mm in diameter from slices 1 mm thick.

as small as possible and the rate of cooling as great as possible.

A suitable liquid for quenching purposes is Isopentane cooled to −160°C with liquid nitrogen. The Isopentane can be kept in a copper calorimeter having fins reaching down into the nitrogen. The fins should allow vaporized nitrogen to escape. Tissues plunged into the Isopentane should be kept constantly on the move for ½ min (Fig. 1).

Freeze-dryers. Freeze-dryers are designed to provide suitable temperature gradients for drying tissues and condensing surfaces for collection of ice. An exception to this is the drying of tissue by means of cryosorption pumping using a molecular sieve. This is a comparatively slow process suitable for drying very small tissue samples (Stumpf and Roth, 1967). Before operating any particular dryer its characteristics should be fully understood.

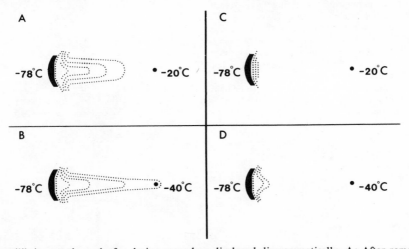

Fig. 2 The equilibrium at the end of a drying procedure displayed diagrammatically. A: After removing a large volume of water from the source by means of a large temperature gradient; B: after removing a large volume of water from the source by means of a smaller temperature gradient; C and D: repeats of A and B respectively with smaller initial volumes of water at the source.

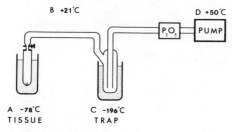

Fig. 3   An inefficient drying system. The cold bath (A) contains CO_2 ind. meth. spirit. The cold trap (C) contains liquid nitrogen.

The dryer should be scrupulously clean, and using as specimens pieces of ice formed from 0.5 ml distilled water, the machine should be calibrated over the complete drying cycle at the selected drying temperature. When the fastest possible results have been achieved the calibration should be repeated using histological specimens; the drying cycle will be somewhat longer in this case. The established pattern of pressures, temperatures, and times should be maintained on all subsequent runs (Middleton, 1967).

Movement of Condensable Vapors along a Temperature Gradient. If a given mass of water is placed in a temperature gradient over a finite length, the molecules will pass down the gradient towards the condensing surface, the system finally reaching equilibrium. At this point not all the molecules will be on the condensing surface, and contours of water concentration will exist along the temperature gradient. Whether the final concentration of water at the original source is zero or not will depend on the various parameters (Fig. 2). To achieve complete removal of water from its original position, both the temperature gradient and condensing surface must be of adequate size.

The speed of establishing equilibrium in the system depends on the magnitude of the temperature gradient, the mean temperature of the gradient, and the amount of water at the source. The process is speeded up considerably if the noncondensable vapors (air, Isopentane) are removed by vacuum pump, thus decreasing the resistance to flow of the water molecules.

Inefficient Drying Systems. With an inefficient drying system the time taken to dry the tissue is excessively long. The most usual cause of this is the use of two opposing temperature gradients which partially cancel each other out. Such a system is shown in Fig. 3.

Water in the tissue situated at point A tends to be held there by reason of the temperature gradient from B down to A, and in isolation this part of the system would not

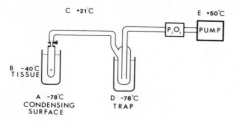

Fig. 4   An efficient drying system (Stowell). The cold bath (A) and cold trap (D) contain CO_2 ind. meth. spirit. The tissue holder at B is maintained at any desired temperature, heat being supplied electrically.

Fig. 5   A vacuum hot plate suitable for embedding frozen dried tissue in paraffin.

Fig. 6   Vacuum embedding chamber with burette and table holding 21 capsules. Suitable for embedding tissues with reasonably viscous methacrylates (Middleton).

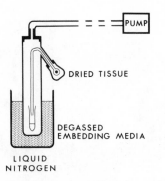

Fig. 7 Equipment for embedding tissues with fairly volatile methacrylates (Sjostrand and Baker, modified by Attramadal).

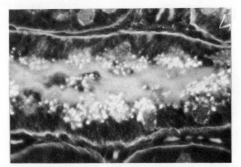

Fig. 9 Microbodies in the proximal tubule, characteristic of normal hamster kidney ($\times$ 800).

dry the tissue. The concentration of water molecules along this gradient decreases on passing from A to B. The more powerful gradient from B down to C is, in effect, the drying force, but is operating under the least favorable conditions, being applied to the region of lowest water concentration in gradient (B/A).

During the drying process, little if any water passes from trap C to the pump. However, oil vapor from the pump readily passes down the gradient from D to C during this procedure, and cold trap C is therefore important to keep the tissue free from oil contamination as well as for drying the tissue.

On completion of drying, the tissue is brought to room temperature by lowering bath A, and dry air is then cautiously bled into the system via a cold trap. There is no problem of rehydration of the tissue.

Efficient Drying Systems. An efficient drying system due to Stowell (1951) is shown in Fig. 4. A heater coil keeps the tissue at any desired temperature in a temperature gradient extending from C down to A.

Water in the tissue situated at B is acted upon by the force of the temperature gradient from B down to the condensing surface (A). There are no opposing forces acting upon the molecules.

During the initial stages of drying, although a good vacuum may exist at point C (0.006 mm Hg), a very poor vacuum exists in the region B to A owing to the transit of water vapor down this gradient.

Trap D ($-78°$C) plays no part in the drying process and during drying remains free of ice. Oil vapor passes down the temperature gradient from pump E to trap D.

On completion of drying, the tissue is brought to room temperature by switching off the heater coil and lowering the cold bath. During the return to room temperature the ice sublimes from the condensing surface (A) and passes to trap D. Although the dried tissue is in the path of this vapor, water molecules are unlikely to enter the tissue because of the superior force exerted upon them by the temperature gradient from B down to D. Design problems make it difficult to remove the ice via an outlet at the base of the drying chamber instead of at the top. Were this done, the vacuum pump would be applied directly to the condensing surface, which might be expected to speed the drying process. Experiment shows this is not so, and clearly the pump deals mostly with the non-condensable vapors, playing little part in the actual drying process.

Choice of Drying Conditions. Up to four pieces of tissue 4 mm in diameter and 1 mm thick can be dried overnight in the same container if held at $-40°$C in a temperature gradient of $38°$C/5 cm. A suitable cold bath consists of a mixture of solid CO_2 and industrial methylated spirit ($-78°$C). A vacuum of 0.006 mm Hg for the noncondensable vapors is low enough for drying to proceed. For maximum efficiency the condensing surface should be hemispherical, concave surface towards the tissue.

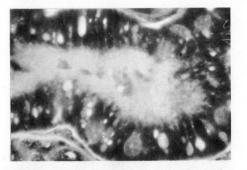

Fig. 8 Microbodies in the proximal tubule, characteristic of normal rat kidney ($\times$ 800). (The sections in Figs. 8–14 were cut at $<2\,\mu$, stained with acriflavine PAS, and viewed with a Leitz fluorescence microscope [exciter filter BG12, barrier filter 510].)

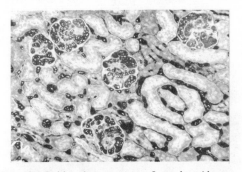

Fig. 10 Golden hamster, age 5 weeks. Absence of capsular space characteristic of freeze-dried normal kidney. Micropuncture techniques do not identify the glomerular space as a separate component (Windhager 1968) ($\times$ 110).

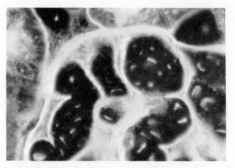

Fig. 11   Golden hamster, age 5 weeks. Capillary system of the glomerulus (×810).

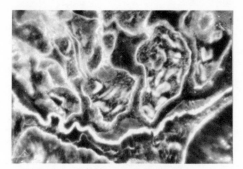

Fig. 13   Golden hamster, age 21 weeks. Further changes in glomerular structure (×810).

This system does not allow 12 such pieces to be dried, and even after prolonged drying, tissue is found still to contain water. For routine drying of large numbers of blocks the problem is effectively solved by transferring the partly dried tissues to a clean drying head for repetiton of the process (Middleton, 1967).

Where it is desired to minimize ice crystal artefact, diffusion of highly mobile molecules, or biomolecular reactions, the drying temperature must be lowered to −70°C, and in this case the cold bath is provided by liquid nitrogen. At the lower drying temperature the process is considerably prolonged.

Embedding Techniques. The necessity of maintaining a high level of cleanliness in drying heads excludes their routine use for embedding purposes. The dried specimens should be quickly transferred to a clean embedding chamber, reevacuated, and then introduced to the embedding media. In this way the conflict between outgoing air and ingoing media is avoided. The dried specimens should not be kept in a high vacuum at room temperature for longer than necessary, for if this period is prolonged the bright red capillary systems visible at the surface of the tissue turn a muddy brown and the specimen takes on a dull appearance. This might be due to the removal of tightly bound water, and although the matter has not been investigated, histologically it may be undesirable to extract this water.

Paraffin. For embedding freeze-dried specimens with paraffin, probably the best form of apparatus is shown in Fig. 5. This consists of a vacuum hot plate upon which individual containers of capacity 2–3 ml can be placed.

Prior to use, the system plus hot wax is thoroughly degassed, and with the wax returned to room temperature, the vacuum is released by bleeding dry air into the system. The dried specimens are then placed on the surface of the solid wax, the system reevacuated, and the wax melted. The specimens then sink and are embedded after a few minutes. If the procedure is a success, bubbling from the tissue ceases within two minutes.

Observations on the sections from wax-embedded freeze-dried specimens suggest that intimate contact between tissue and wax is not complete but that the degree of support is sufficient for cutting purposes (Pearse, 1968).

The difficulties associated with freeze-dried specimens after wax embedding suggest that this method of embedding will be superseded by the use of methacrylic resins.

Resins. Exhaustive comparisons between the properties of the various resins for use in freeze-drying have not been made. However, the embedding of freeze-dried specimens can be easily and satisfactorily achieved by using the following mixture: 65% glycol methacrylate + 35% butyl methacrylate with the addition of 0.5% benzoyl peroxide as accelerator. This mixture is sufficiently viscous to drop from a burette into a system under high vacuum. The dried tissues are transferred to dried gelatin capsules held in a table which can be magnetically rotated so that the capsules pass under a burette from which the methacrylate is dripped (Fig. 6). The system is evacuated and sufficient methacrylate dripped into the capsules to just cover the specimens. This takes about 5 min, after which the vacuum is slowly released. The capsules are then topped up with more methacrylate and the tissues allowed to impregnate at room temperature and pressure for the minimum time necessary to complete

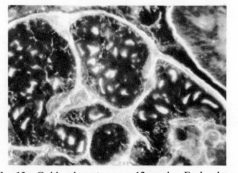

Fig. 12   Golden hamster, age 12 weeks. Early changes in glomerular structure (×810).

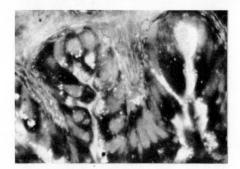

Fig. 14   Golden hamster, age 16 weeks, lung (×700).

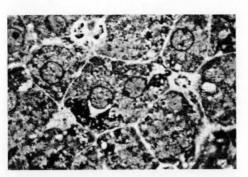

Fig. 15 Golden hamster, age 63 weeks. Liver, stained PAS, viewed phase (×810).

impregnation. This is determined by the type of tissue and the size of the block. For specimens 4 mm diameter and 1 mm thick, brain takes $\frac{1}{2}$ hr and kidney 6 hr to impregnate (Middleton, 1967; Sjostrand and Baker, 1958).

If methacrylates are used which will remain in a vessel placed in a high vacuum but which are too volatile to drop from a burette into such a system, the method of embedding described by Sjostrand and Baker (1958) should be used (Fig .7). In this method the methacrylate is frozen by contact with liquid nitrogen, the dried tissue placed above it, and the system evacuated. The cooling bath is then removed and the tissue allowed to fall into the methacrylate at the moment of melting. The vacuum is maintained until impregnation is complete.

Methacrylates tend to become more viscous under vacuum and this can hinder their passage through the finer membranes. Once the dry tissue has been evacuated and enveloped in methacrylate, there seems little point in maintaining the vacuum (Pearse, 1968).

Polymerization. Polymerization is an exothermic reaction and the heat generated causes acute swelling of freeze-dried tissue. To avoid this, all methacrylate is removed from the surface of the tissue after impregnation and the specimen is then submerged in methacrylate which has been polymerized to within $\frac{1}{2}$ hr of gelling. This methacrylate takes 2 hr at 56°C to polymerize lightly after which it completely restricts any ballooning when the methacrylate within the specimen polymerizes some 4 hr later (Middleton, 1967a,b).

Fixation. Freeze-drying cannot be regarded as a method of fixation. If it is decided to fix the tissue, then no matter what embedding medium is used, vapor fixation prior to impregnation is probably the best approach. This procedure has been fully investigated by Attramadal (1969a,b) with respect to the localization of labeled sex hormones in freeze-dried tissues.

Freeze-dried tissue embedded in methacrylate appears to be fairly stable, and for routine purposes fixation is unnecessary. If the tissue is embedded in wax, however, it is not so stable and fixation is therefore more important (Pearse, 1968).

Methods

The standard of histology to be expected using the routine given below is fairly represented by photomicrographs (Figs. 8–15). Attempts to improve upon these standards would certainly involve drying smaller specimens at −70°C, and also comparisons of results using various methacrylates.

1. Quench tissues 4 mm diameter and 1 mm thick, or smaller, using Isopentane in contact with liquid nitrogen (−160°C).
2. Freeze-dry at −40°C to a gradient of 38°C/5 cm for 48 hr, transferring the tissues to clean drying heads after the first 24 hr.
3. Remove the ice from the condensing surface, or isolate dried tissue from this surface.
4. Bleed dry air into the system extremely slowly; the pressure rises fron 0.006 to 760 mm Hg.
5. Transfer dried tissues quickly to embedding chamber.
6. Submerge evacuated tissues in glycol/butyl methacrylate mixture.
7. Allow to impregnate at room temperature and pressure for the minimum period.
8. Remove the excess methacrylate and place in greatly accelerated mixture.
9. Polymerize at 56°C.

Acknowledgments Concerning the illustrations to this article I would like to thank Mrs. Gail Holland for the histology and Mr. G. D. Leach for the photography.

E. MIDDLETON

References

Attramadal, A., "Preparation of tissues for localization of sex hormones by autoradiography (II). The loss of radioactive material in embedding media after freeze-drying and vapour fixation of the tissue," *Histochemie,* **19**:75–87 (1969a).
Attramadal, A., "Preparation of tissues for localization of sex hormones by autoradiography (III). A critical evaluation of the method using freeze-drying, osmium tetroxide vapour fixation and Epon embedding," *Histochemie,* **19**:110–124 (1969b).
Middleton, E., "Preparation of rodent kidney for light microscopy by freeze-drying and embedding in water-soluble methacrylate," *J. Roy. Microscop. Soc.,* **87**:7–23 (1967a).
Middleton, E., "An impregnating chamber and polymerization oven for preparing blocks of frozen-dried tissue embedded in methacrylate," *Lab. Pract.,* **16**(6): 728–730 (1967b).
Pearse, A. G. E., "Histochemistry, Theoretical and Applied," 3rd ed., London, Churchill 1968 (pp. 27–58).
Sjostrand, F. S., and R. F. Baker, "Fixation by freeze-drying for electron microscopy of tissue cells," *J. Ultrastruct. Res.,* **1**:239–246 (1958).
Stowell, R. E., "A modified freezing-drying approach for tissues," *Stain Technol.,* **2**:105 (1951).
Stumpf, W. E., and L. J. Roth, "Freeze-drying of small tissue samples and thin frozen sections below −60°C. A simple method of cryosorption," *J. Histochem. Cytochem.,* **15**(4):243–251 (1967).
Windhager, E., "Micropuncture Techniques and Nephron Function," in Molecular Biology and Medicine Series, London, Butterworth, 1968.
See also: PARAFFIN SECTIONS, PLASTIC EMBEDDING AND ULTRATHIN SECTION CUTTING.

FREEZE-ETCHING

Although technical improvements in the electron microscope may now be reaching their apogee, this is not so for techniques designed to prepare cells for electron microscopy. One of these techniques, freeze-etching, has infused new interest in established problems of ultrastructure research and has stimulated research in membrane structure at a time when concepts and theories were being accepted as fact.

Freeze-etching as a preparative technique was first introduced in 1957 by Dr. Russell Steere of the U.S. Dept. of Agriculture, although the concept of freezing cells prior to studying with the electron microscope had been suggested years earlier. It was the intention of Steere that the technique might be valuable in the examination of virus-infected plants; specifically to study the structure of tobacco mosaic virus as they were infecting the planter cell. Unfortunately, the technique proved less than ideal, and pursuit of freeze-etching was dropped. Several years later, however, it became apparent that not everyone had considered the technique of freeze-etching a failure. In 1961, a group of biologists at the Swiss Federal Institute of Technology published a paper describing a new apparatus which could successfully and reproducibly perform the technique of freeze-etching as outlined by Steere years earlier.

Freeze-Etching Technique

Prior to the work of Steere in 1957, a low-temperature replica method for electron microscopy had been developed by Dr. Cecil Hall in 1950. In this method, the surface ice of a frozen specimen in an evaporator was made to sublime at low temperature by the application of radiant heat. After sublimation, a replica of the surface was made by metal shadowing while still maintaining the specimen at low temperature. Using this technique, Dr. Hall was able to see individual particles of a frozen suspension of silver halide crystals.

Meryman (1950) was also studying the surface topography of frozen material, and it was his work (Meryman and Kafig, 1954, 1955) that demonstrated the value of examining replicas of new surfaces produced by fracturing frozen specimens in vacuo. This technique provided useful information on ice crystal growth at low temperatures. Since then the freeze-etching technique has undergone considerable modification, although many of the features in the freeze-cleave apparatus of Meryman and Kafig have been incorporated into contemporary devices. Today there are at least four different freeze-etching/fracturing models available. All of these models accomplish essentially the same result and in doing so involve the same 4 or 5 preparational steps. These steps include:

1. Pretreatment and freezing of the object.
2. Chipping or fracturing of the frozen specimen.
3. Etching or subliming.
4. Coating or replicating the object.
5. Cleaning the replica.

Steps 3 and 4 are always carried out *in vacuo*, while steps 1 and 4 are not. Depending on the method being used, step 2 may or may not be done *in vacuo*.

1. Pretreatment and Freezing of the Object. The manner in which specimens are pretreated prior to freezing, as well as the actual freezing, are of utmost importance in terms of survival of the specimen, physical stabilization, and inactivation of the biological specimen. Survival of a specimen through pretreatment and freezing guarantees, *a priori*, that the specimen is not damaged by freeze-fixation. This, of course, is not altogether true but it does indicate that a minimum amount of change has occurred and that the image ultimately obtained reflects the natural arrangement of structures within the cell.

Using the yeast, *Saccharomyces cereviseae*, as a test model, Moor experimented with different freezing methods. He reasoned that the method which yielded the best survival rates coupled with the best results in terms of ultrastructural preservation, would be the best "fixation" method. In all cases the final temperature to which the specimen was lowered was −150°C, which excluded all danger of ice recrystallization. This was done because Meryman (1957) had shown that migratory recrystallization of ice takes place at temperatures as low as −96°C, and it was even postulated that this process might occur continuously to some extent until the temperature is reduced to below −120°C. Furthermore, Moor (1964) has shown that very definite structural artifacts, presumably resulting from cell shrinkage and ice crystallization, occur at slow cooling rates (0.01°–1°C/sec) in yeast cells prepared by freeze-etching.

Since the cost of Helium II is almost prohibitive on any large-scale basis, liquid nitrogen and/or liquid propanes appeared to be the most attractive freezing agents. Freezing in liquid freon (propane) was even more effective when used in combination with a protecting agent such as glycerol, ethylene glycol, and dimethyl sulfoxide (Table 1). Glycerol was generally found to be the most effective, with the final concentration used being very important and varying with each specimen. For a complete evaluation of the effect of protecting agents and freezing rates on the survival of specimens for freeze-etching, the reader is encouraged to consult published literature. It should be pointed out, however, that recent data (Buckingham and Staehelin, 1969) suggest that glycerol does influence the structure of artificial membranes and thus could affect the structure of biological membranes as well.

2. Chipping or Fracturing of the Frozen Specimen. Once the specimen has been frozen, the object is placed in the freeze-etch/cleave apparatus on a stage, precooled

Table 1[a]

	% Survival			
Pretreatment	Slow Dry-Ice −0.010°C/sec	Quick Liq. N₂ −10°C/sec	Snap Liq. Propane −100°C/sec	Fast Liq. He II −10,000°C/sec
None	Variable	1	30	−100
Starvation	—	—	60	—
20% Glycerol	Variable	70	100	100
	Shrinkage from extracellular crystallization	Intracellular crystallization	Intracellular crystallization	Vitrification amorphous solidification of ice

[a]From Moor, 1964.

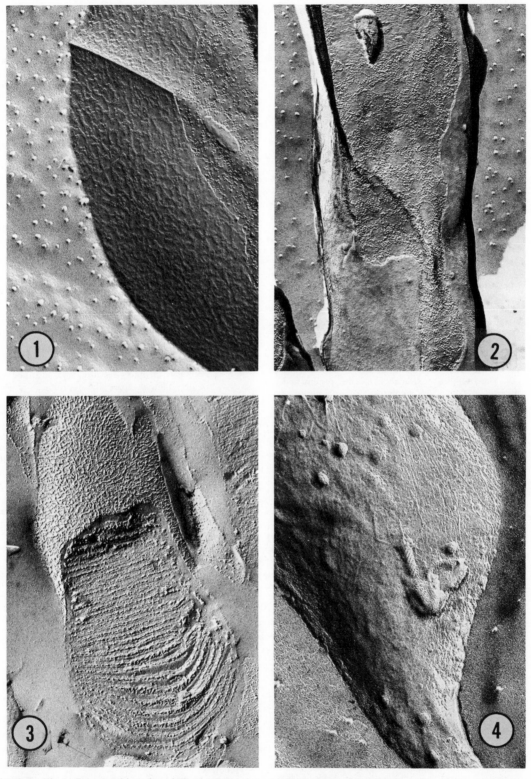

Figs. 1–4 Fig. 1: Freeze-etching of a red blood cell showing an outer surface and an inner face of the plasma membrane (×46,300). Fig. 2: Freeze-etching of the bacterium, *Escherichia coli*, showing portions of the cell wall and the plasma membrane (×48,000). Fig. 3: Freeze-etching of a retinal rod from a rat showing the fracture planes through the retinal membrane (×48,000). Fig. 4: Freeze-etching of a T-2 bacteriophage lying on the cell wall of *Escherichia coli* (×102,500).

to −150°C. A high vacuum is obtained and the specimen is cut, or more accurately, fractured. As Moor and Mühlethaler (1963) have stated, "The production of fracture planes instead of cut faces has two great advantages. (a) A great deal of the object field does not come into contact with the knife and therefore no artifacts are introduced by superficial warming or shifting of material which might cause re-crystallization. (b) The fracture plane may penetrate the structures thereby yielding cross-sections, but also it may follow limiting membranes thereby revealing surface views never seen by the usual sectioning techniques. Even without etching, a replica of the relief produced by 'cutting' may show certain details of the object."

A high vacuum (10^{-6} torr) is recommended during the fracturing phase of the technique in order to get a clean fracture plane and to prevent atmospheric contamination of the fractured specimen. Furthermore, under high vacuum and controlled object temperature, the condensation of oil vapor on the fracture face can be eliminated by using a pumping system which has an extremely low backstreaming of diffusion pump oil. By incorporating a cold trap into this system, virtually all contamination by gases and oil vapors can be eliminated.

3. **Etching or Subliming.** Fracturing alone produces a relief which, when replicated, may show certain structural details; however, many of the finer details stay hidden in the frozen cell sap and can only be demonstrated if part of the ice is removed by sublimation. This freeze-drying, or evaporation of free water from solid, is a physical and chemical process which does not deform structures of the cell. The use of a high-quality vacuum system and an efficient cold trap are prerequisites to obtaining clean surfaces prior to and after sublimation. These are necessary to eliminate any possible contamination from oil and water vapors which are inherent to the vacuum system. Temperature control is also quite important during the etching phase of freeze-etching, since the sublimited rate of ice in a perfect vacuum is a function of the temperature of the sample and the corresponding vapor pressure of ice. In the region of −100°C, where most of the sublimation is carried out, a temperature shift of 1°C can produce a difference in the rate of sublimation of at least 10% and often as much as 25% (Moor, 1966). Thus in a system operating at 1×10^{-6} torr, specimen temperatures of −100°C or lower cannot be expected to produce significant etchings in a reasonable period of time.

4. **Coating or Replicating the Object.** This consists of the evaporation of carbon and heavy metals on the object to produce a suitable replica on the freeze-etched surface. This coating can attack (melt) the specimen with radiation and condensation heating, but if performed within 5–10 sec, the object is sufficiently cooled and these effects are avoided.

Heavy metals normally used are platinum and uranium, although occasionally chromium has been employed. The type of metal used, either solely or in combination with carbon, determines to some extent the limits of resolution of the method. Under optimal conditions, most of these materials produce crystallites up to 30 Å. This is reduced somewhat when the metal is simultaneously evaporated with carbon.

5. **Cleaning the Object.** After the replica has been made, the specimen is removed from the apparatus, and the film removed by floating it off in distilled water. Cellular material that is still present is removed by sequentially replacing the water with a variety of oxidizing solutions.

The concentrations and types of oxidants depend on the specimens being freeze-etched.

Usually the specimen is first treated in a solution of 70% sulfuric acid for periods up to an hour or so. If left too long, the acid will begin to bleach the specimen by removing some of the metal used for contrast. After washing the replica in distilled water, it is treated with a hypochlorite bleaching solution. Commercial bleaches such as Clorox are not suitable as they are generally quite contaminated with other material which tend to "dirty" the replica. Eau de Javelle (Javel wasser) is an exception to the rule and has been found to be quite effective. After a final rinse in distilled water, the replica is then usually ready to be mounted on a grid and examined. If necessary further treatment can be carried out at this time, such as the use of hydrofluoric acid to remove silica diatom frustules.

Artifacts

A major reason for developing freeze-etching as a preparative technique in ultrastructure studies was to provide a true alternative to the purely chemical action of chemical fixtures, dehydration, and embedding in plastic resins. The physical preparation of freeze-etching can be arranged so that cells can remain viable throughout the entire process, and in this way, freeze-etching allows electron microscopists to examine replicas of living specimens, essentially free of artifacts.

This does not mean that the physical stabilization of cells for freeze-etching is inherently free of artifacts. Quite the opposite, artifacts are produced, but they are generally quite different from those produced during chemical fixation. Structural changes can occur and are related to damage initiated during a pretreatment phase, or by mechanical damage from the knife, or by poor temperature control resulting in large ice crystals being formed or in some cases superficial melting. However, since most of these have a physical origin, they can usually be controlled.

Use of Freeze-Etching in Ultrastructural Studies

To date some 250 plus research papers have appeared in the literature in which freeze-etching has been the major preparative technique. Interests have ranged from ultrastructural studies on microbial cells to the characterization of subunits in mitochondrial and chloroplast membranes. Some examples of the types of results that can be obtained with freeze-etching are illustrated in Figs. 1–4.

CHARLES C. REMSEN III

References

Buckingham, J. H., and L. A. Staehelin, *J. Microscopy,* **90**:83–106 (1969).
Hall, C. E., *J. Appl. Phys.,* **21**:61–62 (1950).
Meryman, H. T., *J. Appl. Phys.,* **21**:68 (1950).
Meryman, H. T., *Proc. Roy. Soc.,* **B147**:452 (1957).
Meryman, H. T., and E. Kafig, *Proc. 3rd Int. Conf. Electron Microscopy, London,* pp. 486–488 (1954).
Meryman, H. T., and E. Kafig, *Naval Med. Res. Inst. Project NM* 000 018.01.09 (1955).
Moor, H., *Z. Zellforsch.,* **62**:546–580 (1964).
Moor, H., *Int. Rev. Exp. Pathol.,* **5**:178–216 (1966).
Moor, H., and K. Mühlethaler, *J. Cell Biol.,* **17**:609–628 (1963).
Moor, H., K. Mühlethaler, H. Waldner, and A. Frey-Wyssling, *J. Biophys. Biochem. Cytol.,* **10**:1–13 (1961).
Steere, R. L., *J. Biophys. Biochem. Cytol.,* **3**:45–60 (1957).

FUNGI

The true fungi (Phycomycetes, Ascomycetes, and Basidiomycetes) comprise an assemblage distinct from that of plants and animals (see Moore, 1971). Their cell walls are chitinous, their nutrition lysotrophic, and their cellular dimensions near the limits of resolution of light optics (e.g., fungal nuclei are about 1 μ in diameter). This smallness of fungal cells has had two consequences: (1), a general low-resolution knowledge about the composition of fungal cytoplasm and (2), an extensive controversy about the fungal nucleus.

Much new information has been made available with the introduction of electron microscopy, the development of which has advanced on two fronts—instrumentation (principally electron microscopes and ultramicrotomes) and specimen preparation (fixation, embedding, and staining).

This article concerns itself with specimen preparation for both microscopies and is divided into three sections: the first covers two special techniques (one for handling small and delicate samples for fixation and embedding, the other a protocol for staining Epon sections for light microscopy) that are generally useful and that should be applicable to a variety of materials; the second section includes potassium permanganate fixation and lead staining; the third, and longest, discusses protocols for selected specific structures.

Fibrin Support Medium and Staining Epon Sections for Light Microscopy

Fibrin Support Medium. Furtado (1970) has developed a fibrin clot medium for supporting difficult specimens such as cell suspensions, conidial chains, and delicate fruit bodies. Stock solutions: Thrombin, 1000 units/ml, is stored frozen 1 ml/vial; fibrinogen solution, also stored frozen, contains 300–350 mg fibrinogen, 160 mg Na-citrate, and 850 mg NaCl per 100 ml distilled water. In use, the thrombin is diluted to 10 units/0.1 ml and mixed 1:1 with the fibrinogen at room temperature; in 10–20 sec the fibrinogen begins clotting and by a minute a good clot has formed. Specifically, to cell suspensions that have been fixed, pelleted, and decanted, 0.2–0.3 ml fibrinogen is added, stirred, and recentrifuged; next, an equal volume of thrombin is slowly pipetted into the supernatant; incorporation is effected by, with an applicator stick, first swirling the overlaying activated fibrinogen solution until a gossamer clot appears, then stirring the pellet and entrapping the cell mass on the tip of the stick; the applicator stick carries the specimen through required remaining steps (e.g., for electron microscopy the clot is treated with OsO$_4$, dehydrated, and at the 70% step rendered into pieces the proper size for embedding).

To maintain natural cell configurations, a drop of activated fibrinogen is rapidly spread over a coverslip, allowed to clot 10–20 sec, then impressed on the specimen where, after about 2 min, the clot will be completed and ready for processing for either light or electron microscopy; while in aqueous fixative the fibrin film is partially loosened from the coverslip using either arrowhead needles or scalpels; during dehydration the film can be stained with 0.5% toluidine blue in 70% alcohol to enhance the specimen and to make it easier to see during orientation and sectioning for electron microscopy.

Staining Epon Sections for Light Microscopy. To correlate light and electron microscopic observations it is sometimes desirable to observe sections several microns thick in the light microscope. These can be examined by phase optics or stained. The following is a staining sequence demonstrated by Ghidoni, et al. (1968). Sections on 1 × 3 in. slides are processed in Coplin jars: 2% periodic acid—5 min at room temperature; water rinse; 0.15% basic fuchsin in 50% ethyl alcohol—5 min at 50°C; water rinse; methylene blue/azure II (1:1::1% azure blue in distilled water: 1% methylene blue in 1% borax solution)—5 min at 50°C; rinse in water, dry, and mount.

KMnO$_4$ Fixation and Pb-Staining

In the present context fixation is the critical step. The first fixative to be widely used for preservation of fine structure was a Veronal-acetate buffered 2% OsO$_4$ mixture credited to Palade. This gave good results with animal tissues (if they were minced finely enough) but initially ambiguous results with bacteria in which for several years it was unresolved if all bacteria had a plasma membrane (the modifications of Ryter and Kellenberger in 1958 provided the still used fixative of choice for these organisms; see Nanninga, 1969). Osmium, however, penetrates very slowly and becomes self-blocking. It thus gave very poor results when applied to the walled cells of bacteria, plants, and fungi. In 1956 Luft substituted KMnO$_4$ for the OsO$_4$ in Palade's fixative. His results showed promise, but even at 0°C good results (specifically enhanced membrane contrast) were more the exception than the rule. The reason for this is that KMnO$_4$, while a strong oxidizing agent, is not a true fixative and thus much damage resulted when the embedding medium of that time (methacrylate) polymerized. Subsequent investigators, around 1960, achieved superior results using aqueous 1–2% KMnO$_4$ for shorter periods, as well as sometimes postfixing with OsO$_4$ to reduce the amount of polymerization damage (see Moore, 1963). Several years later methacrylate was supplanted by the epoxy and polyester resins which, because they set by the slow building up of cross bonds rather than rapid linear polymerization, gave good results with permanganate "fixed" material, and further improvement was achieved by eliminating the water rinse between "fixation" and the first dehydration step in 20% solvent.

In the past few years aldehyde/osmium protocols have been found to give the maximum preservation of cell ultrastructure. However, for the student of fungal fine structure there are several caveats. First, the chitinous cell wall often seems to be nearly as much a barrier to glutaraldehyde as it is to osmium, and thus months may be spent tinkering with fixation formulae, times, temperatures, etc. and still produce only mediocre results. Second, the permanganate protocol gives outstanding preservation of membranes and for much of fungal cytology this can still provide a wealth of new information. Third, even if maximum preservation of subcellular components is desired, the simplicity and surety of the permanganate protocol make it a desirable starting point for any new study. The following is a representative KMnO$_4$ schedule: 1.5% (±0.5%) aqueous KMnO$_4$ for 10–15 min; graded acetone dehydration (20%, 35%, 50%, 70%) with 15 min between steps; overnight staining in 1% U-nitrate in 70% acetone; rinse in 70% acetone and transfer to 95% acetone for 15 min; three 10 min changes in 100% acetone; Epon in acetone—25%, 50%, 75%—2 hr per concentration; 100% Epon at 60°C for 24 hr followed by 24 hr at 80°C (NOTES: a) Epon and other embedding plastics are most readily stored and dispensed in 5 or 10 ml plastic syringes (easily plugged with round toothpicks); b) Epon, unlike methacrylates, is non-volatile and need not be covered during curing; c) Epon is best cut with a diamond knife; if glass knives are used an alternative plastic such as an Epon/Araldite mixture [Mollenhauer, 1964] will be more satisfactory).

After sections of aldehyde/osmium prepared material have been picked up on grids they are usually lead stained with either Pb-citrate (Reynolds, 1963; Venable and Coggeshall, 1965) or Pb-tartrate (Peace, 1968). (NOTE: Pre-wetting the grids produces fewer precipitation artifacts. If the drop method is used, 2 rows of drops— one of distilled water and one of stain—can be placed on a piece of Parafilm; the grids are layed inverted on the water drops then transferred to the staining drop. After the desired exposure time (usually 30–60 seconds) they can be rinsed by holding the grid in a pair of forceps and squirting a steady stream of distilled water on to it briefly from a squeeze bottle).

Cell Structure

In the following discussion a structure-by-structure approach has been adopted because techniques that have been successful in one study may or may not yield satisfactory results in the hands of a different investigator working with different materials. As indicated above, potasssium permanganate seldom fails to produce satisfactory results and therefore minor variations in its protocol are not cited.

Walls, Lomasomes, and Plasma Membrane. The walls of fungi while being most noted for their chitinous component also contain a number of other substances such as sugars, glucosamine, and amino acids. Early attempts at determinations of cell wall composition used stains, but the results were seldom clear cut. Today, cell wall studies employ freeze-etching, X-ray diffraction, and chemical analysis, and the interested reader is referred to the papers by Aronson (1965), Bull (1960), Hunsley and Burnett (1970), Kirk (1966), Pfister (1970), Szaniszlo and Mitchell (1971), and Wessels *et al.* (1972).

Lomasomes are porous intumescences occurring between the cell wall and the plasma membrane. In a recent study of their structure and formation in *Saprolegnia ferax* and *Dictyuchus sterile* Heath and Greenwood (1970) employed $KMnO_4$ (2% for 1 hr), OsO_4 (1% for 1 hr), and 2.5% or 5% glutaraldehyde (2 min to 2 hr) followed by OsO_4 (1% for 1 hr) with all three fixatives buffered in M/15 Sorensen's phosphate at pH 7; their sections were stained 20 min in U-acetate followed by 7 min in Reynolds' Pb-citrate. They have also been studied using freeze-etching by Marchant and Moore (1972).

The plasma membrane of fungi reacts ambivalently. In both permanganate and glutaraldehyde prepared material it may be electrolucent or at most about half the density of the endomembranes. Our present knowledge of membrane structure offers no explanation for this phase change, although it does appear to have similarities to the observations of bacterial plasma membranes (see Nanninga, 1971; Stoeckenius and Engelman, 1969).

Septa. Septa are specialized hyphal cross walls that are particularly characteristic of the higher fungi. They are formed by the opposed deposition of wall material in centripetally formed invaginations of the plasma membrane. In both the Ascomycetes and the Basidiomycetes there is a central pore. In the Ascomycetes the septum has the form of a centrally tapered washer while in the Basidiomycetes the pore margin is flared into a dolipore

capped on either side by arced membrane elements termed parenthesomes.

Ascomycete septa have been isolated. In the method of Brenner and Carroll (1968) plugs from growing colonies were first autoclaved to kill the hyphae and to liquefy the agar. The mass of dead mycelium was then washed several times and further autoclaved for 20–30 min in 23M KOH. This was followed by several distilled water rinses and a wash in 2% acetic acid followed by a final distilled water rinse (between each step the material was centrifuged at 30 g for 15 min). Following the final rinse the samples of chemically cleaned hyphae were suspended in 2–3 ml of distilled water and sonicated for 2–3 min. Drops of 1 : 4 and 1 : 8 dilutions were dried down on coated grids and shadowed at a low angle with a platinum-palladium alloy. Although comparable techniques for isolating dolipore septa have not been reported, one would expect that basidiomycete hyphae should be able to be processed in a similar manner.

In a study of mutant strains of *Schizophyllum* Koltin and Flexer (1969) used vapor fixation. They grew their hyphae on cellophane membranes over semisolid media. Material of desired growth was placed in an atmosphere that was in equilibrium with a 25% glutaraldehyde solution at room temperature for 5–10 min and then it was transferred to an aqueous atmosphere to remove residual glutaraldehyde. Next, the mycelial samples were postfixed in vapors of a 2% aqueous solution of OsO_4. Dehydration was effected by transferring the material directly to methyl cellosolve (2-methoxyethanol) at about 0°C and making two solvent changes in 4–24 hr. The dehydrated specimens are moved, sequentially, for 4–24 hr at 0°C to, respectively, 100% ethanol, n-propanol, and n-butanol, and then embedded. The results of this rather elaborate protocol, as evidenced by the electron micrographs, do not appear appreciably better than those available from simpler schedules.

Woronin Bodies, Lipid, Lysosomes, and Glycogen. Both septal types have electropaque inclusions associated with them. Those in the Basidiomycetes are non-membrane bound, champagne cork-like plugs in the dolipore opening (Moore and Marchant, 1972) and have received no special attention but those in the Ascomycetes were early discerned as light microscopically refractile bodies and have long been known as Woronin bodies. In the electron microscope these inclusions may be confused with lipid bodies and lysosomes. Their chemical composition is unknown but they are membrane-bound, crystalline inclusions that appear to be equally well preserved in permanganate and glutaraldehyde, and they are usually found near the septal pore, frequently one on each side.

In contrast, lipid is destroyed by acetone dehydration and its previous presence is frequently noted by membrane-bound clear profiles; if care is taken to preserve lipid by osmium fixation it can be tested for by treating the grids prior to staining with Na-methoxide (Eurenius and Jarskär, 1970). This reagent is prepared by finely subdividing 2.5 g of metallic sodium and dropping it a piece at a time into 25 ml of methanol. When the sodium is completely dissolved 25 ml of benzene are added (if a phase boundary forms more methanol is added until the solution is clear). This stock solution will keep indefinitely in a dark bottle. Sections on formvar- and carbon-coated grids are dipped 1 min into diluted stock solution (3 : 1 of a 1 : 1 : : methanal : benzene mixture), washed 2 min in the methanol/benzene mixture, and finally rinsed 1 min each in two changes of acetone and one of distilled water. The sections are then stained and examined in the electron microscope. (NOTE: this extractive technique will not work if the sections have been previously exposed to the electron beam.) An unusual aspect of fungal lipid is the association on cellular droplets of monolayers of crystalline ferritin; these have been critically studied and analysed by David and Easterbrook (1971).

Lysosomes (or sphaerosomes) in fungi have been studied by Wilson et al. (1970). To demonstrate acid-phosphatase activity electron microscopically they fixed minced blocks of agar containing mycelium in phosphate-buffered (pH 6.8) 3% glutaraldehyde for 3–4 hr. The minced material was then transferred to Gomori medium with Na-glycerophosphate as substrate and incubated at 37°C for 24 hr. Following incubation the material was given a distilled water rinse, postfixed in 2% OsO_4 (phosphate buffered to pH 6.8) for 2 hr and then rinsed in buffer. The embedding sequence included ethanol dehydration followed by propylene oxide before going into Epon. These investigators also studied these minute structures light microscopically using techniques described by Holcomb et al. (1967).

Glycogen can be localized using iodine vapors (Flood, 1970). Grid mounted sections cut from glutaraldehyde/osmium prepared material are treated overnight at 55°C with vapors from a mixture of 2% iodine and 3% K-iodide in 90% ethanol and immediately thereafter Pb-stained at room temperature. CAUTION: The harmful effect of sublimated iodine on objective pole-pieces should be remembered when modifying the method. (Remarks made at the 1971 Mycological Congress, Exeter, indicate that this technique also works well with fungi).

Golgi Apparatus. The Golgi dictyosome is characteristic of the flagellated fungi and the differentiation of its membranes has been specifically studied by Grove et al. (1968). In their study, hyphae of *Pythium* were fixed for 1 hr in 4% glutaraldehyde followed by 8 hr in 1% OsO_4 (both fixatives in 0.1M K-phosphate at pH 7.0 and at room temperature). Prior to examination in the microscope sections were stained for 10 min in 1% aqueous $Ba(MnO_4)_2$.

Nucleus (Figs. 1–5). The minute nuclei of the fungi are near the limits of resolution of light optics. This physical fact plus the paradox that reductive division is satisfactorily meiotic whereas somatic division is ambiguous has engendered an extensive and diverse literature that cannot yet be summed to produce a comprehensive model. The anomaly of somatic division is: (a) The nuclear envelope does not dissociate; (b) there is no evidence, with one exception, of chromosome condensation; and (c) only one class of spindle fibers is formed, those that extend from pole to pole. Further, during both somatic division and meiosis centrioles or centriolar plaques are present. Microscopic studies may be categorized as light optic: still, time lapse, and microspectrophotometric; and electron optic: $KMnO_4$, glutaraldehyde/OsO_4, and freeze-etch protocols.

For light microscopic studies HCl/Giemsa has been used extensively. Knox-Davies and Dickson (1960) studied both types of division in the sexual and asexual stages of *Trichometasphaeria/Helminthosporium* and their technique is representative. Fix material overnight in 3 : 1 : : 100% ethyl alcohol : glacial acetic acid; transfer to 95% alcohol for 10 min; store in 70% alcohol. Hydrolyze by taking stored material down to distilled water (10 min), transfer

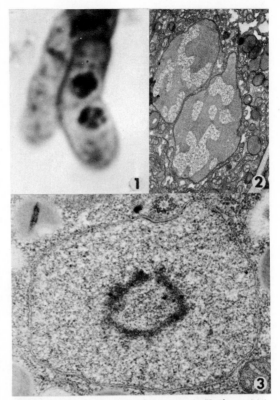

Figs. 1–3   Resting nuclei of fungi. Fig. 1: *Ustilago tritici*, light micrograph, after Lu's fixative. (Courtesy of M. M. S. Malik.) Fig. 2: *Puccinia podophylli*, $KMnO_4$ fixation; note discrete, electrolucent chromatinic areas (compare with lower nucleus in Fig. 1). (Original micrograph.) Fig. 3: *Saprolegnia terrestris*, glutaraldehyde/OsO_4 fixation; note nucleolus and centriole. (From Howard and Moore, 1970.)

to cold 1N HCl (10 min), then immerse in 1N HCl at 60°C (8–12 min), follow with several washes in distilled water, and place in buffer (prepared by adding sufficient 5M NaOH to bring 50 ml of 5M KH_2PO_4 to pH 6.9 and diluting this solution to a final volume of 200 ml with distilled water). Stain overnight in a closed vial of a solution composed of equal volumes of buffer and Giemsa stain. Rinse in a drop of buffer and then mount in another drop of buffer. Permanent mounts can be achieved by ringing the coverslip with a water-soluble mounting medium such as Aquamount, which is also excellent for making permanent preparations for phase microscopy (available from Edward Gurr, Ltd., Michrome Lab's, 42 Upper Richmond Rd. West, London, SW14).

Lu (1967) recommends what he believes to be a superior method. A fixative composed of 9 : 6 : 2–3 : : n-butyl alcohol : glacial acetic acid : 10% aqueous chromic acid is poured directly on actively growing cultures; these are then asperated in a vacuum desiccator and stored at 0–6°C. The fixed material is hydrolyzed at 70°C in either 1 : 1 : : HCl : alcohol for 2.5 min or N-HCl for 35 min and then washed in Carnoy's fluid for 2 min. Staining is effected by using a propionocarmine solution (0.5 gm carmine heated in 100 ml 60% propionic acid) and mordanting with the tip of a clean dissecting needle.

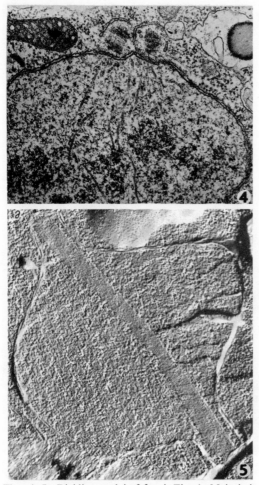

Figs. 4, 5 Dividing nuclei of fungi. Fig. 4: Meiosis in *Saprolegnia terrestris*, glutaraldehyde/OsO₄ fixation; note persistent nuclear envelope, spindle microtubules, and chromosomes. (From Howard and Moore, 1970.) Fig. 5: Somatic division in *Saccharomyces cerevisiae*, freeze-etch preparation; note persistent nuclear envelope, transpolar microtubular bundle, and apparent absence of chromosomes. (From Moor, 1967.)

Another significant paper with outstanding photomicrographs is Robinow's (1961) yeast study using HCl/Giemsa and Feulgen preparations.

Most light microscopic studies use the terms *mitosis* and *chromosomes* and attempt to relate selected configurations to the defined mitosis of higher organisms. However, since the nuclear envelope is persistent so too is the ground nucleoplasm. The consequence is that prolonged staining darkens the whole nucleus and demonstrates primarily the taffy-pull appearance that accompanies final separation, while shorter exposures to stain set out the chromatinic masses. Failure to recognize this differential effect is reflected in the many papers, on the one hand, attempting to prove classic mitosis, and those, on the other hand, presented to refute them (see Moore, 1964). Consequently, there now exists quite a gallery of interpretive diagrams, none of which look like mitosis. In

this connection, it is salutary to view Girbardt's time lapse film of nuclear division in *Polystictus versicolor* and to mark the amoeboid nature of the nucleus as it sunders (see Girbardt, 1968).

The early electron microscope studies of somatic nuclear division used permanganate (see Moore, 1964; 1965). These established the persistence of the nuclear envelope and showed the chromatinic regions to be lighter, discontinuous areas in the darker ground nucleoplasm. Permanganate, however, does not preserve such structures as microtubules and centrioles.

These structures have been elicited in later studies using glutaraldehyde/osmium and freeze-etch techniques. Robinow and Marak (1966) fixed yeast cells in 3% glutaraldehyde in 1/15M Na-K phosphate buffer (pH 6.8) for 8 hr; the cells were subsequently washed several times in the same buffer, then placed in 1 ml snail juice plus 4 ml Michaelis buffer (basic solution: 29.428 g Na-veronal, 19.428 g Na-acetate, 34 g NaCl, distilled water to 1000 ml; mix 5 ml basic solution, 7 ml 10N HCl, 0.25 ml M CaCl₂, and 13 ml distilled water; pH 6.1). Cells were allowed to be digested until only a few shreds of wall remained (1–1.5 hr at 25°C), then postfixed overnight in 1% OsO₄ dissolved in Michaelis buffer. Before dehydration and embedding they were rinsed in Michaelis buffer and soaked in U-nitrate. (See also McCully and Robinow, 1972).

The spindle-like fibers anchored to centriolar plaques set in the nuclear envelope revealed by Robinow were further demonstrated in an adjoining paper by Moor using freeze-etching (see Matile et al. 1969). In this technique (Moor and Muhlethaler, 1963), material that has been suspended in 20% glycerol for 4 hr or more is quick-frozen to −190°C, then fractured using a special microtome in a vacuum evaporator; the cut surface is allowed to freeze-dry to a depth of a few hundred angstroms (the ice sublimation produces the etching effect) and replicated with evaporated carbon/platinum. (The material may also be frozen directly in Freon 22 (Bauer, 1970).) Cleavage planes generally pass along membrane interfaces and the resultant image of cell structure is an intaglio/bas-relief combination of cytoplasmic elements passed over or scooped out. The actual surfaces appear variously roughened or pebbled, but Davy and Branton (1970) have shown that such patterns can be obtained from pure ice, thus caution must be observed in interpreting comparable asperities in biological preparations. (A freeze-etch bibliography, complete to 1970, is available from Balzers, P. O. Box 10816, Santa Ana, Calif. 92711).

Meiosis, in contrast to somatic nuclear division, is characterized by the condensation of chromosomes and their behaving in an orthodox fashion. The light microscopic techniques described above (Knox-Davies and Dickson, 1960; Lu, 1967) are equally applicable here (see also Olive, 1965).

Fine structure studies demand glutaraldehyde/osmium in order to preserve not only microtubular and centriolar structure but also to preserve the signal identifying character of meiosis, the synaptinemal complex. Lu (1967) studied meiosis in the mushroom *Coprinus*. At 4°C, basidia were fixed in 3.4% glutaraldehyde in 0.067M phosphate buffer (+0.7% NaCl in some samples to improve tonicity) at pH 6.5–7 for 2.5–3 hr washed in three buffer changes overnight, postfixed in 1% OsO₄ in buffer for 2 hr; then at room temperature material was washed in three buffer changes and ethanol dehydrated.

Westergaard and von Wettstein (1970) employed a similar protocol for meiosis in the ascomycete *Neottiella* except that they used 6.5 % glutaraldehyde. They and Lu both used Spurr's (1969) low viscosity epoxy medium (available from Polysciences, Inc., Paul Valley Industrial Park, Warrington, Penn. 18976).

Meiosis in the phycomycetes has been more elusive, particularly at the light microscopic level. Somatic nuclei in the fungi are, with a few notable exceptions, haploid, and meiosis is zygotic. However, this interpretation for certain of the Oömycetes, while plausible, came to be questioned by some mycologists. Recently, Bryant and Howard (1969) provided microspectrophotometric evidence that meiosis in *Saprolegnia terrestris* was gametic, and this was confirmed by Howard and Moore's (1970) electron microscope study. One novel aspect of meiosis in this fungus is that at no time does the nuclear envelope dissociate. Howard and Moore fixed entire colonies in 1 : 1 : : 6 % acrolein : 6 % glutaraldehyde buffered with 0.2 *M* cacodylate to pH 7.2 for 1.5 hr in a vacuum desiccator; rinse four times, 15 min each in 0.1 *M* cacodylate and twice in deionized distilled water; stained 4 hr in 0.5 % aqueous U-acetate at room temperature; dehydrated in ethanol and embedded in Epon.

Nuclei can also be isolated intact from cells for concentrated microscopic and biochemical studies (see Bhargava and Halvorson, 1971).

Spores. Spore structures add several special problems for the fungal cytologist. The flagella of motile cells can be light optically observed by inverting a spore suspension

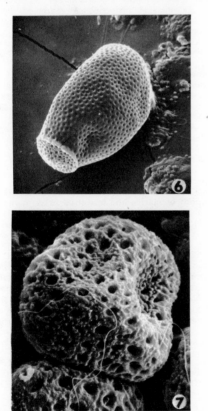

Figs. 6 and 7   Fig. 6: Scanning electron micrograph of resistant sporangium of *Allomyces macrogynus*. (Original micrograph.) Fig. 7: Scanning electron micrograph of spore of *Cintractia junci*. (From Grand and Moore, 1970.)

drop over fumes of 2 % OsO_4 for 1 min; next, near this drop place a drop of 1 % crystal violet and by using a dissecting needle or straightened paper clip add stain to the spore drop until the solution is dark; after mounting and blotting up excess water, ringing with wax will prevent spore movement due to evaporation-caused water currents. Variations on this technique are provided by Koch (1966). For electron microscopy the spore drop can be dried down on a coated grid and shadowed in a vacuum evaporator (developmental stages and internal morphology can be studied using the glutaraldehyde/osmium protocols set forth above).

The surface markings of the variously thickened walls of spores of higher fungi are frequently of taxonomic importance. Though generally discernible light optically, critical visualization of ultratopography has only become practicable with the introduction of scanning electron microscopy (SEM). Spores of any age (including herbarium material from the last century, Fig. 7) can be picked up on double-sticky cellophane tape and lightly mounted on 1 × 3 in. glass slides and stored in dust-proof boxes. Material to be examined is cut out, placed on the SEM stud, and coated in a vacuum evaporator with a monomolecular layer of gold/palladium. Thinner-walled spores frequently show areas of collapse, but this common SEM artifact can be recognized after examining a number of samples (see Grand and Moore, 1970). Other resistant structures such as the meiosporangia of *Allomyces* can be examined in a similar manner (Fig. 6). (The freeze-etch and embedding techniques may also be used; see Hess et al., 1968; Besson, 1971.)

There is now a considerable literature of fungal fine structure encompassing not only many other studies of the structures presented here, but also of other structures, ontogenies, and taxa that are outside the purview of this discussion. Each new material has its own problems and challenges; however, the examples selected should provide the interested student with a starting point—a selection of techniques documented by outstanding results that will help him monitor his own progress with different specimens.

ROYALL T. MOORE

References

Aronson, J. M., "The cell wall," Chap. 3 "The Fungi," Vol. I, Ainsworth, G. C. and A. S. Sussman, eds., New York, Academic Press, 1965.

Bauer, H., "A freeze etch study of membranes in the yeast *Wickerhamia fluorescens*," *Can. J. Microbiol.*, **16**:219–22 (1970).

Besson, M., "Ultrastructure de la paroi sporique des *Laccaria* Berk. et Br. (Agaricales)," *Acad. Sci. Paris Comp. Rend.*, **272**:1078–1081 (1971).

Bhargava, M. M., and H. O. Halvorson, "Isolation of nuclei from yeast," *J. Cell. Biol.*, **49**:423–429 (1971).

Brenner, D. M., and G. C. Carroll, "Fine-structural correlates of growth in hyphae of *Ascodesmis sphaerospora*," *J. Bacteriol.*, **95**:658–671 (1968).

Bryant, T. R., and K. L. Howard, "Meiosis in the Oömycetes. I. A microspectrophotometric analysis of nuclear deoxyribonucleic acid in *Saprolegnia terrestris*," *Amer. J. Bot.*, **56**:1075–1083 (1969).

Bull, A. T., "Chemical composition of wild type and mutant *Aspergillus nidulans* cell walls. The nature of polysaccharide and melanin constituents," *J. Gen. Microbiol.*, **63**:75–94 (1970).

David, C. N., and K. Easterbrook, "Ferritin in the fungus *Phycomyces*." *J. Cell Biol.*, **48**:15–28 (1971).

Davy, J. G., and D. Branton, "Subliming ice surfaces: Freeze-etch electron microscopy," *Science,* **168**:1216–1218 (1970).

Eurenius, L., and R. Jarskär, "A simple method to demonstrate lipids in Epon-embedded ultrathin sections," *Stain Technol.,* **45**:129–132 (1970).

Flood, P. R., "Preliminary experiments with iodine in electron opaque stains for ultrathin sections," Microscopie Électronique 1970. (proc. VI International Cong., Grenoble). Vol. I, pp. 431–432 (1970).

Furtado, J. S., "The fibrin clot: A medium for supporting loose cells and delicate structures during processing for microscopy," *Stain Technol.,* **45**:19–23 (1970).

Ghidoni, J. J., M. M. Campbell, J. G. Adams, H. Thomas, and E. E. Evans, "A new multicolored staining procedure for one micron sections of epoxy embedments," Electron Microscope Soc. Amer. Traveling Show, 1968.

Girbardt, M., "Ultrastructure and dynamics of the moving nucleus," *Soc. Exp. Biol. Symp.,* **22** (Aspects of Cell Motility):249–259 (1968).

Grand, L. F., and R. T. Moore, "Ultracytotaxonomy of Basidiomycetes. I. Scanning electron microscopy of spores," *J. Elisha Mitchell Sci. Soc.,* **86**:106–117 (1970).

Grove, S. N., C. E. Bracker, and D. J. Morré, "Cytomembrane differentiation in the endoplasmic reticulum–Golgi apparatus–vesicle complex," *Science,* **161**:171–173 (1968).

Heath, I. B., and A. D. Greenwood, "The structure and formation of lomasomes," *J. Gen. Microbiol.,* **62**: 129–137 (1970).

Hess, W. M., M. M. A. Sasssen, and C. C. Remsen, "Surface characteristics of *Penicillium* conidia," *Mycologia,* **60**:290–303 (1968).

Holcomb, G. E., A. C. Hildebrandt, and R. F. Evert, "Staining and acid phosphatase reactions of sphaerosomes in plant tissue culture cells," *Amer. J. Bot.,* **54**: 1204–1209 (1967).

Howard, K. L., and R. T. Moore, "Ultrastructure of oögenesis in *Saprolegnia terrestris,*" *Bot. Gaz.,* **131**: 311–336 (1970).

Hunsley, D., and J. H. Burnett, "The ultrastructural architecture of the walls of some hyphal fungi," *J. Gen. Microbiol.* **62**:203–218 (1970).

Kirk, P. W., Jr., "Morphogenesis and microscopic cytochemistry of marine pyrenomycete ascospores," *Nova Hedwigia Beiheft,* **22**:1–128 (1966).

Knox-Davies, P., and J. G. Dickson, "Cytology of *Helminthosporium turcicum* and its ascigerous stage, *Trichometasphaeria turcica,*" *Amer. J. Bot.,* **47**:328–339 (1960).

Koch, W. J., "Fungi in the Laboratory," Chapel Hill, N.C., The Book Exchange, 1966 (113 pp.).

Koltin, Y., and A. S. Flexer, "Alteration of nuclear distribution in *B*-mutant strains of *Schizophyllum commune,*" *J. Cell Science,* **4**:739–749 (1969).

Lu, B. C., "Meiosis in *Coprinus lagopus:* A comparative study with light and electron microscopy," *J. Cell Sci.,* **2**:529–536 (1967).

Marchant, R., and R. T. Moore, "Lomasomes and plasmalemmasomes in fungi," *Protoplasma* **75**: in press (1972).

Matile, P. P., H. H. Moor, and C. F. Robinow, "Yeast cytology," Chap. 6 "The Yeasts," Vol. 1, in A. H. Rose and J. S. Harrison, ed., New York, Academic Press, 1969.

McCully, E. K., and C. F. Robinow, "Mitosis in heterobasidiomycetous yeasts. II. *Rhodosporidium* sp. (*Rhodotorula glutinis*) and *Aessosporon salmonicolor* (*Sporobolomyces salmonicolor*)," *J. Cell Sci.* **11**:1–31 (1972).

Mollenhauer, H. H., "Plastic mixtures for electron microscopy," *Stain Technol.,* **39**:111–114 (1964).

Moor, H., "Der Feinbau der Mikrotubuli in Hefe nach Gefrierätzung," *Protoplasma,* **64**:89–103 (1967).

Moor, H., and K. Mühlethaler, "Fine structure in frozen-etched yeast cells," *J. Cell Biol.,* **17**:609–628 (1963).

Moore, R. T., "Fine structure of mycota. 1. Electron microscopy of the discomycete *Ascodesmis,*" *Nova Hedwigia,* **5**:263–278 (1963).

Moore, R. T., "Fine structure of mycota. 12. Karyochorisis—somatic nuclear division—in *Cordyceps militaris,*" *Z. Zellforsch.,* **63**:921–937 (1964).

Moore, R. T., "The ultrastructure of fungal cells," Chap. 5 in "The Fungi," Vol. I, Ainsworth G. C. and A. S. Sussman, eds., New York, Academic Press, 1965.

Moore, R. T., "An alternative concept of the fungi based on their ultrastructure," in "Rescent Advances in Microbiology," A. Pérez-Miravek and D. Peláez, eds., Mexico City, Libreria Internacional, S.A., 1971.

Moore, R. T., and R. Marchant, "Ultrastructural characterization of the basidiomycete septum of *Polyporus biennis,*" *Can. J. Bot.* **50**: in press (1972).

Nanninga, N., "Preservation of *Bacillus subtilis* by chemical fixation as verified by freeze-etching," *J. Cell Biol.,* **42**:733–744 (1969).

Nanninga, N., "Uniqueness and location of the fracture plane in the plasma membrane of *Bacillus subtilis,*" *J. Cell Biol.,* **49**:564–570 (1971).

Olive, L. S., "Nuclear behavior during meiosis," Chap. 7 in "The Fungi," Vol. 1, G. C. Ainsworth and A. S. Sussman, eds., New York, Academic Press, 1965.

Pease, D. C., "Histological Techniques for Electron Microscopy," 2nd ed. 4th print, New York, Academic Press, 1968 (381 pp.).

Pfister, D. H., "A histochemical study of the composition of spore ornamentations in operculate discomycetes," *Mycologia,* **62**:234–237 (1970).

Reynolds, E. S., "The use of lead citrate at high pH as an electron-opaque stain in electron microscopy," *J. Cell Biol.,* **17**:208–212 (1963).

Robinow, C. F., "Mitosis in the yeast *Lipomyces lipofer,*" *J. Biophys. Biochem. Cytol.,* **9**:879–892 (1961).

Robinow, C. F., and J. Marak, "A fiber apparatus in the nucleus of the yeast cell," *J. Cell Biol.,* **29**:129–151 (1966).

Spurr, A. R., "A low-viscosity epoxy resin embedding medium for electron microscopy," *J. Ultrastruc. Res.,* **26**:31–43 (1969).

Stoeckenius, W., and D. M. Engelman, "Current models for the structure of biological membranes," *J. Cell Biol.,* **42**:613–646 (1969).

Szaniszlo, P. J., and R. Mitchell, "Hyphal wall compositions of marine and terrestrial fungi of the genus *Leptosphaeria,*" *J. Bacteriol.,* **106**:640–645 (1971).

Venable, J. H., and R. Coggeshall, "A simplified lead citrate stain for use in electron microscopy," *J. Cell Biol.,* **25**:407–408 (1965).

Wessels, J. G. H., D. R. Kreger, R. Marchant, B. A. Regensburg, and O. M. H. DeVries, "Chemical and morphological characterization of the hyphal wall surface of the basidiomycete *Schizophyllum cummune,*" *Biochim. Biophys. Acta,* **273**:346–358 (1972).

Westergaard, W., and D. von Wettstein, "Studies on the mechanism of crossing over. IV. The molecular organization of the synaptinemal complex in *Neottiella* (Cooke) Saccardo (Ascomycetes)," *Carlsberg Lab. Compt. Rend. Trav.,* **37**:239–268 (1970).

Wilson, C. L., D. L. Stiers, and G. C. Smith, "Fungal lysosomes or sphaerosomes," *Phytopathology,* **60**:216–227 (1970).

g

GASTROTRICHA

A dissecting microscope equipped with $15\times$ wide field oculars is the best method of finding gastrotrichs in concentrations of moss squeezings or water which was siphoned from the psammon of freshwater lakes or from the intertidal interstitial area of oceans. Transfer of individual animals to a drop of water is best accomplished by the use of a micropipette. Fumes of a 2% solution of osmic acid are a fair fixative, although a percentage of animals die in a contracted condition. A 0.2 M solution of magnesium chloride is an excellent relaxing medium, especially for marine gastrotrichs. The animals can then be fixed in a 4% solution of formalin.

Marine gastrotrichs can be stained in either quadruple stain or Mallory triple. The cuticular covering of freshwater animals does not stain easily, and therefore, the animals are best studied with a phase microscope. The study of the cuticular spines of *Chaetonotus* and others is facilitated by the use of oblique lighting and by the use of dark-field illumination.

Permanent slides of marine animals can be done in the conventional manner by using either balsam or clarite. A special technique should be used with freshwater gastrotrichs. Animals are moved from the preservative solution to a 5% solution of glycerin. This is allowed to evaporate until the individuals are infiltrated with full-strength glycerin. Animals are then transferred to a drop of pure glycerine on a clean slide which has "posts" of wax (wax pencil dots) to support a number one coverslip. Capillary action will draw in the mounting medium to surround the drop of glycerin. The best mounting medium is Murrayite. This can be softened by warming so that the animal can be rotated by a slight push of the coverslip. As many as 25 animals from a colony or clone can be put on a single slide in this manner. The primary disadvantage of using glycerin is that it has a high refractive index and it will draw stain out of the specimens.

Royal Bruce Brunson

GLYCOGEN

Glycogen is the storage form of carbohydrates in animals and man, serving the same purpose as starch in plants. Many cells and various tissues contain glycogen up to a very high content. However, liver and muscle are the most prominent storage centers and of special importance in the intermediary metabolism.

Glycogen is composed of chains of α-1-4' linked D-glucose (α-pyranose) units. The molecules are highly ramified (Fig. 1); the branches are bound by α-1-6-glucosidic linkages.

The statements about the linear chain length of the glucose units vary considerably between 3–5 (72) and 12–18 (32, 86). The same applies to the molecular weight which is given in the range of 1×10^6 (72)–2.9×10^6 (6) for muscle glycogen and 4.4×10^6 (6, 52)–1×10^8 (72) for liver glycogen. For molecular weight determinations, the method used for extracting glycogen has a great influence. It is believed that the popular extraction with 5% trichloroacetic acid (TCA) causes degradation by acid hydrolysis even in the cold and delivers, therefore, lower values than the classical KOH extraction after Pflüger. The "correct" preparation of glycogen is still open to discussion (78).

Külz (39) and later Nerking (65) considered a strong binding of glycogen to proteins which leads to the concept of TCA extractable "lyo-" and strongly to protein bound "desmoglycogen" of Willstätter and Rohdewald (87). In spite of given evidence that desmoglycogen is a mechanical inclusion of glycogen by coagulated proteins (11, 61, 71, 73) the conceptions of "free" and "protein bound" glycogen are still found in textbooks (32) and used. Biologically it was assumed that desmoglycogen is a structural cell component with low metabolic turnover (3, 9, 34), while others found a higher incorporation rate of C^{14}-labeled glucose in this fraction than in the TCA-soluble glycogen (79).

For histological demonstration it has been stated that only the free, labile, and TCA-soluble lyoglycogen can be shown histochemically, while the bound, residual, TCA-insoluble desmoglycogen is undetectable by these methods (36, 37, 38). Although Lillie (46) does not agree with this assumption, the view of Kugler and Wilkinson (36, 37, 38) has been accepted by some histochemical textbooks (4, 68, 82). Leske and Mayersbach (44) confirmed the view of earlier investigators (11, 61, 71, 73) and were able to show conclusively that the desmoglycogen in biochemical assays is an artifact. The amount of trapped desmoglycogen depends mainly on the particle size of the tissue homogenates and the varying coagulability of proteins (Fig. 2). The latter is influenced by the varying biological or experimentally induced states of the tissue proteins (30, 31, 56, 59, 69). In this process of different release of glycogen from bigger tissue particles is certainly involved the variability of the molecular size respective to the glycogen assimilation and dissimilation. Proof is given that high molecular weight glycogen is preferably mechanically included by coagulated proteins (71). However if ultrafine liver homogenates are

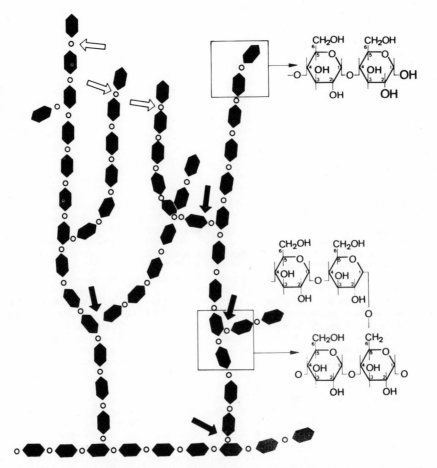

Fig. 1 Scheme of the glycogen configuration (modified after Rapoport [72]). The lower square shows the chemical formula of the linear (1–4) glucosidic bindings of glucose units and the 1–6 linkage of the ramifications. In the upper square are marked the free 1–2 glycol groups, the reaction sites for the histochemical glycogen reaction (PAS). The arrows indicate the points of action, white: phosphorylase; black: amylo-1-6-glucosidase.

submitted to TCA extraction, no discrimination of lyo- and desmoglycogen can therefore be made. There are then practically no quantitative differences between the glycogen yields after TCA and KOH extractions. Furthermore, the authors (44) were also able to show that TCA (2.5– 10 %) neither extracts glycogen from fresh-frozen sections nor impedes the cytochemical demonstration. Even in small liver blocks (1 × 1 mm) glycogen is very well fixed by TCA. In bigger-sized ones there appears a rim of good glycogen fixation on the surface, while glycogen is poorly preserved in the inner parts of these blocks due to the slow penetration of TCA. The interaction of glycogen and glycogensynthetase as shown by Leloir (40) is for quantitative reasons certainly not responsible for the rough differences in TCA extraction and histological fixation.

For microscopical demonstration, as well as for quantitative chemical determinations, two main factors have to be taken into account:

1. biological fluctuations, especially circadian rhythms; and
2. technical factors, which are considerably influenced by the former.

Biological Fluctuations

The values given for the glycogen content in organs are varying in a broad range (73) (Table 1). The reasons for this are rarely considered. Glycogen was among the first substances for which a "Circadian" (24 hr) rhythmicity (25) of quantitative fluctuations was recognized (1, 19, 29). The regularity of time settings of maximal and minimal glycogen contents was only for a short time a matter of discussion (28, 70) and later practically forgotten. The

Table 1 **Glycogen Content of Liver and Muscle in different Animals (after Rauen (73)[a]**

Animal		Muscle	
	Liver	Heart	Skeletal
Dog	800–6700	330–520	670
Rabbit	770–4500	390	280–2200
Guinea pig	2000–9500	240	850–5300
Rat	200–8300	520 (400–650)	570–1230

[a]Values in mg % wet weight.

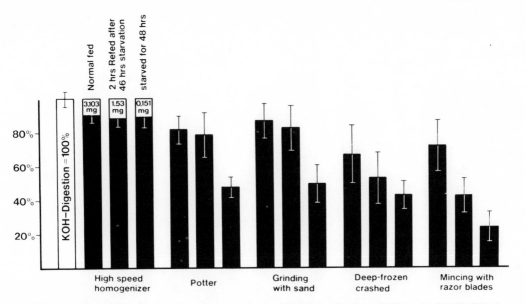

Fig. 2 Influence of particule size for glycogen extraction by TCA (after Leske and Mayersbach [44]). Ultrahomogenization followed by TCA extraction delivers practically the same glycogen amounts as KOH digestion, independently of the biological influences of starvation, while methods yielding coarser particles show irregular differences of extractability (mg/100 mg liver wet weight).

differences in the glycogen contents are ascribed only to the intake of carbohydrate-rich food, a view still held valid by biochemists (72), in spite of the fact that Ekman and Holmgren (17) have already shown that the glycogen phases can be governed by changing the light-dark cycle. Haus and Halberg (26) demonstrated that a glycogen rhythmicity is still present in livers of starved animals until their death by starvation.

Quantitative chemical analyses and morphological demonstrations have given evidence that rats kept under highly standardized environmental conditions (temperature $+21° \pm 1°C$, humidity 40%, free access to standardized food and water ad libit) exhibit a strong 24 hr rhythm with a maximal glycogen content of 4–8% and 0.1–1% of the liver wet weight at the time of minimum.

The pattern, timing of maximum and minimum, mean 24 hr values as well as the magnitude of the amplitudes of the glycogen fluctuations are influenced by sex and seasons (44, 56, 60) (Fig. 3). Within a short period of 1–2 weeks there is a good reproducibility of values when analyzing the animals on consecutive days at the same time of day. However, external factors, such as disturbance of the animals, influences the glycogen content considerably because the glycogen metabolism is governed by the adrenals (42) (Fig. 4).

It must be taken into account that animals of the same strain (e.g. Wistar) but derived from different breeding places may differ strongly in respect of their mean glycogen values, timing of maxima and minima, and magnitude of oscillations (58) (Fig. 5).

In liver, histochemically drastic circadian changes of glycogen distribution can be seen (15). These changes follow a very characteristic pattern within the hepatic lobule (63, 64). In low-power microscopy, the maximum shows a uniform distribution of glycogen over the whole section (Fig. 6a), but cytologically striking differences of the amount and form of distribution in the inner and outer part of the liver lobules are evident (Figs. 7, 8a).

With decreasing glycogen content the lobule structure becomes prominent by diminishing predominantly in the periphery of the lobules (Fig. 6b, c). At the minimum time (Fig. 6d) a small zone of glycogen remains around the central vein, if at all. The ultramorphological aspect of the periphery of the lobule has not only changed in respect to the glycogen content but also in the aspect of endoplasmatic reticulum and its topographic relationship to the mitochondria (Fig. 8a, b).

In addition to these findings Müller (63, 64) found that enzymes related to glycogen metabolism behave similarly in their topographic distribution and activity. Therefore one has to assume that the distribution patterns of glucose-6-phosphatase (85), phosphorylase (81), and glycogen synthetase (77, 81) given in the literature are not static but correspond to a certain circadian state. Circadian changes of glycogen-related enzymes influence nutrition studies more significantly than short-time starvation of several hours (18).

Technical Factors

Chemical Fixation. For histological demonstration one of the most important factors is the fixation or stabilization of glycogen in tissues. Methods for the quantitative and qualitative preservation are still controversial. According to many authors and technical textbooks (4, 14, 18, 21, 22, 23, 24 33, 35, 38, 49, 76, 83) the type and the temperature of fixative used is very important, while others (2, 45, 80, 82, 84) report that common formalin serves as well as any special fixative and the preservation of glycogen is so good that there will be no decrease, even after rinsing the tissue blocks in running tap water. These contrary opinions were elucidated by measuring the glycogen amount in differently fixed tissue blocks (43). Table 2 shows that the effect of fixations for quantitative glycogen preservation is mainly related to the biological states of the times in which the animals had been sacrificed. In one month (March) all fixatives acted equally well,

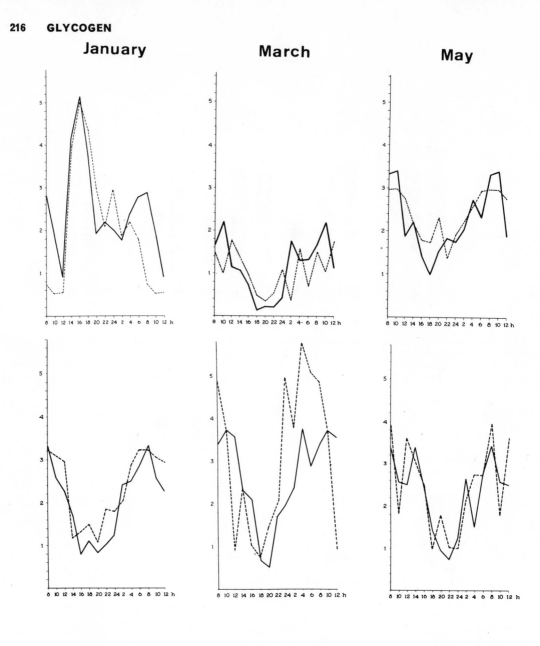

Fig. 3 Seasonal and sexual modifications of the circadian patterns (after Mayersbach [56]). Glycogen in mg/100 mg liver wet weight. All rats were kept under standardized environmental conditions. (——— females; – – – – – males.)

while the same investigation performed in July delivered strong variations in the quantitative glycogen preservation. There is no systematic action. The effect of each fixative is dependent on the biological circadian state of the livers, some delivering better results in the glycogen assimilation, some in the dissimilation phase. Furthermore, it cannot generally be concluded that fixation in the cold will render better quantitative results. These findings favor the view that glycogen fixation is merely a trapping by mechanical inclusion by fixative-coagulated cell proteins. Fixation by glycogen denaturation as dis-

cussed elsewhere (68) seems to be unlikely, at least with common fixatives (100% ethanol, formol, picric acid etc.). Isolated glycogen still dissolves in water after treatments with such reagents, and even ethanol is used for the chemical preparation of glycogen.

The cellular-topographic distribution is strongly influenced by chemical fixatives. If bigger-sized blocks (3 × 3 mm) are used, the typical glycogen polarization occurs. The latter, well known as "*alkoholflucht*" (also caused by ethanol-free fixatives to some extent), is a displacement of glycogen moving in the same direction as

Table 2 Glycogen Determination in Tissue Blocks after Different Fixation[a] (after Leske 43)

Time of Sacrificing	Fixing Temperature	Glycogen Content				
		Original Samples	After Fixation in			
			Carnoy	Formol (10%)	Gendre	Schaffer
10 A.M.[b]	+21°C	100% (5.69 mg)	98.4%	75.2%	93.4%	76.4%
10 A.M.[c]	+21°C	100% (4.64 mg)	74.7%	55.3%	61.6%	51.9%
10 A.M.[c]	− 7°C	100%	63.7%	67.2%	71.7%	77.1%
4 P.M.[c]	+21°C	100% (2.73 mg)	43.9%	30.0%	4.7%	72.1%
4 P.M.[c]	− 7°C	100%	73.6%	56.0%	35.8%	60.8%

[a]The values are percentages of the original samples.
[b]Investigation in March.
[c]Investigation in July.

the penetrating fixative. After block fixation of liver at room temperature or at −7°C there can be seen a "chessboard" pattern caused through the appearance of cells rich in glycogen alternating with cells devoid of it (Fig. 9). This artifact is caused by intracellular glycogen-degrading enzymes, which act postmortally until they are inhibited through the slow-penetrating fixatives (44). Both artifacts can be avoided by (a) using very small tissue blocks in which the fixatives penetrate quickly, (b) fresh-frozen sections (44) and (c) fixation under conditions of freeze-substitution (49, 50).

Stabilization by Freeze-Drying. As for other cell substances, freeze-drying serves as the best tool for revealing the native distribution of glycogen in tissues. The classical artifacts of streaming (glycogen polarization, *alkoholflucht*) as well as the "chess board pattern" are no longer

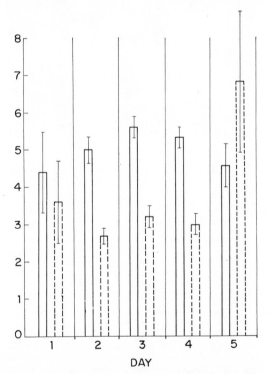

Fig. 4 Glycogen determination on consecutive days (after Leske [42]). Animals were sacrificed daily (Monday–Friday) at 10.00 A.M. (solid bars, elevated glycogen content) and 4 P.M. (stipulated bars, lower glycogen content). Glycogen in mg/100 mg wet weight. Notice the irritation of animals on the first and fifth day of analysis, which was caused by providing the animals with food (Monday) and cleaning the animal quarters (Friday).

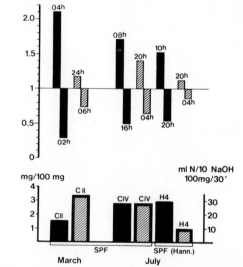

Fig. 5 Time settings of maxima and minima in Wistar rats of different breeds (after Mayersbach [58]). The bars of the upper row indicate the timing and relative amplitudes (variations from the 24 hr means) of maximal and minimal glycogen contents (black bars) and esterase activity (striped bars). The lower row give the 24 hr mean glycogen values (mg/100 mg liver wet weight) and esterase activity (NaOH consumption/100 mg liver wet weight/30 min, necessary to neutralize butyric acid freed from ethylbutyrate hydrolysis by esterase activity). Both animal groups derive from the best national breeding stations. SPF = specific pathogen free animals from "TNO-Netherlands," SPF (Hann) = specific pathogen free animals from "Tierzuchtanstalt der Deutschen Forschungsgemeinschaft" (Hannover), C II = investigations in March, C IV, H 4 = investigations in July.

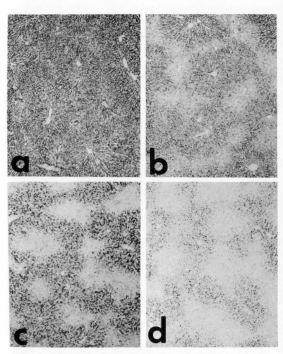

Fig. 6 Morphological appearance of glycogen rhythmicity in rat liver (low-power magnification, Carnoy fixation, PAS). a: Maximal glycogen content; d: minimal glycogen content; b, c: intermediate phases. (Microphotographs by courtesy Dr. O. Müller, Dept. of Anatomy, Medizinische Hochschule Hannover.)

observed after freeze-drying. Furthermore it is stated that glycogen is distributed throughout the cytoplasm only in the form of diffuse, fine granules instead of rough accumulations caused by coagulation through chemical fixatives (52). Recent observations have shown that the even distribution of glycogen in liver cells after freeze-drying is a matter of the circadian glycogen variations. In the hours of maximal contents there occur natural glycogen accumulations (27), which is in agreement with ultramorphological findings (63). As always, freeze-dried material deserves special precautions to avoid substantial losses from tissue sections because the solubility of tissue materials is enhanced by the "lyophilization" process of freeze-drying (10, 54, 55). Embedding in polymerizing resins seems to be superior to other methods proposed for avoiding glycogen loss of freeze-dried materials (27).

Tissue Processing. No precautions are necessary for further processing if the tissues are embedded either in celloidin or polymerizing resins, because the glycogen is protected against extraction through watery solutions. The glycogen extraction from sections of paraffin-embedded materials is amply discussed (48, 49, 76, 80). In our laboratory we were able to show that the glycogen to be visualized by histochemical methods is tremendously influenced by the use of distilled water or different grades of ethanol for section mounting (43). However, there are again the biologically influenced differences in fixation effects causing different resistance to water extraction after chemical fixation; for example, it was noticed that the sections of standardized (Autotechnicon) Carnoy-

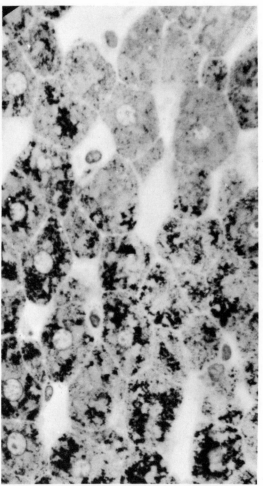

Fig. 7 Glycogen distribution in the liver lobule at time of glycogen maximum (1 μ section, Carnoy fixation, PAS). Left-hand side: periphery with rough glycogen accumulations; Right-hand side: central part, only few and fine granules visible. (Preparation and micrography Dr. O. Müller, Hannover.)

fixed and processed liver blocks of our winter animals exhibited severe losses of glycogen through mounting with distilled water, contrary to those of spring and summer which were very resistant to prolonged water extraction up to 12 hr (41, 42).

The same is valid for the frequently stated necessity of coating paraffin sections with celloidin prior staining.

Threshold Values for Glycogen Demonstration. The minimal amounts necessary in tissues to make glycogen visible by histological means were evaluated by chemical determination of glycogen in the original tissues. The values of critical level differ widely between 0.01 % and 0.6 % of liver wet weight (13, 15, 16, 36, 37, 66). Some of the given values should be taken with certain precautions since the analyses were carried out after various degrees of autolysis or derived from starved and refed animals in order to obtain materials of different glycogen contents. As already pointed out, in different biologic states such as (a) during circadian cycle, (b) seasonal and sexual variations,

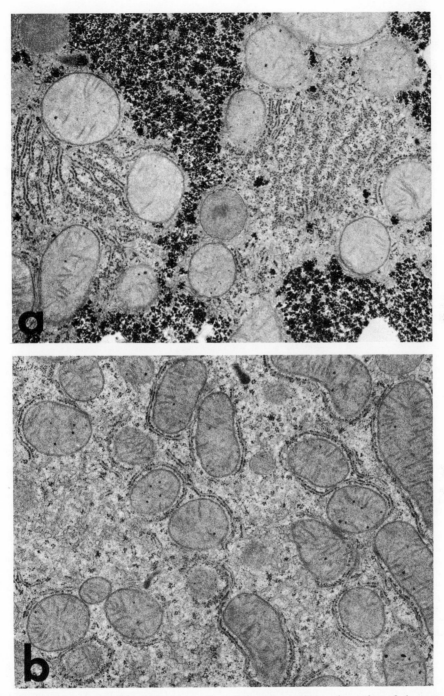

Fig. 8 Ultramorphology of circadian rhythm, a: at time of glycogen maximum; b: at time of glycogen minimum. Both pictures are taken from the peripherial part of a liver lobule (glutaraldehyde-OSO_4). (Preparation Dr. O. Müller, Hannover.)

and (c) experimental treatments, differences occur in the protein precipitation by chemical fixatives and consecutively in the grades of glycogen trapping. Furthermore, naturally and experimentally changed glycolytic enzyme activities must be taken into account as their activity is not immediately stopped by the fixatives, just as fixation does not render glycogen irreversibly insoluble. Histochemical results after chemical block fixation do not necessarily reflect the original amount and distribution of glycogen. A critical level cannot be given,

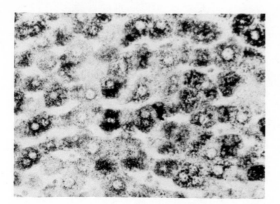

Fig. 9 Glycogen in liver. Chemical fixation in Gendre fixative at −7°C A.M. Note the chess-board pattern.

therefore. In some instances a concentration of 1.1 mg/100 mg liver wet weight may give poor results while occasionally 0.1 mg/100 mg can easily be detected by histochemical means (44). Fig. 10.

Staining methods. The staining methods for glycogen are amply described and discussed in every textbook of histochemistry and histological technologies. No real glycogen-specific histochemical reaction exists. But the respective staining results are in relatively good accordance. In our opinion staining is the smallest problem in glycogen histochemistry, so the methods are outlined only briefly.

Iodine. Staining of glycogen by iodine is one of the oldest histochemical methods (7) and is still in use for staining fresh-frozen sections and freeze-dried material (49). For other purposes iodine is not advisable because of some difficulties with normally processed tissues and in receiving permanent preparations.

Best's Carmine. The carmine solution introduced by Best (8) also stains mucoid substances in salivary glands, fibrin, and mastocyte granules. Several attempts were made to clarify the affinity of this stain for glycogen but no conclusive result has emerged. Nevertheless this empirical method is regarded as more sensitive than any other histological staining for glycogen (37).

Oxydation Methods. All these methods are based on the oxidation of the nonsubstituted, free glycol groups of the glycogen molecule (Fig. 1) producing aldehydes which react with aldehyde reagents, giving a strongly colored reaction product.

These methods were originally created by Bauer (5) using 4% chromic acid for oxidation and Schiff's reagent for the demonstration of the developed polyaldehydes. The main disadvantage of Bauer's method is not the fact that other substances free of glycogen (mucines, galactogen, amyloid, and cellulose) are colored, too, but rather more through the fact that oxidation time, temperature, and chromic acid concentration are very critical. Chromic acid easily causes overoxidations of the glycol groups forming the respective acids which do not react with the Schiff-reagent.

Periodic acid does not carry this danger, because HJO_4 oxidizes with a certain specificity only 1,-2,-glycol groups (Fig. 1) to their respective aldehydes. In combination with the Schiff reagent, the "PAS method" after McManus (51) should be regarded as the most reliable method.

Silver complexes are used as aldehyde reagents (62) following oxidation with $KMnO_4$. This method was criticized (20) for reasons of strong staining of reticuline fibers (which is also the case after PAS) and the variability of results caused by the additional reduction by formalin following the silver bath. It is difficult to understand why preference would be given to this method merely for the ease of microphotography of the deep black silver particles.

Proofs of Specificity. Neither the staining methods nor histochemical reactions are specific for glycogen. Therefore proofs of specificity are inevitable. Specificity can easily be achieved by employing enzymatic extractions. Saliva digestion of glycogen in sections was systematically investigated by Patzelt (67) and is still the most simple and readily available method for routine work. Difficulties may occur by the actions of other enzymes present in saliva (e.g., ribonuclease) or traces of mucins causing unwanted stainings. Therefore malt diastase (47) or the debranching enzyme diazyme (35) is advisable for critical research work. For the correct judging of the enzyme action a blank should be incubated with the respective solvent of the enzymes, a rule valid for every enzyme extraction method. Diastase extraction of glycogen is achieved in paraffin, celloidin, polyacrylmetacrylate and embedded tissues (27), but is impossible in sections of Epon. Proofs of specificity are therefore also possible on the level of electronmicroscopy, to distinguish minute glycogen particles from ribosomes.

H. V. MAYERSBACH

References

1. Argren, W. G., O. Wilander, and O. Jorpes: *Biochem. J.* **25**:777 (1931).
2. Baker, "Cytological Technique," 2nd ed., London, Methuen, 1945.
3. Balzer, H., and D. Palm, *Arch. Exp. Pathol. Pharmacol.*, **243**:65 (1962).
4. Barka, T., and P. I. Anderson, "Histochemistry, Theory, Practice and Bibliography," New York, Hoeber, Medical Division, Harper and Row, 1963.
5. Bauer, H., *Z. mikr. anat. Forsch.*, **33**:143 (1933).
6. Bell, D. J., H. Gutfreund, R. Cecil, and H. G. Ogsten, *Biochem. J.*, **42**:405 (1948).
7. Bernard, M., "Leçon sur le Diabéte," Paris, 1877.
8. Best, F., *Z. Wiss. Mikr.*, **23**:319 (1906).
9. Bloom, W. L., G. T. Lewis, M. Z. Schrumpert, and T. Shen, *J. Biol. Chem.*, **188**:631 (1951).
10. Bruchhausen, D., R. Höfermann, H. Krieg, and H. Mayersbach, *Histochemie*, **20**:215 (1969).
11. Carrol, N. V., R. W. Longley, and I. H. Roe, *J. Biol. Chem.*, **220**:583 (1956).
12. Casellmann, B. W. G., "Histochemical Technique," London, Methuen, 1959.
13. Corring, B., and K. Aterman, *Amer. J. Anat.*, **122**:57 (1968).
14. Culling, C. F. A. "Handbook of Histological Techniques," London, Butterworth, 1968.
15. Deane, H. W., F. B. Nesbett, and A. B. Hastings, *Proc. Soc. Exp. Biol. Med.*, **63**:401 (1946).
16. Eger, W., and H. Ottensheimer, *Virchow Arch. Pathol. Anat.*, **322**:175 (1952).
17. Ekmann, C. A., and H. Holmgren, *Anat. Rec.*, **104**:189 (1949).
18. Fagundes, L. A., and R. B. Cohen, *J. Histochem. Cytochem.*, **13**:553 (1965).
19. Forsgren, E., *Z. Zellforsch.*, **6**:647 (1928).
20. Gomori, G., *Amer. Clin. Pathol.*, **16**:347 (1946).
21. Graumann, W., *Acta Histochem.*, **4**:29 (1957).
22. Graumann, W., *Histochemie*, **1**:97 (1958).

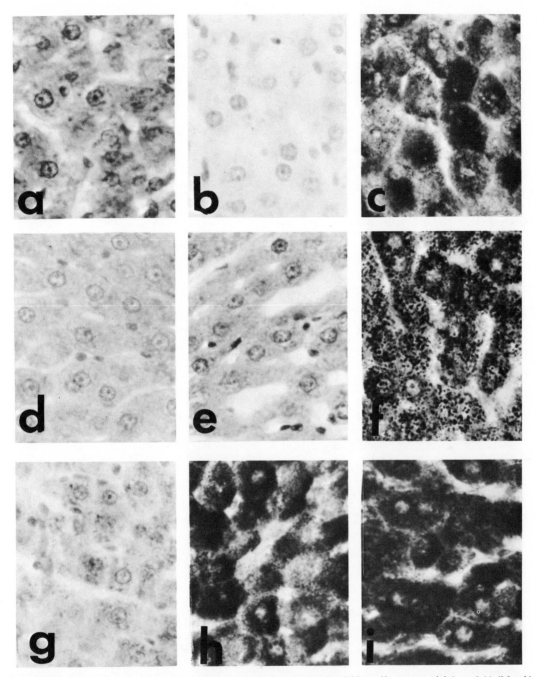

Fig. 10 Glycogen values and histochemical glycogen demonstration (mg/100 mg liver wet weight). a: 0.11 (March); b: 0.5 (January); c: 0.8 (July); d: 0.7 (January); e: 0.8 (March); f: 0.7 (July); g: 1.1 (January); h: 1.2 (July); i: 1.9 (July).

23. Graumann, W., *Acta Histochem.,* **9**:247 (1960).
24. Graumann, W., and W. Claus, *Histochemie,* **1**:241 (1959).
25. Halberg, F., M. B. Vissche, E. G. Flink, K. Berge, and F. Bock, *Z. Vit. Horm. Ferm. Forsch.,* **10**:225, (1959).
26. Haus, E., and F. Halberg, *Experientia,* **22**:113 (1965).
27. Heine, R., H. Mayersbach, W. Arnold and D. Mitrenga, *Acta Histochem.* (in press) (1973).
28. Higgins, G. M., I. Berkson, and E. Flock, *Amer. J. Phys.,* **102**:673 (1932).
29. Holmgren, H., Thesis, Helsingfors, 1936.

30. Horvath, G., *8th Int. Conf. Methodol. Rhythmic Res.,* Hamburg (1963).
31. Horvath, G., *Nature,* **189**:185 (1963).
32. Karlson, P., "Biochemie." Vol. 3, Stuttgart, Thieme, 1962.
33. Kiszely, G., and Z. Posalaky, "Mikrotechnische und Histochemische Untersuchungsmethoden," Budapest, Akademiai Kiado, 1964.
34. Kits van Hijningen, A. J. M., and A. Kemp, *Biochem. J.,* **59**:487 (1955).
35. Kugler, J. H., Master's Thesis, Univ. of Sheffield, 1965.
36. Kugler, J. H., and W. J. C. Wilkinson, *J. Histochem. Cytochem.,* **7**:398 (1959).
37. Kugler, J. H., and W. J. C. Wilkinson, *J. Histochem. Cytochem.,* **8**:195 (1960).
38. Kugler, J. H., and W. J. C. Wilkinson, *J. Histochem. Cytochem.,* **9**:498 (1961).
39. Külz, R., *Z. Biol.,* **22**:161 (1886).
40. Leloir, C. F., and C. E. Cardinie, *J. Amer. Chem. Soc.,* **79**:6340 (1957).
41. Leske, R., "Proceedings of the Second International Congress on Histochemistry and Cytochemistry," Berlin, Springer, 1965 (p. 139).
42. Leske, R., *Int. Symp. Berlin,* **1968**:625 (1968).
43. Leske, R., p. 133 in H. Mayersbach, "The Cellular Aspects of biorhythms," Berlin, Springer, 1967.
44. Leske, R., and H. Mayersbach, *J. Histochem. Cytochem.,* **17**:527 (1969).
45. Lillie, R. D., "Histopathologic Technic and Practical Histochemistry," 2nd ed., New York, Blakiston, 1954.
46. Lillie, R. D., *J. Histochem., Cytochem.,* **10**:763 (1962).
47. Lillie, R. D., and I. Greco, *Stain Technol.,* **22**:67 (1947).
48. Lipp, W., "Histochemische Methoden," Munich, R. Oldenburg, 1954.
49. Lison, L., "Histochimie et Cytochemie Animals," Paris, Gauthier-Vollars, 1960.
50. Lison, L., and R. Vokaer, *Ann. Endocrinol.,* **10**:66 (1949).
51. McManus, J. F. H., *Nature,* **158**:202 (1946).
52. Mancini, R. E., *Anat. Rec.,* **107**:149 (1948).
53. Manners, D. J., and A. Wirgth, *J. Chem. Soc.,* **1961**:1597 (1961).
54. Mayersbach, H., *Acta Anat.,* **30**:487 (1957).
55. Mayersbach, H., *Acta Histochem.,* **8**:524 (1959).
56. Mayersbach, H., "The Cellular Aspects of biorhythms," Berlin, Springer, 1967 (p. 87).
57. Mayersbach, H., ed., "The Cellular Aspects of Biorhythms," Berlin, Springer, 1967.
58. Mayersbach, H., "Proceedings of the Third International Congress on Histochemistry and Cytochemistry," New York, Springer, 1968 (pp. 206, 284).
59. Mayersbach, H., and P. Jap, *Histochemie,* **5**:297 (1965).
60. Mayersbach, H., and R. Leske, *Acta morph., Neerl. Scand.,* **6**:343 (1966).
61. Meyer, K. H., and R. W. Jeanloz, *Helv. Chim. Acta,* **26**:1784 (1943).
62. Mitchell, A. J., and G. B. Wislocki, *Anat. Rec.,* **90**:261 (1944).
63. Müller, O., "Proceedings of the Third International Congress on Histochemistry and Cytochemistry," Berlin, Springer, 1968.
64. Müller, O., *Acta Histochem.,* Suppl. **X**:141 (1971).
65. Nerking, J., *Pflügers Arch.,* **81**:8 (1900).
66. Nielsen, N. A., H. Okkels, and C. C. Stockholm-Borreson, *Acta Pathol. Scand.,* **9**:258 (1932).
67. Patzelt, V., *Wien. Klin. Wochschr.,* **16** (1928).
68. Pearse, A. G. E., "Histochemistry," Vol. 1, 3rd ed. London, Churchill (1968).
69. Philippens, K., "Proceedings of the Third International Congress on Histochemistry and Cyto-

chemistry," New York, Springer, 1968 (p. 206).
70. Pitts, G. C., *Amer. J. Physiol.,* **139**:109 (1943).
71. Prins, P. A., and R. W. Jeanloz, *Ann. Rev. Biochem.,* **17**:67 (1948).
72. Rapoport, S. M., "Medizinische Biochemie," Vol. 5, Berlin, 1969.
73. Rauen, H. M., "Biochemisches Taschenbuch," Vol. 2, Berlin, Springer, 1964.
74. Robinson, J. N., *J. Biol. Chem.,* **236**:1244 (1961).
75. Roe, J. H., J. M. Bailey, R. R. Gray, and J. N. Robinson, *J. Biol. Chem.,* **236**:1244 (1961).
76. Romeis, B., "Mikroskopische Technik," Munich, R. Oldenbourg, 1968.
77. Sasse, D., *Histochemie,* **7**:39 (1966).
78. Stetten, D. Ir., and M. R. Stetten, *Phys. Rev.* **40**:505 (1960).
79. Stetten, M. R., H. M. Katzen, and D. Ir. Stetten, *J. Biol. Chem.,* **232**:475 (1958).
80. Swigart, R. H., C. E. Wagner, and W. B. Atkinson, *J. Histochem. Cytochem.,* **8**:74 (1960).
81. Takeuchi T., and G. G. Glenner, *J. Histochem. Cytochem.,* **8**:227 (1960).
82. Thomson, S. W., "Selected Histochemical and Histopathological Methods," Springfield, Ill., C. C. Thomas, 1966.
83. Trott, J. R., *J. Histochem. Cytochem.,* **9** (1961).
84. Valence-Owen, I., *J. Pathol.-Bacteriol.,* **60**:325 (1948).
85. Wachstein, M., and E. Meisel, *J. Histochem. Cytochem.,* **4**:592 (1956).
86. Whistler, R. C., and C. L. Smart, "Polysaccharide Chemistry," New York, Academic Press, 1953.
87. Willstätter, R., and M. Rohdewald, *Z. Physiol. Chem.,* **225**:103 (1934).

GOLGI APPARATUS

The Golgi apparatus variously referred to as the Golgi complex or dictyosomes was first described in 1898 by the distinguished neurologist, Camillo Golgi, in special silver nitrate preparations of the owl's brain. Golgi considered this body to be a new cellular component and because of its highly reticulated structure termed it the "internal reticular apparatus." It was soon found by numerous investigators, many of whom were the pupils of Golgi and Cajal, that the internal reticular apparatus was present in many different types of cells. However, its reticular form was not so extensively developed in certain types of cells as in the nerve cell and the term was no longer considered appropriate. Partly because of this and partly to honor its discoverer, it has generally been termed the Golgi apparatus. Since the Golgi apparatus could not be convincingly demonstrated in the living cell, and since it could only be demonstrated following metallic impregnated methods, much doubt was cast upon its very existence, notwithstanding the fact that it could be moved about in the cell by ultracentrifugation (Beams). During the long period of controversy concerning whether or not the Golgi apparatus was an artifact, little progress was made concerning its function, although several investigators emphasized its probable role in the formation of secretion. However, new and experimental methods for cytological study such as ultracentrifugation, cyto- and fractionation chemistry, phase contrast and electron microscopy, and autoradiography have all helped to confirm the view that the Golgi apparatus is a real and important cellular organelle. Attention is now focused on its origin, structure, and function, and its relationship to other membrane constituted organelles of the cell.

The Golgi material has been demonstrated in most cells, with the exception of mammalian erythrocytes, bacteria, and blue-green algae. It is often polarized within the cell, occupying a position between the nucleus and the apical end; this is especially characteristic of gland cells. In insect and certain other invertebrate and plant cells the Golgi material is generally unpolarized and distributed as discrete bodies (dictyosomes) throughout the cell (Fig. 3). The fine structure of the Golgi apparatus, whether it be in a reticulated form or as discrete dictyosomes, is much the same, consisting of a series of membrane-bound saccules and associated vesicles. Each membrane is of the unit type and measures about 60–75 Å in diameter. The width of the saccule is highly variable; it may measure from 70 to 500 Å or more, and usually those saccules on the ends of the dictyosome stacks vary in width more than those elsewhere. It is becoming evident that a polarity exists within the dictyosome, and the surface where

material enters the stacks, often adjacent to the smooth-surfaced endoplasmic reticulum, is referred to as the proximal, convex, or forming face, while the surface where the material is budded off from the saccule is usually referred to as the distal, convex, or maturing face. Usually there are about 4–12 membrane-bound saccules within a dictyosome although there may be more, depending in part upon the type and functional state of the cell (Figs. 4 and 5). Dictyosomes often show the saccules to consist of a central flattened portion with a branching network of tubular projections at the surface (Fig. 4). Adjacent to the saccules are present a variable number of vesicles of different size (Fig. 4). This close association of the saccules and vesicles has often been referred to as the Golgi complex. The ultrastructure and function of the Golgi apparatus in plants and animals are similar; they both possess stacks of flattened cisternae or saccules, associated vesicles, and a variable number and complexity

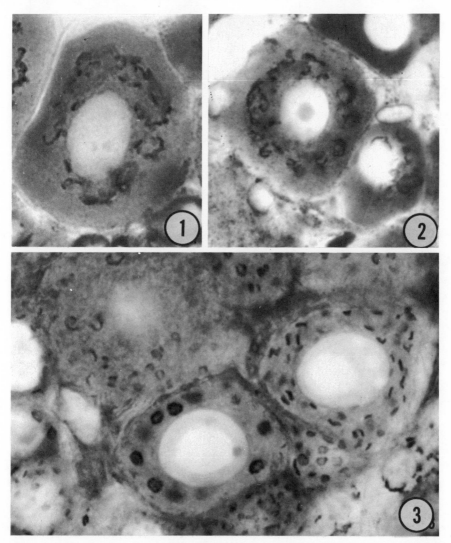

Figs. 1–3   Figs. 1 and 2: Reticulated form of the Golgi apparatus similar to that observed by Golgi in nerve cells. Spinal ganglion cells of the rat. (From Beams and Kessel, 1968.) Fig. 3: Dictyosomes or discrete Golgi bodies in grasshopper nerve cells. (From Beams and Kessel, 1968.)

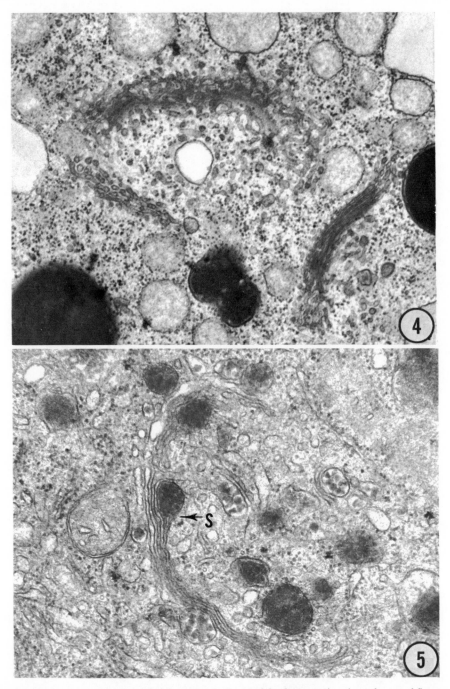

Figs. 4 and 5   Fig. 4: Fine structure of Golgi apparatus. Lower right shows section through central flattened saccules. Upper middle of the picture shows section through anastomosing tubular projections from the central flattened saccules. Observe also the vesicles surrounding the saccules. (Beams and Sekhon, unpublished, from Beams and Kessel, 1968.) Fig. 5: Electron micrograph of trout liver showing secretion within the saccules (S). Nearby are other secretion bodies which have been budded off from the Golgi saccules. (Beams and Sekhon, unpublished, from Beams and Kessel, 1968.)

of tubules which proliferate and branch from the cisternae.

It was noted long ago by Gatenby, Nassonov, Bowen, and others that a close relationship seems to exist between the Golgi apparatus and the forming products of secretion, especially the acrosome of the animal sperm. This view has been extensively confirmed and a close relationship between the endoplasmic reticulum and Golgi apparatus established. Lack of space will not permit mentioning all of the deserving investigators who have worked on this subject; only two well established studies will be used to illustrate the function of the Golgi apparatus: (1) synthesis and packaging of secretion in the pancreas, extensively studied by Palade, Siekewitz, Caro, and Jamieson of the Rockefeller University: and (2) glycoprotein secretion in the goblet cell, investigated by Neutra, Leblond, and associates of McGill University.

Guinea pigs injected with 3H leucene and the pancreas prepared for study by electron microscopical autoradiography revealed the following: after 3–5 min pulse labeling the autoradiographic grains appeared over the rough-surfaced endoplasmic reticulum. After 20–40 min most of the Golgi complex, especially the large condensing vacuoles, were intensively labeled, and after 1 hr most of the autoradiographic grains were over the zymogen granules. Finally the labeled zymogen granules were extruded into the lumen by a kind of reverse micropinocytosis, and concomitantly their bounding membranes fused with the plasma membrane. These results, coupled with certain biochemical studies, seem to prove conclusively that the secretory proteins are synthesized by the ribosomes and transferred through the cisternae of the endoplasmic reticulum to the condensing vacuoles of the Golgi complex, wherein the secretion accumulates and transforms into membrane-bound zymogen granules. In other types of proteinaceous secreting cells, the precursor secretion is blebbed off from the adjacent smooth cisternae of the endoplasmic reticulum and the small vesicles thus formed fuse with the Golgi saccules, where the precursor secretory material is condensed, packaged, and released as membrane-bound granules (Fig. 5, S). However, it should be noted that not all synthesized protein substance, such as certain types of yolk, involves the Golgi apparatus.

Autoradiographic studies of the goblet cells in the intestine of the young rat have clarified the mechanism of synthesis and secretion of "complex carbohydrate" substances. For example, glucose labeled with tritium was injected into young rats and the fate of the sugar was followed in the goblet cells by the method of autoradiography as follows: 15 min after injection it appeared in the Golgi saccules. Here the simple sugars taken up in the Golgi saccules were combined with the incoming protein which was formed by the rough-surfaced endoplasmic reticulum and transported to the Golgi saccules through its cisternae. At 20 min after the injection the globules of mucus containing radioactivity began to appear within the Golgi saccules and at 40 min most of the labeled glycoprotein had been removed from the Golgi saccules. Apparently the saccules on the maturing face simply bud off, carrying with them the secretion; in fact, there may be a complete turnover of Golgi saccules within a period of 40 min. In other cells, carbohydrate-containing secretion may be formed in a similar manner.

Several enzymes have been localized within the Golgi saccules; these are as follows: thiamine pyrophosphatase, alkaline phosphatase, and acid phosphatase. The close cytochemical relationship between the membranes of the Golgi apparatus (G), endoplasmic reticulum (ER) and lyosomes (L) in certain cells has led Novikoff and associates to refer to this association as the GERL system.

The important question confronting investigators dealing with cellular ultrastructure is the extent of interrelation existing among the various components and organelles. It is clear that each organelle performs a more or less specific function which is made possible by the compartmentalization within the cell. However, there is good evidence to show that a "membrane flow" occurs between various organelles, particularly the membranes of the nuclear envelope, endoplasmic reticulum, Golgi apparatus, and plasma membrane. Furthermore, the turnover of the Golgi apparatus membranes must at times be rapid, as evidenced by the condition in the goblet cell. There is at present no convincing evidence that membrane synthesis is limited to any one organelle. It is to be recalled that Chambers in 1924 demonstrated that large areas of *Arbacia* egg membrane could be removed artificially and that a new plasma membrane over the area of injury was immediately formed. To what extent there is a recycling of membranes, i.e., a breaking down into their constituents and reassembly into new membranes, is not clear; this problem constitutes an active field of research.

H. W. BEAMS

References

Beams, H. W., and R. G. Kessel, "The Golgi apparatus: structure and function," in G. H. Bourne and J. F. Danielli, eds., "International Review of Cytology," New York, Academic Press, 1968.

Bowen, R. H., "Cytology of glandular secretion," *Quart. Rev. Biol.,* 4:484 (1929).

Favard, P., "The Golgi apparatus," in Lima-de Faria, ed., "Handbook of Molecular Cytology," Amsterdam, Amsterdam Publishing Co., 1969.

Mollenhauer, H. H., and D. J. Morre, "Golgi apparatus and plant secretion," *Ann. Rev. Plant Physiol.,* 17:27 (1966).

Neutra, M., and C. P. Leblond, "The Golgi apparatus," *Sci. Amer.,* 220:100 (1969).

Novikoff, A. B., and E. Holtzman, "Cells and Organelles," New York, Holt, Rinehart and Winston, Inc., 1970.

Palay, S. E., "Frontiers in Cytology," New Haven, Yale University Press, 1958.

Whaley, G. W., "The Golgi apparatus," in E. E. Bittar and N. Bittar, eds., "The Biological Basis of Medicine," New York, Academic Press, 1968.

GOLGI METHOD, INVERTEBRATE APPLICATIONS

For several years after Golgi's discovery, in the 1870s, anatomists remained ignorant of the technique of impregnating whole single neurons. But, in 1887, the Spanish anatomist, Cajal, chanced to see some of Golgi's original preparations which had been given to Simarro, Cajal's tutor, in Pavia. Cajal immediately siezed upon this technique for the study of the nervous system. He substantially modified the method and, during the next months, used it for an intensive study of the cerebellum. This work finally culminated in Cajal's public renunciation, in 1888, of the widely held opinion that the nervous system was composed of a syncytial arrangement of nerve fibers, one confluent into another. This so-called "reticulum theory," maintained by Golgi and upheld by the majority of anatomists (with the notable exception of His and Forel) was disproven by careful observation of neurons stained by a method that had, ironically, been invented by the theory's most staunch advocate.

Cajal's investigations on the cerebellum, and the independent confirmation of his observations on the retinae of birds, heralded the Golgi technique's meteoric rise in popularity. Apart from a lapse in the 1920s and 1930s, when fear of the technique's enigmatic mechanisms resulted in its loss of favor (although not in the eyes of Herrick and Polyak), this method was recognized as the most rewarding histological procedure for determining the structural organization of the C.N.S.

Surprisingly, the Golgi technique was being applied to insect brain tissue relatively soon after Cajal popularized its use of vertebrate material.

Kenyon's short note on the optic lobes of the bee, and his detailed account of the structure of the bee's brain (1896), demonstrated that this method could be applied equally well on insects and on vertebrates. Cajal, too, used the method on insects, although he did not make his findings public until Vigier (1908) published a short note on single neurons in the optic lobes of the fly. Fired by this article, the Spanish author wrote three papers, the last comprising a mammoth treatise, with Sanchez, on the optic lobes of Diptera, Hymenoptera, and several other orders of insects. Perhaps it is this 1915 account which best illustrates the incredible diversity of neuronal form and the extraordinary complexity of the central nervous system. At least, Cajal's observations inspired him to write the following comment: "Compared with the retina of these apparently humble representatives of life (Hymenoptera, Lepidoptera and Neuroptera), the retina of the birds or the higher mammals appears as something coarse, rudimentary and deplorably elementary. The comparison of a rude wall clock with exquisite and diminutive hunting case watch fails to give an adequate idea of the contrast. . . ."

The advent of the mythelene blue technique for staining single nerve cells in insects and annelids (popularized by Zawarzin [1913] and adopted for many years by the Russian school of invertebrate neuroanatomy) did not detract from the usefulness of the Golgi technique which has since been used on a variety of invertebrates apart from insects (Strausfeld and Blest, 1970; Strausfeld, 1971; Pearson, 1971) and Cephalopods (Young, 1962), the best range of its applicability being, without doubt, portrayed in the works of Hanström (1928).

This short account outlines some of the applications of the Golgi method on insects, with special reference to the visual system, in order to illustrate that it is still probably the most powerful instrument for analyzing the functional anatomy of the C.N.S. Without it neurophysiologists, and students of quantitative behavioral analysis, would be hard put to propose more than a black-box pattern of connectivity for any reaction, or integration, that their techniques might detect.

The insect visual system is a particularly attractive object for anatomical-physiological study; both its retina and underlying neuropil are unequivocally compartmentalized into discrete aggregates of neuronal elements, where a set of receptors is situated in each of several thousand ommatidia, each having its own "private" lenslet. It follows that this system can be exploited for sophisticated studies on visual discrimina-

tion where the stimulus is precisely applied to a particular geometric array of prime receptors (Reichardt, 1970).

Each of the three synaptic regions of the optic lobes consists of a matrix of nerve fibers whose strata of tangentially oriented processes lie parallel to the planar surfaces. Perpendicular processes (the axons, from which arise the tangential fibers) are set at right angles to the surfaces and define neuropil columns (Fig. 1A). It is known that the receptors are mapped onto the most peripheral set of columns in such a way that each column receives an input from a single point in space. These columns are carried through into the next region by interneurons whose projection patterns define an arrangement of subsequent columns that is geometrically the same as that in the first region (Braitenberg, 1970).

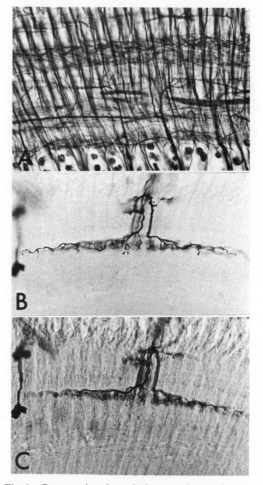

Fig. 1 Cross section through the second synaptic region of the optic lobes of the fly, *Musca domestica*. A: Holmes-Blest (*Quart. J. Micr. Sci.*, **102**:413–417 [1961]) stain showing the stratified and columnar matrix of argyrophilic fibers (axons, axon terminals, and dendrites) (×400) B: A single tangential element "randomly selected" by the Golgi-Colonnier impregnation (bright field illumination). C: The same element as in B, viewed with interference contrast illumination, displayed against the array of neuropil columns and strata (×400.)

This much of the matrix can be derived from reduced silver techniques (Bodian or Holmes-Blest methods) that stain features of whole populations of neurons. Unlike the Golgi method, which impregnates all the processes of *a* single cell, *somewhere* in the brain, reduced silver stains argyrophilic components of *all* the cells of specific populations of neurons. Both stains are selective, the first "randomly so."

Golgi impregnation reveals that the insect visual system is composed of a vast number of cell forms, some of which may eventually be discounted as being involved in visual processing, but none of which can, at present, be ignored in the anatomical analysis. These cell types can be described as a relatively tedious (even though aesthetically appealing) "Linnean" list of neuronal forms. But a far more difficult, and interesting, exercise is to determine the relative populations of cell types, as well as their positions in the visual matrix, their receptive fields (with respect to the efferent pathways arriving at the same neuropil level), and their lateral, and in-depth relationships (Campos-Ortega and Strausfeld, 1972b).

In a few instances it is possible to recognize cell types, identified by the Golgi technique, in preparations stained by reduced silver (Fig. 2A, B). However, these examples of profile correlations are not universal. Mapping the strata relationships of Golgi-stained cells will allow one to exclude certain cell types as interacting together, and needless to say, selective staining of single nerve cells provides invaluable data about the pathways between the optic lobes and the rest of the brain and ventral nerve cord (see Addendum).

The Golgi stain is most eloquent when it can be applied to the most sophisticated type of C.N.S. analysis; namely, that of determining functional contiguity between sets of neurons. To achieve this end a cell is "randomly" labeled by Golgi impregnation (Fig. 2B) and then examined under the electron microscope against the fiber profiles of unimpregnated elements (Fig. 2C). Its position in a neuropil column can be accurately determined, as can its dendritic field through aggregates of columns. In some instances, a proportion of its physiological relationships can be seen: Even though a dense black precipitate

obscures synaptic structures within the impregnated element it does not, however, occlude synaptic structures of cells that are presynaptic to the impregnated neuron.

Fortunately, cells that make up the columns in the optic lobe neuropil have quite precise topographical relationships with one another and can be described as lying along particular geometrical coordinate axes, natural to the columnar neuropil (Braitenberg, 1970). For example, receptor endings in columns of the outermost synaptic region are arranged in a standard cyclic order, and the interneurons post-synaptic to them have characteristic positions with respect to this cyclic arrangement (Strausfeld and Campos-Ortega, 1972). Electron microscopy of Golgi-stained cells deeper in the lobes has shown that, here too, cells have quite precise geometrical arrangements: In fact, these are so specific as to allow the identification of a few unimpregnated profiles of fiber sections (in *normal* E.M. material) as belonging to a certain cell type (Campos-Ortega and Strausfeld, 1972a). Thus it is possible to make estimations of cell populations, as well as identifying synaptic relationships, hence meeting the final objective of anatomical analysis.

To be fair, the analysis of the insect visual system is made easier by its extraordinarily precise geometry. But there is no reason to doubt that combined Golgi-E.M. analysis is just as applicable to the vertebrate brain, even though the coordinates which are natural to the nervous tissue are not nearly so obvious. Indeed, progress was already made in this direction long before the combined Golgi-E.M. technique was applied to insects. The analysis of the neural connectivity in the retinae of fish (Stell, 1967) and primate (Dowling and Boycott, 1966; Boycott and Dowling, 1969) could not have been so satisfyingly carried through without the precise resolution of the shapes, positions, and geometrical relationships of Golgi-impregnated receptors and interneurons.

The combined Golgi-E.M. technique provides the lens, so to speak, between the light and the electron microscope. And it is not too risquant to predict that, in the future, combined studies will become *the* anatomical tool for functional anatomical analysis. It is a prerequisite that the method employed for Golgi impregnation will render

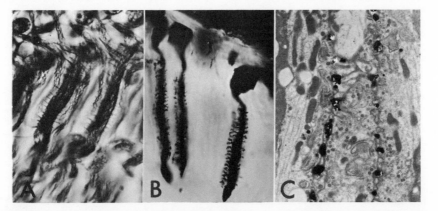

Fig. 2   Reduced silver, Golgi, and Golgi-E.M. of first-order interneurons in the first synaptic region of the fly. A: Bodian preparation showing the axis fibers and dendritic spines of regularly arranged palisades of interneurons. (×550.) B: Three Golgi-impregnated interneurons whose spine arrangement can be correlated to the argyrophilic elements shown in A. (×550.) C: Electron micrograph of the dendritic spines between portions of three receptor terminals from the retina. Even without contrasting, the other unstained profiles and their organelles can be resolved. (Figure 2C by courtesy of Dr J. A. Campos-Ortega, Tübingen.) (×6100.)

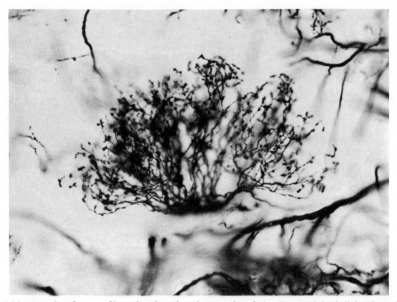

Fig. 3 A dendritic complex from a fiber that invades the matrix of the "central body" in the mid-brain of the butterfly, *Pieris brassicae*. (×510.)

the unimpregnated background tissue amenable for subsequent electron microscopy. Blackstad, in his recent, and erudite, article (1970) recommends fixation in buffered aldehyde prior to immersion in a dichromate–osmium tetroxide mixture; This is followed by impregnation with silver nitrate. Basically the method follows the Golgi rapid procedure, developed by Cajal.

Two methods function well on insects. The first is the Golgi rapid method, used by Pearson (1971) and the second is Colonnier's (1964) modification of the Golgi-Kopsch: Brains are prepared by cutting a small window from the head cuticle and then clearing away as much as possible of the tracheation and fatty tissue about the brain. This should be done as rapidly as possible. The whole heads are immersed in a mixture of 2.5 % potassium dichromate (4 vols) and 25 % "Kodak" glutaraldehyde (1 vol) for up to 7 days in the dark at 20°C. Heads are then washed repeatedly in 0.75 % silver nitrate, until red precipitation stops, and are then immersed in a fresh silver solution for up to 10 days in the dark. Brains are next dehydrated and embedded in araldite (polymerization at 40–60°C). Sections of 60–100 μm are washed in xylol and mounted under coverslips in Permount. Using interference phase contrast microscopy (this will give a good indication of the quality of background fixation) the position of an impregnated cell, with respect to neuropil columns, can be established (Fig. 1B, C). Ideal sections containing a few neurons against a clear background (Figs. 1B, 2B and 3), should be removed for E.M. examination before the Permount sets hard. They are washed in xylol and reembedded in araldite for ultrathin sectioning. No additional contrasting is necessary; in fact any treatment with water involves the risk of removing the silver precipitate which fills the impregnated cell (Fig. 1C). The replacement of silver chromate with lead chromate (Ramón-Moliner and Ferrari, 1972) functions with insect material and may be advantageous in this context.

One need only scan through the classic textbooks of histology to see that innumerable modifications of the Golgi method have been developed since its innovation. Apart from Cajal's own book (1921), those of Kallius (1910), Gatenby and Painter (1937), and Romeis (1969) describe the most important, as well as the strangest Golgi variants. A few variations of this method, as applied to insect material, have been evolved (Strausfeld and Blest, 1970; Strausfeld, 1971), some of which do, in fact, enhance the staining of some types of fibers. Bizarre alchemic modifications, such as adding methylene blue to phosphate buffered glutaraldehyde, prior to staining, or the addition of minute amounts of pyridine to the glutaraldehyde during staining, in no way help to explain the mechanism of Golgi staining even though they seem to induce some regional selectivity of impregnation. It is still true to say that, for all its usefulness, the Golgi stain remains as enigmatic today as it was almost one hundred years ago.

NICHOLAS J. STRAUSFELD

Acknowledgments

I am indebted to my colleague, Dr. J. A. Campos Ortega, for discussing the manuscript and for drawing my attention to many of the historical landmarks mentioned in this account. I also thank Dr. M. Weiss, Department of Zoology, University of Washington, for supplying a detailed description of a Golgi-Cox technique and for valuable recommendations as to its use.

Addendum

Since the manuscript for this article went to press Dr. Weiss (1971, 1972) has developed a Golgi-Cox technique for use on insect brains. This method uses glutaraldehyde-formaldehyde fixation. Fired by his success I have tried Ramon-Moliner's (1970) short Golgi-Cox procedure on flies after immersion with Karnovsky (1969) fixative and subsequently washing the brains in either 2,5 % potassium or sodium dichromate, or in a 0,1 N phosphate or

Fig. 4 Golgi-Cox impregnated fiber tract (to the right) and dendritic tree (to the left) on optic peduncle of the fly *Musca domestica* (×400).

sucrose buffer. The technique functions extremely well and supports Weiss's contention that the Cox method is ideal for the study of mid-brain tracts and neuropil (Fig. 4). In fact the number of elements stained throughout the brain, including optic lobes, is greater than that demonstrated by the Golgi rapid or Golgi-Colonnier procedures.

Weiss recommends alkalinization of sections cut free hand after the mercuric chloride treatment and after washing in water. Whole brains can also be successfully alkalinized. The advantage of sections is that the extent of blackening of the nerve cells can be controlled by microscopical observation. Blackening is extensive after 180 seconds and its onset occurs 40 seconds after application of the alkali. 40–120 minutes is the recommended time for alkalinization of whole brains. Sections or heads are then washed in acidic water, dehydrated in aceton or alcohol and embedded in Spurrs or Araldite.

References

Blackstad, T. W., "Electron microscopy of Golgi preparations for the study of Neuronal relations," in "Contemporary Research Methods in Neuroanatomy," J. H. Nauta and S. O. E. Ebbesson, ed., Berlin, Springer-Verlag, 1970.

Boycott, B. B., and J. E. Dowling, "Organization of the primate retina: light microscopy," *Phil. Trans. Roy. Soc. Lond.* B, **255**:109–176 (1969).

Braitenberg, V., "Ordnung und Orientierung der Elemente im Sehsytem der Fliege," *Kybernetik*, 7:235–242 (1970).

Cajal, S. R., "Elementos de Anatomía Histología Normal y de Técnica Micrográfica," 7th ed., Madrid, 1921.

Cajal, S. R., and D. Sanchez, "Contribución al conocimiento de los centros nerviosos de los insectos. Parte I. Retina y centros opticos," *Trab. del Lab. de Inv. Biol. Madr.*, 13:1–168 (1915).

Campos-Ortega, J. A., and N. J. Strausfeld, "The Columnar Organization of the second synaptic region of the visual system of *Musca domestica* L." *Z. Zellforsch*, 124, 561–585 (1972a).

Campos-Ortega, J. A., and N. J. Strausfeld, "Columns and layers in the second synaptic region of the fly's visual system: The case for two superimposed neuronal architectures." In: "Information processing in the visual system of arthropods" (R. Wehner, ed.), New York, Springer (1972b).

Colonnier, M., "The tangential organization of the visual cortex," *J. Anat. Lond.*, **98**:327–344 (1964).

Dowling, J. E., and B. B. Boycott, "Organization of the primate retina: electron microscopy," *Proc. Roy. Soc. Lond.*, B, **160**:80–111 (1966).

Gatenby, J. B., and T. S. Painter, "The Microtomist's Vade-Mecum," 10th ed., London, Churchill, 1937.

Hanström, B., "Vergleichende Anatomie des Nervensystems der wirbellosen Tiere," Berlin, Springer-Verlag, 1928.

Karnovsky, M. J., "A formaldehyde-gluteraldehyde fixative of high osmolarity for use in electron microscopy," *J. Cell Biol.*, **27**:137A (1965).

Kallius, E., "Golgische Methode," *Enzyk. Mik. Technik J.*, P. Ehrlich, ed., Berlin (1910).

Kenyon, F. C., "The brain of the bee. A preliminary contribution to the morphology of the nervous system of the Arthopoda," *J. Comp. Neurol.*, **6**:133–210 (1896).

Pearson, L., "The corpora pedunculata of *Sphinx ligustri* L and other Lepidoptera: an anatomical study," *Phil. Trans. Roy. Soc. Lond.* B, **259**:477–516 (1971).

Ramón-Moliner, E., "The Golgi-Cox technique." In: "Contemporary Research Methods in Neuroanatomy,' (ed. Nauta, W. J. H. and Ebbesson, S. O. E.) New York: Springer (1970).

Ramón-Moliner, E. and J. Ferrari, "Electron Microscopy of previously identified cells and processes within the central nervous system," *J. Neurocytol.*, 1:85–100 (1972).

Reichardt, W., ed., "Proc. Int. Sch. Phys. 'Enrico Fermi'," New York, Academic Press, 1970 (see articles by K. Kirschfeld [pp. 144–166], W. Reichardt [pp. 465–493], and K. Götz [pp. 495–509]).

Romeis, B., "Mikroskopische Technik," Munich, Oldenbourg, 1969.

Stell, W. K., "The structure and relationships of horizontal cells and photoreceptor-bipolar synaptic complexes in goldfish retina," *Amer. J. Anat.*, **120**:401–414 (1967).

Strausfeld, N. J., "The structural organization of the insect visual system (light microscopy): Part 1. Projections and arrangements of neurons in the Lamina ganglionaris of Diptera," *Z. Zellforsch.*, **121**:377–441 (1971).

Strausfeld, N. J., and A. D. Blest, "Golgi studies on insects. Part 1. The optic lobes of Lepidoptera," *Phil. Trans. Roy. Soc. Lond.* B, **258**:81–134 (1970).

Strausfeld, N. J., and J. A. Campos-Ortega, "Some interrelationships between the first and second synaptic regions of the fly's (*Musca domestica* L.) visual system," In: "Information processing in the visual system of arthropods" (R. Wehner, ed.), New York, Springer (1972).

Vigier, P., "Sur l'existence réelle et le rôle des appendices piriformes des neurons. La neurone perioptique des dipteres," *C. R. Séanc. Soc. Biol.*, **64**:959–961 (1908).

Young, J. Z., "The optic lobes of *Octopus vulgaris*," *Phil. Trans. Roy. Soc. Lond.* B, **245**:19–58 (1962).

Weiss, M. J., "From soma to synapse: the tracings of fibre inputs of a complex brain centre in the American cockroach, *Periplaneta americana* (L)," *Am. Zool.*, **11**:696–697 (1971).

Weiss, M. J., "Golgi-Cox Procedure with Aldehyde Prefixation" (Manuscript), Personal communication (1972, July).

Zawarzin, A., "Histologische Studien über Insekten, IV. Die optischen Ganglien der Aeschna-Larven," *Z. Wiss. Zool.*, **108**:175–257 (1913).

See also: INSECT HISTOLOGY.

GOLGI METHODS

If scientific information could be measured confidently, it is my opinion that the Golgi methods would rate first among all methods of investigation of the nervous system as far as their yield of information about the function of the brain is concerned up to the present. It is also very likely that they will continue in the future to provide the most important information in this field, since the bottle-neck in the process of gaining information through these methods is due to the large number of man-hours which they consume and one feels that very little has been done compared to the complexity of the nervous system which faces us. In fact, the phase of diminishing returns has not been reached yet, as one may easily convince himself by scanning the recent neuroanatomical literature.

The key role of the Golgi method in the task of un-raveling the structure of the nervous system becomes obvious when one imagines the same task to be performed without it. A neuroanatomist who knows how to stain all myelinated fibers, or all neural cell bodies, or all nuclei of nerve cells, or all axons or all dendrites might easily exclaim: give me a midget dissector anytime, doing in the microscopical domain what the gross anatomist does with his blunt dissecting .tools, trained to isolate and follow axonal ramifications in the inextricable neuronal felt-work, and I shall give you all this chemical affinity! In a thicket where the crown of each tree is compenetrated with that of 10,000 other trees, give me a tree shrew rather than a botanist if I have to find out to what stem a par-ticular leaf belongs! The Golgi methods fulfill exactly the role of that micro-dissector, or of the tree shrew running along branches, by virtue of a property which might at first appear as a drawback: their peculiar habit of singling out individual neurons among tens of thousands of other neurons which are left completely unstained.

Golgi staining procedures are unsystematically selec-tive. At least this is what they appear to be in spite of various attempts at finding some rationale to the fact that of many similar constituents of the nervous tissue they will stain only a small proportion. The following ideas have been ventilated to explain this apparently bizarre behavior:

a. Golgi staining may in fact be quite systematic, only it favors certain species of neurons which have not yet been recognized as being distinct from others due to the coarseness of our histological taxonomy. This thought will appear untenable to anybody who has applied Golgi staining repeatedly to the same structure. The set of neurons which are stained at one time will abundantly overlap with the set of neurons which you may stain the next time and so on, so that one could save the idea of a hidden taxonomy underlying the selection performed by the Golgi stain only by making this taxonomy impos-sibly complex.

b. Golgi staining may show neurons in a particular functional state, e.g., neurons that have produced an action potential within a certain time preceding the stop of the circulation, or neurons which are in a state akin to post-tetanic potentiation, or any such idea. This hypo-thesis would add enormous value to the method, making it much more interesting to the physiologists than it ought to be already, but again, it is an idea more likely to be found in the mind of someone who has never used the method systematically, rather than of the specialist. The latter will have noticed such fantastic variations of the overall density, of the patchiness of the distribution of stained neurons, of the dependence of their staining on their distance from the surface of the specimen, that he would soon give up making equally fantastic physiological suppositions to accommodate his histological findings.

c. Golgi staining may show up neurons in a pathological state, or at least, neurons either in their infancy or in a gerontological condition or in any way deviant from the norm as far as their trophic state is concerned. It appears

to be quite certain that in many animals a certain per-centage of neurons may die throughout the whole life, and under normal conditions the distribution of these ailing neurons is probably quite diffuse in the nervous system. It is, in fact, as far as we can judge from the ob-servations of pycnotic or swollen cell bodies in Nissl preparations, too diffuse to serve as an explanation of the selectivity of Golgi staining, which is often patchy, or denser near the surface of the block, etc., as mentioned in the preceding paragraph.

d. Golgi staining may affect a subset of neurons which is characterized by nothing that has anything to do with neurology: In other words, to the neuroanatomist, the selection is totally random. We are ourselves in favor of the latter hypothesis. The concept of random ought not to be taken in an all too abstract sense. Nothing can be farther from a Poisson distribution than the population of stained neurons appearing in a good Golgi preparation of the cerebral cortex: here a convolution with shapeless heaps of precipitate, and no neurons at all, there a dense cluster of neurons stained around such a lump of amor-phous precipitation, all having a process (dendritic or axonal) which ends in that lump, there a convolution, or the bottom of a sulcus, with a modest selection of neurons tenuously stained throughout, making the region appear like a textbook illustration skillfully prepared by the artist. There are obviously explanations to be found for the irregularity of the distribution of the stain, but they probably reside in the process of the formation of the precipitate due to the bichromate-silver interaction, rather than in a specific affinity between certain sites of the tissue and the chemicals involved.

A tentative explanation of the formation of the so called stain, i.e., of the dark deposits of the silver-dichro-mate compound within some cells of the tissue was pro-posed by Braitenberg et al. (1967) on the basis of experi-ments with a modification of the Golgi procedure using formalin and sucrose in the dichromate solution. In a large series of Golgis on frog brains it was observed that the crystals surrounding the tissue after it had been transferred from the dichromate to the silver nitrate solution were of two types, either flaky, planar, jagged crystals deposited as a loose powder around the specimen, or filamentous crystals fixed with one end to the surface of the specimens, forming as it were a thick fur around it. The success of the Golgi procedure was found to be strongly correlated with the type of crystals, since the furry specimens were always much richer in well stained neurons than the others. The supposition seemed then justified that lengthwise growth of crystals may occur also within the tissue in the cases when it is observed on the surface, and that under these conditions a germ of a crystal accidentally arising within a tubular segment of a neuron (an axon or a dendrite, or for that matter, within any tubular object defined by a cell membrane, such as glial fibers or connective tissue fibers, or even capillaries [see Fig. 1]), may become oriented by the very geometry of the structure itself and may thus keep growing within the tubular object, possibly extending also into the branches, and ultimately fill the whole space bounded by the membrane.

It may be that on the basis of the irregular, lumpy appearance of the deposit of dense material within the neurons as it appears at high magnification when the stained block is subsequently embedded and prepared for electronmicroscopy (Blackstad, 1965; Valverde, 1970), the concept of crystals growing within neurons may have to

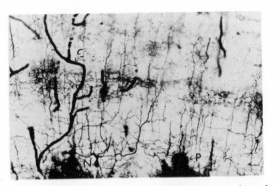

Fig. 1 Photograph of a typical Golgi preparation of the optic tectum of a teleost. The neuron (N) is outlined by the stain with all its widespread dendritic branches. Some capillaries (C) are also filled with the dark substance of the stain and some of the same substance is present in the form of irregular heaps of precipitate (P). The low resolution of the photograph is also typical for the Golgi method, since in order to obtain a sufficient depth of focus to show all the branches of the neuron (N), a lens with a small aperture had to be used.

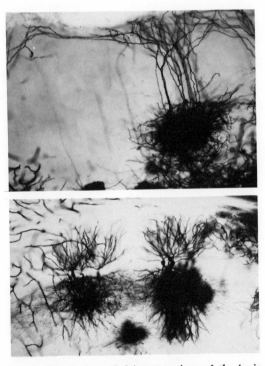

Fig. 2 Details from Golgi preparations of the brain of the mouse. Upper micrograph: cerebral cortex; lower micrograph: hippocampus. These show how the stain affects bunches of neurons which have some processes reaching the same heap of precipitate.

be partially revised. It is, however, quite plausible that the stain occurs as some sort of chain reaction starting at one point of a neuron and proceeding to outline the whole complex structure of its ramifications. The main evidence is the common observation which can easily be made on serially cut Golgi-stained material, that all neurons which have any process reaching a well defined region, such as the surface of the block of tissue, or more frequently, an amorphous deposit of precipitate, will be stained (Fig. 2). A curious isolated observation related by Blackstad (1965) should also be mentioned in this connection. He observed the reaction of the dichromate with the silver nitrate taking place in a thin slice of tissue under the microscope and happened to see "a dendrite with spines becoming stained like a worm creeping into the tissue" (Blackstad, discussion to Ramón-Moliner, 1970).

It should be mentioned at this point that any explanation of the Golgi-type selectivity should not refer exclusively to the dichromate-silver reaction which occurs in the classic Golgi, the rapid Golgi, and in innumerable modifications of the original methods. A very similar type of random, unpredictable selection is performed by the Golgi-Cox method in which the stain is due to a mercury rather than silver compound, and more remarkable still, in the Ehrlich methylene blue stain, where a synthetic organic dye is used. Moreover, in these methods no such dramatic moment occurs as in the original Golgi methods, where the silver in penetrating the tissue reacts violently with the dichromate, with the reaction taking place necessarily in the form of a wave progressing through the tissue.

Fixation in potassium (sodium- or ammonium-) dichromate of blocks of unsectioned tissue which are afterwards transferred into silver nitrate solution and finally cut into very thick (50–100 μ or more) sections constitutes the essence of the Golgi procedure. The silver nitrate is 0.75% in most methods and the time of the silver bath does not seem to be critical, although the longer the specimen stays in the silver, the more intense the strain is said to become. The dichromate, usually in a concentration of 2–5%, is used without any additions only in the original so-called long Golgi method, which, however, was abandoned because of its scant reliability. The principal additive to the bichromate is osmium tetroxide (10 ml of 1% osmium tetroxide in 40 ml of 2.5% potassium dichromate) in the so-called rapid Golgi method which has been used extensively by Cajal and which is still perhaps the most useful of the many variations of the method. Recently the revival of interest in Golgi methods has also led to the publication of some new prescriptions which are recommended because of the constancy of their results such as Colonnier's method (Colonnier, 1964) which adds one volume of 25% glutaraldehyde to 4 volumes of 2.5% potassium dichromate. A method which may be applied to old formalin-fixed material, where the dichromate solution contains varying amounts of the formalin and table sugar, was developed by the author (Braitenberg and Atwood, 1958).

Recently important improvements in the field of the Golgi-Cox methods have been published by Van der Loos (1956) and Ramón-Moliner (1958, 1970). The latter author's review and critical appraisal of this method (1970) should be recommended.

The "Golgi method" is, however, more than just another procedure for staining neurons in blocks of central nervous tissue. The method implies the whole procedure of obtaining information about the structure of brains, which in the case of these unsystematically selective preparations is markedly different from what it is, for instance, in the case of Weigert preparations of myelinated nerve fibers or of Nissl preparations of neural cell bodies, both typical examples of systematically selective methods, both quite predictable in their outcome and therefore useful for quantitative studies. Golgi preparations, on the contrary, are obtained through a procedure which is in itself an experiment with unpredictable outcome every time it is performed. The investigator who uses it preferentially will frequently develop an attitude reminiscent of that of the zoological taxonomist and even that of the collector of zoological oddities. His method being constitutionally unable to give him any indication of the relative frequency of the various neuron types which the Golgi man encounters in his preparations, as complete as possible a list of types of neurons will be the aim of his studies and it is only natural that he will cherish an observation of a strange type of cell the rarer it is, for it might be the only precious example of a species which might constitute in reality who knows how frequent an occurrence.

There is another, technical aspect which ought to be mentioned to round off our sketch of Golgi methods. The most valuable result of a successful Golgi impregnation is a stained neuron with all its ramifications extending over hundreds and even thousands of microns in a very thick histological section. Such an object can rarely be photographed in a satisfactory way and therefore most of the information obtained from Golgi preparations will be presented in the form of tracings made under the microscope with the aid of a camera lucida, a device which optically superimposes the microscopical picture on that of a drawing board. There is, of course, a subjective element in this procedure, which may be in part responsible for the suspicion with which Golgi methods were regarded for many years between the golden period around the turn of the century and the recent renaissance.

Finally, a very promising, powerful extension of the Golgi technique may result from recent experiments in the direction of its use in electron microscopy. The main virtue of the Golgi stain, which is that of establishing the identity of a longer fiber or of a complete neuronal

ramification over a very large region of the nervous tissue, may become even more important in electron-microscopy, where serial sectioning is extremely difficult and the long range relations of neurons among each other are therefore frequently lost. The presupposition for this marriage of two powerful techniques, that they be mutually compatible, i.e., that a mordant which is good for Golgi impregnation should also be a good fixative for electron microscopy, has already been established, e.g., by Blackstad (1970).

VALLENTINO BRAITENBERG

References

Blackstad, T. W., "Mapping of experimental axon degeneration by electron microscopy of Golgi preparations," *Z. Zellforsch.*, **67**:819–834 (1965).

Blackstad, T. W., "Electron microscopy of Golgi preparations for the study of neuronal relations, pp. 186–216 in "Contemporary Research Methods in Neuroanatomy," W. J. H. Nauta and S. O. E. Ebbesson, ed., New York, Springer-Verlag, 1970.

Braitenberg, V., and R. P. Atwood, "Morphological observations on the cerebellar cortex," *J. Comp. Neurol.*, **109**:1–34 (1958).

Braitenberg, V., V. Guglielmotti, and E. Sada, "Correlation of crystal growth with the staining of axons by the Golgi procedure," *Stain Technol.*, **42**(6):277–283 (1967).

Colonnier, M., "The tangential organization of the visual cortex," *J. Anat. Lond.*, **98**:327–344 (1964).

Ramón-Moliner, E., "A tungstate modification of the Golgi-Cox method," *Stain Technol.*, **33**:19–29 (1958).

Ramón-Moliner, E., "The Golgi-Cox technique," pp. 32–55 in "Contemporary Research Methods in Neuroanatomy," W. J. H. Nauta and S. O. E. Ebbesson, ed., New York, Springer-Verlag, 1970.

Valverde, F., "The Golgi method. A tool for comparative structural analysis," pp. 12–31 in "Contemporary Research Methods in Neuroanatomy," W. J. H. Nauta and S. O. E. Ebbesson, ed., New York, Springer-Verlag, 1970.

Van der Loos, H., "Une combinaison de deux vieilles methodes histologiques pour le système nerveux central," *Mschr. Psychiat. Neurol.*, **132**:330–334 (1956).

h

HAIR

Microscopic studies of hair should be made after the hair is cleaned: agitate in ether, ether-alcohol, or some other lipid solvent, dry, wash in water, and dry again. If the hair is too darkly pigmented for distinguishing the structural elements in optical light microscopy it may be bleached in hydrogen peroxide.

Light microscopy is used to distinguish the cuticle, cortex, and medulla of the hair shaft in both whole mounts (5, 18) and cross sections. In a dry mount, a drop or two of glycerin-alcohol mixture applied to the ends of the shaft will bring the free borders of the scale-like cuticular cells into relief for counting with the aid of an ocular micrometer (17); and an enlarged imprint of the surface of the shaft on glycerin jelly (6) or on soft plastic (9) is an aid in the classification of the pattern as well as in the counting and measurement of scales. The phase contrast microscope and, particularly, the Nomarski interference contrast microscope are also suited for studying directly the surface of the shaft and the medullary patterns. The fusi (air vesicles) in the cortex and pigment granules in both the cortex and medulla may be studied under high power after the hair is cleared by immersion in an oil having a refractive index similar to that of the hair (7). Air spaces in the hair shaft may be studied advantageously by employing both episcopic and diascopic illumination in the same microscope. Refractive indices and birefringence of the hair cuticle may be determined under the polarizing microscope by the double-variation method and used as another parameter in identification (4).

The greatest and least diameters of the shaft, for determining the hair form or "index" (15, 16), may be measured on the intact hair with an ocular micrometer when the hair is mounted in a hair-rotating apparatus attached to the stage of the microscope (2). Another simple mechanism which also provides accuracy in measuring the degree of rotation is a capillary tube with one end bent to form a handle; the tube through which the hair is threaded is filled with and immersed in an oil with a refractive index closely matching that of the glass (12). The diameters of hairs can also be measured from cross sections. With the relatively inexpensive instrument known as the "J. I. Hardy Thin Cross-Section Device" (made by A. M. de La Rue, Hyattsville, Maryland), a bundle of 100–200 hairs may be sectioned to a thickness of as little as 10 μ. The hairs are packed with cotton fibers in the slot of the device, its two parts are brought into engagement, the excess material extending from either side of the plate is trimmed off flush with a razor blade, then after locking the plates a turn of a graduated

micrometer screw makes the hairs and fibers protrude from the surface to the desired thickness. The protrusions are coated with collodion, removed as a slice with a razor blade, and are ready for study on a slide. The difficulty of producing thinner cross sections of hair shafts with paraffin microtomy may be minimized by the following technique: Wrap a small bundle of short hair segments from 10 to 50 hairs in tea bag paper, clean and dry as indicated above, place in melted paraffin for 30 min, orient the hair segments in a regular paraffin block, and section on a paraffin microtome with a razor blade or microtome knife in a suitable holder (14). The sections are then ready to be processed and stained by standard histological techniques. Staining the sections with rhodamine B and methylene blue distinguishes the keratins of the hair shaft from those of the inner root sheath of the follicle and from the stratum corneum of the skin (8).

Transmission electron microscopy can be used effectively to study the cuticle, the cortical cells with their constituent keratin filaments, the medullary cells and the melanosomes (melanin granules) of the hair shaft (11). A bundle of clean, dry hairs is tied into 3 mm segments with cotton thread. Any tied segment may be cut out, fixed and stained with 1–2% osmium tetroxide (OsO₄) for 1–3 days, washed in water, and dried. Next, embed in Araldite or similar resin and harden with a slight excess of hardener and prolonged curing at 60°C. Sections may be cut, preferably with a diamond knife, and stained with any of the common EM stains such as lead citrate and uranyl acetate. The filaments and matrix of the cortical cells may be resolved more clearly by pretreating the hair with 0.5M thioglycollic acid at pH 5.6 for 24 hr, washing, and then fixing in 1–2% OsO₄ for at least 72 hr (13). This treatment makes the interfibrillar matrix extremely electron dense, permitting the filaments to be clearly resolved. To study a particular area on a particular hair with the transmission electron microscope, the hair may be flat-embedded in resin between two coverslips which are separated by the thickness of a microscope slide; after the coverslips are broken away, but before trimming and sectioning, the thin, optically flat block can be observed in the light microscope for orientation (12).

Scanning electron microscopy greatly facilitates high resolution study of the surface morphology of the hair shaft (3). Clean, dry hairs should be affixed to the appropriate specimen holder by means of double-sided sticky tape (liquid adhesive tends to obliterate the topography of the hair). The specimen is then made ready for study by coating with chrome or gold in a vacuum evaporator.

Isolation of cells of the cuticle and medulla provides

special material for study with the phase contrast, Nomarski interference contrast, and scanning electron microscopes. Cuticular cells may be isolated by treating hair shafts with formic acid at 100°C for one hour and then agitating gently to detach the cells (10). Medullary cells are isolated by treating hair shafts with keratinolytic agents, such as aqueous sodium hydroxide or peracetic acid, and following with ammonia extraction. A milder method involves immersion of the hair in a 2% aqueous solution of gold chloride, which is deposited preferentially in the medulla; the fiber is disrupted by agitating in formic acid and the heavier medullary cells are separated out by density-gradient centrifugation (1).

MILDRED TROTTER
DAVID MENTON

References

1. Bradbury, J. H., and J. M. O'Shea, "Keratin fibers. II. Separation and analysis of medullary cells," *Australian J. Biol. Sci.,* **22**:1205–1216 (1969).
2. Danforth, C. H., "The hair," *Nat. Hist.,* **26**:75–79 (1926).
3. Dawber, R., and S. Comaish, "Scanning electron microscopy of normal and abnormal hair shafts," *Arch. Dermatol.,* **101**:316–322 (1970).
4. Duggins, O. H., "Age changes in head hair from birth to maturity. IV. Refractive indices and birefringence of the cuticle of hair of children," *Amer. J. Phys. Anthropol.,* **12**:89–114 (1954).
5. Duggins, O. H., and M. Trotter, "Age changes in head hair from birth to maturity. II. Medullation in hair of children," *Amer. J. Phys. Anthropol.,* **8**:399–415 (1950).
6. Eddy, M. W., and J. C. Raring, "Technique in hair, fur, and wool identification," *Proc. Pa. Acad. Sci.,* **15**:164–168 (1941).
7. Hausman, L. A., "Applied microscopy of hair," *Sci. Month.,* **59**:195–202 (1944).
8. Holmes, E. J., "Hot, acidified rhodamine B and methylene blue; a differential stain dichromic for hair follicular keratins, achromatic for epidermal keratin," *Stain Technol.,* **45**:15–18 (1970).
9. Kirk, P. L., S. Magagnose, and D. Salisbury, "Casting of hairs—its technique and application to species and personal identification," *J. Crim. Law and Criminol. Northwestern Univ.,* **40**:236–241 (1949).
10. Leeder, J. D., and J. H. Bradbury, "Conformation of the epicuticle on keratin fibers," *Nature,* **218**:694–695 (1968).
11. Orfanos, C., and H. Ruska, "Die Feinstruktur des menschlichen Haares. I. Die Haar—Cuticula," *Arch. Klin. Exp. Derm.,* **231**:97–110 (1968).
12. Price, V. H., R. S. Thomas, and F. T. Jones, "Pili annulati. Optical and electron microscopic studies," *Arch. Dermatol.,* **98**:640–647 (1968).
13. Rogers, G. E., "Electron microscopy of wool," *J. Ultrastruct. Res.,* **2**:309–330 (1959).
14. Shelley, W. B., and S. Öhman, "Technique for cross sectioning hair specimens," *J. Invest. Dermatol.,* **52**:533–536 (1969).
15. Trotter, M., "The form, size and color of head hair in American Whites," *Amer. J. Phys. Anthropol.,* **14**:433–445 (1930).
16. Trotter, M., and O. H. Duggins, "Age changes in head hair from birth to maturity. I. Index and size of hair of children," *Amer. J. Phys. Anthropol.,* **6**:489–506 (1948).
17. Trotter, M., and O. H. Duggins, "Age changes in head hair from birth to maturity. III. Cuticular scale counts of hair of children," *Amer. J. Phys. Anthropol.,* **8**:467–484 (1950).
18. Wynkoop, E. M., "A study of the age correlations of the cuticular scales, medullas, and shaft diameters of human head hair," *Amer. J. Phys. Anthropol.,* **13**:177–188 (1929).

HEAVY METAL LOCALIZATION

It is generally considered that only zinc, iron, copper, and cobalt exert essential functions in man's metabolism (Prasad, 1966; Thompson, 1966), whereas other heavy elements play a limited role as in for example, intoxications, or are present in trace amounts fortuitously from environmental sources. Heavy metals, when abundant, are easily visualized in tissues by light as well as electron microscopy. This is exemplified by the accumulation of iron pigments in cases of hemosiderosis and hemochromatosis, where the heavy metal appears as brown granules by light microscope and ultrastructurally as electron-dense grains in lysosomal dense bodies. When the metal concentration is low, however, in order to be visualized it must be amplified by means of:

colored reaction products in histochemical reactions,
fluorescent tracing,
radioactive isotopes,
physical development of metal sulfides,
electron diffraction,
electron probe analysis.

Light microscopical methods based upon *colored reaction products* in histochemical heavy metal reactions have been reviewed in several handbooks (Arnold, 1968; Barka and Anderson, 1963; Lillie, 1965; Pearse, 1972; Thompson, 1966). Generally these methods are of a direct-coupling type yielding reaction products of low electron density. Arnold (1967) has calculated that a concentrated amount of 10^{-11}–10^{-12} g of the material sought is the lower limit in any histochemical reaction. This is much greater than the dry mass of most cell organelles (Bahr and Zeitler, 1965). Obviously, reasonable submicroscopical localization can be achieved only when the reactive heavy metal is present in quantities far below the sensitivity limit of direct-coupling methods. Conceivably the organelle-bound heavy metal must be augmented by the binding of other metals or metal complexes.

In the Prussian blue reaction (Thompson, 1966) four ferric ions in the tissue are bound to three ferrocyanide ions in a stoichiometric manner. Dumon and Cone (1970) have successfully applied this method and the Turnbull blue reaction for electron optics (Table 1). It is too early yet to know whether any extreme pH reaction conditions necessary for tracing or excessive crystallization reaction products will seriously impede localization.

So far only little seems to be published on *fluorescent tracing* of tissue-bound heavy metals. The short wavelength of ultraviolet light leads to considerably better resolution than with transmitted light. Generally, more than a hundredfold enhancement could be expected than with nonfluorescent labels (Nairn, 1969). A method for zinc (Smith et al., 1969) is based upon the ability of 8-hydroxyquinoline to form chelate compounds with metals. Only the calcium, magnesium, and zinc chelates are fluorescent (Watanabe et al), 1963), and interference by the two light metals could be avoided by experimental virtuosity (Mahanand and Houck, 1968). The method of

Table 1   Electron Histochemical Methods for Reactive Ferric or Total Iron (Perls' and Turnbull's methods modified by Dumont and Cone, 1969)

Fix in phosphate-buffered glutaraldehyde.
Wash in phosphate buffer overnight.
Wash in distilled water for 10 min.

Perls (Fe^{+++})	*Turnbull* (Fe^{++} + Fe^{+++})
1. Freshly prepared 1 % K$_4$Fe(CN)$_6$ + HCl 3–4 × 30 min.	1. Rinse in distilled water.
	2. Dilute ammonium sulfide 2–4 × 30 min.
2. Distilled water 3 × 5 min.	3. Distilled water until excess sulfide removed.
	4. Freshly prepared 10 % K$_3$ Fe(CN)$_6$ + 0.5 % HCl 2 × 30 min.
	5. Rinse in distilled water.

Conventional EM procedure

Results: Reaction product (Prusian blue and Turnbull blue resp.) appears as electron-dense angular crystals.

Smith et al. (1969) appears to be rapid and highly selective (Table 2) and offers an attractive means of positive zinc localization. Conceivably, it could be combined with the sulfide-silver procedure (Table 3) for electron optics provided hydrosoluble embedding media are used (e.g. glycol methacrylate; Leduc and Bernhard, 1967).

The *radioactive isotope* ^{65}Zn was introduced for autoradiography by McIsaac (1955) and Millar et al. (1960) and has recently also been used at the electron microscopical level by Westmoreland and Koekstra (1969). Suitable heavy metal isotopes such as ^{60}Co, ^{203}Hg, and ^{210}Pb are also available and give satisfactory results by light microscopy (Halbhuber et al., 1970). However,

Table 2   Fluorescent Method for Zinc in Isolated Cells (Smith et al., 1969)

Solutions:
   A. 3 % 8-hydroxyquinoline in absolute ethyl alcohol.
   B. Michaeli's universal buffer, pH 8.0:
      Dissolve 19.428 g sodium acetate and 29.428 g sodium phenobarbital in 1000 ml water. To 100 ml of this solution, add 40 ml physiologic saline. Adjust pH to 8.0 with HCl (about 20 ml 0.1 N HCl).
   C. Add 0.1 ml of A to 25 ml of B.

Staining procedure:
   1. Prepare smear from cell suspension, let air-dry for 1 hr.
   2. Stain for 15 min in solution C.
   3. Drain excess stain, dip twice in distilled water, let air-dry.
   4. Mount in nonfluorescent oil.
   5. Examine at approximately 385 mμ.

Results: Zinc-containing cells and organelles fluoresce a pale greenish color.

Control: Treat control slide with 1 % acetic acid for 5 min prior to treatment with 8-hydroxyquinoline.

Table 3   Electron Histochemical Sulfide-Silver Method for Heavy Metals (Haug, 1967; Pihl, 1967, 1968a)

Fix tissue slices not thicker than 0.1 mm in 3 changes of continuously H$_2$S-gassed, ice-cold glutaraldehyde for 20–30 min.
Continue fixation in fresh aldehyde to make 1 hr.
Wash in tris-maleate buffer.
Dehydrate in ethylene glycol, glycol methacrylate (GMA), or according to routine.
Embed in GMA or Epon.
Cut ultrathin section within 20–24 hr.
Mount sections on thoroughly carbon protected grids or formvar-coated grids for GMA.
Treat grids for 2–30 min at room temperature in darkness with:
      Solution A, 10.0 ml
      Solution B, 0.09 ml

Solution A: Gum arabic 30g/100 ml
            hydroquinone (quinol) 0.17 g/100 ml
            citric acid 0.43g/100 ml
            sucrose 10g/100 ml
   The gum arabic solution must be several days old, pH is adjusted to 3.9–4.0. Hydroquinone should be added immediately before use.
Solution B: 10 % silver nitrate.

Wash grids and stain with uranyl acetate.
Results: Reaction product is distinctly electron-dense.
Control: Omit sulfide treatment.

Note: For maximum preservation of heavy metals ultrathin sections are preferably cut with H$_2$S-saturated 10 % alcohol in the trough, transferred with Marinozzi's plastic loop, washed, and mounted on formvar-coated grids. When the heavy metal is thought to be firmly bound or the tissue very vulnerable treatment with sulfide is omitted during fixation. Semi-thin sections of GMA (or Epon) could be treated with the sulfide-silver developer and will yield high resolution.

Warning: Hydrogen sulfide is poisonous!

Alternative method (Haug, 1967):
   Perfuse for 5–20 min at room temperature with 40 % formaldehyde (or glutaraldehyde—Ibata and Otsuka, 1969) in 0.1 M phosphate buffer, pH 7.5, immediately followed by sulfide solution for 20–30 min.

Sulfide solution: Na$_2$S 11.7 g.
            NaH$_2$PO$_4$ · H$_2$O 11.9 g
            distilled water to make 1000 ml
            adjust pH to 7.5

Cut tissue into 1/2–1 mm slices.
Wash for 2–3 days in phosphate buffer (without Cl ions).
Immerse in sulfide-silver developer until tissue turns light yellow (about 30 min).
Postfixation in OsO$_4$.
Conventional EM procedure.

autoradiography resolution could not be expected to be better than 800 Å even under very favorable conditions (Bachman et al., 1968) and with low-energy isotopes. Thus, heavy metal autoradiography could hardly be recommended as a routine, but would rather be limited to those cases where the heavy metal is present in tissue in such form or quantity that standard methods give equivocal results only, and where positive identification is necessary.

Physical development of metal sulfides in analogy with the common photographic process is used in the sulfide-silver method. The heavy metals of tissue are transformed to insoluble sulfides visible with neither light nor electron optics. In a further step silver ions are reduced to metallic silver, the sulfide molecules serving as "nuclei of condensation." The sulfide silver-method was introduced by Timm (1958) and Voigt (1959). A brief preliminary report on a variant for electron microscopy was given by Müller and Geyer (1965). Detailed methods have been described by Okamoto and Kawanishi (1966), Haug (1967), Pihl (1967, 1968a,c), and Scheuer et al. (1967).

Provided that aldehyde is used for fixation, tissue preservation is satisfactory even at the ultrastructural level. Sensitivity for zinc is in the 10^{-16} g range at least (Pihl et al., 1967; Pihl and Gustafson, 1967; Pihl, 1968c) and probably several hundredfold better.

No single method could yet be recommended for all purposes. In some tissues and cell organelles the heavy metal seems to be bound to diffusible or hydrosoluble compounds necessitating the immediate introduction of the sulfide. On the other hand, the sulfide or rather its SH ions evoke rapid dissolution of some specific granules (Falkmer and Pihl, 1968). The original papers by Haug (1967) and Pihl (1967, 1968a) should be consulted when a specific problem has to be solved, since the amount and distribution of metals depend essentially on the complete precipitation of metal sulfides (Brunk and Sköld, 1967). Consequently the mode of sulfide treatment, fixation, physical development, and embedding medium materially influence the reaction product, resolution, and preservation (Pihl, 1968a,c) (Table 3).

It should be emphasized that the sulfide-silver method is unspecific, and is said to react with Zn, Fe, Co, Ni, Ca, As, Cd, Pt, Au, Hg, Tl, Pb, and U (Voigt, 1959; Falkmer et al., 1964; Timm and Schultz, 1966; Kaltenbach and Eger, 1966; Giusti and Fiori, 1969). Under physiological conditions, however, zinc would constitute the main heavy metal demonstrable in the hippocampal region of the brain (Haug, 1967; Ibata and Otsuka, 1969), the pancreatic islet cells (Fig. 1) (Voigt, 1957, 1958, 1959; Pihl, 1968b), the Paneth cells of the ileum (Müller and Geyer, 1965), the prostate (Müller and Geyer, 1970), and the specific granules of eosinophilic leukocytes (Fig. 2) and mast cells (Pihl et al., Pihl and Gustafson, 1967). For identification purposes, the sulfide-silver method must be combined with conventional methods (Thompson, 1966) or one of those already mentioned. An attractive approach to this problem would be to embed tissue in the hydro-soluble glycol methacrylate (Leduc and Bernhard, 1967) which permits various light microscopical staining methods, fluorescent tracing, and enzymatic digestion (Pihl et al., 1968) on 1 μ thick sections of the same block used for subsequent ultrathin sectioning.

In suitable cases atomic absorption spectrophotometry would be the method of choice (Pihl et al., 1967; Pihl and Gustafson, 1967) permitting detection of 0.03 p.p.m. zinc, i.e., a several hundredfold lower concentration than in most tissues. Neutron activation analysis (Willard et al., 1965; Battistone et al., 1970) would offer still better sensitivity.

Theoretically *electron diffraction* and *electron probe analysis* have to be considered for the localization of

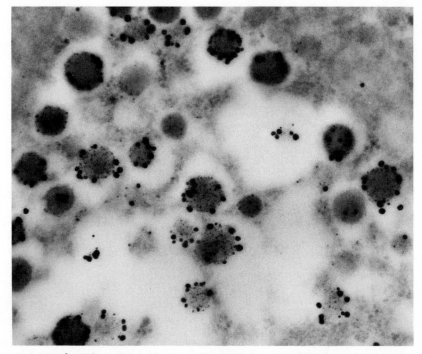

Fig. 1 Pancreatic islet β-cell from Chinese hamster after application of sulfide-silver procedure. Secretion granules show large amounts of silver grains disclosing the distribution of the islet zinc. Contrasted with uranyl acetate ($\times 50,000$)

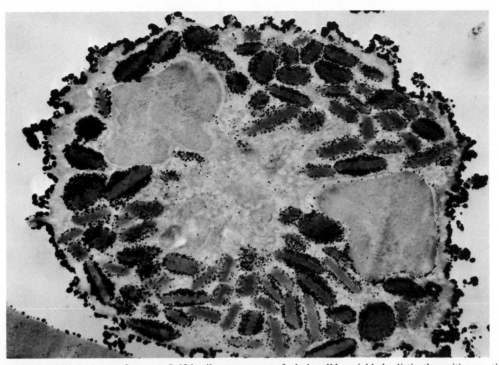

Fig. 2 Eosinophilic leukocyte from rat. Sulfide-silver treatment of whole cell has yielded a distinctly positive reaction, largely attributable to the presence of zinc in the specific granules, the cell surface, and to some degree also in the erythrocyte (bottom left). Contrasted with uranyl acetate and lead. ($\times$ 26,000.)

heavy metals in tissues. The former method, however, would require a minimum amount of some 10^{-14} g of the element (Fuchs, 1958) in solid or crystalline form (Reimer, 1967). These criteria are not fulfilled under physiological conditions, and it should be kept in mind that many cell organelles have substantially less dry mass (Bahr and Zeitler, 1965). Electron probe analysis has about the same sensitivity as electron diffraction but a resolution of only 1 μ (Willard et al., 1965; Zagon et al., 1970) which is apparently too low to be of any advantage compared with conventional methods.

ERIK PIHL

References

Arnold, M., "Die quantitative Erfassungsgrenze in der Histochemie," *Histochemie,* **9**:181–188 (1967).

Arnold, M., "Histochemie," Berlin, Springer, 1968.

Bachman, L., M. M. Salpeter, and E. E. Salpeter, "Das Auflösungsvermögen clektronenmikrokopischer Autoradiographien," *Histochemie,* **15**:234–250 (1968).

Bahr, G. F., and E. Zeitler, "The determination of the dry mass in populations of isolated particles," pp. 217–239 in "Symposium on Quantitative Electron Microscopy," G. F. Bahr and E. Zeitler, ed., Baltimore, Williams & Wilkins, 1965.

Barka, T., and P. J. Anderson, "Histochemistry. Theory, Practice and Bibliography," New York, Hoeber Med. Div., Harper and Row, 1963.

Battistone, G. C., E. Levri, and R. Lofberg, "A simplified ultramicrodetermination of copper in biological specimens by neutron activation analysis," *Clin. Chim. Acta,* **30**:429–438 (1970).

Brunk, U., and G. Sköld, "The oxidation problem in the sulphide-silver method for histochemical demonstration of metals," *Acta Histochem.,* **27**:199–206 (1967).

Dumont, J. N., and Cone, M. N., "Ultrastructural localization of iron by Perls' or by Turnbull's method applied to tissue prior to embedding," *Stain Technol.,* **45**:188–199 (1970).

Falkmer, S., B. Hellman, and G. E. Voigt, "On the agranular cells in the pancreatic islet tissue of the marine teleost, *Cottus scorpius,*" *Acta Pathol. Microbiol. Scand.,* **60**:47–54 (1964).

Falkmer, S., and E. Pihl, "Structural lability of zinc-containing secretion granules of pancreatic β-cells after exposure to hydrogen sulphide," *Diabetol.,* **4**:239–243 (1968).

Fuchs, E., "Identifizierung von Objektverunreinigungen mit Feinbereichsbeugung," *Microchim. Acta,* **5**:674–680 (1958).

Giusti, G. V., and A. Fiori, "The histochemical detection of Thallium in paraffin-embedded tissues by the sulfide-selenium-silver method," *Stain Technol.,* **44**:263–267 (1969).

Halbhuber, K.-J., H.-J. Stibenz, U. Halbhuber, and G. Geyer, "Autoradiographische Untersuchungen über die Verteilung einiger Metallisotope im Darm von Laboratoriumstieren. Ein Beitrag zur Ausscheidungsfunktion der Panetschen Körnerzellen," *Acta Histochem.,* **35**:307–319 (1970).

Haug, F.-M. S., "Electron microscopical localization of the zinc in hippocampal mossy fibre synapses by a modified sulfide-silver procedure," *Histochemie,* **8**:355–368 (1967).

Ibata, Y., and N. Otsuka, "Electron microscopic demonstration of zinc in the hippocampal formation using Timm's sulfide-silver technique," *J. Histochem. Cytochem,* **17**:171–175 (1969).

Kaltenbach, T., and W. Eger, "Beiträge zum histo-chemischen Nachweis von Eisen, Kupfer und Zink in der menschlichen Leber unter besonderer Benücksich-tigung des Silbersulfid-Verfahrens nach Timm," *Acta Histochem.*, **25**:329–354 (1966).

Leduc, E. H., and W. Bernhard, "Recent modifications of the glycol methacrylate embedding procedure," *J. Ultrastruct. Res.*, **19**:196–199 (1967).

Lillie, R. D., "Histopathologic Technic and Practical Histochemistry," New York, McGraw-Hill, 1965.

McIsaac, R. J., "Distribution of Zinc-65 in the rat pan-creas," *Endocrinology* **57**:571–579 (1955).

Mahanand, D., and J. C. Houck, "Fluorometric deter-mination of zinc in biological fluids," *Clin. Chem.*, **14**:6–11 (1968).

Millar, M. J., N. R. Vincent, and C. A. Mawson, "An autoradiographic study of the distribution of zinc-65 in rat tissues," *J. Histochem. Cytochem.*, **9**:111–116 (1960).

Müller, A., and Geyer, G., "Elektronenmikroskopischer Schwermetallnachweis in den Prosekretgranula der Panetschen Zellen," *Acta Histochem.*, **21**:404–405 (1965).

Nairn, R. C., "Fluorescent Protein Tracing," Edinburgh and London, Livingstone, 1969.

Okamoto, K., and H. Kawanishi, "Submicroscopic his-tochemical demonstration of intracellular reactive zinc in β cells of pancreatic islets," *Endocrinol. Jap.*, **13**:305–318 (1966).

Pearse, A. G. E., "Histochemistry. Theoretical and Applied," Edinburgh, London, Churchill Livingstone, 1972.

Pihl, E., "Ultrastructural localization of heavy metals by a modified sulfide-silver method," *Histochemie*, **10**:126–139 (1967).

Pihl, E., "Recent improvements of the sulfide-silver procedure for ultrastructural localization of heavy metals," *J. Microscopie, 7*:509–520 (1968a).

Pihl, E., "An ultrastructural study of the distribution of heavy metals in the pancreatic islets as revealed by the sulfide-silver method," *Acta Pathol. Microbiol. Scand.*, **74**:145–160 (1968b).

Pihl, E., "An electron microscopical sulfide-silver pro-cedure for detecting heavy metal compounds in tissues. A methodological study with particular reference to the occurrence and role of heavy metals in pancreatic islet tissue," Thesis, Umeå, 1968c.

Pihl, E., and G. T. Gustafson, "Heavy metals in rat mast cell granules," *Lab. Invest.*, **17**:588–598 (1967).

Pihl, E., G. T. Gustafson, and S. Falkmer, "Ultrastruc-tural demonstration of cartilage acid glycosamino-glycans," *Histochem. J.*, **1**:26–35 (1968).

Pihl, E., G. T. Gustafson, B. Josefsson, and K.-G. Paul, "Heavy metals in the granules of eosinophilic granulo-cytes," *Scand. J. Haematol.*, **4**:371–379 (1967).

Prasad, A. S., "Zinc Metabolism," Springfield, Ill., Thomas, 1966.

Reimer L., "Elektronenmikroskopische Untersuchungs-und Präparationsmethoden," Berlin, Springer-Verlag, 1967.

Scheuer, P. J., M. E. C. Thorpe, and P. Marriott, "A method for the demonstration of copper under the electron microscope," *J. Histochem. Cytochem.*, **15**:300–301 (1967).

Smith, G. L., R. A. Jenkins, and J. F. Gough, "A fluorescent method for the detection and localization of zinc in human granulocytes," *J. Histochem. Cyto-chem.*, **17**:749–750 (1969).

Thompson, S. W., "Selected Histochemical and Histo-pathological Methods," Springfield, Ill., Thomas, 1966.

Timm, F., "Zur Histochemie der Schwermetalle. Das Sulfid-Silberverfahren," *Dsch. Z. Gerichtl. Med.*, **46**:706–711 (1958).

Timm, F., and G. Schultz, "Hoden und Schwermetalle," *Histochemie, 7*:15–21 (1966).

Voigt, G. E., "Histochemische Untersuchungen über das Zink in den Langerhausschen Inseln bein Alloxan-diabetes," *Acta Pathol. Microbiol. Scand.*, **41**:81–88 (1957).

Voigt, G. E., "Investigation of the human embryonic pancreas for metals (zinc) by the sulphide-silver method," *Acta Pathol. Microbiol. Scand.*, **44**:120–127 (1958).

Voigt, G. E., "Untersuchungen mit der Sulfid-silber-methode an menschlichen und tierischen Bauch-speicheldrüsen unter Berücksichtigung der Diabetes mellitus und experimentellen Metallvergiftungen," *Virchows Arch. Pathol. Anat.*, **332**:295–323 (1959).

Watanabe, S., W. Frantz, and D. Trottier, "Fluorescence of magnesium-, calcium- and zinc-8-quinolinol com-plexes," *Anal. Biochem.*, **5**:345–359 (1963).

Westmoreland, N., and W. G. Hoekstra, "Ultrastructural localization of zinc in rat exocrine pancreas by auto-radiography," *Histochemie*, **18**:261–266 (1969).

Willard, H. H., L. L. Merritt, and J. A. Dean, "Instru-mental Methods of Analysis," Princeton, N.J., D. Van Nostrand, 1965.

Zagon, I. S., J. Vavra, and I. Steele, "Microprobe analysis of protargol stain deposition in two protozoa," *J. Histochem. Cytochem.*, **18**:559–569 (1970).

HEMATOXYLIN

Pure hematoxylin is a colorless powder obtained from *Haematoxylin campechianum L.*, a small leguminous plant. It is the leuco-counterpart of hematein, the reddish weakly acid dye into which hematoxyl is transformed by oxidation, either slowly and spontaneously due to atmos-pheric oxygen ("ripening"), or rapidly with sodium iodate, sodium perborate, mercuric oxide, potassium per-manganate, and potassium periodate.

Hematoxylin and hematein alone are nearly useless in practice. Yet, in conjunction with mordants they yield blue or black or violet lakes which are universally used due to the high contrast (quite convenient in photo-micrography) of their staining, the ease with which dyeing can be controlled, the insolubility of the color in neutral aqueous or alcoholic media after dyeing, as well as its stability in Canada balsam and other mounting resinae.

It is generally agreed that the mordants play a key role in the attachment of hematoxylin lakes to substrates and, thus, in accounting for the selectivity of staining. In fact, stained structures and substrates are subject to major changes according to the nature of the mordant. Most mordants are bivalent or multivalent metals capable of forming covalent bonds; they give basic lakes working not unlike basic dyes. They include iron, aluminum, chromium, and lead salts. Double sulfates (alums) are generally used, but other salts (sulfates, chlorides, nitrates, etc.) of the same cations may be useful. To some degree each metal displays a preferential affinity for given tissue anions. Aluminum hematoxylins (Fig. 1), which are extensively used according to the formulae of Mayer, Ehrlich, Delafield, Harris, Carazzi, and others, and chromium hematoxylins selectively stain nuclear chroma-tin; iron hematoxylins (Fig. 2), as Heidenhain's, Regaud's, Groat's, and Weighert's hematoxylins, besides chroma-tin, heavily stain proteins and phospholipids too, thus providing excellent stainings of cytoplasmic structures like mitochondria or secretory granules. Mucihematein (hematein + aluminum chloride) stains some mucous substances, as do mucicarmin, the lake of aluminum with

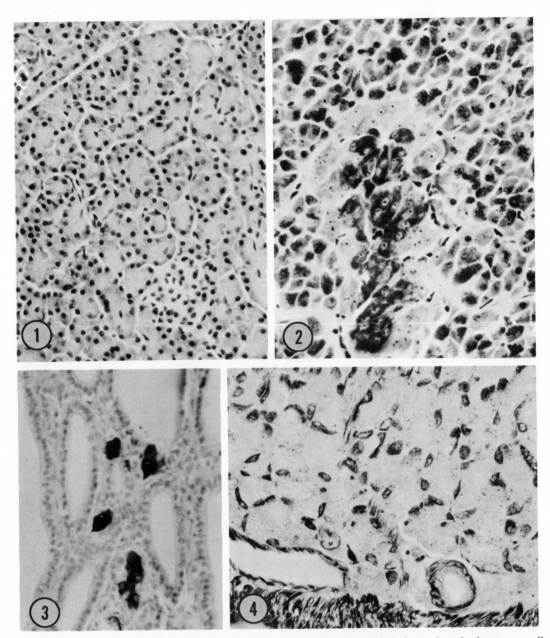

Figs. 1–4   Fig. 1: Carazzi's alum hematoxylin; pig pancreas. Nuclear chromatin is intensely stained; note a small islet at bottom right. (×120.) Fig. 2: Heidenhain's iron hematoxylin, horse pancreas. Note heavily stained glucagon-producing A cells in the center of an islet and zymogen granules in the exocrine parenchyma; nucleoli and red blood corpuscles are also stained. (×450.) Fig. 3: MacConaill's lead hematoxylin; horse thyroid. Heavily stained calcitonin-producing C cells. (×300.) Fig. 4: Mallory's phosphotungstic hematoxylin, pig stomach. Oxyntic cells filled with intensely stained mitochondria; note also reactivity of muscularis mucosae and muscular layer of blood vessels. (×300.)

carminic acid. As a rule, currently used aluminum hematoxylins are combined with red acid dyes (eosin, erythrosin, phloxine) to obtain differential stainings of cytoplasms and intercellular structures. Iron hematoxylins may be used in careful cytological investigations either as such or in polychrome stains in conjunction with other dyes, like Weighert's hematoxylin in Masson's trichrome stain. Gomori's chromium hematoxylin-phloxine technique

has been extensively applied as selective stain for α (red due to phloxine) and β granules (blue due to hematoxylin) of the pancreatic islet cells.

Lakes displaying completely different staining patterns are yielded by hematoxylin with phosphotungstic and phosphomolibdic acids. These acid mordants bind basic groups like amino, guanidine, and imidazole groups; hence structures containing proteins or polypeptides bearing disposable basic groups are preferentially stained. Mallory's phosphotungstic hematoxylin (Fig. 4) gives differential stainings of many structures including nuclei mitochondria, secretory granules, fibrin, neuroglia and muscle fibers (blue), and ground cytoplasm, collagen, elastin, reticulum, cartilage, and bone matrix (yellowish to brownish red).

Aside from the nature of the mordant, the selectivity of staining also depends upon the pH of the lake and the nature of the solvent. For instance, MacConaill's lead hematoxylin (Fig. 3) (pH about 6) stains protein carboxyls much better than alum hematoxylins (with strongly acid pH values, usually of about pH 2–3). In fact, below pH 4 side-chain carboxyls, accounting for most anionic groups of proteins, are practically undissociated. Aqueous or alcoholic solutions are used; their results are mostly comparable. However, the former should be preferable for staining hydrophilic structures made up of nucleic acids, proteins, and some mucosubstances, the latter for staining lipids or lipid-enveloped structures.

The mechanism of staining of hematoxylin lakes is certainly more complex than that of simple acid or basic dyes. In addition to interaction between polar groups, binding of the lake to the substrate seems to be accounted for by covalent links giving stability to the stainings even in nonpolar solvents like ethyl alcohol and xylol. The mordant can be allowed to interact with the substrate independently from the dye (usually before it) with results generally comparable to those of the preformed lake. Preformed lakes are preferred in the case of aluminum and phosphotungstic hematoxylins, which give progressive stainings that are easily controlled. The mordant-dye sequence is frequently employed in the case of iron hematoxylin, for instance in Heidenhain's and Regaud's techniques giving regressive stainings whose selectivity comes from careful differentiation of overstained sections in mordants, oxidizing agents, or acids.

Hematoxylin is scarcely used in modern histochemistry, though it had some application in the past. It was used as a chelating dye for the detection of metals (calcium, copper, iron, lead, beryllium, tin, nickel, etc.). The dye is certainly unable to give exact identification of individual metals, yet it may be useful in routine histopathology as a general indicator of metal deposits. Perhaps, the staining of enterochromaffin, eleidin, keratohyalin, and leucocyte granules as well as of elastic laminae of rodent arteries and hemosiderin by simple hematoxylin solutions (Clara's hematoxylin) is somewhat related to the complexing properties of the dye.

The affinity of bichromate-treated myelin for hematoxylin solutions has been known since 1885, when Weighert introduced his classical myelin method, subsequently modified by Pal and Kultschitzky. Early in this century Smith and Dietrich developed a method for lipid histochemistry based on the same principle; from this method Baker developed his "acid hematein" test for phospholipids which is already extensively used, although its specificity appears to be poor and its mechanism quite disputed.

E. SOLCIA

References

Baker, J. R., "Principles of Biological Microtechnique," London, Methuen, 1958.
Langeron, M., "Précis de Microscopie," Paris, Masson, 1949.
Lillie, R. D., "Histopathologic Technique and Practical Histochemistry," New York, McGraw-Hill, 1965.
Solcia, E., C. Capella, and G. Vassallo, *Histochemie,* **20**:116 (1969).
Terner, J. Y., J. Gurland, and F. Gaer, *Stain Technol.,* **39**:141 (1964).
Thompson, S. W., "Selected Histochemical and Histopathological Methods," Springfield, Ill., Thomas, 1966.
Wigglesworth, V. B., *Quart. J. Microscop. Sci.,* **93**:105 (1952).

HEMATOXYLIN STAINING FORMULAS

Mordant staining with hematoxylin results, in general, in black nuclei, heavily and densely stained, from which the stain is with difficulty removed by subsequent treatment. It is therefore to be recommended in those cases in which it is desired to follow with a complex counterstaining, particularly those which involve the use of an acid rinse. The method most commonly employed is still that of Heidenhain 1892, from which the majority of the other formulas have been derived. These variations are for the most part in the concentration of the solutions employed or in the temperature at which they are used. The technique of Régaud 1910 is very frequently specified in Europe as a prior staining for complex after-staining methods. During the early part of the present century, the substitution of ferric chloride for ferric alum as a mordant was followed in many circles. Mordants other than iron are rarely employed, although the introduction of copper by Faure 1924 has provided histologists with a staining method considered by many people to be a definite improvement. Diamond 1945 (Tech. Bull., **6**:68) finds that the addition of 0.1 % of a wetting agent (Tergitol 7) improves the stain.

After Ferric Alum Mordants

Benda 1893 (*Verh. anat. Ges. Jena.* **7**:161)
REAGENTS REQUIRED: A. Benda's Iron Mordant (see note) B. 1 % hematoxylin; C. Benda's Iron Mordant (see note).
METHOD: [sections] → water → A, 24 hr → rinse → B, until black → C, until differentiated → balsam, via usual reagents.
NOTE: (Benda's Iron Mordant): This is roughly equivalent in iron content to a 70 % solution of ferric sulfate. This salt, however, is so hygroscopic that it is almost impossible to prepare a solution of known concentration from the solid. The additional acids both stabilize the solution and render it more effective as a mordant.
FORMULA: water 40, sulfuric acid 15, ferrous sulfate 80, nitric acid 15.
PREPARATION: Dissolve the sulfate in the sulfuric acid and water. Heat to 50°C. Add nitric acid.

Bütschli (Bütschli, A., "Untersuchungen uber mikroskopische Schaume und das Protoplasma," Leipsig, Engelmann, 1892)
REAGENTS REQUIRED: A. 2 % ferric acetate; B. 0.5 % hematoxylin
METHOD: [sections of protozoans] → A, 24 hr → rinse → B, 3 hr → distilled water → balsam, via usual reagents.

Diamond 1945 (*Amat. Photogr. Lond.,* **15**:68)
REAGENTS REQUIRED: A. 4 % ferric alum; B. water 100, hematoxylin 0.5, Tergitol 7 0.1; C. sat. aq. sol. picric acid

METHOD: [sections or smears] → water → A, 5 min → rinse → B, 5 min → rinse → C. until differentiated, 3–5 min → tap water, until blue → balsam, via carbol-xylene

Dobell 1914 (*Arch. Protistenk*, **34**:139)
REAGENTS REQUIRED: A. 1% ferric alum in 70% alcohol; B. 1% hematin in 70% alcohol; C. 0.1% HCl in 70% alcohol
METHOD: [thin sections or protozoan smears] → 70% alcohol → A, 10 min → B, 10 min → 70% alcohol, quick rinse → C, until differentiated → 70% alcohol, wash → counterstain → balsam, via usual reagents

Freitas 1936 (*Rep. Lond. Homoeop. Hosp.*, **31**:707).
REAGENTS REQUIRED: A. 0.5% ferric alum in 70% alc.; B. 1% hematin in 70% alc. 60; phosphate buffer pH 7.6 30: C. 0.3% picric acid
METHOD: [sections] → A, 1 hr → B, 1 hr → wash → C, until differentiated → wash → balsam, via usual reagents

French 1923 (*Amer. J. Trop. Med.*, **3**:213)
REAGENTS REQUIRED: A. 3.5% ferric alum; B. 95% alc. 98, hematoxylin 1, sat. aq. sol. lithium carbonate 2; C. 1% ferric alum
METHOD: [sections or smears] → A, overnight → rinse → B, overnight → wash → C, until differentiated → wash
NOTE: Solution B is attributed, without reference to Rosenbush

Haggquist 1933 (*Z. wiss. Mikr.*, **50**:77)
REAGENTS REQUIRED: A. 5% ferric chloride; B. 1% hematoxylin; C. 1% ferric chloride
METHOD: [sections] → water → A, 1 hr → quick rinse → B, 1 hr → wash → C, until differentiated → wash → balsam, via usual reagents

Hance 1933 (*Science*, **77**:287)
STOCK FORMULA I: 10% hematoxylin in 95% alcohol
REAGENTS REQUIRED: A. 2.5% ferric alum; B. water 100 ml, stock I 10 ml, 2.5% sodium bicarbonate 0.1
METHOD: [sections] → A, 1/2–2 hr → distilled water, rinse → B, 1/2 hr → tap water, rinse → A, until differentiated → tap water, until blue → balsam, via usual reagents
NOTE: In preparing B solution, the bicarbonate should be added until the color changes from yellow to plum. This color change is of more importance than the quantity of alkali required to produce it.

Heidenhain 1892 (*Festchr. Kolliker*, 118)
REAGENTS REQUIRED: A. 2.5% ferric alum; B. 0.5% hematoxylin, "ripened" at least one month
METHOD: distilled water → A, 30 min–24 hr → distilled water, rinse → B, 30 min–24 hr → tap water, rinse, → A, until differentiated → tap water, until blue → balsam, via usual reagents
NOTE: Murray 1919 (*Rep. Cancer Res. Fd.*, **16**:77) substitutes 3.5% ferric alum for A, above. Masson 1912 (*Bull. Soc. anat. Paris*, **87**:291) substitutes 4% ferric alum for A and 1% hematoxylin for B, mordanting and staining 5–10 min at 50°C. See also Masson 1912.

Kofoid and Swegy 1915 (*Proc. Amer. Ass. Adv. Sci.*, **51**:289)
REAGENTS REQUIRED: A. 0.4% ferric alum in 50% alc.; B. 0.50% hematoxylin in 70% alc.

METHOD: [smears] → 50% alc. → A, 10 min → rinse, 50% alc. → B, 30 min → A, until differentiated → wash thoroughly
RECOMMENDED FOR: protozoan smears

Masson 1912 (Heidenhain 1892 (note)

Murray 1919 Heidenhain 1892 (note)

Régaud 1910 (*Arch. Biol. Paris*, **11**:291)
REAGENTS REQUIRED: A. 5% ferric alum; B. water 80, hematoxylin 1, glycerol 10, alcohol 10; C. sat. sol. picric acid in 95% alc. 65, 95% alc. 35
METHOD: distilled water → A, 30 min, 50°C → distilled water, wash → C, until differentiated → tap water, until blue → balsam, via usual reagents

Shortt 1923 (*Ind. J. Med. Res.*, **10**:836)
REAGENTS REQUIRED: A. 2.5% ferric alum; B. water 95, hematoxylin 1, phenol 5
METHOD: identical with Heidenhain 1892

After Ferric Chloride Mordants

Cole 1926 (*Science*, **64**:452)
STOCK SOLUTIONS: abs. alc. 100, sodium hydrosulfite 1, hematoxylin 5
REAGENTS REQUIRED: A. 95% alc. 50, water 50, ferric chloride 5, acetic acic 10; B. Stock 3, ammonia 0.5, water 100; C. 0.1% hydrochloric acid; D. 0.01% ammonia
METHOD: water → A, on slide, 5 min → rinse → B, on slide, 10 min → C, until differentiated → D, until blue → balsam, via usual reagents

Mallory 1900 (*J. Exp. Med.*, **5**:18)
REAGENTS REQUIRED: A. 10% ferric chloride; B, 1% hematoxylin; C. 0.25% ferric chloride
METHOD: [sections] → distilled water → A, 3–5 min → B, on slide, draining and renewing till precipitate ceases to form, until blue-black, 3–5 min → tap water, wash → C, until differentiated → tap water, wash → balsam via oil of cretan origanum

After Other Mordants

Apáthy 1888 (*Z. wiss. Mikr.*, **5**:47)
REAGENTS REQUIRED: A. 1% hematoxylin; B. 1% potassium dichromate
METHOD: [sections] → water → A, overnight → B, until stain first produced sufficiently differentiated → water, until yellow color removed → balsam, via usual reagents

Bensley and Bensley 1938 (Bensley and Bensley 1938, 79)
REAGENTS REQUIRED: A. sat. sol. copper acetate; B. 5% potassium chromate; C. 1% hematoxylin; D. Weigert's mordant (see note) 20, water 80
METHOD: [sections] → water → A, 5 min → rinse → B, dip → rinse → C, 2 min → A, 1 min → D, until differentiated → balsam, via usual reagents
NOTE: Weigert's (1896) "primary mordant" (*Ergebn. Anat. EntwGesch.*, **6**:14); is prepared by boiling to complete solution 5 pot. dichromate and 2.4 chromium fluoride in 100 water

Heidenhain 1884 (*Arch. mikr. Anat.*, **24**:468)
REAGENTS REQUIRED: A 0.3% hematoxylin; B. 0.5% potassium chromate
METHOD: [whole objects or sections] → water → A, overnight → rinse → B, until dark stain first produced sufficiently differentiated → water, until yellow color removed → balsam, via usual reagents

Knower 1930 (*Science*, **72**:172)

PREPARATION OF STOCK SOLUTION: Dissolve 5 hematoxylin in 100 abs. alc. Add 1 sat. sol. sodium metabisulfite.

REAGENTS REQUIRED: A. (identical with Cole 1926 A. sol. above), B. stock 4, water 8, ammonia 1, 95% alcohol 100; (Mix in order given; leave first three ingredients 30 sec before diluting); C. 0.1% hydrochloric acid; D. sat. sol. lithium carbonate in 70% alcohol

METHOD: [sections from material fixed in any copper-containing fixative] → 95% alc. → A, 5 min → 95% alc. quick rinse → B, 5 min → C, until differentiated → D, 2 min → balsam, via usual reagents

NOTE: Though this was originally intended as a nerve stain, it is far too good a general purpose stain to be omitted from this section. The original calls for "sodium bisulfite" which is not soluble to the extent indicated; however "the bisulphite of commerce usually consists chiefly, or almost entirely, of sodium pyrosulphite" (*Merck Index*, 5th ed., 1940, 506). Sodium pyrosulfite, which is commonly sold as "metabisulfite," is soluble to the required extent and is accordingly specified in the formula given above.

Kulschitzky 1889 (*Anat. Anz.*, **4**:223)

FORMULA: water 98, acetic acid 2, hematoxylin 1

USE: For general purposes as Knower 1930. The formula of Wolters 1890 (*Z. wiss. Mikr.*, 7:466) differs only in containing 2 parts hematoxylin.

Nissl 1894 (*Allg. Z. Psychiat.*, **51**:245)

REAGENTS REQUIRED: A. 2% ferric acetate in 90% alc.; B. 1% hematoxylin in 70% alc.; C. 1% hydrochloric acid in 70% alc.

METHOD: [sections of brain] → A, 30 min → rinse → B, 30 min → rinse → C, until differentiated → balsam, via usual reagents

Wolters 1890 (*Z. wiss. Mikr.*, 7:466)
see Kulschitsky 1889 (note)

Direct Hematoxylin Staining

Hematoxylin cannot be used in direct staining unless some mordant is incorporated with the solution for the purpose of fixing the stain on the material to be colored. The term *direct staining* is used in this case in contrast to *mordant staining* and must not be confused with the term *direct staining* as opposed to *indirect staining*. The former indicates merely that the object is stained in a relatively strong solution for a length of time sufficient to impregnate the whole and that it is subsequently exposed either to an acid solution or to a solution of the mordant with a view to extracting it from those objects which it is desired to bring into contrast. The term *indirect staining*, as used in this same sense, indicates the employment of a very weak solution in order to permit a differential absorption of the stain by those parts of the object to be stained (usually the more dense) which it is intended to bring out. As a generality it may be said that direct staining, in this sense, is usually applied to sections while indirect staining is better for the preparation of whole mounts, provided that one has the leisure to wait for the somewhat lengthy process to finish.

The direct-staining formulas are divided into four classes according to the mordant which is incorporated. The first group, incorporating iron mordants, is used almost exclusively for staining the central nervous system in sections. It is doubtful that these stains could ever be employed for indirect staining of whole mounts, but for staining nuclei they are far better than the other three classes, though less widely employed.

The next two divisions include these formulas containing alum mordants and acid-alum mordants, the separation of these two being necessitated by the large number of formulas to be found in each. Both are employed for sections and for whole mounts, the best known being unquestionably the formula of Delafield (1885). This reagent has the advantage of being almost foolproof, but it has to be ripened for a considerable period before it can be employed: Watson 1945 (*J. R Micr. Soc.*, **63**:21) recommends barium peroxide for ripening these solutions. The formula of Carazzi 1911 is almost identical but may be used as soon as it is prepared. It is strongly recommended to the attention of those whose staining has previously been confined to Delafield. The formula of Mayer 1896 was once very widely employed for staining whole mounts; but it has nowadays fallen somewhat into disuse. It also required ripening for a considerable period before employment.

The formulas incorporating an acid, usually acetic, in addition to the alum mordant are among the best of the general purpose stains. The formula of Ehrlich 1886 is the most widely known, though any of the others can be recommended.

The alum-mordant formulas are the only ones which can be employed in great dilution for indirect staining. It is usually a waste of time to employ acid-alum formulas for this purpose. The diluent to be employed should have the same composition as the formula itself, without the inclusion of hematoxylin. It is a mistake to follow the very wide recommendation that 0.1% hydrochloric acid be employed. This reagent is difficult to remove from the object before its final mounting and leads ultimately to the breakdown of the color.

In all cases hematoxylin stains should be "blued" after they have been differentiated, in some alkaline solution, preferably containing free ions of an alkali metal. Lithium carbonate is widely used, though a weak solution of calcium chloride, adjusted with ammonium hydroxide to a pH of about 8, is more satisfactory. The old exhortation to use tap water originated in Europe where most of the tap waters are alkaline. The majority of city tap waters in the United States are worthless for this purpose.

The chrome-hematoxylins and copper-hematoxylins, which form the fourth class, are of comparatively recent introduction or, at least, of comparatively recent acceptance. The formulas of Hansen 1905 and of Liengme 1930 are, however, excellent reagents and should be tried for sections in those instances in which the more customary formulas do not yield satisfactory results. The phosphomolybdic and phosphotungstic hematoxylins of Mallory, though originally intended for staining nervous structures, are useful for a more general purpose. Mayer's 1891 "haemacalcium" was originally intended for staining whole mounts of small marine invertebrates and is admirable for the purpose.

Formulas Incorporating Iron Mordants

Anderson 1929 (Anderson 1929, 129)

STOCK SOLUTIONS: I. 0.5% hematoxylin in 50% alc. 100, 2% calcium hypochlorite 5; II. water 100, ferric alum 3, sulfuric acid 2.5

WORKING SOLUTION: stock solution I 60, stock solution II 30

NOTE: Overstaining of sections rarely occurs but may be corrected with 0.1 % hydrochloric acid in 70 % alcohol.

Faure 1924 (*C. R. Ass. franç., Trav. publ.*, **90**:87)

REAGENTS REQUIRED: A. 90 % alcohol 100, hematoxylin 3.2; B. water 100, ferric chloride 0.2, cupric acetate 0.1, hydrochloric acid 2; C. 1 % hydrochloric acid; D. sat. aq. sol. lithium carbonate

METHOD: [distilled water] → A + B (equal parts), 5 sec → wash, tap water → C, quick rinse → wash, tap water → D, until blue → wash → [counterstain, if desired] → balsam, via usual reagents

Hansen 1905 (*Z. wiss. Mikr.*, **22**:55)

FORMULA: water 100, ferric alum 4.5, hematoxylin 0.75

PREPARATION: Dissolve the alum in 65 water. Dissolve the dye in 35 water. Mix. Boil. Cool. Filter.

Held (Gatenby and Painter 1937, 374)

REAGENTS REQUIRED: A. 5 % ferric alum; B. Held's phosphotungstic hematoxylin (see below) 5, water 95

METHOD: [water] → A, 24 hr → B, 12–24 hr → A, until differentiated → balsam, via usual reagents

Janssen 1897 (*Cellule*, **14**:207)

FORMULA: water 70, ferric alum 5, hematoxylin 1, abs. alc. 5, glycerol 15, methanol 15

PREPARATION: Dissolve the alum in the water and the hematoxylin in the alc. Mix. Leave 1 week. Filter and add remaining ingredients.

NOTE: This formula is recommended by Lillie and Earle 1939 (*Stain Tech.*, **14**:53) as a substitute for Weigert 1904.

Kefalas 1926 (*J. R. Micr. Soc.*, **46**:277)

FORMULA: acetone 100, ferric chloride 1, hydrochloric acid 0.05, hematoxylin 1

Krajian 1950 (Krajian 1950, 196)

FORMULA: water 50, 95 % alc. 50, hematoxylin, ferric alum, ferric chloride, potassium iodide a.a. 6

PREPARATION: Dissolve salts in water and dye in alc. Mix.

NOTE: This formula, designed for bacteriology is an excellent general-purpose hematoxylin.

La Manna 1937 (*Z. wiss. Mikr.*, **54**:257)

FORMULA: water 100, hematoxylin 1, ferric chloride 3

Lillie and Earle 1939 (*Amer. J. Pathol.*, **15**:765)

STOCK SOLUTIONS: I. 95 % alc. 50, glycerol 50, hematoxylin 1; II. water 100, ferric alum 15, ferrous sulfate 15

WORKING STAIN: stock I, 50, stock II 50

Lillie 1940 (*Arch. Pathol. Anat.*, **29**:705)

FORMULA: water 100, ferric chloride 1.2, hematoxylin 1, hydrochloric acid 1

Morel and Bassal 1909 (*J. Anat. Paris*, **45**:632)

STOCK SOLUTIONS: I. 95 % alc. 100, hematoxylin 1; II. water 100, ferric chloride 2, copper

WORKING SOLUTION: stock I 50, stock II 50

Paquin and Goddard 1947 (*Bull. Int. Ass. Med. Mus.*, **27**:198)

FORMULA: water 75, 95 % alc. 25, glycerol 13, ferric alum 5, ammonium sulfate 0.7, hematoxylin 0.8

PREPARATION: Dissolve the dye in the glycerol and alc. with gentle heat. Cool and add, with constant agitation, the other ingredients dissolved in the water. Leave 24 hr.

Rozas 1935 (*Z. wiss. Mikr.*, **52**:1)

REAGENTS REQUIRED: A. water 74, 95 % alc. 6, glycerol 20, ferric alum 1, aluminum chloride 1.2, hematoxylin 0.6; B. 20 % feric alum

METHOD: [sections] → water → A, 12–24 hr → B, until differentiated → balsam, via usual reagents

Thomas 1943 (*Bull. Histol. Tech. Micr. (= Bull. d Hist. Appl.)*, **20**:212)

FORMULA: water 60, dioxane 40, acetic acid 6, hematoxylin 2.5, ferrous chloride 6, ferric chloride 1.5, ferric alum 3, hydrogen peroxide 1

PREPARATION: Dissolve the hematoxylin in the dioxane and add the hydrogen peroxide. Dissolve the salts in the water and acid. Filter into the hematoxylin solution.

Weigert 1903 cited from 1910 ips. (Ehrlich, Krause, *et al.* 1910, **1**:231)

STOCK SOLUTIONS: I. 0–4 % ferric chloride; II. 1 % hematoxylin in 95 % alc.

REAGENTS REQUIRED: A. stock I 50, stock II 50; B. sat. sol. picric acid in 95 % alc. 65, 95 % alc. 35

METHOD: [distilled water] → A, 1–2 hr → distilled water, rinse → B, until differentiated → tap water, until blue → balsam, via usual reagents

Weigert 1904 (*Z. wiss. Mikr.*, **21**:1)

STOCK SOLUTIONS: I. water 95, ferric chloride 0.6, hydrochloric acid 0.75; II. 1 % hematoxylin in 95 % alc.

REAGENTS REQUIRED: A. stock I 50, stock II 50

METHOD: [distilled water] → stain, until sufficiently colored → distilled water, wash → tap water, until blue → balsam, via usual reagents

Wittmann 1962 (*Stain Tech.*, **37**:27)

MORDANT: 95 % alc. 50, iodic acid 1.7, aluminum alum 1.7. chrome alum 1.7. Immediately before use add 50 hydrochloric acid

STAIN: water 55, acetic acid 45, hematoxylin 4, ferric alum 1

RECOMMENDED FOR: Plant squashes

NOTE: Wittman adapts this stain to animal chromosomes by omitting the mordant and adding 40 chloral hydrate to 20 stain.

Yasvoyn cited from 1946 Roskin (Roskin 1946, 150)

STOCK SOLUTIONS: I. 0.1 % hematoxylin; II. 2.5 % ferric alum

WORKING SOLUTIONS: add II with constant stirring to 20 drops I until solution just remains blue

METHOD: [sections] → stain, 2–5 min → 70 % alc. if differentiation required → balsam, via usual reagents

Formulas Incorporating Alum Mordants

Belloni 1939 cited from 1943 Cowdry (Cowdry 1943, 35)

FORMULA: water 100, potassium alum 3, hematoxylin 0.15, chloral hydrate 0.1, potassium hydroxide 0.01

Böhmer 1868 (*Arch. mikr. Anat.*, **4**:345)

STOCK SOLUTIONS: I. 3.5 % hematoxylin in abs. alc.; II. 0.3 % potassium alum

WORKING FORMULA: stock I 10, stock II 90

Bullard, in verb. Harpst 1951
FORMULA: water 225, 95% alc. 35, glycerol 33, acetic acid 3.5, ammonium alum 6, mercuric oxide 0.8, hematoxylin 0.8
PREPARATION: Dissolve the dye in 15 of 50% alc. with 2 acetic acid. Heat and add 2 ammonium alum in 25 water. Boil, add mercuric oxide, and filter. Add remaining ingredients to cold filtrate.
NOTE: The author has not been able to discover any literature reference for this solution which is widely used in pathological laboratories.

Cajal and de Castro 1933 (Cajal and de Castro 1933, 77)
PREPARATION OF STOCK: To 100 of a 5% solution of hematoxylin add 5 ammonia. Evaporate to dryness.
STAINING SOLUTION: 2% solution of above powder in 50 % alc. 50; 5% ammonium alum 50

Carazzi 1911 (*Z. wiss. Mikr.*, **28**: 273)
REAGENTS REQUIRED: A. water 80, potassium alum 5, hematoxylin 0.1, potassium iodate 0.02, glycerol 20; B. 0.1% hydrochloric acid in 70%
METHOD FOR WHOLE OBJECTS: water → 1A + 10 water, 3–10 min → distilled water → B, if differentiation necessary → tap water until blue → balsam, via usual reagents
METHOD FOR SECTIONS: water → A, until sufficiently stained, 5 min–12 hr → distilled water, wash → tap water, until blue → balsam, via usual reagents

Delafield cited from Prudden 1885 (*Z. wiss. Mikr.*, **2**:288)
REAGENTS REQUIRED: A. water 70, ammonium alum 3, hematoxylin 0.6, abs. alc. 4, glycerol 15, methanol 15
METHOD: as Carazzi 1911
NOTE: The original calls for preparation from sat. aq. sol. ammonia alum and 16% hematoxylin solution.

Friedländer 1882 (Friedländer 1889, 92)
FORMULA: water 30, glycerol 30, 95% alc. 30, hematoxylin 0.6, potassium alum 0.6

Gage 1892 cited from 1896 ips. (Gage 1896, 178)
FORMULA: water 100, potassium alum 4, chloral hydrate 2, 95% alc. 2, hematoxylin 0.1

de Groot 1912 (*Z. wiss. Mikr.*, **29**:182)
FORMULA OF SOLVENT: 95% alc. 65, water 27, glycerol 8
FORMULA OF STAIN: 95% alc. 65, water 27, glycerol 8, hydrogen peroxide 0.75, hematoxylin 0.2, calcium chloride 1.5, sodium bromide 0.75, ammonium alum 2.2, potassium ferricyanide 0.08
PREPARATION OF STAIN: Take 100 solvent. Mix 1.5 solvent with 0.75 hydrogen peroxide and dissolve 0.2 hematoxylin in this. In 25 solvent dissolve 15 calcium chloride and 0.75 sodium bromide. Add this to the hematoxylin solution and dissolve 1.1 ammonium alum in the mixture. In 40 solvent dissolve 0.08 potassium ferricyanide and add this to mixture. In the remaining solvent dissolve 1.1 ammonium alum and add to the mixture.

Harris 1900 (*J. Appl. Micr.*, **3**:777)
FORMULA: 95% alc. 5, hematoxylin 0.5, potassium alum 10, water 100, mercuric oxide 0.25

PREPARATION: Dissolve the hematoxylin in alc. Dissolve alum in water and raise to boiling. Pour hot solution in hematoxylin. Boil and throw mercuric oxide into boiling solution. Cool rapidly. Filter.
NOTE: Mallory 1938, 72 suggests the addition of 5% acetic acid.

Harris and Power cited from 1884 Cole (Cole 1884b, 42)
FORMULA: hematoxylin 20, alum 60, water 100, abs. alc. 6
PREPARATION: Grind the hematoxylin in a mortar with the alum, adding water in small portions while grinding. Filter and add alcohol.

Haug (Pollack 1900, 84)
FORMULA: water 100, aluminum acetate 5, abs. alc. 5, hematoxylin 5.5
PREPARATION: Add the hematoxylin dissolved in the alc. to the acetate dissolved in the water.

Kleinenberg 1876 (Böhm and Oppel 1907, 103)
STOCK SOLUTIONS: I. water 30, 95% alc. 70, calcium chloride to sat., ammonium alum to sat.; Stock I 12, sat. sol. pot. alum in 70% alc. 88; III. sat. alc. sol. hematoxylin
WORKING SOLUTION: stock II 100, stock III 3

Kleinenberg (Cole 1884b, 42)
FORMULA: sat. sol. calcium chloride in 70% alc. 15, sat. sol. potassium alum in 70% alc. 85, sat. sol. hematoxylin in abs. alc. 1

Launcy 1904 cited from 1907 Böhm and Oppel (Böhm and Oppel 1907, 356)
FORMULA: water 100, potassium alum 0.5, hematin 1

Lee 1905 (Lee 1905, 188)
FORMULA: water 100, hematoxylin 0.1, ammonium alum 5, sodium iodate 0.02, chloral hydrate 5

Mallory 1938 (Mallory 1938, 70)
FORMULA: water 100, potassium alum 5, hematoxylin 0.25, thymol 0.25

Martinotti 1910 (*Z. wiss. Mikr.*, **27**:31)
FORMULA: water 70, glycerol 15, methanol 15, hematin 0.2, ammonium alum 1.5
PREPARATION: Dissolve alum in 60 water. Add hematin dissolved in 10 water. Add other ingredients to mixture.

Mayer 1891 (*Mitt. zool. Stat. Neapel.*, **10**:172)
FORMULA: Water 100, 95% alc. 5, ammonium alum 5, hematoxylin 0.1
PREPARATION: Add the hematoxylin dissolved in the alc. to the alum dissolved in the water. Ripen some months.
NOTE: Mayer (*loc. cit.*) also recommended the addition of 2% acetic acid to the above, when used for sections.

Mayer 1896 *Mayer's Glycheaemalum—auct.* (*Mitt. zool. Stat. Neapel.*, **12**:310)
FORMULA: water 70, glycerol 30, ammonium alum 5, hematoxylin 0.4
PREPARATION: Grind the hematoxylin to a stiff paste with a little of the glycerol. Mix the other ingredients and use the solution to wash out the mortar with successive small doses.

Mayer 1901 (*Z. wiss. Mikr.*, **28**:273)
FORMULA: water 100, potassium alum 5, hematoxylin
0.1, sodium iodate 0.02
NOTE: Mayer 1903 (*Z. wiss. Mikr.*, **20**:409) substitutes
ammonium alum for potassium alum.

Prudden 1885 see Delafield (1885)

Rawitz 1895a (Rawitz 1895, 62)

Rawitz 1895b (Rawitz 1895, 63)

Rindfleisch cited from 1877 Frey (Frey 1877, 100)
STOCK SOLUTIONS: I. sat. aq. sol hematoxylin; II.
sat. aq. sol (*circ.* 14%) ammonium alum
WORKING SOLUTION: water 85, stock I 10, stock II 5

Sass 1929 (*Stain Tech.*, **4**:127)
FORMULA: water 100, ammonium alum 5, hematoxylin
0.1, sodium iodate 0.1

Tribondeau, Fichet, and Dubreuil 1916 (*C.R. Soc. Biol.
Paris*, **79**:288)
FORMULA FOR STOCK SOLUTION: 95 alc. 100,
water 22, silver nitrate 2, sodium hydroxide 1, hematoxylin 5
PREPARATION OF STOCK SOLUTION: Add the
hydroxide dissolved in 20 water to the silver nitrate
dissolved in 2 water. Wash ppt. by decantation and
transfer to flask with reflux condenser. Dissolve the
hematoxylin in alc. and add to silver suspension. Raise
to boiling, cool, filter.
WORKING SOLUTION: 5% potassium alum 100,
stock 5

Formulas Incorporating Acid Alum Mordants

Anderson 1923 cited from 1929 ips. (Anderson 1929, 192)
REAGENTS REQUIRED: A. water 90, abs. alc, 5,
hematoxylin 0.25, calcium hypochlorite 0.4, ammonium
alum 2, acetic acid 5; B. 0.1% hydrochloric acid
PREPARATION OF A: Add the hypochlorite to 20
water. Leave 4 hr Filter. Add filtrate to hematoxylin
dissolved in water. Dissolve other ingredients in rest of
water. Add this to dye solution.
METHOD: water → A, 2–3 min → B, until differentiated
→ tap water, until blue
Anderson 1929 (Anderson 1929, 129)
FORMULA: water 70, 95% alc. 5, acetic acid 5, calcium
hypochlorite 4, ammonium alum 3, hematoxylin 0.5
PREPARATION: as Anderson 1923

Apáthy 1897 (*Mitt. zool. Stat. Neapel.*, **12**:712)
FORMULA: water 45, 95% alc. 25, glycerol 34, hematoxylin 0.3, acetic acid 1, salicylic acid 0.03, ammonium
alum 3
PREPARATION: Dissolve the hematoxylin in 10 water,
25 alc. Allow to ripen for some months. Dissolve the
alum and acids in 35 water. Add to dye solution; then
add glycerol.

Cole 1903 (Cross and Cole 1903, 170)
FORMULA: water 32, 95% alc. 32, glycerol 29, acetic
acid 7.5, hematoxylin 0.6, ammonium alum 0.6
PREPARATION: Add alum dissolved in water to
hematoxylin dissolved in alc. Add other ingredients to
mixture.

Conklin (Guyer 1930)
STOCK SOLUTION I Delafield's 1885 alum hematoxylin
(above)
STOCK SOLUTION II water 100, picric acid 0.3,
sulfuric acid 2
WORKING SOLUTION water 80, stock I 20, stock II 4
NOTE: This has been widely used as a combined fixative-
stain for small invertebrates, particularly plankton.
The stock II solution was published as a fixative for
plankton by Kleinenberg (*Quart. J. Microsc. Sci.*,
19:208, 1879)

Ehrlich 1886 (*Z. wiss. Mikr.*, **3**:150)
REAGENTS REQUIRED: A. water 30, 95% alc. 30,
glycerol 30, acetic acid 3, hematoxylin 0.7, ammonium
alum to excess; B. sat 70% alc. sol. lithium carbonate
PREPARATION OF A: Dissolve hematoxylin in the alc.
and acid. Dissolve 1 ammonium alum in water and
glycerol. Mix with dye solution. Allow to ripen for
some months. Add excess (*circ.* 10) ammonium alum
to ripened solution.
METHOD FOR SECTIONS: 90% alc. → A, 1/2–2 min
→ 90% alc. applied from drop bottle, until differentiated → B, until blue → balsam, via usual reagents
NOTE: The passage of sections to water before staining,
or the use of water to differentiate, results in a diffuse
stain.

Harris 1900 see Harris 1900 (note)

Langeron 1942 (Langeron 1942, 523)
REAGENTS REQUIRED: A. Mayers 1901 alum
hematoxylin (see above) 100, chloral hydrate 5, citric
acid 0.1; B. 0.1% HCl in 70% alc.
METHOD: [sections or whole objects] → distilled water
→ A, (sections) 10 min or A, (whole objects) 24–48hr →
B, until differentiated → tap water until blue → balsam,
via usual reagents

Lillie 1941 (*Stain Tech.*, **16**:5)
FORMULA: water 70, glycerol 30, acetic acid 20,
hematoxylin 0.5, sodium iodate 0.1
PREPARATION: Add the iodate to the dye dissolved in
the water. Leave overnight. Add other ingredients.
NOTE: This formula was republished, without any
reference to its previous publication, by Lillie 1942
(*Stain Tech.*, **17**:90)

Mann 1892 cited from 1934 Langeron (Langeron 1934,
475)
FORMULA: water 35, 95% alc. 32, glycerol 25, acetic
acid 3, hematin 0.6; potassium alum 3.5
PREPARATION: Dissolve the dye in the acid. Add
mixed alc. and glycerol. Then add alum dissolved in
water.

Masson cited from 1934 Langeron (Langeron 1934, 475)
FORMULA: water 100, acetic acid 2, hematin 2, po-
tassium alum 6
PREPARATION: Dissolve the alum in boiling water.
Add dye, cool, filter. Add acid to filtrate.

Mayer cited from 1924 Langeron (Langeron 1942, 525)
NOTE: This is Mayer's 1896 alum hematoxylin with the
addition of acetic acid. Mayer, however, recommended
this addition only to his 1891 formula. The present
solution is, therefore, Langeron's variant.

Pearse 1950 (*Stain Tech.*, **25**:77)
INGREDIENTS: water 100, hematoxylin 0.1, sodium
iodate 0.02, potassium alum 5, chloral hydrate 5,
citric acid 0.1
PREPARATION: Dissolve hematoxylin, iodate, and
alum in water. Leave overnight then add remaining
ingredients, boil 5 min and cool.
NOTE: Pearse refers to this as " Mayer's Hemalum."

Sass 1929 (*Stain tech.*, **4**:127)
FORMULA: water 100, ammonium alum to sat., hematoxylin 1, sodium iodate 1, acetic acid 3

Watson 1943 (*J. R. Micr. Soc.* **63**:20)
FORMULA: water 32 abs. alc. 32, glycerol 32, acetic
acid 3, ammonium alum 0.064, hematoxylin 0.64,
potassium permanganate 0.032
PREPARATION: Dissolve the alum and permanganate
in water. Add dye dissolved in alc. and then other
ingredients.
NOTE: This formula stains as well as Ehrlich 1886, from
which Watson developed it, and does not require
ripening.

Formulas Incorporating Other Mordants

Alzheimer 1910 *lithium-hematoxylin* (Nissl and Alzheimer
1910, 411)
FORMULA: water 90, 95% alc. 10, hematoxylin 1,
lithium carbonate 0.03
PREPARATION: Add the alkali dissolved in water to
the dye dissolved in alc.

Bacisch 1937 *lithium-hematoxylin* (*J. Anat. Lond.*, **72**:163)
FORMULA: water 100, hematoxylin 1, lithium carbonate 0.1

Bernbe et. al. 1965 (*Stain Tech.* **40**:165, 1965)
PREPARATION OF DRY STAIN: Dissolve 10 hematoxylin, 10 sodium hydroxide and 70 chrome alumin
600 distilled water, cool to room temperature. Filter,
allowing filtrate to drop into 3500 absolute alcohol.
Filter and air dry precipitate formed in alcohol.
WORKING SOLUTION: water 100, hydrochloric acid 3,
dry stain 3
METHOD: water → stain 20 min–16 hr but routinely 1 hr
→ 3% hydrochloric acid rinse → running water until
blue.

Clara 1923 *molybdic-hematoxylin* (*Z. wiss. Mikr.*, **50**:73)
STOCK SOLUTION: 1% hematoxylin 50, 10% am-
monium molybdate 50, molybdic acid to excess (*circ.* 1)

Cook cited from 1883 Hogg *copper-hematoxylin* (Hogg
1883, 237)
FORMULA: water 100, "extract of logwood" 15, copper
sulfate 2.5 potassium alum 15
PREPARATION: Grind the dry powders in a mortar.
Add enough water to make a paste. Leave 2 days; add
rest of water. Leave 12 hr. Filter.
NOTE: The "extract of logwood" is the result of evaporating an aqueous extract of logwood to dryness; in
addition to hematoxylin and hematin (about 70% of
the whole) it contains tannin, glucosides, and resins.

Crétin 1925 *ferricyanide-hematoxylin* (Langeron 1949, 571)
REAGENTS REQUIRED: A. water 100, ferrous
sulfate 4; B. water 100, potassium ferrocyanide 2,
potassium ferricyanide 1; C. water 100, hematoxylin
0.5; D. water 100, ferric alum 5

METHOD: Distilled water → A, 24 hr → running tap
water, overnight → B, 3–6 hr → distilled water, rinse →
C, overnight → D, until differentiated
RESULTS: nuclei dense, opaque black

Donnaggio 1904 *tin-hematoxylin* (*Ann. Nevrol. Napoli*,
22:192)
FORMULA: water 100, hematoxylin 0.5, stannic
chloride diamine 10
PREPARATION: Dissolve the dry salts each in 50 water.
Mix solutions.
NOTE: Donnaggio 1950 (*Rep. Geol. Trinidad*, **22**:171)
substitutes stannic chloride for the diaminine complex.

Gomori 1941 *chrome-hematoxylin* (*Amer. J. Pathol.*,
17:395)
FORMULA: water 100, sulfuric acid 0.1, potassium
dichromate 0.1, chrome alum 1.5, hematoxylin 0.5
PREPARATION: Dissolve the alum and dye each in 50.
Mix. Add 2.5% potassium dichromate and 2 5%
sulfuric acid. Ripen 2 days.

Hansen 1905 *chrome-hematoxylin* (*Z. wiss. Mikr.*, **22**:64)
FORMULA: water 100, sulfuric acid 0.2, hematoxylin
0.3, chrome alum 3, potassium dichromate 0.2
PREPARATION: boil the chrome alum in 85 water
until green. Dissolve hematoxylin in 5 water. Add to
alum solution. Then add successively, acid in 2 water
and dichromate in 7 water with constant stirring.
Filter.

Heidenhain cited from 1907 Böhm and Oppel *vanadium-
hematoxylin* (Böhm and Oppel 1907, 105)
FORMULA: 0.5% hematoxylin 60, 0.25% ammonium
vanadate 30

Held *phosphomolybdic hematoxylin* (Gatenby and Painter
1937, 158)
FORMULA: water 30, 95% alc. 70 hematoxylin 1,
phosphomolybdic acid 15
PREPARATION: Dissolve hematoxylin in solvents. Add
acid. Leave 1 month. Decant.

Hornyold 1915 cited from Gatenby and Painter 1937
iodine-hematoxylin (Gatenby and Painter 1937, 105)
REAGENTS REQUIRED: A. abs. alc. 25, water 75,
hematoxylin 0.8, ammonium alum 0.5, tincture of
iodine UDP (see note) 0.5; B. 0.1% acetic acid in 70%
alc.
PREPARATION OF A: Add the alum dissolved in the
water to the hematoxylin dissolved in the alc. Ripen
some days. Add iodine.
METHOD: [sections] → water → A, 5–10 min → rinse
→ B, until blue → balsam, via usual reagents
NOTE: The tincture of iodine mentioned in the original
formula is the British, which contains 2 1/2% each of
I_2 and KI in 90% alc. The official American tincture is
double this concentration and should be used in the
preparation above.

Kleinenberg 1876 *iron-copper-hematoxylin* see Squire
1892 (note)

Liengme 1930 *iron-copper-hematoxylin* (*Bull. Histol.
Tech. Micr.* [= *Bull d'Hist Appl.*], **7**:233)
REAGENTS REQUIRED: A. Böhmer 1868 50, Böhmer
1868 50, Morel and Bassal 1909 50; B. 0.1% hydrochloric acid; C. sat. sol. lithium carbonate in 70% alc.

METHOD: 70% alcohol → A, 1–4 days → B, until differentiated → 70% alcohol, wash → C, until blue → balsam, via usual reagents

Loyez cited from 1938 Carleton and Leach *lithium hematoxylin* (Carleton and Leach 1938, 259)
STOCK I: abs. alc. 100, hematoxylin 10; STOCK II: water 100, sat. as. sol. lithium carbonate 4
WORKING SOLUTION: stock I 10, stock II 90

Malllory 1891 *phosphomolybdic hematoxylin (Anat. Anz., 7:375)*
FORMULA: water 100, hematoxylin 1, phosphomolybdic acid 1, chloral hydrate 7.5
PREPARATION: Add the acid dissolved in 10 water to the dye dissolved in 90. Add chloral hydrate to mixture.
METHOD: water → stain, 10 min–1 hr → 30% alc. until differentiated
NOTE: Phenol may be substituted for chloral hydrate. Hueter (cited from Schmorl 1928, 173) substitutes phosphotungstic for phosphomolybdic acid; but see note under Schueninoff 1908.

Mallory 1900 *phosphotungistic hematoxylin (J. Exp. Med., 5:19)*
REAGENTS REQUIRED: A. water 100, hematoxylin 0.1, phosphotungistic acid 2, hydrogen peroxide 0.2
PREPARATION: Add the acid dissolved in 20 water to the dye dissolved in 80. Add hydrogen peroxide to mixture.
METHOD: [sections of material fixed in Zenker or similar fixative] → water, thorough wash → A, 12–24 hr → 95% alcohol, about one minute → abs. alc. until differentiation complete balsam, via xylene
NOTE: This is a polychrome, general-purpose stain. A thorough survey of phosphotungstic acid–hematoxylin lakes has been published by Tuner et al. (*Stain Tech., 39*:141, 1964).

Mallory cited from McClung 1929 *phosphotungstic hematoxylin* (McClung 1929, 298)
REAGENTS REQUIRED: A. 0.25% potassium permanganate; B. 5% oxalic acid; C. water 100, hematoxylin 0.1, phosphotungistic acid 2, potassium permanganate 0.025
METHOD: [sections of material fixed in Zenker or similar fixative] → A, 5–10 min → water, thorough rinse → B, 10–20 min → C, 12–24 hr → 95% alc. quick rinse → abs. alc., least possible time → balsam via xylene
NOTE: This is a polychrome general-purpose stain.

Mayer 1891 *Haemacalcium—auct. (Mitt. zool. Stat. Neapel., 10:182)*
REAGENTS REQURED: A. water 30, alc. 70, acetic acid 1.5, hematoxylin 0.15, aluminum chloride 0.15, calcium chloride 7.5; B. 95% alc. 30, water 70, aluminum chloride 2
PREPARATION OF A: Grind the dye with the aluminum chloride in a beaker. Add solvents and warm to solution. Then add calcium chloride.
METHOD: [whole objects] → 70% alc. → A, until stained, usually overnight → B, until differentiated → balsam, via usual reagents
NOTE: Mayer 1910 "*Haemastrontium*" (*cited* from Gatenby and Cowdry 1937, 160) differs from above only in substitution of strontium chloride for calcium chloride, and of 0.1 citric acid for 1.5 acetic acid.

Mayer 1910 *Haemastrontium—auct.* see Mayer 1891 (note)
Police 1909 *phosphomolybdic-hematoxylin (Arch. zool. (ital.) Napoli., 4:300)*
FORMULA: water 70, alc., 30 chloral hydrate 10, hematoxylin 0.35, phosphomolybdic acid 0.03

Rawitz 1909 *aluminum-hematoxylin (Z. wiss. Mikr., 25:391)*
FORMULA: water 50, glycerol 50, hematin 0.2, aluminum nitrate 2
PREPARATION: Add the aluminum nitrate dissolved in 25 water to the dye dissolved in 25 water. Then add glycerol.

Schröder 1930 *lithium-hematoxylin (Z. ges. Neurol. Psychiat., 166:588)*
FORMULA: water 100, hematoxylin 0.3, lithium carbonate 0.04

Schueninoff 1908 *phosphomolybdic hematoxylin (Zbl. allg. Path. path. Anat., 18:6)*
FORMULA: water 100, hematoxylin 0.9, phenol 2.5, phosphomolybdic acid 0.5
NOTE: Hueter 1911 (Romeis 1948. 351) differs from this only in the substitution of phosphotungistic acid; but see note under Mallory 1891.

Schweitzer 1942 cited from 1946 Roskin *chrome-hematoxylin* (Roskin 1946, 200)
FORMULA: water 125, chrome alum 5, hematoxylin 0.5, 10% sulfuric acid 4, potassium dichromate 0.275
PREPARATION: To the first three ingredients dissolved in 90 water add the dichromate dissolved in 10.

Squire 1892 *calcium-hematoxylin* (Squire 1892, 25)
FORMULA: water 10.5, alc. 96, hematoxylin 1, calcium chloride 8, ammonium alum 1.2.
PREPARATION: Add the alum dissolved in 6.5 water to the calcium chloride dissolved in 4 water. Add alc., leave 1 hr, filter. Dissolve dye in filtrate.
NOTE: In the early 1880s almost any alum-calcium chloride-hematoxylin was referred to as "Kleinenberg," who recommended this method of preparing an aluminum chloride–hematoxylin (which, in effect, this is) without recourse to the very acid salt in commerce in in his time. The method given by Kleinenberg 1876 (Grundsuge der Entwickelungsgeschichte, Leipsig) proved impractical and a revised method (*Quart. J. Micr. Sci., 74*:208) published in 1879, usually erroneously cited as the original, proved litle better.

Thomas 1943 *phosphomolybdic hematoxylin (Bull. Histol. Tech. micr.* [= *Bull d'Hist Appl.*] 20:49)
FORMULA: water 44, dioxane 40, ethylene glycol 11, phosphomolybdic acid 16.5, hematoxylin 2.5, hydrogen peroxide 2
PREPARATION: Dissolve the hematoxylin in dioxane and add the hydrogen peroxide. Dissolve the phosphomolybdic acid in the other solvents and filter into the hematoxylin.

PETER GRAY

References

To save space the undernoted books are referred to in the above article by author and date only. Literature references in the article are in customary form.

Anderson, J. "How to Stain the Nervous System,'. Edinburgh, E. Livingstone, 1929.

Böhm, A., and A. Oppel, "Manuel de Technique Microscopique, Traduit de l'allemand par Etienne de Rouville," 4th ed., Paris, Vigot, 1907.

Ramón y Cajal, S., and F. de Castro, "Elementos de téchnica micrográfica del sistema nerviosa," Madrid, Tipografia Artística, 1933.

Carleton, H. M., and E. H. Leach, "Histological Technique," 2nd ed., London and New York, Oxford University Press, 1938.

Cole, A. C., "The methods of microscopical research, n. d., bound with Cole 1883, vol. 2.

Cowdry, E. V., "Microscopic technique in biology and medicine," Baltimore, Williams and Wilkins, 1943.

Cross, M. I.,* and M. J. Cole, "Modern Microscopy," 3rd ed., London, Bailliere, Tindall and Cox, 1903.

Ehrlich, P., R. Krause, M. Mosse, H. Rosin, and K. Weigert, "Enzyklopädie der mikroskopischen Technik," 2nd ed., 2 vol., Berlin, Urban and Schwarzenberg, 1910.

Frey, H., "Das Mikroskop und die mikroskopische Technik," 6th ed., Leipzig, Wilhelm Engelmann, 1877.

Friedlander, C., "Mikroskopische Technik zum Gebrauch bei medicinischen und pathologisch-anatomischen Untersuchunge," 4th ed., by C. J. Eberth. Berlin, Fischer, 1889.

Gage, S. H., "The Microscope and Microscopical Methods," 6th ed., Ithaca, N.Y., Comstock, 1896.

Gatenby, J. B., and T. S. Painter, "The Microtomist's Vade-mecum (Bolles Lee)." 10th ed., Philadelphia, Blakiston, 1937.

Gray, P., "Microtomist Guide and Formulary," New York and Toronto, The Blakiston Company, Inc., 1954.

Guyer, M. F., "Animal Micrology," 3rd ed., Chicago, University of Chicago Press, 1930.

Hogg, J., "The Microscope," London, Routledge, 1883.

Krajian, A. A., "Histological Technic: Including a Discussion of Botanical Microtechnic," St. Louis, Mosby, 1940.

Langeron, M., "Précis de Microscopie," 5th ed., Paris, Masson, 1934.

Ibid., 6th ed., 1942.

Ibid., 7th ed., 1959.

Lee, A. B., "The Microtomist's Vade-mecum," 8th ed., London, Churchill, 1905.

McClung, C. E., "Handbook of Microscopical Technique," New York, Hoeber, 1929.

Mallory, F. B., "Pathological Technique," Philadelphia, Saunders, 1938.

Nissl, F. von, and A. Alzheimer, "Histologische und histopathologische Arbeiten über die Grosshirnrinde," Jena, Fischer, 1910.

Pollack, B., "Les Méthodes de Préparation et de Coloration du Système Nerveux, traduit de l'allemand par Jean Nicolaide," Paris, Carré et Naud, 1900.

Rawitz, B., "Leitfaden für Histologische Untersuchungen," Jena, Fischer, 1895.

Roskin, G. E., "Mikroskopecheskaya Technika," Moscow, Sovetskaya Nauka, 1946.

Squire, P. W., "Methods and Formulae Used in the Preparation of Animal and Vegetable Tissues for Microscopical Examination," London, Churchill, 1892.

HEMOPOIETIC TISSUES

The components of these tissues, lymph, bone marrow, spleen, and thymus, are probably best demonstrated in three staining methods, the Maximow eosin-azure stain, a combined elastin, reticulum, and collagen stain, and the

---

*Spence (*in litt.* 1955) states that "M. I. Cross is now known to be F. W. Watson Baker the elder."

periodic acid–Schiff method. For special conditions, several other stains are valuable depending upon tissue elements to be studied: hematoxylin and eosin, Masson trichrome, Mallory (Heidenhain's) Azan, iron, fat, Wright's, and Giemsa variations (see Leukocytes), new methylene blue for reticulocytes (Brecher, 1949), pyloxine–methylene blue (Thomas, 1953) and fibrin stains (Humason, 1972).

Maximow's Eosin-Azure Stain. Fix in Zenker-formalin (distilled water, 100.0 ml; potassium dichromate, 2. 5 g; mercuric chloride, 4.0–5.0 g; sodium sulfate, 1.0 g; formalin, 5.0 ml): 30–60 min. For bone marrow wash in running water: 1–24 hr. For tissues larger than 2–3 mm, treat with 3% aqueous potassium dichromate: 6 hr to overnight. Wash, embed, and prepare slides. To stain: deparaffinize and hydrate to water (remove mercuric chloride). Mayer's hematoxylin: 30–45 sec. Running water: 5–10 min. Rinse in distilled water. Stain overnight in eosin-azure (distilled water, 40.0 ml; Wright's buffer, see LEUKOCYTES, 2.0 ml; 0.1% eosin Y, C.I. 45380, in Wright's buffer, 8.0 ml; add with vigorous stirring 0.1% Azure II in Wright's buffer, 4.0 ml). Differentiate in 95% ethanol. Absolute ethanol, 2 changes; xylene, 2 changes; mount. Results: nuclei, dark purple blue; erythrocytes, light pink; eosinophilic granules, red; cytoplasm, pale blue.

Combined Elastin, Reticulum, and Collagen Stain. Fix in any general fixative, embed, and prepare slides. Deparaffinize and hydrate to water (remove mercuric chloride). 2% aqueous silver nitrate: 30 min. Rinse once in distilled water: 2–3 sec. Ammoniacal silver solution: 15 min (to 20 ml 5% aqueous silver nitrate add 20 drops 10% aqueous sodium hydroxide; then concentrated ammonium hydroxide, drop by drop, until only a few grains of precipitate remain. Add distilled water to make 60.0 ml. Use at once). Several quick dips in distilled water and place immediately in 30% formalin: 3 min. Wash in distilled water, gold tone in 0.2% aqueous gold chloride and fix in 5% aqueous sodium thiosulfate. Wash. Rinse in 70% ethanol. Orcein, 37°C: 15 min (orcein, 0.5 g; 70% ethanol, 100.0 ml; concentrated hydrochloric acid, 0.6–1.0 ml). Rinse in 70% ethanol, then in distilled water. 1% aqueous phosphomolybdic acid: 5 min. Wash three changes distilled water: 10 sec each. Sirius blue: 5 min (Sirius supra blue FGL-CF, C.I. 51300, 2 g; distilled water, 100.0 ml; glacial acetic acid, 2.0 ml). Three changes distilled water: 10 sec each. Three changes absolute ethanol or isopropanol, xylene and mount. Results: elastin, red; reticulum, black; collagen, blue. Aniline blue can be substituted for Sirius supra blue, but the color is erratic and not a clear blue.

Periodic Acid–Schiff. Fix in any general fixative, embed, and prepare slides. Deparaffinize and hydrate to water (remove mercuric chloride). Periodic acid: 10 min (periodic acid, 0.6 g; 0.3% aqueous nitric acid, 100.0 ml). Running water: 5 min. Schiff reagent: 10 min (Basic fuchsin, 0.5–1.0 g; sodium metabisulfate, 1.9 g; 0.15N hydrochloric acid, 100.0 ml. Shake at intervals during 2 hr or leave overnight. Add 0.5 activated charcoal. Shake and filter. Store in refrigerator). Three changes 0.5% sodium metabisulfite: 1–2 min each. Running water: 5 min. Mayer's hematoxylin: 5–7 min. Wash and blue. Dehydrate, clear, and mount. Results: acid and neutral mucopolysaccharides, glycolipids, muco- and glycoproteins, phospholipids, glycogen, starch, mucins, reticulum, fibrin, collagen, rose to purplish red. Counterstains other than hematoxylin can be used.

In addition to smears and the commonly used paraffin method for light microscopy, plastic-type embedding and sectioning produces superior sections (Grimley et al., 1965). A simple methacrylate embedding can be used (Humason, 1972) and sectioned on a rotary microtome. Epoxy embedding should be adjusted to a softer medium than used for electron microscopy, 2–3 parts of Luft's (1961) A solution to 1 of B solution is suggested. Zambernard et al (1969) describe a method that is relatively easy, and can be sectioned on a rotary microtome as well as on the sliding microtome as described by the authors. Frozen sectioning for fat is facilitated by embedding in gelatin prior to sectioning.

GRETCHEN HUMASON

References

Brecher, G., "New methylene blue as a reticulocyte stain," *Amer. J. Clin. Pathol.,* **19**:895–896 (1949).

Grimley, P. M., J. M. Albrecht, and H. J. Michelitch, "Preparation of large Epoxy sections for light microscopy as an adjunct to fine-structure studies," *Stain Technol.,* **40**:357–366 (1965).

Humason, G. L., "Animal Tissue Techniques," San Francisco, W. H. Freeman and Company, 1972 (641 pp.).

Luft, J. H., "Improvements in epoxy resin embedding methods," *J. Biophys. Biochem. Cytol.,* **9**:409–414 (1961).

Thomas, J. T., "Phloxine–methylene blue staining of formalin fixed tissue," *Stain Technol.,* **28**:311–312 (1953).

Zambernard, J. M., Block, A. Vatter, and L. Trenner, "An adaptation of methacrylate embedding for routine histopathologic use," *Blood,* **33**:444–450 (1969).

HISTORY OF THE ELECTRON MICROSCOPE

The pioneers of the electron microscope did not at first realize the close connection between Abbé's ideas concerning the imaging process in the optical microscope and that of the electron microscope.

Probably the most difficult step was the development of the concept of electron-optics—that a magnified "image" of an object could be formed in a vacuum by electrons. Moreover, it was also necessary that this concept germinate in the minds of those who could see not only its immense possibilities, but who also had the necessary expertise to put the ideas into practice. The starting point of this process was the iron-shrouded concentrating coil introduced by Gabor in 1926 into his experimental high-voltage oscillograph, shown in Fig. 1. Previously, long iron-free solenoids or "concentrating" coils had been used to bring the "cathode rays" emanating from a source into a small spot on a fluorescent viewing screen. Such a device is *not* analogous to an optical lens and in particular is not capable of magnification. Gabor showed that in practice a short coil can advantageously replace the long concentrating coil as a means of converging a beam of cathode rays.

The explanation for this was given by Busch, who in 1926 showed in a famous paper that a short solenoid converges a beam of electrons in much the same way as a burning glass converges rays of light from the sun. This was a most fruitful idea since the experts in cathode ray techniques had never thought on these lines before. In particular, Ruska and Knoll at the Technische Hochschule at Berlin constructed the simple vertical electron

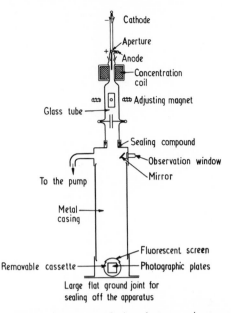

Fig. 1 The forerunner of the electron microscope, Gabor's high-voltage oscillograph (1924–1926) with the short "concentration" coil mounted below the anode. (Courtesy of Professor D. Gabor.)

optical bench shown in Fig. 2 in order to test these ideas. A small aperture illuminated by a beam of electrons formed a well-defined source, and a screen of fluorescent glass was used for observing the image formed by the "lens," the whole system being of course under vacuum. This was the first critical experimental test of Busch's theory since it was possible to define the object and image in a precise manner. Busch's theory emerged unscathed from these searching measurements.

Electron-Optical Imaging. The crucial step towards the realization of an electron microscope took place during the months February to May 1931, when Ruska and Knoll succeeded in magnifying the electron image formed by the first solenoid by means of a second solenoid placed between this intermediate image and the final screen as shown in Fig. 2. The magnification was only 17 times, but the basic principles of electron-optical magnification had been established. No patents were applied for and the results of these experiments together with micrographs of simple objects were described by Knoll on June 4, 1931 at a colloquium at the Technische Hochschule Berlin. Five days previously R. Rüdenberg of Siemens Schuckert-werke at Berlin had filed a comprehensive patent application on the electron microscope. This patent was refused in Germany but granted in France, Switzerland, Austria, and the U.S.A. with priority as from May 31, 1931. Twenty-two years later a German patent (895,635) was granted with the original priority date. Under German Patent Law Knoll and Ruska are co-users of the patent. The curious history of this patent has never been fully documented. This early work of Knoll and Ruska triggered off experiments in electron optics all over the world, none of which, however, held out much hope for the fulfillment of the idea of a practical electron microscope. Knoll left the field but Ruska carried on alone. By 1933 he had produced the electron microscope shown in

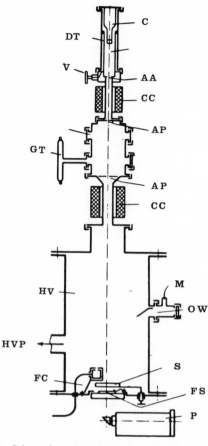

Fig. 2 Schematic arrangement of the two-stage magnifier of Knoll and Ruska (1928–1931). C, cold cathode discharge; V, air-inlet valve; AA, anode aperture; CC, concentration coils; AP, defining apertures; GT, geissler tube; HV, high vacuum; OW, observation window; M, meter for vacuum; FC, Faraday cage; S, observation screen; FS, fluorescent screen (glass); HVP, high vacuum pipe. (Courtesy of *British Journal of Applied Physics*.)

Fig. 3 The first electron microscope (Ruska 1933) to surpass the optical microscope in resolving power. (Courtesy of Professor E. Ruska.)

Fig. 3, the true ancestor of present-day instruments. It was provided with iron-shrouded lenses equipped with pole pieces, an idea developed jointly with von Borries in 1932. The idea behind these pole pieces was to concentrate the magnetic field into as small an axial region as possible. This simple idea has dominated the field of electron microscopy ever since. However, it must be said at once that this new microscope did not inspire confidence among optical microscopists. Organic samples were quickly reduced to ash as had been confidently predicted by the skeptics. The achievement of a resolution of 500 Å, greatly surpassing that of the optical microscope, as shown by the charred remains of the specimen, may have been a success for electron optics but it did not convince potential users.

Specimen Preparation Techniques. Marton, in Belgium, found a way around this difficulty. He showed that much basic structure of interests to biologists could be preserved in skeleton form even after organic material had been removed by the impinging electron beam. After some experimentation with the standard specimen im-

pregnation techniques in optical microscopy he proposed thinner specimens, higher accelerating voltages, and internal photography leading to shorter exposure times. These ideas were incorporated into an improved electron microscope completed in 1935. The maximum operating voltage was increased to 90 Kv and specimen and photographic plate air locks were fitted for the first time in electron microscopy. These air locks made a decisive contribution.

The links between electron and optical microscopes were further strengthened during the latter half of 1935, when H. Boersch showed with remarkably simple apparatus that image formation in the electron microscope takes place in accordance with the Abbe theory of imaging, i.e., electrons form a diffraction pattern of the specimen in the back focal plane of the objective lens before proceeding to form a focused image. As a by-product, dark-field illumination was demonstrated for the first time in electron microscopy simply by stopping

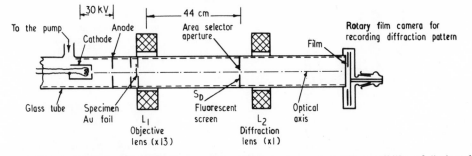

Fig. 4 Schematic arrangement of Boersch's apparatus (1935) to demonstrate the possibility of "selected area" diffraction. (Courtesy of *British Journal of Applied Physics*.)

out the undiffracted beam in the back focal plane. A further important innovation by Boersch was the demonstration of a method of "selected area" electron diffraction, by means of the simple apparatus shown in Fig. 4.

By this time it had become apparent that the electron microscope deserved serious attention in science and medicine and a possible commercial future was envisaged. In 1935 the first commercial electron microscope, the EM1 was built by the Metropolitan Vickers Electrical Company for Professor Martin at Imperial College, London.

The first serial production of electron microscopes took place in Germany in 1938 after two strenuous years of effort by von Borries and Ruska in a special laboratory set up in Berlin by the firm of Siemens and Halske. These were the first production electron microscopes to achieve a higher resolution than that of the optical microscope. During the 1939–45 war, commercial development was largely restricted to Germany and to the United States, where the famous R.C.A. model B with electronically stabilized accelerating voltages and lens current supplies was announced in 1941 by Zworykin, Hillier, and Ramberg. Developments in the field of precision mechanics were made by Von Adenne in Germany, who in 1940 constructed a "universal" electron microscope of advanced design, capable of a resolution of 30 Å, a record for this period. Facilities for dark-field illumination and stereo images were provided. A notable feature of this instrument was the introduction of a *second* condenser lens by means of which a fine electron probe of less than one micron in diameter could be focused onto the specimen. By switching off the objective lens the diffraction pattern from the area selected by the probe was recorded on a photographic plate mounted above the projector lens. An important factor in the rapid spreading of electron microscopy especially in Japan was the publication in 1940 of Von Ardenne's comprehensive book on electron microscopy.

In 1940 Boersch observed Fresnel fringes at the edges of specimens in the electron microscope, emphasizing once more the close connection between optical and electron microscopy. Such fringes were difficult to observe in the electron microscopes of the period because of inadequate electrical stability. Boersch overcame this difficulty by using electrostatic lenses; their focal length does not depend greatly on the stability of the accelerating voltage. Such lenses can in principle be used in high-resolution electron microscopy, but so far, in spite of numerous attempts, such instruments have not succeeded in establishing themselves.

In 1945 Marton constructed at Stanford University

the experimental 100 Kv electron microscope shown in Fig. 5. An important innovation was the three-stage imaging system which greatly increased the range of magnification compared with that of the previous two-stage instruments.

About the same time in the Netherlands, Le Poole, working secretly under wartime conditions and without the knowledge of the occupying military forces, had independently arrived at the principle of the intermediate projector lens as a means of extending the range of magnification. Moreover, he found an elegant way for

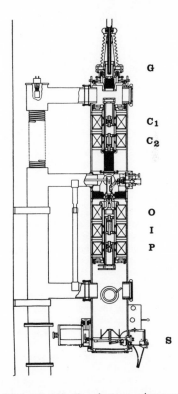

Fig. 5 Marton's 100 Kv electron microscope (1944) with intermediate and final projector lenses giving a large range of magnification. G, electron gun; C_1 and C_2 first and second condenser lenses respectively. I, intermediate lens; P, conventional projector lens; O, objective lens; S, fluorescent screen. (Courtesy of Professor L. Marton.)

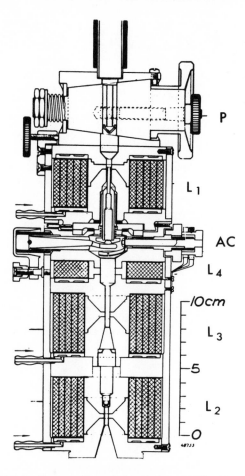

Fig. 6 Cross section of Le Poole's (1944) electron microscope showing objective lens L_1, projector lens L_2, intermediate lens L_3, and diffraction lens L_4. AC, control rod for area selector aperture; P, air lock. (Courtesy of Professor J. B. Le Poole.)

obtaining selected area diffraction, similar in principle to that previously demonstrated by Boersch.

Although Boersch had demonstrated the use of a selector diaphragm he had not had the opportunity of incorporating it into an actual electron microscope. In Le Poole's new electron microscope of very advanced design, shown in Fig. 6, a weak "diffraction lens" L_1 was focused so as to image the diffraction pattern located at the back focal plane of the objective lens onto the object plane of the first projector lens and thence onto the final screen. The area to be imaged could be chosen by means of a small selector aperture D mounted in the image plane of the objective lens, an elegant and practical solution as it enables the microscopist to change the magnification of the diffraction pattern and hence the "camera length" over wide limits, a feature that was not to be fully exploited until many years later.

Thus by the end of 1944 the electron microscope had reached a practical form that truly resembled the original concept that Abbe had once declared impossible, namely a microscope that behaved like an optical microscope but which employed radiation of a much shorter wavelength.

However, these early instruments yielded their results only after much patient effort on the part of the microscopist, and many years were still to pass before the technological near-perfection of today could be approached.

T. MULVEY

HISTORY OF THE MICROSCOPE

A number of histories of the microscope have been written; some limited in scope to specific geographic areas, others limited to scientific accomplishments in specific centuries. This history is based on an exhibit in the Medical Museum of the Armed Forces Institute of Pathology—"The Evolution of the Microscope"—which depicts some 240 historic instruments from the total collection of over 650. These instruments were selected for exhibit to show progressive improvements in design from the late sixteenth to the mid-twentieth centuries.

The word microscope was coined in 1624 by a member of the first Academia dei Lincei, a group of scientists that included Galileo, Cesi, and Stelluti. Galileo, although better known for his work with telescopes, is credited with being the first scientific user of the microscope. However, Hans and Zaccharis Janssen of Middleburg, Holland, constructed an instrument between 1590 and 1610 which is generally recognized as the first compound microscope. It was about 12 in. long and 2 in. in diameter, made of three tubes of tin, and contained two lenses with focusing accomplished by a sliding tube.

The earliest Italian compound microscopes are of the sliding tube and screw-barrel forms and were developed in the late seventeenth century. All of the extant examples conform to the same general appearance and optical structure.

Giuseppe Campani, born in Castel San Felice, Italy, in 1635, was noted for the perfection of his lenses and he invented a special lathe for grinding and polishing them without the use of molds, which he first introduced in 1664. He is credited with having been the first to construct the field lens from a design by Monconys, c. 1660, a form subsequently developed by Robert Hooke in 1665.

In 1662 Campani made an ivory compound microscope on a silver support the optical system of which consisted of an ocular and an object-lens, without a field lens. His 1665 small wooden microscope with a brass stand, also without a field lens, which allowed transparent objects to be seen by directing the instrument toward the light, is considered to be the precursor of the pocket microscopes (Fig. 1). It was also the model from which the Wilson screw-barrel and similar instruments were developed.

Histories of the microscope describe and illustrate the sliding-tube microscope using an example in the collection of the Museo Copernicano in Rome attributed to Eustachio Divini, c. 1668. Divini was an optical instrument maker who established himself in Rome about 1646. A replica of this microscope, made by John Mayall of England from the original, is also in the Billings collection. The socket ring and feet are flat and made of tin, and the cardboard body tubes are covered with gray paper. The lower tube slides within the socket ring for adjustment

---

The opinions or assertions contained herein are the private views of the author and are not to be construed as official or as reflecting the views of the Department of the Army or the Department of Defense.

Fig. 1 Giuseppi Campani, Rome, Italy; before 1665. (Figures 1–7 are Armed Forces Institute of Pathology photographs.)

of the distance between the object lens and the object. The ocular lens, enclosed in a metal holder at the upper end of the body tube, consists of two plano-convex lenses with the convex surfaces in contact.

William Homberg, a Dutch physician and scientist, constructed a microscope in 1715 which utilized a tripod support and a sliding-tube focusing mechanism.

Robert Hooke (1635–1703), London, England, introduced coarse and fine adjustments, devised an illumination system, and used a stage for objects. His early instruments utilized a ball and socket joint at the base of the vertical pillar to allow inclination of the body tube. This was later replaced (1665) by a simpler method in an instrument made for him by Christopher Cock (Fig. 2). Hooke also introduced a field lens. His "Micrographia; or Some Physiological Descriptions of Minute Bodies Made by Magnifying Glasses with Observations and Inquiries Thereupon" was published in London in 1665, and was printed by Jo Martyn and Ja. Allestry, printers to the Royal Society. This was the earliest work devoted to microscopical observations. It included the first reference to cells which were seen in cork when examined under a plano-convex lens. Hooke placed a lens at the lower end of the body tube to the bottom of which the objective was attached; this was first found in a microscope made by Culpeper and adopted by Benjamin Martin about 1759. The objective consisted of a double-convex lens of very short focal length, mounted in a cell with a pin-hole diaphragm close to the lens. The ocular consisted of a large field glass and an eye lens with a cup to control the distance of the eye from the lens.

Fig. 2 Hooke's microscope made by Christopher Cock, London, England; after 1665.

Malpighi of Italy, the discoverer of the anatomy of tissues and the creator of microscopic anatomy, was one of the first biological scientists to use the microscope. He discovered the existence of capillaries in the lungs of frogs in 1661 and he described blood corpuscles in 1665. An instrument in the Billings collection appears identical to one described by Clay and Court as possibly Malpighi's own microscope made by Divini (Fig. 3). The body is bell-shaped and is made of dark wood. The stand consists of a ring fitted with three brass, flat, bent legs and the nose is brass and screws for focusing. There is no adjustment for distance between the lenses. The objective is a biconvex lens 7/16 in. in diameter, and the 1 3/8 in. field lens is plano-convex. The eye lens consists of two biconvex lenses, the lower 1 1/16 in. in diameter and the upper 3/4 in. in diameter. John Mayall said that this instrument was constructed after 1665 by Divini as it has a field lens, and that the upper ocular lens probably was experimental.

Antoni van Leeuwenhoek of Leyden, Holland (1632–1723), one of the greatest early microbiologists, discovered protozoa and bacteria with microscopes he himself made and lenses he ground. In his instrument of 1673 the lens is mounted between two thin brass plates. His instruments were extremely simple and generally poorly

Fig. 3   Eustachio Divini, Bologna, Italy; c. 1670.

finished, but the lenses were of excellent quality. As he used his microscopes for examining only one or two objects, the number he made was quite large, and it is estimated that at his death, there were over 505 microscopes. Of these, only nine are known still to be in existence. The Billings Microscope Collection does not have an original. Johan Joosten van Musschenbroek (1660–1707), a famous instrument maker of Leyden, Holland, made two different models of simple microscopes which were widely imitated. The first form had the objective mounted in a turned cell, pushed tightly on one end of an arm. Hinged to the arm by a ball and socket joint was a second jointed arm for the objects. Accessories included a number of rods of various patterns to which objects of different kinds could be affixed. His second form resembled a fairly complicated compass microscope. Each instrument had six convex lenses, and those with the first form were for low power. Musschenbroek marked his instruments with a small oriental lamp and crossed keys.

A beautifully engraved French simple hand microscope made in 1686 by Depovilly consists of a single lens mounted between two thin brass plates hinged at the bottom, with focusing accomplished by a thin brass wheel containing eight lenses at the top of the instrument.

John Marshall of England, one of the great opticians of the latter part of the seventeenth century, introduced a bull's-eye condenser in an adjustable ring mount below the object holder. Focusing is achieved by a coarse adjustment attached to the upright support arm. The instrument

may be tilted on a ball and socket joint after the design of Hooke.

Edmund Culpeper (1660–1740), a maker of mathematical instruments in London, constructed his first microscope c. 1720. He designed the three-pillar microscope which was produced in three forms: flat wooden stage, flat brass stage, and recessed brass stage. Fixation of the mirror in the optical axis was one of his innovations. The arrangement of the three pillars is peculiar to Culpeper, the upper tier being supported by the stage and not in line with the lower tier. He also made a screw-barrel instrument about 1720, which had been invented many years earlier and was greatly improved by James Wilson in 1702, whose modifications of the optical system popularized the screw-barrel and resulted in the association of his name with the instrument.

Modifications of Culpeper's instrument have been attributable to Edward Scarlett (1677–1743), London, such as a box base with a drawer for accessories and the use of long supporting columns instead of the alternating tiers of Culpeper.

Most of the Culpeper models were made with cardboard draw tubes, but some (1790) were supplied with brass bodies which made possible the use of a rack and pinion for focusing. Continued popularity of the Culpeper model following the development and improvements of the Cuff instrument is somewhat surprising from the scientific point of view. It is likely that most of the later Culpepers were for use by the leisured classes for the amusement of their guests. Science was fashionable and this was an attractive scientific instrument which anyone could easily operate.

John Cuff (1708–1772) of London introduced a newly designed stage in the form of a cross to allow free access in contrast to the obstruction offered by the three pillars of Culpeper. Coarse adjustment was by a sliding bracket on the vertical pillar; fine adjustment was by a screw working in the rectangular block on top of the pillar. He also fitted a Lieberkuhn to a sleeve which slipped over the objective cone. The Cuff model was copied by most makers for almost 100 years.

Nairne and Blunt of London, c. 1760, made a Cuff-type chest microscope, an important development in the hinged inclining pillar. Henry Pyefinch, also of London, made a Cuff-type after 1750 with a screw focusing fine adjustment, a system described by Johannis Hevalius in 1673 and reintroduced by Cuff in 1744.

Benjamin Martin (1704–1782) of London produced cabinets of optical instruments which usually included a telescope, a screw-barrel microscope, and a solar microscope. He later introduced improvements in lens systems and focusing mechanisms.

Benjamin Martin of London was a scientific lecturer before becoming an instrument maker in 1738. He produced several types of instruments and contributed improvements in both the optical and mechanical systems. He adopted the "between-lens" at the top of the nosepiece (originally introduced in the Culpeper-Yarnell microscope) which served as a back lens for the objectives. He replaced the Hevelius clamp and screw focusing with a rack and pinion.

The drum microscope was popularized by Martin, who often included one in his cabinets of optical instruments. Nicolai Bion of Nuremburg had first described the drum base in 1717. The round base with its cut-out side later evolved into the horseshoe base of modern times.

A box microscope utilizing a Cuff model was probably

made by George Brander of Augsburg about 1769. It could be used for opaque objects with direct lighting or for transparent objects with light reflected from a mirror within the box, in which case the condenser would be positioned beneath the stage.

L. F. Dellebarre, Leyden, Holland, made a variety of microscopes. His "Universal" (1777) was very popular in France (Fig. 4). The ocular contained four biconvex lenses which could be used in pairs or altogether. Magnifications from 230 to 1170 diameters were attainable. Focusing was by rack and pinion.

George Adams of London introduced his "Variable" microscope in 1771. Inclination was controlled by a large toothed wheel which was turned by a pinion. The objectives were constructed so that one could be screwed onto another to provide a compound objective lens. It could be used as a simple microscope by replacing the compound body tube with any of the simple lenses and Leiberkuhns[1] which he supplied.

---

[1] An illuminating mirror introduced in 1738 by Johann Nathaniel Lieberkuhn and named after him, which permits observation of objects under reflected light.

Henry Shuttleworth of London constructed a microscope in 1787 after the design of Martin, with a hinged mounting of the substage condenser which permitted it to be moved out of the optical axis. This type of mounting had been introduced by John Bleuler of London about 1780. In 1810 he made an instrument similar to the Adams' and Jones' models but did not include the "aquatic" movement feature. By removal of the body tube, it could be used as a simple microscope.

Earlier, John Bleuler made an instrument with the eyepiece equipped with a third biconvex lens—a feature which appeared several years later in Adams' "Universal."

In 1790 Dollond of London made a Cuff-type, introducing some features previously used by Martin and Adams, with a circular plate over the stage which was perforated with varying-size holes to control the amount of light.

Dollond also made a solar microscope about 1790 with a movable mirror, a Cuff innovation of 1744, to permit the body to remain stationary. The solar microscope, a development from the camera obscura, utilized sunlight to project a magnified image onto a screen. This was used

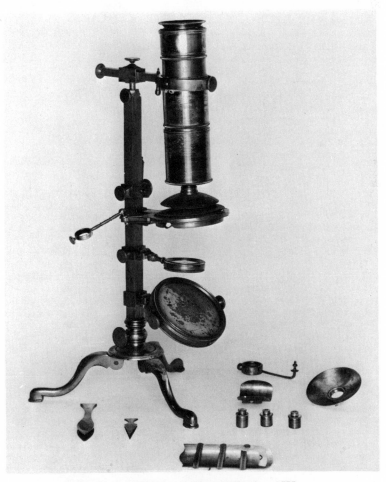

Fig. 4 L. F. Dellebarre, Leyden, Holland; c. 1777.

both for study and entertainment. Most microscope makers produced them from the early eighteenth century to the mid-nineteenth century.

The business of George Adams of London was carried on by W. and S. Jones after 1795; they produced the "Jones Most Improved Microscope" in 1798. It is considered the final development of Cuff's original instrument. It brought into one microscope all the desirable features and improvements made prior to this time. Of particular importance was the positioning of the joint near the center of gravity.

The compass microscope derived its name from the draftsman's tool that it resembles. The compass joint usually carried a rod to which could be attached forceps, live-box, or other means of holding the object to be viewed. T. Harris and Son of London made such an instrument in 1820 and later most other makers produced them.

Jeremiah Sisson of London constructed a pocket (portable) microscope in 1776 after the design of Dr. Demainbray, tutor to the Prince of Wales (later George III), with the objective lenses on a sliding bar and the prepared objects on a revolving disc.

At the beginning of the nineteenth century, chromatic and spherical aberration were unsolved problems which prevented clear, undistorted microscopic images. The lenses being used were little better than those found today in inexpensive toy microscopes. Unsuccessful attempts at correction with various lens combinations and diaphragms had been made by opticians and instrument makers for almost a century. "Achromatic" microscopes were produced but it was nearly 1850 before achromatic lens systems were generally accepted.

In 1825 Vincent Chevalier of Paris made the "Microscope Achromatique Perfectionne" (Fig. 5). He and his son were leading opticians who contributed greatly to the development of achromatic lenses. A revolving disc of diaphragms beneath the stage was a Chevalier innovation, later adopted by most makers. Charles Chevalier of Paris brought out his "Microscope Universal Achromatique" in 1834, furnished with reflecting objectives, condensers, and polarizers. The first instrument maker of the Chevalier family was Louis-Vincent Chevalier (1734–1804), who established a shop in 1765 in Paris. His third son, Vincent Chevalier (1770–1841) became one of the foremost makers of microscopes in the nineteenth century. Vincent Chevalier's son, Charles Chevalier (1804–1859) carried on the tradition of the family.

Jean-Gabriel-Augustin Chevallier of France (1778–1848) may have been related to Vincent and Charles Chevalier but was from another branch of the family. Trained in making optical instruments by his maternal grandfather, François Trochon, he inherited the business in 1796 in Paris and became well known for his optical inventions. He signed his instruments "l'ingenieur Chevallier" and for many years his firm continued producing optical instruments under his name.

In 1825 S. J. Rienks of Friesland, Germany, made a reflecting microscope with concave and convex mirrors in place of objective lenses. The knowledge that a curved mirror did not have chromatic aberration, led to popularity of the reflecting microscope for a number of years.

After 1820 Dollond made a simple microscope using a "Wollaston doublet," two plano-convex lenses in close approximation. This was an early attempt at achromatism. His instrument had a movable stage, the position of which was controlled by a screw on each side.

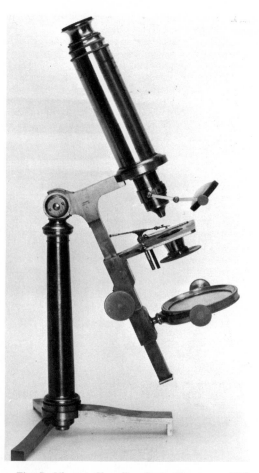

Fig. 5 Vincent Chevalier, Paris, France; c. 1825.

Giovanni Amici (1784–1863) of Modena, Italy, produced both reflecting and achromatic instruments. His achromatic horizontal microscope of 1827 employed an extremely fine micrometer to control movements of the stage.

Amici also made an achromatic instrument about 1833 with a prism which permitted horizontal positioning of the ocular while the stage and objective remained upright. It could also be used as a simple microscope.

John Cuthbert, an English telescope maker, produced a microscope in 1827 which could be used as a reflecting microscope, or, by attaching the achromatic objective lens disc, as a dioptric instrument. A slender tube above the stage contained an elliptic mirror which reflected an enlarged image to the ocular. A series of interchangeable tubes provided a range of magnification.

Dollond made a microscope after 1835, much along the lines popularized by Andrew Pritchard of London, although basically the old Jones design. The pillar telescoped to adjust the height of the instrument and a fine micrometer controlled the mechanical stage.

Froment of Paris constructed a horizontal microscope about 1839, probably in competition with the Chevaliers. Its workmanship was superior but probably its high cost prevented its success.

A drum microscope made by Georges Oberhauser of

Paris before 1840 could be used in a vertical or horizontal position. By attaching the extra tube as a handle it could be used as a demonstration microscope; focusing was by drawtube.

Simon Plössl (1794–1868), an optician and the first maker of microscopes in Vienna, made an instrument before 1840. He used an antiquated type of fine adjustment until the late 1860s. A lever projecting beneath the stage controlled positioning of the condenser. One of his instruments, believed to have been used by Virchow, the father of modern pathology, is in the Billings Collection (Fig. 6).

In 1840 Andrew Pritchard of London made an instrument without a fine adjustment which could not be inclined. One side of the mirror was white plaster for white cloud illumination by sunlight.

Andrew Ross of London made a microscope about 1840 with fine adjustment by a short lever arm controlled by a milled wheel on top of the compass joint. The loosely mounted nosepiece was held in an extended position by an internally located spiral spring, to permit it to telescope into the body tube thereby preventing damage to the object being examined, should the body tube be accidentally moved too far downward.

A dissecting microscope by Georges Oberhauser of Paris after 1840, called "microscope à tambour à disséction," was copied by many makers with modifications mostly in the substage mechanism. It was equipped with a separate simple dissecting microscope.

A microscope of about 1841 by an unknown maker was a modification of an Oberhauser form originally introduced in the 1830s. Earlier models had a rotating stage and a fine adjustment, neither of which appeared on this instrument. Coarse adjustment was by rack and pinion. A second rack and pinion, attached to an inner drawtube that houses the ocular, provided variable magnification (a pancratic ocular).

In 1841 Andrew Ross of London came out with a radical change in construction. The Ross stand, two well-supported, connected pillars, suspended the instrument near its center of gravity. It had a 90° range of inclination.

In 1848 Oberhauser introduced the horseshoe base evolved from the drum which is in general use today.

Before 1848, Powell and Lealand of London constructed a microscope after a design by Cornelius Varley with the fine adjustment mounted on the lower end of the body tube which functioned through a spring-mounted nosepiece. There was a lever action mechanical stage. It was designated the "iron" microscope and was intended as an economical student model.

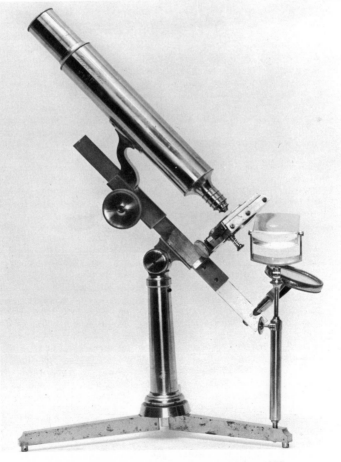

Fig. 6 Simon Plössl, Vienna, Austria; before 1875.

Invention of the inverted microscope is credited to Professor Lawrence Smith of the University of Louisiana, New Orleans. A "chemical" microscope was made in 1850 by C. S. Nachet and Son of Paris. The elder Nachet had worked with the Chevaliers before establishing his own business in 1839.

James Smith, the founding father of the present-day firm of Beck and Beck, London, began constructing microscopes in his own shop in 1829. Richard Beck was apprenticed to Smith about 1843 and in 1847 formed the partnership of Smith and Beck. Joseph, brother of Richard, joined the firm in 1857 and the name was changed to Smith, Beck and Beck. In 1869, Smith retired and the firm then became R. and J. Beck. Conrad Beck, son of Joseph, and William, son of Richard, joined the firm which then became known as Beck and Beck.

Smith and Beck made a student model about 1850. A milled wheel behind the vertical post activated a short lever arm that controlled the fine adjustment. A revolving Brooke's double nosepiece facilitated the changing of objectives.

In 1867–1868, R. and J. Beck made a binocular with the racks and pinions at the upper ends of the body tubes to control adjustment for interpupillary distance. Fine adjustment was by a short lever arm that acted on the nosepiece. Substage equipment included an iris diaphragm and a holder for a condenser.

Beck also made an "economic" model which was popular from 1875 to 1892. An 1878 model had the glass stage held in place by friction from a milled head spring mounted at the base of the limb, which was easily moved in any direction on the polished stage surface.

The drum microscope enjoyed a popularity for more than 100 years. Over 70 years after Martin introduced it, Fraunhofer adopted and stabilized the pattern; Oberhauser, Lerebours, and Hartnack perpetuated it. An 1850 model was quite similar in appearance to those of Martin's day. Refinements included a fine adjustment and a substage disc of diaphragms.

Moreau of Paris introduced a functional compound "monkey" microscope in 1850.

About 1860 Nachet of Paris made a horizontal microscope, a type of demonstration or class instrument, with rack and pinion coarse adjustment. The eyepiece consisted of a single eye lens and the field lens. He also made an Oberhauser-type vertical demonstration instrument, one of which was used at the School of Medicine in Paris. In 1879 he made a binocular instrument in which each body tube prism box unit moved toward or away from the other to adjust to the user's interpupillary distance—a forerunner of the mechanism in modern instruments.

Charles Baker of London made a portable microscope about 1860 which was used in the Army Medical Museum's laboratories in Washington from the late 1860s to 1884, when it was placed on exhibit. It had a sliding coarse adjustment and a milled-head screw fine adjustment.

The main feature of a "Harley" model made after 1865 by Charles Collins of London was the sliding prism box which was designed by Dr. G. Harley of London. It contained both a Nicol and a Wenham prism and permitted the instrument to be used as either a binocular, a monocular, or a polarizing microscope.

Dr. Carl Zeiss (1816–1888) founded a business in Jena, Germany, and began making microscopes in 1847. Today the Zeiss firm produces some of the world's finest instruments.

Père Cherubin D'Orleans of France introduced in 1677 the first binocular microscope with two eyepieces and two objectives; there was no coarse adjustment, and focusing was achieved by moving the stage nearer to or further from the objective, the forerunner of the stage focusing used extensively during the eighteenth century and still used on metallurgical microscopes. J. L. Riddell of the University of Louisiana, New Orleans, was the first to discover (1851) and publish (1854) the optical principle[2] on which depends all the really satisfactory binocular microscopes made prior to 1879. He was also the inventor of two efficient methods of applying that principle—one suitable for the simple or dissecting microscope, the other for the compound. J. and W. Grunow, New Haven, Connecticut, constructed a binocular instrument for Riddell in 1853.

No history of the microscope would be complete without specific mention of the accomplishments of the American microscope builders. One of the first American microscopes, made in 1850 by C. A. and H. R. Spencer of Canastota, New York and modeled after the Pritchard type of instrument, is in the Billings Collection (Fig. 7).

Charles A. Spencer (1813–1881) of Canastota, New York, published a trade circular in 1838 advertising reflecting lens microscopes and telescopes and a second list in 1840 adding compound, achromatic microscopes. By 1850 he was using fluorite and his own improved optical glass to make objectives of greater resolving power. Some of Spencer's objectives received a Gold Medal at the 1878 Paris Exhibition. Spencer's business was destroyed by fire in 1873. In 1875 he and his son Herbert joined the Geneva Optical Works and reestablished their own business in 1877. After Charles' death, the business was continued by Herbert.

Herbert R. Spencer (1849–1900) established his own firm in 1880 at Geneva, New York. In 1889 he formed the H. R. Spencer Optical Company in Cleveland, Ohio. He and Fred R. Smith formed the Spencer and Smith Optical Company in Buffalo, New York, in 1891. In 1895 a group of businessmen in Buffalo bought the Spencer and Smith Optical Company to establish the Spencer Lens Company. Herbert continued as superintendent until his death. In 1935 the American Optical Company purchased the Spencer Lens Company, and in 1945 changed the name to American Optical Company, Instrument Division. Spencer lenses earned a reputation at home and abroad; some experts considered them superior to any European product.

Robert B. Tolles (1821–1883) became an apprentice to Spencer in 1843. He and Spencer corrected spherical aberration from the cover glass by using a ring to move the center lens elements of the immersion objective. He established his own business in Canastota, New York, 1858. The Boston Optical Works was established in 1867 with Tolles as superintendent. He made objectives with two fronts; one for use in air and the other for immersion. Tolles published the optical designs to prove that his lenses were of the high aperture (N.A. 1.25) stated.

Franz Miller, a workman for Tolles, established his business in 1867 in New York City, changed his name to Frank, and was joined by his brother William about 1870. They made microscopes and objectives until 1899. Frank, Jr., was included in 1892.

---

[2]Two prisms placed behind the objective so that the light from the objective was split and deflected up the two parallel eyepiece tubes.

Fig. 7   C. A. and H. R. Spencer, Canastota, New York; 1850.

The Bausch and Lomb Optical Company started in 1853, and a microscope department was added in 1876. Edward Bausch took over the department in 1877. His interest and understanding of production methods soon made available larger numbers of microscopes, more practical in size and construction than the massive models being produced in the United States. They adapted the continental-type horseshoe base and made a concentric stand. Microtomes were added in 1885 and in 1887 petrographic microscopes.

George Wale, a cousin of William Wales, established a business in 1860. His concentric stand, patented in 1879, was copied by both Swift and Ross in England about 1881. He was probably the first to mark his objectives with magnification rather than only focal length and to have made low-power water immersion objectives for use without a cover glass. Bausch and Lomb Optical Company acquired his business in 1880.

Ernest Gundlach made 12 models of microscopes in Berlin in 1866. He designed a glycerin immersion objective which received a special medal at the Paris Exposition. He moved to Hackensack, New Jersey in 1872. In 1876 he became superintendent of Bausch and Lomb Optical Company's microscope department. In 1879 he established the Gundlach Manhattan Optical Company, Rochester, New York, which became the Gundlach Optical Company in 1884.

During 1852 James W. Queen joined with W. Y. McAllister, who started in Philadelphia in 1783. Queen and McAllister, 1853–1854, was taken over by J. McAllister. Queen started his own business in 1854 which became J. W. Queen and Company in 1860. About 1880 Queen retired and Samuel L. and Edward B. Fox continued the business. They acquired the Acme Optical Works in 1881. Eyepieces and objectives were made under the direction of Orford (Gowland). In 1898 the firm became the Queen-Gray Company, makers of electrical instruments.

Joseph Zentmayer (1826–1888) came to the United States and established a shop in Philadelphia in 1848, and began making microscopes in 1853. His "Grand American Microscope" was made in 1859, followed by the "U.S. Army Hospital Microscope," the "Histological Microscope" and, in 1876, the "Centennial Microscope." Zentmayer's larger stands provided flexible lighting for testing objectives with graduated rotating bases, graduated counterable rotating stages, and a substage that rotated about the plane of the specimen for measurement and recording. He made a 1/10 in. objective of N.A. 1.63.

Julius and William Grunow came to New Haven, Connecticut from Berlin in 1849 and began making microscopes in 1851. The Riddell binocular microscope was made in 1853. The J. L. Smith inverted microscope, with an erect image for dissecting was made in 1867 for Major General George H. Thomas. A 128 page treatise and catalogue was published in 1857. Grunow moved to New York City about 1861.

John W. Sidle of Lancaster, Pennsylvania, and Professor J. E. Smith designed a less expensive microscope in 1878 to meet the criticism that students could not afford the microscopes available. Five designs of Acme microscopes, except for the first, were made. Sidle and Poalk became John W. Sidle and Company, 1881, and in 1884, the Acme Optical Works. The Acme microscopes were sold exclusively by Queen & Co., who purchased the Sidle company in 1881.

In 1851 Benjamin Pike was selling imported Beck, Powell and Lealand, and Ross microscopes in New York City. In 1859 Benjamin Pike, Jr. was importing microscopes for sale, mostly of French make, and in 1881 the firm became Benjamin Pike's Son and Company. A two volume catalogue of philosophical and scientific instruments was published in 1856.

L. Schrauer, 50 Chatham Street, New York City, manufactured microscope stands during the 1880s and 1890s, which were equipped with W. Wale's 3/4 and 1/4 in. objectives. Queen listed Schrauer's microscopes in his catalogues.

Walter H. Bulloch came to the United States from England in 1851 and was foreman for B. Pike and Son to about 1864. He became a partner of William Wales; Bulloch making the stands and Wales the optics. He began making microscope stands in Chicago in 1867. After the Chicago fire in 1871 he worked with Tolles in Boston for a short period, then returned to Chicago and made stands until 1890; about 1884 he made microtomes.

By the end of the nineteenth century, the basic design of the microscope was firmly established. The increased demand for reliable high-quality instruments generated by the rapidly developing scientific specialties could not be met by the individual craftsmen, who until then was the principal source.

The twentieth century, therefore, became a period of refinement of the product and improvement of manufacturing methods. Particularly noteworthy were the development of mass production and the establishment of standards which permitted the interchange of components and the introduction of a variety of accessories.

Neither time nor space permit mention of the hundreds of beautifully hand-crafted microscopes in The Billings Microscope Collection of the Medical Museum, AFIP; however, the collection does contain instruments that run the gamut from such artisans as Matthew Loft, George Sterrop, Francis Watkins, and Jan Paauw of the eighteenth century, Hartog van Laun, Hermanus van Deyl, Hendrik Hen, William Ladd, and James White of the nineteenth century, and the electron microscopes of the twentieth century.

H. R. PURTLE

References

Beck, R. and J., Personal Communication to Author.
Bedini, S. A., "Giuseppi Campani, Pioneer Optical Inventor," Ithaca, Cornell University Press, (1962 pp. 26, VIII–2 IX).
Bedini, S. A., "Seventeenth century Italian compound microscopes," *Phys. Riv. Storia Della Sci.,* V(4) (1963).
Bedini, S. A., Personal Communication to the Author, 1966.
Bradbury S., "The Evolution of the Microscope," London, Pergamon Press Ltd., 1967.
Carpenter, W. B., and W. H. Dallinger, "The Microscope," London, J. & A. Churchill, 1901.
Clay, R. S., and T. H. Court, "The History of the Microscope," London, Charles Griffin & Co., Ltd., 1932.
Frison, Ed.: "L'evolution de la partie optique du microscope au cours du XIXe siecle," Communication No. 89 from the National Museum for the History of Science, Leyden, 1954.
Gage, S. H., "Microscopy in America (1830–1945),' Oscar W. Richards, ed., *Trans. Amer. Microscop. Soc.* LXXXIII(4), Supplement (Oct. 1964).
Hooke, R., "Micrographia," London, 1665.
Mayall, J., Jr., "The microscope" (Cantor Lectures, London, 1886), *J. Soc. Arts,* **34**:987 (1886).
Ibid., **34**:1007 (1886).
Ibid., **34**:1031 (1886).
Ibid., **36**:1149 (1888).
Nachet, A., Collection Nachet, Paris, 1929.
Rooseboom, M., "Microscopium," Leiden, 1956.
The Billings Microscope Collection. Washington, D.C.: The American Registry of Pathology, AFIP, 1967.
Van Cittert, P. H., "Descriptive Catalogue of the Collection of Microscopes in Charge of the Utrecht University Museum, Holland," 1934.
Van der Star, P., "Descriptive Catalogue of the Simple Microscopes," National Museum of the History of Science at Leyden, 1953.
Van Heurck, H., "The Microscope," London, 1893. London; Crosby Lockwood & Son; New York: D. van Nostrand Company.
See also: OPTICAL MICROSCOPE.

HISTORY OF MICROTECHNIQUE AND MICROTOMY

No craftsman who has examined the intricate detail on the carved gem stones of antiquity can doubt for a moment that "magnifying glasses" were in use by the ancients: Seneca (1) has left a description of their use in Roman times. Spectacle lenses of molded and polished glass were in use in the thirteenth century (2). The Renaissance reintroduced observation as a tool of science and by the middle of the sixteenth century living things were being studied through simple lenses, a tool brought to such perfection a century later by Leeuwenhoek that he was able to study bacteria. The invention of the compound microscope is usually attributed to the Janssens, a father and son team living in Middleburg, Holland, about 1590. These instruments, however, contributed little or nothing to the progress of either biology or medicine for the next two centuries. As Nordenskiold (3) says, "Microscopy has had, therefore, two periods of brilliant achievement in the course of its history: the seventeenth century and the latter half of the nineteenth century."

This gap of two hundred years, during which there lay fallow an instrument destined to revolutionize the life of man, was not due to imperfections of the instrument or of men who used it. It was due to the lack of materials with which methods could be developed for the preservation of objects for detailed study. The microscope was there: the microscope slide was not.

The origin of the word "slide" is itself of some interest since it is an elision of the "slider" which accompanied the early microscope. This was in two forms, both of which are seen in the illustration (Fig. 1) of "Mr. Culpeper's Improvement of Mr. Marshall's Large Double Microscope," appearing as plate 3 in the first edition of Baker (4).

The flat slider at the upper left was the original form and was usually made of a slip of ivory, through which 1/2 in. holes had been drilled. A ridge of ivory was left in the center of the hole and on each side of the ridge a pair of mica discs were held with spring brass rings. Anything which it was wished to preserve was first dried and then held between the discs. A microscope would usually be provided with a dozen or so of these sliders, the most popular objects for which were small insects, the dried bodies of which could thus be studied in silhouette.

The second, much less usual, form of slider was the circular disc of ivory (just above and to the right of the forceps in Fig. 1), the central axis of which was inserted into one of the holes of the stage.

Early nineteenth century improvements in the production of window glass permitted a glass slip to be substituted for ivory, so that the object could be dried directly on the slip or even in ground cavities. The process is thus described by Pritchard (5):

"Although such exceedingly small creatures as animalcules, when dead, lose many of their characteristic features, especially the soft-bodied ones, yet, for the verification of some parts of their structure, it is absolutely necessary to observe them in a quiescent state; and hence, a method of effectually drying and preserving them must be considered essential. Bacellaria, in this condition, have often been preserved by botanists, in collections of minute Algae, and with very little management; but other families will require more care. Having selected the creature you wish to preserve, remove it with a fine pointed quill, and put it on a slip of glass, or other convenient receptacle. By this means there will be but a small portion of water surrounding it, which may be extracted by some pointed pieces of ragged blotting paper. When you have withdrawn as much of the water as possible from the specimen, the remaining moisture may be readily evaporated, by placing the glass on the palm of the hand. The Hydatinea may be best preserved when destroyed with

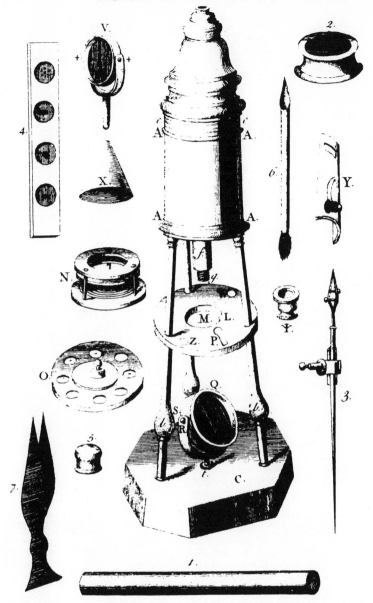

Fig. 1 "Culpepper's Improvement of Marshall's Microscope," showing a straight "slider" at the upper left and a rotating "slider" at 0, just above and to the right of the forceps.

strychinia, and then rapidly dried. By what mode soever life may be taken away, it is absolutely expedient that they should be speedily and carefully dried, otherwise their bodies will be decomposed, gases evolved, and the object will fail."

The object so treated could not be covered with another slip and thus sealed in a preservative fluid. The working distance of the high-power lenses of the period was as little as one-hundredth of an inch. The only thin transparent material which then existed was mica, too porous to retain preservatives.

There was actually very little point in having a second sheet of mica on top of the first, or, indeed, in using a transparent support at all. Many objects were much better examined by reflected than by transmitted light and there therefore developed a complex series of "pill box" mounts, of which a representative collection is seen in Fig. 2, taken from the first edition of Quekett (6). These pill boxes were usually covered at the bottom with black wax into which were fused the dried objects to be examined, and were more in use by botanists than by other workers. The specimens shown in the illustration are all botanical, though only the two moss sporangia are readily recognized. These boxes were either mounted on an arm for insertion into the stage, or more usually (Fig. 3), on long pins which could be inserted into a cork disc cemented to the stage of the microscope.

It must not be thought that only whole objects were

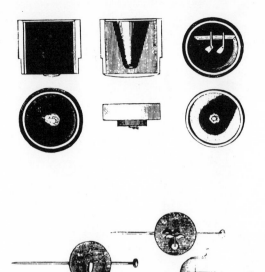

Fig. 2 Pill box mounts from Queckett's "A Practical Treatise on the Microscope," 1848.

Fig. 3 Mounted pill boxes. From Queckett 1848.

Fig. 4 Adam's microtome *circ*. 1770. The figure is from Queckett, 1848.

preserved by these means for examination. Sections of both plant and animal material were quite commonly cut, either in a regular microtome, Fig. 4, or with the aid of a "Valentine's knife," Fig. 5. The microtome shown (reproduced from Quekett *loc. cit.*) is Mr. Custance's improvement of the Adams's microtome which is known to have been in employment in 1775. It consists essentially of a glass plate, along which slides a diagonal razor and in the center of which is mounted an object holder with a fine screw-raising mechanism. The model shown, as will be seen from the illustration, is actually furnished with a ratchet device for advancing the screw automatically with each stroke of the knife. This instrument was commonly used on wood sections, the Valentine's knife (Fig. 5) being used for animal tissues. This device, of which the writer has been unable to establish the date of invention (it is mentioned as a recent invention by Chevalier 1839, (7), was to all intents and purposes a scalpel split down the middle with a mechanism for varying the width of the gap between the two halves. This gap was set to the thickness of the required section and the knife was then plunged with a cutting stroke into the tissue to be sectioned. The two halves were then separated and the section washed off onto a slider, where it was dried for examination.

The reason that all these specimens had to be preserved in the dried condition was not due to any lack of preservative fluids or mountants. Anatomical specimens had been for a considerable period preserved in alcohol, and the use of Canada balsam was well known to anatomical preservers, who used it as a varnish on embalmed preparations of muscle. That these techniques of fluid and resinous preservation could not be utilized for microscopical objects was due to the absence of any thin nonporous material with which the specimen could be covered. The sciences of invertebrate zoology, histology, embryology, pathology, and the like were stalled for more than a century awaiting the invention of the now commonplace coverslip, and it is fascinating that there appears no means to establish at the present time the exact date of the

invention of this revolutionary material. There is little doubt (Quekett *loc. cit.*) that these discs of thin glass were first manufactured by the Chance Brothers of Birmingham in the early 1840s, since the author cited, writing in 1848,

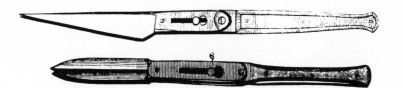

Fig. 5 "Valentine's knives" for sectioning. They are, in effect, split scalpels with an adjustable gap.

regarded them as of recent origin. There is even some collateral evidence for supposing the discovery to be both English and of about that date. In the fifteenth edition of Gould's "Companion to the Microscope" 1848 is the statement "let the piece of thin glass you cover the object with," etc., while Du Jardin, writing from Paris in the same year, still recommends the use of talc, isinglass, or a laboriously hand-polished slip of calcite. The discovery was not widely known even in England at this period, since Pritchard (*loc. cit.*) describes methods of preservation which show him to have been ignorant of the existence of coverslips. Between 1850 and 1860, there are numerous works on microtechnique in English, French, German, and Italian, all of which treat coverslips as a matter of course.

The date of this invention of coverslips would be quite easy to determine if only somebody had preserved copies of Gould's "Companion to the Microscope." This was a little paper-backed volume, apparently produced to be given away free by microscope manufacturers. Several libraries have the first edition (1825), the Royal Microscopical Society in London has the third edition (1828), and the British Museum has the fifth edition (1829); none of these refer to coverslips. There is then no record of any edition before the fifteenth (1848) which mentions coverslips casually.

These glass coverslips were not at first considered capable of being attached to the microscope slide with a mounting medium, so even when balsam was employed, paper covers were used as an additional support. Figure 6, from Martin (1872 (8)), is believed by the authors to be the only illustration showing the manner in which these paper covers were attached. It was customary at the time to lithograph special covers for individual mounts. Whatever may be thought of the aesthetics in these proceedings, mounts of this type were practical and effective.

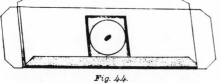

Fig. 42.

Fig. 43.

Fig. 44.

Fig. 45.

Fig. 6 The application of cover papers from Martin 1872.

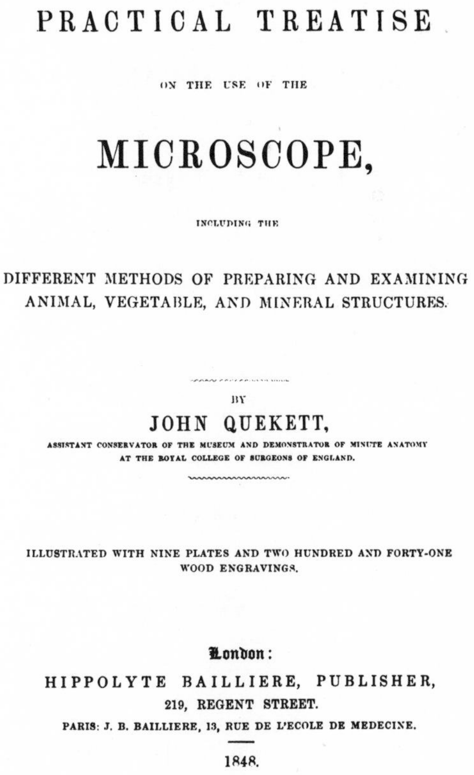

PRACTICAL TREATISE

ON THE USE OF THE

MICROSCOPE,

INCLUDING THE

DIFFERENT METHODS OF PREPARING AND EXAMINING
ANIMAL, VEGETABLE, AND MINERAL STRUCTURES.

BY

JOHN QUEKETT,

ASSISTANT CONSERVATOR OF THE MUSEUM AND DEMONSTRATOR OF MINUTE ANATOMY
AT THE ROYAL COLLEGE OF SURGEONS OF ENGLAND.

ILLUSTRATED WITH NINE PLATES AND TWO HUNDRED AND FORTY-ONE
WOOD ENGRAVINGS.

London:

HIPPOLYTE BAILLIERE, PUBLISHER,
219, REGENT STREET.

PARIS: J. B. BAILLIERE, 13, RUE DE L'ECOLE DE MEDECINE.

1848.

Fig. 7 Reduced facsimile of filter page 1 at edition of Queckett. Original 22 cm × 14 cm.

The junior author has in his collection a slide by Topping, dated 1846, of an injection of the human kidney which is in as good condition as the day it was made. Very few contemporary mounts, in which balsam is the only support to the coverslip, are likely to be available for study 100 years from now.

It cannot be considered a coincidence that no books on the mounting of microscopic objects were published prior to the introduction of the coverslip. It is true that a few earlier works, (e.g., Pritchard *loc. cit.*), incorporate a few directions for preservation, while there are, of course, many previous works on the preservation of microscopic specimens. It is additional evidence for the English invention of the coverslips that the first work on microtechnique came from that country. This was Quekett 1848, of which the title page is reproduced in Fig. 7. At the time of the production of this book, microscopical preparation and microscopical examination of tissues were not taught in most medical schools. The appointment of this 33 year old eccentric to the position of "Assistant Conservator of the Museum" and "Demonstrator of Minute Anatomy" at the Royal College of Surgeons of England was a startling step for so august a body. The microscope was at that time little more than a scientific toy, though it had come to be regarded more highly after the foundation of the Royal Microscopical Society in the development of which the author's elder brother, Edwin John Quekett, a well-known physician, had played a prominent role. Quekett (the author) took hold of his new

THE

MICROSCOPIST;

OR,

A Complete Manual

ON THE

USE OF THE MICROSCOPE:

FOR

PHYSICIANS, STUDENTS,

AND ALL

LOVERS OF NATURAL SCIENCE.

WITH ILLUSTRATIONS.

BY

JOSEPH H. WYTHES, M.D.

PHILADELPHIA:

LINDSAY AND BLAKISTON.

1851.

Fig. 8 Reduced facsimile of title page of Wythes 1851. The original is 18 cm × 12 cm.

THE

MICROSCOPIST'S COMPANION;

A

POPULAR MANUAL

OF

PRACTICAL MICROSCOPY.

DESIGNED FOR THOSE ENGAGED IN MICROSCOPIC INVESTIGATION, SCHOOLS, SEMINARIES, COLLEGES, ETC., AND COMPRISING SELECTIONS FROM THE BEST WRITERS ON THE MICROSCOPE, RELATIVE TO ITS USE, MODE OF MANAGEMENT, PRESERVATION OF OBJECTS, ETC.,

TO WHICH IS ADDED

A GLOSSARY

OF THE PRINCIPAL TERMS USED IN MICROSCOPIC SCIENCE.

BY JOHN KING, M. D.

ILLUSTRATED WITH ONE HUNDRED AND FOURTEEN CUTS.

CINCINNATI:

RICKEY, MALLORY & CO.

1859.

Fig. 9 Reduced facsimile of title page of King 1859. The original is 23 cm × 15 cm.

job with great energy and, by 1855, had persuaded the Royal College of Surgeons to install a microscopical lecture theatre. Quekett's work was translated into German and was followed by a flood of works on microtechnique in many languages, for data on which Gray and Gray 1956 (9), should be consulted. Three works in English are worthy of special mention in the present place.

The science of microscope mounting was apparently brought to the United States by John H. Wythes (or Wythe, as his name is spelled everywhere except on his title), who published in 1851 the first work to appear on this side of the Atlantic. The title page of his work is reproduced as Fig. 8. He was not, however, an indigenous American writer and apparently the first work produced entirely from American sources was King's "The Microscopist's Companion," (1859) of which the title page is reproduced as Fig. 9. This work drew heavily from Quekett, but the appended glossary appears to have been entirely original.

All of these books contain instructions for the use of the microscope itself, in addition to instructions on the preparation of objects for microscopical examination. Apparently, the first work devoted exclusively to the latter

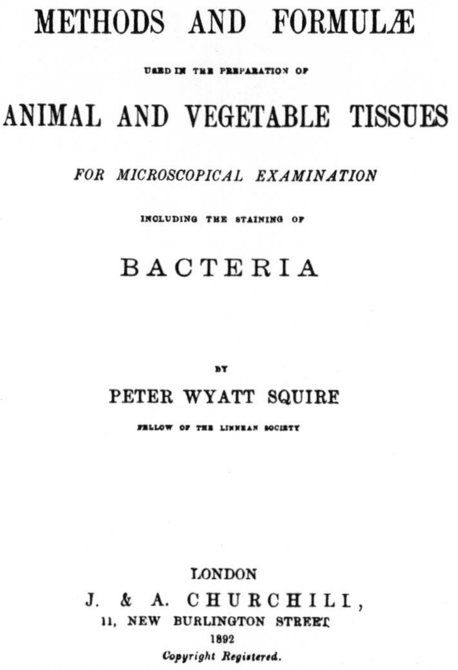

METHODS AND FORMULÆ

USED IN THE PREPARATION OF

ANIMAL AND VEGETABLE TISSUES

FOR MICROSCOPICAL EXAMINATION

INCLUDING THE STAINING OF

BACTERIA

BY

PETER WYATT SQUIRE

FELLOW OF THE LINNEAN SOCIETY

LONDON

J. & A. CHURCHILI,

11, NEW BURLINGTON STREET

1892

Copyright Registered.

Fig. 10 Reduced facsimile of title page of Squire 1892. The original is 18 cm × 12 cm.

subject was that of Squire 1892 (Fig. 10). The voluminous literature which followed has been recorded by the undersigned (9).

FREDA GRAY
PETER GRAY

References

1. "Naturales Quaestiones," Book 1, Chapter 7. Cited from J. Quekett, "A Practical Treatise on the Use of the Microscope," p. 1.
2. "Encyclopedia Britannica," 11th ed., Vol. 25, p. 618.
3. E. Nordenskiold, "The History of Biology," New York, Tudor, 1928.
4. H. Baker, "The Microscope Made Easy," London, R. Dodsley, 1742.
5. A. Pritchard, "A General History of Animalcules," London, Whittaker, 1851.
6. J. Quekett, "A Practical Treatise on the Use of the Microscope," London, Baillière, 1848.
7. C. Chevalier, "Des Microscopes et de leur Usage," Paris, Crochard, 1839 (p. 192).
8. J. H. Martin, "A Manual of Microscopic Mounting with Notes on the Collection and Examination of Objects," London, Churchill, 1872.
9. F. Gray and P. Gray, "Annotated Bibliography of Works in Latin Alphabet Languages on Biological Microtechnique," Dubuque, Iowa, Brown, 1956.

HOLOGRAPHIC MICROSCOPY

I. Introduction. Holography and microscopy are linked together historically, even though many modern applications for holography do not involve microscopy. Holography was invented by Dennis Gabor in 1948 in an effort to improve the imagery of the electron microscope (1). In those early years other workers (2,3)

attempted to extend Gabor's ideas to include X-ray microscopy. Although these early attempts to apply the ideas of holography to microscopy never really came up to expectations, there is still a great deal of promise for holographic microscopy, as modern workers are demonstrating.

The reason for the renewed interest in holographic microscopy is basically the same as that for the renewed interest in holography itself—the improvements to Gabor's early methods introduced by Leith and Upatnieks in 1962 (4). These workers also did some of the first successful holographic microscopy using their new methods (5). Later workers (6,7,8) have shown that a different technique can be employed which allows all of the conventional types of illumination: bright field, dark field, phase contrast, interference contrast, and polarization microscopy.

II. Principles of Holography (9). In order to appreciate fully the interest in holographic microscopy, it is necessary to understand the basic principles of holography, for this method of imagery is unique.

Holography is basically a scheme for recording the information about a wavefront of light, rather than the image formed by that wavefront. Both the amplitude and the phase of the wavefront are recorded in such a way that the wavefront can be faithfully reconstructed. As a result, in principle, a fully three-dimensional, aberration-free image of the original object can be formed. It is these two features that have led to the investigation of holography as a form of microscopy.

The wavefront recording can be described with the aid of Fig. 1. Let O be a complex number representing the amplitude and phase of a monochromatic wave from the object incident on the recording medium H, and let R be a wave coherent with O. The wave O contains information about the object since the object has uniquely determined its amplitude and phase. There are no restrictions on the type of object. To make a hologram, one must record the wave in such a way that the entire wave can later be reconstructed. This is done with a reference beam R. The total light field on the plane H is $O + R$. A square-law recording medium (most often a photographic emulsion) will respond to the irradiance of the incident light, $|O + R|^2$. After processing, the amplitude transmittance of the film, now called a hologram, is given by an expression of the form

$$T_a(x) = f(E_o) + \beta E(x) + \text{higher-order terms} \quad (1)$$

where E_o is the average exposure and x is a linear dimension along the film. The exposure E is the product of the irradiance $|O + R|^2$ and the exposure time t. If we

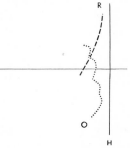

Fig. 1 Illustrating the recording of a general wavefront. The curves O and R are schematic representations of the object and reference wavefronts, respectively.

ignore the constant term $f(E_o)$, the amplitude transmittance of the exposed and processed film is

$$T_a(x) = \beta E(x) = \beta t \, |\, O + R\,|^2$$
$$T_a(x) = \beta t [\,|O|^2 + |R|^2 + OR^* + O^*R\,] \quad (2)$$

where the asterisk denotes a complex conjugate. We now assume that the hologram, having the amplitude transmittance given by Eq. (2), is illuminated with a monochromatic light wave described by the complex number c. The transmitted field at the hologram is then

$$\psi(x) = c \cdot T_a(x) = \beta ct [\,|O|^2 + |R|^2 + OR^*$$
$$+ O^*R\,] \quad (3)$$

If the reconstructing wave c is sufficiently uniform or similar to R, then the third term of Eq. (3) is βctR^*O = constant × O. This term represents the reconstructed wave, which is *identical* with the original object wave O. This is why the whole process is referred to as "wavefront reconstruction." This wave has all of the properties of the original wave and can be used to form an image of the object. This wave can be completely separated from the other waves implied by the other terms of Eq. (3) by means of the improvements introduced by Leith and Upatnieks (4). In the case that $c \equiv R$, the reconstructed wave is unaberrated.

To determine just what the aberrations are and how magnification can be achieved, we refer to the coordinate system shown in Fig. 2. A single object point is situated at (x_O, z_O, y_O) and the reference beam originates from a point located at (x_R, y_R, z_R). The hologram plane is always the x-y plane so that z_O is always negative, but there is no restriction on z_R: $-\infty \leq z_R \leq \infty$. Similarly to the reference wave, the reconstructing wave originates from a point located at (x_C, y_C, z_C). After being exposed to the reference and object waves, the film is processed and illuminated with the reconstructing wave. The object wave O is reconstructed and the coordinates of the center of this wave are

$$Z = \frac{m^2 z_C z_O z_R}{m^2 z_O z_R + \mu z_C z_R - \mu z_C z_O} \quad (4)$$

$$X = \frac{m^2 x_O z_O z_R + \mu m x_O z_C z_R - \mu m x_R z_C z_O}{m^2 z_O z_R + \mu z_C z_R - \mu z_C z_O} \quad (5)$$

$$Y = \frac{m^2 y_O z_O z_R + \mu m y_O z_C z_R - \mu m y_R z_C z_O}{m^2 z_O z_R + \mu z_C z_R - \mu z_C z_O} \quad (6)$$

We have introduced two new quantities here—the scaling

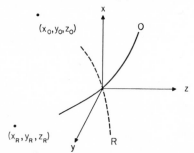

Fig. 2 Illustrating the coordinate system and notation used in describing the recording of a wavefront O from a single object point located at (x_O, y_O, z_O).

factor m, which takes into account a possible scaling up or down of the hologram itself, and μ, which is the ratio of the illuminating wavelength to the recording wavelength, λ_C/λ_O. If the illuminating wave is identical with the original reference wave, then $z_C = z_R$, $x_C = x_R$, and $y_C = y_R$. If in addition $m = \mu = 1$, we find

$$Z = z_O; \quad X = x_O; \quad Y = y_O \quad (7)$$

so that the reconstructed wave exactly corresponds to the original object wave: *There are no aberrations.* This result can be extended to arbitrary, three-dimensional objects. However, no magnification has been achieved.

The magnification is given by

$$M = \frac{dX}{dx_O} \quad (8)$$
$$= \frac{\mu m z_C z_R}{\mu z_C (z_R - z_O) + m^2 z_O z_R}$$

This expression explains the interest in recording the hologram with X-rays or electron waves and reconstructing with light waves. Since the wavelength of X-rays or electron waves is so short, μ can become quite large, resulting in large magnifications. The expression also indicates how magnification can be achieved by changing the relationship between the geometrical positions (z_O, z_R, z_C) of the object, reference, and reconstruction beams. Gabor (1), El Sum (2), and Baez (3) changed the wavelength to achieve high magnification. Leith and Upatnieks (5) and Thompson et al. (10) changed the geometry.

III. Applications to Microscopy. As mentioned earlier, holography was originally invented as an improvement of microscopy—specifically, electron microscopy. There are several features of the wavefront-reconstruction technique that make it potentially more suitable (at least in some areas) for microscopy than conventional imaging techniques. One of these features is the fact that in holography the field of view is a function of the resolution and size of the recording material. We can therefore expect, at least theoretically, to obtain good imagery over much larger fields than are attainable with conventional microscopy at high magnifications. Further, since the complete wavefront has been reconstructed, the hologram contains information about the object through a much greater depth than is possible with conventional microscopy. Still another advantage of holographic microscopy is that the image can be examined at some time after the hologram has been recorded. Thus when a phenomenon has to be studied which changes with time, holographic microscopy can be employed to examine the object as it appeared at one instant. Also, changes can be artificially induced in the object and the changed specimen can be compared with itself before it was changed. Since the hologram reconstructs the actual object wave, the various image-processing techniques that are normally applied to the original object can still be applied. This is a feature unique to holography.

There are basically two distinct methods of holographic microscopy: (1) conventional holography with magnification achieved by changing the scale of the hologram, the illuminating wavelength, or the radius of curvature of the illuminating wavefront; optically magnifying the holographic image; or using any combination of these; or (2) holographically recording an optically magnified wavefront. We shall begin with a discussion of the first method.

Table 1

Spherical:

$$S = \frac{\mu}{m^4}\left[\left(\frac{\mu^2}{m^2} - 1\right)\left(\frac{1}{z_O^3} - \frac{1}{z_R^3}\right) - \frac{3\mu}{z_C}\left(\frac{1}{z_O^2} + \frac{1}{z_R^2}\right) + 3\left(\frac{m^2}{z_C^2} - \frac{\mu}{m^2 z_O z_R}\right)\left(\frac{1}{z_O} - \frac{1}{z_R}\right) + 6\frac{\mu}{z_O z_R z_C}\right]$$

Coma:

$$C_x = \frac{\mu}{m}\frac{1}{z_C^2}\left(\frac{x_O}{z_O} - \frac{x_R}{z_R}\right) - \frac{\mu}{m^3}\frac{1}{z_O^2}\left[\frac{x_O}{z_O}\left(1 - \frac{\mu^2}{m^2}\right) + \frac{\mu}{m}\frac{x_C}{z_C} + \frac{\mu^2}{m^2}\frac{x_R}{z_R}\right] + \frac{\mu}{m^3}\frac{1}{z_R^2}\left[\frac{x_R}{z_R}\left(1 - \frac{\mu^2}{m^2}\right) - \frac{\mu}{m}\frac{x_C}{z_C} + \frac{\mu^2}{m^2}\frac{x_O}{z_O}\right]$$

$$+ \frac{2}{m^2}\frac{\mu}{m^2}\left(\frac{x_C}{z_C} - \frac{\mu}{m}\frac{x_O}{z_O} + \frac{\mu}{m}\frac{x_R}{z_R}\right)\left(\frac{1}{z_O z_C} - \frac{1}{z_C z_R} + \frac{\mu}{m^2}\frac{x}{z_O z_R}\right)$$

Astigmatism:

$$A_x = \frac{\mu}{m^2}\frac{x_C^2}{z_C^2}\left(\frac{1}{z_O} - \frac{1}{z_R}\right) - \frac{\mu}{m^2}\frac{x_O^2}{z_O^2}\left[\frac{1}{z_O}\left(1 - \frac{\mu^2}{m^2}\right) + \frac{\mu}{z_C} + \frac{\mu^2}{m^2}\frac{1}{z_R}\right] + \frac{\mu}{m^2}\frac{x_R^2}{z_R^2}\left[\frac{1}{z_R}\left(1 - \frac{\mu^2}{m^2}\right) - \frac{\mu}{z_C} + \frac{\mu^2}{m^2}\frac{1}{z_O}\right]$$

$$+ 2\frac{\mu}{m}\left(\frac{1}{z_C} - \frac{\mu}{m^2}\frac{1}{z_O} + \frac{\mu}{m^2}\frac{1}{z_R}\right)\left(\frac{x_O x_C}{z_O z_C} - \frac{x_C x_R}{z_C z_R} + \frac{\mu}{m}\frac{x_O x_R}{z_O z_R}\right)$$

Field Curvature:
$$F = A_x + A_y$$

Distortion:

$$D_x = \frac{\mu}{m}\left[\left(\frac{\mu^2}{m^2} - 1\right)\left(\frac{x_O^3}{z_O^3} - \frac{x_R^3}{z_R^3} + \frac{x_O y_O^2}{z_O^3}\right) + \frac{3x_O}{z_O}\left(\frac{x_C}{z_C} + \frac{\mu}{m}\frac{x_R}{z_R}\right)^2 - \frac{\mu}{m}\frac{(3x_O^2 + y_O^2)}{z_O^2}\left(\frac{x_C}{z_C} + \frac{\mu}{m}\frac{x_R}{z_R}\right)\right.$$

$$\left. - 3\frac{x_C x_R}{z_C z_R}\left(\frac{x_C}{z_C} + \frac{\mu}{m}\frac{x_R}{z_R}\right)\right]$$

The main problem associated with this method is that magnification is usually achieved by means of a wavelength and/or radius-of-curvature change between recording and reconstructing. (Few attempts have been made at scaling the hologram.) These changes are always accompanied by the usual wavefront aberrations shown in Table 1. The holographic microscope has to be designed so as to eliminate most of these aberrations. Any magnification due to increasing the illuminating wavelength must be accompanied by a corresponding scaling up of the hologram; otherwise, resolution will be lost by an increase in the aberrations.

One of the earliest and most successful applications of holographic microscopy was particle size determination (10). Conventional methods of physically sampling a dynamic aerosol are unsatisfactory because they are too slow. By recording a hologram of the aerosol using a pulsed ruby laser, one can examine a three-dimensional volume of moving particles in detail at a later time. The useful recorded depth of field is very great compared to that of a conventional photomicrograph.

To illustrate the advantage of the holographic method, consider the problem of trying to record by means of conventional microscopy the image of two particles, 10 μm in diameter, which are separated longitudinally by 1 cm. It is impossible to build a conventional system that will record both particles simultaneously. A microscope that can resolve the 10 μm particles would have a depth of field of the order of 100 μm, not 1 cm. This problem can be solved by holography, a technique which is especially useful in dynamic situations when it is not possible to focus and record each member of a sample of particles separately.

Figure 3 is a schematic of a fog hologram camera used by Thompson et al. (10). There is no separate reference beam since this is a Gabor type of hologram. The sample volume recorded on each hologram is 7 cm³ and the system is capable of 6 μm resolution. Illumination is by a pulsed ruby laser with a pulse duration of 0.5 μsec. Figure 4 is a schematic of the readout system. A He-Ne laser and collimator provide a plane wave to illuminate the hologram and form an image of the volume of particles. The overall magnification is about 300×.

Knox (11) has also demonstrated the usefulness of this technique. He used a Q-switched ruby laser, capable of producing coherent light pulses of 100 nsec duration, to record living marine plankton organisms. He thus recorded, at a single instant of time, a complex, transient, microscopic event, throughout a significant volume, for later analysis with conventional microscopic equipment.

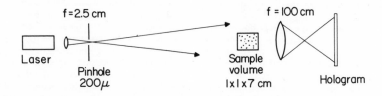

Fig. 3 Schematic of the fog hologram camera of Thompson. After Thomson et al. (10).

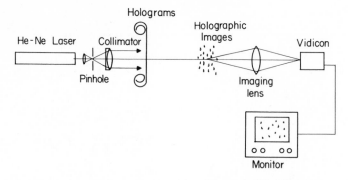

Fig. 4 Schematic of the read-out system used by Thompson. After Thompson et al. (10).

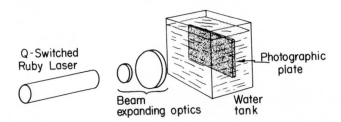

Fig. 5 Schematic of the illumination system used by Knox (11) to record instantaneously a large volume of living marine plankton organisms.

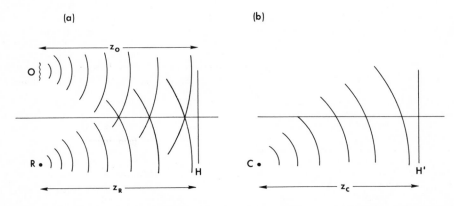

Fig. 6 Schematic of a holographic microscope. (a) A hologram of the object O is recorded with a spherical reference wave derived from a point source a distance Z_R from the hologram plane. (b) The hologram, which may have been scaled, is denoted by H' and is illuminated with a spherical wave derived from a point source a distance Z_C from the hologram.

The total volume that he recorded, using the illumination system shown in Fig. 5, was over 1000 cm³. Using a conventional microscope to view a real image formed with the holograms allowed any point in the volume to be brought into sharp focus.

Figure 6 shows a schematic of the type of holographic microscope used by Leith and Upatnieks (5). Some of their results are shown in Fig. 7. For these photos, $\lambda_R = \lambda_C$, so $\mu = m = 1$. The fly's wing is photographed at a magnification of 60 and the bar chart at $m = 120$. The spacing between the closest bars is about 8μm. Van Ligten (12) reported that the achievable resolution is even less for general objects, a result which he attributes to problems with the photographic emulsion and substrate.

Because of the apparent limitations of this method, most of the current work in holographic microscopy uses involves the second method—that in which an optically magnified wavefront is recorded holographically.

Figure 8 is a schematic of the holographic microscope used by van Ligten and Osterberg (6) to record holographically a magnified wavefront. This is essentially identical with the system used by Ellis (7), who first demonstrated the technique. An unmodified reconstruction yielded an image that was a faithful reproduction of the original microscope image with bright field illumination.

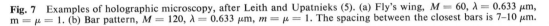

(a)   (b)

Fig. 7 Examples of holographic microscopy, after Leith and Upatnieks (5). (a) Fly's wing, $M = 60$, $\lambda = 0.633$ μm, $m = \mu = 1$. (b) Bar pattern, $M = 120$, $\lambda = 0.633$ μm, $m = \mu = 1$. The spacing between the closest bars is 7–10 μm.

Using this *same hologram*, Ellis was able to obtain:

(a) Dark-field reconstruction. The image is formed by blocking the direct light at the aperture.

(b) Spatial filtering. Fine image detail was removed by reducing the aperture.

(c) Bright-ground positive phase contrast. A lightly darkened mask with a clear central aperture was used as a mask.

(d) Dark-ground negative phase contrast. A lightly darkened spot attenuated and retarded the direct light.

(e) Interference. A small lenticular wedge divided the beam into two parts to produce shear fringes.

Polarization microscopy has been demonstrated by van Ligten (12). The hologram is made with a laser emitting linearly polarized light. Normally, the interference at the hologram plane during recording takes place between the components of the reference and object beams that have the same polarization. This means that the hologram contains information of the specimen as if it were taken between a polarizer and an analyzer set parallel. Introducing a rotator in one of the beams (13) will allow reconstruction of the specimen representing information that would be obtained with the polarizer and analyzer set at an arbitrary angle with each other. This is done during recording and poses no difficulties during reconstruction.

IV. Summary and Conclusions. Holography can circumvent the problems of limited depth of focus, off-axis aberration, and the relatively short working distance of the classical microscope. Theoretically, the hologram records a large volume of object space without loss of resolution. Also, since the hologram contains all of the information about the object, all of the usual image-processing techniques can be applied to the holographically reconstructed wave. These include bright field illumination, dark field illumination, spatial filtering, bright-ground positive phase contrast, dark-ground negative phase contrast, interference microscopy, and polarization microscopy. Holographic microscopy also allows standard microscopic techniques to be applied to dynamic objects throughout a large volume in object space. This feature is unique to the holographic method.

Holographic microscopy at the present time is faced not only with serious problems, but also with unique and exciting prospects for the future. Its principal difficulty, well recognized by workers in the field, consists of the diffraction noise generated by edges, out-of-focus details in the specimen, dust particles in the system, and laser speckle. This last is the noise associated with the random interference patterns in space which occur whenever coherent light is transmitted or reflected in a diffuse fashion. Most investigators believe, however, that these problems will be solved, and advances have already been made.

Prospects for the future in holographic microscopy seem to indicate that conventional microscopy can be fundamentally augmented by the capabilities of the coherent-light microscope. The types of spatial filtering of possible use to the microscopist may be expanded far beyond those presently employed. It has been shown that

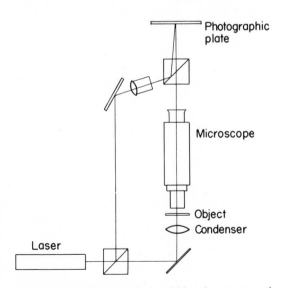

Fig. 8 Schematic of the holographic microscope used by van Ligten and Osterberg (6).

many uses of the microscope can be highly simplified or automated in the performance of such routine tasks as counting and sizing blood cells or other assemblies of small objects (14).

Howard M. Smith

References

1. D. Gabor, *Proc. Roy. Soc.* (London), Ser. A, **197**:454 (1949).
2. H.M.A. E. Sum, "Reconstructed Wavefront Microscopy," Ph.D. Thesis, Stanford Univ., November, 1952.
3. A. V Baez, *J. Opt. Soc. Amer.*, **42**:756 (1952).
4. E. N. Leith and J. Upatnieks, *J. Opt. Soc. Amer.*, **52**:1123 (1962).
5. E. N. Leith and J. Upatnieks, *J. Opt. Soc. Amer.*, **55**:569 (1965).
6. R. F. van Ligten and H. Osterberg, *Nature*, **211**:282 (1966).
7. G. W. Ellis, *Science*, **154**:1195 (1966).
8. W. H. Carter and A. A. Dougal, *IEEE J. Quantum Electronics*, **2**:44 (1966).
9. H. M. Smith, "Principles of Holography," New York, Wiley-Interscience, 1969.
10. B. J. Thompson, J. H. Ward, and R. Zinky, *Appl. Opt.*, **6**:519 (1967).
11. C. Knox, *Science*, **153**:989 (1966).
12. R. F. van Ligten, *Proc. Soc. Phot. Inst. Eng.*, **15**:75 (1968).
13. W. H. Carter, P. D. Engeling, and A. Dougal, *J. Quantum Electronics* Q*E***2**:44 (1966).
14. W. L. Anderson, *Proc. Soc. Phot. Inst. Eng.*, **15**:159 (1968).

HYPOPHYSIS

The variability in the results which are discerned through the staining procedures is well known to anyone who has applied these methods to the hypophysis. These variables depend upon the number of different cell types, and the function which these cells perform in the species which is being used. It is quite apparent that intensity of staining will vary in the same kind of cell in different species as well as the resulting staining reaction, i.e., in most mammals the carminophil increases in number during pregnancy, but in the rat it seems that the orangeophil is the reactive cell during pregnancy.

There are species in which the cells of the pars anterior possess a granulation which will respond to certain staining reactions thus allowing for their immediate identification. However, variations associated with their functional state at the time of staining will become apparent as one compares cell size, granulation, and morphology.

The terms "acidophil" and "basophil" as used by the cytologist who deals with the hypothesis must not be applied to the chemistry of the cells. Merely because a hypophyseal cell stains with a basic dye does not imply that the ergastoplasm, which exhibits a basicity in certain other glandular cells, is being stained. The above term refers to the staining reaction of the granules in the cytoplasm. However, one must realize that a cell which at one time will take a basic dye, can also be stained with an acid dye. Variations in the choice of fixatives will affect the stainability of the cells, and subsequent adjustments in staining techniques must be made if comparable results are to be expected.

Thorough fixation of the gland for a period of several days is thought to be essential for consistent staining. It is also suggested that the gland be bisected prior to fixation. The hypophysis will show evidences of cytolysis unless proper fixation has occurred. The fixatives recommended are formalin-sublimate mixtures, Susa, Helly's, and Zenker's fluids. Fixatives containing dichromate must be washed thoroughly in running water for 24 hr unless one is willing to run the risk of working with hypophyses that will exhibit an alteration of staining properties due to degeneration. Formalin-fixed glands can be stored in 70% alcohol without prior washing in water. The removal of mercuric crystals can take place through the use of iodine once the sections are on the slides.

There are a variety of staining methods available for selectively staining the several cell types which one encounters in the pars distalis of the mammalian hypophysis. I have selected some of those which are recommended because they are tinctorially consistent, and others which exhibit, through the use of histochemical procedures, the presence of hormones within specific cells.

The two types of basophils can best be shown by the use of Halmi's aldehyde-fuchsin (*J. Stain Tech.*, **27**:61–64, 1951) or by Gomori's method (*Am. J. Clin. Pathol.*, **20**:661–664, 1950) in which he introduced the use of aldehyde-fuchsin for elastin as well as for hypophyseal cell granules.

The Gomori procedure is as follows: Basic fuchsin, 0.5 g is dissolved in 100 ml of 70% alcohol. 1 ml HCl (conc.) is added together with 1 ml of U.S.P. paraldehyde. At the end of 24 hr the solution is violet and ready for use. Store in refrigerator. Avoid the use of fixatives that contain chromate, and select either formalin or Bouin's.

1. Decerate and hydrate paraffin sections in the usual way.
2. Transfer sections to 0.5% iodine, 10–60 min.
3. Decolorize with 0.5% $NaHSO_3$, 30 sec.
4. Wash in water, 2 min.
5. Place sections in 70% alcohol.
6. Stain in aldehyde-fuchsin, 1/2–2 hr.
7. Wash in 2–3 changes of 70% alcohol.
8. Counterstain with Masson trichrome or Mallory-Heidenhain method.
9. Dehydrate, clear, and mount in synthetic resin.

The Kerenyi and Taylor (*Stain Tech.*, **36**:169, 1961) method can be utilized for autopsy material following formalin fixation. Sections cut at 5 μ.

1. Decerate and hydrate.
2. Stain in 1.0% aqueous benzo pure blue, 2 min.
3. Water wash, 1 min.
4. Harris' alum hematoxylin, 1 min.
5. Differentiate in 0.125N HCl in 70% alcohol. Use microscope.
6. Blue sections in 1.0% disodium phosphate or 1.0% sodium acetate; rinse in water.
7. Eosin counterstain, if desired, 10–15 sec.
8. Rinse, dehydrate, clear, and mount in synthetic resin.

The Lendrum technique for the hypophysis (*J. Pathol. Bacteriol.*, **57**:267, 270, 1945) utilizes a formalin-sublimate fixative which is followed by an alcohol, chloroform, paraffin embedding procedure. The "carb-acid-fuchsin" stain is prepared as follows: Mix 1 g acid fuchsin with 0.4 g of melted phenol crystals; cool and dissolve in 10 ml of 95% alcohol. Grind to a fine consistency 0.5 g of starch. Add to the above 0.5 g of dextrin and grind finely. The mixture is suspended in 100 ml of water by grinding, with water being added occasionally. Heat to 80°C, cool, filter, and add it to the fuchsin-phenol-alcohol mixture, and make up to a total of 100 ml.

1. Stain sections in alum hematoxylin, 3–5 min.
2. Wash in water.
3. Expose slides to a 1.0% green FCF in a 0.5% acetic acid, 1–2 min.
4. Water rinse.
5. Transfer to Lugol's iodine solution, 2 min. (At this time, of course, mercuric deposits are removed.)
6. Remove color of iodine from tissue in 95% alcohol.
7. Transfer slides to a 2.0% alcoholic phosphotungstic acid solution, 2 min.
8. Water rinse.
9. Stain in "carbacid-fuchsin," 2–6 min. (Use microscope to determine staining intensity of cells.)
10. Water rinse, dehydrate, clear, and mount in synthetic resin.

Lilli (1958) in his modification of the Weil-Weigert method, has shown that in all probability the granules of the acidophils contain a phospholipid component. His technique adapts itself to fixation in formalin mixtures, and can be used successfully with paraffin, celloidin, or frozen sections.

1. Stain sections 1 hr at 60°C in equal parts of the following solutions:
   A. 2 g $FeCl_3$, 0.6 g NH_4Cl, and distilled water to make up to 100 ml; pH should be about 2.
   B. 1 g hematoxylin in 100 ml of either 95% or 100% alcohol.
2. Differentiation is accomplished in 0.5 g borax, 1.25 g potassium ferricyanide, 100 ml distilled water. Continue decolorizing until nuclei are colorless and red corpuscles are brown to black. Myelin is also black.
3. Distilled water, 3–5 min.
4. Counterstain, if desired, in 0.1% safranin in 1% acetic acid.
5. Distilled water rinse, dehydrate, clear in zylene, and mount in synthetic resin.

The Monroe-Frommer method (*J. Stain Tech.*, **41**:248, 1966) utilizes tannic acid as a "trapping" agent for the basic fuchsin used in this technique. In addition, a solution of phosphomolybdic acid is used in order to oppose the action of alcian blue. In this instance, the phosphomolybdic acid should not be looked upon as performing the function of a "mordant" for the alcian blue, but rather with the fact in mind that it opposes its action and thus limits the tendency of alcian blue to color all tissue components. The preferred fixative is Zenker's fluid, and paraffin sections of autopsy tissues are cut at 6 μ. The following solutions are needed:

A. Tannic acid, 10.0 g
   Distilled water, 100.0 ml
Stock solution
B. Basic fuchsin, 1.0 g
   Alcohol, 100%, 20.0 ml
   Distilled water, 80.0 ml
Working Solution (filter before use)

C. Basic fuchsin, 50.0 ml
   Distilled water, 50.0 ml
D. Analine, 1.0 ml
   Alcohol 100%, 90.0 ml
   Distilled water, 10.0 ml
E. Phosphomolybdic acid 1.0 g
   Distilled water, 100.0 ml
F. Alcian blue, 8GX, 1.0 g
   Distilled water, 100.0 ml

1. Decerate and hydrate to distilled water.
2. Remove mercuric crystals with Lugol's or Gram's iodine, and clear with "hypo."
3. Wash in running water, 10 min.
4. Tannic acid solution, 10 min.
5. Wash in running water, 5 min.
6. Working solution of basic fuchsin for 3–5 sec. Agitate slides by dipping in and out of the staining solution.
7. Wash off excess stain in tap water.
8. Differentiate in aniline solution until alpha cells are red and beta cells are pink.
9. Phosphomolybdic acid solution, 30 sec.
10. Distilled water rinse.
11. Stain in alcian blue, 30 sec.
12. Distilled water rinse.
13. Dehydrate, clear in xylene. Two changes in absolute and xylene are recommended.
14. Mount in Permount or other synthetic resin.

Pearse's method (*J. Pathol. Bacteriol.*, **61**:195, 1949) utilizes paraffin sections. Zenker's or Helly's fluids are recommended for fixation.

1. Decerate and hydrate.
2. Remove mercuric precipitates with iodine; follow with "hypo" to remove iodine coloration.
3. Place sections in 0.8 g H_5IO_6, 10 ml 0.2M sodium acetate, 70 ml ethyl alcohol, and 20 ml distilled water, 5 min.
4. A 70% alcohol rinse; 1 min wash in a solution of 1.0 g KI, 1.0 g $Na_2S_2O_3 \cdot 5H_2O$, 20 ml water, 30 ml alcohol, and 0.5 ml 2N/HCl.
5. Rinse in 70% alcohol and subject slides to Schiff's reagent for 15–45 min.
6. Wash in running water, 10–30 min.
7. Stain in 0.5% celestine blue in 5.0% Fe alum, 30 sec. Transfer, without washing, to Mayer's hemalum, 30 sec.
8. Rapid differentiation in 2.0% acid alcohol; blue in water.
9. Alpha granules may be stained in 2.0% orange G in 5.0% phosphotungstic acid, 5–10 sec.
10. Wash in running water.
11. Dehydrate, clear in xylene, mount in polystrene (DPX).

The kresazan method of Romeis (1940) is one of the older methods, and one which requires technical "know-how" to manipulate. This method, according to Harris (1966) necessitates the use of a resorcin-fuchsin which is prepared as follows:

Solution A. Basic fuchsin, 0.5 g
             Resorcinol, 1.0 g (mix in an Erlenmeyer flask)
             Distilled water, 100 ml
Solution B. $FeCl_3$, 0.5g
             Distilled water, 50 ml

The above solutions are prepared in separate flasks. Heat both solutions to boiling. Add B to A and continue boiling with agitation for 5 min. Cool under running water and filter. Do not dry the precipitate but transfer the paper upon which the precipitate is deposited to flask A. Add 70 ml of 95% alcohol and heat to boiling with agitation. Cool and allow to stand for 15 min. Add 1 ml of conc. HCl and filter. One may use the stain immediately, but the life of the stain is relatively short—about three weeks. Stain sections for 1 hr or longer. Remove stain from the slides with 95% alcohol and differentiate to the proper intensity in the same alcohol. Counterstain with either Heidenhain or Romeis methods. The technique has been used on the pars anterior of the cow, dog, horse, rat, cat, and human.

The Wilson-Ezrin method (*Am. J. Pathol.*, **30**:891–899, 1954) uses a 10% formalin-saline fixative, and paraffin sections are cut at 6 μ.

The following solutions are needed:

A. Periodic acid, 1 g
   Distilled water, 100 ml
B. The Schiff reagent which is prepared by dissolving 1.0 g basic fuchsin in 200 ml of hot distilled water. Bring the solution to the boiling point. Cool to 50°C. Filter and add 20.0 ml of N/HCl. Cool further and add 1.0 g of anhydrous $NaHSO_3$ or $Na_2S_2O_5$. Keep in the dark for 48 hr until the solution becomes straw colored. Store in the refrigerator. (To test efficacy of the solution from time to time, proceed as follows: Pour a few drops of the reagent into 10 ml of 37–40% formalin in a small container. The solution, if still usable, will turn reddish purple very quickly. If it is breaking down the color which results from this test will be a deep blue-purple.)
C. $Na_2S_2O_5$, 10% aqueous, 60.0 ml
   N/HCl, 50.0 ml
   Distilled water, 1000.0 ml

D. Orange G, 1.0 g
   Distilled water, 100.0 ml
E. Phosphotungstic acid, 5.0 g
   Distilled water, 100.0 ml
F. Methyl blue, 1.0 g
   Distilled water, 100.0 ml

1. Decerate and hydrate in distilled water.
2. 1% periodic acid, 5 min.
3. Distilled water rinse.
4. Schiff's reagent, 15 min.
5. Three changes of $Na_2S_2O_5$ rinse, 3 min each.
6. Wash in running water, 10 min.
7. Stain in orange G, 1 min.
8. Transfer to phosphotungstic acid solution, 30 sec.
9. Running water, 30 sec.
10. Stain in methyl blue solution, 1 min.
11. Remove excess methyl blue in tap water.
12. Dehydrate, clear in xylene, using two changes.
13. Mount in synthetic resin.

CLYDE W. MONROE

References

Harris, G. W., and B. T. Donovan, "The Pituitary Gland," Vol. 1, Berkeley and Los Angeles, University of California Press, 1966.
Lillie, R. D., "Histopathologic Technic," New York, McGraw-Hill Book Co., 1958.
Lillie, R. D., "Histopathologic Technic and Practical Histochemistry," 3rd ed., New York, McGraw-Hill Book Co., 1965.
Luna, L. G., "Manual of Histologic Staining Methods of the Armed Forces Institute of Pathology," 3rd ed., New York, McGraw-Hill Book Co., 1968.
Romeis, B., "Die Hypophyse," in "Handbuch der Mikroskopichen Anatomie des Menschen," Vol. 6, part 3, von Mollendorf, ed., Berlin, Julius Springer, 1940.

INDIUM STAINING

Indium is a member of the boron family of metals in the periodic table whose atomic number (49) and atomic weight (114.82) are sufficiently greater than that of the aggregate components of most cells to render it useful in "staining" preparations for observation in the electron microscope. Techniques have been developed to utilize chemical properties of compounds of this element in achieving a relatively high selectivity of binding for nucleic acid phosphate. Thereby, nucleic acids may be localized within cells. By coupling with selective extraction techniques (enzymes or cold perchloric acid), the type of nucleic acid may be deduced.

Basic to all presently published procedures are the following seven steps: (1) fixation using organic fixatives (glutaraldehyde or acrolein preferred); (2) dehydration in a graded acetone/water series; (3) reduction of specimen with lithium borohydride ($LiBH_4$) in pyridine; (4) decreasing the tissue ligand spectrum by derivative

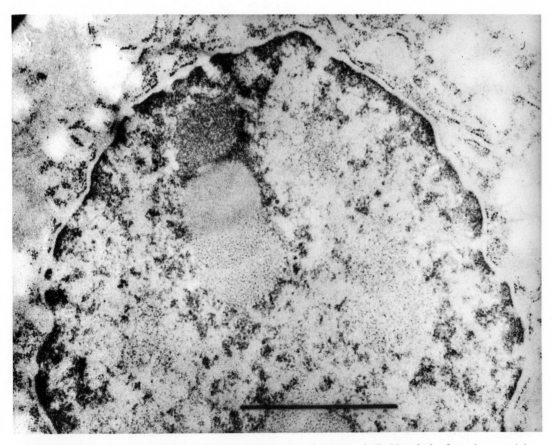

Fig. 1 Indium-stained mouse hepatocyte 15 min after injection of actinomycin D. Note lack of membrane staining and the evident distribution of stain only on chromatin and ribosomes. The nucleolus has separated into three components, a characteristic reaction to cessation of DNA-dependent RNA synthesis. ($\times$ 25,000.)

formation (blocking reactions); (5) infiltration of specimen with indium trichloride dissolved in acetone (staining); (6) embedding in plastic (cross-linked methacrylates, polyesters, or epoxys); (7) sectioning and microscopy.

The following is a typical processing protocol: (1) Fix small (ca. 1 mm³) tissue blocks in 2% unstabilized acrolein containing $0.14M$ NaCl, $0.005M$ Na acetate, and $0.0005M$ $CaCl_2$ at 0–5°C for 15–30 min; (2) dehydrate block with 5 min rinses in each of 25, 50, 75, and 90% acetone/water, and absolute acetone at 0–5°C; (3) in three sequential rinses of 5 min each (25/75, 50/50, 75/25, pyridine/acetone) bring to absolute pyridine at 0–5°C; (4) rinse in pyridine saturated with $LiBH_4$ (2 hr at 0–5°C); (5) wash out unreacted $LiBH_4$ with 3 rinses of pyridine, 10 min each at room temperature; (6) incubate blocks in pyridine/acetic anhydride (60/40) containing a trace of anhydrous sodium acetate for 12–18 hr, room temperature (does not block carboxyl groups); *or* 2% trimethylchlorosilane in absolute pyridine at −10°C for 15 min (blocks carboxyl groups); (7) rinse (×3) in absolute pyridine at −10°C; (8) transfer to absolute acetone through a 50/50 pyridine/acetone mixture (5 min, room temperature); (9) rinse (×3) in absolute acetone 0–5°C, 10 min each; (10) incubate in acetone solution of indium trichloride (25 ml absolute acetone, 625 mg $InCl_3$, 100 λ dist. H_2O) for 2 hr at 0–5°C (make up one day ahead); (11) rinse (×2) in absolute acetone, 15 min each at 0–5°C; (12) begin desired embedding sequence.

In order to secure intense staining with highest selectivity for nucleic acid, several factors must be noted: (1) Metal-containing fixatives and processing solvents (other than sodium, potassium, and calcium) *cannot be used.* (2) Acetylation does not block carboxyl groups of the tissue, and structures not known to contain nucleic acid may stain densely for this reason (sperm tails, mast cell granules, keratohyaline granules). (3) The acetone staining solution must be stored overnight at 0–5°C (before use) to allow sufficient time for indium polymer formation. (4) Embedding resins may contain effective metal chelation agents (such as anhydrides) or metallic compounds such as the activators and initiators of polyester resins. These should be avoided since they slowly remove specimen-bound indium. They may be used if sectioning and observation are promptly carried out (within a week or two). (5) Large amounts of indium are extracted from thin sections by knife-trough water following sectioning on an ultramicrotome. This is controlled by using a 1×10^{-4} M solution of $InCl_3$ in 10% acetone as the solvent in the flotation trough (sections should not be allowed to remain wet any longer than necessary). (6) The density contributed by indium staining is adequate for magnifications up to 100,000×. Detail below 20 or 30Å is difficult and questionable. Unstained tissue, visible at low magnification, cannot be discerned at high magnification where phase contrast permits stained components to be seen distinctly. Therefore stained components will show sharp detail at 20,000× on the viewing screen when viewed with a 10× magnifier.

WILLIAM G. ALDRIDGE

References

Aldridge, W. G., and M. L. Watson, "Perchloric acid extraction as a histochemical technique," *J. Histochem. Cytochem.,* **11**:773–781 (1963).
Coleman, J. R., and M. J. Moses, "DNA and the fine structure of synaptic chomosomes in the domestic rooster," *J. Cell Biol.,* **23**:63–78 (1964).

Watson, M. L., and W. G. Aldridge, "Methods for the use of indium as an electron stain for nucleic acids," *J.B.B.C.,* **11**:257–272 (1961).

INJECTION

For the demonstration of blood, lymphatic, and gall vessel systems vessels are injected with various colored materials which solidify under the action of fixing substances, cold, or chemical reactions, thus remaining in the specimen after preparation.

Rules for Injection. Smaller animals are usually injected *in toto*; with larger ones only a required region or an isolated organ are injected. The vessels must not be damaged (e.g., by cuts) lest the injection material escape. Further, they must not be filled with blood (especially with coagula) and the pressure must not be too great because in either case extravasation might occur. The coloring of the injection material must not diffuse into the neighboring vessel and must not precipitate. When making so-called "corrosive preparations" the color of the injection material must not be lost or changed during the corrosion of the tissue (organ). During the injection the contraction of vessels must be prevented by due regard both to the temperature and to the effect of the injection mass, and fresh tissues obtained from a recently killed animal should be used. Up to now it has not been established unequivocally if the vessels should or should not be washed out prior to the injection; according to the writer washing is not necessary if the organisms or the organ has been well exsanguinated. In any case, when preparing organs after a long postmortem period the blood coagula cannot be removed, and moreover, there is danger of the diffusion of the perfusion solution into the neighboring tissue. If the vessels are washed out the type of the perfusion solution should be selected according to the injection material used. Finally it is necessary to point out that the injection mass must not contain air bubbles.

Injection Method. Smaller animals are injected *in toto* in the following manner: Under the suitable narcosis (ether, thiopental, etc.), the common carotid artery is prepared and three ligatures are put on. First a ligature is made close to the heart and a cannula is then introduced into the artery and tied with a second ligature. The third ligature closes up the artery in the cranial direction. The ligature closest to the heart is then unlaced, and blood flows into a bowl. If the blood-vessel tract is to be washed out the perfusion fluid is warmed up to 40°C. The jugular vein is similarly prepared, another cannula is then introduced into the carotid artery, and the injection material is injected under moderate but continuous pressure. As soon as the material begins to leak out from the canula introduced into the lower part of the artery, both the artery and the vein must be tied up to prevent discharge of the material.

If an organ in a state of postmortem rigidity is to be injected with the warm injection mass, it is desirable to heat the organ in the Ringer's solution or to bathe it with this solution. A 3.3% sodium sulfate solution can also be used. The cannula with the injection material is introduced into the main vessel only; all other vessels, however, must be ligated or at least signed to prevent the outflow of material. If both arterial and venous systems or even several systems (liver) are to be prepared it is desirable to inject them simultaneously. It is possible, particularly in parenchymatous organs (kidneys, liver) to observe the filling of superficial vessels and obtain the optimum connection of both systems in the capillary network. The correct result is indicated by the complete coloration of the organ surface.

Injection Equipment. For injections either an injection syringe or special equipment can be used. Special injection equipment has certain advantages, e.g., it permits precise setting and maintenance of the pressure, but on the other hand, it is rather complicated and is difficult to clean. Here only the simplest equipment will be described. The injection material is stored in a glass funnel transiting into a vertical glass tube connected with a rubber tube equipped with a cannula and a clamp. The pressure is determined by the height of column of the injection material. For injections of gelatin a common injection syringe "Record" can be used. However, in this case a difficulty may occur because, under the effect of warm gelatin, the metal piston expends faster than the glass cylinder and grabs. For injections of plastic materials special syringes with leather or rubber pistons are recommended. The writer has obtained the best results with "Record" and "Jannette" syringes which can be used for all types of injection materials provided that the steel piston ring is replaced with a silicone piston ring. The fully disassemblable injection syringe (Fig. 1, producer Chirana Works, Stará Turá, CSSR) is very serviceable and can be cleaned easily. A further advantage of these syringes is in the mutual replaceability of pistons. If such an injection syringe is not available it is enough to smear a normal injection syringe with oleic acid or liquid paraffin and to put a few drops above the piston. Injection cannulas are most frequently made of glass and are slightly concave at the end to permit the fixation of the vessel with a ligature. They are connected by means of a rubber tube to the injection equipment or syringe. It is convenient to make cannulas from glass tube because the lumen may be

adapted to that of the vessel injected. The simplest method is to make cannulas from metallic injection needles. The vessel can be fixed on them by a thread and they need not be concave.

Injection Materials. There are many kinds of these. The selection of a suitable one is based on the aim of the injection, i.e., it is necessary to decide if rough ramification, terminal ramification, or the total system will be injected. The problem of processing is also important (simple preparation of the vascular bed, histological processing, corrosive casts).

Injection materials can be divided as follows: (1) dyes in the form of suspensions or specially adjusted suspensions; (2) dyes carried in various substances, most frequently gelatin or various proteins; (3) air injections; (4) celluloidine; (5) synthetic plastic materials especially (a) latices of synthetic rubber, (b) polyvinyl chlorides, (c) various kinds of synthetic resins, particularly acrylates and polyester resins, and (d) special injection materials for filling lymphatic vessels.

Synthetic plastic materials have been used extensively in recent years but they have not yet replaced many classical methods. Injections of various dyes and colored gelatin are still used, expecially for histological studies on vascular architectonics. Black drawing ink or India ink are conveniently simple and yield good results. Drawing ink may also be injected in mixture with the blood serum, egg white (in corresponding dilution), and above all, gelatin solutions.

Krause's or Hoyer's methods for the preparation of red injection gelatin with carmine can be recommended if several systems are to be differentiated simultaneously. Carmine must be stabilized in the colloid state; this depends on the pH. Under alkaline conditions the dye diffuses; in the acid state it precipitates. Krause's method using borax carmine is the most reliable. In thin sections, however, the injection material is insufficiently colored. Prussian blue is widely used for the preparation of blue gelatin media. The addition of 2% of chloral hydrate or camphor is recommended for the preservation of gelatin mixtures.

The author has found that gelatin for microscopical purposes can be colored in a simple and easily reproducible way by the paste organic pigments of the Versatine series (Sdružení pro dehtová barviva, Říjnové revoluce 13, Brno ČSSR) dispersed in Kortamol. (See Fig. 2.) They are of excellent fineness and most particles (50%–80%) are smaller than 0.25 μ. Infrequently (2%–5%) particles of the size 2–5 μ can be found. The pH ranges from 7.0 to 9.5. Before mixture with gelatin the pigment is diluted (1 : 3–1 : 5 by volume) with a 1%–2% solution of Kortamol. It colors intensely and 3%–10% by volume is sufficient. A complete color scale can be selected. The best results were obtained with Versatine blue KB, violet RL, green KGN, black KS, red KF4R and E3B, and yellow KG and RT. Yellow RT alone changed to red after staining of histological sections in hematoxylin eosin and became dark when the VG method was used. The properties of pigments mentioned are determined by their chemical character (partly phthalocyanines, derivatives of chinacrodone and dioxazine, partly common or polycondensated azo-dyes).

For injections 6%–20% gelatin is used, according to the injected part of the vascular bed. First more diluted gelatin is injected; thereafter the main trunks are injected with thicker gelatin. For injections this rule is of general validity.

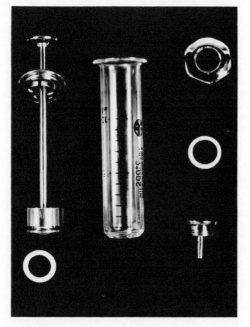

Fig. 1 A fully disassemblable injection "Record" syringe provided with a silicone piston ring.

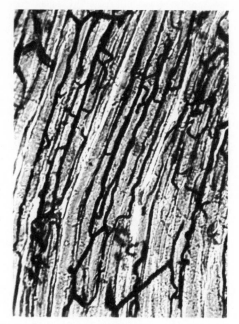

Fig. 2   The capillary network of muscular fibers injected with gelatin colored with Versatine violet RL.

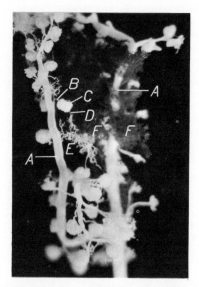

Fig. 3   A corrosion preparation of pig kidney, using rubber latex. A, a. interlobularis; B, vas afferens; C, glomerulus; D, vas efferens; E, arterial capillaries; F, venous capillaries.

Injections of Capillaries with Milk. For this purpose fresh milk is used. This method is suitable for frozen sections. After the injection both afferent and efferent vessels have to be closed. The preparation is fixed in a mixture of formalin and acetic acid.

Synthetic Plastic Injection Materials. For angiological purposes the materials formed by means of polymerization or copolymerization are suitable. At present, above all, latices of synthetic rubbers, acrylates, polyvinyl chlorides (PVC), and polyesters are of interest to anatomists.

Latex of synthetic rubber is a colloidal dispersion in an aqueous medium. The smallest particles (0.15 μ) can be found in chloroprene latices, e.g., in Neopren 601A and Neopren 650, which are of American origin and which are used most frequently. Latex particles are negatively charged, repel each other, and thus prevent spontaneous coagulation. Stability can be affected by bacteria, chemicals, etc., and for this reason latex is preserved with formalin and stabilized with ammonia; pH of the fresh product ranges between 7.0 and 5.8. Later on the acidity increases and a spontaneous coagulation occurs. The rate of coagulation depends on the concentration and may be prevented by adjusting the pH with ammonia (2%–3% by weight). Latex should be stored in a dark and cold room. The coagulation of latex in an acid medium is utilized not only for obtaining corrosion preparations (Fig. 3) but also for the suppression of latex leakage from thinner branches by means of acid.

The injection of the blood vascular system is described in great detail because it represents the most practical injection method: (1) According to the concentration of latex a desirable vessel region or the whole vascular bed can be injected: (2) the object injected can be investigated (a) by means of preparation, (b) histologically (frozen sections), (c) as a corrosion preparation (Fig. 4),

(d) as an angiogram, and (e) as a cleared preparation (Fig. 5).

An X-ray contrast mixture can be prepared from latex in the following manner:

Emulgator	1–1.5
Red lead oxide	10–15
Latex 30%–35%	100

Emulgator is a product of VEB Leuna Werke, GDR. Its general formula can be expressed as $CH_3—(CH)$ CH_2SO_3Na. Lead oxide can be replaced by mercurous

Fig. 4   The terminal ramification of the renal artery. A corrosion preparation using Dipren M-234. A, vas afferens; B, glomerulus; C, vas efferens; D, capillary network.

Fig. 5 Arteriogram of sheep kidney using X-ray contrast latex.

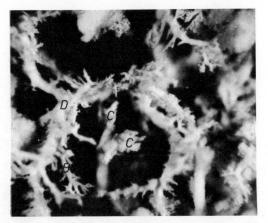

Fig. 7 An analogous synoptical preparation. The sinusoids are not injected. In the picture several lobules can be observed. The key is the same as to Fig. 6.

chloride. The use of the X-ray contrast mixture is the best method for the objective evaluation of angiograms in comparison with corrosion preparations of the same object.

Injection latex can easily be colored with special dyes (e.g., Vulcanosol). When making corrosion preparations these dyes must resist the drastic effects of the corrosive medium and penetrate throughout the capillary network. In this case also the best results were obtained with organic, the pigments mentioned above, which are mixed directly into latex. Microcorrosive preparations are investigated as follows: After the selection of a suitable region by means of a dissecting microscope, the corresponding part of the preparation is cut out and studied in a droplet of water on a depression slide by lateral illumination. Latex corrosion preparations of terminal and preterminal regions retain their natural architectonics only in liquids. (See Figs. 6, 7, 8).

Synthetic Plastic Materials (Resins). These are macromolecular substances which are formed through polymerization of simple molecules (monomers) into polymers. Methacrylates (Fig. 9), polyvinyl chloride, and

epoxide resins are most frequently used. These products are available under various trade names and are suitable for the injection of coarser vessels. Insofar as they retain the topography after solidification it is possible to say that they demonstrate well the mutual functional-topographic relationships between individual systems.

Methacryle consists of a powder and a liquid which are mixed in the ratio 1 : 3–3 : 1 (according to the purpose of the injection). When a homogenous mass appears after the solution of both components, the material is injected or poured in. If the preparation is to be retained *in situ* the cadaver or the organ must be fixed with 5 %–10 % formalin. Otherwise it is put directly into HCl or alkali. After the corrosion of the tissue, the cast is washed out with water. For coloring of the injection material a dye resistant to corrosive effects had to be used, e.g., paste pigments "Plastik" (producer Bayer Company): Plastikweisspaste RA, blau B and rot 5B, gelb GA; or powdered "Mikrolith" pigments (producer Ciba Company): Mikrolith-yellow 2R-K; red R-K; blue 4G-K; black C-K; white R-K. Mikrolith pigments have first to be dispersed in chloroform. An increased elasticity of acrylate casts can be obtained after the addition of 15 %–22 % by volume of dibutyl phtalate.

Polyvinyl chloride (PVC) also has two components, a

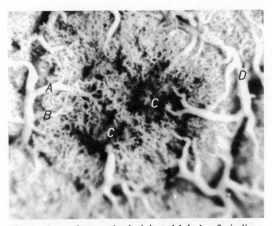

Fig. 6 A nearly completely injected lobule of pig liver. A corrosion preparation. A, B, distribution veins; C, roots of central vein; D, interlobular vein.

Fig. 8 The afferent and the efferent venous system in a one day old pig. A synoptical acrylate preparation.

Fig. 9 A methacrylate corrosion specimen of the afferent and efferent venous systems of the sheep liver from the cranial aspect. "Total" injection.

white powder and a liquid, and it is used in the form of paste, emulsion, or solution. The injection method is similar to that of acrylate. The hardness of PVC can be adjusted by means of various plasticizers and thus may be more suitable than acrylates. In this connection special references should be made to a material based on vinyl, the so-called Plastoid (Schummer), because it gives excellent results even in the field of preterminal and terminal ramifications.

Epoxy resins are substances synthesized on the base of polyesters. Their properties are rather variable and it is necessary to test each before use. The resin is syrupy and injections are carried out after the addition of a hardener. Polymerization takes about 48 hr.

Corrosion preparations are, in point of fact, casts of vessels obtained by means of substances resistant to chemical, bacterial, and other effects after the removal of the tissue by means of chemicals or maceration. Hyrtl (1873), who used celloidin, obtained excellent results in this field. Considerable improvements in this technique have been obtained in recent years through the increased use of plastic materials. The method permits three-dimensional, spatial anatomical studies and has often led to the discovery of entirely new relationships between vessel systems of organs or organisms; this method also allows studies of organ or organism cavities. Substances retaining their normal shape when dry are used for coarse ramifications. For terminal regions, latex casts retaining the natural architectonics in liquids only are used and for that reason they are preserved in a mixture of 0.5%–1% formalin with 0.5% phenol. Smaller objects may be preserved in glycerin.

Cleared Preparations. The angioarchitectonics of some organs (fasciae, joint capsules, very thin muscles, etc.), can be studied by means of the organ clearing method of Spalteholz. Injection of a solution of gelatin with black drawing ink is very suitable for this purpose.

Angiographic Injections. This method was partly described in connection with synthetic rubber latex. X-ray contrast can be obtained through the addition of some X-ray contrast substances (Pb_2PbO_4 or $HgCl$) to materials mentioned, including gelatin; in the case of gelatin the contrasting material can be added directly and mercuric oxide red may also be used.

Injection of the Gall Duct. Gall must be removed prior to the injection, which is made into the bile duct or the hepatic duct. To avoid filling the gall-bladder it is necessary to ligature it in the neck region in such a way that blood vessels are avoided.

The injection of lymphatic vessels calls for a somewhat special technique. In some cases it is possible to inject into the vessel lumen (in intestines or in some organs, e.g., liver) and in other cases it is easier to inject through fine often even glass cannulae, introduced directly into the tissue. For injections, either oil dyes in turpentine, or aqueous media with Prussian blue, black drawing ink, or India ink are used. The cannula is introduced in a flat manner on the surface of the organ; then a slow back stroke is carried out and for the injection only a slight pressure is used. After the injection the organ remains at rest for several hours and then it is fixed in formalin. The vessels injected are investigated as whole mounts or as histological sections.

In recent years many authors have used latices of synthetic rubbers, especially Neopren, for studies on lymphatic vessels. This method gives excellent results, especially after the direct injection into lymphatic vessels, and in this way it is possible to obtain an excellent picture of the complex system of lymphatic vessels in liver.

JIŘÍ KAMAN

References

Hyrtl, J., "Die Corrosions—Anatomie," Wien, W. Braunmüller, 1873.

Kaman, J., "Zum Problem der Herstellung mikrokorrosiver Präparate bei Verwendung von Kautschuklatex," *Mikroskopie,* **19**:303–309 (1964).

Kaman, J., "Simplified preparation of x-ray contrast latex rubber," *Folia Morphol.* (Praha), **13**:131–135 (1965).

Kaman, J., "Beitrag zur Methodik der Gefässinjektion mit künstlichen plastischen Stoffen," *Mikroskopie,* **22**:306–310 (1967).

Kaman, J., "Contribution to the use of methacrylate in the investigation of the vascular architecture," *Folia Morphol.* (Praha), **16**:416–421 (1968).

Kaman, J., "A new type of injection syringe in anatomical practice," *Folia Morphol.* (Praha), **17**:83–85 (1969).

Kaman, J., "Eine neue und einfache Färbemethode der Injektionsgelatine für mikroskopische Zwecke," *Mikroskopie,* **26**:264–267 (1970).

Noble, R. J., "Latex in Industry" [In Russian], Leningrad, Goschimizdat 1962.

Wolf, J., "Mikroskopická technika," Praha, SZN, 1954.

INSECT HISTOLOGY

The most unique feature of insects (and all arthropods) is their impermeable cuticular exoskeleton. This is responsible for most of the problems encountered in the study of insect tissues either at the light or electron microscopical levels.

Fixation. The cuticle has its most important effect on fixation. It is not generally recognized that the cuticle is impermeable to many of the fixatives as well as to other agents used in the preparation of material prior to histological studies. Thus it is not sufficient, in spite of the small size of many insects, simply to immerse the animal in a fixing agent. It is important for proper preservation of the tissues that they be exposed directly, and as quickly as possible, to the fixative. An increasingly popular method of achieving rapid fixation is by injecting the fixative into an animal which has had its turgor reduced by bleeding. The fixed animal is then cut into small (1–2 mm) pieces in fixative. When injection is difficult (as in very small insects or appendages) or impractical (as in the study of an integumental wound), rapid access to fixative can be facilitated by cutting the animal or appendage into small pieces directly in fixative, followed closely by agitation. Penetration of fixative is always aided by agitation.

Blood Cells. Adequate fixation of insect blood cells poses a unique problem. Insect blood cells, called hemocytes, have been studied in the following ways: (1) in fresh smears using phase microscopy; (2) in smears which have been air-dried, vapor-fixed, or heat-fixed; (3) in clots formed by the coagulation of the blood proteins after drops of hemolymph were allowed to fall into fixative; (4) in "pellets" obtained by centrifuging the cells and proteins suspended in fixative which had been agitated to disperse them; (5) *in situ* as they were fixed upon injection of fixative. The limitation of this method lies in the diffi-

These various methods of fixation have resulted in conflicting descriptions and a confused classification of the blood cells of insects. The most satisfactory method, as judged by the fine structural integrity of cell membranes, organelles, inclusions, and matrix, appears to result from injection. The limitation of this method lies in the diffi-

culty of locating cells in sufficient numbers. Centrifugation of cells suspended in fixative overcomes this problem since the cells are then packed lightly together. However, it has been found that, at least in some insects, exposure to the air has a deleterious effect which can be prevented by bleeding the animal under fixative. Another important consideration in this method is the immediate dispersion of the cells in the fixative by agitation. At some stages of their life cycle, e.g., in the last larval instar and in the pupa, the blood of most insects is rich in protein. If the hemocytes are not dispersed quickly in the fixative, they become embedded in the coagulated blood protein, which presumably retards the penetration of the fixative resulting in poor fixation of all but the most peripheral cells in the clot so formed (Lai-Fook and Neuwirth).

The soft tissues present no particular problems except where they are quite thick, e.g., the fat body and muscles, at certain stages of development. These should be cut into small (1–2 mm) pieces, especially for electron microscopy.

Dehydrating and Embedding. The soft tissues present no special problems if they are not too bulky. Well-developed fat body, which usually contains numerous, very large lipid droplets, is known to be poorly fixed and infiltrated with embedding medium if not cut into sufficiently small pieces.

The cuticle presents unique problems to section cutting. This is especially true of very hard, dark cuticles. When sections of insect integument embedded in paraffin are cut, the epidermal layer is most often separated from the cuticle. This can be prevented by using a double-embedding method (Wigglesworth, 1959). The pieces of tissue are embedded first in 5% agar, in which orientation is facilitated, then the agar block is dehydrated, infiltrated, and embedded in ester wax. This technique allows not only undistorted but also fairly thin (0.5–1 μ) sections to be cut.

The embedding media commonly associated with electron microscopy are being used with increasing frequency, usually but not necessarily in conjunction with electron microscopy. Thick (1–2 μ) sections of methacrylate or resin-embedded tissues can be stained either with basic dyes such as Toluidine Blue, Methylene Blue or Azure II at high pH without removing the embedding medium or with many other stains after removing the resin (Hayat, 1970). The limitation to the use of plastics is the necessity of using glass rather than steel knives for sectioning. This severely limits the size of the pieces of tissue that can be sectioned.

Sectioning. The soft tissues present no problem in section cutting regardless of whether they are embedded in paraffin, ester wax, or a plastic. However, the cuticle presents some problem. It is often difficult to obtain thin scratchless sections with conventional steel microtome blades. The answer is provided by "disposable" blades. They can be either safety razor blades, sharpened and mounted in a specially designed blade-holder (Wigglesworth, 1959) for paraffin and ester-wax, or glass knives for cutting sections of plastic-embedded tissues. Diamond knives are too delicate and expensive for making thick sections.

Another problem encountered in sectioning cuticle is due to the lack of penetration of the epicuticle, even in soft-bodied insects, by the embedding media. The embedding medium does not adhere to the epicuticle, presumably from lack of infiltration. This is not too critical, though somewhat bothersome, for the light microscopist since the embedding medium is usually removed before staining and mounting. However, it is much more critical in electron microscopy, especially if the epicuticle itself is under observation. No real remedy has been found for this problem except perseverance. The uninfiltrated epicuticle, acting as the edge of the section and therefore much less stable, can also prove difficult when the epidermis or other parts of the cuticle are under investigation. Then the sections can be supported on a carbon film. It is significant that most of the fine structural studies on insect cuticle have been done on animals which were in a pre-ecdysial stage (Locke, 1966). In the early stages of cuticle deposition the epicuticle is "permeable" to the embedding medium, but shortly before ecdysis, it changes and becomes mostly impermeable. Thus it is recommended that wherever possible, if the age of the animal is not important, pre-ecdysial stages be used.

Whole Mounts. Ideally cells and tissues should be studied live and *in situ*. The best alternative is considered to be the use of freeze-dried sections, especially for the quantitative study of enzyme activity. A much less complex alternative, which has been used widely in the study of many insect tissues, is the use of whole mounts. The various tissues are studied either after dissection, in the whole animal, or after injection, in the small pieces into which the animal has been cut. This technique is possible because most insect organs, e.g., Malpighian tubules, the heart, pericardial cells, nerves, and oenocytes, are not very bulky. It is ideal for the study of organelles, inclusions and enzymes, when such studies are concerned principally with changes in number and size of organelle or inclusion and quantity or activity of enzymes. Whole mounts are particularly useful when enzymes which are very sensitive to embedding procedures are being studied. When the reaction product of enzymic activity will not withstand dehydration and embedding, the tissues can be mounted whole in glycerin jelly. Whole mounts of integument have been used successfully in an autoradiographic study involving the uptake of tritiated thymidine (Lawrence, 1968).

The limitations to the use of whole mounts are the lack of relationship between the cells and cell parts to each other and the superposition of cells in tissues that are more than one cell layer thick. The obvious advantages are the simplicity and rapidity of the method.

Special Techniques. Polarization and fluorescence microscopy have found recent application to the study of cuticle deposition (Neville, 1965). In such studies formalin-fixed, frozen sections have been examined in glycerol or water.

General Comments. The choice of fixative, stain, etc. has not been considered here since this will depend, as in all histology, on the tissue under investigation and the purpose of the study. The importance of choosing the right fixative is fairly generally recognized when we consider different tissues. However, to the uninitiated, it often comes as a surprise to find that it can also be true of different insects. The same techniques, fixatives, etc. do not give equally satisfactory results in different insects. Thus some insects, e.g., *Calpodes*, have become favorite material for fine structural studies while others, such as *Rhodnius*, have not been looked at, to any great extent, with the electron microscope.

J. LAI-FOOK

References

Hayat, M. A., "Principles and Techniques of Electron Microscopy. Biological Applications," Vol. I, New York, Van Nostrand Reinhold Company, 1970.

Lai-Fook, J., and M. Neuwirth, 1972. The importance of methods of fixation in the study of insect blood cells. *Can. J. Zool.*, 50:1011–1013.

Lawrence, P. A., 1968. Mitosis and the cell cycle in the metamorphic moult of the milkweed bug, *Oncopeltus fasciatus*. A radio-autographic study. J. Cell Sci., 3:391–404.

Locke, M., "The structure and formation of the cuticulin layer in the epicuticle of an insect, *Calpodes ethlius* (Lepidoptera, Hesperiidae)," *J. Morphol.*, 118:461–494 (1966).

Neuwirth, M., "The effects of different methods of fixation on insect blood cells" (1971). In preparation.

Neville, A. C., "Chitin lamellogenesis in locust cuticle," *Quart. J. Microscop. Sci.*, 106:269–386 (1965).

Wigglesworth, V. B., "A simple method for cutting sections in the 0.5 to 1μ range, and for sections of chitin," *Quart. J. Microscop. Sci.*, 100:315–320 (1959).

See also: ARTHROPOD SECTIONS, CHITIN.

INSECT WHOLE MOUNTS

The majority of insects may be simply pinned and naturally dried for examination under the low-power dissecting microscope. Very small insects are usually impaled on minute stainless steel pins, "staged" on polyporus, polystyrene, or card strips, secured by a larger pin. The pin should be thrust through the thorax from above to one side of the center line so that one set of the bilaterally symmetrical taxonomic characters is undamaged. Diptera are often pinned through the pleurae so that the insect has its side uppermost. With this method the pin should be inserted near the wing base, care being taken to avoid damaging any obvious characters such as bristles or hairs. Legs and antennae should be drawn out from the body for ease of examination. Direct pinning is especially suitable for larger, strongly sclerotized insects such as Lepidoptera, Trichoptera, Diptera, Hymenoptera, Neuroptera, and large Coleoptera. Stainless steel entomological, preferably headless, pins should be used. Another method of mounting small insects such as Diptera and Hymenoptera is "pointing," i.e. in a tiny blob of water-soluble gum, such as dilute Seccotine, on the point of a small triangle of stiff card or celluloid, then staged on a larger pin. Small to medium sized Coleptera are traditionally mounted in clear gum such as tragacanth or arabic, but in groups where underside characters are important specimens should be card "pointed" or pinned.

Very small or less heavily sclerotized insects need permanent microscopical preparations on glass slides for study under the compound microscope. A very quick method often used for the preparation of insect genitalia but suitable for whole-mounting insects such as Siphonaptera and small Diptera is to heat gently in 10% caustic potash until soft and clear, wash in water then transfer straight to glacial acetic acid, then into beechwood creosote, finally mounting in Canada balsam on a microslide. Another popular method used for small but moderately sclerotized insects is to drop the specimen into pure glacial acetic acid, transfer to a mixture of one part clove oil to two parts glacial acetic acid, then to a mixture of equal parts of clove oil and acetic acid; finally mounting in Euparal. In these methods, if not required for immediate study or mounting, the cleared insects can be stored in the creosote or the last mixture in the Euparal method. Unprocessed insects are best stored in 80% alcohol, if not dry pinned.

For taxonomic research or critical determination

certain groups of insects need very careful preparation for microscopic study.

Coccids should be heated for 5 min in 10% caustic potash, then washed in distilled water and the body contents expelled by gentle pressure. They are then bathed in a mixture of 20 parts glacial acetic acid to 80 parts of 50% alcohol for an hour or so. Then they should be stained in acid fuchsin, washed in 95% alcohol, given three 10 min baths in absolute alcohol, transferred to clove oil, and finally mounted in Canada balsam.

Aphids are very delicate and are best stored in 80% alcohol to which one third volume of 70% lactic acid has been added. They should be cleared in 5% caustic potash for a day or two, then washed in distilled water, and dehydrated carefully in 80%, then to 95%, and finally to absolute alcohol. Finally they are cleared in xylol containing a little phenol. Polyvinyl lactophenol or gum chloral (De Faure's medium) are better mounting media than Canada balsam for aphid research, but less permanent, so the slides should be ringed with a suitable medium. When dry, a ring of Euparal should be put on to seal the mount, then the slide can be finally ringed with balsam or cement.

Thysanoptera should be killed in a mixture of 8 parts of 95% alcohol, 5 parts of distilled water, 1 part of glycerin, and 1 part of glacial acetic acid. Heavily pigmented or brittle specimens should be soaked in 5% caustic potash, washed in water, then dehydrated in alcohol. For final slide-mounting, Canada balsam is preferred, but some workers use gum chloral, polyvinyl alcohol, or Sira mountant.

Collembola should be cleared in warm lactophenol, rinsed in distilled water, and mounted in A. C. S. or other water-soluble mountant.

Lice (Mallophaga and Anopleura) should be soaked in cold 10% caustic potash for 24 hr, then transferred to distilled water and the body contents removed. They are then transferred to 10% acetic acid and left for at least $\frac{1}{2}$ hr or 24 hr if required. They can be stained in acid fuchsin if required, then taken through 40%, 80%, and 95% alcohol, and mounted in Canada balsam.

Insect larvae can be preserved in 80% alcohol, but are better stored in Pampel's fluid (4 cm³ glacial acetic acid, 30 cm³ distilled water, 6 cm³ 40% formaldehyde, and 15 cm³ 95% alcohol). For microscopical examination the methods used for adults can be used. Lepidopterous larvae can be "blown" or dry inflated. Briefly, the gut contents are rolled out of the anus and a drawn-out glass nozzle is inserted. The larva is secured by a small clip on the nozzle and a steady stream of air is gently pumped or blown into the skin while it is held in a small metal oven until dry. Larvae or adults can also be preserved by freeze-drying (*q.v.*), and require special preparation for examination under the electron microscope (*q.v.* scanning electron microscope).

Heavily sclerotized larvae or adults may not need staining. More delicate semitransparent speciments can be stained in a solution of acid fuchsin in 20% alcohol. Tougher specimens can be placed in glacial acetic acid and a few drops of the stain added, but more delicate lightly sclerotized specimens should be taken through a series of alcohols in increasing strengths.

Some small soft-bodied insects can be temporarily mounted directly from spirit or even alive into media such as gum chloral, polyvinyl-lactophenol, polyvinyl alcohol, or shellac gel.

KENNETH G. V. SMITH

References

Beirne, B. P., "Collecting, preparing and preserving insects," Science Service, Entomology Division, Canada Dept. of Agriculture, Publication 932, 1955 (133 pp., 93 figs.).

Eastop, V. F., and H. F. Van Emden, "The insect material," in "Handbook of Aphid Technology," New York, Academic Press, 1972 (344 pp.).

Eltringham, H., "Histological and illustrative methods for entomologists," London, Oxford University Press, 1930.

Hammond, H. E., "The preservation of Lepidopterous larvae using the inflation and heat-drying technique," *J. Lepidop. Soc.,* **14**:67–78 (1960).

Hood, J. D., "Microscopical Whole Mounts of Insects," 2nd mimeographed ed., Ithaca, Cornell University, 1940.

Kosarzevska, E. F., "Methods for making slides of Coccids (Homoptera, Coccoidea) in order to determine them," *Ent. Obozr.* (Rev. Ent. U.R.S.S.), **47**:248–253 [in Russian] (1968).

Mound, L. A., and Pitkin, B. R. "Microscopic Wholemounts of Thrips (Thysanoptera) *Entomol. Gaz.,* **23**:121–125 (1972).

Oldroyd, H., "Collecting, Preserving and Studying Insects," 2nd ed., London, Hutchinson 1970 (336 pp.).

Oman, P. W., and A. D. Cushman, "Collection and preservation of insects," U.S. Dept. of Agriculture Misc. Publications No. 601, 1948 (42 pp.).

Riley, C. V., "Directions for collecting and preserving insects," *Bull. U.S. Nat. Mus.* **39**, 1892 (147 pp.).

See also: INSECT HISTOLOGY WHOLE MOUNTS, ZOOLOGICAL MATERIALS.

INVERTEBRATE LARVAE

The term "larva" has been used to designate a stage or stages in the life history of an animal which differs in its morphology in varying degree from that of the parent. The adult morphological characters are achieved by some sort of metamorphosis on the part of the larvae. Costello (1970) has provided a useful summary of larval types which have been recognized and, in some instances, intensively studied. In this review he has drawn attention to the great variety of larval types and has emphasized the complexity of the morphologies involved. These developmental stages show remarkable adaptation to highly specialized modes of life, often very different from those of the adult. While larval forms may be adapted to the terrestrial and aquatic, fresh or saline environments, they may also be enclosed within protective capsules or membranes or may even be parasitic in habit. Most of the techniques which follow have been developed for use in the studies of larvae of marine invertebrates, but they are often applicable to larvae of other groups.

In studies which seek to provide an understanding of the details of larval morphology and function, the observation of the living specimen is an obvious, but often overlooked, important first step. Bright field microscopic observations are often adequate for such studies. However, the use of dark field and phase contrast microscopy introduces new levels of analysis. Nomarski differential interference microscopy is most excellent in studies requiring high resolution of fine detail. The polarizing microscope can be useful in examination of larvae with skeletal elements such as in the case of the echino or ophiopluteus. The calcareous spicules of these larvae can be illustrated most brilliantly with polarization optics.

Cinematographic as well as still photomicrographic records can be made to good effect with all of these optical systems.

To prepare larvae for examination by any of the several optical means requires preparation of the simple and familiar wet mount. Clean coverslips and microslides, free from detergents and other toxic agents, are required. The living larvae are placed in a drop or two of water of temperature and salinity normal to the conditions under which development occurs. To avoid crushing of the larvae, the coverslip must be supported at an appropriate distance from the slide. This can be accomplished by any of several means: introduction of slivers or fine rods of glass of appropriate thickness; use of fibers of lens paper, algal filaments, or bolting silk; or supplying plasticene or vaseline feet at the corners of the coverslip. The gentle sliding of such a supported coverslip will often result in advantageous rotation of the larvae confined below.

For observations at higher magnification, i.e., where oil immersion is to be used, the substitution of anisole for immersion oil is advantageous. There is less drag on the coverslip with anisole and thus the underlying specimens are not subjected to as much displacement (Strathmann, 1968).

Many marine larvae, especially those from waters of the temperate zone, are highly sensitive to increase in temperature. It therefore becomes important to maintain them at a temperature suitable to their development. Dan and Okazaki (1956) described a simple preparation which may assist, not only in preventing drastic temperature change, but also desiccation where it is desirable to be able to observe a single larva for a prolonged period of time. A piece of filter paper of desired thickness is cut to a smaller scale than the coverslip and a central hole of several millimeters is cut. The filter paper, moistened with sea water, is placed on a clean microslide. A small drop of sea water containing the larva to be observed is placed within the paper borders and a coverslip laid on top. The water droplet should be of such a size that it does not touch the paper at the border. Additional sea water may be added to the edge of the paper, and thus a small moist chamber is prepared. Between periods of observation, the prepared slide may be kept in an appropriately chilled situation to maintain the temperature normal for development. Embryos and larvae may be kept in such a chamber for many hours and undergo normal developmental processes.

More elaborate devices have been described for maintaining either higher or lower temperatures than the room in which one is working. A simple temperature control stage is illustrated in Galigher and Kozloff (1970). Much more precise temperature control for continuous observation or experimentation can be achieved with a stage with thermo-electric control of temperature (Cloney et al., 1970).

The activities of many larvae are characterized by quite rapid movement; thus, for detailed observation and study it often becomes necessary to immobilize the larvae. This may be done by mechanical means or by narcotization.

The activities of the larvae may be limited or confined by such a simple expedient as use of bolting silk of appropriate mesh size or a tangle of lens paper filaments. Certain agents which change the viscosity of the aqueous medium in which the larvae are swimming are sometimes more reliable. A 10% solution of methyl cellulose in sea water is a convenient stock mixture for marine forms. The methyl cellulose is slow to go into solution and it may be necessary to centrifuge the solution after it has been allowed to stand for 24 hr or more. The solution should be kept cold, and a drop or two added to a small container of larvae will increase the viscosity and slow down the swimming activity.

Polyethylene oxide powder produces a highly viscous polymer when added to sea water. The powder can be added in appropriate amounts directly to a preparation on the slide. The material is not known to have a toxic effect (Strathmann, 1968).

Where mechanical means prove not to be satisfactory or reliable, it is often necessary to use agents which narcotize or anesthetize the larvae. A variety of agents have been used with variable success. It is often necessary to select the proper agent in the proper concentration for a specific larval form. This is necessarily determined, in most cases, by experimentation. It is reported (Costello et al., 1957) that one or two drops of a dilute Janus Green B solution (1 : 1000 in sea water) slows the larvae of *Hydroides, Sabellaria, Nereis,* and *Mactra* for a short period. The same source indicates that a small amount of chloral hydrate has a narcotizing effect on the larvae of *Cerebratulus, Crepidula,* and *Cumingia.* Clement (1952) has recommended the use of 1% urethane as an effective agent in slowing down the larvae of *Ilyanassa.*

Chloretone is a commonly used agent to stop muscle contraction without stopping ciliary action. One part saturated, aqueous chloretone to two parts sea water has been used to relax veliger larvae for observation (Strathmann, 1968). Nicotine or other agents, leached from the filter of a smoked cigarette will stop ciliary action. A few fibers from the filter, when added directly to the water in which the larvae are swimming, provide both mechanical and narcotizing immobilization. If one soaks such a filter in sea water for a few hours and then removes it, a solution can be obtained which will be effective for a considerable period of time. This method has been used with success on larvae of echinoderms and of certain polychaetes (P. Dudley, personal communication).

Three agents have proven to be quite useful as ciliary anesthetics because their effects are reversible. MS-222 (tricaine methane sulfonate—Sandoz), "tricaine" (ethyl-m-aminobenzoate methane sulfonic acid—Sigma) and a product by Fisher—ethyl-m-aminobenzoate methane sulfonate (Haarstud, 1967) seem to be equivalent and of comparable effectiveness. Wallace et al. (1967) report that MS-222 and tricaine are both effective as an anesthetic when made up in an isotonic potassium chloride solution. Such a solution is light sensitive and should be stored in the dark or in lightproof bottles. When it turns yellow it becomes ineffective. E. C. Roosen-Runge has used tricaine on gastrulae and planulae of hydrozoans and reports (personal communication) it to be effective and entirely reversible in its effect. A sea water extract from squid egg-string jelly has been useful for slowing down or immobilizing the larval forms of a number of echinoderms (Costello et al., 1957).

The narcotizing or anesthetizing agents have been used extensively to relax larval stages preparatory to fixation for whole mounts and microtomy. A casual survey of the methods used in preparation of larval whole mounts clearly indicates that the technique must vary from one type to the other and no single technqiue is applicable to all. A summary of procedures is provided in Galigher and Kozloff (19 0) for the larvae of echinoderms, polychaetes, and molluscs. It is recommended that fixation in Bouin's fluid be followed by staining according to the dilute acidulated borax carmine method. It is further suggested that polychaete larvae may be fixed in Kleinenberg's picrosulfuric acid mixture or in Bouin's fluid and stained by precipitated borax carmine or in alum cochineal; counterstain lightly with fast green. A number of larvae including tadpoles or tunicates, trochophores and veligers of molluscs, and plutei of echinoids can be fixed in 10% neutral formalin to be followed by staining with Mayer's alcoholic cochineal or with safranin.

Hermans (1966) has used chlorazol black E effectively in the preparation of whole mounts of early and advanced larvae of the polychaete *Armandia*. The larvae were first fixed in 10% formalin in sea water and dehydrated to 70% alcohol. A drop or two of 1% chlorazol black E in 70% alcohol was added to less than 1 ml of 70% alcohol containing the specimens to be stained. The stain penetrates rapidly but can be washed out to the desired level with 85% alcohol. If the larvae are overstained, they can be differentiated by mixing a drop or two of pyridine with 1 ml of 85% alcohol containing the specimens. Differentiations takes from 10 min to as long as a day, depending on the degree of overstaining. The larvae were cleared by transferring directly from 85% alcohol to either clove oil or turpineol and were mounted in piccolyte.

Clement and Cather (1957) have described a technique for veliger larvae of *Ilyanassa* which has proved to be of value for a variety of other forms. Anesthetized larvae were fixed in 10% formalin in sea water, dehydrated in alcohol stained lightly with Orange G, and mounted under supported coverslips in either Euparal or Diaphane. Rice (1967) has reported this to be an effective means of preparing whole mounts of larvae of a variety of Sipuncula. She has also fixed larvae of Sipuncula for two or three minutes in osmium vapor, dehydrating to absolute alcohol, and mounting in Euparal. It becomes clear that a variety of relaxing agents, fixatives, stains, and mounting media can be used to good affect with different larval types.

Where studies of the morphology of the larvae require the use of sectioned materials, many of the conventional techniques for paraffin, celloidin, and Epon embedding are useful. For the paraffin method, fixation in Bouin's fluid has been widely recommended. Ten percent formalin in sea water and Stockard's solution are also recommended. The orientation of the small larvae for sectioning in definite planes has challenged the inventiveness of a number of investigators. Akesson (1961) has provided a limited bibliography relating to this problem and introduced a method which he has found particulary useful for polychaete larvae. Zimmer (1964) provides yet another modification which was used successfully in the study of phoronid actinotrochs.

It is agreed by most workers with paraffin sections that care must be exercised in the selection of the embedding mixture. Cloney (1961) has recommended a mixture which has been used with good results for a variety of larval types. This medium is prepared by melting the following ingredients together at 60–65°C and thoroughly blending them:

Fisher's Tissuemat MP 60–62°C	300 g
Bleached beeswax	35 g
Dry piccolyte	45 g

The sectioning should be done in a cryostat or in a cold room at 5–10°C. This medium is particularly useful where sections of 3–5 μ are to be prepared. Such paraffin sections continue to have real value where serial sections are required for an interpretation of the larval morphology.

The use of Epon embedding following fixation with buffered osmium tetroxide or the double fixation with aldehyde and osmium tetroxide provides a basis for the preparation of thin sections (1 μ) for light microscopy. Preparations superior in quality to those of the classical paraffin technique can be obtained and offer a significant extension of the use of light microscopy for analysis of structure. The utilization of electron microscopy for fine structure analysis can add significant information. Experience would suggest that no single fixative can serve optimally for the preparation of all types of invertebrate larvae for electron microscopy. Rather, it is essential that the investigator be prepared to experiment with a variety of procedures giving particular attention to buffering agents and the osmolality. Suggested combinations and variations are presented elsewhere in this volume as for example in the treatment for larvae and tissues of tunicates by Cloney.

The advantages of the scanning electron microscope for the study of morphology of invertebrate larvae has been little explored. Preliminary application of this technique to a number of pelagosphaera larvae of Sipuncula by Rudolph Scheltema and John Hall (personal communication) suggest real advantages to be gained from this technique.

ROBERT L. FERNALD

References

Akesson, B., *Ark. f. Zool.*, **13**:479 (1961).
Clement, A. C., *J. Exp. Zool.*, **121**:593 (1952).
Clement, A. C., and J. N. Cather, *Biol. Bull.*, **113**:340 (1957).
Cloney, R. A., *Amer. Zool.*, **1**:67 (1961).
Cloney, R. A., J. Schaadt, and J. V. Durcen, *Acta Zool.*, **51**:95 (1970).
Costello, D. P., in "The Encyclopedia of Biological Sciences," 2nd ed., P. Gray, ed., New York, Van Nostrand, Reinhold Co., 1970.
Costello, D. P., M. E. Davidson, A. Eggers, M. H. Fox, and C. Henley, "Methods for Obtaining and Handling Marine Eggs and Embryos," MBL Publ., 1957.
Dan, K., and K. Okazaki, *Biol. Bull.*, **110**:29 (1956).
Galigher, A. E., and E. Kozloff, "Essentials of Practical Microtechnique," Philadelphia, Lea and Febiger, 1970 2nd Ed.
Haarstud, V. B., *Science*, **158**:1524 (1967).
Hermans, C. O., "The Natural History and Larval Anatomy of *Armandia brevis*," Ph.D. Thesis, Univ. of Washington, 1966.
Rice, M. E., *Ophelia*, **4**:143 (1967).
Strathmann, M., "Methods in Developmental Biology," Friday Harbor Labs, 1968. (Unpubl.).
Wallace, R. A., D. W. Jared, and M. E. Toesel, *Science*, **158**:1524 (1967).
Zimmer, R. L , "Reproductive Biology and Development of Phoronida," Ph.D. Thesis, Univ. of Washington, 1964.
See also: WHOLEMOUNTS, ZOOLOGICAL.

k

KIDNEY

The blood pressure is primarily responsible for maintenance of the normal degree of distention of the kidney. In
fact, the normal activity of the kidney depends primarily
upon a normal renal blood flow. The distention is labile
and alteration of any of a number of factors leads to an
immediate and dramatic change in renal morphology.
Although this is now well appreciated, it took many
investigators using a variety of techniques to establish
the factors which could alter renal morphology. In this
chapter, some techniques utilized for studying the morphology of the kidney will be reviewed. The type of
procedure to be employed for preservation of the kidney
should depend on the nature of the investigation to be
performed.

I. Study of the Living Organ

**A. Use of Incident-light Illumination with a Light
Microscope.** One of the best ways to learn about renal
morphology is to look at the living functioning kidney *in
situ* through a light microscope. This technique was first
employed by Ghiron in 1912 (10) when he observed
mouse kidney *in vivo* utilizing incident light. Since then
the utilization of microscopes equipped with incident-
light illuminators has provided valuable information (11,
21, 22, 23). For example, the lumen of the proximal tubule
is normally widely patent (Fig. 1), but a large decrease in
the blood pressure can bring about rapid collapse of the
tubule with disappearance of the patent lumen (Fig. 2).

B. Use of Micropuncture Techniques. The use of delicate micromanipulators which permit the operator to
enter individual renal tubules with small glass micropipettes *in situ* was first utilized by Richards and his
associates (28) to sample fluid from the amphibian nephron. By removing fluid from the renal corpuscle, it was
shown that an ultrafiltrate of blood was produced in this
region. The perfection of micropuncture techniques, and
associated microanalysis of the collected fluid, has allowed
investigators to provide much information about the
functions of the various regions of the nephron. Windhager (29) has recently written a book reviewing the
techniques and the information provided by their use.

C. Use of Isolated Tubules. Methods have been developed to dissect, incubate, perfuse, and analyze viable
fragments from specific sites of single rabbit nephrons (6).
Utilizing modified micropuncture equipment, it has been
found feasible to study transport in these small isolated
segments of individual kidney tubules.

D. Model Systems for Study of Kidney Morphology. One
of the best ways to find out how kidneys function is to
study kidneys during various physiological stresses or
after injury. In addition, the study of nature's own comparative experiments has provided much information on
renal function. From a study of the urine produced by the
aglomerular fish (which lacks a filtering device), the process of urine secretion was elucidated (15). Certain animal
models are available, as well, for the study of specific
problems. For example, the study of the Brattleboro
strain of Long-Evans rats with hereditary hypothalamic
diabetes insipidus has proved useful in the understanding
of the mechanism of urine concentration (27). One can also
administer tracer molecules such as proteins and study
their movement and distribution at various time periods
within the kidney. The study of the distribution of enzymes, or substances such as ions, can be undertaken.
Many of these studies require that the tissue be removed
from the animal to be viewed, so adequate fixation techniques must be considered.

II. Study of chemically fixed tissue for light and transmis-
sion electron microscopy

A. Route of Fixative Application. *1. Immersion.* As
indicated in the introduction, normal kidney morphology
depends upon the maintenance of blood pressure and of
an adequate supply of oxygen. It is no surprise, therefore,
that the route of fixative application has profound effects
on the structures to be visualized. The common practice of
removing an organ from the animal, dicing it into small
pieces, and then immersing it in fixative solution produces
dramatic and reproducible changes in renal architecture.
The cells of the proximal tubule swell, producing large
apical extensions which fill the once patent lumen (Fig. 4).
The cell organelles change in shape and orientation and
may also undergo swelling. Even before the use of the
electron microscope with its requirements for better tissue
preservation, Koenig et al. (13) described an improved
technique for fixation of the kidney which involved the
intravascular perfusion of the fixative solution. Since
immersion fixation does not provide a faithful representation of *in vivo* morphology, its use in experimental studies
should be severely limited. In the case of human renal
disease, however, the only practical way to obtain renal
tissue is through the use of open biopsy or percutaneous
biopsy techniques with subsequent fixation of the tissue by
immersion. It is therefore imperative to know what
changes are produced by the method of procurement and
fixation of this tissue so that these artifacts are not confused with alterations caused by the disease itself. Tisher
et al. (25) have defined many of these preservation artifacts in a study of the kidney of the subhuman primate,
Macaca mulatta.

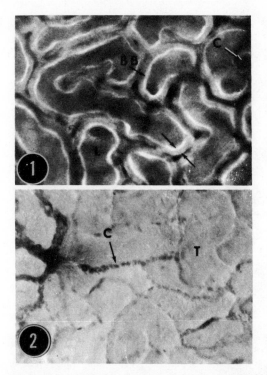

Figs. 1 and 2   Fig. 1: This figure shows the appearance of tubules in a living undisturbed kidney when viewed with incident lighting. Note the widely patent tubular lumina (L), the refractile appearance of the brush border (BB), the low epithelium of the proximal tubule (between arrows), and the patent peritubular capillaries (C).; (Figs. 1–6 from L. D. Griffith, R. E. Bulger, and B. F. Trump, *Lab. Invest.*, 16:220–246, 1967. Reproduced with permission of the International Academy of Pathology.) Fig. 2: This figure shows the appearance of tubules immediately after occlusion of the renal artery when viewed with incident light. The tubules (T) now appear solid and the peritubular capillaries (C) are partially opened.

2. Intravascular Perfusion of Fixative. Several papers have appeared which describe various ways of perfusing intact kidneys with fixative using the intravascular route (Fig. 3) (8, 11, 16, 17). When carefully done, this procedure provides adequate maintenance of the gross morphology of all regions of the kidney. One such procedure includes the following steps: (1) Make a midline ventral incision through the skin and body wall. (2) Expose the descending aorta medial to left adrenal gland and place ligature under the vessel without occluding it. (3) Ligate the vessels to the gastrointestinal tract. (4) Expose the aorta between the renal artery and the aortic bifurcation and place two sutures under the vessel. (5) Temporarily occlude aorta below level of kidney with upper of two sutures from step 4. (6) Nick wall of aorta near bifurcation and insert a cannula filled with heparinized saline. (7) Tie cannula in place with lower of two sutures. (8) Release occlusion of upper suture. (9) Start perfusion of fixative into the cannula making sure not to exceed 180 mm of Hg. (10) After perfusion has begun, occlude aorta at the level of the adrenal, and (11) Cut appropriate vein

to permit outflow of fixative solution. (For detailed description, see reference 11.)

A modification of this basic perfusion technique (11) allows an animal to be perfused without handling at the time of perfusion. This procedure is necessary only if one wishes to prevent the secretion of antidiuretic hormone during the surgery necessary for perfusion.

3. Microperfusion of a Tubule or Its Peritubular Capillary. When a particular renal tubule is being studied by micropuncture techniques, it can be fixed by microinjection of fixative directly into the lumen of the tubule (24). Such a procedure gives rapid fixation of the tubule since the fixative must travel through only a single cell layer for complete fixation. The addition of a marker dye such as lissamine green aids the investigator in localizing the previously studied tubule. Similar procedures can be used to inject microquantities of fixative into the peritubular capillaries of a given nephron providing rapid fixation while causing minimum displacement of the luminal contents of the tubule being fixed (Tisher, personal communication).

4. Applying Fixative to the Surface of the in vivo *Organ.* A procedure was first used by Pease in 1955 (19) in which fixative was dripped on the surface of the kidney *in situ.* This procedure provides adequate fixation for only a small zone of outer cortex but the fixed tubules retain their patent lumens (Fig. 5). Such a procedure is useful only in studies of superficial cortex such as those done in conjunction with micropuncture or in cases where tracer substances have been injected into the vessels (thus eliminating the intravascular route). If, instead of drip fixation, the fixative is simply poured into the abdominal cavity, the animal lives for approximately 15 min and a layer of superficial cortex from the upper half of the kidney (which includes the first row of renal corpuscles) is well fixed (Tyson and Bulger, personal communication). However, if the fixative is dripped *in situ* onto a small portion of superficial cortex and not allowed to enter the abdominal cavity, the animal will live for longer periods with no change in blood pressure (Tisher, personal communication). In this case, the zone of fixed tubules extends deeper into the cortex. The depth of penetration is also increased by adding formaldehyde to the fixative.

B. Chemical Fixatives. Buffered osmium tetroxide solutions were used as the fixative in most of the early investigations on renal ultrastructure (18). Although many useful studies were done utilizing this fixative, it appears that osmium tetroxide may allow swelling of kidney cells and organelles (4). In addition, osmium tetroxide largely inactivates enzymes, which severely limits its use in histochemical studies. The introduction of aldehyde fixatives such as formaldehyde or glutaraldehyde (20) or a combination of these (12) has provided much new information because the fine structure of certain kidney organelles is affected by the fixative used (3, 5, 16, 17, 26) or the concentration of the fixative (17). It is therefore often advantageous to utilize more than one fixative in experimental studies, especially in those involving cell injury because the injury may alter organelles in such a manner as to make them more susceptible to alteration by fixatives.

Maunsbach (17) has shown that the addition of varying amounts of sodium chloride to certain glutaraldehyde fixative solutions can profoundly change the tissue structure. Bohman and Maunsbach (2) have shown that the addition of oncotically active substances such as dextran, or polyvinylpyrrolidone to aldehyde fixatives

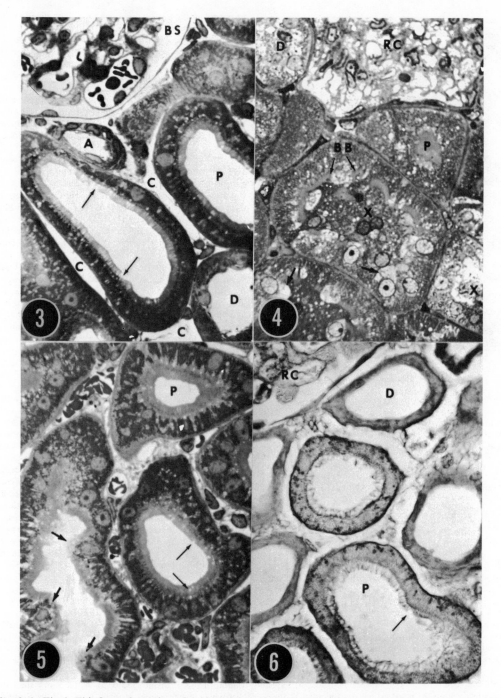

Figs. 3–6   Fig. 3: This figure shows tissue fixed by intravascular perfusion and demonstrates open lumina of proximal (P), distal (D) convoluted tubules, afferent arterioles (A), and peritubular capillaries (C). Bowman's space (BS) is clearly seen. Fig. 4: This figure shows tissue from a kidney fixed by excision and immersion. Apical blebs (thick arrows) can be seen in the occluded lumina of proximal and distal tubules. Disrupted brush border (BB) and debris (X) are also seen in the lumen. The collapse of Bowman's space is seen in the renal corpuscle (RC). Fig. 5: This figure shows renal cortex fixed by dripping of fixative on the *in situ* kidney. The lumina of the proximal tubules (P) are fixed in an open state, although some alteration of the apical cell membranes (thick arrows) can be seen. The thin arrows show some simplification of the apical cell membrane. Fig. 6: This figure shows tissue prepared by *in situ* freezing followed by freeze-drying and vapor phase formaldehyde fixation. The tubules are open and the brush border is intact, although regions of simplification (arrow) can be seen. The renal corpuscle (RC) has a patent Bowman's space. This picture was taken from a section 3 μ thick which was embedded in paraffin, stained by periodic acid–Schiff–Alcian blue.

utilized for intravascular perfusion, prevents expansion of extravascular spaces in tissues including the kidney.

When attempting to obtain adequate fixation of kidney, therefore, one must consider the route of application, as well as the nature, type, and concentrations of fixative, buffer, and other additives.

III. Freeze fixation—alternative to chemical fixation

Because chemicals used in fixation may cause alteration in structure, it is useful to check the adequacy and effect of chemical fixation by fixation using alternate methods such as freezing the tissue *in vivo*. Frozen tissues can be dried at −40°C, fixed in formaldehyde vapor, vacuum-embedded, sectioned, and viewed (Fig. 6). The lumens of the proximal tubules are widely patent, the surface is free of blebs and the peritubular capillaries are patent (11). Alternately, frozen tissues can be fractured (with or without removal of water by etching) and a carbon and platinum replica can be prepared for viewing with an electron microscope (9, 14).

IV. Use of the scanning microscope and the analytical electron microscope for study of kidney

The last decade has increased the availability of new instrumentation to be used in biology. One such new instrument is the scanning microscope. Buss and Kronert (7) and Arakawa (1) studied the surface configuration of glomerular podocytes using a scanning electron microscope. Although this microscope is limited in resolution, much information about surface topography can be gleaned quickly through its use.

A similar technological advance is available in the use of the analytical electron microscope which allows the investigator to carry out high-resolution electron microscopy and X-ray microanalysis (using an electron illumination of 1000 Å diameter) on the same specimen. This instrument should prove of great use in increasing our understanding of the injury produced by metal toxins and perhaps even in localizing the position of substances such as sodium in the kidney. New techniques for the study of kidney morphology must continue to be developed and exploited.

RUTH ELLEN BULGER

(Supported in part by NIH grant number AM 15176.)

References

1. Arakawa, M., "A scanning electron microscopy of the glomerulus of normal and nephrotic rats," *Lab. Invest.,* 23:489–496 (1970).
2. Bohman, S.-O., and A. B. Maunsbach, "Effects on tissue fine structure of variations in colloid osmotic pressure of glutaraldehyde fixatives," *J. Ultrastruct. Res.,* 30:195–208 (1970).
3. Bulger, R. E., "The fine structure of the aglomerular nephron of the toadfish, *Opsanus tau*," *Amer. J. Anat.,* 117:171–192 (1965).
4. Bulger, R. E., "The use of potassium pyroantimonate in the localization of sodium ions in rat kidney tissue," *J. Cell Biol.,* 40:79–94 (1969).
5. Bulger, R. E., and B. F. Trump, "Renal morphology of the English sole (*Parophrys vetulus*)," *Amer. J. Anat.,* 123:195–226 (1968).
6. Burg, M., J. Grantham, M. Abramow, and J. Orloff, "Preparation and study of fragments of single rabbit nephrons," *Amer. J. Physiol.,* 210:1293–1298 (1966).
7. Buss, H., and W. Kronert, "Zur struktur des nierenglomerulum der ratte," *Virchows Arch. Abt. B. Zellpath.,* 4:79–92 (1969).
8. Ericsson, J. L. E., "Glutaraldehyde perfusion of the kidney for preservation of proximal tubules with patent lumens," *J. Microscopie,* 5:97–100 (1966).
9. Friederici, H. H. R., "The surface structure of some renal cell membranes," *Lab. Invest,* 21:459–471 (1969).
10. Ghiron, M., "Uber eine neue methode mikroskopischer unterschung am lebenden organismus," *Zbl. Physiol.,* 26:613–617 (1912). [As cited by Steinhausen (see reference 21).]
11. Griffith, L. D., R. E. Bulger, and B. F. Trump, "The ultrastructure of the functioning kidney," *Lab. Invest.,* 16:220–246 (1967).
12. Karnovsky, M. J., "A formaldehyde-glutaraldehyde fixative of high osmolality for use in electron microscopy," *J. Cell Biol.,* 27:137A (1965).
13. Koenig, H., R. A. Groat, and W. F. Windel, "A physiological approach to perfusion-fixation of tissues with formalin," *Stain Technol.,* 20:13–22 (1945).
14. Leak, L. V., "Ultrastructure of proximal tubule cells in mouse kidney as revealed by freeze etching," *J. Ultrastruct. Res.,* 25:253–270 (1968).
15. Marshall, E. K., Jr., "The aglomerular kidney of the toadfish (*Opsanus tau*)," *Bull. Johns Hopkins Hosp.,* 45:95–102 (1929).
16. Maunsbach, A. B., "The influence of different fixatives and fixation methods on the ultrastructure of rat kidney proximal tubule cells I. Comparison of different perfusion fixation methods and of glutaraldehyde, formaldehyde and osmium tetroxide fixatives," *J. Ultrastruct. Res.,* 15:242–282 (1966)
17. Maunsbach, A. B., "The influence of different fixatives and fixation methods on the ultrastructure of rat kidney proximal tubule cells II. Affects of varying osmolality, ionic strength, buffer system and fixative concentration of glutaraldehyde solutions," *J. Ultrastruct. Res.,* 15:283–309 (1966).
18. Palade, G. E., "A study of fixation for electron microscopy," *J. Exp. Med.,* 95:285–298 (1952).
19. Pease, D. C., "Electron microscopy of the tubular cells of the kidney cortex," *Anat. Rec.,* 121:723–744 (1955).
20. Sabatini, D. D., K. Bensch, and R. J. Barrnett, "Cytochemistry and electron microscopy. The preservation of cellular ultrastructure and enzymatic activity by aldehyde fixation," *J. Cell Biol.,* 17:19–58 (1963).
21. Steinhausen, M., "Microscopy and photomigrography of the living kidney with the ultropak incident-light illuminator," *Sci. Tech. Information* (Leitz), 1(4): 103–106 (1965).
22. Swann, H. G., "The functional distention of the kidney: A review," *Texas Rep. Biol. Med.,* 18:566–595 (1960).
23. Swann, H. G., "Some aspects of renal blood flow and tissue pressure," *Circ. Res.,* 14 & 15 (Supplement 1):115–119 (1964).
24. Thoenes, W., K. Hierholzer, and M. Wiederholt, "Gezeilete fixierung von nierentubuli in vivo durch mikroperfusion zur licht-und elektronenmikroskopischen untersuchung," *Klin. Wschr.,* 43:794–795 (1965).
25. Tisher, C. C., S. Rosen, and G. B. Osborne, "Ultrastructure of the proximal tubule of the Rhesus monkey kidney," *Amer. J. Pathol.,* 56:469–517 (1969).
26. Trump, B. F., and J. L. E. Ericsson, "The effect of the fixative solution on the ultrastructure of cells and tissues," *Lab. Invest.,* 14:507/1245–1323 (1965)
27. Valtin, H., "Hereditary hypothalamic diabetes insipidus in rats (Brattleboro strain). A useful experimental model," *Amer. J. Med.,* 42:814–827 (1967).

28. Wearn, J. T., and A. N. Richards, "Observations on the composition of glomerular urine with particular reference to the problem of reabsorption in the renal tubules," *Amer. J. Physiol.,* **71**:209–228 (1924).

29. Windhager, E. E., "Micropuncture Techniques and Nephron Function," New York, Appleton, Century, Crofts, 1968.

LATICIFERS

Laticifers are specialized plant cells which contain latex. Two types of laticifers are recognized: the articulated, consisting of a series of uninucleated cells joined into a vessel; and the nonarticulated, which is an elongated often branched coenocytic cell. The latex, which is a complex of proteins, carbohydrates, lipids, and frequently alkaloids or rubber, and the positive osmotic pressure of the cellular constituents make this cell particularly difficult to kill and fix for histological examination for either light or electron microscopy.

Picric, chromic, or osmic acid containing fixatives preserve the protein and lipid fractions of the protoplast with excellent rendition. A combination of picric acid and glutaraldehyde in $0.1M$ sodium cacodylate buffer at pH 7.1 (formula 1) can be employed in both light and ultrastructural studies. Substitution of 50% ethyl alcohol for the buffer (formula 2) produces an effective fixative for use with large pieces of tissue in light microscopic histology. The alcohol enhances the rate and depth of penetration of the fixative. Material fixed for 6–12 hr can be dehydrated with alcohol or acetone procedures prior to embedding in paraffin. Compatible stains include hematoxylins or tannic acid–ferric chloride–safranin–fast green.

For electron microscopy the pieces of material must be small, at least one dimension no greater than 1–2 mm, to secure rapid and thorough fixation (6–12 hr) since the aqueous solutions penetrate tissues slowly. Formula (1) followed by a 1% osmic acid vapor post-fixation for 1 hr will fix the lipid fraction that constitutes a considerable volume of the cell. Fixed tissues are prestained in 1% aqueous uranyl acetate (6–24 hr) followed by dehydration in alcohol and embedment in epoxy resins. Uranyl acetate and lead citrate are effective stains for thin sections.

Large segments of the laticifer system can be removed intact by maceration procedures. They are isolated by placing materials in saturated picric acid for several days followed by 0.5% ammonium oxalate for a similar length of time. Maceration is hastened by maintaining material at 40–60°C. Protoplasmic contents are well preserved, and lipids, proteins, carbohydrates, and callose as well as nucleic acids are indentifiable.

Callose, a glucose polymer which serves as a plugging material in the laticifer, is preserved by the techniques described above and can be observed by the aniline blue fluorochrome procedure using a Schott BG-12 exciter filter in conjunction with a Schott GG-9 barrier filter. Callose will appear as a yellow fluorescence after being stained for 10 min in freshly prepared 0.005% water-soluble aniline blue in $0.15M$ K_2HPO_4 at pH 8.5. Precise localization can be made in laticifers isolated by the maceration procedure.

Starch morphology in laticifers frequently differs from that in adjacent cells indicating differences in plastid morphology. Distribution and morphology of starch and other insoluble polysaccharides in either sectioned or macerated tissues can be demonstrated by the periodic acid–Schiff reaction.

Formula
1. Glutaraldehyde (50%): 8 ml
   Picric acid: 0.7 g
   Sodium cacodylate, $0.1M$: to make 100 ml
   pH: 7.1–7.2
(Excess glutaric acid can be reduced by storing stock glutaraldehyde with activated charcoal.)
2. Glutaraldehyde (50%): 8 ml
   Picric acid: 0.7 g
   Ethyl alcohol (50%): to make 100 ml

PAUL MAHLBERG

LEUKOCYTES

The cellular components of leukocytes can be demonstrated in blood and bone marrow smears on slides or cover glasses or in tissue sections. Buffy coat preparation is recommended for a high concentration of leukocytes. Spin heparinized blood in a small tube (11 mm diameter) 2500–3000 rpm, 10 min. With a Pasteur pipette collect from the surface of the packed red cells a white cell buffy coat and smear it as a blood smear (Figs. 1–8). Since stains for blood are often prepared in a fixing agent (methanol), fixation is not always required. With the commonly used Romanowsky (neutral) stains neutrality is essential; dilute with a buffer of known pH.

Wright's stain, the most familiar of the Romanowsky stains, is ground thoroughly in methanol (0.1 g/60.0 ml). Filter before use. The buffer solution, pH 6.4 is: monobasic potassium phosphate, 6.63 g; anhydrous dibasic sodium phosphate, 2.56 g; distilled water, 1000.0 ml. Cover smear with stain: 1–2 min: add equal amount of buffer: 2–4 min. Do not drain off stain; wash it off with distilled water. Blot with two sheets filter paper. Dry thoroughly and examine or apply cover glass with mountant. Results: basophilic granules, deep purple; eosinophilic granules, red to red-orange; neutrophilic granules, reddish brown to lilac; monocytic granules

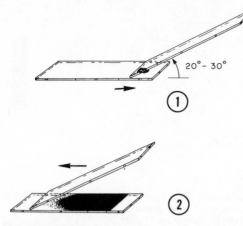

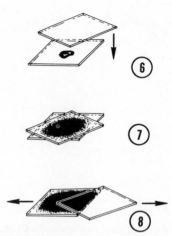

Figs. 1 and 2 Fig. 1: Place drop of buffy coat near one end of slide. Pull to right against drop with second slide and allow buffy coat to spread along edge. (Figs. 1–8 reproduced with permission of W. H. Freeman and Company, San Francisco, California.)

Figs. 6–8 Fig. 6: Place square cover glass on top of drop of buffy coat on another cover glass. Fig. 7: Allow buffy coat to spread. Fig. 8: When spreading stops or has reached edge of glass, pull top cover glass across and off of bottom glass. Air dry.

azure; lymphocytic granules, more reddish than monocytic ones. (For thick smears for parasites see Field [2] or Humason [4].)

Jenner-Giemsa method: Jenner's solution: add 0.2 g Jenner powder to 100.0 ml methanol. Giemsa stock solution: work 3.8 g Giemsa powder into 75.0 ml methanol, warm 2 hr, 60°C. Add 25.0 ml glycerol. Cover tissue sections (deparaffinized and hydrated to water) with

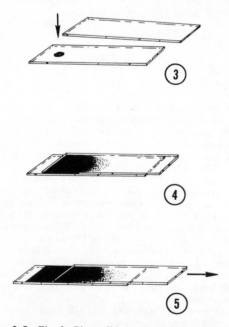

Figs. 3–5 Fig. 3: Place slide on top of drop of buffy coat. Fig. 4: Allow buffy coat to spread between slides. Fig. 5: When spreading stops, pull top slide length of and off end of bottom slide. Air dry. This type of smear is recommended for enzyme preparations; the cells are flatter than in the pull method.

Jenner solution diluted with equal amount of distilled water: 4 min. Cover smears with Jenner solution: 3 min. Dilute with equal amount of distilled water: 1 min. Pour off Jenner from tissues or smears and flood with diluted Giemsa solution (1–10 parts with distilled water): 15–20 min. Rinse off and differentiate with distilled water. If too blue, bring out eosin color by rinsing in 1 % acetic acid, aqueous. Dehydrate in three changes isopropanol, clear in xylene, and mount. Results: similar to those of Wright's stain.

For Maximow eosin-azure stain for bone marrow see HEMOPOIETIC TISSUES.

Neutrophilic sex chromatin is identifiable in Wright's preparations or may be specially stained with pinacyanol (Greenblatt and Manautou [3]). Solution: dissolve 0.5 g pinacyanol in 70% ethanol or methanol. Cover smears with stain: 30 sec. Dilute with equal amount of Wright's buffer (above): 30 sec. Wash, running water: 2–3 min; 50%, 70%, 95%, and absolute ethanol: 30 sec each. Air dry and examine or apply cover glass. Results: chromatin, blue.

Mast cells are easily demonstrated by the metachromatic reaction to several dyes, but the red violet of Toluidine blue O is best. Dissolve 0.2 g Toluidine blue O, C.I. 52040, in 100.0 ml 60% ethanol. Fix smears in methanol-formalin (9 : 1), tissues in any general fixative. Do not use alcohol alone. Deparaffinize and hydrate tissue to 60% alcohol; rinse fixative off smears in 60% alcohol. Stain in Toluidine blue O: 1–2 min. Rinse quickly in tap water. Dehydrate in two changes acetone: 2–3 min each. Xylene and mount. Results: granules, reddish purple.

For alkaline phosphatase (Kaplow [5]), fix buffy coat smears in absolute methanol-formalin (9 : 1), 0°C to −10°C: $\frac{1}{2}$ hr to overnight, no longer. When drawing blood use heparin, not EDTA, as anticoagulant. Buffer: add 25.0 ml propanediol solution (2 amino-2-methyl-1-3-propanediol, 10.5 g; distilled water, 500.0 ml) to 5.0 ml of 0.1N hydrochloric acid and 70.0 ml distilled water. Dissolve 5.0 mg naphthol AS-MX phosphate in 0.2–0.3 ml N-N dimethyl formamide (DMF). Add 60.0 ml buffer and 30.0–40.0 mg fast violet salt LB. Shake 30 sec and filter onto slides which have been washed in distilled water: 20–25 min. Wash in running water: 1 min. Stain in Mayer's hematoxylin: 3 min. Wash in running water: 2 min. Blue in Scott's solution: few seconds. Wash, dehydrate, clear, and mount. Sites: red.

A similar azo-coupling method demonstrates acid phosphatase (Kaplow and Burstone [6]). Fixed or unfixed smears may be used. Fix 20–30 sec, 4°C, in the following: add 300.0 ml acetone to 32.0 ml 0.03M dihydrate sodium citrate and 168.0 ml 0.03M monohydrate citric acid, pH 4.2; or fix in 1.0 g calcium chloride added to 90.0 ml distilled water and 10.0 ml formalin at 4°C. Drain and allow to dry. Dissolve 5.0 mg naphthol AS-MX phosphate in 0.2–0.3 ml DMF. Add 25.0 ml distilled water, 15.0 ml 0.2M acetate buffer (300.0 ml 0.6% acetic acid and 700.0 ml 0.2M sodium acetate, pH 5.0) and 30.0–40.0 mg fast red violet salt LB. Shake 30 sec and filter onto slides. Incubate 37°C: 2 hr, Wash in running water: 1 min. (Fix unfixed smears in 10% formalin: 30 sec. Wash in running water, 2–3 min.) Stain in Mayer's hematoxylin: 3 min. Wash in running water: 5 min. Mount in glycerol jelly. Sites: red. Many prefer the Gomori technique, which produces black sites. See Humason (4) or Wachstein et al. (7).

Peroxidases (1) can be demonstrated in smears fixed in methanol-formalin (9:1): 30 sec. Rinse in diluted Sörenson buffer, pH 7.2 (70.0 ml M/15 potassium sodium phosphate and 30.0 ml M/15 potassium acid phosphate), 1 part; distilled water, 9 parts. Check pH after dilution. Stain in diluted May Grunwald preheated, 35°C: 20 min (May Grunwald stain, National Aniline catalog no 522, 1 part; diluted Sörenson buffer, 8 parts. Wash in diluted buffer. Stain in Giemsa-peroxidase: 40 min. (Sörenson buffer, 15 parts; peroxidase reagent, 0.15 parts; Giemsa stain, National Aniline catalog no. 561, 0.2 parts. Peroxidase reagent: dissolve 0.02–0.03 g benzidine in 6.0 ml 95% ethanol to which 4.0 ml distilled water and 0.02 ml of 3% hydrogen peroxide are added.) Rinse in distilled water and air dry or dehydrate in isopropanol, clear in xylene, and mount. Sites: gray-green to blue-black.

Other enzyme methods can be adapted to leukocyte smears, as can other common methods of staining, such as periodic acid–Schiff, iron, and fat. Frozen sectioning for fat is facilitated by embedding in gelatin prior to sectioning (4).

GRETCHEN HUMASON

References

1. Braunsteiner, H., and D. Zucker-Franklin, "The Physiology and Pathology of Leukocytes," New York, Greene and Stratton, 1962 (293 pp.).
2. Field, J. W., "Further notes on a method for staining malarial parasites in thick blood films," *Trans. Roy. Soc. Trop. Med. Hyg.*, **35**:35–42 (1941).
3. Greenblatt, R. A., and J. M. Manautou, "A simplified staining technique for the study of chromosomal sex in oral mucosal and peripheral blood smears," *Amer. J. Obstet. Gynecol.*, **74**:629–634 (1957).
4. Humason, G. L., "Animal Tissue Techniques," San Francisco, W. H. Freeman and Company, 1972 (641 pp.).
5. Kaplow, L. S., "Cytochemistry of leukocyte alkaline phosphatase," *Amer. J. Clin. Pathol.*, **39**:439–449 (1963).
6. Kaplow, L. S., and M. S. Burstone, "Cytochemical demonstration of acid phosphatase in hematopietic cells in health and various hematological disorders using azo-dye techniques," *J. Histochem. Cytochem.*, **12**:805–811 (1964).
7. Wachstein, M. E., E. Meisel, and J. Ortiz, "Intracellular localization of acid phosphatase as studied in mammalian kidney," *Lab. Invest.*, **11**:1243–1252 (1962).

See also: BLOOD, HEMOPOIETIC TISSUE.

LICHEN

Lichens are composed of two separate plants: a fungus and an alga. The two components are in stable association and form an autonomous organism, which has little external resemblance to either of its components. Understandably, lichens are considered an outstanding example of symbiosis in the plant kingdom. There are about 17,000 species of lichens in the world.

While almost all lichen fungi belong to the Ascomycetes, some are Basidiomycetes. The lichen alga is either a grass-green or a blue-green alga. Morphologically, a distinction is made between three growth forms of the lichen thallus: crustose, foliose, and fruticose. The internal arrangement of the alga and fungus can be unstratified; in this case the symbionts lie in no special order, usually embedded in a gelatinous matrix (Fig. 1). In stratified lichens, the algal cells are restricted to a distinct layer, and protected from above by a cortical layer, comparable to the epidermis of a green leaf. The cortex is composed of compressed, heavily gelatinized hyphae, having in many cases a cellular structure. Medullary tissue lies below the algal layer. The medullary hyphae, less gelatinized, are loosely and irregularly interwoven. The medulla of crustose lichens attaches the thallus directly to the substrate. The foliose (or leafy) thallus has an additional lower cortex and is attached to the substrate by rhizines (Fig. 2). The internal structure of the fruticose lichens is radial; the algal layer is surrounded by a dense cortex on the outside and has an inner lining of medullary tissue. The center may be hollow or may consist of a dense hyphal cord (Fig. 3). The fruiting bodies of lichens are either immersed flask-shaped or superficial discoid ascocarps similar to those of non-lichenized fungi, but having a much longer duration and sporulation capacity. For identification purposes, details achieved from microscopical observations of freehand sections are usually sufficient. The sections are cut with a razor blade, while supporting the object between two halves of a piece of elder pith. Fruiting bodies and crustose and foliose thalli must be cut trans-

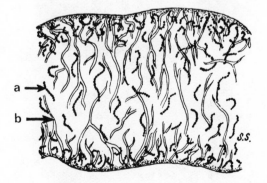

Fig. 1 Schematic drawings of tissue organization in lichen thallus: unstratified (a, alga; b, fungus).

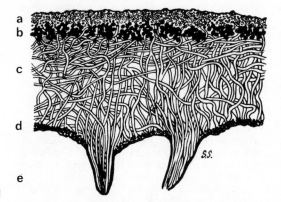

Fig. 2 Schematic drawings of tissue organization in lichen thallus: foliose (a, upper cortex; b, algal layer; c, medulla; d, lower cortex; e, rhizines).

Fig. 4 Usnic acid, recrystallized in glycerin-alcohol-water (1:1:1). (× 168.)

versely; fruticose thalli both transversely and longitudinally. The sections are mounted in lactophenol, lactic blue (lactophenol + cotton blue), or orseillin. For studies of anatomical development and differentiation, and also for the identification of some of the blue-green components, thinner sections are required. Five to ten micron thick sections of undisturbed structure can be prepared using a freezing microtome. Paraffin embedding is unsuitable for lichens.

The composite structure of the lichen causes difficulties in preparing the material for electron microscopy observations. The lichen tissue must be softened by immersion in water for several hours prior to fixation. Each sample should be fixed by three different methods to reveal the details of both components and the contact between them (Fig. 4): (1) Fixation in 5% glutaraldehyde in $0.1M$ phosphate buffer at pH 7.4 for 1 hr at 4°C. The tissue is then washed three times with the same buffer at 15 min intervals and post-fixed for 18 hr with 1% OsO_4. (2) Fixation with 1% OsO_4 in $0.1M$ phosphate buffer, pH 7.2–7.4 for 18 hr at 4°C. (3) Fixation with a mixture of 2 volumes of 4% OsO_4 in $0.1M$ phosphate buffer, pH 7.2–7.4, 1 volume of 50% glutaraldehyde, and 5 volumes of $0.1M$ phosphate buffer, pH 7.2–7.4, for 2 hr at 0°C. The tissue fixed by any of these methods is then dehydrated by 10 min transfer through a graded ethanol series of up to 100%. Propylene oxide serves as a transitional solvent between alcohol and Epon 812, which is the most suitable embedding material. Epon is polymerized at 60°C, for three days. Good sectioning can be achieved with an ultramicrotome diamond or glass knife. The sections are collected on formvar-coated, carbon-reinforced grids. The mounted sections are post-stained with a saturated solution of uranyl acetate in 30% ethanol for 18 hr and then with 0.3% of lead citrate in water. It was found that, as in the case of most tissues, buffered glutaraldehyde gives the best general preservation of fine structure and a little destruction of enzymatic activity. When glutaraldehyde fixation is followed by OsO_4, the quality of preservation is as good as or better than with OsO_4 alone. Lichens contain a unique group of extracellular products, generally named lichen acids, which are either aliphatic (fatty acids, polyols, and triterpenoids) or aromatic (tetronic acid derivatives, depsides, depsidones, quinones, dibenzofuranes, and diketopiperazine derivatives). They can be detected by three complementary techniques: (1) *Spot test*: Both the cortex and the medulla are tested for color reaction employing separately calcium hypochlorite (C) (saturated aqueous solution), potassium hydroxide (K) (10% aqueous solution) and p-phenylenediamine (Pd) (5% alcoholic solution). The last must be prepared freshly every day. C gives a red color with acids having an M-hydroxy configuration and a deep yellow color with xanthons. Pd gives a positive reaction with aldehydic aromatic compounds. Some depsides which do not react with Pd and C give a red color with K. K immediately followed by C gives a positive reaction with depsides and with depsidones which do not react with K, C, and Pd. Quinones give a purple-violet reaction with K. (2) *Crystal test*: The substances are extracted from small fragments of

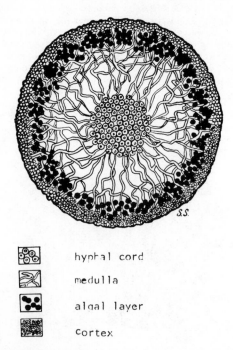

⬚	hyphal cord
⬚	medulla
⬚	algal layer
⬚	cortex

Fig. 3 Schematic drawings of tissue organization in lichen thallus: fruticose.

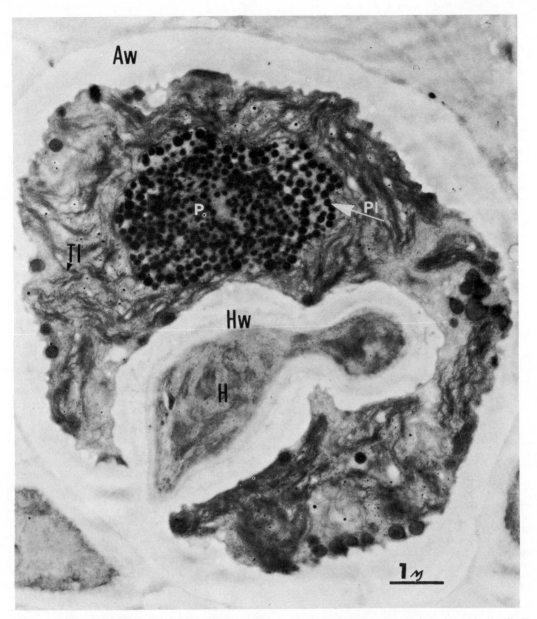

Fig. 5 Section through a mature *Trebouxia* cell of *Lecanora olea* penetrated by haustorium. (Aw, algal wall; H, haustorium; Hw, haustorial wall; P, pyrenoid; Pl, plastoglobuli; Tl, thylakoids). (After Galun et al., 1970.)

the lichen thallus, placed on a microscope slide, with a few drops of an organic solvent (acetone, benzene, or ether). The fragments are discharged after extraction, a drop of crystallizing reagent is added to the dry residue, and it is covered by a coverslip. The substances recrystallize in distinctive shapes and colors which can be identified by comparison with photographs of known crystal structures (Fig. 5). The crystallizing reagents used are: glycerin–acetic acid (1 : 3); glycerin-alcohol-water (1 : 1 :1); glycerin-alcohol-aniline (2 : 2 : 1); and glycerin-alcohol-quinoline (2 : 2 : 1). (3) *Chromatography*: Paper or thin layer chromatography is employed in the identification of lichen acids, using the same techniques as for identification of substances of other plant groups. Rf values of the most common lichen acids, extracted with acetone or benzene, and applied to Eastman Chromagram (K 301 R2) Plates are given in Table 1.

MARGALITH GALUN
YEHUDA BEN-SHAUL

References

Asahina, Y., and S. Shibata, "Chemistry of Lichen Substances," Tokyo: Japan Soc. Promotion Sci., Ueno, 1954.

Table 1

Compound	Solvent	Rf	Detector
lecanoric acid	diethyl ether–glacial acetic acid 50 : 1	.70	bis-diazotized benzidine
erythrin	diethyl ether–glacial acetic acid 50 : 1	.09	bis-diazotized benzidine
gyrophoric acid	diethyl ether–glacial acetic acid 50 : 1	.64	bis-diazotized benzidine
barbatic acid	toluene–butyric acid 19 : 1	.40	bis-diazotized benzidine
evernic acid	toluene–butyric acid 19 : 1	.27	bis–diazotized benzidine
atranorin	toluene–glacial acetic acid 9 : 1	.82	bis-diazotized benzidine
alectoronic acid	toluene–glacial acetic acid 9 : 1	.10	bis-dizaotized benzidine
lobaric acid	toluene–glacial acetic acid 9 : 1	.31	bis-diazotized benzidine
obtusatic acid	toluene–glacial acetic acid 9 : 1	.43	bis-diazotized benzidine
physodic acid	toluene–glacial acetic acid 9 : 1	.06	bis-diazotized benzidine
fumarprotocetraric acid	toluene–glacial acetic acid 9 : 1	.10	bis-diazotized benzidine
norstictic acid	toluene–glacial acetic acid 9 : 1	.34	bis-diazotized benzidine
psoromic acid	toluene–glacial acetic acid 9 : 1	.52	bis-diazotized benzidine
salazinic acid	toluene–glacial acetic acid 9 : 1	.06	bis-diazotized benzidine
stictic acid	toluene–glacial acetic acid 9 : 1	.21	bis-diazotized benzidine
parietin	toluene-cyclohexane 4 : 1	.53	spot: yellow
pulvic acid	toluene–glacial acetic acid 4 : 1	.27	in UV (365 nm)
usnic acid	toluene–glacial acetic acid 4 : 1	.82	in UV (365 nm)

Galun, M., N. Paran, and Y. Ben-Shaul, "The fungus-alga association in the Lecanoraceae. An ultrastructural study," *New Phytol.,* **69**:599-603 (1970).

Hale, M. A., "The Biology of Lichens," London, Edward Arnold Ltd., 1967.

Santesson, J., "Chemical studies on lichens: thin layer chromatography of lichen substances," *Acta Chem. Scand.,* **21**:1162–1172 (1967).

LUNG

Large sections may be cut from whole lungs which have been dried by expansion with compressed air for several days (Ronstrom, 1935; Tobin, 1953) or by passing the compressed air over a solution of formaldehyde (Blumenthal and Boren, 1959). In both methods large sections may be cut and the sections studied unstained or stained with specific dyes. Thinner sections may be cut from whole lungs by hardening them in a fixative, embedding in a gelatin solution, and sectioning the embedded lung on a large sliding microtome. The large sections are then mounted for study on a special paper coated with an adhesive (Gough and Wentworth, 1949; Whimster, 1970), or the sections may be mounted between sheets of glass or plastic, or covered with sheet plastic or plastic spray.

Injection Casts of Pulmonary Vessels or Respiratory Tissue. In addition to the customary injection media, solutions of India ink, gelatin, gum acacia, red lead, woods metal, etc., the use of plastics, such as vinyl acetate solutions, have been found very useful for the study of normal and pathological pulmonary structures (Liebow et al., 1947; Tompsett, 1959; Tobin and Zariquiey, 1953). The solutions of colored plastics can be injected directly into the desired vessel or the respiratory tissue and the lung macerated by hydrochloric acid or pepsin from the injected specimen. Similarly, solutions of liquid latex or latex-silicone preparations (i.e., Microfil) have also been used to demonstrate the bronchopulmonary segments and the blood vessels in normal and diseased lungs. These solutions are injected where it is desired to dissect rather than macerate the specimen (Tobin and Zariquiey, 1950a).

Microscopic sections. Small samples of tissue from the isolated lung should be placed in a container of fixative and vacuum applied to withdraw residual air from the respiratory passageways, so that the fixative can penetrate more readily into the finer structures of the lung. If more normal tissue relationships are desired, the isolated lung should be fixed by gentle perfusion of the fixative via the bronchi until the lung is approximately its normal size at full expansion. However, for the most normal relationship of the pulmonary tissue, the fixative should be perfused into the trachea and/or blood vessels while the lung is still in the chest. After preliminary hardening in the thoracic cavity, the lung should be removed and immersed in a container of fixative for further hardening (see Gough Wentworth, 1949; or Krahl et al., 1958, for details of the latter method).

In addition to the use of microscopic lung sections for normal and/or pathologic studies, sample and serial sections have been used to reconstruct the normal development of the lung (Boyden, 1969).

Electron microscopy of the lung (originally described by Low, 1953) has been used as a means of studying the finer structure of normal and pathologic alveoli and pulmonary capillaries. E. M. methods have also been used to study the developmental origin of various pulmonary tissues as summarized in O'Hare and Sheridan (1970).

Stains. Many different stains may be used on sections of lung tissue, but hematoxylin and eosin have been found to be the best general stains. The elastic tissue may be stained with Unna's Orcein or Verhoff–van Giesen stains. A simple silver impregnation method for reticulum has been described by Gordon and Sweets (1936). After Orth fixation and Masson's Trichrome stain, vacuolated and nonvacuolated cells may be seen in the walls of the smaller respiratory passageways of the lung (Bertalanffy and Leblond, 1955). The basement membrane of the alveoli may be well demonstrated by the McManus–periodic acid–Schiff stain (Reid and Rubino, 1959).

Alveolar pores can be seen in fairly thick (50–200 μ) stained or unstained sections cut from lungs fixed while expanded either by perfusing the respiratory tissue with fixative or by air drying the lung, as well as tissue from lungs with pulmonary edema, or certain types of pneumonia.

Pulmonary Vessels. Various media have been used to inject the pulmonary artery, pulmonary vein, bronchial artery, or bronchial vein to demonstrate their course, relationships, and interconnections: solutions of India ink, colloidal mercury sulfide (8% HgS obtained from Hille Laboratories, Chicago 13, Ill.; see Parks, 1956); solutions of gelatin containing red lead or other pigments, solutions of liquid latex, or solutions of plastics have been used. The latex-injected specimens can be dissected after hardening the specimen in 10% neutral formalin, or the pulmonary tissue can be macerated from the casts of the vessels or respiratory passageways in the specimens injected with vinyl acetate or other plastics.

Pulmonary Lymphatics. The pleural lymphatics and their connections with the deeper lymphatics may be demonstrated by injecting solutions of India ink or liquid latex into the pleural lymphatics (Simer, 1952; Tobin, 1954) or into the pleural lymphatics of edematous adult lungs (MacCallum, 1919).

Dust or Particles. After inhalation of dusts, the finer particles may be visible on the pleural surface of the lung or in the lymph nodes along the course of the lymphatics in sections of the lungs. Heavy metal dust may be seen in roentgenograms of the chest or in isolated lungs, if the concentration of the metal is sufficiently intense.

Inhaled radioactive material may be identified by radioautographs (Belanger and Leblond, 1946) or by recording the radioactivity in the pulmonary lymph channels or nodes with radioactivity recording apparatus (Hahn and Carothers, 1953).

Phagocytes (Heart Failure Cells, Etc.) The phagocytic cells of the lung can be seen in sections of lungs from adult animals or from individuals that have lived in an atmosphere containing particulate contaminates. Phagocytes may be demonstrated by having the animal inhale dusts or dyes, or from cases where pulmonary congestion has allowed red blood cells to enter the alveoli. As Macklin (1951) has pointed out, in sections from lungs which have collapsed from their inherent elasticity or fixation shrinkage, the macrophages are usually within the lumen of the alveoli or of larger respiratory passageways. But if the lungs are fixed in the expanded state, the macrophagees are usually attached to the walls of the alveoli. Action of cilia, mucus, and air currents will propel these free macrophages up the respiratory passages.

Living lung has been studied by Krahl (1953), who used fetal lungs in a small chamber to study the expansion of the lung. His technique has been modified by using the quartz rod technique of transillumination (originally described by Knisely, 1936) to study structure and circulation in living animal lungs (Krahl, 1962).

A-V Shunts. Vessels larger than capillaries connecting branches of the pulmonary artery with those of the pulmonary vein (A-V shunts) have been demonstrated in microscopic sections of lungs and by the passage of glass spheres (Prinzmetal et al., 1948; Tobin and Zariqniey, 1950b) or radiopaque material (Rahn et al., 1952; Tobin, 1966), from the pulmonary artery into the pulmonary vein. Such precapillary connections between branches of the bronchial and pulmonary arteries are best studied from microscopic sections or plastic casts of these vessels.

Biochemistry. Many biochemical studies are being made of pulmonary tissue. Among these are cytochemical studies of Vater et al. (1968) and Holmes and Martin (1967).

CHARLES E. TOBIN

References

Belanger, L. F., and C. P. Leblond, *Endocrinology*, **39**:8 (1946).
Bertalanffy, F. D., and C. P. Leblond, *Lancet,* **2**:1365 (1955).
Blumenthal, B. J., and H. G. Boren, *Amer. Rev. Tuberc, Pulmon. Dis.,* **79**:764 (1959).
Boyden, E. A., *Anat. Rec.,* **163**:158 (1969).
Gordon, H. R., and H. H. Sweets, *Amer. J. Pathol.,* **12**:545 (1936).
Gough, J., and J. C. Wentworth, *J. Roy. Microscop. Soc.,* **69**:231 (1949).
Hahn, P. F., and E. S. Carothers, *J. Th. Surg.,* **25**:265 (1953).
Holmes, R. S., and C. J. Martin, *Biochim. Biophys. Acta,* **146**:138 (1967).
Knisely, M. H., *Anat. Rec.,* **65**:23 (1936).
Krahl, V. E., *Anat. Rec.,* **115**:448 (1953).
Krahl, V. E., *Anat. Rec.,* **142**:350 (1962).
Krahl, V. E., C. E. Tobin, S. Wyatt, and C. Loosli, *Amer. Rev. Resp. Dis.,* **80**:114 (1958).
Liebow, A. A., M. R. Hales, G. E. Lindskog, and W. E. Bloomer, *Bull. Int. Ass. Med. Museums,* **27**:116 (1947).
Low, F. N., *Anat. Rec.,* **117**:241 (1953).
MacCallum, W. G., R.I.M.R. Monograph No. 10, 1919.
Macklin, C. C., *Lancet,* **1**:432 (1951).
O'Hare, K. H., and M. N. Sheridan, *Amer. J. Anat.,* **127**:181 (1970).
Parks, H. F., *Anat. Rec.,* **125**:1 (1956).
Prinzmetal, M., et al., *Amer. J. Physiol.,* **152**:48 (1948).
Rahn, H., R. C. Stroud, and C. E. Tobin, *Proc. Soc. Exp. Biol. Med.,* **80**:239 (1952).
Reid, L., and M. Rubino, *Thorax,* **14**:3 (1959).
Ronstrom, G. N., *Anat. Rec.,* **63**:365 (1935).
Simer, P. H., *Anat. Rec.,* **113**:269 (1952).
Tobin, C. E., *Anat. Rec.,* **114**:453 (1953).
Tobin, C. E., *Anat. Rec.,* **120**:625 (1954).
Tobin, C. E., *Thorax,* **21**:197 (1966).
Tobin, C. E., and M. O. Zariquiey, *Med. Radiog. Photog.,* **26**:38 (1950a).
Tobin, C. E., and M. O. Zariquiey, *Proc. Soc. Exp. Biol. Med.,* **75**:827 (1950b).
Tobin, C. E., and M. O. Zariquiey, *Med. Radiog. Photog.,* **29**:9 (1953).
Tompsett, D. H., *Ann. Roy. Coll. Surg. England,* **24**:110 (1959).
Vater, A. E., et al., *J. Cell Biol.,* **38**:80 (1968).
Whimster, W. F., *Thorax,* **25**:141 (1970).

LYSOSOMES

Through the classical work of De Duve the lysosome, a biochemical concept, was introduced in cytology. Lysosomial is, in fact, termed the cellular fraction which sediments together with the light mitochondria and is characterized by a remarkable acid-hydrolasic activity (mainly due to acid phosphatase, β-glucuronidase, acid ribonuclease and deoxyribonuclease and cathepsine) after treatment with detergent. Such biochemical features are obviously to be primarily taken into account if lysosomes are to be identified in light (L.M.) or electron microscopy (E.M.). In order to understand how some methods for lysosome identification actually work, the origin and the evolution of lysosomal structures is also relevant. In fact, a whole series of such methods is based on intravital administration of substances that label the lysosomes through their own mechanism of formation. A schematic representation of lysosome origin and pathway is given in Fig. 1. From the picture it is apparent that lysosomes may originate either from transformation of phagocytotic or

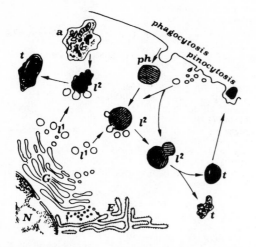

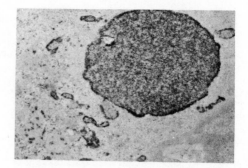

Fig. 2 A lysosome in the human liver shown by means of Miller's methods for acid phosphatase. (E.M. × 10,800; courtesy of Dr. B. Bertolini.)

Fig. 1 Lysosome formation and fate; *a*, autophagic segregation; *E*, granular endoplasmic reticulum (microvesicles arisen from endoplasmic reticulum transfer acid hydrolases to Golgi—small arrows); *G*, Golgi area, *l²*, primary lysosomes; *l²*, secondary lysosomes; *ph*, phagosome; *t*, telolysosomes or residual bodies.

pinocytotic vesicles (heterolysosomes) or from autophagic segregation bodies (autolysosomes). In both instances, their transformation involves fusion with "primary lysosomes" of Golgi origin, containing lysosomal characteristic lytic enzymes. After a primary lysosome has fused with the phagosome or a cluster of pinocytotic vesicles, the structure is termed "secondary lysosome," and can still fuse with more phagosomes or pinocytotic microvescicles.

For the identification of primary and secondary lysosomes L.M. avails itself of these techniques of enzymatic histochemistry that are specific for lysomal enzymes. Among these methods, the most commonly used are the test for acid phosphatase, β-glucuronidase, nonspecific lysosomal ali-esterase, N-acetyl-α-glucosaminidase, and aryl-sulfatases. It is to be stressed, though, that the specificity of these histochemical methods, hence the validity of the identification, is highly dependent on correct fixation and incubation in a suitable medium. Nonetheless other nonlysosomal structures may share similar enzymatic activity, and hence give a deceptive positive test. The most commonly used methods are the acid phosphatases tests by Barka and Anderson (1962) and by Gomori (1952), respectively involving the use of Na-α-naphthyl phosphate and Na-β-glicerophosphate as substrates. The Gomori method was successfully used for E.M. after replacing formalin with glutaraldehyde fixation (Miller, 1962) (Fig. 2). Less frequently used are the tests for β-glucuronidase (Hayashi, 1964) and N-acetyl-α-glucosaminidase (Hayashi, 1965). A very good method for aryl-sulfatase was introduced for L.M. by Goldfischer (1965) and subsequently modified for E.M. A number of methods are also available to test nonspecific esterases in both L.M. and E.M., but their use is objectionable.

Nonenzymatic histochemical methods are also available to identify lysosomes through their glycoprotein matrix (Koenig, 1962). All the specific staining techniques for glyco-lipoprotein complexes can in fact be used: metachromasia after Toluidine blue as well as the PAS

reaction after Hotchkiss. More successful modifications replaced the Schiff solution with ammoniated silver ones (Sandbank and Becker, 1964; Capanna et al., 1966) since the remarkable opacity of the silver precipitate on lysosomal structure results in a high degree of optical contrast (Fig. 3). These techniques were immediately extended to E.M. (Rambourg and Leblond, 1967) due to the high electrondensity of the silver salts.

We have so far pointed out that some histochemical methods, both enzymatic and not, can be used for the E.M. identification of lysosomes. In fact, the validity of such an identification requires that an enzymatic histochemical control be performed. In the E.M. the morphology of lysosomes appears to be extremely varied, the only constant peculiarity being their single limiting membrane. Their size may vary from very small primary lysosomes up to 1 μ or more large autolysosomes. The matrix also may be more or less dense, depending on the functional stage or the kind of substance phagocytosed or pinocytosed. Nonetheless many structures can be presumptively referred to as lysosomal on a morphological basis (dense bodies, multivesicular bodies, etc.).

Dealing with the E.M. identification of lysosomes a few remarks are due on the problem of PTA stainability. In particular conditions phosphotungstic acid (PTA) stains the external layer of the unit membrane, such stainability being retained when the membrane is infolded to

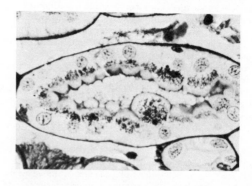

Fig. 3 Lysosomes in convoluted tubule of *Triturus* kidney: silver-methylenamine reaction of Capanna et al. (L.M. × 210.)

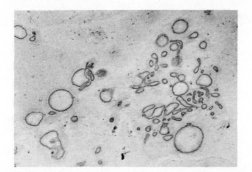

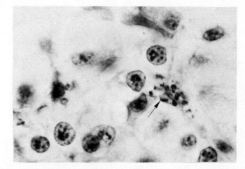

Fig. 4 Lysosomes in a human epatocyte after PTA reaction. (E.M. × 10,500; courtesy of Dr. B. Bertolini.)

Fig. 5 Lysosomes in a Kupffer cell of frog liver after intravital Trypan blue stain. (L.M. × 588.)

form the phagosome and after fusion with the primary lysosome. We shall not discuss this problem here, but the method (Marinozzi and Gautier, 1951) is worth mentioning for use in the E.M. identification of lysosomes (Fig. 4).

A whole series of methods for the identification of lysosomes is suggested by their origin from pinocytotic or phagocytotic vesicles; a number of suitable substances, to be intravitally administered, can in fact be pinocytosed and identified in both light and electron microscopical preparations. To this aim the ancient trypan blue vital stain, or other similar anionic dyes, can be successfully used to label heterolysosomes. (Fig. 5). The rapidly sequestred vital dyes are not degraded within the secondary lysosomes and are to be found still in the telolysosomes. Though trypan blue is electron-dense enough to be recognized in E.M. preparations, other substances, such as ferritin, are more convenient owing to their high electron density and macromolecular peculiar conformation. A variety of substances can be also administred intravitally, provided that some devices are available to recognize them in the preparations. Radioactively labeled macromolecules can be used as markers and radioautographically traced down. Protected colloidal solution of heavy metals can be seen directly in the E.M. preparations or visualized by the Prussian blue staining reaction (Kent, 1963), for the L.M.

Fluorescence microscopy can also be of some advantage in the identification of lysosomes both for their own autofluorescence or after intravital administration of fluorescent compounds such as Acridine orange or Quinacrina (Allison and Young, 1969).

Ernesto Capanna

References

Allison, A. C., and M. R Young, p. 600 in "Lysosomes," Vol. 2, Dingle and Fell, ed., Amsterdam, North-Holland Publ. Co., 1969.
Barka, T., and P. J. Anderson, *J. Histochem. Cytochem.*, **10**:741 (1962).
Beck, F., and J. B. Lloyd, p. 567 in "Lysosomes," Vol. 2, Dingle and Fell, ed., Amsterdam, North-Holland Publ. Co., 1969.
Capanna, E., et al., *Experientia,* **22**:114 (1966).
Goldfischer, S., *J. Histochem. Cytochem.,* **13**:520 (1965).
Gomori, G., "Microscopic Histochemistry," Chicago, Chicago Univ. Press, 1952.
Hayashi, M., *J. Histochem. Cytochem.,* **12**:659 (1964); *Ibid.,* **13**:355 (1965).
Kent, G., et al., *Lab. Invest.,* **12**:1094 and 1102 (1963).
Koenig, H., *Nature,* **195**:782 (1962).
Marinozzi, V., and A. Gautier, *C.R. Acad. Sci.* (Paris) **253**:1180 (1951).
Miller, F., "Proc. 5th Congr. Electron Microscopy, Q-2," New York, Academic Press, 1962.
Rambourg, A., and C. P. Leblond, *J. Cell Biol., ***32**:27 (1967).
Sandbank, U., and N. H. Becker, *Stain Technol.,* **39**:27 (1964).

m

MAGENTA (BASIC FUCHSIN)

According to Conn 1961 ("Biological Stains," 7th ed., Baltimore, Williams & Wilkins Company) the dye commonly sold as magenta (basic fuchsin) is a mixture of pararosanilin, rosanilin, and magenta I. The writer prefers the term magenta to basic fuchsin since it avoids confusion with acid fuchsin which is a sulfonated derivative of magenta. It may be added that fuchsin is the German word for magenta and there is no reason to retain it in English.

Magenta is rarely used alone as a dye at the present time. Except in the form of "aldehyde fuchsin" (see Gomori 1950 and Gabe 1952 below); or as the leucobase in the Fuelgen reaction, the formulas and methods of which are given elsewhere (see FUELGEN STAINS).

Albrecht cited from 1943 Cowdry (Cowdry 1943, 17)
FORMULA: water 100, 95% alc. 20, phenol 8, magenta 4
PREPARATION: Dissolve the dye in phenol and alc., then add water.

Auguste 1932 see Ziehl 1882 (note)

Biot 1901 see Gallego 1919 (note)

Davalos cited from 1894 Kahlden and Laurent (Kahlden and Laurent 1894, 89)
FORMULA: water 100, abs. alc. 1, phenol 5, magenta 10
PREPARATION: Grind the dye with the alc., wash out mortar with 10 successive portions of 5% phenol.

Duprès 1935 (*J. Oxf. Univ. Jr. Sci. Cl.*, **46**: 77)
STOCK FORMULA: water 80, 95% alc. 20, magenta 0.75
PREPARATION OF STOCK: Digest 48 hr at 40°C. Cool. Filter.
REAGENTS REQUIRED: A. sat. aq. sol. aniline 100, stock 5, acetic acid 1.5; B. Duprès 1936.
METHOD: [sections] → water → A, 1–10 mins → B, until differentiated, 5–10 mins → counterstain → balsam, via usual reagents

Gabe 1953 (*Bull. Micro. Appl.*, **3**(2): 153)
PREPARATION OF DRY STAIN: Dissolve 0.5 magenta in 100 boiling water. Cool filter and add 1 hydrochloric acid and 1 paraldehyde. Test drops on filter paper at daily intervals until no red stain appears. Filter and dry ppt.
STOCK SOLUTION: 0.75% solution in 70% alc.

Gallego 1919 (*Trab. Lab. Invest. biol. Univ. Madr.*, **17**: 95)
REAGENTS REQUIRED: A. Ziehl 1882 2, water 98, B. 0.5% formaldehyde
METHOD: [sections of formaldehyde material] → A, 5 min → wash → B, 5 min → wash → balsam, via usual reagents
RESULTS: nuclei, dark violet, cytoplasmic structures, polychrome
NOTE: This method of converting magenta to a blue-black stain was originally proposed for bacteriological purposes by Biot 1910 (*C.R. Ass. Anat.*, Congrès de Lyon, 234).

Gomori 1950 (*Amer. J. Clin. Pathol.* (**20**: 665)
PREPARATION OF DRY STAIN: Boil for 1 min. 100 0.5% magenta. Cool, filter, and add 1 hydrochloric acid and 1 paraldehyde. Leave until no further ppt forms. Filter, dry ppt.
STOCK SOLUTION: Sat. sol. ppt from above in 70% alc.
WORKING SOLUTION: 70% alc. 75, stock 15, acetic acid 0.6
NOTE: This is the original "aldehyde fuchsin." The method of preparation of Gabe 1953 (above) is now almost universal.

Goodpasture and Burnett 1919 (*U.S. Nav. Med. Bull.*, **13**: 177)
FORMULA: water 80, 95% alc. 20, phenol 1, aniline 1, magenta 0.5
PREPARATION: Dissolve the dye in the mixture of alc. phenol and aniline. Add water.
NOTE: This solution was republished by Goodpasture 1925 (*Amer. J. Pathol.*, **1**: 550) to which paper and date reference are frequently made.

Henneguy 1891 (*J. Anat. Parist.*, **27**: 397)
REAGENTS REQUIRED: A. 1% potassium permanganate; B. 1% magenta
METHOD: [sections or smears from Flemming 1882 fixed material] → water → A, 5 min → water, wash → B, 5 min → water, rinse → 95% alc., until partly differentiated → clove oil, until differentiation complete → balsam, via xylene
NOTE: Schneider 1922, 113 applies this method to any section by mordanting in 1% chromic acid.

Huntoon 1931 (*Amer. J. Clin. Pathol.*, **1**: 317)
FORMULA: water 75, glycerol 25, phenol 3.75, magenta 0.6

Kinyoun cited from 1946 Conn et al. (Conn et al. 1946, 1 V, 6)

FORMULA: water 100, 95% alc. 20, phenol 8, magenta 4

Krajian 1943 (*J. Lab. Clin. Med.*, **28** : 1602)
FORMULA: xylene 65, creosote 35, 95% alc. 5, magenta 0.3
PREPARATION: Dissolve dye in alc. Add to other solvents.

Maneval 1928 (*Stain Technol.*, **4** : 21)
FORMULA: water 90, aniline 2.5, 95% alc. 7.5, acetic acid 0.2, magenta 1

Muller and Chermock 1945 (*J. Lab. Clin. Med.*, **30** :169)
FORMULA: water 100, 95% alc. 20, phenol 8 magenta 4, wetting agent 0.1
NOTE: The original specifies "Tergitol 7" as the wetting agent.

Pottenger 1942 cited from Farber 1942 (*Stain Technol.*, **17** : 183)
FORMULA: water 80, 95% alc. 15, phenol 5, magenta 1.6

Schneider 1922 see Henneguy 1891

Tilden and Tanaka 1945 (*Tech. Bull.*, **9** : 95)
FORMULA: water 90, methanol 10, phenol 5, magenta III 0.5

Ziehl 1882 (*Deuts. med. Wschr.*, **8** : 451 and 1890 *Z. wiss. Mikr.*, **7** : 39)
REAGENTS REQUIRED: A. water 100, 90% alc. 10, magenta 1, phenol 5; B. 1% acetic acid
PREPARATION OF A: Grind the dye with phenol in a mortar. When dissolved add the alc. in 10 successive lots while grinding. Then use water in 10 successive lots to wash out mortar. Filter accumulated washings.
METHOD: [sections or smears] → water → A, 10–20 min → B, until differentiated → counterstain → balsam, via usual reagents
NOTE: The solution of Auguste 1932 (*C.R. Soc. Biol. Paris*, **111** : 719) contains 2 magenta but is otherwise identical.

PETER GRAY

Books Cited

Journal references in the above formulas are given in the standard form. The following books are cited by author and date only.

Conn, H. J., "Biological Stains," 7th ed., Baltimore, The Williams and Wilkins Company, 1961.
Cowdry, E. V., "Microscopic Technique in Biology and Medicine," Baltimore, Williams and Wilkins, 1943.
Gray, P., "The Microtomist's Formulatory and Guide," New York and Toronto, The Blakiston Company, 1954.
Kahlden, C. von, and O. Laurent, "Technique Microscopique," Paris, Carré, 1896.
Schneider, H., "Die botanische Mikrotechnik". Jena, Fischer, 1922.

MESOGLOEA

Two general approaches have been taken to isolate mesoglea for histological and biochemical possibilities, enzymatic and mechanical. Normandin (1962), in an extensive study of various enzymes and times of treatment deduced that the best approach for isolating the mesoglea

of *Hydra pseudoligactus* involved hyaluronidase and papain. The mesoglea retained the shape of the entire hydra as could be seen in whole-mounted preparations.

A mechanical method has been described (Shostak et al., 1965) based on the use of a mild acid treatment which Lenhoff and Muscatine (1963) first used to separate the epithelia of *Hydra viridis*. Hydra culture medium (Muscatine, 1961) less bicarbonate and tris, is adjusted to pH 2.5 by the addition of HCl. If done in bulk, cellulose nitrate test tubes should be used. However, individual hydra or small numbers of animals can be managed in glass petri dishes, particularly if a gelatin lining is prepared in advance. It is essential that the solution be ice cold; the epidermis will roll back into an annulus near the foot and may be removed either by rapidly shaking when the animals are in tubes or with the use of forceps. This solution can be poured off at that point or within 10 min. The mesoglea will adhere to the gastrodermis, which will be elongated in the form of a cigar with tentacles at one end. The preparation is rinsed three times in normal culture medium and may be centrifuged in a clinical centrifuge to facilitate rinsing. The final pellet with minimum fluid remaining is then freeze-thawed three times following which distilled water is added and the contents of the

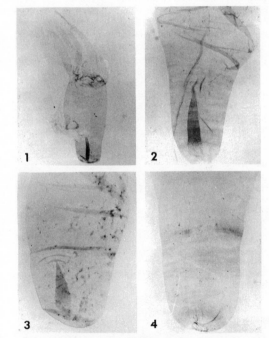

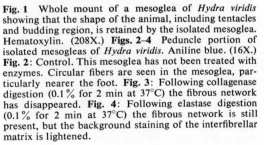

Fig. 1 Whole mount of a mesoglea of *Hydra viridis* showing that the shape of the animal, including tentacles and budding region, is retained by the isolated mesoglea. Hematoxylin. (208X.) **Figs. 2–4** Peduncle portion of isolated mesogleas of *Hydra viridis*. Aniline blue. (16X.) **Fig. 2**: Control. This mesoglea has not been treated with enzymes. Circular fibers are seen in the mesoglea, particularly nearer the foot. **Fig. 3**: Following collagenase digestion (0.1% for 2 min at 37°C) the fibrous network has disappeared. **Fig. 4**: Following elastase digestion (0.1% for 2 min at 37°C) the fibrous network is still present, but the background staining of the interfibrellar matrix is lightened.

tubes spun on a Vortex mixer for 10 sec. The purpose of this operation is to compress the mesogleas so as to force out their liquid contents which were the gastrodermal cells. When individual animals are prepared the same can be accomplished by repeatedly forcing the material through the surface of a dish of water, either through simply dropping the material through a medicine dropper onto the water or by holding one end of it to a slide with forceps and pulling the slide in and out of the water. The mesogleas can then be collected in the centrifuge tube by centrifugation at approximately 2000 rpm for three minutes. A "plug" of aqueous of 100% sucrose at the bottom of the tube serves to prevent adhesion of mesoglea to the bottom of the tube.

After drying, these mesoglea are stained with aniline blue (Fig. 1) and other basophilic stains, revealing prominent circular fibers in the peduncle (stalk) (Fig. 2) and less prominent longitudinal fibers throughout the length of the animal. A diffuse analine blue positive staining appears in addition to the fibrous material. The fibers are rapidly collagenase (Worthington, highly purified) digestible (Fig. 3). The diffuse staining materials are elastase (Worthington) digestible (Fig. 4).

Another mechanical method which works for the brown hydra and the giant hydra is like the above but skips the mild acid bath, beginning instead with freeze-thawing (Hauseman and Burnett, 1969). Mesogleas prepared in this way resembled those just described, except in these species the fibrous systems are more nearly diagonal and are prominent in the mouth and foot regions. The fibrous system, moreover, seems to change as a function of feeding, budding, and regeneration.

Electronmicroscopy of glutaraldehyde-fixed, post-osmicated araldite or Maraglas embedded hydras have likewise shown the presence of fibrous material in the mesoglea (Davis et al., 1968). These fibers, however, are entirely too thin to correspond directly to those seen with the light microscope.

STANLEY SHOSTAK

References

Davis, et al., J. E. Z., 167:167, 295 (1968).
Hauseman, R. E., and A. L. Burnett, J. E. Z., 171:7 (1969).
Lenhoff and Muscatine, Science, 142:956 (1963).
Muscatine, p. 255 in "The Biology of Hydra," W. F. Loomis and H. M. Lenhoff, ed. Miami, University of Miami Press, 1961.
Normandin, D., Ph.D. Thesis, University of Illinois, Urbana, 1962.
Shostak, S., et al., Developmental Biol., 12:434 (1965).
See also: COELENTERATA.

MESOZOA

The so-called "Mesozoa," often treated as a phylum, consists of two very different groups of microscopic organisms, the Orthonectida and Dicyemida. The resemblance of male and female orthonectids to the nematogen and rhombogen stages of dicyemids is superficial, for the latter reproduce asexually; the reduced sexual generation of dicyemids is unlike any stage in the life cycle of orthonectids. The few known species of orthonectids are tissue parasites of diverse marine invertebrates: turbellarians, nemerteans, polychaetes, gastropods, pelecypods, and ophiuroids. Smears of parasitized tissue fixed in Bouin's fluid respond well to staining with iron

hematoxylin and to impregnation with Protargol; the latter method favorably demonstrates the distribution of kinetosomes, and may bring out ciliary rootlets and nuclei as well. After fixation in Champy's fluid, Protargol impregnation may differentiate myocytes and elastic cells, as well as kinetosomes and rootlets. Impregnation of smears (without prior fixation) in 2% silver nitrate, followed by reduction in sunlight, brings out cell boundaries and therefore shows the very precise arrangement of cells in the outer jacket. Parasites in small pieces of infected tissue fixed in a mixture of 2 parts of 4% osmium tetroxide and 1 part of $0.2M$ s-collidine (pH 7.5) are suitable for studies on ultrastructure. (For preparation of material for both light microscopy and electron microscopy, see Kozloff, 1969.) Dicyemids live in the renal organ and branchial heart coelom of many cephalopod molluscs. Among the better fixatives for smears to be stained by iron hematoxylin or alum hematoxylin are Bouin's fluid and Sanfelice's fluid. Protargol and silver nitrate impregnations do not, as a rule, give particularly useful results. For fixation of material for electron microscopy, fairly good preservation has been achieved with 1% osmium tetroxide in sea water, and 1% osmium tetroxide in $0.4M$ sucrose buffered to pH 7.5 with sodium cacodylate (see Ridley, 1968).

EUGENE N. KOZLOFF

References

Kozloff, E. N., J. Parasitol., 55:171–195 (1969).
Ridley, R. K., J. Parasitol., 54:975–998 (1968).

METALLOGRAPHIC MICROSCOPES

Principle. Metallographic microscopes are used to study polished opaque metallurigical specimens. Their design, therefore, is slightly different from a transmitted light microscope. A thin plane glass beam splitter is mounted behind the objective at an angle of 45° to the optic axis of the microscope. The illuminating light is directed through the objective to the specimen by this beam splitter. This means that the objective also serves as condenser. The light reflected by the specimen is collected by the objective, transmits the beam splitter, and the image is formed in the intermediary image plane in the eyepiece. (Fig. 1.)

The beam splitter should be located in a collimated beam. Therefore, metallographic objectives of some manufacturers are corrected for infinity—the objective projects the image to infinity. A tube lens positioned behind the beam splitter brings the image back to the proper image plane, inside the eyepiece. The latter serves as a secondary magnifying system and projects the intermediary image in the focal plane of the eyelens to infinity.

Classification of Metallographic Microscopes. The two principal types of metallographic microscopes are the upright or bench microscope and the inverted metallograph.

In *upright* microscopes the specimen is illuminated from above, through the objective positioned over the specimen surface. The microscope stand is similar to or the same as one used for transmitted light microscopy. Its advantages are: It can easily be converted for transmitted light work. Furthermore, the operator of the upright microscope can see the whole specimen area, so particular structures are easily located. Accessories such as photomicrographic attachments are readily adapted mechanically and

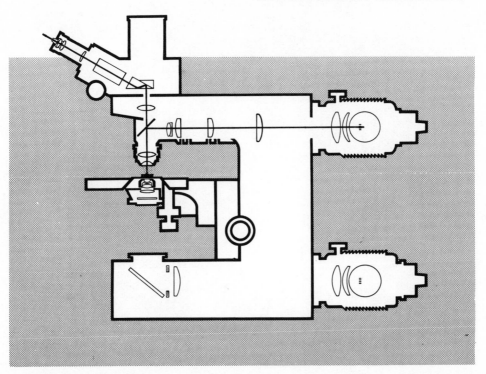

Fig. 1   Schematic of metallographic microscope, also equipped for transmitted light work. Abstract beam path and optical components for reflected light illumination.

optically. Samples are easy to mount in heating stages, and if they melt they do not drop toward the objective.

A disadvantage, however, is the accurate leveling necessary to align the polished specimen surface perpendicular to the optical axis. If the surface is not fragile, a simple handpress can be used to level the specimen by pressing it onto a slide with putty. This is time consuming when large numbers of specimens have to be studied. With sensitive surfaces or brittle material, an attachable tilting stage has to be used. Such stages are mounted on a ball and socket joint and can be very accurately aligned. In another arrangement, the specimen surface is pressed from below against the aperture of a well-aligned auxiliary stage. Another disadvantage of the upright microscope is the fact that dust settles on the polished surface.

Upright microscopes range from small stands to very sophisticated research instruments versatile enough to allow all possible illumination techniques and to accept a variety of accessories. Two such stands are shown in Fig. 2.

Inverted metallographs (LeChatelier) are designed to be used exclusively for study of metallurgical specimens. The polished specimen is placed upside down on the mechanical stage. Therefore, the specimen is automatically leveled and no dust can fall onto the surface. The specimen is illuminated through an objective from beneath, and the image projected downward into a viewing head inclined toward the observer. The larger inverted Metallographs usually can be equipped with:

1. illuminators for bright field, dark field, polarized light, phase contrast, differential interference contrast, etc.;

2. special objectives corrected to eliminate curvature of field;

3. a high-intensity light source for photography and projection;

4. Projection screen—often of large size—for demonstrating and teaching;

5. camera system with bellows or zoom optic to match standard magnifications used in metallography, often with automatic exposure control;

6. accessories for the metallograph such as hardness testers, heating stages, special reticles (grain size, etc.) just to mention a few. (See Fig. 3.)

Metallographs historically originate from an actual optical bench arrangement. Therefore, some models are still elongated; others consist of compact, heavy stands mounted on a large base plate or table. Photomicrography is essential in studies with the metallograph. It should therefore be mounted absolutely vibration free. Vibrations become a problem when the instrument is used in a steel mill or other industrial plants. Rubber or plastic shock absorbing feet placed under the base plate are most helpful. More sophisticated absorbers are matched to the frequency of the vibration.

A few factors about metallographs deserve careful consideration. They are:

1. High-power light source: In many cases, it is necessary to utilize a high-power light source for photomicrography as well as for projection. The light source should emit a continuous spectrum in the visible range for true color reproduction in photomicrographs and should be intense enough to produce a bright image on large projection screens. The previously used carbon arc lamps have been replaced by powerful xenon arc lamps (150, 450 watts) which meet all requirements.

2. Lamp housing: The importance of a proper lamp housing for the high-intensity light source is widely

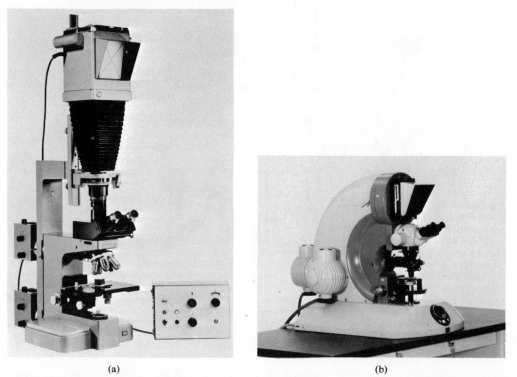

(a)   (b)

Fig. 2 Upright microscope. (a) Metallographic microscope and attachable 4 × 5 in. bellows camera with automatic exposure control (b) Metallographic microscope with built-in 4 × 5 in. camera with automatic exposure control. (Courtesy of Carl Zeiss, Inc.)

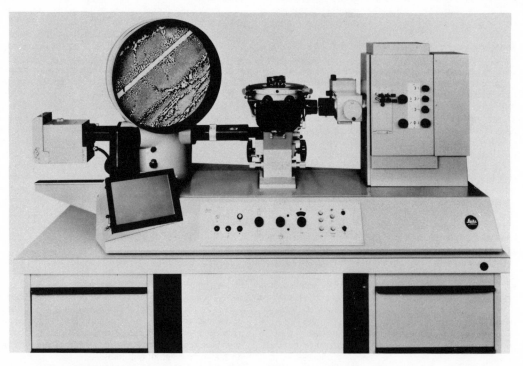

Fig. 3 Inverted Metallograph with high intensity light source, large projection screen, large format and 35 mm camera with automatic exposure control and zoom system.

underestimated. It is important that the lamp housing permit accurate centration of the arc lamp when it is in operation. Such centration should be possible with the lamp housing closed. The lamp housing should have an independently adjustable concave mirror mounted behind the light source in order to increase the illumination intensity. A properly designed adjustable lamp condenser as well as filter holders for heat absorption— , neutral density— , contrast filters and a diffusion disc should be part of the housing. It should permit the attachment of an incandescent lamp for visual observation. Quick change from the low-voltage to high-power source should be possible.

3. A Fresnell lens screen is superior to a ground glass for projection because of its high intensity and better image quality.

4. Parfocality between the image in the microscope tube, the projection screen, and the film plane of the photographic attachment, is of great convenience to the operator.

Illumination Techniques and Microscope Optics. Even with small microscope stands it is advantageous to have Köhler illumination. This gives the most homogeneous illumination of even large fields, as well as control of contrast and resolution in the image. With this technique the light source is projected into the rear focal plane of the objective so that parallel light rays reach the specimen. In order to eliminate stray light, a field diaphragm is imaged into the specimen plane to restrict the illumination to that area which is observed. A second diaphragm, the aperture diaphragm, is placed in a plane which is conjugated to the rear focal plane of the objective. It regulates the illumination aperture which effects resolution and contrast. (Fig. 4.)

Most of the metallographic illuminators allow use of oblique illumination by insertion of a half stop into the aperture diaphragm plane or by an arrangement which decenters the aperture diaphragm. Coarse surface topography shows up three-dimensionally as if shadow cast. In bright field illumination two thirds to the full aperture of the objective are generally used to give best image quality. In dark field illumination the central portion of the illuminating beam is blocked. The illuminating light bypasses the objective and is deflected towards the speci-

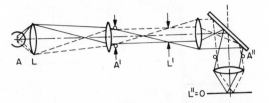

Fig. 4 Schematic of Köhler illumination in reflected light bright field. A, light source; A′, aperture; L′, field diaphragm; A″, focal plane of objective; L″, specimen plane.

men by a circular cone-shaped metal mirror. A small amount of light is scattered by very fine surface structures. The surface detail appears bright on a black background. With some systems the central stop can be removed to allow use of the same objective for bright field and dark field. (Fig. 5.)

Illumination with polarized light reveals the often weak birefringent nature of certain metals such as titanium, zirconium, and beryllium. For good contrast, it is essential that the objective be strain free. A filter or prism polarizer is part of the illuminator. A second polarizer— called analyzer—is mounted in the tube.

A plane glass beam splitter mounted in 45° position depolarizes linearly vibrating light coming from the polarizer. Different illumination arrangements have been designed which overcome depolarization. The most common types are:

1. Foster prism: This is a calcite prism which, in addition to the 90° deflection, acts as polarizer and analyzer. The illuminating beam is reflected by the metal-coated surface (A in Fig. 6/2) onto a prism interface. The angle of incidence is such that the ordinary ray is totally reflected toward the specimen. The extraordinary ray, however, is transmitted and eliminated at the blackened side (B in Fig. 6/2) of the prism. The ordinary ray, after reflection by the specimen surface, is again totally reflected at the interface and does not participate in the formation of the image. The extraordinary ray introduced by the specimen reaches the eye, the prism acting as a permanently crossed polarizer and analyzer. It is a shortcoming of this prism

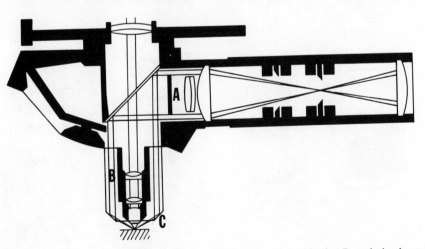

Fig. 5 Dark field illuminator. Note central stop A, illumination bypassing objective B, and circular cone-shaped reflector C.

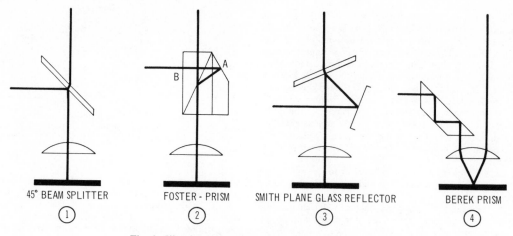

Fig. 6 Illumination arrangements for polarized light.

that analyzer and polarizer cannot be uncrossed. Especially with weakly birefringent metal surfaces, the best contrast is often obtained with an analyzer setting slightly off crossed position.

2. In the Smith illumination arrangement (Fig. 6/3), the illuminating beam first strikes a front surface mirror from which it is reflected upward onto a thin glass beam splitter which is tilted at 22° and reflects the beam toward the specimen, undergoing only negligible depolarization. The light is reflected back from the metal specimen surface and passes through the beam splitter. Therefore, with the Smith illumination arrangement, the extinction is hardly impaired and remains quite uniform over the field of view. However, in comparison to a 45° beam splitter, the intensity is reduced.

3. The Berek type prism covers half the illumination aperture; the other half is used for imaging. Therefore, the illumination is slightly oblique so that the light reflected by the specimen bypasses the prism. The image brightness is considerably higher than it is with plane glass reflectors. The Berek prism is made of glass with refractive index $n = \sqrt{3}$. The illumination beam undergoes three internal total reflections before being directed to-

ward the specimen. The linear polarization of the incident light bundle is ideally preserved.

Compensator plates such as first-order red, also called sensitive tint or gypsum plate, can be inserted behind the objective introducing a well-defined retardation which is superimposed over the one—usually small—introduced by the specimen. The first-order red permits the detection of very small birefringence which is translated into pronounced color change.

Other illumination techniques make finest details of the surface, differing in height or refractive index, visible by converting the phase shift which is produced by such details into an amplitude difference in the image.

Phase contrast is such an illumination technique. It requires a special set of objectives and an illuminator which contains annular diaphragms.

Differential interference contrast (systems according to Nomarski, Francon, Smith, etc.) translates the phase image into an image of real or apparent surface relief. Two laterally shifted images of the same specimen detail are brought to interference. The distance of shear is in the order of the resolution limit of the microscope objective. Edges, slopes, refractive index boundaries, appear as if

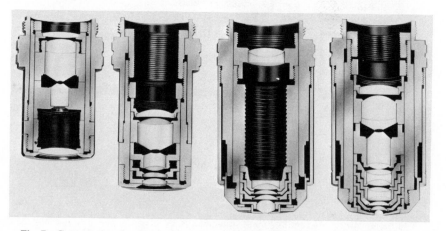

Fig. 7 Cross section showing a set of objectives corrected especially for flatness of field.

shadows had been cast in a specific direction. The interpretation of such differential interference contrast images is, however, very difficult. In spite of this, the system is a very powerful tool for qualitative studies of the finest polished surfaces.

Objectives for the metallographic microscope are designed to be used without cover glass. They span a magnification range from low-power 2.5 to 160× dry or oil immersion with maximum aperture of 1.4. The total microscopical magnification should stay within 500–1000 times the numerical aperture of the objective, which is sometimes surpassed for measuring purposes only. The extreme microscopical magnification in this case should be limited to about 2500×. The magnification values of objectives and eyepieces as well as the aperture are computed by the manufacturer, taking into account the standard magnifications for photomicrography. The ASTM standard magnifications are listed under I; the European standard, II, differs slightly.

I		II
		100×
250×	and multiples in	125×
	powers of ten	160×
500×		200×
		250×
750×		320×
		400×
1000×		500×
		630×
1500×		800×
ASTM		European
Standard		Standard

Metallographers prefer to use dry objectives because of the convenience of working without immersion oil. However, for highest resolution or with low-reflective material, oil immersion systems are used. A set of parfocal microscope lenses on a turret allows instant magnification change.

One lens shortcoming, curvature of field, is especially undesirable with flat, polished metallurgical specimens.

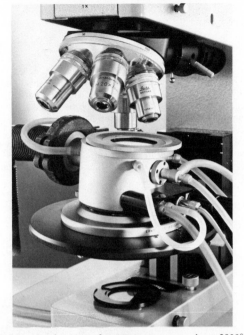

Fig. 8 Heating stage for temperatures to about 2800°C attached to a microscope. To the left, connection with a vacuum pump; to the right, inert gas and cooling tubing.

Objectives have, therefore, been designed with considerable expenditure, requiring in some cases up to 16 individual lens elements to achieve flat fields of view as large as 28 mm. (Fig. 7.)

The advantage of such objectives is that large areas of the specimen can readily be observed, and photomicrographs are obtained which are sharp to the corner.

The tube lens in a microscope corrects for the fact that the mechanical tube length for which the objective is designed is not necessarily the same as the one of the microscope. The tube lens can carry a magnification factor other than 1× which has to be taken into account

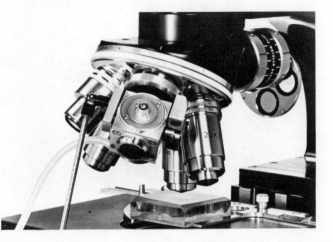

Fig. 9 Pneumatic hardness tester mounted to nosepiece turret. Diamond indentor (center of rings on cube) and parcentered measuring objective on swivel mount.

for the determination of the total microscopical magnification.

$$M_{total} = M_{obj.} \times M_{tubelens} \times M_{eyepiece}$$

For the exact determination of the magnification in the intermediary image, an eyepiece micrometer is introduced in this plane. A stage micrometer is used to calibrate the eyepiece micrometer. Such measuring eyepieces can be equipped with suitable reticles for lateral measurement, grain size determination and point count methods, etc. They should contain a focusable eyelens which insures that the reticle is seen in sharp focus.

Use of optical components of different manufacturers on one microscope is not recommended. The microscope optics of each manufacturer are computed as a unit. This means that an optical element of a second manufacturer may not necessarily carry the proper correction for the microscope of the first. As a result mixing of optics almost always leads to an image quality loss.

For heating stages, long working distance objectives are required, corrected for quartz or glass plates which seal the heating space.

Strain-free objectives are necessary for polarized light microscopy and differential interference contrast. Special objectives must be used for illumination techniques like phase contrast and dark field.

Accessories for the Metallographic Microscope. The

(a)

(b)

Fig. 10   (a) Remote controlled metallograph extending into "hot cell" for work on radioactive metals. (b) Side view.

metallographic microscope is very seldom used without photomicrographic equipment operated either manually or automatically. Attachable 35 mm or 4 × 5 in. cameras for plates, sheet film, or Polaroid convert any bench metallograph into a camera microscope. In more sophisticated metallographs, the camera is usually integrated.

The study of metals and other materials at elevated temperatures has vastly increased. Many microscope manufacturers now offer heating and cooling stages for various temperature ranges as attachments to microscopes. Temperatures up to about 1800°C are recorded with the help of thermocouples; above that pyrometric methods have to be used. It is essential that high-temperature heating stages be equipped for use with vacuum or inert gas atmosphere. (Fig. 8.)

In micro hardness testing, the hardness of individual metal grains, inclusions, and other microstructures is studied. For this purpose hardness test attachments are available for the metallographic microscope. The force of small weights is applied to a pyramid-shaped diamond, which is pressed into the metal. The diagonals of the indentation are measured and the hardness value of the test material deducted. Two different indenters are commonly used in micro hardness tests: The Vickers indenter has a square base; the Knoop type a rhombic one. The Knoop pyramid produces an indentation diagonal about three times larger, and a depth one-third smaller than the Vickers pyramid under identical test conditions. Because of the higher reading accuracy longer indentations are preferable with hard materials, shallower ones with thin layers of test material. After the indentation is performed a precentered objective is swung into position to measure the indentation diagonals. This has the advantage that the specimen does not have to be moved from the metallographic microscope to a special hardness tester. (Fig. 9.)

Besides the previously mentioned differential interference contrast systems, quantitative interference attachments for the measurement of surface elevations and depressions are increasingly utilized. Double-beam interferometers permit these measurements without contacting the specimen. The fringe pattern is visible in polychromatic light; however, the fringes are relatively wide even in monochromatic light. Multiple-beam interferometers have to be brought into contact with the specimen surface. They are used with highly monochromatic light only, and depending on surface reflectivity, produce very narrow fringes for high measuring accuracy. Most simply, these interference systems are attached to the metallographic microscope objectives or are interference eyepieces. There are also a number of single-purpose interference microscopes.

Special-Purpose Metallographic Microscopes. The investigation of radioactive metallurgical materials requires the special attention of microscope designers. With α-radioactive material, it is sufficient to place the entire microscope in a transparent plastic box, the hot box. The microscope controls are accessible by means of integral rubber gloves. The light source, as well as the viewing head, is usually placed outside the box. This requires optical extensions which have to be gas sealed. Neutron or γ-emitting materials are more dangerous.

Regular glass would become brown under their radiation and rendered useless. Therefore, special nonbrowning glasses had to be found and new lenses computed. In metallographs for this type of work, the motor-driven specimen stage and the interchangeable objectives are colated in the hot cell. The rest of the metallograph and the operator are outside, shielded by a concrete wall of considerable thickness. This means the optical image has to be relayed from inside the hot cell through sealed apertures, and the operation of the microscope is controlled remotely. Figure 10 shows such an instrument.

Automatic quantitative image analysis is a new field of increasing interest in the metals industry. Due to their fast speeds, television systems seem to have the commercial edge over flying spot scanners and photometric devices. In the last two years not fewer than four new instruments have appeared on the market, all of them consisting of a metallographic research microscope which projects an image onto a vidicon or plumbicon camera

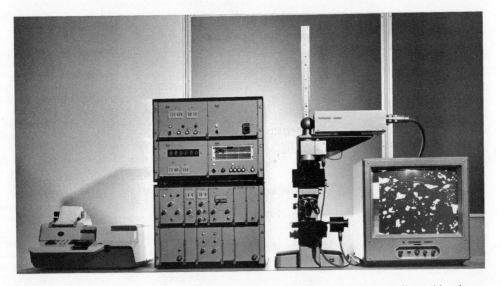

Fig. 11   TV scanning device for automatic quantitative image analysis connected to metallographic microscope.

tube. Here the image is converted in a fraction of a second to electronic signals which are analyzed and processed for specific image parameters. The more advanced systems incorporate a step scanning stage which moves the specimen in a programmed pattern from one measuring field to the next. With these instruments, the operator can preset his own measuring programs (sequence of data and function flow). On-line computing equipment is used to calculate a large number of related image parameters by point counting or lineal analysis methods if the TV system works on that principle. In this way percentage areas of metal phases, size and area distribution of nonmetallic inclusions (stringers), average grain size, and many more are accurately found over a large number of measuring fields, at high magnifications, in a very short time. For example, it takes the equipment shown in Fig. 11 anywhere from about 8 to 80 min to analyze 1000 measuring fields on a specimen, depending on the number of parameters asked from each field.

Conclusion. In spite of the importance of the mechanical parts of a metallographic microscope, its usefulness is determined primarily by the quality of the optics. Any flaw in optical design, or manufacture, will seriously impair the image quality. This will be especially apparent in photomicrography.

The metallographic microscope reaches even beyond the application for which it was designed. In many cases in research or production control, microstructures of opaque materials have to be studied, and the unmodified metallographic microscope can be utilized. A typical example is the semiconductor industry, where the metallographic microscope is used extensively for microscopic control of different production steps in integrated circuit manufacturing. This is just one example, which shows the importance of the metallographic microscope outside the metals field.

WERNER HUNN

References

1. R. C. Gifkins, "Optical Microscopy of Metals," New York, American Elsevier, 1970.
2. E. N. Cameron, "Ore Microscopy," New York, John Wiley & Sons, Inc., 1961.
3. H. Freund, "Handbuch der Mikroskopie in der Technik," Band 1, Teil 2 Auflicht-Mikroskopie, Frankfurt, Umschau Verlag, 1960.
4. M. Francon, "Progress in Microscopy," New York, Pergamon Press, 1961.
5. G. K. Kehl, "Principles of Metallographic Laboratory Practice," 3rd ed., New York, McGraw-Hill, 1949.
6. "Light Microscopy," Philadelphia, Amer. Soc. Test. Mat., 1953.
7. R. H. Greaves and H. Wrighton, "Practical Microscopical Metallography," 4th ed., London, Chapman and Hall, 1956.
8. B. Chalmers and A. S. Quarrel, "Physical Examination of Metals: Vol. 1 Optical Methods," London, Arnold, 1960.
9. W. Rostoker and J. R. Dvorak, "The Interpretation of Metallographic Structures," New York, Academic Press, 1956.
10. R. Smallman and K. H. G. Ashbee, "Modern Techniques in Metallography," Oxford, Pergamon, 1966.
11. H. C. Claussen, "Mikroskope, Handbuch der Physik," Band XXIX, Optische Instrumente, New York, Springer, 1967.
See also: OPTICAL MICROSCOPE, POLARIZATION MICROSCOPE.

MICROFOSSILS

Microfossils are an arbitrary grouping of small fossils that are most commonly studied with a microscope. They represent a vast array of mostly unrelated organisms ranging from bacteria through most of the plant and animal phyla including the tiny bones and teeth of rather large vertebrates. There is no natural grouping associated with microfossils or micropaleontology as there is, for example, with vertebrate paleontology or paleobotany, other than the limits of resolution of the unaided human eye. Microfossils are composed of vastly different materials and may range in size from less than one micron to several centimeters in greatest dimension (Table 1). This diversity of materials and size requires that many different preparatory and microscopical methods be used to study microfossils.

Preparation techniques vary depending on the nature and size of the microfossils, the kind of enclosing matrix, the objectives of the study, and the instruments with which the study will be made. Each situation is unique and best results often may be obtained by experimentation rather than by rigidly following a set of instructions. Nevertheless, there are techniques with which broad classes of microfossils may be processed that will generally produce satisfactory results. More detailed information is given in the references cited.

In general the techniques avoid mechanical treatments, such as crushing, grinding, and ultrasonic cleaning, because these methods damage specimens. They can always be used during the preparation process should difficult situations arise, but they should be reserved as a last resort in most cases. For rapid examination of numerous samples, as in certain stratigraphic studies where the detailed morphologic study of microfossils is not necessary, mechanical techniques may significantly speed the recovery.

Free Specimens

Disaggregation and Dispersion. These procedures are used to free calcium carbonate microfossils or others that might be damaged by acids. Initially the bulk samples should be thoroughly dried in an oven. Three procedures may be used. (1) The sample is covered with sodium pyrophosphate and dilute hydrogen peroxide (15%) which disaggregates the rock particles. The process may be speeded by heating. (2) The sample is covered with a solution containing a commercial surface active agent (particularly Quaternary O, manufactured by Geigy Industrial Chemical Corporation, Ardsley, New York) diluted according to the manufacturer's recommendations. The sample will break down at varying rates depending on the degree of its consolidation. (3) The sample is covered with kerosene or a similar solvent and allowed to soak for several hours or more depending on the rate at which the solvent penetrates the rock. After saturation, the solvent is poured off and the sample is covered with water, to which 50–100 g of sodium bicarbonate per quart of water have been added, and then boiled. If the sample does not break down, it is again dried and the process repeated until it is completely disaggregated.

Once the sample is disaggregated, the particles are wet sieved using a 200 or 250 mesh Tyler screen (openings 74 or 61 micra). Larger microfossils, such as foraminifera, ostracodes, conodonts, teeth, and otoliths, will be retained by the screen, while the smaller nannofossils will pass through it.

Table 1

Microfossil	Description	Size	Composition	Habitat	Geologic Range	Recovery Techniques
Bacteria	unicells, filaments	< 10 μ	organic	all	PreCambrian–Recent	acids, sectioning
Cyanophytes	unicells, filaments	< 50 μ	organic	all	PreCambrian–Recent	acids, sectioning
Acritarchs	cysts of various eucaryotic algae?	< 100 μ	cellulose	marine	PreCambrian–Recent	acids, sectioning
Chlorophytes	unicells	100–400 μ	silica, calcite, aragonite	all	PreCambrian–Recent	acids, sectioning
Charophytes	spheres, ovoids	.5–3 mm	calcite	aquatic	Silurian–Recent	disaggregation, sectioning
Euglenophytes	unicells	< 200 μ	organic	aquatic	Cretaceous, Eocene–Recent	acids, sectioning
Pyrrhophytes (Dino-flagellates)	cysts	5–150 μ	organic	aquatic	Silurian, Permian–Recent	acids
Ebriaceans	mesh of solid rods	< 100 μ	silica	marine	Paleocene–Recent	acids
Diatoms	circular, elongate, or irregular valves	< 200 μ	silica	aquatic	Jurassic?, Cretaceous–Recent	disaggregation, acids
Silicoflagellates	mesh of hollow rods	< 100 μ	silica	marine	Cretaceous–Recent	disaggregation, acids
Coccolitho-phorids (in-cluding discoasters, nannoconids, and similar forms)	plates, asters, rods	< 50 μ	calcite	marine	Triassic?, Jurassic–Recent	disaggregation, sectioning
Fungi	coccoid, filaments	< 10 μ	organic, calcite	all	PreCambrian–Recent	acids, sectioning
Gymnosperm pollen	unicells, masses	< 150 μ	cellulose	terrestrial-marine	Devonian–Recent	acids
Angiosperm pollen	unicells, masses	< 150 μ	cellulose	terrestrial-marine	Jurassic–Recent	acids
Spores	spheres	1–900 μ	cellulose	terrestrial-marine	Silurian–Recent	acids
Amoebas	cysts	< 150 μ	organic	parasitic	Quaternary	reconstitution of coprolites, saline solution
"Thecamo-bians"	single-chambered tests	10–500 μ	organic, agglutinated	mainly fresh water or terrestrial in damp places	Mississippian–Recent	disaggregation and dispersal
Foraminiferans	chambered tests	.01–100 mm	organic, agglutinated, CaCO₃	marine	Cambrian–Recent	disaggregation, acids, sections
Radiolarians	spherical or conical mesh-work	< 1000 μ	silica	marine	Cambrian–Recent	disaggregation, acids, sections
Tintinnids	single-chambered lorica	< 300 μ	organic, agglutinated, CaCO₃	marine	Ordovician–Recent	disaggregation, sections
Poriferans	spicules	.05–3 mm	siliceous, CaCO₃	fresh, marine	Cambrian–Recent	disaggregation, sections
Octocorallians	spicules	1–3 mm	CaCO₃	marine	Silurian?–Recent	disaggregation, sections
Acantho-cephalans	eggs, cysts	< 100 μ	organic	parasitic	Quaternary	reconstitution of caprolites, saline solution

Table 1 (continued)

Microfossil	Description	Size	Composition	Habitat	Geologic Range	Recovery Techniques
Nematodes	eggs, cysts	< 100 μ	organic	parasitic	Quaternary	reconstitution of caprolites, saline solution
Trematodes	eggs	< 100 μ	organic	parasitic	Quaternary	reconstitution of caprolites, saline solution
Cestodes	eggs	< 100 μ	organic	parasitic	Quaternary	reconstitution of caprolites, saline solution
Ostracods	bivalved carapace	.1–5 mm	$CaCO_3$	freshwater- marine	Cambrian– Recent	disaggregation and dispersal
Holothuroids	spicules	.05–1 mm	$CaCO_3$	marine	Ordovician– Recent	disaggregation and dispersal
Scolecodonts	tooth or toothed plate	.05–5 mm	chitin	marine	Cambrian– Recent	disaggregation, acid, sections
Conodonts	toothed or flattened plates (of agnathan fish?)	.1–3 mm	$CaPO_4$	marine	Ordovician– Recent	weak acids, disaggregation, and dispersal
Chitinozoa	flask shape	75–700 μ	chitin	marine	Ordovician – Devonian	acids

The smaller-sized fraction may be prepared for study by dispersing the material in water on a coverslip, and mounting with Canada balsam or Permount on a standard biological or petrographic slide. In cases where the clay and silt dominate over nannofossils, repeated short centrifugation may separate the fossils. In this process the sample is placed in a water solution of Calgon which disperses the fine material. The solution containing the sample is centrifuged at about 3500 rpm for 30–50 sec, and braked slowly. Experimentation with the rpm and time is necessary, for each type of sample varies considerably. Too much centrifugation results in the retention of undesired silt, clay, and organic particles, while too little results in loss of finer-sized nannofossils. After each centrifugation, the sample is decanted, distilled water is mixed with the centrifugate, and the process is repeated until the water is clear. Calcareous nannofossils will generally be concentrated in the lighter-colored material. This material is dispersed and mounted on a slide for optical view, or used for electron microscopy.

The larger-size fraction is dried, separated into different-size fractions with screens for convenience of study, and spread in trays (usually black in color with ruled guidelines on the bottom) for separation under a low-power dissecting microscope. The microfossils are removed by manually picking them out with a moistened fine brush (in 0–000 sizes) and mounted on cardboard microfaunal slides (scientific supply houses sell many varieties suitable for mounting individual specimens, several specimens, or entire assemblages).

Some samples may contain relatively few microfossils. In these cases the microfossils may be concentrated by separation in a magnetic separator (Frantz Isodynamic Magnetic Separator) which separates magnetic from nonmagnetic materials, including most microfossils. Experimentation with the settings for field strength, vibration, slope, and feeding rate is necessary, again because samples are variable in composition. Some microfossils (i.e., agglutinated foraminifera) may also be magnetic, and those residues always should be inspected for microfossils.

A second separation technique involves flotation of the microfossils in a heavy liquid. The dried sample is shaken into the liquid and the microfossils float while the heavier clastic particles sink. The liquid and light particles are decanted through filter paper, which retains the microfossils. Although carbon tetrachloride is commonly used for this purpose, bromoform provides a more efficient separation and is less toxic to humans. Neither of these separation techniques is completely successful. In all cases, both fractions should be examined to determine whether or not the separation is adequate.

Acid Macerations. For siliceous and organic fossils preserved in rocks of differing composition, specimens may be freed by dissolving or oxidizing the unwanted fraction in concentrated acids. About 5–10 g of disaggregated sample are placed in a beaker under a fume hood, and 40 ml HCl and 20 ml HNO_3 are added. After the initial reaction ceases, the mixture is heated until the brown fumes disappear; 40 ml concentrated H_2SO_4 may be added and boiled for more complete removal of organic material. In some cases, this could destroy desirable fossils; experimentation is necessary. The beaker is filled with water and allowed to stand. Repeated decantation will remove the acid. Slides may be prepared for siliceous microfossils. An additional boiling in 40 ml concentrated HF will remove the silica and clay leaving only resistant organic microfossils. The sample is rinsed with water until neutral. The organic fossils may be stained for better viewing optically with Safranin O or methyl green. Single-grain mounts may be prepared by micropipette. Strewn slides are prepared by dispersing a water-specimen mixture on a coverslip, drying, and mounting with Canada balsam or Permount. Radiolaria in some cherts may be freed by immersion of sample fragments in 10% HF.

Embedded Specimens

Thin Sections. Microfossils too commonly occur in indurated rocks of the same chemical composition as themselves; for example, calcareous foraminifera and tintinnids in limestone, or radiolarians in chert. Such fossils cannot be freed except by laborious, if not impossible, mechanical grinding, chipping, or scraping with sharpened probes and needles. In most cases it is simpler and more practical to make petrographic thin sections.

As thin sections are simply narrow slices through a rock, care should be taken in selecting the plane of slicing. The rock sample should be inspected to determine whether or not particular parts of the rock contain more fossils than others, or whether or not there is a common orientation of the fossils. From these observations, sections can be selected that best reveal the detailed morphology of the specimens. In cases where determinations such as these cannot be made, several sections at different angles through the rock should reveal the fossils. Ordinarily the section should be as thick as possible so that many complete secimens will be included, but this depends in part on the transparency of the rock, the kinds of fossils, and their composition.

The rock is first cut to a thin slab. One side is carefully polished on a lapidary wheel or ground glass plate with fine abrasive, and mounted polished-side down on a petrographic (or biologic) glass slide with a thermoplastic. The other side is then ground to the desired thickness, polished, and covered with a coverslip attached with thermoplastic. The slide can then be viewed with an ordinary light or petrographic microscope.

An alternative method is to make acetate peels from any polished section or from a fracture surface. After polishing, the surface is etched in suitable acid (10% HCl or EDTA for limestones, 10% HF for chert) for about 30 sec, usually long enough to produce suitable microrelief. A few drops of replicating fluid are placed on the surface, to which a piece of acetate replicating tape is then pressed (materials available from electron microscopy supply houses). After 10 min the peel is stripped off. The process should be repeated several times because loose particles often are retained on the initial peels. The final peels may be viewed directly in optical microscopes or carbon-metal shadowed for transmission electron microscopy (see below).

Oriented Thin Sections. The internal ultrastructure of microfossils usually requires making oriented thin sections of free specimens. The specimens should be picked from the sample and mounted with the desired plane of section parallel to the surface of a clean glass coverslip. The specimen(s) is covered with epoxy, cured, and ground nearly to the desired plane of section. Should air remain inside the specimen, some epoxy can be placed in the holes at this time. The specimen is carefully ground and polished at the section plane. The specimen and glass coverslip are mounted with thermoplastic on a slide, and ground through the coverslip and the specimen until the desired details can be observed. A drop of water on the ground surface will permit viewing the specimen as grinding proceeds. For detailed, fragile work grinding may be done with a ground glass slide under a binocular microscope. When satisfactory results are visible, the section is polished and covered with a coverslip. For certain optical

observations and transmission electron microscopy, acetate peels can be made at any stage in the process.

Instrumental Techniques

Optical Microscopy. The larger microfossils are best studied with low-power binocular dissecting microscopes. Thin sections and most calcareous nannofossils should be studied with ordinary and polarized light. Phase and interference contrast reveal much detail of nannofossils not otherwise observable with optical microscopes. Likewise, infrared study often reveals details of some organic microfossils.

Transmission Electron Microscopy (TEM). Images are formed by passing an electron beam through a specimen to a phosphor screen or photographic plate. Because even the smallest microfossils are too thick for the electron beam to penetrate, TEM requires special shadowing or thinning techniques.

Nannofossils freed from the rock and cleaned by centrifugation are sprayed with a spray mounter onto a freshly cleaved piece of mica. The distance of spraying and amount of material are best determined by experimentation, as samples and the mounters are variable. The mica, usually attached by tape to a slide, is then shadowed simultaneously or sequentially with platinum or platinum-palladium (use about 2–4 cm metal wire) and carbon in a vacuum evaporator. The metal should be evaporated first at an angle of 30–45°, followed by carbon from directly above the specimens. After the evaporation process, the mica is dipped slowly at a very low angle into a 10% HCl solution. The liquid's surface tension will strip the shadowed film from the mica and at the same time the HCl will dissolve the $CaCO_3$. For siliceous fossils or samples with much extraneous siliceous or clay material, the film can be transferred to dilute HF. The acid order should not be reversed, for HF converts $CaCO_3$ to CaF (fluorite), insoluble in HCl. The film is next transferred through several distilled water baths. Electron microscope grids are held with tweezers and lifted carefully through the floating film, thus emplacing a piece of film on the grid. After drying, the grid may be stored or viewed on the TEM.

Acetate peels, prepared as described above, allow TEM of larger microfossil ultrastructure. The peel is placed in a vacuum evaporator, replica side up, and shadowed with metal-carbon. A selected piece of the peel is cut to fit on a TEM grid. Separation of the peel from the film can be done by condensing acetone vapor onto the peel in a reflex unit, obtainable from electron microscopy supply houses. After the acetate is completely removed the grid may be stored or viewed.

TEM is possible through very thin fossils. This requires thinning the specimens in an ion beam thinning machine (Ion Micro Milling Instrument, Commonwealth Scientific Corp.), a technique currently under development. In this method microfossils are embedded in epoxy and thin sectioned by grinding. The specimen-epoxy wafer is removed from the slide and placed in the receptacle in the ion beam thinning machine. In this procedure, the wafer is inclined at a low angle between two argon ion beams and rotated in its own plane. The beams erode the specimen molecule by molecule. When a small hole has been made in the wafer, it is transferred to a TEM grid. Viewing may be done with a high-voltage TEM, and is possible only in the vicinity of the hole made by the thinning machine.

Scanning Electron Microscopy (SEM). The SEM is

probably the single most important advance in modern micropaleontology. It permits high-resolution, high-magnification three-dimensional micrographs of specimens and details impossible to view otherwise. The instrument focuses a scanning electron beam on the specimen, producing the emission of low-energy secondary electrons. These, picked up by a detector, are displayed on a cathode ray tube for viewing or photographing.

Preparation techniques are simple: Mount a specimen or spray material on a SEM specimen plug, coat it with a thin layer ($\simeq 30$ Å) of gold, aluminum, platinum-palladium, gold-palladium, or carbon in a vacuum evaporator, place in the instrument, and view. One difficulty comes in choosing a suitable adhesive to mount specimens to the plugs. Double-stick tape or gum tragacanth or other glue are satisfactory, although the tape may crack in vacuum and the glue may accidentally cover parts of the specimen. If mounting with glue directly on the plug, the final pictures are less distracting if the ridges and grooves normally present on plugs are polished off. Some specimens, especially particulate ones like agglutinated foraminifera, tend to have charging effects because of poor electron conductivity. Commonly a thicker or multiple coating of metal diminishes this effect.

X-ray Microscopy. Three methods of X-ray microscopy of microfossils are possible—reflection X-ray microscopy (RXM), contact microradiography (CM), and point-projection X-ray microscopy or projection microradiography (PXM). RXM has not proven very useful in micropaleontology. CM and PXM hold considerable promise for future studies.

In CM, the specimen is placed very close to a maximum-resolution photographic plate in a special brass camera which lies in the path of an X-ray beam. X-ray diffraction equipment may be used. The exposed plate is developed, and examined microscopically or enlarged.

In PXM, an electron beam is focused on a metal target from which X-rays are emitted. The X-rays penetrate the specimen, thus producing an enlarged image on a phosphor screen or photographic plate. Advantages are depth of focus, resolution, contrast, and visibility of internal structures. Specimens are glued on a formvar film (2 μ thick) or on tape and placed in the specimen holder. PXM may be performed on specially converted transmission electron microscopes.

Electron-Probe Microanalyzer (EPM). The EPM focuses an electron beam to a spot about 1 μ in diameter on a target specimen. X-rays characteristic of the elements present are emitted and monitored by crystal spectrometers. The chemical composition of microfossils can thus be determined without the problems of contamination inherent in individual or groups of small specimens. Most specimens must be mounted with epoxy, sectioned, and polished. In some cases, semiquantitative results can be obtained on very thin specimens without sectioning.

JERE H. LIPPS

References

Bé, A. W. H., W. L. Jongebloed, and A. McIntyre, "X-ray microscopy of recent planktonic foraminifera," *J. Paleontol.*, **43**:1384–1396 (1969).

Camp, C. L., and G. D. Hanna, "Methods in Paleontology," Berkeley, Univ. California Press, 1937.

Gibson, T. G., and W. M. Walker, "Flotation methods for obtaining foraminifera from sediment samples," *J. Paleontol.*, **41**:1294–1297 (1967).

Hay, W. W., and P. A. Sandberg, "The scanning electron microscope, a major break-through for micropaleontology," *Micropaleontology*, **13**:407–418 (1967).

Hooper, K., "Electron-probe X-ray microanalysis of foraminifera: an exploratory study," *J. Paleontol.*, **38**:1082–1092 (1964).

Kummel, B., and D. Raup, ed., "Handbook of Paleontological Techniques," San Francisco, W. H. Freeman and Co., 1965.

Lipps, J. H., and P. H. Ribbe, "Electron-probe microanalysis of planktonic foraminifera," *J. Paleontol.*, **41**:492–496 (1967).

McLean, J. D., Jr., "Manual of Micropaleontological Techniques," Alexandria, Va., McLean Paleontol. Lab., 1959–date.

Pessagno, E. A., Jr., and R. L. Newport, "A technique for extracting Radiolaria from radiolarian cherts," *Micropaleontology*, **18**:231–234 (1972).

See also: ROCK SECTION GRINDING AND POLISHING.

MICROINCINERATION

Introduction

Microincineration is a histochemical technique in which the mineral components of biological materials are freed of organic matter so that their distribution can be examined microscopically. The technique has been known for more than a century, but Policard (1923a,b) was the first to appreciate and develop its possibilities.

Spodography, the study of the localization of mineral in biological materials after microincineration, has an extensive literature (Hintzsche, 1956; Kruszynski, 1966) and the method has recently been extended to the level of the electron microscope (Boothroyd, 1968; Thomas, 1969).

I. Specimen Preparation

Ideally, specimen preservation with neither morphological nor chemical change is required. Since this is not possible, a technique should be chosen which best suits the circumstances of the investigation.

a. Unfixed Tissue. Whole mounts of viruses, bacteria or isolated cells, thin smears and spread membranes may be used. For high resolution, tissue more than about 20 μm thick should be sectioned before incineration. In frozen sections, the processes of freezing and thawing generally produce shifting of cell organelles and chemical substances. Recent methods of microtomy at very low temperatures (about $-150°$C) followed by freeze-drying are less damaging than the more common freeze-sectioning techniques. However, pieces of tissue greater than about 2 mm³ suffer badly from ice-crystal growth damage during cooling.

Freezing techniques retain virtually all minerals, but good cytological preservation is difficult to maintain.

b. Fixed Tissue. Fixation usually improves the morphological preservation, but aqueous fixatives cannot preserve soluble minerals. Formaldehyde or osmium tetroxide vapor improves the preservation of thin smears, isolated cells and other thin whole mounts. Aqueous fixatives must be buffered by volatile salts. Large specimens may be fixed with 4% formaldehyde buffered with ammonium acetate or s-collidine for light microscope spodography. Similarly, 2—5% glutaraldehyde, or a mixture of these, e.g., 2.5% glutaraldehyde with 2% formaldehyde in a 0.1 M buffer, gives good cytological preservation both for light and electron microscopic spodography of the less soluble minerals. Buffered osmium tetroxide gives excellent cytological preservation, but reduced osmium may remain to contaminate the ashed residue.

For light microscopic spodography of the more soluble minerals, a mixture of absolute ethanol and formalin (9 : 1) (Scott, 1933) has been widely used. Carnoy's fixative is also useful.

c. Microtomy. Fixed tissues sectioned on the freezing microtome show better cytological preservation when compared with unfixed tissue due to the increased stability of tissue components, but usually such material is embedded in waxes, epoxy resins or plastics. Dehydration into the alcohol or solvent phase should be done quickly to avoid further mineral loss. Paraffin wax was the conventional embedding medium. Polyester and ester waxes have also been used. Epoxy resins are excellent embedding compounds (Kruszynski and Boothroyd, 1967) and show virtually no shrinkage during polymerization. Thin (0.3–1.0 μm) serial sections can easily be cut so that adjacent sections may be used, one for microincineration and the other for histological comparison.

d. Section Mounting. Wax sections may be mounted on mica or hard glass slides. Avoid water for section flattening. Use liquid paraffin or 50–80 % ethanol solutions instead. Mineral-free albumen may be used as an adhesive, if necessary.

Epoxy-resin embedded sections cut onto water or aqueous acetone should be transferred quickly to a slide or coverslip. These sections may be examined and photographed by phase contrast microscopy (Kruszynski, 1966) before incineration. For electron microscopy sections are mounted on uncoated nickel grids and supported by a silicon monoxide film (Boothroyd, 1968). Specimen localization after incineration is difficult because little ash remains and special mapping procedures are needed (Boothroyd, 1968; Thomas, 1969). More ash is present in sections for light microscopy so location of the residue is easier.

II. Incinerators and Incineration

a. The Furnace Required for incineration is not complex. Basically a quartz tube large enough to contain the specimen support is wrapped with resistance wire and covered with asbestos to promote heating efficiency. Mains electricity supplied via a variable transformer provides controlled power and a simple iron/constantan thermocouple directly connected to a sensitive millivolt-meter is sufficiently accurate as a temperature gauge.

Thomas (1969) ashed specimens for electron microscope spodography in atomic oxygen at low temperatures (less than 100°C). Volatile salts are not lost with this system which works well on very thin specimens.

b. Incineration for Light Microscopy. The first rise of temperature to 80°C causes most damage in whole mounts and wax-embedded specimens. Shrinkage and distortion often occur quite unpredictably. Epoxy-embedded specimens are much more stable. This initial temperature rise should take 10–15 min. From 80°C to 600°C, 20–30 min should be allowed. For very thin sections, 2 μm or less, a temperature of 550°C is sufficient, particularly with low levels of phosphate or potassium.

Slow cooling, to avoid specimen shrinkage or cracking, is required, e.g., one hour from 600°C to room temperature. The ash requires some form of protection to avoid spontaneous hydration and carbonation; usually a coverslip may be sealed over the warm preparation with wax or shellac cement.

Qualitative analysis based on solubility, ash color, crystal formation and spot reagent tests is possible with sufficient ash. Topographic detail, however, may easily be lost.

c. Specimen Examination. Usually the spodogram is not visible by transmitted light microscopy, and dark field, incident light or phase contrast methods are necessary. Incident light illumination gives good results and color differentiation is good, but Kruszynski (1966) favored phase contrast microscopy. Very small traces of ash show by this technique although the halo effect may apparently increase the size of residues. With dark field illumination fine deposits are often obscured by the brighter dense particles.

A photographic record should be taken at the earliest opportunity since preparations lose sharpness by hydration.

d. Incineration for Electron Microscopy. Sectioned specimens may be 100–300 nm thick. The latter show poorer resolution while the former are difficult to observe and record. Compared with classical incineration less care is required in heating. Boothroyd (1968) heated to 550°C in 5 min and maintained this for a further 15 min or so. The heater was then switched off and allowed to cool. The mounted specimens were immediately examined by electron microscopy. Examination prior to incineration using marker support grids enables the same area to be examined before and after ashing. The ash is often invisible and particles in the area of choice are used as focusing aids. High-contrast photography is required to obtain a usable image. A light shadowing with chromium or uranium may improve the visibility of ashed specimens, but this is not always effective (Boothroyd, 1968).

III. Future Developments

Electron microscopic microincineration of normal biological material is hindered by lack of contrast in the spodogram. Where abnormally high mineral deposition occurs it is easy to localize and record. This technique has an inherent advantage over classic microincineration because of the available higher resolution. With improvements in specimen preparation and the use of additional techniques such as electron probe microanalysis it should become possible to obtain a complete and accurate analysis of ash deposits down to the limit of the microanalyzer.

B. BOOTHROYD

References

Boothroyd, B., "The adaptation of the technique of microincineration to electron microscopy," *J. Roy. Microscop. Soc.*, **88**:529–544 (1968).

Hintzsche, E., Das Aschenbild Tierischer Gewebe und Organe. Berlin, Springer, 1956 (pp. 1–140).

Kruszynski, J., pp. 96–187 in "Handbuch der Histochemie," Vol. 1, pt. 2, W. Graumann and K. Neumann, eds., Stuttgart, Gustav Fischer Verlag, 1966.

Kruszynski, J., and B. Boothroyd, "A cytochemical study of mitochondria, Golgi structure and other cellular components by microincineration," *Acta Anat.*, **68**: 400–411 (1967).

Policard, A., "La minéralisation des coupes histologiques par calcination et son intérêt comme méthode histochimique générale," *C.R. Acad. Sci.*, **176**:1012–1014 (1923a).

Policard, A., "Sur un méthode de microincineration applicable aux recherches histochimiques," *Bull. Soc. Chim.*, **IV**(33):1551–1558 (1923b).

Scott, G. H., "The localisation of mineral salts in cells of some mammalian tissues by microincineration," *Amer. J. Anat.,* **53**:243–288 (1933).

Thomas, R. S., "Microincineration techniques for electron-microscope localisation of biological minerals," pp. 99–154 in "Advances in Optical and Electron Microscopy," Vol. 3, R. Barer and V. E. Cosslett, eds., New York, Academic Press, 1969.

MICROMANIPULATION

Micromanipulation, microsurgery, or micrurgy (*mikros,* minute; *ourgos,* working) deals with the instrumentation, techniques and applications of microdissection, microvivisection, microisolation, and microinjection. Many mechanical operations can be performed under high magnifications by micromanipulation when properly instrumented. Applications of microsurgery are many, especially in cancer research, embryology, cell physiology, cytogenetics, somatic cell hybridization, cytochemistry, microchemistry, biophysics, microelectrophysiology, colloid technology, subminiature assembly, microcircuitry, and metallurgy.

The applications of microsurgery to the study of living cells include: single cell isolation, especially bacteria, yeast, spores, and ascites tumor cells; microdissection of virus inclusion bodies in plant and animal cells; microdissection of the neuromotor system in ciliated protozoa; measuring adhesiveness of cells; enucleating cells; stretching and cutting chromosomes; dissecting polytene chromosomes; production of translocations in chromosomes; microinjecting pH or redox indicators, salt solutions, enzymes, drugs, and oils into cells; bioelectrical studies involving action and membrane potentials with microelectrodes; electrical resistance of protoplasm; physical properties of extraneous coats; mechanisms or cell division; cytochemistry with the isolation of subcellular structures; and transplantation of subcellular structures.

Subcellular components that have been successfully transplanted from cell to cell include: micronuclei of ciliated protozoa; normal and irradiated cytoplasm; nuclei of amebas; nuclei of differentiated frog cells into the cytoplasm of unfertilized eggs; inclusion bodies of viral origin; and normal or malignant nucleoli usually with attached chromosomes.

Most recent applications of subcellular transplantation techniques include: transplantation of isolated and washed mitochondria (*Neurospora*) into the hyphae of *Neurospora* with indications that the transplanted mitochondria (genetically marked) can reproduce in other strains of *Neurospora*. Other exciting applications involve the reassembly of amebas by both nuclear and cytoplasmic transplantation. The zoochlorellae from *Paramecium bursaria* have been successfully transplanted into the cytoplasm of *Amoeba proteus* and *A. dubia.* When these transplantations are properly done, the zoochlorellae will survive and reproduce in the ameba cytoplasm. Chromosomes, including Barr bodies (inactive × chromosome), have been transplanted into mouse cells. Openings through the zona pellucida of mouse ova can be made by the local application of the polyvalent proteolytic enzyme, pronase, with a micropipette.

Nonbiological applications include: measuring consistency and elasticity of colloidal systems; demonstration of elasticity of metallic crystals, whiskers, and synthetic or natural fibers; manufacture, assembly, and testing of fine mechanical devices and electronic components such as transistors and integrated microcircuits; orienting fine particles and preparing specimens for electron microscopy; isolating and determining chemical properties or rare elements such as plutonium.

The basic instruments required for microsurgery are: (1) micromanipulators or micropositions which convert the crude hand movements to delicate fine motion of the microtools; (2) microtools, such as needles, pipettes, hooks, loops, electrodes, scalpels or forceps; (3) microinjectors, for controlling small volumes of fluids; and (4) microscope and accessories for recording optical data. Currently, records from the microscope can be made rapidly by utilizing video techniques, including video tape recording.

Functionally, the micromanipulator is a micropositioner with which one can place the tip of a microneedle or micropipette into any position (three axes of space, namely, X, Y, and Z) covered by the optical chain. At any instant, the tip of a microneedle or micropipette (or any other microtool) must be either where it is needed or out of the way. Over 200 different types of micromanipulators have been described since Schmidt (1859) published a description of his "microscopic dissector." Of these, perhaps 20 have reached the stage of commercial production. These instruments range from simple rack-and-pinion assemblies to massive, accurately fitted ball-bearing slides actuated by precision feed screws or other driving mechanisms. Many commendable adaptations of this principle have been engineered in the excellent fine adjustments incorporated into modern research microscopes. Through the inclusion of spring-loaded ball-bearing slides and various actuating mechanisms, the movements are smooth and without backlash.

Although precision positioning of microtools can be accomplished with feed screw type micromanipulators, there is a serious drawback in such instruments. They are inconvenient to operate, especially if the microtool needs to be moved diagonally. Undoubtedly, this handicap was responsible for the development of several varieties of lever or joystick actuated micromanipulators. With these instruments, the tip of a microtool can be positioned fairly rapidly—a great advantage if the object is motile, such as ciliates. Movements may be transmitted from the control lever to the slides by pneumatic or hydraulic means; in others, the movements are transmitted directly through mechanical couplings. One lever-controlled micromanipulator generates motion for the microtools through the expansion of spring-loaded, electrically heated wires.

The new Leitz (Wetzlar) micromanipulator, one of the better commercially available instruments, consists of three massive ball-bearing slides mounted at right angles to each other to produce movements in three directions of space. A joystick controls the two horizontal fine motions. Vertical movement is provided by a coarse and fine adjustment operated by turning two coaxial knobs. There are also two horizontal controls for the preliminary positioning of the microtools. Movement is transmitted from the joystick to the two horizontal slides by a ball-segment. By raising or lowering the eccentrically mounted ball-segment, the ratio between lever movement and needle motion is continuiusly varied from 16 : 1 (coarse) to 800 : 1 (fine). The gigantic Kulicke and Soffa Model 411-VXYZ micropositioner provides an excellent facility for very fine positioning, achieved by the chessman-joystick mechanically coupled to ball-bearing slides.

A new instrument has been built that combines the precision of feed screws with the convenience of joystick controls. This can be achieved by actuating the feed screw drives by servo motor mechanisms. Such drives can be operated by push button-controls. Furthermore, through added electronic feedback devices, such micropositioners can be programmed for performing certain operations automatically.

More recent advances in microsurgical instrumentation include the construction of micropositioners actuated by step-motor driven hydraulic feeds. The X-Y-Z positions are controlled by an assembly of 12 push buttons for each micropositioner to provide single-step (approximately 1 μ increments) or multi-step (variable speed) in any direction in space. Although each step drives the micropositioner approximately 1 μ, it does so at the rate of approximately 800 μ/sec and thus provides a rapid thrust motion.

Beaudouin (Paris) is now manufacturing a new version (designed by M. Bessis) of the original de Fonbrune micromanipulator. The micropositioners (model C) have been miniaturized so that as many as seven can be mounted on a special stage which can be added to any microscope. Each micropositioner is joystick-controlled and pneumatically coupled. Leverage ratios (up to 2500 : 1) may be continuously altered so that incredibly fine movements can be obtained. A miniaturized, electromagnetically actuated "sub-micron positioner" was developed. This instrument also can be directly mounted on the microscope stage.

Each microsurgical problem generally requires special instrumentation. Obviously, the construction or testing of a transistor or integrated microcircuit presents entirely different requirements from those needed to transplant a nucleus, nucleolus, or even a chromosome. The transplantation of a nucleus from one ameba to another requires one approach. The transplantation of a frog embryonic nuclei into an unfertilized frog egg needs an entirely different procedure. Microsurgical investigations of cells in tissue culture present strikingly different problems from those encountered with sea urchin eggs.

Many microsurgical procedures can be satisfactorily performed with a single micropositioner. On the other hand, subcellular transplantations can best be performed with at least four micropositioners. The principle here is one of redundancy—or back-up systems—so that if one micropipette becomes plugged, there is another one immediately available to take its place. In the Robert Chambers Laboratory for Microsurgery at New York University, there are now eight micromanipulators each with from four to eight micropositioners. All other micromanipulators have at least two micropositioners. (See Fig. 1.)

The microtools, such as needles, hooks, loops, and pipettes, can be fabricated by hand using a simple gas microburner and glass rods or tubing. Glass is the best substance for making microtools since this is the only material that consistently has ample rigidity along with flexibility even when reduced to micro or submicro dimensions. Several mechanical devices, usually with electric heating elements, are available for pulling microneedles and micropipettes. Micropipettes with special tips can be constructed in the microforge. Microforges are optical electromechanical devices for controlling the position of needles or pipettes relative to an electric filament in the field of a low-power microscope. Several versions of the microforge are now available. With these

devices, where the heating element is generally a V-shaped platinum wire, microhooks and microloops can be made; micropipettes can be bent near the tip and their minute openings can be beautifully fire-polished. The tip of a microneedle may be submicroscopic in size (<0.2 μ). Microneedles can be made with functional tips less than 0.5 μ in diameter. Microelectrodes with tips of the order of 1 μ are routinely made and used in biophysical laboratories.

The glass microtools are mounted in metal holders which, in turn, are held by clamps on the micropositioners. The better instruments provide controllable coarse adjustments for the preliminary centering and orientation of the microtools in the field of the microscope.

Any conventional microscope equipped with a good mechanical stage and a long working distance substage condenser will suffice for most microsurgical studies. Ordinarily, biological specimens, such as living cells suspended in a fluid medium, are mounted on cover glasses and supported by a box-shaped moist chamber. The moist chamber with its moist filter paper lining which prevents excessive evaporation of the preparation is positioned by the mechanical stage. With conventional microscopes, the cells are suspended in a hanging drop from the bottom surface of a cover glass. With inverted microscopes, however, lying drops are used so that the cells can rest on the upper surface of the cover glass. Microscopes which achieve focusing by moving the stage up or down are generally undesirable for microsurgery unless the micropositioners are mounted directly onto the stage of the microscope. This can be done, for example, with the Beaudouin Model C receivers.

The basic microsurgical procedures include: setting up micromanipulator, microscope, and light source; preparation of moist chamber; centering of microtools in optical field; proper vertical placement of microtools in relation to depth of hanging drop or height of lying drop; preparation and placement of cells or other material on cover glasses and subsequent mounting on the moist chamber; practice in manipulating needles or pipettes in the field of the microscope by familiarizing oneself with all the adjustments provided by the micromanipulators; practice in manipulating the microinjectors and skill in making microtools.

Many parts of a cell such as nucleoli, chromosomes, and mitochondria can be transplanted into another cell with a micropipette by microinjection techniques. The removal of a nucleolus from a nucleus and its insertion into the cytoplasm of the same or a different cell presents special and rather difficult technical problems. Specifically, the orifice of the micropipette must be correctly positioned towards the nucleolus. Then, suction must be carefully applied to start pulling the nucleolus, usually with its chromosome(s) into the orifice of the micropipette. At the right instant, the suction must be increased vigorously in order to dislodge and to extract the nucleolus from its site while avoiding the entry of extraneous material into the micropipette.

Such microsurgery requires exceptionally fine microsurgical and microinjection equipment and skill. The proper positioning of the tip of a micropipette towards the nucleolus with minimal injury to the cell can be done only with superb micropositioners. The transplantation procedures and facilities should be designed so that the entire operation can be performed in a matter of seconds; otherwise there is the danger of overexposing the nucleolus to the environment and thus damaging it. Siliconized

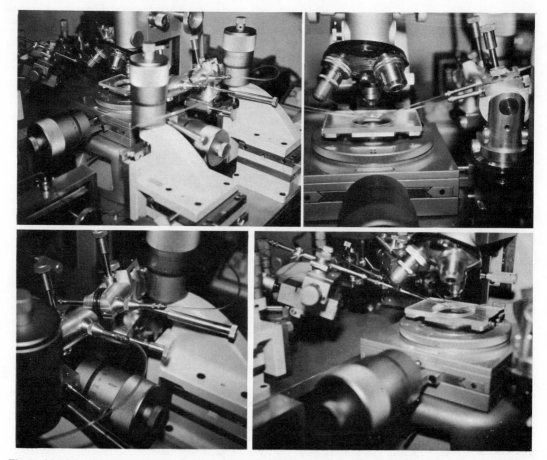

Fig. 1 Multiple micromanipulator consisting of six micropositioners surrounding a toolmaker's microscope, especially designed for transplanting embryonic nuclei into unfertilized frog egg cytoplasm. Upper right and lower left pictures show a special yaw-pitch-roll movement mounted on a large three-way micropositioner with ball-bearing slides actuated by micrometer screws with large micrometer heads.

micropipettes also minimize damage to subcellular structures.

Spherical cells may be conveniently and harmlessly held during removal and implantation of a nucleolus or other subcellular structures by a modification of the microelastimeter. The "cell-holding micropipettes" are made from thick-walled tubing. These micropipettes may have an orifice diameter of 10–15 μ, approaching one-fourth to one-third the diameter of the cell to be held. The tips of such micropipettes are fire-polished to avoid tearing the cell surface. The cell is picked up with the cell-holding micropipette by bringing its tip against the surface of the cell and applying sufficient negative pressure with the microinjector to aspirate a small segment of the cell into the micropipette. Such cells are moved into position with the micromanipulator. The "transplanting micropipette" is made from thin-walled tubing. Its orifice generally has a diameter of approximately 2 μ.

Transplantation of subcellular structures by micro-injection techniques can be executed properly only if volume displacement in the microinjector is under control. Heretofore, such volumetric controls were provided by the fine piston (0.005 in. in diameter) actuated by complex,

carefully fitted, worm gear–feed screw mechanisms. A new method of controlling minute volumes was developed by employing the principle of the differential piston. (See Fig. 2.) Two pistons, about 1 mm in diameter, with one piston slightly larger than the other, are mounted on a common barrier bracket. Each piston enters the same volume chamber but from opposite ends. As one piston is moved into the chamber, the other piston moves out. Accordingly, the volume displaced per unit length of travel is the difference in volume displaced by each of the two pistons. The piston actuator is a slightly modified coarse-line focusing mechanism as used on microscopes. This unit can be driven by a servo or step motor, thereby establishing the basis for a programmed or semiautomatic microinjector, so that the different rates of aspiration needed for removing nucleoli from a nucleus or other subcellular structures from the cytoplasm can be automatically programmed.

In the Robert Chambers Laboratory for Microsurgery, four Leitz micromanipulators, each equipped with a microinjector, are generally used for transplanting nucleoli. Each one carries a micropipette, a micropipette holder, and a micropositioner clamp. Two microinjectors,

Fig. 2   Microinjectors. (1) Microinjector consisting of a micrometer-driven syringe coupled with a micrometer that drives a small piston. The differential piston microinjector (2) is shown with the dual pistons and chamber assembly (3) and coarse volume control (4) consisting of a steel volume chamber and a micrometer. The two pistons (3) are moved by manipulating the coarse-fine adjustment as used in microscopes. The new Leitz micromanipulator is shown by (5).

mounted behind the micromanipulators, grouped around a microscope, are connected to the micropipette holders with Kel-F or Teflon tubing. These microinjectors consist of glass syringes, the spring-loaded pistons being moved by micrometer screws. A modified micrometer with a small piston provides the necessary fine volume control. The micropipettes are large, with fire-polished orifices, and are used for holding the cells. In front of the micromanipulators are mounted the differential piston microinjectors. These are also connected to the micropipette holders with Kel-F or Teflon tubing and are used for transplanting nucleoli or other subcellular structures. These microinjectors include a large piston moved by a micrometer screw for coarse volume adjustments in addition to the differential piston capability.

A new procedure was designed to permit the transplantation of a nucleolus (or chromosome) into the nucleus. Generally, a nucleus is damaged, sometimes irreversibly, if punctured. To avoid this possibility, the transplantation is done when the nuclear membrane does not exist. A nucleolus or chromosome is placed behind one of two anaphase sets of chromosomes in a cell undergoing mitosis. Both donor and host cells are held by large micropipettes which function as microsuction cups and are moved into place with micropositioners. The donor cell should be in late interphase or early prophase with the nucleolus prominently displayed. Similarly, mammalian cells of female origin should be in the interphase, if the inactive X chromosome (Barr body) is to be transplanted. The host cell should be in the mid-anaphase stage of mitosis. The nucleolus, with its attached chromosome(s), is gently extracted by a fine micropipette, using the differential piston microinjector. Then, the micropipette, preferably siliconized, is removed from the donor cell with the nucleolar complex inside. The micropipette is inserted into the host cell so that the nucleolus (or chromosome) is deposited just behind one of the anaphase sets of chromosomes. The nucleolus is then gently expelled and the micropipette is carefully removed. This is an extremely delicate operation. During telophase, when the new nuclear membrane is formed, the transplanted nucleolus (or chromosome) will be enclosed, and thereby becomes a member of the new nucleus.

What are the limitations of microsurgery? As far as delicacy of micropositioning is concerned, there is no limit. Micromanipulators are now available that permit micropositioning of a microtip to within the limits of resolution of the light microscope. Microinjections can now be performed on a quantitative basis using either the micropiston or differential piston principle. Volumes of the order of 1–10–100 micro-microliters can be routinely measured.

There are certain practical limitations in the size of the micropipete. If the micropipette is too small, great difficulties are encountered in causing flow of liquid (even water) through the orifice. Generally, the rate of flow varies as the fourth power of the radius of the orifice, other factors being constant. The second limitation in micropipette size is determined by the toughness of the extraneous coats which cover most cells. Obviously, a superfine glass micropipette would lack the strength needed to penetrate a tough cell wall. For this reason, microsurgery on plant cells whose cell walls are intact has been seriously restricted. There are many animal cells with vitelline membranes or other extraneous coats too tough to permit entry of a micropipette (*Urechis* eggs, for example).

Since the minimum size of a micropipette is limited for practical considerations to a diameter of 0.5 μ, this will also limit the minimum size of a cell. Based on the author's experience, the minimum *cell/micropipette* (*c/m*) diameter ratio should be not less than 10. Accordingly, the minimum diameter of a cell for a 0.5 μ micropipette would need to be at least 5 μ. Such a *c/m* ratio obviously excludes bacteria since in these cells the diameter is rarely greater than 1 μ. The second limitation that more or less excludes bacteria is the cell wall, which unfortunately is a tough structure. Whether this factor can be overcome by the use of enzymes needs to be investigated. It is unfortunate that microinjections into bacteria cannot, as yet, be satisfactorily performed. There are many problems open for investigation such as the obvious injections of viral DNA or RNA into bacteria. In this connection, the bacterial viruses have solved, and quite admirably, the problem of microinjecting nucleic acids into the interior of a bacterial cell. This is probably one of the most elegant examples of nature's microinjection techniques!

The transplantation of nucleoprotein-rich subcellular structures, such as chloroplasts, mitochondria, centrioles, and chromosomes, with the induction of various cell changes, can add new horizons to the study of somatic cell genetics. No longer is one dependent on the chance inclusion into cells of, or infection by, some exogenous subcellular particle. This is especially true with the recent advances in somatic cell fusion and hybridization. One may now deliberately take out a selected subcellular structure from one cell and place it precisely into the cytoplasm or nucleus of another cell. Virus-cell interactions can be approached through new experiments since inclusion bodies or structures associated with viruses can also be transplanted into uninfected cells. The interaction of viruses and cells, especially where striking cellular effects may be induced, such as transformation, interference with normal processes of differentiation, or gene location in chromosomes, could very well be one of the most fruitful areas of future work.

Much of the success of survival and propagation of the cells following transplantation will be enhanced by more precise procedures as well as by the most modern instrumentation.

M. J. KOPAC

References

Chambers, R., and E. L. Chambers, "Explorations into the Nature of the Living Cell," Cambridge, Harvard University Press, 1961.

Chambers, R., and M. J. Kopac, "Micrurgical techniques for the study of cellular phenomena," in Jones, R. M., ed., "McClung's Handbook of Microscopical Technique," 3rd ed., New Hork, Hoeber, 1950.

El-Badry, H. M., "Micromanipulators and Micromanipulation," New York, Academic Press, 1963.

Kopac, M. J., "Cytochemical micrurgy," *Int. Rev. Cytol.,* **4**:1–29 (1955).

Kopac, M. J., "Micrurgical studies on living cells," Chapter 6 in J. Brachet and A. E. Mirsky, eds., "The Cell," Vol. I, New York, Academic Press, 1956.

Kopac, M. J., "Structure and microsurgery of chromosomes," Chapter 10 in F. J. Kallman, ed., "Expanding Goals of Genetics in Psychiatry," New York, Grune and Stratten, 1962.

Kopac, M. J., "Micromanipulators: Principles of design, operation, and application," Chapter 5 in W. L. Nastuk, ed., "Physical Techniques in Biological Research," Vol. 5, New York, Academic Press, 1964.

Kopac, M. J., and J. Harris, "Microsurgery and visible light television," *Ann. New York. Acad. Sci.,* **97**(3):31–345 (1962).

MICRO-MELTING POINT DETERMINATION

The desire to observe the melting process under the microscope is of long standing and this was first successfully accomplished in 1877 by Otto Lehmann. A series of commercial microheating stages has been known since then among which two deserve special notice: the Kofler Hot Stage, because it has the widest distribution, and the Mettler Hot Stage, because it conforms to the modern development of equipment.

The Kofler Hot Stage (Fig. 1) is a simple device which

Fig. 1   Kofler Hot Stage. 1, thermometer; 2, frame of the object guide; 3, glass cover plate; 4, object guide; 5, glass bridge.

is electrically heated on the underside and by which the temperature reading is obtained on a mercury-calibrated thermometer. Its advantage is that each position of the object slide can be conveniently and quickly brought into the field of vision with the help of an object guide (4). The glass cover plate (3) makes it possible to have the whole sample under control. The heat is regulated by a transformer, which has a scale graduated in centigrade. The heating rate is rapid from the beginning by the appropriate setting of the regulating transformer; it takes about 2°/min during the rise through the last 10° before the desired melting point. The thermometers, one with range 20–230°C, a second from 120 to 350°C, show 1° markings—a half degree can be guessed.

In the Mettler Hot Stage (Fig. 2) there is an oven which is heated above and below. The temperature reading is obtained through a platinum resistance thermometer, which is brought near to the observation aperture in the lower part of the oven. By an electric control device the linear temperature rise is regulated, and by this means there is a choice of several rates of increase. Digital readings of the temperature rise with data storage make possible the uninterrupted observation of the sample during the critical phases. With a temperature rise of 0.2°/min the standard deviation shows only ±1°C, so this apparatus is particularly designed for exact temperature measurements. Also there is the possibility at the end of photoelectric recording. The equipment can also be used as a cooling stage to −20°.

Both hot stages can be set up on any kind of microscope that has a suitable object stage and an objective with a free working distance of more than 6 mm. In addition the optical firm of C. Reichert, Vienna, manufactures a special microscope which contains the necessary adjustments, but no superfluous components. The Thermopan (Fig. 3) is a microscope with a fixed Kofler Hot Stage, which meets all the requirements.

Thermomicroscopic melting point determination has the advantage over all other methods that the melting point can be determined in the microgram range. A further advantage is the fact that not only the melting point is measured, but that all the preliminary temperature readings, during the heating, can be observed and used to distinguish substances.

Melting Point. The melting point under the microscope can be determined in various ways. Kofler has suggested determining the equilibrium between the melt and the tiny residue of the crystals, a method which certainly should be used more. In general it is sufficient if the temperature reading is taken at the moment that the bulk of the crystals has melted, and the last remains of the crystals are just melting. The supposition, of course, is that the heating rate (2°/min for the Kofler Hot Stage) will be maintained.

Sublimation. In many substances, during the heating there are various changes conditioned by sublimation to observe. Many substances alter so radically that before the melting temperature has been reached, they have a completely new appearance. Sublimation can just begin directly before melting temperature has been reached or considerably before this. For many substances form and appearance are characteristic of the sublimate. Sublimation can also be used for isolation and purification. A coverslip serves as a recipient, and this, coupled with the melting point determination, adapts this method to pharmacy and toxicology.

Crystal Solvates. Under the microscope the presence of

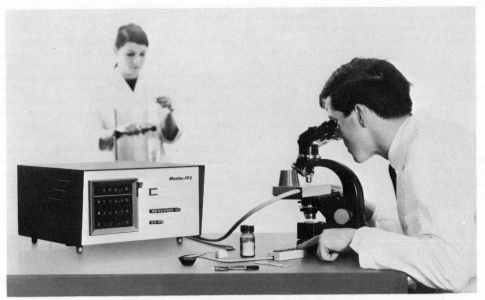

Fig. 2 Mettler Hot Stage.

a crystal solvent can be recognized—frequently it behaves like crystal water—in many cases even with normal heating. The crystals containing the solvent medium under the protection of their outer form change into microcrystalline aggregates of the solvent-free substance, so that by transmitted light one sees the formerly clear, transparent crystals become cloudy, brown, or black (Fig. 4). In some hydrates the crystals remain unchanged up to melting temperature, and from the melting of the hydrate, on further heating, the anhdyrous substance crystallizes. Since even polymorphous transition can proceed under decomposition in an aggregate, in doubtful cases the sample is treated with one drop of paraffin oil (Nujol), covered with the coverslip, and heated. At the decomposition temperature of the solvate one sees bubbles of gas rising from the crystals, which often produce foam visible to the naked eye.

Melting with Decomposition. In decomposing substances one usually finds there is a long melting interval period which depends on the rate of heating and the grain size. A higher starting temperature in the case of strongly decomposing materials as a rule leads to higher melting intervals. Carefully controlled observation of the heating rate (4°/min), and regard for the starting temperature are absolutely essential for achieving reproducible data.

Polymorphous Crystal Modifications. Many organic substances have the property of crystallizing in various forms. This leads to the fact that in many cases from a

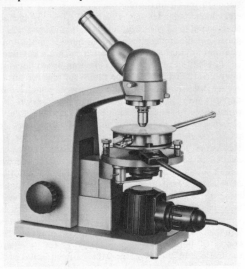

Fig. 3 Thermopan.

Fig. 4 Loss of crystal water.

specific solvent a different form crystallizes out, as would happen if another solvent were used. Unstable crystals can change during heating, which in polarized light most often results in a change of colour; in this case, if the crystals are metastabilized, there are complications, because another melting point occurs. In many cases a part of the crystals melts at the melting point of the unstable form, while another part changes and only melts at the melting point of the stable form. For example, sulfathiazole melts partly at 175°, the remaining crystals at 202°. Investigation of polymorphs is a special chapter of thermomicroscopy which cannot be pursued in this article.

Mixed Melting Points. For determining a mixed melting point it is sufficient to grind a few small crystals of the sample with a like amount of the authentic sample between two slides. From this mixture the melting interval is determined. When the substances are not identical, a part of the mixture begins to melt at the eutectic temperature, while the remaining crystals gradually dissolve in the liquid. This results in a longer melting period than in the case of a single substance. If, on the other hand, the samples are identical, then the melting period of the mixture is the same as with a single component.

Thermomicroscopy. The microscopic melting point determination is only one part of the whole field of thermomicroscopy. By identifying organic materials according to L. Kofler, the microscopic determination of the refractive index of the melt and determination of the eutectic temperature are also obtained. In addition, the heatable microscope is indispensable for investigating mixtures of organic and inorganic compounds (there are also high-temperature microscopes), and also for investigating crystallization procedures and polymorphs.

M. KUHNERT-BRANDSTÄTTER
(trans. from German)

References

1. M. Kuhnert-Brandstätter: "Thermomicroscopy in the Analysis of Pharmaceuticals," Pergamon Press, Oxford, 1971.
2. W. C. McCrone: "Fusion Methods in Chemical Microscopy," Interscience Publishers, Inc., New York, 1957.

MICROPHOTOGRAPHY

As originally defined in 1857, microphotography is the process of producing minute images on a relatively grainless photographic emulsion.[1] The literature of the period described document copies so minute that they could be hidden in an ink blot or period, and telescope eyepiece reticles (or graticules) with lines "so fine that the image of a star positively leaped across the gap." In addition to the use of wet-collodion emulsions which were "grainless" at the instant of exposure, success depended on employment of a well-corrected microscope objective and special procedures for focusing on the light-sensitive layer. This technique may be called "extreme-resolution photography."

The history of the applications of extreme-resolution

---

[1] Readers should, however, note that *Mikrophotographie* in German denotes photomicrography, and many English-speaking authors erroneously use microphotography in this sense.

photography is a story of initial promise and long-delayed but most rich fruition. Early examples were the production of minute pictures mounted on Stanhope lenses as curios; operation of a postal service in which photographically reduced messages were flown by pigeons into the besieged city of Paris (1870), and the manufacture of stage micrometers and reticles for surveying and military optical instruments. Until 1940, the art was confined to a small number of workshops by the necessity for users to sensitize their own glass blanks, and in the absence of organized literature, to train their own labor.

In 1940, this situation was transformed by the commercial production of dry-plate emulsions with mean grain diameters of 60 nm, and by dissemination of information about their handling. Although these plates gave high-density and contrast, the undeveloped emulsion was transparent, scattered very little light, and with an appropriate microscope objective gave images resolving well over 1,000 lines/mm. With condenser illumination of a transparency, their speed was sufficient to permit very brief exposures. These properties were immediately exploited by the Allies for the production of reticles in World War II.

Although these plates produced images of very high information content, the limited field of microscope objectives meant that a single exposure could cover only about a square millimeter with a micron-line image. Larger images were produced by mounting the unexposed plates on the table of a dividing engine, exposing successive areas after exactly controlled movements. The beam of the microphotographic projection head was substituted for the engraving tool. Lines of any width or shape or whole groups of lines were thus produced by a device which neither wore out nor scattered debris on the workpiece. Series of hundreds or thousands of sequential exposures were used to build large images such as circles for surveying instruments, binary coded scales for computer systems, and gratings for metrology.

Extreme-resolution photography entered another dimension when engineers transferred the information in their negatives by printing on layers of photoresist applied by simple methods to any impervious support. By plating or etching, material was added or removed through the stencil formed from the photoresist. Layers only a micrometer thick had adequate resistance to withstand solutions capable of attacking all metals and ceramics and glass, and the stencils recorded all the information in the microphotographic negatives. Such methods are used for the fabrication of microsieves, electron microscope specimen grids, and interference filters for the far infrared. Above all, they control the area-wise penetration of dopants into ultrapure silicon: a process essential for production of microelectronic integrated circuits. Repeated operations produce multilayer structures comprising within a square millimeter a dozen or more transistors connected to all the capacitors and resistors needed to constitute a completely functioning electronic circuit smaller than a pinhead. The economic importance of this work has stimulated the evolution of a range of new diffraction-limited monochromatic objectives of unprecedentedly large flat fields. Such objectives have valuable potential for special photomicrographic tasks.

Early suggestions for the production of greatly reduced copies of books had to wait 70 years for commercial realization. In 1927, George McCarthy in collaboration with the Eastman Kodak Co. founded the Recordak Corporation for recording information from checks and invoices on 16 mm film as a precaution against fraud and loss. The requirement that multitudes of documents be economically recorded in a short time was incompatible with the two-stage reduction used for extreme-resolution photography. The process was therefore based on direct reduction by factors of 15 (and upwards) from the opaque documents. Because only a small proportion of the light diffusely reflected from an opaque document can ever enter the lens, the brightness of the aerial image was limited, and it proved necessary to use much faster emulsions similar to a fine-grain motion-picture film. These had grain sizes in the range of 100–400 nm, scattered light quite strongly, and gave resolution figures in the range of 200–500 lines/mm.

Once a viable process was marketed, users perceived that it gave the further advantages of easier duplication and transportation, preservation of file integrity, and incorporation into a variety of office routines. Librarians and research workers rapidly exploited its potential for filling gaps in literature collections, publications of doctoral theses, and dissemination of technical reports. To comply with different users, adaptations in format and packaging were introduced. Roll microfilms were equipped with photoelectrically sensed coded information for automating the location of known documents or for searching of the file by information content. Another method has been to divide the film into convenient units. Engineering drawings are filed as short lengths of 35 mm film mounted over an aperture in a machine-sortable punched card, and copies of American patents are available on short lengths of 16 mm film. Much technical literature is available as up to 100 pages on a filing-card sized film, and this concept has been applied to publishing much scholarly literature.

For a century, the only proven application of extreme-resolution document copies has been in furtive communication and espionage, but in developing integrated circuit technology, the electronics industry proved the practicability of filling a rectangular area with microdot-sized images. Extension of the concept is giving document copies with up to 3000 pages on a 6×4 in. film. This technique has already been applied to the production of mail-order catalogues and automobile parts lists, and to the republication of collections of 20,000 reference books for inexpensive stocking of libraries in university departments.

What could ultimately prove to be the biggest application of microfilming may be just at the start of an exponential growth. Computer output microfilming (COM) aims to solve the retardation of computer operations by the relatively sluggish output of impact printers. Output an order of magnitude faster can be achieved by microfilming of pages of text displayed on a cathode-ray tube. Because the information is displayed at a rate of several pages a second it cannot be read directly. But the microfilm camera more nearly keeps pace with the computer and simultaneously reduces the physical mass of the output to a more easily handled size.

Conclusion. This curious activity of using microscope objectives to produce minute but sharp images (microscopy in reverse) has proved a major growth point for modern technology.

G. W. W. STEVENS

References

Eastman Kodak Co., "Techniques of Microphotography," Kodak Publication No. P52, Rochester, New York, 1970.

Eastman Kodak Co., "An Introduction to Photofabrication," Kodak Publication No. P79, Rochester, New York, 1970.

Nelson, C. E., "Microfilm Technology (Engineering and Related Fields)," New York, McGraw-Hill Book Co., 1965.

Stevens, G. W. W., "Microphotography: Photography and Photofabrication at Extreme Resolution," 2nd ed., London, Chapman and Hall, and New York, John Wiley & Sons, Inc., 1968.

See also: PHOTOMICROGRAPHY.

MICROPROBE

1. Introduction and Physical Description. The electron-beam (also the ion-beam) microanalyzer, commonly called the electron (or ion) microprobe consists of a group of closely related systems: gun, lenses, and diaphragms to produce a concentrated electron (or ion) beam of very small diameter (in some instruments electrostatic or other means are provided for sweeping this beam either on a line or in a two-dimensional raster over a chosen area); an optical microscope to observe the sample and control the progress of the analysis; and one or more X-ray spectrometers, detectors, and readout devices to measure the quality (wavelength) and intensity of the X-rays produced where the electron (ion) beam strikes the specimen. The first practical instrument was developed by R. Castaing and described in a doctoral dissertation at the University of Paris (1951). There are now several commercial models, and many such instruments are finding wide application in metallurgy, mineralogy, petrology, chemistry, and biology.

Briefly, the operation consists of bombarding the surface of the sample to be analyzed with the beam of energetic particles and measuring the X-ray spectrum given off by the sample; then to compare this spectrum with that of a known standard measured under conditions as nearly identical as possible. If the sample and the standard have the same composition, their spectral distributions will be alike; if the compositions differ only a little, the compositions by weight are proportional to the X-ray intensities; if the compositions differ significantly, as when pure elements are used as standards, then correction factors must be applied to the intensity ratios before applying the rules of proportionality mentioned above. These correction factors are themselves functions of the (unknown) composition of the sample, but iterative computation methods are available that converge with only a few iterations.

The electron-optical system consists of a gun which accelerates electrons emitted from a hot filament, and aims them in the general direction of a condenser lens. Diaphragms limit the beam at certain focal points. The condenser focuses a reduced image of the first diaphragm upon an objective diaphragm; the second condenser, commonly called the objective lens, further reduces the size of the beam and focuses it upon the sample. In some instruments, a third lens between these is used to further reduce the diameter of the beam.

Deflection plates are provided in most instruments which deflect the beam periodically so as to scan along a line on the sample; a second set of deflection plates then moves this line periodically at right angles to itself, forming a square raster which can be synchronized with similar deflections of the beam in a cathode ray oscillo-scope. Oscilloscope brightness may be modulated in several ways: (1) with the electron current through the sample, (2) with the back-scattered electron current, or (3) with the output from an X-ray detector tuned to the wavelength of characteristic X-radiation from an element present in the sample, forming a profile or image of the sample surface.

A microscope is arranged in most probes to permit observation of the sample during bombardment with the beam. This microscope is commonly used to adjust the sample in the focal position for both the electron beam and the X-ray spectrometer(s).

One or more Bragg spectrometers are commonly arranged to receive the X-rays given off by the sample where it is excited by the beam. After selection of a particular wavelength by reflection in a crystal used as diffraction grating, the X-rays are detected by a quantum counter. The pulses from the counter are filtered through a pulse-height discriminator. The filtered pulses then actuate ratemeter, scaler, and in some instruments, the brightness circuits of the oscilloscope.

Nondispersive readout systems do not require the crystal spectrometer, so the quantum counter may be placed very much closer to the source of the X-rays, gaining many orders of magnitude in counting rate, and also eliminating the critical spectrometer-focusing opera-tion, but not the less critical electron beam focusing, which both depend upon the accuracy of optical focusing with the microscope. The counter signals then pass through multichannel pulse-height discriminators which activate corresponding counters, rate meters, etc. The term "non-dispersive" for such systems is a misnomer, for dispersion is in fact achieved electronically instead of by means of the spectrometer.

The axis of the visible-light microscope may be physically separate from the electron- and X-ray-optical paths. Beam alignment, spectrometer tuning, and speci-men positioning must then be done indirectly. In most instruments, the microscope objective and the electron-optical path are coaxial. A hole along the central axis of the objective permits the beam to strike the sample at the center of the field of view, luminescence excited by the beam is a valuable tool for beam alignment. The sample may be observed by reflected or transmitted light. Sample preparation consists of mounting in a solid block, or on a microscopical object-slide, and polishing a flat surface. Because the sample must be located accurately in three dimensions, the final focusing operation with the micro-scope is always by reflected light. This step alone may account for half or more of the statistical variance of the analytical readings.

Readout systems generally consist of scalers or rate-meters, or both; the scalers are arranged to permit counting for a preset time, or to a preset count limit. The results should be printed out, including counts on each channel and the length of time of counting. It is convenient for the output devices to produce records, such as punched cards or paper tape, that can be used with high-speed computers. These data will be similar whether the system uses spectrometers or a multichannel "non-dispersive" system.

Counting pulses sent to oscilloscopic readout devices during scanning operations with the beam covering a small area in raster scanning mode produce light spots or flashes on the oscilloscope screen, which gradually build up a picture resembling a halftone print. Low resolution in this mode need not be a serious limitation, because even if the beam diameter is a fraction of a micron, the X-ray excitation probably originates through-out a volume of at least $100 \, \mu^3$.

Oscilloscope presentation of specimen current, or of backscattered electron current may be used in either line-scanning mode or raster mode, yielding in the one case a record of the variation of current along a selected line on the sample, or in the other case a brightness-modulated image; these tend to show grain boundaries, scratches in the polish, and variations in the mean atomic number of material in the sample.

Ratemeter data may be presented in useful ways as a meter deflection or as pen deflection on a strip-chart recorder varying either (1) as the sample is moved linearly, giving a profile of the chemical element content along some chosen line, or (2) as the spectrometer setting is changed through positions corresponding to the X-ray wavelengths for the several elements.

2. Data Reduction. As with many other kinds of automatic data-collection systems, there tends in micro-probe analysis to be much replication of observations. This is useful for improved estimates of the composition, but more important, it permits making estimates of the statistical variance associated with the composition num-bers. This aspect of microprobe data reduction can be handled by well-known, standard statistical procedures and will not be considered further here.

In microprobe analyses one attempts to eliminate as many variables as possible by the simple expedient of comparing an unknown sample with a known one under the same conditions. One part of "same conditions" that is particularly desirable is that the sample and the standard for comparison have as nearly as possible the same composition! At one extreme, this leads to the necessity for almost infinite numbers of standards, and the problem is then how to select the right standard. At the other extreme, one pure-element standard should be sufficient for each element. In that case, one computes the counting rate that would be obtained from a theoretical standard with the same composition as the sample.

Given the counting rate for some element A from a compound standard of known composition, similar computations permit us to estimate the counting rates that would be obtained from the pure elements, which in turn may be applied as above. Because the same algor-ithms are used forward and backward in this process, it is well to have a standard that is fairly similar to the un-known sample in composition so as to eliminate most of the error caused by possible errors in theory or in empirical constants used in the algorithm.

Corrections on Counting Rates. When ionizing radiation is absorbed in the quantum detector, the first ionizing photon causes further ionizations in the gas in the detector that tend to deaden its sensitivity for a short time. If the X-ray intensity is low, very few interferences occur and the effect is quite negligible. However, when the counting rate exceeds a few thousand counts per second, the deadtime, τ, is no longer negligible, and it can be shown that the true number of counts per second, N, can be calculated from the observed number, N', and the deadtime as follows

$$N = N'/(1 - N'\tau) \qquad (1)$$

This correction is routinely made in computations by computer; for deadtimes τ of 5 –10 μ-sec, the correction is significant only with N' larger than about 5000 counts per second.

Cosmic rays, electrical noise, and X-ray noise picked up from various sources contribute a background counting rate that is nearly independent of any small variations in sample composition. This rate can therefore be measured a few times for a typical sample in a group of related ones and applied to all of them. For each element, ideally, the background should be measured at the exact wavelength setting used for the element, while the beam is striking a "same" sample without the element concerned. This is impractical, but measurements can be made at wavelengths above and below the analytical one, or, in the case of very simple compounds or elements, background measurements can be made on samples with elements of adjacent atomic number substituted for the element analyzed. In the latter technique, for example, the background for a pure metal standard with atomic number Z can be obtained by substituting a different pure metal with atomic number $(Z-1)$, or $(Z-2)$. The same sort of device can be used with pure oxide standards, as by taking background for SiO_2 on a pure Al_2O_3 standard. Background may be interpolated on a linear, weight-composition basis in cases where the sample has a simple composition, as in binary alloys, or in simple solid-solution series like olivine $(Mg_{1-x}Fe_x)_2SiO_4$, $0 \leqq x \leqq 1$.

Background and all significant deadtime corrections must always be made on every X-ray intensity before any other calculations are performed. In all further discussion it will be assumed that these corrections have been made, and we shall henceforth use corrected intensity data as if they were the data obtained directly from the probe.

Drift. Electronic systems tend to drift slowly; if remeasurement of standard signals indicates that the system drift has been more than about 4% or 5% per hour during the course of an analysis, the instrument should be realigned and the analysis repeated. A standard signal should be monitored before, during, and after any analysis that takes more than about 1 hr. If drift is small, it may be ignored; better, it can be corrected by assuming a constant rate of drift throughout the time of the analysis.

Relative Intensity. The intensity should be expressed relative to that of the pure element standard after making corrections for background, deadtime, and drift.

$$K_A = I_{A,x}/I_{A,S} \qquad (2)$$

where K_A is the corrected intensity for element A; $I_{A,s}$ and $I_{A,x}$ are corrected intensities for the element A as read on the pure-element standard and on the unknown sample, respectively.

Calibration. Experimental calibration of the analysis using three or more standards for a binary alloy, and correspondingly more for more complicated compounds, is the most accurate, and in some cases, the most convenient method; it requires preparation of many standards, and careful microprobe tests on these to establish calibration curves. Naturally, standards must be pure and should be homogeneous on a submicron scale. A standard may be traversed with an enlarged beam to average the composition over regions as large as 50 μ if it contains finely divided inhomogeneities.

Many calibration curves have been measured; they may be expressed in the linear form,

$$(1 - K_A)/K_A = \alpha_{AB}(1 - C_A)/C_A \qquad (3)$$

where K is the intensity ratio defined by (2), C_A is the weight fraction of element A in the sample, and α_{AB} is the calibration constant for element A in alloy A_mB_n (or $A_{1-x}B_x$).

Equation (3) can be rearranged so as to read α_{AB} from the intercept of a linear plot of C_A/K_A against C_A:

$$C_A/K_A = \alpha_{AB} + (1 - \alpha_{AB})/C_A \qquad (4)$$

In principle, only one calibration standard is required because there is only one unknown variable α_{AB}, in (4). For improved accuracy, several standards should be used.

The same kind of reduction can be extended to componds of several elements. In principle, one element at a time is added to the system by taking α_{AB} as known from the binary alloy A_mB_n, considering this alloy as the end-point of a series from it to element C, and finding $\alpha_{A,ABC}$, and so forth. Following Ziebold and Ogilvie (1966) we have

$$(1 - K_A)/K_A = \bar{\alpha}_{A,(N)}(1 - C_A)/C_A \qquad (5)$$

where the average parameter $\bar{\alpha}_{A,(N)}$ for element A in the compound containing elements $A, B, \cdots, N$ is given by

$$\alpha_{A,(N)} = \frac{\alpha_{AB}C_B + \alpha_{AC}C_C + \cdots + \alpha_{AN}C_N}{C_B + C_C + \cdots + C_N} \qquad (6)$$

If the composition of multielement compound $A_mB_nC_pD_q \cdots N_z$ is unknown, equation (6) may be applied as follows: Assume that all $C_i = K_i$ and compute a zero-th approximation of the unknown composition. From this the $\bar{\alpha}_{i,(N)}$ can be estimated for the first approximation, whence a new set of C_i. Convergence is usually rapid; the second estimate of $\bar{\alpha}_{i,(N)}$ will ordinarily differ but little from the first, and the third, even less from the second.

Theoretical Correction from Physical Principles. Theoretical relations between the ratios K_i and the composition C_i have been studied extensively. Available solutions depend upon empirical data such as the well-established mass-absorption constants of the elements, backscatter and ionization data for elements under electron bombardment, and enhancement of X-ray intensities due to fluorescence. These factors are known, or can be estimated, with varying precision. The theoretical reduction takes the form of an adjustment to each K_A (K for element A)

$$K_A = \left(\frac{R^*p^*}{R_Ap_A}\right)\left(\frac{f(\chi^*)}{f(\chi_i)}\right)\left(1 + K_f\right)C_A \qquad (7)$$

where R^*/R_A accounts for the differences of electron backscatter between the sample and the standard, p^*/p_A similarly for differences in ionization efficiency, and the asterisk (*) indicates weight-fraction average over the sample composition; both R and p are somewhat controversial, but they tend to compensate one another, and the first factor, called the *backscatter-ionization factor*, is commonly only slightly different from 1.0 (Duncumb and Reed, 1968). The second factor, involving $f(\chi)$, is due to absorption of the X-rays in the specimen and the standard. It is the best known of the three factors, and also the most important factor in the whole correction. The basic treatment of absorption corrections is due to Philibert (1963), and subsequent refinements have been mainly to the mass-absorption constants (Heinrich, 1966) and to details of certain approximations used. Extensive tables have been prepared by Adler and Goldstein (1966), and by

Colby (1966), and the treatment is readily amenable to computer programming.

The third factor is due to enhancement of X-ray intensity by fluorescence after absorption of X-rays produced in surrounding atoms in the sample. This is likely to be significant only for absorption of characteristic radiation from the surrounding atoms, and then only if the fluorescing element has very strong absorption, as it does just below an absorption edge, for that radiation. Thus in an alloy of Fe and Ni, fluorescent enhancement of the $FeK\alpha$ X-ray line is high if there is much Ni present, producing much Ni radiation for the Fe to absorb and reemit by fluorescence. The fluorescing element must be just a few atomic numbers lower than the element whose characteristic radiation is causing the fluorescence. Depending, however, upon the accuracy desired, a fluorescence enhancement correction is usually not necessary in oxides, or in samples generally that contain major amounts of many different elements.

3. **Uses and Limitations of the Microprobe.** The primary uses of the microprobe are at once evident, clearly depending upon its ability to obtain a complete chemical analysis from a microscopic area of material, and only to a microscopic depth, in the polished flat surface of a sample (Marton (1969) showed that the ultimate sensitivity of the microprobe, of the order of 10^{-15} g depending upon the element sought and upon the matrix, etc., compares very favorably with sensitivities of most other analytical tools such as emission spectrography, fluorescence microscopy, and mass spectrometry; to put this number in perspective, he noted that the olfactory sense of some animals permits detection of amounts of 10^{-18} g, that of bees permits detection of a sexual attractant at the level of 10^{-20} g, while chemical methods such as wet chemical analysis, X-ray fluorescence, and absorption spectroscopy can detect elements at levels of 10^{-7}–10^{-9} g. Some of these limits may be optimistic, but they are certainly impressive. Practical limitations of the microprobe include the very great difficulty of obtaining detectors for X-rays characteristic of the lightest elements: H, He, Li, and Be are impossible at present; B, C, N, O, F, Ne, and Na may be arranged along a scale of decreasing difficulty ranging from research-level experiments to the limit of routine capabilities; Mg, Al, Si, P, S, and Cl are "light elements" measurable routinely with only a little specialization of the equipment, and the heavier elements are mostly truly routine. Na and K are rather easily volatilized, yielding a rapid drop in intensity during the first few seconds of an analysis; the noble gases are of course difficult to prepare in solid form that would be stable in the vacuum of the electron path. Spatial resolution is fairly good if qualitative or semiquantitative results are sufficient. The electron beam, focused initially to an area less than $1\ \mu$ in diameter, diffuses out in all directions from the point of impact, and excites atoms that are certainly as far as several microns from that point. This diffusion, both laterally and in depth, tends to make quantitative analysis of elements inaccurate in the neighborhood of steep gradients such as grain boundaries, even though spectacular differentiations can be made at scales approaching a few microns or less (Rapperport, 1969, p. 136, concludes that a resolution of 1/5 to 1/10 of the beam size should be attainable by mathematical deconvolution techniques that he describes and applies to a specimen-current profile. This is about an order of magnitude better than the "beam-diameter" resolution to be expected as the ultimate in normal, present-day

work). Spatial resolution is probably much better in specimen-current profiles than in X-ray analytical modes.

HORACE WINCHELL

References

Adler, I., and J. Goldstein, "Absorption Tables for Electron Probe Microanalysis," NASA, Washington 1965, NASA TN D-2984, (276 pp.).

Castaing, R., "Application des sondes electroniques à une methode d'analyse ponctuelle chimique et cristallographique," Thesis, University of Paris, 1951.

Colby, J. W., "The applicability of theoretically calculated intensity corrections in microprobe analysis," pp. 95–188 in T. D. McKinley, K. F. J. Heinrich, and D. B. Wittry, ed., "The Electron Microprobe," New York, Wiley, 1966.

Duncumb, P., and S. J. B. Reed, "The calculation of stopping power and backscatter effects in electron probe microanalysis," pp. 133–154 in K. F. J. Heinrich, ed., "Quantitative electron probe microanalysis," U.S. NBS Spec Pub Washington (1968).

Heinrich, K. F. J., "X-ray absorption uncertainty," pp. 296–377 in T. D. McKinley, K. F. J. Heinrich, and D. B. Wittry, ed., "The Electron Microprobe," New York, Wiley, 1966.

Philibert, J., "A method for calculating the absorption correction," pp. 379–410 in A. J. Tousimis and L. Marton, ed., "Electron Probe Microanalysis," New York, Academic Press, 1969.

Rapperport, E. J., "Deconvolution: a technique to increase electron probe resolution," pp. 117–136 in A. J. Tousimis and L. Marton, ed., "Electron Probe Microanalysis," New York, Academic Press, 1969.

Ziebold, T. O., and R. E. Ogilvie, "Correlations of empirical calibrations for electron microanalysis," pp. 378–389 in T. D. McKinley, K. F. J. Heinrich, and D. B. Wittry, ed., "The Electron Microprobe," New York, Wiley, 1966.

MICROPROJECTION

1. **The Usefulness and Limitations of Microprojection.** The direct observation of an object with the aid of a magnifier or a microscope engages the attention of the observer through his involvement in the production of the image to such an extent that his personal attention produces the strongest impulse for his intellectual absorption in the object. This applies to both research workers and students. Both learn and discover mainly because they let themselves be captivated by the object and therefore by the questions the problem to be investigated poses; in other words, the act of acquaintance with a new object becomes a personal experience.

This directness of the microscopic investigation of the object cannot be adequately replaced by any teaching aid. But in schools, during lessons or lectures, direct microscopic observation raises many technical difficulties, so that the use of teaching aids becomes essential whenever it is impossible to issue each pupil or student his own microscope and his own specimen as practical instruction in the use of the microscope.

The modern aids allowing the transmission of the experience of personal observation are mainly the photograph (still transparency or cine-film), microprojection, and microtelevision) The transparency as an already existing means of reproduction which, at the moment of showing, has already been abstracted from the original object, can, by the impact of this abstraction, in its context or through aesthetical composition of line and color attract the interest of the viewer; the microfilm can, in addition, not only interpret the moment of motion, but also use it as a stimulus of attention; but the original object is remote.

From a tutorial point of view television transmission from a microscope, particularly when it is in color, is superior. But here the equipment required is, at least for the time being, so expensive and the correct color rendering on several monitor screens or television projection, for instance with the Eidophore methods, requires such special training and experience that the use of television in teaching is out of the question unless it is based on programmed, long-term, careful preparation, which is the only way to ensure success. But, on the other hand, microtelevision offers facilities in teaching, such as the demonstration of infectious material, which no other method of demonstration makes available.

The gap between the individual microscope and the universally suitable but complicated television method is filled by microprojection, which is a technique of direct demonstration of microscopic objects to a large number of people. The extent of equipment this requires is modest —even small schools and teaching establishments will find the financial outlay within their means—and the operation calls for hardly any knowledge and skill that is not already part of the teacher's basic training in microscopy.

2. Principle and Technique of Microprojection. In principle the optical aspect of microprojection is identical with that of photomicrography:

a. With single-stage magnification (by magnifying lens) a short-focal-length optical system is used to project a microscope specimen placed closely in front of the lens at a large image distance. But the image distance and therefore the image size vastly exceed the dimensions of photographic reproduction.

b. Also as in photomicrography two-stage magnification by means of objective and eyepiece is used for projection, with the eyepiece taking over the function of a projection eyepiece for the real aerial image formed by the objective. The higher magnifications and increased resolution obtainable with the two-stage method based on a "genuine" microscope result in reduced brightness of the projected image, which can however, be adequately compensated by the introduction of high luminous densities for the very small object space.

These conditions of space and lighting technique which exceed the requirements of photomicrography, have created some characteristic features in the microprojection devices.

For use in microprojection ordinary microscopes require a high-quality light source of sufficient brightness, a total radiation suitable for the reproduction of all the colors of the spectrum, and carefully screening to prevent the radiation of stray light into the space outside the object. This has led to the use of halogen or xenon lamps in lamp housings of special technical and optical design. In addition the normally vertical light emerging from the microscope must be deflected so that it is horizontal.

For this purpose deflecting devices such as mirrors, or better still, prisms, are available, and with the projection eyepiece in position the screen size is adapted to the usually predetermined projection distance.

Basically, microprojection devices, designed to meet these conditions, are easier to handle and optimum projection can be achieved more quickly and readily than with ordinary microscopes extended for microprojection if it is adequately equipped for this purpose.

Horizontal position of the object stage and therefore vertical position of the microscope tube is preferable to the horizontal position of the entire microscope because there the object is placed vertically, which mostly excludes a projection of a specimen in vivo in an aqueous medium. A horizontal position of the object stage is preferable particularly when ordinary 35 mm projectors are used in conjunction with single-stage magnification or with a microscope attachment. For this purpose microattachments with deflecting mirror are available.

3. Some Practical Hints. The most important preparation for effective and successful microprojection consists in the choice and production of instructive, durable, contrasty, transparent specimens of sufficient thinness. In addition, the microscope and its light source must be well adjusted and the degree of magnification must be chosen in accordance with the purpose of the demonstration.

It is important that the viewer witness the search for an instructive area of the object, the focusing, and the ascent through the various stages of magnification; this enhances the essential direct relationship to the object. Naturally this requires a smooth, precise object movement which at increased magnification can be obtained only by means of a mechanical stage.

The magnification should always be made known, since without it the viewer will be unable to form a correct idea of the scale of what he sees. The additional magnification produced by the projection distance, which for every meter should be multiplied by 4, can be neglected. The magnification to be given is that of the visual observation in the microscope, which is the product of the objective magnification and the magnification of a normal observation eyepiece (usually $10\times$).

The viewers should be seated, if spatial conditions permit, near the projection axis. In conjunction with the reflecting power of the projection screen this improves the utilization of the available image brightness.

The entire light quantity required for a projected image of sufficient brightness must be passed through the object area, which decreases in inverse proportion as the magnification increases. This object area measures only a few tenths of a square millimeter at high magnification, and is magnified to areas of the order of square meters on the projected screen. The brightness therefore decreases at a ratio of more than $1:1,000,000$. The required high intensities of illumination in the object involve a considerable amount of light energy.

It can, however, be taken for granted that good-quality microprojectors incorporate heat filters to remove infrared radiation (heat) from the beam path. The ultraviolet portion is likewise absorbed by a lens made of UV-absorbing glass included in the lamp condenser. These safety measures in conjunction with suitable histological stains and a compromise between the choice of object and the degree of magnification, which is based on experience, eliminate the risk of damage to the microscope specimen owing to heat.

Hematoxylin and light-fast azo dyes have been found most reliable for use in suitable microprojection specimens, even when the very powerful xenon microlight sources were used.

J. GREHN

See also: OPTICAL MICROSCOPE, PHOTOMICROGRAPHY.

MICROSPECTROPHOTOMETRY

Introduction. New microanalytical instrumentation is necessary to study the chemistry of a living cell in its dynamic state. Living cells contain cellular organelles, e.g., a nucleus, mitochondria, chloroplasts, and other microbodies, for which direct chemical analyses are lacking. To begin to attack this problem sensitive spectroscopic methods began to be applied.

In the late 1930s, Caspersson (1) initiated the development of microspectrophotometry, by combining a microscope with a monochromator. By this method, he was able to measure the ultraviolet absorption spectrum of the cell nucleus for nucleic acids.

A rapid advance in microspectrophotometer instrumentation was made in the 1960s for obtaining absorption spectra of living cells. These instruments are of various types, from relatively simple microspectrophotometer (cytophotometers, cytodensitometers) to very complex ones, in which spectral data are rapidly recorded and stored by computer techniques.

Microspectrophotometers are now becoming a useful analytical tool in the investigations of a variety of biological, biochemical, and chemical processes. Descriptions of the designs of some of these instruments and their application to biological studies will be briefly examined.

Microspectrophotometers of Various Design. The microspectrophotometers presently in use are laboratory-built instruments. There are, though, several commercial recording microspectrophotometers that have recently become available (Carl Zeiss, Incorporated; Canal Industrial Corporation; Olympus Corporation of America); however, they still lack features which have been designed into laboratory-built instruments.

The essential components of all microspectrophotometers are a microscope, a monochromatic light source, a light chopper, a sensitive photocell (usually a photomultiplier tube), an amplifier, and an output device for recording the absorption.

Simplified microspectrophotometers, as illustrated in Fig. 1, can be easily assembled. The components are a microscope, monochromator, light source, and voltmeter (2). The photocell is usually a photomultiplier tube. A cadmium selenide crystal cell can also be adapted as the photocell, for it can be used for spectral measurements from the ultraviolet through the infrared and even for X-ray spectroscopy. The cadmium selenide cells do not require high voltages and have a photosensitive surface of 0.5×1 mm with a radiation sensitivity per unit area

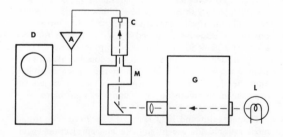

Fig. 1 Diagram of a simplified microspectrophotometer. D, oscilloscope or recorder (voltmeter); A, dc amplifier, C, photomultiplier tube; M, microscope; G, monochromator; L, light source.

roughly equivalent to a photomultiplier tube. The peak response is in the visible region of the spectrum at 720 nm, and at high light intensity the time constant of the cell is of the order of milliseconds. At extremely low light levels, the time constant is of the order of seconds. The instrument is usable over the wavelength range from 200 to 990 nm, depending on the monochromator and the light source; using a tungsten ribbon light source, the useful range is from 370 to 900 nm. The maximum magnification employed is $500\times$, and at this magnification, absorption spectra from areas of $1\ \mu^2$ are obtained. No limiting apertures are used in this simplified instrument, since the entire photosensitive surface is used for measurements. Optical alignment is not critical, and no light chopping is necessary. The noise level as determined by photocell current fluctuations is about 2×10^{-3} optical density units. Since the entire photosurface of the cell is exposed to the light beam, no effects due to variation in the sensitivity of the photosurface are observed.

For the study of living cells and photosensitive pigments, better resolution, faster time constants, low light levels, and rapidly recorded absorption spectra are necessary.

To meet these requirements, a recording spectrophotometer can be adapted for microspectrophotometry. Brown (3) converted a Cary Model 14 dual-beam recording spectrophotometer into a microspectrophotometer by replacing the sample cell compartment with a microscope for the sample light beam and equivalent optics for the reference beam. This instrument can be used for solution spectroscopy or for microspectrophotometry by simply interchanging components. For sample areas of 0.1–1 mm in diameter, the usable wavelength range is 300–700 nm. Additional provisions include increasing the sensitivity of the recorder from 0–1 optical density units to 0.01 optical density units full scale and the addition of a monochromator light source for locating the specimen. This modified Cary 14 microspectrophotometer can sweep the visible spectrum in less than 2 min and has been successful in obtaining recorded absorption spectra with a peak-to-peak noise level of about 10^{-3} optical density units. A disadvantage of this system is that the reference beam does not pass through the microscope slide chamber containing the experimental specimen, and unless a pure solution is used for the suspending medium, the absorption of other cellular constituents surrounding the sample area under investigation must be corrected for.

Recording microspectrophotometers, which have been constructed in several laboratories, are elaborate instruments (4–9). These instruments use either split or dual-beam optical systems with mechanical beam chopping devices, voltage-regulated photomultiplier tubes with feedback mechanisms, and automatic recording outputs. Quantitative measurements of the optical density of a given specimen, with an instrumental error of less than 5%, can be obtained. The response times of the instruments may be of the order of milliseconds; absorption measurements from the ultraviolet through the visible are recorded in seconds.

One example of such instruments is that of a single-beam recording microspectrophotometer, designed and constructed by Chance et al. (6). This instrument covers the spectral range from 240 to 600 nm, with a noise level of the order of 10^{-4} optical density units to detect concentrations of 10^{-18} mole at a signal-to-noise ratio of 50 to 1. A modified design of this microspectrophotometer was made by Liebman (7) to investigate absorption and dichroism in isolated single retinal rods and cones. In these studies, the microscope was fitted with Zeiss ultrafluar optics (320×) and was illuminated by a Bausch and Lomb 250 mm grating monochromator with 1200 lines/mm. To fill the condenser back aperture (N.A. = 0.4), the monochromator exit slit-width was set at 1.8 mm for a spectral bandwidth of $5\ \mu$. A beam splitter above the ocular permits viewing of the specimen and provides a light path for the signal, which is imaged in the plane of a pair of circular apertures (in the object plane). A mechanical shutter alternately selects the object or the reference area at 60 Hz for illumination of the photomultiplier. The output of the photomultiplier is controlled by a dynode feedback circuit. Integration of the measure-reference signal is performed by an RC network, and the output is recorded on a Varian dc recorder in percent absorption, proportional within 10 % to an optical density of 0.1 or less. A time constant of 2 sec and a noise level of about 4×10^{-4} optical density units peak-to-peak are possible with this instrument.

Another recording microspectrophotometer was designed and constructed in our Biophysical Research Laboratory (8). The instrument is shown in Fig. 2, and a block diagram is illustrated in Fig. 3. The main characteristics of this microspectrophotometer are improved optical resolution accompanied by greater sensitivity from the ultraviolet through the visible.

In operation, light from a tungsten or a xenon lamp passes through a Canalco grating monochromator with variable slits to a specially designed chopper. The chopper divides the beam of light into two beams which pulse alternately in time. Each beam is on 40 % of the time and is pulsed 480 times/sec. The two beams of light are directed into a condenser of a microscope, which is an ultrafluar objective lens of 100× or less. Its purpose is to focus the two light beams as two spots of light on the specimen slide. The slide is moved to locate the specimen area to be investigated over one of the spots while the other spot is used as a reference. The light is then collected and focused by the objective, and a simple quartz lens (focused on the back diaphragm of the objective) directs all the light on the photomultiplier tube. The photomultiplier tube used in this instrument is an EM19558QA.

For viewing the specimen and locating it with respect to the spots of light, a swingout prism directs the light to the binocular eyepieces. A rotating reflecting blade serves to direct light from the microscope illuminator into the light path to illuminate the background which can then be viewed simultaneously with the monochromatic spots of light. When the swingout prism is in position for a measurement, the rotating reflecting blade is automatically stopped so as not to interrupt the monochromatic light path and only the monochromatic light goes to the photomultiplier.

The photomultiplier output is fed into the ac-coupled amplifiers. The dc level is restored by a combination of a clamping circuit and a pulse-shaping circuit. The signal is then synchronously switched into separate RC smoothing circuits. The synchronization signals are supplied by a light source and phototransistor arrangement at the chopper. Now that the signal has been separated the sample signal is fed into a recorder. The reference signal

Fig. 2 Recording microspectrophotometer M-5. h, light source; m, monochromator; c, chopper; cs, cuvette chamber for solution spectroscopy; b, background illuminator; p, photomultiplier tube housing; r, chart recorder; e, electronics; pw, power supply.

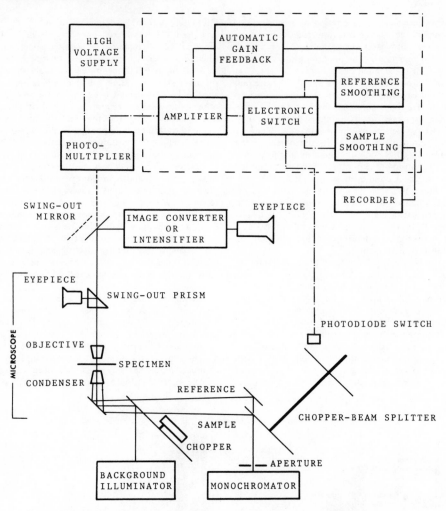

Fig. 3 Electronic and optical block diagram of recording microspectrophotometer M-5.

or AGC is then amplified and the output is applied to the gate of a load resistance field effect transistor located at the photomultiplier output. This keeps the average level of the reference signal at the amplifier output a constant. The average level of the sample signal is then directly proportional to the absorption of the specimen.

Living and stained tissue cells have been studied by measuring the absorption spectrum of specimen areas as small as $0.5\,\mu$ in diameter. With quartz, optics throughout a spectral range from the ultraviolet through the visible to the infrared can be recorded. This instrument is capable of scanning a specimen at low light levels (0.9 $\times$ 10^6 photons) from 225 to 800 nm. The spectrum from 350 to 750 nm can be scanned within 2 sec. For kinetic studies, a temperature-controlled specimen chamber (with thermoelectric elements for temperatures from $-20°$ to $+100°$C) has been adapted to the microscope stage. For specimen location when the material under investigation is photosensitive to visible light, an image tube intensifier or IR image converter can be fitted to the eyepiece of the microscope. Incorporated in the design is a chamber to hold cuvettes, which permits the instrument to be used for solution spectroscopy as well.

Instrumentation for microspectrophotometry, in addition to those described here, is being developed and constructed (4,5) in many laboratories. As a result of these developments, instruments for *microfluorospectrophotometry* have also been constructed to record the emitted fluorescence spectrum from cellular organelles, e.g., porphyrins and flavines (10).

Examples of Spectra. Microspectrophotometers are designed to rapidly record absorption spectra of living cells or cell organelles *in situ*. It is especially useful in determining cellular pigments e.g., chlorophyll, porphyrins, hemes, cytochromes, carotenoids, and flavines. Microspectrophotometry also provides information on cellular DNA, RNA, and protein concentration.

In research related to photosynthesis, the absorption spectrum of chlorophyll and the pigments of the chloroplast is of considerable importance in elucidating the process. In Fig. 4, the absorption spectrum of the *Euglena* chloroplast from the ultraviolet through the visible was obtained within 2 sec with a single sweep. The spectrum is similar in peak position to the spectrum of chlorophyll *a*, which is known to account for 80–90% of the *Euglena* chloroplast chlorophylls. The peak near

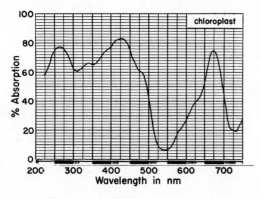

Fig. 4 Continuous absorption spectrum of *Euglena* chloroplast. The area sampled was 2.5 μ × 3 μ at a half bandwidth of 8 nm.

480 nm is that of carotenoid, which is associated with the chlorophylls in the chloroplast.

Microspectrophotometry has immediate application to photochemistry and to the understanding of how the retinal photoreceptors function in vision. Studies of the retinal photoreceptors, the rods and cones of the eye which are known to be sensitive to visible light, indicate that microspectrophotometry is most useful in studying the visual pigments (i.e., the photochemistry of the visual process during light ↔ dark reactions). The principal component of the retinal rods is the visual pigment rhodopsin. It accounts for about 40% of the dry weight of the frog retinal rod outer segment. Figure 5 is a typical absorption spectrum from 250 to 650 nm for a frog retinal rod outer segment. Curve 1 of the retinal rod shows the spectrum of rhodopsin and that the major absorption is near 500 nm; in addition, there are peaks near 280 and 350 nm, which are associated with rhodopsins. The light-bleached spectra, curves 2, 3, and 4, show the release of retinal from the rhodopsin complex with a shift in the spectral peak towards 380 nm, that of retinal[1]. The protein opsin absorption peak near 280 nm does not change with light bleaching. The search for the identification of the pigments in the retinal cones responsible for color vision can now be determined.

The number of chlorophyll molecules in a chloroplast or in a retinal rod outer segment can be calculated from

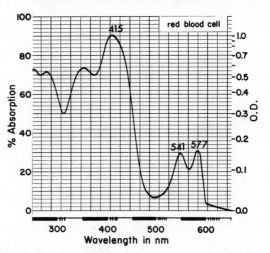

Fig. 6 Continuous absorption spectrum of a single freshly isolated human red blood cell. Area sampled was 1.5 μ × 1.5 μ at a half bandwidth of 8 nm.

such spectra (Figs. 4 and 5), if the thickness of the specimen and their extinction coefficients are known. For example, calculations from the absorption spectrum of the *Euglena* chloroplast indicate that there would be 1.3×10^9 molecules of chlorophyll in the chloroplast, and from the absorption spectrum of the frog retinal rod outer segment there would be 3×10^9 rhodopsin molecules in the retinal rod.

Another example is that of the red blood cell. Hemoglobin is present in large quantities in the cytoplasm of the red blood cell. The hemoglobin molecule contains four hemes which are bound to the protein globin. The heme pigments can be identified by their absorption peaks. Using the microspectrophotometer, the absorption spectrum of a single human red blood cell is shown in Fig. 6, which shows the absorption peaks near 415, 541, and 577 nm.

These few examples are given to illustrate the uniqueness of the instrumentation for obtaining information about the chemistry of the cell and cell organelles by direct spectroscopy.

In addition, the instrument has application to the study of the spectra of single crystals, especially those of biological origin. It also can be useful for studies in the identification of organic molecules in oceanic sediments, fossils, meteorites, and extraterrestrial matter.

Therefore, microspectrophotometry is a versatile, sensitive, and quantitative microanalytical tool for performing a variety of spectroscopic analyses of living cells, cell organelles, crystals, and particles, and in minute and dilute solutions which would be difficult if not impossible to obtain by other methods.

JEROME J. WOLKEN

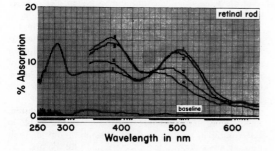

Fig. 5 Absorption spectrum of frog retinal rod. Curve 1, rhodopsin spectrum; curves 2, 3, and 4, after bleaching with white light for 15 sec intervals. This illustrates the conversion of the visual pigment rhodopsin to *retinal* (Vitamin A aldehyde) and the protein *opsin*. Area of sample was 5 μ × 20 μ with a half bandwidth of 8 nm.

References

1. Caspersson, T., "Cell Growth and Cell Function," New York, W. W. Norton and Company, Inc., 1950.
2. Wolken, J. J., and G. K. Strother, "Microspectrophotometry," *Appl. Opt.,* **2**:899 (1963).
3. Brown, P., "A system for microspectrophotometry employing a commercial recording spectrophotometer," *J. Opt. Soc. Amer.,* **51**:1000 (1961).

4. Barnes, J. C., and A. J. Thomson, "The design and construction of a microspectrophotometer," *J. Sci. Instr.,* **44**:577 (1967).

5. Engle, J. L., and J. J. Freed, "Double-beam vibrating mirror flying spot scanning-integrating microspectrophotometer," *Rev. Sci. Instr.,* **39**:307 (1968).

6. Chance, B., R. Perry, L. Akerman, and B. Thorell, "Highly sensitive recording microspectrophotometer," *Rev. Sci. Instr.,* **30**:735 (1959).

7. Liebman, P. A., "*In situ* microspectrophotometric studies on the pigments of single retinal rods," *Biophys. J.,* **2**:161 (1962). Liebman, P. A., "Microspectrophotometry of retinal cells," *Ann. N.Y. Acad. Sci.,* **57**:250 (1969). P. A. Liebman, "Microspectrophotometry of Photoreceptors" Sensory Physiology VII/1 Photochemistry of Vision, (H. J. A. Dartnall, ed.), New York, Springer-Verlag, (1972) pp 481–528.

8. Wolken, J. J., R. Forsberg, G. J. Gallik, and R. G. Florida, "Rapid recording microspectrophotometer," *Rev. Sci. Instr.,* **39**:1734 (1968).

9. Montgomery, P. O'B., ed., "Scanning techniques in biology and medicine," *Ann. N.Y. Acad. Sci.,* **97**:329–526 (1962).

10. Runge, W. R., "A recording microfluorospectrophotometer," *Science,* **151**:1499 (1966).

MILK

Freshly expressed milk contains protein casein granules 0.3–3 μ in size, fat globules ranging from 0.5 μ to 20 μ in diameter, white blood cells, lactose, various soluble salts, and vitamins in an aqueous phase forming 50–90 % by volume in different species. The two major problems in the preservation of milk for electron microscopy are the lack of any structural matrix connecting the two particulate components of the secretion and the difficulty of preservation of the saturated lipid which makes up the bulk of the fat globule.

Separation of the particulate components from the liquid phase is necessary at some stage and this destroys the relationship of the two major constituents *in vivo*. However, unless the milk has been homogenized the two components are maintained as separate entities after centrifugation, the bulk of the casein granules in a pellet, and the lipid droplets in a floating cream layer. This concentration of material is convenient in that far more examples of lipid or protein particles can be examined in a single sample.

Sectioning. It makes little difference whether the initial fixative is added to the milk prior to centrifugation or to the pellet and cream separately after spinning. Centrifugation at 900 *g* for 5 min is sufficient to separate most of the fat and protein. The blood and other cells in normal milk are found concentrated at the bottom of the protein granule pellet. If colostrum is used it is better to spin before fixation since the high serum protein concentration results in gel formation on addition of the fixative and the lipid and protein constituents cannot be concentrated by centrifugation. Pellet and cream are processed according to the following schedule:

4 % glutaraldehyde in 0.1 *M* phosphate buffer pH 7.2, 30–60 min

1 % osmic acid in 0.1 *M* Veronal-HCl buffer pH 7.2, 30–60 min

5 % uranyl acetate aq, 1–2 hr

dehydration in alcohol series to absolute, 1 hr

propylene oxide (PO), 30 min

3 : 1 PO : Araldite mixture, 2 hr

1 : 1 Araldite mixture, overnight

1 : 3 Araldite mixture, overnight

Araldite mixture, overnight

Embed in fresh Araldite, cure at 60°C for 6–7 hr.

This procedure gives good preservation of the casein granules and cells in the pellet and round fat globules in the center of the pieces of the original cream layer. The globules at the edge of cream pieces tend to be fragmented by manipulation of the sample during processing.

Instead of centrifugation, milk can be freeze-dried (Roelofsen and Salome, 1961; Board et al., 1970) or sealed into a narrow gelatin microcapsule prior to processing (Henstra and Schmidt, 1970a,b). The results seem to be comparable to those produced by centrifugation as far as can be judged from the low magnification of the published micrographs. Early work with milk using methacrylate embedding procedures produced very crumpled lipid droplet fragments (Knoop et al., 1958). Since in an aqueous environment a lipid droplet would be expected to have a spherical shape, it seems reasonable to assume that a smooth circular contour would indicate good fixation, and any corrugations would point to a failure in processing method.

Extraction of the lipid of the fat globule during processing may be considerable. Alcohol dehydration, infiltration with propylene oxide and monomer plastic all result in losses of neutral fat after osmium or glutaraldehyde-osmium fixation (Cope and Williams, 1968). However, the total amount lost varies considerably depending upon the exact conditions used (Stein and Stein, 1967; Korn and Weisman, 1966). Most electron micrographs of milk so far published show a membrane more or less crumpled around an electron-transparent area representing the lipid. Electron micrographs of mammary tissue range from the homogeneously electron-dense fat droplets in Bargmann and Knoop (1961) (osmium-methacrylate) and Murad (1970) (glutaraldehyde-osmium-maraglas) through droplets with a dense zone around the periphery (Pier et al., 1970) (glutaraldehyde-osmium-maraglas) to droplets with no greater electron density than is found with pure plastic (Helminen and Ericsson, 1968; Wooding et al., 1970).

Since there is good evidence that only unsaturated neutral fat reacts with the glutaraldehyde and/or osmium tetroxide used for electron microscope processing, the electron density of the interior of a fat droplet is more likely to reflect the degree of unsaturation of the bulk fat rather than the total amount of fat remaining in the section after processing. This would agree with Stewart and Irvine's (1970) finding that intravenous infusion of unsaturated fat into a cow increases the unsaturation of the milk fat by 5 % and that this "improves the fixation for electron microscopy producing fat droplets which are more electron dense than usual."

It is difficult to establish criteria for "good" fixation, since the normal organelles are present only in the blood cells found in milk, and these have been out of their normal environment for indeterminate times prior to fixation. However, at least some of the cells should have well-fixed organelles of conventional structure, and where the milk fat globules are cut in true transverse section, remnants of the unit membrane which envelops them completely on secretion (Fig. 1) should be visible. Between the unit membrane and the lipid there is a zone of dense material 1–200 Å wide. Occasionally this may contain ribosomes and organelles originating from the cytoplasm of the mammary secretory cell, in which case the zone may widen locally to up to 1 μ or more (Wooding et al., 1970). In view of this fact and the method by which the zone is produced on secretion I consider that this dense material is cytoplasmic in origin. After the fat droplet has been secreted, the originally continuous unit membrane and dense material splits and blebs off spheres of dense material surrounded by unit membrane into the milk serum. This process starts while the fat globule is still in the alveolus and continues during the life of the milk. The loss of the membrane and dense material leaves a thin 20–30 Å dense line around the fat droplet (Fig. 2).

Other Techniques. A recent negative staining study has claimed that the membrane around the fat globule differs significantly in appearance from plasmalemma membrane, indicating a change of state in the membrane on secretion (Keenan et al., 1970). Another method used in early studies was metal shadowing of whole milk preparations, on the basis of which Morton (1954) suggested that as well as casein and fat globules, "milk microsomes" equivalent to microsomes from tissue homogenates, were present in milk. This has never been confirmed by other structural investigations.

Freeze-etching of native milk has not so far been very informative because of the high water content (Buchheim, 1969). However Eggmann (1969) has produced good micrographs of freeze-etched milk powders which he mixed with glycerol to inhibit ice crystal formation.

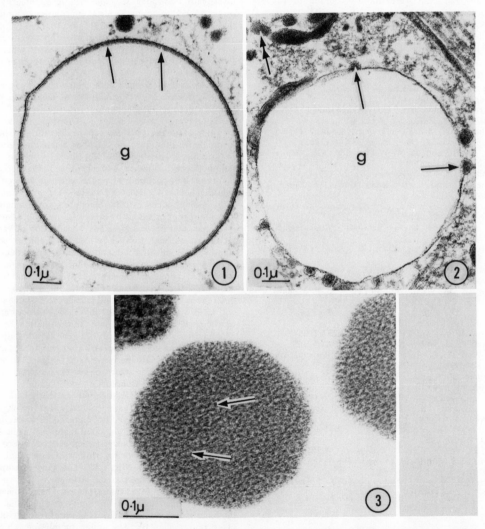

Fig. 1   Alveolar milk fat globule from goat, just after secretion. Note the continuous membrane consisting of a plasmalemma (arrow) and a zone of dense material (double arrow) around the structureless fat globule (g). Such a profile could come from goat, cow, or guinea pig; the morphology is identical.

Fig. 2   Milk fat globule in freshly expressed Jersey cow milk. Remnants of the plasmalemma plus dense material can be seen as small blebs (single arrows) on the thin dense line around the globule (g). Similar material can be seen free in the milk plasma (double arrow). Such a profile is typical of Holstein, Jersey, or Friesian cow, or goat milk.

Fig. 3   Casein micelles from goat milk. Note the 80–100 Å granules whic h make up the micelle (arrows).

Using this technique he showed that nonhomogenized milk fat droplets have an ordered lamellar structure within the bulk lipid. In homogenized milk powder, the casein granules are closely associated with the membrane of the fat droplets which are much smaller after homogenization, as has been shown also by Henstra and Schmidt (1970b) and Board et al., (1970) using sectioned material.

Casein Granule Substructures. Freeze-etching demonstrated the subunit structure of the casein micelle very clearly, cow casein having a 90 Å particle size and human 70 Å. The particles are not arranged in any particular order. This 90 Å "unit particle" size for the cow casein micelle agrees with the results from section (Fig. 3), also Shimmin and Hall, 1964) and negative stain (Calapaj, 1968) work. This would seem to eliminate the suggestion by Rose and Colvin (1966) that the particles seen by Shimmin and Hall were due to staining of calcium phosphate and not the casein unit particle. Theoretical discussions of the internal organization of the casein unit particle have no basis as yet in known structure (Garnier and Ribadeau-Dumas, 1970; Rose, 1969).

F. B. P. WOODING

References

Bargmann, W., and A. Knoop, *Z. Zellforschung. Mikrosk. Anat.,* **49**:344–388 (1959).
Board, P. W., J. M. Bain, D. W. Gove, and J. T. Mullett, *J. Dairy Res.,* **37**:513–519 (1970).
Buchheim, W., *Milchwissenschaft.,* **24**:6–11 (1969).
Calapaj, G. G., *J. Dairy Res.,* **35**:1–6 (1968).
Cope, G. H., and M. A. Williams, *J. Roy. Micr. Soc.* **88**:259–277 (1968).
Eggmann, H., *Milchwissenschaft.,* **24**:479–483 (1969).
Garnier, J., and B. Ribadeau-Dumas, *J. Dairy Res.,* **37**:493–504 (1970).
Helminen, H. J., and J. L. E. Ericsson, *J. Ultrastruct Res.,* **35**:193–213 (1968).
Henstra, S., and D. G. Schmidt, *Naturwiss,* **57**:247 (1970a).
Henstra, S., and D. G. Schmidt, *Neth. Milk Dairy J.,* **24**:45–51 (1970b).
Keenan, T. W., D. J. Morré, D. E. Olson, W. N. Yunghans, and S. Patton, *J. Cell Biol.,* **44**:80–93 (1970).
Knoop, W., A. Wortmann, and A. Knoop, *Milchwissenschaft.,* **13**:154–159 (1958).
Korn, E. D., and R. A. Weisman, *Biochem. Biophys. Acta,* **116**:309–316 (1966).
Morton, R. K., *Biochem. J.,* **57**:231–237 (1954).
Murad, T. M., *Anat. Rec.,* **167**:17–36 (1970).
Pier, W. J., J. C. Garnacis, and J. F. Kuzma, *Amer. J. Pathol.,* **60**:119–130 (1970).
Roelofsen, P. A., and M. M. Salome, *Neth. Milk Dairy J.,* **15**:392–394 (1961).
Rose, D., *Dairy Sci. Abstr.,* **31**:171–175 (1969).
Rose, D., and J. R. Colvin, *J. Dairy Sci.,* **49**:351–355 (1966).
Shimmin, P. D., and R. D. Hall, *J. Dairy Res.,* **31**:121–123 (1964).
Stein, O., and Y. Stein, *J. Cell Biol.,* **34**:251–264 (1967).
Stewart, P. S., and D. M. Irvine, *J. Dairy Sci.,* **53**:279–288 (1970).
Wooding, F. B. P., M. Peaker, and J. L. Linzell, *Nature,* **226**:762–764 (1970).

MITOCHONDRIA

The structure and function of the subcellular particles known as mitochondria have been the subject of investigation since the latter half of the nineteenth century. It is only since 1948, however, when intact mitochondria were first isolated in a relatively pure suspension from rat liver, that knowledge of their function has been obtained. Improvement in specimen preparation techniques for electron microscopy led shortly afterwards to the demonstration of mitochondrial ultrastructure and it was found that mitochondria, whatever their source, have a common basic structure.

Broadly speaking, the structure and function of mitochondria may be studied under two conditions, that is, with mitochondria *in situ* or in an isolated state. While the former would appear to approximate the *in vivo* condition and therefore be the more ideal state for investigation, it is readily apparent that, in functional studies, difficulties would arise due to activities of other cellular components. For this reason work correlating structure and function has been done mainly on isolated mitochondria. Most investigators consider desirable criteria of integrity and purity to be: (a) mitochondrial morphology similar to that observed in the cell of origin; (b) ability of the isolated mitochondria to perform their function of oxidative phosphorylation with a high degree of efficiency; (c) absence from the mitochondrial suspension of other cell organelles.

The technique of differential centrifugation in sucrose medium, described by Hogeboom et al. (1), for the isolation of rat liver mitochondria forms the basis of many present-day methods of isolation. The steps involved are as follows: (a) the cells are ruptured in the suspending medium; (b) the unbroken cells and nuclei are removed by low-speed centrifugation; (c) the supernatant, containing the mitochondria, is recentrifuged at higher speed to obtain a mitochondrial pellet which is washed and resuspended in the isolated medium. The mitochondrial suspension obtained by this procedure is contaminated to a greater or lesser extent by other cell organelles, such as lysosomes and microsomes. The introduction of various modifications, e.g., the use of density gradient sucrose solutions, has allowed the separation of mitochondria in a relatively homogeneous suspension. Since the properties of isolated mitochondria are greatly influenced by the composition of the suspending medium (2), various different media have been tried in order to find one which will yield a sample of mitochondria with structural and biochemical properties closely approximating *in vivo* characteristics. The most frequently employed medium is a 0.25 M sucrose solution at pH of 7.0–7.5. All the manipulative procedures are performed at 0°C–4°C to avoid deleterious changes in structure and enzymic function of mitochondria during isolation.

The ultrastructure of mitochondria, whether *in situ* or in isolated form, is usually studied either in ultrathin sections, positively stained to enhance contrast, or in preparations of negatively stained material. In ultrathin sections of material fixed and embedded by routine methods, the mitochondrion consists of two membrane-bound compartments (Fig. 1A). The inner membrane which encloses the matrix compartment is thrown into folds which project into the matrix area. These projections are known as cristae. The continuity of the cristae with the inner membrane can be observed in favorable planes of section (Fig. 1B). The outer membrane which is unfolded is separated from the inner one by a narrow space of constant dimension. The outer compartment consists of the intracristal space and the space between the inner and outer membranes. This basic design has minor variations, e.g., cristal form and number, from one type of tissue to another.

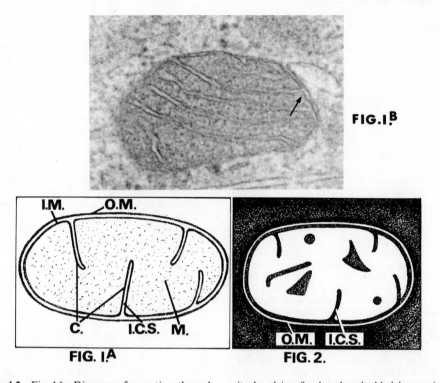

FIG.1.B

FIG. 1.A

FIG. 2.

Figs. 1 and 2   Fig. 1A: Diagram of a section through a mitochondrion fixed and embedded by routine methods. I.M., inner membrane; O.M., outer membrane; C., crista; I.C.S., intracristal space; M., matrix compartment. Fig. 1B: Electron micrograph of a section through a mitochondrion of a monkey kidney cell in tissue culture showing continuity between inner membrane and crist; (arrow). Fig. 2: Diagrammatic representation of an intact mitochondrion negatively stained. The negative stain surrounds the mitochondrion and is present in the outer compartment, i.e., between the outer and inner membranes and in the intracristal space.

At the conclusion of the usual isolation procedure the mitochondrial pellet is large and loosely packed so that it disintegrates easily on handling. The procedures of fixation, dehydration, and embedding for ultrathin sectioning are facilitated if the material remains compact throughout the manipulations. Furthermore, since mitochondrial suspensions may consist of heterogeneous populations of light and heavy mitochondria and some nonmitochondrial contaminants which sediment at different rates, a method of obtaining sections representative of the entire pellet population is necessary. Malamed (3) has described a method of obtaining a mitochondrial pellet that satisfies both of these requirements. The isolated mitochondria are suspended in fixative and very small volumes of the suspension transferred to polyethylene conical microcentrifuge tubes. The tubes are centrifuged at approximately 10,000 g for 5 min, the fixative withdrawn with a pasteur pipette, and the tubes then cut across above and just below the pellet, which can be pushed out with a blunt needle into a container of dehydrating fluid. Throughout the remaining steps of dehydration and embedding the pellet retains its shape and firmness and may be handled in the same way as a tissue block. Suitable microcentrifuges are produced by Coleman Instruments Inc. and Beckman Instruments Inc., Spinco Division.

The application of the negative staining technique to suspensions of isolated mitochondria has provided rewarding results. The principle of the method is that a solution of a negative stain is added to the mitochondrial suspension which is then applied to a specimen grid. When dry, the solution forms an electron-dense cast and any structures surrounded by it are defined but in a negative sense (Fig. 2). As stated by Valentine and Horne (4), a good negative stain should have a high weight density, high solubility, and a high melting point so that it does not melt in the electron beam. Until recently, the most frequently used negative stain was phosphotungstate. It now apears (5,6) that, of the negative stains tried, a 2%–4% solution of ammonium molybdate at pH 7.0 is best for study of mitochondrial morphology. Prior fixation is not required when ammonium molybdate is used and this may be due, at least in part, to the fact that at these concentrations the ammonium molybdate solutions are isotonic with the sucrose solutions usually employed in the isolation of mitochondria. Negative staining has revealed the presence of small particles on the inner membrane of many mitochondria. Each particle is attached to the membrane by a stalk so that it projects into the matrix compartment. The demonstration of the presence of these particles (which have been named elementary particles), and the number seen depends on the stain used and whether fixation is done before staining.

The techniques used in the study of mitochondrial ultrastructure have led to a "static" image of mitochondria whereas phase contrast microscopy and time lapse microcinematography have shown that *in vivo* these organelles are constantly moving and changing shape.

Their primary function is that of providing a readily available supply of energy, in the form of adenosine triphosphate, for the metabolic needs of the cell. The production of high-energy phosphate is coupled to oxidation of substrate and the process is known as oxidative phosphorylation. It was found that the morphology of mitochondria alters with change in metabolic state and also that the addition of certain substances to suspensions of mitochondria could cause them to swell or shrink while at the same time affecting their metabolic activity. The volume alterations have been studied by various techniques, e.g., particle diameter measurements, packed volume, dry weight, and most frequently, optical density measurements. This last method is very rapid and convenient but does not by itself always give an accurate indication of volume changes and should be performed in conjunction with one of the other techniques, such as packed volume or dry weight measurements.

The separation of the inner and outer mitochondrial membranes for biochemical analysis has permitted confirmation that certain of the mitochondrial enzymes are bound to one membrane or the other, e.g., the inner membrane contains the respiratory chain enzymes. With perfection of cryo-ultramicrotomy techniques more precise localization of mitochondrial enzymes should be possible.

The existence of specific mitochondrial DNA and RNA which have many properties in common with the nucleic acids of chloroplasts and bacteria has been demonstrated. It is hoped that the elucidation of the precise functions of these mitochondrial nucleic acids will provide a clue to the origin of mitochondria.

MARY T. O'HEGARTY

References

1. Hogeboom, G. H., W. C. Schneider, and G. E. Pallade, "Cytochemical studies of mammalian tissues. 1. Isolation of intact mitochondria from rat liver; some biochemical properties of mitochondria and submicroscopic particulate material," *J. Biol. Chem.*, **172**:619–636 (1948).
2. Chappell, J. B., and G. D. Greville, "The influence of the composition of the suspending medium on the properties of mitochondria," p. 39 in "Methods of separation of subcellular structural components," *Biochem. Soc. Symp.*, No. 23 (1963).
3. Malamed, S., "Use of a microcentrifuge for preparation of isolated mitochondria and cell suspensions for electron microscopy," *J. Cell Biol.*, **18**:696–700 (1963).
4. Valentine, R. C., and R. W. Horne, "An assessment of negative staining techniques for revealing ultrastructure," in "The Interpretation of Ultrastructure," R. J. C. Harris, ed., New York, Academic Press, 1962.
5. Munn, E. A., "On the structure of mitochondria and the value of ammonium molybdate as a negative stain for osmotically sensitive structures." *J. Ultrastruct. Res.*, **25**:362–380 (1968).
6. Muscatello, U., and R. W. Horne, "Effect of the tonicity of some negative-staining solutions on the elementary structure of membrane-bounded systems," *J. Ultrastruct. Res.*, **25**:73–83 (1968).

MOLLUSCA

Light microscopy remains important despite a current emphasis on ultrastructural research. Historically, light microscopy enabled investigators to describe basic tissue and organ morphology. Currently, the application of light microscopy has been widened to the fields of cytochemistry and cellular physiology. Conventional light microscopy, phase contrast, immunofluorescence, polarizing microscopy, autoradiography, and multifarious histochemical techniques are applied in the delineation of cellular structure and function. A detailed exposition of these methods is not possible within the limits of this paper, but fortunately their application to molluscan tissues requires little or no modification from the standard procedures described in several volumes (8,12,15).

Ultrastructural techniques, unlike those of light microscopy, must frequently be adapted for application to molluscan tissues. The physiological individuality of molluscs and the specific organ chosen for study frequently necessitate diverse solutions to the fundamental problems of microscopy. Approaches to these problems are presented in the following discussion of whole mount, replication, and ultramicrotomy techniques. General references on ultrastructural techniques can be consulted for introductions to these methodologies (16,20). Briefly, these techniques involve viewing the specimen either in its entirety with whole mounts, or by means of a mold cast from a surface by replication, or by sections of less than 1000 Å thickness.

Whole Mounts. Whole mounts are most functional in topographical investigations and can be used in conjunction with negative staining, shadow casting, and scanning electron microscopy. The use of the scanning electron microscope is analogous to that of the dissecting microscope; however the former microscope features resolution to about 150 Å as well as maintenance of focus in a wide depth of field. This rather simple technology has been applied to the calcified organs of molluscs, the shell, and the radula. Routine preparation for scanning electron microscopy involves mounting the specimen on a stub and vacuum coating it with a thin (100–200 Å) metallic layer. This metallic layer prevents image distortion that results from an accumulation of charge on the poorly conducting specimen during electron bombardment (19). Additional metal shadowing and/or etching with weak acids can enhance the geographical contrast of natural or fractured shell surfaces. In studies of a gastropod shell, Wise and Hay (26) noted that etching rates vary as the orientation of the anisotropic calcium carbonate crystals changes. The variation results in differential decalcification of the nacreous and prismatic layers. The high relief of the radular surfaces makes etching unnecessary; however, this flexible organ must be hardened for scanning electron microscopy. Air drying (19) and lyophilization (18) produce adequate preservation.

References to negative staining and shadow casting techniques are not abundant in the literature of Mollusca; yet the high resolving power gained by these techniques might facilitate topographical investigations of isolated cells or intracellular structures (10,16).

Replication. Before the genesis of the scanning electron microscope, knowledge of the ultrastructure of molluscan shells has been gained by viewing formvar (21), collodion (1), and carbon (7,24,26) replicas with the transmission electron microscope. Also, the radula of *Cryptochiton stelleri* has been studied by replication (23). Again, preliminary induration of the radula is required and can be effected by either of two standard fixatives, osmium tetroxide (buffered with veronyl acetate) or glutaraldehyde. For one-stage replication a direct casting, or preliminary replica, is made of the surface, then separated from it, and examined. Two stage replication utilizes the preliminary replica as a mold to prepare a second casting, or final replica. Both replication techniques are common, but the latter is sometimes preferred because (a) the final replica is a positive duplication of the surface topography, thus facilitating interpretation of the image seen with the electron microscope and (b) removal of the preliminary replica does not destroy the surface, thus permitting additional experimental treatments and repeated replication of the same area. For two-stage replication Grégoire (7) recommended the use of a water-soluble plastic for preliminary replication to minimize desiccation artifacts that could occur with organic solvents. Metal shadowing of the replica itself or of the surface before replication and limited etching of the surface will augment the contrast of the replica. Detailed discussions of the methodology are available (16). Replication of soft tissues by vacuum coating the exposed surface of a fractured, deep-frozen specimen is possible. These freeze-etching techniques were reviewed by Moor (14). The use of freeze-etching is limited by the difficulty of preparing watery molluscan tissues. Ice crystal formation must be minimized by displacing some of the free cellular water with glycerin.

Ultramicrotomy. Although the soft organs of molluscs can conceivably be scrutinized with whole mounts, most researchers have relied upon ultramicrotomy. To a lesser extent, shells have been similarly studied. The procedures for the thin sectioning of tissues include chemical fixation with consequent preservation of fine subcellular detail and distortion-free embedding in a medium which survives the rigors of ultramicrotomy.

Soft Tissues. The pliability of soft tissues may generate problems for the electron microscopist. With such tissues fixation is critical. The standard fixatives are osmium tetroxide, glutaraldehyde, or some combination of these two. The osmolality and pH of the fixative solution are important. Saline (2), sucrose (3,17), molluscan Ringer's (11), artificial sea water (4), and sea water or squid physiological solution (13) have been used to maintain suitable osmolality. Control of pH has been achieved with numerous buffering systems; e.g., veronyl acetate (27), phosphate (9), sodium bicarbonate (2), cacodylate (17), and S-collidine (6). Experimentation will reveal the combination of factors which produce optimum preservation of a particular mollusc. Even when good fixation is obtained subtle artifacts may have been introduced (5,13). Embedding in such standard media as Epon, Araldite, Vestopol W, and Durcupan follows fixation. The sectioning qualities of the tissue and the extractions occurring during embedding are considerations which may influence the choice of embedding media.

Shell. Difficulties in the preparation of shells for thin sectioning originate from the dense crystalline structure. First, it hinders the infiltration of embedding media into the shell. Complete infiltration of entire *Mytilus edulis* valves required three months (24). Acceleration by heating was avoided to prevent the introduction of artifacts into the crystalline architecture. Vacuum embedding is an alternative approach (22). Secondly, arduous ultramicrotomy is anticipated after embedding. Thirdly, the opaqueness of the shell must be reduced before the specimen can be visualized. Severe etching either before (22) or after (24,25) ultramicrotomy produces the needed translucence. During this decalcification, disruption of the shell organic matrix is lessened by including a buffer and fixative in the etching solution (24).

Ultrastructural Cytochemistry. The growth of this em-

bryonic subspeciality of electron microscopy hinges on the development of mild preparative techniques. Glutaraldehyde (9) is presently the fixative of choice, and water-soluble resins show promise as nondesiccative embedding media (16).

The microscopist who is interested in studying molluscan tissues can choose from an array of techniques. A choice of several techniques is preferable because (a) many techniques are complementary—one compensates for the theoretical and procedural limitations of another; (b) the examination of a specimen at different levels of magnification results in a synergistic accumulation of observations; and (c) by a correct coordination of techniques the researcher can expand a morphological investigation into one which deals with structural physiology.

LOUISA SCHMID

References

1. Bevelander, G., and H. Nakahara, "An electron microscope study of the formation of the periostracum of *Macrocallista maculata*," *Calc. Tiss. Res.,* **1**:55 (1967).
2. Clony, R. A., and E. Florey, "Ultrastructure of cephalopod chromatophore organs," *Z. Zellforsch Mikroskop Anat.,* **89**:250 (1968).
3. Coggeshal, R. E., "A fine structure analysis of the statocyst in *Aplysia californica*," *J. Morphol.,* **127**:113 (1969).
4. Dilly, P. N., E. G. Gray, and J. Z. Young, "Electron microscopy of optic nerves and optic lobes of *Octopus* and *Eledone*," *Proc. Roy. Soc. London,* **158**:446. (1963).
5. Doggenweiler, C. F., and J. E. Heuser, "Ultrastructure of prawn nerve sheaths. Role of fixation and osmotic pressure in vesiculation of thin cytoplasmic membranes," *J. Cell Biol.,* **24**:407 (1967).
6. Gray, E. G., "Electron microscopy on the gleovascular organization of the brain of *Octopus*," *Phil. Trans. Roy. Soc. London,* **255**:13 (1968).
7. Grégoire, C., "Structure of the conchiolin cases of the prisms in *Mytilus edulis* Linne," *J. Biophys. Biochem. Cytol.,* **9**:395 (1961).
8. Humason, G. L., "Animal Tissue Techniques," San Francisco, W. H. Freeman and Co., 1967.
9. Japha, J. L., and A. W. Wachtel, "Transmission in the visceral ganglion of the freshwater pelecypod *Elliptio complanatus*, II. esterase histochemistry," *Comp. Biochem. Physiol.,* **29**:571 (1969).
10. Kay, D. H., ed., "Techniques for Electron Microscopy," Philadelphia, F. A. Davis Co., 1965.
11. Kelly, R. Z., and R. L. Hayes, "The ultrastructure of smooth cardiac muscle in the clam *Venus mercenaria*," *J. Morphol.,* **127**:163 (1969).
12. Lillie, R. D., "Histopathologic Technique and Practical Histochemistry," New York, McGraw-Hill Book Co., 1965.
13. Martin, R., and P. Rosenburg, "Fine structure alterations associated with venom action on the giant squid nerve fibers," *J. Cell Biol.,* **36**:34 (1968).
14. Moor, H., "Freeze-etching," in "International Review of Cytology," Vol. 25, G. H. Bourne and J. F. Danielli, ed., New York, Academic Press, 1969.
15. Pearse, A. G. E. "Histochemistry, Theoretical and Applied," Boston, Little, Brown and Co., 1968.
16. Pease, D. C., "Histological Techniques for Electron Microscopy," New York, Academic Press, 1964.
17. Personne, P., and W. Anderson, "Compartmentalization of enzymatic activities in the spermatozoon of certain gastropod molluscs. I. Localization of dehydrogenases," *J. Cell Sci.,* **4**:693 (1969).

18. Runham, N. W., "The use of the scanning electron microscope in the study of the gastropod radula: the radulae of *Agriolimax reticulatus* and *Nucella lapillus*," *Malacologia*, **9**:179 (1969).

19. Runham, N. W., and P. R. Thornton, "A scanning electron microscopic study on mechanical wear of the gastropod radula," *J. Zool.*, **153**:445 (1967).

20. Sjöstrand, F. S., "Electron Microscopy of Cells and Tissue," Vol. I, Instrumentation and Techniques, London, Academic Press, 1967.

21. Taylor, J. D., "The influence of the periostracum on the shell structure of bivalve molluscs," *Calc. Tiss. Res.*, **3**:274 (1969).

22. Towe, K. M., and G. H. Hamilton, "Ultrastructure and inferred calcification of the mature and developing nacre in bivalve mollusks," *Calc. Tiss. Res.*, **1**:306 (1968).

23. Towe, K. M., and H. A. Lowenstam, "Ultrastructure and development of iron mineralization in the radular teeth of *Cryptochiton stelleri* (Mollusca)," *J. Ulstrastruct. Res.*, **17**:1 (1967).

24. Travis, D. F., "The structure and organization of, and the relationship between, the inorganic crystals and the organic matrix of the prismatic region of *Mytilus edulis*," *J. Ultrastruct. Res.*, **23**:183 (1968).

25. Watanabe, N., "Studies on shell formation. XI Crystal-matrix relationships in the inner layers of mollusc shells," *J. Ultrastruct. Res.*, **12**:351 (1965).

26. Wise, S. W., Jr., and W. W. Hay, "Scanning electron microscopy of molluscan shell ultrastructures. I. Techniques for polished and etched sections" (p. 411); "II. Observations of growth surfaces" (p. 419), *Trans. Amer. Microscop. Soc.*, **87** (1968).

27. Yasuzumi, G., "Spermatogenesis in animals as revealed by electron microscopy. I. The fine structure and function of endoplasmic reticulum and of peculiar bodies appearing in a typical maturing spermatids and nutritive cells of *Cipangpaludinae malleata* (Rieve)," *Amer. J. Anat.*, **45**:431 (1964).

MOLYBDENUM

1. Introduction

Molybdenum is a refractory metal, the industrial applications of which are becoming more and more important. The properties of molybdenum are directly correlated with its structure, and principally with its purity when it is used at high temperatures, so that metallographic examination is often necessary to establish its character.

Preparation of the surface of a specimen of molybdenum and its micrographic study are carried out in the following ways.

2. A Preparation of the Specimen

2.1. Cutting. The specimen to be examined is cut with the aid of a saw or of a slicing disk. Diamond impregnated or carbide-fritted disks are best for this cutting.

2.2. Embedding. This operation is often indispensible to avoid rounding the edges of the specimen in the course of polishing. Cold polymerizable resins are at present used since they allow rapid embedding. Certain resins are destroyed by immersion in liquid nitrogen, which permits the ready recovery of the specimen after polishing.

2.3. Mechanical Polishing. This is a very important operation since defective polishing can cause a mistaken interpretation of a structure. The following operations are carried out in order to bring the surface to be examined to a mirror polish:

Smoothing with a lap or a file.

Smoothing with emery paper of successively finer grain, the normal succession is the following: 80, 120, 180, 240, 400, 600.

(This operation is best carried out under running water. Each paper should be applied at right angles to the direction of the previous one.)

Polishing with the aid of a diamond dust pastes used on the surface of felt covered laps.

(Pastes with diamond particle dimensions of 6, 3, 1, and 0.1 microns are used successively. In the course of polishing it is necessary to rotate the specimen to avoid the polishing defect known as "comets.")

A final polish with the aid of ultrafine magnesium oxide or aluminum oxide is carried out in some cases to eliminate the very fine scratches of the 0.1μ diamond.

2.4. Polishing by Vibration. By this technique it is possible to obtain polished surfaces lacking scratches, comets, or orange peel effects entirely automatically.

The specimens to be polished are placed on an abrasive vibrating surface. Complete polishing requires long hours of vibration.

2.5. Electrolytic Polishing. Specimens treated in this manner yield a metallic surface without strains and without contamination. It is necessary to conduct a preliminary mechanical polishing to the level of 600 emery paper. Numerous formulae for polishing baths have been described in the literature but the following is the most frequently used:

methyl alcohol, 600cc
concentrated sulfuric acid, 20cc
concentrated hydrochloric acid, 20cc
Method of use
temperature, 20°C
time, 15 min
potential, 18 v
current density, 5 a/cm^2
molybdenum cathode

The potential must be maintained on the specimen while it is rinsed.

2.6. Electrolytic Polishing with a Plug. This is principally used for the control of localized polishing on a finished piece.

The specimen is subsequently etched micrographically and a celluloid varnish replica is made which is examined by reflected or transmitted light.

2.7. Micrographic-Polish Etching. In this technique micrographic etching is combined with polishing.

After polishing to the level of 6 μ diamond paste, the surface is manually polished with a suspension of ultrafine aluminum mixed with a 10% solution of potassium ferrocyanide in water.

After three minutes the structure is brought out and the surface appears to have a mirror polish. This method of micrographic etching does not allow the examination of structure to be pushed as far as can be done after the etching techniques described below.

3. Micrographic Etching

3.1. The Purpose of Micrographic Etching. Etching reagents by selectively attacking crystal boundaries or particular constituents principally bring out:

the methods of production or manufacture

defects or accidents in production

mechanical, chemical, or thermal treatments

crystalline structure

physical and chemical impurities

In general, micrographic etching should be proportionally lighter as the magnification used for examination becomes higher, in order that it may be compatible with the depth of field of the objective used.

3.2. Chemical Etching

3.2.1 ferrocyanide of potassium, 360 g

sodium carbonate, 36 g

distilled water, 1000 cm^2

immersion time, 5–20 sec

3.2.2 ferrocyanide of potassium, 100 g

sodium carbonate, 100 g

distilled water, 1000 cm^2

immersion time, 2–3 min

3.3. Electrolytic Etching

3.3.1 oxalic acid, 5 g

distilled water, 1000 cm^2

potential, 5 v

current density, 0.3 a/cm^2

time, 10–20 sec

3.3.2 sodium carbonate, 100 g

distilled water, 1000 cm^2

potential, 2 v

current density, 0.4 a/cm^2

time, 5 sec

3.4. Color-Producing Chemical Agents

3.4.1 Slow Reagent (1, 2, 3)

concentrated hydrochloric acid, 25 cm^2

95 % ethanol, 75 cm^2

ferric chloride, 6.5 g

emersion time, 2–3 min, rinse lightly dry with compressed air

3.4.2 Rapid Reagent

concentrated hydrochloric acid, 25 cm^2

distilled water, 75 cm^2

ferric chloride, 6.5 g

emersion time, 25–30 sec, rinse lightly dry with compressed air

Note: The ferric chloride in the above two formulas is usually added in the form of 50 cm^2 of a solution of 130 g FeCl$_3$ per liter.

3.4.3 Coloration of Molybdenum

With these two reagents perfectly uniform interference films are developed of which the color is a function of the crystal orientation of the underlying grains of molybdenum.

Orientation	Color
Plane (100)	golden yellow
Plane (110)	dark blue
Plane (111)	violet blue
Plane (112)	light blue

These reagents color only pure molybdenum and thus permit its ready identification in complex mixtures.

R. HASSON

References

Alleau, T., M. Clemot, and R. Hasson, "Post mortem examinations of thermoionic emitters," Thermoionic Conversion Specialist Conference, Palo Alto, California, October 1967.

R. Hasson, "Sur la métallographie en couleurs du molybdène," Symposium de Microscopie, Cambridge, August 1967.

R. Hasson, "Sur les applications aux études des émetteurs thermoioniques d'un nouveau réactif du molybdène," Deuxième Conférence Internationale sur la Production Thermoionique, d'Energie Electrique. Stresa, Italie, May 1968.

MOUNTANT FORMULAS

Introduction

Mountants, as pointed out in the following article, may be divided into three groups according to the treatment which the object must receive before mounting. The first group (Mountants Miscible with Water) contains those mountants to which the object may be directly transferred either from water or from glycerol. The second class (Mountants Miscible with Alcohol) comprises those mountants to which objects may be transferred after dehydration but without having passed through a clearing agent. The third class of mountants (Mountants not Miscible with Either Alcohol or Water) are the conventional resins and balsams to which objects can be transferred only after they have been dehydrated and cleared in the usual manner.

There are at present on the market many proprietary compounds of secret composition which cannot be noticed in this place for reasons elsewhere given by the present writer (Gray 1954, "The Microtomist's Formulary and Guide," Philadelphia, Blakiston, p. 3.) and endorsed by Conn (1954, *Stain Technol.* 29 : 222). The physical characteristics of many of these are given by Lillie *et al.*, 1953 (*Stain Technol.*, 28 : 57).

Mountants Miscible With Water

This section may be divided into three groups. The first is gum arabic media (of which Farrants' is the type) to which objects may be transferred either directly or from water. These media should be far more widely used than is commonly the case, for a great deal of time is wasted in dehydrating and in transferring to balsam objects which were better mounted in gum arabic. The second group (gelatin media) contains glycerol jellies which are widely used by botanists and to a lesser extent by zoologists for the preparation of Crustacea. These media have the disadvantage that they must be melted before use and are, therefore, not nearly as simple to use as the gum arabic media. The third group, other media, which will probably become more numerous as time goes on, employs water-thickening agents other than gum arabic or gelatin. Any of these three groups may be mixed with stains for special purposes and the best-known media of this type are those of Zirkle, some of which will be found in each of the three sections.

Gum Arabic Media

Formulas

Allen cited from 1937 Gatenby and Cowdry (Gatenby and Cowdry 1937, 221)

FORMULA: water 45, glycerol 11, 40 % formaldehyde 4.5, gum arabic 45
PREPARATION: Dissolve gum arabic in water. Mix formaldehyde in the glycerin and add slowly, with constant stirring, to gum.

André cited from 1942 Langeron (Langeron 1942, 930)
FORMULA: water 50, glycerol 20, gum arabic 30, chloral hydrate 200
PREPARATION: Dissolve gum arabic in water. Mix glycerol in gum and add chloral hydrate.

Apáthy 1892 (*Z. wiss. Mikr.*, **89** : 1065)
FORMULA: water 30, gum arabic 30, levulose 30
PREPARATION: dissolve gum arabic in water. Add levulose to solution.
NOTE: Much grief in the prevention of bubbles may be avoided by reducing the water to 20 and using commercial levulose syrup in place of the dry sugar.

Berlese cited from 1929 Imms (*Bull. Entomol. Res.*, **20** : 165)
FORMULA: water 10, acetic acid 3, dextrose syrup 5, gum arabic 8, chloral hydrate 75
PREPARATION: Dissolve the acid in the water with the syrup and gum arabic. Add chloral hydrate to the solution.
NOTE: Swan 1936 (*Bull. Entomol. Res.*, **27** : 389) states that Berlese first disclosed the formula to Davidson in 1919, who communicated it to Lee by whom it was published in 1921. Doestschman 1944 (*Trans. Amer. Micr. Soc.*, **63** : 175) uses three times as much water in his formula for "Berlese."

Chevalier 1882 (Chevalier 1882, 319)
FORMULA: water 60, gum arabic 20, glycerol 20

Dahl 1951 (*Stain Technol.*, **26** : 99)
PREPARATION OF GUM ARABIC: Dissolve 100 gm arabic in 100 water and heat to 100°C. Add 3 sodium carbonate dissolved in 10 water and add to hot solution. After a few days decant from precipitate.
FORMULA: Gum arabic solution 100, Amman 1896, lactophenol 50 (see below under "other water miscible mountants")

Davies *circ.* 1865 (Davies 1865, 82)

Doetschman 1944 (*Trans. Amer. Micr. Soc.*, **63** : 175)
FORMULA: water 35, glycerol 20, dextrose syrup 20, gum arabic 20, chloral hydrate 20, sat. aq. sol. magenta 0.3

Ewig cited from 1944 Doetschman (*Trans. Amer. Micr. Soc.*, **63** : 175)
FORMULA: water 35, glycerol 12, dextrose syrup 3, gum arabic 20, chloral hydrate 30

Farrants cited from 1880 Beale (Beale 1880, 68)
FORMULA: water 40, glycerol 20, gum arabic 40
NOTE: This mixture requires a preservative. Arsenic, camphor, and phenol have all been recommended. If only dirty gum arabic is available the mixture may be filtered through glass wool. The author's name is almost always misspelled "Farrant."

Faure 1910 (*Ann. Bot. Roma.*, **8** : 25)
FORMULA: water 50, chloral hydrate 50, gum arabic 30, glycerol 20

Gater 1929 (*Bull. Entomol. Res.*, **19** : 367)
FORMULA: water 10, acetic acid 2.7, gum arabic 8, chloral hydrate 74, glucose syrup 5, cocaine hydrochloride 0.3

Highman 1946 (*Arch. Pathol. Anat.*, **41** : 559)
FORMULA: water 50, gum arabic 25, sucrose 25, potassium acetate 25

Hogg 1883 (Hogg 1883, 237)
FORMULA: water 75, gum arabic 25, phenol 5
PREPARATION: Dissolve gum arabic in 25 water. Dissolve phenol in 50 water and mix with gum.

Hoyer 1882 (*Biol. Zbl. L.*, **2** : 23)
FORMULA: water 50, gum arabic 50, chloral hydrate 2
PREPARATION: Use water and chloral hydrate to dissolve gum.

Landau 1940 (*Bull. Histol. Tech. Micr.*, = *Bull. d'Hist. appl.*, **17** : 65)
FORMULA: water 30, gum arabic 30, dextrose 30, glucose 5

Langerhaus 1879 (*Zool. Anz.*, **2** : 575)
FORMULA: water 20, glycerol 25, gum arabic 60, phenol 1
PREPARATION: Dissolve gum in 20 water and filter. Add glycerol and phenol to filtrate.

Lieb 1947 Abopon mountant—auct (*Amer. J. Clim. Pathol.*, **17** : 413)
This involves a proprietary product of secret composition and cannot, therefore, be further noticed.

Lillie and Ashburn 1943 (*Arch. path. Anat.*, **36** : 432)
FORMULA: water 100, gum arabic 50, sucrose 50, thymol 0.1

Marshall 1937 (*Lab. J.*, **7** : 565)
FORMULA: water 50, gum arabic 0.5, gum tragacanth 1.5, Archibald and Marshall 1931 50
PREPARATION: Dissolve gum arabic in water with gum tragacanth with boiling. Cool. Mix Archibald and Marshall 1931 (see below under other water miscible mountants) with gums. Filter.

Martin 1872 (Martin 1872, 169)
FORMULA: water 50, gum arabic 50, glycerol 25, camphor 0.2

Morrison 1942 (*Turtox News*, **20** : 157)
FORMULA: water 50, glycerol 20, acetic acid 3, gum arabic 40, chloral hydrate 50

Robin 1871a (Robin 1871, 372)
FORMULA: water 45, gum arabic 15, glycerol 30

Robin 1871b (Robin 1871, 372)
FORMULA: water 100, gum arabic 50, glycerol 50

Robin 1871c (Robin 1871, 372)
FORMULA: water 60, gum arabic 20, glycerol 20

Schweitzer 1942 cited from 1946 Roskin (Roskin 1946 200)
FORMULA: water 65, gum arabic 20, chloral hydrate 8, glycerol 7

Semmens 1938a (*Microscope*, **2** : 120)
FORMULA: water 40, gum arabic 20, Belling 1921

Semmens 1938b (*Microscope*, **2** : 120)
FORMULA: water 45, gum arabic 10, chloral hydrate 25, acetic acid 37.5, carmine 0.5
PREPARATION: Dissolve the chloral hydrate in 25 acetic acid with 25 water, raise to boiling, stir in carmine, cool, and filter. Dissolve the gum in 20 water and 12.5 acetic acid. Mix the solutions.

Swan 1936 (*Bull. Entomol. Res.*, **27** : 389)
FORMULA: water 20, gum arabic 15, chloral hydrate 60, glucose syrup 10, acetic acid 5

Womersley 1943 (*Trans. Roy. Soc. S. Aust.*, **67** : 181)
FORMULA: water 100, 95% alc. 50, gum arabic (powder) 40, phenol 50, chloral hydrate 50, glucose syrup 10, lactic acid 20
PREPARATION: Mix alc. and powdered gum to a smooth paste. Flood 100 water onto paste. Stir rapidly. Leave 2 hr, then filter. Evaporate until volume 100. Grind phenol and chloral hydrate in a mortar until solution complete. Add to solution. Add syrup and lactic acid to mixture.

Zirkle 1940 (*Stain Technol.*, **15** : 144)
FORMULA: water 65, formic acid 41, gum arabic 10, sorbitol 10, ferric nitrate 0.5, carmine 0.5
PREPARATION: Dissolve the gum in the solvents. Incorporate the iron and then the dye.
NOTE: See also under Gelatin Media, and Other Water-Miscible Mountants, Zirkle 1940.

Gelatin Media. It is presumed, in all the formulas that follow, that a gelatin is employed which will give a crystal-clear solution in water. Such purified gelatins are today available on the market for bacteriological use. If commercial gelatin is being used, it is necessary that it should first be clarified, and directions for doing this are given in all the older formulas. Soak the gelatin overnight, drain it carefully, and then melt it on a water bath at about 40°C. Then add, for each 100 ml of the fluid so produced, the whites of two fresh eggs. These are mixed thoroughly with the molten gelatin and the temperature of the water bath is then raised to boiling and left until the whole of the egg white is coagulated. The medium must not be stirred during this time. The egg white coagulates in large lumps, which may readily be strained out through cheesecloth, and which retain, attached to them, all the fine particles which cause cloudiness of the gelatin.

Baker 1945 (Baker 1945, 173)
FORMULA: water 65, gelatin 5, glycerol 35, cresol, 0.25
PREPARATION: Soak gelatin in 25 water for 1 hr, then melt at 60°C. Mix glycerol in 40 water with cresol, then heat to 60°C and mix with gelatin.

Beale 1880 (Beale 1880, 67)
FORMULA: clarified gelatin 50, glycerol 50

Brandt 1880 (*Z. wiss. Mikr.*, **2** : 69)
FORMULA: gelatin 40, glycerol 60, phenol 0.5
PREPARATION: Soak gelatin in water for 24 hr. Drain and melt. Mix glycerol and phenol with molten gelatin. Clarify *s.a.*

Bruere and Kaufmann 1907 (*Bull. Int. Ass. Med. Mus.*, **2** : 11)
FORMULA: gelatin about 25, glycerol about 50, water *q.s.*, 40% formaldehyde 0.1

PREPARATION: Soak the gelatin overnight. Drain, melt, and add an equal volume of glycerol. Clarify *s.a.* filter, and add formaldehyde.

Carleton and Leach 1938 (Carleton and Leach 1938, 115)
FORMULA: water 60, gelatin 10, glycerol 70, phenol 0.25
PREPARATION: Melt gelatin in water at 80°C. Raise glycerol and phenol to 80°C. and add.

Chevalier 1882 (Chevalier 1882, 297)
FORMULA: water *q.s.*, gelatin 25, pyroligneous acid 10
PREPARATION: Soak the gelatin in water, drain, and melt. Add pyroligneous acid to molten gelatin.

Deane cited from 1877 Frey (Frey 1877, 135)
FORMULA: water 30, gelatin 15, glycerol 55
PREPARATION: Dissolve gelatin in water with heat and add glycerol.

Dean cited from 1880 Beale (Beale 1880, 67)
FORMULA: gelatin 30, honey 120, 95% alc. 15, creosote 0.2
PREPARATION: Soak gelatin overnight. Drain. Melt on water bath. Heat honey on water bath. Mix with gelatin. Mix creosote in alc. and add to mixture when cooled to about 35°C. Filter.

Delépine 1915 (*Bull. Int. Ass. Med. Mus.*, **5** : 71)
FORMULA: gelatin 5.2, sat. sol. arsenic trioxide 19, glycerol 71
PREPARATION: Dissolve gelatin in hot arsenic solution, add glycerol.

Fischer 1912 (*Z. wiss. Mikr.*, **28** : 65)
FORMULA: water 100, sodium borate 2, gelatin 10 glycerol 17
PREPARATION: Dissolve ingredients with heat and maintain at 40°C until the medium remains liquid on cooling to room temperature.

Forbes 1943 (*Trans. Amer. Micr. Soc.*, **62** : 352)
FORMULA: water 40, gelatin 9, glycerol 50, phenol 0.6

Geoffroy 1893 (*J. Bot. Paris.*, **7** : 55)
FORMULA: water 100, chloral hydrate 10, gelatin 4

Gilson cited from 1905 Lee (Lee 1905, 273)
PREPARATION: Soak gelatin in water overnight, drain, and melt. To 50 of this add 50 glycerol and enough chloral hydrate to bring the total volume to 50.

Guyer 1930 (Guyer 1930, 96)
FORMULA: water 50, gelatin 8, glycerol 50, egg white (fresh) about 10, phenol 0.25
PREPARATION: Soak gelatin in water overnight. Dissolve with gentle heat. Add egg white to warm gelatin. Autoclave 15 min at 15 lb. and filter. Add glycerol and phenol to filtrate.

Heidenhain 1905 (*Z. wiss. Mikr.*, **20** : 328)
FORMULA: water 60, gelatin 13, glycerol 10, 95% alc. 20. Dissolve gelatin in water with glycerol. Add alc. drop by drop to solution.

Kaiser 1880 (*Bot. Zbl.*, **1** : 25)
FORMULA: water 40, gelatin 7, glycerol 50, phenol 1

Kisser 1935 (*Z. wiss. Mikr.*, **51** : 372)
FORMULA: water 60, glycerol 50, gelatin 16, phenol 1

Klebs cited from 1877 Frey (Frey 1877, 135)
PREPARATION: Soak isinglass in water, drain, and melt. Add enough glycerol to increase volume by one half.

Legros 1871a cited from Robin (Robin 1871, 371)
FORMULA: water 20, gelatin 10, glycerol 30, sat. aq. sol. arsenic trioxide 30
PREPARATION: Melt gelatin in water. Mix glycerol and arsenic with molten gelatin.

Legros 1871b cited from Robin (Robin 1871, 372)
FORMULA: water 50, gelatin 10, arsenic trioxide 0.3, glycerol 30, phenol 0.1
PREPARATION: Soak gelatin in 20 water some hours and melt. Dissolve arsenic in 30 water and add to molten gelatin. Add glycerol and phenol to mixture.

Martindale cited from 1884 Cole (Cole 1884b, 49)
FORMULA: water 50, gelatin 5, 95% alc. 3, egg white 3, glycerol 50, salicylic acid 0.25
PREPARATION: Soak gelatin in water and melt. Add alc. to molten gelatin. Add remaining ingredients to mixture at 30°C. Mix well and heat to 100°C for 5 min and filter.

Moreau 1918 (*Bull. Soc. Mycol. Fr.*, **34** : 164)
FORMULA: water 42, gelatin, 7, glycerol 50, phenol 1
PREPARATION: Soak gelatin in water and melt. Add glycerol and phenol to molten gelatin.

Muir cited from *circ.* 1938 Wellings (Wellings *circ.* 1938, 146)
FORMULA: sat. aq. sol. thymol 100, glycerol 5, gelatin 10, potassium acetate 0.5

Nieuwenhuyse 1912 cited from 1915 Kappers (*Bull. Int. Ass. Med. Mus.*, **5** : 116)
REAGENTS REQUIRED: A. 30% gelatin; B. 4% formaldehyde
METHOD: Sections on slide are covered with a fairly thick layer of A and, after chilling, placed in B for 30 min. Air dry at 30°C until transparent.

Nordstedt 1876 cited from 1883 Behrens (Behrens 1883, 180)
FORMULA: water 42, gelatin 14, glycerol 56

Roskin 1946 (Roskin 1946, 123)
FORMULA: water 42, gelatin 7, glycerol 50, phenol 0.5
PREPARATION: Soak gelatin in water 2 hr. Melt. Incorporate glycerol. Filter.

Roudanowski 1865 (*J. Anat. Paris*, **2** : 227)
FORMULA: water *q.s.*, gelatin 20, glycerol 50
PREPARATION: Soak gelatin in water overnight. Drain and melt. Mix glycerol with molten gelatin.

Schact cited from 1883 Behrens (Behrens 1883, 181)
FORMULA: water 36, gelatin 12, glycerol 48

Squire 1892 (Squire 1892, 84)
FORMULA: water 25, gelatin 6.5, glycerol 50, chloroform 0.6, egg white 5
PREPARATION: Soak gelatin in water 24 hr, drain. Mix glycerol in 25 water with 0.1 chloroform, heat, and add to soaked gelatin. Heat to solution. Add egg white and clarify *s.a.* Filter. Add enough water to make filtrate 100. Add 0.5 chloroform and stir well.

Wood 1897 (*Bot. Gaz.*, **24** : 208)
FORMULA: water 100, clarified gelatin 20, glycerol 10, 40% formaldehyde 1
PREPARATION: Dissolve gelatin in water with glycerol on water bath. Add formaldehyde immediately before use.

Wotton and Zwemer 1935, see Zwemer 1933 (note)

Yetwin 1944 (*J. Parasitol.*, **30** : 201)
FORMULA: water 83, gelatin 5, glycerol 17, chrome alum 0.3, phenol 0.3

Zirkle 1937 (*Science*, **85** : 528)
FORMULA: water 50, acetic acid 50, glycerol 1, gelatin 10, dextrose 4, ferric chloride 0.05, carmine to sat.

Zirkle 1940a (*Stain Technol.*, **14** : 143)
FORMULA: water 60, acetic acid, gelatin 10, sorbitol 10, ferric nitrate 0.5, carmine 0.5
PREPARATION: Mix all ingredients except dye. Bring to boil and add dye. Boil 5 min. Do not filter.
NOTE: See also Gum Arabic Media Formulas and Other Water-Miscible Mountants Formulas Zirkle 1940.

Zirkle 1940b (*Stain Technol.*, **15** : 143)
FORMULA: water 55, acetic acid 45, gelatin 10, gluconic acid 15, ferric nitrate 0.5, carmine 0.5
PREPARATION: As Zirkle 1940a.

Zirkle 1947 (*Stain Technol.*, **22** : 87)
FORMULA: water 45, acetic acid 30, lactic acid 15, gelatin 10, orcein to sat.
PREPARATION: Dissolve the gelatin in water. Add other ingredients, boil 2–3 min.

Zwemer 1933 Glychrogel—auct (*Anat. Rec.*, **57** : 41)
FORMULA: water 80, gelatin 3, glycerol 20, chrome alum 0.2 camphor 0.1
PREPARATION: Dissolve gelatin in 50 water with heat. Add glycerol to hot solution. Dissolve chrome alum in 30 water and add to hot mixture. Filter, then add camphor.
NOTE: This formula was republished by Wotton and Zwemer 1935 (*Stain Technol.*, **10** : 21)

Other Water-Miscible Mountants
Amman 1896 (*Z. wiss. Mikr.*, **13**, 18)
FORMULA: Lactic acid 20, phenol 20, glycerol 20, water 20
NOTE: This is popularly called "lactophenol." See Gray 1954 for several modifications and Dahl 1951 above.

Archibald and Marshall 1931 (*Parasitology*, **23** : 272)
FORMULA: water 60, gum tragacanth 0.5, alc. *q.s.*, gum acacia 1.5, lactic acid 12, glycerol 12, phenol 12
PREPARATION: Add enough alcohol to gum tragacanth to make thin paste. Flood 10 water on paste with constant stirring. Dissolve gum acacia in 50 water. Mix with tragacanth mucilage. Add acid, glycerol, and phenol to mixed gums. Filter.
RECOMMENDED FOR: whole mounts of invertebrate larvae.

Berhnardt 1943 (*Arch. Derm. Syph. Wien.* [See also *Arch. Derm. Syph. N.Y.*, **48** : 533])
FORMULA: water 1, glycerol 2, phenol 1, lactic acid 1

Downs 1943 (*Science*, **97** : 639)
STOCK SOLUTION: dissolve 15 polyvinyl alc. in 100 water at 89°C
WORKING FORMULA: stock 56, lactic acid 22, phenol 22
NOTE: The same formula was republished, with proper acknowledgment, by Huber and Caplin 1947 (*Arch. Derm. Syph. N. Y.*, **56** : 763) to whom the medium is often attributed.

Jones 1946 (*Proc. Roy. Entomol. Soc. London*, **21** : 85)
FORMULA: water 35, polyvinyl alc. 6.3, sat. sol. picric in abs. alc. 18, lactophenol 45
PREPARATION: mix polyvinyl alc. to a paste with the picric solution. Add water to paste and stir to jelly. Add lactophenol to jelly and heat to transparency on water bath.

Gray and Wess 1950 (*J. Roy. Micr. Soc.*, **70** : 290)
FORMULA: water 30, lactic acid 15, glycerol 15, 70% acetone 20, polyvinyl alc. 6
PREPARATION: Add the acetone slowly and with constant stirring to the dry resin. Mix half the water with the lactic acid and glycerol and add to resin mixture. Add remaining water slowly and with constant stirring. Heat on a water bath until clear.

Monk 1938 (*Science*, **88** : 174)
FORMULA: levulose syrup 35, pectin gel 35, water 20, thymol 0.1

Monk 1941 (*Trans. Amer. Micr. Soc.*, **60** : 75)
FORMULA: dextrose syrup 30, pectin gel 30, water 30
NOTE: *Karo* brand dextrose and *Certo* brand pectin are specified in the original of both of Monk's formulas.

Roudanowski 1865 (*J. Anat. Paris*, **2** : 227)
FORMULA: water *q.s.*, isinglass 5, glycerol 8
PREPARATION: Leave isinglass in water and soak overnight. Drain and melt. Add glycerol to molten material.

Salmon 1954 (*Microscope*, **10** : 66)
FORMULA: Prepare with heat a 15% sol. of polyvinyl alc. Mix 10 of this with 10 lactic acid and 1 glycerol
NOTE: Salmon and Ralph 1955 (*Microscope*, **10** : 141) recommend staining with dyes dissolved in lactophenol before this mountant.

Zirkle 1937 (*Science*, **85** : 528)
FORMULA: Belling 1921, levulose syrup 10, pectin jelly 10
NOTE: The original specifies *Karo* brand levulose syrup and *Certo* brand pectin jelly.

Zirkle 1940a (*Stain Technol.*, **15** : 142)
FORMULA: water 60, acetic acid 50, dextrin 10, sorbitol 10, ferric nitrate 0.5, carmine 0.5
PREPARATION: Dissolve dextrin in water. Add other ingredients in order given. Boil, cool, and filter.
NOTE: See also Gum Arabic Media formulas and Gelatin Media formulas, Zirkle 1940.

Zirkle 1940b (*Stain Technol.*, **15** : 144)
FORMULA: water 55, acetic acid 55, sorbitol 5, pectin gel 10, levulose syrup 10, ferric nitrate 0.5, carmine 0.5
PREPARATION: Mix all ingredients except pectin. Leave some days. Filter. Incorporate pectin.
NOTE: The original calls for *Certo* brand pectin and *Karo* brand levulose syrup.

Mountants Miscible With Alcohol

Media of this type, into which objects may be mounted directly from alcohol without the necessity of clearing, fall into three classes. In the first class (Gum Mastic Media) are the media based on gum mastic; in the second (Venice Turpentine Media) are the media based on Venice turpentine. Both these media are regularly used for objects which are considered too delicate to withstand the action of a clearing agent. The third class (Gum Sandarac Media) containing the gum sandarac media, comprise those formulas which are usually referred to as *neutral mountants*. These are widely used both for substances which are considered too delicate to preserve in Canada balsam, which involves prior clearing, or for sections which have been stained in material which fades rapidly under the influence of acid balsam. Many are derived from the original *euparal* of Gilson, the formula for which has never been disclosed and which is a preparatory substance of secret composition. These media have a refractive index much lower than the gum mastic or gum turpentine media so that they cannot satisfactorily be used for thick objects. It may be pointed out that "dry" Canada balsam is soluble in absolute alcohol and has from time to time been recommended. It is the writer's experience that this solution is not satisfactory, for mounts made with it darken more rapidly than those made from balsam which has been dissolved in hydrocarbons.

Gum Mastic Media

Artigas 1935 (*Mem. Inst. Butantan*, **10** : 71)
FORMULA: 95% alc. 100, gum mastic 30, beechwood creosote 100
PREPARATION: Dissolve gum in alc. Centrifuge or filter. Add creosote and evaporate until no alc. remains.
RECOMMENDED FOR: nematode worms after clearing in creosote.

Hoyer 1921 (*C. R.Soc. Biol. Paris*, **84** : 814)
PREPARATION: Suspend mastic in a cloth bag in a considerable volume of 95% alc. Withdraw bag in which remain gross impurities. Shake solution thoroughly, allow to settle, and decant clear solution. Evaporate to required consistency.

Venice Turpentine Media

Langeron 1942 Venice turpentine-alcohol (Langeron 1942, 654)
PREPARATION: Dilute crude Venice turpentine with an equal volume of 95% alc. Mix well and allow impurities to settle. Decant and reevaporate to convenient consistency.

Vosseler 1889 (*Z. wiss. Mikr.*, **6** : 292)
FORMULA: 95% alc. 50, Venice turpentine 50
PREPARATION: Mix ingredients. Allow to settle. Decant.

Wilson 1945 (*Stain Technol.*, **20** : 133)
FORMULA: Venice turpentine 25, phenol 50, proprionic acid 35, acetic acid 10, water 20
PREPARATION: As Venice Turpentine Media formulas.

Zirkle 1940 (*Stain Technol.*, **15** : 147)
FORMULA: water 25, acetic acid 15, proprionic acid 35, phenol 55, Venice turpentine 20, ferric nitrate 0.5, carmine 0.5

PREPARATION: Mix the Venice turpentine with the proprionic acid. Add the phenol and acetic acid. Then add the water, in which the ferric nitrate has been dissolved, slowly and with constant stirring. Incorporate the dye in this mixture.

NOTE: See also Canada Balsam Media, Zirkle 1940.

Gum Sandarac Media

Armitage 1939 (*Microscope*, **3**: 215)
PREPARATION: To 100 of a syrupy filtered solution of gum sandarac in dioxane add the fluid produced by the mutual solution of 3 salol and 2 camphor.

Buchholz 1938 (*Stain Technol.*, **13**: 53)
FORMULA: oil of eucalyptus 60, paraldehyde 30, gum sandarac to give required consistency

Cox 1891 (*Arch. Mikr. Anat.*, **37**: 16)
FORMULA: alcohol 150, sandarac 150, turpentine 60, camphor 30, lavender oil 45, castor oil 0.5
PREPARATION: Dissolve sandarac in alc. Dissolve camphor in turpentine and mix in alc. solution. Add oils to mixed solutions.

Denham 1923 Camphoral—auct (*J. Roy. Micr. Soc.*, **43**: 190)
STOCK I: chloral hydrate 50, camphor 50
PREPARATION OF STOCK I: Grind ingredients in a mortar until solution complete.
STOCK II: gum sandarac *q.s.*, isobutyl alc. *q.s.*
PREPARATION OF STOCK II: Make a thin solution of the ingredients. Shake with activated charcoal. Filter. Evaporate to a thick syrup.
WORKING MEDIUM: stock I 60, stock II 30

Gilson 1906 Euparal—compl. script. (*Cellule*, **23**: 427)
NOTE: This is a proprietary mixture of secret composition and cannot be further noticed. The reference cited does not disclose the composition.

Mohr and Wehrle 1942 (*Stain Technol.*, **17**: 157)
FORMULA: camsal 10, gum sandarac 40, eucalyptol 20 dioxane 20, paraldehyde 10
NOTE: Camsal is produced by the mutual solution of equal quantities of camphor and phenyl salicylate (salol). This medium may be diluted with dioxane. It may be colored green (in imitation of green euparal) by adding a solution of copper oleate in eucalyptol.

Shephered 1918 (*Trans. Amer. Micr. Soc.*, **37**: 131)
FORMULA: sandarac 30, eucalyptol 20, paraldehyde 10 camsal 10
NOTE: Camsal is a mixture of equal parts camphor and phenyl salicylate. Shephered (*loc. cit.*) recommends dissolving the sandarac in 150 abs. alc., altering under anhydrous condition and reevaporating to dryness.

Other Alcohol-Miscible Media

Hanna 1949 (*J. Roy. Micr. Soc.*, **69**: 25)
FORMULA: sulfur 40, phenol 100, sodium sulfide 2

Seiler 1881 (*Trans. Amer. Micr. Soc.* [Vol 1 to 16 as "Proc"], **3**: 60)
FORMULA: Canada balsam 40, abs, alc. 60

Mountants not Miscible With Either Alcohol or Water

The great majority of all mounts are prepared in these media of which Canada balsam (mixtures containing which form the first class, Canada Balsam Media), is the best known. Gum damar (Gum Damar Media) is a substance which is so variable in composition that it is difficult to recommend it. It has less tendency to become either yellow or acid with age than has balsam, provided that one secures a good specimen. But there are numerous accounts in the literature of mounts which have become granular within a few years of having been made. It would appear probable that these mounts were made from an impure sample of the gum, and the worker who wishes to use damar media is recommended to be very particular as to his source of supply. The next class (Other Resins and Mixed Resins) covers the few other natural resins which have from time to time been proposed for mounting media as well as mixtures of these resins with Canada balsam and with gum damar. The last class (synthetic resins below) is likely to increase very rapidly with time. It includes numerous synthetic resins which have been proposed as a substitute for the natural resins usually employed. No methcrylate mixtures are included since there is abundant evidence in the literature (Richards and Smith 1938: *Science*, **87**: 374) that they are worthless. Unfortunately, many authors have proposed media based on resins of which only a trade name is quoted. These have not been included since they are almost impossible to duplicate. Formulas using trade names have, however, been included if a reasonable chemical identification of the resin is given in the original description.

Canada Balsam Media. Canada balsam is the natural exudate of *Abies balsamea*. It consists of a resin (*Canada resin* in the formulas given below) dissolved in a variety of hydrocarbons. The material sold on the market as *dried balsam* has had the lower boiling-point natural hydrocarbons driven off with heat but the higher boiling-point fractions, which act as natural plasticizers, remain. This dried balsam is commonly used as a 40% solution in xylene or benzene. If true Canada resin is used, a plasticizer must be added. A method of purifying Canada balsam is given by Bensley and Bensley 1938, 38. Bender 1967 (Departmental Publication No. 1182, Canadian Department of Forestry and Rural Development) also gives some useful details. *Neutral balsams* are a delusion and some synthetic resins formula should be used in their place.

Apáthy 1909 (*Fauna u. Flora Neapel*, **22**: 18)
FORMULA: Canada balsam 50, cedarwood oil 25, chloroform 25

Becher and Demoll 1913 (Becher and Demoll 1913, 107)
FORMULA: Canada resin 40, abs. alc. 40, terpineol 20

Curtis 1905 salicylic balsam—compl. script. (*Arch. Med. Exp.*, **17**: 603)
FORMULA: dried Canada balsam 30, sat. sol. salicylic acid in xylene, 70

Hays 1865 (*Trans. Amer. Micr. Soc.* [Vol 1 to 16 as "Proc"], **1**: 16)
FORMULA: Canada balsam 50, chloroform 50
PREPARATION: Mix ingredients and allow to stand 1 month. Decant. Evaporate to required consistency.

Sahli 1885 (*Z. wiss. Mikr.*, **2**: 5)
NOTE: This paper is often quoted as recommending a solution of balsam in cedar oil. Sahli recommends only that sufficient of the oil used for clearing be left to soften the balsam.

Zirkle 1940 (*Stain Technol.*, **15**: 149)
FORMULA: water 20, acetic acid 40, phenol 65, oleic acid 10, dried Canada balsam 10, ferric nitrate 0.5, carmine 0.5
PREPARATION: Mix the balsam with the acetic and proprionic acids. Add the oleic acid and then the phenol. Incorporate the water in which the ferric nitrate has been dissolved. Then mix in the dye.

Gum Damar Media. Gum damar is the natural exudate of *Shorea wiesneri*, but it is almost always adulterated and usually contains solid impurities. The raw gum should be dissolved in chloroform, filtered, and evaporated until the chloroform is driven off. The purified gum is then dissolved in benzene or toluene to a suitable consistency.

Cooke *circ.* 1920 (Cooke *circ.* 1920, 49)
PREPARATION: Warm together till dissolved equal parts of gum damar, benzene, and turpentine. Filter. Evaporate to desired consistency.

Vögt and Jung cited from *circ.* 1890 Francotte (Francotte, 248)
FORMULA: Canada balsam 30, benzene 100, gum damar 30
PREPARATION: Dissolve ingredients separately, then mix solutions.

Other Resins and Mixed Resins

Artigas 1935 see Artigas 1935 (under Gum mastic media, above)

Chevalier 1882 (Chevalier 1882, 327)
FORMULA: chloroform 90, rubber 3, gum mastic *q.s.* to give required consistency.

Frémineau cited from 1883 Chevalier (Chevalier 1882, 297)
FORMULA: Canada balsam 60, gum mastic 20, chloroform *q.s.* to give required fluidity

Lacoste and de Lachand 1943 (*Bull. Histol. Tech. micr.* = *Bull d' Hist appl.*, **20**: 159)
FORMULA: toluene 100, rosin 120

Rehm 1893 (*Z. wiss. Mikr.*, **9**: 387)
FORMULA: benzene 100, rosin 10

Seiler 1881 (Seiler 1881, 90)
FORMULA: naphtha 17, turpentine 15, Canada balsam 45, damar 23
NOTE: Seiler (*loc. cit.*) recommended this either as a mountant or as a ringing cement.

Southgate 1923 (*Brit. J. Exp. Pathol.*, **4**: 44)
PREPARATION: Digest 200 crude Yucatan elemi in the cold with 200 95% alc. Filter and evaporate filtrate to dryness and redissolve in benzene to a suitable consistency.
NOTE: The solution of amyrin-free gum elemi thus obtained is stated to preserve Giemsa or other stains indefinitely.

Synthetic Resins

Deflandre 1933 (*Bull. soc. franc. microsc.*, **2**: 67)
FORMULA: coumarone resins 20, xylene 80, monobromonaphthalene 1

Deflandre 1947 Kumadex—auct (Deflandre 1947, 96)
PREPARATION: Mix 3 parts of a syrupy solution of a coumarone resin in xylene with 1 part of Canada balsam.

Fleming 1943 (*J. Roy. Micr. Soc.*, **63**: 34)
FORMULA: Naphrax 40, xylene 59, dibutyl phthallate 1
NOTE: The resin mentioned is a high refractive index (1.7–1.8) naphthalene derivative, the synthesis of which is fully described in the reference cited. Fleming 1954 (*J. Roy. Micr. Soc.*, **74**: 427) gives a simplified method for the synthesis of Naphrax.

Gray and Wess 1951a (*J. Roy. Micr. Soc.*, **71**: 197)
FORMULA: ethyl cellosolve 68, isoamyl phthalate 12, polyvinyl acetate 20
RECOMMEND FOR: sections.

Gray and Wess 1951b (*J. Roy. Micr. Soc.*, **71**: 197)
FORMULA: ethyl cellosolve 50, isoamyl phthalate 10, polyvinyl acetate 40
RECOMMENDED FOR: whole mounts.

Groat 1939 (*Anat. Rec.*, **74**: 1)
FORMULA: toluene 40, "nevillite 1" or "V" 60
NOTE: The nevillites are mostly hydrogenated coumarones of which clairite (nevillite 1) is the best known. Nevillite V is a naphthaline polymer.

Kirkpatrick and Lendrum 1939 DPX—auct. (*J. Pathol. Bacteriol.*, **49**: 592 4)
FORMULA: xylene 80, tricresyl phosphate 15, "Distrene-80" 10
NOTE: Distrene-80 is a polystyrene with a molecular weight of about 80,000.

Lillie and Henson 1955 (*Stain Technol.*, **30**: 133)
FORMULA: Cellulose caprate 50, xylene 50

Skiles and Georgi 1937 (*Science*, **85**: 367)
FORMULA: vinylite 20, xylene 80
METHOD: used to varnish bacterial films

Wicks, Carruthers, and Ritchey 1946 (*Stain Technol.*, **21**: 121)
FORMULA: "Piccolyte" 60, xylene 40
NOTE: There are a whole series of beta-pinene polymers marketed under the general name "piccolyte." The authors cited found "WW-85," "WW-100," "S-85," and "S-100" the most suitable.

PETER GRAY

References

Journal references in the above formulae are given in the standard form. The following books are cited by author and date only.

Baker, J. R., "Cytological Technique," 2nd ed., London, Methuen, 1945.
Beale, L. S., "How to Work with the Microscope," 5th ed., London, Harrison, 1800.
Becher, S., and R. Demoll, "Einführung in die mikroskopische Technik," Leipzig, Quelle und Meyer, 1913.
Behrens, W., "Hilfsbuch zur Aüsfuhrung mikroskopischer Untersuchungen im botanischen Laboratorium," Braunschweig, Harald Bruhn, 1883.
Carleton, H. M., and E. H. Leach, "Histological Technique," 2nd ed., London and New York, Oxford University Press, 1938.
Chevalier, A., "L'Etudiant Micrographe," 3rd ed., Paris, Chevalier, 1882.

Cole, A. C., "The Methods of Microscopical Research," n.d., bound with Cole 1883, vol. 2.

Cooke, M. C., "One Thousand Objects for the Microscope with a Few Hints on Mounting," Popular ed., London, Warne, n.d. (1920).

Davies, T., "The Preparation and Mounting of Microscopic Objects," New York, William Wood, circ. 1865.

Deflandre, G., "Microscopie Pratique," 2nd ed., Paris, Lechevalier, 1947.

Francotte, P., "Manuel de Technique Microscopique," Paris, Lebègue, circ. 1890.

Frey, H., "Das Mikroskop und die mikroskopische Technik," 6th ed., Leipzig, Wilhelm Engelmann, 1877.

Gatenby, J. B., and T. S. Painter, "The Microtomist's Vade-Mecum (Bolles Lee)," 10th ed., Philadelphia, Blakiston, 1937.

Gray, P., "The Microtomist's Formulary and Guide," New York and Toronto, The Blakiston Company, 1954.

Guyer, M. F., "Animal Micrology, 3rd ed., Chicago, University of Chicago Press, 1930.

Hogg, J., "The Microscope," London, Routledge, 1883.

Langeron, M., "Précis de Microscopie," 6th ed., Paris, Masson, 1942.

Lee, A. Bolles, "The Microtomist's Vade-Mecum," 8th ed., London, Churchill, 1905.

Martin, J. H., "A Manual of Microscopic Mounting," London, Churchill, 1872.

Robin, C. P., "Traité du Microscope; son mode d'emploi: ses applications à l'étude des injections; à l'anatomie humaine et comparée: à la pathologie médico-chirurgicale; à l'histoire naturelle animale et végétale; et à l'économie agricole," Paris, Baillière, 1871.

Roskin, G. E., "Mikroskopecheskaya Technika," Moscow Sovetskaya Nauka, 1946.

Seiler, C., "Compendium of Microscopical Technology," Philadelphia, Brinton, 1881.

Squire, P. W., "Methods and Formulae Used in the Preparation of Animal and Vegetable Tissues for Microscopical Examination," London, Churchill, 1892.

Wellings, A. W., "Practical Microscopy of the Teeth and Associated Parts," New York, Staples, n.d. (circ. 1938).

MOUNTANTS

Mountants are substances in which specimens are placed for microscopical observation. They may be temporary or permanent and may be solid, semisolid, liquid, or gaseous, i.e., air. In biomedical usage, mountants not only permanently preserve the specimen but hold the coverslip in place, as distinct from preservatives in which objects may also be permanently preserved, but which do not inherently retain the coverslip. Preservatives must be sealed under the coverslip with some kind of cement, usually by ringing with asphaltum or one of the newer plastic enamels. Industrial usage does not make this distinction between mountants and preservatives; any medium in which a specimen is placed is its mountant.

Biomedical Mountants. Mountants for tissues, microorganisms, and other biologicals are selected primarily on the basis of the treatment the specimen must receive prior to mounting. Other considerations in choosing a mountant are solubility of specimen elements in the solvent and dye or pigment extraction by the solvent. Three major groups are recognized: (1) mountants miscible with water such as the glycerin jellies, lactic acid–polyvinyl alcohol mixtures, and gum arabic media; (2) mountants miscible with alcohol such as the gum mastic, Venice turpentine, and gum sandarac media; (3) mountants miscible with solvents other than water or alcohol—usually aromatic hydrocarbons—such as natural and synthetic resins, of which Canada balsam is the classic example. Synthetic resin mountants may be proprietary or nonproprietary; examples include Permount, Euparal, Harleco Synthetic Resin (HSR), Technicon Medium, Histoclad, and Piccolyte.

Mounting media vary in several important respects, including chemical type, solvent, concentration, viscosity, refractive index, color, melting point, acid number, iodine number, drying time, tendency to form air bubbles, tendency to fade, bleed, or bleach stains, and tendency to crystallize. Mountants must therefore be carefully selected for compatibility with the specimen and stains, and for the anticipated storage time.

Phase contrast objectives are computed to give optimal contrast with the majority of objects prepared in the conventional manner. For special cases, there are commercially available sets of mountants for achieving the refractive index relationships necessary for satisfactory contrast.

Industrial Mountants. Industrial microscopists, such as those in air pollution, criminalistics, microelectronics, and general product failure analysis, are almost always dealing with "unknowns" which must be identified. These may be biologicals, wind erosion products, combustion products, industrial dusts—literally any particulate substance regardless of nature or origin. In these cases, optical properties are of paramount importance for identification, and mountants are selected primarily for their refractive index. If a particle and the mountant have the same or nearly the same refractive index, the particle is "invisible." Only when the particle and mountant have different refractive indices is a degree of contrast possible and the particle boundary visible. It usually does not matter if the refractive index of the medium is higher or lower than that of the particle, as long as the indices differ sufficiently. Thus, refractive index differences between mountant and sample are the basis for contrast enhancement, and reversing the principle, sameness of refractive indices is the basis for determining the refractive indices of specimens with a microscope: If the refractive index of the mountant is known, and that optical property is the same for the particle, a step toward identifying the particle has been made. Accordingly, the most useful mountants are sets of refractive index liquids ranging from about 1.35 to 2.00, most of which differ by intervals of 0.002 refractive index units. Such sets are commercially available (R. P. Cargille Labs., Cedar Grove, New Jersey) or can be made up; sets of organic liquids are most common, but sets of aqueous liquids are sometimes needed, and silicone fluids are also frequently used. Of course it is possible to have liquids below 1.35 and above 2.00; at the upper end of the scale, realgar and melts of sulfur and selenium extend the refractive index range up to as much as 3.17. Great caution must be observed with the high-index nonresinous mountants, because all of the media containing arsenic, selenium, and tellurium are very toxic, especially when heated.

For permanent preparations, the industrial microscopist seeks a mountant which requires virtually no time for drying or hardening, is nonreactive with most solid substances, and allows for dispersing the sample or removing a portion at any time. Furthermore, since no stains are used, a high refractive index is desirable for visibility. All of these requirements are met by the

Aroclors, a series of chlorinated biphenyl and polyphenyl compounds made by the Monsanto Company. Aroclor 5442, with a refractive index of 1.66, is the most generally useful. It is thermoplastic and must be used on a hot plate. The slide with specimen and cover slip is placed on the hot plate, and the Aroclor 5442 is run under the cover slip by capillarity. Another Aroclor, 1260, has the consistency of very thick molasses. With it, the microscopist can roll or tumble a dust particle into the best crystallographic orientation by pushing on one side of the coverslip. Other high-index resinous media include Hyrax, Naphrax, and Pleurax.

Feed microscopists favor one mountant which they use almost exclusively—a 1:1:1 mixture of glycerin, chloral hydrate, and water.

JOHN GUSTAV DELLY

References

Delly, J. G., "Mounting media for particle identification," *Microscope,* **17**:205–212 (1969).
Gray, P., "The Microtomist's Formulary and Guide," Philadelphia, Blakiston, 1954.
Greco, J. P., "Refractive indices of currently used mounting media," *Stain Technol.,* **25**:11–12 (1950).
Lillie, R. D., et al., "Final report of the Committee on Histologic Mounting Media," *Stain Technol.,* **28**:57–80 (1953).
White, G. W., "A survey of refractive indices," *Microscopy,* **31**:257–266 (1970).

MYELIN (MYELIN SHEATH)

The myelin, or myelin sheath, is a lipid and protein sheath that segmentally encases large axons (the long cytoplasmic processes of neurons). Myelin is part of a living nerve-sheath cell, not an extracellular material. In the peripheral nervous system myelin is the part of a Schwann cell wrapped spirally around an axon. In the central nervous system oligodendroglial cells form the spiral myelin sheath.

The spiral pattern of myelin was recognized first in peripheral nerves by use of electron microscopy. In a peripheral nerve one Schwann cell encases a longitudinal segment of an axon, and only one axon, in myelinated (medullated) nerves. The myelin sheath is interrupted where the Schwann cell wrap terminates at the node of Ranvier, the gap between neighboring Schwann cells (Fig. 1). The most superficial Schwann cell wrap extends nearest the node, the successively deeper wraps terminate progressively further from the node. In most parts of the spiral wrap Schwann cell cytoplasm is not identifiable because the inner and outer cell (plasma) membranes of the layer have been compacted together forming a major dense line or myelin membrane (Fig. 2). Each wrap contains recognizable cytoplasm in the paranodal region, where the wraps terminate. V-shaped internodal interruptions in the compacted layers, called Schmidt-Lanterman incisures, are foci in which the wrap is uncompacted and contains cytoplasm, but does not terminate (Fig. 3). The most superficial and least superficial Schwann cell wraps do not compact. Myelin (compact myelin) recognized by light microscopy corresponds to that part of the nerve sheath where the spiral cellular layers, identified by electron microscopy, are compacted; and also to the concentric lamellar structure recognized by X-ray diffraction methods. The major dense lines create a repeating

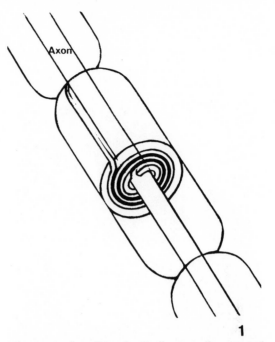

Fig. 1 A three-dimensional diagram of axon and Schwann cell sheath of a myelinated nerve. A cross section of the spiral Schwann cell myelin sheath is illustrated. The outermost layer contains Schwann cell cytoplasm as does the innermost layer. In the other layers of the spiral sheath plasma membranes of each layer are compacted creating a myelin membrane. The outermost wrap extends nearest the node of Ranvier.

pattern with intervals of approximately 120 Å, with a less dense intraperiod line occurring approximately half-way between neighboring major dense lines. The lamellae of the myelin sheath are composed of layers of mixed lipids (oriented with paraffin chains of the molecules extending radially) with polar groups loosely bonded to the protein layers at the aqueous interfaces (X-ray diffraction studies by Schmitt, Bear, and Palmer). The structure of the myelin sheath is relatively insensitive to the action of temperature, electrolytes, and detergents.

The term loose myelin has been used for a thick sheath around axons that contains lipid but fails to stain as intensely as compacted myelin. By electron microscopy loose myelin corresponds to uncompacted spiral wraps, the only pattern present in some invertebrates such as the earthworm, and also is seen in the uncompacted segments of myelinated mammalian nerves.

Small nonmyelinated axons are sheathed in a simple tunnel, not surrounded by a spiral wrap. Generally axons over 2 μ in diameter are myelinated and those under 1 μ are not. In the central nervous system nodes of Ranvier are found, but Schmidt-Lanterman incisures are rare.

Techniques for preparing specimens for examination by light or electron microscopy generally use lipid solvents which result in loss of part of the fatty component of myelin, but do not destroy its architecture. Specimens for light microscopy are usually fixed by immersion in neutral buffered formalin, dehydrated and embedded in paraffin for sectioning. If exceptionally large specimens are needed embedding is done in celloidin. For electron

1st Line
2nd Line
3rd Line

Major
period
Lines

AXON

Fig. 2 Electron micrograph of part of the thick myelin sheath of a large axon. The axon (upper left) has a cell membrane, seen as the first line (or lines since it is a unit membrane) from the top of the picture. The second and third lines are the cell membranes of the cytoplasmic (innermost) wrap of the Schwann cell. The other wraps in the picture are compacted and each is seen as a dense major line or myelin membrane with an intraperiod minor line between ($\times$ 140,000.)

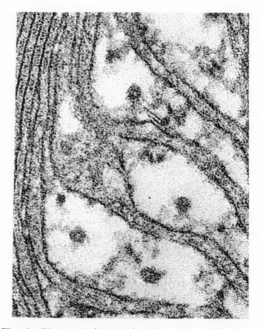

Fig. 3 Electron micrograph of a section through a Schmidt-Lanterman incisure of the myelin sheath. The Schwann cell layers do not terminate at Schmidt-Lanterman incisures. ($\times$ 160,000.)

microscopy fixation of tissue in buffered glutaraldehyde (by immersion or perfusion) provides considerably better preservation of myelin than fixation with osmium tetroxide. A trace of calcium improves preservation of membranes, such as myelin. Specimens are rapidly dehydrated in a few changes of ethanol. Different parts of the myelin sheath (major and minor dense lines and the unit membranes) are demonstrated in different fashions by staining of the sections with different electron-dense compounds, or by using such compounds in the earlier stages of specimen preparation.

In tissue sections stained with hematoxylin and eosin for light microscopy the myelin barely stains and is often described as a clear ring surrounding an axon, containing only a few eosinophilic filaments, the neurokeratin net-

work. The myelin sheath can be demonstrated in tissue processed in the same fashion and stained with Weil, Weigert, or other preparations. The so-called myelin stains are not specific for myelin, but do accentuate the myelin sheath so that it is easily seen. Probably the simplest and most reproducible method is the Luxol Fast Blue stain, often used with a P.A.S. (periodicacid–Schiff) counterstain. The disadvantage of the Luxol Fast Blue-P.A.S. stain is that black and white photographs of such preparations are disappointing in comparison to the color.

Other methods of specimen preparation for light microscopy include nerves teased apart, stained, and placed on a slide without sectioning; and frozen sections of undehydrated specimens with or without fixation in which Oil red O or osmium tetroxide intensely stain myelin and other lipid-rich structures. A modified Sudan IV method used on frozen sections stains broken myelin (free fat) differently from the fat in the myelin sheath; the technique is somewhat more reproducible than the erratic Marchi stain for demonstration of damaged myelin.

The commonly used stains (Luxol Fast Blue, Weil) demonstrate normal myelin and are useful in demonstrating areas of missing myelin in disease. Absence of myelin can be difficult to appreciate unless sections are compared with normal controls. Changes in pattern of myelin distribution are studied at relatively low magnification, and can frequently be recognized in a tissue section with the naked eye. For comparison to controls the usual method of sectioning specimens is to cut peripheral nerves longitudinally, the spinal cord and brain stem transversely, and the cerebral hemispheres and basal ganglia in a coronal (frontal) plane.

Because myelin sheath stains fail to adequately demonstrate nuclei, axons, and supporting cells (glia and Schwann cells) different stains are required on other sections to demonstrate these other parts of the nervous system. Electron microscopy is the one technique that clearly illustrates all the structures at one time, but it has the disadvantage of confining its field of vision to a fragment less than 1 mm in diameter.

JAMES C. HARKIN

MYXOMYCETES

Even though the Myxomycetes have been subjected to a variety of investigations in recent years, they are still not generally well known to biologists. For this reason a diagram of their life history is provided. Figure 1 shows the sequence of events in a heterothallic form. In many species, however, no mating types are identifiable and such homothallic forms are capable of completing their life histories without benefit of bringing together genetically unlike amoebae. Otherwise, the patterns are apparently the same.

Techniques for handling the various stages in the laboratory will be emphasized, since this is likely to be an important obstacle for most microscopists.

Laboratory stock material may be available as spores, myxamoebae, or plasmodia. These are likely to carry a bacterial associate since few species have been cultivated axenically. When kept in a dry cool place, spores remain viable for months, or even years. Maintenance of the two other stages is more involved, but usually not difficult. If the strain is heterothallic, single-spore-derived cultures do not ordinarily yield plasmodia, but instead produce only a clonal population of microscopic myxamoebae. The cells remain active for up to a few weeks, after which most of them die, but some encyst. In the microcyst stage, they may survive in wet agar culture for a few months to a year or more, depending on particular clones and culture conditions. Alternatively, long-term storage of myxamoebal or spore populations can be achieved by lyophilization. The plasmodium apparently cannot survive the lyophil process. Maintenance is accomplished either by periodic transfers of the active stage, or, in some species, in the dormant sclerotial state. Sclerotia can often be induced by allowing the plasmodium to dry gradually after letting it migrate onto moist filter paper. Storage under room conditions for months or longer is then possible.

Activation of any stage may be brought about by

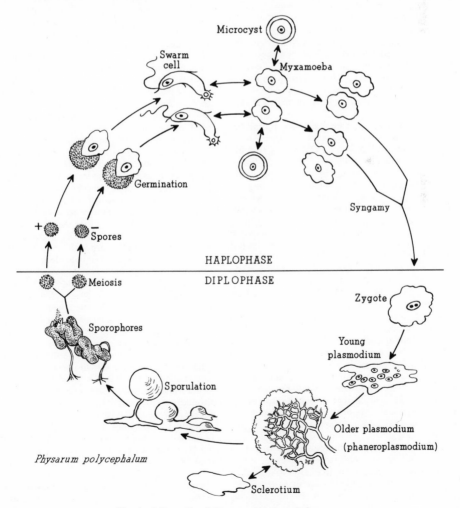

Fig. 1 Life cycle of *Physarum polycephalum*.

transfer to fresh agar medium, followed by addition of a film of sterile distilled water. One commonly used agar medium is Difco's corn meal agar, modified as follows: 12.5 g corn meal agar, 1000 ml distilled water. This will be referred to as CMA/2.

The Myxomycetes have been utilized in certain kinds of work where it is desirable to isolate new clones, beginning with single haploid spores. One very straightforward method is to make a dilute water suspension of the cells to be isolated. Then a few drops are placed at one edge of a Petri dish containing CMA/2. By tilting the plate to the opposite side, the suspension is distributed in a straight line across the surface of the medium. The process may be repeated one or two times, starting at different points on the same plate. In about one-half hour the water is absorbed, and small blocks of agar, each containing a single cell, may be plucked up with a fine-pointed instrument. This is done under a dissecting microscope set at magnifications of ×60 or higher and with substage lighting adjusted to give a fairly dark field. When adjustment is satisfactory, each spore appears brighter than the background, so that they are readily identifiable as individuals. Each isolated agar block containing a spore is placed on CMA/2 and the agar surface is thinly covered with a dilute water suspension of a bacterium, such as *Aerobacter aerogenes* or *Escherichia coli*. Bacteria provide food for the holotrophic cells. Inspection after about one or two weeks with a dissecting microscope (× 60 or higher) reveals whether a population, or clone, has developed. If so, the surface may be covered with active cells and microcysts, which can be transferred and stored on CMA/2 in screw-cap tubes. Addition of enough water to cover about a third of the agar slant allows for longer storage.

Crosses between clones can be made at any time that sufficient cells are available. This is accomplished simply by mixing cells from two clones on CMA/2. A thin film of water may be added to achieve good distribution of myxamoebae and swarm cells, but this is not usually necessary. No bacteria need be added, since they will already be present. Fertile crosses yield macroscopic plasmodia in about one to two weeks. These too feed on available bacteria. If kept in a room where ordinary lighting is available, plasmodia of several species will sporulate in the plate where they develop. Others do not, but instead eventually sclerotize or die. To prevent sporulation and encourage growth to a large size, the small macroscopic plasmodia may be fed sterilized Quaker Oats. Either pulverized or whole flakes may be best, depending on the strain or species. Once a vigorous, large plasmodium is available, the agar on which it is growing may be cut into several pieces to provide plasmodial inoculum. Each piece is transferred to a new agar plate and fed once again. Plasmodia of certain kinds may be perpetuated almost indefinitely in this fashion, but others lose their vigor after a while and require periodic rejuvenation. This can be done by recrossing their parental clones. Thus, it is possible either to directly perpetuate genetic duplicates or to regenerate them. This is one truly unusual and outstanding feature of the Myxomycetes.

Aside from the technique mentioned above, sporulation can be induced in some species by transfer of a piece of an old (past its peak of vigor, often 10 days or older) plasmodium to 4% water agar. Results are often dramatic, fruiting occurring overnight. Sometimes transfers on

several successive days from the old stock are necessary before sporangia are obtained.

Although most species cultivated in the laboratory require a bacterial associate, the plasmodial stage of a few species have been grown axenically on killed bacteria, and fewer still have also been grown on chemically defined media. Because of a plasmodium's capacity to migrate over solid surfaces it is sometimes easy to free it from adhering bacteria, or other microbiol contaminants. The plasmodium is transferred to nonnutrient agar, allowed to migrate some distance from the point of inoculation, transferred again, and so on. Antibiotics, such as penicillin and streptomycin, may be incorporated into the nonnutrient medium, but this is not always necessary. Finally, the plasmodium is transferred to chemically defined or semidefined media (see references). Often the medium is kept in a liquid state and the plasmodium is inoculated into it. To provide for good aeration, the culture flasks are constantly agitated on a mechanical shaker. The inoculum grows and then breaks up into smaller pieces. These in turn grow and fragment, and so the process continues until a large amount of plasmodium is obtained. The hundreds of microplasmodia may then be transferred by means of a pipette to a solid surface such as agar or filter paper supported by glass beads immersed in a liquid medium. There they fuse with one another and form one large plasmodium. Characteristically, the nuclei of each plasmodium divide synchronously, and a new synchrony becomes established when two to several plasmodia are allowed to fuse with one another.

Plasmodia of *P. polycephalum* grown in chemically defined media can be induced to sporulate if transferred to a solid starvation medium containing niacin and salts, kept at least four days in darkness, and exposed to white light for 4 hr or more.

Microscopic studies of Myxomycetes have been made on all stages in their life cycle. Many of these have utilized the light microscope. One convenient method for the study of spore-bearing stage is to mount a portion of the sporangium, or if small enough, a whole one, on a slide, wet with absolute alcohol, add a drop of 3 % KOH. Excess KOH is then blotted off with filter paper before adding a drop of 8 % glycerin. After placing a coverslip on the drop, allow the glycerin to dry for a few days before sealing with nail polish or some other suitable agent. The mount may last for years. To study spore sculpturing and ornamentation, viewing under oil immersion objectives is recommended, since the spores are very small, often less than 10 μ in diameter. Other structures which may be associated with mature sporangia, such as capillitia, lime deposits, and other granules, can also be studied in this fashion.

Myxamoebae and swarm cells can be observed in temporary water mounts under either phase or ordinary light. Under phase, organelles such as nuclei, vacuoles, mitochondria, and flagella can be seen quite readily. In addition, stages in the life cycle, from spore germination through development of small plasmodia, can be observed directly in microcultures. These are made by coating a coverslip with a suitable agar medium followed by inoculation with cells in a water drop. The coverslip is then inverted over a polyethylene ring. Vaseline or paraffin may be used to seal the apparatus for prevention of rapid evaporation. Microcultures, of course, may be started from spores, clonal myxamoebae, or crosses between sexually compatible clones. Permanent slides can be made directly from the microcultures by removing the coverslips and immediately placing them over 2 % osmium tetroxide for fixation. After allowing them to dry at room temperature, they may be stained with Heidenhain's hematoxylin, then dehydrated in a graded series of alcohol and to xylene. Coverslips then are mounted in Clarite on a slide. A slightly different method for making permanent slides has also been used successfully. Cells at various stages in ordinary cultures may be harvested by differential centrifugation and placed on albumen-coated coverslips. They are fixed in Schaudinn's fixative and stained by the long method for Heidenhain's iron hematoxylin. Both methods highlight nuclei and nuclear activity.

Electron microscope studies are becoming prevalent and basically the techniques are those commonly used. Samples of myxamoebae, plasmodia, and immature sporangia have often responded well to glutaraldehyde-osmium fixation. Embedding in Epon-Araldite mixture, and post-staining with uranyl acetate and lead citrate have yielded excellent results. It has been possible to effectively study mitotic and meiotic nuclear division, myxamoebal–swarm cell transformations, and ultra-structural changes during differentiation associated with sporangial and sclerotial development.

O'NEIL RAY COLLINS

References

Aldrich, H. C., "The ultrastructure of meiosis in three species of *Physarum*," *Mycologia*, **59**:127–148 (1967).

Aldrich, H. C., "The ultrastructure of mitosis in myxamoebae and plasmodia of *Physarum flavicomum*," *Amer. J. Bot.*, **56**:290–299 (1969).

Collins, O. R., "Multiple alleles at the incompatibility locus in the myxomycete *Didymium iridis*," *Amer. J. Bot.*, **50**:477–480 (1963).

Goodman, E. M., and H. P. Rusch, "Ultrastructural changes during spherule formation in *Physarum polycephalum*," *J. Ultrastruct. Res.*, **30**:172–183 (1970).

Gray, W. D., and C. J. Alexopoulos, "Biology of the Myxomycetes," New York, The Ronald Press Co., 1968.

Henney, H. R., and T. Lynch, "Growth of *Physarum flavicomum* and *Physarum rigidum* in chemically defined minimal media," *J. Bacteriol.*, **99**:531–534 (1969).

Kerr, S. J., "A comparative study of mitosis in amoebae and plasmodia of the true slime mold *Didymium nigripes*," *J. Protozool.*, **14**:439–445 (1967).

Martin, G. W., and C. J. Alexopoulos, "The Myxomycetes," Iowa City, University of Iowa Press, 1969.

Ross, I. K., "Syngamy and plasmodium formation in the Myxogastres," *Amer. J. Bot.*, **44**:843–850 (1957).

Rusch, H. P., "Some biochemical events in the life cycle of *Physarum polycephalum*," in "Advances in Cell Biology," Vol. 1, D. M. Prescott, ed., New York, Appleton-Century-Crofts. [In press.]

n

NEGATIVE STAINING

Negative staining of materials for electron microscopy is not true staining, but instead involves the deposition of essentially amorphous electron-dense material around the surfaces and within the penetrable cavities of certain relatively stable cell structures. Thus, the biological material stands out as white against a neutral or dark (depending in part on the material and the stain) background. The method has been used successfully to study such organelles as microtubules (Figs. 1 and 2), ciliary rootlets, mitochondria, nuclear envelopes, muscle fibrils, and ribosomes. Other applications of negative staining include observations on viruses, insect tracheoles, acrosomes of spermatozoa (Figs. 3–5), glycogen (Fig. 6), blood pigment molecules (Figs. 7 and 8), bacteria (Fig. 11), mycoplasma-like organisms, enzyme molecules, vitreous body fibrils, and fibrin. Probably any reasonably stable structure with internal cavities and/or irregularities of the surface could successfully be negatively stained, with a judicious adjustment of such parameters as timing, pH, possibly temperature, prefixation, and concentration of the stain. The possibilities of the technique were first pointed out by Hall (in 1955) and by Huxley (in 1956). Brenner and Horne (1959) refined the method further, in a study of tobacco mosaic and turnip yellow mosaic viruses, and it has been used extensively since then, particularly for the study of viruses, bacteria, microtubules, and muscle. The method does not involve sectioning and is a simple and rapid one. Brenner and Horne's original technique involved using an ordinary atomizer and spraying a mixture of virus and negative stain onto carbon-coated electron microscope grids; they stated that Formvar-coated grids were unsatisfactory because of thermal drift in the microscope. Subsequently, however, it has been found that a combination of Formvar and carbon coatings is stable and quite readily penetrable by the electron beam. It is not feasible to use completely uncoated grids, because the support the coatings afford to the specimen is essential.

A number of substances have been used for negative staining, with varying degrees of success; this is apparently associated with the nature of the material one wishes to study, to some extent at least. Among the reagents which have been successfully used for negative staining are sodium and potassium phosphotungstate (PTA), ammonium molybdate, uranium nitrate and acetate, sodium dihydrogen phosphate, and cadmium iodide (Sjöstrand, 1967). Most workers have used PTA or uranyl acetate, but there is some evidence that ammonium molybdate may be an even more satisfactory stain. Addition of small

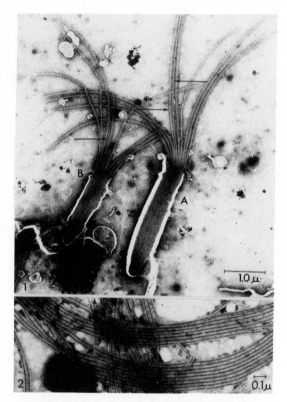

Figs. 1 and 2 Fig. 1: Two surface cilia (A and B) and a portion of a third (C) from the freshwater rhynchocoel *Prostoma rubrum*. Arrows designate the central singlet microtubules. For further explanation, see text. (Courtesy of W. D. Russell-Hunter, Managing Editor, *The Biological Bulletin*. From Henley, 1970.) Figure 1 and all other illustrations in this article are electron micrographs [made with a Zeiss 9A microscope except for Figs. 7 and 8, which were made with a JEM 100-B instrument], of material negatively stained with phosphotungstic acid. Fig. 2: Spermatozoa of the marine polyclad flatworm *Notoplana* have singlet cortical microtubules arranged in a row just beneath the plasma membrane. These have a striking helical substructure in their walls. The lumen of each singlet microtubule is marked by a dark line, representing an accumulation of electron-dense PTA.

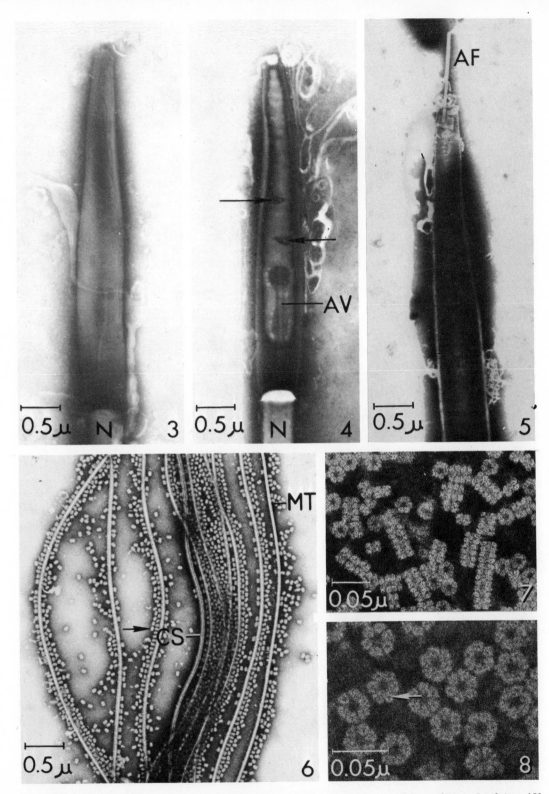

Figs. 3–8 Unreacted (Figs. 3 and 4) and reacted (Fig. 5) acrosomes of spermatozoa of the earthworm *Lumbricus*. AV, acrosome vesicle; N, nucleus; AF, acrosome filament. Arrows designate regions where the PTA has eroded the surface of the acrosome. Fig. 6: Glycogen granules (arrow) in intimate association with the nine doublet microtubules (MT) in spermatozoa of the earthworm *Lumbricus*. The central singlets (CS) of this spermatozoon are attached to one another and appear as a single unit, although they are two separate entities. Figs. 7 and 8: Erythrocruorin molecules from the blood of the earthworm. When seen in side view (Fig. 7) these are stacked in rouleaux; in flat view (Fig. 8) a petal-like six-part arrangement is apparent, with a relatively large central cavity and a smaller cavity (arrow) in each "petal."

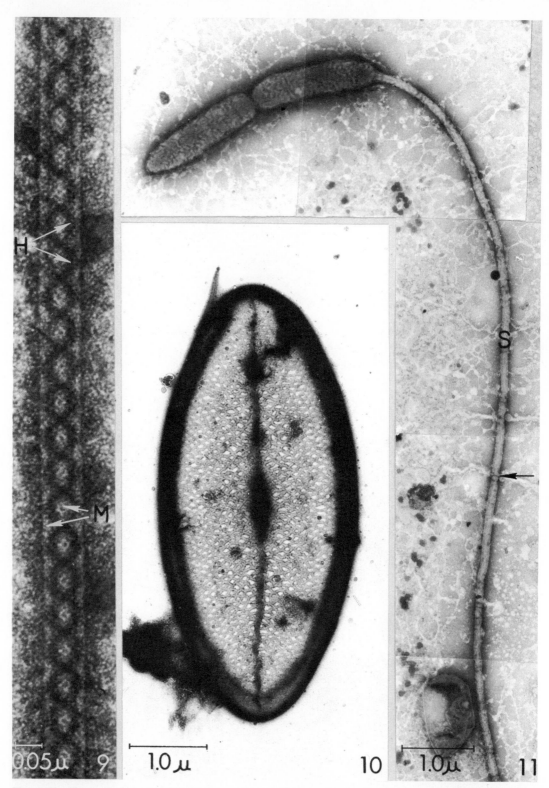

Figs. 9–11   Fig. 9: Central core of spermatozoon from a flatworm having the "9 + 1" pattern of microtubules. Instead of the two central singlet microtubules found in the more familiar 9 + 2 pattern, these spermatozoa have a large single core, of very complex substructure. There are apparently two hollow structures wound in a double helix (H) in and around matrix material (M). Fig. 10: Unidentified diatom-like structure from fresh water. Fig. 11: Certain bacteria give off a long stalk (S), not comparable to true bacterial flagella; this stalk is interrupted at irregular intervals by node-like structures (arrow), possibly indicative of spurts of growth.

amounts of a wetting agent, such as bovine serum albumin, to the stain greatly improves its spreading and staining qualities. The molecules of both stain and wetting agent must be sufficiently small as to remain unresolved by the electron microscope, and they should not volatilize in the electron beam.

If the biological material is small, it can merely be mixed with the stain; a drop of the mixture is then picked up with a narrow-bore pipette and transferred to the coated grid. In the case of PTA, the stain itself has the capacity of macerating small organisms differentially if they are dropped into it. For staining with uranium acetate, a premaceration, cytolysis, or fractionation is necessary. Some investigators prefer to use uranium acetate to avoid this macerating action of PTA. Alternatively, the desired region or material may be dissected out in a suitable physiological medium, and a drop of this then placed on a Formvar–carbon-coated grid, the stain being added to the grid. In both cases, after the requisite staining period, a strip of filter paper is then applied to the under-surface of the grid, drawing off all excess stain solution as rapidly as possible. Following a period of drying, the specimen is ready for examination with the electron microscope. Usually it is advisable to study preparations as soon as possible after they have been made, because the stain (in the case of PTA, at least) has a tendency to continue to macerate and digest away proteinaceous material, particularly in the presence of atmospheric moisture. If grids are to be stored or saved for later examination, they should be kept in a desiccator, away from dust. Most workers prefer to use freshly prepared stain, although in our experience this is not essential if one is careful to add the bovine serum albumin immediately before use.

We find it necessary to determine empirically, for each new biological material, the optimum staining time. Some forms (protozoa, for example) and some organelles (notably ciliary rootlets) are very sensitive to the macerating action of the PTA we routinely use, and the liquid stain should therefore be left in contact with such biological material only for a few seconds. Other materials (particularly spermatozoa of insects) are much more resistant to the lysing action of PTA, and considerably longer periods of staining are necessary, sometimes in conjunction with sonication and/or treatment with detergents. The concentration of the stain solution is apparently not very critical; most workers use 1% or 2% aqueous solutions, at a pH approximating neutrality or slightly below it (6.8). Higher pH levels result in artifactual changes in the appearance of microtubules, and very likely cause undesirable effects in other structures as well.

One of the most baffling (and exasperating) features of phosphotungstate as a negative stain is the unpredictability of its action, particularly insofar as its macerating effect, alluded to above, is concerned. Figure 1 exemplifies this well; three cilia, in varying stages of degeneration, are in close juxtaposition to one another on a single grid square, and all, presumably, were exposed to PTA (1%, pH 6.8, for 4 min) under exactly the same circumstances. In cilium A, however, the binding matrix is still present for a considerable distance; all nine doublet microtubules and both central singlet microtubules (arrows) can readily be identified in higher-magnification micrographs. By contrast, in cilium B (left), there is less matrix present; all nine doublets can be counted, but only one of the pair of central singlets (arrow) is present and degeneration has begun in it. The matrix has completely disappeared

from around cilium C (lower left corner), and higher-magnification micrographs indicate clearly that both central singlets have disappeared, leaving only the nine peripheral doublets. For further discussion of this topic, see Henley (1970).

It is obvious that if one drops an entire small organism, such as a flatworm, into negative stain, there is the strong likelihood that many cell organelles will be present, in varying stages of degeneration. Included among these would be mitochondria, Golgi, endoplasmic reticulum, and fragments of membrane from other sources. Identification of such structures is quite difficult, however, because their usual appearance, familiar from sectioned material, may be greatly changed by the action of the negative stain.

CATHERINE HENLEY
DONALD P. COSTELLO

References

Brenner, S., and R. W. Horne, "A negative staining method for high resolution electron microscopy of viruses," *Biochim. Biophys. Acta*, **34**:103–110 (1959).

Henley, C., "Changes in microtubules of cilia and flagella following negative staining with phosphotungstic acid," *Biol. Bull.,* **139**:265–276 (1970).

Pease, D. C., "Histological Techniques for Electron Microscopy," New York, Academic Press, 1964.

Sjöstrand, F. S., "Electron Microscopy of Cells and Tissues," Vol. I, New York, Academic Press, 1967.

NEMATOCYST

Nematocysts are important in the systematics of Cnidaria. The spination of discharged threads has been used in a classification of nematocysts. Nematocysts occur abundantly on the tentacles of coelenterates such as *Hydra* and jellyfish and in the acontia of certain sea anemones such as *Metridium*. Pieces of tentacles or acontia may be placed on a glass slide with a drop of distilled water and a coverslip and the discharged threads may be observed using a standard light microscope or a phase contrast microscope. For best results, however, it is necessary to isolate the nematocysts from the animal by one of the methods given below. Discharge of isolated nematocysts on a fresh slide preparation may be studied by introducing a drop of a 1% methylene blue solution at one edge of the coverslip. The blue dye will diffuse into undischarged capsules and may be seen to pass out the everted thread at the time of discharge. A drop of 1N HCl or 1N NaOH may be used to induce discharge of nematocysts.

1. Isolated nematocysts

Small Number of Cnidae. Nematocysts or stinging capsules may be isolated from tentacles or small whole specimens of Cnidaria such as *Hydra*, sea anemones, and jellyfish by treatment with 1.0M sodium citrate. A clean preparation of nematocysts of *Hydra* for photographic or other purposes may be obtained by the following procedure: Pipette approximately 100 whole *Hydra* from the culture medium into a centrifuge tube. Centrifuge lightly for approximately 3 min. Remove supernatant and replace with 1.0M sodium citrate. Shake violently by hand or on a vortex mixer for 1–2 min. Store overnight or longer in a refrigerator to allow time for the nematocysts to be expelled from the tissues. Shake 1 min and centrifuge at 1400 g for 15 min. Remove supernatant and resuspend

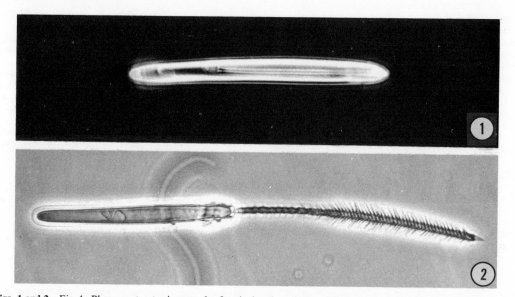

Figs. 1 and 2 Fig. 1: Phase contrast micrograph of an isolated undischarged nematocyst. ($\times$ 980.) Fig. 2: Phase contrast micrograph of a discharged nematocyst with a spined everted tube. ($\times$ 770.)

pellet in approximately 3 ml of $1.0M$ sodium citrate. Shake on vortex mixer for 2 min and centrifuge once more for 15–20 min at 1400 g. If storage of capsules is desired, remove supernatant, add 3 ml $1.0M$ sodium citrate, and resuspend pellet by pipetting. Store at 0–4°C. If nematocysts will be used within 24 hr, remove all but 0.5 ml of supernatant and add 4 ml of distilled water. Break up pellet by repeated pipetting of the preparation and again centrifuge at 1400 g for 15 min. Remove all but 0.5 ml of supernatant and resuspend pellet by pipetting. The preparation can be examined using a phase contrast or standard light microscope. Approximately 25% of the nematocysts will be discharged but 75% of the capsules will remain undischarged.

Large Number of Cnidae. Large numbers of nematocysts for experimental purposes may be isolated from whole sea anemones by homogenizing the animals in 3.85% NaCl buffered with 0.01M NaPO$_4$ at pH 7.4. Homogenize for 10 sec in a Waring blender at low speed; then for 10 sec at high speed. Centrifuge in a large test tube at 10,000 g for 5 min. Decant supernatant and resuspend pellet in 3.85% phosphate-buffered saline. Centrifuge again, decant supernatant, and resuspend material using same process. Filter suspension through glass wool previously rinsed with buffered saline. Centrifuge, decant supernatant, and resuspend nematocysts in saline. Use nematocyst preparation immediately or store in refrigerator.

Quick Method for Electron Microscopy. Place acontia of sea anemone *Metridium* in a corked plastic test tube in a sonicator set at maximum voltage for 2 min. Centrifuge, decant supernatant, and resuspend isolated nematocysts in 95% ethanol where they may be kept until needed.

Transmission Electron Microscopy. The threads of discharged nematocysts may be prepared for transmission electron microscopy by the following procedure: Place a drop of distilled water on a Formvar-covered specimen screen and, using an oral micropipette, inject several nematocysts into the drop while observing the process

with a stereomicroscope. The nematocysts tend to discharge in the distilled water and may be dried by evaporation onto the film. If discharge does not occur, add a microdrop of 2% phosphotungstic acid to discharge the capsules before they have settled on the film. Rinse the grids of dried nematocysts in distilled water to remove any remaining salts. The nematocysts may be viewed in a transmission electron misroscope directly or after shadow casting with metal.

Scanning Electron Microscopy. Isolated undischarged and discharged nematocysts may be prepared for scanning electron microscopy by the following procedure: With a micropipette inject several capsules into a drop of distilled water on an 8 mm square piece of coverslip. After the capsules have settled on the glass, use a pointed slip of filter paper to draw off most of the remaining fluid. Plunge the coverslip into liquid nitrogen and freeze-dry the specimen in a high-vacuum evaporator. Coat the entire surface of the nematocysts by *in vacuo* evaporation of a length of 60% gold–40% palladium wire. Mount coverslip on specimen stub and view in a scanning electron microscope.

2. Sectioned Tissue

Undischarged nematocysts within their cells (nematocytes) may be studied by means of light microscopy and electron microscopy of fixed and sectioned pieces of tissue.

Light Microscopy. Best results are obtained with Epon-embedded tissue rather than paraffin-embedded tissue because of the minute size of nematocysts and the difficulty of preserving their internal structure. Fix and embed as for electron microscopy (see below), but cut 0.5–1.0 μ thick sections on an ultramicrotome using either a glass or diamond knife. Spread the sections by waving a wooden stick dipped in xylene above the ribbon of sections. Lift one or more sections from the water trough by dipping a camel's hair brush under the sections and transfer them to a drop of distilled water confined within a circle etched on a glass slide using a carborundum pencil. Heat the slide on a hot plate until the water has evaporated and the sections have adhered to the glass. Stain the sections by covering them with one drop each of 2% sodium borate and 2% toluidine blue solutions. Shake the slide gently to mix and heat until the border of the stain turns green. Rinse the slide in distilled water and dry on a hot plate. The sections may be studied directly or covered with a drop of Permount and a cover-slip for permanent preservation.

Electron Microscopy. This method may be applied to pieces of tentacle extirpated from large coelenterates, or to small whole hydras. Put four or five specimens of *Hydra*, for example, into a small vial containing about 3 ml of culture medium. Pipette off the excess fluid leaving the specimens in about 0.5 ml solution for about 5 min until the animals extend their tentacles completely. Pour ice cold fixative into the vial. A suitable fixative for *Hydra* is one part 8% glutaraldehyde and one part $0.1 M$ cacodylate buffer (pH 7.2). For marine coelenterates it is best to add 6% sucrose to the fixative. Fix at 0–4°C for 2 hr. Rinse in $0.05 M$ cacodylate buffer (pH 7.2) for five changes or overnight. Trim off tentacles in buffer and postfix in a cold solution of one part 4% osmium tetroxide and three parts $0.1 M$ cacodylate buffer (pH 7.2) at 0–4°C for 2 hr. Dehydrate tissues 5 min each in 50%, 70%, and 90% ethanol and ten min each in three changes of absolute ethanol followed by two changes of propylene oxide. Leave the tissues in a 1 : 1 mixture of propylene oxide and Epon (Luft, 1961) for 2–10 hr, and then in a 1 : 3 mixture of the same substances for 10–12 hr. Transfer the tentacles to fresh Epon in a flat embedding dish for infiltration. Use microknives made with pieces of razor blades soldered to a needle to trim the tentacles into smaller pieces after they have infiltrated from 5 to 10 hr. Observe the trimming with a stereomicroscope in order to avoid excessive damage to the tiny pieces of tissue. Keep the dish or plastic cap containing the specimens on a layer of calcium chloride in a covered Petri dish to avoid hydration of the Epon. The final embedding dishes are prepared by adding a thin layer of Epon mixture to plastic caps which are placed in a 60°C oven for 6 hr or overnight to cure. Add an equal amount of uncured Epon to this substrate. Transfer small pieces of trimmed tissue to these dishes and orient for cross or longitudinal sectioning. Cure the Epon by placing the Petri dish for approximately one day each in 35°C, 45°C, and 60°C ovens. The cured blocks are trimmed into pyramids of tissue which are sectioned on an ultramicrotome for either light or electron microscopy.

JANE A. WESTFALL

References

Luft, J. H., *J. Biophys. Biochem. Cytol.*, **9**:409 (1961).
Pease, D. C., "Histological Techniques for Electron Microscopy," New York, Academic Press, 1964.
See also: COELENTERATA.

NERVE AND GLIA CELLS IN SECTIONS

Histological sections through tissue reveal a complicated structure, composed of nerve and glial cells and their processes. The identification of individual elements is not always easy.

Staining of Nissl substance visualizes the nerve cell perikarya, with the exception of very small neurons, i.e., granular cells of the cerebellum. Various basic dyes can be used for the identification of nucleo-protein particles in the cell. For quantitative determination of nucleic acids using cytophotometry, gallocyanin staining has been used: progressive staining at low pH, 1.5–1.7. With basic dyes, in the case of glial cells, it is the nucleus which takes up the strain.

Perikarya of nerve cells and their processes are visible after Golgi's impregnation method modified by Cajal (rapid Golgi method). A small piece of nervous tissue is immersed in an osmium-dichromate solution followed by impregnation with silver nitrate. This method was later modified by Bubeinat and Cox. About 10% of nerve cells are impregnated and this method has been applied for demonstrating the dendrites.

Methods demonstrating neurofibril are used for the identification of nerve fibers. Mostly commonly applied is the method proposed by Cajal (impregnation by silver nitrate and subsequent reduction by hydroquinone) and impregnation after Bielschowsky, either in the original version or modified (Barr, 1939; Bodian, 1936). These methods demonstrate neurofilaments of cell bodies and of bigger axons. In this way sometimes the dark fibrils of cell bodies can be distinguished from surrounding light axoplasma. It is possible to see in some regions, for instance, spinal cord axon terminals in the form of bouton-like structures.

Special methods have been applied for the demonstration of degenerated axons as described Nauta and Gygax (1951), Fink and Heimer (1967). The staining of normal fibers is suppressed and degenerating axons with some of their terminals are stained.

To visualize glial cells with their processes in the optical microscope, impregnation techniques were developed. A technique for nonspecifically impregnating glia is Cajal's gold-sublimate method. Glial cells are stained red and neurons purple.

The demonstration of microglia and macroglia is facilitated either by the bromformol impregnation (according to Cajal), or by the method described by Rio del Hortega (tissue is saturated with bromine, and then impregnated with silver). Without these specific methods it would be difficult to distinguish glial cells even on stained sections, except in cases where the glia has a typical localization, e.g., the satellite glia surrounding spinal neurons and neurons of dorsal root glanglia, glia from white matter, and Schwann cells.

Quantitative evaluation of sections enables computation of the ratio of glial cells to neurons and neuropil. The measurement can be performed by counting the single elements in the field of the microscope according to Chalkley (1943), where the volume ratio is obtained by taking random samples of points, determined by four pointers in the eyepiece. Where the elements are equally stained, the flying-spot microscope may be used to count them. Fixation artifacts influence quantitative evaluation. The space of neuropil relatively increases after perfusion fixation with 1% OsO_4 and embedding in Epon. During the histological treatment of the tissue nonuniform shrinkage of glia, neurons, and the neuropil appears, while maximal shrinkage of 40–60% was observed in astrocytes. Most shrinkage took place during embedding as was revealed by the optimal results obtained in cryostat post-fixed sections (Lodin et al., 1969).

The type of fixation is important for the preservation of the ultrastructure of the nervous tissue. Fixation by perfusion of the brain is considered to give optimal results.

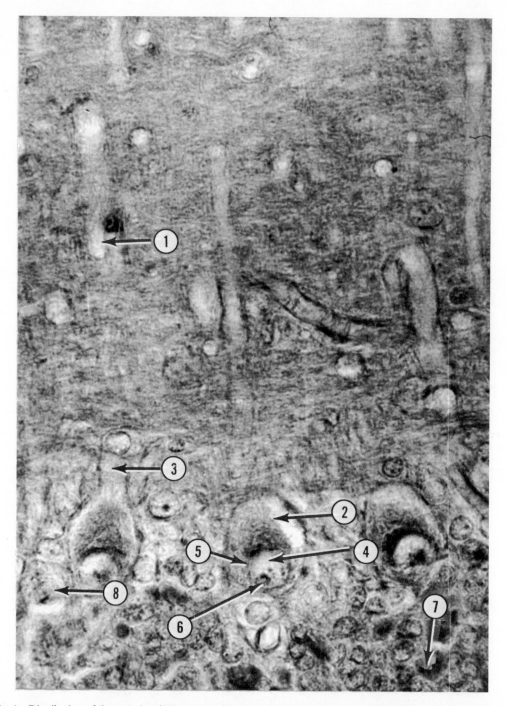

Fig. 1 Distribution of dry mass/cerebellar cortex, homogenous filed of interference microscope (Lodin et al., 1967). 1, Distal dendrite—0.18 pg/μ^3; 2, cytopiasm—0.37 pg/μ^3; 3, proximal dendrite—0.21 pg/μ^3; 4, nucleus—0.20 pg/μ^3; 5, membrane of the nucleus—0.43 pg/μ^3; 6, nucleolus—0.45 pg/μ^3; 7, glomerulus cerebell—0.43 pg/μ^3; 8, body of Bergmann's astrocyte—0.20 pg/μ^3.

The method proposed by Palay et al. (1962) used 1% of buffered osmic acid (animals were maintained on artificial respiration, vasodilatation was produced by 1% sodium nitrite, and heparin was injected). The composition of the fixative and the time of perfusion seem to play a decisive role in the quality of the fixation. Osmium tetroxide was substituted by glutaraldehyde (it is necessary that the latter be pure) or by paraformaldehyde, also in combination. Post-fixation involved immersing 1–2 mm³ blocks in buffered 1% osmic acid. Here fixation and not embedding is the major cause of artifacts. The success of fixation was judged by Palay (1962) according to the absence of the following features: discontinuity of cell membranes, swelling of mitochondria, shrinkage, retraction spaces, explosions, vacuoles, and dehiscences.

Medium-size and large neurons are easily identified by the characteristic size of cell body and nucleus, high density of rough endoplasmic reticulum, and typical processes. The majority of glial cells can be identified by their typical nuclear chromatin distribution, density of cytoplasm, and pale processes. Identification of neuropil structures, predominantly of small axons, dendrites, and glial processes, is much more difficult, however, and sometimes impossible.

In unstained histological sections the distribution of the dry mass can be determined by means of interference microscopy (see Fig. 1) and X-ray historadiography. A disadvantage of these methods is the fact that the thickness of the section fluctuates considerably (see Fig. 2) and this is a source of artifacts.

Using histochemical methods, the distribution of lipids, nucleic acids, proteins, and various enzymes can be studied. Accumulating evidence demonstrates changes during development of the nervous tissue, accompanying pathological processes, characterizing topographical differences, and also metabolical differences between glial cells and neurons. Histochemical methods allow easy morphological interpretation of the data obtained. Insufficient specificity of reactions and the limited possibility of quantitative evaluation are disadvantages of these methods. Evidence seems to indicate that oligodendroglial cells have a relatively higher activity of enzymes coupled with glycolysis, hexose-monophosphate shunt and citric acid cycle than astrocytes (Friede, 1965). The activity of most enzymes is generally lower in nerve and glial processes; weak enzymatic activity occurs in the body of astrocytes. In white matter glial cells show higher enzymatic activity than myelinated fibers. White matter

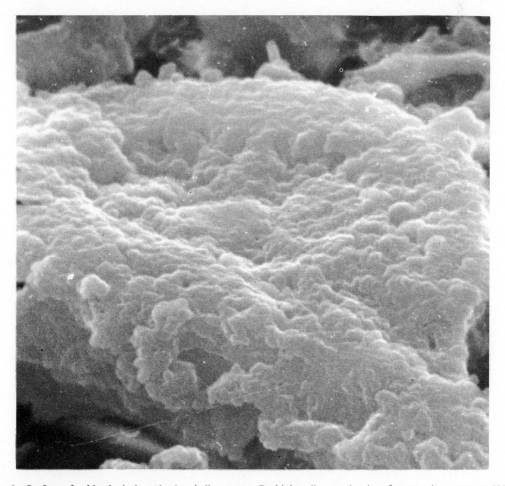

Fig. 2 Surface of a histological section/cerebellar cortex. Purkinje cell, scanning interference microscope, × 8000.

is characterized by much lower enzymatic activity than the gray matter.

In the interpretation of all the above data it is essential to bear in mind the parameters of histological treatment, as these have a profound effect on both the quantity and quality of the histochemical response.

Z. LODIN

See also: SILVER STAINS

NERVOUS SYSTEMS

Introduction

The diversity of nervous systems makes it impossible to do justice to the numerous technical approaches to their study. The decision was made to present some of the more widely used and reliable methods which can serve as a reference and guide. The omission of many methods does not reflect on their reliability, but indicates the necessity for some selectivity with the hope that those presented may form a basis for further reference.

Tissue Preparation

Microscopic investigation of the nervous system can be achieved through a wide variety of methods which have been modified, adapted, or developed specifically for this tissue. Of primary importance for success is the choice of a method of fixation and a fixing solution which will preserve the structural and spatial relationships of cells and cell organelles as faithfully as possible to their occurrence in the living animal. Although this is the goal of all tissue preservation, it is especially important for the nervous system due to the susceptibility of this tissue to anoxia. The slow rate of penetration of fixative into the tissue during tissue preparation may lead to artifacts. The chemical constituents of the tissue structures must be stabilized and yet retain their active binding sites for dyes and histochemical reagents.

One of the problems in preparation of nervous tissue for study obtains from the fact that fixation and stainability of neuronal structures and cellular elements vary from one part of the brain to another.

Preservation may be accomplished by both physical and chemical fixatives, each with its own limitations. Therefore, special recognition must be given not only to the specific details of the method and its qualities of preservation, but also to the goals of the investigation. This is especially true of tissue for enzymatic stains.

Probably the best general cytologic fixative for nervous tissue has proven to be 10 % neutral formalin (4, 9, 13, 43, 64, 75). Optimal preservation at the light and the electron microscopic level may be achieved when the animal is anesthetized and the tissue fixed by vascular perfusion. The demands of fixation for the electron microscopist are more exacting than for the light microscopist. Well-fixed material should exhibit plasma and cytoplasmic membranes free of discontinuities with an absence of swelling on the part of organelles (59, 70). If it is not possible to fix the tissue by perfusion, the nervous structure can be removed by careful dissection with the aid of the dissecting microscope and then immersed in the fixing solution. To prevent distortion and compression of the delicate neural structures the specimen is suspended by a thread or placed on a bed of cotton or glass wool.

For rapid microscopic study fresh unfixed or fixed tissue may be frozen, sectioned, and stained, thereby avoiding infiltration and embedment otherwise necessary in tissue preparation. One of the simplest methods is to place the tissue on a brass freezing platform and to freeze it with dry ice or short bursts of carbon dioxide (13). A variety of cold knife, cryostat, and freeze-drying and substitution methods have been developed and are described by Pearse (58). The choice of freezing method with or without fixation and the choice of fixative will be dictated by the nature of the components to be studied. No one method can be recommended as best for all purposes. The parameters of temperature and rapidity of freezing tissues are critical to control the formation of ice crystals from unbound extra and intracellular water. The use of freezing methods to avoid degradation of chemical components, especially enzymes, has been established in histochemistry and cytochemistry (2, 43, 58). The diffusion of products into the incubating medium or from the original locus of the cell can lead to artifacts of false localization of cytochemical reaction products. Often brief fixation in alcohol or acetone (1–2 hr) is sufficient to reduce the diffusion and permit preservation of even labile enzymes without interfering with their function. The higher resolution of electron microscopic methods has necessitated refinements to further eliminate distortions and artifacts of freezing. A detailed discussion of fixation methods may be found elsewhere in this text as well as standard references of microtechnique (9, 13, 22, 43, 58, 59, 75).

The diverse forms and composition of invertebrate nervous tissue make it especially difficult to recommend general methods. Most vertebrate techniques can be applied with appropriate modifications to the species under investigation. Reference may be made to Pantin (57), Gatenby and Beams (22), Wiesner (76), and Galigher and Kozloff (21) for details of invertebrate methods utilized by other investigators.

Whole body contractions of aquatic forms due to immersion fixation may be a serious problem. Distortion can be partially prevented by the use of anesthetics such as ethyl alcohol, magnesium sulfate, chloroform, or ethyl urethane which may be added to the natural medium of the animal. The exact anesthetic dose must be determined for the individual species. For terrestrial invertebrates fumes of ether, chloroform, or cyanide may be introduced into a closed container. The effect of anesthetics on tissue preparation and staining remains unknown.

Fixation in 10 % formalin, 10 % isotonic saline-formalin, Bouin's, or Zenker's solutions are recommended. Although perfusion methods are not usually applicable, the general fixation methods discussed above are useful.

Various embedment methods can be found in texts of general histologic techniques (4, 9, 12, 13, 15, 43, 64). Again no one procedure is unique for the nervous system and the ultimate choice will be dictated by the goals of the investigation. Recent interest has centered on plastic-embedded tissue adapted for examination with the light microscope and for correlation with the electron microscope. This subject will be covered below in its application to the nervous system.

Cytologic Staining Methods

Perhaps the most widely used general staining procedure utilizes hematoxylin and eosin because of their properties as a broad-spectrum, simple, orientational method routinely applied to normal and pathological nervous tissue. Although individual processes of glia and neurons remain unstained in satisfactory preparations, nuclear details of all cells and Nissl bodies are clearly distinguished. The choice of hematoxylin solution from the many successful formulae used is a matter of personal preference. It may be made on the basis of stability and chromatic intensity preferred. In addition to the routinely used eosin Y, hematoxylin may be counterstained with safranin, phloxine B, eosin B, or erythrosin B and certain acid azo dyes.

The preparation of hematoxylin solutions is affected by the partial oxidation of hematoxylin, either by slow, spontaneous, natural "ripening" or by chemical means through the addition of oxidizing agents such as mercuric oxide, hydrogen peroxide, potassium permanganate, potassium periodate, or sodium iodate to the dye hematein. The mordant used in combination with the hematein to form a basic "lake" helps to create the affinity of the dye for tissue components. Ammonium aluminum sulfate or potassium alum is a mordant for selective nuclear staining, as is iron for myelin and phosphotungstic acid for glia.

Lillie (33) describes the "azure eosin" methods and recommends their use for routine procedures. Excellent neural tissue preparations can be obtained from Mallory's phloxine methylene blue, Maximow's hematoxylin azure II eosin, and Lillie's azure A–eosin B method applied to 10% formalin-fixed tissue. Of particular value is the procedure of Lillie for its ability to regulate the blue-red color balance of tissue by variations of stain pH with buffers.

Intra-vitam staining with methylene blue for entire nerve cells and their fibers has been utilized for many animals and can give spectacular results. Insects and crustaceans have been stained by intra-abdominal or intrathoracic injections of 0.4% methylene blue at pH 5 (73). Tinctorial characterization of the invertebrate tissues may be achieved by application of Masson's trichrome and its many variants.

Most recently the introduction of intracellular injection methods for neurons and large azons has been demonstrated in studies of neuron geometry (14, 66), axon transport (45, 71), and synaptic terminals (33) to mention only a few. The fluorescent dye, procion yellow, which has the property of diffusing throughout the neuron by axoplasmic flow, and can withstand histologic procedures, has been used in crayfish (66) and lobster (14). This method provides combined electrophysiological and cytoarchitectural studies of selected neurons. This is in contrast to the Golgi methods which also stain entire cells, but do so by random selection.

It has been my intent to limit any discussion of Golgi methods, since it would be impossible to do justice to such a classic and elegant tool which has broad application today. A very complete and practical analysis of their capacities and limitations can be found in Nauta and Ebbeson (52), with reference to Golgi methods and their relevance to recent disciplines.

Myelin and Nissl Methods

In 1885 Nissl noted that basic aniline dyes stained clumps of material in the cytoplasm. These clumps have been called Nissl granules and have been shown with the electron microscope to be rough endoplasmic reticulum. Thionin is the most dependable dye and gives excellent results following the procedure of Conn et al. (12) which utilizes progressive or regressive staining in buffer solutions.

The Kluver and Barrera (41) method after fixation in 10% formalin followed by infiltration in paraffin is perhaps the best choice for a relatively simple technique providing details of myelinated fibers, nuclei, and Nissl substance of cells, on one slide. A 0.1% Luxol fast blue solution, followed by differentiation in alcohol and 0.05% lithium carbonate, and counterstaining in 0.1% cresyl fast violet is commonly used. A wide range of tinctorial differentiation can be obtained by combinations of Nissl staining with basic fuchsin, Darrow red, safranin, Gallocyanin chrome alum, or toluidine blue, and with myelin stains, Luxol fast blue or hematoxylin variants. Sections first stained with Bodian's method for neurofibrils followed by the Luxol fast blue method often provide a more complete picture of myelinated and unmyelinated fibers, cell bodies, and neurofibrillar components of terminal knobs than is available with other methods. The Luxol fast blue method has been used by Margolis and Pickett (47) in combination with periodic acid–Schiff reaction for complex carbohydrates, phosphotungstic acid hematoxylin for glial filaments, oil red O for lipids, and Holmes silver nitrate for axons with good results.

The staining of normal myelin sheaths by the method of Weigert in 1884 (43) and its many modifications has been widely used to trace myelinated pathways of the nervous system. The method may also be used in neuropathology and in neuroanatomy to map degenerated myelinated pathways due to their loss of myelin staining. Hematoxylin is used with a mordant in this method and its modifications. The mordant, frequently a chromium or ferrous salt, is used either before tissue embedding or after sectioning. Myelin is composed of lipid complexes which are partially extracted by the solvents used in tissue preparation. Methods, therefore, will differ in their ability to stain myelin and to demonstrate finely myelinated fibers. However, a number of techniques may be used with reliable results: (a) the Pal-Weigert modifications (75); (b) Lillie's (43) method which utilizes a mordant before embedding; and (c) the Weil method employing a combination mordant-hematoxylin stain (4).

Page (56) found the solochrome dye for myelin sheaths superior to Loyez's or Weil's method because of its simplicity, reliability, and clarity. This method is valuable for paraffin and celloidin sections of the brain and spinal cord, and cryostat sections of unfixed peripheral nerves.

Staining of myelin by osmium vapors or immersion in dilute osmic acid solutions (0.5–1.0%) is particularly recommended for peripheral nerve. It is possible to use formalin-fixed or fresh unfixed tissues. For fresh tissue the osmium tetroxide serves as both a fixative and a stain and is more satisfactory than using formalin-fixed tissue. The procedure of Bruesch cited in Conn et al. (12) has been reliable.

The Bridge Between Light and Electron Microscopy

Improved methods of nervous tissue preservation for electron microscopy have stimulated interest in the histologic examination of semithin sections (0.5–3 μ) by light microscopic methods in an effort to "bridge the gap" between high-resolution electron microscopy and the view provided by the light microscope. The application of conventional procedures for neuronal staining is not always possible in tissue sections which have been prepared by fixation and plastic embedments for electron microscopy. Such attempts have been limited by a variety of factors such as: (a) the poor penetration of some dyes into plastic; (b) the staining of the plastic itself; (c) the altered availability of bonding groups resulting from the chemical effects due to the variety, complexity, and combinations of reagents for fixation; (d) the use of osmium tetroxide post-fixation. Some cationic dyes (methylene blue, basic fuchsin, axure B, thionin, toluidine blue, alcian blue, nile blue B, and safranin) have been successful in achieving monochromatic contrast of the cells and their organelles with reliability and simplicity. Color distinctions have been gained with combinations of these stains (26, 38, 42).

The oxidation or "bleaching" of the plastic sections prior to the application of staining solutions by oxidizing agents such as periodic acid (49), oxone (67), potassium permanganate followed by oxalic acid (68), acidified hydrogen peroxide (62), and performic acid (30) has been effective in producing available reactive groups for acid stains. These stains include orange G, eosin, and phloxine B, which may then be used with hematoxylin and eosin and PAS methods. In contrast to the need for oxidation of osmium tetroxide for acid stains, unmodified osmium fixation has been essential for successful silver impregnation of epoxy-embedded tissues in the procedures developed by Goldblatt and Trump (23), Berkowitz et al.(5), and Heimer (31). A suppressive silver method (31) for Epon-Araldite sections has been developed and its advantageous approach in a combined light and electron microscope study has been demonstrated.

The stain paraphenylenediamine, which was introduced by Estable-Puig in 1965 (18), has more recently been used by Hollander and Vaaland (36). They suggested it as a possible method for identification of degeneration in axons and axon terminals.

The application of an epoxy solvent (a ripened saturated solution of NaOH in ethyl alcohol) has made possible the differential staining of neurons and glia in brain tissue (5). Use was made of a hematoxylin "lake" and the dyes malachite green, azure B, and basic fuchsin. The chromatic differentiation of neurons (pink), astrocytes (red), and oligodendrocytes (green) permitted identification of profiles where size and shape alone could be misleading criteria. A second method using a combined methylene blue, basic fuchsin stain followed by differentiation in the malachite green–azure B solution achieved identification of terminal endings (red), myelin (dark blue-green), and axoplasm (bluish-pink). The identification of neurons and glia was also possible as a result of differences in color intensity. Cell organelles such as mitochondria (red), lipid inclusions (dark blue-green), and endoplasmic reticulum (pink) can be clearly shown. In addition a "reduced silver" method was presented in which the osmium tetroxide is suggested as the possible "reducing agent" for the silver with resultant impregnation of neuronal structures (5). (See Figs 1 and 2).

Comparison of adjacent "thick" and "thin" sections for correlation of light and electron microscopic patterns in the same cell can provide a basis for evaluation of organelles and topographic organization difficult to obtain with either method alone.

Fiber Methods

The various silver impregnation methods designed for demonstration of nerve fibers, cells, endings, and neuro-fibrils can yield rewarding results when great care is taken in technical detail and manipulation. A few general prin-

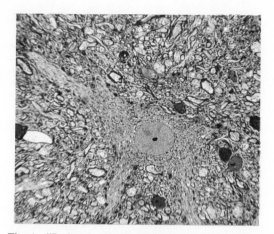

Fig. 1 "Reduced silver" method (see reference 5) anterior horn—macaca mulatta. With this method neurons, glia cells, and cytoplasmic components are stained in varying intensities of gold to black. "Axonal knobs" and mitochondria are especially prominent. Tissue in this figure and Fig. 2 were prepared for electron microscopy, embedded in araldite, sectioned at 2 μ, and stained after removal of the embedding medium ($\times$ 630).

ciples apply to all silver methods: (a) Contamination by metallic ions from instruments and vessels must be avoided; therefore, preference is given to glass containers and utensils; (b) all glassware and utensils must be chemically clean and are best reserved for these methods; (c) chemicals should be reagent grade and distilled or double-distilled water used unless otherwise specified. Latitude in the implementation of the method to take into account variations in individual tissues and in the solutions is necessary.

Observations made by electron microscopic study have greatly enhanced the understanding and interpretation of the variety of silver methods and the structures impreg-

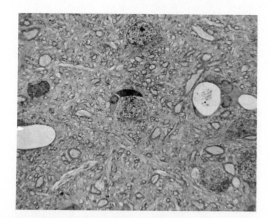

Fig. 2 Hemataxylin "lake" modification of Grimley stain (see reference 5) cerebral cortex—macaca specioso. This method permits the identification of oligodendro-cytes and astrocytes on the basis of differential staining of both cytoplasm and nuclei. Nissl bodies, myelin, mitochondria nucleoli, and lipofuscin are green as distinct from red cytoplasmic matrix and nucleoplasm ($\times$ 630).

nated through their application. Recent experimentation combining light microscopic silver methods with electron microscopic observations have shown that silver granules are deposited upon neurofilaments (44, 32). These studies and the views of Gray and Guillery (24) have been given further support by cytoarchitectural electron microscopic correlations of the neurofibrillar silver impregnations using the methods of Glees, Holmes, and paraffin Nauta. Argyrophilia of normal axons and terminal knobs are shown to be dependent upon the presence of neurofilaments (63). Laminae I and II of the cortex in which microtubules are abundant are poorly stained. Recent findings are reviewed by Guillery (27) and Heimer (31) of normal and degenerating axonal and terminal structures in which the nature of "silver and structure" is elucidated and interpreted in detail.

When a nerve cell body is separated from the axon, that axon will undergo degenerative changes. Degenerating myelin contains many unsaturated fatty acids which can be demonstrated by osmium tetroxide. When the tissue is treated with potassium dichromate the normal myelin is oxidized and becomes unreactive, but the degenerative myelin retains sufficient reactive groups to reduce the osmium tetroxide and is stained black. The reactive groups have been shown to be cholesterol esters (2). The selective staining of these groups forms the basis of the Marchi method and is useful in tracing myelinated pathways in central and peripheral nervous tissue. Modifications of these methods include the substitution of other oxidizing agents, e.g., sodium iodate (12) or potassium chlorate (12). A histochemical method for the simultaneous demonstration of normal and degenerating myelin has been developed by Adams (1) called the OTAN method. This technique utilizes the Swank-Davenport method of osmium tetroxide for degenerating myelin with the resultant black Marchi reaction, while normal myelin appears red-brown due to chelation of a-naphthylamine by the normal myelin lipids.

Efforts to clearly define the fiber tracts and cytoarchitecture of nervous systems of nonmammalian and invertebrate forms have been successful primarily by modifications of Cajal, Bielschowsky, and selective silver impregnations. These animals are characterized by a greater abundance of nonmyelinated and thinly myelinated axons than occur in mammalian forms.

The Holmes silver technique (37) is recommended for lower forms because of its versatility which is derived by utilizing a sequential pH series of buffered silver solutions. The optimal argyrophilia for each individual specimen may then be determined. The Bielschowsky methods also offer versatility for diverse forms by controlling various aspects of the impregnation procedure, in particular, the amount of ammonia in the ammonical silver solution, the concentration and time for the silver nitrate solution, and the chemical components of the ammonical silver. Double silver impregnation methods have been useful for "difficult to stain" tissue. Such a modification has been applied to several insects with intense staining of nerves (11).

Silver methods selective for degenerating axoplasm and terminals with suppression of the argyrophilia of normal nerve fibers, although recent in application, have been significant in their contributions to the studies of a variety of animal nervous systems. Suppression has been achieved by pretreatment of sections with phosphomolybdic acid or uranyl nitrate and potassium permanganate. The following references are only a few in the development and application of these methods (3, 16, 31, 40, 51, 53, 54).

Glial Methods

Ramón y Cajal's formalin–ammonium bromide is the recommended fixative for the demonstration of glial cells utilizing silver methods. Mention is made of a few procedures which may be used with a fair degree of reliability. The silver carbonate method of Del Rio-Hortega developed in 1917 (12, 75) demonstrates oligodendroglia and possibly microglia after fixation of 12–48 hr. Longer fixation, as much as 30 days, will favor the impregnation of astrocytes. The gold sublimate method of Cajal is especially recommended for staining of astrocytes (12, 15). Following fixation of 2–5 days, staining of protoplasmic astrocytes will be best demonstrated; longer fixation, up to 25 days, will favor staining of fibrous astrocytes. These methods for glia are designed for frozen sections cut at 15–30 μ.

Holzer's and Mallory's phosphotungstic acid hematoxylin methods are usually successful for glial fibers after formalin fixation and paraffin infiltration.

Recent electron microscopic studies have evaluated the identity of various glial elements. The interpretations suggest that we may have to modify some of our concepts of glial morphology in the future.

Interaction of Method

For a greater appreciation of the nervous system it is of increasing value to integrate a variety of relevant methods. The interaction of techniques has been demonstrated in most fields including histochemistry, cytochemistry, autoradiography, fluorescence, microscopy, and tissue culture. No effort has been made to discuss completely these techniques and their contributions in relation to the nervous system. The methods cited are merely an adaptation which is specific for its application to neural investigations.

Autoradiography. The methods of autoradiography have been reviewed by Caro (10), Salpeter and Bachmann (65), Sidman (69), and Joftes and Kury (39). The development of fine-grained, sensitive, dependable emulsions and simpler, refined techniques for applying tissue sample has made this method available and useful for light and electron microscopy. For the nervous system the effective dose of thymidine-H^3 is 2–10 times higher than that used for study of other organs. This dose, according to Sidman, is 5 μc/g body weight.

Fluorescence Microscopy. Fluorescence methods have recently been applied to the study of nervous tissue. The demonstration of storage sites of tissue amines in neurons such as the monamines (19), histamine (17), and tryptamine (6) are some of the contributions resulting from application of these methods. Briefly the procedure outlined by Fuxe et al. (20) involves the condensation of active monamines by formaldehyde to yield fluorescent compounds when activated by light of short wavelengths used in the fluorescent microscope. The combined use of the fluorescent and electron microscope has permitted a study of degeneration of unmyelinated axons and axon collaterals. These fibers, which are too small to be identified with silver methods, may be identified by fluorescence of their dopamine granules resulting from the degenerative process. In skilled hands one can make identification of types as well as localization of various monamines.

Tissue Culture. The first observations of the developing nerve fiber in tissue culture were made in 1907 by Ross G. Harrison when he explanted a portion of neural tube from a frog embryo and described the growing tip. Since that time the role of neural tissue in tissue culture has evolved through improvements of the culture media and substrate which maintained nerve cells (34, 61) for several months. In 1955 Peterson and Murray (60) observed development of myelin in cultures of chick spinal ganglia. The technical achievements which included the addition of glucose to the medium (7, 61) have led to greater organization of the tissue *in vitro* to mirror *in vivo* organization, differentiation, and development thus permitting continuous observations of fibers, neurons, supporting cells, and their organelles (8, 28, 46, 60, 72).

LILLIAN R. BERKOWITZ

References

1. Adams, C. W. M., "A histochemical method for the simultaneous demonstration of normal and degenerating myelin," *J. Pathol. Bacteriol., 77*:648 (1959).
2. Adams, C. W. M., "Neurohistochemistry," New York, Elsevier Publishing Co., 1965.
3. Albrecht, M. H., and R. C. Fernstrom, "A modified Nauta-Gygax method for human brain and spinal cord," *Stain Technol., 34*:91–94 (1959).
4. Armed Forces Institute of Pathology, "Manual of Histologic and Special Staining Technics," 2nd ed., New York, McGraw-Hill Book Co., Inc., 1960.
5. Berkowitz, L. R., O. Fiorello, L. Kruger, and D. S. Maxwell, "Selective staining of nervous tissue for light microscopy following preparation for electron microscopy," *J. Histochem. Cytochem., 16*:808–814 (1968).
6. Bjorklund, A., B. Falck, and R. Hakansson, "Histochemical demonstration of trytamine. Properties of the formaldehyde-induced fluorophores of trytamine and related indole compounds in models," *Acta physiol. scand., 74*(Suppl. 318): 1–31 (1968).
7. Bornstein, M. B., "Reconstituted rat-tail collagen used as sub-strate for tissue cultures on coverslips," *Lab. Invest., 7*:134–140 (1958).
8. Bunge, M. B., R. P. Bunge, and E. R. Peterson, "The onset of synapse formation in spinal cord cultures as studied by electron microscopy," *Brain Res., 6*:728–749 (1967).
9. Carleton, H. M., and R. A. B. Drury, "Histological Technique," 3rd ed., London, Oxford University Press, 1957.
10. Caro, L. G., "High resolution autoradiography," pp. 327–363 in "Methods in Cell Physiology," Vol. 1, D. M. Prescott, ed., New York, Academic Press, 1964.
11. Chen, J. S., and M. G. M. Chen, "Modifications of the Bodian Technique applied to insect nerves," *Stain Technol., 44*:50–51 (1969).
12. Conn, H. J., M. A. Darrow, and V. M. Emmel, "Staining Procedures," 2nd ed., Baltimore, The Williams & Wilkins Co., 1960.
13. Davenport, H. A., "Histological and Histochemical Technics" Philadelphia, Saunders, 1960.
14. Davis, W. J., "Motoneuron morphology and synaptic contacts: determination by intracellular dye injection," *Science, 168*:1358–1360 (1970).
15. Disbrey, B. D., and J. H. Rack, "Histological Laboratory Methods," London, L. & S. Livingstone, 1970.
16. Ebbesson, S. O. E., "The selective silver-impregnation of degenerating axons and their synaptic ending in nonmammalian species," in "Contemporary Research Methods in Neuroanatomy," W. S. H. Nauta, and S. O. E. Ebbesson, New York, Springer-Verlag, 1970.
17. Ehringer, B., and R. Thumberg, "Induction of fluorescence in histamine-containing cells," *Exp. Cell Res., 47*:116–122 (1967).
18. Estable-Puig, J. F., W. C. Bauer, and J. M. Blumberg, "Paraphenylene diamine staining of osmium-fixed plastic embedded tissue for light and phase microscopy," *J. Neuropath. Exp. Neurol., 24*:531–535 (1965).
19. Falck, B., N.-A. Hillarp, G. Thieme, and A. Torp, "Fluorescence of catecholamines and related compounds condensed with formaldehyde," *J. Histochem. Cytochem., 10*:348–354 (1962).
20. Fuxe, T., T. Hokfelt, G. Jonsson, and O. Ungerstedt, "Fluorescence microscopy in neuroanatomy," pp. 275–314 in "Contemporary Research Methods in Neuroanatomy," W. J. H. Nauta, and S. O. E. Ebbesson, ed., New York, Springer-Verlag, 1970.
21. Galigher, A. E., and E. N. Kozloff, "Essentials of Practical Microtechnique," Philadelphia, Lea & Febiger, 1964.
22. Gatenby, J. B., and H. W. Beams, "Bolles Lee's The Microtomist's Vade Mecum," 11th ed., New York, McGraw-Hill Book Co., 1950.
23. Goldblatt, P. J., and B. F. Trump, "The application of del Rio Hortega's silver method to epon-embedded tissue," *Stain Technol., 40*:105–115 (1965).
24. Gray, E. G., and R. W. Guillery, "Synaptic morphology in the normal and degenerating nervous system," *Int. Rev. Cytol., 19*:111–182 (1966).
25. Gray, P., "The Microtomists' Formulary and Guide," New York, Blakiston Co., 1954.
26. Grimley, P. M., J. M. Albrecht, and H. J. Michelitch, "Preparation of large epoxy sections for light microscopy as an adjunct to fine-structure studies," *Stain Technol., 40*:357–366 (1965).
27. Guillery, R. W., "Light and electron microscopical studies of normal and degenerating axons," pp. 77–105 in "Contemporary Research Methods in Neuroanatomy," W. J. H. Nauta, and S. O. E. Ebbesson, ed., New York, Springer-Verlag, 1970.
28. Guillery, R. W., H. M. Sobkowicz, and G. L. Scott, "Relationships between glial and neuronal elements in the development of long-term cultures of the spinal cord of the fetal mouse," *J. Comp. Neurol., 140*:1–34 (1970).
29. Hakansson, R., and C. Owman, "Concomitant histochemical demonstration histamine and catecholamines in enterochromaffin-like cells of gastric mucosa," *Life Sci., 6*:759–766 (1967).
30. Heath., E., "The use of performic acid oxidation to facilitate differential staining of epoxy-embedded adenohypophysis," *Z. Zellforsch. Mikro. Anat., 107*:1–5 (1970).
31. Heimer, L., "Bridging the gap between light and electron microscopy in the experimental tracing of fiber connections," pp. 162–172 in "Contemporary Research Methods in Neuroanatomy," W. J. H. Nauta and S. O. E. Ebbesson, New York, Springer-Verlag, 1970.
32. Heimer, L., and R. Ekholm. "Neuronal argyrophilia in early degeneration states: A light and electron microscopic study of Glees and Nauta techniques," *Experientia* (Basel), *23*:237–239 (1967).
33. Hendrickson, A., "Electron microscopic radioautography. Identification of origin of synaptic terminals in normal nervous tissue," *Science, 165*:194–196 (1969).
34. Hogue, M. S., "A study of adult brain cells grown in tissue cultures," *Amer. J. Anat., 93*:397–416 (1953).
35. Hokfelt, T., and K. Fuxe, "Cerebellar monoamine nerve terminals, a new type of afferent fibers to the cortex cerebelli," *Exp. Brain Res., 9*:63–72 (1969).
36. Hollander, H., and J. L. Vaaland, "A reliable staining method for semi-thin sections in experimental neuroanatomy," *Brain Res., 10*:120–126 (1968).

37. Holmes, W., "Silver staining of nerve axons in paraffin sections," *Anat. Rec.,* **86**:157–187 (1943).

38. Huber, S. D., F. Parker, and G. F. Odland, "A basic fuchsin and alkalinized methylene blue rapid stain for epoxy-embedded tissue," *Stain Technol.,* **43**:83–87 (1968).

39. Joftes, D. L., and G. Kury, "Radioautography in neuropathology," p. 401 in "Neuropathology: Methods and Diagnosis," C. G. Tedeschi, ed., Boston, Little, Brown and Co., 1970.

40. Johnstone, G., and D. Bowsher, "A new method for the selective impregnation of degenerating axon terminals," *Brain Res.,* **12**:47–53 (1969).

41. Kluver, H., and E. Barrera, "A method for the combined staining of cells and fibers in the nervous system," *J. Neuropath. Exp. Neurol.,* **12**:400–403 (1953).

42. Leeson, C. R., and T. S. Leeson, "Staining methods for sections of epon-embedded tissues for light microscopy," *Can. J. Zool.,* **48**:189–190 (1970).

43. Lillie, R. P., "Histopathologic Technic and Practical Histochemistry," 3rd ed., New York, McGraw-Hill Book Co., 1965.

44. Lund, R. D., and L. E. Westrum, "Neurofibrils and the Nauta method," *Science,* **151**:1397–1399 (1966).

45. Lux, H. D., P. Schubert, G. W. Kreutzberg and A. Globus, "Excitation and axonal flow: Autoradiographic study on motoneurons intracellularly injected with a ³H amino acid," *Exp. Brain Res.,* **10**:197–204 (1970).

46. Lyser, K. M., "Early differentiation of the chick embryo spinal cord in organ culture: light and electron microscopy," *Anat. Rec.,* **169**:45–64 (1971).

47. Margolis, G., and J. P. Pickett, "New applications of the luxol fast blue myelin stain," *Lab. Invest.,* **5**:459–474 (1956).

48. Marsland, T. A., P. Glees, and L. B. Erikson, "Modification of the Glees silver impregnation for paraffin sections," *J. Neuropath. Exp. Neurol.,* **13**:587 (1954).

49. Munger, B. L., "Staining methods applicable to sections of osmium-fixed tissue for high resolution light microscope," *J. Biophys. Biochem. Cytol.,* **11**:502–505 (1961).

50. Murray, M. R., "Nervous tissues in vitro," pp. 373–455 in "Cells and Tissue in Culture," Vol. 2, New York, Academic Press, 1965.

51. Nauta, W. J. H., "Silver impregnation of degenerating axons," pp. 17–26 in "New Research Techniques of Neuroanatomy," H. F. Windle, ed., Springfield, Illinois, Thomas, 1957.

52. Nauta, W. J. H., and S. O. E. Ebbesson, eds., "Contemporary Research Methods in Neuroanatomy," New York, Springer-Verlag, 1970.

53. Nauta, W. J. H., and P. A. Gygax, "Silver impregnation of degenerating axon terminals in the central nervous system: (1) Technic (2) Chemical Notes," *Stain Technol.,* **26**:5–11 (1951).

54. Nauta, W. J. H., and P. A. Gygax, "Silver impregnation of degenerating axons in the central nervous system: A modified technique," *Stain Technol.,* **29**:91–93 (1954).

55. Nayyar, R. P., and M. L. Barr, "Histochemical studies on the accessory body of Cajal in neurones of the cat," *J. Comp. Neurol.,* **132**:125–134 (1968).

56. Page, K. M., "Histological methods for peripheral nerves. Part I," *J. Med. Lab. Technol.,* **27**:1–17 (1970).

57. Pantin, C. F. A., "Notes on Microscopical Technique for Zoologists," London, Cambridge University Press, 1948.

58. Pearse, A. G. F., "Histochemistry: Theoretical and Applied," 3rd ed., Vol. 1, Boston, Little, Brown & Co., 1968.

59. Pease, D. C., "Histological Techniques for Electron Microscopy," New York, Academic Press, 1964.

60. Peterson, E. R., and M. R. Murray, "Modification of development in isolated dorsal root ganglia by nutritional and physical factors," *Develop. Biol.,* **2**:461–476 (1960).

61. Pomerat, C. M., and I. Costero, "Tissue cultures of cat cerebellum," *Amer. J. Anat.,* **99**:211–247 (1956).

62. Pool, C. R., "Hematoxylin-eosin staining of OsO_4 fixed epon-embedded tissue; prestaining oxidation by acidified H_2O_2," *Stain Technol.,* **44**:75–79 (1969).

63. Ralston, H. J., III, "The fine structure of neurons in the dorsal horn of the cat spinal cord," *J. Comp. Neurol.,* **132**:275–301 (1968).

64. Ruthmann, A., "Methods in Cell Research," London, G. Bell & Sons, Ltd., 1970.

65. Salpeter, M. M., and L. Bachmann, "Autoradiography with the electron microscope," *J. Cell Biol.,* **22**:469–477 (1964).

66. Selverston, A. I., and D. Kennedy, "Structure and function of identified nerve cells in the crayfish," *Endeavor,* **28**:107–113 (1969).

67. Sevier, A. C., and B. L. Munger, "The use of oxone to facilitate specific tissue stainability following osmium fixation," *Anat. Rec.,* **162**:43–52 (1968).

68. Shires, T. K., M. Johnson, and K. M. Richter, "Hematoxylin staining of tissues embedded in epoxy and resins," *Stain Technol.,* **44**:21–25 (1969).

69. Sidman, R. L., "Autoradiographic methods and principles for study of the nervous system with thymidine-H³," pp. 252–274 in "Contemporary Research Methods in Neuroanatomy," W. J. H. Nauta and S. O. E. Ebbesson, ed., New York, Springer-Verlag, 1970.

70. Sjöstrand, F. S., "Electron Microscopy of Cells and Tissues," Vol. 1, New York, Academic Press, 1967.

71. Sjöstrand, J., and J. O. Karlsson, "Axoplasmic transport in the optic nerve and tract of the rabbit: a biochemical and radioautographic study," *J. Neurochem.,* **16**:833–844 (1969).

72. Sobkowicz, H. M., R. W. Guillery, and M. B. Bornstein, "Neuronal organization in long term cultures of the spinal cord of the fetal mouse," *J. Comp. Neurol.,* **132**:365–396 (1968).

73. Stark, M. J., K. N. Smalley, and E. C. Rome, "Methylene blue staining of axons in the ventral nerve cord of insects," *Stain Technol.,* **44**:97–102 (1969).

74. Stevens, A. R., "High resolution autoradiography," p. 255 in "Methods in Cell Physiology," Vol. II, D. Prescott, ed., New York, Academic Press, 1966.

75. Tedesch, C. G., ed., "Neuropathology: Methods and Diagnosis," Boston, Little, Brown & Co., 1970.

76. Wiesner, F. M., "General Zoological Microtechniques," Baltimore, The Williams and Wilkins Co., 1960.

NEURONS AND GLIA CELLS IN CULTURE

I. Introduction

Several researchers have observed that neurons separated from glial cells and from other constituents of the nervous system can survive and grow *in vitro*. Many studies have been directed toward embryonic development of neural tissue from the chick central or peripheral (including the sympathetic) nervous system dissociated in culture (Nakai, 1956; Levi-Montalcini and Angeletti, 1963; Cohen et al., 1964; Nakajima, 1965; Shimizu, 1965; Utakoji and Hsu, 1965; Rieske, 1969; Scott et al., 1969;

Sensenbrenner et al., 1969; Lodin et al., 1970a, 1970b). Some of these investigations have involved the cultivation of dissociated cells from the spinal cord (Cavanaugh, 1955; Courtey and Bassleer, 1967a; Meller et al., 1969) and brain (Varon and Raiborn, 1969; Booher et al., 1969; Grosse and Lindner 1970; Sensenbrenner and Mandel, 1971; Sensenbrenner et al., 1971). Some studies have been made on young and adult nerve tissue from rats and rabbits (Hillman, 1966, Hillman and Sheikh, 1967, 1968; Hillman and Khalawan, 1970).

The cultivation of dissociated nerve cells and the methods and procedures for obtaining these kinds of preparations have become of great importance in the study of the neuron-glial relationship, myelination, cellular behavior and differentiation, and metabolic and physiological activities, as well as the effects of growth factors.

II. Preparation of Cell Suspensions

Most of the procedures employed today involve mechanical or enzymatic dissociation or the combination of the two. In tissue culture it is always considered that each step in the procedures involved must be performed under sterile conditions.

A. Dissociation of Neurons and Satellite Cells from Peripheral Nerve Tissue. The dorsal ganglia are the most commonly employed tissue for studies of the peripheral nervous system in tissue culture. The ganglia are aseptically dissected under a dissecting stereomicroscope and transferred to sterile nutrient solution. The connective tissue is then carefully separated from the periphery of each ganglion. The ganglia are placed in a depression slide and covered with nutrient medium (Eagle's basal medium supplemented with 20% calf serum and antibiotics: penicillin 50 units/ml and streptomycin 50 μg/ml).

1. *Mechanical Dissociation.* a. Hand-Free Dissection Method. With the use of finely sharpened needles the cells contained within each ganglion are gently teased away from the capsule into the nutrient medium. The dissecting stereomicroscope is necessary for best results in this procedure. The capsular material from the ganglia is removed and discarded (Hillman, 1966; Sensenbrenner et al., 1969; Lodin et al., 1970a).

Some authors (Lodin et al., 1970a) have recommended a light centrifugation to increase the amount of separation of the nondispersed cellular aggregates and clumps. The cell suspension in the nutrient medium is then transferred to a 10 ml centrifuge tube in which gentle aspiration with a Pasteur pipette increases the amount of cellular separation. This cell suspension is then centrifuged (800–1000 rpm for 3–5 min) to concentrate the neuron-rich fraction as a pellet. The lighter cellular material is decanted and the pellet resuspended in fresh nutrient medium. For best results this procedure is repeated three times, which washes the cell preparation and further increases dissociation.

b. Sieving Technique. It has been demonstrated (Sensenbrenner et al., 1969) that by gently passing the ganglia through a nylon sieve of 82 μ pore size into nutrient medium, cellular dissociation is achieved. This portion of the procedure is accomplished with the aid of a glass rod to gently press the nicked ganglia through the sieve. Further dissociation of the cell aggregates is accomplished with repeated aspiration using a Pasteur pipette.

c. Enzymatic Dissociation. Enzymes and other dispersing agents have often been employed for dissecting cells from embryonic chick ganglia (Nakai, 1956; Levi-

Montalcini and Angeletti, 1963; Cohen et al., 1964; Shimizu, 1965; Utakoji and Hsu, 1965; Rieske, 1969; Scott et al., 1969). The effectiveness of pronase for this technique was described by Banks et al. (1970). However, trypsin digestion has been most frequently employed.

The ganglia are incubated in a 0.25% trypsin solution for 15 hr at 4°C, as described by Utakoji and Hsu (1965). After two washings in Hanks' or Tyrode solution, and gentle pipetting of the cell suspensions, the cells are again lightly centrifuged and resuspended in nutrient medium (Sensenbrenner et al., 1969). This procedure in which tissues can be digested under refrigeration temperature for as long as 48 hr renders excellent cellular dissociation and recovery.

All of these procedures (a,b,c) produce well-isolated neurons, satellite cells, and nonneuronal fibroblasts and differentiated mesenchymal cells, freely dispersed in nutrient medium. With the sieving technique and the chemical dissociation method the cells of the capsule remain in the suspension, while with the hand-free dissociation procedure, most of the capsular material can be removed and discarded. Recently, a fractionation procedure was reported which provides cultures largely, and sometimes entirely, free of non-neuronal cells (Okun et al., 1972).

B. Dissociation of Neurons and Glial Cells from Brain and Spinal Cord. To obtain dissociated neurons and glial cells from the central nervous system various techniques can be employed. The brain or spinal cord is first dissected and transferred to a sterile balanced salt solution and then into the nutrient medium (Eagle's basal medium fortified with 20% calf serum and antibiotics: penicillin 50 units/ml and streptomycin 50 μg/ml) where the connective tissue coverings or meningeal membranes are removed.

1. *Mechanical Dissociation.* The tissue is gently passed with the aid of a glass rod through a sterile nylon sieve of 45 μ, 58 μ, or 82 μ pore size into nutrient medium. Repeated gentle aspirations of the cell suspension with a Pasteur pipette further increase the dissociation process. This method has been described by Varon and Raiborn (1969) and Booher et al. (1971b) to dissociate neurons and glial cells from embryonic cerebral hemispheres.

2. *Enzymatic Dissociation.* Cavanaugh (1955), Courtey and Bassleer (1967a), and Meller et al. (1969), employed trypsin digestion for the dissociation of chick embryo spinal cord. Spinal cord was cut into small fragments in Hanks' salt solution. These tissue fragments were incubated at 37°C in a calcium and magnesium free balanced salt solution containing 0.25% trypsin for five to six 8 min intervals. At the end of each interval the supernate solution was removed and put into a centrifuge tube containing a trypsin inhibitor (85% nutrient medium Difco TC 1066 with 15% horse serum). This solution was then centrifuged 10 min at 500 rpm; fresh medium was then replaced and a second centrifugation produced the cellular pellet which was then resuspended in 3 ml of nutrient medium. The cells were then resuspended with repeated aspiration against the wall of the centrifuge tube with a Pasteur pipette (Meller et al., 1969).

Both of these procedures (1,2) produce well isolated neurons, various glial cells, and nondispersed cellular clumps in suspension. Following dissociation of the heterogeneous nerve cell tissue, it has often been desirable to separate these cells into homogeneous cell types and culture them. Varon and Raiborn (1969) have described their technique for separation of three distinct cell types from 11-day-old chick embryo cerebral hemispheres. This cellular purification technique could perhaps ideally be accomplished by differential centrifugation procedures using sucrose, ficoll, or albumin in the appropriate density or by sedimentation in a gradient of increasing concentrations of these substances. Recently, mechanical dissociation of brain and ganglia from embryos of the cockroach *Periplaneta americana* and the culture of the dissociated cells results in the survival of nerve, but not of glial cells (Chen and Levi-Montalcini 1970a; 1970b).

III. Culture Techniques and Media for Dissociated Nerve and Glial Cells

For the first 6–12 hr of cultivation, neurons and glial cells following dissociation either mechanically or enzymatically are spherical in their morphology and identical to each other. Generally, all cell processes have been destroyed by the dissociation procedure. However, when such cells, suspended in nutrient medium, are cultivated for more than 12 hr, cellular structure and process regeneration can be expected to resume. Of the various tissue culture techniques commonly employed, the hanging-drop (Maximow chamber) and the multipurpose chamber (Rose chamber) methods have been extensively utilized and well described in the literature, for the cultivation of nerve tissue, as well as dissociated nerve cells. The "flying coverslip" method of Costero and Pomerat (1951) was excellent for maintaining the tissue from the central nervous system, but it could not be visualized unless transferred to a bridge-mount slide. This lasted only a few hours.

The purpose of this section is to describe technical developments which have permitted maintenance and normal function of dissociated neuronal tissue in chambers in which morphological and behavioral patterns can be visualized with phase contrast, bright field, or Nomarski interference optics.

A. Hanging-Drop Culture in Maximow's Double Coverslip Assembly. This culture procedure has been employed for dissociated nerve cell cultivation by Cavanaugh (1955), Nakai (1956), Courtey and Bassleer (1967a), Rieske (1969), Sensenbrenner et al. (1969).

For this type of preparation one drop of cell suspension is placed onto a coverslip (22 × 22 mm) which has been coated with either a matrix of reconstituted collagen or rooster plasma (Bornstein, 1958). Perhaps the most commonly employed substrate for this type of preparation has been the so-called "plasma clot" in conjunction with the collagen substrate matrix. This culture matrix consists of the underlining collagen substrate to which is added one drop of heparinized rooster plasma, one drop of nutrient medium containing the cell suspension, plus one drop of freshly prepared chick embryo extract from 8–10-day-old chick embryos at a concentration of 10–20% in Tyrode solution. Following coagulation of the plasma clot, the 22 × 22 mm coverslip, which has previously been cohered to a larger coverslip (24 × 40 mm) is placed centrally over the depression of the Maximow chamber (Maximow, 1925). This assembly is then sealed with paraffin wax so as to permit O_2-CO_2 diffusion. The cultures are incubated at 37°C. Every 3–4 days the cultures are washed with Tyrode solution and replenished with one drop of nutrient medium (Eagle's basal medium fortified with 20% fetal bovine serum plus embryo extract in the proportion 70 : 20 : 10, and containing penicillin 50 units/ml and streptomycin 50 μg/ml).

Along with the many advantages advanced by the Maximow double coverslip assembly, two limiting factors are inherent in this type of culture preparation:

(1) The hollow-ground parabolic depression of the Maximow chamber can present difficulties when phase-contrast observations are desired, and (2) the wax closed system and limited quantity of medium can be restrictive to studies in which drug perfusion is required.

B. Culture in Multipurpose Chamber (Rose type). Varon and Raiborn (1969) employed the Rose chamber for the cultivation of dissociated cells from 11-day-old chick embryo cerebral hemispheres. Lodin and coworkers (1970a, 1970b) demonstrated that optimum conditions for long-term cultivation of dissociated nerve tissue can be obtained in the Rose chamber (Rose, 1954) employing the cellophane strip technique (Rose et al., 1958). This culture system affords not only optimum optical advantages for phase contrast or time-lapse cinematography observation, but is also easily adapted for studies involving perfusion.

Cultivation of neural tissue in the Rose multipurpose chamber can be accomplished in a variety of ways. Many modifications of this procedure have been adapted in recent years for neurological tissue culture. However, the basic procedure for the cultivation of both central and peripheral dissociated neurons and glia has involved the following method: One drop of rooster plasma is applied onto the collagen-coated coverslip (40 × 40 mm). One drop of the concentrated cell suspension in medium is then inoculated and combined with one drop of freshly prepared chick embryo extract (from 8–10-day-old chick embryo at a concentration of 10% in Tyrode solution) to constitute a plasma extract collagen matrix. The preparation is then covered with a 30 mm wide strip of Visking dialysis cellophane membrane (Ref. 36/32) which has been presterilized in 70% alcohol (20 min) and equilibrated to medium. The chamber is then completed, in the usual manner, with the addition of a rubber gasket which inhibits O_2-CO_2 diffusion (Hendelman and Booher, 1966), and the top coverslip. The chamber is then secured with the final stainless steel plate assembly. By applying two small hypodermic needles (25 gauge) through the gasket, on opposite sides of the chamber, the nutrient medium (Eagle basal medium fortified with 20% calf serum) can be introduced or in some cases perfused, if desired, into the chamber assembly. The cultures are incubated at 37°C and the nutrient medium is renewed every 2–3 days of cultivation.

It should be noted from recent studies involving the cultivation of neural tissue in the Rose chamber that by allowing a substantial air phase to remain within the chamber, i.e., introducing only 1 ml of nutrient medium, neurons and glial cells can be maintained for long periods of time and retain normal metabolic, physiologic, and morphological character (Booher and Kasten, 1966; Booher et al., 1971a).

For normal incubation of these preparations in Rose chambers rubber gaskets are required to control the pH which must be maintained between 7.0 and 7.2. However, with silicone gaskets these preparations can be maintained in an atmosphere of 95% O_2 and 5% CO_2 incubation (Hendelman and Booher, 1966).

With strict adherence to the previously described environmental conditions, neurons and glial cells can be maintained *in vitro* as long-term cultures. It has been demonstrated that, with collagen as a matrix and fetal bovine serum-enriched Eagle's medium as the only nutritional substrate, central and peripheral nuerons and glial cells can be maintained as long-term (more than one month) cultures with no change in their morphology or metabolism (Booher et al., 1971a). It can be noted that the addition of the embryo extract–plasma matrix had relatively no effect upon the development of dissociated neurons from the dorsal ganglia of 14-day-old chick embryos (Booher et al., 1971a). However, marked morphological differences have been observed in cultures of dissociated neurons and glial cells from 5–7-day-old chick embryos when in the presence or absence of the plasma substrate matrix (Sensenbrenner et al., 1970).

C. Culture in plastic Falcon flasks or petri dishes. Werner et al. (1971) were successful in the cultivation of dissociated brain cells from chick embryos in Falcon plastic flasks which had been previously coated with collagen. Booher and Sensenbrenner (1972) obtained successful cultivation of dissociated nerve cells from both young chick and rat embryos in Falcon plastic flasks without collagen as substrate.

One ml of the cell suspension obtained after mechanical or trypsin dissociation ($2 - 2.5 \times 10^5$ cells) is introduced into a flask or a petri dish and the nutrient medium is then added (Eagle's basal medium fortified with 20% fetal calf serum). The cell cultures are incubated at 37°C in an atmosphere of 95% oxygen or air and 5% CO_2.

The flask culture system provides a high cell yield and thus can be more useful for biochemical investigations then the other culture methods described above.

IV. Effects of Various Nutrients and Metabolites on the Growth and Development of Dissociated Neurons in Culture

A. Glucose. All chemically defined tissue culture media employed today contain glucose at a concentration of 100 mg %. In studies involving myelin formation and maintenance in explant cultures and dissociated cultures, it has been shown that the glucose concentration must be increased to 600 mg/100 ml (Murray et al., 1962).

B. NGF. Several workers (Levi-Montalcini and Angeletti, 1963; Cohen et al., 1964; Rieske, 1969, Sensenbrenner et al., 1969; Shahar and Saar, 1970) demonstrated that nerve growth factor (NGF), a specific protein extracted from the salivary gland of the mouse, promotes a stimulatory effect on the outgrowth of nerve fibers from isolated peripheral and sympathetic neurons in tissue culture. The most effective concentration of NGF (Wellcome Co.) is approximately 1–2 units/ml.

C. Brain Extract. Treska et al. (1968b), Sensenbrenner et al. (1969) and Sensenbrenner et al. (1972) have demonstrated that, with the addition of homogenized brain extract, favorable outgrowth and differentiation in neurons from embryonic chick central and peripheral nervous system was enhanced. This "growth promoting" extract is prepared from brains of 8, 10 or 12-day-old chick embryos at a concentration of 10-20% in Tyrode solution.

D. Potassium. Scott and Fisher (1970) demonstrated that a potassium concentration of approximately 40.8 mM is necessary for the survival of dissociated neurons from the peripheral nervous system of chick embryos, whereas for human material a concentration of 20.8 mM only is necessary (Scott 1971). This cationic concentration is standard for most nutrient media employed in neurological tissue culture today.

V. Conclusion

The nervous system is composed of a wide variety of cell types and complex interrelationships. Tissue culture and the cultivation of dissociated neurons and glial cells

have advanced our knowledge and ability to investigate the morphologic, metabolic, and physiologic behavior of these individual cells. Many questions and challenges are left to be encountered by the neurochemist or neurobiologist. However, with the technical skills and tools at our disposal, concomitant advances in our knowledge of the nervous system are, indeed, to be expected.

In the past, the cultivation of dissociated neurons and glial cells has revealed much information concerning morphological differentiation and function of these cells. It was at one time postulated that the neurons had not the ability to survive and were dependent upon their glial counterparts. However, it has been demonstrated in recent years that neurons, in tissue culture, are not only able to survive but have the capacity to regenerate new processes and retain a normal metabolic and physiological character.

In recent years, increasing attention has been paid to the advances achieved in neurological tissue culture. It has been demonstrated that dissociated neurons in tissue culture synthesize actively nucleic acids (Utakoji and Hsu, 1965; Courtey and Bassleer, 1967b; Sensenbrenner et al., 1968; Sensenbrenner et al., 1970; Amaldi and Rusca, 1970), retain their biochemical integrity (Burdman, 1967, 1969), demonstrate a normal enzymatic function (Hermetet et al., 1968, 1970; Treska et al., 1968a; Ciesielski-Treska et al., 1970) and are physiologically normal in their oxidative metabolism and ATP content (Schousboe et al., 1970). Electrophysiological characteristics of dissociated chick embryonic spinal ganglion neurons have been reported and have demonstrated that action potentials can be recorded in these cells (Scott et al., 1969; Crain, 1971; Crain and Bornstein, 1972; Okun, 1972).

Neurological tissue culture and the various techniques which have been developed in this highly specialized field supply to the neurochemist, the neurophysiologist, the neuropathologist, and the morphologist a viable and dynamic system for investigations. It has been demonstrated in tissue culture that neurons and glial cells are able to regenerate and reorganize a normally functioning model of cell-to-cell relationships. Technical advances are ever increasing in this field which afford new information and understanding to our knowledge of the nervous system.

M. Sensenbrenner

References

Amaldi, P., and G. Rusca, "Autoradiographic study of RNA in nerve fibers of embryonic sensory ganglia cultured *in vitro* under NGF stimulation," *J. Neurochem.*, **17**:767 (1970).

Banks, B. E. C., D. V. Banthorpe, D. M. Lamont, F. L. Pearce, K. A. Redding, and C. A. Vernon, "Dissociation of sensory ganglia from the embryonic chick by pronase and other dispersing agents," *J. Embryol. Exp. Morph.*, **23**:519 (1970).

Booher, J., L. Hertz, and Z. Lodin, "A simplified technique for cultivation of dissociated central and peripheral neurons and glial cells in the Rose chamber," *Neurobiology* **1**:27 (1971a).

Booher, J., and F. H. Kasten, "A modified technique for cultivating tissues from the central nervous system in the Rose chamber," Abstracts of the 7th Annual Meeting of the Tissue Culture Association 95, San Francisco, 1966.

Booher, J., M. Sensenbrenner, J. C. Hermetet, and P. Mandel, "Différenciation de cellules d'hémisphères cérébraux d'embryon de Poulet en culture," *C.R. Acad. Sci.*, (Paris), **272**:272 (1971b).

Booher, J., and M. Sensenbrenner, "Growth and cultivation of dissociated neurons and glial cells from embryonic chick, rat and human brain in flask cultures." *Neurobiology*, in press (1972).

Bornstein, M. B., "Reconstituted rat-tail collagen used as substrate for tissue cultures on cover slips in Maximow slides and roller tubes," *Lab. Invest.*, 7:134 (1958).

Burdman, J. A., "Early effects of a nerve growth factor on the RNA content and base ratios of isolated chick embryo sensory ganglia neuroblasts in tissue culture," *J. Neurochem.*, 14:367 (1967).

Burdman, J. A., "Conversion of adenine to guanine and cytosine to uracil by chick embryo sensory ganglia in tissue culture," *Brain Res.*, 15:515 (1969).

Cavanaugh, M. W., "Neuron development from trypsin-dissociated cells of differentiated spinal cord of the chick embryo," *Exp. Cell Res.*, 9:42 (1955).

Chen, J. S., and R. Levi-Montatcini, "Axonal growth from insect neurons in glia-free cultures". *Proc. nat. Acad. Sci., Wash.*, 66:32, (1970a).

Chen, J. S., and R. Levi-Montalcini, "Long-term cultures of dissociated nerve cells from the embryonic nervous system of the cockroach *Periplaneta americana*". *Arch. ital. Biol.*, 108:503, (1970b).

Ciesielski-Treska, J., J. C. Hermetet, and P. Mandel, "Histochemical study of isolated neurons in culture from chick embryo spinal ganglia," *Histochem.*, 23:36 (1970).

Cohen, A. I., E. C. Nicol, and W. Richter, "Nerve growth factor requirement for development of dissociated embryonic sensory and sympathetic ganglia in culture," *Proc. Soc. Exp. Biol. (N.Y.)*, 116:784 (1964).

Costero, I., and C. M. Pomerat, "Cultivation of neurons from the adult human cerebral and cerebellar cortex," *Amer. J. Anat.*, 89:405 (1951).

Courtey, B., and R. Bassleer, "Culture in vitro de cellules nerveuses isolées d'embryon de Poulet," *C.R. Acad. Sci.* (Paris), Ser. D, 264:389 (1967a).

Courtey, B., and R. Bassleer, "Etude histoautoradiographique de l'incorporation de thymidine, d'uridine et de leucine tritiées dans des cellules nerveuses d'embryon de Poulet isolées et cultivées in vitro," *C.R. Acad. Sci.* (Paris), Ser. D, 264:497 (1967b).

Crain, S. M., "Intracellular recordings suggesting synaptic functions in chick embryo spinal sensory ganglion cells isolated in vitro." *Brain Res.* 26:188, (1971).

Crain, S. M., and Bornstein, M. B., "Organotypic bioelectric activity in cultured reaggregates of dissociated rodent brain cells." *Science*, 176:182 (1972).

Grosse, G., and G. Linder, "Untersuchungen zur Differenzierung isolierter Nerven- und Gliazellen des Zentralnervösen Gewebes von Hühnerembryonen" *Zellkultur. 4. Hirnforsch.* 12:207 (1970).

Hendelman, W. J., and J. Booher, "Factors involved in the culturing of chick embryo dorsal root ganglia in the Rose chamber," *Tex. Rep. Biol. Med.*, 24:83 (1966).

Hermetet, J. C., J. Treska, and P. Mandel, "Histochemical study of isolated neurons in culture from chick embryo sympathetic ganglia," *Histochem.*, 22:177 (1970).

Hermetet, J. C., J. Treska, M. Sensenbrenner, and P. Mandel, "Catécholamines et activité monoamine-oxydasique dans des ganglions sympathetiques et des neurones isolés en culture." *C.R. Soc. Biol.*, 162:2287 (1968).

Hillman, H., "Growth of processes from single isolated dorsal root ganglion cells of young rats," *Nature*, 209:102 (1966).

Hillman, H., and S. A. Khalawan, "The growth of new processes from isolated dorsal and sympathetic cell bodies of rat and rabbit," *Tiss. Cell.*, 2:249 (1970).

Hillman, H., and K. Sheikh, "Growth of new processes from isolated rabbit vestibular neurone cell bodies," *J. Physiol.*, 194:69P (1967).

Hillman, H., and K. Sheikh, "The growth in vitro of new processes from vestibular neurons isolated from adult and young rabbits," *Exp. Cell Res.*, 50:315 (1968).

Levi-Montalcini, R., and P. U. Angeletti, "Essential role of nerve growth factor in the survival and maintenance of dissociated sensory and sympathetic embryonic nerve cells in vitro." *Develop. Biol.* 7:653 (1963).

Lodin, Z., J. Booher, and F. H. Kasten, "Long-term cultivation of dissociated neurons from embryonic chick dorsal root ganglia in the Rose chamber," *Exp. Cell Res.*, 59:291 (1970a).

Lodin, Z., J. Booher, and F. H. Kasten, "Phase-contrast cinematographic study of dissociated neurons from embryonic chick dorsal root ganglia cultured in the Rose chamber," *Exp. Cell Res.*, 60:27 (1970b).

Maximow, A., "Tissue cultures of young mammalian embryos," *Contrib. Embryol. Carnegie Inst.*, 16:47 (1925).

Meller, K., W. Breipohl, H. H. Wagner, and A. Knuth, "Die Differenzierung isolierter Nerven-und Gliazellen aus trypsiniertem Rückenmark von Hühnerembryonen in Gewebekulturen," *Z. Zellforsch.*, 101:135 (1969).

Murray, M. R., E. R. Peterson, and R. P. Bunge, "Some nutritional aspects of myelin sheath formation in cultures of central and peripheral nervous system," *4th Int. Congr. Neuropathol. Proc.* (Stuttgart) II:267 (1962).

Nakai, J., "Dissociated dorsal root ganglia in tissue culture," *Amer. J. Anat.*, 99:81 (1956).

Nakajima, S., "Selectivity in fasciculation of nerve fibers in vitro," *J. Comp. Neurol.*, 125:193 (1965).

Okun, L. M., "Isolated dorsal root ganglion neurons in culture: cytological maturation and extension of electrically active processes." *J. Neurobiol.* 3:111 (1972).

Okun, L. M., F. K. Ontkean, and C. A. Thomas, "Removal of non-neuronal cells from suspensions of dissociated embryonic dorsal root ganglia." *Exptl. Cell. Res.* 73:226 (1972).

Rieske, E., "Einfluss eines spezifischen Nervenwachstumsfaktors (NGF) auf Zellkulturen des Ganglion trigeminale," *Z. Zellforsch.*, 95:546 (1969).

Rose, G. G., "A separable and multipurpose tissue culture chamber," *Tex. Rep. Biol. Med.*, 12:1074 (1954).

Rose, G. G., C. M. Pomerat, T. O. Shindler, and J. B. Trunnell, "A cellophane-strip technique for culturing tissue in multipurpose culture chambers," *J. Biophys. Biochem. Cytol.*, 4:761 (1958).

Schousboe, A., J. Booher, and L. Hertz, "Content of ATP in cultivated neurons and astrocytes exposed to balanced and potassium-rich media," *J. Neurochem.*, 17:1501 (1970).

Scott, B. S., V. E. Engelbert, and K. C. Fisher, "Morphological and electrophysiological characteristics of dissociated chick embryonic spinal ganglion cells in culture," *Exp. Neurol.*, 23:230 (1969).

Scott, B. S., and K. C. Fisher, "Potassium concentration and number of neurons in cultures of dissociated ganglia," *Exp. Neurol.*, 27:16 (1970).

Scott, B. S., "Effect of potassium on neuron survival in cultures of dissociated human nervous tissue." *Exptl. Neurol.* 30:297 (1971).

Sensenbrenner, M., Z. Lodin, J. Treska, M. Jacob, M. P. Kage, and P. Mandel, "The cultivation of isolated neurons from spinal ganglia of chick embryo," *Z. Zellforsch.*, 98:538 (1969).

Sensenbrenner, M., Z. Lodin, J. Treska, M. Jacob, and P. Mandel, "Metabolism of RNA in isolated, cultured neurons. Histoautoradiographic study," Tagung der Tschechoslowakischen Gesellschaft für Cyto-und Histochemie. Brno 17–19 Juni, 1968.

Sensenbrenner, M., J. Treska-Ciesielski, Z. Lodin, and P. Mandel, "Autoradiographic study of RNA synthesis in isolated cells in culture from chick embryo spinal ganglia," *Z. Zellforsch.*, 106:615 (1970).

Sensenbrenner, M., and P. Mandel, "Differentiation of chick embryo brain cells in culture". *Experientia.*, **27**:830 (1971).

Sensenbrenner, M., J. Booher, and P. Mandel, "Cultivation and growth of dissociated neurons from chick embryo cerebral cortex in the presence of different substrates." *Z. Zellforsch.* **117**:559, (1971).

Sensenbrenner, M., N. Springer, J. Booher, and P. Mandel, "Histochemical studies during the differentiation of dissociated nerve cells cultivated in the presence of brain extracts." *Neurobiology.*, **2**:49 (1972).

Shahar, A., and M. Saar, "Cultivation of isolated nerve cells in a perfusion chamber and the early effects of nerve growth factor on them," *Brain Res.*, **23**:315 (1970).

Shimizu, Y., "The satellite cells in cultures of dissociated spinal ganglia," *Z. Zellforsch.*, **67**:185 (1965).

Treska, J., Z. Lodin, M. Sensenbrenner, and P. Mandel, "Metabolism of RNA and proteins in isolated, cultured neurons. Histochemical study," Tagung der Tschechoslowakischen Gesellschaft für Cyto-und Histochemie. Brno 17–19 Juni, 1968a.

Treska, J., M. Sensenbrenner, Z. Lodin, M. Jacob, and P. Mandel, "Action d'extraits embryonnaires de cerveau sur la différenciation morphologique des cellules nerveuses en culture *in vitro*," *C.R. Acad. Sci.*, (Paris), Ser. D, **267**:2034 (1968b).

Utakoji, T., and T. C. Hsu, "Nucleic acids and protein synthesis of isolated cells from chick embryonic spinal ganglia in culture," *J. Exp. Zool.*, **158**:181 (1965).

Varon, S., and C. W. Raiborn, "Dissociation, fractionation, and culture of embryonic brain cells," *Brain Res.*, **12**:180 (1969).

Werner, I., G. R. Peterson, and L. Shuster, "Choline acetyltransferase and acetylcholinesterase in cultured brain cells from chick embryos." *J. Neurochem.* **18**:141 (1971).

NEUROSECRETORY STRUCTURES

The investigation of the neurosecretory structures may be performed by applying both light and electron microscopical techniques. These two methodologies must be regarded as complementary since practically all the histochemical techniques appear to demonstrate biologically active principles and their carrier molecules, and therefore, they are very useful in evaluating the functional stage of a given neurosecretory structure. On the other hand, the conventional electron microscopical techniques provide information regarding the subcellular organization of the neurosecretory formations as well as the necessary evidence for the elucidation of fundamental questions such as site of synthesis, packaging and storage of the neurosecretory products, and mechanism of their transport and release.

Light Microscopy

Most of the classical techniques used for the demonstration of neurosecretory material (n.s.m.) may be applied to sections of paraffin-embedded material or to staining the whole mass of tissue containing the neurosecretory structure.

Histochemical Techniques Suitable for Paraffin Sections.
1. Chrome-alum hematoxylin-phloxine (Gomori, 1941)
2. Aldehyde-fuchsin (Gabe, 1953)
3. Alcian blue (Adams and Sloper, 1956)
4. Alcian blue–alcian yellow (Peute and Van de Kamer, 1967)
5. Chrome-alum gallocyanine (Bock, 1966)

6. Pseudoisocyaninchloride (Sterba, 1961)
7. Crotonaldehyde-diaminobenzophenone (Bock and Ockenfels, 1970)

The last two techniques are fluorescence methods and seem to be more sensitive and specific than the classical methods of Gomori. However, one of the simplest and most reliable methods is the aldehyde-fuchsin according to Gabe (1953) (Figs. 1 and 4). The material is processed as follows:

Fixation: 4–24 hr in the Halmi fluid or 7 days in diluted Bouin (Baker, 1946)
Dehydration: graded alcohols
Embedding: paraffin
Staining procedure:
1. Removal of the paraffin and hydration of the sections
2. Application of the oxidizing Gomori's mixture* for 1–2 min
3. Rinsing in distilled water
4. Bleaching with 2.5% sodium bisulfite (about 10–30 sec)
5. Rinsing in tap water for 30 sec
6. Staining with the aldehyde-fuchsin solution** for 1–2 min
7. Differentiation in acidic alcohol (usually a few seconds are enough)
8. Dehydration and mounting

In Block Techniques. The main advantage of this technique is that a whole neurosecretory structure may be demonstrated in a single preparation, which allows its precise localization and distribution of the neurosecretory pathways. The most reliable methods are:
1. Victoria blue (Braak, 1962)
2. Aldehyde-fuchsin (Oksche et al., 1964)
3. Aldehyde-thionin (Dogra and Tandan, 1964)

Oxidizing solution of Gomori:
Potassium permanganate, 2.5%, 10 ml
Hydrochloric acid, 5%, 10 ml
Distilled water, 60 ml
**Preparation of the aldehyde-fuchsin solution:*
A. Stock Solution
1. Heat 200 ml of distilled water to about 50°C, add 1 g of basic fuchsin, and *then* heat to boiling point, allowing to boil for 2 min.
2. Allow the solution to cool to room temperature, and then add 2 ml of concentrated HCl and 2 ml of paraldehyde.
3. Leave the solution to stand at room temperature and control the ripeness of it by dripping a few drops (one to three) onto a filter paper. At the beginning a blue spot surrounded by a reddish halo is obtained. When the solution is ripe (after about four days), the red halo disappears.
4. Filter the ripe solution discarding the filtrate and keeping the precipitate obtained on the filter paper. Dry the filter paper with the precipitate still on it in an oven, at about 80°C.
5. Remove the dry powder of aldehyde-fuchsin from the filter paper and place it in a bottle, then add 70% alcohol to obtain a saturated solution of aldehyde-fuchsin (usually about 100–150 ml of alcohol are enough). An alternative is to weigh the dry precipitate of aldehyde-fuchsin and dissolve it in 70% alcohol in a proportion of 0.5 g of precipitate per 100 ml of alcohol.
B. Staining Solution
Stock solution, 25 ml
70% alcohol, 75 ml
Acetic acid, 1 ml

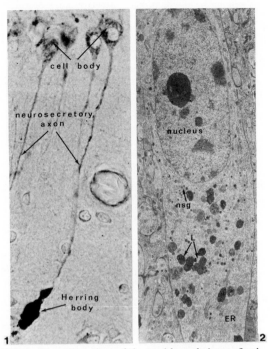

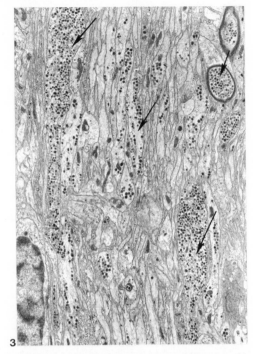

Fig. 1 Preoptic nucleus of the toad hypothalamus fixed in diluted Bouin and stained with aldehyde-fuchsin according to Gabe (1953).

Fig. 2 Electron micrograph of a neurosecretory neuron of the toad preoptic nucleus. nsg, neurosecretory granules; L, lysosomes; ER, rough endoplasmic reticulum. Fixation by *ventricular perfusion* with a threefold aldehyde mixture (Rodríguez, 1969).

Fig. 3 Neurosecretory axons (arrows) of the preoptic-hypophysial tract of the frog fixed by *ventricular injection* of a threefold aldehyde mixture (Rodríguez, 1969).

B. Electron Microscopy

There are two main considerations which must be taken into account when processing neurosecretory structures for electron microscopy. One is the way in which the fixative will reach the tissue and the other is the composition of the fixative itself.

Ways of Fixation:

1. *Vascular perfusion*: This is the selected method of fixation for those neurosecretory structures lying deep in the central nervous system, such as the supraoptic nucleus. For a description of this method see Karlsson and Schultz (1965).

2. *Ventricular perfusion*: This is the most suitable way of fixing neurosecretory structures underlying the ventricular walls, such as the paraventricular nucleus or median eminence (Rodríguez, 1969) (Figs. 2 and 3).

3. *Immersion*: This way of fixation is most convenient for those areas in which the neurosecretory material is concentrated and where the first two methods generally fail to produce a good fixation (i.e., neural lobe of the hypophysis (Fig. 5), urophysis, or neurohaemal areas of invertebrates).

Preparation of Fixative. Whatever the way of fixation, the best fixative for the ultrastructural study of neurosecretory formations is a threefold aldehyde mixture containing glutaraldehyde, formaldehyde, and acrolein (Rodríguez, 1969). This fixative is prepared in the following way:

1. One or two grams of paraformaldehyde are added to

25 ml of distilled water and heated to about 60°C whereupon a cloudy solution is obtained. 0.1N sodium hydroxide solution is then added dropwise (10–15 drops) until the solution clears.

2. When the solution has cooled to room temperature, 5 ml of 25% glutaraldehyde are added, either fresh or purified with an ion exchange resin. A purified and stabilized glutaraldehyde is now available. Any glutaraldehyde solution with a pH less than 3 has to be discarded.

3. Half, 1, or 2 ml of acrolein are then added.

4. A solution of $0.2M$ phosphate buffer pH 7.6 is added to make up a total volume of 50 ml of fixative.

The osmolarity of this fixative is in the range of 1,300 mOsM.

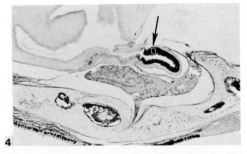

Fig. 4 Hypophysial region of the lizard fixed in diluted Bouin and stained with aldehyde-fuchsin according to Gabe (1953). The neural lobe is deeply stained. An area similar to that framed in square (arrow) is shown in Fig. 5.

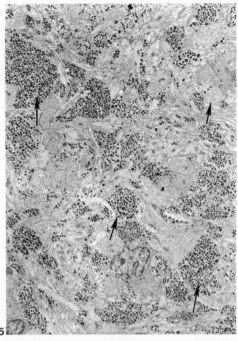

Fig. 5 Electron micrograph of an area of the lizard neural lobe similar to that shown in square of Fig. 4. The arrows point out neurosecretory endings. Fixation by *immersion* in a threefold aldehyde mixture (Rodríguez, 1969).

Fixation Procedure

1. Keep the tissue in contact with the threefold aldehyde mixture, either by perfusion or immersion, for about 2–3 hr.
2. Wash the tissue in $0.1M$ phosphate buffer pH 7.6, three washes of 20 min each.
3. Fix the tissue in 1 % OsO_4 buffered to pH 7.5 with $0.1M$ phosphate buffer, for 2 hr.
4. Wash the blocks with double-distilled water to remove any traces of phosphate left in the tissue, three washes of 10 min each.
5. Dehydrate and embed.

ESTEBAN MARTIN RODRÍGUEZ

References

Adams, C. W. M., and J. C. Sloper, "The hypothalamic elaboration of posterior pituitary principles in man, the rat and dog. Histochemical evidence derived from a performic acid–alcian-blue reaction for cystine," *J. Endocrinol.*, 13:221–228 (1956).

Baker, J. R., "Cytological Technique," London, Methuen, 1946.

Bock, R., Über die Darstellbarkeit neurosekretorischer Substanz mit Chromalaun-Gallocyanin im supraoptico-hypophysären System beim Hund," *Histochemie*, 6:362–369 (1966).

Bock, R., and H. Ockenfels, "Fluoreszenzmikroskopische Darstellung Aldehydfuchsin-positiver Substanzen mit Crotonaldehyd-Diaminobenzophenon," *Histochemie*, 21:181–188 (1970).

Braak, H., "Eine Methode zur räumlichen Darstellung des neurosekretorischen Zwischenhirn-hypophysensystems," *Mikroskopie*, 17:344–346 (1962).

Dogra, G. S., and B. K. Tandan, "Adaptation of certain histological techniques for *in situ* demonstration of the neuro-endocrine system of insects and other animals," *Quart. J. Micr. Sci.*, 105:455–466 (1964).

Gabe, M., "Quelques applications de la coloration par la fuchsine-paraldehyde," *Bull. Micr. Appl.*, 3:153–162 (1953).

Gomori, G., "Observations with differential stains on human islets of Langerhans," *Amer. J. Pathol.*, 17:395–406 (1941).

Karlsson, U., and R. L. Schultz, "Fixation of the central nervous system for electron microscopy by aldehyde perfusion. I. Preservation with aldehyde perfusates versus direct perfusion with osmium tetroxide with special reference to membranes and the extracellular space," *J. Ultrastruct. Res.*, 12:160–186 (1965).

Oksche, A., W. Mautner, and D. S. Farner, "Das räumliche bild des neurosekretorischen System der Vögel unter normalen und experimentellen Bedingungen," *Z. Zellforsch.*, 64:83–100 (1964).

Peute, J., and J. C. Van de Kamer, "On the histochemical differences of aldehyde-fuchsin positive material in the fibres of the hypothalamo-hypophyseal tract of *Rana temporaria*," *Z. Zellforsch.*, 83:441–448 (1967).

Rodríguez, E. M., "Fixation of the central nervous system by perfusion of the cerebral ventricles with a threefold aldehyde mixture," *Brain Res.*, 15:395–412 (1969).

Sterba, G., "Fluoreszenzmikroskopische Untersuchungen über die Neurosekretion beim Bachneunauge (Lampetra Planeri Bloch)," *Z. Zellforsch.*, 55:763–789 (1961).

NITROCELLULOSE EMBEDDING AND CUTTING

Embedding of specimens in paraffin is the easiest method of preparing soft tissues for cutting and staining. However, large specimens and tough or hard tissues such as decalcified bone, embryo, and brain require low-viscosity nitrocellulose (LVN) or Parlodion as the embedding medium. This method does not use heat, and therefore, shrinkage and distortion of the tissue are avoided. Since the density of nitrocellulose approaches that of the specimen, the cutting of thin sections is facilitated. The chief disadvantage of this technique is the long time required for the preparation.

Embedding at body temperature is ideal; therefore LVN is normally employed at 18–25°C. Walls (1) in 1932 and Koneff and Lyons(2) in 1937 reported the use of a temperature of 56°C for more rapid penetration. However, such temperature has proved dangerous since the LVN solution can cause a hot and intense fire. For this reason the use of the higher temperature has been discontinued in many laboratories. Instead, to speed up the time of infiltration low vacuum rather than heat is employed. With the use of negative pressure of approximately 16–20 in. of mercury, the time of LVN embedding has been reduced by half and the quality of the sections improved.

Preparation of LVN Solutions. Two solutions, a thick and a thin, of LVN are made for the embedding process. Because LVN is flammable, it comes moistened with approximately 30 % alcohol[1] for safety in shipping.

---

[1] *Ethyl* alcohol should be specified when LVN is ordered.

Narrow-mouthed, glass-stoppered bottles with a five-pint capacity are used for making and storing the solutions.

Thick Stock Solution. The thick solution (3) is made by placing 700 g of LVN in a bottle. To this is added 640 ml of absolute ethyl alcohol. The solution is left to stand, without being stirred, for 24 hr. This allows the LVN to soften and speeds up the dissolving process. Finally, 960 ml of anhydrous ether is added to the bottle. The solution is agitated several times a day until the LVN is completely dissolved.

Thin Solution. The thin solution is made from the thick stock solution (3). To 100 ml of the thick stock LVN solution is added 450 ml of absolute ethyl alcohol and 450 ml of anhydrous ether.

Parlodion. Parlodion may also be used as the embedding medium. The thick solution is prepared as follows: To 100 g of dry Parlodion 750 ml of equal parts of absolute alcohol and anhydrous ether (A and E) are added. The flask must be shaken frequently to dissolve the Parlodion. The thin solution is made in the same manner, but the amount of A and E is doubled.

Both LVN and Parlodion are subject to deterioration in the presence of light, so storage should be in a dark place or in dark glass bottles (4).

Embedding Dishes. Generally, the specimen dishes used have straight sides. Particularly after the last step of embedding with the thick solution, the straight sides facilitate the removal of the solid block of tissue. Also the height of the dish should not exceed the diameter; otherwise the block will not harden well. Before a specimen is placed in a dish, the sides and bottom of the dish are coated with a very thin layer of silicone grease. Usually the LVN solution (thin or thick) is then poured into the dish and the specimen placed in the solution.

Embedding. After the last stage of dehydration and before it is placed in embedding solution, the specimen is soaked in a dish containing a 1 : 1 solution of A and E. The amount of time that the specimen should soak depends on its size.

The specimen is then transferred to a dish containing the thin solution of LVN. A small amount of A and E is poured on the top and the dish is placed, uncovered, in a desiccator. Vacuum grease is applied to the rim of the desiccator cover, which should have a ground-in stopcock. The desiccator is then covered and a vacuum of approximately 16–20 in. of mercury is applied.

The specimen is maintained under this vacuum in the covered desiccator for 10 min. It may be necessary to tap the desiccator in order to start the solution bubbling ("boiling process"); the pressure should be carefully controlled to prevent excessive bubbling. After the 10 min period, the vacuum is turned off and the specimen left in the desiccator overnight. (See Fig. 1.)

On the following day the specimen is transferred to a dish containing the thick solution of LVN. A thin layer of A and E is put on the surface of the solution. The specimen dish is placed, uncovered, in the desiccator which, in turn, is covered, and the vacuum process is begun again. After 10 min, the vacuum is turned off and the specimen left overnight, uncovered, in the closed desiccator.

The process of maintaining a vacuum for approximately 10 min, once every 24 hr, is continued until the LVN solution has become very thick. Occasionally A and E is added to the surface in order to prevent over-hardening. Thus the LVN is "cured" very slowly. The total amount of time for preparation of the specimen depends on the size and type of the tissue to be processed. For example, a

Fig. 1 Desiccator with specimen under vacuum. Note bubbling ("boiling") effect.

block 1 cm in thickness and 1.5 cm in diameter from a femur of a one-year-old monkey should be completely infiltrated in 72 hr.

Several times a day the desiccator cover is removed so that a vacuum hose may be used to suck out the ether and alcohol vapors. During the time that the vacuum hose is being applied, the embedding dish must be covered tightly or the top of the thickening block will become too hard.

When the block is solid, i.e., when it can be just slightly dented on the edge by a thumbnail, it is taken from the desiccator (or left in with the lid of the desiccator removed) and placed where it can harden further at room temperature. The block should be turned over in the dish once a day to insure even hardening. Finally, the hardened block is stored in 80% alcohol.

Mounting. The hardened LVN blocks are mounted on fiberboard blocks which have deeply grooved grids (Fig. 2). First, the LVN blocks are trimmed to the shape desired; several millimeters of the LVN mass should be left surrounding the specimen. The side of the LVN block to be cemented to the fiberboard block must be perfectly flat so that it will make good contact. To obtain a flat surface, a very coarse sandpaper is placed on an even area and the

Fig. 2 LVN block mounted on grid of fiberboard block.

Fig. 3   Microtome with specimen in place.

LVN block is then rubbed across the sandpaper in a circular motion.

Next, the fiberboard block is placed in a dish containing enough 1 : 1 alcohol and ether mixture to submerge it 0.5 cm. About 0.5 cm of the specimen block is also placed in a dish of 1 : 1 alcohol and ether. The bottom of the specimen block will soften in about 1 or 2 min.

The fiberboard block is then placed with the grooves up, and thick LVN solution is immediately poured over it. The specimen block is put on top, its soft side in contact with the fiberboard block, and held firmly in place for a few minutes. The two pieces are left to stand together until a dry crust forms around the fiberboard block. This takes 5–10 min. The large block is then put in chloroform so that the junction between the fiberboard block and the specimen block is covered. Small blocks, 2 cm square, should be left for $\frac{1}{2}$ hr, large ones overnight.

Cutting. Large blocks of tissue are cut by a heavy, sturdy, sliding microtome, with a knife of suitable length set at 30° (Fig. 3). The block is mounted in the microtome holder; the knife is drawn across the surface; and the section is picked up with a brush and dropped into a dish of 70–80% alcohol, where it slowly spreads out to its full size. Care must be taken to keep the edge along the junction of the LVN block and the fiberboard block moist with 80% alcohol, to prevent loosening.

There are three methods of cutting LVN blocks. Thinner and better sections are obtained by a semidry method using 80% alcohol.

Wet Method. Both knife and specimen are kept wet with 70% or 80% alcohol.

Semidry Method. An alcohol sponge is rubbed over the surface of the specimen and then the excess alcohol is rubbed off with the side of the palm of the hand.

Dry Method. The block is impregnated with cedar oil and cut on a rotary microtome.

Judging the preparation time of a well-cured block is an important part of laboratory planning. It is difficult to determine this length of time because it differs with the size and the type of tissue being processed. For the preparation of a 2 × 7 × 1.5 cm slice of the head of the human femur, the processes of dehydration and embedding require about 20 days. A monkey metacarpal measuring 0.3 × 0.8 × 0.5 cm can be prepared for cutting in 48 hr if necessary, although a period of 96 hr is preferable.

Soft tissue requires more dehydration time and less vacuum pressure to prevent shrinkage and distortion of the tissue. A whole monkey brain is fixed for 7 days, dehydrated for 10 days, and embedded in about 14 days. Human embryos of 40 mm are fixed in 3 days, dehydrated for 9 days, and embedded in 10–11 days. In a tissue with many interior air spaces, the embedding time should be increased to remove all possible air holes in the finished block.

LEO SAKOVICH

References

General References
Gray, P., "The Microtomist's Formulary and Guide," New York, Blakiston, 1954 (pp. 142–148).
Sakovich L. and H. F. Burns, "Vacuum embedding of tissues in nitrocellulose," *Amer. J. Clin. Pathol.,* **43**:396–398 (1965).
Text
1. G. L. Walls, "The hot celloidin technic for animal tissues," *Stain Technol.,* **7**:135–145 (1932).
2. A. A. Koneff and W. R. Lyons, "Rapid embedding with hot low-viscosity nitrocellulose," *Stain Technol.* **12**:57–59 (1937).
3. A. A. Koneff, Personal communication.
4. H. A. Davenport, "Histological and Histochemical Technics," Philadelphia, Saunders, 1960 (p. 71).
See also: PLASTIC EMBEDDING.

NONFERROUS METALS

The mechanical properties of metals and alloys have been investigated at three levels: macroscopic, microscopic, and submicroscopic. Although the information revealed by each of them is at first instance different as in scale concerns, the correlation extensively made between them permits at the present time a unified but not complete understanding of the physical mechanism of deformation. Each level mentioned above provides valuable information to the metallurgist and to the materials engineer. The latter must use the results of the physical approach, together with the fundamental equations of continuum mechanics, in order to give a good approximate description of the constitutive relations for a real metal or alloy.

The macroscopic level is essentially concerned with the measurement of the resistance to plastic deformation and fracture. Many useful properties such as elastic modulus, yield strength, tensile strength, and ductility can be determined from the simple tension test. Other tests give further information essential to the design engineer; among them can be mentioned the fatigue test, which concerns the premature fracture of metals under repeated applied low stresses; the creep test, which measures the plastic flow under constant stress conditions; the impact test, which permits determination of the suitability of the metal to shock loading conditions; the hardness test, which measures the metal resistance to penetration, giving a rapid indication of its degree of deformation; and the stress-rupture test, which measures the long-time resistance to fracture at elevated temperatures. These tests are not only useful to physical metallurgy and design requirements but also for checking material quality against standard specifications. But macroscopic measurements provide little direct information about the relation between mechanical behavior and microstructure.

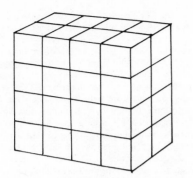
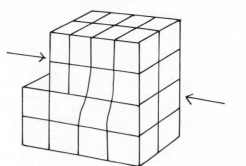
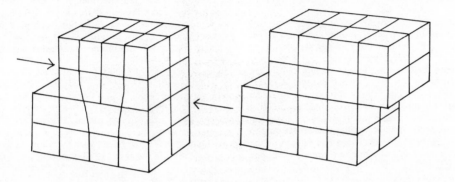

Fig. 1 Sequence of diagrams showing an edge dislocation. Under stress, the perfect crystal (upper left) is squeezed one atomic spacing (upper right). The extra half plane of atoms which appears in this figure produces at its base an edge dislocation. The dislocation moves across the crystal producing a slip step.

The microscopic level works at a magnification that permits one to obtain most of the structure results by observing surface defects. Deformation often alters the local structure and this level allows the study of the associated microscopic processes of slip, twinning, kinking, deformation bands, crack nucleation, and modes of fracture. Detailed descriptions of these microscopic aspects of deformation can be found in some general references (1–3).

The first steps in exploring the submicroscopic features of plastic deformation were made 30 years ago with the introduction of the dislocation concept by Taylor, Orowan, and Polanyi in order to explain slip in metals. The concept of dislocation primarily arose from the study of plastic deformation in crystalline materials. The books listed in the references (4–6) give account of the early work on dislocations.

The basic geometry of an "edge" dislocation is shown in Fig. 1 in a simple cubic crystal. The upper left drawing is a simplified perfect crystal; the application of a shear stress produces a region of mismatch in which the atoms are not lined up in a perfect array. In this way, the upper half crystal has slipped one atomic space distance over the lower half. Along the bottom edge of this plane runs what is called an edge dislocation; the movement of the dislocation deforms the crystal in a series of discrete steps.

A second kind of dislocation, the "screw" dislocation, is shown in Fig. 2. In this dislocation the two parts of the crystal have slipped parallel to the line of the dislocation rather than perpendicular to it, as in the case of an edge

dislocation. Usually, dislocations do not have a single screw or edge character but they can blend from one type to another forming a dislocation line.

In the last decade a great extent of theoretical and experimental work has been performed in order to obtain a better understanding of dislocation mechanics, and through that, of many mechanical properties of metals and alloys. In recent years it has been possible to observe and study many submicroscopic defects and their incidence in the mechanical behavior of materials, due to the rapid development of microscopic techniques.

Light, X-rays, electrons, and ions can be used as optical means, visible light and electrons being the most important for metallographic research. Visible light has a wavelength of 4000–8000 Å, corresponding approxi-

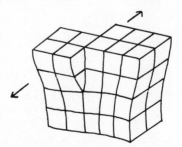

Fig. 2 Screw dislocation occurring when a plane of atoms slip each other along the dislocation line.

mately to 3000 Å resolution, and electrons have a wavelength of 0.039 Å when accelerated under a potential of 100 Kv, giving a resolution of 4 Å.

Figure 3 illustrates the successes reached by metallography. The several methods of observation are shown in connection with the magnification and resolution range achieved. The dimensions of the observed defects are also shown in three types of specimens: polished surfaces, thin foils, and hairpins. It can be noticed that many metallographic instruments overlap respecting magnification, although their scope is somewhat different regarding the type of observation to be performed.

Surface Effects. Surface effects produced by plastic deformation in nonferrous metals depend strongly upon the crystal structure, the nature of the deforming process itself, and the conditions under which this process is performed.

Plastic deformation in metals bascially occurs by the slide of one plane of atoms over another, within a grain: the slip planes. As pointed out before, the motion of dislocations on a slip plane results in a surface step: the slip line. Since further deformation is possible in the neighborhood of an original slip line, parallel slip lines group themselves, producing optically observable traces on the surface: the slip bands, which can be detected if the surface is polished before plastic deformation. The optical microscope has been the most familiar instrument used to study slip bands. In Fig. 4 slip bands can be seen on the surface of a single crystal of aluminum slightly deformed

under tension. The nature of the deformation process has an important influence on the slip character. As Fig. 5 shows, grooves and ridges can be observed in fatigued pure aluminum revealing one of the differences existing between static and cyclic loading surface topography.

Many other microscopic surface effects arising from plastic deformation, such as deformation bands, kink bands, and deformation twin bands, can be directly studied with the aid of the optical microscope. Structures commonly produced by large plastic deformation processes, such as drawing and rolling, in which the grains are distorted, can also be studied to some extent by these means.

Many models representing different features of plastic deformation were based upon the direct observation of the surface of the deformed metals or alloys with the optical microscope. Among these can be mentioned the dislocation model of void formation in copper during fatigue (9), and recently, the coarse slip fatigue model that explains crack propagation and many other well-known features of fatigue (10).

The observation of etched surfaces is extensively used in studying surface effects due to plastic deformation, since the emerging points of dislocations can be revealed. When a polished specimen surface is etched by a chemical reagent, many structural details are detected. Of primary importance is the fact that in a metal subjected to an environment which removes atoms from the surface, the rate of removal in the points of emergence of a dislocation

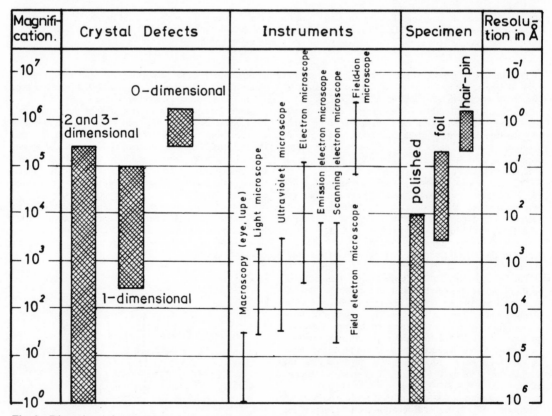

Fig. 3 Dimensions of crystal defects, specimens, and instruments utilized for their examination. The magnification and resolutions achieved are also indicated (7).

at the surface is different from that of the surroundings. The principal methods of etching are: (a) chemical etching; (b) electrolytic etching, successfully used for instance to reveal pits in Cu and α-brass (11); (c) thermal etching, consisting in etching the metal in an atmosphere that contains some oxygen, which produces etch pits at dislocation sites (12); and (d) preferential oxidation, also employed in Cu (13) where dislocation sites oxidize preferentially. A complete table of etchants appears in the book by Amelinckx (14).

Etch pit studies give information about dislocation density and distribution and also about dislocation arrays in slip traces and deformation bands. A great amount of work has been performed in recent years, based upon the development of etch pits in surfaces of plastically deformed metals. As a matter of illustration, Fig. 6 shows an isolated dislocation pileup in Cu–5 at . % Al single crystal; etch pits were produced by using Livingstone's etching reagent (16). These isolated pileups were used to calculate frictional stresses that move dislocations in face-centered cubic metals and alloys, and these calculations have an enormous importance upon the yield strength determina-

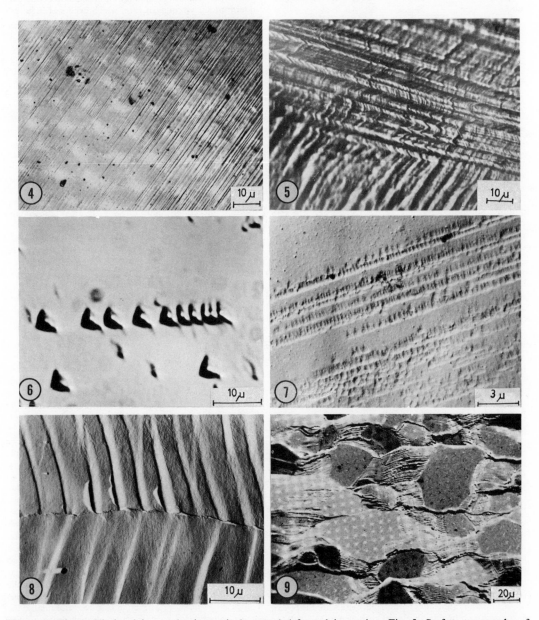

Figs. 4–9 Fig. 4: Slip band in an aluminum single crystal deformed in tension. Fig. 5: Surface topography of fatigued pure aluminum (8). Fig. 6: A dislocation pile up in Cu–5 at . % Al (15). Fig. 7: Etch pits distribution in Cu–15 at . % Al deformed in tension at a shear strain of 0.088 (21). Fig. 8: Formation of cracks in fatigued polycrystalline molybdenum showing slip band and grain boundary cracking (22). Fig. 9: Surface structure of deformed superplastic alloy (23).

tion. Recent studies on the variation of etch pit distribution near crystal surfaces have led to important results regarding dislocation multiplication phenomena (17).

The fine surface effects of plastic deformation are not usually visible under the light microscope because of resolution restrictions, e.g., the slip lines in the surface of a metal or an alloy, which cannot be seen because the relative translation on a single slip plane is out of its resolution limit, as well as a high density of etch pits. Surface replicas examined under the electron microscope are one of the available means of studying such effects, although they are leaving place to more modern techniques.

We shall not give a detailed reference of the several kinds of replica techniques, but it is worthwhile to mention the plastic replicas (18) that give a resolution of 200 Å, the widely used carbon replica (19) with a resolution of Å, and the oxide replica (20), successfully applied in studying aluminium and its alloys, which can resolve between 20 and 50 Å.

Figure 7 illustrates the distribution of etch pits along slip lines in a deformed Cu–15 at . % Al single crystal. It is an electron micrograph of a Ge-shadowed carbon replica.

Another example of the use of replicas in the study of surface topography is shown in Fig. 8, which illustrates a combination of slip band and grain boundary cracking in fatigued polycrystalline molybdenum in an axial tension-compression test.

We can sinterize the preceding by establishing that within resolutions of about 20 Å, replicas are still very useful in the observation of surface effects due to plastic deformation, considering the limitations inherent in the stripping process.

Surface features developed during plastic deformation can now be directly explored by the scanning electron microscope with its large depth of focus and wide range of available magnifications. Figure 9 illustrates the surface structure of a deformed Mg-Al alloy observed with the scanning electron microscope. Such an alloy is called superplastic, meaning that it is capable of undergoing extensive plastic deformation. It has a high dependence of stress on strain in a narrow range of strain rates. The individual grains are very fine and noncoplanar, so direct observation of the surface by optical microscopy is difficult. Replica techniques can be used, but the magnification range of interest is between 500 and 3000×, which is at the lower end of the electron microscope magnification.

Another instrument that has become very useful in the study of deformation structures, due to its pronounced orientation contrast, is the emission electron microscope. Figure 10 shows fine deformation twins in a deformed brass.

Fracture Surfaces. The most general classification of fracture describes two kinds of fracture processes: brittle and ductile fracture. Brittle fracture occurs in completely brittle materials (there are some exceptions in ductile materials) in which the only work required is that needed for overcoming the cohesive forces between the atoms. Brittle failures may occur in some refractory nonferrous metals because of their inherent low ductility, principally at low temperatures. From this point of view, the most important and interesting nonferrous metals are the body-centered cubic (BCC) refractories such as Nb, Ta, Cr, Mo, and W; the high-strength aluminium alloys; and, because of their low crystal symmetry, the hexagonal close-packed (HCP) refractories Mg, Be, and Ti.

Ductile fracture, in which plastic slip, mechanical twinning, and kinking play an important role, is particularly typical of face-centered cubic (FCC) metals such as Al, Cu, and Ni base alloy. Among the different types of ductile fracture, can be mentioned: the fibrous fracture, formed at the bottom of the cup in the cup-and-cone fracture; and shear fracture, typical of ductile metals in shear and torsion as well as in tension tests of sheets; the fatigue fracture, produced by mechanical fatigue under cyclic stressing, which has many similarities with the brittle fracture, although it cannot occur without local plastic deformation at the tip of a propagating crack; the intergranular viscous fracture, which occurs at high temperatures and low rates of deformation; and finally, rupture, which cannot be regarded as fracture in the narrow sense of the word, and which occurs by localization of plastic deformation, particularly in single crystals with one preferred set of slip planes such as Zn, Cd, and Mg.

The microscopic examination of fracture surfaces provides valuable information about the mechanism of fracture. The replication method is useful when the fracture surface is not too rough, the most convenient being the direct carbon replica already mentioned. The aluminium oxide replication is suitable for light alloys. Figure 11 shows a microfractograph of the fracture surface of an Al–2.5% Mg alloy subjected to a tension-compression fatigue test. Fatigue grooves can be seen in this picture. The method for obtaining the replica was more sophisticated than those mentioned above, and proceeded in two stages: First a primary surface replica of plastic was obtained over which carbon was deposited afterwards; then the plastic was dissolved and the remaining carbon replica examined in the electron microscope.

However, the replication technique allows an indirect examination of fractures surfaces. The scanning electron microscope, which does not require any sample preparation, is at present a useful tool in studying fracture surfaces when the needed resolution is not greater than 200 Å. It also gives a three-dimensional nature of images, which facilitates the interpretation. In Figure 12 is shown a scanning electron micrograph of the central portion of the fracture surface in cup-cone fracture of polycrystalline copper-containing inclusions; numerous points can be seen that correspond to the necking down of the metal between the inclusions, which are surrounded by relatively smooth areas.

The principal advantages of the scanning electron microscope in this field are summarized as follows: the depth of focus, which is 300 times that of the optical microscope, and, even if this fact does not represent an advantage regarding the electron microscope, rough fracture surfaces are very difficult to replicate; the sample size, which can be 10 mm diameter and 10 mm height, while in the electron microscope the maximum diameter is 3 mm and also considerable area is lost in the grid that supports the replica; its range of magnification, which goes from 20 to 140,000×, the low magnification end being very useful in studying the initiation and propagation of cracks; the several possibilities in manipulating the specimen under examination inside the microscope, this being one of its more attractive features; and finally, the image characteristics, which give a much better understanding of the fracture surface topography.

The scanning electron microscope has no facilities for making diffraction, and this appears to be its only disadvantage, but many attempts are being made to overcome this problem.

Bulk Effects. It is a well-known fact that metal samples become transparent to electrons when they are a few thousand angstroms thick, making the thin film method the most versatile one for studying submicroscopic effects due to plastic deformation, since it permits a direct understanding of the dislocation behavior. The transmission electron microscopy method was developed principally by Hirsch, Whelan, and Howie (27) at the Cavendish Laboratory in Cambridge.

In transmission electron microscopy, dislocations are observed by diffraction contrast, instead of absorption contrast, usual in surface replication. The observation of dislocations arises from the scattering of electrons in a strained region around the dislocation. The necessary theory to understand the origin of the contrast is very well developed in the book by Hirsch et al. (28). Diffraction contrast theory allows the determination of the role played by crystalline defects in plastic deformation and the achievement of a better understanding of the nature and properties of the defects themselves; thus it offers a powerful tool to study dislocations, stacking faults, twins, grain boundaries, etc., within a resolution at best of 4 Å.

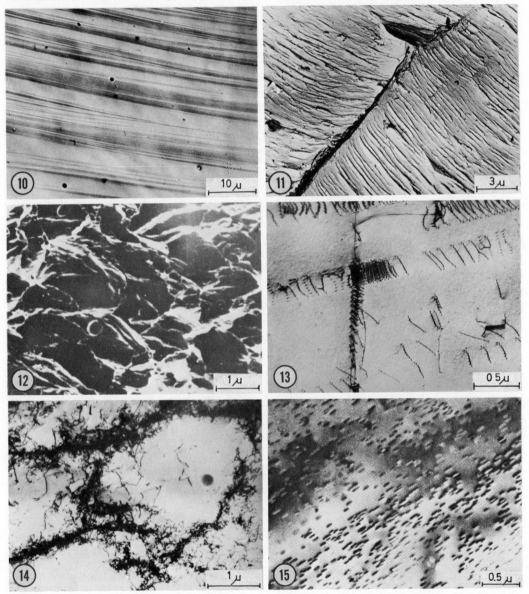

Figs. 10–15 Fig. 10: Photoemission micrograph of fine twins in a deformed brass (24). Fig. 11: Fracture surface of a fatigued Al–2.5% Mg alloy (25). Fig. 12: Scanning electron micrograph of the fracture surface of deformed copper containing inclusions. The effect of inclusions is to initiate voids in the internal cavity of the "cup and cone" fracture (26). Fig. 13: Dislocation distribution in deformed α-brass (30). Fig. 14: Dislocation cell in fatigued polycrystalline copper. Fig. 15: Dislocation loops in Be subjected to hydrostatic pressure (31).

One of the main advantages of transmission electron microscopy is the fact that it gives a three-dimensional picture of dislocations in crystals, contrasting with the etch pit technique that traces the distribution and movement of dislocations in two dimensions, i.e., along the surface. It also makes it possible to perform quantitative measurements with the aid of the selected area diffraction pattern which can be obtained for each image.

Plastic deformation studies must be representative of the bulk material. In order to accomplish this, it is convenient to start with a specimen 100 μ thick, which can be obtained by any method that avoids further deformation. One of the modern available techniques among these methods is that based upon spark erosion, which seems to introduce little surface damage, although the dislocation distribution present in the initial mechanical state can to some extent be affected, mostly in soft metals. Another available method is the chemical slicing of the specimen, e.g., by means of an acid saw; in this method surface damage almost completely disappears, but it is very slow. Subsequent thinning to a thickness of about 2000 Å is more frequently made by chemical or electric polishing. An excellent review of metal thinning techniques can be found in the paper by Kelley and Nutting (29).

The sequence of photographs in the figures illustrates submicroscopical effects due to some plastic deformation processes. Slip lines in α-brass deformed 5% in tension are shown in Fig. 13. It can be noticed that in the transmission electron micrograph the individual dislocations lying along the slip plane can be resolved; the dislocations pile up in the intersection of the two slip lines. This micrograph can be compared with Fig. 8, where dislocations appearing in a slip line are revealed through etch pits.

A typical dislocation configuration of a copper single crystal subjected to cyclic loading is shown in Fig. 14, where dislocations are heavily tangled. Tangles harden the crystal by obstructing subsequent dislocation motions. Dislocations are concentrated in certain areas which constitute a system of small dislocation-free "cells."

Finally, a bulk effect arising when single-phase polycrystals of high anisotropic compressibility are subjected to external hydrostatic pressure is illustrated. Shear stresses build up across the boundaries of adjacent grains, causing plastic deformation. Figure 15 shows that in Be subjected to this type of loading, a high density of small defects identified as vacancy loops formed by vacancy clusters is developed.

The dislocation distribution in thin foils prepared by thinning bulk material is representative of the prior distribution in the original thick specimen, this fact being based upon two points. First, the dislocation distribution in the foil could be altered by heating with the electron beam; but it has been demonstrated that this effect is negligible, since under suitable illuminating conditions the temperature rise does not exceed 10–20°C. Second, the dislocation distribution may be altered by the thinning process as a consequence of the relief of internal stresses, or of the re-motion of slip obstacles, but several experiments have confirmed that relatively minor arrangements take place during thinning, it having been proved that the distributions were the same as those inferred by X-ray diffraction studies in bulk specimens (32). This can be attributed to the locking of dislocations due to their interaction with surface films, which becomes very important in thin foils, and also to the fact that as the foil becomes thinner, the range of the elastic-strain field, and therefore the dislocation interactions, becomes smaller.

Transmission electron microscopy allows correlation between mechanical properties after deformation and the actual distribution of the defects present.

Modern high-voltage electron microscopes allow a more realistic picture of dislocation interaction after plastic deformation, since thicker foils can be examined. This fact also permits the extension of the observation of features to bigger areas.

Plastic deformation depends on the dislocation motions through the metal and it has not yet proved possible to solve the equations of wave mechanics to obtain approximate solutions for the distribution of atoms around the dislocations and the forces required to move them. But the extension of continuum mechanics has made it possible to establish a dislocation theory that explains how they move, interact, and multiply under the influence of applied stresses. This fact allows application of the theory to the interaction of few dislocations, but actually there may be a million miles of dislocations involved in a standard specimen of a cold-worked metal, and besides, these dislocations are not in an ordered array, but are rather heavily tangled, which makes it difficult to apply the theoretical solutions obtained from the dislocation theory. The mechanical behavior of metals is not fully developed, and the improvement of metallographic techniques is particularly important in this area. Present experimental knowledge is far ahead of the capabilities to obtain theoretical solutions, and more insight into plastic deformation mechanism will be made as metallographic techniques become more quantitative.

Acknowledgments

Thanks are given to all the authors who have kindly permitted the reproduction of their photographs in this paper.

ARI VARSCHAVSKY

References

1. G. E. Dieter, "Mechanical Metallurgy," New York, McGraw-Hill Book Co., Inc., 1961.
2. F. A. McClintock and A. S. Argon, "Mechanical Behavior of Materials," Reading, Mass., Addison-Wesley Publ. Co., Inc., 1966.
3. A. S. Tetelman and A. J. McEvely, Jr., "Fracture of Structural Materials," New York, John Wiley & Sons, Inc., 1967.
4. W. T. Read, Jr., "Dislocations in Crystals," New York, McGraw-Hill Book Co., Inc., 1953.
5. A. H. Cotrell, "Dislocations and Plastic Flow of Crystals," The Clarendon Press, 1953.
6. Ch. Kittel, "Introduction to Solid State Physics," New York, John Wiley & Sons, Inc., 1956.
7. E. Hornbogen and G. Petzow, *Metallkünde,* **61**:81 (1970).
8. P. J. Forsyth, *Proc. Roy. Soc.,* **A242**:198 (1957).
9. W. A. Wood, S. Cousland, and K. R. Sargant, *Acta Met.,* **11**:643 (1963).
10. P. Neumann, *Acta Met.,* **17**:1219 (1969).
11. P. A. Jacquet, *Acta Met.,* **2**:725 (1954).
12. A. A. Hendrickson and E. S. Machlin, *Acta Met.,* **3**:64 (1955).
13. F. W. Young and A. T. Gwathmey, *J. Appl. Phys.,* **31**:225 (1960).
14. S. Amelinckx, "The Direct Observation of Dislocations," New York, Academic Press, 1964.
15. S. Yoshioka, Y. Nakayama, and T. Itô, *Trans. Jap. Inst. Met.,* **10**:383 (1969).
16. J. D. Livinstone, *J. Appl. Phys.,* **31**:1070 (1960).

17. S. Kitajima, H. Tanaka, and H. Kaieda, *Trans. Jap. Inst. Met.,* **10**:12 (1969).
18. *Proc. Amer. Soc. Test. Mater.,* **50**:444 (1950).
19. E. Smith and J. Nutting, *Brit. J. Appl. Phys.,* 7:214 (1959).
20. G. Thomas and J. Nutting, *J. Inst. Met.,* **85**:1 (1956).
21. S. Yoshioka, Y. Nakayama, and T. Itô, *Trans. Jap. Inst. Met.,* **10**:390 (1969).
22. P. Beardmore and P. H. Thorton, *Acta Met.,* **18**:109 (1970).
23. E. Lifshin, W. G. Morris, and R. B. Bolon, *J. Met.,* **21**:1 (1969).
24. L. Wegman, *Balzers Rep.,* **19** (1970).
25. H. Nordberg, S. Karlsson, and B. Aronsson, *Rev. de Mét.,* **66**:861 (1969).
26. C. Baker and G. C. Smith, *Trans. Met. Soc. AIME,* **242**:1991 (1968).
27. P. B. Hirsch, A. Howie, and M. J. Whelan, *Phil. Trans. Roy. Soc.,* A **252**:499 (1960).
28. P. B. Hirsch et al., "Electron Microscopy of Thin Crystals," London, Butterworths, 1965.
29. P. M. Kelly and J. Nutting, *J. Inst. Met.,* **87**:385 (1959).
30. R. L. Segall and J. M. Finney, *Acta Met.,* **11**:685 (1963).
31. S. V. Radcliffe and C. W. Andrews, Sixth International Congress for Electron Microscopy, Kyoto, Japan, 1966, (p. 319).
32. P. Gay, P. B. Hirsch, and A. Kelly, *Acta Cryst.,* **7**:41 (1954).

O

ONYCHOPHORA

The Onychophora are the remains of a very ancient, wide-spread fauna, and therefore of great interest. In the present time they are restricted to the area between 22 degrees north latitude (America) and 45 degrees south latitude (New Zealand), prevalent to Central and South America, equatorial and South Africa, Indonesia, Australia, New Guinea, New Zealand. Near relatives are found as fossils in the Middle-Cambrium of British Columbia (*Aysheaia pedunculata*-Prot-onychophora) and in the Old Cambrium of Scandinavia (*Xenusionauerswaldae*). This is striking evidence of a widespread ancient distribution.

Technique of Keeping

The extant species prefer wet habitats underneath a litter of leaves or moldy wood in the tropics. Most of them may be kept in the laboratory, if the natural conditions are observed. It is worth noting that the Onychophora require a high humidity but they avoid wetness and mustiness. They are best kept in unglazed earthenware moistened on the outside. The subsoil should be filtering gravel, sandy earth, and above it moldy wood and some stones. They are fed with small arthropods, e.g., termite, apterygota, isopods, and centipedes. The optimal temperature is 26°C in the daytime, 20°C during the night. Fluctuating temperatures are endured.

Technique of Research

I. General Body Plan. For the better study of the organization, preparation of single organs is necessary, e.g., of the legs, the mandibulae with the muscles (Pflugfelder, 1968). The inner organization of specimens treated with ether, chloroformium, or carbon dioxide may be studied after post-mortem examination.

II. Histological Methods. *1. General Methods.* To get good histological sections, fixatives used are: Susa, Bouin's, or Flemming's fluid; for frozen sections (Baker, (1944), formaldehydecalcium and osmium ethylgallate (Wigglesworth, 1957, 1959). The best staining techniques are: Azan (Romeis, 1968), iron hematoxylin (Heidenhain), Mallory, Masson (Richards, 1951).

2. The Cuticle. a. Robson (1964) prepared replicas of the cuticle by shadowing with carbon, and then with chromium, backing with a layer of 0.5% formvar or collodium, and dissolving away the specimen with 20% sodium hypochlorite. Replicas were washed with distilled water and cut into pieces which were picked up on grids for examination. Replicas examined with the electron microscope confirm the details of light microscope observations: cell outlines, tracheal openings (at least 50 in every segment), papillae, blunt spines.

b. Chemical Research. Kunike (1925) showed that *Peripatus broelmanni* gives a positive Schultze-chitin reaction (α-Naphtol). Richards confirmed the presence of chitin. X-ray analysis indicated that it occurs as α-chitin (Lotmar and Picken, 1949; Rudall, 1955, 1963). It is dissolved slowly by 20% chromic acid, more rapidly by 10% sodium hypochlorite in a warm solution of potassium nitrate in nitric acid.

c. Electron Microscopy. Material for electron microscopy was fixed by Robson (1964) in 1% OsO_4 in veronal acetate buffer at pH 7.2, usually for an hour at 0°C. It was taken through acetone and propylene oxide (Luft, 1961) into araldite (Glauert, 1962), and embedded in specially hard mixture (Luft, 1961) containing araldite resin:

araldite resin, 10 ml
dodecenyl succinic anhydride, 9 ml
methyl nadic anhydride, 1 ml
accelator, 0.4 ml

Staining was with uranyl acetate (Gibbons and Grimstone, 1960).

According to Lavallard (1965) OsO_4 is not suitable for demonstrations of the five different zones of the cuticle. These appear with better contrast when fixed in permanganate.

3. The Eye. The general organization of the eye is similar to that of a polychaete, but on the basis of the ultrastructure it shows affinities to the arthropods, namely in the arrangement of the microvilli. Eakin and Westfall (1965) got excellent results by the following fixatives: Dalton's solution (Dalton, 1955) at pH 7.2–7.4, phosphate-buffered 1% OsO_4 in 1.25% bicarbonate at pH 7.4 (Cloney, 1964) and 6.25% glutaraldehyde (Sabatini et al., 1963) in 0.1 M phosphate buffer at pH 7.2, postfixation in Dalton's solution. Rapid dehydration in ethanol, transfer to Epon. Staining: saturated aqueous uranyl acetate (Watson, 1958) followed by lead citrate (Reynolds, 1963).

4. Neurosecretory System. Gabe (1954) found by the method of Gomori's paraldehyde-fuchsin five groups of neurosecretory cells in the brain of *Peripatopsis, Opisthopatus* and *Ooperipatus*: one group in front of the *corpus centrale* and two pairs of groups lateral behind the *nervus antennae*. In the ventral chain of the nervous system the secretory cells are very small.

5. Muscles. In the body wall there are three layers of muscles. In the outer layer the fibers follow a circular course, in the eight bundles of the inner layer they run longitudinally. Between these two layers is a third layer of diagonal muscles.

Lavallard (1965, 1966) has investigated the fine structure of the somatic muscles of *Peripatus acacioi*. The appropriate fixative was glutaraldehyde. The fibrils do not show any striae. There are two sorts of myofilaments, some of 180 Å diameter, and others of 60 Å diameter, the latter being the more numerous. The sarcoplasma is rich in glycogen and in ATP. The activity of ATPase could be shown by application of lead precipitation.

6. Circulation System. Rajulu and Singh (1969) investigated the rate of heartbeat of *Eoperipatus* under normal and experimental conditions. When first isolated the heartbeat was slow and irregular. On recovery it was rhythmic, at a rate 38–42 beats/min/25°C. The pH of freshly drawn blood was found to be 7.5. Both increasing and decreasing the pH with phosphate buffers depressed the heart-rate. Increasing temperature accelerates the heart-rate (7–10 at 5°C, 61–64 at 30°C). Faradic stimulation of the brain and ventral cord (up to 2 v) accelerates the rate. Acetylcholine (10^{-4}) atropine (10^{-6}), and adrenaline (10^{-4}) accelerate, histamine (10^{-5}) and Chloroform (1/4–1/2 saturated) depress the heartrate.

The composition of the blood in *Peripatopsis* has been studied by Robson et al. (1966). The blood consists of a fluid plasma within which five types of corpuscles are suspended (5000 cells/mm³): (1) roundish hemocytes with a big nucleus, (2) leucocytes with an irregular nucleus, (3) hemocytes with a fibrillary plasma, (4) big cells with many vacuoles, and (5) nephrocytes. Two counts by Tuzet and Manier (1958) disclosed 2000 cells/mm³.

Picken (1936) found, for species of *Peripatopsis*, that osmotic pressure approximated 110 m mole/liter NaCl. This blood concentration is very low in comparison with other terrestrial arthropods and is accurately regulated. Robson et al. suppose that the low tonicity may be of advantage if desiccation reduces the volume. They used the following methods:

Na+ and K+ by flame photometer or Unicam flame the legs, the mandibula with the muscles (Pflugfelder, spectrophotometer SP900
Ca++ by ethylendiaminetetraacetic titration
Cl– by electronic titration
The tonicity is equal to the composition of *Peripatus* Ringer's solution.

Stock solutions:

	M	g/liter
NaCl	0.54	31.6
KCl	0.54	40.2
CaCl$_2$·6H$_2$O	0.36	78.2
MgCl$_2$·6H$_2$O	0.36	36.6
NaHCO$_3$		2.0
NaH$_2$PO$_4$ anh.		0.1

Distilled water to make 1000 ml
Final pH 7.3
Glucose 0.4 g added to 100 ml before use

7. Digestive System. Gabe (1954) studied the digestive system by the following methods:

Fixation: Bouin's fluid or Dubosq-Brazil

Embedding: Celloidin-paraffin
            Hemalum-indigocarmine
Staining: Triple staining Prenant
            Triple staining Gomori (1950)
Iron hematoxylin Groat
Counter staining: Eosin Y lg
Lightgreen, 0.2 g
Phosphotungstic acid, 0.5 g
Distilled water, 100 ml
Acetic acid, 0.5 ml
Histochemical methods:
    Chondriom: After post-chroming Altman's Fuchsin
        Gabe's Picro–methyl green
        Groat's iron hematoxylin
    Nucleic acids: Feulgen and Rossenbeck
    RNA: Brachet
    Glucides: Hotchkiss-McManus
        Bauer
        Best's Carmine
    Calcium: Method of Kossa and Stoelzer
    Iron: Tirmann and Schmeltzer
    Alc. phosphatases: Gomori (1950)

8. Excretory System. Gabe (1957) used the same methods as for his investigations on the digestive system. Eakin (1964) fixed with glutaraldehyde and embedded in Epon.

9. Reproductive System. Gabe (1959) has studied the reproductive system of three species of *Peripatopsis* by special methods: The plasma of the spermatogonia contain pyrenophilic ribonucleides and Hotchkiss-positive glucides. The *ductuli efferentes* are rich in mucopolysaccharides. The *ductus ejaculatorius* contains glucides and SH-proteids. The oocytes are rich in Hotchkiss-positive glucides and SH-proteids, the oviducts in mucopolysaccharides.

10. Development. The study of the development is easiest in viviparous species, the uteri of which are often filled by many embryos and various stages. The following fixatives may be used: Bouin, Susa, or Flemming; staining: iron hematoxylin–Fuchsin, Mallory.

OTTO PFLUGFELDER

References

Baker, J. R., *Quart. J. Micr. Sci.*, **85**:1 (1944).
Cloney, R. A., *Acta Embr. Morph. Exp.*, **7**:111–130 (1964).
Dalton, A. J., *Anat. Rec.*, **121**:281 (1955).
Eakin, R. M., *Amer. Zool.*, **4**:433 (1964).
Eakin, R. M., and J. A. Westfall, *Z. Zellforsch.*, **68**:278–300 (1965).
Gabe, M., *C. R. Acad. Sci.*, **238**:272–274 (1954).
Gabe, M., *Bull. Soc. France*, **79**:141–150 (1954).
Gabe, M., *Bull. Soc. France*, **81**:170 (1956).
Gabe, M., *Arch. Anat. Micr.*, **46**:283–306 (1957).
Gabe, M., *Bull. Biol. France et Belg.*, **93**:140–154 (1959).
Gatenby, J. B., *Trans. Roy. Soc. New Zealand*, **87**:51–53 (1959).
Gibbons, J. R., and A. V. Grimstone, *J. Biophys. Biochem.*, **7**:697 (1960).
Glauert, A. M., *J. Roy. Micr. Soc.*, **80**:269 (1962).
Gomori, G., *Amer. J. Clin. Pathol.*, **20**:661–664 (1950).
Graumann, W., and K. Neumann, "Handbuch der Histochemie," Fischer, Stuttgart, 1966.
Kunike, B., *Z. vergl. Physiol.*, **2**:233 (1925).
Lavallard, R., *C. R. Acad. Sci. Paris*, **260**:965–968 (1965).
Lavallard, R., *C. R. Acad. Sci. Paris*, **263**:148–151 (1966).
Lavallard, R., and S. Campiglia-Reimann, *C. R. Acad. Sci. Paris*, **263**:1728–1731 (1966).

Lotmar, W., and L. E. R. Picken, *Experientia,* **6**:58 (1949).

Luft, J., *J. Biophys. Biochem. Cytol.,* **9**:409 (1961).

Millonig, G. V., *Int. Congr. Electr. Micr.,* **2**:P8 (1962).

Manton, S. M., *Trans. Roy. Soc. London,* **B228**:421–441 (1938).

Manton, S. M., *Ann. Mag. Nat. Hist.,* **1**:515 (1938).

Pflugfelder, O., *Zool. Jahrb. Anat.,* **69**:443–493 (1948).

Pflugfelder, O., *Mikrokosmos,* **1955**:169–171.

Pflugfelder, O., "Onychophora," Stuttgart, Fischer, 1968.

Picken, L. E. R., *J. Exp. Biol.,* **13**:309 (1936).

Rajulu, G. S., and M. Singh, *Naturwiss.,* **56**:38 (1969).

Reynolds, E. S., *J. Cell Biol.,* **17**:208–212 (1963).

Richards, A. G., "The Integument of Arthropods," Minneapolis, Univ. of Minnesota Press, 1951.

Robson, E. A., *Quart. J. Micr. Sci.,* **105**, (pt. 3):281–299 (1964).

Robson, E. A., A. P. M. Lockwood, and R. Ralph, *Nature* (London), **209**:533 (1966).

Romies, B., "Mikroskopische Technik," Ed. 6, München, 1968.

Rudall, K. M., *Soc. Exp. Biol. Symp.,* **9**:49 (1955).

Rudall, K. M., *Adv. Insect Physiol.,* **1**:257 (1963).

Sabatini, D. D., K. Bensch, and J. R. Barrnett, *J. Cell Biol.,* **17**:19–58 (1963).

Tuzet, O., and J. Manier, *Bull. Biol. France et Belg.,* **92**:7 (1958).

Watson, M., *J. Biochem. Cytol.,* **4**:475–478 (1958).

Wigglesworth, V. B., *Proc. Roy. Soc. B.,* **147**:185 (1957).

Wigglesworth, V. B., *Quart. J. Micr. Sci.,* **100**:315 (1959).

OPTICAL MICROSCOPE

Elements of the compound "research" microscope, which is the type of instrument primarily considered here, are illustrated diagrammatically in Fig. 1. Illumination is provided by a lamp, which may be equipped with a lens and variable iris diaphragm. Light is usually reflected by a mirror at the base of the microscope stand, and is focused at or near the specimen by the condenser lens system. A second iris diaphragm is located below the condenser at its first focal plane. The specimen rests on the stage, which may be equipped for calibrated translation in two mutually perpendicular directions ("mechanical stage") or, principally in polarizing microscopes, for calibrated rotation about the microscope axis. Imaging is performed by the objective and ocular lenses, each of which is a system of lens elements designed for the correction of aberrations. The objective forms a real intermediate image within the microscope tube. Further focusing by the ocular (eyepiece) forms a virtual image located near the specimen plane (in the case of viewing by eye) or a real image on a photographic emulsion placed above the microscope tube.

Focusing of the image is achieved by adjusting the distance between specimen and objective lens. Magnification is determined by the choice of objective and ocular, and by the distance between these two lenses.

The general plan just described may be modified in a number of ways. For example, a continuously varying

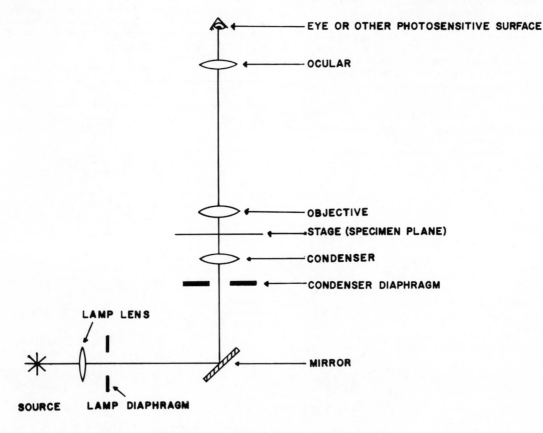

Fig. 1 Compound optical microscope (schematic).

range of magnification can be obtained with "zoom" optics; i.e., by adjusting the spacing of elements within the objective. This feature is particularly convenient in dissecting microscopes.

Magnification, Resolution, and Contrast. The purpose of any microscope is to produce an image in which more detail can be discerned than is visible by direct viewing. Obviously, this requires enlargment of the image. Magnification alone is not sufficient, however; the resolving power of the microscope and the contrast produced in the image must both be sufficient to render detail visible. Magnifications greater than about 1250 or 1550× (depending upon the visual acuity of the micro-

scopist) are therefore of no avail in the light microscope.

Resolving power specifies the ability of an optical system to render two closely spaced points as distinct. This quantity is discussed in detail later; here, it may simply be said that the wave-like properties of light impose a definite limit on resolving power. Expressions specifying this limit are of the form:

$$d_{min} = K \frac{\lambda}{n \sin \alpha} \qquad (1)$$

where d_{min} is the minimum resolvable separation of points, λ is the wavelength of the imaging light, n is the refractive index of the space surrounding the object, α is the aperture angle of the lens (i.e., the half angle of the

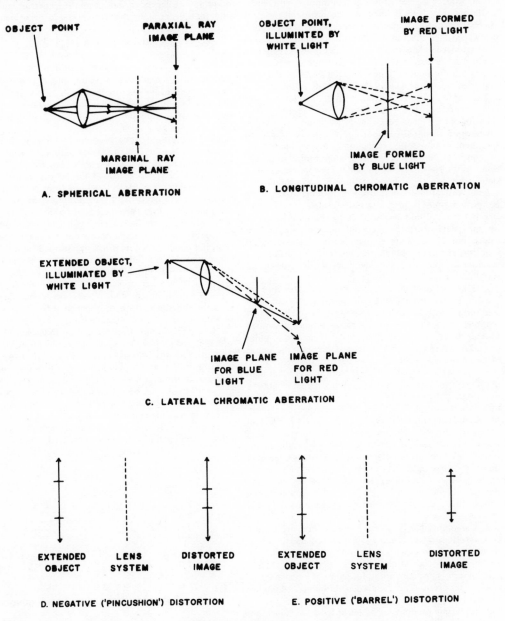

Fig. 2 Lens aberrations.

cone of illumination) and K is a constant of the order of unity, the exact value of which is determined by the physical nature of the illuminating beam. The quantity ($n \sin \alpha$) is the *numerical aperture* and is clearly a measure of the resolving power of the lens. Optimal resolving powers, from Eq. (1), correspond to values of d_{min} slightly greater than 0.1 μ. (Note that this value refers strictly to discerning a separation of points; the light microscope may succeed in detecting particles somewhat less than 0.1 μ in diameter, or in resolving smaller crystal spacings.)

Contrast is a term which refers to differences in intensity between images of specimen points and background. In the ordinary light microscope, differential absorption is the primary source of contrast, although differential scattering at surfaces is important also. Modifications of the instrument (see "Special Types of Light Microscope" below) have the effect of introducing or enhancing certain contrast mechanisms. Since differences in light scattering produce more contrast when the lens aperture is small, contrast unfortunately tends to deteriorate at the high numerical apertures which favor resolving power.

The terms "resolving power" and "resolution" are often confused. The former refers to the theoretical capacity of an instrument, whereas the latter refers to the extent of detail actually obtained in the image of a given specimen. Paradoxically, reduction of microscope resolving power, by stopping down the lens aperture, can actually result in improved resolution under appropriate circumstances because of the improved contrast so obtained.

Lens Aberrations. Apart from limitations of resolving power, lenses are subject to a series of defects which can largely but not totally be corrected by skillful microscope design. These aberrations are an inherent property of lenses, not to be confused with such purely technical problems as uneven grinding of lens surfaces or inhomogeneities of optical glasses. The defects will be discussed very briefly here.

1. Spherical aberration results from the fact that refraction by the outer (marginal) portions of a spherical lens surface focuses light more powerfully than that by the inner (axial) region, as suggested (to an exaggerated degree) by Fig. 2A. Consequently, light from point objects is spread out into extended discs in the image plane producing a softness in image definition.

2. Chromatic aberration results from the fact that light of different wavelengths is refracted to different extents by any single optical glass. The effect of longitudinal chromatic aberration, shown in Fig. 2B, is obviously analogous to spherical aberration, except that the light is of different colors in different zones of the extended image. Lateral chromatic aberration, shown in Fig. 2C, results in the formation of images of different magnifications by light of different wavelengths. (The latter effect, which also has its analogy in spherical aberration, accompanies the longitudinal aberration in images formed by uncorrected lenses, but tends to persist after correction of the other effect.)

3. Curvature of field, as its name implies, produces a curved focal surface. The image cannot be made to appear simultaneously in exact focus at any microscope setting. During visual observation, the defect can be substantially overcome by continuous fine focusing, but curvature of field creates a serious problem when photomicrographs are to be recorded.

4. Distortion produces a greater or lesser magnification at the edges of the field of view than at the axis, as shown in

Fig. 2D and E. The effect, which is associated with lens systems of limited aperture, is particularly undesirable in recording or measuring extended areas of a specimen.

5. Coma, the imaging of off-axis points as a comet-shaped blur, is due to differences in magnification by different zones of the lens suface. Depending upon lens shape, magnification by peripheral zones may be either greater or less than that at the lens axis.

6. Astigmatism is the imaging of off-axis points as two separate mutually perpendicular line images which are formed at different distances from the lens.

Illumination: The Condenser Lens System. The illuminating system of a microscope should provide even and adequately intense lighting of the specimen. Since image brightness is dependent both upon the aperture angle of illumination and upon the area observed, it is evident that

$$\text{Image brightness} = K \left[\frac{N.A._{\text{obj}}}{M_{\text{total}}} \right]^2 \tag{2}$$

where $N.A._{\text{obj}}$ is the numerical aperture of the objective, M_{total} is the overall magnification, and K is a constant characteristic of the microscope and its light source. Since the N.A. is limited to values not much greater than 1.4, very intense illumination is required at high magnifications.

The widely used Abbe condenser, shown in Fig. 3A, provides an N.A. of 1.30. A shortcoming of this lens is that spherical and chromatic aberrations affect focusing by the two uncorrected lens elements. These defects are corrected, and a higher N.A. (1.40) achieved in multi-element "achromatic" condensers, an example of which is diagrammed in Fig. 3B. When extended fields of view are to be observed, "variable focus" condensers are suitable.

Apart from its function of providing adequate illumination, the condenser system has important if more subtle effects upon the resolution and/or contrast achieved by the microscope. It is through skilled use of the illuminating system that the microscopist can fully exploit the resolving power inherent in lens and microscope design. The two systems which have been recommended for high-resolution microscopy are described below.

"Critical" illumination focuses an image of the light source in the plane of the specimen, as shown in Fig. 4A. This method was believed in fact to be critical for optimum resolution at a time when extended but not highly brilliant kerosine lamps were used. More recently, however, the method has been shown to possess no theoretical advantages over the method of Köhler illumination (Fig. 4B)(1). In that method, the lamp lens forms an image of the source at the condenser substage diaphragm; i.e. at the first focal plane of the condenser. That lens collimates

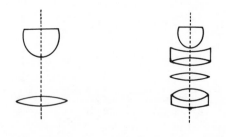

A. ABBE CONDENSER   B. ACHROMATIC CONDENSER

Fig. 3 Types of condenser lens.

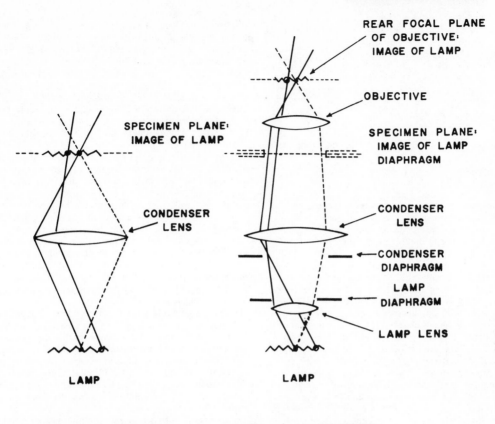

Fig. 4 Illumination systems for high-resolution microscopy.

the beam, so that each specimen point receives light from an extended area on the source. Local variations of source intensity, which produce unevenness in critical illumination, are therefore without influence on Köhler illumination. Furthermore, since an image of the lamp (field) diaphragm is formed in the plane of the specimen, there is direct control of the size of the illuminated area in the latter method. Focused imaging of the diaphragm is also helpful in establishing that a condition of Köhler illumination exists.

The Objective Lens. The objective effects the first stage of imaging, and any defects of the intermediate image must subsequently be enlarged by the ocular. Consequently the objective is the most critical optical element of the microscope, and is the lens system which must be most extensively corrected for aberrations. In general, corrections are applied to a degree which renders N.A., rather than residual aberrations, the factor which determines the resolving power of a microscope.

Objectives are characterized by the following parameters: (a) N.A., (b) magnification, (c) focal length (the distance from the lens at which parallel light rays are brought to a point focus. The focal length is thus a measure of lens strength; the shorter the focal length, the stronger the lens), (d) the working distance (the distance from the top of the specimen coverslip to the surface of the objective when the microscope is in focus.) Some typical values of these quantities are given in Table 1.

Table 1   Characteristics of Typical Objective Lenses

Type	Magnification	f	N.A.	Working Distance
Low power	4×	40 mm	.10	15 mm
Medium power	10×	17 mm	.25	5 mm
High dry	40×	4 mm	.70	.4 mm
Oil immersion	100×	1.8 mm	1.30	.1 mm

Highest resolution requires the use of oil immersion lenses of maximum numerical aperture. In the oil immersion method, the space between specimen and objective is filled with a material of high refractive index, increasing the value of the quantity $(n \sin \alpha) = N.A.$ As shown in Fig. 5A, there is an air space between coverslip and lens in the absence of immersion oil. Ray 1 from a specimen point is refracted away from the normal to the coverslip as it enters the air space, but nevertheless enters the aperture of the objective. Ray 2 is similarly refracted, but at a larger angle, and passes outside the lens. Ray 3 meets the coverslip-to-air interface at an angle greater than the critical angle for total internal reflection, and is also lost to the system. Figure 5B shows the course of the same rays when the space between coverslip and lens is filled with a medium of the same index as that of the

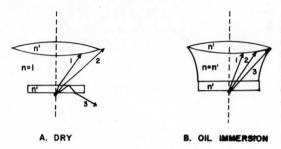

A. DRY   **B. OIL IMMERSION**

Fig. 5   The oil immersion principle.

coverslip glass. All three rays now enter the lens without undergoing previous refractions. The angular size of the cone of light from the specimen which reaches the lens is thus increased substantially by oil immersion.

Ideally, the refractive index of the oil should exactly match that of the coverslip and lens ("condition of homogeneous immersion"). In recent years, however, some designers have found it advantageous to depart somewhat from this condition. In either case, it is clear that use of a "dry" immersion lens must result in serious deterioration of image quality. The ideal immersion oil is transparent, chemically inert, and free from tendency to spread or creep. Its optical properties must be highly stable, and the material should not harden on exposure to air. Cedarwood oil ($n = 1.52$) has been widely used.

Optimal resolution imaging is dependent on the use of a coverslip of appropriate thickness and refractive index. Usually, objectives are designed for use with coverslips 0.18 mm thick, whereas actual thicknesses vary between 0.15 and 0.22 mm. As shown in Fig. 6, the cover glass has the effect of displacing rays from each specimen point, thereby introducing aberrations which must be accounted for in designing the objective. Some objectives can be adjusted to compensate for differences in coverslip thickness by means of a "correction collar"; with other lenses coverslips of correct thickness must be selected for high-resolution work. Variations in the refractive index of coverslips are usually not sufficient to affect image quality.

Correction of spherical aberration in objective and other lenses is based upon suitably designed doublet systems in which a positive (converging) lens is combined with a negative (diverging) lens of different index. The members of the combination produce aberrations in opposite senses which may nullify each other. While the focusing power of the combination is less than that of one lens singly, a net focusing power can be retained. With a

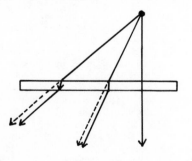

Fig. 6   Deviation of rays by refraction at coverslip.

two-element system, the images formed by any two selected wavelengths can be made to coincide, but a surrounding halo of intermediate colors remains. Chromatic aberration can be corrected at a third wavelength by the use of a third lens element. A lens which corrects the (axial) chromatic aberration for two colors only is called an *achromat*, while *apochromats* correct for three colors and in addition, are more highly corrected for spherical aberration. Commercially, lenses described as "achromatic," "semi-apochromatic," and "apochromatic" are available.

Perhaps the most important innovation in lens design in recent years has been the development of high-quality *flat-field* lenses, which largely eliminate curvature of field without sacrificing correction of spherical and chromatic aberrations. These lenses are particularly important on account of increased exploitation of photomicrographic techniques. (Note that the flat-field lenses are not the same as "aplanats"; the latter is a term applied to systems simultaneously free of spherical aberration and coma.)

Quality objectives designed for use with any given microscope are *parfocal* and *parcentric*; i.e., upon exchange of objectives, the field of view remains centered without shift of focus.

The real intermediate image formed by the objective lens can be viewed as a concentration of light within the microscope tube. Special telescopic eyepieces are made for convenient viewing of this image.

The Ocular Lens. The ocular (eyepiece) lens serves to magnify the intermediate image formed by the objective, and also to correct for any residual aberrations present in that image. Ocular magnifications usually range from $5\times$ to $20\times$.

In the case of viewing by eye, the intermediate image lies at or near the first focal point of the ocular, so that a virtual image of the specimen is formed at some distance below the lens, as shown in Fig. 7. (Note that when a virtual image is formed, the eye interprets the rays it receives as having traveled in straight lines from the image plane, and is unaffected by the fact that there is no physical concentration of light at that position.) Ultimately, the rays leaving the microscope are focused by the eye, forming a real image on the retina. In photomicrography, the focusing elements of the eye are replaced by a camera lens which forms a real image on the photographic emulsion. Photomicrography can be carried out simply by placing a suitably focused camera above the microscope tube or, more conveniently, by using a special ocular which combines the functions of eyepiece and camera lens.

Oculars are designed to complete the correction of aberrations begun at the objective. It is therefore important to use an ocular corrected to the same level as the objective; i.e., achromatic, apochromatic etc. Many objectives are deliberately overcorrected with respect to chromatic aberration; i.e., they introduce an aberration of the opposite sense to that which would be produced by a single lens. The overcorrection is then balanced by a "compensating ocular." For ultimate resolution it is necessary to make use of objective-ocular pairs which are specifically designed to complement each other. In practice, however, the corrections effected by different high-quality lenses are so similar that very nearly optimal resolution can be obtained without attention to this more stringent condition.

The Huygenian is the basic and still widely used type of ocular lens, consisting of two simple lens elements. Field

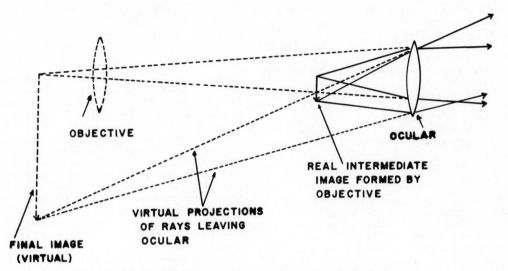

OBJECTIVE

OCULAR

REAL INTERMEDIATE
IMAGE FORMED BY
OBJECTIVE

VIRTUAL PROJECTIONS
OF RAYS LEAVING
OCULAR

FINAL IMAGE
(VIRTUAL)

Fig. 7 Imaging by the microscope ocular (visual observation).

of view and eye relief are improved by inclusion of additional lens elements in the "wide field" ocular, while "hyperplane" and "compensating" oculars contain lens elements added for the purpose of correcting aberrations.

Parameters which characterize ocular lenses are: (a) magnification, (b) type of correction, (c) diameter of field of view, and (d) eye relief (the distance between lens and eye during observation). Characteristics of some typical ocular lenses, as used in conjuction with a 40× objective, are listed in Table 2.

Table 2

Type	Magnification	Eye Relief	Diameter of Field of View
Huygenian	5×	8 mm	0.5 mm
Huygenian	10×	8 mm	0.4 mm
Wide field	10×	21 mm	0.5 mm
Wide field	15×	14 mm	0.4 mm
Wide field	20×	10 mm	0.3 mm

Near or farsightedness of the microscopist can be compensated by the focal setting of a microscope, but astigmatic vision requires the use of eyeglasses (note, however, that "astigmatism" of the eye results from asymmetry and is not analogous to the astigmatism of lenses mentioned previously). At lower powers, special "high point" oculars are available for use with glasses, but at high magnifications the only adequate provision for astigmatic vision is the use of custom-designed corrected oculars.

Both monocular and binocular eyepiece lenses are used; the latter are more comfortable for prolonged viewing. Binoculars use a prism and mirrors for even division of light intensity between the two lenses. Other modified oculars in which the light beam is divided by a prism include eyepieces for simultaneous direct observation and photomicrography, and for display of the final image on a relatively large viewing screen. Most oculars also make provision for insertion of a measuring reticule.

Lens Testing. Precision testing of optical components requires specialized equipment, and must be carried out by the manufacturer. However, a good indication of lens quality can be obtained quite simply with appropriate test specimens (2).

Resolution can be tested by observing structures, such as *Pleurosigma angulatum* and other species of diatoms, in which regular fine spacings are known to be present.

An "Abbe test plate" can be used to estimate adequacy of correction for spherical and chromatic aberrations. This device is a slide ruled with several series of fine parallel lines, which are first focused upon under direct illumination. As the lighting is shifted to an oblique angle, no deterioration in image sharpness should be detected if the lens is corrected for spherical aberration. If chromatic aberration is present, colored fringes appear at the edges of the lines under oblique illumination. A good achromat produces narrow fringes only, while fringes should be entirely undetectable with an apochromat. The Abbe test plate also incorporates a calibrated wedge-shaped cover glass which allows for determination of the optimum optical thickness of the specimen coverslip.

An approximate measure of curvature of field can be obtained by observing any thin flat specimen, such as an object micrometer. The difference in exact focal setting for a point at the center of the field and for a peripheral point can be read from the fine focus control of the microscope. For meaningful comparison of objectives producing different magnifications, the peripheral point chosen must be the same distance from the axis in each case (i.e., the distance in object space must be constant).

Depths of Field of Focus. Ideally, a single plane of a specimen would form an image in focus at a single plane only. Actually, however, each image point is of finite extent in three dimensions, so that the image is observed to be "in focus" throughout a finite range. The extent of this range, measured along the microscope axis, is the *depth of focus*. Correspondingly, a finite depth within the specimen appears in focus at any chosen image plane; the depth of this region, in the specimen, is the *depth of field*. While depths of field and focus are large for lenses of small angular aperture, high-magnification images formed by objectives of large N.A. correspond approximately to a single specimen plane, and are said to constitute an "optical section." The limited depth of field obtained at high magnifications requires continuous adjustment of fine focus in order to observe the full depth even of very thin sections.

Depths of field for any lens may be computed by calculating the distance over which rays from a point subtend a disc of diameter no greater than the minimum resolvable distance. In this way, depths of both field and focus are computed to be given by (3):

$$D = \frac{K\lambda \cos \alpha}{n \sin^2 \alpha} = \frac{\lambda \sqrt{n^2 - (N.A.)^2}}{(N.A.)^2} \qquad (3)$$

where K is a constant of the order of unity, as given in Eq. (1), n is the refractive index of the medium surrounding the specimen, λ is the wavelength (usually taken as an average value of about 550 mμ for white light), α is the aperture angle of the lens, and $N.A.$ its numerical aperture.

Values of D are of the order of 10 μ for low-power objectives ($N.A. = 0.25$), 1 μ for high dry lenses ($N.A. = 0.65$) and 0.2 μ for oil immersion objectives ($N.A. = 1.3$). These values apply strictly for photomicrography, but are increased in visual observation to the extent to which accommodation of the eye occurs.

Magnification, Calibration, and Measurement. The magnification of a compound microscope is the product of magnifications by objective and ocular. In general, magnification by any lens is:

$$M = \frac{\text{image distance}}{\text{object distance}} \qquad (4)$$

In the case of the objective, (4) becomes, specifically:

$$m_{obj} = \frac{\text{mechanical tube length}}{f_{obj}} \qquad (5)$$

Assuming that the intermediate image lies very close to the focal point of the ocular, and that the final image is formed at 25 cm (the distance of most comfortable vision for the average observer), the magnification of the ocular is

$$m_{oe} = 25/f_{oc} \qquad (6)$$

The total magnification for a standard mechanical tube length setting of 16.0 cm is thus:

$$M_{total} = m_{obj}m_{oe} = \frac{-16 \times 25}{f_{oc}f_{obj}} = -\frac{400}{f_{obj}f_{oc}} \qquad (7)$$

The negative sign in (7) refers to the fact that the image is inverted by the objective; the inversion is preserved by the ocular.

A desirable level of magnification is one which renders detail resolved by the objective comfortably visible to the eye. Magnification in excess of that level is termed "empty," since no further detail is revealed. At excessive magnifications, the image appears "soft," and may contain diffraction effects which are not directly related to the structure of the specimen. With normal vision, 0.1 mm spacings can be resolved, but viewing of 0.2 mm is somewhat more comfortable. Under oil immersion, ultimate resolution is approximately 0.15–0.2 μ. Maximum useful magnification in light microscopy is thus:

$$M_{max} = \frac{\text{size of spacing resolved by eye}}{\text{size of spacing resolved by objective}} \qquad (8)$$
$$= \frac{1 - 2 \times 10^{-2} \text{ cm}}{1.5 \times 10^{-5} \text{ cm}} \cong 1300 \times$$

Similarly, maximum magnification for lower-powered objectives should also be about 1000 N.A. For example, M_{max} for a 10× objective, $N.A. = 0.25$, would be 250×, requiring the use of a 25× ocular.

Frequently the light microscopist is unconcerned by the precise numerical level of magnification obtained, since the order of magnitude of object dimensions is obvious or at least well known. The instrument is nevertheless suited to exact measurements also. A ruled graticule can be inserted into the ocular at such a level that its in-focus image is superimposed upon that of the specimen. The graticule is then calibrated with respect to a ruled slide placed on the microscope stage. (Calibration slides, ruled at 0.01 mm intervals, are obtainable commercially.) If the images of spacings on graticule and calibration slide are similar in size, the mechanical tube length of the microscope can be adjusted to bring the two into coincidence. Otherwise, a numerical factor is determined from which conversion of graticule readings can be made.

Accuracy of microscope measurements can be improved by use of an image-splitting ocular (4). In this device, twin prisms form images which may either superimpose or be moved some distance apart, according to the mutual rotation of the prisms. Rotation can be calibrated so that object dimensions are read off at the setting for which the two images just touch. The advantage of the system is that contact between the images can be detected with greater accuracy than can the location of a graticule line on the image.

Special Types of Light Microscope

The design of the compound microscope can be altered in a number of ways which greatly increase contrast in images of appropriate specimens. Specialized instruments are discussed at length elsewhere in this volume, but will be summarized briefly here.

Dark field microscopy is a method used for viewing specimens of high scattering power, usually suspensions of small particles. The material is illuminated obliquely, with the direct transmitted beam falling outside the aperture of the objective lens. Only light scattered by the specimen can then enter the microscope, and the resulting image is bright against a dark background. The difference in light intensity between specimen and background is then much greater than would be the case in normal bright field microscopy, where light scattering reduces image point intensity only slightly with respect to background. As shown schematically in Fig. 8, the specimen is illuminated by an annulus of light which falls outside the objective. An annular aperture may be placed below the condenser for this purpose, or else a special dark field condenser may be used. Immersion oil contact is maintained between condenser and specimen slide in order to minimize extraneous scattering of light at interfaces.

It is sometimes possible to detect structures of dimensions less than d_{min} in dark field images, e.g., intensely illuminated polystyrene spheres of 1000 Å (0.1 μ) diameter. However, since two or more adjacent particles of this size cannot be distinguished as separate, such objects cannot be said to be "resolved."

TO IMAGE

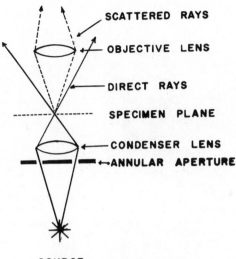

SCATTERED RAYS

OBJECTIVE LENS

DIRECT RAYS

SPECIMEN PLANE

CONDENSER LENS

ANNULAR APERTURE

SOURCE

Fig. 8 Dark field microscopy.

Ultraviolet microscopy is used for the observation of structures which absorb strongly in the wavelength range 250–400 mμ, including, principally, cellular nucleic acids and proteins. Despite an approximately twofold improvement in resolving power in the UV (cf. Eq. (1)), the principal advantage of the method is the high inherent contrast of these important biological structures.

Ordinary optical glasses are opaque to wavelengths shorter than about 3100 Å, so quartz optical parts must be used for UV microscopy. Since the dispersive power of optical materials is much greater in the ultraviolet than in the visible range, it was necessary, until recently, to use monochromatic light for UV microscopy. "Monochromat" lenses were employed which, since they required no correction for chromatic aberrations, could be highly corrected in other respects. In recent years, however, improvement of optical materials and design has resulted in the development of lenses which can be used at all wavelengths and which, it is claimed, perform as well in all respects as the older monochromat lenses (5).

The UV microscope image can be detected on a fluorescent screen and recorded photographically. Precautions must of course be taken against damage to the eye by UV radiation.

Fluorescence microscopy is used for the observation of structures which either are inherently fluorescent or, more commonly, have an affinity for a specific fluorescent dye. The specimen is illuminated by an appropriate UV wavelength, the "exciting radiation." A filter subsequently removes this wavelength, and fluorescent structures appear as characteristically colored images against a dark background.

Requirements for fluorescence microscopy include a highly intense source (since the efficiency of fluorescence is often low), optical parts transparent to the wavelengths employed, and the use of slides, coverslips, and reagents free from traces of fluorescing materials.

Phase Contrast and Interference Microscopes. Unstained biological objects normally produce minimal contrast in the ordinary light microscope since they do not appreciably absorb or scatter light. Such specimens can be rendered visible by exploiting differences in optical path (i.e., of the product; refractive index × thickness) with respect to background in order to produce contrast in the image. This is achieved in the phase contrast and interference microscopes. The phase contrast microscope is an essentially qualitative instrument which makes use of interference between diffracted and undiffracted light from the specimen. In the interference microscope, which is essentially a quantitative tool, interference between specimen and reference beams produces a series of fringes across the field of view. These fringes are displaced by the specimen in proportion to its optical thickness. With careful adjustment of the instrument, either the refractive index or the thickness of minute quantities of material can be measured precisely.

The polarizing microscope is used for the observation of birefringent specimens; i.e., for specimens in which refractive index varies as a function of the direction of transmission of light. The degree of birefringence can be measured quantitatively.

Acknowledgement

I wish to thank American Optical, Bausch & Lomb, Metropolitan-Vickers, Eric Sobotka, and Carl Zeiss companies for making technical literature available to me, and Dr. B. G. Uzman for reviewing the manuscript.

ELIZABETH M. SLAYTER

References

1. M. Berek, "Zur Theorie der Abbildung im Mikroskop," parts 1–8, *Optik,* **1**:475 (1946); **3**:289 (1948); **4**:377 (1949); **5**:1, 144, 329 (1949); **6**:1, 219 (1950).
2. "Is This a Good Microscope Objective? Advice on Ascertaining the Quality of Microscope Objectives," Eric Sobotka Co. Inc., 110 Finn Court, Farmingdale, N.Y.
3. J. R. Benford, "The Theory of the Microscope," Rochester, New York, Bausch & Lomb, 1965 (p. 6).
4. J. Dyson, *J. Opt. Soc. Amer.,* **50**:754 (1960) "Precise measurement by image splitting"; Vickers bulletin "A.E.I. Image Splitting Eyepiece."
5. Carl Zeiss Co. "Optical Systems for the Microscope," 1967 (p. 27).

A number of commonly cited works on microscopy are difficult to obtain or are not available in English. The following general references dealing with aspects of microscopy have either been published recently in the U.S.A. or are available from the respective instrument manufacturers:

Bausch & Lomb Co. "Glossary of Optical Terms" (21 pp.).

J. R. Benford, "The Theory of the Microscope," Rochester, New York, Bausch & Lomb Co., 1965 (24 pp.).

L. C. Martin, "The Theory of the Microscope," New York, American Elsevier, 1966 (488 pp.).

F. K. Möllring, "Microscopy from the Very Beginning," New York, Carl Zeiss Co. (64 pp.).

O. W. Richards, "The Effective Use and Proper Care of the Microscope," Buffalo, N.Y., American Optical Co., 1958 (63 pp.).

E. M. Slayter, "Optical Methods in Biology," New York, John Wiley, 1970 (Ch. 8–12).

See also: DISSECTING MICROSCOPES, HISTORY OF MICROSCOPE METALLOGRAPHIC MICROSCOPE, POLARIZATION MICROSCOPE, RESOLUTION, ULTRA-VIOLET MICROSCOPE, ZOOM MICROSCOPE.

OSMIUM STAINING

Over the past decade the use of osmium staining in biology has focused upon enzyme cytochemical techniques for the electron microscope. The selection of osmium compounds as staining reagents for tissue and cellular components is based upon unique properties of opacity to the electron beam, high density, amorphous character, and insolubility. No other element combines a unique molecular image imprint in the electron micrograph with characteristic reactivity as well as osmium (1). This element incorporated into specific reactive moieties allows the essential demonstration of enzymatic activity and functional groups of tissue components delineating *specificity*, *localization*, and *chemical identity of reacting groups*.

The study of the chemical reactions of osmium compounds utilized in cytochemical methods is sparsely begun (11). This is due to a number of reasons. Certainly the fact that many of these reactions with *tissue* and *cellular* substances were unknown a decade ago points up that this is a frontier science. Perhaps even more cogent is that enzyme cytochemistry is an interdisciplinary study. It borrows much from histology and cytology while its emphasis is the biochemical machinery of the living cell and its ultimate goals are the correlation of biochemical activity, morphology, and function. Yet, without the physics of the electron microscope and chemistry of the specifically designed osmium compounds the science could not move ahead. Another impediment to clarity has been the lack of distinction between the characteristics of osmium chemical reactions in cytochemical work as opposed to the osmium chemistry of organic reactions.

In all tissue staining for enzymes the chemical reactions are mild and must be performed under rigorous incubator conditions often narrowly limited by optimum temperature of enzyme activity and denaturation effects. Osmium reactions with tissues are usually performed in the cold or no higher than 55°C, whereas reactions performed with organic materials under a wide variety of conditions may not be at all applicable to cytochemical techniques. Thus, a relatively small body of literature exists characterizing the osmium reactions and reagents common to enzyme cytochemical reactions and products.

Historically, osmium tetroxide was used to stain for fat, degenerating myelin, and later as a fixative for electron microscopy (18, 19, 20). Osmium staining in cytochemical reactions was initiated analytically with the work of Criegee (5, 6). Classically, the reaction that Criegee studied, that of the alkene linkage with osmium tetroxide, has been the dominant reaction of osmium with tissue and cellular substances. As studied by Criegee, the reaction of an olefin with osmium tetroxide in an inert solvent (ether or dioxane) occurs at room temperature over a period of several days. Since these conditions are mild they have been generally interpreted to be comparable to the conditions of cytochemical reactions. Under the above conditions a black osmic ester is formed and can be isolated and characterized either by precipitation from the reaction mixture or by evaporation of the solvent. The reaction may be represented as follows:

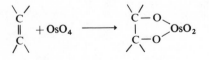

The isolation of the above osmic ester is actually an intermediate step in the overall reaction of an alkene with osmium tetroxide. The overall reaction or the completion of the reaction if water is present is the hydrolysis of the above osmic ester which yields the final product a 1,2 glycol. It must be emphasized that the reaction of many different kinds of alkenes with osmium tetroxide goes to completion as a 1,2 glycol. The osmic ester intermediate only serves to identify one stable intermediate state. The fact that the reaction occurs in a reducing medium pinpoints the fact that this is an oxidation-reduction reaction. However, in describing this reaction authors write the product of OsO_4 in different ways. A survey indicates that products listed range through OsO_3, OsO_2, OsO, and Os. To clarify this point, it is useful to turn to the organic chemist and consider the characteristics of osmium in this reaction as defined by Royals (22). He points out that all alkene linkage plus osmium tetroxide can react to completion to yield a 1,2 glycol and osmium (zero state); and that this osmium may be recovered and reoxidized to osmium tetroxide stoichiometrically. He also points out that the identical reaction can be obtained (i.e., hydroxylation of the alkene linkage) by using osmium tetroxide as a catalyst rather than in stoichiometric quantities. It is postulated that the OsO_4 is still forming the osmic ester which is then hydrolyzed continuously to form the 1,2 glycol and then reoxidized in the medium to again start the cycle. Although this mechanism has never been proved as such, other oxides (e.g., vanadium pentoxide and selenium dioxide) can perform the same function as catalysts lending credence to this point of view.

Let us now summarize our understanding of Criegee's work:

1. The reaction of *all* alkenes with OsO_4 *when allowed to go to completion* is:

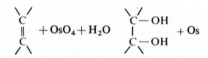

2. *In some* reactions, when water is carefully excluded, the osmic ester can be identified and characterized:

3. It is postulated that the same above osmic ester intermediate is the mechanism by which OsO_4 acts as a catalyst with other oxidizing agents. Further evidence that this is so is that other metal oxides (V_2O_5, SeO_2) with similar chemical properties act also as a catalyst for the same reaction.

4. Finally, in conclusion we may infer from the above evidence that the reduction of OsO_4 in reaction with an alkene most probably proceeds through the osmic ester where Os (+6) has been identified and characterized. Other than Os (zero valence state) and Os (+6) no other valence states were identified prior to this time although many have been postulated. It is reasonable to assume from this evidence alone that in the reduction of osmium (+8) in OsO_4 osmium passes through all of the oxidation states. The question is, which of these states is sufficiently

stable so that compounds can be formed and isolated, particularly with carbon. The work from Criegee is clear cut in that regard: OsO_4 ($+8$) ⟶ Os ($+6$) ⟶ Os (zero) if OsO_4 is not used in excess.

Of more recent vintage, the most substantive contribution to our knowledge of what occurs chemically in the osmium tetroxide staining of lipid tissue components is the work of Korn outlined in his review (12). Again, the primary reaction is osmium tetroxide reacting with an alkene in an oxidation-reduction type reaction. Korn was able to show that under certain conditions dimers and polymers could be formed of osmic esters. Thus, starting with tetramethyl dipotassium osmate a dimer could be synthesized:

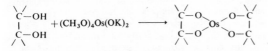

He then proceeded to produce the same product from the reaction from osmium tetroxide using a specific alkene under specific conditions in which the alkene was present in excess and water was excluded. Thus, he was able to isolate an Os ($+4$) intermediate osmic ester. However, it must be pointed out that these conditions do not exist when osmium tetroxide is used as a fixative of tissue or is used for cytochemical staining. It probably does not represent a stable compound from the more general reactions of OsO_4 with alkenes. This work confirms the previous view that osmium compounds proceed through several oxidation states and that intermediates can be isolated. It is noted that only the Os (zero) is completely stable for *all* reduction products of alkenes although this zero state may not be reached as osmium tetroxide is used biologically. In summary, Korn (a) has extended the osmium tetroxide + alkene reaction to include isolation of Os ($+4$) intermediates; (b) as far as the reduced form of osmium isolated in his reactions, he states: "In the model reactions, an approximately equal amount of osmium was recovered as uncharacterized products, presumably lower oxides" (13). However, it is more likely that the reaction did not go to completion and that there was a mixture of incompletely oxidized osmium esters which are more likely to be osmium-containing polymers than osmium oxides (10, 11).

It is of further interest to consider products of excess OsO_4 with cellular and tissue components particularly with regard to the final form in which osmium appears. In cytochemical terms, this is often described as the formation of "osmium black," a characteristic color common to most osmium tetroxide staining reactions with cellular and tissue components as well as products of enzyme reactions. The term "osmium black" refers more to the appearance of a generic class of substances than any specific compound resulting from reactions of osmium tetroxide with tissues or products of enzymatic reactions. Although various types of brown reduction products are also found from these reactions, most of the resulting substances are black in color, hence the term "osmium black" (10, 11). Since all products of osmium, including the oxides below valence state ($+8$), are black or brown, it indicates to the biologist that the reaction has occurred. Early views on the nature of osmium black often refer to this product either as finely divided osmium metal formed by the reduction of OsO_4 in solution or an incompletely reduced $OsO_2 \cdot nH_2O$ by product. In recent studies the term osmium black is used by Seligman and

his associates to refer to the end products for the identification of specific enzyme systems, and they have characterized these substances in some detail (10, 11). Some of the characteristic properties of "osmium black" such as its amorphous, noncrystalline state, insolubility in tissue constituents and nonconductivity point to a nonmetallic or nonosmium metal interpretation (10, 11). However, other properties such as insolubility in organic solvents used for dehydrating tissue, insolubility in acrylic and epoxy monomers used in preparation of sections for electron microscopy, formation of colloidal suspensions in pyridine, and variation in properties depending upon the sulfur content of the reaction products did not offer any alternative explanation. It was not until elemental and infrared analyses showed that "osmium black" consisted of coordination compounds of osmium with organic sulfur ligands, that an approach to the problem was first achieved. However, the concept of "osmium black" as a coordination polymer was not new, but strong evidence for that point of view had been lacking. Many investigations with the products of OsO_4 with thiosemicarbazide, thiocarbohydrazide, thiophenol, p-chlorothiopenol and 2-naphthalenethiol yielded insolubility of resultant black materials (2, 9, 11, 21, 24, 25, 27, 29, 32). Thus, it was difficult to determine molecular weights of these "osmium black" substances by colligative methods. A breakthrough was achieved by Seligman and his associates (10) by the preparation of a *water soluble* "osmium black" by use of a hydrophilic ligand, (3-mercapto-1,2-propanediol). The molecular weight of 2555 in water was determined by a vapor pressure osmiometric measurement. Elemental analysis indicated a "tetramer"-like formula. (Calculated m.w. = 2475, empirical = 2555.) Osmometric molecular weight measurements with a blacker product obtained by varying the water content in the above reaction showed a molecular weight of 3040 (cf. theoretical: 3093) for formation of the pentamer.

Verification that "osmium blacks" were acting as bridging agents between osmium stain and tissue components was demonstrated by treating an aqueous solution of the pentamer with excess OsO_4. In so doing a further black precipitate was obtained, insoluble in organic or aqueous solvents. It is to be noted that NMR spectra studies of the tetramer and pentamer above confirm the point of view that they are in true aqueous solution. Further evidence to this unusual solubility effect of this particular "osmium black" in water was obtained by intravenous injection of this dark solution made isotonic with sucrose into each of several albino mice. The dark brown solution colored the blood, skin areas, and eyes (those areas rich in blood vessels) of these animals. Within 45 min these areas returned to normal color and dark material was found in the urine. This is additional biological evidence that the osmium-containing polymer can traverse the glomerular membrane and is in true solution. No acute toxicity was noted (10).

As we have seen above, the chemistry of osmium compounds is quite involved, and our interest here is to understand only what happens when OsO_4 in excess reacts with tissue components. What are the *possible* products of these reactions? The answer is simple: They cannot be products that are not stable. In order to exist they must be formed from osmium (in any form) through some known chemical reaction. As a very simple example, osmium in the $+5$ or $+7$ state is extremely rare where only a few known fluoride compounds have ever been isolated or identified. Thus, to postulate OsO_4 + tissue compounds ──────→ Os (+5 or +7) oxides would be contrary to the known facts of osmium chemistry. Let us now extend this reasoning further.

Sjöstrand (35) points out that OsO_4 (and equally true for all osmium compounds) reacts by either oxidation-reduction reactions or by "complex" formation in which there is no change in oxidation state. Since OsO_4 is a powerful oxidizing agent most of its reactions center about the oxidation-reduction products. Rarely does one find coordination or complex formation of interest to OsO_4 in cytochemistry. This may occur in part with protein or fixation of tissue with osmium tetroxide and probably there is no significant color change when the valence state is not changed. To what extent protein-bound osmium (+8) is washed out in embedding procedures is not known. Suggestive evidence of some removal by thiocarbohydrazide wash was shown in the matrix of mitochondria in the OTO staining reaction (32). Furthermore, it is known that choline and lecithin, as two examples, do form coordination compounds with OsO_4 such that there is no change in valence state. This is the exception rather than the rule, and we shall confine ourselves here *only* to oxidation-reduction products.

The first question that must be asked is what osmium compounds are known to form stable bonds with carbon compounds. Or, putting the question a different way, if osmium tetroxide reacts with an organic compound, the osmium must be reduced and either forms a stable carbon-osmium compound or forms one or more of the stable osmium oxides. Let us examine these two classes of compounds.

Griffith in his review of osmium and its compounds points out that "there is some doubt about the existence of (OsO_3) osmium (+6) oxide" (8). The only reference to it is to Criegee's original work, where it is a by-product of the decomposition of the cyclic osmium esters. Current thinking is that this is a transient hydrated trioxide since no one else has been able to synthesize a pure OsO_3 species similar to it. The stable dioxide (OsO_2) is normally made by heating osmium metal in a stream of nitric oxide. Osmium disulfide can also be made from its elements at red heat. Nowhere in the literature is any record of these compounds being identified from oxidation-reduction products of osmium tetroxide plus an organic compound. Quoting Griffith's review article: "Nitrogen and carbon compounds are somewhat rare for this (+4) oxidation state of the element." On the other hand, a large number of bivalent osmium-carbon compounds are known. It is generally thought that "carbon compounds and good π acceptors stabilize the low (+2, 0) states while oxygen and halides stabilize the high (+8, +6) states. The intermediate states (+4, +3) are stabilized by σ (sigma) bonds with NH_3 or some halides" (8). Therefore, one may conclude that if the products of OsO_4 + organic compounds are osmium-bound organic material, then osmium must be in the +2 or zero state in these substances. If polymers are formed as with sulfhydryl-containing compounds then the +2 state is the more common form to expect.

Let us take some specific examples:

1. Osmium tetroxide (OsO_4) reacts with thiourea to yield $[Os (thiourea)_6]^{+2}$ with osmium in the +2 state (23).

2. Osmium tetroxide (OsO_4) reacts with acetylacetone, 2'2 bipyridyl and 1, 10-phenanthroline to form osmium +2 compounds (3).

3. Treatment of OsO_4 with glycols and Na_2SO_3 yields $Na_4[Os(SO_3)_3] \cdot 6H_2O$ which characterizes osmium in the +2 state (5, 6).

4. Sulfur compounds in general appear to reduce the OsO_4 to +2 Os compounds. Thus $OsO_4 + SO_2$ ──────→ $OsSO_3$ (4).

Other carbon compounds that stabilize the +2 state could be listed. The above four examples should suffice. Let us remember the fact that "almost all osmium compounds are very easily oxidized to the tetroxide" (8). Thus to stabilize lower oxidation states (+4, +6) oxygen and halides are necessary while carbon, sulfur, and nitrogen compounds stabilize +2 and zero states. Of course in the oxidation-reduction process of OsO_4 + organic material, as Criegee and Korn have been able to accomplish, intermediate oxidation states *can* be isolated and identified. However, this is the exception and as such it is quite difficult to find conditions in biological material to allow such states to be stabilized. On the other hand, as both Criegee and Korn have indicated, when allowed to go to completion OsO_4 will go to osmium metal when reducing an alkene. That this does not occur in histological or cytochemical use of OsO_4 is because OsO_4 is always present in great excess.

Finally, no review of contributions of osmium staining to ultrastructural cytochemistry and ultimately to biology would be complete without mention of the *osmiophilic staining principle*. The osmiophilic staining principle consists of taking well-established enzyme cytochemical procedures and altering them by utilizing reagents which contain a group capable of reacting selectively with OsO_4 after the cytochemical reaction has been completed (11). This allows for the demonstration of specific enzymes by the light and electron microscope. Thus, the "osmium black" described earlier is an ultimate product of such enzyme reactions. The preparation of ligands such as thiocarbohydrazide (TCH) that will react with osmium (11, 24) or other metals (37) and the synthesis of osmiophilic reagents that will selectively bind with products from enzyme reactions, is a continuing process early in its development.

Osmiophilic reagents fall into four categories:

1. A product of enzyme-substrate reaction in a subsequent step reacts with osmium tetroxide to yield osmium black. In this case the osmiophilic group or ligand is incorporated into the enzyme reaction product. Upon incubation and reaction with the enzyme the reaction product must be captured or rendered insoluble not only in water but in lipid as well. The best reaction products are therefore large polymers (15, 25). The osmiophilic group or ligand is thereby precisely localized in the tissue. Later it reacts with the osmium tetroxide to form osmium black.

2. A bridging process can also be used. In this case osmium bound in the tissues by fixation in osmium tetroxide can react with an osmiophilic ligand, thiocarbohydrazide (TCH) which can in turn react with additional osmium tetroxide or silver proteinate (37). This bridging process, called the OTO method (32) (osmium bridged via TCH to tissue-bound osmium) (9, 11, 24) results in enhancing the contrast of osmiophilic lipid components of membranes in the electron microscope. Silver proteinate may be used in the last step with equally good results (37). This is shown diagrammatically in reference 9. Some enhancement is also due to removal from tissue, such as mitochondrial matrix of loosely bound osmium tetroxide by the incubation in the solution of TCH and subsequent wash.

3. A recent extension of the osmiophilic principle which has been used to advantage with the oxidases (7, 17, 25, 33) has been the production of an osmiophilic

polymer originating from enzymatic *hydrolysis* of a substrate containing its own diazonium coupling group (15). After enzymatic hydrolysis the product polymerizes. Thus, the *polymer* is expected to localize the specific enzyme involved more accurately than formerly. Whereupon after osmication the very insoluble polymeric deposits appear as black stain in the light microscope and opaque deposits on synaptic membranes of the motor endplate in the electron microscope (15).

4. A further refinement of osmiophilic ligands has been the synthesis of substances which contain *specific* properties to make them lipophobic as well as insoluble in water that will enhance the localization and resolution of enzymes under the electron microscope. The most important advance in this area has been the development of a reagent, diazotized 4-amino-phthalhydrazide, that yields a new *lipophobic osmiophilic* substance after coupling into 2-naphthylamine (34). The electron microscope preparations for amino peptidase utilizing this substance are distinctly membraneous in location in contrast to the droplet distribution obtained with earlier osmiophilic products.

Some examples of methods developed with the osmiophilic principle for the electron microscope are the OTO staining reaction for enhancing membranes (32), the demonstration of horseradish peroxidase (7), cytochrome oxidase (25), dehydrogenases (25, 29), esterase (30), lipase (28), phosphatases (26), amino peptidase (34), acetylcholinesterase (2, 15), macromolecules with adjacent glycol groups (24), osmium-containing acid and basic stains (31), and detection of antigens and antibodies by peroxidase-labeled antibodies and antigens (14, 16, 36).

ROBERT FRIEDENBERG
ARNOLD M. SELIGMAN

References

1. Adams, C. W., *Histochemie,* **9**:68 (1967).
2. Bergman, R. A., H. Ueno, Y. Morizono, J. S. Hanker, and A. M. Seligman, *Histochemie,* **11**:1 (1967).
3. Buckingham, D. A., F. D. Dwyer, H. A. Goodwin, and A. M. Sargeson, *Austral. J. Chem.,* **17**:315, 325 (1964).
4. Claus, C. J., *Pract. Chem.,* **90**:80 (1863).
5. Criegee, R., *Annalen,* **522**:75 (1936).
6. Criegee, R., B. Marchand, and H. Wannowius, *Annalen,* **550**:99 (1942).
7. Graham, R. C., and M. J. Karnovsky, *J. Histochem. Cytochem.,* **14**:291 (1966)
8. Griffith, W. P., *Quart. Rev.,* **254** (1966).
9. Hanker, J. S., C. Deb, H. L. Wasserkrug, and A. M. Seligman, *Science,* **152**:1631 (1966).
10. Hanker, J. S., F. Kasler, M. G. Bloom, J. S. Copeland, and A. M. Seligman, *Science,* **156**:1750 (1967).
11. Hanker, J. S., A. R. Seaman, L. P. Weiss, H. Ueno, R. A. Bergman, and A. M. Seligman, *Science,* **146**:1039 (1964).
12. Korn, E. D., *Science,* **153**:3743 (1966).
13. Korn, E. D., *Science,* **153**:1495 (1966).
14. Mason, T. E., R. F. Phifer, S. S. Spicer, R. A. Swallow, and R. B. Dreskin, *J. Histochem. Cytochem.,* **17**:563 (1969).
15. Mednick, M. L., J. P. Petrali, N. C. Thomas, L. A. Sternberger, R. E. Plapinger, D. A. Davis, H. L. Wasserkrug, and A. M. Seligman, *J. Histochem. Cytochem.,* **19**:155 (1971).
16. Nakane, P. K., and G. B. Pierce, Jr., *J. Histochem. Cytochem.,* **14**:929 (1966).
17. Nir, I., and A. M. Seligman, *J. Cell Biol.,* **46**:617 (1970).
18. Palade, G. E., *J. Exp. Med.,* **95**:285 (1952).
19. Pearse, A. G. E., "Histochemistry. Theoretical and Applied," 3rd ed., Vol. 1, London, Churchill, 1968, pp. 475–494.
20. Pease, D. C., "Histological Techniques for Electron Microscopy," 2nd ed., New York, Academic Press, 1964.
21. Plapinger, R. E., S. Linas, T. Kawashima, C. Deb, and A. M. Seligman, *Histochemie,* **14**:1 (1968).
22. Royals, H. E., "Advanced Organic Chemistry," p. 333, Englewood Cliffs, N.J., Prentice Hall, 1960 (p. 333).
23. Sauerbrann, R. D., E. B. Sandell, *J. Amer. Chem. Soc.,* **75**:3554 (1953); *Analyt. Chim. Acta,* **9**:86 (1953).
24. Seligman, A. M., J. S. Hanker, H. L. Wasserkrug, H. Dmochowski, and L. Katzoff, *J. Histochem. Cytochem.,* **13**:629 (1965).
25. Seligman, A. M., M. J. Karnovsky, H. L. Wasserkrug, and J. S. Hanker, *J. Cell Biol.,* **38**:1 (1968).
26. Seligman, A. M., T. Kawashima, H. Ueno, L. Katzoff, and J. S. Hanker, *Acta Histochem. Cytochem.,* **3**:29 (1970).
27. Seligman, A. M., R. E. Plapinger, H. L. Wasserkrug, C. Deb, and J. S. Hanker, *J. Cell Biol.,* **34**:787 (1967).
28. Seligman, M. L., H. Ueno, J. S. Hanker, S. P. Kramer, H. L. Wasserkrug, and A. M. Seligman, *Exp. Mol. Pathol.,* sup. **3**:21 (1966).
29. Seligman, A. M., H. Ueno, Y. Morizono, H. L. Wasserkrug, L. Katzoff, and J. S. Hanker, *J. Histochem. Cytochem.,* **15**:1 (1967).
30. Seligman, A. M., H. Ueno, H. L. Wasserkrug, and J. S. Hanker, *Ann. Histochimie,* **11**:115 (1966).
31. Seligman, A. M., H. L. Wasserkrug, C. Deb, and J. S. Hanker, *J. Histochem. Cytochem.,* **16**:87 (1968).
32. Seligman, A. M., H. L. Wasserkrug, and J. S. Hanker, *J. Cell Biol.,* **30**:424 (1966).
33. Seligman, A. M., H. L. Wasserkrug, and R. E. Plapinger, *Histochemie,* **23**:63 (1970).
34. Seligman, A. M., H. L. Wasserkrug, R. E. Plapinger, T. Seito, and J. S. Hanker, *J. Histochem. Cytochem.,* **18**:542 (1970).
35. Sjöstrand, F. S., "Electron Microscopy of Cells and Tissues," Vol. 1, New York, Academic Press, 1967 (p. 379).
36. Sternberger, L. A., P. H. Hardy, Jr., J. J. Cuculis, and H. G. Meyer, *J. Histochem. Cytochem.,* **18**:315 (1970).
37. Thiery, J. P., *J. Microsc.,* **6**:987 (1967).

OVARY

The ovaries of all vertebrates have two main constituents, the *soma* and *germ cells.* The soma comprises a variety of cell and tissue types including a covering epithelium, a *cortex* of supporting cells with a little connective tissue and aggregations of glandular, hormone-secreting cells, and a *medulla* of variable composition with a basis of connective tissue and blood vessels. The extent to which these regions are developed varies with age and reproductive state and shows marked differences according to the class of vertebrate examined. For example, in many fishes the medulla is repressed during development, and in Amphibia and Reptilia it is largely cavernous (see Franchi, 1962). In mammals, on the other hand, the medulla is a tissue-filled zone which increases in density and vascularity during life.

The cortex of the ovary is also variable in that the

female germ cells which reside in it are largely responsible, directly or indirectly, for the extreme changes in ovarian volume at various times during the life of the animal, particularly in those species with well-defined breeding cycles. In the majority of vertebrates the germ cells (oocytes) elaborate appreciable volumes of yolky reserves in the cytoplasm as they mature, and such cells may be said to constitute the bulk of the cortex, the reserve stock of germ cells occupying only small interstices of the cortex in comparison. In mammals little or no yolk as such is present in the cytoplasm of the oocyte; immediately surrounding cortical cells become organized into relatively complex "follicles." During each breeding cycle, a proportion of these follicles enlarge and develop fluid-filled antra; they are then termed Graafian follicles. The enclosed oocyte itself remains microscopic in size.

Of the several factors influencing the composition of vertebrate ovaries the features outlined above emphasize the important structural differences which exist between the organ in mammals, on the one hand, and in non-mammals on the other. They could influence the choice of technical approach to microscopic study, especially with regard to the penetration of fixatives and other media, and subsequent microtomy. For instance, the mammalian ovary has a consistency which is similar to many other "compact" tissues, but that of most other vertebrates is "saccular" and, therefore, more prone to distortion during various stages of histological processing, especially during microtomy. In birds, because of the deposition of yolk, a number of developing oocytes may exceed 1 cm in diameter. The large areas of yolky cytoplasm in the latter may have become relatively brittle and be inclined to "chatter" during sectioning. Furthermore, being unsupported by other tissues, such areas may become dislodged when the sections are floated out. Should such difficulties arise, consideration should be given to prior dissection of the ovary so that regions of widely differing consistency may be processed separately.

Light Microscopy. With the possible exceptions mentioned above, the vertebrate ovary presents no special problems with regard to its preparation for histology, for histochemistry, or for any of the conventional methods by which cells and tissues in other situations are studied. Certain preferences are, however, apparent. For example, compound fixatives containing formaldehyde seem generally to be recommended. Satisfactory results are reported for representatives of all classes of vertebrates after the use of Bouin's fluid (Picro-formol-acetic). For some applications several other standard fixatives, e.g., Carnoy, Formol-saline, mercuric-formol, and Zenker have been employed, especially where the picric acid interferes with subsequent staining reactions (see Appendices in Pearse, 1968). For histochemical studies of lipids formaldehyde-calcium fixatives may be used, followed by frozen sectioning or gelatin embedding (see Nath, 1960; Guraya, 1959).

Studies on the ovaries of various mammals in the author's laboratory have employed the standard aqueous solution of Bouin's fixative. Fixation time is kept as short as is consistent with complete penetration of the tissue. Fetal ovaries thus require a relatively short immersion time compared with those of mature animals. Subsequent processing is conventional for this fixative and embedding in paraffin wax at 56°C is routine. For both qualitative examination and quantitative estimation of oocyte numbers (see Mandl and Zuckerman, 1951) serial sections are cut at 5 μ and stained in Weigert's iron hematoxylin, and counterstained lightly with Chromotrop 2R or with eosin. Nuclear chromatin in oocytes is clearly shown and the cytoplasm and zona pellucida of growing oocytes, glandular cells in the follicular theca and the corpus luteum, and other cell types, are well shown with minimal distortion.

Examination of Chromosomes in Oocytes. In most female mammals the primordial germ cells (oogonia) pass through the characteristic stages of meiotic prophase in fetal or early postnatal ovaries. Examination of these in sections may be supplemented by squash preparations, which are procedures specifically designed for chromosome studies. The techniques of Ford and Hamerton (1956) have been used by Beaumont and Mandl (1962) and Baker (1963), among others, on the fetal ovary. The methods do not work as well for postnatal ovaries since the development of connective tissue prevents adequate dispersion of the cells following treatment with trypsin. It is nevertheless possible to examine in this way oocytes at later stages of meiosis after aspirating them from preovulatory Graafian follicles. "Pronase" (Calbiochem) may be employed to remove the zona pellucida and adherent follicle cells. The same squash techniques have been employed to study chick oocytes (Hughes, 1963). Other techniques may work equally well, and the reader is referred to Sharma and Sharma (1965) for a full treatment of this subject.

The ovarian oocytes of a number of species among lower vertebrates present particularly suitable material for the study of chromosomes in the living, unfixed state. The oocytes are large and their nuclei may be dissected out for examination under phase contrast illumination. Upon rupturing the nuclear envelope, the diplotene chromosomes, which are often very large and are of the "lampbrush" type, are clearly depicted. They are amenable to enzyme treatment, cytochemical and autoradiographic study, as well as being eminently suitable objects for genetic analysis. They have also been examined as whole mount preparations with the electron microscope (Miller, 1965). The techniques for the manipulation of lampbrush chromosomes are described by Callan and Lloyd (1961) and Gall (1966). It is beyond the scope of this article to describe the particular osmotic and ionic conditions of the medium, etc., which are required for each species. Similar techniques for the study of chromosomes mammalian oocytes at this stage, also believed to be of the lampbrush form (Baker and Franchi, 1967a), await development.

Electron Microscopy. The techniques which are currently used for the study of the fine structure of ovarian tissues in vertebrates are those which, with minor variations, have been applied to most other tissues. Fixatives which are now commonplace, e.g., glutaraldehyde, formaldehyde, acrolein, and osmium tetroxide, suitably buffered, are employed. Variations in fixative strength, buffers, and additives (saline, sucrose, etc.) are not related to features peculiar to the ovary but have usually been adjusted to match the ionic or osmotic properties of the blood or tissue fluids of particular species. For routine electron microscopy of the ovary various laboratories have devised particular procedures, one of which, in the author's experience, provides consistently good results on mammalian ovaries.

We prefer to use osmium tetroxide as a primary fixative, made up as given below. It is also used following initial fixation in 6.25% glutaraldehyde buffered with cacodylate, although the latter tends to increase the general density of the cytoplasmic and nuclear matrix to a variable extent. The osmium fixative consists of the following:

Solution 1:
Micháelis buffer:
   19.428 g sodium acetate–3H_2O
   29.428 g sodium veronal
   500 ml distilled water
Solution 2:
N/5 HCl
Solution 3:
Balanced salts solution:
   a. sodium chloride 96.8 g/liter
   b. calcium chloride 3.42 g/liter
   c. potassium chloride 2.85 g/liter
These separate solutions should be mixed in the proportions: a. 1 liter: b. 21.7 ml: c. 17.3 ml.
Solution 4:
2% aqueous OsO_4
Solution 5:
CO_2-free distilled water
All stock solutions should be stored at 0–4°C.
The fixative is made by mixing the stock solutions in the proportions and in the order:
   5 ml of Solution 1 (Michaelis buffer)
   13 ml of Solution 5 (dist. water)
   2 ml of Solution 3 (balanced salts)
   approx. 5 ml of Solution 2 (HCl) to make a pH of 7.4–7.6.

For convenience approximately half the above may be mixed with an equal amount of OsO_4 (Solution 4) to make the final fixative. The remainder should be diluted with water in equal quantity to provide the post-fixation wash. Fixation is carried out at 0–4°C for 3–4 hr. Fetal and early postnatal rodent ovaries are small enough to fix entire, but larger pieces of tissue should be trimmed carefully to no thicker than 1 mm. The majority of oocytes in postnatal primate ovaries are present in the superficial cortex; thin slices are cut from the surface of the ovary with a sharp blade, and may be subdivided further if required.

Fixation may be followed by any of the standard methods of dehydration and infiltration with embedding resin. For the latter the author prefers Vestopal W (Martin Jaeger), particularly for primate ovaries, where the dense stromal connective tissue is better supported during sectioning than with Araldite. Bulk-staining of tissue blocks in 1% PTA at the 100% alcohol stage provides added contrast, although sections may be stained in uranyl or lead salt solutions as an alternative (Baker and Franchi, 1967a,b).

The staging of oocytes in meiotic prophase as seen under the electron microscope is made considerably easier if adjacent 1–2 μ sections are cut from the same block and examined with an optical microscope. The landmarks visible in the latter act as a guide to the identification of specific cell stages under the electron microscope. The thick sections are stained by a very rapid method which does not interrupt normal operations as it can be performed while thin sections are drying out on grids: Sections are picked up from the trough and deposited in a small drop of 10% acetone on a microscope slide. The slide is dried out fairly rapidly over a small flame and while still hot it is flooded with a few drops of toluidine blue (1% in 50% methanol, containing 0.5% borax). Sections are usually adequately stained in 5–10 sec and

the stain is washed off with tap water, the water evaporated with gentle heating and the preparation examined as a temporary mount, using immersion oil. Suitable specimens can be permanently mounted in D.P.X., or the coverslip may be sealed down with lacquer.

L. L. Franchi

References

Baker, T. G., *Proc. Roy. Soc. B.,* **158**:417 (1963).
Baker, T. G., and L. L. Franchi, *Chromosoma* (Berl.), **22**:358 (1967a).
Baker, T. G., and L. L. Franchi, *J. Cell Sci.,* **2**:213 (1967b).
Beaumont, H. M., and A. M. Mandl, *Proc. Roy. Soc. B.,* **155**:557 (1962).
Callan, H. G., and L. Lloyd, *Phil. Trans. Roy. Soc. B.,* **243**:135 (1961).
Ford, C. E., and J. L. Hamerton, *Stain Technol.* **31**:247; *Nature,* **178**:1020 (1956).
Franchi, L. L., "The structure of the ovary: vertebrates," Chap. 2B in "The Ovary," Vol. I, S. Zuckerman, ed., New York and London, Academic Press, 1962.
Gall, J. G., "Techniques for the study of lampbrush chromosomes," Chap. 2 in "Methods in Cell Physiology," Vol. II, D. M. Prescott, ed., New York and London, Academic Press, 1966.
Guraya, S. S., *Res. Bull. Panjab. Univ.,* **10**:81 (1959).
Hughes, G. C., *J. Embryol. Exp. Morph.,* **11**:513 (1963).
Mandl, A. M., and S. Zuckerman, *J. Endocrinol.,* **7**:190 (1951).
Miller, O. L., Jr. *Nat. Cancer Inst. Monogr.,* **18**:79 (1965).
Nath, V., *Int. Rev. Cytol.,* **9**:305 (1960).
Pearse, A. G. E., "Histochemistry: Theoretical and Applied," Vol. I, London, J. & A. Churchill, 1968.
Sharma, A. K., and A. Sharma, "Chromosome Techniques; Theory and Practice," London, Butterworths, 1965.

OVIDUCT

The improved morphological techniques of glutaraldehyde perfusion fixation and Epon embedding make it possible to study whole female genital organs of smaller animals with the gametes and zygotes preserved *in situ.* Therefore, the various processes in the oviduct, e.g., fertilization and the passage of spermatozoa, ova, or early zygotes can be examined. This article deals with a preparatory technique for this purpose. Certain requirements have to be set up with reference to the mode of fixation, embedding, and sectioning.

The *fixation* should be by the perfusion method using a rapidly penetrating fixative (Nilsson and Reinius, 1969; Reinius, 1969, 1970). The advantage of the perfusion fixation lies in the rapidity of the process and the uniform preservation of the entire organ, which is fixed *in situ* and without preceding manipulation; the hardening of the tissues by the fixative further facilitates dissection of the organ (cf. Forssmann et al., 1967). *In situ* preservation is necessary when one wishes to establish the width of the oviductal lumen or evaluate the interrelationship between the epithelium of the reproductive tract and the gametes or zygotes. However, the requirement of perfusion for fixation cannot always be fulfilled, and ordinary immersion fixation must be used. This, however, increases the risk of artifacts and makes it difficult to compare specimens obtained by various techniques of fixation. The fixation by perfusion is preferred for smaller animals, e.g., from the mouse up to the rhesus monkey. For larger animals and man the oviductal tissue, rapidly cut into small pieces with a razor blade, is fixed by immersion.

The fixative is a 2.5% solution of glutaraldehyde in Soerensen's phosphate buffer, pH 7.4, osmolality about 450 mOsm (Sabatini et al., 1963). As a postfixative 1% osmium tetroxide in Soerensen's phosphate buffer, pH 7.4 is used, but the osmium tetroxide can also be used for immersion fixation directly. The fixatives are used at room temperature or at a temperature of about 4°C.

The perfusion aggregate is preferably an ordinary flask and an administration set with a cannula for intravenous infusion. Excess pressure can be achieved either from a nitrogen cylinder, a hand pump (see Fig. 1), or simply by raising the flask to a proper level above the animal.

The procedure for fixation by perfusion is as follows:

1. Anesthetize the animal by an i.p. or i.v. injection of Nembutal (Abott) and fasten the animal to a cork-plate or a table.
2. Open the abdomen by a midline incision with lateral extensions, and move the intestines carefully aside to avoid muscular contractions of the reproductive tract by mechanical irritation.
3. Expose the abdominal aorta and the inferior vena cava in the retroperitoneal space by careful splitting and retracting the peritoneum covering them.
4. Clamp the aorta just caudal to the renal vessels and then insert a cannula of appropriate size or a polythen tubing,[1] connected with the apparatus for injection under pressure, into the abdominal aorta just distal to the clamp, aided by a magnifying glass if necessary.
5. Cut the vena cava about the level of the renal veins, and immediately inject the fixative into the aorta under pressure of 100–200 mm Hg for a period of about 5 min. With this setup the fixative will perfuse through the vessels of the lower half of the body only. (Perfu-

[1] The outer diameter of the cannula or polythen tubing used in different animals (the tubing is preferred in larger animals): mouse, 0.5 mm; rat, 0.8 mm; rabbit, 1.8 mm; and rhesus monkey, 3.0–3.5 mm.

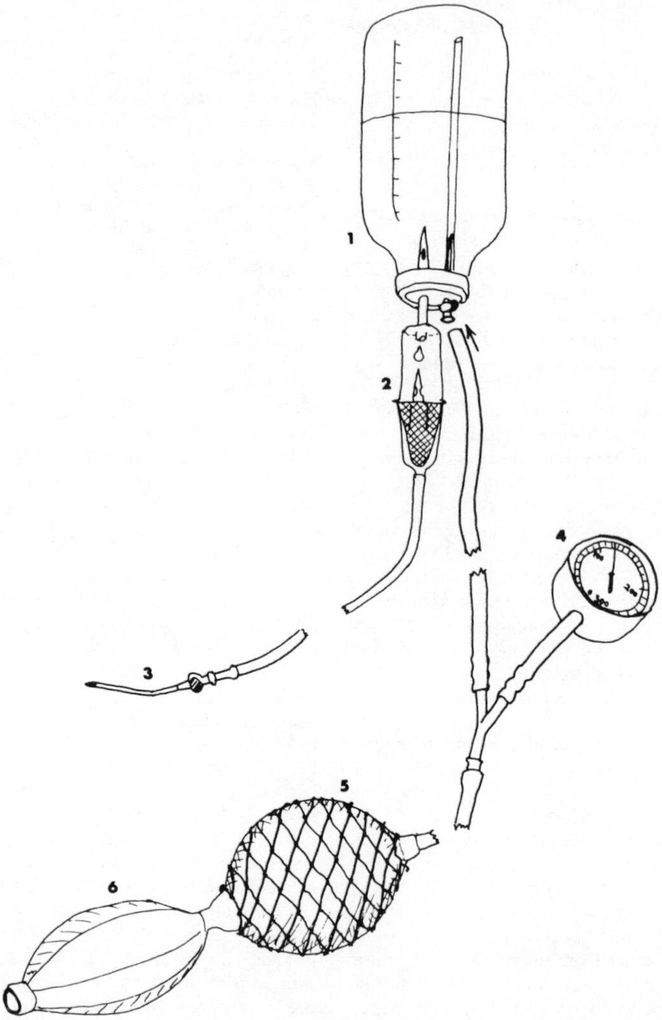

Fig. 1 The perfusion setup: 1, flask; 2, administration set for intravenous infusion; 3, cannula; 4, manometer; 5, rubber balloon; 6, rubber hand-pump.

sion can also be accomplished through the thoracic aorta or directly from the heart). The amount of fixative that is used varies from about 100 ml in the mouse to 1–2 liters in the monkey.

A successful fixation is indicated by an immediate disappearance of blood from small vessels and a stiffening and yellowing of the perfused tissues (Hopwood, 1967). Diffusion of the fixative from the capillaries quickly fixes the epithelial linings, and the gametes or zygotes, if present.

When the perfusion is completed the oviduct is removed carefully, avoiding any pinching or tearing and place in the fixative for some hours up to several days.

Dissection of the oviduct into appropriate sizes is preferably made of the different regions that have been

suggested according to the distribution and ultrastructure of the nonciliated cells: the preampulla, the ampulla, the isthmus, and the junctura (Nilsson and Reinius, 1969; Reinius, 1969, 1970). (See Fig. 2.) The specimens are rinsed in Soerensen's phosphate buffer, and postfixed for 3–5 hr in the osmium tetroxide. A similar postfixation is used also after immersion fixation in glutaraldehyde. The purpose of the postfixation is to facilitate later orientation and trimming of the specimen embedded in Epon, and to contribute to the contrast of the section when studied with the electron microscope.

The *embedding* in some of the Epoxy resins is preferred. Epon 812 (Luft, 1961) after dehydration in ethanol gives a minimum of distortion of tissues, although the cutting area of specimens has to be relatively small. However, there is the advantage that the same specimen may be

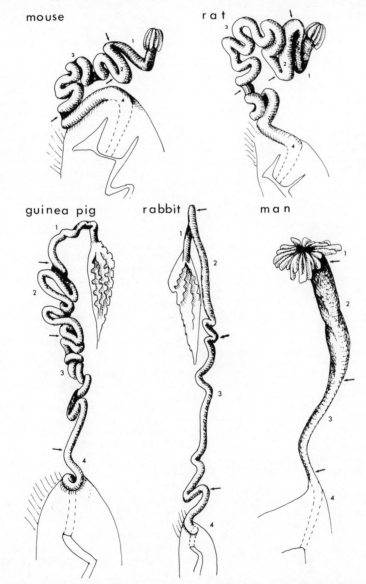

Fig. 2 Drawings of the oviducts of the mouse, rat, guinea pig, rabbit, and man: 1, preampulla; 2, ampulla; 3, isthmus; 4, junctura. (From Nilsson and Reinius 1969, by permission of the University of Chicago Press).

used for both light and electron microscopy. This is important when searching for eggs in the oviduct. The use of sections of about 1 μ thickness for light microscopy also gives a better result than can be obtained with paraffin embeddings.

The *sectioning* for light microscopy is done with a glass knife without a trough on an ultramicrotome or an LKB Pyramitome. The sections are then carefully transferred with forceps to a slide. This method makes it possible to get Epon sections of a maximum size of 5 $\times$ 8 mm, with a thickness of 1–10 μ. A fairly good view of the tissue in the sections is obtained by observing them through the stereoscope of the ultramicrotome. This is a great advantage, making possible a rapid search for eggs, without mounting, staining, and then observing each section under a microscope. The sections intended for further study are stained by heating, on the slide, in a few drops of an aqueous solution of 1 % toluidine blue with 1 % borax or sodium bicarbonate for about 1/2 min (Reinius, 1966).

For electron microscopy blocks are sectioned with an ordinary trough-mounted glass knife on a convenient ultramicrotome; the sections are preferably collected on one-hole grids (Galey and Nilsson, 1966). For staining, an ordinary solution of uranyl acetate (Watson, 1958) followed by lead citrate (Reynolds, 1963) can be used.

S. REINIUS

References

Forssmann, W. G., G. Siegrist, L. Orci, L. Girardier, R. Pictet, and C. Rouiller, "Fixation par perfusion pour la microscopic électronique essai de généralisation," *J. Micr.,* **6**:279 (1967).

Galey, F. R., and S. E. G. Nilsson, "A new method for transferring sections from the liquid surface of the trough through staining solutions to the supporting film of a grid," *J. Ultrastruct. Res.,* **14**:405 (1966).

Hopwood, D., "Some aspects of fixation with glutaraldehyde," *J. Anat.,* **101**:83 (1967).

Luft, J. H., "Improvements in epoxy resin embedding methods," *J. Biophys. Biochem. Cytol.,* **9**:409 (1961).

Nilsson, O., and S. Reinius, "Light and electron microscopic structure of the oviduct," pp. 57–83 in "The Mammalian Oviduct," E. S. E. Hafez and R. J. Blandau, ed., Chicago, University of Chicago Press, 1969.

Reinius, S., "Sectioning tissue for light microscopy with the Ultrotome ultramicrotome," *Sci. Tools,* **13**:10 (1966).

Reinius, S., "Morphology of oviduct, gametes and zygotes as a basis of oviductal function in the mouse," Thesis, University of Uppsala, Uppsala, 1969.

Reinius, S., "Morphology of oviduct, gametes, and zygotes as a basis of oviductal function in the mouse. I. Secretory activity of oviductal epithelium," *Int. J. Fertil.,* **15**:191 (1970).

Reynolds, E. S., "The use of lead citrate at high pH as an electron opaque stain in electron microscopy," *J. Cell Biol.,* **17**:208 (1963).

Sabatini, D. D., K. Bensch, and R. J. Barrnett, "Cytochemistry and electron microscopy," *J. Cell Biol.,* **17**:19 (1963).

Watson, M. L., "Staining of tissue sections for electron microscopy with heavy metals. I," *J. Biophys. Biochem. Cytol.,* **4**:475 (1958).

OXAZINE DYES

Oxazine dyes are analogous to the thiazins, with a sulfur replacing the oxygen. The oxazine dyes have been used industrially for calico printing, usually after chromium mordanting. They were introduced to biological technic by Becher (1921), who wrote extensively on nuclear staining with solutions of celestin blue B, gallamin blue, gallocyanin, and some of the alizarines, in solutions incorporating aluminium, chromium, and iron mordants. He reported that celestin blue B in combination with ferric alum was a particularly good nuclear stain. His results were checked, and though logical, the conclusion was made that this stain was good for amphibian but not mammalian tissues. The question was reopened by Proescher and Arkush (1928) who reported good results with three oxazine dyes, including celestine blue with ferric alum. This work was confirmed by Lendrum (1935) who introduced a modified solution containing glycerol. All these objections to the use of oxazine dyes were discussed by Clark and Powers 1953 (*Stain Technol.,* **29**: 93). Several methods of preparing a staining solution were reported by Gray, 1954 ("*The Microtomist's Formulary and Guide*"). This publication of these methods led to considerable correspondence, from which it became apparent that the stain was extremely variable in its results even though it was apparently prepared according to standardized methods from standard ingredients.

In consequence of this Gray (1956) published a new method of preparation (see below) which resulted in a staining technique absolutely specific to nuclei, independent of the time of staining, and requiring no differentiation beyond a simple wash in water. Moreover, nuclei stained in this manner are not affected by subsequent acid staining. With the possible exception of Berube et al. (1965) chrome hematoxylin (see HEMATOXYLIN STAINING SOLUTIONS), this is the simplest and most effective method of staining the nuclei yet devised. In the note appended to the formula given below, further applications of the dye are recorded.

Anonymous 1936 Catalogue of Vector Mrf. Co., London, n.d. (received 1936)
FORMULA: water 100, ferric alum 2.5, celestin blue B 0.5, glycerol 14, sulfuric acid 2
PREPARATION: Boil the dye 5 min in alum solution. Cool. Filter. Add other ingredients.
METHOD: [sections] → water → stain, 5 min to 1 hr → water, wash → [counterstain] → balsam, via usual reagents
NOTE: This is probably derived from Lendrum 1935 (*q.v.*).

Becher 1921a (Becher 1921, 46)
FORMULA: water 100, ferric alum 5, napthopurpurin 0.5
PREPARATION: Boil 5 min. Cool. Filter.
METHOD: [sections] → water → stain, 2 hr → water, wash → [counterstain] → balsam, via usual reagents
RESULT: nuclei black

Becher 1921b (Becher 1921, 72)
FORMULA: water 100, chrome alum 5, gallocyanin 0.5
PREPARATION: as Becher 1921a
METHOD: as Becher 1921a save that 24 hr staining is recommended
RESULT: nuclei deep blue
NOTE: Buzaglo 1934 (*Bull. Histol. Tech. micr.* [= *Bull d'Hist. appl.,* **11**:40]) diminishes the gallocyanin to 0.1 and adds 1 % formaldehyde to the filtrate.

Becher 1921c (Becher 1921, 40)
FORMULA: water 100, aluminum chloride 5, napthazarin 0.5

PREPARATION: as Becher 1921a

METHOD: as Becher 1921a save that 24 hr staining is recommended

RESULT: nuclei, deep blue violet

NOTES: Becher (*loc. cit.*) recommends also naptho-purpurin (dark red nuclei), purpurin (scarlet nuclei), and galloflavin (yellow nuclei) in solutions of aluminum chloride.

Berube et al. 1966 (*Stain Technol.*, **41**:73)

PREPARATION OF DRY STAIN: Boil 15 chrome alum and 0.15 gallocyanin in 100 water to 20 min. Filter and restore filtrate to 100 with filter washings. Cool, adjust to pH 8–8.5 with ammonium hydroxide. Filter, wash ppt with ether, dry.

WORKING SOLUTION: ppt from above 3, $1N$ H_2SO_4 100

Cole 1947 (*Stain Technol.*, **22**:103)

FORMULA: water 100, chrome alum 5, gallocyanin 1.5

PREPARATION: Boil the dye 5 min in the alum solution.

Demke 1952 (*Stain Technol.*, **27**:135)

STOCK SOLUTION: 2% ferric alum 100, sulfuric acid 2, celestin blue B 1, methyl alc. 10, glycerol 10

PREPARATION OF STOCK: Add acid to ferric alum sol. and heat to boiling. Add dye, boil 5 min, cool, and add alc. and glycerol.

WORKING SOLUTION: stock 25, water 75

METHOD: [fixed, flattened, helminths] → water → stain, 5–30 min → wash, 2 changes, 30 min each → [balsam via isopropyl alc. and creosote]

Einarson 1932 (*Amer. J. Pathol.*, **8**:295)

FORMULA: water 100, chrome alum 5, gallocyanin 0.15

PREPARATION: Boil 20 min. Cool. Filter.

METHOD: [50 μ celloidin sections of Zenker 1894 fixed material] → water → stain, 12–24 hr → water, wash → 80% alc., wash → 95% alc., 1 hr → abs. alc., until dehydrated → abs. alc. and ether, till celloidin removed → abs. alc., wash → balsam, via oil of Cretan thyme

NOTE: Bowie and Edmunson 1960 (*Stain Technol.*, **35**:1) recommend this method for staining autoradiographs.

Einarson 1935 (*J. Comp. Neurol.*, **61**:105)

FORMULA: water 100, gallamin blue 0.2

PREPARATION: Boil 5 min. Cool. Filter.

METHOD: [sections] → water → stain 12–24 hr → water, wash → 50% alc., until differentiated → 95% alc., → balsam, via usual reagents

Gray et al. 1965 (*Stain Technol.*, **31**:141)

PREPARATION OF STAIN: Mix 1 celestin blue B to a paste with 0.5 conc. sulfuric acid. Flood on a solution, at 50°C, of 2.5% ferric alum and 14% glycerol. Cool and adjust to pH 0.8 with conc. sulfuric acid.

METHOD: [sections] → water → stain 1 min or longer → wash → balsam, via usual reagents

NOTE: Overstaining is impossible and differentiation unnecessary. Nuclei are specifically stained and the stain is not removed by acid after stains. It has the disadvantage that the shelf life of the solution rarely exceeds a few weeks. The method was adapted to plant tissues by Gray and Pickle 1956 (*Phytomorph*, **6**:156 see POLYCHROME STAINING FORMULAS), to microorganisms by Pickle and Gray 1957 (*Mikroskopie*, **12**:27) to frozen sections by Sisca 1958 (*Proc. Iowa Acad. Sci.*, **65**:450) and an improved technique for chromosomes was published by Gray et al. 1960 (*J. Roy. Microsc. Soc.*, **78**:85). Cooke 1962 (*Stain Technol.*, **37**:317) Combines this stain with hematoxylin and toluidine blue.

Lendrum 1935 (*J. Pathol. Bacteriol.*, **40**:415)

FORMULA: water 84, glycerol 14, sulfuric acid 2, ferric alum 4.2, celestin blue 0.42

PREPARATION: Boil the dye in the alum 5 min. Cool. Filter. Add glycerol and acid.

Petersen 1926 (*Z. wiss. Mikr.*, **43**:355)

FORMULA: water 100, aluminum sulfate 10, gallocyanin 0.05

PREPARATION: Boil 10 min. Cool. Filter. Make up to 100.

Proescher and Arkush 1928 (*Stain Technol.*, **3**:36)

FORMULA: water 100, ferric alum 5, celestin blue B 0.5 0.5

PREPARATION: as Becher 1921a

METHOD: [sections] → water → stain, until nuclei deep blue-black, 3 min to 2 hr → water, wash → [counterstain] → balsam, via usual reagents

NOTE: Proescher and Arkush (*loc. cit.*) also recommended gallocyanin blue and gallocyanin in ferric alum solutions; but for the latter see Becher 1921. These solutions are less stable than celestin B solutions. For celestin blue B as a polychrome stain see Becher 1921, Lendrum and McFarlane 1940 (*J. Pathol. Bacteriol.*, **50**:381) add 14 glycerol to this solution.

PETER GRAY

Books Cited

Becher, S., "Untersuchungen über die Echfarbung der Zellkerne mit Künstlichen Beizen-farbstoffen und die Theorie des histologische Färbprozesses mit gelösten Laken." Berlin, Verlag von Gebrüder Bornträger, 1921.

Gray, P., "Handbook of Basic Microtechnique," Philadelphia, The Blakiston Company, 1952.

Gray, P., "The Microtomist's Formulary and Guide," Philadelphia, The Blakiston Company, 1954.

p

PAINT MICROSCOPY

Paints consist of hiding pigments such as titanium dioxide, extender pigments such as calcium carbonate, talc, and clay, and the color pigments such as Hansa yellow and iron oxide. The pigment is colloidally dispersed in a vehicle which consists of a resin or several resins and solvents. Various minor but essential ingredients such as catalysts, flow control agents, thickeners, and mildewcides are also present. The resin vehicle may be a true solution, or as in the case of latex base paints, consist of a colloidal dispersion of polymer in water (solvent).

The paint is applied to a surface by brushing, spraying, rolling, dipping, etc. The coating forms a solid film by one or more processes such as solvent evaporation, air oxidation, and chemical reaction of functional groups. The curing may take place at room temperature, or in most industrial processes, by baking in forced-air convection ovens.

The primary functions of paints are protection and decoration. Most industrial applications of microscopy to paints involve the appearance of the coating and factors which influence appearance.

Some factors which influence coating appearances are the nature of the substrate to be coated, the composition of the liquid paint, the method of application and curing, and the action of weather or environment on the painted surface or article.

Since the paint technologist has direct control over the composition of the liquid coating, his efforts to solve an appearance problem will frequently involve microscopic investigation of the paint or individual raw materials as well as the finished article.

The variety of substrates (metal, wood, plastic, etc.), coatings types (alkyd, epoxy, latex, etc.), raw materials, and methods of application of coatings require the adaptation of a variety of microscopic techniques.

Examination of Coated Panels or Coated Articles

The most frequent use of the microscope in coating laboratories is for the examination of cured coatings on wood or metal test panels to observe the presence of film defects such as cracks, craters, fisheyes, Bernard cells, foreign inclusions (dust, etc.), and other blemishes. The presence of these defects can often be observed visually and they detract from the decorative or protective functions of the paint. A $5\times-10\times$ magnifier such as a linen tester or doublet or triplet hand lens is useful for routine examination of the defects. If blemishes cannot be observed at $10\times$ magnification, they are considered to be visually absent.

Stereomicroscope. To examine the structure of defects, such as craters, higher magnifications are necessary. The stereomicroscope is very useful for the general examination of the panel surface. The variable magnification models which have a range of magnification from $8\times$ to $40\times$ are convenient since the surface defect can be examined continuously as the magnification is varied.

Control of the illumination is necessary for observing the structure of the painted surface. A light source illuminating the surface at grazing incidence will frequently define the surface structure more sharply than direct illumination. Control of the light intensity is also important. Too little light will result in loss of important details but too much light will produce glare obscuring the surface. Variation of the angle of incidence and light intensity while viewing the sample through the stereomicroscope will often permit discerning surface details that would otherwise be overlooked.

Research Microscope. For higher magnifications a research microscope with a dark field reflected light system gives the best results. The Leitz Ultropak objectives are especially suited for the examination of coated surfaces and unpainted substrates.

Control of the illumination is again crucial for satisfactory results. The artificial shadows which can be controlled with the sector diaphragms are useful as is the ability to vary the angle of illumination.

Occasionally an external light source focused on the sample at a grazing angle can be helpful. Variation of the

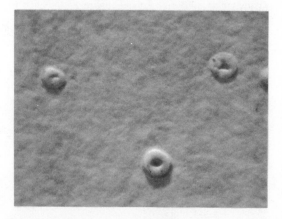

Fig 1 Surface defects produced by release of solvent during the baking of an industrial enamel. This phenomenon is called solvent popping. Reflected light. ($\times$ 48.)

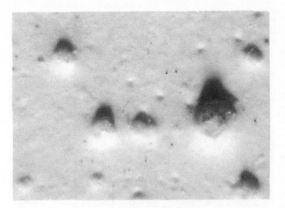

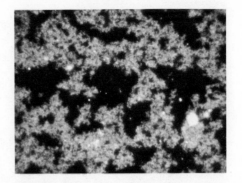

Fig 2 Surface defects caused by reaction of zinc oxide pigment with the alkyd paint vehicle. The zinc soap seeds which cause the defect are rich in palmitic acid. Reflected light. ($\times$ 30)

Fig 4 A sample of clear varnish with a flocculated fine particle silica flatting agent. The almost transparent varnish was photographed with dark field transmitted light illumination. ($\times$ 115.)

illumination while viewing the sample is again very effective.

Polarized light is useful with glossy or very reflective surfaces. Troublesome stray reflections which cannot be removed with the sector diaphragms can sometimes be eliminated using the polarizing filter and analyzer slide.

Colored filters improve the quality of observation and increase the sharpness of photomicrographs.

In general the nature and structure of the blemishes in coated panels can best be observed at lower powers of magnification (50$\times$, 81$\times$, 137$\times$). Because of the irregularity of most painted substrates and size of paint defects observation at higher magnifications is generally less rewarding.

Cross Sections. Many paint defects have their origin below the surface of the film. Some may involve the substrate or the paint-substrate interface. Cross-sectioning of the painted article or panel can provide valuable clues as to the origin of the defect. Additional information such as the number of coats of paint, the thickness of primer or filler coats, and size of irregularities in the substrate can frequently solve otherwise difficult problems.

The easiest method for preparing a cross section con-

sists of sawing the sample into suitable sections followed by sanding the section smooth with abrasive paper. The wet abrasive papers generally give a cleaner cross section for metal articles. Dry papers give better results with wood and wood-like materials. The section is usually finished with a 500 grit paper. This method frequently fails if the coating or substrate is weak or if there is a great disparity in the hardness or strength of the substrate and film. The most common types of failure are the tearing of the coating from the substrate or wearing of the sample edges obscuring the film structure.

Embedding of the sample in a suitable material will avoid damage to the coating and result in a superior cross section. The embedding material should be chosen to have a hardness approximating the embedded object. An embedding compound which does not attack the coating should be chosen. The acrylic resin embedding compounds are very good but they will attack some coatings. The epoxy resin embedding compounds are compatible with more types of coatings but are less convenient because of their longer cure times.

The sample can be precut prior to embedding and placed so that the sample face is in the plane of the casting. A glass plate with a metal ring is a suitable mold. The glass plate and ring should be coated with a layer of mold release agent. A silicone stop cock grease is a suitable release agent. Disposable aluminum dishes are also convenient molds.

The cured casting is sanded with coarse abrasive paper (180–220 grit) to remove the embedding resin from the surface of the cross section. Wet sanding with a 500 grit paper will finish the cross section for some purposes but additional wet polishing with an alumina abrasive and a metallurgical polishing wheel will generally give superior results.

Soft films on soft substrates can be sectioned by embedding in wax media followed by microtoming. A hand microtome is sufficient for most purposes. Razor blades are useful for cutting sections and have the advantage that they are disposable. Most pigmented paints are abrasive and will quickly dull conventional microtome knives.

Scanning Electron Microscope. The scanning electron microscope is a useful tool for examining coated surfaces and substrates. Some advantages to the method for paints are ease of sample preparation, range of magnification from the lower optical range (100$\times$) to above the optical

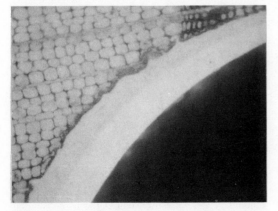

Fig 3 A cross section of a piece of wood coated with a wood primer and topcoat. Reflected light. ($\times$ 82.)

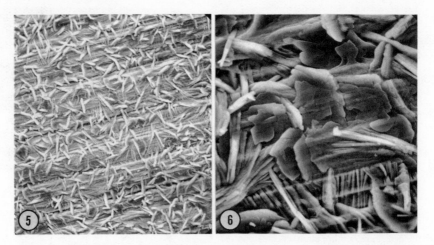

Figs. 5 and 6   Fig. 5:   A phosphate-pretreated piece of steel observed with the scanning electron microscope. ($\times$ 235.)
Fig. 6:   The phosphate pretreated steel panel shown in Fig. 5 observed at $\times$ 1,150 with the scanning electron microscope revealing the details of the phosphate treatment.

range (10,000$\times$), and ability to handle irregular surface, because of the high depth of focus. The main disadvantages are the high cost of the available instrumentation as compared to good optical microscopes. Also the method is not suited for observation of liquid samples.

The scanning electron microscope has been used to examine such paint problems as the surface treatment of metals, the changes of the painted surface during weathering, factors which influence gloss, and the structure of surface defects.

Examination of Liquid Coatings

The origin of appearance problems in the cured paint is sometimes traceable to the liquid coating. Gel particles can lead to defects that have the appearance of craters. Dust particles can produce cratering. Phase separation of the vehicle or flocculation of the pigment will give loss of gloss or lack of gloss uniformity on the panel surface. Color pigment separation can result in a shift in color of the cured coating.

Paints. The first attempts at microscope examination of the paint should be made on an unthinned sample or a sample thinned with the recommended solvent to application viscosity. A drop of a paint is placed on a microscope slide. A coverslip is placed over the paint and the drop allowed to flow out under the weight of the coverslip. In the case of high-viscosity coatings, some pressure on the coverslip is necessary.

The paint is usually examined with transmitted light. A thin film is required because of the high opacity conferred to the paint by the titanium dioxide hiding pigment present in most pigmented coatings. Koehler illumination is required for the best results.

Gel particles or dust of sufficient size to cause surface defects in the cured film can usually be detected in the liquid paint. Dilution of the sample with solvent can be used to isolate the gel or dust particles. Dilution must be done with the proper solvent or the resin will precipitate. Sometimes on dilution the gel particles seem to disappear. This is because the solvent-swollen gel particles are transparent with refractive index close to the solvent. Dark field illumination is very effective for observing gel

particles and some types of dust. A Cardiod dark field condenser is quite satisfactory and is more convenient than the oil immersion types.

Pigment flocculation problems can also be diagnosed by observing the liquid paint. Dilution of the sample should be avoided because of the danger of changing the condition of the sample. Although pigments range in particle size from less than 1 μ to 40 μ in size, relatively low magnifications (100–500$\times$) will generally reveal pigment flocculation. This flocculation can be caused by incompatibility of solvent and vehicle, insufficient or incompatible pigment dispersents, poor dispersion technique, etc.

Paints with large opaque pigment particles such as the polychromatic aluminum coatings which consist of clear vehicle, aluminum pigments, and a tinting color are best examined with a combination of transmitted and reflected light. The Leitz Ortholux microscope and the Zeiss

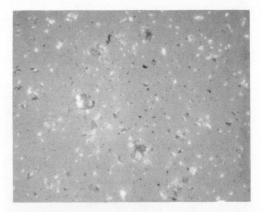

Fig 7   A sample of liquid paint containing clear vehicle, aluminum pigment, and a blue color pigment observed with a combination of transmitted and reflected light. The aluminum plates reflect the overhead illumination and can be distinguished from the blue color pigment which appears as dark particles. ($\times$ 215.)

Universal microscope are examples of instruments that provide this flexibility.

Resins and Clear Coatings. The resin and vehicle are import constituents of the coating. Gel particles and dust in resins can give rise to various defects, such as craters and eyeholes. Since these particles are frequently transparent and have refractive indices close to the refractive index of the vehicle, they are easily overlooked if ordinary transmitted light illumination is used. A drop of resin on a microscope slide with a cover glass can be effectively examined for gel particles and dust using transmitted dark field illumination. A Cardioid air dark field condenser is suitable for routine work. The gel particles, fibers, and dust appear as light bodies or pinpoints of light on a dark background.

Certain vehicles such as latexes, emulsions, and some water-dispersable resins are colloids. The structure of these vehicles can also be effectively observed with dark field transmitted light. Changes in colloidal state such as flocculation and change in size distribution are typical determinations.

Some clear coatings contain fine particle size pigments to control the gloss of the coating or control the viscosity characteristics. These pigments have refractive indices close to the refractive index of the cured film. The state of dispersion of these additives can also be investigated by dark field microscopy.

Pigments. Most industrial coating laboratories rely on macroscopic or instrumental methods to analyze or identify the pigments in a paint sample. This is a result of the availability of large samples in most situations. The microscopic identification of pigments is, however, possible by studying the morphology, optical properties, or in some cases microchemical reactions of the pigments.

Morphological features such as size, shape, and color are most frequently used. Optical properties such as refractive index may also be used but preliminary separation of the pigment from the vehicle is required. Cargille refractive index liquids can be used as dispersing liquids for the separated pigments. Microchemical techniques can be used, but since ample samples are generally available in industrial laboratories, conventional methods of chemical and physical analysis are usually employed to analyze the pigment content of unknown samples.

The chemical or petrographic microscope is most suitable for systematic identification of pigments. The small particle size of many pigments makes application of microscopic methods difficult. Extender pigments can be most easily identified.

Proper dispersion of the pigment or pigment mixture is important for best results. The medium chosen for dispersion should have a refractive index different than the materials under study. Most extender pigments have refractive indices in the range of 1.4–1.6. The viscosity of the dispersion medium should be high enough to immobilize the pigment particles. Glycerin jelly or polyvinyl alcohol solutions having refractive indices near 1.4 and Aroclor resins having refractive indices of 1.66, higher than most extenders, have been suggested. Commercial mounting resins or Duco cement can also be used for some pigments.

A drop of resin solution is placed on a microscope slide. An amount of pigment about the size of a pinhead is added and dispersed by rubbing under the tip of a flat-bladed spatula. If the mixture sets too rapidly, a drop of solvent will keep the sample fluid. The dispersed pigment in resin is spread to a thin film and a cover slide is placed

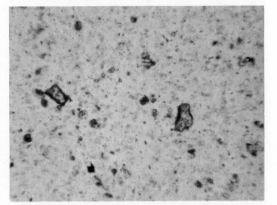

Fig 8   A basic lead silicate extender pigment dispersed in glycerol. Transmitted light. (× 240.)

on the dispersion. The cover glass is squeezed and rotated gently. If the preparation is allowed to set before examination, particle motion which occurs as the resin dried will be avoided. Some trial and error may be necessary to obtain the proper dilution of the pigment.

Other Methods

Because of the varied nature of coatings and coating applications, many methods have promise for the solution of particular problems. The variety of these applications are suggested by the following suggestive listing.

Hot stage microscopy has been used to observe the curing film during baking. Microelectrophoresis has been applied to determine the sign and magnitude of the surface charge on pigments which is related to paint stability. Ultraviolet fluorescence microscopy has provided useful information on the wood coating interface and penetration of the paint into the wood structure. X-ray diffraction has been used to identify foreign inclusions in paint films. The electron microprobe has provided information on the distribution of metal ions in coating films. Conventional electron microscopy has been used to observe the distribution of pigment in baked paint films.

PERCY E. PIERCE

PAINTINGS AND OTHER WORKS OF ART

George Stout, one of the foremost persons in the field of conservation of paintings, observed in 1935 that "a technical examination consists of study, by physical, optical and chemical means of the materials, construction and present condition of a particular painting." The various magnifiers probably play a greater part than anything else in these examinations. And *hunch*, although often hard to justify and for the most part frowned upon by serious scholars, is frequently the springboard suggesting the way in examination and research. The use of magnification (from the single lens to the most elaborate microscope) is not new. As long ago as 1870 von Pettenkofer introduced the use of the microscope in the examination and treatment of paintings. But it was not until about 1932 that technical evidence, particularly that resulting from magnification and photographic evidence so made, carried more weight in a court of law than did stylistic evidence. And it is only much more recently that it has really come into its own. Great advances have been made in the last few years. In 1914 A. P. Laurie began to use paint cross sections in his studies, this being carried further at the Fogg Museum in the 1930s. The use of cross sections in the examination and treatment of paintings at the National Gallery, London, has for the most part dated from 1951. Kögel demonstrated the value of ultraviolet about 1914 and between 1924 and 1932 Bayle began to develop infrared photography, with Maché introducing spectrography about the same time as a usable technique in such examinations.

The purpose of examination of paintings and other works of art is to secure information on (a) present condition, (b) the technique of the artist, and (c) other facts to guide the critic and conservator. A general scrutiny in the bright sunlight is preferably the first step and may go a long way toward pointing the direction of the entire investigation. Except for the general inspection of the surface, the examination begins with the support, following from there through the ground and paint to the surface film or varnish. The paint layer usually comes in for the most critical examination and the microscope is probably the most useful tool, whether it be in microchemical procedures or in viewing physical conditions. Microchemical tests and photomicrographic records of the result or of physical facts tell what to expect before starting treatment. The fibers most generally found in fabric supports of older paintings are cotton, linen, and sometimes wool. These three are quite easy to recognize. Cotton resembles a twisted noodle, linen resembles bamboo, and the less frequently found wool looks like asparagus. In addition, close observation of the scale pattern on the wool fiber will give a clue to the animal from which it came. Characteristics of natural fibers persist even though the fibers may be centuries old. Other fibers, particularly the synthetics, are much more difficult to determine and frequently require some chemical analysis in addition to observation under the microscope at about 100×. Paper fibers likewise are often difficult to pin down and may require some chemical tests in addition to the microscope. Any single test, whether it be of the pigment, the support, the ground or medium, taken alone may not mean much, but several taken together may tip the scales to confirm or condemn. Interpretation of the facts in many cases may mean more than the facts themselves. Photomicrographs are invaluable for comparison and for records.

Often the first step in an examination, and certainly the simplest after the naked-eye observation, is viewing with a 4×–6× single lens held nearly against the eyebrow and perfectly parallel to the surface being examined. This manner of using a single lens is highly essential to success but it is often overlooked or not appreciated by even the otherwise most knowledgeable of operators. A 2×–5× binocular loupe made up as a headset is very useful for general observation and for compensation of minute losses. The more precise low-power (7×) binocular microscope is invaluable for noting structure and condition as well as for taking samples to be used in chemical analysis and in making sections. For simple chemical tests and observation of fiber structure about 100× magnification is much used. Elaborate chemical tests are usually carried out under much higher magnification and frequently with polarized light. However, pigment particles must have about 250× magnification and paint sections anywhere between 30× and 150×.

The single lens is hand-held and the binocular loupe attached to a headband, so neither presents any particular problem of support, although they are obviously unsteady. The observation of sections and chemical tests, as well as photomicrography of these, is carried out with much heavier table-supported microscopes which are quite steady. The binocular microscope as used in examination of the surface of a painting, however, presents a considerable problem of safe support. It must be perfectly steady, be as close as 13 cm for the 7× magnification (and much closer for the higher power) and yet run absolutely no risk of touching the surface of the painting, particularly when the microscope is being moved into position or focused. The most obvious means of support, a long adjustable arm with the microscope at on end and attached to a weighted column at the other, is probably the least safe. This form, if the utmost care is not exercised, can swing into the painting and in addition is almost impossible to keep from vibrating. It can, however, be used for examination of a painting in a horizontal position on a table, or vertically against the wall. But in either case, if it must be used, it is better to have the weighted column standing on the floor. A much better approach to the problem is to set the painting against the wall on an easel that is attached to the wall and which will permit the painting to be moved up and down. Totally independent of this, at an appropriate distance from the painting, and not more than 5 or 6 feet apart, are two pipe columns anchored firmly between the floor and ceiling. Running between these columns are two bars a few inches apart secured to the columns by an attachment which permits them to move up and down and yet to be locked firmly at any selected height. Fine adjustment of this is desirable. The distance between the horizontal bars is such that a binocular microscope may be mounted between them on a carriage that will permit horizontal movement with fine adjustment. Focusing of the microscope is accomplished in the usual way. The convenience of having the microscope at a constant height (for either standing or sitting) may suggest vertical adjustment of the painting rather than of the microscope. An arrangement somewhat related to this has been used in the horizontal examination of small paintings on a table, but it has not proved entirely satisfactory. The mounting of a microscope on a camera tripod, with the adjustable column removed, has recently been suggested as a simple arrangement for the examination of pictures in a vertical position on an adjustable easel. It would, however, seem more advisable not to remove the central column, but to make use of it for minor vertical adjustment rather than depending on the movement of the painting for such adjustments. The use of a tripod for support has many advantages over the more elaborate systems, including portability. However, the very fact that it is portable and not fastened down would make it somewhat hazardous to use. If a low-power (7×) binocular microscope is to be used and it is light enough in weight, it can sometimes be set directly on the painted surface providing the painting is well supported from the reverse and there appears to be no loose paint. By whatever means they are examined, the crackle and general condition of the paint and varnish give clues as to the history and origin of the painting to almost the same extent as does the chemical makeup of the ground and paint film.

Some years ago samples, for determining the sequence of paint layers and for chemical analysis, were taken by pushing a squarely sharpened hypodermic needle vertically through the entire structure of a painting. However, it was found that by using this method the materials and stratification tended to become crushed and confused. More recently it has been found desirable to pry loose a small sample from the edge of a damaged area or from a crack. This can be done with a scalpel or harpoon and need not be much larger than a millimeter square, often much smaller, so it cannot be said to damage the painting or be visible to the ordinary observer. The taking of such a sample can best be done under a low-power microscope with adequate light. The sample, whose location must be recorded, can then be mounted as a section for observation as such, or reserved for chemical analysis. In some cases much of the analysis can be carried out directly on the mounted section itself. In earlier days sections were mounted using paraffin wax as the support. However, the hard, brittle paint tended to break up as it was shaved or cut. This led to using various synthetic resins as the support. Such resins are much tougher and can be polished to a fine degree without danger of confusing the sample. The laboratory at Brussels uses methacrylate in preparing their sections, while the National Gallery at London used cold-setting polyester resin. The methacrylate has certain advantages, in some cases, in that it has a very feeble fluorescence and thus it is possible to observe more clearly fluorescent media, if they be present.

Measurement of crackle, thickness of paint layers, or size of pigment particles can be done by placing a measuring reticle at the eyepiece diaphragm of the microscope. Although this will appear to float on the object viewed, it must be calibrated to indicate the true size of the object. A scale can also be placed in the field of view, but this is often inconvenient, if not impossible, at relatively high magnifications.

Photomicrographs, whether they be in black and white, color, under UV or IR radiation, must have on them the location, date, and other conditions at the time, if they are to be of any value and/or used for comparison. Photomicrographs are usually made with a 35 mm camera mounted on the microscope by means of an adapter fitted over the tube. The adapter is screwed directly into the camera body in place of the camera lens, which is not used. Considerable variation in the magnification of the object can be obtained by eliminating the microscope eyepiece or inserting, with or without the eyepiece, an additional tube or bellows between the adapter and camera body. The most desirable combination for a certain instance will obviously have to be determined on the basis of trial. The exposure can be determined by taking a meter reading through the lenses and must also be based on a certain amount of trial. It is most convenient to use a 35 mm single-lens reflex camera for this sort of work, as with it one can get an exact focus and field of view. Stereo photomicrographs have been used to advantage in many cases.

F. DUPONT CORNELIUS

References

Bachmann, K. W., "Conservation and technique of the Herlin Altarpiece," *Studies in Conservation*, **XV**(4): 370–400 (1970).

Bradbury, S., "The Microscope Past and Present," London, Pergamon Press, 1968.

Buck, R. D., and G. L. Stout, "Original and later paint in pictures," *Technical Studies in the Field of Fine Arts*, **VIII**:123–150 (1940).

Butler, M. H., "Polarized light microscopy in the conservation of painting," *Centennial Volume State Microscopical Society of Illinois*, 1960.

Cellerier, J. F., "Scientific methods in the examination of paintings," *Mouseion*, **XIII–XIV**(1–2):3–21 (1931).

Coremans, P., "La technique des 'Primitifs Flamande,' III," *Studies in Conservation*, **I**:41–46 (1954).

Coremans, P., "Scientific research and the restoration of pictures," *Bulletin de l'Institut Royal du Patrimoine Artistique*, Brussels, **IV**:109–116 (1961).

Coremans, P., R. J. Gettens, and J. Thissen, "The technique of the Flemish 'Primitives,'" *Studies in Conservation*, **I**:1–29 (1952–3).

Coremans, P., A. Philippot, and R. Sneyers, "Van Eyck —The Adoration of the Lamb," Antwerp, De Sikkel, 1951 (6 pp., 4 plates).

Coremans, P. and J. Thissen, "The introduction of thin sections to the examination of paintings," *Bulletin de l'Institut Royal du Patrimoine Artistique*, **II**:41–46 (1959).

Corrington, J. D., "Exploring with Your Microscope," New York, McGraw-Hill, 1957.

Delbourgo, S., "Colour macro- and microphotography at the Laboratory of the Louvre Museum," *Bulletin du Musée du Louvre*, **I**:13–15 (1956).

Delbourgo, S., and J. Petit, "Application de l'analyse microscopique et chimique a quelques tableaux de Poussin," *Bulletin du Laboratoire du Musée du Louvre*, Paris, **V**:40–54 (1960) (illustrated in color).

DeWild, A. M., "The Scientific Examination of Pictures," London, G. Bell & Sons Ltd., 1929 (106 pp., illus.).

Eastman Kodak Co., "Infra-red and Ultra-violet Photography," 4th ed., Rochester, N.Y., 1951 (40 pp., illus., tables).

Eibner, A., "The microchemical examination of binding materials," *Museion*, **XX**(4):5–23 (1932).

Elskens, I, "The introduction of thin sections to the examination of paintings. Spectrophotometric study," Brussels, **III**:20–34 (1960).

Flieder, F., "Mise au point des techniques d'identification des pigments et des liants inclus dans la couche picturale des enluminures de manuscrits," *Studies in Conservation*, **13**:49–86 (1968).

Gettens, R. J., "A microsectioner for paint films," *Technical Studies in the Field of the Fine Arts*, **I**:20–28 (1932).

Gettens, R. J., "An equipment for the microchemical examination of pictures and other works of art," *Technical Studies in the Field of Fine Arts*, **II**:185–202 (1934).

Gettens, R. J., and G. L. Stout, "The stage microscope in the routine examination of homogeneous binding mediums," *Technical Studies in the Field of Fine Arts*, **V**:18–22 (1937).

Goulinat, J. G., "The application of scientific methods in the restoration of paintings," *Mouseion*, **XV**:(3):47–54 (1931).

Graeff, W., "The examination of paintings and optical means," *Mouseion*, **XIII–XIV**(1–2): 21–42 (1931).

Gray, P. "Handbook of Basic Microtechniques," 3rd ed., New York, McGraw-Hill, 1964.

de Henau, P., and B. Tint, "Contribution a l'étude des peintures murales de Pagan en Birmanie," *Bulletin de l'Institut Royal du Patrimoine Artistique*, Brussels, **XI**:82–90 (1969).

Hood, W., "A simple microscope stand for picture examination," *Studies in Conservation*, **16**:24–28 (1971).

Keck, S., "The technical examination of paintings," *Brooklyn Museum Journal*, Brooklyn, N.Y., 1942 (pp. 68–82, illus, by 36 photographs).

Keck, S., "Mechanical alteration of the paint film," *Studies in Conservation*, **14**:9–30 (1969).

Lank, H., "Simple stereomicroscope stand for picture examination," *Studies in Conservation*, **4**:152–155 (1959).

Laurie, A. P. "Pigments and Mediums of the Old Masters," London, Macmillan and Company, 1914.

Laurie, A. P., "Photomicrography applied to the study of the technique of Rembrandt and his school," *Mouseion*, **XV**(3):5–8 (1931).

Laurie, A. P., "Microscopic examination of paint film," *Technical Studies in the Field of Fine Arts*, **III** (1935).

Lyon, R. A., "Ultra-violet rays as aids to restorers," *Technical Studies in the Field of Fine Arts*, **II**:153–157 (1934).

Marette, J., "The Identification of Primitives by the Study of the Wood, from the Thirteenth to the Sixteenth Centuries," Paris, Picard, 1959 (383 pp., 41 plates).

McGraw-Hill, "Encyclopedia of Science and Technology," Vol. 8, New York, 1960.

Plenderleith, H. J., "Notes on the technique in the examination of panel paintings," *Technical Studies in the Field of Fine Arts*, **I**:2–7 (1932).

Plesters, J., "Cross-sections and chemical analysis of paint samples," *Studies in Conservation*, **II**:101–157 (1956).

Plesters, R. J., "The preparation and study of paint cross-sections," *The Museums Journal*, London, **LIV**: 97–101 (1954).

Rawlins, F. I. G. "Evidence: its nature and place in the scientific examination of paintings," *Technical Studies in the Field of Fine Arts*, **VII**:75–82 (1939).

Ruggles, M., "Stereomicrography using the binocular microscope," *Studies in Conservation*, **14**:31–35 (1969).

Ruhemann, H., "Criteria for distinguishing additions from original paint," *Studies in Conservation*, **III**:145–161 (1958).

Ruhemann, H., "The Cleaning of Paintings, Problems and Potentialities," New York, Praeger, 1968 (508 pp., bibliography, illustrations, appendix).

Stotlow, N., J. F. Hanlon, and R. Boyen, "Element distribution in cross-sections of paintings," *Studies in Conservation*, **14**:139–151 (1969).

Stout, G. L., "A museum record of the condition of paintings," *Technical Studies in the Field of Fine Arts*, **III**:200–216 (1935).

Stout, G. L., "One aspect of the so-called 'mixed-technique'," *Technical Studies in the Field of Fine Arts*, **VII**:58–72 (1938).

Straub, R. E., "An apparatus for the surface examination of paintings," *Studies in Conservation*, **VI**:46–48 (1961).

Thissen, J., and J. Vynckie, "A note from the laboratory concerning the works of Juan de Flandes and his school at Palencia and Cervera," *Bulletin de l'Institut Royal du Patrimoine Artistique*, Brussels, **VII**:196–218 (1964).

West, E. H., "A ring-mount for micro-cross-sections of paint and other materials," *Studies in Conservation*, **IV**:27–31 (1959).

PANCREAS

Material. For the cytological study of the *exocrine elements*, the pancreas of amphibia and fish is favorable material as the acinar cells and their secretory or zymogen granules generally are very large in these animals. Among mammals, dog and guinea pig pancreas is easy to handle because it forms a well-definable and consistent organ and the exocrine elements are easily fixed.

For the study of the *endocrine elements*, the variable distribution of this portion of the pancreas should be kept in mind. In the cartilage fishes (Selachii and Holocephali) the endocrine pancreas is either the outer layer of the double epithelium of the exocrine ducts or its outgrowth into islands. In the bone fishes the endocrine pancreas forms an independent organ called Brockmann's body which is generally found adjacent to the gall bladder but may be separated into several corpuscles.

The endocrine pancreas of the higher animals is represented by the islets of Langerhans disseminated in the pancreatic parenchyme. In some reptiles the islets are condensed into a few large bodies in the splenic part of the pancreas. The islets of birds are divided into large "dark" and small "light" islets, the former occurring mainly in the small splenic lobe (1). As for the mammalian endocrine pancreas, the rabbit and guinea pig are favorite materials as the islets and their cells are large and easily stained.

Fixation. For the cytological examination of the exocrine and endocrine pancreas, Helly's fixative and GPA by Solcia et al. (2) (1 part 25% glutaraldehyde, 3 parts saturated picric acid and acetic acid to 1%) are highly recommended. The cytoplasm and secretory granules, both zymogen and endocrine, are beautifully preserved in species from fish to mammals. Well stainable ergastoplasm in the acinar cells is also preserved. Bouin's fluid has been widely used especially for the examination of endocrine cells and is still the best fixative for large-sized tissue blocks as in pathological laboratory work, but this fixative causes unfavorable shrinkage of the cytoplasm and sometimes the disappearance of zymogen granules.

Staining. For general staining of the *exocrine elements*, Masson-Goldner's trichrome method is recommended. This method is superior to the azan method because of its rapidity and because the basophily of the ergastoplasm can be demonstrated by Weigert's iron hematoxylin included in this technique. Crossmon's method is also recommended in which Masson-Goldner's light green is replaced by anilin blue. The contrast of the acinar and centroacino-ductular elements is very good after these stainings. Dominici's orange-eosin-toluidine blue stain gives good contrast of the ergastopasm, zymogen granules, and cytoplasmic matrix. For the simultaneous demonstration of the Golgi apparatus, ergastoplasm, and zymogen granules, the pancreas (dog is favored) fixed and silver-impregnated according to Aoyama is stained either with hematoxylin and acid fuchsin or with eosin and toluidine blue.

The *endocrine elements of the pancreas*, A, B, and D cells respectively corresponding to glucagon, insulin, and an unidentified third hormone, occur in every vertebrate species. The C cell, which has long been believed to be characteristic of the guinea pig is identical with the D cell of other animals (3).

For the differential staining of the endocrine cells of the pancreas (Table 1), azan and Masson-Goldner's trichrome are available. The color contrast especially of the A cells is enhanced by the preoxidation of the sections in an acidified (0.5% H_2SO_4) solution of $KMnO_4$ (0.25%) for 10–60 sec.

Table 1   Coloration of Islet Cells in Some Staining Methods

	A	B	D
HE	pale to pink	pale	pale
Azan	red	pale brownish	blue
Trichrome	red	pale reddish	green
AF-trichrome	red	violet	green
Phosphotungstic acid hematoxylin	dark gray	unstained	pale lilac
Impregnation	yellow	yellow	black

Phosphotungstic acid hematoxylin staining after pre-oxidation and treatment with iron alum as mordant was proposed by Gomori and by Levene (4) for the selective staining of the A cells. The A cells stain dark gray.

The B cell granules are stained by chrome-hematoxylin and aldehyde-fuchsin (AF) both originated by Gomori. The latter is more favored because of the higher selectivity of its staining and of the facility of the counterstaining. We recommend the following largely revised formula for the preparation of AF solution. Dissolve 2 g basic fuchsin (or 1 g pararosanilin) in 100 cm³ of 60% ethanol, add 2 cm³ paraldehyde and 2 cm³ conc. HCl, and leave the solution 3 days at room temperature. The solution may be used only for about two weeks. Gabe recommends therefore his "durable AF" (5). After the above-mentioned preoxidation, the sections are dipped in the AF solution until the desired coloration of the B cell granules is attained (usually from several minutes to one hour). For counterstaining Masson-Goldner's trichrome (usually omitting the first procedure with iron-hematoxylin) is recommended, as not only the A but also the D cells can be differentiated by this method. For the AF-trichrome staining, fixation in GPA is by far superior to the widely used Bouin fixation both in the preservation of cellular structures and in the stainability of the islet cell granules.

For the demonstration of the D cell, whose clear identification may be difficult with the above methods because of the pale coloration of its cytoplasm Davenport's silver impregnation (slightly modified by Hellman and Hellerström [6]) is the most reliable and the most widely applicable at present. The D cell granules are selectively blackened in the materials fixed in formalin, Bouin, or GPA (3). (See Fig. 1.)

Staining of the islet elements in basic anilin dyes such as toluidine blue and azure A was originated by Manocchio (7) for the selective demonstration of D cells. These dyes are, however, now known to stain (more or less metachromatically) also B and, in some conditions, A cells. The effects of pH values, of previous methylation and demethylation of sections, and of previous hydrolysis in warm $0.2N$ HCl have been reported (8, 9, 10).

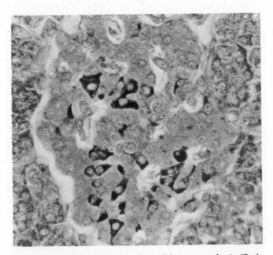

Fig 1   An islet of the monkey, *Macaca cyclopis* (Swinhoe). D cells are darkly demonstrated by silver impregnation. ($\times$ 440.)

The metachromatic reaction of preoxidized B cells in the solution of pseudoisocyanin, which has been proposed by Schiebler and Schiessler as the histochemical reaction of insulin, is now suggested to be of identical value as the staining in toluidine blue above described. The stained substances in this dye and in AF seem not to be the hormones themselves but their carriers in the secretory granules of the islet cells (10, 11).

TSUNEO FUJITA

References

1. S. Mikami and K. Ono, *Endocrinology,* **71**:464–473 (1962).
2. E. Solcia, G. Vassallo, and C. Capella, *Stain Technol.,* **43**:257–263 (1968).
3. T. Fujita, *Arch. Histol. Jap.,* **29**:1–40 (1968).
4. C. Levene, *J. Anat.,* **97**:647 (1963).
5. M. Gabe, *Bull. Microsc. Appl.,* **2**:153–162 (1953).
6. B. Hellman and C. Hellerström, *Z. Zellforsch.,* **52**:278–290 (1960).
7. I. Manocchio, *Zentralbl. allgem. pathol. Anat.,* **101**:1–4 (1960).
8. E. Solcia and R. Sampietro, *Z. Zellforsch.,* **65**:131–138 (1965).
9. E. Solcia, G. Vassallo, and C. Capella, *Stain Technol.,* **43**:257–263 (1968).
10. T. Fujita and K. Takaya, *Stain Technol.,* **43**:329–332 (1968).
11. T. Fujita, *Arch. Histol. Jap.,* **29**:313–325 (1968).

PARAFFIN SECTIONS

Introduction

The article HISTOLOGICAL TECHNIQUES, AUTOMATED describes in great detail the procedures commonly used for the routine preparation of blocks of mammalian tissue. This is much the best way of preparing such material but is ill adapted to preparing sections of anything else, since the machines' schedules must be readjusted for each type of material. Moreover, the majority of such objects as small invertebrates require individual judgment in the rate in which they are transferred from one fluid to another. Little difficulty is experienced in cutting normal histological material by any method, and indeed, frozen sections of a quality equal to paraffin sections are nowadays obtainable. However, most objects must be supported in a matrix which will itself section well, and those containing cavities must be impregnated throughout their whole substance with the embedding medium. Wax, nitrocellulose, and a variety of water-soluble materials have from time to time been suggested as impregnating and supporting agents, but the use of wax is so convenient and simple that only in special cases should any other material be employed.

The advantage of wax is not only that it readily passes from a solid to a molten state at temperatures which do not damage the material, but also that it is somewhat sticky, so that ribbons of sections may be prepared, each section being in the ribbon in the same order as it was cut from the object. Thus, if a rectangular block of wax is mounted in some kind of holder and then brought sharply down on a horizontal knife, the thin slice of wax which is cut off will adhere by its edge to the edge of the knife. If the block is then advanced by some mechanical device—such as a microtome—a small distance and again brought down on the knife, a second section will be cut off which will displace the first section, to which it will adhere on one edge, while the other edge remains attached to the knife. By the repetition of these movements a long ribbon may be produced. Preparation of paraffin sections is quite a complex operation and involves the following stages:

1. Fixation of the material.
2. Dehydration, in order that the material may be impregnated with a fluid capable of dissolving wax.
3. The removal of the dehydrating agent with a material solvent of, or miscible with, molten wax.
4. The soaking of the cleared specimen in a molten wax for sufficiently long to insure that it shall become completely impregnated.
5. Casting the now impregnated specimen into a rectangular block of wax.
6. Attaching this block of wax to some holder which itself may be inserted into a suitable microtome.
7. The actual cutting of the sections of the block into ribbons.
8. The placing of these ribbons on a glass slide in such a manner that they will lie flat and that the contained section will be adherent after the wax has been dissolved away.
9. The removal of the wax solvent.
10. Staining and mounting.

Each of these operations will be dealt with in due order.

Fixation

Selection of a Fixative. The theories involved in the process of fixation are dealt with in the article FIXATION and several hundred formulas for fixatives are given in the article FIXATIVE FORMULAS. Many other articles in this encyclopedia dealing with specific organs or organisms also suggest fixatives.

When one intends to section a small invertebrate, with the primary function of preserving its parts in as natural as possible a relation to each other, the same fixative should be employed as is recommended for those invertebrates intended to be made into wholemounts. The purpose in each case is to preserve the object in as natural a shape as possible without special regard to the preservation of the fine details of the cells themselves.

Something of the same consideration applies to blocks of tissue which are to be fixed in such a manner that their general structure or histology will be displayed. In this case, however, there is no problem of contraction of parts, so fixatives which would be quite useless for a whole animal may safely be applied to a block of tissue. The following notes are written only for the benefit of the beginner who, presented with this bewildering display, lacks the experience on which to base his choice.

A. RECOMMENDED FIXATIVES FOR EMBRYOS OR WHOLE ORGANS EXCEEDING 5 MM IN THICKNESS
1. FOR USE WHEN THE PRESERVATION OF SHAPE IS OF PRIMARY IMPORTANCE
   Bensley 1915
   Erlitzky 1877
   Hoyer 1899
   Lavdowski 1894
   Maximov 1909
   Müller 1859
   Orth 1896
   Régaud 1910

2. WHEN IT IS DESIRED, AS FAR AS POSSIBLE, TO PRESERVE BOTH SHAPE AND PROTOPLASMIC DETAIL
   a. *When shape is of greater importance*
      Helly 1903
      Petrunkewitsch 1933
      Rawitz 1895
      Smith 1902
      Zenker 1894
   b. *When protoplasmic detail is of greater importance*
      Fol 1896
      Gilson 1898
      Kohn 1907
      Mayer 1880
      Rabl 1894
      Tellysniczky 1898

B. RECOMMENDED FIXATIVES FOR SMALL PORTIONS OF ORGANS OR WHOLE ORGANS OR EMBRYOS NOT EXCEEDING 5 MM IN THICKNESS
1. WHEN A GENERAL-PURPOSE FIXATIVE IS REQUIRED
   Carleton and Leach 1938
   Gatenby 1937
   Gerhardt 1901
   Schaudinn 1900
2. WHEN PROTOPLASMIC DETAIL IS OF GREATER IMPORTANCE
   a. *When nuclear fixation is especially required*
      Allen 1929
      van Beneden 1905
      Carnoy 1887
      Carnoy and Lebrun 1887
      Sanson 1928
   b. *When cytoplasmic detail is especially required*
      Champy 1911
      Flemming 1884
      Kultschitzky 1887
      Mann 1894
      Smith 1935

There are only two general precautions to be observed in the practical application of fixatives: first, that adequate volumes (at least 100 times the volume of the part to be fixed) be employed; second, that mixtures containing either chromic acid or potassium dichromate with formaldehyde be used in the dark. After fixation, tissues should be thoroughly washed in water if this is the solvent for the fixative, or in alcohol if the fixative is based on the latter. Objects are usually stored, after fixation and washing, in 70% alcohol; though if they are to be kept a long time before dehydration, it is recommended that 5% of glycerol be added to the alcohol. This glycerol must, however, be very thoroughly washed out before dehydration commences.

Choice of a Dehydrating Agent. The classic method of dehydration for animal material is to soak the object in a graded series of alcohols, usually 10 or 15% apart. Dehydration through gradually increasing strengths of alcohol may be vital when one is dealing with delicate objects containing easily collapsible cavities, such as chicken and pig embryos, but a block of tissue may be taken from water to 95% alcohol without any apparent damage. Even though one uses increasing strengths of alcohol, the series normally in employment at the present time is by no means satisfactory. It is customary, for example, to pass from water to 30% alcohol at one end of the series and to pass from 85% to 95% alcohol at the other. The diffusion currents between water and 30%

alcohol are far greater and far more intense than those between 85% and 95%, and an intelligently graded series for delicate objects should run from water to 10%–20%–50%–95% alcohol rather than through the conventionally spaced gradations. This is not at all in accordance with the recommendations in most textbooks but is based on the author's experience over a long time. In using this classic method of dehydration, it is not necessary to confine the technique to ethanol. Methanol or acetone will dehydrate just as effectively, though they are rather more volatile.

There is a considerable vogue nowadays for the substitution for a straight dehydrating agent of some solvent which is both miscible with water and also with molten wax. The best known of these is dioxane, though n-butanol has also been recommended. The writer is not fond of these methods for, though the solvents involved are excellent dehydrating agents, they are relatively poor solvents of paraffin and frequently cause great shrinkage of delicate objects in the final transition between the solvent and the wax. For such purposes as the routine examination of the tissue blocks in a pathological laboratory, or for sectioning relatively sturdy plant materials, they may justifiably be employed. For sections intended, however, to retain intact structures on which research is subsequently to be conducted, it is most strongly recommended that the standard routine of passing from a dehydrating to a clearing reagent be retained.

Plant materials are much more difficult to deal with because of the tendency of the cytoplasmic materials to contract away from the cell walls. Most botanists start with the series 5%–10%–20%–30% ethanol and then continue dehydration with a water-ethanol-tertiary butyl alcohol series. The first of these mixtures is usually 10% tertiary butyl alcohol in 90% ethanol (i.e., 10 cm³ of tertiary butyl alcohol added to 90 cm³ of 70% alcohol), the second is 20% tertiary butyl alcohol in 65% ethanol, the third 35% butyl alcohol in 80% ethanol, the fourth 55% butyl alcohol in 95% alcohol, and the fifth 75% butyl alcohol in absolute ethanol. This is commonly followed by either one or two changes in pure tertiary butyl alcohol from which the specimen is transferred directly to the molten wax.

Selection of a Clearing Agent. The choice of a clearing agent in section cutting is of far more importance than the choice of a dehydrant, for there is not the slightest doubt that prolonged immersion in some of the volatile hydrocarbons, particularly xylene, leads to a hardening of the tissue with subsequent difficulty in sectioning. The classic method is to pass from alcohol to xylene, but there is no logical reason for the choice of xylene over toluene or benzene. There is little choice in the solvent power of any of these three hydrocarbons on wax; the writer's preference is for benzene, though it seems impossible to shake the faith of the conventional that the more expensive xylene is a necessity both as a solvent for embedding media and as a clearing agent before them. These three hydrocarbons are so cheap, and are obtainable in such a pure form, that there seems no necessity to use any other clearing agent, unless one prefers the reagents which are supposed to combine the functions of both dehydration and clearing.

Surprisingly little use has been made by histologists of "coupling agents" that permit mixing of relatively dilute alcohols with hydrocarbons. An excellent example of this is the phenol-xylol mixture usually attributed to Weigert under the name of "carbol-xylol." This is prepared by dissolving 25% phenol in xylol and is a useful intermediary between 95% ethanol and xylol.

It is still occasionally recommended that essential oils, such as cedar oil, be used for clearing objects for embedding. There is no justification for this unless it is vital that the object be rendered transparent (rather than alcohol-free) in order that some feature of its internal anatomy may be oriented in relation to the knife. Essential oils are excellent for wholemounts, but they are not readily removed from the specimen by molten wax; therefore, if they must be used, they should always be washed out with a hydrocarbon before the wax bath. Relatively small traces of any essential oil will destroy the cutting properties of any wax mixture, and as they are nonvolatile, there is no chance of getting rid of them in the embedding oven.

Choice of an Embedding Medium. The choice of an embedding medium should be dictated less by the nature of the specimen than by the conditions under which it should be cut. If pure paraffin is to be employed, it should be selected with such a melting point that the hardened wax will give a crisp section at the required room temperature. In the Europe of 20 years ago, when many writers were recommending a wax with a melting point of 52°C., the average laboratory temperature in winter was between 50° and 60°F. A wax of 52°C melting point, in an American laboratory kept between 70° and 80°F, is far too soft to cut any but the thickest sections. The use of waxes of 58°C, which are quite hard enough for cutting sections in an American laboratory is unfortunate, since such use requires an oven temperature of at least 60°C which results in many tissues becoming hard and brittle. As the introduction of any foreign substance automatically lowers the melting point of the wax, it is obviously desirable to use mixtures rather than the pure material. The advantage of the mixtures is that they have a relatively low melting point but soften very little before reaching the melting point. For ordinary routine preparations the writer's preference is for any of the paraffin-rubber-bayberry-wax mixtures or for one of the many proprietary products of this general composition. The introduction of rubber undoubtedly increases the stickiness of the wax and makes it easier to secure continuous ribbons, while the bayberry wax not only prevents the crystallization of the paraffin but also lowers its melting point. The beginner is strongly recommended to experiment with several of the rubber-bayberry-wax compositions and to select after experiment that which gives him uniformly successful results in his own laboratory. The following formulas are all worth trying:

Altmann cited from 1942 Langeron (Langeron 1942, 422)
FORMULA: 60°C paraffin 85, tristearin 10, beeswax 5

Beyer 1938 (*Amer. J. Clin. Pathol.*, **2**:173)
FORMULA: paraffin 100, rubber 2, beeswax 0.5

Gray 1941 (U.S. Patent 2,267,151)
FORMULA: paraffin (MP 58°C) 70, rubber 5, beeswax 5, spermaceti 5, nevillite "5" ("clarite") 15
NOTE: This composition melts at about 50°C but will cut 5 μ ribbons at a room temperature of 85°F. By increasing the resin to 30 it is possible to cut 1 μ ribbons but impregnation is very slow.

Hance 1933 (*Science*, **77**:353)
STOCK: Dissolve about 20 crude rubber, in small pieces, in 100 paraffin heated to smoking.

WORKING FORMULA: paraffin 100, stock 4–5, beeswax 1

Johnston 1903 (*J. Appl. Micr.*, **6**:2662)
FORMULA: paraffin 99, asphalt 0.1, para rubber 1
PREPARATION: Heat ingredients to 100°C with occasional stirring for 48 hr. Decant.

Maxwell 1938 (*Stain Technol.*, **13**:93)
FORMULA: paraffin 56–58° 100, rubber paraffin (see Hance 1933 above) 4–5, bayberry wax 5–10, beeswax 1

Ruffini 1927 (Ruffini 1927, 28)
FORMULA: paraffin (MP 52°–54°C) 100, beeswax 10, lard 15

Steedman 1947 (*Quart. J. Micr. Sci.*, **88**:123)
FORMULA: ethylene glycol monostearate 10, ethylene glycol distearate 73, stearic acid 5, ethyl cellulose 4, castor oil 8

Steedman 1949 (*Nature, Lond.*, **164**:1084)
FORMULA: diethylene glycol distearate 80, ethyl cellulose (low viscosity) 4, stearic acid 5, castor oil 4, diethylene glycol monostearate 5
NOTE: This composition melts at 53°C but cuts well at 89°–90°F.

Waterman 1939 (*Stain Technol.*, **14**:55)
FORMULA: paraffin 80, stearic acid 16, spermaceti 3, bayberry wax 1
NOTE: This mixture melts about 3°C below the melting point of the paraffin base, but it is sufficiently hard to cut good sections at room temperatures.

Technique of Dehydrating, Clearing, and Embedding. The techniques of dehydration and de-alcoholization do not differ materially from those used in the preparation of WHOLEMOUNTS which are described elsewhere. The whole process could, however, be much simplified if people would only remember that water is heavier than the majority of dehydrating agents, and that the majority of dehydrating agents are lighter than most clearing agents. Translating this theory into practice it must be obvious that the object to be dehydrated should be suspended toward the top of a tall cylinder of dehydrant in order that the water extracted from it may fall toward the bottom of the vessel, and that an object for clearing should be held at the bottom of the vessel for the reverse reason. It is, indeed, practically impossible to dehydrate a large object unless it is so suspended. The process of impregnating the tissues with wax has not, however, previously been discussed and will be dealt with fully.

The first prerequisite is some device which will maintain wax just at its melting point. Most people employ complex thermostatically controlled ovens for this purpose. And these are certainly necessary when any great volume of work is to be done. For the occasional embedder a very simple device may be made by placing a glass vial under an incandescent lamp which may be raised or lowered. The vial is filled with molten wax and the lamp adjusted at hourly intervals until the upper half of the wax in the vial is liquid and the lower half solid. This distance remains invariable and there is always a layer of wax at exactly the melting point at the interphase of the two states. A slightly more complex arrangement of the same kind is shown in Gray 1954 ("The Microtomist's Formulary and Guide," 98).

Vacuum ovens are occasionally required for the impregnation of the most difficult material but should be avoided whenever possible. If a vacuum oven is to be employed, moreover, it is necessary that all volatile solvents be removed from the material before it is placed in the vacuum, so it is always desirable to precede exposure in a vacuum oven by a considerable period of embedding in an ordinary oven.

Assuming that the material has been passed through dehydrating and clearing agents, and is now awaiting embedding, there are two main methods by which this may be done. Either the object may be transferred directly to a bath of molten wax, or it may be passed through a graded series of wax-solvent mixtures. The writer is strongly in favor of the latter course. Let us suppose benzene has been selected as the clearing agent and that the object is in a vial containing a few milliliters of this solvent. Chips are then shaved from the block of embedding agent and added to the vial. These usually dissolve very slowly and form a thickened layer at the bottom of the tube through which the object to be embedded sinks. The average object will be satisfactory if left overnight. The tube is then placed in the embedding oven, maintained at a temperature slightly above the melting point of the wax, and as many further shavings as possible are added into the tube. When these are completely molten, and most of the volatile solvent has evaporated, the object is removed with a pipet, or forceps, and placed in a dish of pure wax for an hour or two before being transferred to a second dish of pure wax for the time necessary to secure complete impregnation.

There is no method of forecasting how long an object will take to become completely impregnated with wax. It is very easy to find out, when one has started to cut sections, that the impregnation is not complete; but there is no basis save experience on which to base the timing in the different baths. If the object is to be transferred directly from solvent to wax, at least three baths should

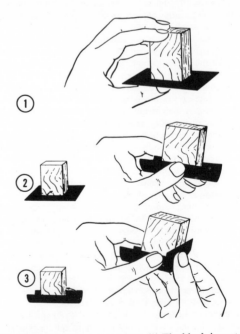

Figs 1–3 Folding a paper box. (1) The block is centered on the sheet; (2) the sides are folded up; (3) the end is folded up.

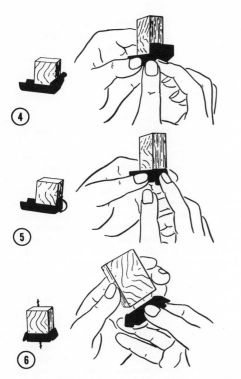

height. Center the sheet between the finger and the thumb (Fig. 1) and then fold up the sides (Fig. 2) creasing the paper where it is in contact with the edges of the block. Push up the end with the forefinger (Fig. 3) creasing both the paper in contact with the block and the flaps. Fold the flaps to the center (Fig. 4), being careful to get them straight and creasing them up the sides. Fold down the projecting flap (Fig. 5) and crease it firmly. Repeat these operations with the other side of the block and then slide the box off the end of the block (Fig. 6). It is well to have a series of these blocks made in both square and rectangular shapes. An additional advantage of this type of box is that one can put the data about the block on the flaps. Boxes cannot be made by this method much larger than 1 in. × ½ in.

After the box has been prepared we come to the actual process of embedding, which is shown in detail in Figs. 7–10. Before starting it is necessary to make sure that the following items are available: (1) a dish of water of sufficient size that the finished block may be immersed in it (in the illustration an ordinary laboratory fingerbowl is in use); (2) some form of heat, an alcohol lamp being just as effective as a bunsen burner; (3) a slab of plate glass; (4) a wide-mouth, eye-dropper type pipette. It is presumed that the object itself is in the oven, which also contains a supply of molten medium. It must be emphasized that an object cannot successfully be impregnated with one kind of wax and embedded in another. Next wet the *underside* of the bottom of the paper box and press it into contact with the plate-glass slab. Then take from the oven (Fig. 7) a beaker of molten embedding material and fill the little paper box to the brim. The eye dropper is then heated in the flame to a temperature well above that at which the wax will melt, and is used to pick up the object from its own dish (Fig. 8) and to transfer it to the paper box. By the time this has been done, a layer of hardened wax will have been formed at

Figs 4–6 Folding a paper box (continued). (4) The flaps are folded in; (5) the end is folded down and creased; (6) the cycle is repeated with the other end and the box removed.

be employed, for nothing is more destructive to a good section than the presence of a small quantity of the clearing agent in the embedding medium. To a beginner seeking a rough guess, it may be said that a block of liver tissue of 3–5 mm side will be satisfactorily impregnated with wax after 30 min in each of three baths, while a 96 hr chicken embryo will require at least two hours in each of three baths for its successful impregnation.

While the object is being impregnated with the wax it is necessary to decide what type of vessel will be used to cast the final block. This will depend more on the size of the object than on the preference of the worker. Very small objects may be most satisfactorily embedded in ordinary watch glasses (i.e., ordinary thin-walled watch glasses, not Syracuse watch glasses of the laboratory type) or in any other thin-walled glass vessel. Very large objects are often embedded with the aid of two thick L-shaped pieces of metal, which by being slid against each other may be caused to form a rectangular mold of varying dimensions. The writer prefers to prepare a cardboard or paper box than to endeavor to maneuver metal molds which are always getting jarred out of place at the wrong moment.

There is a very convenient method of folding small boxes which requires a series of wooden blocks of cross section equal to that of the boxes required. Take such a block (Figs. 1–6) and cut as many sheets of bond paper as there are boxes to be made. The length of the sheet should be the length of the box plus twice the height of the box plus twice the length of the flaps; the width of the sheet should be the width of the block plus twice the

Fig 7 Filling with wax an embedding box which has been attached with water to a glass slide.

Fig 8 Transferring the object from the embedding dish to the wax-filled paper box.

the bottom of the paper box, so the object will rest on the layer of solidified wax with a molten layer above. It will almost invariably happen that the surface has also cooled, so a crust of cool wax will have been carried down with the object in the box. It is essential to get rid of this if the wax is to adhere through section cutting, and the pipette is again heated, used to melt the entire surface of the wax (Fig. 9), and to maneuver the object into the approximate position in which it is required to lie in the finished block. Then blow on the surface until the wax is sufficiently solidified to enable you to pick up the box carefully and to hold it on the surface of the water used for cooling. With most wax media it is desirable to cool the block as rapidly as possible; it should never be permitted to cool in air. It cannot, however, be pushed under the surface of the water, or the molten center is liable to break through the surface crust and thus destroy the block. After it has been held in position until it is fairly firm throughout, it may be pushed under the surface to complete the cooling.

A necessity for careful cooling is greatly diminished if one of the wax rubber media, or their proprietary equivalence, is employed. Many of these will yield a grain fine enough for cutting if left to cool at room temperature, though it is usually better to place them in the refrigerator nowadays available in most laboratories.

The boxes may be kept in air for any reasonable length of time, but if they are to be kept for months or years it is better to immerse them in a 5% solution of glycerol and 70% alcohol.

It is widely believed that since the block is perfectly dehydrated before being impregnated with wax, it must subsequently be kept out of contact with fluids. Nothing could be further from the truth. As will be discussed later, when dealing with the actual technique of sectioning, it is often desirable to expose a portion of the object to be sectioned and leave it under the surface of water for some days, in order to get rid of the brittleness which has been imparted through the embedding process. Blocks which have been stored dry for a long period of time should always be soaked in a glycerol-alcohol mixture for at least a day before sectioning.

It is, in any case, undesirable to section a block as

Fig 9 Remelting the wax around the object with a heated pipette.

soon as it has been made, for it is necessary for successful sectioning that the block should be the same temperature throughout. If a block is made in the evening, it is better to take it out of the water and to leave it lying on the bench overnight in order that the temperature may be stabilized. Assuming, however, that we have such a block at hand, the next thing to do is to mount it in whatever holder is to be used.

Microtomes and Knives. The subject of capitalized microtomes is dealt with in a separate article. The type most commonly employed is that referred to as the American Optical, originally invented by Minot. Almost all other existing microtomes are modifications of these.

It is impossible to cut a good paraffin section without a properly sharpened knife. Numerous devices have from time to time appeared on the market for the purpose of using interchangeable razor-type blades, but these are rarely employed save in elementary instruction where the cost of providing enough microtome knives for an entire class is prohibitive. In times past it was customary for the operator to sharpen his own microtome knives, and those interested in this process may find a full description of it in Gray 1954 ("The Microtomist's Formulary and Guide"). Fortunately it is today possible either to send knives out to firms engaged in commercially sharpening them or to purchase one of the many excellent microtome knife sharpeners at present available. Of these the writer much prefers the model made by the American Optical Company. It is, however, necessary to understand the theory involved if such machines are to be properly used.

Three types of solid blade are available: first, those which are *square-ground*, i.e., in which the main portion of the knife is a straight wedge; second, those which are hollow-ground, i.e., in which both sides of the knife have been ground away to a concave surface, which results in a relatively long region of thin metal towards the edge; third, knives which are *half-ground*, i.e., knives of which one side is square- or flat-ground and the other side hollow-ground. This last type of knife, which the writer prefers, is a compromise. There is no doubt that a square-ground knife is sturdier than a hollow-ground knife, a point of some importance when cutting large areas of relatively hard tissues; but there is equally no doubt that a hollow-ground knife can be brought more readily to a fine edge. Microtome knives must be sharpened frequently; but it is necessary, before discussing how to do this, to give a clear understanding of the nature of the cutting edge itself.

If a wedge of hardened steel were to be ground continuously to a fine edge, as in Fig. 10A, it would be utterly worthless for cutting. After only a few strokes the fine feather-edge, which would be produced by this type of grinding, would break down into a series of jagged saw teeth. A microtome knife, or for that matter any other cutting tool, requires to have ground on its cutting edge a facet of a relatively obtuse angle, whether it be a square-ground knife, as in Fig. 10B, or a hollow-ground knife as Fig. 10C. The process of applying this cutting facet to the tip is known as *setting*. The nature and purpose of this cutting facet is best explained by reference to the mechanism of cutting shown in Fig. 10D. Notice first that the knife blade itself must be inclined at such an angle to the block that the cutting facet is not quite parallel to the face of the block. There must be left a *clearance angle* to prevent the knife from scraping the surface every time that it removes a section. This clearance angle should, in cutting wax, be as small as possible, and it is for this reason that the blade holder of a microtome is furnished with a device for setting the knife angle. The knife angle should not be set with reference to any theoretical consideration, but with regard only to securing this small clearance angle. The only way to judge whether or not a satisfactory clearance angle has been obtained is to observe the sections as they come from the knife. If the clearance angle is too large, so that the section is not being cut from the block but is being scraped from it, the section will have a wrinkled appearance and will also usually roll up into a small cylinder. If the clearance angle is too small, so that the lower angle of the facet is scraping the block after the tip has passed, the whole ribbon of sections will be picked up on the top of the block, which will itself crack off when the knife point reaches it. It is obvious that the knife angle will be changed as the angle of the cutting facet is changed, so it is desirable to maintain the cutting facet of as uniform an angle as possible.

Mounting the Block. A sharp knife being available and the microtome selected, it now remains to trim the block to the correct shape and to attach it to the object holder of the microtome. The rough block of wax containing the object must be first removed from the mold, or if a paper box was used, the box cut away roughly with a knife. The block should now be held against a light so that the outlines of the contained object can be seen clearly. The block is then trimmed until the object lies in the center of a perfect rectangle, with the major axis of the object exactly parallel to the long sides. This is best achieved by finding first the major axis, at right angles to which the sections are to be cut, and trimming down one side of the block with a sharp safety-razor blade, taking off only a little wax at a time. If one tries to remove a large quantity of wax there is danger of cracking the block. When one side has been shaved to a flat surface, the other side is shaved parallel to it. The top and bottom surfaces of the block may now be shaved,

and it is essential that these be exactly parallel to each other. A skilled microtomist can cut these edges parallel with a safety-razor blade without very much difficulty, but numerous devices have been described from time to time in the literature to enable one to do this mechanically. It does not matter if these two edges are exactly parallel with the plane of the object; it is only essential that they be parallel with each other. At this stage plenty of wax should be left both in front of, and behind, the object.

This trimmed block has now to be attached to some holder which can itself be inserted into the microtome. Two kinds of block holder are commonly available on contemporary microtomes. The first of these, used in the illustrations that follow, is a disc of metal firmly attached to a rod. The advantage of this is that it can easily be rotated to adjust to the edge of the blocks squared to the knife.

The other type is a rectangular block of plastic one surface of which is scored with grooves at right angles to each other. These blocks have the advantage that it is easier to attach the object and that they are available in a variety of sizes.

Whether the metal holder or the block is used, the technique is essentially the same. A layer of molten wax is built up on the surface and allowed to cool. The block (see Fig. 11) is then pressed lightly onto this hardened wax and fused with it with the aid of a piece of heated metal. Some people use old scalpels but the writer prefers the homemade brass tool shown in the figure. Care must be taken to press only very lightly with the forefinger and to perform the whole operation as speedily as possible to avoid softening the wax in which the object is embedded. The metal tool should be heated to a relatively high temperature and touched lightly to the base of the block. If the block is very long, it is also desirable to build up small buttresses of wax against each side, being careful not to bring these buttresses so far up the block that they reach the tip of the object to be cut. The metal should now be placed on one side and allowed to reach room temperature. Many people at this point throw the block and holder into a fingerbowl of water, which is all right provided the water is at room temperature. But there is no more fruitful source of trouble in cutting sections than to have the knife, the block, and the microtome at different temperatures. It is much better to mount the blocks the day before one intends to cut them and to leave them on the bench to await treatment. A final inspection is then made of the block to make certain that its upper and lower surfaces are flat, smooth, and parallel. Many people do not make the final cuts on these surfaces until after the block has been mounted in the block holder. The block and the block holder, after insertion in the jaws of the microtome, are seen in Fig. 12

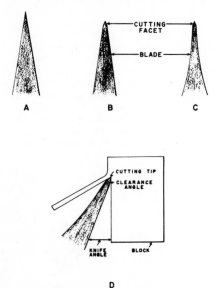

A   B   C

D

Fig 10 Types of cutting edge and cutting action. A, Simple wedge without cutting facet; B, Flat-ground razor showing cutting facet; C, hollow-ground razor; D, cutting action of knife on wax block.

Fig. 11 Mounting the wax block on the block holder.

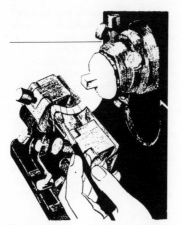

Fig. 12   Starting the paraffin ribbon.

and it will be noticed that setscrews on the apparatus permit universal motion to be imparted to the block so that it can be correctly orientated in relation to the knife. It is easy to discover whether or not the edges are parallel by lowering the block until it does not quite touch the edge of the knife, adjusting it until the lower edge is parallel, then lowering the block again and comparing the relation of the upper edge with the edge of the knife.

Cutting Paraffin Ribbons. The first step in cutting sections on this type of microtome is to make sure that every one of the setscrews seen in Fig. 12 is fully tight. The setscrews holding the block holder may be tightened in any order, provided that the result leaves the block correctly oriented, but those connected with the knife must be done in the correct order. First the knife is inserted into the holder and fixed firmly, but not tightly, in place by the two bearings at each end. The tightening of these screws causes the two movable holding arms to hold the knife near its edge. The two original setscrews, which hold the knife in place, are now screwed up as tightly as the thumb can bear. This leaves two setscrews which come through the inclinable hemicylinders and bear on the bottom edge of the knife. These two setscrews should then be tightened simultaneously and uniformly. The effect of this is to force the knife upward and thus wedge it with extreme firmness in the knife holder. The knife is now held in a pair of hemicylinders which may be moved so as to adjust the knife angle. The knife should be set at that angle which experience has shown to be desirable—no guide other than experience can be used—and the two setscrews which lock these inclinable hemicylinders in place then tightened.

Now that everything is tight the handle on the back of the microtome is turned until the block is as far back as possible, and the entire knife is moved on its carriage until the edge of the blade is about $\frac{1}{4}$ inch in front of the block. A last-minute check is now made to make sure that the divisions of the setting device exactly coincide with the thickness desired; then the handle is rapidly rotated until the block starts cutting. The front face will rarely be parallel to the blade of the knife, therefore a considerable number of sections will have to be cut until the entire width of the block is coming against the knife. No particular attention need be paid to the quality of this initial ribbon, which may be thrown away.

We will assume that all is going well and that the ribbon is coming off in a perfect condition; if it is not,

refer to Table 1. The remaining operations of preparing and mounting the ribbon are far more clearly seen in illustration than by description. As soon as the ribbon is the width of the knife in length a dry soft brush, held in the left hand, is slipped under the ribbon which is then raised in the manner shown in Fig. 12. Care should be taken that a few sections always remain in contact with the blade of the knife, for if the ribbon is lifted until only the edge of the section lies on the edge of the knife, the ribbon will usually break. As the handle is turned, the brush in the left hand is moved away until the ribbon is the same length as the sheet of paper on which it is to be received. Legal size (foolscap) paper is quite commonly employed and is shown in Fig. 13. Notice that the left-hand edge of the ribbon has been laid flat some distance from the edge of the paper and that a loop, sufficiently large to avoid strain on the ribbon attached to the knife, is retained with the brush, while the ribbon is cut with a rocking motion with an ordinary scalpel or cartilage knife. The larger and colder this scalpel is, the less likelihood there will be of the section adhering to it. The purpose of leaving a good margin around the edge of the paper is that it may be desirable to interrupt ribbon cutting for some time and to continue later. In this case the worker should furnish himself with a little glass-topped frame which is laid over the paper to prevent the sections from being blown about. As the inexperienced worker will soon find out, the least draft of air, particularly the explosive draft occasioned by someone opening the door, is quite sufficient to scatter the ribbons all over the room. These operations of carrying the ribbon out with the left hand, transferring the brush to the right hand, and cutting the ribbon off, are continued until the whole of the required portion of the block has been cut and lies on the paper.

The ribbon must then be divided into suitable lengths for mounting on a slide (Fig. 14). Though in theory a section should be of the same size as the block from which it came, this practically never occurs in practice and it is usually safe to allow at least 10% and sometimes 20% for expansion when the sections are finally flattened. The ribbon should not be cut up until a sample has been flattened on a slide in order to determine the degree of expansion. Though the sections shown in the illustration are being mounted on an ordinary 3 in. × 1 in. slide, it would be more practical (for a ribbon as wide as this) to use a 3 in. × 1¼ in. or even a 3 in. × 2 in. slide. The sections should never occupy the whole area of the slide, but at least $\frac{1}{4}$ in. should be left at one end for subsequent labeling. When the decision has been made as to how many sections shall be left in each segment of ribbon,

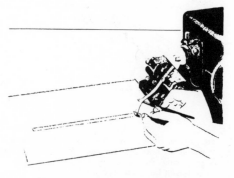

Fig. 13   Laying out the ribbon.

Table 1   Defects Appearing in Ribbons While Being Cut

Defect	Possible Causes	Remedies
Sections alternately thick and thin, usually with compression of thin sections	1. Block, or wax holding block to holder, still warm from mounting	1. Cool block and holder to room temperature.
	2. Block, or wax holding block to holder, cracked or loose	2. Check all holding screws. Remove block from holder and holder from microtome. Melt wax off holder and make sure holder is dry. Recoat holder and remount block. Cool to room temperature.
	3. Knife loose	3. Release all holding screws and check for dirt, grit, or soft wax. Check knife carriage for wax chips on bearing.
	4. Knife cracked	4. Throw knife away.
	5. Microtome faulty	5. Return microtome to maker for overhaul.
Sections bulge in middle	1. Wax cool in center, warm on outside	1. Let block adjust to room temperature. This is the frequent result of cooling blocks in ice water.
	2. Only sharp portion of knife is that which cuts center of block	2. Try another portion of knife-edge or resharpen knife.
	3. Object impregnated with hard wax and embedded in soft, or some clearing agent remains in object	3. Re-embed object.
Object breaks away from wax or is shattered by knife	1. If object appears chalky and shatters under knife blade, it is not impregnated	1. Throw block away and start again. If object irreplaceable, try dissolving off wax, redehydrating, reclearing and re-embedding.
	2. If object shatters under knife but is not chalky, it is too hard for wax sectioning	2. Soak block overnight in phenoglycerol mixture, rinse thoroughly, and dry, or spray section between each cut with celloidin, or dissolve wax and re-embed in nitrocellulose.
	3. If object pulls away from wax but does not shatter, the wrong dehydrant, clearing agent, or wax has been used	3. Re-embed in suitable medium, preferably a wax-rubber-resin mixture. Avoid xylene in clearing muscular structures.
Ribbon splits	1. Nick in blade of knife	1. Try another portion of knife-edge.
	2. Grit in object	2. Examine cut edge of block. If face is grooved to top, grit has probably been pushed out. Try another portion of knife-edge. If grit still in place, dissect out with needles. If much grit, throw block away.
Block lifts ribbon	1. Ribbon electrified. (Check by testing whether or not ribbon sticks to everything else.)	1. Increase room humidity. Ionize air, either with high frequency discharge or bunsen flame a short distance from knife.
	2. No clearance angle	2. Alter knife angle to give clearance angle.
	3. Upper edge of block has fragments of wax on it (a common result of 2)	3. Scrape upper surface of block with safety-razor blade.
	4. Edge of knife (either front or back) has fragments of wax on it	4. Clean knife with xylene.
No ribbon forms: 1. Because wax crumbles	1. Wax contaminated with clearing agent	1. Re-embed. (*Note:* Wax very readily absorbs hydrocarbon vapors.)
2. Because sections, though individually perfect, do not adhere	2. Very hard, pure paraffin used for embedding	2. Dip block in soft wax or wax-rubber medium. Trim off sides before cutting.
3. Because sections roll into cylinders	3a. Wax too hard at room temperature for sections of thickness required	3a. Re-embed in suitable wax. If the section is cut very slowly, and the edge of the section held flat with a brush, ribbons may sometimes be formed.
	3b. Knife angle wrong	3b. Adjust knife angle.

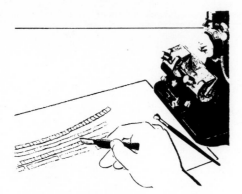

Fig. 14 Cutting the ribbon in lengths.

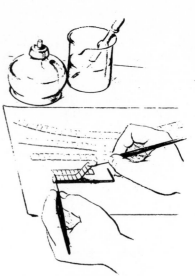

Fig. 15 Mounting the dry ribbon.

the first row of ribbons is then cut into the required lengths (Fig. 14). Then the worker must decide what shall be used to make them adhere to the slide.

Flattening and Attaching Ribbons to the Slide. It has long been customary to use some form of adhesive to attach the flattened ribbon to the glass slide. The best known of these materials is that of Mayer 1884 (*J. Roy. Micros. Soc.,* **4**:317) which is made by mixing equal quantities of glycerol and fresh egg white and then adding some preservative. The original calls for 1 % of sodium salicylate; the formula of Heidenhain 1905 (*Z. wiss. Mikr.,* **22**:331) is made by dispersing fresh egg white in 25 % ethanol. The former is best diluted at the rate of one drop per 50 cm³ with the water which is to be used for flattening. The latter is smeared similarly on the clean slide and allowed to dry. Various proprietary compounds serving the same purpose are available.

It will have been apparent to the worker from the moment that he started cutting the sections that they are not absolutely flat. They may be slightly crinkled, or slightly distorted, and must be flattened by being warmed on water heated just below the melting point of the wax. If only one or two sections are to be mounted, rather than an entire series of ribbons, it is convenient to have available an electrostatically controlled water bath to which the sections are transferred to expand. They are then transferred to the slide by slipping the latter diagonally under the surface of the water and raising it so that the section or sections are stranded in the right position.

If a series of ribbons are to be mounted, it is better to place them on the slide before flattening them, due allowance being made for their expansion. Some people place the water on the slide and then add the sections to it, but the writer prefers to lay the ribbons on the slide as shown in Fig. 15. This is not nearly so easy as it looks. Two brushes must be moistened with the tongue just enough to bring the hairs to a point. The two moist points are then delicately touched down (too much pressure will cause the ribbon to adhere to the paper) on each end of the selected piece of section. This piece is then lifted as shown in the illustration and placed on the slide. When a sufficient number have been accumulated the slide is then picked up carefully, reversed, and laid on top of the last three fingers of the left hand as shown in Fig. 16. It is fatal to grasp the slide by the sides; if this is done, when the water is flooded on from the pipette, the meniscus coming to the edge of the slides will break against the fingers, to which the sections will

permanently adhere. The technique shown is quite safe and the water containing the adhesive (if none has been applied to the slide) is then flooded on from a pipette in the manner shown. Enough fluid should be applied to raise a sharp meniscus at the edge of the slide.

The sections must now be flattened, and this is better done rapidly with a flame than slowly on a hot plate. Fig. 17 shows the slide being held over a small alcohol lamp, but a micro-bunsen can be employed equally well. The slide should be exposed to heat for a moment, withdrawn to give time for the heat to pass from the glass to the fluid, warmed again, and so on, until the sections are observed to be flat. The utmost care must be taken at this point for, if the paraffin is permitted to melt, the sections will not stick to the glass. As soon as the sections are flattened, the slide is gently tilted backward towards the hand so as to run off the excess water against the thumb, leaving the sections stranded in place. The slide is now usually placed on a thermostatically controlled hot plate and dried. Most people leave their slides overnight but frequently an hour would be sufficient. Dryness can be gauged without the least trouble by the fact that a moist slide shows the wax to be more or less opalescent, while on a properly dried slide it is almost glass-clear.

Fig. 16 Flooding the ribbon.

Fig. 17 Warming the flooded ribbons in order to flatten them.

The method just described is susceptible of several variations which may be briefly noticed. Some people do not drain the water from the slide, nor do they heat the slide over the lamp; they merely place the slide, as soon as the water has been added to it, on the thermostatically controlled hot plate so that the sections dry and flatten at the same time. The objection to this procedure is that dissolved air in the water used for flattening usually comes out in the form of bubbles which accumulate under the section, either causing it to fall off or at least making it very difficult to observe properly when

mounted. There is also the risk in this procedure that the water will not stop at the edge of the slide, but will unexpectedly flood off, carrying the sections with it onto the surface of the hot plate.

Another procedure, frequently used by the author but not recommended for the inexperienced, is to blot the sections before putting them on the hot plate. A *water-saturated* piece of *coarse* filter paper is placed on the drained slide and pressed hard with a rubber roller, which squeezes much of the water out of both the paper and the sections. This makes sure that the sections are perfectly flattened in contact with the slide, but requires a strong nerve to try for the first time, because most people fear that the sections will stick to the paper. This has never happened in a good many thousands of slides which the author has made by this means. Slides so prepared are always free of air bubbles. The appearance, cause and cure of the more common defects are shown in the tables.

Staining and Mounting Sections. Assuming that all difficulties have been overcome, and that one now has a series of slides bearing consecutive ribbons, the paraffin must next be removed in order that the sections may be stained. It is conventional, though probably not necessary, to warm each slide over a flame until the paraffin is molten. The slide is then dropped into a jar containing xylene, benzene, or some other suitable paraffin solvent. Large numbers of small slides are more conveniently

Table 2 Defects Appearing in Sections During Course of Mounting

Defect	Cause	Remedy	Method of Prevention
Sections appear wrinkled	1. Blunt knife used for cutting	1. None	1. Sharpen knife and cut new sections.
	2. Water used for flattening too hot, so that folds in sections fused into position	2. None	2. Watch temperature of water used for flattening.
	3. Sections unable to expand sufficiently: (a) because water used for flattening too cold (b) because area of water too small	3. None	3(a) Watch temperature of water used for flattening. (b) Make sure that slide is clean, so that water flows uniformly over it.
Sections have bubbles under them	1. Sections insufficiently flattened, so that air is trapped	1. If sections still wet, reflood slide with water and reheat to complete flattening.	1. Check flatness of sections before draining slide.
	2. Air dissolved in water used for flattening has come out and is trapped under sections in drying	2. If sections still wet, reflood slide with water, work out bubbles, and reheat to complete flattening.	2. Use air-free (boiled) water for flattening. Drain slide thoroughly and blot off excess moisture. Squeeze sections to slide.
Sections fall off slide	1. Wax melted in flattening	1. None	1. Watch temperature of water used for flattening.
	2. Slide greasy	2. None	2. Use clean slides.
	3. Alkaline reagents dissolve albumen adhesive. (Sections start to work loose in course of staining or dehydrating)	3. Treat slides with collodion.	3. Use another adhesive.
	4. Sections not flattened into perfect contact with slide	4. None	4. Sometimes caused by swelling of sections which causes center to lift. Squeeze sections to slide and dry as rapidly as possible.

Table 3   Defects Appearing in Sections After Staining and Mounting

Defect	Cause	Remedy	Method of Prevention
Sections distorted	1. Blunt knife and soft wax	1. None, though prolonged flattening on warm water may help.	1. Use suitable knife and embedding medium.
	2. Ribbon stretched when picked up on hot day	2. As (1) above	2. Handle ribbons in short lengths or use harder wax.
	3. Tissues not properly hardened before embedding	3. None	3. Use more suitable fixative or fix longer. Take extra care in dehydrating, clearing, and embedding.
Sections appear opaque or have highly refractive lines outlining cells and tissues	1. Clearing agent evaporated before mountant added	1. None	1. Obvious
	2. Sections insufficiently cleared or cleared in agent not miscible with mountant	2. Soak off cover. Clear properly.	2. Check quality and nature of clearing agents and mountants.
Sections will not take stain, or stain irregularly	1. Wax not properly removed before staining	1. Return sections through proper sequence of reagents to xylene. Leave until wax removed. Restain.	1. Change first jar of xylene frequently.
	2. Section not uniform thickness	2. None	2. See Table 1.
	3. Tissue "old" (has been stored for a long time in alcohol or, worse still, fixative)	3. Return sections through proper sequence of reagents to water. Wash overnight.	3. Store all tissues embedded in paraffin blocks—never liquids.
	4. Fixative not suitable before staining technique employed	4. Try mordanting sections in recommended fixative	4. Obvious.
	5. Fixative not fully removed		5. Treat tissues as indicated.
Sections contain fine opaque needles or granules	1. Imperfect removal of mercuric fixatives	1. Return sections through proper sequence of reagents to water. Treat 30 min with Lugol's iodine, rinse, and bleach in 5% sodium thiosulfate. Restain.	1. Treat tissues with Lugol before sectioning.
	2. Long storage in formaldehyde	2. None	2. Never store tissues in formaldehyde—always in paraffin blocks.

handled by being placed in racks, which may be moved from one rectangular jar to another.

It is necessary through the subsequent proceedings to be able to recognize instantly on which side of the slide the section lies. This is not nearly as easy as it sounds; a lot of good slides have been lost by having the sections rubbed off. The simplest thing to do is to incline the slide at such an angle to the light that, if the section is on top, a reflection of the section is seen on the lower side of the slide. A diamond scratch placed in the corner is of little use because it becomes invisible when the slide is in xylene. The greatest care should be taken to remove the whole of the wax from the slide before proceeding further. It is usually a wise precaution to have two successive jars of xylene, passing the second jar to the position of the first, and replacing it with fresh xylene, after about ten or a dozen slides have passed through. It must be remembered that paraffin is insoluble in the alcohol which is used to remove the xylene, so it is no use soaking a slide in a solution of xylene in wax and imagining that it will be sufficiently free from wax for subsequent staining. Some people go further than this and have the first two jars containing xylene, and then a third containing a mixture of equal parts of absolute alcohol and xylene, to make sure that the whole of the

wax is removed. If even a small trace of wax remains, it will prevent the penetration of stains. Assuming that one is proceeding along the classic xylene-alcohol series, the slide is transferred from either the fresh xylene or the xylene–absolute alcohol mixture, to a coplin jar of absolute alcohol. It is unfortunate that nobody seems yet to have placed on the market a coplin jar, or slide-staining dish, the lid of which is satisfactorily ground into position so that absolute alcohol, which is very hygroscopic, remains uncontaminated. It does not matter if xylene is carried over into the absolute alcohol, but as soon as the first trace of a white flocculent precipitate appears in the alcohol—indicating that some wax is being carried over—the alcohol must be replaced.

The writer never bothers to use a series of graded alcohols between absolute alcohol and water. These graded series are necessary, of course, when one is dealing with the dehydration of whole objects which may be distorted, but the author has never been able to find the slightest difference between thin sections which have been passed from absolute alcohol to water, and those which have laboriously been downgraded through a series. As soon, however, as the slide has been in water long enough to remove the alcohol, it should be withdrawn and examined carefully to make sure that it has been sufficiently dewaxed. If the water flows freely over the whole surface, including the sections, it is safe to proceed to staining by what ever manner is desired. If, however, the sections appear to repel the water, or if there is even a meniscus formed round the edge of the section, it is an indication that the wax has not been removed, and that the slide must again be dehydrated in absolute alcohol, passed back into a xylene-alcohol mixture, and thence again into pure xylene. The slides are now in condition, though they may be passed through alcohol, or through alcohol to water and then subjected to any of the numerous staining methods suggested elsewhere in this volume. It is by no means necessary to confine staining to the conventional hematoxylin-eosin, and the beginner who is tired of this combination is recommended to refer to the articles POLYCHROME STAINING and POLY-CHROME STAINING FORMULAE. The reference to the index will also disclose methods applicable to specific organs or specific organisms.

PETER GRAY

References

Clayden, E. C., "Practical Sectioning, Cutting and Staining," Brooklyn, Chemical Publishing Company, 1948.

Gray, P., "Microtomist's Formulary and Guide," New York, Blakiston Company, 1954.

Gray, P., "Handbook of Basic Microtechnique," 3rd ed., New York, McGraw-Hill Book Company, 1964.

See also: AUTOMATED HISTOLOGICAL EQUIPMENT AND TECHNIQUE, NITROCELLULOSE EMBEDDING AND SECTION-ING, PLASTIC EMBEDDING AND ULTRA THIN SECTIONING.

PHASE MICROSCOPY

The study of living cells has long been a difficult problem in biological microscopy. Although methods such as polarizing and fluorescence microscopy have been available for observing specimens with special physical or chemical properties the more general methods remained unsatisfactory until the development of phase contrast microscopy. The basic problem is that the living cell is an essentially transparent object and has very little effect on the intensity of light passing through it. A photo-receptor such as the human eye or a photoelectric cell is sensitive only to intensity changes in the incident light. A perfectly transparent object absorbs no light, so the transmitted intensity is unchanged and therefore un-detected. Nevertheless the transmitted light is affected by its passage through even a transparent object. By definition, the refractive index, n, of a material is the ratio of the velocity of light *in vacuo* to the velocity of light in the material. Thus in passing through a slab of glass the incident waves would be delayed and arrive at the eye a tiny fraction of a second later than in the absence of the glass. The delay depends on the product of the refractive index n_o of the glass and its thickness t, i.e., $n_o t$. If the glass were removed the light would pass through the same thickness of surrounding medium (e.g., air) of refractive index n_m and would undergo a delay depending on $n_m t$. Thus in general the relative phase change or optical path difference for an object of re-fractive index n_o and thickness t, surrounded by a medium of refractive index n_m is given by

$$\phi = (n_o - n_m)t$$

For an object of heterogeneous composition such prod-ucts have to be integrated throughout the thickness of the specimen so that

$$\phi = \int (n_o - n_m)dt$$

Living cells or other transparent objects may vary in refractive index or thickness, or both. In any case phase delays are impressed on the incident wave which thus carries the information about the object structure. Since ordinary photodetectors are insensitive to phase changes some means has to be found of converting such changes to intensity changes which can be detected. The Dutch physicist Zernike developed such a method in the early 1930s but commercial development was delayed by the Second World War and major progress was made after 1945. Today the technique is widely used and is the method of choice for the routine study of living cells.

Theory. Phase contrast depends upon the phenomenon of diffraction. Whenever light waves encounter an object some of the energy is scattered or diffracted by edges and irregularities in the object. The diffracted light can be regarded as waves originating at the object and spreading out as a set of spherical wavefronts. Thus in any micro-scope the objective lens will collect both diffracted and direct (background) light. It is the interaction between these two sets of waves that results in the formation of the final image. The position is summarized diagram-matically in Fig. 1. C and O represent the condenser and objective lenses respectively. D is the substage condenser iris diaphragm, S the specimen plane, and I the image plane, which is normally inspected through an eyepiece E. Consider any point on the plane of D; for simplicity the point P on the optical axis is chosen. The plane of D is usually at or very close to the focal plane of the con-denser. Thus a pencil of rays emanating from P would emerge from the condenser as a pencil of parallel rays. After passing through the specimen these rays fall on the objective O, and since they are parallel, are brought to a focus at P′ on the rear focal plane of O. Thus the point P on the front focal plane of the condenser is imaged at P′ on the rear focal plane of the objective. If we consider any other (nonaxial) point in the plane of D it can be seen that it too will be imaged at the rear focal plane of O. The specimen plane will be illuminated by a number of pencils of parallel rays varying in their obliquity to the optical axis. After being focused at P′ the rays from P diverge to cover a large area of the image plane I. The overlapping of all the rays from various points of the plane D forms a uniform back-ground of light over the image plane. This is true whether a specimen is present or not and the illumination is usually referred to as the direct or background light. Now let us consider the effect of a specimen at S. Each parallel pencil passing through the specimen can be regarded, in terms of wave optics, as a set of plane waves. These waves are diffracted by the object, which can be

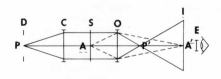

Fig. 1

regarded as a source of spherical waves which spread out over the front lens of the objective. This light is brought to a focus by the objective O on the image plane I. The broken lines in Fig. 1 show how the light from the axial point of the specimen A is focussed at A′ where it interacts with the unfocused direct light. The state of affairs at the rear focal plane of the objective requires special study. This plane can be inspected by removing the eyepiece. In the absence of an object only the image of the diaphragm plane D will appear there. Thus if D were closed down to a pinhole so that only P is illuminated, a bright image of P will be seen at P′ but the rest of the rear. focal plane will be dark. If, however, an irregular specimen is present the light diffracted by it will pass through a broad area of the rear focal plane of the objective as shown in Fig. 1. In a practical case the energy distribution of the light in this plane will depend on the nature of the specimen. If it contains a lot of very fine detail (i.e., high spatial frequencies) light is diffracted over a wide angle so that the peripheral parts of the rear focal plane are fairly well illuminated. Broad details (of low spatial frequency), on the other hand, are diffracted over a narrow angle so that only the region around P′ will be illuminated. If the object has a regular periodic structure such as a ruled grating or a diatom, light will be diffracted at a number of definite angles so that several discrete images of the point P will appear at the rear focal plane. In his well-known theory of the microscope Abbé considered such periodic objects and showed that for resolution to occur the direct image and at least one diffracted image must pass through the rear focal plane.

Let us now consider the nature of the specimen and its effect on the diffracted light. A specimen detail may differ from its surroundings in refractive index, thickness, or light absorption, or in a mixture of all of these properties. The case of a partially absorbing object immersed in a medium having the same refractive index as itself is relatively simple. The final effect of such an object on an incident wave is to reduce its amplitude without change of phase. Thus in Fig. 2 if wave A represents the incident wave, B represents the transmitted wave which differs from A only in amplitude. Now as we have seen, the transmitted wave is the resultant of the incident wave A and a wave diffracted by the specimen. We must therefore ask "What wave, added to wave A will give wave B as the resultant?" The answer is clearly the wave shown by the broken line in Fig. 2 which, by definition, represents the diffracted wave. The two waves must be added with due regard to sign, i.e., downward displacements must be subtracted from upward ones. The diffracted wave is thus seen to be exactly half a wavelength out of phase with the incident wave, so that its troughs coincide with the peaks of the latter.

The case of a transparent object is rather more complicated. In Fig. 3a A represents the incident wave. The transmitted wave B has the same amplitude, since no light is absorbed. However it is changed in phase. This can be represented by a lateral displacement. We shall assume that the phase delay is small and wave B is shifted slightly to the left of A (i.e., delayed). As far as our detector is concerned the two waves have the same intensity (proportional to the square of the wave amplitude) and are therefore indistinguishable, unlike the situation in Fig. 2. Let us now ask the same question as before, namely "What wave added to wave A will give wave B as the resultant?" The answer is the wave

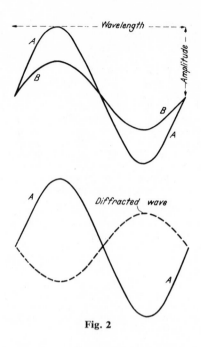

Fig. 2

shown by the broken line in Fig. 3b. It does not seem to bear any obvious relationship to wave A but in fact it can be shown that if the phase change is small the two waves are approximately one-quarter of a wavelength out of step. Thus from the point of view of wave optics the main difference between an absorbing and a refractile object resides in the different phase relationship between the incident and diffracted waves.

Since the diffracted wave is displaced in phase by about one-quarter of a wavelength from the incident wave, it is clear that if we could somehow separate these two waves physically and increase the displacement to half a wavelength, we should reach the situation shown in Fig. 3c. The diffracted wave would now be exactly out of phase with the incident wave and the new resultant wave would be indistinguishable from wave B in Fig. 2. This is the essential step needed to produce phase contrast and if it could be achieved the image of a transparent object would resemble that of an absorbing one.

Practical Design. Figure 4 shows how this problem can be solved in practice. In a conventional microscope with a circular substage iris diaphragm an image of the plane of this diaphragm is formed at the rear focal plane of the objective (Fig. 1). If a specimen is present there will be diffracted light also present at this plane, but if the substage iris is fairly widely opened there will be no way of distinguishing between the direct and diffracted light as normally the rather faint diffracted light will be swamped by the much brighter direct light. If, however, the diaphragm D (Fig. 1) is replaced by an aperture of a definite shape, such as the clear annulus shown in Fig. 4, an image of that annulus will be formed at the rear focal plane of the objective. This image will occupy only a relatively small part of the rear focal plane, the rest of which will be illuminated by diffracted light only. Thus, in practice, the direct and diffracted light can be separated at this plane except for the relatively small area of overlap superimposed on the image of the annulus. The phase difference between these two sets of waves can be altered

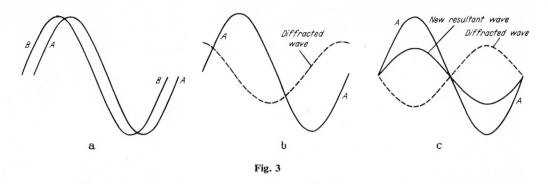

Fig. 3

by inserting a phase plate *P* at the rear focal plane of the objective. This can take the form of a transparent plate with an annular groove which coincides with the image of the substage annulus. The direct light will now pass through the groove whereas the diffracted light will be spread out over all or most of the area of the phase plate. Since the diffracted light has to pass through a greater thickness of material it will be delayed relative to the direct light. By proper selection of the depth of the groove and refractive index of the phase plate the phase difference can be made about one-quarter of a wavelength so that the conditions shown in Fig. 3c are fulfilled. Since the amplitude of the direct wave is usually much higher than that of the diffracted wave it is common practice to deposit a thin layer of metal or dye on the groove of the phase plate. This reduces the intensity of the direct light so that object details tend to look blacker, i.e., contrast is improved. It is impossible to match the characteristics of a fixed phase plate to those of every type of object likely to be encountered. It is not necessary for the phase plate to introduce a phase change of exactly one-quarter of a wavelength though this is favored by most manufacturers. The degree of absorption of direct light by the phase plate can also vary widely. For highly refractile objects low-absorption plates may be desirable, but to obtain high contrast from small details of low refractility, absorptions of between 75% and 95% are frequently used. Phase plates are usually built into the objective and a set of interchangeable annuli are provided in the substage condenser, usually mounted on a rotating turret. The substage annulus corresponding to the phase plate of a given objective can then be selected rapidly and a centering device is usually provided to enable the image of the substage annulus to be superimposed exactly

on the groove of the phase plate. Apart from this adjustment, the procedure for setting up a phase contrast microscope is very similar to that adopted in conventional microscopy. Köhler illumination, in which an image of the light source is focused on the plane of the substage condenser diaphragm, is generally used and it is important to ensure that the size of this image is adequate to cover the largest annulus likely to be used.

A number of more elaborate phase contrast systems have been proposed from time to time. In some of these the rear focal plane of the objective is reimaged by means of an auxiliary optical system, so that phase plates can be inserted at some position outside the objective. This enables an ordinary objective to be used with phase plates of different characteristics. Other systems in which polarizing elements and birefringent crystals are used allow the phase and amplitude characteristics of the phase plate to be varied continuously. However, though such systems are interesting in principle, it cannot be said that they have found great favor in practice. For the majority of biological work fixed phase plates and perhaps a number of alternative objectives are usually preferred.

In the system discussed above the image of a refractile detail is made to appear darker than its surroundings. This is generally known as positive phase contrast. However, if the diffracted wave is displaced by the phase plate by such an amount as to make its peaks coincide with those of the incident wave, the new resultant wave will have an amplitude greater than that of the background wave, so that object details will appear brighter than the background. This is known as negative phase contrast and it occasionally has certain advantages, particularly when the object contains a mixture of details of

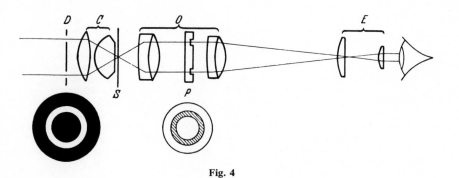

Fig. 4

low and high refractility. The system known as Anoptral contrast is simply a negative phase contrast method in which a layer of soot was originally used to provide both phase change and absorption. Negative phase contrast in some ways resembles the image seen by dark ground illumination and, indeed, approaches the latter as the absorption of the phase plate increases. If all the direct light is completely absorbed by the phase plate, the object will only be visible by virtue of the light it diffracts, giving one form of dark ground illumination.

Practical phase contrast microscopy is capable of yielding images of living cells and other transparent objects with excellent optical quality and a resolving power of about the same order as that achieved in conventional microscopy. However, there is one defect that must be clearly appreciated. This is the presence of a halo around all object details. If the object appears dark the halo will be bright and vice versa. The halo is the result of diffraction at the phase plate. Perhaps a simpler way of looking at it is to regard it as the redistribution of light inherent in any phase contrast image. Since the object is transparent no light is absorbed, so that if one part of the image field is brighter or darker than the rest some other part of the field must show compensatory changes in illumination. In practice, the object is surrounded by a more or less conspicuous halo, though this is usually obtrusive only if the phase annulus is very broad or heavily absorbing. A related defect is what is known as the "zone of action" effect. If the object is large and uniform, light will be diffracted over a relatively narrow angle, so that most of the diffracted light will pass through the phase annulus. Such objects exhibit the full phase contrast effect only at their edges. The intensity at the center of the object may approach that of the background. This may actually be helpful when examining a large cell containing many fine granules. The latter diffract light over a wide angle and show full phase contrast, whereas the broad details of the cell appear with lower contrast. Thus the internal contrast of the fine details may be heightened. This effect is often very noticeable when comparing phase contrast with interference contrast images where internal contrast tends to be low.

Quantitative Applications. Simple theoretical considerations show that the relationship between image intensity I and phase change ϕ is given by a formula of the type

$$I = a - b \cos \phi + c \sin \phi$$

This ignores the halo and zone of action effects, which are extremely difficult to take into account. In general, in positive phase contrast the intensity falls to a minimum for a certain value of ϕ and then increases with ϕ until at a certain stage reversal of contrast may occur. This nonlinear relationship makes it difficult to estimate ϕ visually or from photometric measurements. In any case the presence of the zone of action and halo effects would vitiate such estimates. Phase contrast is not therefore a suitable method for determining optical path differences. However, it can be used as a very delicate null method for determining refractive index. As we have seen, ϕ is defined by the relationship $\phi = (n_o - n_m)t$. Thus, when $n_o = n_m$, $\phi = 0$ for all values of t. At the same time the image intensity would equal that of the background, so the image would disappear. If the refractive index of the medium is slightly less than that of the object, ϕ is positive but if it exceeds that of the object, ϕ becomes negative.

In other words, the contrast reverses through the null point. The high sensitivity of phase contrast makes it an excellent method for determining the refractive index of a transparent object by means of immersion refractometry. The object can be observed while mounted in different media until a close match in refractive index is obtained. This is a relatively simple procedure in the case of inert objects such as many crystals or fixed biological tissues. In such cases a wide range of organic media can be used as immersion fluids. Measurements of this sort have proved of some value in studying tissue sections. Quite apart from the quantitative applications, variation of refractive index of the mounting medium can be a very valuable procedure as it enables the observer to change the relative contrast of different parts of the specimen and to bring out the particular features in which he is interested. The use of immersion refractometry for studying living cells is more valuable as it provides quantitative cytochemical data. The problem here is to find a suitable immersion medium that is nontoxic, will not penetrate the cell, and whose refractive index can be varied over a fairly wide range. In addition, the osmotic effects of the medium should be minimal, so that it can be brought to isotonicity with the cell by the addition of salt. Changes in cell volume must be avoided as they affect the refractive index. Protein solutions may fulfill these conditions. Bovine plasma albumin is most widely used as it can be obtained commercially and is soluble up to a concentration of about 55%. For measurements on single cells the concentration of the surrounding medium has to be changed until the cell or at least its periphery becomes virtually invisible. However, in many cases a knowledge of the distribution of refractive indices among different members of a cell population is more useful. Here it is easier to make cell suspensions in media of various concentrations and to estimate the relative number of bright and dark cells, i.e., those with a refractive index above or below that of the medium. The importance of the refractive index is that it can be directly related to the concentration of organic solids within the cell. This is because there is a linear relationship between the refractive index of a solution and the concentration of solute, i.e.

$$n = n_s + \alpha C.$$

Where n is the refractive index of the solution, n_s that of the solvent (usually water or saline), α is a characteristic constant known as the specific refraction increment, and C is the concentration of solute in grams per 100 ml. The value of α for proteins is approximately .00184 with relatively little variation. Other substances of biological importance have values of α that do not deviate much from this and it is generally accepted that an average value for "protoplasm" can be taken as .0018. Thus, once the refractive index of the cell has been determined the value of the solid concentration C can be calculated.

The application of this method is limited to those parts of the cell that are in contact with the medium. However, if the refractive index of the cytoplasm is matched to that of the medium it is possible to judge whether an intracellular inclusion has a higher or lower refractive index than that of the cytoplasm. For cells of simple geometry or for which the cell volume can otherwise be measured the volume of solid concentration can be combined with cell volume to calculate both dry and wet mass.

R. BARER

References

Barer, R., "Phase, interference and polarizing microscopy," in R. C. Mellors, ed., "Analytical Cytology," 2nd ed., New York, McGraw-Hill, 1959.

Barer, R., "Phase contrast and interference microscopy in cytology," in A. W. Pollister, ed., "Physical Techniques in Biological Research," Vol. 3A, New York, Academic Press, 1966.

Bennett, A. H., et al., "Phase Microscopy," New York, Wiley, 1951.

Ross, K. F. A., "Phase Contrast and Interference Microscopy for Cell Biologists," London, Arnold, 1967.

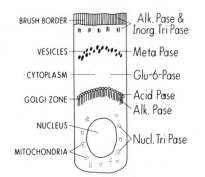

Fig. 2 Regional distribution of seven phosphatases in a columnar absorbing cell of human intestine.

PHOSPHATASES

1. Scope of Technique. Phosphatase is the common name for a phosphohydrolase, an enzyme that catalyzes the hydrolysis of phosphate esters. The site of hydrolysis can be marked in a tissue section by a suitable staining reaction. This provides a histochemical localization of the enzyme, which is identified biochemically by the substrate, cofactors, and hydrogen ion concentration in the medium chosen for the reaction. For example, a section of rat jejunum viewed at low power is described as having alkaline phosphatase in the columnar epithelium of villi, acid metaphosphatase in the *lamina propria mucosae*, nucleoside polyphosphatase in *lamina propium mucosae*, smooth muscle, and arterial endothelium, and a neutral nucleoside triphosphatase (lead-sensitive) in the ganglia of the myenteric plexus (Fig. 1). The picture is, of course, a composite. With good technique, four consecutive frozen sections can be differentially fixed and stained so that each of the first three enzyme reactions appears alone in one section, while both nucleoside phosphatase reactions appear in the fourth one.

When the section is to be viewed at higher power, similar procedures serve to localize various phosphatases differentially within a cell (Fig. 2). When stained sections are post-fixed and processed for electron microscopy, differential localization can be extended to organelles (Fig. 3) and portions of organelles (Fig. 4).

2. First Level of Resolution. Phosphatase histochemistry is a spectrum of procedures of increasing biochemical specificity and microscopic precision. Routinely used reactions represent the lowest level of microscopic and chemical resolution. They have to be interpreted carefully,

because the site of reaction need not use the same substrate *in vivo* (the enzyme hydrolyzing adenosine triphosphate may not be an ATPase, or may not even be a phosphohydrolase under physiological conditions), and the sites that gave no reaction in the section are not necessarily free of the enzyme *in vivo*. Experience with staining patterns found after different fixing procedures and after incubation with different substrates and cofactors led to the general acceptance of eleven classes of phosphatase reactions. They are listed here in order of increasing specificity (key substrates are shown in parentheses). The last four reactions may be specific for one substrate each.

1. nonspecific alkaline phosphatase (organic esters of mono and polyphosphate; inorganic pyrophosphate)
2. nonspecific acid phosphatase (organic esters of mono- and polyphosphate)
3. nucleoside polyphosphatase (nucleoside tri-, di-, and occasionally monophosphates)
4. nucleoside diphosphatase
5. nucleoside triphosphatase (ATP, other nucleoside triphosphates)
6. nucleotide phosphatase (5′ nucleotides)
7. metaphosphatase (inorganic tri- and tetrametaphosphate)
8. glucose-6-phosphatase
9. thiamine pyrophosphatase
10. triphosphatase (inorganic triphosphate)
11. pyrophosphatase (inorganic pyrophosphate)

Each reaction is obtained by a standard procedure (14). The differences between procedures are not limited to the incubation medium, but may include every step of technique. The four examples given below illustrate all the techniques of fixing, sectioning, localizing the reaction, staining, and mounting that are of general use in phosphatase histochemistry.

The names of the standard phosphatase reactions are a key to a wealth of descriptive literature that has been accumulating since 1939 (1, 16). Each reaction produces a distinctive anatomical pattern (Figs. 1 and 2) which remains remarkably constant from species to species. Microscopists used both the regularities of distribution and the departures from norm to solve problems in comparative anatomy, histology, embryology, and pathology (Examples 1, 2, and 3 below). In sum, the routinely performed phosphatase reactions serve as highly selective differential stains.

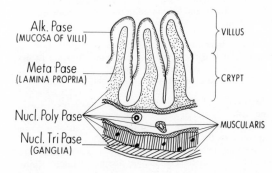

Fig. 1 Tissue distribution of four phosphatases in rat jejunum.

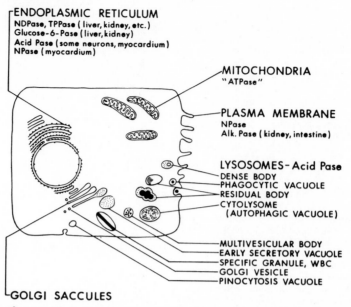

Fig. 3 Distribution of seven phosphatases among organelles of mammalian cells. (By courtesy of Goldfischer et al. and *J. Histochem. Cytochem.* [6].)

Example 1. Trimetaphosphatase Reaction in Comparative Anatomy and Embryology. Method. Small specimens of living tissue are fixed in ice-cold neutral 4% formaldehyde, rinsed, dehydrated in acetone, cleared in benzene, and embedded in paraffin at no more than 56°C. Sections are spread on a warm solution of .02M magnesium acetate pH 6, which is blotted during mounting on albumenized slides. Slides are processed through benzene, dioxane, and water, for incubation in the following reaction mixture:

substrate: sodium trimetaphosphate (Victor Chem.) 0.6 mM
buffer: acetate and acetic acid 50 mM
capturing ion: lead II (as lead acetate) 4.6 mM
pH: adjusted with KOH to pH 4.8

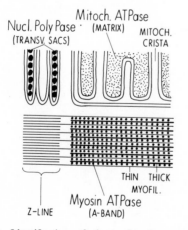

Fig. 4 Identification of three phosphatases at high resolution within three organelles of a mammalian heart muscle fiber.

The reaction is stopped by a thorough rinse in water, the color is developed in 0.3% ammonium sulfide. In this, as in all other lead sulfide stains, care must be taken not to bleach the sulfide during counterstaining, dehydrating, and final mounting. A good mounting medium is obtained by dissolving a dry resin (H.S.R. or Bioloid) in high-purity tetrachloroethylene.

Application. The trimetaphosphatase reaction was found in digestive tract epithelia of a wide variety of animals from flatworm through frog to man (not in the rat, however, where the reaction was in *lamina propria*, Fig. 1). In nearly all species, the reaction was positive in the epithelia derived from endoderm (Fig. 2), and stopped at their junction with epidermis. In frog tadpoles, the epidermis of external gills also had a positive reaction. This pattern of distribution provided support for the theory, that gill epidermis of these tadpoles was derived from endoderm rather than from ectoderm.

Example 2. Nucleoside Triphosphatase Reaction in Muscle Histology. Method. A specimen of living muscle is quick-frozen (quenched) and freeze-sectioned across the muscle fibers. Sections are finger-mounted on coverslips, air-dried, and incubated in the freshly made medium of Padykula and Herman (14).

buffer: barbiturate (sodium) 20 mM
cofactor and capturing ion: calcium (chloride) 18 mM
substrate: ATP (sodium) 5 mM
pH: adjusted with NaOH to pH 9.4

The reaction is stopped with repeated rinses in 1% CaCl$_2$, and color is developed in two steps: exposure to 2% CoCl$_2$, followed by a thorough rinse in water and exposure to 0.3% ammonium sulfide.

Application. Vertebrate skeletal muscles have three anatomically distinct kinds of fibers: "white," "intermediate," and "red." The nucleoside triphosphatase reaction of unfixed sections is selectively weak in one fiber type, which belongs to the "intermediate" group, and is strong in all other fibers, i.e., in all the "white" and "red" and in a few "intermediate" ones. (If the section had been fixed in formaldehyde, the reaction of "white" fibers would have also turned weak; if it had been pretreated in a mildly acid solution, only the "white" fibers would have given a reaction.)

The same muscles have fibers with two distinct mechanisms of contraction, "slow" and "fast" ones, and there is recent evidence that the "weak" staining reaction in unfixed fibers is a selective histological marker for physiologically "slow" fibers (5).

Such patterns of staining may be given a rational explanation in the future, but are purely empirical in present use. Hydrolysis of trimetaphosphate, for example, belongs to no known biochemical pathway, and there is no explanation for the presence of the trimetaphosphatase reaction in some cells. By contrast, hydrolysis of ATP occurs in many biochemical pathways. It is consequently expected in every cell, and there was no theoretical basis for predicting in advance of experiment that a particular histochemical technique leading to the hydrolysis of ATP would give a selectively weak stain in a particular muscle fiber and not another. This did not detract from the use of a phosphatase reaction as a selective stain, which is illustrated again in the third example.

Example 3. Acid Phosphatase Reaction in Pathology. Method (3). Small biopsy specimens are fixed overnight in refrigerator (4°C) in:

formaldehyde: 4%
CaCl₂: 1%
NaOH, to pH 7

Frozen sections are cut and floated on water and incubated in Burstone's reaction mixture, freshly prepared from four stock solutions:

5 ml A. buffer: 2% barbiturate, 3% acetate with Na⁺ to pH 5.0
1 ml B. substrate: 1% Naphthol AS-TR phosphate in N,N-dimethylformamide
0.8 ml C. dye stabilizer: 4% sodium nitrate
0.8 ml D. capturing dye: 4% pararosanilin-HCl in 2N HCl
12 ml water

Solutions C and D are mixed together before adding; special precautions are required in preparing both the stock solution and the mixture. Reaction is stopped by a rinse in water, and it is complete at that point, with the color fully developed. Sections are counterstained and mounted in the usual way.

Application. Acinar cells of the prostate gland have a distinctively strong acid phosphatase reaction in mammals. The function of the prostatic enzyme is not known. Lymph node cells have a weak acid phosphatase reaction. The appearance of a strong reaction in lymph node biopsies of a patient with a prostatic tumor serves to diagnose a carcinoma with metastases to lymph nodes (16). Diagnosis is made faster by incubating frozen sections taken directly from an unfixed biopsy specimen. The histological reaction pattern is the same but cytological detail is poorly preserved.

3. Higher Levels of Resolution. The eleven varieties of phosphatase activity are subdivided further by characterizing the reactions in terms of differential denaturation, differential activation and inhibition, and substrate specificity. The goal is to assign the phosphatase reaction

found at some selected site to an enzyme (or a set of enzymes) that is defined in terms of (a) biochemistry, (b) physiology, and (c) cell anatomy.

a. The biochemical name and number of the enzyme is found by referring to the official handbook of enzyme nomenclature (10). For example, the histochemical nucleoside triphosphatase reaction is assigned to a "nucleoside triphosphate phosphydrolase, E.C.3.6.1.3." The official classification necessarily lags a few years behind published research, and enzymes that are biochemically distinct may still share one name. The microscopist may need to work out his own, more precise definition of the "individual enzyme" he works with.

b. The physiological definition specifies the contribution made by the phosphatase reaction to a known metabolic pathway and a known physiologic function. This again is a first approximation; in more precise studies each enzyme is regarded as a transducer which alters the rate of reaction in response to selected stimuli, and distinctions are made between enzymes which control the same reaction differently in different types of cells.

c. The cytological definition relates an enzyme to a structure within the cell. Localizations resolved in a light microscope are a first approximation. Ultimately, localization must be done at the level of ultrastructure and correlated with cytochemical fractionation.

In spite of the vast amount of published observations to date, only four sites of phosphatase reactions can be assigned with any precision to a known enzyme performing a known function. They are:

1. Glucose-6-phosphatase, E.C.3.1.3.9, found in the cisterna of endoplasmic reticulum and functioning in gluconeogenesis (8).
2. Mitochondrial nucleoside triphosphatase, E.C.3.6.1.3, found in the matrix between mitochondrial cristae, and functioning in respiratory synthesis of ATP (7).
3. Actomyosin, E.C.3.6.1.3, found at the crosslinks of myofibrils in striated muscle, and functioning in contraction (15).
4. Bicarbonate-activated ATPase, E.C.3.6.1.3, found on the inner face of microvilli of oxyntic-peptic cells in amphibian stomachs, and functioning in the secretion of chloride ions (11).

Example 4. Definition of Phosphohydrolase Activities of Contractile Apparatus and Mitochondria of a Muscle Fiber. The experiment depends on the comparison of several procedures at each step in the histochemical reaction sequence.

Preparation of Specimen. Muscle samples are mechanically shredded into fibers, or else longitudinally freeze-sectioned in thin sections (8 μ) and thick sections (20 μ).

Fixation. Unfixed preparations are standard, but they are routinely compared with shreds or sections that were fixed for 15 min at 0°C in neutral buffered solutions of formaldehyde with MnCl₂ and CaCO₃ (9), or glutaraldehyde (9) or hydroxyadipaldehyde (15).

Incubation. Two standard reaction mixtures are used: (a) Padykula-Herman formula found in Example 2, (b) Wachstein-Meisel formula (9).

buffer: tris-maleate 80 mM
cofactor: mangesium (sulfate) 10 mM
substrate: ATP (sodium) 1 mM
capturing ion: lead (nitrate) to opalescence
pH adjusted with NaOH to pH 7.2.

Several variants are prepared of each formula, using alternate substrates, alternate cofactors, and additional activators and inhibitors (2). Incubation is stopped by a rinse in water or in fixative.

Preparation for Light Microscopy. Color is developed in shredded specimens and in thin sections as described in Example 2 (for a calcium precipitate) and in Example 1 (for a lead precipitate), and the specimens are mounted between slide and coverslip in a suitable medium, such as aqueous polyvinyl pyrrolidone.

Preparation for Electron Microscopy. Thick sections are taken directly from the post-incubation rinse into a strong fixative (buffered osmium tetroxide) and processed like a standard specimen for electron microscopy (9). Shredded specimens are centrifuged into a pellet in the rinse fluid, and the pellets processed similarly to thick sections.

Results. The properties of phosphatases at locations shown in Fig. 4 are as follows. The reactions in myofibrils and mitochondria are specific for nucleoside triphosphate substrates, and inactivated by exposure to formaldehyde fixatives. This distinguishes the two enzymes from nucleoside polyphosphatase, E.C.3.6.1.5, which is found in the same muscle fibers at a site in the T-complex (6).

At the unfixed myofibril the reaction requires calcium salts for activity; is activated less strongly by Ba^{++} and Sr^{++}, and not at all by Mg^{++}; is inhibited by dinitrophenol, but not by azide; and is specific for ATP as substrate at a neutral pH (but uses both ATP and inosine triphosphate at an alkaline pH). In all these properties it behaves as would the enzyme myosin of striated muscle in the presence of actin. This definition of the enzyme is confirmed by other histochemical methods (differential extraction, immunofluorescent labeling) which identify the crosslink as myosin, and the coating of the thin myofilament as actin (4).

In the mitochondrion, the reaction requires magnesium salts for activity; is activated less strongly by Mn^{++} and Co^{++}, and not at all by Ca^{++}; is activated by dinitrophenol and inhibited by azide; and uses both ATP and ITP as substrates at a neutral pH. In all these properties the enzyme behaves as would the nucleoside triphosphatase of mammalian mitochondria. The location agrees with the ultrastructural location of "headpieces" which are thought to contain the coupling mechanism for the respiratory synthesis of ATP (7).

The histochemical phosphatase reaction carried out at a high level of resolution, as in Example 4, becomes a rational test for the detection of a known enzyme. When the specific reaction is present at one site and absent at another in the microscopic preparation of a cell, this is a reliable indication of the distribution of the enzyme in the living cell.

4. Highest Resolution. The ultimate in resolution is to define a phosphatase in the biophysical sense, i.e., to identify the isoenzyme molecules and their structural relation to other molecules at the site of phosphatase activity. Work at this level is still scarce. Isozymes which may correspond to unique amino acid sequences are known for an acid phosphatase of human white cells, and an alkaline phosphatase of mouse intestine.

Example 5. Differential Location of Isoenzymes. In white cells of human blood, seven isoenzymes of acid phosphatase have been identified by acrylamide gel electrophoresis, molecular weight, and substrate specificity. Different combinations of these isoenzymes were found in each of six types of white cells. For example, only the neutrophilic granulocytes contained isozyme 2 (12).

Example 6. Differential Ratios of Isoenzymes. Alkaline phosphatase is a zinc metalloenzyme activated by magnesium ions. In the mucous epithelium of intestine it is located on the outer face of the plasma membrane of microvilli, and at anatomically contiguous sites in vesicles and Golgi apparatus (Figs. 1 and 2). The intestinal enzyme is biochemically distinct from alkaline phosphatases of other organs. Intestinal alkaline phosphatase of mice has, in turn, been separated into two isoenzymes by chromatography, electrophoresis, immune reactions, and affinity to different substrates. The ratio of the two isoenzymes varies from site to site, and the differences have been localized with some precision by means of a microdissection technique based on freeze-sectioning. For example, the isoenzyme with high affinity to phenyl phosphate is the predominant form in the duodenum; it is inactive in crypt cells of mucous epithelium, becomes active when cells are at the base of the villus, and increases at least fourfold in activity in cells toward the tip. By contrast, the activity of the other isoenzyme increases very little from base to tip. As the columnar cells are produced in the crypt and glide along the villus to the tip, these regional changes show that the molecular structure of the membrane keeps changing with the age of the cell (13).

5. Principles of Microtechnique. The standard procedures of fixing, sectioning, incubating, clearing, and mounting, illustrated in Examples 1 through 4, represent an optimum compromise between the conflicting requirements of preservation of structure and preservation of enzyme activity, and a detail of any step may prove critical. The histochemical incubation mixtures have a substrate in solution, together with a capturing ion. The phosphatase reaction always breaks a phosphate ester bond of the substrate, and a product of the reaction forms an insoluble precipitate with the capturing ion, but there are three mechanisms for producing this precipitate.

a. Dye Coupling. The substrate is a substituted naphthyl phosphate, and the dephosphorylated residue is insoluble. The capturing ion is a selective dye (Example 3).

b. Phosphate Precipitation. A capturing cation is present in excess in solution, to make the phosphate residue of the reaction insoluble. The cation is calcium in alkaline media (Example 2), and lead in acid media (Example 1). The insoluble residue can be orthophosphate (Example 2), pyrophosphate, or triphosphate (Example 1). The precipitate forms apatite-like crystals and can be viewed directly in the electron microscope; it has to be stained for viewing in the light microscope.

c. Chelate Removal. In neutral solution containing polyphosphate esters and lead, the substrate (such as ATP) would form an insoluble salt if lead were added in excess, as in the previous method. If substrate is added in excess, it binds all the lead ions in a chelate, so free lead ions are not available for precipitation. Reaction mixtures must consequently be titrated so that they are close to saturation with lead (Example 4). The phosphatase reaction contributes to precipitation in such a mixture in three ways: by releasing two insoluble products of hydrolysis (for example ADP and orthophosphate), and by releasing lead ions from chelate bonds. These three compounds form mixed precipitates, and precipitation can be enhanced further by including lead-capturing anions such as chloride in the reaction mixture. The sediment is stained with sulfide for viewing in a light microscope, but can be seen directly in the electron microscope, and is also visible without staining when the phosphatase reaction is done in an acrylamide gel electropherogram.

Any technique chosen for a phosphatase reaction brings its own assortment of artifacts, both negative (loss of activity from sites containing the enzyme) and positive (a stain at a site that has no enzyme). Artifacts can arise at every step, from killing (autolysis and leakage of enzyme) to final staining (lead ions, for example, may act as a mordant for a stain at a nonenzymatic site). Most phosphatase reactions have been thoroughly validated and require only routine controls. Some, like the nucleoside triphosphatase (neutral ATPase) reaction of cell nuclei, still have to be validated for each new site, by tests that compare the kinetics of histochemical reaction to biochemical properties of the enzyme. Two kinds of artifacts which were supposed to be most damaging to the neutral ATPase reaction on theoretical grounds proved to be trivial in practice. The first was the binding of lead to protein. In practice, this produces only a diffuse background stain, because ATP has the fortuitous property of inhibiting much of the direct binding of lead to tissue. The second was the nonenzymatic breakdown of ATP, catalyzed by lead ions in the incubation medium. In practice, the fraction of total sediment which is contributed by this artifact can be made insignificantly small, by exposing many enzymatically active sections to a small volume of medium, and by keeping incubation time short. Microscopists can now look forward to localizing a wide variety of cell functions (active transport, mechanical work, anabolism) by means of the ATPase reactions coupled to each kind of work. Thirty years after Gomori's discovery of the histochemical reaction for phosphatase, we are beginning to fill in the gaps between the microscopically observed patterns of reaction and their biochemical and physiological meaning.

Acknowledgment

Dr. Berg's work is supported by the U.S. Atomic Energy Commission contract at the University of Rochester Atomic Energy Project. This article has been assigned Report No. UR-49-1397.

GEORGE G. BERG

References

1. Arvy, L., pp. 154–303 in "Handbuch der Histochemie," Vol. II, W. Graumann and K. Neumann, ed., Stuttgart, G. Fischer, 1962; *Int. Rev. Cytol.*, **25**:333–362 (1969).
2. Barden, H., and S. S. Lazarus, *Lab. Invest.*, **13**:1345–1358 (1964).
3. Barka, T., and P. J. Anderson, "Histochemistry: Theory, Practice, and Bibliography," New York, Hoeber Division, Harper and Row, 1963.
4. Bendall, J. R., "Muscles, Molecules and Movement," New York, American Elsevier, 1969.
5. Eversole, L. R., and S. M. Standish, *J. Histochem. Cytochem.*, **18**:591–593 (1970).
6. Goldfischer, S., E. Essner, and A. B. Novikoff, *J. Histochem. Cytochem.*, **12**:72–95 (1964).
7. Grossman, I. W., and D. H. Heitkanp, *J. Histochem. Cytochem.*, **16**:645–653 (1968).
8. Hugon, J. S., M. Borgers, and D. Maestracci, *J. Histochem. Cytochem.*, **18**:361–363 (1970).
9. Hunt, R. D., Chap. 10 in "Selected Histochemical and Histopathological Methods," S. W. Thompson, ed., Springfield, Ill., C. S. Thomas, 1966.
10. International Union of Biochemistry, "Enzyme Nomenclature," New York, Elsevier and Co., 1965.
11. Koenig, S., C. and J. D. Vial, *J. Histochem. Cytochem.*, **18**:340–353 (1970).
12. Li, E. Y., L. T. Yam, and K. W. Lam, *J. Histochem. Cytochem.*, **18**:901–910 (1970).
13. Moog, F., and R. D. Gray, *J. Cell Biol.*, **32**:C1–C6 (1967).
14. Pearse, A. G. E., "Histochemistry, Theoretical and Applied," 3rd ed., Boston, Little Brown Co., 1968.
15. Tice, L., and R. Barrnett, *J. Cell Biol.*, **15**:401–416 (1962).
16. Wachstein, M., pp. 73–153 in "Handbuch der Histochemie," Vol. II, W. Graumann and K. Neumann, ed., Stuttgart, G. Fischer, 1962.

PHOTOGRAPHIC MATERIALS

For both light and electron microscopy of photographic materials, the amount of information obtainable depends strongly on sample preparation. The most common types of preparations are sections, smears, and whole mounts. In light microscopy, the examination is, with but few exceptions, by conventional transmitted light. Both transmission and scanning techniques can be used in electron microscopy.

A systematic study of an unfamiliar photographic material normally begins with the examination of preliminary cross sections of the material cut to a thickness of 2 μm. This examination will indicate the optimum section thickness, lamellar structure, size, shape, and concentration of the particulates. It will also indicate whether other types of preparations may be needed to complete the investigation.

Sections for photomicrography require the use of a good rotary microtome, a special sample holder, a stereomicroscope, and a sharp dissecting needle. The brass sample holder, shown in Fig. 1, eliminates the time-consuming procedure of sample embedding. It is the only item that is not commercially available.

No more than 1 mm of the film must project beyond the end of the holder. If a piece of the film (5 × 10 mm is a convenient size) is placed between the jaws of the sample holder and the projecting material is trimmed with a pair of sharp dissecting scissors, not only will the proper amount of material project from the jaws, but the edge of the sample will be correctly aligned for cutting in the microtome.

Photographic materials coated on a paper base require an additional step for sectioning. If the paper base is of no interest, most of it can be removed either by peeling it off with pressure-sensitive tape or by scraping with a razor blade. A piece of film base can then be cemented to the remaining paper with Eastman 910 Adhesive (Eastman Chemical Products, Inc.) and the sample sectioned in the same manner as film. Sections of the entire paper base can be obtained by painting the edge projecting beyond the jaws with collodion or a 25% solution of Duco cement in acetone. The cement will dry in seconds and a section can be cut.

The stereomicroscope is an essential visual aid to the microtomist. It enables him to place the tip of the dissecting needle on the start of the section where it forms

Fig. 1   Brass sample holder.

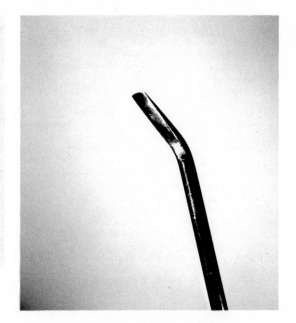

Fig. 2   Section lifter (4 ×).

on the bevel of the knife and yet avoid touching the tip of the needle to the edge of the knife. Observation through the microscope enables him to coordinate the action of his right hand, which is rotating the wheel of the microtome, with that of his left, which is holding the needle and is drawing the cut section down the knife. Tension on the section must be sufficient to keep the section from curling, but not great enough to keep it taut and thus stretch it.

It is possible to transfer the section from the knife to a microscope slide with the dissecting needle. A more satisfactory method is to use a special section lifter made from a bent dissecting needle, as shown in Fig. 2. After the needle has been shaped with a file, its surfaces are smoothed with a pocket-hone. The same hone is used to maintain a sharp point on the needle used in the sectioning.

A mineral oil, such as Nujol, is an excellent mounting medium for the preliminary section. A drop of oil is placed on the slide near the section, a cover glass is placed gently over the oil, the oil is permitted to flow around the section, and light pressure is applied to the center of the cover glass with the eraser tip of a pencil.

A number of different mounting media may be used for the final sections. The only restriction is that the medium must not cause any unwanted change in the section. Synthetic resin solutions, such as Hyrax or Permount, are good for films containing only silver or silver halide, but they may cause fading of the dyes of color films. For these films, Nujol or Aroclor 1254 is satisfactory.

Examination of a film section can provide information in addition to the obvious mechanical structure. The presence of an ultraviolet absorber can be detected by photographing the section with near-ultraviolet radiation. Brighteners can be detected by fluorescence photomicrography. Nujol is an excellent mounting medium for both of these techniques. Detail in thin layers can be made visible by intentionally swelling the emulsion. An aqueous mountant, such as glycerin jelly, will cause moderate swelling and provide a semipermanent preparation. Water

will cause much greater swelling but will limit the life of the slide.

Sections of processed film are useful for assessing lateral image diffusion and depth of image within the coating. They can also be a valuable adjunct to a microdensitometer trace. However, such sections can lead to erroneous conclusions concerning the size and shape of crystals if the observer forgets that the act of sectioning will cut through many of the crystals, thus altering the shape and decreasing the size.

A photomicrographic record of crystal morphology and size distribution of photographic grains is best made from a smear of the sample. If the film is single-layered, the procedure is very simple. A portion of the film is soaked in warm water until the coating is softened. The softened coating is then scraped off the base and transferred to a small beaker held in a 40°C water bath. An unhardened emulsion will soon melt. Hardened emulsions will require the addition of a few drops of concentrated nitric acid. The sample must be stirred gently until uniformly melted. Then a drop is placed near one end of a clean microscope slide. A small glass rod held parallel to the surface of the slide is lowered into the drop and pushed to the opposite end of the slide. The rod should almost, but not quite, touch the slide. Drying of the smear may be expedited by the use of a small hair dryer.

If there are too few crystals per field on the slide, a larger drop of the melt should be used and the coating rod held slightly higher and moved more slowly. Too great a concentration of crystals can be corrected by diluting the melt with distilled water or with a 1% solution of gelatin. If the crystals are not well dispersed, either the sample was not completely melted or too much of the gelatin was destroyed by an excessive amount of acid. If the crystals are not in one plane, the viscosity of the melt was too high; further dilution of the melt with water is indicated. Some films and many papers that have been hardened to a high degree may not soften in warm water. These materials may be softened for removal from their bases by soaking in a warm 0.1 % solution of a proteolytic enzyme, such as Takamine.

Smears of grains from multilayered films can be made in the same manner, but extreme care is necessary to remove only one layer at a time. Some workers prefer to harden the film completely and then soften one layer at a time by soaking for experimentally determined times in an enzyme solution, a differential softening technique.

Whole mounts offer the best method for studying the distribution and morphology of developed silver in the planes parallel to the support. Although smears can be made (nitric acid must NOT be used!) the silver particles may bear no resemblance in size or shape to those in the original gelatin matrix. Very low power microscopy will show the relative graininess of the material, medium powers the size of the silver aggregates, and high powers the individual silver particles and filaments. However, as the magnification increases, the depth of field decreases. A photographic record at high magnification can require several pictures at different focus levels. Examination of both sections and whole mounts of the same material is advisable for complete characterization.

Although gross changes in surface profile are readily shown by a section, small changes are difficult to detect. The examination of a shadowed replica, prepared as described later, is preferable for the study of such small changes. In some instances the replication stage can be omitted and an opaque aluminum coating evaporated normal to the surface of the sample and directly upon it. The sample must then be examined by reflected light.

When the size of silver halide grains or developed silver particles is less than a micron, it may be advantageous to use the electron microscope, which provides better definition of the shape of the particles. When studying light-sensitive materials with an electron beam, the reduction of silver halide to silver (print-out) is a more serious difficulty than in light microscopy. Over the years, sample preparation techniques have been developed to minimize this problem.

The carbon replica technique (1) avoids print-out. Briefly, the technique involves removing the emulsion grains from the sensitized coating with a 0.1% proteolytic enzyme solution when the gelatin binder has been hardened. In the case of an uncoated emulsion or an unhardened coating, the gelatin can be melted in water at 40°C and diluted to a suitable concentration as described above. Because gelatin has a high electron-stopping power, centrifuging is required to remove the grains from the excess gelatin. After the grains have been resuspended in distilled water, a drop is deposited on a Formvar-covered stainless steel specimen screen and dried. To make the replicas, the specimen screens are transferred to a vacuum evaporator set up with a sample rotator, a tungsten filament for evaporating gold-palladium (18° is a suitable angle), and carbon electrodes for making the replica. Gold-palladium is first evaporated to produce the shadow effect, and then the carbon is evaporated in short bursts while the samples rotate at 60 rpm. Complete coverage of the grains with carbon prevents the replicas from collapsing during the dissolution of the silver halide particles in a fixing bath. Finally, the specimen screens are washed in distilled water and dried. Figure 3 shows a typical carbon replica of emulsion grains.

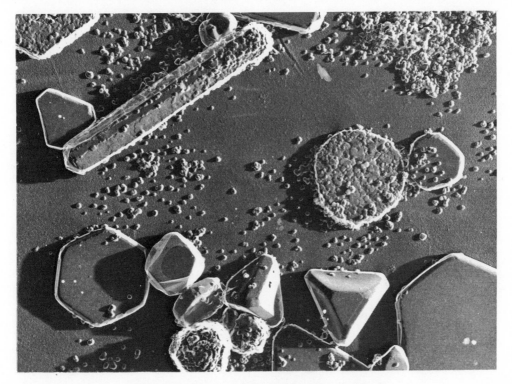

Fig. 3 Carbon replica of silver halide grains (12,500 ×).

The carbon replica technique is not useful with silver halide grains less than 0.5 μm in diameter, because the thickness of the carbon film adds appreciably to the size of the small particles. For example, a 0.5 μm grain would be increased 10% in size by a carbon film 250 Å thick. Since it is difficult to measure with any certainty the thickness of the carbon film, replicas are unsuitable for highly accurate particle-size determination. For particles smaller than 0.5 μm, replica techniques also fail to resolve some surface detail. For instance, a cube with sharp corners will appear to have rounded corners, or an octahedral grain will appear spherical.

When replicas cannot be used, the particles must be examined directly. Sample preparation is similar to that just described without the carbon evaporation and fixing steps. One must focus and take the picture as rapidly as possible to keep the print-out effect to a minimum. Mounting the suspended grains on carbon substrates instead of on Formvar will reduce thermal drift in the microscope, allow faster photography, and give sharper micrographs. Figure 4 shows silver halide grains for direct examination. If the print-out (evident as the small protuberances from the grain surface) goes too far, it is almost impossible to determine the size and shape of the grains.

Developed silver particles, which are important in image studies, are also examined in the electron microscope. Figure 5 shows filaments of silver which are characteristic of chemical development. Since silver particles are more stable in the electron beam than are silver halides, the sample preparation techniques are quite simple. The image silver is removed from the coatings by techniques described for grains. When examining these silver particles in the electron microscope, one must avoid use of too intense a beam of electrons, which may cause the small silver particles to melt and change shape.

Another method for examining developed silver is the gel-capsule technique, used for studying the relationship between the developed silver and the original silver halide grain (2). This technique depends on the fact that chemical interaction between the gelatin and silver halide changes the solubility of the gelatin associated with the surface of the grain, and the resulting shell, or capsule, can be observed in the microscope. For the technique to be successful, (a) the coating must be unhardened so that the grains can be removed by dissolving the gelatin in water at 40°C and (b) the sample must be exposed and developed, but not fixed in hypo, so that the silver halide-developed silver particles can be removed together from the coating. The resulting suspension is centrifuged to remove the soluble gelatin, and then resuspended in distilled water. A drop of the sample is placed on a carbon-covered stainless steel grid and dried. The sample is then fixed by placing the grid in a fixing bath such as KODAK Fixer (Formula F-5) for not more than 15 sec; longer times or a more active fixing bath will dissolve some of the image silver. A 1 min wash followed by drying completes the preparation. As shown in Fig. 6, the gelatin capsule indicates in outline where a silver halide grain was before fixing. The capsule is distorted as the sample dries after fixing, and no longer shows the original shape of the grain. However, the black specks shown are the developed silver particles and the number per grain can be counted. It is possible to study filamentary silver and its relation to the original grain surfaces by this method, but it is a little more difficult

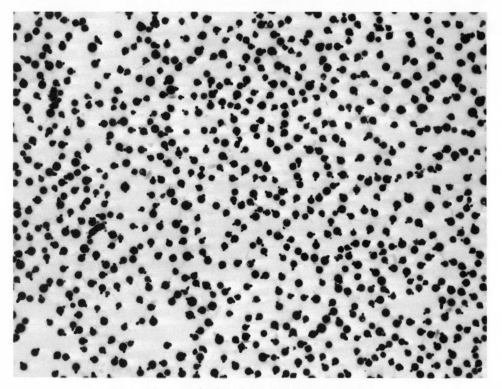

Fig. 4 Direct examination of silver halide grains (40,000 ×).

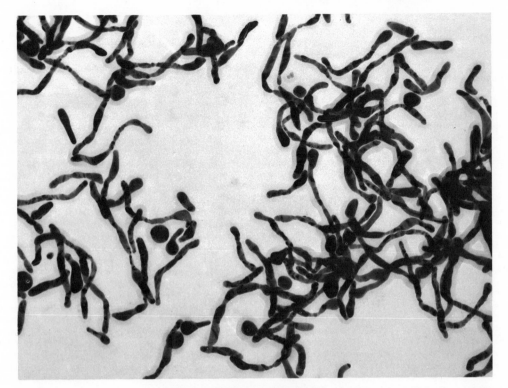

Fig. 5   Filamentary silver (80,000 ×).

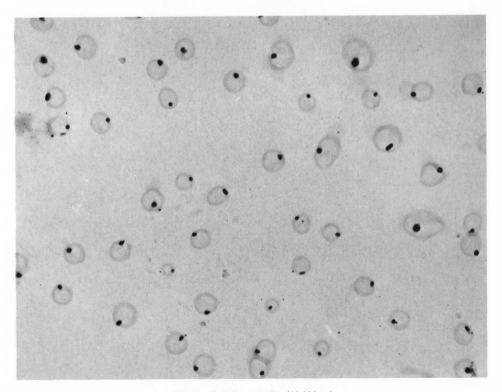

Fig. 6   Gelatin capsules (40,000 ×).

to keep the filaments from breaking off at the grain surface during the preparation of the sample.

For electron microscopy of the distribution of photographic emulsion grains from top to bottom in an emulsion layer, ultrathin sectioning techniques are useful. Ultramicrotomes equipped with diamond knives are capable of cutting good sections from 100 Å to 1000 Å in thickness. For adequate transparency to electrons, most sections should be 500 Å or less. Figure 7 is a cross section of unprocessed film showing the support, the adhesion layer, and the emulsion layer. The emulsion grains are evenly distributed through the layer. Figure 8 is a section through a piece of film that has been physically developed. Again the same layers are apparent, but this type of development causes a variation in particle size from the top of the coating to the bottom. Without ultrathin sectioning techniques, this size distribution would be difficult to measure.

Samples are usually embedded to support them during the sectioning process, but it is better to cut photographic film samples directly. However, the samples should be treated in a post-hardening bath (3% formaldehyde—1% sodium carbonate) for 5 min and rinsed in distilled water for 30 sec to harden the emulsion layer, and then dried. Since this layer will swell as ultrathin sections float on the water surface, there will be less folding of the gelatin layer if there is no embedding material on each side of the sample. This is not the case for materials coated on paper support. The fibrous nature of paper makes it difficult to cut ultrathin sections and it is better to embed the paper for maximum support during the sectioning process. One should remove as much of the paper support as possible before embedding.

The most common embedding materials for ultramicrotomy are methacrylates and epoxy polymers. Two major factors govern the choice of the embedding material: (a) The monomer should not be a solvent for either the support or the emulsion layer of the sample, and (b) the hardness should be controllable to match the sample hardness by varying the percentages of monomers and hardeners. Butyl methacrylate with varying amounts of methyl methacrylate is a good embedding material, but cannot be used with film that has a cellulose acetate support. In general, epoxy resins such as Epon 812, Araldite, and Maraglas are satisfactory for all photographic materials.

Ultramicrotomes are fitted preferably with diamond knives mounted in a trough. This trough is filled with a 0.05% aqueous solution of a wetting agent, such as Triton X-100 to collect the sections for subsequent mounting on the carbon-covered sample grid. The thickness of the sections can be estimated by comparing their interference colors to a color chart indicating thickness.

Scanning electron microscopy is a relatively new tool and it appears to have a number of important advantages for studying photographic materials. In the scanning electron microscope (SEM) the electron beam does not have to pass through the sample so there is no limitation on sample thickness. Also, the depth of field is much larger than in transmission microscopy and extremely rough surfaces can be studied in greater detail than is possible with replicas. However, present scanning microscopes have a resolution limit of about 200 Å when secondary or back-scattered electrons are detected.

With the SEM, one can examine large silver halide

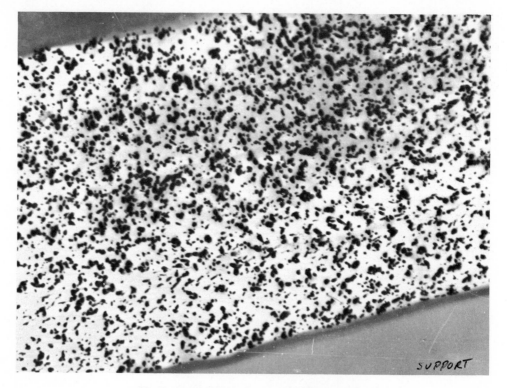

Fig. 7   Cross section of unprocessed film (3,000×).

Fig. 8 Cross section of physically developed film (20,000×).

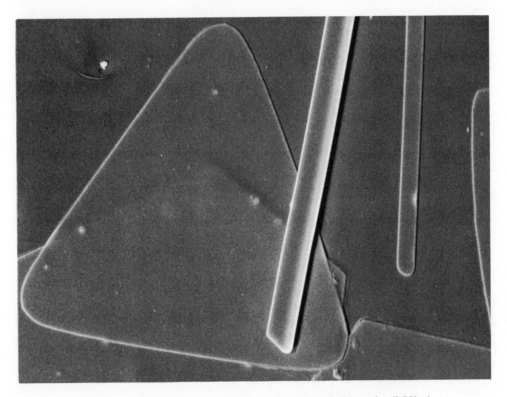

Fig. 9 Scanning electron micrograph of tabular silver halide grains (3,750×).

grains without making replicas and without undue print-out. Since the beam is continuously scanning in a raster, the rate of print-out is reduced enough to allow time for taking the micrograph. The only added requirement for examining the sample in the SEM is that its surface be conducting. In the case of emulsion grains, the sample is dried on a polished metal sample stub and is vacuum-coated with a layer of gold about 200 Å thick. To ensure a proper coating, the samples are simultaneously rotated and tilted during the evaporation. Figure 9 shows a scanning micrograph of large tabular grains of silver halide. Since these grains are quite soft, they deform on drying, as indicated in the micrograph.

Surfaces of photographic coatings can also be studied *in situ* using the SEM after a conductive layer has been vacuum coated. For example, the protrusion of grains from an emulsion surface that has not been overcoated can be observed, as can the details of matte surfaces.

BERNARD M. SPINELL
CARL F. OSTER, JR.

References

1. Bradley, D. E., *Brit. J. Appl. Phys.,* **5**:65 (1954).
2. Hamm, F. A., and J. J. Comer, *J. Appl. Phys.,* **24**:1495 (1953).

PHOTOMICROGRAPHY

Before we embark upon the subject of the apparatus used for photomicrography, we must first give some account of the capabilities and limitations of the human eye. Since, however, we are talking here about a variable and complex system, it is not such a simple matter to explain in an easily comprehensible manner the manifold functions of the eye.

We shall therefore limit ourselves to say that at the distance of distinct vision (250 mm) we can, with the naked eye, distinguish, or resolve, three to five lines per millimeter, or details which are 0.2–0.3 mm apart. We further want to try to comprehend the *main rules for the formation of an optical image*, using a few simple examples. For this purpose we need a magnifier or a reading glass.

The focal point of this lens is that point at which parallel rays from an object at infinity intersect after passing through the lens. This is illustrated by the use of a lens to burn a hole in a piece of paper by focusing an image of the sun. The distance from the center of the lens to its focal point is called the focal length. If we measure this, we are then immediately in a position to calculate the "magnification" of the lens used. The formula for a magnifier states quite simply:

magnification of a simple lens

$$= \frac{\text{conventional viewing distance 250 mm}}{\text{focal length of lens}}$$

If the focal length was 100 mm, for example, then the magnification of the lens would be

$$\frac{250}{100} = 2.5 \times$$

(A 50 mm focal length objective for a 35 mm camera has a "magnification" of 250/50 = 5×.)

The distinct or *conventional viewing distance of 250 mm* therefore gives us a *means of comparison* and thus enables us to define magnifications, which are otherwise a purely relative concept. We shall return to this question later when discussing eyepiece magnification.

Once the focal length of a lens is known, there is a straightforward construction which permits us to determine the image planes under various optical conditions. (So doing, we are naturally greatly simplifying things and must disregard the optical quality of the image thus formed which would be quite unsatisfactory with simple lenses.)

Case 1: Very distant object, at "infinity." (See Fig. 1.) (Comparable to the setting ∞ on a camera in photography)
Result: Image in the focal plane on the image side.
Case 2: Object closer, but still outside twice the focal length. (See Fig. 2.) (Comparable to setting the camera on "close" in photography)

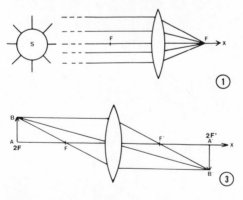

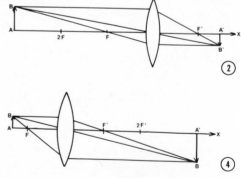

Fig. 1 S = sun; F = object-side focal point; F′ = image-side focal point; X = optical axis.
Fig. 3 AB = object at 2F, twice the focal length; F,F′ = focal points; A′B′ = image at 2F, twice the focal length.

Fig. 2 AB = object; A′B′ = image; 2F = twice the object-side focal length; F = object-side focal length; F′ = image-side focal length.
Fig. 4 AB = object just outside F, the focal length; A′B′ = image outside 2F, twice the focal length.

The following constructions are based on only two rules:

a. Parallel rays pass through the rear focal point after passing through the lens, and focal rays, so called because they pass through the focal point on the object side, leave the other side of the lens parallel to the X axis.

b. For the purpose of this simplified construction, principal rays passing through the center of the lens are considered not to be refracted.

Result: The inverted image is reduced in size and is located between the focal length and twice the focal length on the image side.

Case 3: Object at *twice the focal length*. This is a *special case* as the 1:1 image formation is of importance in *photomacrography* and *slide reproduction*. (See Fig. 3.)

Result: The image is the same size as the object, inverted, and is at twice the focal length on the image side.

Case 4: Using the same construction, it can easily be seen that the closer we bring the object to the focal plane on the object side, the further away from the lens will be the image plane. (See Fig. 4.)

Such relationships, where the image formed is magnified, are valid for magnifiers and also for photomacrography using a "camera extension" (image scale 1:1

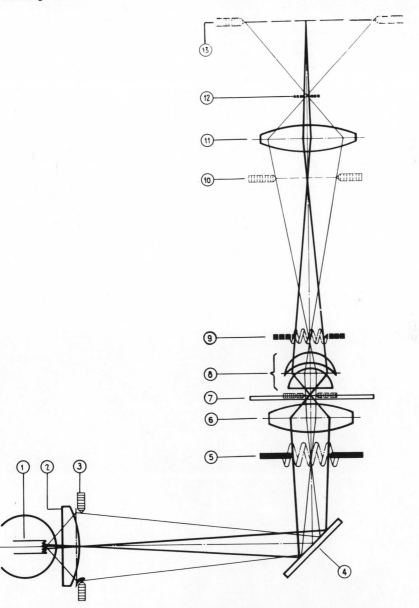

Fig. 5   1 = lamp with filament;   2 = lamp collector;   3 = lamp (field) diaphragm;   4 = mirror;   5 = condenser diaphragm;   6 = condenser;   7 = specimen plane;   8 = objective;   9 = rear focal plane of objective;   10 = intermediate image plane; 11 = eyelens of eyepiece; 12 = exit pupil of eyepiece; 13 = image plane.

to 10:1), but it is also the basis of single-stage image formation in the microscope.

Case 5: is basically the same as case 1. We must simply interchange the positions of the image and the object, or, alternatively, read the drawing from right to left. That is to say, if the object is in the focal plane of a lens, then the image will be at "infinity."

Following these basic explanations, we are now ready to discuss the light path in the *compound microscope.*

This is certainly somewhat more complicated since with microscope magnification we are dealing with *image formation in two stages.* It is further complicated due to the fact that the light path begins not at the object, the microscopy specimen, but at the *light source.* This is very important for photomacrography and photomicrography.

As a result of the increased amount of light which is needed for higher magnifications, simple lamps are completely out of the question. Indeed, special low-voltage lamps and arc lamps are used in conjunction with collectors and field (lamp) diaphragms.

Let us now look at the complete light path from lamp filament to film plane as used in photomicrography. (See Fig. 5.)

1. The lamp filament is imaged in the plane of the condenser diaphragm. This means that as much of the light flux supplied by the lamp as possible enters the condenser. It can also be seen from Fig. 5 that it is above all the illumination intensity of the filament which is the determining factor for the total light flux.

2. The field diaphragm of the lamp is imaged in the specimen plane via the collector and the condenser. This therefore ensures that only that field of view is illuminated which is to be observed with the particular optics chosen. Another advantage of this field diaphragm is that it can also be used as a *reference* for centering so that major decentering errors can be avoided. Thus the condenser is adjusted to such a height that the field (lamp) diaphragm is imaged in the specimen plane. (See Fig. 6.)

The next feature in the light path is the presence of the *specimen.* It cannot be overemphasized just how important are well-prepared, clean microscopy specimens which should be mounted, if at all possible, between standard slides (thickness 1.1 ± 0.1 mm) and standard cover glasses (0.17, +0.01, −0.03 mm), or in the form of perfect smears or polished specimens. Since it is beyond the scope of this article to deal with all possible specimens, we suggest the use of the following two auxiliary specimens: a stage micrometer (Fig. 7) for transmitted light, and a piece of printed material (Fig. 8) for reproduction work and low-power magnifications in reflected light.

If we consider suitable eyepiece-objective combinations as forming the "optical soul" of the microscope, then the *coarse and fine focusing mechanisms* are its "mechanical soul," and are of no lesser importance.

Accepting as a fact the depth of field available is very limited, we can say that the best optics can be no better than the mechanical system used for focusing (see Fig. 9). In other words, if with high-power magnifications the depth of field in the microscope is only fractions of a micron, then the mechanism must permit focusing operations within this extremely small range.

Assuming that the correct combination of optics has been selected (pay attention to manufacturers' advice), the first thing to do in every case is to observe the specimen and decide whether or not it is worth photographing.

Great care was taken by the manufacturers to ensure that the optics of the microscope are so computed and matched that the objective produces at the upper end of the microscope tube an image of the specimen situated just outside the focal length of the objective on the object side. This *intermediate image* is now observed through one or two eyepieces (termed monocular and binocular respectively), just as with a magnifying glass. By defini-

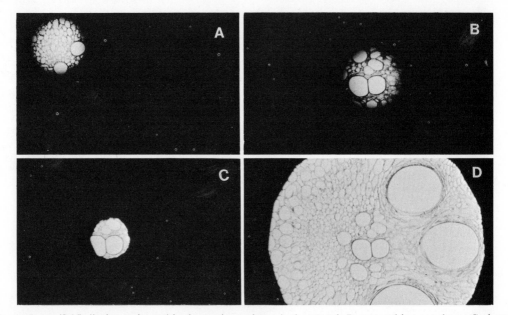

Fig. 6   Lamp (field) diaphragm imaged in the specimen plane: A, decentered; B, centered but not sharp; C, sharply focused; D, opened.

Fig. 7 Stage micrometer.

ie right of

society of

Fig. 8 Printed specimen.

tion, the intermediate image should be in the focal plane of the microscope eyelens, so that rays originating from a point in the intermediate image are parallel on leaving the eyepiece and the image is at infinity. If another optical system is now used in conjunction with the eyepiece, e.g., the eye, or a photographic, television, or cine camera,

then the objective focused on infinity (relaxed eye) will form an image of the intermediate image on the retina, or on the light-sensitive layer of whatever type of camera is employed (see Fig. 10).

Now we are just left with the problem of how to check picture composition and sharpness.

In principle there are three paths open to us:

1. *Direct checking of composition and sharpness in the plane of the film or plate* (Fig. 10).

   For this we simply use a bellows camera, with or without objective, which is fixed at a suitable distance above the eyepiece. Advantage: Allows a sure check on composition and sharpness with correspondingly careful focusing on a frosted-glass screen, with or without a clear-glass spot. Disadvantage: complicated and inconvenient.

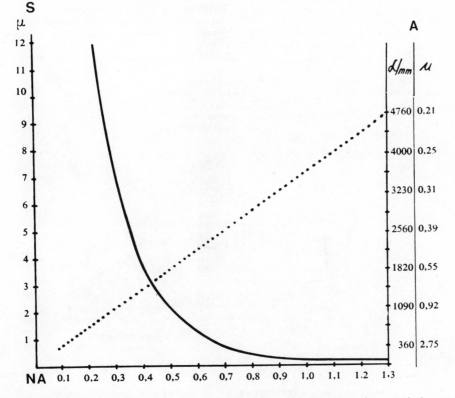

Fig. 9 Depth of field (S) and resolving power (A) of the microscope dependent upon the numerical aperture (NA).

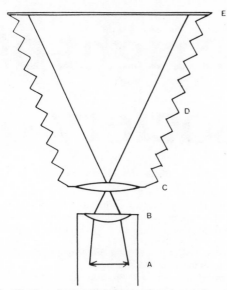

Fig. 10 A = intermediate image plane in eyepiece; B = eyelens of eyepiece; C = camera objective; D = camera extension; E = image–film plane.

2. Checking of composition and sharpness *in a plane conjugate to the film plane* (Fig. 11).

This is the case with single-lens reflex cameras and attachable cameras having a frosted-glass screen for focusing or a focusing telescope. Advantage: With sufficient care, convenient and sure method of focusing. Disadvantage: Binocular observation cannot be made during exposure.

3. Checking of composition and sharpness *in a plane conjugate to that of the intermediate image in the phototube* (Fig. 12).

Advantage: Binocular observation possible during exposure. Disadvantage: Focusing in the intermediate image plane requires special care.

So it is left to the individual to choose from the whole range of possibilities the combination which best suits his needs. It follows from what has been said that, using a single-lens reflex camera, everything from photography through a telescope, reproduction work, photomacrography, and photography with a simple magnifier to real stereophotography and photomicrography at high magnifications can, in principle, be carried out.

But with difficult, low-contrast specimens and cases on the border line of resolution, both frosted-glass screens and focal plane shutters are decidedly disadvantageous. In such cases it is best to use a focusing telescope.

It is often not satisfactory to use only 35 mm film, as it is pointless to use this for color reproductions when the very practical 4 × 5 in. format, which is becoming far more common, can do the job even better. The *three formats* 24 × 36 mm, 60 × 60 mm and 4 × 5 in. are those, therefore, which can be used to best advantage, according to the purpose intended for the film.

Black and White Exposures. Depending on contrast conditions, one chooses:

With very *weak contrast*, e.g., test diatoms, black and white emulsions with steep gradation, strong developer, e.g., Agepe FF (Agfa) developed with Rodinal 1 + 10 or 1 + 20 for $5\frac{1}{2}$ min at 20°C. Specimens with poor contrast and small differences in brightness are *not* suitable for color exposure.

With anything from *weak* to *strong contrast*, black and white single emulsion of medium gradation can be used, e.g., Isopan FF (Agfa) or Panatomic X (Kodak). According to conditions, the contrast can be somewhat increased (e.g., concentrated developer Rodinal 1 + 10 or 1 + 20) or decreased (e.g., Rodinal 1 + 50 or 1 + 100). It is best to try to obtain black and white negatives which, at the thinnest point still of importance for the picture, have a density of about 0.2–0.3 above the fog, and a corresponding maximum density of picture detail of 1.3 –1.5.

Color Exposures. With colored specimens whose contrast leaves something to be desired, a didymium filter, which accentuates red and blue, may be useful for making color exposures.

Fig. 11 Attachable camera (Wild Mka 1) with frosted screen or focusing telescope.

Fig. 12 Attachable camera (Wild Mka 2) on phototube and with format-indicating eyepiece in binocular tube.

Since low-voltage lamps used often have a too-low color temperature of around 2600–3000°K, even when operated above their rated voltage, a suitable *color correction filter* is to be used; so, for example, Kodachrome A, artificial light film 3400° K, 40 ASA, should be used with color correction filter 82 A if a low-voltage lamp is employed. No correction is necessary if a halogen lamp (100 w) is used.

Ektachrome X *daylight film*, 19 DIN, should be used in conjunction with color correction filter 80A (Kodak), whereas xenon lamps and electronic flash do not need correction.

If there seems to be insufficient contrast in the specimen, then microscopy methods such as phase contrast, polarization, interference, or dark field may be used to increase contrast.

With *dark field* especially, additional structural colors appear (interference phenomena). These can make a

ND 0,4	CC 50 R
CC 50 B	CC 50 G

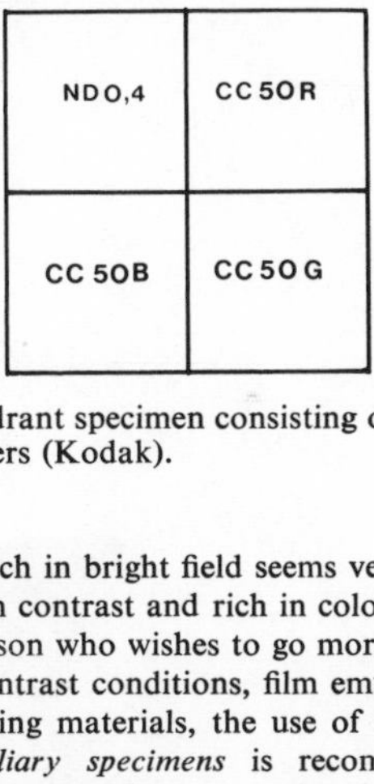

Fig. 14 Quadrant specimen consisting of gray and color correction filters (Kodak).

specimen, which in bright field seems very uninteresting, appear high in contrast and rich in color.

For the person who wishes to go more deeply into the subjects of contrast conditions, film emulsions, developers, and printing materials, the use of two simply constructed *auxiliary specimens* is recommended. These consist of quadrants of gray or colored filters of known characteristics. (See Figs. 13 and 14.)

A carefully maintained *record* of all technical data employed is an integral part of the whole procedure, thereby guaranteeing that the same mistake is never made twice. In view of the many factors involved, it would be quite wrong to try to be sparing with materials; on the contrary, a certain amount of generosity in the use of materials is well worthwhile as the chances of obtaining usable results are then greatly increased.

R. GANDER

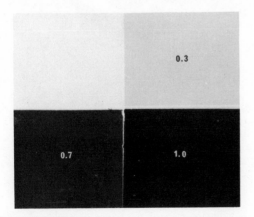

Fig. 13 Quadrant specimen with crossed gray filters, density 0.3 and 0.7.

References

Gander, R., "Photomicrographic Technique," New York, Hafner Publishing Company, 1969.
"Photography through the Microscope," Rochester, N.Y., Eastman Kodak Company, 1970.
Traber, H. A., "The Microscope as a Camera," London and New York, Focal Press, 1971.
See also: CAMERA ATTACHMENTS, MICROPHOTOGRAPHY.

PINEAL

The pineal (pineal gland, pineal body, pineal organ, epiphysis cerebri) belongs to the parietal organs. In mammalian brains it is part of the neurohumoral system, producing different neurohormones (melatonin, serotonin). It develops as an evagination of the diencephalic roof (epithalamus) and is connected to the habenular and posterior commissure (Fig. 1). In nearly all vertebrates a pineal complex is present. In most of the nonmammalian, cold-blooded species, it consists of two separated systems: a photosensorial component, which receives photostimuli, and a secretional component with neurohumoral activity. In accordance with this functional separation the pineal complex can be divided into two distinct organs (parapineal organ—pineal organ in Lampetra [Fig. 1a], frontal organ—epiphysis in amphibians, parietal eye—epiphysis in lizards [Fig. 1b]). Birds and mammals exhibit only one single pineal organ (Fig. 1c and d). Beside this duality the pineal organ seems to change from a saccular system to a solid, parenchymal system (Fig. 1). There is some evidence of parallelisms between saccular architecture and photoreceptor function on one hand and parenchymal structure and neurohumoral activity on the other (Wurtman et al., 1968). However, up to the present day a problem is unsolved,

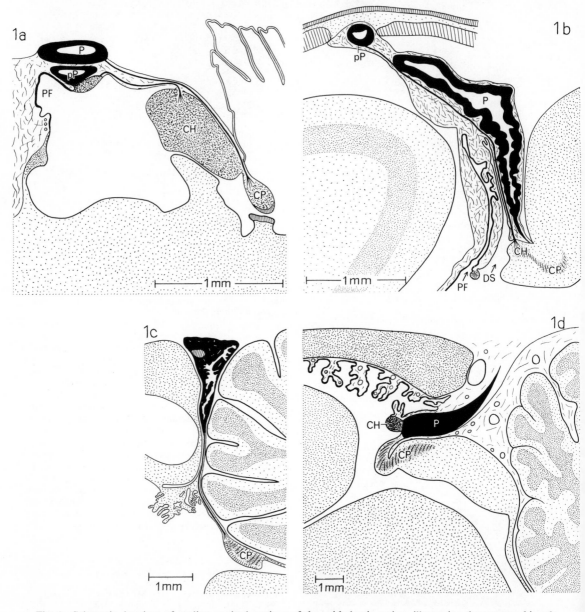

Fig. 1 Schematic drawings of median, sagittal sections of the epithalamic region, illustrating the topographic relationship of the pineal organ (P. black) parapineal organ (pP = parietal eye), dorsal sac (DS) and paraphysis (PF) to the habenular commissure (CH) and the posterior commissure (CP). a, lamprey (*Petromyzon fluviatilis*); b, lizard (*Lacerta viridis*); c, blackbird (*Turdus merula*) (after Breucker, 1967); d, rhesus monkey (*Macaca mulatta*).

i.e., whether photoreceptor cells change to cells with secretorial activity during evolutionary development (Collin, 1969, 1971) or whether pinealocytes with sensorial and secretional activities exist side by side, decreasing and increasing in number while the significance of the first system diminishes to the credit of the second.

Cytologically photoreceptor cells and/or secretory pinealocytes, glial cells (supportive cells) and the vascular apparatus constitute the pineal structure (Fig. 2).

Since morphologically the pineal organ occupies a position between brain and hormonal organs, many microscopical methods have been employed:

a. In order to stain cell processes of mammalian pinealocytes, Rio-Hortega (1923) developed a silver-carbonate method. (1) Fix in 10% formalin 48 hr. (2) Rinse frozen sections in dist. water several times. (3) Incubate sections in 2% $AgNO_3$ solution, mixed with 3 drops of pure pyridine per 10 ml. (4) Warm up to 50°C for 5–10 min and keep jar at room temperature for 24 hr or at 37°C for a few hours: Sections stain dark ocherbrown. (5) Rinse in dist. water plus 2 drops of pyridine per 10 ml. (6) Treat in 10 ml of silver-carbonate solution (20 ml 5% Na_2CO_3 + 5 ml 10% $AgNO_3$, add ammonia until yellow precipitate is dissolved, add 15 ml dist. water) mixed with 3 drops of pyridine at 50°C: Sections should become sepia-brownish. (7) Rinse in tap or dist. water. (8) Reduce in 10% formalin. (9) Treat in 0.2% $H(AuCl_4)$ solution, warm up until sections become violet. (10) Treat in 5% sodiumthiosulfate, rinse in dist. water, dehydrate, clear, and mount in Canada balsam.

b. Information on course and distribution pattern of neuroglial fibers (= processes of fibrous astrocytes) (Hülsemann, 1967) can be obtained by selective staining with the Luxol fast blue procedure (Klüver-Barrera, 1953) combined with Goldner's trichrome method (Fleischhauer, 1960) or with the chrome alum-hematoxylin-phloxin method (Gomori, 1941): In sections stained with the latter method glial fibers give a brilliant yellow fluorescence, excited with UV light at 350 nm (excitation filter: Schott BG 3; 4 mm, barrier filter: Schott 53 and 47, absorbing light up to 500 nm) (Fleischhauer, 1961).

For a review of classical histological techniques used in pineal studies, see: Benoit (1935), Bargmann (1943), and Romeis (1968).

Modern trends in pineal research are concerned with the following problems and methods:

a. Fine Structural Analysis of Mammalian and Nonmammalian Pineal Architecture (See, e.g., Anderson, 1965; Arstila, 1967; Collin, 1969; Collin and Ariens Kappers, 1968; Duncan and Micheletti, 1966; Gusek, 1968; Kelly, 1965; Kelly and Smith, 1964; Oksche and v. Harnack, 1963; Oksche and Kirschstein, 1968; Rüdeberg, 1969; Sano and Mashimo, 1966; Wartenberg, 1968; Vivien-Roels, 1969, 1970; Wartenberg and Baumgarten, 1968; Wurtman et al., 1968).

Fixation for EM-studies: (1) Immersion in buffered OsO_4 or OsO_4-glutaraldehyde mixture at 4°C (for methods see Sjöstrand, 1967; Trump and Bulger, 1966). (2) Vascular perfusion of 2.3–6% phosphate-buffered glutaraldehyde (which is highly effective as a vasodilatator) retrograded through the thoracal aorta or through the (left) heart ventricle. Postfixation for 2–4 hr in phosphate-buffered 1% OsO_4 (Sjöstrand, 1967; Palay et al., 1962; Karlson and Schulz, 1965, 1966; Wartenberg, 1968; Wartenberg and Baumgarten, 1968, 1969a). Embedding in Epon 812 (Luft, 1961; Sjöstrand, 1967).

Results: Basically pineal and brain tissues show common features in architecture. Pinealocytes and their processes are surrounded by numerous glial processes (Fig. 2). Terminals of the pinealocyte processes get access to the perivascular space. In mammalian pineal organs one or more types of pinealocytes have been described. Most of them show secretory activities. In reptiles photoreceptor cells with outer segments can be demonstrated beside secretory cells (Fig. 2), whereas in lower vertebrates photoreceptor cells predominate. Principally, ' pineal organs of all vertebrates contain special synaptic structures: synaptic ribbons (vesicle-crowned rodlets) which are characteristic for sensory transmission.

b. Studies on Innervation of the Pineal Organ. Main problems are: distribution of sympathetic, aminergic fibers and neuronal connections to the brain (Ariens Kappers, 1960, 1965, 1967, 1969; Collin, 1969; Collin and Ariens Kappers, 1968; Hülsemann, 1971; Oksche and Kirschstein, 1969, 1971; Oksche and Vaupel-v.Harnack, 1965; Pellegrino de Iraldi et al., 1965; Scharenberg and Liss, 1965; Wartenberg and Baumgarten, 1969b).

Technique for demonstration of adrenergic nerve fibers: Falck-Hillarp fluorescence technique (Falck and Owman, 1965; Baumgarten, 1967). Freeze-dried tissue should be treated with vapor developed from paraformaldehyde in vacuum, in order to produce a condensation product between aldehyde and monoamine (noradrenaline, dopamine, serotonin). After embedding in paraffin and sectioning, noradrenaline-containing structures exhibit a brilliant greenish fluorescence caused by UV light.

Further techniques employed for pineal innervation studies are: Silver impregnation methods for neurofibrils.

Results: A broad innervation of mammalian and nonmammalian pineal organs with autonomic, postganglionic fibers is well established (Fig. 2). In mammals these neurons have their origin in the superior cervical ganglion and enter the organ as Nervus conarii or via perivascular bundles. The pineal seems to be the only part of the brain where peripheral neurons cross the barrier of brain parenchyma, contacting derivatives of the central nervous system. Biochemical studies of tissue cultures of rat pineal organ give good evidence that "noradrenaline liberated from sympathetic nerves stimulates the formation of melatonin" (Axelrod et al., 1969). Noradrenaline liberation will be influenced by environmental lighting (e.g., continuous darkness). This stimulus will be transmitted through the following neural pathways: retina—inferior accessory optic tract—medial forebrain bundle—medial terminal nucleus of the accessory optic system—preganglionic sympathetic tract in spinal cord—superior cervical ganglia—postganglionic fibers—parenchymal cell of the pineal (Moore et al., 1968; Axelrod, 1970a).

Further neural connections to the brain can be demonstrated morphologically; however, their functional significance in mammals is questionable. In nonmammals efferent nerves leave the pineal as Tractus pinealis sive epiphyseos (pineal nerve).

c. Histo- and Cytochemical Studies. Main subjects are biogenic amines, lipids (Ariens Kappers et al., 1964; Prop, 1965; Prop and Ariens Kappers, 1961; Quay, 1957; Zweens, 1965). (1) Melatonin: There has not been developed so far a useful cytochemical method for localization of melatonin. A native, primary blue fluorescence caused by melatonin under in vitro conditions may be due to an emission peak at 430–420 nm seen at the microspectrofluorometric analysis under standard conditions in the Falck-Hillarp preparation (Owman and

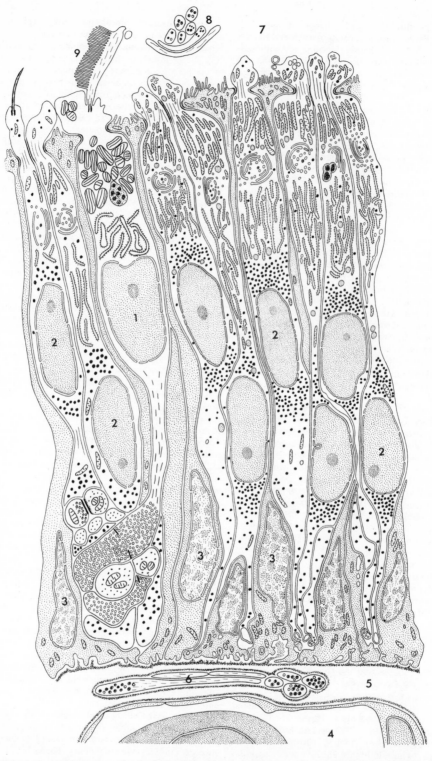

Fig. 2 Schematic representation of cell types of the pineal organ (part of an epithelial arrangement in a saccular pineal system = lacertilians). 1, photoreceptor cell; 2, secretory pinealocyte; 3, supportive (glial) cell; 4, capillary; 5, perivascular space; 6, sympathetic nerves inside perivascular space; 7, lumen of pineal organ; 8, sympathetic nerves inside lumen; 9, outer segment of photoreceptor cell (after Wartenberg and Baumgarten, 1968, supplemented by author).

Rüdeberg, 1970; Owman et al., 1970). (2) Serotonin (5-hydroxytryptamine), noradrenaline, and other indoles and catecholamines have been located by the Falck-Hillarp fluorescence technique (see section b) (Bertler et al., 1964; Collin, 1969, 1971; Oksche et al., 1969; Owman, 1964, 1965; Owman and Rüdeberg, 1970; Owman et al., 1970; Quay et al., 1968; Ueck, 1970; Wartenberg and Baumgarten, 1969a). Histoautoradiographic studies have confirmed the results obtained by the Falck-Hillarp method (Wolfe et al., 1962; Taxi and Droz, 1966). (3) On the electronmicroscopical level the use of "false transmitters" is very promising: Endogenous amines can be replaced by exogenous analogous compounds with additional OH groups (e.g., 5-OH-dopa, 5-OH-dopamine, Tranzer and Thoenen, 1967; Richards and Tranzer, 1969; Thoenen, 1969). Since glutaraldehyde precipitates these substances very well and OsO_4 will be reduced by them very strongly, a localization succeeds in "dense-core vesicles" by reason of their high density. Sites of concentration of these "false transmitters" are aminergic nerve fibers and their terminals in the pineal (see section b). (4) Similar experimental electron microscopical studies have been done by pretreatment with intermediate compounds of the melatonin and 5-HT synthesis (e.g., 5-hydroxytryptophan), simultaneously using inhibitory substances (e.g., reserpine, monoaminooxidase inhibitors).

Most of the important studies on identification of monoamines and enzymatic activities in pinealocytes have been done on the biochemical and pharmacological level (see Wurtman et al., 1968).

Results: 5-HT can be localized in pinealocytes of most species either inside membrane-bounded granules (intravesicular pool) or in an extravesicular, cytoplasmic pool (Wartenberg and Baumgarten, 1969a; Owman and Rüdeberg, 1970).

d. Developmental and Comparative Studies. (See e.g., Kelly, 1963, 1971; Collin, 1971; Hülsemann, 1971).

Reviews and summarizing articles including morphological, physiological, biochemical, and pharmacological aspects of the pineal: Studnička, 1905; Rio-Hortega, 1932; Bargmann, 1943, 1958; Kitay and Altschule, 1954; Kelly, 1962; Owman, 1964. Structure and function of the epiphysis cerebri: Ariens Kappers and Schadé, 1965; Quay, 1965; Wurtman and Axelrod, 1965; Arvy, 1966; Gusek, 1968; Wurtman et al., 1968; Ariens Kappers, 1969; Axelrod, 1970a; symposium with contributions of Hoffmann, 1970; Bagnara and Hadley, 1970; Ralph, 1970; Quay, 1970; Reiter and Sorrentino, 1970; Axelrod, 1970b. The pineal gland: Wolstenholme and Knight, 1971.

HUBERT WARTENBERG

References

Anderson, E., "The anatomy of bovine and ovine pineals. Light and electron microscopic studies," *J. Ultrastr. Res.*, Suppl. 8, 1–80 (1965).

Ariens Kappers, J., "The development, topographical relations and innervation of the epiphysis cerebri in the albino rat," *Z. Zellforsch.*, 52:163–215 (1960).

Ariens Kappers, J., "Survey of the innervation of the epiphysis cerebri and the accessory pineal organs of vertebrates," *see* Ariens Kappers and Schadé (1965), pp. 87–151.

Ariens Kappers, J., "The sensory innervation of the pineal organ in the lizard, *Lacerta viridis*, with remarks on its position in the trend of pineal phylogenetic structural and functional evolution," *Z. Zellforsch.*, 81:581–618 (1967).

Ariens Kappers, J., "The mammalian pineal organ," *J. Neuro-visceral Rel.*, Suppl. IX,, 140–184 (1969).

Ariens Kappers, J., N. Prop, and J. Zweens, "Qualitative, evaluation of pineal fats in the albino rat by histochemical methods and paper chromatography and the changes in pineal fat contents under physiological and experimental conditions," *Progr. Brain Res.*, 5:191–199 (1964).

Ariens Kappers, J., and J. P. Schadé, ed., "Structure and function of the epiphysis cerebri," *Progr. Brain Res.*, 10 (1965).

Arstila, A. U., "Electron microscopic studies on the structure and histochemistry of the pineal gland of the rat," *Neuroendocrinology*, Suppl. to vol. 2:1–101 (1967).

Arstila, A. U., and V. K. Hopsu, "Studies on the rat pineal gland. I. Ultrastructure," *Ann. Acad. Sci. Fenn.*, Ser. A, V 113:1–21 (1964).

Arvy, L., "La glande pinéale, centre régulateur des amines biogènes," *Ann. Biol., Paris*, 5:565–594 (1966).

Axelrod, J., "The pineal gland," *Endeavour*, 29:144–148 (1970a).

Axelrod, J., "Comparative biochemistry of the pineal gland," *Amer. Zool.*, 10:259–267 (1970b).

Axelrod, J., H. M. Shein, and R. J. Wurtman, "Stimulation of C^{14}-melatonin synthesis from C^{14}-tryptophan by noradrenaline in rat pineal in organ culture," *Proc. Nat. Acad. Sci.*, 62:544–549 (1969).

Bagnara, J. T., and M. E. Hadley, "Endocrinology of the amphibian pineal," *Amer. Zool.*, 10:201–216 (1970).

Bargmann, W., "Die Epiphysis cerebri," in W. v. Möllendorff, ed., "Handb. Mikroskop. Anat. Menschen," Vol. VI, Berlin, Springer, 1943.

Bargmann, W., "Epiphysis cerebri," pp. 459–466 in P. Cohrs, R. Jaffé, and H. Meessen, ed., "Pathologie der Laboratoriumstiere," Vol. 1, Berlin-Göttingen-Heidelberg, Springer, 1958.

Baumgarten, H. G., "Vorkommen und Verteilung adrenerger Nervenfasern im Darm der Schleie (*Tinca vulgaris* Cuv.)," *Z. Zellforsch.*, 76:248–259 (1967).

Benoit, W., "Über die histologischen Färbemethoden der Zirbel," pp. 1575–1586 in E. Abderhalden, ed., "Handbuch der biologischen Arbeitsmethoden," Vol. VIII, Berlin and Wien, Urban & Schwarzenberg, 1935.

Bertler, A., B. Falck, and Ch. Owman, "Studies on 5-hydroxytryptamine stores in pineal gland of rat," *Acta Physiol. Scand.*, Suppl. 239, 63:1–18 (1964).

Breucker, H., *Verh. Anat. Ges., Anat. Anz.*, 120 (1967).

Collin, J.-P., "Contribution a l'étude de l'organe pinéal. De l'épiphyse sensorielle a la glande pinéale: Modalités de transformation et implications fonctionnelles," *Ann. Stat. Biol. Besse-en-Chandesse*, Suppl. 1, 1–359 (1969).

Collin, J.-P., "Differentiation and regression of the cells of the sensory line in the epiphysis cerebri," *see* Wolstenholme and Knight (1971).

Collin, J.-P., and J. Ariens Kappers, "Electron microscopic study of pineal innervation in lactertilians," *Brain Res.*, 11:85–106 (1968).

Duncan, D. and G. Micheletti, "Notes on the fine structure of the pineal organ of cats," *Texas Rep. Biol. Med.*, 24:576–587 (1966).

Falck, B. and Ch. Owman, "A detailed methodological description of the fluorescence method for the cellular demonstration of biogenic amines," *Acta Univ. Lund.*, Sect. II, 1–23 (1965).

Fleischhauer, K., "Fluoreszenzmikroskopische Untersuchungen an der Faserglia," *Z. Zellforsch.*, 51:467–496 (1960).

Fleischhauer, K., "Regional differences in the structure of the ependyma and subependymal layers of the cerebral ventricles of the cat," pp. 279–283 in Kety and Elkes, ed., "Regional Neurochemistry" London, Pergamon Press, 1961.

Gomori, G., "Observations with differential stains on human islets of Langerhans," *Amer. J. Pathol.,* **17**:395–406 (1941).

Gusek, W., "Neue Befunde zur Morphologie und Funktion der Epiphysis cerebri," *Ergebn. allg. Path. path. Anat.,* **50**:104–148 (1968).

Hoffman, R. A., "The epiphyseal complex in fish and reptiles," *Amer. Zool.,* **10**:191–199 (1970).

Hülsemann, M., "Vergleichende histologische Untersuchungen über das Vorkommen von Gliafasern in der Epiphysis cerebri von Säugetieren," *Acta anat.,* **66**:249–278 (1967).

Hülsemann, M., "Development of the innervation in the human pineal organ. Light and electron microscopic investigations," *Z. Zellforsch.,* **115**:396–415 (1971).

Karlsson, U., and R. L. Schultz, "Fixation of the central nervous system for electron microscopy by aldehyde perfusion. I. Preservation with aldehyde perfusates versus direct perfusion with osmium tetroxide with special reference to membranes and the extracellular space," *J. Ultrastr. Res.,* **12**:160–186 (1965); "III. Structural changes after exsanguination and delayed perfusion," *J. Ultrastr. Res.,* **14**:47–63 (1966).

Kelly, D. E., "Pineal organs: Photoreception secretion, and development," *Amer. Sci.,* **50**:597–625 (1962).

Kelly, D. E., "The pineal organ of the newt; a developmental study," *Z. Zellforsch.,* **58**:693–713 (1963).

Kelly, D. E., "Ultrastructure and development of amphibian pineal organs," *see* Ariens Kappers and Schadé (1965).

Kelly, D. E., "Developmental aspects of amphibian pineal systems," *see* Wolstenholme and Knight (1971).

Kelly, D. E., and St. W. Smith, "Fine structure of the pineal organs of the adult frog, *Rana pipiens,*" *J. Cell Biol.,* **22**:653–674 (1964).

Kitay, J. I., and M. D. Altschule, "The Pineal Gland," Cambridge, Mass., Harvard Univ. Press, 1954.

Klüver H., and E. Barrera, "A method for combined staining of cells and fibres in the nervous system," *J. Neuropath. Exp. Neurol.,* **12**:400–403 (1953).

Luft, J. H., "Improvements in epoxy resin embedding methods," *J. Biophys. Biochem. Cytol.,* **9**:409–414 (1961).

Moore, R. Y., A. Heller, R. K. Bhatnager, R. J. Wurtman, and J. Axelrod, "Central control of the pineal gland: visual pathways," *Arch. Neurol.,* **18**:208–218 (1968).

Oksche, A., and H. Kirschstein, "Unterchiedlicher elektronenmikroskopischer Feinbau der Sinneszellen im Parietalauge und im Pinealorgan (Epiphysis cerebri) der Lacertilia. Ein Beitrag zum Epiphysenproblem," *Z. Zellforsch.,* **87**:159–192 (1968).

Oksche, A., and H. Kirschstein, "Elektronenmikroskopische Untersuchungen am Pinealorgan von *Passer domesticus,*" *Z. Zellforsch.,* **102**:214–241 (1969).

Oksche, A., and H. Kirschstein, "Weitere elektronenmikroskopische Untersuchungen am Pinealorgan von *Phoxinus laevis* (Teleostei, Cyprinidae)," *Z. Zellforsch.,* **112**:572–588 (1971).

Oksche, A., Y. Morita, and M. Vaupel-v. Harnack, "Zur Feinstruktur und Funktion des Pinealorgans der Taube (*Columba livia*)," *Z. Zellforsch.,* **102**:1–30 (1969).

Oksche, A., and M. Vaupel-v. Harnack, "Elektronenmikroskopische Untersuchungen an der Epiphysis cerebri von *Rana esculenta* L.," *Z. Zellforsch.,* **59**:582–614 (1963).

Oksche, A., and M. Vaupel-v. Harnack. "Elektronenmikroskopische Untersuchungen an den Nervenbahnen des Pinealkomplexes von *Rana esculenta* L.," *Z. Zellforsch.,* **68**:389–426 (1965).

Owman, Ch., "New aspects of the mammalian pineal gland," *Acta Physiol. Scand.,* **63**, Suppl. 240, 1–40 (1964).

Owman, Ch., "Localization of neuronal and parenchymal monoamines under normal and experimental conditions in the mammalian pineal gland," *see* Ariens Kappers and Schadé (1965).

Owman, Ch., and C. Rüdeberg, "Light, fluorescence, and electron microscopic studies on the pineal organ of the pike, *Esox lucius* L., with special regard to 5-hydroxytryptamine," *Z. Zellforsch.,* **107**:522–550 (1970).

Owman, Ch., C. Rüdeberg, and M. Ueck, "Fluoreszenzmikroskopischer Nachweis biogener Monoamine in der Epiphysis cerebri von *Rana esculenta* und *Rana pipiens,*" *Z. Zellforsch.,* **111**:550–558 (1970).

Palay, S. L., S. M. McGee-Russel, S. Gordon, and M. A. Grillo, "Fixation of neural tissues for electron microscopy by perfusion with solutions of osmium tetroxide," *J. Cell Biol.,* **12**:385–410 (1962).

Pellegrino de Iraldi, A., L. M. Zieher, and E. De Robertis, "Ultrastructure and pharmacological studies of nerve endings in the pineal organ," *see* Ariens Kappers and Schadé (1965).

Prop, N., "Lipids in the pineal body of the rat," *see* Ariens Kappers and Schadé (1965).

Prop, N., and J. Ariens Kappers, "Demonstration of some compounds present in the pineal organ of the albino rat by histochemical methods and paper chromatography," *Acta anat.,* **45**:90–109 (1961).

Quay, W. B., "Cytochemistry of pineal lipids in rat and man," *J. Histochem. Cytochem.,* **5**:145–153 (1957).

Quay, W. B., "Indole derivatives of pineal and related neural and retinal tissues," *Pharmacol. Rev.,* **17**:321–345 (1965).

Quay, W. B., "Endocrine effects of the mammalian pineal," *Amer. Zool.,* **10**:237–246 (1970).

Quay, W. B., J. Ariens Kappers, and J. F. Jongkind, "Innervation and fluorescence histochemistry of monoamines in the pineal organ of a snake (*Natrix natrix*)," *J. Neuro-Visceral Rel.,* **31**:11–25 (1968).

Ralph, Ch. L., "Structure and alleged functions of avian pineals," *Amer. Zool.,* **10**:217–235 (1970).

Reiter, R. J., and S. Sorrentino, "Reproductive effects of the mammalian pineal," *Amer. Zool.,* **10**:247–258 (1970).

Richards, J. G., and J. P. Tranzer, "Electron microscopic localization of 5-hydroxydopamine, a 'false' adrenergic neurotransmitter, in the autonomic nerve endings of the rat pineal gland," *Experientia,* **25**:53 (1969).

Rio-Hortega, P. del., "Constitution histologique de la glande pinéale. I. Cellules parenchymateuses," *Trav. Lab. Rech. Biol. Univ. Madrid,* **21**:95–141 (1923).

Rio-Hortega, P. del, "The pineal gland," in W. Penfield, ed., "Cytology and Cellular Pathology of the Nervous System," Vol. II, New York, Hoeber, 1932.

Romeis, B., "Mikroskopische Technik," Munich, Leibniz, 1968.

Rüdeberg, C., "Light and electron microscopic studies on the pineal organ of the dogfish, *Scyliorhinus canicula* L.," *Z. Zellforsch.,* **96**:548–581 (1969).

Sano, Y., and T. Mashimo, "Elektronenmikroskopische Untersuchungen an der Epiphysis cerebri beim Hund," *Z. Zellforsch.,* **69**:129–139 (1966).

Scharenberg, K., and L. Liss, "The histologic structure of the human pineal body," *see* Ariens Kappers and Schadé (1965).

Sjöstrand, F. S., "Electron Microscopy of Cells and Tissues," New York and London, Academic Press, 1967.

Studnička, F. K., "Die Parietalorgane," in A. Oppel ed., "Lehrbuch der vergleichenden mikroskopischen Anatomie der Wirbeltiere," Jena, Fischer, 1905.

Taxi, J., and B. Droz, "Etude de l'incorporation de noradrénaline-³H (NA-³H) et de 5-hydroxytryptophane-³H (5-HTP-³H) dans l'épiphyse et le ganglion cervical supérieur," *C. R. Acad. Sci., Paris,* **263**:1326–1329 (1966).

Thoenen, H., "Bildung und funktionelle Bedeutung adrenerger Ersatztransmitter," Berlin-Heidelberg-New York, Springer, 1969.

Tranzer, J. P., and H. Thoenen, "Electronmicroscopic localization of 5-hydroxydopamine (3,4,5-trihydroxy-phenyl-ethylamine), a new 'false' sympathetic transmitter," *Experientia*, 23:743 (1967).

Trump, B. F., and R. E. Bulger, "New ultrastructural characteristics of cells fixed in a glutaraldehyde-osmium tetroxide mixture," *Lab. Invest.*, 15:368–379 (1966).

Ueck, M., "Weitere Untersuchungen zur Feinstruktur und Innervation des Pinealorgans von *Passer domsticus* L." *Z. Zellforsch.*, 105:276–302 (1970).

Vivien-Roels, B., "Etude structurale et ultrastructurale de l'épiphyse d'un Reptile: *Pseudemys scripta elegans*," *Z. Zellforsch.*, 94:352–390 (1969).

Vivien-Roels, B., "Ultrastructure, innervation et fonction de l'épiphyse chez les Chéloniens," *Z. Zellforsch.*, 104:429–448 (1970).

Wartenberg, H., "The mammalian pineal organ: Electron microscopic studies on the fine structure of pinealocytes, glial cells and on the perivascular compartment," *Z. Zellforsch.*, 86:74–97 (1968).

Wartenberg, H., and H. G. Baumgarten, "Elektronenmikroskopische Untersuchungen zur Frage der photosensorischen und sekretorischen Funktion des Pinealorgans von *Lacerta viridis* und *L. muralis*," *Z. Anat. Entwickl.-Gesch.*, 127:99–120 (1968).

Wartenberg, H., and H. G. Baumgarten, "Untersuchungen zur fluorescenz- und elektronenmikroskopischen Darstellung von 5-Hydroxytryptamin (5-HT) im Pineal-Organ von *Lacerta viridis* und *L. muralis*," *Z. Anat. Entwickl.-Gesch.*, 128:185–210 (1969a).

Wartenberg, H., and H. G. Baumgarten, "Über die elektronenmikroskopische Identifizierung von nor-adrenergen Nervenfasern durch 5-Hydroxydopamin und 5-Hydroxydopa im Pinealorgan der Eidechse (*Lacerta muralis*)," *Z. Zellforsch.*, 94:252–260 (1969b).

Wolfe, D. E., "The epiphyseal cell: an electron-microscopic study of its intercellular relationships and intracellular morphology in pineal body of the albino rat," *see* Ariens Kappers and Schadé (1965).

Wolfe, D. E., L. T. Potter, K. C. Richardson, and J. Axelrod, "Localizing tritiated norepinephrine in sympathetic axons by electron microscopic autoradiography," *Science*, 138:440–442 (1962).

Wolstenholme, G. E. W., and J. Knight, ed., "The Pineal Gland," A Ciba Foundation Volume, London and Edinburgh, Churchill Livingstone, 1971.

Wurtman, R. C., and J. Axelrod, "The pineal gland," *Sci. Amer.*, 213:50–60 (1965).

Wurtman, R. C., J. Axelrod, and D. E. Kelly, "The Pineal," New York–London, Academic Press, 1968.

Zweens, J., "Alterations of the pineal lipid content in the rat under hormonal influences," *see* Ariens Kappers and Schadé (1965).

PLACENTA

The placenta, broadly defined as a union of fetal and maternal tissues for physiological exchange, represents the principal means of fetal homeostasis in viviparous organisms (1). There is, however, great morphological diversity from species to species in the arrangement and complexity of the maternal and fetal tissues. The need to detail this morphological diversity in order to better understand placental function has largely been the impetus behind many electron microscopic investigations of placenta. Placenta is also unique in that placental components undergo a complete life cycle in a time span limited at its maximum extent by a species-specific period of gestation. Thus, a full cycle, consisting of differentiation, functional maturation, and degeneration, is available for cytological study. Further impetus to fine structural study of placenta is the juxtaposition of genetically dissimilar tissue for periods of time beyond which a normal homograft is usually rejected (1).

Electron microscopic techniques applied to the study of placentas from various species have generally followed a routine pattern. Placental tissue is removed from anesthetized females or at autopsy, immediately immersed in a chemical fixer, dehydrated, passed through a transitional fluid, and finally embedded in plastic. Thin sections are mounted on grids and stained with uranyl and/or lead salts.

The two primary chemical fixatives most utilized in the study of placental fine structure, as they are in the study of most tissues, are buffered solutions of osmium tetroxide or glutaraldehyde. Pieces of placental tissue destined for immersion in slowly penetrating solutions of osmium tetroxide should be 1 mm^3 or less in size while those to be immersed in glutaraldehyde may be somewhat larger. Whole body vascular perfusion of fixative has largely been unsuccessful as a means of fixing placentas although Enders[2] has used vascular perfusion of fixative in studies of early implantation.

Various osmium tetroxide fixatives have been used on placental tissue with good results (3–10). The present writer has had the most consistent success in rat and mouse placentas with the Millonig (11) fixative. Placental tissue is cut with a clean razor blade into small cubes (1 mm^3 or less) under a drop of fixative and quickly immersed for 2 hr in 3–5 ml of cold Millonig's osmic acid at pH 7.4. The fixative is kept chilled by packing the fixing vials in crushed ice. The vials are occasionally swirled to dislodge any tissue cubes that may stick to the bottom. After fixation, the tissue blocks are washed quickly (3–5 min) in a buffer solution (pH 7.4) containing all components of the Millonig fixative except the osmium tetroxide and then passed to the dehydration series to be embedded, usually in Epon 812. Figure 1 illustrates the type of preparation that can be consistently obtained with this technique.

The potential benefits of glutaraldehyde as a primary fixative for electron microscopy were first suggested in 1963 by Sabatini, Bensch, and Barrnett (12). Satisfactory fixation of rat liver was obtained in 0.5–2 hours with 4–6.5% glutaraldehyde in 0.1M phosphate buffer at pH 7.2. Some shrinkage was noted with the 6.5% solutions, especially in free cells of tissue culture suspensions. In the latter case, the aldehyde concentration was reduced to 2–3%. Postosmication of tissue blocks, even after long storage in cold buffer (0.1M phosphate) with sucrose added (0.22M) produced results equivalent to optimal fixation with osmium tetroxide alone. Since that time numerous investigators have applied glutaraldehyde fixation with postosmication to ultrastructural studies of placenta, but not all have used the same glutaraldehyde formulations. Wynn (6, 13) used 5% glutaraldehyde in phosphate buffer at pH 7.4 and 0°C to fix tissue from human, macaque, and guinea pig placentas. Tighe, Garrod, and Curran (5) have also fixed human placental tissue in 5% glutaraldehyde. Carpenter and Ferm (14), at the other extreme, applied cold 2% glutaraldehyde in 0.05M phosphate buffer at pH 7.4 for 1–3 hr to yolk sac placenta of golden hamsters with good results. Glutaraldehyde fixatives of different pH and ionic

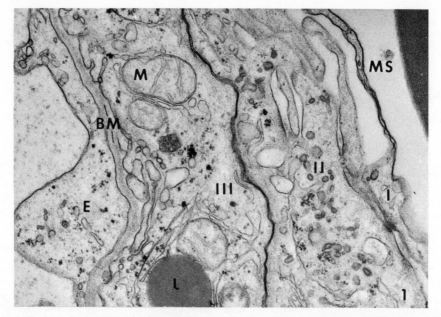

Fig. 1 Labyrinth of rat chorioallantoic placenta at 20 days' gestation. Three layers of trophoblast (I, II, III) separate the maternal blood space (MS) from the fetal capillary endothelium (E). The inner layer of trophoblast (III) is separated from the fetal capillary endothelium by a basement membrane (BM). A mitochondrion (M) and a lipid droplet (L) are evident in the cytoplasm of the inner layer of trophoblast. The tissue was fixed by immersion in Millonig's osmic acid, embedded in Epon 812, and double stained with 3% aqueous uranyl acetate and Venable's lead citrate. (17,500×.)

strength were tested by Anderson (15) as fixing agents for dog placenta. The observations that he reported were obtained with cold (0°C) 2.5% glutaraldehyde in 0.075 M or 0.1 M phosphate buffer at pH 7.4. Tissue cubes were cut under fixative and then immersed for 30–40 min longer. Many investigators, including the present writer, have utilized 3% glutaraldehyde in 0.1 M phosphate buffer (generally cold) at pH 7.3–7.4 for periods ranging from 1 to 3 hr on placental tissue from mouse (8, 9), cat (10), rat (8, 9, 16–19), rabbit (9), guinea pig (9), chipmunk (9), and human (9, 20). In some cases the authors stated that tissue blocks were rinsed after glutaraldehyde fixation in either phosphate buffer (15, 20) or in phosphate buffer with sucrose added (14, 17). Postosmication of the various placental tissue was carried out in 1% or 2% buffered osmium tetroxide for 1–2 hr. Some authors did not detail the buffer system used in their osmium solutions while others indicated that the vehicle for osmium tetroxide was identical to that for glutaraldehyde. Such practice would seem rational since transfer from glutaraldehyde to osmium tetroxide would be accomplished with the least disturbance to the tissue. Subsequent techniques utilized by the investigators referenced above were largely conventional and included dehydration in ascending concentrations of ethanol. Transitional fluid, when stated, was propylene oxide and embedding was done in Epon 812 (5, 6, 8, 13–15, 18), Araldite (6, 9, 13, 16), Durcupan (17, 19, 20), or Maraglass (6, 14). The methods by which glutaraldehyde fixation and subsequent techniques are employed in the present author's laboratory are described below.

Placenta are singly removed from animals under sodium pentobarbital anesthesia and immersed in cold 3% glutaraldehyde in a petri dish where they are cut into strips measuring approximately 1 × 2 × 4 mm. The

tissue strips are transferred to fixing vials containing fresh 3% glutaraldehyde at ice bath temperatures. After the harvest of placentas is finished, the tissue strips are trimmed into smaller blocks suitable for further processing. The 3% glutaraldehyde fixative is made up in a 0.1 M phosphate buffer at pH 7.4. The phosphate buffer contains 1 ml of calcium chloride per 200 ml of buffer. The tissue blocks are fixed for at least 2 hr in glutaraldehyde and subsequently rinsed for 15–20 min in 0.1 M phosphate buffer with calcium chloride and 5% sucrose added. Tissue blocks of rat and mouse placenta have occasionally been stored overnight in buffer-sucrose solution. Postosmication is carried out for 2 hr in 1% Millonig's osmic acid at pH 7.4. Tissue blocks are again briefly rinsed in buffer and then dehydrated by passing through single changes of 50%, 70%, 80%, and 95% ethanol, each for 3–5 min. Two 15 min changes of absolute ethanol are used before passing the tissue through 2 changes of propylene oxide. The last change of propylene oxide is poured off the tissue blocks and 2–3 ml of fresh solvent and an equal amount of the complete mixed resin with accelerator (Epon 812, Luft [21]) are added, mixed by swirling and allowed to stand at room temperature for 1–1.5 hr. Tissue blocks are then transferred to plastic capsules containing the complete resin mixture and cured for at least 24 hr at 60°C. Thin sections mounted on 200 mesh copper grids without support membranes are double stained with 3% uranyl acetate and Venable's lead citrate (22). Each grid is placed tissue-side down on a drop of uranyl acetate in a wax-lined, covered petri dish and permitted to stain for 1–2 hr. After a thorough rinse in distilled water the grids are allowed to dry before transfer to drops of lead citrate in another covered petri dish for up to 5 min. The grids are then very carefully rinsed with 0.02 N

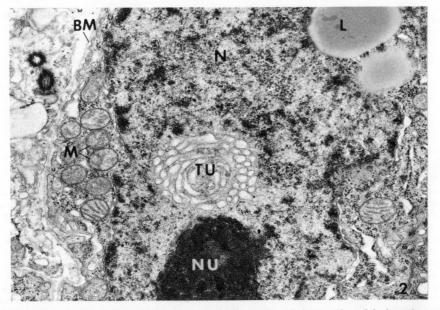

Fig 2 Labyrinth of mouse chorioallantoic placenta at 19 days' gestation. Only a portion of the inner layer of trophoblast is shown. A lipid droplet (L) and a cluster of mitochondria (M) are closely adjacent to the nucleus (N) of this syncitial layer of trophoblast. The nucleus contains a nucleolus (NU) and a tubular unit (TU). The latter structure represents an intranuclear system of tubules that are found in this layer of trophoblast in mouse and rat placenta[8]. A basement membrane (BM) separates the trophoblast from a tissue space that sometimes intervenes between it and the fetal capillary endothelium. The tissue was fixed in 3% glutaraldehyde, postosmicated with Millonig's osmic acid, embedded in Epon 812, and double stained with 3% aqueous uranyl acetate and Venable's lead citrate. (14,000×.)

NaOH and distilled water and allowed to dry in a covered petri dish lined with a fine filter paper. When proper care is taken, tissues stained in this manner are reasonably free of contamination. The results that one can expect from techniques outlined above are illustrated in Fig. 2.

DWAYNE A. OLLERICH

References

1. Wynn, R. M., "Morphology of the placenta," pp. 94–184 in "Biology of Gestation," Vol. 1, N. S. Assali, ed., New York and London, Academic Press, 1968.
2. Enders, A .C., and S. Schlafke, "Cytological aspects of trophoblastic-uterine interaction in early implantation," *Amer. J. Anat.*, **125**:1–30 (1969).
3. Jollie, W. P., "Fine structural changes in placental labyrinth of the rat with increasing gestational age," *J. Ultrastruct. Res.*, **10**:27–47 (1964).
4. Bjorkman, N. H., "On the fine structure of the porcine placental barrier," *Acta Anat.*, **62**:334–342 (1965).
5. Tighe, J. R., P. R. Garrod, and R. C. Curran, "The trophoblast of the human chorionic villus," *J. Pathol. Bacteriol.*, **93**:559–567 (1967).
6. Wynn, R. M., "Comparative electron microscopy of the placental junctional zone," *Obstet. Gynecol.*, **29**:644–661 (1967).
7. Bjorkman, N. H., "Light and electron microscopic studies on cellular alterations in the normal bovine placentome," *Anat. Rec.*, **163**:17–30 (1969).
8. Carlson, E. C., and D. A. Ollerich, "Intracellular tubules in trophoblast III of rat and mouse chorioallantoic placenta," *J. Ultrastruct. Res.*, **28**:150–160 (1969).
9. Enders, A. C., "A comparative study of the fine structure of trophoblast in several hemochorial placentas," *Amer. J. Anat.*, **116**:29–68 (1965).
10. Wynn, R. W., and N. Bjorkman, "Ultrastructure of the feline placental membrane," *Amer. J. Obstet. Gynecol.*, **102**:34–43 (1968).
11. Millonig, G., "Advantages of a phosphate buffer for osmium tetroxide solutions in fixation," *J. Appl. Phys.*, **32**:1637 (1961) [abstract].
12. Sabatini, D. D., K. Bensch, and R. J. Barrnett, "Cytochemistry and electron microscopy: the preservation of cellular ultrastructure and enzymatic activity by aldehyde fixation," *J. Cell Biol.*, **17**:19–58 (1963).
13. Wynn, R. M., "Fetomaternal cellular relationships in the human basal plate: an ultrastructural study of the placenta," *Amer. J. Obstet. Gynecol.*, **97**:832–850 (1967).
14. Carpenter, S. J., and V. H. Ferm, "Uptake and storage of thorotrast by the rodent yolk sac: an electron microscopic study," *Amer. J. Anat.*, **125**:429–456 (1969).
15. Anderson, J. W., "Ultrastructure of the placenta and fetal membranes of the dog. 1. The placental labyrinth," *Anat. Rec.*, **165**:15–36 (1969).
16. Jollie, W. P., "Fine structural changes in the junctional zone of the rat placenta with increasing gestational age," *J. Ultrastruct. Res.*, **12**:420–438 (1965).
17. Martinek, J. J., "Fibrinoid and the fetal-maternal interface of the rat placenta," *Anat. Rec.*, **166**:587–603 (1970).
18. Ollerich, D. A., and E. C. Carlson, "Ultrastructure of intranuclear annulate lamellae in giant cells of rat placenta," *J. Ultrastruct. Res.*, **30**:411–422 (1970).
19. Jollie, W. P., "Nuclear and cytoplasmic annulate lamellae in trophoblast giant cells of rat placenta," *Anat. Rec.*, **165**:1–14 (1969).
20. Enders, A. C., "Fine structure of anchoring villi of the human placenta," *Amer. J. Anat.*, **122**:419–451 (1968).

21. Luft, J. H., "Improvements in epoxy resin embedding methods," *J. Biophys. Biochem. Cytol.,* **9**:409–414 (1961).
22. Venable, J. H., and R. Coggeshall, "A simplified lead citrate stain for use in electron microscopy," *J. Cell Biol.,* **25**:407–408 (1965).

PLASTIC EMBEDDING

Ordinary electron microscope specimens must be 0.1 μ or less in thickness in order to produce useful images. Waxes and other substances used to embed tissues for light microscopic procedures, which require much thicker sections, are impossible, in practice, to cut thinly enough for electron microscopy; in addition, waxes are thermally unstable in the electron beam and they degrade the fine structural preservation of the tissue. Several classes of plastics have properties suitable for embedding electron microscope specimens; in fact, the advantages of plastics have resulted in their widespread use in many light microscopic procedures which formerly employed waxes. Each of the commonly used embedding plastics will be discussed in turn.

Methacrylates. Mixtures of methyl and butyl methacrylates were described in 1951 as the first suitable plastic embedding media for electron microscopy. The hardness of the polymerized plastic is controlled by the properties of the two components, pure methyl methacrylate being extremely hard. The monomers are unstable and inconvenient to prepare, but they infiltrate tissue rapidly and generally are considered easy to section. A major disadvantage of methacrylates is their shrinkage during polymerization. The loss in volume may approach 20% and is inhomogeneous, resulting in submicroscopic pockets of distortion. Special precautions, such as infiltrating with partially polymerized mixtures, polymerizing with ultraviolet light rather than heat, and very slow polymerizing may reduce the polymerization damage to acceptable levels. The major disadvantage of methacrylates is their thermal instability in the electron microscope. Polymerized methacrylates are not highly crosslinked and, consequently, tend to flow when heated by the electron beam. A substantial portion of the methacrylate matrix of a thin section evaporates in the electron microscope, leaving some tissue structures unsupported. These conditions can be remedied by sandwiching the section between thin supporting films, but the films themselves add noise to the image.

In general, methacrylates are inferior to epoxy resins as embedding media for general morphological studies. However, for special purposes, particularly investigations of cytochemical reactions and of lipid soluble substances, some methacrylates are advantageous. Some glycol methacrylates are water soluble; tissues embedded in them do not require dehydration in polar solvents. Thus, certain types of important tissue components, which would be lost or rendered chemically inert by standard embedding, can be retained in glycol methacrylate.

Epoxy Resins. Epoxy plastics were introduced in the mid 1950s as embedding media for electron microscopy. Several suitable formulations have been described, the most commonly used of which are Epon 812 (Luft, 1961) and Araldite (Glauert and Glauert, 1958), or mixtures of the two.

Epoxy resins are less advantageous than methacrylates in ease of sectioning, infiltration, and subsequent staining. However, their great superiority in thermal stability and low polymerization damage make epoxy resins the embedding media of choice for morphological study. The properties of the polymerized resin can be controlled by adjusting the properties of added acid anhydrides and accelerators. The anhydrides act as "plasticizers" or "hardeners" by inhibiting or enhancing the amount of crosslinking during polymerization. The monomer mixtures are polymerized by mild heat, usually within 48 hr.

The extreme complexity of the reactions that occur during polymerization, the potential interference of trace impurities with these reactions, and the imprecise composition of the resins combine to require, in practice, the strict following of previously successful recipes. Unexplained periods of unsuccessful embedding constitute a pandemic plague even in the best of electron microscopy laboratories.

Polyesters. Vestopal-W was described by Ryter and Kellenberger (1958) as an embedding medium in their fine structural studies of bacteria. The superiority of their morphological results over previous investigations of bacteria resulted in the widespread use of Vestopal-W, especially among microbiologists. It appears, in retrospect, that this success was due not so much to the plastic as to its users. Polyesters generally are less convenient to handle, infiltrate, and section than are the other plastics, and there seems to be no inherent advantage in their use.

General Considerations. As a rule, harder plastics are more difficult to section than softer ones, particularly with glass knives. Yet, it is preferable to use hard mixtures because of their greater thermal stability; when tissue which contains hard components is embedded with soft plastic, hardness inhomogeneities are produced which may interfere with sectioning and distort tissue structures. Thus, the first approximation in formulating the hardness of an embedding medium is to match the hardness of the plastic with the hardest component of the specimen.

Plastics also are used as embedding media for nonbiological specimens, such as paper and textiles. For less porous specimens, as metals and minerals, the plastic can serve as a convenient means to hold the sample firmly in the microtome, even if infiltration does not occur.

M. D. Maser

References

Erlandson, R. A., "A new Maraglas, D. E. R. 732 embedment for electron microscopy," *J. Cell Biol.,* **22**:704 (1964).
Glauert, A. M., and R. H. Glauert, "Araldite as an embedding medium for electron microscopy," *J. Biophys. Biochem. Cytol.,* **4**:291 (1958).
Leduc, E. H., and W. Bernard, "Recent modifications of the glycol methacrylate embedding procedure," *J. Ultrastruct. Res.,* **19**:196 (1967).
Leduc, E., V. Marinozzi, and W. Bernard, "The use of water soluble glycol methacrylate in ultrastructural cytochemistry," *J. Roy. Microscop. Soc.,* **81**:119 (1963).
Luft, H. H., "Improvements in epoxy resin embedding methods," *J. Biophys. Biochem. Cytol.,* **9**:409 (1961).
Mollenhauer, H. H., "Plastic embedding mixtures for use in electron microscopy," *Stain Technol.,* **39**:111 (1964).
Ryter, A., and E. Kellenberger, "L'inclusion au polyester pour l'ultramicrotomie," *J. Ultrastruct. Res.,* **2**:200 (1958).

Spurlock, B. O., V. C. Kattine, and J. Freeman, "Technical modifications in Maraglas embedding," *J. Cell Biol.,* **17**:203 (1963).

Spurr, A. R., "A low-viscosity epoxy resin embedding medium for electron microscopy," *J. Ultrastruct. Res.,* **26**:31 (1969).

PLASTID

Plastid is the general term applied to a group of intracellular organelles, of which the best known is the chloroplast. The principal types of plastid which are to some (often undefined) extent interconvertible are as follows:

Proplastids: small, poorly differentiated organelles, occurring in plant meristems and embryonic tissues. In higher plants, all other types of plastid are thought to originate from proplastids.

Chloroplasts: photosynthetic, chlorophyll-containing plastids possessing a complex internal membrane system. Found in green cells of all eukaryotic plants (except fungi).

Etioplasts: chloroplast precursors found in the leaves of some dark-grown higher plants, and converted to chloroplasts on illumination.

Amyloplasts: plastids packed with starch, and found chiefly in roots and storage organs.

Chromoplasts: plastids containing much carotenoid material, often formed from chloroplasts in, e.g., ripening fruits.

The microscopy of plastids has been concerned chiefly with the study of internal structure and organization, and to a smaller extent with their composition. Chloroplasts of higher plants range in size from 3 μ to 10 μ, and are thus easily seen with the light microscope. A regular lenticular shape is common in higher plants, but in the algae the chloroplasts are frequently irregular in shape, and not resolved from the rest of the cell with the light microscope. *In vivo* observations on higher-plant chloroplasts suggest that they are plastic and readily deformed, a property seen well in cells showing cytoplasmic streaming. In phase contrast, chloroplasts appear highly refractive; this seems to be associated with the presence of an intact outer envelope, since *in vitro*, both intactness (as seen with the electron microscope) and refractivity decrease with time. Consequently, phase contrast observations are frequently used to check the quality of chloroplast preparations.

The fluorescence microscope may be used to detect chloroplasts, since the chlorophylls fluoresce a brilliant red when excited by, e.g., blue light. The site(s) of the chlorophyll precursor (protochlorophyll) in etioplasts may also be detected in this manner, though the fluorescence is much weaker.

A simple method of detecting starch grains in plastids is to use the blue color reaction with iodine, either *in vivo* or after fixation.

The fixation of plastids for structural studies with the light microscope is not common since most internal detail is beyond the resolution of the light microscope. Moreover, most fixatives for light microscopical work, particularly those containing ethanol or other organic solvents, cause distortion of shape and disruption of the internal structure of plastids. The best structural preservation is obtained using the glutaraldehyde method for

electron microscopy outlined below. Also satisfactory is buffered formaldehyde at a concentration of about 5% and suitably adjusted osmolarity.

More frequently, light microscopy has been used for the histochemical and autoradiographic detection and localization of nucleic acids, proteins, and other components. Fixatives suitable for such purposes have included ethanol (100% or 70% aq), ethanol/acetic acid (Carnoy's fixative), formaldehyde, formaldehyde/acetic acid, and many others. Freeze-substitution and freeze-drying have also been employed. DNA stains such as Feulgen, methyl green, and acridine orange fluorescence have been successfully used to detect DNA in plastids in some cases; unsuccessfully in others. Such differences probably reflect variation in the amount and distribution of plastid DNA between species. RNA (Azur B, Pyronin) and protein (Naphthalene yellow S, Millon) stains have mostly given positive results when applied to plastids.

The electron microscope has been widely used for studies on plastid structure and development. Early work used material fixed in potassium permanganate which revealed clearly the complex internal membrane system of chloroplasts, but in recent years, most workers have preferred to use glutaraldehyde followed by an osmium tetroxide postfixation since this method preserves more of the internal components. In designing a fixation schedule, it is important (particularly for isolated plastids) to match the osmolarity of the glutaraldehyde solution with that of the organelle, and to provide adequate buffering capacity. Typically, glutaraldehyde (2–6%) in phosphate or veronal acetate (barbital) buffer at about 0.1M, pH 6.5–7.5, and containing 0.25M sucrose is used as the initial fixative. (Tris type buffers give inadequate preservation.) The glutaraldehyde is then washed out with at least three changes of a buffered, osmotically adjusted solution before postfixation in buffered osmium tetroxide (about 1%). The criteria for a well-preserved chloroplast are an intact two-layered outer envelope, fairly electron-dense stroma (osmotically damaged chloroplasts have a balloon-like appearance and little stroma material), and internal membranes with tightly stacked regions. It should also be possible to resolve ribosome-like particles in the stroma, and fibrillar material (probably DNA) may be seen. Among the algae, chloroplast morphology varies widely, and regular stacking patterns of the thylakoids may not be seen. In other types of plastid, the thylakoid system is less prominent, but other components such as starch grains (amyloplasts), para-crystalline prolamellar bodies (etioplasts) or osmiophilic globules (chromoplasts) are conspicuous. (See Figs. 1 and 2.)

Recently, it has been shown that freeze-fracturing and freeze-etching reveal further components of the chloroplast membrane system. Isolated membranes frozen in water are commonly used; for whole chloroplasts or tissues, it is necessary first to infiltrate with glycerol (5–15%) in order to minimize ice-crystal damage. It is generally thought that chloroplast (and other) membranes are split into two halves by the fracturing process, and hence the particles (ranging in size from 80 Å to 180 Å) seen on the fracture faces represent some internal component(s). (See Fig. 3.)

Negative staining of chloroplast fragments reveals yet another set of components. The most commonly used stains are potassium phosphotungstate at neutral pH and a concentration of 1–2% aq, and ammonium molybdate at 2–4% aq. Apart from the membranes themselves, two

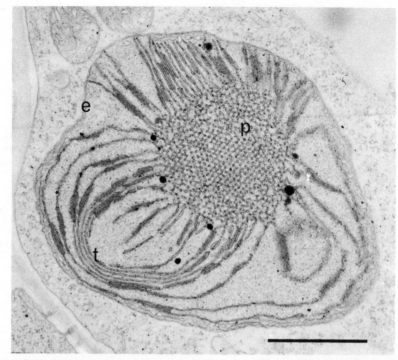

Fig. 1 Etioplast from dark-grown leaf of *Zea mays*. Material was fixed in glutaraldehyde, postfixed in osmium tetroxide, dehydrated, and embedded in epoxy resin. Sections were stained with aqueous uranyl acetate and lead citrate. p, prolamellar body; e, etioplast envelope; t, thylakoid membranes. (Scale = 1 μ.)

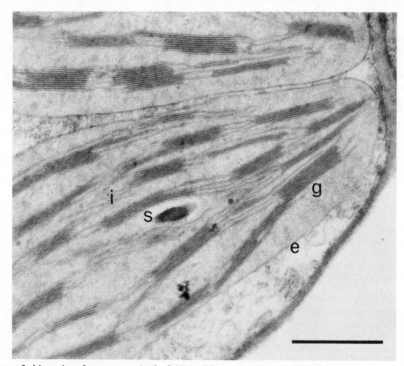

Fig. 2 Portion of chloroplast from young leaf of *Vicia faba* prepared as in Fig. 1. e, envelope; g, grana stacks; i, intergranal membranes; s, starch grain (Scale = 1 μ) (*Courtesy of Dr. Joanne Rosinski.*)

Fig. 3 Freeze-fractured chloroplast membranes from *Chlamydomonas reinhardi*. Membranes were frozen rapidly, fractured at about −100°C, shadowed with platinum, and replicated with carbon. The inner surfaces of the membranes appear covered with particles of varying size with this technique. Shadow direction, bottom to top. (Scale = 0.5 μ.) (*Courtesy of Dr. L. Andrew Staehlin.*)

classes of regularly shaped particles are commonly seen in negatively stained preparations: a cubic particle, about 120 Å in width, and a more rounded particle about 90 Å in diameter. Considerable evidence suggests that the former can be equated with the enzyme ribulose 1.5 diphosphate carboxylase, and the latter with a calcium-dependent ATPase enzyme. These enzymes appear to be localized in the stroma (RUDP carboxylase) or attached to the external surface of the membranes (ATPase), and are thus unlikely to be related to the particles seen by freeze-etching. Indeed, particles are still present on the fracture faces of thylakoids after both enzymes have been removed.

C. L. F. WOODCOCK

References

Branton, D., and R. B. Park, "Subunits in chloroplasit lamellae," *J. Ultrastruct. Res.,* **19**:283–303 (1967).

Jensen, W. A., "Botanical Histochemistry," San Francisco, W. H. Freeman and Co., 1962.

Kirk, J. T. O., and R. A. E. Tilney-Bassett, "The Plastids," San Francisco, W. H. Freeman and Co., 1967.

PLATYHELMINTHES

About 30 years ago the essential part of the following technique was found most appropriate for preparing stained whole mounts of parasitic flatworms. This basic technique has been further improved with some recent modifications.

I. Fixation. Fixatives to be used vary according to the type of stains chosen for the specimen. For example, alcohol-fixed material is best stained with carmin, whereas acetic Schaudinn's is a prerequisite for Heidenhain's hematoxylin. Of all the fixatives ever used by helminthologists, I prefer concentrated formalin (commercial formalin diluted with equal volume of water) for the material that is not fresh enough. For freshly collected material, however, acetic Schaudinn's (equal volumes of saturated solution of sublimate in pure saline water and absolute alcohol plus glacial acetic acid 5% by volume) is used. The fixative is added by a dropper on the slide where the specimen is mounted, or on the surface of the cover glass which is to be applied to the specimen on a slide. Fixative may also be added by drops on one side of the cover glass and drained away slowly, with small pieces of thick blotting paper, from the specimen at the opposite end of the cover glass in a manner permitting repeatedly added fresh fixative to come constantly into contact with the specimen. After proper fixation, the worm should appear uniformly opaque. If it is found insufficiently fixed, raise the cover glass gently to allow more fixative to enter the area where the specimen lies.

The cover glass used in flattening the worm must not be too large nor too small with reference to the size and freshness of the worm. Should the cover glass be too large for the worm, the pressure alone due to it may ruin the specimen. When the worm is large and massive, it must be subjected to a stronger cover glass pressure by a wire compressorium as shown in Fig. 1.

Fixation under a cover glass alone is usually inadequate.

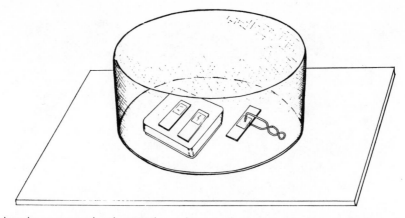

Fig. 1   Showing wire compressorium in operation and square plastic dish under large upsidedown glass dish.

After the specimen has become uniformly opaque, the compressorium, when used, is released, and the slide is placed in a covered plastic dish which in turn is left overnight under a large glass dish turned upside-down on a plate glass. When the specimen is thick, the mounted slide is placed, without being released from the compressorium, under a large glass dish similar to the above-mentioned plastic dish (Fig. 1) to prevent desiccation.

For the purpose of staining, the cover glass has to be removed. The process of removing the cover glass from the slide requires utmost care. Through experience I find the best way is to soak the slide gently in water and to wait until the cover glass loosens itself from the slide. Another method would be to raise the cover glass slowly in water by inserting the point of a fine Japanese drawing hair brush (the type which has a bamboo handle without any metal parts) under one edge of the cover glass, while the opposite edge is held down with the fingertip. When the cover glass is released from the slide, the flatworm usually remains entirely or partly attached to either the cover glass or the slide, or becomes completely detached. If it remains firmly attached, leave it as it stands and put it into acetic Schaudinn's for refixation. When the worm is completely detached from the cover glass or slide, refix it in acetic Schaudinn's like the attached one. At times when the specimen may be only partially adhered to the slide or about to be detached, a very thin film of celloidin dissolved in alcohol-ether may be used to cover the specimen. This technique is very useful for worms which are small and delicate. For most of the average-size specimens, refixation does not require more than 3 hr, but thicker specimens will need a longer time of refixation. After refixation the sublimate in the tissue of the worm must be removed completely by iodine alcohol as in the routine procedure.

II. Staining. Commercial carmin solution (alcohol-Mayer) is used for alcohol-fixed material; commercial Delafield's hematoxylin solution with glacial acetic acid added in 4% for formalin-fixed material (this is recommendable for didymozoid trematodes); and Heidenhain's hematoxylin for the material fixed in acetic Schaudinn's. During the entire course of the Heidenhain's staining, it is desirable to use distilled water for washing instead of tap water. If the Heidenhain-stained material cannot be bleached adequately with ferric ammonium sulfate, then treat it with 1% oxalic acid for a few minutes or longer until the whole worm becomes bleached properly.

III. Dehydration, Clearing, and Mounting. Whenever the stained specimen is distorted and does not lie flat on the slide, it may be straightened out on the slide while in water with a fine Japanese hair brush before being covered with a cover glass. Use a compressorium, if necessary, to make sure that the specimen is nicely flattened.

Dehydration is achieved by placing the mounted slide (with or without the compressorium) into absolute alcohol. It is also possible to drop absolute alcohol repeatedly at one end of the cover glass and blot it at the other end. For massive worms, however, the cover glass used in water for flattening must be removed for complete dehydration. There is no risk of shrinkage of the worm due to direct use of absolute alcohol. Preliminary dehydration by the use of alcohol in graded percentages is superfluous.

Complete dehydration and subsequent clearing in creosote can be accomplished by repeated passage through absolute alcohol and creosote, even under a cover glass for delicate worms. Clearing in creosote instead of xylol is absolutely necessary not only for preventing desiccation of the specimen during mounting, but also for the differentiation of the internal structure of the worm which will take place slowly after mounting in inadequately bleached specimens. This "post-differentiation" by creosote increases the clarity of the structures of the worm with age. The excess hematoxylin dissolved into the mounting medium in and around the stained worm in aged mounts can be removed by remounting after the cover glass is released in creosote. Any unevenness of the specimen due to handling may also be corrected even after mounting by the use of the compressorium.

Satyu Yamaguti

POGONOPHORA

Morphological and histological studies using techniques for light microscopy have been carried out on numerous Pogonophora ranging from some of the smallest known species, e.g., *Siboglinum ekmani*, to the largest, *Lamellibrachia barhami*. Electron microscopic techniques have been employed on *Nereilinum punctatum*, *Oligobrachia floridanum*, *Siphonobrachia ilyophora*, and *Siboglinum ekmani*. Histochemical and physiological techniques have been used mainly on species of *Siboglinum*, and *Oligobrachia ivanovi*, *Polybrachia canadensis*, and *Lamellisabella* sp.

A. Light and electron microscopic techniques

1. **Whole Mounts.** Smaller pogonophore species, e.g., the genera *Siboglinum* and *Sclerolinum*, can be studied within their tubes as long as the tubes are semitransparent by treating with Gaultheria (Wintergreen) oil, or with Euparal Essence, then mounting in Euparal. However, it is preferable to remove the animal from its tube by squeezing it out before fixation/preservation or *immediately* after narcotization. Narcotization by means of a 7–10% magnesium salt solution in seawater, and fixation for 1–2 min in 2% formaldehyde in seawater proved satisfactory. Removal of the animal before narcotization or brief immersion in formaldehyde results in the animal throwing itself into tight coils which are impossible to straighten before fixation.

If specimens are preserved in any of the common preservatives/fixatives, e.g., Bouin, 70% alcohol, 2–4% formaldehyde in seawater, or calcium formaldehyde, the animal cannot be removed by squeezing, so the tube can be split lengthwise or slit by using a razor-sharp finely pointed knife, or in large species it can be cut away.

Removed animals can be studied as whole mounts without prior staining, or by bulk staining in, e.g., Delafield's hematoxylin, alcoholic borax carmine, or chlorazol black E, dehydrated, cleared, and mounted. Good results are obtained using Euparal Essence and Euparal for clearing and mounting respectively.

Larvae can be kept alive for 2–3 weeks in small dishes containing seawater of the animal's environmental temperature. Colloids are ideal in studying ciliary movements and behavior of individual larvae. Whole-mount preparations are similar to those used for adults.

2. **Sectioning.** *a. For Light Microscopy.* Larvae and pieces of adults of small species are best prepared by using the agar-ester wax double embedding method. In large species, the animal, after removal from the tube, is cut into equal lengths of ca. 10 mm and each piece is bulk stained, embedded in paraffin wax, and sectioned. The resultant sections, using either of the embedding methods, are stained. Good results have been achieved with Delafield's hematoxylin and Eosin or with Ponceau 2R-phosphomolybdic acid-light green. Euparal Essence and Euparal for clearing and mounting give good results.

Material prepared for electron microscopy and embedded in either Epon 812 or Araldite is equally suitable for light microscopy as thinner sections can be cut than with the use of wax-embedded material. Such sections are stained either in 4 parts 1% toluidine blue in 1% borax and 1 part 1% pyronine, or in 1% methylene blue in 1% borax. The sections are air-dried and mounted in neutral immersion oil.

b. For Electron Microscopy. Freshly collected material is kept in seawater at the animal's environmental temperature. The removed animals are fixed:

(1) in 2% OsO₄ in seawater for 15–90 min, or
(2) by immersing them in a primary fixative of a 2.5–5% solution of glutaraldehyde buffered with $0.1M$ sodium cacodylate at pH 7.2. The osmolarity of the final solution is adjusted to 1100 m-osmoles/l by adding $0.31M$ sucrose. Alternative buffers used with the primary fixative are either $0.05M$ sodium cacodylate at pH 7.8 containing $0.8M$ sucrose, or $0.1M$ Sörensen phosphate buffer at pH 7.8 with $0.7M$

sucrose. The material is fixed for 2 hr at 4–8°C, and washed for 24 hr in either ice-cold seawater or in the respective buffer solution containing $1.1M$ sucrose; thereafter it is postfixed for 1 hr in ice-cold 1% solution OsO₄ in veronal acetate buffer at pH 7.4 containing $1M$ sucrose. The material is dehydrated in ethanol and embedded either in Epon 812 or Araldite.

B. Amino acid and nitrogen determination of tubes

Each of two equal-weight samples of powdered tube is processed identically through the following procedure:

The sample is placed in a long-necked Pyrex hydrolyzate flask containing 15 ml $6N$ HCL. Nitrogen is passed through for 15 min; thereafter the flask is heated in boiling water to expel excess gas, the neck is sealed, and the flask is placed in an oven for 24 hr at 105°C.

After 24 hr the flask is cooled to room temperature and the contents filtered into an Erlenmeyer flask through Whatman's No. 4 filter paper. The flask is covered, placed in a deep freeze until the solution is frozen, after which the flask is uncovered and the contents dried completely in a freeze-drier. The resultant powder is dissolved in a little pH 2.2 sodium citrate buffer. Filter through Whatman's No. 1 filter paper into a 10 ml volumetric flask and wash the filter paper carefully with buffer to make up to 10 ml. Take 1 ml of hydrolyzate and dilute to 10 ml with pH 2.2 sodium citrate buffer and add one drop of Brij. The hydrolyzate is now ready to be put through the column of an automatic amino acid analyzer.

Nitrogen determination is by the Kjeldahl method using 1 ml of the hydrolyzate to which is added 5 ml 70% NaOH. The resultant product is distilled into an Erlenmeyer flask containing ca. 5 ml 1% boric acid with ca. 2 drops of methyl orange indicator. The distillate is titrated against $0.01N$ H₂SO₄.

C. Physiological and Histochemical techniques

1. **Physiological Techniques.** *Salinity Tolerance.* Tubes containing animals are placed individually in dishes with 5–10 ml "Millipore"-filtered seawater (pore size 0.45 μ) (35°/₀₀) and kept in the dark at 5°C. Equal numbers are treated differently, e.g., dishes covered to maintain salinity; dishes open permitting an increase in salinity; and a daily addition of 0.3 ml distilled water to decrease salinity. In all cases the salinity is determined by measuring chloride content of the water.

Osmotic Pressure of Blood. Blood sampling is done only after the animals have been maintained in a constant salinity medium for 24 hr. Removed animals are placed under liquid paraffin, excess water is removed with a micropipette, and blood is collected with a micropipette by puncturing a longitudinal vessel. The osmotic pressure of the blood and seawater is determined by the freezing point method and expressed as mM/literNaCl.

Uptake of Nutrients and Autoradiography. For measurement of uptake rate of amino acids, animals, in tubes, are placed in the dark at 5–7°C in 1–10 ml "Millipore"-filtered seawater containing ¹⁴C-phenylalanine (specific activity 504 mCi/mmole) at concentrations from 2×10^{-6} to $2 \times 10^{-8}M$, or tritium-labeled glycine (S.A. 1.6 or 0.16 μCi/12 ml) at concentrations from 10^{-6} to $10^{-7}M$.

After incubation for varying periods the animals are removed, weighed on an electrobalance, and placed for 3 hr in 100 μl of 80% ethanol. 50–75 μl of the extract is evaporated on a planchet and C¹⁴ counted on a gas flow counter. To show that the label is absorbed and not adsorbed blood samples are taken, under oil, after exposure to $2 \times 10^{-5}M$ ^{14}C-phenylalanine. The animal is rinsed in distilled seawater, macerated on a planchet, dried under an infrared lamp, and the C[14] counted.

For autoradiography the labeled amino acid-treated material is, after 1–20 hr incubation, rinsed in filtered seawater, fixed, either in cold calcium formaldehyde or for 2 hr in 4% glutaraldehyde in $0.1M$ sodium cacodylate buffer at pH 7.4 with added sucrose giving a final measured concentration of 1100 m-osmoles/liter. After glutaraldehyde fixation the material is rinsed in sucrose cacodylate buffer and then postfixed in buffered OsO_4, dehydrated, and embedded in Araldite. The calcium formaldehyde-fixed material is dehydrated, cleared in benzene, and embedded in hard wax.

The sectioned material in both cases is coated with either diluted Ilford L4 liquid emulsion, or with Ilford K2 nuclear emulsion. The sections are exposed for 8–16 weeks at 2°C after which they are developed in D19b for 4–7 min at 20°C.

Uptake of Macromolecules. Specimens are placed for 10–60 min in 5 ml "Millipore"-filtered seawater containing 10 µg/liter denatured ^{14}C-labeled *Chlorella* protein. To this is added 2 mg% streptomycin sulfate with 2 mg% penicillin G or 0.01% chloramphenicol. After this treatment the animals are rinsed, macerated on planchets, dried, and counted for C[14].

2. Histochemical Techniques. Histochemical studies can be carried out successfully on cold calcium formaldehyde-fixed material for 2–24 hr at 4–6°C, subsequently washed in distilled water for 2–24 hr, dehydrated in acetone at 4–6°C, and embedded in paraffin wax (42°C mp) or ester wax (35–40°C mp). A technique recommended for enzyme studies involves fixation of material in calcium formaldehyde for 2–24 hr followed by an enzyme localization procedure. The material is returned to the fixative to be eventually embedded in gelatin or paraffin wax.

Enzyme Localization Procedures. *Esterase* is localized in whole specimens by two methods:

a. a two-solution modification of the $\propto$-naphthyl acetate method using Brentamine fast red TR as the coupling azo dye, incubating for 1 hr in the dye solution and 1 hr in the full medium at 15–20°C;

b. a modification of the indoxyl acetate method with O-acetyl-bromoindoxyl as the substrate, incubating for 8 hr at 15–20°C, using copper sulfate ($10^{-3}M$) as the oxidation catalyst.

Material treated by method (a) must be embedded in gelatin, and the frozen sections mounted in glycerin or PVP. The (b) treated material can be wax-embedded and the sections mounted in glycerin or PVP.

Alkaline Phosphatase. Whole specimens and paraffin sections are treated with a modification of the $\propto$-naphthyl phosphate-azo dye reaction or by applying the Gomori-cobalt technique to paraffin sections.

Polysaccharides. Gelatin sections, either pre- or post-treated in acid ethanol are handled with short PAS and alcian blue at pH 2.5 and 1.0 neutral red lake, and aqueous or alcoholic toluidine blue. Chitin can be demonstrated by using the chitosan test, and with Diaphanol followed by Schulze's chlor-zinc-iodide solution.

Lipids. Whole mounts and frozen sections are stained with oil red O and sudan black.

Phospholipids. Paraffin sections are stained with luxol blue MBS in ethanol.

Protein. Tests include Millon's reagent as modified by Baker; acid solochrome cyanine; diazonium reaction; performic acid/alcian blue (PFA/AB); the DDD reagent; and tetrazolium blue.

Protease. Tests are best done by the gelatin-silver method. Medium-speed, medium-contrast photographic plates are fogged by exposing to light for 10–15 sec, processed in MQ developer, fixed, washed, and dried. Animals taken from their tubes are placed on small pieces of photographic plate, dampened with seawater, and placed in Petri dishes containing wetted filter paper. The extract of macerated animals is treated identically, and the dishes are incubated for 1–20 hr at 4–20°C.

MICHAEL WEBB

References

Baker, J. R., "The structure and chemical composition of the Golgi element," *Q. J. Microsc. Sci.,* **85**:1–71 (1944).

Baker, J. R., "The histochemical recognition of phenols, especially tyrosine," *Q. J. Microsc. Sci.,* **97**:161–174 (1956).

Blackwell, J., K. D. Parker, and K. M. Rudall, "Chitin in pogonophore tubes," *J. Mar. Biol. Ass. U.K.,* **45**:659–661 (1965).

Brunet, P. C. J., and D. B. Carlisle, "Chitin in Pogonophora," *Nature (London),* **182**:1689 (1958).

Gupta, B. L., and C. Little, "Studies on Pogonophora II: Ultrastructure of the tentacular crown of *Siphonobrachia,*" *J. Mar. Biol. Ass. U.K.,* **49**:717–741 (1969).

Gupta, B. L., and C. Little, "Studies on Pogonophora III: Uptake of nutrients," *J. Exp. Biol.,* **51**:759–773 (1969).

Gupta, B. L., C. Little, and A. M. Philips, "Studies on Pogonophora. Fine structure of the tentacles," *J. Mar. Biol. Ass. U.K.,* **46**:351–372 (1966).

Holt, S. J., and R. F. J. Withers, "Cytochemical localization of esterases using indoxyl derivatives," *Nature (London),* **170**:1012–1014 (1950).

Ito, S., and R. J. Winchester, "The fine structure of the gastric mucosa in the bat," *J. Cell Biol.,* **16**:541–577 (1963).

Jägersten, G., "Investigations on *Siboglinum ekmani* n.sp., encountered in the Skagerak with some general remarks on the group Pogonophora," *Zool. Bidrag Uppsala,* **31**:211–252 (1956).

Lewis, P. R., "A simultaneous coupling azo dye technique suitable for whole mounts," *Q. J. Microsc. Sci.,* **99**:67–71 (1958).

Little, C., "A note on salinity tolerance in *Siboglinum ekmani* (Pogonophora)," *Sarsia,* **38**:87–90 (1969).

Little, C., and B. L. Gupta, "Pogonophora. Uptake of dissolved nutrients," *Nature (London),* **218**:873–874 (1968).

Manwell, C., E. C. Southward, and A. J. Southward, "Preliminary studies on haemoglobin and other proteins of the Pogonophora," *J. Mar. Biol. Ass. U.K.,* **46**:115–124 (1966).

Nørrevang, A., "Structure and function of the tentacle-and pinnules of *Siboglinum ekmani* Jägersten (Pogonophora)," *Sarsia,* **21**:37–47 (1965).

Pearse, A. G. E., "Histochemistry. Theoretical and Applied," 2nd ed., London, Churchill, 1960.

Southward, A. J., and E. C. Southward, "Notes on the biology of some Pogonophora," *J. Mar. Biol. Ass. U.K.,* **43**:57–64 (1963).

Southward, A. J., and E. C. Southward, "Uptake and incorporation of labelled glycine by pogonophores," *Nature (London),* **218**:875–876 (1968).

Southward, E. C., and A. J. Southward, "A preliminary account of the general and enzyme histochemistry of *Siboglinum atlanticum* and other Pogonophora," *J. Mar. Biol. Ass. U.K.,* **46**:579–616 (1966).

Steedman, H. F., "Section Cutting in Microscopy," London, Blackwell Scientific Publications, 1960.

Webb, M., "The larvae of *Siboglinum fiordicum* and a reconsideration of the adult body regions. (Pogonophora)," *Sarsia,* **15**:57–66 (1964).

POLARIZATION MICROSCOPE

The *polarization microscope*, also called a *polarizing microscope*, differs in several essential details from an ordinary microscope. In addition to the features found in conventional microscopes a typical polarizing microscope has a rotatable stage, a lower linear polarizer in the substage condensing system, an upper linear removable polarizer above the objective lens system, and a removable Amici-Bertrand lens above the upper polarizer and below the ocular lens system.

Each linear polarizer is constructed of clear calcite or Polaroid and transmits plane-polarized light which vibrates in a single direction. Provisions are made for rotating either or both of the polarizers in their mounts, but for routine purposes the orientations of the two polarizers are adjusted so that their directions of vibration are mutually perpendicular ("crossed"), and, in the absence of a birefringent object on the microscope stage, no light is transmitted beyond the upper polarizer.

With the Amici-Bertrand lens removed from the optical system, the microscope serves as an orthoscope, which enables examination of magnified images of objects on the microscope stage, either in linearly polarized light from the substage polarizer or between "crossed" polarizers.

The polarizing microscope serves its most useful function in the study of birefringent (optically anisotropic) substances in which incident light is resolved into two linearly polarized components which vibrate in mutually perpendicular planes, and are transmitted with different velocities. Upon emergence, and because of the different velocities, a phase difference has been produced between the two components.

Birefringence is present in strained noncrystalline substances and in all crystalline substances except those in the cubic (isometric) system. The manner of passage of light through nonopaque substances is determined by the distribution of strain or by the crystal structure. Each birefringent crystalline substance is characterized by two or three principal refractive indices, fixed directions of vibration of light corresponding to these indices, and in colored substances, a geometrically identifiable pattern of selective differential absorption of transmitted light.

Refractive indices are measured routinely by the immersion method in which fragments or small crystals are compared with immersion liquids of known refractive indices (Wahlstrom, 1969). Other properties such as birefringence, differential absorption of light, and optic orientation are observed or calculated with or without the aid of a variety of accessory devices.

Uniaxial substances, including tetragonal and hexagonal crystals, are characterized by two principal refractive indices, and biaxial substances, including orthorhombic, monoclinic, and triclinic crystals, have three principal refractive indices. The vibration directions of light corresponding to the principal refractive indices as related to crystal structure defines the *optic orientation*, and the optic orientation is characteristic and diagnostic property of each crystalline substance.

The phase difference produced by a birefringent substance on the stage of the microscope depends on the orientation, the thickness, and the difference between the refractive indices of the crystal for the two transmitted components (the birefringence). Depending on the orientation, the birefringence may vary between zero and a maximum value equal to the difference between the maximum and minimum principal indices. Only one component is transmitted by a crystal when one of its vibration directions coincides with the vibration direction of either of the polarizers, a statement which leads to the conclusion that a crystal produces a phase difference only when the vibration directions of crystals do not coincide with the vibration directions of the polarizers.

The mutually perpendicular linearly polarized components exiting from the crystal do not interact because they are not coherent. The upper linear polarizer allows transmission of light vibrating in one direction only and converts the noncoherent light from the crystal into quasi-coherent light capable of interfering. Depending on the phase difference produced by the crystal and the wavelength of the light (color) an interference color is produced, which, during a 360° rotation of the microscope stage, passes through four positions of maximum intensity and through four positions of zero intensity (extinction positions).

In illumination from an omnichromatic white-light source, a crystal may produce phase differences which are 0, 360°, or whole-number multiples of 360° for a certain wavelength and fractional multiples of 360° for other wavelengths. Accordingly, certain interference colors are transmitted by the upper polarizer with maximum intensity, and others are transmitted with intensities ranging from zero to a maximum. Interference colors observed under the orthoscope in white light, then, are composite colors which in some respects are analogous to and resemble interference colors in Newton's rings or produced by diffraction gratings.

Insertion of the Amici-Bertrand lens into the microscope tube together with introduction of a high numerical aperture lens into the substage condenser converts the orthoscope into a *conoscope*. Commonly a high numerical aperture objective lens also is employed. The Amici-Bertrand lens brings the ocular lens system into focus at the upper curved focal plane of the objective lens system and enables observation of interference patterns in this plane. These patterns, which consist of symmetrically or nonsymmetrically disposed color curves (isochromates) and black or gray extinction brushes (isogyres) are called *interference figures* and reveal much concerning the optical properties and orientation of a birefringent object on the microscope stage.

Many accessory measuring devices are available for measurement or estimation of ortical properties. Especially useful are calibrated *compensators*, which are plates of anisotropic substances which produce known phasal differences, or wedges which produce phasal differences ranging from zero to some maximum. Typical compensators are the quarter wavelength plate, which produces a phasal difference corresponding to a quarter wavelength of yellow light; the sensitive tint plate, which yields a violet or magenta interference color; and the quartz wedge. Other kinds of compensators are constructed of birefringent plates mounted in rotating calibrated mounts and yield a range of phase differences under controlled conditions.

Compensators generally are introduced into an accessory slot in the microscope tube and, when a birefringent object on the microscope stage is rotated into an opposing position, reduce or nullify the phase difference produced by the stage object.

The utility of the polarizing microscope is enhanced by attaching a multi-axis universal stage to the microscope stage. The universal stage enables controlled rotation of stage objects mounted between glass hemispheres into critical positions about a combination of vertical and horizontal rotation axes, and can be adapted for either orthoscopic or conoscopic observation.

ERNEST E. WAHLSTROM

Reference

Wahlstrom, E. E., "Optical Crystallography," 4th ed., New York, John Wiley & Sons, 1969.

See also: METALLOGRAPHIC MICROSCOPE, OPTICAL MICROSCOPE.

POLLEN

Introduction

Pollen may be examined with the electron microscope from a number of aspects and in a variety of ways. The recently developed scanning electron microscope permits nearly complete morphological observation of the surfaces of pollen grain walls (exines), while the transmission electron microscope, although also permitting exine surface studies, is primarily used for examination of internal morphology: both of the exine layers and of the pollen cytoplasm. This article will consider the various preparative techniques necessary for electron microscopic study of pollen; however, it will emphasize the more important aspect of each rather than giving step-by-step descriptions of individual manipulations. A number of technique-oriented papers are listed in the references.

I. Surface Morphology

A. Scanning Electron Microscope Techniques. Sample preparation for the scanning electron microscope is relatively simple and certainly the least time-consuming of all pollen techniques. Fresh, dried, herbarium, acetolyzed, and fossil pollen can be used directly. Thin sections (Echlin, 1968) and developmental stages (Heslop-Harrison, 1969) can also be examined. The primary requirements are that the pollen be suitably affixed to specimen stubs and that it be dry. Several commercial adhesives are available for attachment: those reported for pollen include polyurethane adhesive (Leffingwell et al., 1970), the adhesive made from dissolving Sellotape with chloroform (Echlin, 1968), and double-stick tape. With the latter it is possible to press one side of the tape against open anthers and remove pollen (Martin and Drew, 1969). It is also possible to omit an adhesive if the pollen is dried directly on specimen stubs.

Although drying from absolute ethanol (Pilcher, 1968) or simple air-drying over a chemical desiccant or under vacuum are methods commonly used in preparing pollen for scanning electron microscopy, freeze-drying has been suggested for pollen with fragile walls (Echlin, 1967). It also appears that the various "critical point" drying methods (Anderson, 1951, 1952, 1956; Cohen, Marlow, and Garner, 1968; Horridge and Tamm, 1969) offer the potential of drying pollen without the artifacts caused by declining surface tensions.

Coating of the pollen surface is usually necessary in order to minimize distortional effects resulting from electron bombardment. This is accomplished with conducting materials such as commercial antistatic aerosols (Echlin, 1968), or more frequently with metal alloys. A number of heavy metal alloys are available for specimen coating, but gold-palladium is most commonly used. (In some cases it is helpful to precede the metal coating with a thin layer of evaporated carbon.) The thickness of the coat has been variously reported to be from 200 Å to 500 Å. Rotation of the sample is necessary when applying the coating. In addition, a more even deposition of coat is obtained if the angle of the specimen is continuously varied. Holders to accomplish these tasks are available commercially or can be easily constructed.

B. Replication Techniques. Since the adoption of the scanning electron microscope for pollen work, replica techniques, which employ the transmission electron microscope, are receiving less emphasis. Nevertheless, if a scanning electron microscope is inaccessible, replication may be the only means of examining exine surfaces. Furthermore, with present scanning electron microscopes the resolution obtained is approximately one-tenth that of replicas.

Perhaps the most significant advance in the preparation of replicas has been the reduction of time and the minimizing of unpredictable results attained by making single-stage replicas in lieu of double-stage replicas (Rowley and Flynn, 1966). This method depends upon the chemical destruction of the exine. The use of hot 2-aminoethanol (ethanolamine) results in the removal of pollen grains from the carbon film coating and the direct examination of the latter with the electron microscope. 2-aminoethanol is more satisfactory than chromic acid (Muhlethaler, 1955; Rowley and Flynn, 1966) since when it is used the pollen has less of a tendency to break out of the replica film.

Another technique, similar to scanning electron microscopy in that the pollen wall is directly examined, has been described by Flynn and Rowley (1967). This consists of treatment with hot 2–3N potassium or sodium hydroxide followed by a coating of carbon prior to examination.

II. Internal Morphology—Thin Sectioning Techniques

In preparing pollen for thin sectioning one must first decide whether primary interest is in the pollen wall, the cytoplasm, or both. However, the decision is governed by the condition of the available pollen. Only freshly collected pollen properly preserved can be readily examined from both standpoints. While it is obviously best to use fresh pollen, this is not always possible. For example, taxonomic and morphological studies may necessitate the use of herbarium specimens; fossilized pollen is self-restricting. Since the initial condition of available pollen is variable, differences in processing for electron microscopy occur in the initial stages of fixation and staining; subsequent steps, with few exceptions, are similar.

A. Fixation. *1. Freshly Collected Pollen.* The most consistent results with fresh pollen are obtained by fixation with glutaraldehyde. The concentration can vary from 2% to 6% and must be determined empirically for the particular pollen. Since stock solutions of glutaraldehyde deteriorate autolytically to form glutaric acid, it is essential to use a highly purified grade. This can be achieved by the addition of 3 g/ml of barium carbonate (Maser, in *Fisher Scientific Technical Bulletin—Biological*

Grade Glutaraldehyde: Fast-acting Tissue Fixative) or 0.16 g/ml of activated coconut charcoal (Morre et al., 1965). It is also possible to obtain purified glutaraldehyde in sealed vials which can then be made up just prior to use.

A choice of buffers including phosphate, cacodylate, veronal acetate, S-collidine, and maleate (see Gomori, 1955) is available; phosphate and cacodylate are most frequently used. As with the concentration of glutaraldehyde, the buffer of preference must be determined empirically. The pH should be in the range of 7.0–7.4. Molarity of the buffer is in the range of 0.01–0.1M.

Although osmolarity of the final fixing solution is considered to be important (Maser, Powell, and Philpott, 1967), little information is available for pollen. Rowley (personal communication) uses glutaraldehyde in a cacodylate-HCL buffer at about 310 m-osmols.

Fixation time in glutaraldehyde is not critical and may vary from a few hours to several days. This is usually determined by the thickness of the pollen walls. The temperature of fixation may also vary from room temperature to 0°C.

Since most pollen is powder-like, fixation is best done in tapered glass or plastic test tubes. This facilitates the removal of solutions and the concentration of pollen by centrifugation.

After fixation the pollen should be washed several times in the same buffer that was used with glutaraldehyde. Usually this requires changing washes every 15–30 min for a few hours, or until all apparent traces of fixative are removed. The buffer also can be used to store the pollen for future secondary fixation. Rowley (personal communication) feels that buffer-washing is unnecessary and that after removal of glutaraldehyde, OsO_4 can be immediately added. He also has had excellent results with glutaraldehyde and OsO_4 mixed.

Following buffer washes, OsO_4 is used for staining and secondary fixation. Stock solutions of 2–4% OsO_4 are diluted to 1–2% with the buffer of the primary fixative. Recommended fixation time may vary from 2 to 12 hr and can be accomplished either at room temperature or in ice baths.

If desired a uranyl acetate secondary stain can be applied after the OsO_4 has been removed and the pollen thoroughly washed in distilled water. However, since uranyl acetate can be applied after microtome sectioning, the choice is subjective. If used immediately following staining and secondary fixation, aqueous solutions of 0.5% or 1% should be at room temperature for 1–2 hr, after which time the pollen should be washed in distilled water. At this time the pollen is ready for dehydration.

Aqueous $KMnO_4$ is a second less common fixative. Because permanganate corrodes the pollen wall and destroys most of the cytoplasmic components except lipoprotein complexes, it is infrequently employed. Nevertheless, it does provide sharp membrane definition and it usually fixes pollen cytoplasm which, on some occasions, does not respond to other fixatives. The fixative is primarily 2% aqueous with fixing time 1–2 hr either at room temperature or in ice. Fixation time should not exceed 2 hr since $KMnO_4$ is highly corrosive. After the fixative is removed and the pollen rinsed to eliminate fixing solution, it can be either secondarily stained in uranyl acetate or dehydrated.

If only the morphology of fresh exines is of interest, fixation may be omitted, and the loose exines dehydrated in alcohol and embedded.

2. Developmental Stages. When anthers are fixed to provide the entire sequence of microsporogenesis it is essential that they are not mechanically injured. Individual locules are separated from each other by longitudinal cuts through the connecting tissue. This can be done by making "scalpels" or "spears" from double-edged razor blades. If desired some of the locules may be squashed and stained to determine the developmental stage. The remaining locule(s) are immersed in fixative. Fixation can present a problem: If the anthers are young (stages from archesporial cells through early meiosis) the fixatives will usually penetrate the locules and preserve the sporogenous tissue. With older anthers fixation is often erratic. In such cases it is necessary to puncture the anther or remove an end. A standard procedure is to fix at least two locules per anther, one whole, the other punctured.

Glutaraldehyde fixative, prepared and applied exactly as given above, has been found to provide the most consistent results. However, a second aldehyde fixative, paraformaldehyde, has also been found to provide adequate to excellent fixation (Glauert, 1965). Because paraformaldehyde will decompose in stock solutions, it should not be made up more than 24 hr before use. A stock solution is prepared by dissolving powdered paraformaldehyde in distilled and demineralized water heated to about 80°C. Clearing of the solution is done with N NaOH, after which it is filtered. Phosphate is usually recommended as the most stable buffer vehicle with paraformaldehyde. Subsequent procedures are the same as those used for loose pollen.

$KMnO_4$ is an especially good fixative for developmental stages since in anthers with exceptionally thick epidermal walls it is sometimes the only one which appears to adequately penetrate and preserve the sporogenous tissues.

3. Herbarium Pollen. Dried pollen from herbarium sheets can be treated in two ways for subsequent examination with the electron microscope: either intact or with acetolysis (Erdtman, 1960).

a. Intact Pollen. After removal from herbarium sheets, dried buds or anthers are placed in solutions of either 50–70% ethyl alcohol or in Vatsol (Pohl, 1965). After the flower parts have softened (usually 4–34 hr), the pollen may ordinarily be separated by washing through screens of appropriate mesh. If separation proves difficult, gentle sonication is helpful. The free pollen is then rinsed in distilled water to remove softening solutions. At this point the pollen can be dehydrated or stained in OsO_4.

The decision to stain in OsO_4 is left to the investigator. Although this stain is not considered effective by all investigators, it nevertheless appears to contribute to exine contrast under the electron microscope. It is not unusual to find that without OsO_4 exine contrast is inconsistent, even with the application of a variety of section stains. Therefore, exines are usually stained in solutions of 1–2% OsO_4 with any of the previously mentioned buffers for 2–4 hr at room temperature. If OsO_4 is used, an optional treatment with uranyl acetate may be employed.

b. Acetolyzed Pollen. Herbarium pollen is frequently treated by acetolysis (boiling pollen in a mixture of 1 part concentrated H_2SO_4 and 9 parts acetic anhydride for 5–10 min to remove cytoplasm and most extraneous bud materials. This technique has been outlined elsewhere (Erdtman, 1960); its application to electron microscopy has been discussed by Skvarla (1966a).

After acetolysis there is again the option of staining with OsO₄ and uranyl acetate. Following acetolysis the exine generally appears more electron dense than unacetolyzed exines from similar sources, but contrast is still low and it is not unusual to observe a lack of differential stainability of the individual exine layers.

Treatment of acetolyzed exines (either after OsO₄ or without it) differs from that of unacetolyzed exines in that the exines are next concentrated in 0.9–2% agar pellets (Skvarla, 1966a). Such a concentration permits a large number of pollen walls to be subsequently sectioned and examined by electron microscopy. Concentration in agar is employed only after acetolysis, since neither agar nor embedding resins penetrate pollen containing cytoplasm very well.

4. Fossil Pollen. One of the most effective ways of preparing pollen from rock samples for electron microscopy is described by Leffingwell et al. (1970). Pollen is removed from chemically digested rock samples with a micromanipulator and placed on the surface of a 2% agar cube. The specimens are then covered with an additional drop of agar. Staining of fossil pollen is not necessary, but as recommended for acetolyzed exines, OsO₄ can sometimes contribute to overall contrast.

B. Dehydration. All of the above preparations must be dehydrated. Graded solutions of ethyl alcohol are most frequently used, but some investigators employ acetone. Although dehydration usually involves increments of 10%, it is possible in some cases to confine it to a single step by going directly to absolute alcohol (Skvarla and Kelley, 1968).

Following dehydration, transitional solvents can be propylene oxide, absolute alcohol, or reagent grade acetone.

C. Embedding. Any number of embedding resins may be used, but Epon (Luft, 1961) and Araldite-Epon (Mollenhauer, 1964) are by far those most employed by investigators. With the former a large number of steps requiring 2–3 days is necessary; with the latter, the steps are considerably shortened and time is reduced to 12–24 hours. Araldite-Epon embedding (Mollenhauer, 1964) for pollen (Skvarla, 1966a) requires 1.8 ml of a stock mixture consisting of 25 ml Epon 812, 15 ml Araldite, and 4 ml dibutyl phthalate, and 2.2 ml of dodecenyl succinic anhydride. The mixture is catalyzed by 1–2% tridimethylaminomethylphenol (DMP-30). Since this catalyst has a tendency to deteriorate, the more stable benzyldimethylamine (BDMA) (Skvarla, 1966b), is sometimes used (6–12 drops of BDMA per 4 ml of resin).

Pollen containing cytoplasm should be bathed in a mixture of Araldite-Epon and transitional solvent (1:1) for at least an hour prior to embedding pure resin. This procedure can be followed with acetolyzed and fossil pollen, but direct embedding in pure resin is also acceptable (Skvarla and Kelley, 1968).

After the addition of final resin, impregnation in a heated vacuum desiccator is strongly recommended in order to aid infiltration of the exine and cytoplasm.

D. Section Staining. The preparation of polymerized blocks for ultramicrotomy, including trimming, sectioning, section collection, and the coating of screens with films, has been discussed elsewhere (Skvarla and Pyle, 1968) and will therefore be omitted here. After sections are obtained, the application of stains, which will enhance exine contrast, is very important. For pollen initially treated with OsO₄, staining in 0.5% aqueous uranyl acetate for a few minutes followed by lead citrate,

(either Venable and Coggeshall, 1965; Reynolds, 1963; or Karnovsky, 1961), is most commonly employed and usually provides excellent contrast. If pollen has been prestained with uranyl acetate (as discussed earlier), it is only necessary to use lead citrate as a section stain. Aqueous KMnO₄, KMnO₄ in acetone, or the above-mentioned stains can be used with varying degrees of success for pollen not previously stained. It is also possible to float thin sections of unstained exines which provide difficult contrast problems on solutions of OsO₄ and subsequently stain them with uranyl acetate and lead citrate.

JOHN J. SKVARLA

References

Anderson, T. F., "Techniques for the preservation of three-dimensional structure in preparing specimens for the electron microscope," *Trans. N. Y. Acad. Sci.,* **13**:130–133 (1951).

Anderson, T. F., "The structures of certain biological specimens prepared by the critical point method," in "C. R. Congrés International Microscopie Electronique," Paris, 1952.

Anderson, T. F., "Electron microscopy of microorganisms," pp. 178–181 in "Physical Techniques in Biological Research," Vol. III, 1956.

Bradley, D. E., "The study of pollen grain surfaces in the electron microscope," *New Phytol.,* **57**:226–229 (1958).

Cohen, A. L., D. P. Marlow, and G. E. Garner, "A rapid critical point method using fluorocarbons (freons) as intermediate and transitional fluids," *J. Microscopie,* **7**:331–342 (1968).

Echlin, P., "The use of the scanning reflection electron microscope in the study of plant and microbial material," *J. Roy. Micr. Soc.,* **87**:407–418 (1968).

Erdtman, G., "The acetolysis method, a revised description," *Svensk Bot. Tidskr.,* **54**:561–564 (1960).

Flynn, J. J., and J. R. Rowley, "Methods for direct observation and single-stage surface replication of pollen exines," *Rev. Paleobot. Palynol.,* **3**:227–236 (1967).

Glauert, A., "The fixation and embedding of biological specimens," in D. Kay, ed., "Techniques for Electron Microscopy," Philadelphia, F. A. Davis Company, 1965.

Gomori, G., "Preparation of buffers for use in enzyme studies," pp. 138–146 in S. P. Colowick and N. O. Kaplan, ed., "Methods in Enzymology," Vol. 1, New York, Academic Press, Inc., 1955.

Graham, A., and S. A. Graham, "Palynology and systematics of *Cuphea* (Lythraceae) I. Morphology and ultrastructure of the pollen wall," *Amer. J. Bot.,* **55**:1080–1088 (1968).

Heslop-Harrison, J., "An ultrastructural study of pollen wall ontogeny in *Silene pendula,*" *Grana Palynol.,* **4**:7–24 (1963).

Heslop-Harrison, J., "The origin of surface features of the pollen wall of *Tagetes patula* as observed by scanning electron microscopy," *Cytobios,* **2**:177–186 (1969).

Horridge, G. A., and S. L. Tamm, "Critical point drying for scanning electron microscopic study of ciliary motion," *Science,* **163**:817–818 (1969).

Karnovsky, M. J., "Simple methods for staining with lead at high pH in electron microscopy," *J. Cell Biol.,* **11**:729–732 (1961).

Kay, D., "Techniques for Electron Microscopy," Philadelphia, F. A. Davis Company, 1965.

Larson, D. A., "Processing pollen and spores for electron microscopy," *Stain Technol.,* **39**:237–243 (1964).

Leffingwell, H. A., D. A. Larson, and M. J. Valencia, "A study of the fossil pollen *Wodehousia spinata.* I. Ultrastructure and comparisons to selected modern taxa. II. Optical microscopic recognition of foot layers in differentially stained fossil pollen and their significance," *Bull. Can. Petrol. Geol.,* **18**:238–262 (1970).

Luft, J. H., "Improvements in epoxy resin embedding methods," *J. Biophys. Biochem. Cytol.,* **9**: 409–414 (1961).

Martin, P. A., and C. M. Drew, "Scanning electron micrographs of southwestern pollen grains," *J. Ariz. Acad. Sci.,* **5**:147–176 (1969).

Maser, M. D., T. E. Powell, and C. W. Philpott, "Relationships among pH, osmolality, and concentration of fixative solutions," *Stain Technol.,* **42**:175–182 (1967).

Mollenhauer, H. H., "Permanganate fixation in plant cells," *J. Biophys. Biochem. Cytol.,* **6**:431–435 (1959).

Mollenhauer, H. H., "Plastic embedding mixtures for use in electron microscopy," *Stain Technol.,* **39**:110–114 (1964).

Morre, D. J., H. H. Mollenhauer, and J. E. Chambers, "Glutaraldehyde stabilization as an aid to Golgi apparatus isolation," *Exp. Cell Res.,* **38**:672–675 (1963).

Muhlethaler, K., "Die struktur einiger pollenmembranen," *Planta,* **46**:1–13 (1955).

Pettitt, J. M., "A new interpretation of the structure of the megaspore membrane in some gymnospermous ovules," *J. Linn. Soc. (Bot.),* **59**:253–263 (1966).

Pettitt, J. M., "Exine structure in some fossil and recent spores and pollen as revealed by light and electron microscopy," *Bull. Brit. Mus. (Nat. Hist.) Geol.,* **13**:223–257 (1966).

Pilcher, J. R., "Some applications of scanning electron microscopy to the study of modern and fossil pollen," *Ulster J. Archaeol.,* **31**:87–91 (1968).

Pohl, R. W., "Dissecting equipment and materials for the study of minute plant structures," *Rhodora,* **67**:95–96 (1965).

Reynolds, E. S., "The use of lead citrate at high pH as an electron-opaque stain in electron microscopy," *J. Cell Biol.,* **17**:208–212 (1963).

Ridgway, J. R., and J. J. Skvarla, "Scanning electron microscopy as an aid to pollen taxonomy," *Ann. Mo. Bot. Gard.,* **56**:121–124 (1969).

Rowley, J. R., and J. J. Flynn, "Single-stage carbon replicas of microspores," *Stain Technol.,* **41**:287–290 (1966).

Skvarla, J. J., "Techniques of pollen and spore electron microscopy. Part I. Staining, dehydration and embedding," *Okla. Geol. Notes,* **26**:179–186 (1966a).

Skvarla, J. J., "Addendum to techniques of pollen and spore electron microscopy, Part I, Staining, dehydration and embedding," *Okla. Geol. Notes,* **26**:285 (1966b).

Skvarla, J. J., and A. G. Kelley, "Rapid preparation of pollen and spore exines for electron microscopy," *Stain Technol.,* **43**:139–144 (1968).

Skvarla, J. J., and C. C. Pyle, "Techniques of pollen and spore electron microscopy. Part II. Ultramicrotomy and associated techniques," *Grana Palynol.,* **8**:255–270 (1968).

Thornhill, J. W., R. K. Matta, and W. H. Wood, "Examining three-dimensional microstructures with the scanning electron microscope," *Grana Palynol.,* **6**:3–6 (1965).

Venable, J. H., and R. Coggeshall, "A simplified lead citrate stain for use in electron microscopy," *J. Cell Biol.,* **25**:407–409 (1965).

POLYCHROME STAINING FORMULAS

The formulas given below are arranged according to the undernoted scheme which has the advantage of placing together similar methods. Those who know the specific

method for which they are looking should consult the index.

Stains for sections the nuclei of which are prior stained:
  Single contrast formulas
    Aqueous solutions
    Weak alcohol solutions
    Strong alcohol solutions
    Clove-oil solutions
    Phenol solutions
    Other solutions and mixtures
  Double contrasts from one solution
    Contrasts for red nuclei
      Formulas containing picric acid
      Other formulas
    Contrasts for blue nuclei
      Formulas containing picric acid
      Other formulas
  Complex contrast formulas
    Techniques employing the phosphotungstic (molybdic) reaction with fuchsin
    Techniques employing the phosphotungstic (molybdic) reaction with other dyes
    Other complex contrasts
Complex techniques involving both nuclear and plasma staining:
  Techniques employing the "eosinates" of the thiazins without other admixture
    Methylene blue eosinates
    Polychrome methylene blue eosinates
    Other thiazin eosinates
  Techniques employing thiazins and their eosinates in combination with other dyes
    In combination with orange G
    In combination with other dyes
  Techniques employing methyl green as the nuclear stain
    In combination with pyronin
    In combination with other dyes
  Techniques employing acid fuchsin as the nuclear stain
    Methods employing the acid fuchsin–phosphomolybdic reaction
    Methods employing the acid fuchsin–phosphotungstic reaction
    Methods using neither phosphotungstic nor phosphomolybdic acid
  Techniques employing safranin as the nuclear stain
  Techniques employing hematoxylin as the nuclear stain
  Other polychrome staining methods

Double Contrasts From One Solution

Double contrasts from one solution are just as easy to use as are single contrasts. In those cases in which there is no histological or cytological difference between the elements which it is desired to stain, they present no improvement over the more conventional techniques, but unless one is dealing with homogeneous organs, or with a very young embryo, there seems to be no possible reason why the double-contrast material should not be employed. The division of these stains into two groups is based entirely on the color of the nucleus which they are designed to set off. The first group, contrasting with red nuclei, can therefore be used either after magenta, carmine, or safranin, the picro-contrasts being by convention more commonly employed after carmine than after the other reagents.

Contrasts for blue nuclei present more difficulties than do those for red since nuclei for these techniques are customarily stained by hematoxylin, which is very sensitive to acid. Though many of the picro-contrasts can be employed with hematoxylin, it is strongly recommended that one of the oxazine nuclear stains be employed in its place.

Contrasts for Red Nuclei
Formulas Containing Picric Acid
Borrel 1901, see Cajal 1895 (note)

Cajal 1895 cited from 1905 Lee, *picro-indigo-carmine* (Lee 1905, 20)
FORMULA: water 100, picric acid 1, indigo-carmine 0.25
METHOD: [sections with red nuclei] → water → stain, 3–5 min → water, quick rinse → abs. alc., until connective tissue clear blue
RESULT: muscle, green; most connective tissues, blue
NOTE: This stain is referred to "Borrel" (without reference) by Besson 1904, 751.

Curtis, 1905, *picro–naphthol black* (*C.R. Soc. Biol. Paris.*, **57**:1038)
STOCK SOLUTIONS: I. sat. aq. sol. picric acid; II. water 80, glycerol 20, naphthol blue black 1
WORKING SOLUTION: stock I 90, stock II 10
METHOD: [red, preferably safranin-stained, nuclei] → stain freshly prepared, flooded on slide, 10–15 min → abs. alc. until differentiated → toluene to stop differentiation → balsam
RESULT: nuclei, red; cartilage, blue; other structures, yellow

Curtis 1905b, *picro–naphthol black* (*Arch. med. exp.*, **17**:603)
FORMULA: water 100, glycerol 2, acetic acid 0.01, picric acid 0.9, naphthol blue black 0.1
METHOD, ETC.: as Curtis 1905a

Domagk cited from 1948 Romeis, *picro–thiazin red* (Romeis 1948, 168)
FORMULA: water 100, picric acid 1, thiazin red 0.01

Dubreuil 1904, *picro–methyl blue* (*C.R. Ass. Anat.*, **6**:62)
FORMULA: water 100, methyl blue 0.1, picric acid 0.9

Grosso 1914, *picro–methyl green* (*Folia haematol.*, **18**:71)
PREPARATION OF DRY STOCK: Add a sat. aq. sol. picric acid to a sat. aq. sol. methyl green until no further ppt. is formed. Wash and dry ppt.
PREPARATION OF STOCK SOLUTION: methanol 100, dry stock 0.5
PREPARATION OF WORKING SOLUTION: stock solution 15, water 65
METHOD: [sections with red nuclei] → stain, 5–10 min → water, quick rinse → abs. alc., until red color clouds cease → balsam, via xylene

Krause 1911 cited from 1948 Romeis, *picro–indigo-carmine* (Romeis 1948, 169)
FORMULA: water 100, picric acid 1, indigo-carmine 0.3

Lillie 1945, *picro–blue* (*J. Tech. Meth.*, **25**:1)
FORMULA: water 100, picric acid to sat., methyl blue 0.1 or anilin blue 0.1
NOTE: Lillie 1948 recommends this as a contrast for blue nuclei.

Lillie 1948, *picro-naphthol black* (Lille 1948, 191)
FORMULA: water 100, picric acid q.s. to sat., naphthol blue black 0.02 to 0.04

Masson cited from 1934 Langeron, *picro-indigo-carmine* (Langeron 1934, 552)
REAGENTS REQUIRED: A. 1% acetic acid; B. sat. aq. sol. picric acid 100, indigo-carmine 0.25; C. 0.2% acetic acid
METHOD: [nuclei red, preferably by some magenta method] → A, thorough rinse → B, 10 min → water, quick rinse → C, until connective tissue clear blue, about 2 min → abs. alc. shortest time to complete dehydration → balsam, via xylene
RESULT: very like Cajal 1895 but with a greater range of shades

Minchin cited from 1928 Goodrich, *picro–light green* (Gatenby and Cowdry 1928, 432)
FORMULA: 90% alc. 100, picric acid 5, light green 1
METHOD: [red nuclei] → 90% alc. → stain, 10 min → abs. alc. until differentiated → balsam, via xylene
RESULT: similar to Smith 1912

Neubert 1922, *picro–thiazin red* (*Z. Anat. EntwGesch*, **66**:424)
REAGENTS REQUIRED: A. water 100, 95% alc. 10, thiazin red 0.15, picric acid 0.03; B. sat. 95% alc. sol. picric acid
METHOD: [sections with hematoxylin-stained nuclei] → water A, until collagen deeply stained → rinse → B, thorough wash → balsam, via usual reagents

Pfitzer 1883, *picro-nigrosin* (*Ber. deuts. bot. Ges.*, **1**:44)
FORMULA: sat. sol. picric acid 100, nigrosin 0.2

Pol 1908 cited from 1948 Romeis, *picro-indigo-carmine* (Romeis 1948, 169)
FORMULA: water 100, picric acid 1, indigo-carmine 0.4

Shumway 1926, *picro-indigo-carmine* see SAFRANIN (Shumway 1926)

Smith 1912, *picro–spirit blue* (*J. Morphol.*, **23**:94)
FORMULA: sat. alc. sol. (*circ.* 1%) spirit blue 100, picric acid 1
METHOD: [sections of material bulk stained in carmine] → abs. alc. → stain, 2 min → abs. alc., until differentiated → balsam, via xylene
RESULT: nuclei, red; yolk, yellow green; yolk-free cytoplasm, clear blue

White cited from 1905 Hall and Herxheimer, *picro-erythrosin* (Hall and Herxheimer 1905, 63)
FORMULA: water 100, sat. alc. sol. erythrosin 4, picric acid 0.6, calcium carbonate to excess

Other Formulas
Chatton 1920, *eosin Y–light green* (*Arch. Zool. exp. gén.*, **59**:21)
REAGENTS REQUIRED: A. 95% alc. 100. light green 1, eosin Y 2; B. 5% acetic acid in abs. alc.
PREPARATION OF A: Dissolve with occasional agitation over period of some days. Filter.
METHOD: [sections with red nuclei] → 95% alc. → A, 5 min → B, until connective tissue clear green → balsam, via usual reagents
RESULT: On arthropod material, for which the stain was designed, chitin is green on a red background. On vertebrate material the picture is similar to Patay 1934.

Kostowiecki 1932, *orange G–anilin blue* (*Z. wiss. Mikr.*, **49**:337)
FORMULA: water 100, anilin blue 0.06, orange G 0.2, phosphomolybdic acid 1
PREPARATION: Boil dyes with water 3 min. Add acid to hot solution. Cool. Filter.
METHOD: [sections with red nuclei] → water → stain, until dark-colored, $\frac{1}{2}$–12 hr → water, rinse → 95% alc., 1 min → balsam, via usual reagents
RESULT: procartilage, light blue; cartilage, dark blue; muscle, orange; other connective tissues, blue-green

Roux 1894, *dahlia–methyl green* (*Anat. Anz.*, **9**:248)
FORMULA: water 90, abs. alc. 20, dahlia violet 0.5, methyl green 0.5
PREPARATION: Grind each dye separately in 10 abs. alc. Wash out each mortar with 50 water in small successive doses. Collect washings; leave 24 hr; filter. Mix filtrates; leave 24 hr filter
METHOD: [red nuclei] → water → stain, 5–15 min → blot → abs. alc., until differentiated → balsam, via usual reagents
NOTE: The "dahlia violet" used by Roux may have been almost any mixture of pararosaniline derivatives. Poor quality samples of "gentian violet" work admirably. The result is an excellent polychrome counterstain wherever reproducibility of research results is of less importance than classroom clarity of demonstration. The working solution has been evaporated to dryness and sold as Roux's blue.

Unna cited from 1928 Hill, *anilin blue–orcein* (Gatenby and Cowdry 1928, 280)
REAGENTS REQUIRED: A. 0.1% acetic acid; B. water 50, abs. alc. 25, acetic acid 2.5, glycerol 10, orcein 0.5, anilin blue 0.5
PREPARATION OF B: Dissolve blue in water with gentle heat. Filter. Dissolve orcein in alc. and add to it the acid and glycerol. Add this mixture to the blue.
METHOD: [red nuclei] → A, few minutes → B, 1–10 hr → A, until differentiated → balsam, via usual reagents
RESULT: bone and elastic fibers red-brown against a blue background
NOTE: This reaction is more usually applied by the method of Pasini (1928), given under "Complex Contrasts" below.

Contrasts for Blue Nuclei
Formulas Containing Picric Acid

Curtis 1905, *picro-ponceau* (*Arch. med. exp.*, **17**:603)
FORMULA: water 100, ponceau S 0.1, picric acid 1, acetic acid 0.04

Fite 1939, *picro-fuchsin* (*J. Lab. Clin. Med.*, **25**:344)
FORMULA: water 100, picric acid 0.5, acid fuchsin 0.1

van Gieson 1896, *picro-fuchsin* (*Z. wiss. Mikr.*, **13**:344)
FORMULA: sat. sol. (*circ.* 1.2%) picric acid 100, acid fuchsin 0.05
METHOD: [blue nuclei] → water → stain, 2–10 min → water, quick rinse → balsam, via usual reagents

Gnanamuthu 1931, *picro–congo red* (*J. Roy. Micr. Soc.*, **51**:401)
FORMULA: sat. sol. (*circ.* 1.2%) picric acid 50, ammonia 50, congo red 2
PREPARATION: Add the ammonia to the picric solution. Dissolve the dye in mixture and boil until no odor

of ammonia is apparent. Cool. Add sufficient water to redissolve ppt. formed on cooling.
METHOD: [blue nuclei (Ehrlich 1886 acid alum hematoxylin specified in original)] → water → stain 1–2 min → blot → abs. alc., minimum time for dehydration → balsam, via usual reagents
RESULT: Muscle, red; other tissues, yellow and orange. Good for most heavily muscularized tissues.

Hansen 1898, *picro-fuchsin* (*Anat. Anz.*, **15**:152)
REAGENTS REQUIRED: A. sat. sol. (*circ.* 1.2%) picric acid 100, acetic acid 0.3, acid fuchsin 0.1; B. water 98, A. 2
METHOD: [blue nuclei] → water → A, some hours → B, wash → balsam, via xylene
RESULT: selective red stain on white fibrous connective tissue

Lillie 1948, *picro-fuchsin* (Lillie 1948)
FORMULA: water 100, picric acid 1, acid fuchsin 0.1, hydrochloric acid 0.25
NOTE: Lillie (*loc. cit.*) also recommends his picro–naphthol black solution as a contrast for hematoxylin-stained nuclei.

Ohlmacher 1897, *picro-fuchsin* (*J. Exp. Med.*, **2**:675)
FORMULA: sat. sol. (*circ.* 1.2%) picric 50, water 50, acid fuchsin 0.5

Schaffer 1899, *picro-fuchsin* (*Z. wiss. Zool.*, **66**:214)
FORMULA: sat. aq. sol. picric acid, acid fuchsin 0.15, acetic acid 0.05

Unna cited from Lillie 1948, *picro-fuchsin* (Lillie 1948, 190)
FORMULA: water 90, acid fuchsin 0.25, nitric acid 0.5, glycerol, 10, picric acid *q.s.* to sat.

Weigert 1904, *picro-fuchsin* (*Z. wiss. Mikr.*, **21**:3)
FORMULA: sat. sol. (*circ.* 1.2%) picric acid 100, acid fuchsin 0.1
METHOD: as van Gieson above

Wilhelmini 1909, *picro-fuchsin* (*Fauna u. Flora Neapel.*, **22**:18)
FORMULA: water 90, 95% alc. 10, ammonium picrate 0.8, acid fuchsin 0.2
METHOD: as van Gieson above
Other Formulas
Delèphine cited from *circ.* 1938 Wellings, *fuchsin-orange* (Wellings 1938, 104)
FORMULA: water 100, acid fuchsin 0.04, orange G 0.2

Gray 1952, *ponceau-orange* (Gray 1952, 24)
FORMULA: water 100, orange II 0.6, ponceau 2R 0.4
METHOD: [blue nuclei] → water → stain, 1–2 min → blot → abs. alc., until differentiated → balsam, via usual reagents

Gregg and Puckett 1943, *eosin-orange* (*Stain Technol.*, **18**:179)
FORMULA: 95% alc. 100, eosin 0.2, orange G 0.01

Guyler 1932, *indigo-carmine–eosin Y* (*Lab. J.*, **18**:314)
FORMULA: water 100, indigo-carmine 0.25, eosin Y 1, thymol trace
METHOD: [blue nuclei (original specifies Delafield 1885)] → water → stain, overnight → water, quick rinse → abs. alc., until differentiated → balsam, via usual reagent

Hayem cited from 1896 Kahlden and Laurent, *eosin-aurantia* (Kahlden and Laurent 1896, 117)
FORMULA: 1% eosin W, 1% aurantia, *a.a. q.s.* to give rose-colored solution

Kingsbury and Johannsen 1927, *orange–acid fuchsin* (Kingsbury and Johannsen 1927, 76)
FORMULA: water 100, glycerol 7, orange G 1, acid fuchsin 2

Male 1924, *fuchsin–martius yellow* (*R.A.M(C. J.*, **42**:455)
FORMULA: water 80, 95% alc. 20, acid fuchsin 0.6, martius yellow 0.8

Masson 1911, *saffron-erythrosin* (*C.R. Soc. Biol. Paris*, **70**:573)
REAGENTS REQUIRED: A. 1% erythrosin; B. water 100, saffron 2.5% tannin 1, 40% formaldehyde 1
PREPARATION OF B: Extract the saffron in the water 1 hr 90°C. Filter. Add other ingredients to filtrate.
METHOD: [blue nuclei] → water → A, 5 min → water, quick rinse → 70% alc., few seconds → water, thorough wash → B, 5 min → blot → abs. alc., flooded over slide, until dehydrated → balsam via xylene
NOTE: Langeron 1942 substitutes eosin B for erythrosin in the above.

Semichon 1920, *methyl blue–eosin–victoria yellow* (*Bull. Soc. zool. Fr.*, **45**:73)
REAGENTS REQUIRED: A. water 100, methyl blue 0.04, eosin Y 0.2, victoria yellow 0.1
METHOD: [blue nuclei] → water → A, overnight → drain → abs. alc., until differentiated → xylene → balsam
RESULT: horn, hair, chitin, yellow; cartilage, blue; other tissues, orange

Squire 1892, *fuchsin-orange* (Squire 1892, 42)
FORMULA: water 80, 95% alc. 20, acid fuchsin 0.3, orange G 2.0
METHOD: as Gray 1952 above

Szütz 1912, *polychrome alizarin* (*Z. wiss. Micr.*, **28**:289)
REAGENTS REQUIRED: A. 5% aluminum acetate; B. sat. alc. sol. alizarin red S 1, water 100
METHOD: [hematoxylin-stained sections of Szütz 1912 fixed material] → water → A, 5 hr → rinse → B, 5 hr → wash → balsam via usual reagents
RESULT: nuclei, blue; cytoplasm, varying shades of red, cytoplasmic inclusions being generally very darkly stained

Complex Contrast Formulas

Techniques Employing the Phosphotungstic (Molybdic) Reaction with Acid Fuchsin

Brillmeyer 1929, *acid fuchsin–anilin blue–orange G* (*J. Tech. Meth.*, **12**:122)
REAGENTS REQUIRED: A. 0.2% acid fuchsin; B. water 100, phosphomolybdic acid 1, anilin blue 0.5, orange G 2.0
METHOD: [blue nuclei (original specifies Delafield 1885)] → A, 1 min → drain → B, 2–3 hr → water, quick wash → balsam, via usual reagents
NOTE: Weiss 1932 (*Stain Technol.*, **7**:131) differs only in the dilution of A to 0.04% and in the substitution

of 4 min for 3 hr immersion in B. Mayer 1901 is recommended for prior staining of sections from picric fixed or mordanted material.

Crossman 1937, *acid fuchsin–orange G–light green (or –anilin blue)* (*Anat. Rec.*, **69**:33)
REAGENTS REQUIRED: A. water 100, acetic acid 1, acid fuchsin 0.3, orange G 0.13, thymol 0.06, B. 1% phosphomolybdic acid; C. either water 100, acetic acid 1, light green 1 or water 100, acetic acid 2, anilin blue 2; D. 1% acetic acid
METHOD: [sections, nuclei hematoxylin-stained] → water → A. 1 min → tinse → B. until collagen decolorized → quick rinse → C, 5 min → rinse → D, until differentiated → rinse → abs. alc. → balsam, via xylene

Goldner 1938, *acid fuchsin–ponceau 2R–orange G–light green* (*Amer. J. Pathol.*, **14**:237)
REAGENTS REQUIRED: A. water 100, ponceau 2R 0.07, acid fuchsin 0.03, acetic acid 0.2; B. 1% acetic acid; C. water 100, phosphomolybdic acid 4, orange G 2; D. water 100, light green 0.2, acetic acid 0.2
METHOD: [sections with nuclei stained by Weigert 1904] → thorough wash → A, 5 min → B, wash → C, until collagen decolorized → B, rinse → D, 5 min → B, 5 min → blot → abs. alc., least possible time → balsam, via xylene
RESULT: general cytoplasm, red; erythrocytes, orange; collagen, green
NOTE: Other color combinations may be obtained by substituting for A above either water 100, azophloxin 0.5, acetic acid 0.2 or water 100, acid fuchsin 0.025, ponceau 2R 0.075, azoploxine 0.01, acetic acid 0.2

Haythorne 1916, *acid-fuchsin–orange G–anilin blue* (*Bull. Int. Ass. Med. Mus.*, **6**:61)
REAGENTS REQUIRED: A. water 100, hydrochloric acid 0.05, 95% alc. 4, orange G 0.8, ferric alum 5; B. 0.5% acid fuchsin; C. sat. aq. sol. phosphomolybdic acid 100, anilin blue 2.5, orange G 2.5
PREPARATION OF A: Dissolve the orange G in 70 water with the alc. and acid. Dissolve the alum in 25 water and add to the dye solution. Filter.
METHOD: [sections of Zenker 1894 fixed material, after 30 min staining in Bohmer 1868 alum hematoxylin] → water → A, 2 min → water, 5 min → B, 3 min → blot → C, 20 min → blot → 95% alc., quick rinse → abs. alc., from drop bottle, until differentiated → balsam, via xylene
RESULT: nuclei, reddish black; cartilage, white fibrous tissue, blue; keratin, chitin, erythrocytes, bright orange; muscle, red

Lendrum and McFarland 1940, *picro–orange G–acid fuchsin–ponceau 2R–anilin blue* (*J. Pathol. Bact.*, **50**:381)
REAGENTS REQUIRED: A. water 20, 95% alc. 80, picric acid 1, orange G 0.2; B. water 99, acetic acid 1, acid fuchsin 0.5, ponceau 2R 0.5, sodium sulfate 0.25; C. 1% acetic acid; D. 1% phosphomolybdic acid; E. water 99, acetic acid 1, anilin blue 2
METHOD: [sections with nuclei stained by an oxazine technique] → water → A, 2 min → overnight → rinse → B, 1–5 min → C, rinse → D, until collagen not quite decolorized → E, 2–10 min → C, rinse → balsam, via usual reagents
NOTE: Fast green FCF may be substituted for anilin blue in E above.

Lillie 1940, *Biebrich scarlet–fast green* (*Stain Technol.*, **15**:21)

REAGENTS REQUIRED: A. water 99, acetic acid 1, Biebrich scarlet 1; B. water 100, phosphotungstic acid 2.5, phosphomolybdic acid 2.5; C. water 97.75, acetic acid 2.5, fast green FCF 2.5; D. 1% acetic acid

METHOD: [sections with blue nuclei (original specifies Weigert 1903)] → water → A, 2 min → rinse → B, 1 min → C, 2 min → D, 1 min or until differentiated → balsam, via acetone and xylene

Masson 1912a, *acid fuchsin–anilin blue* (*Bull. Soc. anat. Paris*, **87**:290)

REAGENTS REQUIRED: A. water 100, acetic acid 0.5, acid fuchsin 0.5; B. 1% phosphomolybdic acid; C. water 100, acetic acid 2.5, anilin blue to sat; D. 1% acetic acid; E. 0.1% acetic acid in abs. alc.

METHOD: [sections (original required prior staining in Régaud 1910)] → water → A, 5 min → water, quick rinse → B, 5 min → drain → C, poured on slide, 2–5 min → D, until differentiated 5–30 min → E, until dehydrated → salicylic xylene → salicylic balsam

RESULT: nuclei, black (if prior stained in Régaud) or deep red; collagens, light blue; bone, dark blue; epithelia, muscle, some glands, light red; erythrocytes, orange; nervous tissues, violet

Masson 1912b, *acid fuchsin–ponceau 2R–anilin blue* (*Bull. Soc. anat. Paris*, **87**:290)

REAGENTS REQUIRED: A. water 100, acetic acid 1, acid fuchsin 0.35, ponceau 2R 0.65; B. C. D. E. as Masson 1912a above

METHOD: as Masson 1912a above

Masson 1912c, *acid fuchsin–metanil yellow* (*Bull. Soc. anat. Paris*, **87**:290)

REAGENTS REQUIRED: A. water 1, acetic acid 1, acid fuchsin 100; B. 1% phosphomolybdic acid; C. sat. sol. (*circ.* 6%) metanil yellow; D. 1% acetic acid

METHOD: [sections (original required prior staining in Régaud (1910) iron hematoxylin] → water → A, 5 min → water, quick rinse → B, 5 min → drain → C, poured on slide, 5 min → D, 5 min → salicylic balsam, via usual reagents

McFarlane 1944a, *picro–acid fuchsin–anilin blue* (*Stain Technol.*, **19**:29)

REAGENTS REQUIRED: A. water 98, acetic acid 2, picric acid 0.2, phosphotungstic acid 1, acid fuchsin 1, anilin blue 2; B. 2% acetic acid; C. water 90, 95% alc. 10, picric acid 0.25, phosphotungstic acid 2.5

METHOD: [sections with blue nuclei] → water → A, 5 min → B, rinse → C, until differentiated → D, 1 min → B, wash → balsam, via usual reagents

McFarlane 1944b, *picro–acid fuchsin–anilin blue* (*Bull. Soc. anat. Paris*, **19**:23)

REAGENTS REQUIRED: A. water 98, acetic acid 2, acid fuchsin 0.8, picric acid 0.2; B. 2% acetic acid; C. water 60, 95% alc. 40, picric acid 1, phosphotungstic acid 10; D. water 97.5, acetic acid 2.5, anilin blue 2.5; E. water 90, 95% alc. 10, picric acid 0.25, phosphotungstic acid 2.5

METHOD: [sections with iron hematoxylin nuclei] → water → A, 5 min → B, rinse → C, 5 min → rinse → D, 5–10 min → B, rinse → E, 5 min → B, wash → balsam, via usual reagents

McFarlane 1944c, *picro–orange G–acid fuchsin–ponceau 2R–anilin blue* (*Bull. Soc. anat. Paris*, **19**:23)

REAGENTS REQUIRED: A. water 20, 95% alc. 80, picric acid 1, orange G 0.25; B. water 99, acetic acid 1, acid fuchsin 0.25, ponceau 2R 0.25; C. 2% acetic acid; D. water 20, 95% alc. 80, picric acid 1, phospho-tungstic acid 10; E. water 97.5, acetic acid 2.5, anilin blue 2.5; F. water 80, 95% alc. 20, picric acid 0.5, phosphotungstic acid 5

METHOD: [sections, stained but not differentiated, in Régaud 1910 iron hematoxylin] → rinse → A, until nuclei differentiated → wash, until only erythrocytes yellow → B, 5–10 min → C, rinse → D, 5 min, until differentiated → C, rinse → E, 10 min → C, rinse → F, until differentiated → C, wash → balsam, via usual reagents

Pasini cited from 1928 Hill, *eosin B–acid fuchsin–anilin blue–orcein* (Gatenby and Cowdry 1928, 280)

REAGENTS REQUIRED: A. 2% phosphotungstic acid; B. 50% alc. 35, glycerol 40, eosin B 0.7, Unna (1928) anilin blue–orcein (see above, "Contrasts for Red Nuclei") 35, sat. aq. sol. acid fuchsin 4

PREPARATION OF B: Dissolve the eosin in 35 50% alc. Add, in order, the Unna's stain, fuchsin solution, and glycerol.

METHOD: [sections with blue nuclei (original specifies Ehrlich 1886)] → water → A, 10 min → water, quick rinse → B, 15–20 min → 70% alc. quick rinse → abs. alc., 20 sec → A, 5 sec → abs. alc. until differentiated → balsam, via xylene

RESULT: collagen, blue; elastic fibers; purple; erythrocytes, bright orange

NOTE: See also Water 1930.

Pollak 1944, *orange G–light green–ponceau 2R–acid fuchsin* (*Arch. Pathol.* (*Lab. Med.*). **37**:294)

REAGENTS REQUIRED: A. water 50, 95% alc. 50, acetic acid 1, phosphotungstic acid 0.5, phosphomolyb-dic acid 0.5, orange G 0.25, light green 0.15, ponceau 2R 0.33, acid fuchsin 0.17; B. 0.2% acetic acid

PREPARATION OF A: Mix water, alc., and acetic acid. Divide into four portions. In first dissolve phosphomolybdic acid with heat; in second dissolve phosphotungstic acid and Orange G; in third dissolve light green, in fourth dissolve acid fuchsin and ponceau 2R. Mix and filter.

METHOD: [sections with blue nuclei] → water → A, 3–7 min → B, until differentiated → 95% alc., until dehydrated → balsam, via usual reagents

Wallart and Honette 1934, *acid fuchsin–fast yellow* (*Bull. Histol. Tech. Micr.* (*Bull. d'Hist appl.*), **10**:404)

REAGENTS REQUIRED: A. 1% acid fuchsin in 1% acetic acid 30, 3% fast yellow in 1% acetic acid 30, 1% phosphomolybdic acid 30; B. 1% acetic acid; C. 1% acetic acid in abs. alc.

METHOD: [sections with nuclei stained in Weigert 1903] → tap water → A, 5 min → quick rinse → B, 5 min → C, dropped on from pipet, 30 sec → abs. alc. → xylene → balsam

RESULT: black nuclei, red cytoplasm, yellow collagen, pink elastin

Weiss 1932 see Brillmeyer 1929 (note)

Techniques Employing the Phosphotungstic (Molybdic) Reaction with Other Dyes

Dupres 1935, *toluidine blue–orange G* (*Monit. zool. ital.*, **46**:46)

REAGENTS REQUIRED: A. 1% phosphomolybdic acid; B. water 100, toluidine blue 0.25, orange G 4, oxalic acid 4

METHOD: [red nuclei (original specifies Duprès 1935) magenta] → A, 10 min → water, thorough wash → B, 2–5 min → drain → 95% alc. until differentiated → balsam, via usual reagents

RESULT: nuclei, red; collagens, blue; bone, dark green, muscle, light green; erythrocytes, bright orange; epidermis, blue; layer of Malpighi, orange; hair, etc., bright red

NOTE: Duprès 1935 (*loc. cit.*) also recommends methyl green in place of toluidine blue in B above.

Gomori 1950, *chromotrope–fast green* (*Tech. Bull.*, **20**:77)

REAGENTS REQUIRED: A. water 100, acetic acid 1, chromotrope 2R 0.6, fast green FCF 0.3, phosphotungstic acid 0.6; B. 0.2 acetic acid

METHOD: [smears, or sections not more than 5 μ thick, prior stained in hematoxylin] → water → A, 5–20 min → rinse → balsam, via usual reagents

NOTE: Wheatley 1951 (*Tech. Bull.*, **21**:92) recommends this technique for protozoans in intestinal smears.

Hollande 1912, *magenta–orange G–light green* (*Arch. Zool. exp. gen.*, **10**:62)

REAGENTS REQUIRED: A. 1% magenta in 70% alc.; B. 0.1% hydrochloric acid in 70% alc.; C. 1% phosphomolybdic acid; D. sat. sol. (*circ.* 11%) orange G; E. 0.2% light green

METHOD: [blue nuclei (Langeron 1942, 606 alum hematoxylin)] → water → A, 6–12 hr → water, 5 min → B, until color clouds cease, few seconds → water, thorough wash → C, 5 min → water, rinse → D, 5 min → E, poured on slide ½–1 min → 95% alc., few seconds, until differentiated → balsam, via amyl alc. and benzene

RESULT: resting nuclei, blue; mitotic figures, red; cartilage, purple; fibrous tissue, light green; erythrocytes and keratin, bright orange

Masson 1912, *ponceau 2R–anilin blue* (*Bull. Soc. anat. Paris*, **87**:290)

REAGENTS REQUIRED: A. water 100, acetic acid 1, ponceau 2R 1; B; C; D; E, as Masson 1912a

METHOD: as Masson 1912a

Patay 1934, *ponceau 2R–light green* (*Bull. Histol. Tech. Micr.* (*Bull. d'Hist. appl.*,) **11**:408

REAGENTS REQUIRED: A. 1% ponceau 2R; B. 1% phosphomolybdic acid; C. 0.5% light green in 90% alc.

METHOD: [blue nuclei (original recommends Masson (1934) insufficiently differentiated)] → water → A, 2 min → water, brief rinse → B, 2 min → water, brief rinse → C, 30 sec → balsam, via usual reagents

RESULT: cartilage, blue (from hematoxylin); other collagens, light green; bone, brilliant green; epithelia and muscle, orange; erythrocytes, yellow; nervous tissue, gray

Other Complex Contrasts

Flemming 1891, *gentian violet–orange G* (*Arch. mikr. Anat.*, **37**:249)

REAGENTS REQUIRED: A. 1% gentian violet; B. sat. sol. (*circ.* 11%) orange G

METHOD: [sections of Flemming 1882 fixed material, nuclei stained by safranin method] → water → A, 3 hr

→ quick wash → B, few moments → abs. alc., until no more color comes away → balsam, via clove oil

NOTE: Johansen (Johansen 1940, 84) substitutes a saturated solution of orange G in clove oil for B, above.

Hubin 1928, *eosin Y–orange G–safranin* (*Arch. Bio-. Paris*, **37**:25)

REAGENTS REQUIRED: A. 0.1% acetic acid; B. abs. alc. 70, water 30, eosin Y 0.1, orange II 0.2, safranin 0.2; C. 0.1% hydrochloric acid in 70% alc.

PREPARATION OF B: Dissolve the eosin in 50 alc. and 10 water. Add to this the orange dissolved in 10 water. To this add the safranin dissolved in 20 alc. and 10 water.

METHOD: [blue nuclei (original specifies Carazzi 1911 alum hematoxylin)] → A, 4–5 sec → 70% alc., 5 min → B, 1–3 min → C, until differentiated, 2–10 sec → balsam, via usual reagents

RESULT: nuclei, cartilage, bone, blue; nerves and blood vessels, yellow; muscles and ganglia, brown

Lendrum 1939, *GEEP stain—auct.* (*J. Pathol. Bact.*, **49**:590)

REAGENTS REQUIRED: A. water 50, 95% alc. 50, eosin Y 0.2, erythrosin 0.2, phloxine 0.2, gallic acid 0.5, sodium salicylate 0.5; B. sat. sol. tartrazine N.S. in ethylene glycol monoethyl ether

PREPARATION OF A: Add the dyes dissolved in the alc. to the other incredients dissolved in water.

METHOD: [sections with hematoxylin-stained nuclei] → A, 2 hr → 95% alc., rinse → B on slide, until color balance satisfactory

Lendrum 1947 (*J. Pathol. Bact.*, **59**:394)

REAGENTS REQUIRED: A. water 100, calcium chloride 0.5, phloxine 0.5; B. sat. sol. tartrazine in ethylene glycol monoethyl ether

METHOD: [sections with blue nuclei] → A, 30 min → rinse → B, from drop bottle, until differentiated → balsam, via usual reagents

Lillie 1940, *Biebrich scarlet–picro–anilin blue* (*Arch. Orthop. MechTher.*, **29**:705)

REAGENTS REQUIRED: A. water 99, acetic acid 1, Biebrich scarlet 0.1; B. water 100, picric acid 1, anilin blue 0.1; C. 1% acetic acid

METHOD: [sections, nuclei stained in hematoxylin] → water → A, 4 min → rinse → B, 4 min → C, 3 min → salicylic balsam, via usual reagents

Margolena 1933, *phloxine–orange G* (*Stain Technol.*, **8**:157)

REAGENTS REQUIRED: A. 0.5% phloxine in 20% alc.; B. 0.5% orange G in 95% alc.

METHOD: [blue nuclei] → water → A, 1–5 min → 70% alc., thorough rinse → B, dropped on slide, ½–1 min → abs. alc., until no more color comes away → balsam, via usual reagents

Masson, 1929, *metanil yellow–picro–fuchsin* (*J. Tech. Meth.*, **12**:75)

REAGENTS REQUIRED: A. water 100, acetic acid 0.5, metanil yellow 0.5; B. 0.2% acetic acid; C. 3% potassium dichromate; D. van Gieson 1896 picrofuchsin; E. 1% acetic acid

METHOD: [sections with blue nuclei (original requires Regaud 1910)] → water → A, 5 min → B, rinse → C, 5 min → D, poured on slide still wet with C, 2 min →

E, until yellow clouds cease → salicylic balsam, via usual reagents
RESULT: collagens, clear red

Masson 1911, *erythrosin-saffron* (*C.R. Soc. Biol. Paris,* **70**:573)
REAGENTS REQUIRED: A. water 100, erythrosin 1, 40% formaldehyde 0.25; B. water 100, saffron 2, 40% formaldehyde 1, 5% tannic acid 1
PREPARATION OF B: Boil saffron 1 hr in water. Cool. Filter. Add other ingredients.
METHOD: [blue nuclei] → A, 5 min → water, quick rinse → 70% alc., until collagens colorless → water, quick rinse → B, 5 min → water, rapid rinse → abs. alc., minimum time possible → balsam, via xylene
RESULT: nuclei, blue; collagens, yellow; muscle, red

Maximow 1909, *eosin Y–azur II* (*Z. wiss. Mikr.,* **26**:177)
STOCK FORMULAS: I. 0.1% eosin Y; II. 0.1% azur II
WORKING SOLUTION: water 100, stock I 10, stock II 10
METHOD: [sections with nuclei stained in very dilute alum hematoxylin → water, 24 hr → stain, 12–24 hr → 95% alc., quick rinse → abs. alc., until differentiated → neutral balsam, via xylene

Millot 1926, *acid fuchsin–martius yellow* (*Bull. Histol. Tech. Micr.* (= *Bull. d'Hist. appl.,* **3**:2)
REAGENTS REQUIRED: A. 5% acid fuchsin in 40% alc.; B. 5% martius yellow in 40% alc.
METHOD: [blue nuclei] → A, 5 min at 30°C → 40% alc., until no more color comes away → B, 5 min → balsam, via usual reagents
NOTE: This method was developed for insect histology, for which it is excellent.

Pianese 1896, *malachite green–acid fuchsin–martius yellow* (*Beitr. pathol. Anat.,* **1**:193)
FORMULA: water 75, 95% alc. 25, malachite green 0.25, acid fuchsin 0.05, martius yellow 0.005

Reinke 1894, *gentian violet–orange G* (*Arch. mikr. Anat.,* **44**:262)
FORMULA: A. sat. sol. (*circ.* 1%) gentian violet 25, sat. sol. (*circ.* 11%) orange G 0.2, water 75
NOTE: Use after safranin nuclear staining and differentiate with clove oil.

Scriban 1924, *picro-fuchsin–brilliant green* (*C.R. Soc. Biol. Paris,* **90**:531)
REAGENTS REQUIRED: A. water 100, picric acid 0.5, acid fuchsin 0.1; B. 60% alc. 100, picric acid 0.2, brilliant green 0.1
METHOD: [sections with iron hematoxylin stained nuclei] → thorough wash → A, 3–4 sec → abs. alc., wash → B, 3–4 sec → abs. alc., wash → balsam via usual reagents

de Winiwarter and Sainmont 1908, *crystal violet–orange G* (*Z. wiss. Mikr.,* **25**:157)
REAGENTS REQUIRED: A. 1% crystal violet; B. sat. sol. (*circ.* 11%) orange G; C. 0.1% HCl in abs. alc.
METHOD: [sections of osmic-chromic and osmic-chromic-acetic fixed material; nuclei stained by safranin method] → water → A, 24 hr → brief rinse → B, 1 min → C, 2–3 hr → clove oil, until differentiated → balsam
NOTE: de Winiwarter 1923 (1825, 32,329) recommends a process which differs only in the substitution of 0.2% orange G for B, above.

Complex Techniques Involving Both Nuclear and Plasma Staining

This is a large class of staining techniques, involving those methods in which, by a series of successive and interlocking operations, both the nuclei and all the elements of the background are differentially stained. They are divided broadly, for purposes of this work, into seven classes, of which the first two (Techniques Employing the "Eosinates" of the Thiazins without Other Admixture and Techniques Employing the Thiazins and Their Eosinates in Combination with Other Dyes) employ the thiazins and their related compounds either in combination with eosin, or in the next class, with such other dyes as have been employed. The next class (Techniques Employing Methyl Green as the Nuclear Stain) takes up the methyl green combinations and is followed by the group (Techniques Employing Acid Fuchsin as the Nuclear Stain) of complex formulas in which the Mallory reaction is employed. This is followed by a small group (Techniques Employing Safranin as the Nuclear Stain), largely of French origin, in which safranin is employed as the nuclear stain, and another small group (Techniques Employing Hematoxylin as the Nuclear Stain) in which hematoxylin is employed. This still leaves for the last class (Other Polychrome Staining Methods) a considerable miscellaneous group.

Techniques Employing the "Eosinates" of the Thiazins without Other Admixture
Methylene Blue Eosinates

Assmann 1906a cited from 1928 Schmorl (Schmorl 1928, 241)
REAGENTS REQUIRED: A. May-Grünwald 1902 (working sol.) [below]; B. water 100, Unna 1892
METHOD: [dried smear] → A, 1 ml poured on slide lying in petri dish, 3 min → B, 15 ml poured into dish around slide, 3–4 min → wash → dry

Assmann 1906b cited from 1928 Schmorl (Schmorl 1928, 246)
REAGENTS REQUIRED: A. May-Grünwald 1902 (working sol.) (see below); B. 0.001% acetic acid
METHOD: [sections of "dichromate alone" Müller 1859 or "mercuric-dichromate-acetic" Zenker 1894 fixed material] → A, several hours → B, until color changes to clear eosin → wash → balsam, via usual reagents

Chenzinsky 1894 (*Z. wiss. Mikr.,* **11**:269)
FORMULA: sat. sol. (*circ.* 4.5%) methylene blue 40, 0.5% eosin Y in 70% alc. 20, water 20
METHOD: [fresh smear] → stain, 5 min → water, rinse → dry

Ellerman 1919 (*Z. wiss. Mikr.,* **36**:56)
REAGENTS REQUIRED: A. water 100, 40% formaldehyde 5, eosin Y 1; B. May-Grünwald 1902 (working sol.) (see below) 50, water 50
METHOD: [5 μ paraffin sections of "mercuric-dichromate-formaldehyde" Ellerman 1919 fixed material] → water → A, 15 min → wash, 2–4 min 45°C → B, 30 min → wash, 5–10 min → blot → abs. alc. until differentiated → mountant via usual reagents

Greenstein 1947 (*Stain Technol.,* **32**:75)
STOCK I: Add 1 HCl to 100 1% phloxine B. Wash ppt. by decantation. Filter out washed ppt. and dissolve 0.5 in 100 95% alc.

STOCK II: 1% azur II
STOCK III: 1% methylene blue in 1% borax
REAGENTS: A. 95% alc. 60, stock I 40, 28% acetic acid 1; B. 95% alc. 100, sat. aq. sol. lithium carbonate 0.5; C. stock II 50, stock III 50 D 1% rosin in 95% alc.
METHOD: [sections] → water → 80% alc. 2 min → A, 3–4 min → B, rinse → wash → C, 5 min → D, 2 min in each of 2 changes → [balsam via usual reagents]

Jenner 1899 (*Lancet, Lond.*, **6**:370)
PREPARATION OF DRY EOSINATE: Mix equal parts 1.25% eosin Y and 1% methylene blue. Leave 24 hr. Filter. Wash and dry filtrate.
WORKING SOLUTION: dry powder 0.5, methanol 100
METHOD: [fresh smear] → stain, 3 min → water, rinse → dry
NOTE: The most usual employment of this formula is as a fixative before such methods as Slider and Downey (1929).

Jenner cited from 1905 Lee (Lee 1905, 385)
FORMULA: 0.5% methylene blue 50, 0.5% eosin Y in methanol 62.5
METHOD: as Jenner 1899

Lim 1919 (*Quart. J. Micr. Sci.*, **63**:542)
REAGENTS REQUIRED: A. 1% ethyl eosin in abs. alc.; B. 1% methylene blue
METHOD: [sections or smears] → abs. alc. → A, on slide, 1 min → water, thorough wash → B, on slide, 1 min → blot → abs. alc., flooded on slide, 2–3 sec → benzene → neutral mountant *or* → dry

May-Grünwald 1902 (*Zbl. inn. Med.*, **11**:265)
PREPARATION OF DRY EOSINATE: Mix equal parts of 0.5% eosin Y and 0.5% methylene blue. Filter. Dry filtrate. Wash dried filtrate and redry.
WORKING SOLUTION: methanol 100, dry powder from above to sat.
METHOD: [air-dried smear] → stain, on slide, 3 min → add equal volume distilled water, leave 1 min → wash → dry
NOTE: See note under Jenner 1899 with which this technique is nearly, but not quite, identical. Zieler (cited from Schmorl 1928, 246) stains paraffin sections 2–3 min and dehydrates in acetone; for another application of this stain to sections see Assmann 1906b above.

Michaelis 1901 (*Deuts. med. Wschr.*, **27**:127)
FORMULA: water 40, abs. alc. 25, acetone 35, methylene blue 0.25, eosin 0.15
PREPARATION: Dissolve the methylene blue in 25 water, 25 alc. Dissolve the eosin in 15 water, 35 acetone. Mix. Filter.

Müller cited from 1928 Schmorl (Schmorl 1928, 239)
REAGENTS REQUIRED: A. 0.5% eosin Y in 70% alc.; B. A 60, 0.25% methylene blue 30
METHOD: [blood smear fixed 3 min in methanol] → A, 3–5 min → distilled water, wash → blot → B, ½–1 min → distilled water, wash → dry

Nocht cited from 1903 Ehrlich (Ehrlich, Krause et al. 1903, 784)
FORMULA: water 65, acetone 17, 0.1% thiazin 10, 0.1% eosin 10, buffer pH 4.6
NOTE: This is not a specific technique, though it is often quoted as such; it is a general direction for experiments with any thiazins and any eosin.

Sabrazès 1911 (*C.R. Soc. Biol. Paris*, **70**:247)
REAGENTS REQUIRED: A. 30% eosin Y in 95% alc.; B. 0.2% methylene blue
METHOD: [smear] → place 1 drop A on smear and 1 drop B on coverslip → seal with wax for temporary examination → *or* water, quick rinse → dry

Thomas 1953 (*Stain Technol.*, **28**:311)
REAGENTS: A. water 100, acetic acid 0.2, phloxine 0.5 B. water 100, methylene blue 0.25, azur B 0.25, borax 0.25 C. 0.2% acetic acid
METHOD: [sections] → water → A, 1–2 min → rinse → B, ½–1 min → C, until almost differentiated → 95%, until differentiation complete → [balsam, via usual reagents]

Willebrand 1901 (*Deuts. med. Wschr.*, **27**:57)
FORMULA: water 65, abs. alc. 35, eosin 0.25, methylene blue 0.5, 1% acetic acid *q.s.*
PREPARATION: Dissolve the eosin in 35 alc., 15 water. Dissolve the methylene blue in 50 water. Mix the solutions and add acid drop by drop until solution turns red. Filter.
METHOD: [fresh smear] → stain 5 min → rinse → dry

Polychrome Methylene Blue Eosinates

Diercks and Tibbs 1947 (*J. Bact.*, **53**:479)
REAGENTS REQUIRED: A. water 100, MacNeal 1922 (see below) 6
METHOD: [moist smear] → methanol 3–5 min → A, 15–20 min → acetone, until no more color comes away → blot → dry

Gordon 1939 (*J. Lab. Clin. Med.*, **24**:405)
REAGENTS REQUIRED: A. 10% ammonia; B. water 100, copper sulfate 12, mercuric chloride 2.5, potassium dichromate 1, sodium sulfate 0.5; C. water 100, eosin Y 0.25, phloxine 0.75; Loffler 1890 methylene blue (see THIAZINE DYES)
METHOD: [sections of formaldehyde material] → A, 1 min → wash → B, 3 min → wash → C, 2 min → wash → D, 4 min, 37°C → wash → abs. alc., until differentiated → balsam, via xylene

Leishman 1901 (*Brit. Med. J.*, **2**:757)
PREPARATION OF DRY EOSINATE: To 100 0.5% methylene blue add 0.25 sodium carbonate. Digest 12 hr at 65°C followed by 10 days at 15°C. Filter. Add 50 0.5% eosin B to filtrate. Leave 12 hr. Filter. Wash and dry filtrate.
WORKING SOLUTION: methanol 100, dry stain 0.15
METHOD: [air-dried smear] → stain, on slide, 5–10 min → water, 1 min → blot dry
NOTE: This method is stated by Leishman (*loc. cit.*) to be a modification of a stain proposed by Rowmanowski 1891 (*St. Petersburg Med. Wschr.*, **16**:297) to the original of which the writer has never had access.

Lillie and Pasternack (*Bull. Int. Ass. Med. Mus.*, **15**:65)
PREPARATION OF DRY STOCK: Dissolve 0.25 silver nitrate in 12.5 water. Add 1–2 5% sodium hydroxide, wash ppt. five times by decantation. Add 50 1% methylene blue to wet ppt. Shake at intervals for 11 days. Filter and add 45 1% eosin. Leave overnight, filter, and dry ppt.
FORMULA OF STOCK SOLUTION: methanol (acetone-free) 75, glycerol 25, dry stock 0.6 6, methanol 6, stock solution 4

WORKING SOLUTION: citric acid–sodium phosphate, dibasic buffer (see note) 70, acetone 6, methanol 6, stock solution 4

NOTE: The buffer must be so adjusted that after the addition of the other ingredients the pH is 5.3.

Mallory cited from Langeron 1942 (Langeron 1942, 614)
REAGENTS REQUIRED: A. water 100, eosin Y 5; B. water 85, Sahli methylene blue [see THIAZINE DYES] 1885 15; C. 0.5% rosin in 95% alc.
METHOD: [water] → A, 20 min → rinse → B, 5 min → B, fresh solution, 15 min → B, fresh solution, 30 min → wash → C, until differentiated → abs. alc. least possible time → balsam, via cedar oil
RESULT: nuclei, blue; other structures, polychrome

Manwell 1945 (J. Lab. Clin. Med., 30:1078)
REAGENTS REQUIRED: A. Manwell 1945 methylene blue [see THIAZINE DYES]; B. water, adjusted to pH 6.5 with acetic acid; C. 0.2% eosin Y
METHOD: [methanol-fixed smear] → dry → A, 30 sec → B, wash → A, 30 sec → dry → neutral mountant

Marie and Raleigh 1924 (J. Lab. Clin. Med., 10:250)
PREPARATION OF DRY STOCK: Dissolve 1 methylene blue in 100 0.5 sodium bicarbonate and expose to UV arc 30 min in shallow dish. Cool and add 500 0.1% eosin Y. Filter, wash, and dry ppt.
WORKING SOLUTION: methanol 100, dry stock 0.16

Michelson 1942 (J. Lab. Clin. Med., 27:551)
PREPARATION OF DRY STOCK: Dissolve 1 methylene blue in 100 N/100 sodium hydroxide. Heat 2½ hr, 55°C shaking for 1 min at half-hour intervals. Add 1 sodium bromide, continue heating 2½ hr. Cool. Filter. Add 60 1% eosin Y and mix. Add further eosin in doses of 5 until solution is reddish. Leave 24 hr. Filter. Dry filtrate.
WORKING SOLUTION: 0.017% dry stock in methanol

Raadt 1912 (Z. wiss. Mikr., 29:236)
STOCK SOLUTION: water 100, methylene blue 1, lithium carbonate 0.5
REAGENTS REQUIRED: A. stock 10, water 90; B. Jenner 1899 methylene blue–eosinate 25, water 75
METHOD: [methanol-fixed smears] → A, on slide, 5–10 min → rinse → blot → B, 5–10 min → wash → dry

Roques and Jude 1940 cited from 1949 Langeron (Langeron 1949, 621)
FORMULA OF STOCK SOLUTION: methanol 50, glycerol 50, eosin B 0.2, methylene blue 0.2, Roques and Jude 1940 methylene blue [see THIAZINE DYES] 0.5
PREPARATION OF STOCK SOLUTION: Dissolve dyes in methanol, leave 24 hr, filter, and add glycerol.
WORKING SOLUTION: water 95, stock 5

Senevet 1917 (Bull. Soc. Pathol. exotique., 10:540)
STOCK SOLUTIONS: I. methylene blue 1, sodium borate 3, water 100; II. eosin Y 1, water 100
WORKING SOLUTION: water 100, stock II 0.2, stock I 0.25 to 0.4
METHOD: [methanol-fixed smears] → A, 2–3 hr → wash → dry

Stafford 1934 (Johns Hopkins Hosp. Bull., 55:229)
REAGENTS REQUIRED: A. water 100, potassium dichromate 1, eosin 1; B. Goodpasture (1934) methylene blue [see THIAZINE DYES]; acetone 100, abs. alc. 10

METHOD: [frozen sections of formaldehyde-alc. material] → water → A, 1 sec → wash → B, 30 sec → wash → blot on slide → C, dropped on section until dehydrated → balsam, via xylene

Wright 1910 (Rep. Mass. Gen. Hosp., 3:1)
FORMULA OF METHYLENE BLUE STOCK: water 100, methylene blue 1, sodium bicarbonate 0.5
PREPARATION OF STOCK SOLUTION: Digest at 100°C 1½ hr.
WORKING SOLUTION: methanol 80, eosin B, 0.16, stock blue 24
METHOD: [smear] → stain, on slide 2 min → add water, drop by drop, until green scum forms on surface, leave 2 min → water, thorough wash → [dry] → or → neutral mountant, via acetone
NOTE: Wright 1910 also recommended (Quart. J. Micr. Sci., 57:783) the dilution of the working solution with an equal volume of water to be used for 10 min. It is nowadays universal practice to substitute a phosphate buffer at pH 6.4 for the water used to dilute the stain on the slide.

Other Thiazin Eosinates

Böhm and Oppel 1907, *methylene blue–thionin–eosin* (Böhm and Oppel 1907, 114)
REAGENTS REQUIRED: A. water 96, 40% formaldehyde 4, methylene blue 0.3, thionin 0.15; B. 1% acetic acid; C. 1% eosin B
METHOD: sections → A, some minutes → B, until no more color comes away → C, few seconds → balsam, via usual reagents

Boye 1940 (Bull. Soc. Pathol., exotique., 33:248)
REAGENTS REQUIRED: A. water 100, eosin Y 0.1; B. Stevenel 1918 methylene blue [see THIAZINE DYES]
METHOD: [methanol-fixed smears] → A, 15–20 sec → B, flooded on slide, 45 sec → A, until differentiated → rinse → dry

Endicott 1945 (Stain Technol., 20:5)
PREPARATION OF DRY STOCK: To 250 ml 0.8% toluidine blue add 250 0.4% eosin B. Filter, wash, and dry ppt.
WORKING SOLUTION: methanol 50, glycerol 50, dry stock 0.3

Geschickter, Walker, Hjort, and Moulton 1931 (Stain Technol., 6:3)
REAGENTS REQUIRED: A. water 60, glycerol 20, 95% alc. 20, sodium hydroxide 1.47, potassium acid phosphate 0.675; B. ethylene glycol 75, 95% alc. 25, acetic acid 0.2, thionine-eosinate 0.75, barium-eosinate 0.25, azur A 0.25; C. 20% glycerol in 95% alc.; D. diethylene glycol monobutyl ether; E. n-butyl phthalate
PREPARATION OF B: Dissolve the barium-eosinate with heat; raise solution to boiling, add thionine-eosinate. Filter solution hot.
METHOD: [frozen sections of fresh tissue] → A, until required → B, 20–30 sec → C, 10 sec → C, fresh solution, 3 sec → D, 10–15 sec → E. 20 sec → dammar
NOTE: Solution A should be adjusted, if necessary, to pH 7.

Haynes 1926a (Stain Technol., 1:68)
REAGENTS REQUIRED: A. 1.5% azur II or azur C; B. sat. sol. ethyl eosin in clove oil
METHOD [water] → A, 5 min → abs. alc. quick rinse → B, 30 sec → balsam, via xylene

Haynes 1926b (*Stain Technol.*, **1**:107)
REAGENTS REQUIRED: A. 2.5% phloxine; B. 0.1% azur I; C. 0.5% rosin in 95% alc.
METHOD: [sections] → water → A, 15 min → water, thorough wash → B, 30 min → water, thorough wash → C, from drop bottle, until differentiated → abs. alc., minimum possible time → balsam, via xylene

Kingsley 1935 (*Stain Technol.*, **10**:127)
STOCK SOLUTIONS: I. water 50, methanol 10, glycerol 10, buffer pH 6.9 30, methylene blue 0.130, azur A 0.020; II. acetone 70, methanol 20, glycerol 10, methylene violet 0.026, eosin Y 0.090
REAGENTS REQUIRED: A. stock I 50, stock II 50; B. 0.015% acetic acid; C. acetone 60, acetic acid 0.005, eosin Y 0.0005; D. n-butyl alc. 60, eosin Y 0.001
METHOD OF SMEARS: [air-dried smear] → methanol ½–1 min → dry → A, 5–8 min → water, thorough wash → dry
METHOD FOR PARAFFIN SECTIONS: water → A, flooded on slide, 9–10 min → water, thorough wash → C, rinse → D, rinse → balsam, via xylene
TEMPORARY METHOD FOR FROZEN SECTIONS: [blot section to slide] → A, flooded on slide, 2–3 min → wash → [examine]
PERMANENT METHOD FOR FROZEN SECTIONS: [blot section to slide] → A, 4–5 min (fresh) or 8–10 min (fixed) → [thence as for paraffin sections]
NOTE: Ritchie 1941 (*Tech. Bull.*, **2**:157) prefers, for smears, to dilute the A stain 8:5 with water. Berlanger 1961 (*Stain Technol.*, **36**:313) recommends a modification of this technique for staining autoradiographs.

Langeron 1942a, *Giemsa for wet-smears—auct* (Langeron 1942, 583)
REAGENTS REQUIRED: A. water 100, potassium iodide 2, iodine 0.12 3; B. 0.5% sodium thiosulfate; C. water 97, Giemsa 1902 (above) 3; D. 1% monosodium phosphate
METHOD: [smears fixed in Schaudinn 1893 or Zenker 1894, 24 hr] → rinse → A, 5–10 min → rinse → B, 10 min → running water 5 min → distilled water → C, 30 min → C, fresh solution, 12 hr → D, if differentiation required → neutral mountant, via graded acetone-xylene mixtures

Langeron 1942b, *Giemsa for sections—auct* (Langeron 1942, 583)
REAGENTS REQUIRED: A. 70% alc. 97, iodine 0.12; B. 0.5% sodium thiosulfate; C. water 97, Giemsa 1902 (see above) 3
METHOD [5 μ paraffin section of Schaudinn 1893 or Zenker 1894 material, dewaxed and brought to water] → A, 20–30 min → rinse → B, 10 min → tap water 5 min → distilled water → C, 30 min → C, fresh solution, 2–12 hr → balsam, via graded acetone-xylene mixtures

Groat 1936 (*J. Lab. Clin. Med.*, **21**:978)
PREPARATION OF DRY STOCK: Dissolve 1.2 eosin Y in 100 water. Dissolve 1 methylene blue, 0.2 methyl violet 2 B, 0.04 thionine in 10 water. Mix solutions, heat to 50°C and hold at 37°C, 24 hr. Filter. Wash and dry filtrate.
FORMULA OF WORKING SOLUTION: methanol 100, dry stock 0.5
METHOD: [blood smears] → stain, on slide, 5 min → plunge in distilled water until smear rosy-pink → dry

Kingsley 1937 (*J. Lab. Clin. Med.*, **22**:524)
STOCK SOLUTIONS: I. water 80, phosphate buffer pH 6.920, methylene blue 0.070, methylene azur 0.025, II. acetone 70, methanol 20, glycerol 10, methylene violet 0.018, eosin Y 0.065
WORKING SOLUTION: stock I 50, stock II 50

Kühn 1933 (*Sbor. Stud. biol. Kruzh. Odessa.*, **7**:758)
PREPARATION OF DRY STOCK: Mix 9 ammonia with 6 2% copper sulfate. Add this to 200 0.5% methylene blue and leave 24 hr at 18°–20°C. Then add, in small amounts and with continuous agitation, 40 2% eosin Y. Filter, wash ppt. with 4% ammonia, and dry.
STOCK SOLUTION: methanol 60, glycerol 30, dry stock 0.36
WORKING SOLUTION: stock solution 25, methanol 75

Lillie 1948, *azur-eosin* (Lillie 1948, 82)
PREPARATION OF DRY STAIN: Dissolve 1 azur A or C in 60 water. Add 0.8 eosin B or Y dissolved in 10 water. Mix solutions. Filter. Wash and dry ppt.
PREPARATION OF STOCK SOLUTION: glycerol 50, methanol 50, dry stock 1
PREPARATION OF STOCK BUFFERS: I. water 75, methanol 25, sodium phosphate, dibasic 2.84; II. water 75, methanol 25, citric acid 9.12
PREPARATION OF WORKING SOLUTION: stock stain 0.5, acetone 5, mixed buffers I and II 2, water to make 40
METHOD: [sections] → water → stain, 1 hr → rinse → acetone, until dehydrated → balsam, via xylene
NOTE: The ratio of buffers I and II in the working stain must be adjusted to suit the fixative used and the color balance desired. Lillie, *loc. cit.*, recommends from 0.7 I 1:3 II to 1:1. This is a slight modification of Lillie 1941 (*Stain Technol.*, **16**:1).

MacNeal 1922, *azur I–methylene violet–eosin Y* (*J. Amer. Med. Ass.*, **78**:1122)
FORMULA: methanol 100, eosin Y 1, azur I 0.6, methylene violet 0.2
METHOD: [air-dried smears] → flood with stain, 2 min → add water, drop by drop, until green scum forms on surface → leave 1 min → water, thorough wash → neutral mountant, via acetone
NOTE: The dry dyes, mixed in the proportion indicated, have appeared in commerce under the name *MacNeal's Tetrachrome.*

Maximow 1924, *azur-eosin* (*J. Infect. Dis.*, **34**:549)
REAGENTS REQUIRED: A. water 100, Delafield 1885 0.1; B. 0.1% eosin Y 10, water 100, 0.1% azur II 10
METHOD: [sections from mercuric-dichromate-formaldehyde-acetic fixed material] → water → A, 24 hr → water → B, 24 hr → 95% alc., until differentiated → balsam, via xylene

McNamara 1933 (*J. Lab. Clin. Med.*, **18**:752)
REAGENTS REQUIRED: A. water 90, iodine 0.4; B. 0.5% sodium thiosulfate; C. water 100, Slider and Downey 1929 (see below) 10, acetone 10, methyl alc. 10, 0.5% sodium carbonate 2; D. 0.5% rosin in 95% alc.
METHOD: [4 μ sections of Zenker 1894] → A, 30 min → 95% alc. wash → water, wash → B, 10 min → wash → C, 15 min → D, until differentiated → balsam, via acetone and xylene

Pappenheim 1908, *panoptic stain*—compl. script. (*Med Klinik.*, **4**:1244)

REAGENTS REQUIRED: A. May-Grünwald 1902 (see above); B. water 97, Giemsa 1902 3

METHOD: [dry smear] → A, flooded on slide, 3 min → add 10 drops distilled water, mix well, leave 1 min → drain → B, 10–15 min → wash off with jet of distilled water → wash, until differentiated → drain, dry

NOTE: Agulhon and Chavannes 1919 (*C.R. Soc. Biol. Paris*, **82**:149) differentiate with 1% boric acid or 1% monosodium phosphate. Otherwise the technique is identical. Wolbach 1911 (*J. Amer. Med. Ass.*, **56**:345) uses his "rosin alcohol" (0.5% in 95% alc.) for differentiation. Langeron 1942 refers to his "method for sections" as *Wolbach's technique.*

Pappenheim 1912a (*Anat. Anz.*, **42**:525)

REAGENTS REQUIRED: A. water 75, May-Grünwald 1902 (see above) 25; B. water 95, Giemsa 1902; C. 0.2% acetic acid

METHOD: [5 μ paraffin sections of Orth 1896 Helly 1903 material, dewaxed and brought to water] → A, 15 min, 37°C → B, 30 min at 37°C → wash → C, until differentiated → balsam, via graded acetone-xylene mixtures

Shortt 1918 (*Ind. J. Med. Res.*, **6**:124)

REAGENTS REQUIRED: A. Langeron 1908 methylene blue [see THIAZINE DYES] 10, water 90; B. eosin Y 0.01, water 100

METHOD: [methanol-fixed smears] → 1 A mixed with 1, 2, or 3 B (proportion established by trial) → quick rinse → abs. alc., until differentiated → water, to stop differentiation → dry → balsam

NOTE: The original calls for a 0.1% solution of *Borrel's blue*. A above is identical in composition to such a solution.

Slider and Downey 1929, *azur-eosin* (McClung 1929, 246)

FORMULA: glycerol 50, azur II 0.16, azur II–eosin 0.6, methanol 50

METHOD: [smears, treated with methanol or Jenner 1899 or May-Grünwald 1902] → equal parts stain and water (or pH 6.4 buffer), 15 min → water, until differentiated → blot → dry

NOTE: Gatenby and Cowdry 1928, 490 specify 75 methanol to 25 glycerol.

Svihla 1924 (*J. Amer. Med. Ass.*, **83**:2093)

PREPARATION OF DRY STOCK: To 50 water add 0.69 methylene blue, 0.44 silver oxide, 0.38 sodium bicarbonate. Boil 1 hr and decant supernatant liquid which is added to 50 0.62% eosin. Mix thoroughly, filter, wash and dry ppt.

WORKING SOLUTION: methanol 100, sodium phosphate, dibasic 0.125, potassium dihydrogen phosphate 0.080, potassium hydroxide 0.004

Techniques Employing the Thiazins and Their Eosinates in Combination with Other Dyes. The combination of the thiazin eosinates with orange G is almost as old as the utilization of the thiazin-eosinates themselves. The first and still the best known formula is that of Mann 1894, though in French literature it has largely been replaced by the method of Dominici 1902. The preoccupation of the French school with the work of Dominici has led to the production of such almost incredibly complex methods as those of Gausen 1929 and Houcke 1928. Complex methods of this type have proved of some value in the investigation of pathology, but are not to be recommended to the routine worker.

The combination of the thiazin eosinates with other dyes than orange G (In Combination with Other Dyes) have not been very successful, the main exception being the technique of Kull 1914 which, though originally designed for the demonstration of mitochondria, is one of the best and surest triple staining methods yet developed. The formula of Rhamyl 1930 is of great interest in the staining of blood smears, and yields pictures which are not only of as great diagnostic value but also far more readily preserved than are the standard eosin-methylene blue mixtures.

In Combination with Orange G

Arnold 1909, *methylene blue–safranin–orange G* (*Arch. Zellforsch.*, 3:434)

REAGENTS REQUIRED: A. 6% iodine in 4% potassium iodide; B. sat. sol. safranin in 70% alc.; C. water 100, methylene blue 7, sodium carbonate 0.5; 1% orange G

METHOD: [sections of chromic or dichromate material] → A, 5 min → wash → B, 4 hr → wash → C, 4 hr → abs. alc., until dehydrated → D, until differentiated → balsam, via usual reagents

RESULT: nucleoli, centrosomes, red; many cell inclusions, blue; other structures, orange

Cowdry 1943, *phloxine–orange G–azur A* (Cowdry 1943, 69)

REAGENTS REQUIRED: A. water 100, phloxine 0.12, orange G 0.3; B. 0.1% azur A

METHOD: As Dominici 1902 below

NOTE: See also note on Cowdry 1943 modification under Dominici 1902.

Dominici 1902, *eosin Y–orange G–toluidine blue* (*C.R. Soc. Biol. Paris*, **54**:221)

REAGENTS REQUIRED: A. water 100, eosin Y 0.5, orange G 0.5; B. 0.5% toluidine blue

METHOD: [water] → A, 5–10 min → water, quick rinse → B, 20–30 sec → water, quick rinse → 95% alc., until differentiated → balsam, via xylene

NOTE: Cowdry 1943 recommends the substitution of 0.5% acid fuchsin for the eosin Y in A above. See also Cowdry 1943 above and Mann 1894 below.

Gausen 1929, *methylene blue–orange G–magenta* (*C.R. Soc. Biol. Paris*, 97:1658)

FORMULA OF METHYLENE BLUE SOLUTIONS: 80% alc. 150, lactic acid 3, methylene blue 2.5

FORMULA OF ORANGE G SOLUTION: 80% alc. 100, lactic acid 2, orange G 2

PREPARATION OF COMPLEX: Add blue to orange. Heat to 80°C. Filter. Save both ppt. and filtrate.

PURIFICATION OF PRECIPITATE: Dissolve ppt. from above in 30 80% alc. Heat to 80°C. Cool. Filter. Save filtrate. Reject ppt.

PREPARATION OF WORKING SOLUTION: Mix both filtrates with 50 methylene blue solution. Filter.

REAGENTS REQUIRED: A. Ziehl 1890 magenta [see MAGENTA]; B. working solution

METHOD: sections → water → A, 5 min → water, thorough wash → B, 4 min → abs. alc., until differentiated; → neutral mountant, via usual reagents

RESULT: nuclei, red; cartilage, blue; muscle, yellow; bone, green

Houcke 1928, *toluidine blue–orange G–thionin-eosin-azur II* (*C.R. Soc. Biol. Paris*, 99:784)

PREPARATION OF STOCK SOLUTIONS: I. Mix 10 1% toluidine blue with 5 1% orange G. Dilute to 100. Leave 24 hr. Decant and leave ppt. dry. Dissolve dried ppt. in 10 methanol. II. Add 17 1% eosin Y to 100 sat. sol. (circ. 25%) thionine. Leave 24 hr. Decant. Dry ppt. Prepare 0.5% solution ppt. in methanol. III. Add 11 1% eosin B to 10 1% methylene blue. Thence as in II. IV. Add 1.25 1% eosin Y to 40 0.08% azur-eosin. Filter. Dry ppt. on filter paper. Cut in strips and extract with 10 methanol. V. Add 8 1% eosin Y to 10 1% toluidine blue.
WORKING SOLUTION: water 100, stock I 0.5, stock II 1.5, stock III 1.5, stock IV 1.5, stock V 1.5, 0.1% acetic acid 1.0
METHOD: [water] → stain, 24 hr → abs. alc., minimum possible time → balsam, via xylene
NOTE: Houcke (loc. cit.) recommends varying the acidity of the working solution by experiment to adapt it to use after various fixatives.

Holmes and French 1926, azur C–eosin Y–orange II (Stain Technol., 1:25)
REAGENTS REQUIRED: A. 1.5% azur C; B. abs. alc. 99, acetic acid 1, eosin Y 0.025, orange II 0.025
METHOD: [water] → A, 5 min → methanol, until color clouds cease, 5–10 sec → B, until no more blue comes away → abs. alc., quick rinse → balsam, via xylene
RESULT: nuclei, and some bacteria, blue; cell inclusions and blood, as Giemsa; collagens, bright orange; muscle, pink

Mann 1894 cited from 1942 Langeron cit. Masson, erythrosin–orange G–toluidine blue (Langeron 1942, 613)
REAGENTS REQUIRED: A. 6% iodine in 4% potassium iodide; B. 5% sodium thisulfate; C. water 100, orange G 1, erythrosin 0.2; D. 1% toluidine blue; E. 0.2% acetic acid
METHOD: [sections] → water → A, ½ hr → water, rinse → B, until bleached → water, thorough wash → C, 15 min → water, rinse → D, on slide, 1–2 min → water, rinse → E, until differentiated → balsam, via usual reagents
NOTE: Langeron (loc. cit.) says "Ce procédé, dit de Dominici, jouit en France d'une assez grande vogue—etc." See, however, Dominici 1902 above.

In Combination with Other Dyes
Bauer and Leriche 1934, methylene blue–eosin Y–cresyl blue (Pr. med., 42:1385)
REAGENTS REQUIRED: A. water 100, brilliant cresyl blue 0.25; B. Jenner 1899 (working sol. [see above])
METHOD: Mix 4 parts blood with 1A, as drop on slide, 2 min → smear → dry → B, 2 min → water, rinse → dry → neutral mountant

Blank 1942, mercurochrome-azur-eosin (J. Lab. Clin. Med., 27:1342)
STOCK SOLUTIONS: I. water 82, 40% formaldehyde 9, abs. alc. 5, Giemsa 1902 2.25, methylene blue 0.25, sodium borate 0.5
REAGENTS REQUIRED: A. 0.1% mercurochrome 220 in 25% methanol; B. stock I 5, water 95
METHOD: [sections (nuclei may be hematoxylin-stained)] → water → A, 1 min → wash → B, 2 min → 90% alc., until no more color comes away
Geschickter 1930, azur-erie garnet (Stain Technol., 5:81)
FORMULA: water 100, azur A 0.8, erie garnet B 0.1

PREPARATION: To the azur dissolved in 80 water add, very rapidly, the garnet dissolved in 20. Filter.
METHOD: [frozen sections of fresh tissue] → stain, 15–20 sec → wash → M 10 mountant

Houcke 1928a, methylene blue–toluidine-thionin-fuchsin (C.R. Soc. Biol. Paris, 99:786)
PREPARATION OF STOCK FORMULAS: I. Mix, without agitation, in a small graduate, 14 1% acid fuchsin with 22 1% methylene blue. Leave 24 hr. Pour liquid from viscous ppt. which adheres to side of graduate. Dry ppt. and dissolve in 20 methanol. II. Add 0.5 1% acid fuchsin to 10 sat. sol. thionine. Leave 1 hr. Centrifuge. Decant and drain. Dry ppt. in tube. Dissolve in 10 methanol. III. Mix 11 1% toluidine blue with 5 1% acid fuchsin. Then treat as II.
WORKING FORMULA: water 100, stock I 2.5, stock II 2.5, stock III 2.5, 1% acetic acid 1
METHOD: [sections] → water → stain ½–2 hr → abs. alc., shortest possible time → balsam, via xylene
NOTE: Houcke (loc. cit.) comments that the pH of the working solution is critical, but the optimum, which varies with both types of tissue and fixatives employed, can only be established empirically.

Houcke 1928b (C.R. Soc. Biol. Paris, 99:788)
FORMULA: sat. sol. (circ. 5%) methylene blue in 95% alc. 3, sat. sol. aniline 27, 0.5% rhodamine B 70
METHOD: [section] → water → stain, 2–3 hr → abs. alc., minimum possible time → dammar via xylene
RESULT: chromatin, blue violet; nucleoli, red; collagens, orange; muscle, deep orange; erythrocytes, bright red
NOTE: For tissues recently fixed Houcke recommends that the aniline solution be diluted 1:1 with water.

Langeron 1942a, polychrome methylene blue–orcein (Langeron 1942, 612)
REAGENTS REQUIRED: A. water 100, polychrome methylene blue 1; B. 70% alcohol 100, orcein 0.25
METHOD: Distilled water → A, 5 min → wash → B, 5 min → abs. alc., least possible time → balsam, via xylene
RESULT: nuclei, blue; connective tissue, red; other structures, polychrome

Langeron 1942b, polychrome methylene–blue tannin-orange (Langeron 1942, 612)
REAGENTS REQUIRED: A. water 100, polychrome methylene blue 1; B. Unna 1892
METHOD: Distilled water → A, 5–15 min → wash → B, until no further blue removed → rinse → balsam via usual reagents
RESULT: nuclei in mitosis, bacteria, some connective tissues, blue; resting nuclei, most other structures, orange

Kull 1913, toluidine blue–aurantia–acid fuchsin (Anat. Anz., 45:153)
REAGENTS REQUIRED: A. sat. sol. aniline 100, acid fuchsin 20; B. 0.5% toluidine blue; C. 0.5% aurantia in 70% alc.
METHOD: [sections] → water → A, 1 min warmed to steaming → cool → water, quick rinse → B, 1–2 min → water, quick rinse → C, until sufficient red extracted, 20–30 sec → 95% alc., until differentiation complete → balsam, via usual reagents

NOTE: Though originally intended for the demonstration of mitochondria, this is an excellent general-purpose stain. For the mitochondria technique see Kull 1914.

Maldonado and José 1967 (*Stain Technol.*, **42**:11)
REAGENTS REQUIRED: A. 1% phloxine B. 3% phosphotungstic acid C. 0.05% azur II D. Weigert 1904 iron hematoxylin
METHOD: [sections] → water → A, 10 min → rinse → B, 1 min → rinse → C, ½ min → wash → D, 1 min → wash → [balsam, via usual reagents]

Masson cited from Langeron 1942, *thionin–picric acid* (Langeron, 1942, 613)
REAGENTS REQUIRED: A. Nicolle (1942) methylene blue (see THIAZINE DYES); B. 0.2% acetic acid; C. sat. sol. picric acid in toluene
METHOD: water → A, 15 min → wash → B, until differentiated → abs. alc. until dehydrated → toluene → C, until green balsam, via toluene
RESULT: nuclei, bacteria blue; other tissues, yellow, blue, or green

Rhamy 1930, *methylene blue–eosin Y–magenta* (*J. Lab. Clin. Med.*, **15**:490)
STOCK FORMULAS: I. sat. sol. (*circ.* 6%) magenta in abs. alc.; II. sat. sol. (*circ.* 45%) eosin Y in water; III. sat. sol. (*circ.* 1.5%) methylene blue in abs. alc.
WORKING FORMULA: 30% alc. 100, stock I 4, stock II 5, stock III 15
METHOD: [sections] → 70% alc. → stain, 5 min → abs alc., until differentiated → balsam, via usual reagents

Unna cited from 1928 Schmorl, *methylene blue–orcein* (Schmorl 1928, 76)
REAGENTS REQUIRED: A. Unna 1892 methylene blue (see THIAZINE DYES); B. 1% orcein in 95% alc.
METHOD: [celloidin sections of alc.-fixed material] → water → A, 10 min → thorough wash → blot → B, 15 min → balsam, via bergamot oil

Techniques Employing Methyl Green as the Nuclear Stain. Methyl green in combination with pyronin has long been recognized as an excellent stain for smears, but, like all other methyl green combinations, it has the disadvantage of being very sensitive to alkalies—so much so that extreme care must be taken either to provide a perfectly neutral mounting or to use some acid mountant such as salicylic balsam. The formula of Pappenheim 1901 has been adapted for use with bacteria by Saathof 1905, and is now widely used for this purpose. The combination of methyl green with other dyes is best known from the Ehrlich 1898 (or Heidenhein 1888) "triacid" mixtures. These mixtures and their variations continue to occur from time to time in the literature, but it is difficult to justify their employment today. Far better methyl green–acid fuchsin combinations are the two triple stains of Foley 1930 and 1931, which will give all the staining reactions of the earlier mixtures without their manifest disadvantages. Most of the other formulas in this class are various modifications of the "triacid" mixtures.

In Combination with Pyronin
Grosso 1912 (*Pathologica.*, **4**:41)
FORMULA: water 100, sat. sol. (*circ.* 9%) pyronin 5, sat. sol. (*circ.* 5%) methyl green 3.5, sat. sol. (*circ.* 11%) orange G 3.5

METHOD: water → stain, 30 sec → abs. alc., until differentiated → n-propyl alc., if dehydration insufficient → neutral mountant
RESULT: nuclei, green; plasma, red, orange, and yellow

Kay 1953 (*Stain Technol.*, **28**:41)
FORMULA: 0.37% methyl green and 0.11% pyronin B in glycerol, 20; 2% phenol 100, 95% alc. 25

Kurnick 1952 (*Stain Technol.*, **27**:233)
REAGENTS: A. 0.2% methyl green; B. n-butyl alc.; C. sat. col. pyronin in alc.
METHOD: [sections] → water → A, 6 min → B, until differentiated → C, ½–1½ min → [balsam, via cedar oil]

Kurnick 1955 (*Stain Technol.*, **30**:213)
FORMULA: 2% pyronin Y 25, 2% methyl green 15, water 60
METHOD: [sections] → water → stain, 6 min → n-butyl alc. 5 min in each of 2 changes → [balsam via xylol and cedar oil].

Langeron 1942 (Langeron 1942, 615)
STOCK SOLUTIONS: I. water 100, methyl green 4, phenol 5; II. water 100, pyronin 4, phenol 5
REAGENTS REQUIRED: A. stock I 50, stock II 50; B. alc. 50, acetone 50; C. amyl alcohol
METHOD: distilled water → A, 15 min at 50°C → distilled water, rinse → B, until differentiated → C, until dehydrated → balsam, via toluene
RESULT: good differential staining of blood cells and glandular cell inclusions

Lipp 1940, see Lipp 1940c

Pappenheim 1901 (*Arch. pathol. Anat.*, **155**:427)
STOCK SOLUTIONS: I. water 100, phenol 0.25, methyl green 1; II. water 100, phenol 0.25, pyronin 1
WORKING FORMULA: stock A 30, stock B 70
METHOD: water → stain, 5–10 min → water, quick rinse → neutral mountant, via acetone
RESULT: nuclei, violet; lymphocytes and plasma cells, red; other tissues, orange and green

Sandiford 1937 (*J. Pathol. Bact.*, **45**:467)
FORMULA: water 75, glycerol 20, 95% alc. 5, phenol 1.5, methyl green 0.15, pyronin 0.5

Taft 1951 (*Stain Technol.*, **26**:205)
PREPARATION OF STAIN: Dissolve 0.5 methyl green in 100 m/10 acetate buffer pH 4.4. Extract repeatedly with chloroform to remove methyl violet. Add 0.2 pyronin B.
METHOD: as Pappenheim 1901

Scott and French 1924 (*Milt. Surg.*, **55**:337)
FORMULA: water 80, glycerol 16, abs. alc. 4, phenol 1.6, methyl green 0.8, pyronin 0.2

Unna 1910 cited from 1928 Gatenby and Cowdry cit. Gandletz (Gatenby and Cowdry 1928, 176)
FORMULA: water 100, phenol 0.5, glycerol 20, abs. alc. 2.5, pyronin 0.25, methyl green 0.15
PREPARATION: Grind the dyes with the alc. in a mortar. Heat glycerol to 50°C and add to mortar in small portions while grinding. Dissolve phenol in water and use this to wash out mortar with small successive doses.
METHOD: water → stain, 10 min 30°C → water, rinse → abs. alc. until differentiated → balsam, via usual reagents

NOTE: Rosa 1950 (*Stain Technol.* **25**:166) recommends dehydrating and differentiating in a 50:50 mixture of triethyl phosphate and xylol.

In Combination with Other Dyes
Auerbach cited from 1930 Guyer, *acid fuchsin–methyl green* (Guyer 1930, 230)
STOCK SOLUTIONS: I. 0.1% acid fuchsin; II. 0.1% methyl green
WORKING SOLUTION: stock I 40, acetic acid 0.005, stock II 60
METHOD: [3 μ sections of mercuric-fixed material] → water → stain, 15 min → 95% alc., until green clouds cease → abs. alc. rinse → balsam, via xylene

Biondi cited from 1888 Heidenhain, *orange G–acid fuchsin–methyl green* (*Pfluegers Arch.*, **63** (suppl.):40
STOCK SOLUTION: Mix 100 sat. sol. (*circ.* 11%) orange G with 20 sat. sol. (*circ.* 13%) acid fuchsin. Add slowly and with constant agitation 50 sat. sol. (*circ.* 5%) methyl green.
WORKING FORMULA: water 100, stock 1–2
METHOD: [sections] → water → stain, 6–24 hr → balsam, via usual reagents
NOTES: This stain is variously attributed to Ehrlich, Ehrlich-Biondi, and Ehrlich-Biondi-Heidenhain. Heidenhain (*loc. cit.*) explains quite clearly "—*welche dieselben Ingredientien enthalt, wie die von Babes empfohlene Ehrlich'sche Mischung (Berhend's Ztschr f. Mikroskopie, Bd IV, S 232) aber in anderen Verhaltnissen.*" This would seem to dispose of the "Ehrlich" in the name of the mixture. These new proportions, however, were stated (*loc. cit.*) as "*nach Versuchen von Biondi*" and the mixture was later referred to as "*Biondi'sche Flussigkeit.*"

Ehrlich 1898 cited from Lee 1905, *orange G–acid fuchsin–methyl green* (Lee 1905, 212)
PREPARATION: Mix 16 sat. sol. (*circ.* 11%) orange G with 7.5 sat. sol. (*circ.* 13%) acid fuchsin. Dilute mixture with 40 50% alc. Add slowly and with constant agitation 15 sat. sol. (*circ.* 5%) methyl green. Then add 10 each 95% alc. and glycerol.

Foley 1930, *methyl green–acid fuchsin* (*Anat. Rec.*, **45**:340)
REAGENTS REQUIRED: A. 2% methyl green 80, 0.1% acid fuchsin 20; B. 0.1% hydrochloric acid
METHOD: [sections of osmic-fixed, or mordanted, material] → running water, overnight → A, 24 hr → blot → B, until differentiated → balsam, via carbolxylene
RESULT: chromatin, green; muscle, brick red; other tissues, pink

Foley 1931, *methyl green–acidf–uchsin–orange G* (*Anat. Rec.*, **49**:15)
PREPARATION: To 10 glycerol and 20 0.1% acid fuchsin and 30 0.1% orange G, add drop by drop, with constant agitation, 30 0.25% methyl green.
METHOD: [water] → stain, 12–24 hr → blot → 95% alc., until differentiated, 5–30 sec → balsam, via carbolxylene

Guinard 1889, *methyl green–acid fuchsin* (*Rev. gén. Bot.*, **1**:19)
PREPARATION: Mix 1% methyl green with 1% acid fuchsin in such proportion as give a violet solution. Add acetic acid to pH 3.

Korson 1951 (*Stain Technol.*, **26**:265)
REAGENTS: A. water 100, orange G 4; B. water 100, methyl green 0.15; C. water 100, toluidin blue 0.05, methyl green 0.075
METHOD: [sections or smears] → water → A, 2 min → rinse → B, 15 min → C, 5 min → tertiary butyl alc., overnight → [balsam via xylene]

Krause 1893, *methyl green–acid fuchsin–orange G* (*Arch. mikr. Anat.*, **42**:59)
FORMULA: sat. sol. (*circ.* 13%), acid fuchsin 0.4, sat. sol. (*circ.* 11%) orange G 0.7, sat. sol. (*circ.* 5%) methyl green 0.8, water 100

Krause 1911 cited from 1948 Romeis, *methyl green–acid fuchsin–orange G* (Romeis 1948, 169)
FORMULA: water 100, methyl green 3.4, acid fuchsin 4.2, orange G 3
PREPARATION: Grind the dry dyes together to a fine powder. Dissolve in water.
METHOD: As Biondi (1888)

Lower 1955 (*Stain Technol.*, **30**:209)
REAGENTS: A. 1% acetic acid; B. 1% azocarmine G in 1% acetic acid; C. Sat. aq. sol. orange G; D. 5% phosphotungstic acid; E. 0.1% methyl green in 1% acetic acid
METHOD: [sections] → water → A, rinse → B, 15 min or longer → A, rinse → C, 20 sec → A, wash → [repeat A → C → A until endoarticle yellow] → rinse → D, 3 min → A, rinse → D, 10 sec → rinse → differentiate in 95% alc. → Mohr and Wehrle 1942 mountant via "camsal"]
RECOMMENDED FOR: insect histology

Mayer 1901 cited from 1901 Lee and Mayer, *methyl green–acid fuchsin–orange G* (Lee and Mayer 1910, 197)
FORMULA: water 60, glycerol 15, 95% alc. 25, orange G 2.6, acid fuchsin 4, methyl green 1.3
PREPARATION: Grind dyes together and dissolve in mixed solvents.

Morel and Doleris 1902 (*C.R. Soc. Biol. Paris*, **54**:1255)
FORMULA: Ehrlich 1898 50, 8% formaldehyde 50, acetic acid 0.1

Oppell cited from 1895 Rawitz, *methyl green–acid fuchsin–picric acid*
REAGENTS REQUIRED: A. 1% methyl green 60, 1% eosin 1, 1% acid fuchsin 20, abs. alc. 20; B. sat. sol. picric acid 80, abs. alc. 20
METHOD: [sections] → A, 15 min → B, 30 sec → abs. alc., minimum possible time → balsam, via usual reagents

Squire 1892, *methyl green–acid fuchsin* (Squire 1892, 37)
PREPARATION: Mix 30 0.5% methyl green with 10 1.5% acid fuchsin.

Stropeni 1912, *methyl green–acridine red* (*Z. wiss. Mikr.*, **29**:302)
REAGENTS REQUIRED: A. water 100, sodium borate 1; B. water 100, glycerol 20, methanol 30, phenol 2, methyl green 0.05, acridine red 0.25
PREPARATION OF B: Grind each dye with 1 phenol and wash out each mortar with 50 water. Mix washings and add other ingredients.
METHOD: water → A, 10 min → rinse → B, 30 min → abs. alc. until differentiated → balsam via xylene

Thome 1898, *methyl green–acid fuchsin–orange G* (*Arch. mikr. Anat.*, **52**:820)

FORMULA: sat. sol. (*circ.* 13%) acid fuchsin 0.15, sat. sol. (*circ.* 11%) orange G 0.35, sat. sol. (*circ.* 5%) methyl green 0.6, water 100

Techniques Employing Acid Fuchsin as the Nuclear Stain. Within this subdivision of the complex staining techniques lie the majority of the methods which are understood today whenever the term *triple stain* is used. Originated by Mallory in 1901, they depend for the most part upon the fact that phosphomolybdic acid will extract acid fuchsin from collagens and leave it in muscle and nuclei. Various mixtures are then used differentially to stain the decolorized tissues. The original method of Mallory used methyl blue and orange G and has been widely copied. So numerous have these formulas become that it is necessary in the present instance to divide them into those using phosphomolybdic acid (Methods Employing the Acid Fuchsin–Phosphomolybdic Reaction) and those using phosphotungstic acid (Methods Employing the Acid Fuchsin–Phosphotungstic Reaction), though in point of fact the results of the stains can scarcely be distinguished. Mallory himself (1936; *Stain Technol.*, **11**:101) prefers phosphotungstic acid. No one formula in either of these two groups can be singled out as better than another, and only one of them (Heidenhain 1905) has become sufficiently well known to acquire a popular name. This stain is frequently referred to as *Heidenhain's azan* because azocarmine is used as the first solution.

Methods Employing the Acid Fuchsin–Phosphomolybdic Reaction. Waterman 1937 (*Stain Technol.*, **12**:21) recommends that dioxane be substituted for alc. in the dehydration of sections stained by these techniques.

Bensley cited from 1938 Mallory cit. Warren (Mallory 1938, 210)
REAGENTS REQUIRED: A. 20% acid fuchsin in sat. aq. sol. aniline; B. 1% phosphomolybdic acid; C. water 100, orange G 2, anilin blue 0.5
METHOD: [sections] → water → A, 10 min → rinse → B, 10 min → quick rinse → C, 1 hr → 95% alc. until color clouds cease → balsam, via usual reagents.

Duprès 1935 (*Monit. zool. ital.*, **46**:77)
REAGENTS REQUIRED: A. Gallego 1919 magenta (see MAGENTA); B. water 50, acetid acid 25, 40% formaldehyde 20; C. 1% phosphomolybdic acid; D. water 100, oxalic acid 4, toluidine blue 0.25, orange G 4
METHOD: [sections of Ruffini 1927 fixed material] → A, 1–10 min → B, wash → rinse → C, 10 min → wash → D, 1–2 min → blot → 95% alc., until differentiated → balsam, via usual reagents
NOTE: For toluidine blue in D above there may be substituted *either* methylene blue 0.3 *or* malachite green 0.2 *or* methyl green 0.3.

Kricheski 1931 (*Stain Technol.*, **6**:97)
REAGENTS REQUIRED: A. 0.25% acid fuchsin; B. 2% methyl blue 30, 1% orange G 30, 1% phosphomolybdic acid 30
METHOD: water → A, 1–3 min → water, thorough rinse → B, 3–5 min → water, quick rinse → 70% alc., 2 or 3 dips → 95% alc., 2 or 3 dips → abs. alc. until differentiated 1–3 min → balsam, via xylene
RESULT: as Mallory 1901

Mallory 1901 (*J. Exp. Med.*, **5**:15)
REAGENTS REQUIRED: A. 1% acid fuchsin; B. 1% phosphomolybdic acid; C. water 100, methyl blue 0.5, orange G 2, oxalic acid 2
METHOD: [water] → A, 2 min → water, thorough rinse → B, 2 min → water, quick rinse → C, 15 min → water, thorough wash → abs. alc., until differentiated → balsam, via xylene
RESULTS: nuclei, red; collagens, blue; nerves and glands, violet; muscle, red; erythrocytes and keratin, orange
NOTE: Mallory 1936 (*Stain Technol.*, **11**:101) substituted 1% phosphotungstic acid in B. Lee-Brown 1936 (*J. Urol.*, **21**:259) differs only in using sol. C before sol. B. Rexed and Wohlfart 1939 (*Z. wiss. Mikr.*, **56**:212) recommend that A above be buffered to pH 3.3 with citrate.

Masson 1913 (*Bull. Soc. anat. Paris*, **14**:291)
REAGENTS REQUIRED: A. 4% ferric alum at 50°C; C. 2% ferric alum; D. 0.1% acid fuchsin; E. 1% phosphomolybdic acid; F. 1% anilin blue 50, 1% phosphomolybdic acid 50
METHOD: [water] → A. 5 min, 50°C → water, 50°C rapid rinse → B, 10–15 min, 50°C → C, until nuclei alone colored → running water, 15 min → D, 10 min → tap water, if overstaining has occurred → E, 5–10 min → F, 20 min to 1 hr → water, quick rinse → 95% alc., quick rinse → abs. alc. until dehydrated → balsam, via xylene

Maxwell 1938 see Maxwell 1938

Milligan 1946 (*Tech. Bull.*, **7**:57)
REAGENTS REQUIRED: A. water 75, 95% alc. 25, potassium dichromate 2.5, hydrochloric acid 2.5; B. 0.1% acid fuchsin; C. 1% phosphomolybdic acid; D. 2% orange G in 1% phosphomolybdic acid; E. 1; acetic acid; F. 0.1% *either* fast green FCF *or* anilin blue in 0.2% acetic acid
METHOD: [sections of formaldehyde material] → water → A, 5 min rinse → B, 5 min → rinse → C, 1–5 min → D, 5–10 min → rinse → E, 2 min → F, 5–10 min → E, 3 min → 95% alc. → balsam, via usual reagents

Schneidau 1937 (*Trans. Amer. Micr. Soc.*, **56**:260)
REAGENTS REQUIRED: A. 1% acid fuchsin; B. 10% phosphomolybdic acid; C. 0.1% thionine
METHOD: [whole objects] → A, 2–4 min → wash → B, 1 min → rinse → C, 2–4 min → wash → balsam, via usual reagents
RECOMMENDED FOR: double-stained whole mounts

Methods Employing the Acid Fuchsin–Phosphotungstic Reaction
Cason 1950 (*Stain Technol.*, **25**:225)
FORMULA: water 100, phosphotungstic acid 0.5, orange G 1, anilin blue 0.5, acid fuchsin 1.5
METHOD: [6 μ sections] → water → stain, 5 min → water, rinse → balsam, via usual reagents

Heidenhain 1905, Heidenhain's Azan—compl. script. (*Z. wiss. Mikr.*, **22**:339)
REAGENTS REQUIRED: A. water 100, azocarmine 2, acetic acid 1; B. 0.1% aniline in 95% alc.; C. 0.1% hydrochloric acid in abs. alc.; D. 5% phosphomolybdic acid; E. water 100, orange G 2, anilin blue 0.5, acetic acid 7.5

METHOD: water → A, 1 hr, 50°C → water, quick rinse → B, until nuclei well marked, dip in C, before examining → C, thorough rinse → D, 2 hr → E, 2–3 hr → water, rinse → balsam, via usual reagents
RESULT: nuclei, scarlet; muscle, orange; collagens, blue

Kohashi 1937 (*Folia. Anat. Jap.*, **15**:175)
REAGENTS REQUIRED: A. water 100, azocarmine 0.1, acetic acid 1; B. 90% alc. 100, aniline 0.1; C. 1% acetic acid in 95% alc.; D. 5% phosphotungstic acid; E. Pasini (1928) (sol. B)
METHOD: [sections] → water → A, 12–15 min 60°C → wash → B, until nuclei differentiated → C, ½–1 min → rinse → D, ½–1 hr → rinse → E, 15–20 min → 95% alc. until differentiated → balsam, via carbol-xylene

Walter 1930 (*Z. wiss. Mikr.*, **46**:458)
REAGENTS REQUIRED: A. 2.5% ferric alum; B. 2% phosphotungstic acid; C. Pasini (1928) sol. B
METHOD: water → A, 24 hr → water, quick rinse → B, 10 min → water, quick rinse → C, 15–20 min → 95% alc., until color clouds cease → abs. alc.; 1 min → balsam, via usual reagents
RESULT: nuclei and elastic fibers, purple; collagens, blue; erythrocytes, orange

Methods Using Neither Phosphotungstic Nor Phosphomolybdic Acid
Böhm and Oppel 1907 (Böhm and Oppel 1907, 115)
STOCK FORMULAS: I. water 100, orange G 0.05, acid fuchsin 0.05, acetic acid 1, 40% formaldehyde 1; II. water 90, methanol 10, 0.25% brilliant cresyl blue 0.25, 40% formaldehyde 1
WORKING SOLUTION: stock I 50, stock II 50
METHOD: water → stain, 20–30 min → abs. alc.; until differentiated → balsam, via xylene
RESULT: nuclei, red-purple; cartilage, blue; bone, orange; muscle, red

Maresch 1905 (*Zbl. allg. Pathol. pathol. Anat.*, **16**:41)
REAGENTS REQUIRED: A. sat. methanol sol. methyl green 50, sat. methanol sol. picric acid 50; B. 0.75% acid fuchsin
METHOD: [sections] → abs. alc. → A, 10 min → water, rinse → B, 5–10 sec → blot → abs. alc. dropped on slide, until color changes from dark violet to blue-gray → balsam, via xylene

Roskin 1946 (Roskin 1946, 154)
REAGENTS REQUIRED: A. 0.1% indigocarmine in sat. sol. picric acid; B. sat. aq. sol. magenta
METHOD: [sections] → A, 10–30 min → B, poured on slide, left till greenish scum appears → abs. alc., until differentiated

Waterman 1937 (*Stain Technol.*, **12**:21)
PREPARATION OF STOCK SOLUTION: water 100, phenol 5, magenta 3
REAGENTS REQUIRED: A. water 100, acetic acid 0.6, 40% formaldehyde 0.6, stock 10; B. water 100, picric acid 1, acetic acid 2, indigo-carmine 0.25
METHOD: [sections of material fixed in cupric-nitric Waterman 1937] → water → A, 5 min → rinse, 2 sec → B, 90 sec → rinse, 2 sec → 50% dioxane, 2 sec → balsam, via dioxane and xylene
RESULT: nuclei, red; collagen, blue; other tissues, yellow and green

Techniques Employing Safranin as the Nuclear Stain
Bryan 1955 (*Stain Technol.*, **30**:153)
REAGENTS: Mixed 50 0.1% safranin O with 50 0.1% fast green FCF
METHOD: [sections] → water → stain → wash, 5 min → abs. alc. until differentiated → [balsam, via xylene]

Conant cited from 1940 Johansen, *safranin–crystal violet–fast green–orange II* (Johansen 1940, 87)
REAGENTS REQUIRED: A. 1% safranin in 50% alc.; B. sat. sol. (*circ.* 1%) crystal violet; C. 1% fast green in abs. alc.; D. sat. sol. orange II in clove oil
METHOD: [sections] → 70% alc. → A, 2–24 hr → water, rinse → B. 1 min → water, rinse → abs. alc. until dehydrated → C, 5–10 dips → D, until alcohol removed → D, fresh solution, until differentiated → balsam, via xylene

Dionne and Spicer 1958 (*Stain Technol.*, **33**:15)
FORMULA: water 55, acetic acid 45, safranin O 0.6, aniline blue 0.8
RECOMMENDED FOR: pollen tube squashes

Foley 1929, *safranin–orange G–crystal violet* (*Anat. Rec.*, **43**:171)
REAGENTS REQUIRED: A. 1% safranin O; B. N/40 hydrochloric acid; C. 0.3% crystal violet in 70% alc.; D. Lugol (1905); E. 1% mercuric chloride; F. sat. sol. orange G in clove oil
METHOD: [sections of osmic-fixed, or mordanted, material] → wash → A, overnight → B, until outline of nuclei distinct → C, 20 min → B, until outline of nuclei distinct → D, until sections deep b!ack, 1–3 min → running water, until iodine removed → E, until sections bright blue, 1–3 min → water, thorough wash → blot → 95%, rinse 5 sec → carbol-xylene until differentiated, 1/4–5 min → xylene, 2 min → F. 1–2 min → clove oil wash → balsam, via xylene

Henneguy 1898 cited from 1907 Böhm and Oppel, *safranin–methyl violet–orange G* (Böhm and Oppel 1907, 111)
REAGENTS REQUIRED: A. water 100, ammonium thicyanate 1, methyl violet 0.1, orange G 0.1; B. Zwaademaker 1887 safranin (see SAFRANIN)
METHOD: [sections] → A, 10 min → rinse → B, 15 min → rinse → A, 15 min → abs. alc., minimum possible time → balsam, via clove oil
RESULT: nuclei, red; cytoplasm, various shades of blue and blue-gray

Johansen 1940, *safranin–methyl violet–fast green–orange G* (Johansen 1940, 88)
REAGENTS REQUIRED: A. Johansen 1940 safranin (see SAFRANIN); B. 1% methyl violet; C. 95% alc. 30, ethylene glycol monomethyl ether 6, fast green *q.s.* to sat., 95% alc. 36, tert. butyl alc. 36, acetic acid 12; E. 95% alc. 50 tert. butyl alc. 50, acetic acid 0.5; F. sat. sol. orange G in ethylene glycol monomethyl ether 30, ethylene glycol monomethyl ether 30, 95% alc. 30; G. clove oil 30, ethylene glycol monomethyl ether 30, 95% alc. 30; H. clove oil 30, abs. alc. 30, xylene 30
METHOD: [sections] → 70% alc. → A, 1–2 days → water, rinse → B, 10–15 min → water, rinse → C, 15 sec → D, 10–15 min → E, quick rinse → F, 3 min → G, rinse → H. rinse → balsam, via xylene
RESULT: (plant tissues) dividing chromatin, red; resting nuclei, purple; lignified and suberized tissues, red;

cellulose, green–orange; cytoplasm, bright orange; starch grains, purple; fungal mycelia, green.

Kalter 1943, *safranin–crystal violet–fast green–orange II* (*J. Lab. Clin. Med.*, **29**:995)
REAGENTS REQUIRED: A. water 50, 95% alc. 50, 40% formaldehyde 4, sodium acetate 0.5, safranin 0.2; B. 0.5% crystal violet; C. clove oil 100, fast green FCF and orange II, each to sat.; D. sat. sol. orange II in clove oil
METHOD: [sections] → water → A, 24 hr → rinse → B, 1–2 min → wash → 95% alc. → C. 5 min → clove oil, until connective tissue green → D, 10 min → clove oil → xylene, thorough wash → balsam
RESULT: nuclei, red; cytoplasm, pink or light green; muscle, tan; collagen, bright green

Laguesse 1901 cited from 1907 Böhm and Oppel, *safranin–crystal violet–orange G* (Böhm and Oppel 1907, 154)
REAGENTS REQUIRED: A. 2% potassium sulfite; B. Babes 1887 safranin (see SAFRANIN); C. sat. aq. sol. crystal violet 50, sat. aq. sol. orange G 2
PREPARATION OF C: Mix the solution. Add just enough water to redissolve ppt.
METHOD: [sections] → A, 12–24 hr → wash → B, 6–12 hr → wash → C, 24 hr → blot → abs. alc. minimum possible time → clove oil, until differentiated → balsam, via xylene

Maácz and Vágás 1961 (*Mikroscopie*, **16**:40)
REAGENTS REQUIRED: A. water 100, "astrablau" 0.5, tartaric acid 2; B. sat. aq. sol. auramine; C. 1% safranin O
METHOD: [sections] → water → A, 1–5 min → rinse → B, 1–5 min → rinse → C, 1–5 min → [balsam via acetone and carbol-xylol]
RECOMMENDED FOR: plant histology

Stockwell 1934, *safranin–gentian violet–orange G* (*Science*, **80**:121)
REAGENTS REQUIRED: A*a*. water 90, chromic acid 1, potassium dichromate 1, acetic acid 10 1934 *or* A*b*. 1% chromic acid; B. 1% crystal violet 20, 1% safranin O 40, water 40; C. 80% alc. 100, iodine 1, potassium iodide 1; D. 1% picric acid in 95% alc.; E. 95% alc. 100, ammon. hydroxide 0.3; F. 0.2% orange G in clove oil
METHOD: [sections] → water → A*a*, if bleaching required (or A*b* if not chromic fixed) overnight → water, thorough wash → B, 1–6 hr → water, rinse → C, 30 sec → 70% alc., quick rinse → D, few seconds → E, few seconds → abs. alc., few seconds → F, few seconds → balsam, via xylene

Unna 1928 cited from 1928 Schmorl, *safranin–orcein–anilin blue–eosin* (Schmorl 1928, 343)
STOCK SOLUTIONS: I. water 50, abs. alc. 25, glycerol 10, acetic acid 2.5, orcein 0.5, anilin blue 0.5; II. 80% alc. 100, ethyl eosin 1; III. 1% hydroquinone
REAGENTS REQUIRED: A. stock I 50, stock II 15, stock III 15; B. 1% safranin; C. 0.5% potassium dichromate
METHOD: [sections] → water → A, 10 min → wash → B, 10 min → thorough wash → C, 10–30 min → wash → balsam, via usual reagents
RESULT: nuclei, black with red granules; protoplasm violet; collagen, blue; elastic fibers, red
RECOMMENDED FOR: skin

Unna cited from 1928 Schmorl, *safranin–anilin blue* (Schmorl 1928, 176)
REAGENTS REQUIRED: A. 1% safranin; B. water 100, tannin 15, anilin blue 0.5
METHOD: [sections] → A, 10 min → thorough wash → B, 10–15 min → thorough wash → abs. alc. minimum possible time balsam, via xylene

Techniques Employing Hematoxylin As the Nuclear Stain. Hematoxylin is, of course, the commonest nuclear stain to be employed before any of the plasma techniques. The formulas given here are those in which the nuclear staining is an integral portion of a complex technique which cannot be employed in combination with any other nuclear stain.

Barbrow 1937 (*J. Lab. Clin. Med.*, **22**:1175)
STOCK SOLUTIONS: I. 1% hematoxylin in 95% alc. (well "ripened"); II. water 99, ferric chloride 2, hydrochloric acid 1; III. water 100, picric acid 1, acid fuchsin 1
WORKING SOLUTION: stock I 25, stock II 10; stock III 50
METHOD: [frozen sections of unfixed tissues] → stain, 1 min → wash → balsam, via usual reagents

Delamare 1905 cited from 1907 Böhm and Oppel (Böhm and Oppel 1907, 131)
STOCK SOLUTIONS: I. abs. alc. 100, hydrochloric acid 2, orcein 2; II. sat. aq. sol. picric acid 100, sat. aq. sol. acid fuchsin 0.5, Ehrlich 1886 1
REAGENTS REQUIRED: A. stock I 50, stock II 50; B. 0.1% hydrochloric acid
METHOD: [sections] → water → A, 30 min → B, rinse → tap water, to "blue" hematoxylin → balsam, via usual reagents
RESULT: nuclei, violet; muscle and general cytoplasm, yellow; collagens, red; elastic fibers, black

Friedländer 1889 (Friedländer 1889, 94)
FORMULA: abs. alc. 30, hematoxylin 0.5, glycerol 30, sat. sol. potassium alum 30, 1% eosin 10
PREPARATION: Dissolve the hematoxylin in the alc. Add the glycerol and alum. Leave 1 week. Filter. Add the eosin sol. to filtrate.

Fullmer 1959 (*Stain Technol.*, **34**:81)
REAGENTS: A. To 26.8 acetic acid add 72.5 3% hydrogen peroxide and 0.8 sulfuric acid. Leave 2 days then add 0.012 disodium phosphate; B. 70% alc. 100, hydrochloric acid 1, orcein 1; C. Mayer 1891 hemalum; D. 0.1% hydrochloric acid in 70% alc.; E. water 100, light green SF 0.2, orange G 1, phosphotungstic 0.5, acetic acid 1 F 0.2% acetic acid in 70% alc.
METHOD: [sections] → abs. alc. → A, 30 min → wash 2 min → B, 15 min 37°C → 70% alc. until differentiated → C, 4 min → D, until differentiated → wash, until blue → E, 20 sec → F, brief rinse → [balsam, via usual reagents]

Gabe 1953 (*Bull. Micr. Appl.* Series 2, **3**:153)
REAGENTS: A. water 100, sulfuric acid 1, potassium permanganate 0.5; B. 2% sodium bisulfite; C. Masson (1934) acid alum hematoxylin; D. Gabe 1952 paraldehyde fuchsin (see MAGENTA)
STOCK SOLUTION: 25, 70% alc. 75, acetic acid 1
METHOD: [sections] → water → A, ½ min → rinse → B, until colorless → wash → C, until nuclei stained → rinse → D, 2 min → rinse → E, 20 sec → [balsam, via usual reagents]

Galiano 1928 cited from 1928 Findlay (*J. Roy. Micr. Soc.*, **48**:314)

REAGENTS REQUIRED: A. 3% ferric alum; B. water 80, acetic acid 20, hematoxylin 0.2; C. 95% alc. 75 m acetic acid 25, eosin 1.5; D. 0.1% ammonia in 95% alc.

METHOD: [sections] → water → A, 15 min → B, until nuclei darkly stained → wash, 15 min → 70% alc., 1 min → C, until differentiated → D, wash → balsam, via usual reagents

Gude 1953 (*Stain Technol.*, **28**:161)

REAGENTS: A. Harris 1900 alum hematoxylin; B. sat. aq. sol. lithium carbonate; C. 0.5% acid fuchsin; D. 1% phosphomolybdic acid; E. water 100, methyl blue 0.5, orange G 2, oxalic acid 2

METHOD: [sections] → water → A, 2–3 min → rinse → B, 1 min → rinse → C, 1–2 min → rinse → D, 2–5 min → E, 1–2 hr → 95% alc., until differentiated → [balsam via carbol-xylol]

Halmi 1952 (*Stain Technol.*, **271**:61)

REAGENTS REQUIRED: A. 60% alc. 100, magenta 0.5, paraldehyde 1, HCl 1.5: B. Ehrlich's acid alum hematoxylin; C. water 100, light green SF 0.2, orange G 1, chromotrope 2R 0.5, phosphotungstic acid 0.5, acetic acid 1; D. 0.25% acetic in 95% alc.

METHOD: [sections] → water → A, 2–10 min → 95% alc., rinse → 70% alc. rinse → water → B, 3,4 min → differentiate in acid alc. → wash until blue → C, 45 sec → E, rinse → [balsam via usual reagents]

NOTE: The A stain requires "ripening" from 2 to 4 days.

Isaac and Aron 1952 (*Bull. Micr. Appl.* Series 2, **2**:99)

REAGENTS: A. water 100, potassium permanganate 1, sulfuric acid 1; B. Harris 1900 alum hematoxylin; C. 0.2% HCl; D. 1% acid fuchsin; E. 1% phosphomolybdic acid; F. water 100, methyl blue 0.5, orange G 2, oxalic acid 2

METHOD: [sections of Bouin 1897 fixed material] → water → A. until rusty brown → rinse → B, 5 min → C, until differentiated → wash → D, 2 min → C, rinse → E, 1 min → rinse → F, 1–12 hr → 95% alc. until differentiated → balsam via usual reagents]

Kefalas 1926 (*J. Roy. Micr. Soc.*, **46**:277)

REAGENTS REQUIRED: A. Kefalas 1926 iron hematoxylin; B. sat. sol. Biebrich scarlet in acetone

METHOD: [sections] → acetone → A, until slightly overstained → B, until counterstained and differentiated → acetone → balsam, via xylene

Koneff 1936 (*Anat. Rec.*, **66**:173)

REAGENTS REQUIRED: A. 5% ferric alum; B. Harris 1900 alum hematoxylin; C. water 100, anilin blue 0.03, phosphomolybdic acid 5, oxalic acid 0.6

Ladewig 1938 (*Z. wiss. Mikr.*, **55**:215)

REAGENTS REQUIRED: A. Weigert 1903 iron hematoxylin; B. 1% phosphotungstic acid; C. water 100, methyl blue 0.5, orange G 2, oxalic acid 2, acid fuchsin 1

METHOD: [sections of formaldehyde material] → water → A, 3–5 min rinse → B, 2 min → rinse → C, 4 min → quick rinse → 95% alc., until dehydrated → balsam, via xylene

Lillie 1945 (*Bull. Int. Ass. Med. Mus.*, **25**:33)

REAGENTS REQUIRED: A. sat. 95% alc. sol. picric acid; B. Weigert 1903 iron hematoxylin; C. 1%

Biebrich scarlet in 1% acetic acid; D. 3% ferric chloride; E. water 99, acetic acid 1 *and either* anilin blue 1 *or* methyl blue 1 *or* wool green S 1; F. 1% acetic acid

METHOD: [sections] → 95% alc. → A, 2 min → thorough wash → B, 6 min → wash → C, 4 min → D, 2 min → E, 3–5 min → rinse → F, 2 min → balsam, via acetone and xylene

RESULT: nuclei, brown black; muscle and cytoplasm, red; connective tissue, blue or green

Lillie 1948 (Lillie 1948, 149)

REAGENTS REQUIRED: A. Weigert 1904 iron hematoxylin; B. 0.02% fast green FCF; C. 1% acetic acid; D. 0.1% Bismarck brown in 1% acetic acid

METHOD: [sections] → water → A, 6 min → wash → B, 3 min → C, wash → D, 4–6 min → balsam, via usual reagents

RESULT: nuclei, black; general cytoplasm, gray-green; mucus, cartilage, cell granules, brown

NOTE: Magenta, or new magenta, may be substituted for Bismarck brown in D. Eosin Y may be substituted in B, in which case crystal violet or malachite green should be substituted in D.

Löwenthal 1892 cited from 1907 Böhm and Oppel (Böhm and Oppel 1907, 119)

PREPARATION OF STOCK SOLUTIONS: I. Dissolve 0.4 carmine in 100 0.05% sodium hydroxide. Add 0.25 picric acid; II. To 0.5 hematoxylin dissolved in 50 abs. alc. add 50 1: ammonium alum.

WORKING SOLUTION: stock I 100, stock II 20, acetic acid 2.5

PREPARATION OF WORKING SOLUTION: Mix the stock solutions slowly and with constant agitation. Leave 24 hr. Filter. Ripen 4 weeks.

METHOD: [sections] → water → stain, 24 hr → wash → balsam, via usual reagents

Mollier 1938 (*Z. wiss. Mikr.*, **55**:472)

REAGENTS REQUIRED: A. 50% alc. 100, orcein 0.8, hydrochloric acid 0.5; B. Weigert 1904 iron hematoxylin; C. 1% hydrochloric acid in 70% alc.; D. water 100, azocarmine 2, acetic acid 1; E. 5% phosphotungstic acid; F. water 100, naphthol green B 1, acetic acid 1

METHOD : [sections] → 70% alc. → A, 12 hr → wash, until no more color comes away → B, 1–3 min → rinse → C, until nuclei well differentiated → wash, 15 min → D. 15–30 min rinse → E, 3 changes, 2–6 hr, until collagen decolorized rinse → F, 15–30 min → 95% alc., 30 sec with constant agitation → balsam, via usual reagents

RESULT: nuclei, deep blue; general cytoplasm, purple; elastic fibers, black; collagen, green; erythrocytes, scarlet

Möllendorf cited from 1946 Roskin (Roskin 1946, 158)

REAGENTS REQUIRED: A. Hansen 1905 iron hematoxylin; B. 1% eosin in 0.3% acetic acid; C. 2% phosphomolybdic acid; D. 1% methyl blue

METHOD: [sections] → water → A, 5 min → distilled water, rinse → running water, wash → B, 20 min → rinse → C, 10 sec → rinse → D, 1–2 min → rinse → 95% alc. until color clouds cease → abs. alc. balsam, via xylene

Monroe and Spector 1963 (*Stain Technol.*, **38**:187)

REAGENTS REQUIRED: A. 20% tannic acid; B. 0.5% magenta in 10% alc.; C. 1% anilin in 90% alc.; D.

1% phosphomolybdic acid; E. 10% hematoxylin in 95% alc. 50, 4% ferric alum 50; F. 1% alcian blue 8 GX

METHOD: [sections of Bouin 1900 fixed material with picric "removed" with lithium chloride sol.] → A, 15 min → B, wash 5 min → C, 2–4 sec → rinse → C, until differentiated → D; 1 min → wash → E, 0.5 min → wash, 1–2 min → wash 1–2 min → F, 15 sec → [balsam via usual reagents]

Paquin and Goddard 1947 (*Bull. Int. Ass. Med. Mus.*, **27**:198)

REAGENTS REQUIRED: A. Paquin and Goddard 1947 iron hematoxylin; B. 0.5% picric acid in 95% alc.; C. water 100, phosphotungstic acid 0.1; eosin 0.07, phloxine 0.03, orange G 0.1; D. 0.2% phosphotungstic acid; E. 0.4% acetic acid; F. 0.04% anilin blue in 1% acetic acid

METHOD: [sections of picric-chromic-formaldehyde-acetic Masson (1947) material] → water → A, 5 min → wash, 5 min → B, 15–20 sec → wash → C, 5 min → D, 5 min → E, double rinse → F, 5 min → E, double rinse → D, 5 min → E, 30 min → 95% alc., 3 dips → balsam, via isoamyl alc. and toluene

RESULT: nuclei, black; elastic tissue, cherry red; other connective tissues, blue; general cytoplasm, pink

Rénaut cited from 1889 Friedländer (Friedländer 1889, 94)

FORMULA: sat. sol. potassium alum in glycerol 65, sat. sol. eosin Y 15, sat. alc. sol. hematoxylin 20

Reeve 1948 (*Stain Technol.*, **23**:13)

REAGENTS REQUIRED: A. water 90, Delafield (1885) alum hematoxylin 10; B. water 40, 95% alc. 60, safranin 0.01, sodium acetate 0.01; C. xylene 75, abs. alc. 25, sat. sol. fast green FCF in 50:50 clove oil—abs. alc. 2–5

METHOD: [sections] → water → A, 5–15 min → wash → B, 5–15 min → rinse → 95% alc., until no more color comes away → C, 1–3 min → balsam, via xylene

RECOMMENDED FOR: general plant histology

Romeis 1948 (Romeis 1948, 364)

REAGENTS REQUIRED: A. water 100, potassium ferricyanide 2.5, borax 1; B. Weigert 1903 iron hematoxylin; C. water 100, azophloxine 0.5, acetic acid 0.2; D. 1% acetic acid; E. water 100, phosphomolybdic acid 4, orange G 2; F. water 100, light green 0.2, acetic acid 0.2

METHOD: [sections] → 80% alc. → A, 15 min → wash → B, 2–3 min → thorough wash → C, 5 min → D, wash → E, until collagen decolorized → D, rinse → F, 5 min → D, 5 min → abs. alc., least possible time → balsam, via xylene

RESULT: nuclei, black; muscles and general cytoplasm, red; collagen, green; elastic fibers, black

NOTE: The C, D, E, F, solutions are from Goldner 1938.

Slater and Dornfeld 1939 (*Stain Technol.*, **14**:103)

REAGENTS REQUIRED: A. Harris 1900 alum hematoxylin; B. 1% safranin O in sat. aq. sol. aniline; C. 0.5% fast green in 95% alc.

METHOD: [sections by dioxane technique of amphibian embryos from picric-formaldehyde-acetic Puckett 1937] → A, 5 min → C, 2–5 min → C, 2–5 min, until yolk granules red on green cytoplasm → balsam, via usual reagents

Other Polychrome Staining Methods

Aoki and Gutierrez 1967 (*Stain Technol.*, **42**:307)

REAGENTS REQUIRED: A. 1% toluidine blue in cacodylate buffer (pH6); B. 1% magenta in 50% alc.; D. 1 light green, 1 acetic acid 50% alc. 100

METHOD: [epoxy sections] → dry → A, 3 min 90°C → wash → [repeat A, if necessary] → B, 3–5 min → 70% alc., rinse → C, 2 min 20–25°C → rinse → D, 3–5 min 20–25°C → dry → mount

RECOMMENDED FOR: epoxy sections

Becher 1921a, *polychrome gallamin blue* (Becher 1921, 70)

FORMULA: water 100, sodium alum 5, gallamin blue 0.5

PREPARATION: Boil 5 min. Cool. Filter.

METHOD: [sections] → water → stain, 24 hr → water, wash → balsam, via usual reagents

RESULT: nuclei, black; cartilage, violet

Becher 1921b, *polychrome quinalizarine* (Becher, 1921, 55)

FORMULA: water 100, chrome alum 5, quinalizarin 0.5

PREPARATION: Boil 5 min. Cool. Filter.

RESULT: nuclei, blue; muscle and nerves, red

Becher 1921c, *polychrome coelestin blue* (Becher, 1921, 73)

FORMULA: water 100, chrome alum 5, coelestin blue 0.5

PREPARATION: Boil 5 min. Cool. Filter.

METHOD: [sections] → water → stain, 24 hr → water, wash → balsam, via usual reagents

RESULT: blue-black nuclei, with a violet to red polychrome staining of connective tissues

Bensley and Bensley 1938, *acid fuchsin–crystal violet* (Bensley and Bensley 1938, 97)

PREPARATION OF DRY STAIN: Add a sat. sol. acid fuchsin to a sat. sol. crystal violet until no further ppt. produced. Filter. Wash and dry ppt.

PREPARATION OF STOCK SOLUTION: abs. alc. 100, dry stain from above to sat.

REAGENTS REQUIRED: A. water 72, abs. alc. 18, stock solution 10; B. clove oil 75, abs. alc. 25

METHOD: [sections] → water → A, 5 min → blot → acetone, until dehydrated → benzene → B, until differentiated → benzene → balsam

Böhm and Oppel 1907, *bismarck brown–dahlia violet–methyl green* (Böhm and Oppel 1907, 127)

REAGENTS REQUIRED: A. Sat. sol. (*circ.* 1.5%) bismarck brown; B. Roux 1894

METHOD: water → A, 10 min → water, quick rinse → B, 1 min → abs. alc. until differentiated, few seconds → carbol-xylene → balsam, via xylene

RESULT: as Roux 1894 but with black nuclei and brown cartilage

Bonney 1908, *methyl violet–orange G–pyronin* (*Virchows Arch.*, **193**:547)

REAGENTS REQUIRED: A. water 100, methyl violet 0.25, pyronin 1; B. acetone 100 2% orange G *q.s.*

PREPARATION OF B: To 100 acetone add 2% orange G until ppt. first formed just redissolves.

METHOD: [sections, mercuric-fixed, or mordanted] → water → A, 2 min → B, flooded on drained slide, 1 min → balsam, via acetone and xylene

RESULT: nuclei, purple; plasma, red and yellow

Buzaglo 1934, *quinalizarin–acid alizarin blue–alizarin viridine* (*Bull. Histol. Tech. Micr.* (= *Bull. d'Hist. app.*), **11**:40)

REAGENTS REQUIRED: A. Becher 1921b (above); B. 70% alc. 100, hydrochloric acid 1, orcein 1; C. water 100, aluminum sulfate 10, acid alizarin blue 5; D. 5% phosphomolybdic acid; E. water 100, hydrochloric acid q.s., to make pH 5.8, alizarin viridine 0.2

PREPARATION OF C: Boil 10 min. Cool. Filter.

METHOD: [sections] → water → A, 24 hr → rinse → B, 3 changes, 5 min in each → rinse → C, 7 min → rinse → D, until muscles differentiated → wash → E, 7 min → blot → abs. alc. minimum possible time → balsam, via carbol-xylene and xylene

RESULT: nuclei, dark blue; elastic fibers, brown; muscle, blue; cartilage, green

Calleja 1897, *carmine-picro-indigocarmine* (*Z. wiss. Mikr.*, **15**:323)

REAGENTS REQUIRED: A. sat. sol. lithium carbonate 100, carmine 2; B. 0.1% hydrochloric acid; C. sat. sol. picric acid 100, indigo-carmine 0.25; D. 0.2% acetic acid

METHOD: [sections] → A, 5–10 min → B, 20–30 sec → wash → C, 5–10 min → D, few seconds → balsam, via usual reagents

Canon 1937 (*Nature, Lond.*, **139**:549)

FORMULA: 70% alc. 100, chlorazol black E 1

METHOD: [sections, plant or animal tissue] → stain, 15–30 min → wash → balsam, via usual reagents

RESULT: a very pleasing, and well-differentiated series of gray tones with some green

NOTE: This stain is often attributed to Darrow (*Stain Technol.*, **15**:67) who introduced it into American literature.

Castroviejo 1932, *magenta-picro–indigo carmine* (*Amer. J. Clin. Pathol.*, **2**:135)

REAGENTS REQUIRED: A. water 100, 40% formaldehyde 0.6, Ziehl 1882 magenta (see MAGENTA) 5; B. Cajal 1895 picro-indigo-carmine 100, acetic acid 0.6

METHOD: [sections of formaldehyde material] → water → A, until nuclei well stained → wash → B, until stained → wash → balsam, via usual reagents

Dobell 1919, *methylene blue–eosin Y–orange G* (Dobell 1919, 7)

REAGENTS REQUIRED: A. Mann 1894 methylene blue; B. 70% alc. 90, sat. sol. (*circ.* 0.2%) orange G in abs. alc. 10

METHOD: [smears] → water → A, 12 hr → B, applied from drop bottle, until differentiated → abs. alc., minimum possible time → balsam, via xylene

Drew-Murray 1919, *picro–acid fuchsin–nile blue* (*Rep. Cancer Res. Rd.*, **6**:77)

REAGENTS REQUIRED: A. van Gieson 1896 picro fuchsin; B. 2% nile blue sulfate

METHOD: [water] → A, 1–3 min → water, thorough rinse → B, 2–24 hr → water, wash → A, 1–5 min → wash → abs. alc., minimum possible time → xylene → clove oil until differentiated → xylene → balsam

RESULT: nuclei and some cell inclusions, blue-black; keratin, orange-yellow; collagens, red

Ehrlich cited from 1905 Lee cit. Grubler, *indulin-aurantia–eosin Y* (Lee 1905, 218)

FORMULA: glycerol 90, aurantia 6, eosin Y 6

PREPARATION: Digest at 40°C until completely dissolved.

METHOD: [smears] → stain, 4–5 hr, 40°C → water, thorough rinse → balsam, via acetone and xylene

RESULT: nuclei, deep blue; some cytoplasmic inclusions, violet; plasma in general, orange

Ewen 1962 (*Stain Technol.*, **37**:94)

REAGENTS REQUIRED: A. 0.3% potassium permanganate in 0.2% sulfuric acid; B. 2.5% sodium bisulfite; C. Gabe 1953 (see MAGENTA) stock aldehyde fuchsin sol. 2.5, 70% alc. 75, acetic acid 1; D. 0.5% hydrochloric acid in abs. alc.; E. water 100, phosphomolybdic acid 1, phosphotungstic acid 4; F. water 100, light green SFY 0.4, orange G 1, Chromotrope 2R 0.5, acetic acid 1; G. 0.2% acetic acid in 95% alc.

METHOD: [sections] → water → A, 1 min → rinse → B, until colorless → rinse → 30% alc. → 30% alc → C, 2–10 min → wash, 95% alc. → D, until differentiated (10–30 sec) → 70% alc. rinse → 30% alc. → water → E, 10 min → rinse → F, 1 hr → G, rinse → [balsam, via usual reagents]

Freeborn 1888, *picro-migrosin* (*Amer. Mon. Micr. J.*, **9**:231)

FORMULA: water 100, nigrosin 0.05, picric acid 0.4

METHOD: [sections] → stains, 5–10 min → wash → balsam, via usual reagents

RESULT: nuclei, black; collagens, blue; other tissues, yellow

Glenner and Lillie 1957 (*Stain Technol.*, **32**:187)

FORMULA: water 70, 0.1m citric acid 2.75, 0.2m disodium phosphate 2.25, 1% anilin blue 5, 1% eosin B 5

METHOD: [sections of formaldehyde or Heidenhain 1916 ("Zenka-formol") fixed material] → water → stain 1 hr at 60°C → wash → [balsam, via acetone and xylol]

Gray and Pickle 1956 (*Phytomorph.*, **6**:197)

PREPARATION OF STAIN: Dissolve successively 14 glycerol and 5 ferric alum in 100 water at 50°C. Mix 3 safranin O with 1 celestin blue B and mix to a paste with 2 sulfuric acid. Add glycerol-alum solution little by little with constant stirring and dissolve with further heating.

METHOD: [sections of plant tissues] → water → stain 5 min or more → wash → [balsam via usual reagents]

RECOMMENDED FOR: plant histology

NOTE: Many plasma stains may be used after this method which stains cellulose black and lignin red. Tolbert 1962 (*Stain Technol.*, **37**:165) recommends bulk staining pieces of plant material by this method before sectioning.

Green and Wood 1959 (*Stain Technol.*, **34**:313)

REAGENTS REQUIRED: A. 70% alc. 100, 0.5 magenta, hydrochloric acid 1, paraldehyde 1; B. 1% pontacyl blue black SX 25, 2% potassium dichromate 75; C. 2% fast yellow TN in 95% alc.

METHOD: [sections] → 70% alc. → A, 30 min → 70% alc., rinse → water → B, 15 min → rinse → 70% alc. until no more blue comes out → C, 5 min → 95% alc. rinse → [balsam via usual reagents]

NOTE: A requires to be ripened overnight.

Hall 1970 (*Stain Technol.*, **45**:49)

REAGENTS REQUIRED: A. sat. aq. sol. magenta; B. sat. aq. sol. picric acid 50, sat. 70% alc. sol. indigo-carmine 50

METHOD: [sections] → water → A, 15 min → wash until no more color comes away → B, 1–1½ min → 70% rinse → 70% until red color washed out → [balsam via usual reagents]

Huber et al. 1968 (*Stain Technol.*, **43**:83)
REAGENTS REQUIRED: A. 4% magenta; B. 2% methylene blue adjusted to pH 12.5 with 1*N* NaOH
METHOD: [sections] → dry → A, 1 min 70°C → blot → dry → B, 2 min → rinse → dry → [immersion oil]
RECOMMENDED FOR: epoxy sections

Kornhauser 1943, Quad Stain—auct (*Stain Technol.*, **18**:95)
REAGENTS REQUIRED: A. 95% alc. 90, water 10, nitric acid 0.4, orcein 0.4; B. water 100, aluminum sulfate (cryst.) 10, ferric chloride 0.8, acid alizarin blue 2B 0.35; C. 5% phosphotungstic acid; D. water 100, acetic acid 2, orange G 2, fast green FCF 0.2
PREPARATION OF B: Boil the dye in the sulfate solution 10 min. Add the ferric chloride dissolved in little water.
METHOD: [sections of mercuric, or mercuric-chromic, fixed material after removal of mercury by iodine treatment] → 85% alc. → A, 2–24 hr → 85% alc., thorough wash → water, via graded alcs. → B, 5–10 min → rinse → C, until collagen destained → rinse → D, 10 min → 50% alc., wash → balsam, via usual reagents
RESULT: elastic fibers, brown; nuclei, blue; cytoplasm and muscle, violet; collagen, green; erythrocytes, orange

Kornhauser 1945 (*Stain Technol.*, **20**:33)
REAGENTS REQUIRED: A. as Kornhauser 1943 above; B. water 100, acetic acid 0.475, sodium acetate 0.023, acid alizarin blue 2 B 0.35, ammonium alum 5; C. water 100, phosphotungstic acid 4, phosphomolybdic acid 1; D. as Kornhauser 1943 above
METHOD: [material as for Kornhauser 1943 above] → 85% alc., → A, 2–24 hr → 85% alc., thorough wash → water, via graded alcs. → B, 5–10 min → rinse → C, 10–30 min → quick rinse → D, 10 min → 50% alc. wash → balsam, via usual reagents
RESULT: as Kornhauser 1943 above

Krugenberg and Thielman 1917 (*Z. wiss. Mikr.*, **34**:234)
REAGENTS REQUIRED: A. water 108, anilin blue 0.45, eosin B 0.22, phloxine 0.45
PREPARATION OF A: The eosins are each dissolved in 45 water and mixed with the blue dissolved in 18.
METHOD: [sections of alcohol-fixed material] → A, 2–10 min rinse → abs. alc., minimum possible time → balsam, via usual reagents

Lillie 1945a (*Bull. Int. Ass. Med. Mus.*, **25**:27)
REAGENTS REQUIRED: A. water 99, acetic acid 1, brilliant purpurin R 0.6, azofuchsin G 0.4; B. 1% acetic acid; C. water 100, picric acid 1, naphthol blue black 1
METHOD: [sections, nuclei stained by Weigert 1903 hematoxylin (see HEMATOXYLIN STAINING FORMULAS] → water → A, 5 min → B, rinse → C, 5 min → B, 2 min → balsam, via usual reagents
RESULT: collagen, reticulum, basement membranes, dark green; muscle and glands, brown; erythrocytes, brownish red

Lillie 1945b (*Bull. Int. Ass. Med. Mus.*, **25**:28)
REAGENTS REQUIRED: A. either 0.1% fast green FCG in 1% acetic acid or 0.3% wool green FCF in 1% acetic acid; B. 1% acetic acid; C. either 0.2% acid fuchsin or 0.2% violamine R

METHOD: [sections, nuclei stained in Lillie 1948] → water → A, 4 min → B, wash → C, 10–15 min → B, 2 min → balsam, via usual reagents
RESULT: connective tissue, red; erythrocytes, green; muscle and cytoplasm, gray-green

Lillie 1945c (*Bull. Int. Ass. Med. Mus.*, **25**:32)
REAGENTS REQUIRED: A. 1% eosin Y; B. 3% ferric chloride; C. 1% naphthol green B; D. 1% acetic acid
METHOD: [sections, nuclei stained by Weigert 1903] → water → A, 3 min → rinse → B, 4 min → rinse → C, 5 min → D, 2 min → balsam, via acetone and xylene
RESULT: general cytoplasm, pink; collagen, green

Lillie 1945d (*Bull. Int. Ass. Med. Mus.*, **25**:32)
REAGENTS REQUIRED: A. 1% Biebrich scarlet in 1% acetic acid; B. 1% acetic acid; C. 0.5% methyl blue in 0.3% hydrochloric acid
METHOD: [sections, nuclei stained by Weigert 1903] → water → A, 5 min → B, 2 min → C, 5 min → balsam, via usual reagents

Lillie 1945e (*Bull. Int. Ass. Med. Mus.*, **23**:41)
REAGENTS REQUIRED: A. 1% phloxine B; B. 1% acetic acid; C. water 100, hydrochloric acid 0.25, methyl blue 0.1, orange G 0.6
METHOD: [sections with nuclei stained by Weigert 1903] → water → A, 10 min → B, 2 min → C, 10 min → B, 5 min → balsam, via acetone and xylene

Lonnberg 1891, *carmine–spirit blue* (*Svenska Akad. Handl.*, **6**:1)
REAGENTS REQUIRED: A. Any formula; B. 0.1% hydrochloric acid in 70% alc.; C. sat. 60% alc. sol. spirit blue 100, hydrochloric acid 1 drop; D. 85% alc. adjusted with ammonia to pH 8
METHOD: [whole objects] → A, until stained → B, until differentiated → 70% alc., wash → C, 15 min → D, until violet → balsam, via usual reagents
RECOMMENDED FOR: double-staining whole mounts

Lopez 1946 (*Tech. Bull.*, **7**:53)
REAGENTS REQUIRED: A. water 100, Ziehl 1882 10, acetic acid 0.2; B. water 100, 40% formaldehyde 4, acetic acid 0.2; C. water 100, phosphomolybdic acid 1, anilin blue WS 0.5, methyl orange to sat.
PREPARATION OF C: Dissolve the acid with gentle heat. Add the anilin blue. After complete solution, add an excess of orange. Warm a few minutes. Filter.
METHOD: [sections of formaldehyde material] → water → A, 1 min → wash → B, 3 min → wash → C, ½–1 min → quick rinse → abs. alc., least possible time → balsam, via xylene

Lynch 1930, *carmine-indulin* (*Z. wiss. Mikr.*, **46**:465)
REAGENTS REQUIRED: A. Grenacher 1879; B. hydrochloric acid; C. 0.5% hydrochloric acid in 70% alc.; D. sat. sol. indulin in 80% alc.
METHOD: [whole objects, preferably mercuric-fixed] → A, until thoroughly saturated → add B, drop by drop, until brick-red ppt. formed; leave 12 hr → C, until object clear pink → add D drop by drop until solution faint blue, leave until blue round edges → 95% alc., 1 hr → balsam, via usual reagents
RESULT: thick or dense, structures, red; thin or diffuse structures, blue
RECOMMENDED FOR: double-staining whole mounts

Mann 1894a, *methyl blue–eosin* (*Z. wiss. Mikr.* **11**:490)
REAGENTS REQUIRED: A. water 55, 1% methyl blue 20, 1% eosin 25; B. 0.005% potassium hydroxide in abs. alc.; C. 1% eosin
METHOD: [smears or sections] → water → A, 24–48 hr → water, rinse → abs. alc., until dehydrated → B, until red → abs. alc., thorough wash → water → C, until differentiated → balsam, via usual reagents
NOTE: Alzheimer 1910 (Nissl and Alzheimer 1910, 409) gives A as water 150 to 35 of each of the dye solutions. Langeron 1942, 598, attributes this formula, without reference, to "1892." Perdrau 1939 (*J. Pathol. Bact.*, **48**:609) recommends mordanting in 2% ammonium molybdate before this method.

Mann 1894b, *toluidine blue–erythosin* (*Z. wiss. Mikr.*, **11**:489)
REAGENTS REQUIRED: A. water 100, erythrosin 0.1; B. water 100, toluidine blue 1.0; C. 0.2% acetic acid
METHOD: [water] → A, 1–2 min → rinse → B, on slide, 1–2 min → C, until differentiated → balsam, via usual reagents
RESULT: nuclei, cartilage, blue; other structures, polychrome red and violet

Masson cited from 1942 Langeron, *erythrosin–toluidine blue–orange G* (Langeron 1942, 613)
REAGENTS REQUIRED: A. 6% iodine in 4% potassium iodide; B. 5% sodium thiosulfate; C. water 100, erythrosin 0.2, orange G 1; water 100, toluidine blue 1; E. 0.2% acetic acid
METHOD: [sections of Helly 1903, Orth 1896, or Zenker 1894 fixed material] → water → A, 30 min → B, 2 min → thorough wash → C, 1–2 min → rinse → D, on slide, 1–2 min → E, until differentiated → balsam, via usual reagents
RESULT: as Mann 1894b but with bacteria deep blue

Matsura 1925, *polychrome neutral red* (*Folia. Anat. Jap.*, **3**:107)
REAGENTS REQUIRED: A. 1% congo red in 95% alc.; B. 1% phosphomolybdic acid in abs. alc.
METHOD: [sections] → A, 12–24 hr → abs. alc., rinse → B, 5 min → abs. alc., until differentiated → neutral mountant, via oil of thyrne
RESULT: nuclei, red; elastic fibers, red violet; collagen, green; white blood cells, violet; other tissues, brown

Merbel 1877, *carmine–indigo carmine* (*Amer. Mon. Micr. J.*, **1**:242)
STOCK SOLUTIONS: I. water 100, sodium borate 7, carmine 1.7; II. water 100, sodium vorate 7, indigocarmine 7
PREPARATION OF STOCK SOLUTIONS: Boil 15 min. Cool. Filter.
REAGENTS REQUIRED: A. stock I 50, stock II 50; B. sat. sol. oxalic acid
METHOD: [sections of Muller 1859 material] → water → A, 15–20 min → B, until differentiated → wash → balsam, via usual reagents

Norris and Shakespeare cited from 1895 Rawitz, *carmine–indigo carmine* (Rawitz 1895, 67)
FORMULA: water 130, sodium borate 8, carmine 1, indigo carmine 4
PREPARATION: Boil the carmine in 65 water with 4 borax for 15 min. Cool. Filter. Boil indigo carmine in 65 water with 4 borax 5 min. Cool. Filter. Mix filtrates. Filter.

Pearse 1950 (*Stain Technol.*, **25**:97)
REAGENTS REQUIRED: A. Lendrum and McFarlane 1940 celestin blue B solution [see OXAZINE DYES]; B. Pearse 1950 alum hematoxylin [see HEMATOXYLIN FORMULAS]; C. 2% HCl in 70% alc.; D. 2% orange G in 5% phosphotungstic acid
METHOD: [sections of Helly 1903 fixed material, prior stained by Feulgen technique] → A, ½–3 min → B, ½–3 min → C, rinse → wash, 5 min → D, 10 sec → wash until differentiated → balsam, via usual reagents

Petersen 1924 cited from 1946 Roskin (Roskin 1946, 230)
REAGENTS REQUIRED: A. water 100, aluminum sulfate 10, acid alizarin blue 0.5; B. 5% phosphotungstic acid; C. water 100, acetic acid 8, orange G 2, anilin blue 0.5
METHOD: [sections] → water → A, 5 min → rinse → B, several minutes → distilled water, rinse → C, 2 min → rinse abs. alc., until dehydrated → balsam, via xylene

Plehn cited from Kahlden and Laurent, *methyl blue–eosin* (Kahlden and Laurent 1896, 126)
FORMULA: water 20, sat. aq. sol., methyl blue 60, 0.5% eosin in 70% alc. 20, 20% sodium hydroxide 0.5

Proescher, Zapata, and McNaught 1946 (*Tech. Bull.*, **7**:50)
STOCK SOLUTIONS: I. 0.1% azophloxine; II. celestine blue B (anonymous 1936) see OXAZINE DYES]
WORKING SOLUTION: stock I 60, stock II 30
METHOD: [frozen sections of hot-formaldehyde-fixed materials] → stain, 10–30 sec → abs. alc., least possible time for dehydration → mountant
RECOMMENDED FOR: rapid staining for diagnosis

Roskin 1946 (Roskin 1946, 231)
REAGENTS REQUIRED: A. 70% alc. 100, hydrochloric acid 1, orcein 1; B. any alum hematoxylin formula; C. van Gieson 1896 picrofuschin 100, acetic acid 0.3; D. water 100, C (preceding) 4
METHOD: [sections] → water → A, 1 hr → wash, 3 min → B, until thoroughly overstained → distilled water, 3 min → running water, 3 min → 3 min → distilled water, 3 min → C, 5 min → D, quick rinse → blot → abs. alc., minimum possible time → balsam, via xylene.

Schleicher 1943 (*Tech. Bull.*, **4**:35)
REAGENTS REQUIRED: A. 0.1% azocarmine B in 5% acetic acid; B. 5% phosphotungstic acid; C. water 100, acetic acid 0.3, orange G 0.13, anilin blue 0.07
METHOD: [sections of Helly 1903 fixed material with mercury removed by iodine treatment] → A, 15–30 min → 56°C → rinse → B, 15–30 min → rinse → 95% alc., until color clouds cease → balsam, via usual reagents

Shumway 1926, *magenta-picro–indigocarmine* (*Stain Technol.*, **1**:37)
REAGENTS REQUIRED: A. sat. sol. (*circ.* 1%) magenta; B. sat. sol. (*circ.* 0.2%) indigocarmine 50, sat. sol. (*circ.* 1.2%) picric acid 50
METHOD: [sections] → water → A, 20 min → B, 5 min → 70% alc. until pink, few seconds → abs. alc., until blue-green, several seconds → balsam, via xylene

RESULT: resting nuclei, dark blue; mitotic figures, dark red; cartilage, pink; procartilage, light blue; bone, dark blue; muscle, bright green; nerves, purple

Twort cited from Minchin 1909, *neutral red–light green* (*Quart. J. Micr. Sci.*, **53**:755)
PREPARATION OF DRY STOCK: Dilute 50 sat. sol. (*circ.* 3%) neutral red to 100. Dilute 50 sat. sol. (*circ.* 20%) light green to 100. Mix at 50°C. Cool. Filter. Wash and dry ppt.
WORKING SOLUTION: water 70, propanol 30, dry stock 0.5
METHOD: [sections from acetic alone or dichromate-formaldehyde fixed material] → water → A, 40°C, 10 min → distilled water, rinse → abs. alc., until differentiated → balsam, via xylene
RESULT: nuclei, purple; blood, green; connective tissues, blue-green
NOTE: see also Twort 1924. Langeron 1942, 565 recommends the addition of 1% phenol to the working solution.

Volkman and Strauss 1934, *azocarmine–naphthol green–crystal violet* (*Z. wiss. Mikr.*, **51**:244)
REAGENTS REQUIRED: A. water 100, chrystal violet 1, resorcinol 2, 30% ferric chloride 12.5; B. 0.1% azocarmine in 1% acetic acid; C. 0.1% aniline in 95% alc.; D. 5% phosphotungstic acid; E. 1% naphthol green B in 1% acetic acid
METHOD:]sections] → water → A, 1–2 hr → 70% alc., rinse wash → E, 15 min → wash → abs. alc., least possible time → balsam, via xylene
RESULT: nuclei and muscles, red; collagen, green; elastic fibers, black

Westphal 1880 cited from Böhm and Oppel, *carmine–crystal violet* (Böhm and Oppel 1907, 193)
FORMULA: water 40, abs. alc. 40, glycerol 20, acetic acid 4, ammonium alum 0.4, carmine 0.4, phenol 0.4, crystal violet 5
PREPARATION: Boil the carmine and alum, in the water 15 min. Filter. Dissolve the crystal violet in the alc. Add to cooled filtrate. Leave overnight. Filer; add the glycerol and acid.

Williams 1935 (*J. Lab. Clin. Med.*, **20**:1185)
PREPARATION OF STAIN: Dissolve 1 cresyl violet and 1 potassium carbonate (anhyd.) in 95 water and 5 40% formaldehyde. Shake for 30 min. Add slowly, and with constant agitation, 3 acetic acid. Shake 30 min. Filter and add 5 isopropyl alc.
METHOD: [frozen sections of unfixed tissues] → stain, 6 sec → wash → examine
RESULT: nuclei and muscle, blue; fat, yellow; other tissues, pink

PETER GRAY

Books Cited

To save space the undernoted books are referred to in the above article by author and date only. Literature references in the article are in customary form.

Becher, S., "Untersuchungen über Echtfarbung der Zellkerne mit künstlichen Beizenfarbstoffen," Berlin, Borntraeger, 1921.
Bensley, R. R., and S. H. Bensley, "Handbook of Histological and Cytological Technique," Chicago, University of Chicago Press, 1938.
Böhm, A., and A. Oppel, "Manuel de Technique Microscopique," Traduit de l'Allemand par Etienne de Rouville, 4th ed., Paris, Vigot, 1907.
Cowdry, E. V., "Microscopic Technique in Biology and Medicine," Baltimore, Williams and Wilkins, 1943.
Dobell, C., "The Amoebae Living in Man," London, John Bale, Sons and Danielsson, 1919.
Ehrlich, P., R. Krause, M. Mosse, H. Rosin, and K. Weigert, "Encyklopädie der mikroskopischen Technik mit besondere Berücksichtigung der Farbelehre," Berlin, Urban and Schwarzenberg, 1903.
Friedlaender, C., "Technik sum Gebrauch bei medicinischen und pathologischantomischen Untersuchungen," 4th ed., by C. J. Eberth, Berlin, Fischer, 1889.
Gatenby. J. B., and E. V. Cowdry, "Bolles Lee's Microtomist's Vade-Mecum," Philadelphia, Blakiston, 1928.
Gray, P., "Handbook of Basic Microtechnique," Philadelphia, Blakiston, 1952.
Guyer, M. F., "Animal Micrology," 3rd ed., Chicago, University of Chicago Press, 1930.
Hall, W., and G. Herscheimer, "Methods of Morbid Histology and Clinical Pathology," Philadelphia, Lippincott, 1905.
Johansen, D. A., "Plant Microtechnique," New York, McGraw-Hill, 1940.
Kahlden, C. von, and O. Laurent, "Technique Microscopique," Paris, Carré, 1896.
Kingsbury, B. F., and O. A. Johannsen, "Histological Technique," New York, Wiley, 1927.
Langeron, M., "Précis de Microscopie," 5th ed., Paris, Masson, 1934.
Idem. 6th ed., 1942.
Lee, A. B., "The Microtomist's Vade-Mecum," 8th ed., London, Churchill, 1905.
Lillie, R. D., "Histopathologic Technic," Philadelphia, Blakiston, 1948.
Mallory, F. B., "Pathological Technique," Philadelphia, Saunders, 1938.
Rawitz, B., "Leitfaden für histologische Untersuchungen," Jena, Fischer, 1895.
Romeis, B., "Mikroskopische Technik," 15th ed., Munich, Leibniz, 1948.
Roskin, G. E., "Mikroskopercheskaya Technika," Moscow, Sovetskaya Nauka, 1946.
Schmorl, G., "Die pathologisch-histologischen Untersuchungs-methoden," 15th ed., Leipzig, Vogel, 1928.
Squire, P. W., "Methods and Formulae Used in the Preparation of Animal and Vegetable Tissues for Microscopical Examination," London, Churchill, 1892.

PORIFERA

The Porifera can be considered the simplest multicellular animals, set apart by their features from all other metazoans. The recent resurgence of interest in the use of sponges as experimental material came from the desire to know why, and to what extent, these multicellular animals are really set apart from the others.

Whatever may be our understanding of a sponge, either as a population of cells lacking consistent general organization, or as an organism built with true tissue layers, it is apparent that both biophysical and biochemical mechanisms and structural abilities in such an animal are adequate to carry out the integration of cell behavior and differentiation (1). But this integration appears to lead to the development of local and versatile cellular systems, able to change according to the needs of morphogenesis and homeostasis.

It follows that, in virtue of the plasticity of their cell organization, the Porifera offer a good model for studies of the fundamental attributes of cells.

The techniques available correspond, in general, to some peculiar adaptations of general microtechnical methods utilized in cell and tissue biology. Nevertheless, some special methods have been developed to deal with the special functions of sponges, in wound healing, in the restorative processes following transplantation (2), in the reconstitution of lost structures, and in the redevelopment from small fragments of the body, or from aggregates of somatic cells after dissociation.

These phenomena of regeneration proper and of "somatic embryogenesis" (3) are analyzed with microtechniques quoted below, but the first step depends on what might be called, broadly speaking, cell and tissue cultures.

Gross or quite delicate excision can be undertaken with success, even on larvae (4), in order to obtain one cell-type culture. In general, however, it is more usual to secure cell dissociation and culture by rubbing a piece of sponge through the fine meshes of bolting cloth (3, 5). The resulting suspension is transferred either into small dishes where cells reaggregate (6), or into shaker flasks to avoid that aggregation (7). One other of the most spectacular of these techniques is the "sandwich culture" of hatching gemmules (which are rounded masses of cells giving asexual reproductive bodies in all freshwater and some marine sponges) the tissues of which are compressed between two cover glasses and thus are obliged to organize themselves in a thin plan (8). Excellent observations *in vitro* can be made by this procedure.

In any case, microscopy is the basic technique used to examine living or dead samples of sponges, for cyto- and histomorphology. No special techniques are necessary for standard light microscopy; except after fixation and embedding in paraffin, in regard to sectioning some species the tissues of which contain organic ("spongin fibers") or mineral (spicules) components as skeletal structures. After dissolving calcareous spicules with various acids the results are often histologically correct but this is almost never the case with siliceous spicules treated with fluorhydric acid.

Still in regard to light microscopy, data obtained with timelapse cinematography of living material (cultures or thin pieces of tissue) contribute to the understanding of the movements of spicules (9), cell displacements, and contractile activities (10).

For electron microscopy, standard methods are to be used with caution; in regard to securing first, the correct iso-osmolarity of the fixative, second a good embedding of small pieces of tissue in polymerized resin to facilitate cutting ultrathin sections in spite of the skeletal components of many species.

Furthermore scanning electron microscopy permits the investigation of the detailed shape of spicules or the surface of epithelia (11). In the same way, spicules can be analyzed with replicas (12) and the connective tissue with freeze-etching (13).

In the field of cyto and histochemistry, many data can be collected with classical techniques of light microscopy specific for: nucleic acids, mucoproteins, pigments, polysaccharides (15), monoamines, and also several enzymes such as phosphatases, cholinesterases, mono-amines oxydases (16). The use of labeled compounds (uridine ^{3}H, thymidine ^{3}H, proline ^{3}H) (6) is also possible. In electron microscopy, cytochemical methods give data dealing with proteinpolysaccharides (17), monoamines, glycogen, phosphatases, etc. (18). Some enzymatic digestions with RNase, pepsine, hyaluronidase, etc. (19) have also yielded interesting results.

Finally, numerous biochemical and biophysical micro-

analyses complement earlier techniques available for sponges. This is the case: for "antigenic substance" related to the sponge cell surface, composed of heteropolysaccharides and peptides, and extracted with trichloracetic acid (20); for "aggregation factor" containing also polysaccharides and proteins, and coming from the supernatant of the medium where gemmules are growing (21). In such studies chromatography and electrophoresis are often utilized (22).

In the case of ions, measurements are made with flame photometers, chloride titrimeters or specific dyes, while inulin space is calculated with labeled carbon and scintillation counter (23). Warburg apparatus measures respiration rate of living tissue (8). X-ray diffraction and arc spectrography are useful methods for the study of the crystalline substance of spicules (24), as are X-ray and infrared spectrography for collagen and organic components of skeletons (19, 25).

At last, one may note myographic records using electronic force transducer displayed on a polygraph, in the investigation of contractile activities of cells *in situ* on the whole sponge (23). Similarly electrophysiological methods with intracellular glass microelectrodes allow a voltage clamp to be applied and thus record the electrotonic coupling of living cells in culture, displayed on an oscilloscope (26).

M. Pavans de Ceccatty

References

1. W. G. Fry, *Symp. Zool. Soc. London*, **25**:XI–XIV (1970).
2. H. Mergner, *Symp. Zool. Soc. London*, **25**:365–397 (1970).
3. G. P. Korotkova, *Symp. Zool. Soc. London*, **25**:423–436 (1970).
4. R. Borojevic, *Devl. Biol.*, **14**:130–153 (1966).
5. H. V. Wilson, *J. Exp. Zool.*, **5**:245–258 (1907).
6. S. M. Efremova, *Symp. Zool. Soc. London*, **25**:399–413 (1970).
7. T. Humphreys, *Devl. Biol.*, **8**:27–49 (1963).
8. R. Rasmont, *Ann. Soc. Roy. Zool. Belg.*, **91**:147–156 (1961).
9. W. C. Jones, *J. Mar. Biol. Ass. U.K.*, **44**:67–85, 311–331 (1964).
10. M. Pavans de Ceccatty, *Experientia*, **27**:57–59 (1971).
11. R. Rasmont (personal communication).
12. R. W. Drum, *J. Ultrastr. Res.*, **22**:12–21 (1968).
13. R. Garrone and J. Pottu, (personal communication).
15. T. L. Simpson, *Peabody Mus. Nat. Hist., Yale Univ. Bull.*, **25**:1–141 (1968).
16. T. L. Lentz, *J. Exp. Zool.*, **162**:171–180 (1966).
17. Y. Thiney and R. Garrone, *7th Int. Cong. Elec. Micr., Grenoble*, 1:639–640 (1970).
18. Y. Thiney (personal communication).
19. R. Garrone, *J. Micr. France*, **8**(5):581–598 (1969).
20. A. P. MacLennan, *Symp. Zool. Soc. London*, **25**:299–323 (1970).
21. F. Rozenfeld, *Arch. Biol. Liege*, **81**:193–214 (1970).
22. P. R. Bergquist and W. D. Hartman, *Mar. Biol.*, **3**:247–268 (1969).
23. C. L. Prosser, *Zeit. verg. Physiol.*, **54**:109–120 (1967).
24. W. C. Jones, *Symp. Zool. Soc. London*, **25**:91–123 (1970).
25. J. Gross, Z. Sokal, and M. Rougvie, *J. Histochem. Cytochem.*, **4**(3):227–246 (1956).
26. W. R. Loewenstein, *Devl. Biol.*, **15**:502–520 (1967).

PROTOZOA

Even if the protozoa are considered simply as cells (and some protozoologists would quarrel with such a definition), there are such numbers of them (at least 50,000 described and myriad unknown species) and they demonstrate such diversity in size, form, habitat, and morphological complexity that it is difficult to present even an adequate introduction to the many microscopic methods and microtechniques of their study in a single short article. As the celebrated English cytologist-protozoologist E. A. Minchin pointed out some 60 years ago, concerning solely the fixing and staining of protozoa, every species or strain "requires its own special technique, which must be established empirically by trial, and can be discovered only to a very limited extent and with great uncertainty by analogy."

No attempt is made in the present account, though attention is here directed to their tremendous importance, to cover *physiological* or *biochemical* techniques, including methods of cultivation often indispensable to (subsequent) thorough investigation of the cytology or morphology of a protozoan species. Along with culture media, collecting methods and narcotizing agents also will not be discussed.

In spite of their diversification, most protozoa fall in a size range requiring a high degree of magnification for their proper observation; thus the whole science of protozoology has developed in parallel with that of microscopy (indication of this is borne out in many of the references cited below).

Living protozoa may be studied to great advantage by methods beyond the standard light or bright field microscopy. Such techniques include phase contrast microscopy, dark field illumination, and polarizing, fluorescence, ultraviolet, and Nomarski interference microscopy. Generally, no prestaining (vital) is required. Even the centrifuge microscope has been employed, although to date to a limited extent only. Microcinematography should also be mentioned; and living specimens may be subjected to various micrugical techniques to advantage in selected kinds of studies.

Common temporary stains used on living or otherwise nonfixed Protozoa include such basic "vital" dyes as neutral red, Janus green B, brilliant cresyl blue, methylene blue, and Bismarck brown. "Negative" staining with such relief "stains" as Nigrosine and Aniline, Toluidin, and other Blues often provides a quick and reliable technique for demonstration of surface sculpturing, etc. Lugol's solution or D'Antoni's iodine are widely used, especially with parasitic protozoa of various groups (routinely, e.g., with fecal smears). Several aceto-carmine methods are popular. Even writing fluids (e.g., red and blue inks) have been used with success. Following osmic acid fixation, methyl green is excellent for temporary demonstration of nuclei, even chromosomes within nuclei in some material.

Fixed, and generally *stained*, material, sectioned or as whole mounts, in temporary or in permanent preparations, may be investigated in three major ways: by light microscopy, by electron microscopy, or by various biochemical or cytochemical tests (often involving microscopy of some sort as well). Methods of fixation and of staining can be mentioned only briefly: often they parallel those commonly known to cytologists and histologists working with metazoan or metaphytan material, and references to them abound in most of the citations given below.

Embedding and sectioning of material, operations often unnecessary with protozoan specimens, require approaches also quite standard—thus they are not treated further here. Still the single most indispensable manual of microtechniques (for the protozoa) available today is the aging but excellent work of Kirby (1950). Minchin's warning (quoted above), when attempting to adapt general cytological methods to usage with protozoa, should be kept constantly in mind. Even among the protozoa, parasitic and nonparasitic forms and forms belonging to vastly different systematic groups (e.g., amoebae as contrasted with ciliates, hemoflagellates with foraminifera, malarial organisms with phytoflagellates, microsporidians with radiolarians) require quite different approaches, even if the chemical reagents used are the same. Methods of *handling* specialized material differ, too, and the original works of the experts, often scattered research papers or notes, must be consulted for specific directions.

Kind and state of the material, temperature of the reagents, concentration of the specimens, and purpose of the study are all factors in the choice of fixative. Desiccation alone is sometimes the most appropriate killing agent; formalin or alcohol alone are also often employed, though they are widely used much more effectively as constituents of more elaborate fixing fluids. Of the sublimate fixatives, Schaudinn's fluid is generally acknowledged to be the best standard, routine protozoological fixative for many kinds of material. Bouin's, Carnoy's, and Flemming's represent other classes of penetrating fixatives. There are several osmium tetroxide (commonly known as osmic acid) solutions which are excellent: a straight 1–2% solution is often used as fumes; or the OsO_4 may be mixed with other chemicals (such as with potassium dichromate and chromic acid to yield Champy's fluid).

Stains and staining techniques are numerous. Most popular with protozoan material are the several hematoxylin stains (e.g., the long-known Heidenhain's iron hematoxylin); Giemsa's, Wright's, and other blood stains; the Feulgen nucleal reaction; and two or three methods of silver impregnation (primarily for flagellates and ciliates). Even more versatile than the Chatton-Lwoff silver technique (in its turn, far more sophisticated than the Klein "dry" method), the Protargol technique reveals intact ciliature, infraciliature, and nuclear material as well as basal bodies (kinetosomes) and other elements of the "silverline system" proper. In recent years, parallel studies of (samples of) the same organisms (in the same and in different groups) using methods of both silver impregnation and electron microscopy have thrown considerable light on alleged "homologous" structures and thus also on the possible taxonomic interrelationships of the ciliate or flagellate groups involved.

Utilizing light microscopy (in all of its variations), fixed and generally stained protozoan material may be studied in various conventional ways. Subcellular cytoplasmic organelles are common objects of such study, though also body coverings, structures of locomotion and attachment, feeding organelles, cystic membranes, secreted tests, spore envelopes, shells, and loricae may be examined equally well by the same or similar approaches. Chromosomes, nucleoli, etc. in the nucleus also may be so investigated.

Transmission electron microscopy (TEM) generally requires thin-sectioned material (or sonic fragments), since protozoa are rarely small or thin enough to be examined whole. (Methods of fixation, embedding, sectioning, etc.

for electron microscopy are beyond the scope of this brief article: see appropriate sections elsewhere in the volume.) The ability to study ultrastructural details of the protozoan cytoplasm and nuclear material, of pseudopodia, cilia, flagella, etc. has been a great boon to protozoology and has brought about significant advances in our overall knowledge of these microscopic animalcules. In very recent years, scanning electron microscopy (SEM) has allowed examination of protozoan whole mounts at magnifications (e.g., 2000–20,000) conveniently intermediate between those typical of light microscopy and TEM. Since cortical movements, de- and redifferentiation of somatic and buccal ciliature, and the like are essentially surface phenomena, SEM studies should prove to be particularly valuable in investigations of morphogenetic processes (such as binary fission and stomatogenesis) and thus, secondarily, in work in such fields as systematics and phylogeny.

Discussion of various biochemical, biophysical, or cytochemical tests is beyond the scope of this contribution, but many such techniques may be usefully employed in cytological or comparative morphological or taxonomic study of protozoan material: e.g., autoradiography, cytophotometry, X-ray analysis, radioactive-isotope labeling, fluorescent antibody techniques, enzyme assays (a widely used example: the Gomori method for alkaline phosphatase), base ratio DNA analyses, immunological comparisons, and nucleic acid hybridization studies. Genetic tests may be considered to fall into the same broad category and thus are also not treated here.

JOHN O. CORLISS

References

A relatively large number of pertinent citations, of varying age, length, and language, are purposely included to give the interested reader an opportunity to find both (1) first-hand sources of directions for handling various kinds of protozoa, and (2) indirect references to much of the vast literature containing specialized information by the experts on the many individual taxonomic groups impossible to treat separately in the above account.

Bělař, K., "Methoden zur Untersuchung der Protozoen," pp. 735–826 in Peterfi, ed., "Methodik der Wissenschaftlichen Biologie," Vol. 1, Berlin, Springer, 1928.

Buhse, H. E., Jr., J. O. Corliss, and R. C. Holsen, Jr., "*Tetrahymena vorax*: analysis of stomatogenesis by scanning electron and light microscopy," *Trans. Amer. Micros. Soc.,* **89**:328–336 (1970).

Chatton, E., and A. Lwoff, "Technique pour l'étude des protozoaires, spécialement de leurs structures superficielles (cinétome et argyrome)," *Bull. Soc. franç. Micros.,* **5**:25–39 (1936).

Chen, T. T., ed., "Research in Protozoology," Vols. I–IV, New York and London, Pergamon Press, 1967–1972.

Corliss, J. O., "Silver impregnation of ciliated protozoa by the Chatton-Lwoff technic," *Stain Technol.,* **28**:97–100 (1953).

Corliss, J. O., "Application of modern techniques to problems in the systematics of the protozoa," *Proc. XVIth Int. Cong. Zool.,* Washington, D.C., Aug. 1963, **4**:97–102 (1963).

Corliss, J. O., "Newer trends in the systematics of the protozoa," *Bull. Nat. Inst. Sci. India,* (34):26–33 (1967).

Dragesco, J., "L'orientation actuelle de la systématique des ciliés et la technique d'imprégnation au protéinate d'argent," *Bull. Micros. Appl.,* **11**:49–58 (1962).

Frankel, J., and K. Heckmann, "A simplified Chatton-Lwoff silver impregnation procedure for use in experimental studies with ciliates," *Trans. Amer. Micros. Soc.,* **87**:317–321 (1968).

Galigher, A. E., and E. N. Kozloff, "Essentials of Practical Microtechnique," 2nd ed., Philadelphia, Lea & Febiger, 1971.

Garnham, P. C. C., "Malaria Parasites and Other Haemosporidia," Oxford, Blackwell, 1966.

Goodrich, H. P., "Protozoa," in J. B. Gatenby and H. W. Beams, ed., "The Microtomist's Vade-Mecum" (Bolles Lee), 11th ed., London, Churchill, 1950.

Hoare, C. A., "Handbook of Medical Protozoology," London, Baillière, Tindall & Cox, 1949.

Honigberg, B. M., "Developments in microscopy in relation to our understanding of protozoa," *Trans. Amer. Micros. Soc.,* **86**:101–112 (1967).

Kidder, G. W., ed., "Protozoa," in M. Florkin and B. T. Scheer, ed., "Chemical Zoology," Vol. I, New York and London, Academic Press, 1967.

Kirby, H., "Materials and Methods in the Study of Protozoa," Berkeley, University of California Press, 1950.

Klein, B. M., "The 'dry' silver method and its proper use," *J. Protozool.,* **5**:99–103 (1958).

Kudo, R. R., "Protozoology," 5th ed., Springfield, Ill., Thomas, 1966.

Lom, J., and J. O. Corliss, "Morphogenesis and cortical ultrastructure of *Brooklynella hostilis*, a dysteriid ciliate ectoparasitic on marine fishes," *J. Protozool.,* **18**:261–281 (1971).

Lom, J., J. O. Corliss, and C. Noirot-Timothée, "Observations on the ultrastructure of the buccal apparatus in thigmotrich ciliates and their bearing on thigmotrich-peritrich affinities," *J. Protozool.,* **15**:824–840 (1968).

Mackinnon, D. L., and R. S. J. Hawes, "An Introduction to the Study of Protozoa," London and New York, Oxford University Press (Clarendon), 1961.

Manwell, R. D., "Introduction to Protozoology," New York, St. Martin's Press, 1961.

Minchin, E. A., "An Introduction to the Study of the Protozoa," London, Arnold, 1912.

Pitelka, D. R., "Electron-Microscopic Structure of Protozoa," London and New York, Pergamon Press, 1963.

Small, E. B., and D. S. Marszalek, "Scanning electron microscopy of fixed, frozen, and dried protozoa," *Science,* **163**:1064–1065 (1969).

Taylor, A. E. R., ed., "Techniques in Parasitology," Ist Symp. Brit. Soc. Parasitol., Oxford, Blackwell, 1963.

Tuffrau, M., "Perfectionnements et pratique de la technique d'imprégnation au protargol des infusoires ciliés," *Protistologica,* **3**:91–98 (1967).

Wenrich, D. H., and W. F. Diller, "Methods of protozoology," in R. McC. Jones, ed., "McClung's Handbook of Microscopical Technique," 3rd ed., New York, Hoeber, 1950.

PSAMMON

The tremendous number and variety of microorganisms that live in the interstitial spaces between sand grains both intertidally (thalassopsammon) and subtidally (meiobenthos), their unique morphology and especially their many adaptations for adhesion, and the relatively short time this rich, vigorous, and in great part undescribed biocoenose has been known, have made it necessary to develop special techniques both for collecting these organisms and for removing them from their lacunar environment. The interstitial flora is virtually unknown, and there is very little currently in the literature concerning their extraction. On the other hand, relatively much is known about sampling the protozoa and the micrometazoa inhabiting the interstitial milieu and removing them from their habitats.

Collecting, Maintaining, and Extracting Live Material. The easiest and often a preliminary method of intertidal collecting in the field (qualitative sampling), is to dip a hand net with a mesh size of 50 μ into the water at the bottom of a hole 60–90 cm in diameter dug to the water table, and sweeping it about while disturbing the bottom. The net is then inverted and washed into a small jar with seawater, and the collecting resumed. Several sweeps later, the bottom of the hole is redug, and the crumbling walls of sand removed to allow renewed working space. This system is generally adopted when specific collecting is desired (i.e., searching for the presence of a certain taxon), or for the preliminary investigation of a new beach. A variety of beach sampling methods utilizing heavy-walled plastic tubes of known diameters and lengths, forced vertically or horizontally into the beach, with samples capable of being subsectioned in the laboratory, have been developed and are widely used.

In the shallow subtidal and in deeper waters, a suction corer (Riedl and Ott, 1971) is the most useful instrument developed thus far. Additional apparatus to use in these areas include modified dredges and grab samplers, as well as an improved box sampler (Reineck, 1963) that can be used for investigation of the vertical distribution of organisms. In the laboratory, both intertidal and deeper subtidal samples should be both stored and extracted in cool rooms. Refrigeration at 45° is helpful in holding sand samples for several weeks until they can be examined.

There are many recently developed methods for extraction. Because of the variety and numbers of adhesive structures, morphological adaptations and varying granulometry, no single extraction method is equally efficient for all forms. In addition to the mechanical stirrer, elutriation with seawater, air bubbler (Zinn, 1968), bicycle pump, and other mechanical methods, and the novel but efficient sea ice method (Uhlig, 1964), a generalized technique has been developed by Riedl and his associates that apparently is quite satisfactory.

A solution of $MgCl_2$ isotonic to seawater is mixed into the sediment from which the interstitial fauna is to be extracted. The sample is then shaken for several minutes and abruptly stopped. While the heavier sediment settles, the wash water is decanted, and the suspended organisms are filtered through a small fish tank net frame with plankton netting. Both the filter and the deposited filtered material are then placed in a petri dish with filtered seawater. The animals migrate through the filter and then can easily be collected from the bottom of the dish with camel's hair paintbrush, fine drawn pipette, filed dentist's root canal drill, or dissecting needle made from "minuten nadeln" (insect pins). This should provide a rich diversity of interstitial fauna groups. For quantitative studies, the whole filter deposit should be washed into a small plastic petri dish scribed with 10 cm squares; the organisms are then counted. After separation from the sedimentary substratum, specimens can be studied alive by means of phase contrast microscopy, regular light microscopy, or squeeze preparations, and anesthesia and live staining can be applied at this time for morphological or ethological observations. Effective anesthetics are 5–10% $MgCl_2$, eucain hydrochloride, low-percentage ethanols, chlorotone, etc.; useful vitual stains include fast green, neutral red, nile blue sulfate, and methylene blue (Hulings and Gray, 1971).

Preserving and Microdissection. Fixation fluids such as Bouin, Worcester, Flemming, Zenker, glutaraldehyde and osmium tetroxide have been proved efficient with most interstitial forms (Gray, 1954; Galigher and Kozloff, 1964). For light microscopy, a method developed by Antonius (1965), combining embedding in celloidin and paraffin with scale photography of the transparent animal has proven very satisfactory. With subsequent serial sectioning and staining, this method allows three-dimensional reconstruction; oriented specimens are sectioned with a diamond knife, every 10–30 thin sections alternating with a thick section. By counting the sections mechanically, and by simultaneously controlling the progress of cutting, through using micrographs of the block, it becomes possible to reconstruct the whole organism on two levels: The thick sections are mounted on a slide, serially stained, and furnish the material for classical reconstruction, while the thin sections can be used for electron micrographs or for scanning micrography. This is especially useful for the study of cuticularized structures.

Culture Methods. The laboratory culture of interstitial thalassopsammon has met with indifferent success thus far; certain basic problems apparently as yet unisolated have halted progress. Sporadic positive results have been obtained with ciliates, nematodes, and certain arthropods. The basic thigmotactic requirements of most of the interstitial fauna require that they be kept in containers (petri dishes) with sediment samples of their natural substratum, or reasonable artificial substitutes. Micro-glass beads inoculated with sand bacteria, or coated with artificial nutrients together with natural or synthetic interstitial lacunar water variously composed and containing potential food as cyanophyceae, fungi, bacteria, or absorbed molecular organic material, are now being tried experimentally in an effort to solve the vexing culturing and rearing problems.

Quantitative Evaluation of Interstitial Fauna Populations. Riedl (1953) has developed a so-called "homogeneity calculation" for the comparison of different faunal samples for their degree of conformity: it expresses the boundary between biotopes as a more or less gradual shifting of both factors and organisms. In addition, a statistical program has been developed for most of the currently used diversity indices, including those of Brillouin (1963), Simpson (1964), Sanders (1968), Margalef (1968), and Shannon and Weaver (1969).

Macro- and Microstratification Measurements of the Physical and Chemical Environment. Apparatus for these measurements is constantly being developed both here and abroad, but unfortunately most of it is not yet available commercially. Most investigators favor equipment utilizing probes and sensors for measuring temperature, oxygen, pH, Eh, carbon dioxide, salinity, and so on. It is felt that this methodology least disturbs the physiocochemical system, is best replicated, is most advantageously used worldwide for parameter comparisons, and is most easily employed in the field. Instrument packages designed for continuous recording for periods up to several weeks are presently being tested. For example, Eh has been measured over a tidal cycle by means of a glass fiber rod equipped with platinum microelectrodes connected to a microammeter.

Sediment Analysis. Analysis of sediments are customarily made accordng to the standard procedures outlined by Krumbein and Pettijohn (1938). This includes dispersion with Calgon (glassy sodium hexametaphosphate buffered with sodium phosphate), sieve analysis of sand fraction using standard sieve sizes at regular phi intervals, and pipette analysis of silt and clay fractions.

The sediment should be sieved through a nest of seven U.S. Standard Sieves, each separated by one phi unit. The sieves suggested have mesh openings of 4 mm, 2 mm, 1 mm, 0.5 mm, 0.25 mm, 0.125 mm, 0.0625 mm. The silt-clay fraction that passes through the 0.0625 mm sieve should be collected in a bucket of known volume. The fractions collected on the sieves as well as a 100 ml aliquot of the silt-clay fraction should be put into 250 ml tared beakers, dried at 80°C, and weighed. The silt-clay weight is then extrapolated to the total volume of the bucket. All of the grain size analyses are most easily determined by wet-sieving the sediment in fresh water, thereby avoiding the aggregation that often accompanies drying of carbonate sediments.

Organic content is ordinarily determined by digestion with chromic acid (Walkley, 1947).

Dissolved Organic Matter. The problem of dissolved organic matter as a source of organic energy for interstitial thalassopsammon has been successfully attacked by measuring the uptake of labeled organic substances by sterile natural sands, or by techniques that have been developed for measuring the uptake of organic materials by microorganisms by Strickland (1965), Wright and Hobbie (1966), and Voccaro (1969).

DONALD J. ZINN

References

Antonius, A., "Methodischer Beitrag zur mikroskopichen Anatomie und graphischen Darstellung sehr kleiner zoologische Objekte," *Mikroskopie,* **20**:145–153 (1965).

Brillouin, L., "Science and Information Theory," New York, Academic Press, 1962.

Galigher, A. E., and E. N. Kozloff, "Essentials of Practical Microtechnique," Philadelphia, Lea and Febiger, 1964.

Gray, P., "The Microtomist's Formulary and Guide," New York, The Blakiston Company, 1954.

Hulings, N. C. and J. S. Gray, "A Manual for the study of Meiofauna," Smithsonian Contributions to Zoology No. 78, Washington, D.C., 1971.

Krumbein, W. C., and F. J. Pettijohn, "Manual of Sedimentary Petrography," New York, Appleton-Century-Crofts, 1938.

Reineck, H. E., "Der Kostengreifer," *Natur und Museum,* **93**:102–108 (1963).

Riedl, R., "Quantitativ oekologische Methoden mariner Turbellarien-Forschung," *Oesterr. Zool. Z.,* **4**:108–145 (1953).

Riedl, R., and J. Ott, "The suction corer, a device to yield elastic potentials in coastal sediment layers," *Senckenb. Marit.,* **2**:67–84 (1971).

Sanders, H. L., "Marine benthic diversity: a comparative study," *Amer. Natural.,* **102**:243–282 (1968).

Shannon, C. E., and W. Weaver, "The Mathematical Theory of Communication," Urbana, University of Illinois Press, 1969.

Simpson, G. G., "Species diversity of North American recent mammals," *Sept. Zool.,* **13**:57–73 (1961).

Strickland, J. S. H., "Production of organic matter in the primary stages of marine food chains," in J. P. Riley and G. Skirrow, ed., "Chemical Oceanography," Vol. 1, New York, Academic Press, 1965.

Uhlig, G., "Eine einfache Methode zur Extraction der vagilen mesopsammalen Mikrofauna," *Helgol. wiss Meeresunters.,* **11**(3/4):178–185 (1964).

Vaccaro, R. F., "The response of natural microbial populations in seawater to organic environment," *Limnol. Oceanogr.,* **14**:726–735 (1969).

Walkley, A., "A critical examination of a rapid method for determining organic carbon in soils," *Soil Sci.,* **63**:251–264 (1947).

Wright, R. T., and J. E. Hobbie, "Use of glucose and acetate in aquatic ecosystems," *Ecol.,* **47**:447–464 (1966).

Zinn, D. J., "A brief consideration of the current terminology and sampling procedures used by investigators of marine interstitial fauna," *Trans. Amer. Micr. Soc.,* **87**:219–225 (1968).

q

QUANTIMET IMAGE ANALYSER

Quantimet Image Analysing Computers[1] are modular systems for deriving, at very high speeds, quantitative data—numerical, geometrical, and densitometric—from images produced by optical and electron microscopes, photographs, and other imaging systems. They have been shown to have many applications in biology and in the other life sciences where precise quantitative analysis of features in an image of a specimen is required.

Initial applications were, for technical reasons, confined to measurement of features with good optical, or

---

[1]Developed and manufactured by Image Analysing Computers Ltd. (Imanco), a subsidiary Company of Metals Research Ltd, of Melbourn, Royston, Hertfordshire, England.

gray level, contrast from other features and from the background of an image. However, the recent introduction of the Imanco Quantimet 720 (Fig. 1), using more advanced techniques of discrimination and computation, means that accurate measurements can now be made even where optical contrast is low and also where there is inadequate optical contrast, but only a difference in shape criteria.

The Quantimet 720 is shown schematically in Fig. 2. An input peripheral, typically an optical or electron microscope, an epidiascope, or a 16 mm or 35 mm projector, produces an image which is projected into the scanning system. The Quantimet 720 uses slow-speed vidicon and plumbicon scanners, which have been specially designed for precision image analysis of static images. The scan standard is 720 lines, with no interlace of successive scans, and the scanners are controlled by a

Fig. 1 Quantimet 720 Image Analysing Computer.

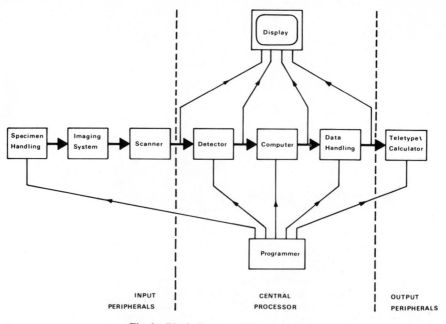

Fig. 2   Block diagram of Quantimet 720.

digital clock which eliminates the positional uncertainty due to analog drift, associated with conventional T.V. scanning systems, thus achieving not only extremely high, but also very uniform resolution.

The signal from the scanner is then detected above a variable threshold in a detector module and is then digitized to produce a digital representation in the time domain of the selected features in the image. All subsequent computation to produce the required data is digital, and the advanced logic used eliminates measurement errors due to features which are only partially in the field of view, counts features of complex shape accurately, and allows preselection of features by a number of criteria.

Results are shown on the display monitor screen, which also shows exactly which features are being measured. Alternatively, results can be printed, or punched out on paper tape for further processing by computer, or fed directly into a desk-top mini-computer, in which further computation of derived parameters, such as standard deviations, can be made—and this desk-top computer can further feed the data to a plotter to give a graphical representation of results.

The operation of a Quantimet 720 can be automatically controlled and directed by a programmer module, which automates the movement of the specimen and specifies the measurements performed on each field of view and the mode of data output.

The introduction of quantitative image analysis into biology has been aided by recent improvements in histochemical techniques, which can now explicitly select a particular cell or tissue component. Most current applications involve comparisons between test and control situations and the Quantimet 720 is enabling experiments to be carried out which, because of the magnitude of work involved, were, until recently, considered to be impossible, using nonautomated, manual methods.

As an example, a bio-assay of irritation of the respiratory tract has been evolved in the Department of Path-

ology at Huntingdon Research Centre, England, under Dr. Lionel Mawdesley-Thomas. This involves the evaluation of areas covered by PAS positive material in rat tracheal and bronchiolar epithelium. This test requires the evaluation of 2000 fields of view, each covering 300 μ of epithelium, from each of 10 animals per test and control group. Provided that all the groups are processed and stained simultaneously and measured by the same operator, using the Quantimet, it is possible to produce a dose-related response to almost any respiratory irritant, with sufficient sensitivity to obtain differences between cigarette smokes and 50 ppm increments of sulfur dioxide.

The features counted were PAS-positive droplets in goblet cells of rat bronchiolar epithelium (Fig. 3). Figure 4 shows the differences in absolute numbers when machine counts are compared with those of an experienced technician. With low overall counts, the instrument and human operator roughly parallel each other, but as the number of particles increases, the superiority of the

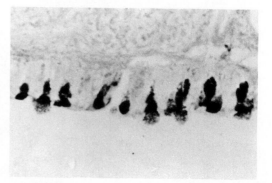

Fig. 3   Goblet cells in rat bronchiolar epithelium stained by collodial iron/PAS technique.

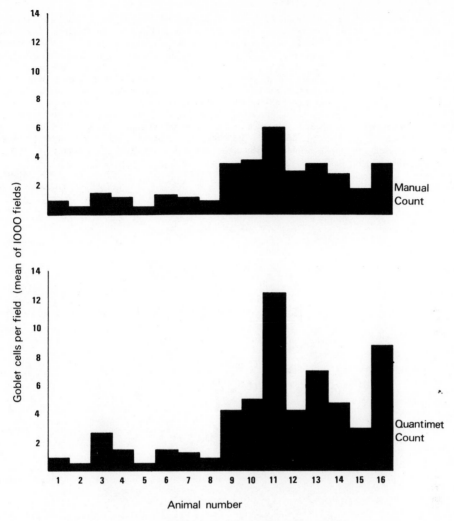

Fig. 4 Comparison of manual and machine counts.

machine becomes evident, producing results which are more accurate and much faster. Indeed, on high counts, the Quantimet proved to be approximately 25 times faster than the human operator and this with a maximum of only 14 cells per field of view.

As a second example, in the evaluation of nonspecific esterase activity in mouse skin sebaceous glands, it has been found that the most sensitive parameter to measure is the area covered by a reaction product at a set density threshold (Fig. 5). By evaluating sufficient material ($68 \times 10^6 \mu^2$) from groups of animals, the skins of which have been painted with various potentially carcinogenic agents, the Quantimet has enabled a screening test to be evolved to evaluate the potential activity of tobacco condensates, etc.

To date, applications have included measurements of specific features of the central nervous system, thyroid colloid, cell counts, pancreas islet volume, fat spaces in bone marrow, emphysema in lungs, carbon particles in liver, separate measurements of rods and cocci in one specimen, and many others.

Although the new Quantimet 720 has been able to

widen considerably the scope of image analysis applications, there are certain areas where inherently slow scanning microdensitometers are still used. Already, however, the Quantimet is being used for densitometric measure-

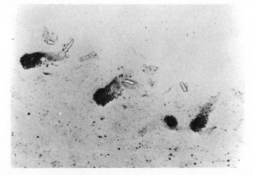

Fig. 5 Nonspecific esterase reaction products on mouse skin.

ments in quantitative histopathology, where its high speed is very valuable, and it will not be long before the Quantimet takes over even more rigorous densitometric tasks. A densitometric module for the Quantimet 720 is currently under development and promises to make accurate measurements, comparable to those obtained by computer controlled scanning densitometric systems, but at speeds nearly 100 times as fast, yet at a capital cost an order of magnitude less. Moreover, the recent introduction of a light integrating tube has opened a new field of applications in fluorescence microscopy, and with the addition of a monochromator it is possible to select a particular wavelength and thus only to analyze features of a specific color.

Quantitative image analysis techniques have been used in biology for only a relatively short time and the biologist himself is still reorienting specimen preparation techniques for this new technology. Quantimet systems have, however, already found numerous successful applications where their very high speed has made possible the collection of more data, more accurately, reproducibly, and rapidly than has hitherto been possible by traditional manual methods.

In short, the Quantimet 720 is opening new fields of study to the biologist due to its speed and greatly improving the accuracy with which a biologist can state his conclusions, based on a solid ground of quantitative data.

MICHAEL COLE
GEOFFREY L. STURGESS

r

RADIOAUTOGRAPHY

Introduction

Radioautography is a method used to identify the sites of localization of radioactive substances within biological specimens. This is done by placing the radioactive specimen in intimate contact with a photographic emulsion, exposing for a suitable time interval, and subsequently processing as in ordinary photography. The rays emitted by the isotope provide the energy necessary to transform the silver halide crystals of the emulsion into photolytic silver (latent image), which is then photographically developed into visible, black silver grains. Together, the grains, which overlie a radioactive site in the specimen, constitute the radioautographic image and give a permanent record of the localization of the radioactive substance.

The increasing availability of radioactive compounds opens new fields for investigation of living systems, thus extending the possibilities of application of radioautography. Accordingly, many variations of the radioautographic method have been developed to suit particular experimental situations. Furthermore, different types of nuclear track emulsions, which are specially sensitized to radiation, have become commercially available. These emulsions, which consist of silver halide crystals suspended in a gelatin matrix, vary in initial background fog, sensitivity, size of the silver halide crystals, and properties of the gelatin.

Radioautographs can be prepared for three main levels of resolution:

1. Macroscopic radioautographs, where large specimens, like paper chromatograms (1), frozen sections of whole animals (2), large sections of ground bone (3) are brought into close contact with X-ray emulsion. Because the resolution of these radioautographs is measured in millimeters, they can be viewed with the naked eye. The distribution of radioactivity between relatively large areas may thus be determined.

2. Light microscope radioautographs, which are examined at a cellular level with the light microscope. The resolution depends largely on the experimental system, i.e., section and emulsion thickness, distance between radioactive source and emulsion, type of emulsion and isotope (4). The resolution can be 5–10 μ for ^{32}P, 2–5 μ for ^{14}C and ^{35}S, 0.5–1 μ for ^{3}H and ^{125}I. Light microscope radioautographs can be prepared from a large variety of specimens: Histological sections of tissues embedded in paraffin or plastics, frozen histological sections, blood and bone marrow smears, or whole mounts of small histological structures (5), are most commonly used. The

principal liquid emulsions are Kodak NTB2, NTB3 (0.25 μ),[1] Ilford G5 (0.27 μ), K2, K5 (0.2 μ), and L4 (0.14 μ). The Kodak stripping film plates AR.10 (0.2 μ) are also widely used (6).

3. High-resolution electron microscope radioautographs, where the high resolution of the electron microscope is combined with very fine grain nuclear track emulsions. These radioautographs are analyzed on a subcellular level. The requirements for best results are very thin specimens (0.05–0.1 μ thick sections), isotopes of low maximum energy (^{3}H, ^{125}I), reproducible emulsion coats of uniform monolayers of tightly packed silver halide crystals with small diameter and high sensitivity. The problems of resolution and efficiency have been discussed by several authors (7–11). The available emulsions are Ilford L4 (0.14 μ), Gevaert NUC 307 (0.07 μ), Kodak NTE (0.05 μ), Sakura (Japan) NR-M1 (0.14 μ) NR-H1, NR-H2 (0.08 μ), Fuji (Japan) ER-29M (0.07 μ).

Methods for light microscope and electron microscope radioautography are described in detail in the following sections.

A Coating Technique for Light Microscope Radioautography (12)

This method is applicable to a large variety of specimens, provided that the labeled material is tightly bound to the components of the specimen and is insoluble in a wide range of solvents, such as the histological processing fluids and the liquid emulsion.

1. Histological Preparation of the Specimen: Tissue Sections. The experimental animals are injected with a predetermined dose of the radioactive precursor and sacrificed at appropriate times thereafter. In the case of paraffin embedding, tissues fixed in Bouin, Carnoy, alcohol etc. are sectioned at 2–4 μ and mounted on precleaned microscope slides with albumin fixative. In the case of plastic embedding, tissues fixed in OsO_4, glutaraldehyde or paraformaldehyde, are sectioned at 0.5–1 μ and mounted on clean slides. The preparations may be stained before or after radioautography. For *prestaining*, staining procedures must be selected which do not remove radioactive material from the specimen, do not produce chemical fogging, do not desensitize the emulsion, and are not lost during photographic processing. For instance, modified Harris hematoxylin, eosin, or the PAS technique are suitable for prestaining (13).

---

[1]Mean diameter of silver halide crystals is given in parentheses.

2. Emulsion Coating. The emulsions are stored in light-proof containers in a refrigerator at 4°C. Every batch of emulsion is tested before use to estimate the number of spontaneously developable silver grains (background fog). This is done by coating and processing clean slides as indicated below for ordinary radioautographs. A satisfactory batch of Kodak NTB2 emulsion has 1–3 initial background fog grains /1000 μ^2 counted with the light microscope.

The emulsion coating is carried out in an entirely light-proof darkroom maintained at atmospheric conditions of 28°C and about 80% relative humidity, at 3 ft from a safe light with filter Wratten Series 2 (14). The flask in which the emulsion is supplied is immersed in a water bath at 40°C. After 30 min the melted emulsion is poured slowly into a cylindrical staining jar (Borrel tube). The jar is filled with emulsion up to a level where the tissues on the slides will be covered during dipping, but the slide label will not. The jar is then covered and kept in the water bath for 30 min, in order to allow the air bubbles in the emulsion to disappear. Because air bubbles cause an irregular emulsion coat, their absence must be confirmed by dipping a clean glass slide and examining it under the safe light. If bubbles are found, they can either be scooped out with a porcelain spoon or allowed to disappear with time.

The specimen-bearing slides are held at the label end and are dipped vertically into the emulsion without tilting. Then the slides are withdrawn vertically from the emulsion, the excess emulsion is drained onto a tissue paper held against the lower portion of the slides, and the back of the slides is wiped clean. The emulsion coat is dried by standing the slides vertically on moist tissue paper on plastic racks. Under the atmospheric conditions of the darkroom the approximately 2 μ thick coat obtained with undiluted NTB2 emulsion dries within an hour. By slow drying at high humidity, stress artifacts in the form of grain accumulations are minimized. The safe light is turned off during the drying period. After dipping, the emulsion is solidified at 4°C and can be remelted several times.

3. Exposure. After drying, the coated slides are stored in lightproof, black plastic boxes. In the case of ³H labeled specimens, slides can be placed in consecutive slots in the box, but the distance between the slides must be increased with higher-energy isotopes. To minimize the *fading of the latent image*, which is believed to be caused by oxidation of the latent image by atmospheric oxygen in the presence of water vapor, the exposure box contains a tissue paper bag with 20 g of the desiccant Indicating Drierite. The box, sealed with adhesive tape, is stored in a refrigerator at 4°C. The optimal duration of exposure is determined by developing test slides at different time intervals. During prolonged exposure times the Drierite bag must be changed each month. No chemicals or radio-isotopes should be stored in the refrigerator where the radioautographs are exposing.

4. Photographic Processing. After a suitable exposure time, the preparations are processed in a darkroom maintained at 18°C. The photographic development must be of high efficiency, i.e., it must produce a large number of grains over the radioactive sites (intensity of the radioautographic reaction) as compared with the background fog. The developed silver grains must be mainly single and easily countable under the oil immersion objective of the microscope. Of a variety of processing procedures, the following has been adapted for routine use with Kodak NTB2 emulsion. The slides are placed horizontally in developing racks with the emulsion side facing upwards. Development is in freshly prepared Kodak D-170 developer (15, p. 62) for 6 min. After a 30 sec rinse in distilled water the preparations are fixed for 3 min in 24% sodium thiosulfate. The pH values of these solutions (pH 7.1, pH 6.7 respectively) allow better preservation of prestained radioautographs when stains like hematoxylin are used. The slides are then thoroughly washed in slowly running filtered tap water of 18°C for 10 min and rinsed in distilled water.

If sections have not been prestained, *poststaining* may now be done. Hematoxylin, eosin, toluidine blue, methyl green–pyronin, Giemsa, and many other methods are suitable (13, 16, 17). Poststaining involves the danger of altering the radioautograph; the emulsion could be displaced or the silver grains dissolved; the emulsion gelatin is often heavily stained, but can usually be destained by a fast dedifferentiation.

For mounting radioautographs of paraffin sections in a permanent form, the preparations are dehydrated after washing in 50%, 75%, 95%, and two changes of absolute alcohol, 2 min each. They are then placed in a cedar wood oil–absolute alcohol mixture (1:1) for 1 hr—a procedure which prevents or removes air lock artifacts. Then they are transferred to a Canada balsam–xylene mixture (1:1), for 1 hr and finally they are mounted in balsam under a coverslip.

Radioautographs of plastic sections are air-dried after washing. They may be poststained by covering them in horizontal position with a toluidine blue solution (1% in saturated borax) for 15–40 min. After washing and drying they are mounted with Permount. The arrangement of the final light microscope radioautograph is seen in Fig. 1.

The liquid emulsion technique can also be used to prepare radioautographs of material which has been *sim-*

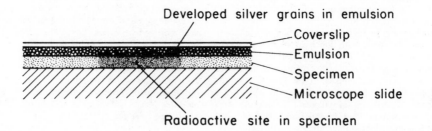

Fig. 1 Schematic diagram showing the arrangement of layers in a light microscope radioautograph: ○, unexposed silver halide crystals of the emulsion dissolved in the photographic fixer; ●, visible silver grains developed from exposed silver halide crystals.

ultaneously labeled with two isotopes of different energy ranges. In this case, a single, thick layer (30 μ) of Ilford G5 emulsion is applied over a specimen labeled with ^{3}H and ^{14}C. The majority of grains produced by ^{3}H would lie within a 1 μ distance from the source, while most of the grains caused by ^{14}C are found within 10 μ of the source. Another possibility is to coat the specimen with two different emulsions, whose different grain size and sensitivity help to distinguish between grains produced by the two isotopes (18, pp. 195–197).

Liquid emulsion methods, however, are not usable for *radioautography with soluble radioisotopes.* Special techniques are required in order to avoid extraction or displacement of the labeled substance during the preparation of the specimen and application of the emulsion. In these methods, freeze-dried specimens are brought in contact with dry emulsion (2, 19, 20, 21).

A Method for Electron Microscope Radioautography

1. **Preparation of Specimen.** Precleaned microscope slides (25 × 75 mm) with frosted ends are cleaned by wiping with tissue paper and lens paper. They are then celloidin-coated by dipping them vertically into freshly diluted 0.8% celloidin in isoamylacetate. The coated slides are dried vertically in a dustfree cabinet overnight. EM sections routinely cut at a thickness of silver to pale gold interference color are transferred from the boat of the microtome knife with a 3 mm diameter platinum wire loop to the celloidin-coated slides. Several groups of sections are placed within an area of 1.5 × 1.5 cm, which starts at about 2 cm from the nonfrosted end of the slides. The location of the groups of sections is marked on the back of the slides with a circle of masking tape. Care must be taken not to damage the celloidin coat while placing the sections and to immediately remove the excess of distilled water around the sections with a wedge of filter paper. Otherwise, the background fog might increase and the final stripping become more difficult. Sections may be prestained with uranyl acetate and Reynolds lead citrate, in which case they must be carbon-coated before radioautography.

2. **Emulsion Coating.** The emulsions are diluted on a volume basis with double-distilled water, placed in a water bath at 40°C, and well stirred. The dilution is 1:3 (emulsion:distilled water) for Ilford L4, 3:2 for Gevaert NUC 307, 1:1 for Kodak NTE and 2:3 for Sakura NH_2. The emulsion is then transferred to the water bath of the semiautomatic coating instrument (22; Fig. 2), where it is kept covered for 20 min at 32°C with the safelight turned off. Eventually air bubbles evaporate from the emulsion during the cooling period.

For coating, the specimen-bearing slides are placed in the clips of the slide holder of the coating instrument. (The instrument has two types of interchangeable slide holders, to dip two or five slides simultaneously). The slides are manually lowered into the emulsion and then mechanically withdrawn from it at the slow and constant speed of the instrument (64 mm/min). The speed of removal of slides can be precisely adjusted; this speed, combined with the dilution and temperature of the emulsion, determines the thickness of the emulsion coat. Thus, a densely packed monolayer of homogeneously distributed silver halide crystals can be achieved reproducibly for any type of emulsion. The drying of the emulsion coat is completed by keeping the slides in a vertical position on plastic racks.

3. **Exposure.** The radioautographs are exposed either in the presence of a drying agent as described for light microscope radioautography, or in special containers, where the air is replaced by an inert gas. The elimination of atmospheric oxygen is necessary for the very fine grain emulsions NTE and 307, which are more subject to latent image fading.

4. **Photographic Processing and Transfer to EM Grids.** After exposure, the radioautographs are processed vertically with the frosted end uppermost in the developing racks. The type of developer and the developing time and temperature determine the number, size, and shape of the visible silver grains (23). Thus, a strong reaction intensity is obtained with a 1 min development in the potent, undiluted Kodak D19b developer (15, p. 58) at 20°C. The silver grains are large and consist of heavily coiled silver filaments. If the D19b developer is diluted 1:5 or 1:10, the still numerous silver grains are smaller and the silver filaments thinner. For better resolution, a fine-grain development is preferable; however, this results in a decreased number of silver grains. In order to obtain a strong reaction intensity, the fine-grain development is combined with gold latensification (24), i.e., metallic gold is deposited on the latent images before development. The development is then arrested before the silver grains are fully developed, and only the development centers, which are located on the sites of the latent images, are visible.

After development, the preparations are placed for 30 sec in distilled water, fixed for 2 min in 24% sodium thiosulfate, and washed in five changes of distilled water, 1 min each. After the radioautographs have dried, a strip of the celloidin-emulsion complex, which carries the sections, is cut and floated off the glass onto the surface of distilled water. While the films are floating, Athen type EM grids (200–300 mesh) are placed with forceps on the sections. Moistened discs of filter paper (4.25 cm diameter) are placed on an aspirating device. The aspirator is then placed over the floating grids and the grids with the radioautographs are sucked up onto the filter paper. The latter is removed from the aspirator and the preparation is dried face up (Fig. 3). Poststaining may then be done with uranyl acetate (saturated in 50% alcohol) for approximately 5 min and Reynolds lead citrate for 20–40 min. Preparations must be well washed after staining.

Evaluation of Radioautographs

In order to assess radioautographs with the *light microscope,* the structural details of the specimen must be clearly distinguishable and the overlying silver grains must be precisely identifiable. The radioautographs are evaluated either *qualitatively* by studying the distribution of silver grains over the tissue sections or *quantitatively* by grain counting. The grain counts are proportional to the radioactivity in the specimen, provided there has been no loss of radioactivity, no desensitization of the emulsion, and no artifactually produced silver grains. In visual grain counting, the individual silver grains are counted directly under the oil immersion objective of the microscope, under bright or dark field conditions. The area to be counted is subdivided by an ocular grid. Visual grain counting is facilitated by projecting the field of view onto a screen, by the use of closed-circuit television, or by counting grains on microphotographic prints. Grain counting can also be automated by means of photometric estimation of grain densities (18, pp. 144–160).

Electron microscope radioautographs are assessed by

Fig. 2 The semiautomatic coating instrument for radioautography: 1, motor box; 2, pulleys to adjust withdrawal speed of slides; 3, motor switch to withdraw the slides automatically from the emulsion; 4, emulsion jars; 5, thermostatically controlled water bath; 6, drain for water bath; 7, temperature adjustment for water bath; 8, thermostat switch for water bath; 9, plexiglas frame to hold emulsion jars in place; 10, interchangeable slide holder; 11, manually operated knob to dip the slides into the emulsion.

analyzing the distribution of developed silver grains over subcellular structures in the electron microscope or on electron microscope prints. With present radioautographic methods the resolution of the radioautographic image (silver grains) is relatively poor as compared to the resolving power of the electron microscope. The interpretation of electron microscope radioautographs is based on various considerations concerning the relation between the location of the silver grain and the source of radioactivity which caused it (7, 9, 10, 11). The object is to assign each silver grain to a structure visible in the electron microscope or on the print. Apart from the difficulty that silver grains can conceal underlying structural details, two problems are encountered. The first is

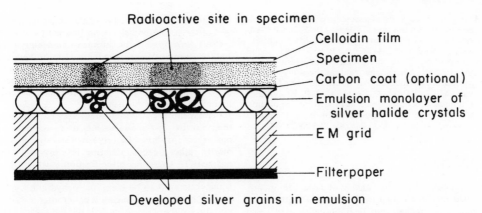

Radioactive site in specimen

Celloidin film

Specimen

Carbon coat (optional)

Emulsion monolayer of silver halide crystals

E M grid

Filterpaper

Developed silver grains in emulsion

Fig. 3 Schematic representation depicting the sequence of layers in an electron microscope radioautograph.

to establish boundary limits within which would lie the source of radioactivity responsible for each grain. In practice a circle calculated to correspond to a designated probability of enclosing the source is drawn about each silver grain in the print (10, 25, 26). The second problem arises if more than one structure is visible within the circle and it becomes necessary to decide which structure is the most probable source of radioactivity. For this purpose statistical methods are available to analyze the results and deduce the probable distribution of radioactive label among the visible structures (25, 26).

Both light microscope and electron microscope radio-autography is widely used in studying the synthesis and metabolism of biochemical compounds in living cells and in analyzing the kinetic behavior of cell populations.

BEATRIX M. KOPRIWA

References

1. J. Chamberlain, A. Hughes, A. W. Rogers, and G. H. Thomas, "An evaluation of the available techniques for the autoradiography of tritium in chromatograms," *Nature,* **201**:774–775 (1964).
2. S. Ullberg, "Autoradiographic localization in the tissues of drugs and metabolites," *Biochem. Pharmacol.,* **9**:29–38 (1962).
3. C. P. Leblond, G. W. Wilkinson, L. F. Belanger, and J. Robichon, "Radio-autographic visualization of bone formation in the rat," *Amer. J. Anat.,* **86**:289–341 (1950).
4. J. Gross, H. R. Bogoroch, N. J. Nadler, and C. P. Leblond, "The theory and methods of the radio-autographic localization of radioelements in tissues," *Amer. J. Roentgenology and Radium Therapy,* **65**:420–458 (1951).
5. C. Huckins and B. M. Kopriwa, "A technique for the radioautography of germ cells in whole mounts of seminiferous tubules," *J. Histochem. Cytochem.,* **17**:848–851 (1969).
6. S. R. Pelc, "The stripping-film technique of auto-radiography," *Int. J. Appl. Radiation and Isotopes,* **1**:172–177 (1956).
7. L. G.. Caro, "High resolution autoradiography," *J. Cell Biol.,* **15**:189–199 (1962).
8. S. R. Pelc, "Theory of electron autoradiography," *J. Roy. Microsc. Soc.,* **81**:131–139 (1963).
9. P. Granboulan, "Comparison of emulsions and techniques in electron microscope radioautography," pp. 43–63 in "The Use of Radioautography in Investigating Protein Synthesis. Symposium of the International Society of Cell Biology," Vol. 4,

C. P. Leblond and K. B. Warren, ed., New York, Academic Press Inc., 1965.
10. M. M. Salpeter, L. Bachmann, and E. E. Salpeter, "Resolution in electron microscope radioautography," *J. Cell Biol.,* **41**:1–20 (1969).
11. K. Uchida and V. Mizuhira, "Electron microscope autoradiography with special reference to the problem of resolution," *Arch. Histol. Jap.,* **31**:291–320 (1970).
12. B. M. Kopriwa and C. P. Leblond, "Improvements in the coating technique of radioautography," *J. Histochem. Cytochem.,* **10**:269–284 (1962).
13. C. P. Leblond, B. Kopriwa, and B. Messier, "Radio-autography as a histochemical tool," in "First International Congress of Histochemistry and Cytochemistry Paris, "London, Pergamon Press, 1963 (pp. 1–31).
14. B. M. Kopriwa, "A model dark room unit for radio-autography," *J. Histochem. Cytochem.,* **11**:553–555 (1963).
15. K. M. Hornsby, "Basic Photographic Chemistry," The Fountain Press in conjunction with the British Journal of Photography, 1958.
16. J. M. Thurston and D. L. Joftes, "Stains compatible with dipping radioautography," *Stain Technol.,* **38**:231–235 (1963).
17. W. Sawicki and J. Rowinski, "Periodic acid-Schiff reaction combined with quantitative autoradiography of ³H-thymidine or ³⁵S-sulfate-labeled epithelial cells of colon," *Histochemie,* **19**:288–294 (1969).
18. A. W. Rogers, "Techniques of Autoradiography," Amsterdam, Elsevier Publishing Company 1967.
19. O. L. Miller, G. E. Stone, and D. M. Prescott, "Autoradiography of soluble materials," *J. Cell Biol.,* **23**:654–658 (1964).
20. W. E. Stumpf and L. J. Roth, "High resolution autoradiography with dry mounted, freeze dried frozen sections," *J. Histochem. Cytochem.,* **14**:274–287 (1966).
21. T. C. Appleton, "The application of autoradiography to the study of soluble compounds," *Acta Histochem.,* Suppl. 8, 115–133 (1968).
22. B. M. Kopriwa, "A semiautomatic instrument for the radioautographic coating technique," *J. Histochem. Cytochem.,* **14**:923–928 (1967).
23. B. M. Kopriwa, "The influence of development on the number and appearance of silver grains in electron microscope radioautography," *J. Histochem. Cytochem.,* **15**:501–515 (1967).
24. M. M. Salpeter and L. Bachmann, "Autoradiography with the electron microscope," *J. Cell. Biol.,* **22**:469–477 (1964).

25. M. A. Williams, "The assessment of electron microscopic autoradiographs," pp. 219–272 in "Advances in Optical and Electron Microscopy," Vol. 3, New York, Academic Press, 1969.

26. N. J. Nadler, "The interpretation of grain counts in electron microscope radioautography in the appendix by Haddad, A., Smith, M. D., Herscovics, A., Nadler, N. J., Leblond, C. P. Radioautographic study of in vivo and in vitro incorporation of ³H-fucose into thyroglobulin by rat thyroid follicular cells," *J. Cell Biol.* **49**:856–882 (1971).

RADIOLARIA

Introduction. Acantharia and Radiolaria are mainly planktonic and strictly marine Protozoa. Together with Heliozoa, they form the group the Actinopods, all of which are characterized by possessing, among an important layer of jelly, radiated pseudopods, thin and straight, some of them being supple (filopods), or rigid (axopods). The skeleton is composed of either amorphous silica (Radiolaria) or of strontium sulfate (Acantharia). The more primitive forms have spines dispersed in their cytoplasm, simple or branched, but the majority have their skeletons composed of one or more latticed concentric shells. Others have either 10 diametral or 20 radial spines, showing a definite orientation according to Muller's law. The constitution and the molecular arrangement of these spines has been elucidated by the use of the polarizing microscope, spectrography, and microdiffraction with X-rays. The polarizing microscope demonstrates that the spines have a crystalline structure, that the crystal axes either of the vegetative skeleton or of the cystic membrane have definite orientations. X-ray spectrography indicates that the skeleton is composed of strontium sulfate and microdiffraction shows that each spine is a monocrystal.

The cytoplasm is differentiated into two regions: first the ectoplasm, which is cortical and mucilaginous, constituting what is called "the calymma," and from where arise the filopods and the plasmatic coating of the axopods (rheoplasm); second the endoplasm, which contains the nucleus, the different organelles, and the axoplast. In Radiolaria alone the endoplasm and the ectoplasm are separated from each other by a particularly thick membrane, the "capsular membrane." This differentiates the cortex of the cell and is composed of skeletal plates made of glycoproteinic substance lying under the cell membrane. It is pierced by a varying number of pores, "the fusules," which allow passage to the axes of the axopods. The fusules can be numerous and uniformly distributed on the surface of the capsular membrane (*Spumellaria*) or gathered at a single pole (*Nassellaria*); only three of them are observed in *Phaeodaria* (a tubular main opening called "astropyle," and two smaller ones, the parapyles). The axoplast is usually located in the center of the central capsule (*Acantharia*, *Sphaerellaria Centroaxoplastidida* and *Nassellaria Apoaxoplastidida*); but is occasionally transported to the side, and there lodged in a cavity of the nucleus (*Sphaerellaria Periaxoplastidida*, *Nassellaria Proaxoplastidida*). The axoplast is made up of a microfibrillar substance; the microfibrils are organized to make the wall of microtubules, which themselves form parallel rows constituting prismatic meshes, or spiral palisades which are at the origin of the stereoplasmic rods of the axopods.

Techniques. Observations on living Protozoa permit only the study of peripheral regions of the skeleton and of the cytoplasm—the structure of the calymma, the evolution of the pseudopodial system (filopods, axopods), the cyclosis of rheoplasm, the capture of alimentary particles, and the rejection of waste products. The whole endocapsular region is generally very opaque and its study can be made only after it has been cleared or sectioned. The destruction of the organic substance makes the observation of the skeleton much easier.

Permanent preparations are obtained by mounting these protists in Canada balsam or in synthetic resins; but, except for those of a very small size, the cells remain mostly rather opaque. Clearing in clove oil is recommended. This requires prior fixation (such as Bouin in most cases, or in the case of the Acantharia, 95% alcohol if the skeleton is to be kept) and dehydration. Observation of the various organelles is easier after a short staining with eosin or light-green. The refraction index of clove oil makes both the opal of radiolarian skeletons and the strontium of acantharian spines extremely transparent. l-bromonaphtalene can be substituted for clove oil when the delicate features of the spines or of the shells have to be studied; the very high refraction index of this reagent makes the skeleton much more conspicuous while the cytoplasm remains clear. These methods for clearing specimens in a liquid medium have the additional advantage that the protists may be oriented while being observed. Radiolaria so treated can be used later either for light microscopy sections or for the production of clean-skeleton preparations.

Thin sections are prepared after paraffin or resin (Epon) embeddings. They are often scratched or torn by the spines, mainly when these are made of silica. Striae when produced by small or scattered spines are not too inconvenient, but it is impossible to secure useful preparations from forms with compact skeletons.

The removal of Acantharia spines is a minor problem requiring no more than a light acid treatment, during or after fixation. In the case of Radiolaria, only a fluoride ion can be used, but hydrofluoric acid is unsuitable when good preservation of the cytoplasmic structures is required. However, a saturated solution of ammonium hydrogen fluoride (NH_4HF_2), mixed with an equal quantity of 95% alcohol, can be recommended unless fixation and dehydration are unduly prolonged. The time required to dissolve the skeleton varies, according to the importance or the thickness of the siliceous elements, from 30 min to several hours and must be controlled under a microscope. As soon as possible protists are rinsed several times in 95% alcohol.

This technique can also be applied to electron microscopy, the fixative being Karnowsky's method slightly modified: a prefixation with glutaraldehyde and paraformaldehyde followed by osmic acid—but the ionic concentration and the pH of the phosphate buffer as well as of the fixative and the rinsing media, must be adapted to those of seawater (monobasic and dibasic sodium phosphate about 0.4 M; pH 7.4). The organic sheath and the pellicle which surrounds the spines are of course preserved. These are not very permeable to embedding materials, so after being passed through propylene oxide, the embedding resin must be by slow stages. For this reason, the embedding media have to be very fluid, with relatively small molecules such as those which are proposed by Spur.

The preparation of radiolarian skeletons offers no difficulty. The protists are plunged in a solution of potassium dichromate mixed with concentrated sulphuric acid. This method presents one inconvenience: The mixed solutions have the same refractive inlet as the opal, so skeletons are not conspicuous after a good distilled water rinsing. Skeletons so prepared can be used, after washing in distilled water, either for light microscopy or for scanning electron microscopy; the latter apparatus permits great field depth and thus the opportunity to secure three-dimensional pictures. For optical microscopy skeletons can be studied in any suitable mounting medium; but for electron microscopy the preparations must be dried. Forms with spines unusually thin or long must be freeze-dried; otherwise surface tension effects during liquid-water evaporation could cause distortion. It is not necessary to use special equipment: The specimens, surrounded by a microdrop of water, may be kept in an atmosphere dehydrated by sulfuric acid or by phosphorus pentoxide, a few degrees above the melting point. When the specimens are sufficiently dry (about 24 hr), they are placed in a vacuum evaporator and coated with gold or gold : palladium.

Similar preparations of Acantharia are more difficult. Strontium sulfate is not soluble in the potassium dichromate and sulfuric acid mixture; however, the spines and the plates of the cystes, which are united only by an organic cement, become dissociated. This does not occur with the order Symphyacantha, the different species of which have all their spines strongly united in the center. Moreover the elimination of the acid solution should be very fast to prevent the disintegration of the spines through the action of the wash water: Freeze-drying has to be used as soon as possible. If the different elements of the acantharian skeleton have to be kept in a good order, calcination is required; specimens freshly taken from plankton samples, without any previous fixation, are plunged into distilled water and rinsed quickly to avoid salt precipitations. They are then passed through alcohol and dried, if the nature of their skeleton is not too delicate, or rapidly freeze-dried. The calcination of the specimen requires a quartz vessel, since glass makes fusible combinations with strontium. In any case the temperature should not exceed 600°C (dark red) to avoid combinations with the quartz itself. This technique is adequate for the light microscopy; but the ultrastructures of the cell surface are somewhat hidden by cytoplasmic remains when observed in the scanning electron microscope. In this case, this technique may be improved by a previous treatment using proteolytic enzymes on living Protozoa. These enzymes should belong to the trypsin family to act in a medium adjusted to an alkaline condition with ammonium acetate. The preparations must also be washed in ammonium acetate solution, this salt offering the advantage of being completely evaporated during the course of the calcination.

JEAN CACHON
MONIQUE CACHON

References

Cachon, J., and M. Cachon, "Le système axopodial des Nassellaires. Origine, organisation et rapports avec les autres organites cellulaires. Considérations générales sur l'organisation macromoléculaire du stéréoplasme chez les Actinopodes, *Arch. f. Protist.,* **113**(1):80–97 (1971).
Cachon, J., and M. Cachon, "Acantharia and radiolaria," in "Illustrated Guide for Protozoa,"
Cachon, J., M. Cachon, "Les modalités du dépôt de la Silice chez les Radiolaires," *Arch. f. Protist.,* **114**:1–13 (1972).
Hollande, A., and M. Cachon-Enjumet, "Cytologie, évolution et systématique des Sphaeroïdés (Radiolaires)," *Arch. Mus. Paris,* **VII**(7):1–134 (1960).
Hollande, A., J. Cachon, and M. Cachon, "Les modalités de l'enkystement présporogénétique chez les Acanthaires," *Protistologica,* **I**(2):91–104 (1966).
Hollande, A., J. Cachon, and M. Cachon, "La signification de la membrane capsulaire des Radiolaires, et ses rapports avec le plasmalemme et les membranes du reticulum endoplasmique. Affinités entre Radiolaires, Héliozoaires et Péridiniens," *Protistologica,* **VI**(3):311–318 (1971).

RECONSTRUCTION

Reconstruction from serial sections is a means of studying the anatomy of structures that are too small for normal dissection. Only two methods—graphical and model reconstruction—are widely used, though other methods have been devised for specialized cases. Graphical reconstructions are best used for tracing the shape and course of simple structures such as muscle blocks, nerves, and blood vessels, whereas model reconstructions are preferable for more complicated structures such as toothgerms and ear ossicles.

In all reconstructions correct alignment of the sections is essential, and this is usually achieved by choosing a natural reference point in the material. This reference point, which is often referred to as a guideline, must appear in every section and have a known course through the original wax block. For example, for a dorsal-view graphical reconstruction of a transversely sectioned whole animal, a suitable reference point would be the midline of the body. It is easily recognized in each section and its course through the wax block should be known. For a side-view reconstruction of the same animal the central nerve cord or a major longitudinal blood vessel or a combination of two such structures could be used. For model reconstructions two reference points in planes at right angles to each other are needed, unless the shape of the model is known beforehand.

To make a graphical reconstruction, the course of the chosen reference point or guideline is drawn to scale on graph paper and the section numbers marked off so that each division of the graph paper represents a set number of sections (if the sections are 10 μ thick a convenient scale is 1:100 so that each section is represented by one millimeter division on centimeter squared graph paper). Then by means of a microprojector or camera lucida the image of each section in turn is projected at the correct magnification onto the graph paper, and the reference point on the paper is aligned with the reference point in the material. The outer limits of the structures to be reconstructed are then marked on the line assigned to that section (Fig. 1). In this way a dotted outline is built up which can be completed by drawing a smooth line through the greatest number of points. This "smoothing" eliminates the irregularities caused by uneven shrinkage during fixation and uneven stretching of the wax ribbon during mounting.

Model reconstructions can be made from sheets of any suitable material, such as thick paper or cardboard, but dental wax is recommended, particularly Ash's grade 5 wax and Ash's silicone-toughened wax. In wax-model reconstruction the outline of the structure in each section

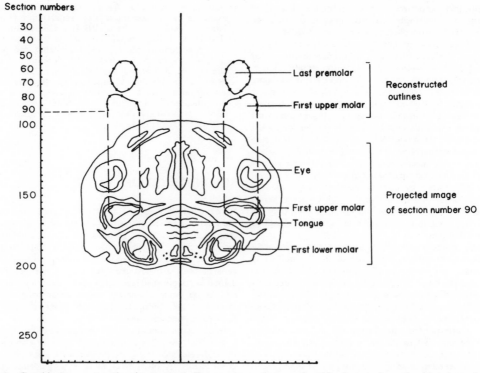

Fig. 1 Graphical reconstruction in progress. Transverse section number 94 of an embryonic chiropteran head is projected onto graph paper and oriented so that the midline coincides with the midline guideline. The outlines of the upper toothgerms are then marked onto the appropriate line on the graph paper.

or every second or nth section is traced onto a wax sheet with the point of a needle. These wax outlines are then cut out with a scalpel and assembled. Ash's grade 5 wax is sold in sheets that are 1.2 mm thick, so that if the material has been sectioned at 10 μ a magnification of 60× will enable each wax cutout to represent two sections. If the shape of the finished model is known, the wax cutouts can be assembled by eye, but if it is not known, aids for correct alignment are needed. One such aid is to make dorsal-view and side-view graphical reconstructions of the model outlines on the same scale. The model can then be built up within these outlines. A second method is to include in the wax cutout extra strips which extend to reference points in the material. For example, in wax models of toothgerms one can include a strip of wax extending to and along the midline of the head (or to a line parallel to the midline), and another strip extending to the mandibular bone (Fig. 2). The midline strips should be aligned in a straight row (if the block was cut exactly transversely) and the mandibular bone outline should be aligned to coincide with a side-view graphical reconstruction of the mandible. When assembled the wax cutouts can be fused by running a hot scalpel blade between them, after which the superfluous guide strips can be carved away. To make the model somewhat more durable it can be painted with three coats of vinyl plastic (Vinalek 5909, Vinyl Products Ltd.).

If the material to be sectioned contains no natural reference point then an artificial one may be introduced by embedding the material alongside a plate of stained tissue or elder pith. Sections are then cut at right angles

to this plate and its face seen in each section provides the reference point for alignment. Alternatively the side or sides of the wax block can be used as a reference point. This is made easier if the surroundings of the wax ribbon are painted before the wax is removed. However, if this method is used the block should be trimmed very close to the material and have straight sides, the cutting should be exactly at right angles to the block, and the sections should be cut at a thickness of not less than 13 μ, otherwise errors due to uneven expansion of the wax become too great. As a last resort, if rare material has already been sectioned thinly and has no apparent natural or artificial reference point, it may still be possible to find a reference point using the edges of the wax block, traces of which usually remain as faint broken lines of stain or dust particles in at least some sections. These traces can be emphasized by marking them with a fine felt pen. By using only those sections in which it is visible, the edge of the wax block (or corners) can be used as a reference point to reconstruct any convenient large smooth structure in the material. This first outline will be very jagged due to the uneven expansion of the wax in the thin sections, but since only smooth structures are involved the outline is easily smoothed over. This initial reconstruction can then be drawn on fresh graph paper and can itself be used as a reference point for reconstructing other organs in the material. In most cases it will be found more convenient to make the initial reconstruction on a small scale (e.g., 10× or 20×) using approximately every 10th or 20th section. This can then be enlarged for the final reconstruction.

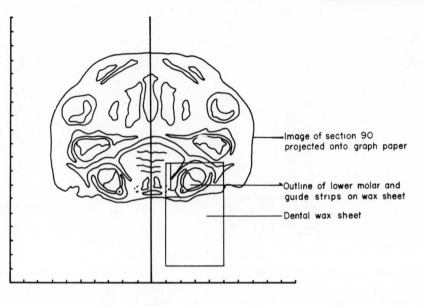

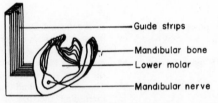

Fig. 2 Upper: Wax model reconstruction in progress. The outline of the toothgerm and guide strips in every *n*th section are traced onto dental wax. Lower: The wax cutouts are assembled.

Reconstructions are best made from exactly transversely cut material. However, if material has been sectioned obliquely, accurate reconstructions can be made, provided the orientation of the material, and hence the true course of the reference point, in the original wax block is known. Where the material has not been cut exactly transversely and there are no artificial guidelines with known courses, inaccurate alignment of sections will result. Obliquity of cut in the transverse plane itself (causing the material to roll to one side) will not affect dorsal-view or side-view reconstructions as this error is automatically eliminated during alignment of the projected image. Obliquity of cut in the lateral plane (where the material veers to right or left through the block) or in the sagittal plane (where the material dips to front or rear) will, however, cause "slewing" of the sections during alignment and foreshortening of structures, so that shapes and measurements will be inaccurate.

In bilaterally symmetrical material the error due to obliquity of cut in the lateral plane can be reduced as it is possible to find the *approximate* course of the material through the original wax block. This can be done by preparing a simple, dorsal-view graphical reconstruction of symmetrical structures, using the midline of the graph paper to represent the midline of the animal. A "slewed" reconstruction will result with the structures appearing asymetrically about the midline (Fig. 3 [left]). If points known to be symmetrically opposite each other in the material are then joined, the perpendicular to this line will give the approximate orientation of the midline in the original wax block at that point. The angle between this line and the middle of the graph paper approximates the angle between the true midline of the material and the midline of the original wax block (Fig. 3 [right]).[1] A midline guideline can then be drawn on fresh graph paper and a more accurate reconstruction made.

The degree of obliquity of cut in the sagittal plane is more difficult to assess unless there are structures known to be directly opposite each other above and below the reference point used. These can then be reconstructed as the bilaterally opposing structures above and the approximate orientation of the reference point in the wax block can be found. Alternatively a side-view reconstruction of the gross smooth structures or even of the reference point itself can be made, using the traces of the edge of the wax block as a reference point. This reconstruction will give the true orientation of the whole animal in the wax block

---

[1] From this approximate orientation it is geometrically possible to calculate the exact orientation of the midline as the error is constant for any given angle. However, the error is small, being less than 10%, which means less than 0.4° for material deviating up to 4° from the longitudinal axis of the wax block.

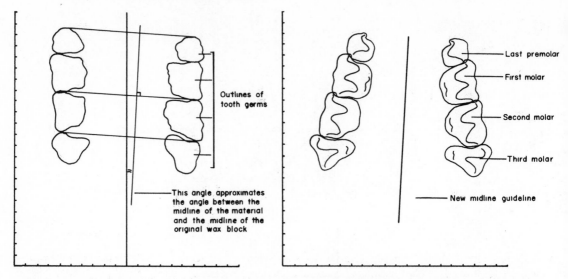

Outlines of tooth germs

This angle approximates the angle between the midline of the material and the midline of the original wax block

Last premolar

First molar

Second molar

Third molar

New midline guideline

Fig. 3 Left: A dorsal-view reconstruction of obliquely cut toothgerms, using the midline of the graph paper as the midline guideline. The toothgerms appear misshapen. Lines are drawn joining structures known to be symmetrically opposed. The perpendicular to these lines gives the approximate orientation of the midline of the material in the original wax block. Right: A new reconstruction is made using the midline guideline found in the figure at left. The orientation of this midline is accurate within less than 10% of the true angle of obliquity.

provided that the edges of the wax block were parallel. The true course of the reference point can then be drawn on fresh graph paper.

Correction for obliquity of cut must be made in assembling model reconstructions.

P. M. FOGDEN

REFRACTIVE INDEX DETERMINATION

The measurement of light refraction indices of microscopic particles of solids is based on the possibility of an accurate determination of the equality of refractive indices of a solid and a liquid in an immersion preparation under a microscope. This determination is always indirect and includes on one hand the attained balancing of the refractive index of the measured material and the immersion liquid, on the other hand the measurement of the refractive index of the same liquid under the same conditions under which balancing has been attained.

On the interface of a grain of the solid and the surrounding liquid occurs the refraction of light rays which makes even colorless and transparent grains visible under the microscope. Only with a mutual equality of the refractive indices does the colorless grain in the viewing field of the microscope disappear, and the colored grain is distinguishable only due to a different light absorption. For a more accurate determination of the moment of balancing of the refractive indices, the contrast in the viewing field of the microscope is increased by a suitable method of illumination or with the aid of special facilities for enhancing the contrast. Arranged in the order of sensitivity, the following four methods can be used: oblique illumination, Becke lines, double screening, and phase contrast (Figs. 1–4). The Becke line is formed by the refraction of light on the grain periphery, and when the microscope tube is raised, it is shifted into an optically more dense medium. The contrast with oblique illumination and double screening is obtained by circling out the

rays deflected by refraction, and phase contrast utilizes phase differences of the light wave in media of a different optical density. For colorless objects the phase contrast method is the most sensitive one, and for colored objects the employment of the Becke line is most advantageous. The sensitivity of all the above methods decreases with an increasing aperture of the objective and is in practical cases influenced considerably by the shape and size of the measured grains, by the number of inclusions, the degree of transformation, etc. When using phase contrast, the grain possesses contrast with respect to its surroundings on the entire cross section. This can be utilized for the photoelectric determination of the equality of refractive indices which removes the subjective factor and is therefore most sensitive. It is based on the microphotometric comparison of the light intensity of the grain and its immediate surroundings.

Accurate Determination of Refractive Indices of Optically Anisotropic Materials. This requires facilities for adjusting the space orientation of the measured grain in the immersion liquid or the adjustment of the entire preparation with a stationary grain. For this purpose are used spindle stages in which the grain is cemented onto the tip of a rotary needle or a glass fiber oriented perpendicularly to the optical axis of the microscope. This permits the rotation of the grain around two mutually perpendicular axes. Spindle stages are of significance especially for the measurement of refractive indices of perfectly cleavable materials, in which it is not possible to orient suitably all the optical directions in the universal Fjodorov stage. In this stage is carried out the orientation of the entire preparation with stationary grains.

Balancing of the Refractive Index of the Measured Grain and the Immersion. This is attained by mixing two fluids directly on the underlying glass under the microscope after the previous application of a set of fluids with known refractive indices. Substantially faster and more

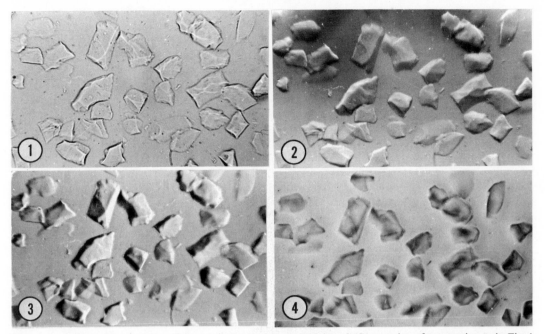

Figs. 1–4 1: Becke line on glass fragment in immersion preparation. 2: Same point of preparation as in Fig. 1, oblique illumination. 3: Same preparation as in Fig. 1, double screening. 4: Same point of preparation as in Fig. 1, phase contrast.

accurate are variance methods which utilize differences in light and thermal dispersion of solids and fluids. With the thermal variation method (Gaubert, 1922) the preparation is heated in which the measured material is immersed into a fluid with a somewhat higher refractive index. The refractive index of the solid is determined by measuring the refractive index of the same liquid at the same wavelength of illumination and temperature at which the balance of indices was obtained under the microscope. Though the change of the refractive index of a solid within the temperature range of the determination is very small, it has to be taken into account in order to obtain accurate results and should be determined by measurements in different liquids at different balancing temperatures.

The chromatic variation method (Tsuboy, 1923) attains the balance by a continuous change of the wavelength of light and utilizes the differences in light dispersion of liquids and solids. The light dispersion of solids is, however, not negligible, and it should therefore be determined by measuring refractive indices for several wavelengths of light and the refractive index for a certain wave length should be determined from the dispersion curve.

The common disadvantage of the two previous methods, i.e., the necessity of changing the immersion liquid several times during each determination, has been overcome by the dual variation method (Emmons, 1928). By a combination of both the change of the preparation temperature and the change of the wavelength of light it is possible to balance the refractive indices of the measured material and the fluid several times, and thus the dispersion curve of the measured material can be attained during a single determination. A further advantage of this method is the possibility of employing monochromatic filters instead of a monochromator, and since in the Hartman dispersion network the dispersion curve has the shape of a straight

line, it is sufficient for the plotting of the latter to achieve the balance for two or at most for three wavelengths of light (Fig. 5). The possibility of obtaining dispersion curves without handling the preparation is of considerable significance especially in the measurement of optically anisotropic materials in the universal Fjodorov stage, which permits the determination of two or three refractive indices of the same grain.

The Measurement of Refractive Indices of Liquids. This is carried out with the aid of refractometers, by the prismatic method on the single-circle goniometer or with a somewhat lower accuracy by a polarization interferometer on the microscope. Without the use of those instruments it is possible to determine the refractive index of liquids by comparison with glass standards directly in the immersion preparation, but the accuracy of this method, as well as the utilization of the universal Fjodorov stage as a refractometer directly under the miroscope, is substantially lower. The range of refractive index values that can be

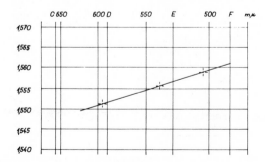

Fig. 5 Curve of light dispersion of measured material in the Hartman dispersion network.

determined on refractometers is limited by the power of the glass employed for their manufacture and by their sensitivity towards certain liquids. By suitable adaptations of prismatic stages it is possible to overcome those drawbacks of the prismatic method and with the application of electric heating this method can be employed to measure also the refractive indices of highly refracting melts.

Immersion Liquids. Organic compounds with a high boiling point are used; they can be mutually mixed and do not react either with the measured material or with one another. For the measurement of highly refracting materials, solutions of yellow phosphorus and sulfur in methylene iodide and low-melting mixes of piperine with arsenic and antimony iodides and mixes of thalium halogenides are employed.

For the variation methods, the immersion liquids must have a high light and thermal dispersion, and the use of the mixes should be limited to a minimum. Within the temperature range employed for the determinations, the dispersions of the set of fluids should cover continuously the entire field between the marginal members. An example is a set of immersion fluids proposed by R. C. Emmons (1943) (Table A1).

Instrumentation. Variation methods require certain special instruments and adaptations of the microscope accessories. For the chromatic variation method a sufficiently strong monochromator is needed, and for the thermal and dual variation method a heating chamber for the microscope stage and for the universal Fjodorov stage. Several water-tempered chambers have been proposed, e.g., the Hipple chamber and the Emmons chamber for the universal stage (Figs. 6 and 7). Also the refractometer or the prismatic stage must be tempered within the same temperature range, with facilities for the exact reproduction of the chamber temperature. With the last two methods the monochromator can be replaced by monochromatic discharge tubes and filters. The previously employed method of tempering with circulating water has been replaced more recently by electric heating, and the use of thermistors has facilitated the construction of a special "underlying glass" (Fig. 8, 1) which has the

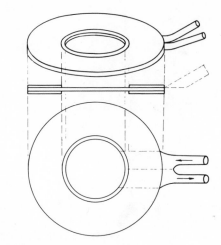

Fig. 6 Water-tempered Hipple chamber (Emmons, 1943).

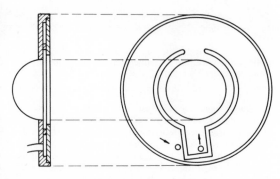

Fig. 7 Water-tempered Emmons chamber for universal Fjodorov stage (Emmons, 1929).

Table 1   Set of Immersion Fluids for the Dual Variation Method (Emmons 1943)

	b.p. (°C)	N_c 10°	N_c 50°	N_F 10°	N_F 50°	dn/dt	$N_F - N_c$
00. ethyldiiodinearsene		1.808	1.777	1.834[a]	1.800[a]	0.00081	0.049$\pm$
0. methylene iodide + sulfur		1.775[b]					
1. methylene iodide	180	1.737	1.711	1.774	1.747	0.00068	0.0369
2. α-iodinenaphtalene	305	1.698	1.678	1.734	1.714	0.00047	0.0368
3. α-iodinenaphtalene + α-bromonaphtalene		1.675	1.652	1.711	1.689	0.00055	0.0365
4. o-bromoiodinebenzene	257	1.660	1.640	1.687	1.665	0.00052	0.0261
5. phenylisothiocyanate	220	1.646	1.623	1.682	1.658	0.00058	0.0353
6. α-chloronaphtalene	259	1.629	1.611	1.660	1.640	0.00047	0.0300
7. iodobenzene	188	1.618	1.596	1.644	1.620	0.00055	0.0247
8. α-naphthylethylether	276	1.599	1.579	1.628	1.607	0.00050	0.0290
9. glycerotribromohydrine	220	1.583	1.560	1.608	1.586	0.00055	0.0250
10. m-chlorobromobenzene	196 $\pm$	1.570	1.550	1.590	1.568	0.00055	0.0188
11. o-bromotoluene	181	1.555	1.534	1.573	1.552	0.00050	0.0181
12. o-nitrotoluene	222	1.544	1.525	1.568	1.547	0.00049	0.0228
13. ethylsalicylate	233	1.519	1.501	1.551	1.533	0.00046	0.0326
14. ethylbenzoate	211	1.502	1.483	1.531	1.512	0.00047	0.0287
15. cymene	176	1.490	1.471	1.503	1.483	0.00049	0.0126
16. methylthiocyanate	130	1.469	1.449	1.481	1.459	0.00054	0.0111

[a] $\lambda = 550$ m$_u$.

[b] n_D at 25°C.

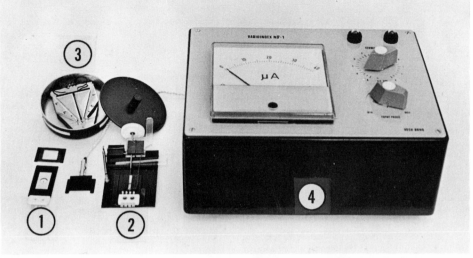

Fig. 8   Accessories of thermistor apparatus for the determination of refractive indices by variation methods Chromý, 1965): 1, special "underlying glass"; 2, stage with rotary glass fiber and adapted cell; 3, heated prismatic stage for goniometer; 4, thermistor thermometer and heating current source with electric thermostat.

usual dimensions, is provided with an electric resistance winding and a thermistor, and can be employed also for determinations in the universal Fjodorov stage. Similarly equipped is the small cell for the spindle stage and the prismatic stage for the goniometer (Fig. 8, 2, 3). Those accessories (Chromy, 1965) permit, in conjunction with a thermistor thermometer and a heating current source (Fig. 8, 4), prompt measurements by all methods with an accuracy limited by the sensitivity of the effect for the determination of the quality of the refractive index of the measured material and the immersion liquid.

STANISLAV CHROMY

References

Chromy, S., "Thermistor device for refractive index determination by means of variation methods," *Cas. Miner. Geol.,* **10**:157–164 (1965) (in Czech).

Emmons, R. C., "The double dispersion method of mineral determination," *Amer. Mineral.,* **13**:504 (1928).

Emmons, R. C., "A modified universal stage," *Amer. Mineral.,* **14**:441 (1929).

Emmons, R. C., "The universal stage," Geological Society of America, Memoir 8 (1943).

Gaubert, M. P., "Mesure des indices de refraction d'un solide par immersion dans un liquide porté à une temperature determinée," *Bull. Soc. Franc. Min.,* **45**:89–94 (1922).

Tsuboy, S. A., "A dispersion method for determining plagioclase in cleavage flakes," *Min. Mag.,* **20**:108–122 (1923).

REPRODUCTIVE ORGANS—MALE

Few, if any, techniques in microscopy have been developed solely for material from the male reproductive organs, but it is worthwhile considering the various stages in the preparation of material for examination under the microscope.

Sampling

Testis. In some species this organ can be heterogeneous in its composition. Therefore a number of samples should be taken from as many areas as possible. The material immediately beneath the tunica albuginea should perhaps be avoided, partly because it may be difficult to cut good sections after fixation, but in some species it is essential to examine sections from this location. For the purpose of obtaining human testicular tissue, the technique of biopsy is used.

Epididymis. Basically this is a highly convoluted tubule which can be divided into three broad anatomical regions (caput, corpus, and cauda); according to several workers a number of histological regions can be distinguished principally by the differences in height of the columnar cells of the epithelium and by the presence or absence of pseudo stratified epithelium. The number of spermatozoa in a section of the tubule depends upon the position of the section within the tubule; areas of high and low sperm density can be reliably selected within a species and may lie very close together, but the named anatomical regions are not necessarily homologous between species. Sampling must be very carefully done if comparisons between individuals are to be made.

Vas Deferens. The lumen varies in diameter along its length: It tends to be wider near the epididymis end and again towards the ampulla.

Accessory Organs of Reproduction. In man and the domestic animals the principal organs are the seminal vesicles, prostate, Cowpers gland, and bulbo-urethral glands. The relative sizes of the glands vary from species to species; e.g., the seminal vesicle is absent in the dog, in the bull and ram the prostate is small compared with the seminal vesicle, and in the boar the Cowpers gland is especially prominent. In the laboratory animals the nomenclature and homology are confused but from the point of view of histology these are all ducted glands, often lobed with varying amounts of muscle and connective tissue between the lobes.

Fixation

Preparation of material from the male reproductive tract for light microscopy presents no particular problems in fixation. The choice of fixation method for electron microscopy depends on the particular aspect of the cell which is under investigation. However, the best results are obtained when the fixative is perfused into the tissue (Christensen, 1965), provided the pH is correct for that particular organ of that species.

Staining for Light Microscopy

A broad range of stains must be applied to any organ to obtain as complete a picture as possible of its composition, structure, and physiological state. Some of the stains require the material to be examined by normal transmitted light and some require a light source which will excite fluorescence.

Transmitted Light

The particular stain used to delineate the general appearance of any of the male reproductive organs is a matter of personal preference but hematoxylin with or without eosin is commonly used. However, each organ has a function or functions and these are best seen after treatment with the appropriate histological or histochemical techniques.

Testis. The testis has two functions, to produce spermatozoa, and to secrete testosterone.

The stages of spermatogenesis have been carefully worked out on sections of testis stained with PAS (Leblond and Clermont, 1952). In such sections the cells of the spermatogenic tissue can be classified according to the stage of division, and the stage of the cycle can be most closely defined by the state of development of the acrosome in the developing spermatid. The distribution of lipid and of acid phosphatase granules within the seminiferous tubule also change during the spermatogenic cycle.

The steroidogenic activity of the testis can be assessed by treating sections to reveal steroid dehydrogenase activity and by using Threadgold's technique (Threadgold, 1957) which shows lipid in the Leydig cells of the interstitium.

Epididymis. Spermatozoa pass through the epididymis when they leave the testis and their passage is facilitated by the circular muscle which surrounds the epididymal tubule. In addition the epididymis has two other functions, namely absorption and secretion. The epididymis is a target organ for testicular testosterone but it is also capable of producing its own steroids. Thus a wide range of treatment is required for a study of the epididymis.

Vas Deferens. The muscle surrounding the seminal duct is the most prominent feature of this tissue. The epithelium is capable not only of secreting, but also of absorbing material from the lumen.

Accessory Glands. These are all organs which secrete various substances into ducts, and the staining must therefore be appropriate to the metabolic process likely to be involved.

Fluorescent Microscopy

The adrenergic nerve supply to the organs of the male reproductive tract can be demonstrated by causing the monoamines to fluoresce (Falck and Owman, 1965) by exposing sections to paraformaldyde before mounting.

Testis. Acridine orange and similar fluorescent dyes have been used to study the distribution of lysosomal material and the condensation of the nucleus during spermatogenesis.

Epididymis. Some cells have a characteristic white autofluorescence which differs from the more yellow fluorescence produced by lysosomal material (Risley, 1970).

Electronmicrography

The organs of the male reproductive tract have been used as a source of material for the study of endoplasmic reticulum (epididymis, Hamilton et al., 1969; prostate and seminal vesicles, Brandes, 1965). The techniques applied to these organs are generally similar to those used for other animal tissues.

H. M. Dott

References

Brandes, D., "Observations on the apparent mode of formation of 'pure' lysosomes," *J. Ultrastruct. Res.,* **12**:63–80 (1965).

Christensen, A. K., "The fine structure of testicular interstitial cell in guinea pigs," *J. Cell Biol.,* **26**:911–935 (1965).

Falck, B., and C. Owman, "A detailed methodological description of the fluorescence method for the cellular demonstration of biogenic amines," *Acta Univ. Lund.,* Sect II, No. 7 (1965).

Hamilton, D. W., A. L. Jones, and D. W. Fawcett, "Cholesterol biosynthesis in the mouse epididymis and ductus deferens: a biochemical and morphological study," *Biol. Reprod.,* **1**:167–184 (1969).

Leblond, C. P., and Y. Clermont, "Definition of the stages of seminiferous epithelium in the rat," *Ann. N.Y. Acad. Sci.,* **55**:548–573 (1952).

Risley, P. L., "Fluorescence of holocrine epithelial cells of the epididymis," *Biol. Reprod.,* **3**:67–75 (1970).

Threadgold, L. T., "Sudan black and osmic acid as staining agents for testicular interstitial cells," *Stain Technol.,* **32**:267–270 (1957).

RESOLUTION

Limit of Resolution of an Optical System With a Diaphragm of Circular Aperture

1. The optical system L is geometrically arranged to form in the focal plane ψ', a stigmatic image M' of a luminous point M situated at infinity in the direction zz' (Fig. 1).

If a diaphragm D, having the form of a slit of width Δ, is placed in front of L it can be stated that the more Δ is diminished the more the image M' spreads as an illuminated line parallel to Δ and cut by several equidistant black fringes m'_1, $m_2 \cdots$ (Fig 2). The curve C in Fig. 1 represents the light received in the plane ψ' along this line. Owing to the wave nature of light, this phenomenon is always present, although it may become imperceptible when the width Δ is sufficiently great.

2. The point M emits from infinity plane waves normal to zz'. The principle of Huyghens applies to each plane wave W_0 which passes D: This means that all points of the free part of the wave act as primary in phase oscillating sources all in phase (Fig. 3).

Suppose, beyond D, a plane W_θ making an angle θ with W_0. Two of these primary sources, i and j, transmit in i' and j' their oscillations which are subjected to a relative phase shift dependent on the difference $\overline{jj}' - \overline{ii}'$. When this difference is $\lambda/2$ (λ being the wavelength), the

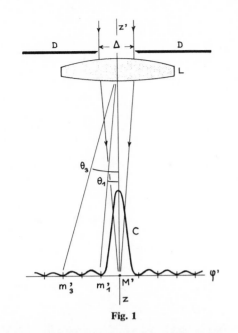

Fig. 1

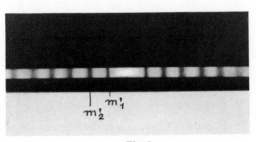

Fig. 2

two oscillations, opposite in phase, carry in sum no energy to the plane W_θ. One part of the source present on W_θ can thus form couples of opposite phase; the remainder produce a certain energy on W_θ which is, from then on, a luminous defracted wave.

But when θ is such that, from one side to the other of

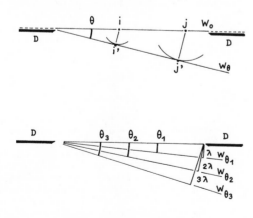

Figs. 3 (upper) and 4 (lower)

the diaphragm, the difference in path length varies as λ, or more generally as $n\lambda$ where n is an integer (Fig. 4), the formation of couples of opposite phase absorbs all the primary sources and there is no energy on the corresponding plane waves W_{θ_1} $W_{\theta_2}\cdots$. Along the system L (Fig. 1), these directions correspond to the black fringes $m_1'\, m_1'\cdots$.

If the intervening angles are small ($\lambda \ll \Delta$), then:

$$\theta_1 = \frac{\lambda}{\Delta}, \cdots \theta_n = \frac{n\lambda}{\Delta}$$

and, if f' is the focal length of L;

$$\overline{M'm_1'} = f'\frac{\lambda}{\Delta} \tag{1}$$

3. The case of a slit diaphragm allows a simple study of these phenomena. But optical instruments are generally based on circular diaphragms.

When the opening of D is a circle around the axis zz', the phenomena are those of revolution, and there results on ψ' a circular luminous spot surrounded by rings (Fig. 5). The illumination is represented by a surface on the axis zz' of which the meridian is analogous to the curve C in Fig. 1.

The comparison of the two cases, a slit of size Δ and circle of diameter Δ diaphragming the same system L, is experimentally easy. If ε' be the first angular black ring of Fig. 5, the result is:

$$\varepsilon' = 1, 2, \overline{M'm^1} = 1, 2f'\frac{\lambda}{\Delta} \tag{2}$$

If a point N, at infinity, is angularly close to M, their images M' and N' on ψ' are two spots analogous to Fig. 5 which can encroach on each other; admittedly the two spots are more visually and photographically distinct when the center of one coincides with the first black ring of the other (Fig. 6). The distance which then separates the two centers is the limit of resolution in the plane ψ'; it is the length ε' of formula (2).

The angle:

$$\theta_m = 1, 2\frac{\lambda}{\Delta} \tag{3}$$

is the angular limit, independent of system L.

4. The eye is a system L and the iris is D. The examination of an object placed 30 cm away from the eye reveals on the object a limit $\varepsilon = 0.1$ mm. The limit θ_m is thus 1/3000 of a radian. This value is fixed, even though the aperture of the iris is variable. The limit is not therefore derived from the diaphragm but is due to the discontinuous structure of the retina.

But let us calculate the diameter Δ of the iris which corresponds to this limit when $\lambda = 0.55\ \mu$:

$$\Delta = 1, 2\frac{\lambda}{\theta_m} = 1.2 \times 0.55 \times 3000 = 2000\ \mu$$

or 2 mm. This is roughly a minimum diameter that can be reached by the pupil. The limit due to the "grain" of the retina receptor and that due to refraction are equal when the iris is closed as far as possible.

Limit of Resolution for a Microscopic Objective

5. A microscope objective, supposedly corrected for infinity, is a convergent stigmatic and aplanatic system with regard to its object focus M (Fig. 7) here considered as a luminous point. A ray MI, at any angle U, has thus

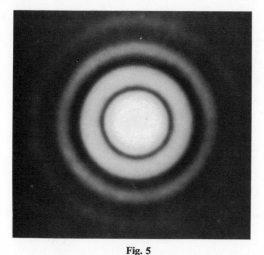

Fig. 5

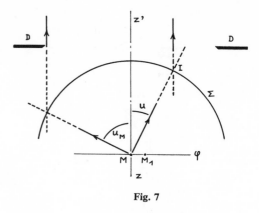

Fig. 7

for conjugate a ray *IJ*, parallel to the axis *zz'*: this constitutes stigmatism.

The point path Σ of point *I*, when *u* varies, is the principal object surface of the objective, which is aplanatic when Σ is, as in the figure, a sphere with center *M*. Aplanatism means that the stigmatic conjugate with infinite, obtained for *M*, continues for the points of plane ψ, such as M_1, which remains close to *M*. Aplanatism is essential for a microscope objective; the maximum angular aperture U_M, determined by the circular diaphragm *D*, can reach 70°; it is no longer at all possible to argue along the lines of Gauss' approximation that that which pertains to a point on the axis applies also, more or less, to its surroundings.

6. The objective is identical to the preceding system *L*; the only change is in the direction of the light. Allowing for this, the limit of resolution in the object plane will be called ε, and, following formula (2):

$$\varepsilon = 1,2\,\lambda\frac{f}{\Delta}$$

f, focal length of the objective, is the ray from the sphere Σ; Δ is the diameter of *D*. Also there derives from Fig. 7:

$$\sin U_M = \frac{\Delta}{2f}$$

On the other hand, if λ is the wavelength in a vacuum, the wavelength in an environment of refractive index *n* (the case with immersion objectives) is λ/n. The limit ε is thus in general:

$$\varepsilon = 1,2\frac{\lambda}{2n\sin U_M}$$

Finally, if $n\sin U_M$ is the numerical aperture *A* of the objective:

$$\varepsilon = 1,2\frac{\lambda}{2A} \tag{4}$$

7. The relation (4) applies only to the objective: resolution is, in point of fact, only dependent on it.

An objective corrected for infinity gives a real image at a finite distance when there is added to it an aplanatic telescope objective with the principal surface image Σ (Fig. 8) and a focal length F', such that $\gamma = F'/f$ is the desired magnification.

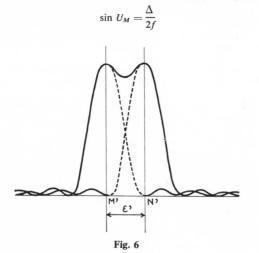

Fig. 6

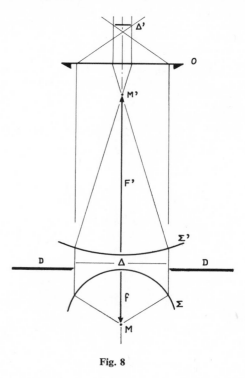

Fig. 8

On condition that D is the limiting aperture diaphragm of one and the other objective, the value (4) of ε does not change.

In the case of visual examination, the ocular O gives an image Δ' on the diameter Δ (diameter of the exit pupil) that must coincide with the pupil of the eye. If Δ' is greater than 2 mm, it is the retina, and not the objective, that limits the resolving power x (paragraph **4**). It is necessary therefore to choose an ocular of such power that Δ' becomes less than 2 mm.

For photography Δ' must be the only aperture diaphragm of the projector. It is therefore necessary to choose a magnification such that the image of ε, obtained on the emulsion, matches the grain of this emulsion.

8. In the preceding discussion, the point-objects have been considered to be luminous. Dark field omnidirectional illumination applied to two very small objects M and N meets this condition. M and N radiate in all directions, and in particular to the objective, part of the light that they receive. The whole aperture of the objective is thus used.

Let us suppose that M and N are each of a size significantly less than the limit ε. If the light that they receive is sufficiently intense, M and N are nevertheless visible as two luminous points. This is the reason that, for example, Brownian movement of "ultramicroscopical particles" can be seen. In fact, under these conditions, dark field has been named an "ultramicroscope," which should not rise to the belief that there is an increase in resolution: The images M and N are only the circular defraction spots described in paragraph **3**. They do no more than indicate the presence of these objects without giving any information about them.

If M and N approach each other their images are no longer separated when the distance $\overline{MN}$ becomes less than ε, and it is no longer possible to say whether the observed spot corresponds to one or several objects. The images of points M and N are comparable to two stars observed at night. The formulas (2) and (3), from which (4) derives, are in good agreement with the actual resolving power of astronomical instruments. Microscopic observation in dark field accords well with formula (4).

9. Bright field observation corresponds only to the relations given in paragraph **6** if the full aperture of the objective is used. The condenser, a system in every way comparable to the objective, and of the same numerical aperture, permits this condition to be fulfilled (Fig. 9: Σ_c is the principal image surface of the condenser, D_0 its iris diaphragm). But all users of the microscope know that this situation is, in fact, not satisfactory for viewing small isolated objects and that it is scarcely possible to find, in the image of any object at all, detail that can be usefully compared to ε.

It is also customary to determine experimentally the resolution achieved by an objective, by observing objects of periodic structure, comparable to gratings in the optical sense of this word. These test objects can be assembled in a collection of sizes covering the useful range of values of ε. Such test objects have been manufactured by several firms; but various natural objects, among which are the siliceous frustules of certain diatoms, furnish excellent test material.

The existence of natural objects is not surprising. The morphology of living creatures shows innumerable examples of periodic structures and their resolution is an important problem for the microscopist.

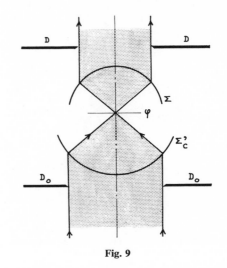

Fig. 9

An idea has been developed that the limit ε of formula (4) is the minimal spatial periodicity of structures that can be resolved by an objective. This is only roughly true. It is good enough, for example, when choosing test objects appropriate to objectives to be tested. But the resolution of periodic structures is a process that must be studied independently.

Resolution of Periodic Structures

10. The object here considered is a grating of equidistant transparent lines traced on a plane opaque surface. These lines t are normal to the plane of Fig. 10, their spatial period P is of the order of λ, or greater.

In a luminous incident beam s_0 normal to the plane of the wave, W_0 and s_0 are two "rays" which encounter two successive slits.

Above the grating, the defracted rays, as S_n, are those for which the difference in the passage $\overline{\delta} = \overline{\delta n} + \overline{\delta}_0$ is an integer n of wavelength λ: Thus a plane wave parallel to W_n is formed giving rise to the beam S_n. There are thus $n - 1$ other defracted rays (Fig. 11) between S_0 and S_n. It is apparent from Fig. 10:

$$\overline{\delta}_n + \overline{\delta}_0 = P (\sin U_n - \sin U_0) = n\lambda$$

Thus, for two consecutive rays of the orders i and $i + 1$:

$$\sin U_{i+1} - \sin U_i = \frac{\lambda}{P} \tag{5}$$

11. The objective is represented in Fig. 11 by its principal-object surface Σ (see paragraph **5**). The central rays of the beam S_i correspond, in the space image, to rays s'_i parallel to the axis zz'. The radius of Σ being f, the distance between two consecutive rays s' is written:

$$f(\sin U_{i+1} - \sin U_i) = \frac{f\lambda}{P} = a$$

these rays are thus equidistant.

For each parallel beam of the space-object, the objective produces a focus situated in the focal-image plane ψ'. There are thus produced in this plane equidistant foci f_1 (from f'_{-1} to f_3 in the figure). These points are coherent sources since all are derived from the incident beam S_0, and concordant because of the constancy of the optical paths from M to ψ'.

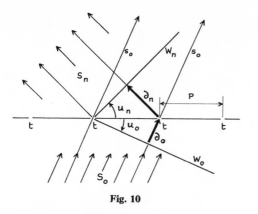

Fig. 10

On a plane normal to zz' at infinity (the objective being corrected for infinity), the sources f'_1 give interference fringes. In each direction of the angle θ with zz' (θ is postulated to be small and in the plane of the figure), there are added as many waves, with their equivalent phases, as there are sources f'_1.

The amplitude received by each point of the plane at infinity is thus the vector sum of an initial amplitude and of dephased amplitudes on it of the angles α, 2α, $\cdots$ $n\alpha$ which, insofar as they are functions of θ, constitute a harmonic series:

The first harmonic is due to the deviation a of the two sources. Its special period is the angular interfringe θ_a:

$$\theta_a = \frac{\lambda}{a} = \lambda \frac{P}{f\lambda} = \frac{P}{f}$$

It is the angle subtended by the image of the period P at infinity.

The subsequent harmonics correspond to the deviations $2a$, $3a$, $\cdots$ na of the sources, then to the angular periods 2, 3, $\cdots$ n times smaller.

The effect of the diffraction is a harmonic analysis of the periodic structure of the object. The reconstitution of the

image is better in proportion as the number of harmonics admitted to the objective becomes greater.

E. Abbe, author of this theory, has given a simple experimental form to it: The period P being such that one has several foci f'_i one arranges, as close as possible to the plane ψ', a diaphragm so cut as to mask one of the two foci. The first harmonic is then due to the distance $2a$ of the sources, its angular period is $\theta_a/2$: The observed image is that of a grating on which the lines would be twice as close.

12. The relation (5), written in the form;

$$\lambda = P (\sin U_{i+1} - \sin U_i)$$

shows that since $\sin U_{i+1} - \sin U_i \leq 2$, there is no defraction except when $P > \lambda/2$. But we are dealing with a limited case where $U_n = -U_0 = \pi/2$. For an actual objective this value is replaced by the angular aperture U_M. Let us call ε_p the smallest spatial period that can be resolved by the objective:

$$\varepsilon_p = \frac{\lambda}{2 \sin U_M}$$

a formula illustrated in Fig. 12 and which must be generalized in the case of immersion; when diffraction takes place in a medium of index refraction n, the wavelength is γ/n. Finally:

$$\varepsilon_p = \frac{\lambda}{2 n \sin U_M} = \frac{\lambda}{2A} \qquad (6)$$

The plane ψ' then only contains the two sources f'_0 and f'_1. The harmonic analysis is reduced to fundamentals. The grating described in paragraph **10** is thus translated into the image plane, if its period is ε_p, by sinusoidal variation of the illumination that is shown in the top of Fig. 13. The base of the same figure is relative to the same type of grating but with a periodicity $P = 5\,\varepsilon_p$: The same objective gives interference image through the interaction of five harmonics.

Formulas (4) and (6) are not contradictory. If one considers two points, separated by the distance P, on adjacent slits of the grating, and applies formula (4) to them, one ignores that the diffraction phenomena are dependent upon two or more foci, f', the lines of the grating, and consequently the points which comprise them are the sources to which the phases are linked. Formula (4) applies to points independent of each other in respect to phase.

In dark field, the interference process is also involved in the case of periodic structures. The incident ray S_0

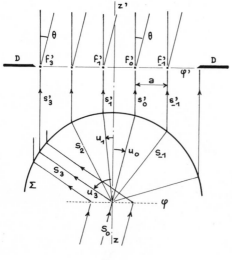

Fig. 11

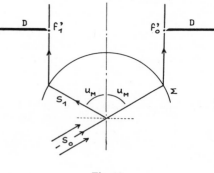

Fig. 12

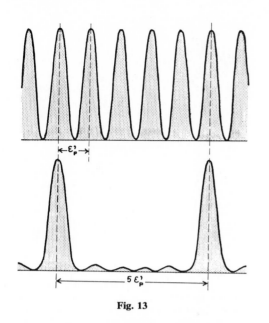

Fig. 13

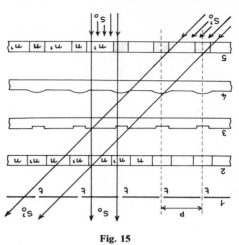

Fig. 15

(Fig. 14), which does not enter the objective, produces defracted rays S_1 and S_2 through the grating R. But the figure shows how, in the case of large numerical apertures the narrowness of the angular region that S_0 can occupy causes the diffracted rays to lose an area ω which becomes greater as the structure to be resolved becomes smaller. The limit continues to be established by the value ε (paragraph 6).

Resolution Under the Customary Observation Conditions

13. The grating described in paragraph 10, and indicated as 1 on Fig. 15, is called an amplitude grating: In the slits, luminous amplitude is undiminished but becomes zero in the opaque parts. The gratings, 2, 3, and 4 seen, as 1, in section, are phase gratings; being of transparent material they do not in practice modify the amplitude but, in passing through the sheet, the phase of a luminous wave is modified in periodicity according as the waves encounter the index n or n', or even varying thickness. Finally 5, where it is assumed that the medium of index n is partially absorbent, is at the same time a phase and an amplitude grating.

These diagrams, more or less approximating the actual structure of objects, allow certain conclusions:

—The existence of a spatial period $P > \varepsilon_p$ always

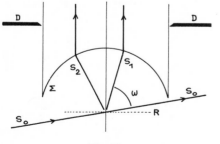

Fig. 14

involves the process of image formation which has just been described.

—When P is very close to ε_p, only the very strongly inclined incident rays permit resolution (Fig. 12).

—When P equals several times ε_p, it seems better to revert to axial illumination. But the incident defracted rays S_0 normal to the plane of the object are, in detail, different from those that give an oblique incident S_0' (Fig. 15). As it is not only a matter of observing the period P, but of securing a better understanding of the object, oblique illumination should be employed. Finally, a general remark can be made: The image better conforms to the object to the extent that it contains information derived from the object; this information is in the form of defracted rays; all possible defracted rays are obtained only if all possible incident rays are utilized, that is to say, all those which are contained in the cone U_M (Fig. 12). The conditions of paragraph 9, thus reconstituted, appear theoretically optimal.

But the description of defraction phenomon has been made in paragraphs **10, 11,** and **12** for only a single direction S_0 of the incident light: It is necessary to re-examine the case of a very open beam.

14. The entry pupil D_0 (iris diaphragm) of the condenser (Fig. 16), situated in the focal plane of the object, is uniformly lit from a source not shown.

A parallel beam S_0 crossing the object plane ψ corresponds first to a point f_0 in this pupil, then to a focus f_0' in the image focal plane ψ' of the objective. The exit pupil of the objective, the diaphragm D, is in the plane ψ'. There is therefore concentric to D an image D_0' of the diaphragm D_0 (Fig. 17). When the iris is opened, D_0' becomes equal to D at the moment when the illuminating cone has the same aperture as the objective.

A grating having been placed in the plane ψ, the diffraction causes the foci $f_1' \cdots f_1'$ to correspond to f_0' by the perpendicular shift a of the slits of grating (paragraph 11). The homologous points of these images, at a distance a are coherent sources. The diffraction images of D_1', $D_2' \cdots D_{-1}'$, D_{-2}' are thus added to the illuminated circle D_0: The interference effect at infinity of the circles D_1' is the same as that of the foci f_1' in paragraph 11.

15. As in Fig. 17, Fig. 18 represents the exit pupil of the objective but at full aperture: D_0' is thus confused with D. Three cases are considered, relative to the examination

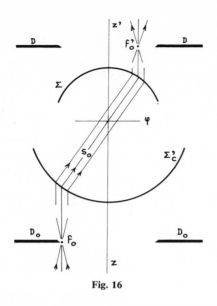

Fig. 16

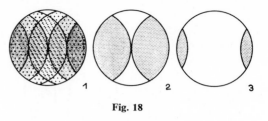

Fig. 18

of gratings of which the periods are $P_1 = 4\varepsilon_p$; $P_2 = 2\varepsilon_p$; P_3 is slightly greater than ε_p. The drawings are to be interpreted from the aspect that the incident light is present over the whole pupil, also in the shaded parts, which represent the defracted light.

—With P_1, the six defracted rays very nearly fill the whole pupil and are for the most part superimposed.

—With P_2, the two defracted rays occupy about 3/4 of the pupil.

—With P_3, they occupy only a very small part of it.

On the other hand, the illuminous energy defracted by a transparent object is proportional to the phase shift when the shift is small (in the sense that $\sin X = X$). The smallest structures, in general, produce the least phase shift. The defracted rays in case 3 thus contain only a very small part of the incident light. The interference that they produce, superimposed on a background illuminated by almost all the light, becomes less and less visible in proportion as it approaches the limit ε_p.

This fact almost always prevents observation at full aperture, by transmitted light, with powerful objectives. One obtains sufficient contrast only by partially closing the substage iris; the necessity of so doing has given birth to an empirical rule: A being the numerical aperture of

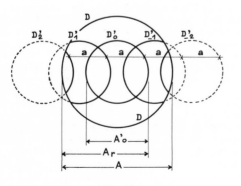

Fig. 17

the objective, the substage condenser is set to between $A/2$ and $2A/3$.

It is thus necessary to lose a part of the resolving power. In Fig. 17, A measures the numerical aperture of the objective, A_0' that of the diaphragmed substage condenser. A_r, which is, in the pupil, the greatest possible distance between the incident and the defracted rays, measures what is known as the "resolving aperture" of the objective-substage combination:

$$A_r = \frac{A + A_0'}{2} \tag{7}$$

The practical, and classical, method of reaching ε_p with adequate contrast is the device of oblique lumination which corresponds to Fig. 12: The resolution of some structures is improved but the illumination for the whole of the object is poor. Annular illumination, less efficient for any given structure, is better for the whole because it does not interfere with the axial symmetry of the phenomena.

Increase in Resolving Power

16. There are two methods of endeavoring to increase resolution:

a. To influence the factors in formula (6).
b. To seek to fully utilize the characteristics of objectives. Actually, under the practical conditions of observation described in paragraph 15, this fails to utilize the full potential of high-power objectives.

By method a, modifying λ led to photomicrographic objectives for ultraviolet light. The improvement is significant since ε_p becomes twice as small when one changes, at equal aperture, from $\lambda = 550$ nm to λ uv $= 275$ nm. But the restrictions of paragraph 15 remain, and the use of ultraviolet is technically difficult. Modifications of A consist, on one hand, in making U_M as close as possible to 90°; this ultimately results in $\sin U_M \simeq 0.95$; it is not possible to conceive of a greater angle. On the other hand, the index of refraction n of the medium can be increased. The operative indices are more of the frontal lenses of the condenser and objective, and of all that is between them, in particular the mounting medium and the immersion fluid. With monobromnaphthalene Abbe reached $A = 1.60$.

Formula (7) under the condition $A_0' = 2A/3$ gives the following table:

Numerical Aperture A of the Objective	Resolving Aperture A_r
0.65	0.55
0.95	0.80
1.30	1.08
1.40	1.17
1.60	1.33

The augmentation of A_r is slow and scarcely justified the difficulties in producing the required increases of A. In fact, an aperture of 1.60 is no longer manufactured and the aperture 1.40 is little used.

In method b above, the purpose is to make $A_r = A = A_0'$, while at the same time maintaining a high contrast. In this case very large apertures again become of interest. The search for maximum resolution has long excited suspicion based on the fact that the "images" of fine structures are proportionately more schematic as ε_p is more closely approached (see Fig. 13); there is thus a risk of inaccurate interpretations. The suspicion is legitimate when the unknown begins immediately beyond the limit of the optical microscope. Nowadays electron images can be used a control. All work on the limit of resolution must at times use objects which can also be observed in the electron microscope with a view to evaluating the fidelity of the optical images.

In seeking a solution to problem b, polarized light can be considered:

17. When a beam rectilinear polarized light reaches a grating, the defracted light is rectilinear if the incident wave plane is parallel or rectangular to the slits of the grating. Otherwise, as is usually the case, the defracted light is elliptical.

When a microscope is being illuminated by polarized light, the image receives:

—Rectilinear polarized light (a slight degree of depolarization is inherent in the objective) which is the direct light, i.e., the incident light from which the defracted portion is removed.

—The defracted light, usually elliptical.

When the observed structures are extremely fine, the direct portion is very much more intense that the defracted portion (paragraph **15**).

In Fig. 19 V represents the amplitude of the direct light and the ellipse E represents the defracted light. The figure shows how observation through an analyzer An can produce the same order of magnitude as the transmitted light amplitude: v for direct light and e for the ellipse. The angle β being close to $\pi/2$, experiment confirms that, for $A_0' = A$, the spatial periodicies close to ε_p are resolved with good contrast.

The conditions regarded an optimal in paragraphs **9** and **13** would be satisfied if the specific orientation of the amplitude V in relation to the object did not destroy the symmetry of rotation of the illuminating beam.

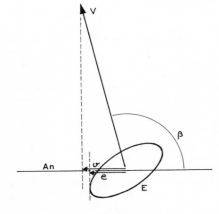

Fig. 19

If the polarizer and the analyzer are rotated each in its plane, at a constant speed, about a general axis with maintenance of a given adjustment β, the symmetry of rotation is reestablished as soon as an integral number of half turns results in the formation of the image. For visual observation a continuous rotation is necessary at a speed that does not produce fluttering.

18. The practical application only involves some difficulties in the construction of the apparatus. The use of the microscope furnished with two rotating polarizers is the same as that of an ordinary microscope. One comment: The use of an immersion objective requires also that the condenser be oiled and that the object be mounted in a medium of refractive index $n \geq 1.52$; otherwise it is not possible to make $A_0' = A$.

The possible birefringences of the object do not cause orientation effects when the polarizers revolve; they act as colorants.

It is obviously possible to use white light, and the method does not diminish the interest of stained preparations. But when the colors are not of importance a monochromatic light is preferable.

This method was used for the photographs of test diatoms in Figs. 20, 21, and 22 made in green light (546 nm), each matched by an electronic image of the same object. These figures show that the limit of resolution:

$$\varepsilon_p = \frac{0.546}{2.8} = 0.20\,\mu$$

is almost reached. An object without periodic structure (Fig. 22) was photographed under the same conditions.

The method can be used with ultraviolet. Available quartz optics provide an aperture of 1.25 for the wavelength 275 or 250 nm. The limit ε_p is thus in the neighborhood of 0.1 μ.

19. The production of maximum resolution is not the only problem in the production of a microscopic image. The need to demonstrate very slight phase shifts, for example, has given rise to methods of very great interest; phase contrast and interference microscopy.

How is it possible to forecast the effect on resolving power of a particular method of study? There is a universal response; all that modifies the uniform distribution of the light across the full extent of the exit pupil of the objective, or an image of this pupil in the absence of the object, bears on the optimal conditions of resolution.

It would be a misunderstanding to read into this remark a prohibition against modifying the pupil. Optimum resolution means the simultaneous restitution of all the structures of an object up to the limiting period ε_p. But if there is no interest in spatial periods close to ε_p, Fig. 18 suggests that a central circular diaphragm stopping all the unwanted direct illumination can improve the contrast of the image: this is the angular illumination indicated at the end of paragraph **15**.

In the region of visible light the order of magnitude of the limit of resolution is immutable and the leap forward taken by the electronic microscope has been admirable. But examination *in vivo* can be made only in visible light and the attempts to increase still further the possibilities of the optical microscope are thus justified.

The electron microscope aids this by the comparisons that it makes possible. Its practical limit of resolution is in the neighborhood of a few Ångströms and depends above all on the quality that it is possible to produce and to maintain in the lenses.

Fig. 20 Diatom—test *Pleurosigma angulatum*. A: ×3200; resolving aperture 1.40; objective and condenser in oil. B: ×3700 O.N. objective 1.40, O.N. condenser 0.8. C: ×7400; electronic image.

Technical Remarks

20. Qualities of the objective. Figure 12 brings the resolution of a spatial period into the scheme of an interference experiment. The two beams that form the foci f_0' and f_1' only pass through narrow regions of the objective and it is sufficient that f_0' and f_1' be fairly exact in order to produce interference. The properties of the objective are scarcely important: It is a well-known fact that a test object close to the limit of ε_p can be resolved in oblique light by a badly corrected objective.

It is an altogether different matter when pairs of coherent points, spread over a major part of the area of the exit pupil (Fig. 18), must interact in the image plane. The correction of the objective is then paramount, and simple manipulations, e.g., little variations in the objective ocular distance, may be desirable.

21. Simplicity of the Optical Setup. This follows from the preceding paragraph since, even though the objective alone is taken in the discussion of these phenomena all the optical elements are involved in the geometric correction of the microscope. It is therefore desirable to retain only the indispensable ones: condenser, objective, and ocular in the straight tube. When observing without immersion oil it is best to use an uncovered object and an objective corrected for this case because the coverslip of uncertain optical quality introduces serious errors when numerical aperture is large.

22. Light Source, Critical Illumination, Köhler Illumination. The source must be of uniform intensity, a ribbon filament incandescent lamp being typical. Under this condition critical illumination and Köhler illumination

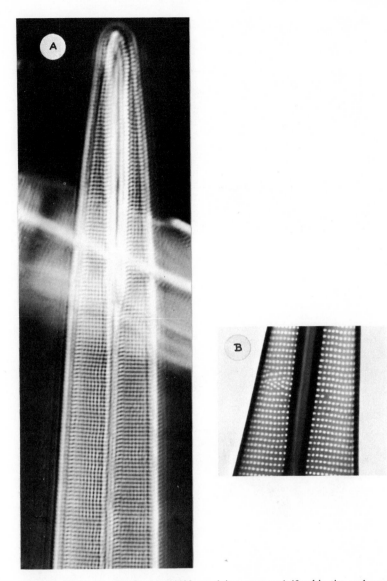

Fig. 21 Diatom—test *Amphipleura pellucida*. A: ×4400, resolving aperture 1.40; objective and condenser in oil; fixed polarizer. B: ×6300 electronic image.

are the same. The photographs of test objects illustrating this article were made with critical illumination.

Köhler illumination has the advantage of a field iris in a useful position. Under critical illumination it is necessary to make the image of the source, superimposed on the object, just sufficiently large to cover the field. Köhler illumination produces a coherent beam in all parts of the object-space. But it would be a mistake to believe that coherence is not possible without this method of illumination.

Forming the image of the light source in the focal plane of the condenser, the method of Köhler seems to allow the use of any light source whatever, even a more or less loose spiral filament, since the object plane is in any case uniformly illuminated. But an image of the light source is produced in the exit pupil of the objective and a spiral

in this position is arbitrarily cut into illuminated and nonilluminated areas, which is altogether contrary to the conditions that have been established above. The only acceptable filaments are those that are flattened and closely wound.

Arcs require that the regions used be carefully chosen; the examination next described is particularly important for those sources.

23. Examination of the Exit Pupil of the Objective. This is made by removing the ocular and observing the back of the objective either by the naked eye or with a telescope (of the type used to center the rings of a phase microscope). The purpose is to adjust the illumination to the point that the exit pupil is completely and uniformly illuminated. Many other things can be noticed by this

Fig. 22 (upper) Diatom—test *Surirella gemma*. A: ×4000; resolving aperture 1.40; objective and condenser in oil. B: ×7400; electronic image.

Fig. 23 (lower) *Trypanosoma equiperdum*. ×2600, resolving aperture 1.40; objective and condenser in oil.

examination, particularly the presence of air bubbles in immersion liquid. It also allows the observation of those aspects of the exit pupil that have been described in paragraphs **10–15**.

24. Magnification—Focusing. A rule due to Abbe fixes the maximum magnification at between 500 and 1000 times the numerical aperture. This rule was suggested for the examination of objects under the conditions of paragraph **15**. When the observation takes place with $A_r = A$, the value of $1000A$ can be passed. A magnification of 1500 to 2000 can easily be obtained with an aperture of 1.40.

At high magnifications the image changes rapidly in terms of moving into focus and the decision to cease at a given point is arbitrary, which is one of the reasons for

the suspicions pointed out in paragraph **16**. The frustule of PLEUROSIGMA is in this regard interesting: rather than hesitating between the various appearances of the image A (Fig. 20) it was usual to stop on the image B, a classic and prudent example of "good" resolution up to the time when the electronic image C proved that any part of A was in point of fact better than B.

The fine adjustment must permit one to stop exactly on, and remain stationary at a given point; any looseness in the transmission can produce spontaneous changes in focusing. The viscosity of immersion liquid acts in an analogous manner and it is necessary to have several trials before the image is stabilized.

25. Depth of Field. The formulas purposed as an expression of axial resolution and of the depth of field add nothing to the primordial problem of the choice of the focal point. But it shows that at equal lateral resolution the field depth varies as a function of magnification; the "slice" of the object that forms the image becomes proportionally thinner as the magnification increases. These formulas are also a means of determining the thickness that is desirable for sections, the rule being that the thickness of the sections should not be greater than the depth of the field. Berek has given the following formula:

$$\rho = 0.125 \frac{4n\lambda}{A^2} + 0.0014 \frac{nF}{A}$$

where: $\rho =$ depth of field; $A =$ numerical aperture; $n =$ refractive index of the object (and not of the immersion medium); $\lambda =$ the wavelength in air; $F =$ focal length of the complete microscope, if the magnification is G, $F(\text{nm}) = 250$: G. The same units of length for λ and F must also be the units for ρ.

R. BOUYER

References

1. E. Abbee, *J. Roy. Microsc. Soc., London,* **2**(1):812–824 (1879); **2**(2):300–309, 460–473 (1882); **3**:790–812 (1883); **9**(2):21–724 (1882).
2. M. Berek, *Z. f. wiss. Mikros.,* **41**:1–15 (1925); *Z. f. Physik,* **40**:420–450 (1927); **53**:483–493 (1929).
3. R. Bouyer, *J. Microsc.,* Paris, **3**(2):225–228 (1964); **4**(1):1–8 (1965); **7**(4):459–464 (1968).
4. A. Köhler, "Mikrophoto mit U.V. Licht," *Z. f. wiss. Mikros.,* **21**:129–165, 273–304 (1904).
5. K. Michel, "Die Grundlagen der Theorie des Mikroskops Wissensch," Stuttgart, Verlagsges, 1950.

RHYNCHOCOELA

The phylum Rhynchocoela (also referred to as Nemertea, Nemertinea, nemertean worms, nemertines, and ribbon worms) is an interesting group of acoelomate worms characterized by the presence of a unique proboscis contained in a sheath, the rhynchocoel. Most of these animals are marine and can be found inhabiting the littoral zone of temperate oceans; however, a few species have adapted to the rigors of the freshwater environment and a lesser number still to a semiterrestrial existence in moist tropical areas (Hyman, 1951). Certain species occur as symbionts in the mantle cavities of clams and snails and in the branchial chambers of tunicates. According to Cheng (1969) none of the nemerteans are truly parasitic, but are rather obligate commensals; however, at least one genus (*Carcinonemeters*) feeds on the eggs of its host.

The major problem with preparing many species of nemertean for microscopic studies is their fragility and tendency to fragment upon fixation, and narcotization is usually necessary. Generally, the smaller, more delicate forms can be narcotized by placing them in shallow finger bowls containing water from their environment and adding to this water a few small crystals of chlorotone. This drug will dissolve fairly rapidly and its effect upon the worms will likewise be rapid. In addition to this method, a number of other techniques can be used. A dilute solution (1.0–1.5) of chloral hydrate or a warmed (37°C) solution of magnesium chloride (8%) have been used. An alternative method that can be used with smaller specimens is to position the animal on a microscope slide under a weighted coverslip or in an apparatus designed to immobilize them (Poluhowich, 1965). Fixative can then be flooded under the coverslip with little fear of fragmentation. In some instances, this latter technique can be used without narcotization.

The choice of a killing and fixing reagent will vary according to the requirements of the staining procedure. A good general fixative is FAAG (24 ml 95% ethanol, 5 ml glacial acetic acid, 1.5 ml formalin, 10 ml glycerin, and 46 ml water). Marine forms can be fixed with marine Bouin's (75 ml seawater saturated with picric acid, 25 ml formaldehyde, 5 ml glacial acetic acid) or Susa fixatives or a 4% solution of formaldehyde in seawater.

For observation of general anatomy, hematoxylin and eosin staining is sufficient. Whole mounts of small specimens can be stained with Mayer's paracarmine and counterstained with fast green; care should be taken not to overstain with latter. Methyl salicylate is used for clearing. The stylet region of hoplonemerteans is best studied with the smaller forms, by applying light pressure to a coverslip that covers the animal prior to fixation. Once the proboscis is extended, extreme care has to be taken during the subsequent handling of the worm. The larger forms, particularly the marine nemerteans, require embedding and sectioning to view their interal structures. Paraffin embedding of short segments of these animals will give excellent results for 8–10 μ sections.

A number of histochemical studies have been conducted on nemertean tissues. The digestive physiology of representatives from three of the four orders has been extensively studied by Jennings and Gibson (1969). Enzyme activity in the gut was characterized by a variety of histochemical procedures which included carbonic anhydrase, endopeptidase of the cathepsin-C type, exopeptidase of the leucine amino-peptidase type, esterase, lipase, acid phosphatase, alkaline phosphatase α-glucosidase, and β-glucuronidase determinations. Ohuye (1942) has studied the blood corpuscles and hemopoietic processes in the marine *Lineus fuscoviridis*, and Gibson and Jennings (1967) have observed the leucine amino-peptidase activity in the circulatory systems of a variety of nemertean species. The structure and function of the basement membrane and muscular system in *Amphiporus lactifloreus* has also been studied through the use of a silver impregnation technique (Cowey, 1952). The stylet region of the freshwater hoplonemertean *Prostoma rubrum* is presently being examined (Poulous and Poluhowich, unpublished). The main stylet (Fig. 1) is mounted on a bulb that has been presumed to contain a venom. Preliminary examinations indicate that the bulb contains a substance that stains positively for protein and PAS while the cells below the bulb show a positive indol reaction. This suggests that the bulb contains a mucoprotein which may act in conjunction with an alkaloid of some type.

In addition to these studies, Iwata (1960) has studied the embryology of a variety of nemerteans. Sexually mature worms were placed in shallow finger bowls and allowed to shed their gametes. Fertilized and unfertilized eggs and the subsequent stages of development were fixed in Bouin's and placed in the lower portion of dahlias in 70% alcohol prior to sectioning. Eight-micron sections of the embryos were stained with Delafields' hematoxylin and eosin or Heidenhain's azan for study. The collection of gametes from common marine nemerteans from the Atlantic and Pacific coasts of North America has been described by Coe (1937).

Smaller nemerteans (e.g., *Prostoma rubrum*) can be observed and photographed in the living condition using phase microscopy. With this technique, the animals can

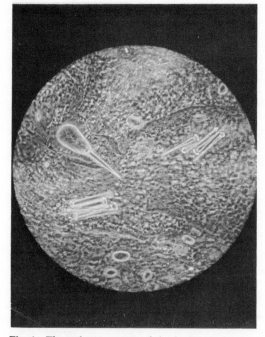

Fig. 1 The stylet apparatus of the freshwater hoplone-mertean *Prostoma rubrum* showing the main stylet mounted on the bulb and the pouches containing the reserve stylets. (Phase microscopy, 450×.)

be maintained for some time and demonstrate nicely flame cell and amoeboid cell activity in addition to the structure of the proboscis and stylet apparatus.

JOHN J. POLUHOWICH

References

Cheng, T. C., "The Biology of Animal Parasites," Philadelphia, W. B. Saunders Co., 1969 (pp. 665–668).

Coe, W. R., "Methods for the laboratory culture of nemerteans," pp. 162–165 in J. G. Needham, F. E. Lutz, P. S. Welch, and P. S. Galtsoff, "Culture Methods for Invertebrate Animals," New York, Dover Publications, Inc., 1959.

Cowey, J. B., "The structure and function of the basement membrane and muscle system in *Amphiporus lactifloreus* (Nemertea)," *Quart. J. Micro. Sci.,* **93**:1–15 (1952).

Gibson, R., and J. B. Jennings, "Leucine aminopeptidase activity in the blood system of rhynchocoetan worms," *Comp. Biochem. Physiol.,* **23**:645–651 (1967).

Hyman, L. H., "The Invertebrates: Platyhelminthes and Rhynchocoela, The Aceolornate Bilateria," Vol. 2, New York, McGraw-Hill Book Co., Inc., 1951 (pp. 459–531).

Iwata, F., "Studies on the comparative embryology of nemerteans with special reference to their relationships," *Pub. Akkeshi Marine Biol. Sta.,* No. 10, pp. 1–51 (1960).

Jennings, J. B., and R. Gibson, "Observations on the nutrition of seven species of rhynchocoelan worms," *Biol. Bull.,* **136**:405–433 (1969).

Ohuye, T., "On the blood corpuscles and the hemopoiesis of a nemertean, *Lineus fuscoviridis,* and of a sipunculus, *Dendrostoma minor,*" *Sci. Rep. Tohoku Imp. Univ.,* Ser. 4, Biol. **17**:187–196 (1942).

Poluhowich, J. J., "On flattening flatworms," *Turtox News,* **43**(2):76–77 (1965).

ROCK SECTION GRINDING AND POLISHING

The thin section method of studying rocks, minerals, and fossils is over 140 years old. It is not known who actually made the first thin section but certainly J. Nicol, inventor of the polarizing prism, was one of the very early investigators. He is reported to have given a paper on the preparation of thin sections as early as 1829. Other pioneers in the field, though somewhat later than Nicol, were Sorby, Oschatz, Zirkel, and Vorgelsand. All of these early technicians used rather primitive equipment and the making of good thin sections must have been a very slow and laborious process indeed. This situation of lack of adequate equipment persisted for many decades. While some technicians, through the years, have devised their own special pieces of equipment, or adapted and modified some existing machine as an aid, the fact remains that the use of the time-honored, hand method continues to some extent to this day. It was not until 1960 that equipment specifically engineered and designed for the rapid and precision production of thin sections became commercially available. In addition to more sophisticated equipment becoming available, great progress has been made in improving cementing media. Canada balsam was probably the only cement used for mounting sections for more than 100 years and is still one of the very best for cementing the coverslip. The use of Canada balsam as a mounting medium finally gave way to Lakeside 70 Cement, which continues to be widely used. Fairly recently, epoxy resins with a reasonably good index (near 1.54) have become available and are excellent in many respects, especially because of their adhesive properties. This has greatly contributed to the increasing ease of preparing thin sections.

Statement of Problem. Before attempting to make a thin section it is essential to have a thorough understanding of the problems involved and a precise knowledge of the essential requirements that must be met in order to produce a thin section of top quality. The perfect thin section may be described very simply as consisting of five different layers. They are, from top to bottom:

1. A thin glass plate (coverslip).
2. A thin layer of Canada balsam (or some suitable adhesive material with an index of 1.54) completely free of air bubbles.
3. A layer of specimen material 30 μ (.0011811 in.) thick, the surfaces of which are smooth, optically flat, and absolutely plane parallel.
4. A layer of cementing material with an index of refraction of 1.54 which is as thin as practical, of absolutely uniform thickness and completely free of air bubbles.
5. A glass plate (microscope slide) with optically flat surfaces which are perfectly plane parallel.

The combination of these five layers as described above constitute an ideal thin section, and the specifications set forth must be met within very narrow limits in order to produce a good thin section.

Preparation. There are essentially eight different steps involved in converting a piece of rough sample material into a completed thin section ready for microscopic examination.

Step 1. The rough material must be cut into slabs. The actual thickness of the slab is of no great importance; however, it should be at least 1/8 in. thick. Slabbing is best accomplished with a diamond saw with adequate blade diameter to accommodate the size of the material being cut.

Step 2. The slab must be dimensioned to a size which will fit conveniently on the microscope slide. For standard petrographic (27 mm × 46 mm) slides a satisfactory size is 3/4 × 1 1/2 in. It is well to cut the sample to such a dimension so that when it is mounted on the slide there will be sufficient uncovered space at one end of the slide for an identification number. A push-through type diamond saw with a 4–6 in. blade is excellent for this operation. This dimensioned sample is usually called a chip.

Step 3. The dimensioned sample or chip must be prepared for cementing to the microscope slide. The down surface of the chip, i.e. the surface which is to be cemented to the slide, must be lapped smooth and flat. The importance of correctly preparing the down surface of the chip cannot be overemphasized. It must be lapped as nearly optically flat as possible (See Fig. 1.) An 8 or 12 in. diameter cast iron lap plate rotating at approximately 450 rpm used with a thin slurry of silicon carbide abrasive grains and water is best for this procedure. It is convenient to have two such laps, one for initial, coarse lapping using 400 grit abrasive or coarser to quickly remove the saw marks from the surface of the chip and another lap for final grinding using 600 grit abrasive or finer for finishing. Remember that it is impossible to grind a flat surface on the chip if the lap plate is not itself flat. As the lapping procedure nears its final stage, it is most helpful to check the degree of flatness being generated on the chip's surface by means of a toolmaker's knife edge. This is accomplished by placing the knife edge against the surface of the chip and holding toward a light source. Any deviation of the chip's surface from the straight surface of the knife edge is readily apparent. Lapping should be continued until no light is observed along the entire length of the junction of the knife edge and the surface of the chip. This test should be made along both length and width dimensions and across the diagonals of the chip. The most common undesirable condition to watch for is a slightly convex surface of the chip. As a final preparation of the down surface of the chip, many technicians prefer to hand lap the chip on a sheet of plate glass with a water slurry of 1200 grit aluminum oxide.

Step 4. For best results the microscope slide requires some preparation. Very few microscope slides are truly flat. Many are slightly wedged and most have slightly concave or convex areas on the surface. The surface to which the chip is to be mounted should be lightly lapped on a cast iron lap or plate glass to remove these irregularities. The frosted surface developed on the slide by this procedure is by no means objectionable, but indeed an advantage, since it assures a much better bonding surface for the cementing medium. A good-grade one inch micrometer graduated in ten thousandths is most helpful in checking flatness and parallelism of the microscope slide.

Step 5. The chip must be mounted to the slide. Before mounting, it is essential that both the chip and the slide be completely clean and free from all foreign matter and grease. This is best accomplished by a thorough washing in soap and water followed by a wash in some solvent such as acetone. An ultrasonic cleaner may also be used. After cleaning, slide and chip should be placed, mounting surfaces up, on a hot plate at 200°F to dry. As to mounting media, there are several excellent epoxy resins available for this purpose. While the mixing ratio of plastic and hardener and the exact curing temperature may vary somewhat with each brand, the mixing and mounting procedure is basically the same for all. It should be noted that only epoxy cements described by the manufacturer as being suitable for microscopy techniques, i.e., cements with an index very near 1.54, should be used. The ordinary household epoxy kits sold in hardware stores are in no way suitable for thin section mounting. The mounting procedure here described, so far as mixing ratio and curing temperature is concerned, is for epoxy resins of the Araldite AY-1;05 type.

A. Place clean sheet of paper on top of hot plate preheated to 200°F. The temperature should be held within as narrow limits as possible. Place chip and slide, mounting surfaces up, on hot plate. Allow to remain on hot plate until thoroughly dry and heated (10–20 min).

B. Measuring with extreme care, place equal amounts of plastic and hardener in a small, disposable aluminum foil dish. Mix gently but thoroughly with disposable wooden spatula, stirring until mixture becomes clear (30–40 sec.).

C. With chip and slide on the hot plate, apply a thin, even coat of the epoxy on the mounting surface of the chip, not on the slide. The slide is now carefully lowered on the chip from a 45° angle so as to minimize the trapping of any air bubbles. Immediately work out any bubbles by rubbing and pressing back side of slide with eraser end of pencil or tapered wooden dowel. When all bubbles have been removed, quickly position slide on chip. Leave on hot plate at 200°F for at least 30 min to completely cure. Remove from hot plate and allow to cool. Any epoxy on back surface and sides of slide should be carefully removed with a razor blade. The mounted chip is then ready for subsequent processing into a thin section.

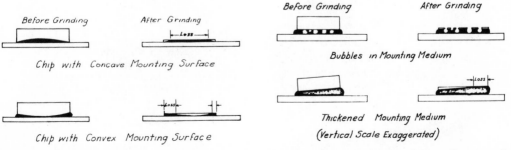

Before Grinding   *After Grinding*

Chip with Concave Mounting Surface

Chip with Convex Mounting Surface

Before Grinding   *After Grinding*

Bubbles in Mounting Medium

Thickened Mounting Medium
(Vertical Scale Exaggerated)

Fig. 1 Loss of section resulting from nonflatness of chip.   **Fig. 2** Loss of section resulting from careless mounting.

Step 6. Chip thickness must be reduced leaving a layer approximately .020 in. for final grinding. This may be done by the hand method of grinding on rotating cast iron and slurry as in Step 3 above or by equipment designed especially for this purpose. There are several thin sections machines presently available. The type most widely used will be described here. It consists of two companion machines (Figs. 3 and 4), a thin section cut-off saw and thin section grinder. The grinder will be described briefly in Step 7 below. The thin section cut-off saw consists of a sintered, diamond blade 5 in. in diameter and .025 in. thick mounted vertically on the shaft of a horizontally mounted motor. The mounted chip to be sawed is held by a vacuum to face plate of chuck attached to the rocking arm assembly which is mounted on a micrometer slide. The amount of material to be cut off by the saw blade is controlled by advancing or retracting the micrometer slide. The micrometer slide is positioned so as to remove the desired amount (thickness) of material from the chip. A cut is made by moving the rocking arm into the diamond blade. Excess material (chip) is quickly sawed off in a minute or less leaving a thin layer (.020 in. thick or as desired) of the chip for final grinding.

Step 7. The sawed chip must be reduced in thickness to 30 μ. For this procedure the thin section grinder (Fig. 4) is used. This machine consists of a 6 in., vitrified bond, is the type of cup wheel mounted vertically on the shaft of a horizontally mounted high-precision spindle driven by an electric motor. The vacuum chuck and rocking arm are identical with those on the saw; however, the rocking arm spindle and micrometer slide are of far greater accuracy than those used on the saw. The sawed chip from Step 6 is transferred to the vacuum face plate of the grinder and

may be moved back and forth across the face of the cup wheel by means of the rocking arm. To reduce thickness of the chip it is advanced into the cup wheel by means of a geared micro adjustment mechanism on the micrometer slide, and is rapidly reduced in thickness to 40–50 μ. During this operation it is well to examine the semifinished section from time to time under the microscope to determine the exact thickness. If the chip has been prepared and mounted correctly, as described above, it may be finished to standard, 30 μ thickness directly on the grinder. If, however, the two surfaces of the microscope slide are not plane parallel and/or if the cement layer is not uniform in thickness, a slight wedging may develop. Should this situation occur it is necessary to hand lap on a glass plate with a water slurry and 600 aluminum oxide to correct the wedging and reduce the thickness to standard (30 μ). Also, if a finer finish than that produced by the cup wheel (equivalent to water slurry and 600 abrasive) is required, the last 10 μ should be removed by hand lapping on a glass plate with a water slurry and 1500 aluminum oxide. When a thickness of 30 μ has been achieved the thin section is thoroughly washed in soap and water and placed specimen side up on a hot plate at 200°F to dry.

Step 8. The last procedure in completing the thin section is cementing the coverslip. Although there are several cements available for this purpose, unrectified, paper-filtered Canada balsam is still about the most satisfactory. While the clean, dry section is on the 200°F hot plate, apply a thin even layer of balsam on its upper surface and allow to remain heating for approximately 3 min. This is usually sufficient time for the balsam to cook to the point that it will become hard on cooling. Lower the coverslip

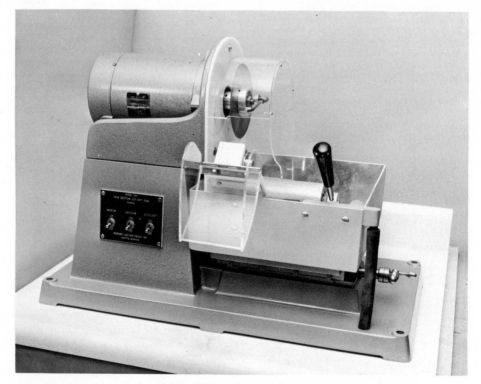

Fig. 3   Thin section cut-off saw.

Fig. 4 Thin section grinder.

onto the section and quickly work out any trapped air bubbles. This must be done with considerable care to avoid cracking the thin glass slip. Remove it from the hot plate and allow to cool. Balsam on edges of the slide may be moved with a heated knife or razor blade. Wash in alcohol, removing excess balsam on edges and upper surface of coverslip. The thin section is now completed.

W. FRANK INGRAM

References

Anderson, J. L., "Determination of thin section thickness by the U-stage," *Amer. Mineral.,* **36**:622–624 (1951).

Kennedy, G. C., "The preparation of polished thin sections," *Econ. Geol.,* **40**:353–360 (1945).

Moreland, G. C., "Preparation of polished thin sections," *Amer. Mineral.,* **53**:2070–2074 (1968).

Reed, F. S., and J. L. Mergner, "Preparation of rock thin sections," *Amer. Mineral.,* **38**:1184–1203 (1953).

Rogers, A. F., and P. E. Kerr, "Optical Mineralogy," 2nd ed., New York, McGraw-Hill, 1942 (pp. 3–7).

Weymouth, A. A., "Simple method of making thin sections," *Econ. Geol.,* **23**:323–330 (1928).

Woodbury, J. L., and T. A. Vogel, "A rapid, economical method for polishing thin sections for microprobe and petrographic analyses," *Amer. Mineral.,* **55**:2095–2102 (1970).

ROTIFERS

Introduction. The Rotifer was one of the first of the little animals to attract the attention of the early microscopists and it has been a favorite with naturalists who go hunting under the microscope ever since. One of the smallest of the Metazoa, it has a highly organized internal structure revealed by its transparent integument, which has been the subject of many years of speculation among anatomists and taxonomists from the days of Ehrenberg (1795–1876). By way of introduction to this phylum it is as well to point out that the literature deals at great length with the comparative morphology, physiology, taxonomy, and ecology of rotifers, which may be found in all works on invertebrate zoology, but information on methods of collection, culture, and investigation on the stage of the microscope is usually omitted from textbooks with a few notable exceptions, e.g., Hudson and Gosse's classic "The Rotifera," and Prof. Williamson's monograph in Prichard's "History of The Infusoria." To some extent, therefore, the ingenious novice will devise his own techniques, and the simpler the better.

Habitat. Rotifers live in great numbers in natural sources of fresh water, in water storage, in the mud of dried-up pools, and in the water film on the leaves of bog mosses like *Sphagnum Cuspidatum* and *S. plumulosum.* Many genera are found in ponds by field and farmyard, in wayside tarns, and on the shores of lakes. The genus *Hydatina,* a large and common rotifer, abounds in water green with *Euglena* and algae, and the leaves of water lilies and the pondweed *Potamageton* yield a variety of species if the surfaces are washed off into shallow dishes. The bdelloids become desiccated when deprived of moisture and continue their existence in a state of suspended animation. These rotifers are found in old water channels among the dust, mud, and decayed plant life; a supply of animals is always available if the detritus is stored in jars near to hand. This dust is sprinkled on a petri dish and tap or distilled water added. It should be left for several hours at room temperature, 60–70°F, and then placed on the stage of the microscope and examined under the low-power objective when the cysts will be seen gradually resuming an active state. A beautiful colored rotifer of this order, *Philodina roseola*, encysts in this manner and it is visible to the naked eye as a pink speck in gutters and ornamental bird-baths from which the water has evaporated. To identify any but the commoner species requires the experience won by long attention to the habits and structural minutiae of these curious animals, and it narrows the field of search if the enthusiast has a fair idea where genera and species are most likely to be found. Literature about this is extensive and to be found in books on invertebrate zoology, and reference to these is to be found at the end of this article. Information about the more specialized aspects of the physiology and genetics of rotifers is scattered and infrequent, appearing unexpectedly in obscure journals of experimental biology. Rotifers are selective in their choice of environment, some exclusively so. *Asplancha*, for example, is pelagic and swims all its life in the open waters of lakes, while *Bdelloids* and the sessile *Floscularia* inhabit the littoral and benthic zones of ponds, large and small alike. Others living on leaves seem to have a preference for different plants and none prosper in a situation where active decomposition is taking place.

Collection. To collect rotifers with the plankton of open waters a plankton net is a useful piece of apparatus. It is made of fine bolting silk or nylon and fitted with a handle, extensible in sections, and stout enough to make wide sweeps deeply below the surface. A clear glass bottle is screwed to the apex of the net into which the catch is washed when a preliminary inspection with a hand lens can be made on the spot. Small bottles with metal caps known to pathologists as "universals" make very good dippers in shallow water among the stones and rocks of the pondside. The cap is removed and two fingers are used to close the mouth of the bottle, which is then submerged at the place desired. The fingers are removed and the inrush of water carries with it a representative sample of the microscopic flora and fauna of the immediate vicinity. The dipping should be examined at once or the caps removed, as the larger animals will die from lack of the dissolved oxygen in the water. Leaves and stems of plants are collected in jars or polythene bags thus retaining the surface moisture.

Cultures. The supply of rotifers is everywhere so abundant that elaborate methods of culture, except for the maintenance of pure strains for research purposes, need not be undertaken by the naturalist studying this phylum. Microscopists who need a supply of the commoner genera for demonstration can grow flourishing colonies in shallow glass dishes if they feed them with the cultured *Euglena* or the motile green algae *Chlamydomonas*. Rotifers like *Hydatina* will thrive in cultures of soil amoebae grown as follows. Boil a few wheat grains in 100 cm³ tap water or glass distilled water if the local water supply is too acid, for 10 min. Pour the water into petri dishes, two kernels to each dish, and allow to cool. Cover the cultures and keep out of bright sunlight for a week, when a rich growth of bacteria will be observed. Sprinkle a little garden soil onto each plate and after some days numbers of a small species of amoeba will be available as a food supply. Inoculate with 5 cm³ of pond water containing rotifers and this will furnish a healthy culture indefinitely. Subcultures should be made if the numbers of individuals are seen to diminish in a marked degree. A wide range of temperature is tolerated, but the stock are most vigorous at 60–70°F. Contrary to some opinions infusions of chopped hay do not make suitable food for rotifers. A few kinds are to be found in fresh infusions, but during the process of decomposition they disappear almost completely.

Examination. *The Microscope.* However obtained, the rotifer should be studied alive on the stage of the microscope. Most instruments deal adequately with the larger members of the Infusoria if suitable objectives are chosen. High powers are not essential and require much more careful focusing; they shorten the working distance, a great disadvantage for the manipulation of single specimens, when these have to be placed on a slide from a watch glass or other source. Three objectives should be used, a 24 mm, 16 mm, and 4 mm, and for a preliminary search in a shallow watch glass or on a large glass slide (4 × 8 cm) a 36 mm gives a wide field and a useful depth of focus. As it may be necessary to spend long periods at the microscope, a binocular head is to be preferred to the single eyepiece as it causes less fatigue and strain to the eyes.

Illumination. A lamp fitted with a lamp condenser and iris diaphragm using a 75 w opal bulb is an adequate source of light, but where more critical work requires the use of the oil immersion lens, e.g., in the study of the mastax which contains the parts of the jaws, a high-intensity lamp with low-voltage lamp is needed. Dark ground is particularly rewarding with living specimens as it shows up the finer details by reflection more clearly than by transmitted light. A "spider" of patch stops cut from black card or thin metal sheet is fitted into the filter holder below the substage condenser, and using the 16 mm objective adjustments are made with the condensers and diaphragms in the optical train until the best result is arrived at with a stop of the right size.

Handling. The microscopist will have his own preference for a particular slide or cell. If these are made of glass slips of regulation size, the mechanical stage, if fitted, can be used for searching the field, but if out-size slides, watch glasses, or small petri dishes are chosen, it is better removed. Actively swimming rotifers must be restrained if any details of structure are to be observed and a Rousselet compressor is a useful part of the equipment. To transfer a single specimen from the watch glass to the slide use a fine pipette. A platinum loop or a nylon bristle stuck to a toothpick may be found to be more positive when conveying the object selected to the stage. A cover is then placed over the rotifer on the slide and pressed gently down, the excess water being removed with filter paper from one edge of the coverslip. This method is adopted when chemicals or dyes are introduced into the water on the slide.

Slowing Down Movement. Pressure by the coverslip on the rotifer must be applied very gently as it frequently distorts the animal beyond recognition: Even the top glass of the compressor which is gradually brought down upon the object can have the same effect. Some form of narcotic, skilfully applied, is of considerable assistance in the process of retarding the movements of the more restless swimmers and loopers and holding them in the field of vision. Rousselet's solution is specially recommended for rotifers:

2% cocaine (or eucaine), 3 cm³
90% alcohol, 1 cm³
distilled water, 6 cm³

This should be freshly prepared and experiment made as to the quantity to be added to the water in a watch glass. Alternatively a 1 % solution of formalin can be added one drop at a time to a fair amount of water as for the above solution. Rapid movement will cease in 10 min or so and kill the rotifers within an hour.

Vital Staining. Solutions of Neutral Red or Methylene Blue into which the rotifers are pipetted show up the finer details of structure.

Permanent Preparations. Narcotize the rotifers in a solution of eucaine hydrochloride and fix in an aqueous solution of osmic acid. Stain with Delafield's hematoxylin so diluted as to be almost clear, or with weakly colored (palest pink) alum carmine. Dehydrate specimens and mount in Canada balsam.   H. W. Denyer

References

Ehrenberg, "Die Infusionthiere als Volkkommenere Organismen," 1838.

F. Dujardin, "Histoire Naturelle des Zoophyte," 1841.

Hudson and Gosse, "The Rotifera," 1861.

Williamson, "The rotifera," in Prichard's "History of the Infusoria," 1861.

Marcus Hartog, "Rotifera," in "Cambridge Natural History," 1901.

Jennings, "The rotifera," in "Synopses of North American Invertebrates.

J. Donner, "Radertiere (Rotatorien)," 1956.

S

SAFRANIN

Introduction

According to Ehrlich et al. (1910), safranin was introduced into microtechnique by Hermann (1893). It is commonly employed either as the strong alcohol solution introduced by Blanc (1883), in the aniline water solution introduced by Babes (1888) on the suggestion of Zwardemaker (1887), or the mixed solvents solution of Johannsen (1940).

The safranins are a fairly large group of azin dyes but in microtechnique the stain specified is almost invariably "Safranin O." This is a mixture of trimethyl phenosafranins, the optimum mixture for microscopy, according to Conn 1961, having an absorption maximum at 530 mμ. So many mixtures have been sold as "safranin" that it is difficult to know what earlier writers may have been using.

Safranin is a "nuclear stain" but is rarely used in animal microtechnique. It is widely used in plant microtechnique for staining both nuclei and lignified tissues.

Formulas and Methods

Babes 1887 (*Z. wiss. Mikr.*, **4**:470)
FORMULA: water 100, aniline 2, safranin 7
PREPARATION: Heat to 60°C for 1 hr stirring frequently. Cool. Filter.
METHOD: [sections of osmic-fixed material]→ water→ stain, until nuclei bright red, 1 hr to 10 days→ water, wash→ balsam, via usual reagents
NOTE: This is what is usually meant by *Babes Safranin* even though Babes had previously (*Arch. Ophthal.*; *Arch. mikr. Anat.*, **21**:356) recommended both alcoholic and aqueous solutions. A slight modification of this is given by Langeron 1934, (100) as "Babes-Langeron-Dubosq." This method with a light green counterstain is sometimes called *Benda's stain*.

Böhm and Oppel 1907 (Böhm and Oppel 1907, 113)
FORMULA: water 85, 40% formaldehyde 5, 95% alc. 10, phenosafranin 1

Chamberlain 1915 (Chamberlain 1915, 51)
FORMULA: water 100, 95% alc. 50, water soluble safranin 0.5, alc. soluble safranin 0.5
PREPARATION: Dissolve each dye in the appropriate solvent and mix the solutions.
NOTE: For a rationalization of this procedure see Conn 1961:120. It is nowadays customary to substitute Safranin O for this mixture.

Gray and Pickle 1956 (*Phytomorph.*, **6**:196)
FORMULA: water 100, glycerol 14, ferric alum 5, safranin O 2
PREPARATION: Heat water to 50°C, add the glycerol and ferric alum. Stir to complete solution. Pour the hot solution, with rapid stirring, onto the dye. Cool and filter.
METHOD: [sections]→ water→ dye solution, 5 min→ water, until no more color comes away→ [counterstain]→ balsam, via usual reagents

Johansen 1940 (Johansen 1940, 62)
REAGENTS REQUIRED: A. 95% alc. 25, water 25, methyl cellosolve 50, safranin 0.1, sodium acetate 1, 40% formaldehyde 2; B. sat. sol. picric acid in 95% alc. 65, 95% alc. 35.
PREPARATION OF A: Dissolve the dye in the cellosolve. Add first alc., then water. Then add remaining ingredients.
METHOD: [sections]→ water→ A, 24–48 hr→ B, until differentiated→ [counterstain]→ balsam, via usual reagents

Pfitzner 1881 (*Morph. Jahrb.*, **7**:289)
FORMULA: water 60, 95% alc. 40, safranin 0.3

Rawitz 1895 (Rawitz 1895, 76)
REAGENTS REQUIRED: A. 20% tannin; B. 2% potassium antimony tartrate; C. sat. sol. safranin; D. 2.5% tannin
METHOD: [sections of chrome-fixed or mordanted material]→ water→ A, 24 hr→ B, 24 hr→ rinse, D, 24 hr→ water, until no more color comes away→ D, if further differentiation required→ balsam, via usual reagents

Roskin 1946 (Roskin 1946, 152)
STOCK SOLUTION: water 50, 95% alc. 50, safranin 3.3
PREPARATION: Dissolve the dye in alc., mix with water.
WORKING SOLUTION: stock 20, 50% alc. 80

Sémichon 1920 cited from 1934 Langeron (Langeron 1934, 500)
FORMULA: water 50, 90% alc. 50, safranin 0.5, 40% formaldehyde 1
PREPARATION: Dissolve dye in alc. then add other ingredients.

METHOD: [sections]→ 70% alc.→ stain, 30 min to 2 days→ absolute alcohol, until differentiated→ balsam, via xylene

Zwaademaker 1887 (*Z. wiss. Mikr.*, **4**:212)
REAGENTS REQUIRED: A. 95% alcohol 50, safranin
1.5, sat. aq. sol. aniline 50; B. absolute alc.
PREPARATION OF A: Dissolve dye in alc., then add
aniline.
METHOD: [sections of osmic-chromic-acetic material]
→ 95% alc.→ A, overnight→ B, until differentiated→
balsam, via xyelene

Mordanting has not commonly been recommended
before safranin staining. The use of formaldehyde for
this purpose was recorded by Rouville (1907) and is
erroneously attributed to Sémichon by Langeron (1942).
Federici (1906) used a 1% solution of iron alum to
differentiate materials stained in hematoxylin and safranin
but records no effect on the safranin. Popham (1948),
in a general discussion of mordanting plant tissues, deals
only with safranin as a nuclear stain. Foster (1934) and
Sharman (1943) mention the use of safranin for staining
lignified tissues before staining meristematic tissues with
iron alum and tannic acid. Northern (1936), discussing
the same type of staining, says, "If the safranin washes
out easily, it is desirable to stain with the tannic-acid–
ferric-chloride combination before staining with the
safranin." This appears to be the only suggestion in the
literature that iron alum could hold safranin in lignified
tissues and there is no record of the use of safranin in
iron alum solutions except Gray and Pickle 1956.

Neither celestine blue B, nor any other oxazine dye,
appears to have been used for plant histology.

PETER GRAY

References

To save space the undernoted books are referred to in
the above article by author and date only. Literature
references in the formulas are in customary form.

Blanc, H., "Structure des cupules membraneux an
'calcioli' chezles amphipodes," *Zool. Anz.*, **6**:370–372
(1883).
Böhm, A., and A. Oppel, "Manuel de Technique Micro-
scopique, traduit de l'allemand par Etienne de
Rouville," 4th ed., Paris, Vigot, 1907.
Chamberlain, C. J., "Methods in Plant Histology," 3rd
ed., Chicago, University of Chicago Press, 1915.
Conn, "Biological Stains," Baltimore, Williams and
Wilkins, 1961.
Federici, F., "Un nuovo metodo per la colorazione
specifica delle Mastzellen," *Anat. Anz.*, **29**:357–361
(1906).
Foster, A. S., "The use of tannic acid and iron chloride
for staining cell walls in meristematic tissues," *Stain
Technol.*, **9**:91–92 (1934).
Hermann, F., "Methoden zem Studium des Archiplas-
mas und der Centrosomen tierischer und pflanzlicher
Zellen," *Ergebn Anat. Entw-Gesch.*, **2**:23–36 (1893).
Johansen, D. A., "Plant Microtechnique," New York,
McGraw-Hill, 1940.
Langeron, M., "Précis de Microscopie," 5th ed., Paris,
Masson, 1934.
Northern, H. T., "Histological applications of tannic
acid and ferric chloride," *Stain Technol.*, **11**:23–24
(1936).
Popham, R. A., "Mordanting plant tissues," *Stain
Technol.*, **23**:49–54 (1948).
Rawitz, B., "Leitfaden für Histologische Untersuchun-
gen," Jena, Fischer, 1895.
Roskin, G. E., "Mikroskopecheskaya Technika,"
Moscow, Sovetskaya Nauka, 1946.
Rouville, E. de, "Manuel de Technique Microscopique,"
Paris, 1907.
Sharman, B. C., "Tannic acid and iron alum with
safranin and orange G in studies of the shoot apex,"
Stain Technol., **18**:105–111 (1943).

SAND

Sand can be defined in a number of ways, perhaps most
commonly as an aggregate of mineral or rock grains
greater than 1/16 mm and less than 2 mm in diameter
(Fig. 1). Most sand is composed mainly of quartz and
may also include opal, siliceous sinter, silica glass, and
especially chert, with very small amounts of other
minerals possible. Garnet, illmenite, and feldspar sands
are among others found occasionally in both the modern
environment and the fossil record; these frequently contain some quartz (Pettijohn et al., 1965).

Upon weathering of granitic rocks, individual quartz
grains are released and may be transported away from
their point of origin across the face of the land by running
water, wind, gravity, or glacial ice and deposited. Some
of this material is carried across the continental shelf
and into the deep ocean by waves and currents.

Thin sections of sands and sandstones have been used
to determine such diverse parameters in ancient sedimentary deposits as the source rock from which a given
sediment originated, the transportation distance to the
site of deposition, the physical and chemical environment in which the grains were finally deposited, burial
temperature, type of water movement which carried the
grains to their depositional site, and recent outcrop
weathering history (Pettijohn et al., 1965).

Thin sections of sands are prepared for light microscopy as follows. The loose sand is washed in acid and
distilled water to remove adventitious material; an ultra-

Fig. 1 A group of sand grains from a Sahara Desert
dune sand near Cairo, Egypt. Light micrograph, 5.5.

sonic bath is used if necessary. The grains are then mounted with Lakeside 70 cement on a microscope slide and a standard thin section is prepared for light microscopic examination (Krumbein and Pettijohn, 1938); the slide may be viewed both in plane polarized light and under crossed nicols.

Luminescent petrography is a new technique that has recently been used to distinguish solution and overgrowth phenomena in sands and sandstones where other petrographic techniques have failed. Luminescence reveals the detrital fabric so that size, shape, and degree of rounding of the detrital grains can be seen (Sipple, 1968). A luminescent attachment to a petrographic microscope is used (Sipple, 1965); electron bombardment of a thin section produces visible light. Preparation of samples is simple, as ordinary unpolished and uncoated thin sections can be used. However, they must be epoxymounted: Lakeside 70 is not satisfactory since heat from the electron beam will produce boiling in a few minutes. Polished thin sections are used so that light scattering is kept to a minimum; here magnifications of $10 \times -20 \times$ are best, but even at $200 \times$ luminescent effects can be observed in great detail using special long-distance objectives.

All of the above techniques are concerned with cross sections of sand grains; another meaningful parameter is the surface texture (roughness) of individual sand particles. This type of study is difficult with the techniques indicated above because of the limited resolving power of the instruments concerned.

The transmission electron microscope (TEM) has been used since about 1960 and the scanning electron microscope (SEM) since 1967 to study the surface textures of sand grains. These instruments have revolutionized the examination of sand as well as other sedimentary particles; in particular the recent use of scanning electron microscopy has permitted the distinction of river grains, littoral grains, eolian grains, glacial grains, and those subjected to weathering. It is therefore possible to determine the history of certain sand deposits in the geological record in great detail (Krinsley and Margolis, 1969).

Some idea of the differences between the light, TEM, and SEM can be obtained from Table 1, which has been adapted and modified from Hay and Sandberg (1967). The transmission instrument, of course, has the highest resolution, but specimen preparation is somewhat involved. The sample is first described with the binocular microscope; then the grains are cleaned as indicated above. After cleaning, the grains must be replicated using an acetate peel technique, as they are much too thick to be examined directly with the TEM. The peel is plated and shadowed in a vacuum using a heavy metal and carbon; the original acetate peel is removed with acetone leaving the metal-carbon replica supported on a 200 mesh metal grid. The thickness of this replica is only about 200 Å and it is placed directly in the transmission instrument for viewing (Krinsley and Takahashi, 1964).

Many of the problems involved in transmission electron microscopy are due to the necessity of using the peel technique; these can mostly be eliminated by using the SEM (Figs. 2 and 4). Grains are observed directly without the need for replication, thus avoiding most artifacts and distortion. More detail can be seen with the SEM, indicating that many small features are not observed or are somewhat distorted with the replica method. A number of grains can be directly observed at one time with the SEM before selecting a particular grain to be zoomed in on and photographed at high magnification. Most microscopes are supplied with a tilting stage so that considerably more than half of a given grain can

Table 1[a,b]

	Stereoscopic Binocular Microscope	Compound Light Microscope	Transmission Electron Microscope	Scanning Electron Microscope
Useful Magnification Range	$8 \times -212 \times$	$8 \times -2500 \times$	$200 \times -1,000,000 \times$	$15 \times -130,000 \times$
Resolution	20,000 Å	2500 Å	2 Å	200 Å
Depth of Field at:				
50 ×	200 μ	20 μ	—	10,000 μ (=1 cm)
500 ×	—	2 μ	800 μ	1000 μ
5000 ×	—	—	80 μ	100 μ
50,000 ×	—	—	8 μ	10 μ
Working Distance at:				
50 ×	40 mm	16 mm	—	20 mm
500 ×	—	0.4 mm	2 mm	20 mm
5000 ×	—	—	2 mm	20 mm
50,000 ×	—	—	2 mm	20 mm
Field of View at:				
50 ×	4000 μ	2100 μ	—	2000 μ
500 ×	—	210 μ	200 μ	200 μ
5000 ×	—	—	20 μ	20 μ
50,000 ×	—	—	2 μ	2 μ
Illumination Ordinarily Used	Incident dark field	Transmitted bright field	Resembles transmitted bright field	Resembles incident dark field
Image Formation System	Light optical	Light optical	Electron optical	Nonoptical [e]

[a]Data have been determined by experiment or calculated from information provided by manufacturers of the best instruments available.

[b]Adapted from Hay and Sandberg, 1967.

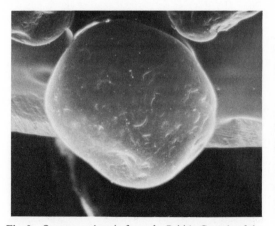

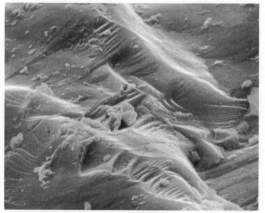

Fig. 2 Quartz sand grain from the Pebble Gravels of the Thames Valley, England. Photograph taken with the scanning electron microscope in the emissive mode. The surface contains crescent-shaped breakage patterns which probably originated mechanically in a high-velocity stream. (36 ×.)

Fig. 4 Portion of a quartz sand grain from a core off the Argentine coast showing glacial abrasion. Photograph taken with the scanning electron microscope in the emissive mode. Note the three-dimensional appearance (212 ×.)

be observed. Finally, depth of focus and certain other SEM features give a simulated three-dimensional photograph which simplifies interpretation.

Preparation for SEM observation is quite simple. The sand grains are cleaned, mounted on a specimen stub, and a heavy metal such as platinum-palladium or gold is evaporated in a vacuum upon the specimen while it is rotated and tilted on a special stage. About 300 Å of heavy metal is deposited on the specimen to prevent charging; the stub is then placed in the microscope for observation. The specimen is bombarded with electrons from a tungsten filament, and emitted electrons are picked up by a detector and displayed on a cathode ray screen (for details see Nixon, 1969).

Another technique which has been used in SEM work is cathodoluminescence (CL). This technique involves the bombardment of the specimen by the primary beam of

electrons as in the above mode; luminescence emitted naturally by the specimen or as a result of excitation by the primary beam is detected (Fig. 3). More detail can be observed with this technique than with the luminescent microscope, as resolution here is considerably better, perhaps better than 1000 Å. Additionally the same grain area may be studied with a nondispersive X-ray attachment and the chemical composition related to the luminescence and the topography.

When the surfaces of the grains are observed in the cathodoluminescent mode, a thin layer below the surface can frequently be seen which would otherwise be invisible; thus more of the previous history of the grain can be studied than was heretofore possible. New dimensions have thus been added to the study of sands and sandstones through the use of current advances in microscopy.

DAVID KRINSLEY

Fig. 3 Same grain as in Fig. 2, but taken with the scanning electron microscope in the cathodoluminescent mode. A series of healed cracks can be discerned below the surface, probably at a depth of about 1 μ. (36 ×.)

References

Hay, W. H., and P. A. Sandberg, "The scanning electron microscope, a major breakthrough for micropaleontology," *Micropal.,* **13**:407–418 (1967).

Krinsley, D., and S. Margolis, "A study of quartz grain surface textures with the scanning electron microscope," *Trans. N.Y. Acad. Sci.,* **31**:457–477 (1969).

Krinsley, D., and T. Takahashi, "A technique for the study of surface textures of sand grains with electron microscopy," *J. Sed. Petr.,* **34**:423–426 (1964).

Krumbein, W. C., and F. J. Pettijohn, "Manual of Sedimentary Petrography," New York, Appleton-Century-Crofts, 1938, (549 pp.).

Nixon, W. C., "Scanning electron microscopy," *Contemp. Phy.,* **10**:71–96 (1969).

Pettijohn, F. J., P. E. Potter, and R. Siever, "Geology of Sand and Sandstone," Ind. Geol. Sur. and Dept. of Geol., Univ. of Ind., 1965, (205 pp.).

Sipple, R. F., "A simple device for luminescence petrography," *Rev. Sci. Instr.,* **36**:1556–1558 (1965).

Sipple, R. F., "Sandstone petrology, evidence from luminescence petrography," *J. Sed. Petr.,* **38**:530–554 (1968).

SAPROLEGNIALES

Light Microscopy. The majority of descriptive, developmental, and morphological studies of saprolegniaceous fungi, as well as their identification, require microscopic examination of living material. Temporary wet mounts are routinely used for this purpose and are prepared by placing either a whole specimen or a various portion of a specimen in 1–2 drops of distilled water on a clean glass slide and adding a coverslip. Care should be taken in positioning filamentous forms to avoid excessive tangling of the filaments. Hanging-drop mounts are particularly useful in following the development of reproductive structures, germlings, parasitic species, and spore motility. These are prepared by placing a drop of water containing material to be examined on a coverslip, which is then inverted over the well in a depression slide. The rate of evaporation can be reduced by rubbing petroleum jelly around the edge of the well before adding the coverslip. The few naturally occurring parasites of aquatic plants and animals are handled in a similar manner. Parasites of algae, microscopic animals, etc. can be mounted directly in water and a coverslip added to the preparation. Species associated with larger animals, such as fish, must be removed from the host and thoroughly washed with distilled water before making a wet mount.

Permanent and semipermanent mounts are used primarily for preserving type specimens, voucher species, and reference material on glass slides; these do not, however, replace the use of live specimens. Dilute Amman's (lactophenol) [1 part phenol in 39 parts glycerin, 1 part lactic acid, 9 parts distilled water, with or without 0.01 g of either acid fuchsin or cotton blue] is an excellent mounting medium and if tightly sealed will last for several years. If a wet mount is to be preserved, the *excess* water is blotted off, 1–2 drops of mounting medium and a round coverslip are added, and the preparation allowed to dehydrate for 1 week. The coverslip is then sealed to the slide with clear fingernail polish. Alternatively, living material can be placed directly into Amman's on a clean glass slide and subsequently treated as above. More complex techniques for making permanent mounts have been described by Koch (1966) and Johnson (1956). Special techniques and schedules for staining various cell organelles, zoospores, and flagella are also described in the Koch paper.

Electron Microscopy. Very few saprolegniaceous fungi have been subjected to transmitted electron microscopy and insofar as is known there are no reports of studies of these fungi using the scanning microscope. Except for a report by Moore and Howard (1968), describing the ultrastructure of oosporogeny in a species of *Saprolegnia*, previous workers have dealt primarily with the fine structure of zoospore cysts and flagella. Procedures used in preparing the latter are briefly outlined by Meier and Webster (1954).

ROLAND SEYMOUR

References

Johnson, T. W., Jr., "The Genus *Achlya*," Ann Arbor, University of Michigan Press, 1956 (180 pp.).

Koch, W. J., "Fungi in the Laboratory," Chapel Hill, N.C., Book Exchange, 1966 (113 pp.).

Meier, H., and J. Webster, "An electron microscope study of cysts in the Saprolegniaceae," *J. Exp. Bot.,* **5**:401 (1954).

Moore, R. T., and L. K. Howard," Ultrastructure of oosporogeny in *Saprolegnia terrestris*," *J. Cell Biol.,* **39**:94a (1968).

SCANNING ELECTRON MICROSCOPE

Historical Introduction. The science of electron optics grew up in the 1930s, and the German physicists, who played such a large part in laying the foundations of the subject, were well aware that their researches might lead to new electron optical instruments of various types. However, their development of the conventional transmission electron microscope was so successful that work on other possible lines received less attention and was finally brought to an end by the outbreak of war.

In the conventional microscope, electrons pass through the object and are then focused by two or more electron lenses to form a greatly magnified image on a fluorescent screen or photographic film. The process is exactly analogous to that which occurs in an ordinary optical microscope when a transparent object is being viewed; information is collected simultaneously from each point of the object and is displayed simultaneously on the screen or film. There is, however, a quite different method of producing an electron-optical image which is illustrated diagrammatically in Fig. 1.

Electrons from a hairpin tungsten filament F are formed into a very fine electron probe by successive demagnification by lenses L_1, L_2, and L_3 so that when the probe strikes the object P, it may have a diameter of about 0.01 μm. The probe is caused to move over the surface of P in a zigzag raster by two pairs of deflecting coils (one pair A_1, A_2, is shown) which carry currents from a saw-tooth generator G. The same saw-tooth currents also traverse the deflecting coils of a cathode-ray tube T, to produce on the face of this tube an identical, but very much larger zig-zag raster. Finally, electrons leaving P are collected by C and the resulting current is amplified and used to control the brightness of the cathode-ray tube. Thus there is a point-by-point correspondence between the raster on P and that on the face of T. Moreover, the brightness of each point of the latter is governed by the number of electrons leaving the corresponding point of the surface of P, so the picture built up on the cathode-ray tube must, in some sense, be an image of this surface. The magnification is determined by the relative sizes of the two rasters and this can be varied at will by controlling the magnitudes of the currents flowing in the two sets of deflecting coils.

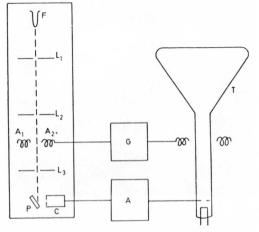

Fig. 1   Schematic diagram of scanning electron microscope.

An instrument based on the above principles has become known as a "scanning electron microscope." As described, the formation of the image depends on secondary electrons or electrons reflected from the object, but it would obviously be possible to arrange matters so that the brightness of the cathode-ray rube was controlled by electrons which had passed through a transparent object. In fact, the first true scanning electron microscope, constructed by von Ardenne in 1938, made use of transmitted electrons, though the recording system was quite different from that described above.

When a transparent object is to be examined, the conventional transmission electron microscope will, as a rule, be the most convenient instrument for the purpose. However, recent work by Crewe and his colleagues has shown that, in particular cases, important advantages can be secured by the application of the scanning principle. In the present article we shall restrict ourselves to the more common use of the scanning microscope, in which opaque specimens are examined and in which the formation of the image depends on secondary electrons or electrons reflected from the specimen.

In 1942, Zworykin, Hillier, and Snyder reported the construction of a scanning electron microscope in which thick objects could be examined with a resolution of about 0.05 μm. In this instrument the ancillary electronic equipment was very complicated and the published micrographs show a good deal of background noise. Presumably for these reasons, development of this instrument was not followed up.

Research on the scanning electron microscope was begun in the Engineering Laboratory of the University of Cambridge in 1948 and has been actively pursued since that date. As a result of this work, the first commercial scanning electron microscope was marketed by the Cambridge Instrument Company in 1965, under the trade name Stereoscan.

Factors Limiting Resolution. The resolution of a scanning electron microscope is limited by two quite different sets of circumstances. One of these is concerned with the fundamental laws of electron optics, while the other depends on the penetration of electrons into the object and each, independently, suggests a practical limit to resolution of about 0.01 μm with techniques available at the present time.

For an investigation of the electron-optical limitations, the starting point is an arbitrary choice for the number of lines N in the final picture on the cathode-ray tube. Assuming this picture to be about 10 cm², N might be made equal to 1000 for reproduction of high quality. To make full use of the detail that can be displayed with so large a value of N, the electron probe in the microscope must be sufficiently fine to permit discrete resolution of an equal number of lines in the area scanned on the object and the probe will then be providing information about N^2 separate picture elements.

It is sufficiently accurate for our purpose to assume the probe to have square cross section of side d_0 so that if j is the current density in the electron spot, t the time taken to scan the object and e the electronic charge, the number n of electrons falling on each picture element is given by

$$n = \frac{jd_0^2 t}{N^2 e} \qquad (1)$$

The emission of electrons from the cathode is, how-

ever, a random process, so n will be subject to statistical fluctuations of r.m.s. value $(n)^{\frac{1}{2}}$. This fluctuation causes noise in the final picture and we may take the ratio $n/(n)^{\frac{1}{2}} = (n)^{\frac{1}{2}}$ to be the basic signal/noise ratio for the element being scanned.

The effect of noise of this kind has been exhaustively studied in connection with television reproduction and it is found that an area of brightness B in the final image will not be distinguishable from an adjacent area of brightness $B \pm \Delta B$, unless the empirical relation

$$(n)^{\frac{1}{2}} \geq \frac{5B}{\Delta B} \qquad (2)$$

is satisfied. Substituting from equation (1), we have

$$25\left(\frac{B}{\Delta B}\right)^2 \leq \frac{jd_0^2 t}{N^2 e} \qquad (3)$$

It thus appears that the minimum value of d_0 is limited not only by the values of N and $B/\Delta B$ that we wish to achieve in the image, but also by j and t. However, j itself is limited by the well known Langmuir relation

$$j = j_0 \{1 + (eV/kT)\} \sin^2\alpha \qquad (4)$$

where j_0 is the current density emitted by the cathode, V the accelerating voltage, k Boltzmann's constant, T the absolute temperature of the cathode and α the semi-angle of the cone of electrons which converge to form the final spot. If, as is usual, a tungsten hairpin cathode is used, j_0 and T are fixed and, to avoid excessive penetration of electrons into the specimen, V should not greatly exceed 30 kV.

We are thus left with α as the only variable that can be adjusted to make j as large as possible but, once again, there is a strict limit to what can be done. The aberrations of the final lens increase rapidly with α and these aberrations must be kept small in comparison with d_0.

Returning to equation (3), the only quantity that we have not yet considered is t, the time required to build up the picture. Here, a practical limit of about 5 min suggests itself; a longer time is very tedious and can cause trouble from movement of the object.

Taking all these factors into consideration, it is possible to determine the minimum value of d, the effective probe size when lens aberrations are taken into account, for any given set of operating conditions. Details of the calculation have been given elsewhere (Oatley et al., 1965) and a typical set of results is plotted in Fig. 2. These relate to a working voltage of 20 kV, $B/\Delta B = 10$, and magnetic lens performance which can be achieved without too much difficulty.

It would clearly be of great advantage to reduce the time t corresponding to any particular value of d and the best hope of doing this lies in the development of cathodes giving greater current densities of emission. Some success has been obtained by the use of field-emission cathodes or of thermionic cathodes in which the active emitter is lanthanum hexaboride; oxide cathodes of the dispenser type have also been used. Nevertheless, the tungsten hairpin cathode is the source of electrons in the great majority of scanning microscopes at the present time.

Assuming an accelerating voltage of 20 kV, the primary electron beam will penetrate an average specimen to a depth of the order of 1 μm producing secondaries along

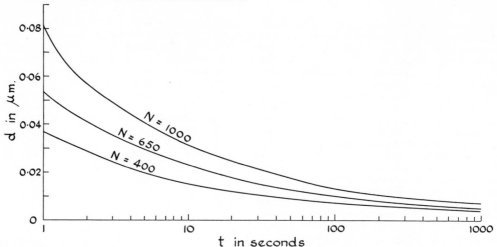

Fig 2 Relation between effective spot diameter d and scanning time t (probe size d, taking lens aberrations into account, as a function of t).

its path. These secondaries move off in random directions, but few of them reach the surface. In fact, most of the secondary electrons which escape through the surface will have originated within about 0.01 μm of the surface, and calculations based on a simple model suggest that more than half of them will emerge within a distance of 0.005 μm from the point where the primary electron entered. Since the number of secondaries reaching the collector is likely to be determined, at any rate partly, by conditions at the point of emergence, the spreading of the secondaries must set a limit to the resolution that can be achieved. In the light of the above calculations this limit should be about 0.01 μm, i.e., of the same order as that set by the laws of electron optics.

In practice a resolution better than 0.01 μm has been obtained with a small number of specimens. For the average specimen a more realistic figure might be 0.02 μm.

Practical Details. The foregoing discussion indicates that, in the design of a scanning electron microscope, provision should be made for operation of the instrument with an electron spot size of 0.01 μm or less. If a tungsten hairpin cathode is used, the "crossover" of the electron gun will have a diameter of about 100 μm so a total demagnification of, say, 2×10^4 must be provided. In a column of reasonable length it is difficult to achieve so large a value with two lenses, and three lenses are commonly used. Magnetic lenses are generally preferred to the electrostatic variety, since the former have lower aberration coefficients and are easier to clean. With a three-lens system, the working distance between the final lens and the object can be about 0.5 cm when the highest resolution is required; at lower magnifications the distance can be much larger.

The column of the microscope and the specimen chamber must, of course, be evacuated. So far as operation of the instrument is concerned, a pressure of about 10^{-5} torr is low enough, but for many applications, a much lower pressure may be needed to prevent contamination of the object and there appears to be no difficulty in making provision for this.

The secondary electron current leaving the object is of the order of 10^{-12} A and, so far as possible, this must be amplified without degradation of the signal/noise ratio. In a convenient form of detector the electrons enter a brass box through a metal gauze window, having been attracted to the box by a potential difference of a few hundred volts applied between box and object. Once inside, the electrons are accelerated to raise their energy to about 10 keV and are then focused on to a plastic scintillator. The light thus produced is conveyed by a light pipe to a photomultiplier mounted outside the evacuated specimen chamber, and the multiplier provides most of the required amplification. A convenient amplifier, with variable gain, is connected between the photomultiplier and the display cathode-ray rube. For special purposes, where signal/noise ratio is not of the first importance, other detectors using channel multipliers, semiconductors, or field-effect transistors have been found convenient.

Reference to Fig. 2 shows that, for the highest possible resolution, the object must be scanned for a time t of the order of 1000 sec. Theoretically it is immaterial whether the object is scanned once, taking a total time t, or p times, with each scan occupying t/p sec. In practice, the former procedure is greatly to be preferred since, with multiple scanning, any drift in the position of the object will blur the final picture. With a single scan, drift may cause slight distortion of the picture but will not impair its sharpness.

Although a high-definition cathode-ray tube with a slow single scan is used for photographic recording, provision must also be made for visual display, so that the appropriate area of the specimen can be selected and brought into focus. For this purpose a second cathode-ray tube with afterglow screen is generally provided. With this tube the object may be scanned at a rate of about once per second.

In the scanning electron microscope the magnification is the ratio of the length of the side of the area scanned on the cathode-ray tube to the length of the corresponding side of the area scanned on the object. Since both scans are brought about by current from the same source, magnification can be controlled by varying the ratio of the currents flowing through the two sets of scanning coils respectively. It is a very valuable feature of the instrument that the magnification can be rapidly changed over a range of, say, 20 to 10^5.

The Nature of the Image. Since there is a one-to-one

correspondence between the scanning of points on the object and points on the screen of the cathode-ray tube respectively, the image built up on the screen must represent a map of some property of the object. The nature of this map will depend on the factors which control the number of electrons leaving any point of the object and these must now be considered.

Electrons leaving the object may be divided into two groups, which we shall term secondary electrons and reflected electrons, respectively. The true secondaries leave the surface of the specimen with energies of only a few electron volts. With the arrangement of the apparatus described above, these electrons will be attracted towards the collector and will provide the greater part of the signal which ultimately controls the brightness of the cathode-ray tube. The reflected electrons, on the other hand, are primaries which, passing near an atomic nucleus, have been turned back with little loss of energy. They therefore leave the specimen with energies of several thousand electron volts and their subsequent trajectories are relatively little affected by the electric field between specimen and collector. Thus, only a few of them, which happen to be traveling in the right direction, will enter the collector and contribute to the output signal.

Occasionally it is advantageous to modify the apparatus to cause the bulk of the output signal to be provided by the reflected electrons. For example, this might be worthwhile when examining an object at such a high temperature that it was emitting thermionically. A weak retarding field between object and collector would then reject both thermionic and secondary electrons, while allowing the reflected electrons to enter the collector. In general, however, the most useful results are obtained when the output signal results from secondary electrons, and this condition will be assumed in what follows.

One obvious factor causing variation in the number of electrons leaving different points of the object is variation in chemical constitution or surface condition, leading to change in the secondary emission coefficient. If this were the only factor, the image would be a map of the relevant chemical or physical properties of the object—useful for certain specialized purposes, but not necessarily bearing any resemblance to an optical micrograph of the same object.

Fortunately, a much more potent cause of variation in the secondary emission current reaching the detector is the change, from point to point, in the angle between the incident electron beam and the surface. Experience shows that a change of as little as 1° or 2° is sufficient to produce a detectable change of brightness in the image. Other major variations in brightness result from changes in the extent to which electrons leaving any particular point are prevented by adjacent topographical features from reaching the detector. The great importance of these mechanisms lie in the fact that they are closely analogous to those which operate to produce contrast when an illuminated object is observed by the eye; light or shade in the image formed on the retina depends on the quantity of light reflected from the object and this, in turn, is governed by the inclination of the relevant portion of the object to the line drawn to it from the eye, and to the extent to which light from any portion is intercepted by adjacent portions. Because the two processes are analogous, the images formed in the two cases are of a similar nature—hollows appear dark, projecting regions cast shadows, and so forth. In consequence, the eye

which is accustomed to the interpretation of a two-dimensional picture, in terms of a three-dimensional world, has little difficulty in interpreting the image formed in a scanning electron microscope.

Another factor which renders the image particularly lifelike is the very large depth of field of the scanning instrument. Aberrations in present-day electron lenses are very large in comparison with those found in optical microscope objectives, so it is possible to use the former only by restricting the angle of the electron beam by means of very small apertures. It thus comes about that the depth of field in the scanning electron microscope is some hundreds of times as great as that in an optical microscope used under comparable conditions. The marked three-dimensional effect and the large depth of focus are well illustrated in the micrograph shown in Fig. 3.

When necessary these effects can be enhanced by the use of stereo-pairs of micrographs. Two pictures are taken, corresponding to slightly different views of the object, and are then examined in an ordinary stereo-viewer.

Advantages of the Scanning Electron Microscope. The scanning electron microscope is necessarily a fairly expensive instrument to manufacture. The electron-optical components and stabilized power supplies must be of a quality comparable with that needed for an ordinary transmission electron microscope, and the scanning instrument requires more electronic equipment. It is thus likely to cost rather more than the transmission microscope and many times as much as a high-quality optical microscope. It can therefore hope to compete with these well-established instruments only when it can offer special advantages in performance.

The resolution that can be obtained with the scanning microscope is about 20 times as good as that provided by an optical microscope, but for the examination of transparent specimens, it falls short of the performance of a conventional transmission electron microscope by another factor of 20. When specimens which are opaque to electrons have to be examined, the transmission microscope can operate only by way of replicas but,

Fig. 3   Aluminum recrystallized from the melt.

even so, its resolution is still better than that which can be attained with the scanning instrument. However, there are many opaque specimens of which replicas cannot be made satisfactorily, and here the scanning microscope comes into its own. For example, replication is difficult with objects which are heavily indented (e.g., metallurgical specimens resulting from brittle fracture) and quite impossible with objects, such as thermionic cathodes, which cannot be exposed to air and which may have to be examined at high temperatures.

For many purposes the outstanding advantage of the scanning electron microscope is its very large depth of field. The comparison with optical microscopes in this respect has been discussed previously, but the superiority of the scanning instrument over the transmission electron microscope is perhaps less obvious. It arises from the fact that, in the latter, the object must necessarily be thin enough to transmit the electrons, so the inherently great depth of field of the lenses is largely wasted. Experience has shown that, for a very wide range of specimens, the scanning microscope gives micrographs which are greatly superior to those produced by any other instrument. Moreover, some of the most striking results have been obtained at quite low magnification, where resolving power is unimportant.

It has been pointed out that the process of contrast formation in the scanning electron microscope is very similar to that which operates when an object is viewed directly by eye. There is, however, one important difference. With visual examination, the object is generally illuminated by light falling on it from a variety of directions, while the reflected light entering the eye is limited to a narrow cone of rays. With the scanning microscope, exactly the opposite is the case; the object is "illuminated" by the very narrow electron probe, but the secondary electrons can travel to the wide-angle collector by a variety of paths. One result of this difference is that the image represents the magnified view of the object that would be seen by looking along the incident electron beam. A more important result is that the image reveals details of structure inside relatively deep fissures or holes on the surface of the object. The very narrow incident electron probe is able to penetrate such apertures to produce secondaries from the interior walls. The fact that the secondaries may reach the collector by devious paths, perhaps by repeated reflection from the walls of the cavity, makes no difference to the formation of the image. Of course, many of them get trapped, but if the magnification is not too high, enough survive to yield a satisfactory output signal. The micrograph shown in Fig. 4 illustrates this effect.

A quite different advantage of the scanning electron microscope arises from the way in which electrons travel from the specimens to the collector. Most of the secondaries have energies of only a few electron volts, and since they are emitted in all directions, it is usual to attract them to the collector by maintaining the latter at a positive potential of a few hundred volts with respect to the specimen. In general, therefore, the secondaries travel to the collector along curved paths, and moreover, their trajectories are greatly affected by any variation in the electric field near the specimen, where the electrons are traveling rather slowly. It follows that, if there are any changes in potential from point to point of the surface of the specimen, the resultant fields will influence the number of electrons reaching the collector in such a way as to produce additional contrast in the image. If the

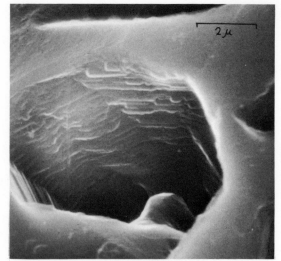

Fig. 4 Crystal growth of graphite with negative whiskers due to *lamphanen*.

aperture of the collector is reduced to a slot about 0.5 cm in width, the effect can be very marked and potential variations of as little as 1 V can cause readily observable changes in brightness of the corresponding areas of the image. The effect has been found particularly valuable in the study of p-n junctions in semiconductors, and a micrograph illustrating this use is shown in Fig. 5.

In a scanning electron microscope the distance between the object and the face of the final lens need never be less than about 0.5 cm and when the highest magnification is not needed, it can be very much greater. This large working distance, coupled with the almost unlimited size of the specimen chamber, makes the instrument an ideal one when it is desired to carry out experiments on an object while it is under observation. The following brief list of examples will serve to illustrate the potentialities of the scanning microscope in this direction. The process of sputtering of metals by positive-ion bombardment has been studied, a beam of ions from a high-frequency source being allowed to fall on the specimen while the latter was mounted in the microscope. The activation and, subsequently, the poisoning of thermionic cathodes of the dispenser type have been investigated and, on occasion, micrographs have been taken of the cathode at its normal operating temperature. Useful information has been obtained about the thermal decomposition of unstable chemical substances such as azides, and the microscope has proved a most useful tool for the study of the electron-beam machining of thin films. In this instance, the electron beam of the microscope was used also for machining the films.

Acknowledgment

The present article is based on an earlier one written by the author and published in *Science Progress*, **54**, 1966. The author is indebted to Blackwell Scientific Publications (Oxford) for kindly allowing him to reproduce portions of the earlier article.

C. W. OATLEY

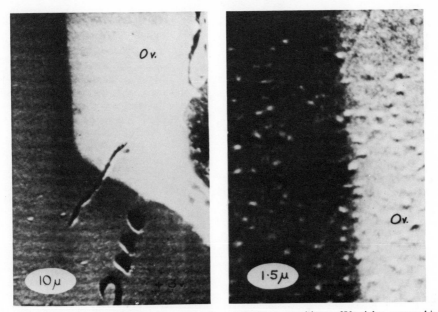

Fig. 5 Germanium-indium p–n junctions in a semiconductor. Left: reverse bias = 3V; right: reverse bias = 1V.

References

Oatley, C. W., W. C. Nixon, and R. F. W. Pease, "Advances in Electronics and Electron Physics," Vol. 21, New York, Academic Press, 1965 (p. 181).

Thornton, P. R., "Scanning Electron Microscopy," London, Chapman and Hall, 1968.

Oatley, C. W., "The Scanning Electron Microscope," London, Cambridge University Press, 1972.

See also: TRANSMISSION MICROSCOPY.

SCANNING MICROSPECTROPHOTOMETRY

The Bouguer-Beer Law and the Need for Scanning in Microspectrophotometry. To grasp the essentials of scanning microspectrophotometry, or indeed any spectrophotometry, some knowledge of the Bouguer-Beer (or Lambert-Beer) absorption law is essential. The law states that if a parallel beam of monochromatic light passes through a homogeneous absorbing medium, then

$$\log(I_0/I_t) = \log(I/T) = E = abc \qquad (1)$$

Here I_0 and I_t are respectively the intensities of the light on entering and leaving the medium, T is the transmittance, E is the absorbance (optical density, extinction), a (the absorbtivity) is constant for a given substance and wavelength, b is the path length through the medium, and c is the concentration. Where the law holds, the mass of light-absorbing material per unit area, measured in a plane perpendicular to the light beam, is $bc = E/a$, and the total amount of material in an area A is AE/a. The *integrated absorbance* of a discrete, uniform specimen is the product of its area and absorbance, AE, and is proportional to its content of light-absorbing substance.

In the application of the Bouguer-Beer law to non-uniform microscopic specimens, an initial difficulty arises from the fact that the concentration of light-absorbing material in a beam is proportional to the absorbance E, not the transmittance T. If the object is optically heterogeneous, the mean concentration is accordingly proportional to the mean absorbance, but not to the readily measurable mean intensity of the transmitted light. This is the origin of the so-called *distribution error* (Ornstein, 1952) affecting the spectrophotometry of nonuniform objects.

Consider, for example, a parallel beam of light of 100 μm^2 cross-section, passing through a microscopic object of projected area 100 μm^2 and uniform transmittance 0.5 (Fig. 1A). No light passes through the empty background. The absorbance of the specimen is log $(1/0.5) = 0.3010$, and its integrated absorbance is $100 \times 0.3010 = 30.10$ units. If a similar object has a projected area of only 25 μm^2, but contains the same total amount of material (Fig. 1B), its transmittance will be 6.25%, and its true integrated absorbance will be $25 \times$ log $(1/0.0625) = 30.10$ units, as before. If, however, the light beam is still 100 μm^2 in cross section, the mean transmittance of object plus background will be $(1.0 \times 75/100)$ $+ (0.0625 \times 25/100) = 76.6\%$, giving an apparent absorbance of log $(1/0.766) = 0.116$, and an apparent integrated absorbance of only $100 \times 0.116 = 11.6$ units.

To generalize, we may say that the amount of light-absorbing material in a heterogeneous object is proportional to its true integrated absorbance, defined as the sum of the integrated absorbances of its constituent homogeneous subunits. Alternatively, the true integrated absorbance can be regarded as the product of the total area A and the mean absorbance $\bar{E}$. The mean absorbance is *not*, however, equal to the apparent absorbance calculated from the mean transmittance $\bar{T}$. Simple measurement of the total light passing through an object ("plug"

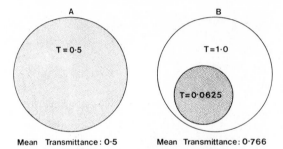

Mean Transmittance: 0·5   Mean Transmittance: 0·766

Fig. 1 Effect of distribution error on apparent absorbance. The specimen in Fig. 1A has an area of 100 μm^2 and a transmittance of 50%. Its absorbance is 0.3010. The specimen in Fig. 1B contains the same amount of light-absorbing material, but has an area of only 25 μm^2. Traversed by a light beam of 100 μm^2 area, its apparent absorbance is only 0.116 (see text).

photometry) is therefore not suitable for accurate photometric work on nonuniform histological and cytological specimens.

Distribution error can be at least partially eliminated in several ways, including a two-wavelength procedure, a two-area method, photographic cytophotometry, and fluorescence techniques (see articles by Mendelsohn, Garcia and Iorio, Kelly, and Ruch, all in Wied, 1966). The most direct and generally applicable approach is by scanning, integrating microspectrophotometry: The integrated absorbance of the whole specimen is obtained by summating the absorbances of subunits so small that they can be regarded as homogeneous.

Alternative Modes of Scanning in Microspectrophotometry. Scanning can be carried out on a photomicrograph, but difficulties associated with the linearity and reproducibility of the photographic response are avoided if the microscopic specimen itself is scanned. In scanning microspectrophotometers, light passes through a minute area of a specimen and eventually falls on the window of a photomultiplier tube, the electrical output of which is proportional to the transmittance of the object point measured. An electronic device ("analog computer") automatically transforms the output of the photomultiplier tube into an electrical signal proportional to the instantaneous absorbance, and the measuring spot scans over the specimen in a systematic pattern ("raster"). The integrated absorbance of the whole area measured is obtained by adding ("integrating") the electrical signals corresponding to the instantaneous absorbances occurring during the scan.

Several different methods for moving the measuring spot relative to the specimen are used in commercially available scanning microspectrophotometers. Thus in the Carl Zeiss "UMSP" instrument, based on a design by Caspersson, the spot measured is defined by a small fixed aperture in an image plane of the microscope, and the specimen itself is moved by a special stage. In the Deeley (1955) microdensitometer produced by Barr & Stroud, both specimen and image remain stationary, and the image is scanned by mechanical movement of an image-plane aperture. In "flying-spot" instruments such as the Vickers M85 designed by Smith, a reduced image of an illuminated aperture is formed in the object plane by the microscope objective, and moves over the specimen.

Optically the various systems are fundamentally equivalent, but each has characteristic practical advantages and disadvantages. Moving-specimen instruments are relatively slow and cumbersome, but make it possible to scan a specimen larger than the field of view of the objective, and require illumination of an area of specimen only slightly larger than that actually measured at any given moment (see the discussion on glare, below). Flying-spot and image-scanning instruments, on the other hand, can be much more rapid. This is advantageous because of the saving in time, and because errors due to mechanical or electronic instability are thereby reduced.

In selecting an instrument for a particular application, in addition to the points just described relevant factors which should be considered include availability, cost, ease and convenience of operation, and flexibility of control. As will be discussed later, for the most exacting work variability of the measuring-spot size is valuable, as is the ability to offset the dark current of the photomultiplier tube. It goes without saying that mechanical, optical, and electronic components should be of high quality. The relative weight accorded to the various considerations will depend on the program of research envisaged; a machine which is unsuitable for the most critical work may nevertheless be almost ideal for relatively crude measurements on a large number of specimens.

Potentialities and Pitfalls. In principle, scanning microspectrophotometry at a given wavelength makes it possible to estimate microscopic quantities of biological substances absorbing either visible light (e.g., chlorophyll, hemoglobin) or ultraviolet light (e.g., protein, nucleic acid). Colored histochemical reaction products or dyes in a stained preparation are also measurable, and if it can be shown that under defined conditions the concentration of dye in a section or smear is proportional to the concentration of a particular substrate, the latter can be measured indirectly.

The absorbance spectrum of a specimen (i.e., the plot of absorbance or integrated absorbance vs. wavelength of the light used) may be very revealing, since it is affected by and hence yields information on the chemical identity and state of aggregation of the light-absorbing substance, and its interaction with neighboring molecules.

In addition to the measurement of absorbance spectra and integrated absorbances, scanning microscopes are increasingly being used for automatic or semiautomatic image analysis. This can range from simple measurement of the specimen area having an absorbance greater than an arbitrarily set threshold, to object counting, recognition, and classification using sophisticated digital-computer techniques. Advances in this field are too rapid for a useful review at the present time.

Before the various potentialities of scanning microspectrophotometry can be realized, certain difficulties must be overcome. Among the most intractable of these are factors relating to the specimen itself. Thus the Bouguer-Beer law deals only with *absorption* of light, and light losses due to scattering, refraction, or reflection must be eliminated or allowed for. This is not always easy, especially with ultraviolet light. The very factors which make changes in the absorbance spectrum of a substance interesting may complicate the interpretation of the integrated absorbance at a given wavelength. The absorptivity of a substance remains constant with increasing concentration only if no interaction of molecules takes place; gross deviations from the Bouguer-Beer law are often found in solutions, and are sometimes revealed by metachromasia in microscopic preparations. The absorption of light by a substance can be affected by other molecules in the vicinity, by the pH, dielectric constant, and temperature of the environment, and by a host of other factors. The uptake of dyes may be inhibited or facilitated in various ways. Questions connected with dye uptake and stoichiometry are in fact among the most controversial problems in microspectrophotometry. Even in the most-studied cases, such as the Feulgen reaction, proportionality between dye concentration and substrate concentration is not universally accepted.

The attention of the reader is drawn to these rather contentious issues, but they will not be further discussed here. The remainder of the present article deals only with selected topics of instrumentation and methodology in scanning microspectrophotometry, and is intended to aid the beginner to avoid, or at least to recognize, some of the technical pitfalls which face him.

Calculation of Results in Absolute Units. Scanning microspectrophotometers in general give results in arbitrary machine units, and conversion to absolute units is necessary if data obtained with different instruments, or even with a given instrument at different times, are to be compared.

A simple method of calibration is to measure the apparent integrated absorbance of an empty field of known area, with and without a neutral-density filter of known absorbance in the optical path. The difference between the two readings is equal to the integrated absorbance in arbitrary units of a specimen having an area equal to the field scanned and a uniform absorbance equal to that of the filter. If the radius of the scanned area is R μm, the absorbance of the filter F, and the apparent integrated absorbance in machine units AE_{arb}, the calibration constant K is defined as

$$K = \frac{\pi R^2 F}{AE_{arb}} \qquad (2)$$

Experimental results in arbitrary units, when multiplied by K, are obtained in absolute units with the dimensions of μm^2.

If very accurate calibration is required it should be recognized that the apparent area scanned may not be identical with that actually measured. In the Vickers M85, for example, the area measured is controlled by an auxiliary photomultiplier tube and masking system, and incorrect setting of the "gating level" results in the area scanned being slightly larger or smaller than it should be. An analogous error occurs with the Barr & Stroud instrument, even though the masking system here functions quite differently.

A corrected value for the calibration constant is obtainable if the calibration procedure is carried out using two different areas. Let the apparent radii of the measured areas be R and r um, respectively, and the corresponding integrated absorbances in arbitrary units be AE_R and AE_r. The effective radii as "seen" by the machine will be $(R + x)$ and $(r + x)$ microns, where x is the difference between the apparent and the effective radii, and the corrected calibration constant is given by

$$K = \frac{\pi(R + x)^2 F}{AE_R} = \frac{\pi(r + x)^2 F}{AE_r} \qquad (3)$$

whence

$$x = \frac{AE_r R - AE_R r + (R - r)(AE_r \cdot AR_R)^{\frac{1}{2}}}{AE_R - AE_r} \qquad (4)$$

Uniformity of Illumination. Uneven illumination results in some parts of the specimen appearing darker, and others brighter. The resultant error in the apparent integrated absorbance can be eliminated by measuring the apparent integrated absorbance of a field containing only empty background, and subtracting this as a blank from the experimental reading.

Some machines do not integrate negative absorbances (corresponding to transmittances higher than the nominal 100% transmittance level). With such instruments the intensity of illumination should be reduced enough to ensure that all parts of the measured blank area have a positive absorbance, i.e., have a transmittance of less than 100%.

Spot Size and Focus. Ideally, the measuring spot in scanning microspectrophotometry should be so small that the specimen area measured at any given moment is essentially homogeneous. Even ignoring the possibility that the light-absorbing molecules are distributed in a nonrandom fashion at a submicroscopic level (Ornstein, 1952), this would imply that the measuring spot is small relative to the resolving power of the lens system. If no detail is to be lost in a scanning system, the diameter of the spot should not in fact be greater than half the limit of resolution, i.e., about 0.15 μm with green light and an oil-immersion objective.

The amount of light reaching the photomultiplier tube is, however, proportional to the area of the measuring spot, and a relatively large current from the tube (signal/noise ratio) is necessary if reproducible results are to be obtained. In almost all currently available scanning microspectrophotometers the spot is therefore considerably larger than ideal, and some residual distribution error is to be expected, especially with very small or very dense objects.

Provided the measuring spot is not too large, the deficit in the apparent integrated absorbance is approximately proportional to the spot width (Fig. 2). By measuring the apparent integrated absorbance with spots of two different diameters it is therefore possible to calculate the true integrated absorbance corresponding to zero spot size. In the special case where the large spot

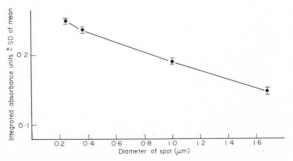

Fig. 2 Effect of measuring-spot diameter on apparent integrated absorbance of a single Gram-stained staphylococcus. The deficit in the result is directly proportional to the spot width.

is exactly twice the width of the smaller, the true integrated absorbance is given by the approximate expression

$$AE_0 = 2AE_1 - AE_2 \qquad (5)$$

where AE_1 and AE_2 are the apparent integrated absorbances with the smaller and larger spot, respectively. With very large spots the error increases more slowly, and the apparent integrated absorbance eventually approaches a plateau value. Extrapolating to zero spot size is therefore a conservative correction, and cannot give too high a result.

Incorrect focus is to a first approximation equivalent to an increase in effective size of the measuring spot in the plane of the specimen. Careful focusing is essential, especially if (as with many flying-spot and ultraviolet microscopes) the specimen is focused with light of a different wavelength from that used for the actual measurement.

The specimen thickness should ideally be less than the depth of focus of the objective. A "crushing condenser" (Davies et al., 1954) may be used for flattening thick or dense objects specially mounted under "Cellophane," but the method is not universally applicable.

Glare (Stray Light). Multiple reflections of light in the optical system, mainly off glass-air surfaces in the microscope objective, lead to a redistribution of energy in the image plane of the microscope and an apparent increase in the transmittance of heavily absorbing objects. Stray light or glare, as measured by the apparent transmittance of a completely opaque specimen, is seldom less than 2–4% unless the field illuminated by the microscope condenser is severely restricted. It is often much higher, and can lead to serious errors in the microspectrophotometry of dense objects (Fig. 3).

Glare can be somewhat reduced by meticulous attention to cleanliness, and by special selection of the objectives used. The most effective single measure is to reduce the area A_{ill} illuminated: If A_{spec} is the area of the specimen itself, the amount of glare is approximately proportional to $(A_{ill} - A_{spec})/A_{ill}$. A small field stop between the

specimen and the microscope lamp should therefore be imaged into the object plane by a good-quality, preferably achromatic condenser. In flying-spot microscopes the equivalent glare-reducing field stop is situated between the light-collecting lens and the photomultiplier tube.

Glare is a source of error only if the specimen incompletely fills the area illuminated. With flying-spot and image-scanning systems the area illuminated cannot, however, be smaller than the specimen, and even with moving-specimen instruments there must be moments during the scan when the illuminated area overlies the edge of the object and therefore includes some empty background.

In the presence of glare (expressed as a fraction F of the intensity of the incident light I_0), the *true* absorbance E_t of a specimen is related to the *apparent* transmittance I_p as follows:

$$E_t = \log \frac{I_0 - F}{I_p - F} \qquad (6)$$

Glare would be corrected if it were possible to subtract an amount of light equal to F from both the light passing through empty background (I_0) and that which has apparently passed through the specimen (I_p). Although light itself cannot be subtracted, with some instruments an electrical current proportional to F can be subtracted from the output of the photomultiplier tube, using the "set zero" control intended for balancing out the dark current. In this way unavoidable glare can be compensated for electronically. The correction should not be used during calibration, since no glare is present while the neutral-density filter ("virtual object") completely fills the illuminated field.

Monochromator Errors. Absorption of light by a medium in general depends on the wavelength λ of the light. At a particular wavelength, λ_{max}, the absorbance shows a maximum and the transmittance a minimum, while secondary, smaller absorbance peaks may be present at other wavelengths.

For high-accuracy spectrophotometry the specimen absorbance should be neither very low nor very high, the optimum value being about 0.43 (Hiskey, 1955). It may be necessary to measure particularly dense specimens at a wavelength some distance from the absorbance maximum, but this situation seldom arises in microspectrophotometry, where specimens are more commonly too pale than too dense. For most purposes it is best to set the wavelength empirically to λ_{max}, i.e., the wavelength is chosen which gives the highest apparent integrated absorbance of a scanned specimen. This procedure avoids errors due to incorrect calibration of the wavelength scale of the monochromator, and makes results obtained with different instruments directly comparable. Measurement at the absorbance peak is also desirable if the light used is imperfectly monochromatic (see below).

If a specimen contains two light-absorbing substances with different absorbance spectra, microspectrophotometry can readily be carried out at a wavelength where only the substance of interest has a significant absorbance. If no such wavelength exists it is necessary to measure the absorbance (or integrated absorbance) at two different wavelengths, and obtain the desired value by calculation (Hiskey, 1955). This method has been used for measurement in cells of nucleic acid and protein, which have overlapping ultraviolet absorbance spectra. In principle the procedure could be extended to three or even more

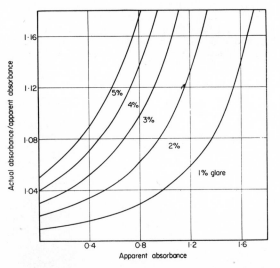

Fig. 3 Ratio of actual to apparent absorbance in the presence of varying amounts of glare.

different substances, but in practice the necessary accuracy of measurement is probably unattainable, and where possible, measurement at two wavelengths should also be avoided.

If the light used in spectrophotometry consists of a mixture of beams of different wavelengths, the concentration of light-absorbing substance may not be proportional to the apparent absorbance calculated from the observed mean transmittance. The Bouguer-Beer law is valid only for strictly monochromatic light, or for light of a wavelength range so small that the absorptivity of the medium is essentially identical for all wavelengths present.

Almost perfectly monochromatic light is emitted by some lasers and by low-pressure vapor lamps used with filters to isolate the emission band required. In spectrophotometry it is, however, highly desirable for the wavelength to be continuously variable. Most microspectrophotometers are accordingly equipped with a continuous-spectrum light source such as a tungsten lamp or xenon arc, and the required wavelength is selected by a prism, grating, or interference-filter monochromator. The resulting light has a finite bandwidth (range of wavelengths) determined by the design and quality of the monochromator and by the setting of its entrance and exit slits. With some instruments an admixture of white light results from scattering of light off edges of the monochromator slits, or imperfections in the optical system, or in the case of interference filters, from pin-hole defects in the film.

A wide monochromator bandwidth results in the apparent absorbance being too low at peaks and too high at minima of the absorbance spectrum. At some intermediate wavelength the error will be zero, at least for a given absorbance. In theory this wavelength could be used for relatively accurate microspectrophotometry even with an imperfect monochromator, but in practice it would be extremely difficult to find and set the correct wavelength.

Since the rate of change of absorptivity with wavelength is minimal over relatively flat parts of the absorbance spectrum, imperfect monochromaticity is less harmful near the middle of a wide absorbance maximum or minimum than on the "shoulders" of the curve. A relatively wide monochromator bandwidth is therefore tolerable if the absorbance maximum is fairly wide, as with many histological dyes. The error is more serious with narrow peaks such as the hemoglobin absorption peak (the Soret band) at about 415 nm.

Measured at λ_{max}, the percentage deficit in the apparent absorbance, due to the use of imperfectly monochromatic light, increases with the actual absorbance of the specimen. The resulting systematic error is cumulative with that caused by glare and by the use of a finite scanning spot. In all three cases the error may be regarded as being due to the photomultiplier tube receiving light not subjected to absorption by the specimen. As already discussed, glare can be compensated for by an electronic biasing of the photomultiplier tube output, and in principle it should be possible to correct the monochromator bandwidth error in a similar way.

Numerical Aperture of Objective and Condenser. According to the Bouguer-Beer law, absorbance is directly proportional to the path length through the medium. If a plane-parallel object lying normal to the optic axis of the microscope is illuminated by a solid cone of light of high numerical aperture, the mean path length through the specimen and hence the apparent absorbance will be greater than if the specimen is traversed by a parallel beam. With a dense object, oblique rays will be more heavily absorbed than vertical ones and contribute less to the final intensity of light reaching the photomultiplier tube. The mean path length will therefore be slightly shorter than in the case of a pale specimen, and the apparent absorbance will be increased slightly less. This leads to a small systematic error in the comparison of objects of different densities.

Several workers have accordingly recommended relatively low condenser apertures, of the order of 0.3, in microspectrophotometry. The objective aperture should, on the other hand, be kept high, in order to collect as much diffracted light as possible and to give good resolution. In flying-spot systems this would be equivalent to reducing the aperture of the collecting lens, while maintaining that of the (objective) lens which forms the reduced image of the measuring spot in the plane of the specimen.

It is, however, doubtful whether in practice the condenser aperture is quite as important as has sometimes been suggested. If the specimen is spherical, the path lengths of the oblique and direct rays are identical, and even in the extreme plane-parallel case the error is not great. Thus with a specimen of refractive index 1.53 and true absorbance 1.0, the apparent absorbance is about 4% too high with an illuminating aperture of unity (Walker, 1956). The error is admittedly greater with living cells, of lower refractive index, and with this sort of specimen it appears prudent to limit the condenser aperture somewhat. In any event, the condenser aperture should be kept constant throughout a given series of observations.

Precision and Accuracy in Scanning Microspectrophotometry. The *precision* (reproducibility of results) of modern, scanning microspectrophotometers is usually satisfactory, and can readily be checked by making replicate measurements on a given specimen. Precision should not, however, be confused with *accuracy*, i.e., an absence of systematic errors. These are difficult to detect, and yet harder to correct.

For many types of investigation the utmost technical refinements are neither necessary nor desirable. In experimental work aiming at the highest accuracy, however, many apparently trivial technical factors, some of which have been briefly discussed in this article, must be carefully controlled. If this is not done, the evident precision of the results may lead to a quite unwarranted belief in their accuracy.

D. J. GOLDSTEIN

References

General:

Caspersson, T. O., "Cell Growth and Cell Function," New York, Norton & Co., 1950.

Caspersson, T. O., and G. M. Lomakka, "Scanning microscopy techniques for high resolution quantitative cytochemistry," *Ann. N.Y. Acad. Sci.,* **97**:449–463 (1962).

Davies, H. G., and P. M. B. Walker, "Microspectrophotometry of living and fixed cells," pp. 195–236 in "Progress in Biophysics and Biophysical Chemistry," Vol. 3, J. A. V. Butler and J. T. Randall, ed., New York, Pergamon Press, 1953.

Hiskey, C. F., "Absorption spectroscopy," pp. 73–130 in "Physical Techniques in Biological Research," Vol. 1, G. Oster and A. W. Pollister, ed., New York, Academic Press, 1956.

Leuchtenberger, C., "Quantitative determination of DNA in cells by Feulgen microspectrophotometry," pp. 219–278 in "General Cytochemical Methods," Vol. 1, J. Danielli, ed., New York, Academic Press, 1958.

Mayall, B. H., and M. L. Mendelsohn, "Errors in absorption cytophotometry: some theoretical and practical considerations," pp. 171–197 in "Introduction to Quantitative Cytochemistry," Vol. 2, G. L. Wied and G. F. Bahr, ed., New York, Academic Press, 1970.

Mayall, B. H., and M. L. Mendelsohn, "Deoxyribonucleic acid cytophotometry of stained human leukocytes. II. The mechanical scanner of CYDAC, the theory of scanning photometry and the magnitude of residual errors," *J. Histochem. Cytochem.*, **18**:383–406 (1970).

Ornstein, L., "The distributional error in microspectrophotometry," *Lab. Invest.*, **1**:250–265 (1952).

Swift, H., and E. Rasch, "Microphotometry with visible light," pp. 354–400 in "Physical Techniques in Biological Research," Vol. 3, G. Oster and A. W. Pollister, ed., New York, Academic Press, 1956.

Walker, P. M. B., "Ultraviolet absorption techniques," pp. 402–487 in "Physical Techniques in Biological Research," Vol. 3, G. Oster and A. W. Pollister, ed., New York, Academic Press, 1956.

Walker, P. M. B., "Ultraviolet microspectrophotometry," pp. 163–217 in "General Cytochemical Methods," Vol. 1, J. F. Danielli, ed., New York, Academic Press, 1958.

Wied, G. L., "Introduction to Quantitative Cytochemistry," New York, Academic Press, 1966.

Wied, G. L., and G. F. Bahr, ed., "Introduction to Quantitative Cytochemistry," Vol. 2, New York, Academic Press, 1970.

Spot size and focus:

Davies, H. G., M. H. F. Wilkins, and R. G. H. B. Boddy, "Cell crushing: a technique for greatly reducing errors in microspectrophotometry," *Exp. Cell Res.*, **6**:550–553 (1954).

Goldstein, D. J., "Aspects of scanning microdensitometry. II. Spot size, focus and resolution," *J. Microsc.*, **93**:15–42 (1971).

Glare:

Goldstein, D. J., "Aspects of scanning microdensitometry. I. Stray light (glare)," *J. Miscrosc.*, **92**:1–16 (1970).

Lison, L., "Schwarzschild-Villiger effect in microspectrophotometry," *Science*, **118**:382–383 (1953).

Naora, H., "Microspectrophotometry of cell nucleus stained by Feulgen reaction. I. Microspectrophotometric apparatus without Schwarzschild-Villiger effect," *Exp. Cell Res.*, **8**:259–278 (1955).

Condenser aperture:

Pillat, G., "Zum Einfluss der Kondensorapertur auf die mikrophotometrische Absorptionsmessung biologischer Objekte," *Acta Histochem.*, **9**:169–173 (1960).

SILVER PROTEINATE STAINS

Silver proteinate (protargol) staining technique was devised by Bodian (1) for histologic demonstration of myelinated and unmyelinated fibers in the central (Figs. 1, 2) and peripheral (Fig. 3) nervous system, and of other nervous tissue elements of vertebrates and invertebrates. Striated muscles (Fig. 4), spermatozoa (Fig. 5), and various structures in flagellate (Figs. 6, 7, 9), ciliate (Fig. 8), and sporozoan protozoa also stain well by this technique. The standard method involves activation of the silver proteinates by metallic copper (1–7), but with some tissues good results can be obtained without the activator (7–9).

For most elements of the adult mammalian nervous system, an alcoholic solution of formalin and glacial acetic acid is the fixative of choice (2, 6, 7). Numerous other fixing fluids also give good results with various nervous tissue elements, Bouin's and other picric acid-containing mixtures, which are superior by far for protozoa (3–6), often being very suitable; alkaline fixatives and those containing heavy metal salts generally are useless (2, 6, 7). The most satisfactory impregnation is obtained after perfusion of the animal or organ with the fixative (1).

Paraffin sections of tissues are used customarily for silver proteinate staining. Celloidin sections also can be employed, but celloidin must be removed before staining. Protozoa and spermatozoa are impregnated successfully in smears prepared on cover glasses (3–6, 8) and fixed by dropping them onto the surface of appropriate fluids (3–6).

Impregnation of protozoa requires in most instances potassium permanganate–oxalic acid treatment (Mallory's bleach) before staining (4, 5), and this treatment is useful also for spermatozoa. Bleaching insures satisfactory staining of the argentophilic organelles in all protozoa, and in many species such structures fail to impregnate unless the organisms are bleached. The following schedule (4, 5) is recommended for this treatment (all solutions are aqueous):

1. Place in a 0.25–0.5% solution of potassium permanganate for 3–5 min.

2. Wash in several changes of distilled water.

3. Place in a 2–5% solution of oxalic acid for 3–5 min.

4. Wash as in step 2.

Bleaching often renders the preparations quite brittle and causes them to peel off, especially in the hydroquinone reducing bath (see below); however, adjustment of the strength of the reagents and of the times of treatments (4), and careful handling of the preparations usually offset this drawback.

Figs. 1–9 [Figs. 1, 2, 5–7, and 9—preparations fixed in Bouin's fluid, bleached by the potassium permanganate(0.25%)–oxalic acid(2%) treatment, and stained with copper-activated prewar protargol of German manufacture. Figs. 3 and 4—preparations fixed in a mixture of formalin (5 ml), glacial acetic acid (5 ml), and ethyl alcohol (90 ml), and stained with copper-activated Protargol S. Fig. 8—preparation fixed in Bouin's fluid, bleached by the potassium permanganate (5%)–oxalic acid(5%) treatment, and stained with Roques' silver proteinate without copper according to a significantly modified procedure (8). Figs. 1–7 and 9 are original photomicrographs; Fig. 8 was reproduced from Jerka-Dziadosz and Frankel (8)].

Fig. 1: Cross section of a spinal cord from the laboratory rat showing a junction between an area of motor neurons and an area of myelinated nerve fibers. (260 ×). Fig. 2: Enlarged part of the motor neuron area represented in Fig. 1. (500 ×.) Fig. 3: Cross section of a spinal nerve from the laboratory rat. (550 ×.) Fig. 4: Longitudinal section of a skeletal muscle from the laboratory rat. (1900 ×.) Fig. 5: Spermatozoa from a bull. (1900 ×). Fig. 6: *Giardia duodenalis* (Davaine) from a deer mouse (*Peromyscus californicus*). (2900 ×.) Fig. 7: *Trichomitopsis* (*Trichomonas*) *termopsidis* (Cleveland) from the termite, *Zootermopsis angusticollis*. (700 ×.) Fig. 8: *Urostyla weissei* Stein. (390 ×.) Fig. 9: Mastigont system of *Trichonympha sphaerica* (Kofoid and Swezy) from the termite, *Z. angusticollis*. (460 ×.)

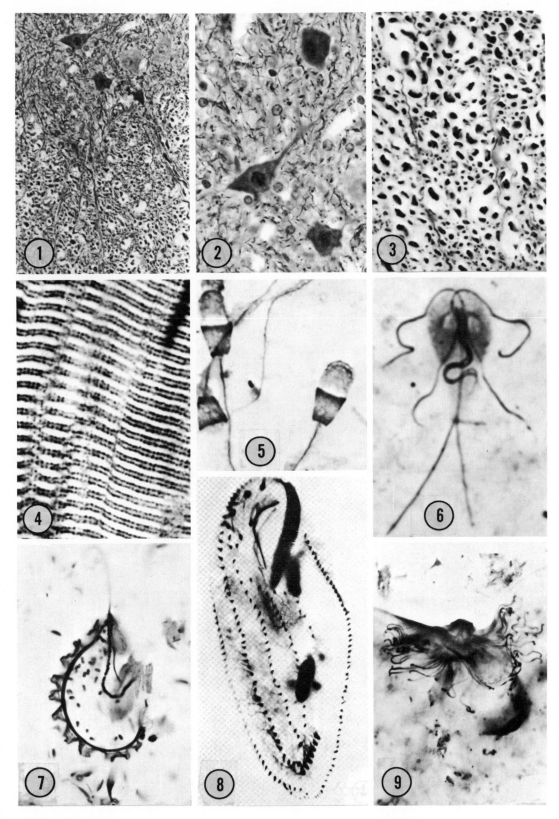

A schedule essentially like that suggested by Bodian (1, 2) has been used successfully for staining of vertebrate and invertebrate tissues (6, 7) as well as of protozoa (3–6) and spermatozoa.

1. Stain in a 1% silver proteinate solution (this and other solutions in this schedule are aqueous) with 4–6% (weight/volume) of metallic copper (thin wire or foil) for 12–48 hr at 37°C. (The solution can be used only once.)

Variation in the concentration of silver proteinates from 0.2 to 2.0% and in the exposure time may improve staining (4, 7). Optimum pH is 8.0–8.5 (but see 9), with little or no impregnation at pH below 5.0 and very intense but nonspecific staining at pH above 9.0.

2. Wash in several changes of distilled water.

3. Reduce in 1% hydroquinone in a 5% solution of sodium sulfite for 10 min. (Add hydroquinone just before use.)

4. Wash firmly adhering preparations for 5 min or longer in running tap water; then rinse in distilled water. If the material tends to peel off, wash gently in several changes of distilled water.

5. Subject to gold toning as follows:

a. Place in a 1% solution of gold chloride for 5 min. [The often recommended acidification of the solution is superfluous (7), but, on occasion, variation in the gold chloride concentration (0.2–2.0%) and in the time of treatment improves staining.]

b. Rinse briefly in distilled water. (Insufficient rinsing results in a nonspecific precipitate during the next step; excessive washing causes weak staining.)

c. Reduce gold chloride in a 2% solution of oxalic acid for 5 min or until the preparations appear purple or bluish. [Variation in the concentration of oxalic acid from 0.1 to 5.0% and in the time of treatment is indicated in some instances (7).]

d. Wash as in step 4.

6. Remove residual silver and gold salts in a 5% solution of sodium thiosulfate: 5–10 min.

7. Wash as in step 4.

8. Dehydrate, clear, and mount.

With certain ciliate protozoa, excellent results have been obtained by significantly modified methods (see 8 for pertinent information).

Silver-impregnated preparations may be counterstained with acridine red, methyl green, or other dyes; combinations with other staining methods also are possible. In staining certain nervous tissues, double impregnation, involving two successive silver proteinate treatments, improves the results (2).

From the chemical viewpoint, the "silver proteinates" used for staining actually represent mixtures of silver "proteosates" and "peptonates" (7). The mechanism of staining by these compounds is not fully understood. It has been suggested that impregnation is determined by the amount of ionized silver in solution which affects the speed of reduction and the size of particles available for deposition (2). Inasmuch as silver is deposited upon the copper, addition of this latter metal to the staining solution further decreases the silver ion levels. Copper has been proved to go into solution and to be deposited together with silver on the nerve fibers. This may explain the slow and only partial reduction of silver in the hydroquinone bath as well as the weak impregnation of metazoan nervous tissue and protozoan organelles in preparations not subjected to gold toning. The toning causes replacement of the silver by gold deposits and

intensifies the stain by increasing these latter deposits upon reduction of gold chloride by oxalic acid (7). Copper appears to aid also in differentiation between the nervous and connective tissues (7).

Protargol S (Winthrop Chemical Co., New York), certified by the Biological Stain Commission, is the most common commercially available silver proteinate. All batches of this compound are satisfactory for staining of vertebrate, especially mammalian, tissues. Only occasional batches give acceptable results with protozoa. The only compound that has produced uniformly superior staining of protozoa is the prewar protargol of German manufacture. "Gurr's Protargol" (Pfalz & Bauer, Inc., New York) and Roques' "Protéïnate d'argent" (Roboz Surgical Instrument Co., Washington, D.C.), the commercially available compounds of foreign manufacture, vary from one batch to another in their suitability for staining protozoa. Some success in impregnation of nervous tissue and of protozoa has been achieved by using silver proteinates prepared according to the methods recommended by several workers (5, 9–12).

B. M. HONIGBERG

References

1. Bodian, D., "A new method for staining nerve fibers and nerve endings in mounted paraffin sections," *Anat. Rec.*, **65**:89–97 (1936).
2. Bodian, D., "The staining of paraffin sections of nervous tissues with activated protargol. The role of fixatives," *Anat. Rec.*, **69**:153–162 (1937).
3. Kirby, H., "The structure of the common intestinal trichomonad of man," *J. Parasitol.*, **31**:163–175 (1945).
4. Honigberg, B. M., "Structure and morphogenesis of *Trichomonas prowazeki* Alexeieff and *Trichomonas brumpti* Alexeieff," *Univ. Calif. Pub. Zool.*, **55**:337–394 (1951).
5. Honigberg, B. M., and H. A. Davenport, "Staining flagellate protozoa by various silver-protein compounds," *Stain Technol.*, **29**:241–246 (1954).
6. Galigher, A. E., and E. N. Kozloff, "Essentials of Practical Microtechnique," Philadelphia, Lea & Febiger, 1964.
7. Davenport, H. A., "Histological and Histochemical Technics," Philadelphia, W. B. Saunders Co., 1960.
8. Jerka-Dziadosz, M., and J. Frankel, "An analysis of the formation of ciliary primordia in the hypotrich ciliate *Urostyla weissei*," *J. Protozool.*, **16**:612–637 (1969).
9. Moskowitz, N., "The use of protein silver for staining protozoa," *Stain Technol.*, **25**:17–20 (1950).
10. Porter, R. W., and H. A. Davenport, "Synthesis of silver proteinates for neurological staining," *Stain Technol.*, **26**:1–4 (1951).
11. Davenport, H. A., R. W. Porter, and B. A. Myhre, "Preparation and testing of silver-protein compounds," *Stain Technol.*, **27**:243–248 (1952).
12. Tengler, M., "Un metodo rapido de impregnacion de plata para protozoarios enterozoicos," [in Spanish, English summary], *Acta Cientifica Venezolana*, **8**:131–133 (1957).

See also: SILVER STAINS.

SILVER STAINS

Solutions of silver compounds have been used as histological stains since 1843. Since the late part of the nineteenth century a wealth of techniques have been developed by Cajal, Golgi, del Rio-Hortega, Bodian,

Davenport, and others for the demonstration of various components of the nervous system. By appropriate choice of reagents and methods with silver one may selectively stain axis cylinders, degenerating axons, glial cells, neurons, and boutons of nerve endings. The Golgi method with dichromate oxidation followed by silver impregnation is remarkable in that one may stain an individual neuron with all of its ramifying dendrites and axon while adjacent neurons are unstained. Ramon y Cajal extended this silver method, and with del Rio-Hortega, they were able to comprehensively investigate the architecture of the nervous system. Much of present understanding of neural architecture has depended on these silver techniques. The earlier silver procedures generally used small tissue blocks or frozen sections. More recently many procedures have been developed to stain nervous tissue in paraffin sections because of the need to correlate conventional stains and silver stains. This is of particular importance in studying diseased tissues.

Most of the methods for silver staining depend on exposure of tissue to silver compounds (silver nitrate, ammoniacal silver, or silver proteinate) followed by development in a reducing solution containing formaldehyde, pyrogallol, or similar agents. The initial exposure to the silver solution results in the production of minute metallic silver nuclei of a few silver atoms formed by the reaction of tissue substances of high redox potential with silver ions. Subsequent development in an alkaline solution of a developer such as pyrogallol results in deposition of much additional silver on these nuclei so that as in photography a latent image becomes a visual image. Carbonyl and sulfhydryl groups are thought to be particularly responsible for formation of silver nuclei. The formation of silver nuclei during silver impregnation is affected by choice and concentration of silver compounds, time, temperature, the presence of myelin, and optimal pH. The commonly used ammoniacal silver solutions depend largely on a complex silver diamine cation $[Ag(NH_3)_2]^+$ which is a very reactive oxidant at high pH. Gold chloride is often used to tone the silver image from brown-black to a lavender-black. The differentiating effect is thought to be due to a deposition of gold plating on large silver aggregates and a bleaching of fine silver deposits to colorless silver chloride with a resulting improvement in contrast. While these silver techniques for neural tissues are empiric and capricious their remarkable ability to selectively stain structures with great contrast renders them irreplaceable.

Silver impregnation and reduction will stain reticular connective tissue fibers more accurately than will aniline dyes. Not only is this reticulum silver stain useful in histology, but the pathologist often uses such stains as an aid to diagnosis of diseased tissues. The demonstration of the treponema of syphilis and similar organisms in fixed diseased tissues is possible only with modifications of the silver impregnation techniques (Levaditi, Warthin-Starry). Such stains have been of great importance in the diagnosis and study of these diseases.

Silver ions may be directly reduced to black silver deposits without development by substances such as vitamin C, melanin, adrenal medullary catecholamines, and enterochromaffin substance (5-hydroxytryptamine).

Silver reactions of this type may be useful histochemical procedure if appropriate controls are used, i.e., staining concurrently tissues known to contain these substances or staining concurrently tissues in which the sought-for substances have been removed or chemically blocked. Histochemical quantitation of a substance with silver reactions is unreliable. Reactions which directly reduce much silver have been called argentaffin reactions while those requiring subsequent development have been designated argyrophilic reactions. The Von Kossa stain for calcium deposits in tissue which actually demonstrates phosphates, depends on the photochemical blackening of these deposits by silver nitrate which is reduced to silver by exposure to daylight. While not entirely specific it is a useful stain.

Certain carbohydrate moieties in tissues may be oxidized to form aldehyde groups by treatment with chromic acid or periodic acid. Such aldehydes will reduce silver methenamine (Gomori) and result in a black silver deposit at the aldehyde site. This reaction may be useful in demonstrating mucus, glycogen, basement membranes, and the capsule of yeasts and fungi. The chromic acid methenamine silver stain is extremely useful in demonstrating pathogenic yeasts and fungi in tissues. The periodic acid silver methenamine stain is widely used as a precise stain of the basement membrane of renal glomeruli and will show subtle changes in disease not revealed by other techniques.

Recently there has been increasing use of periodic acid silver methenamine stains in electron microscopy. Histochemical stains for electron microscopy are limited to procedures where an electron opaque product is deposited, thus aniline dyes are generally useless. Although the silver image produced has the disadvantage of being granular, it will stain nuclear chromatin, ribosomes, glycogen, mucus, reticular and collagen fibers, basement membrane, lysosomes, and the carbohydrate-rich cell coat (sialic acid) which lies outside the plasma membrane of cells. With appropriate controls such silver stains provide useful morphologic and histochemical information to the electron microscopist.

While the specificity of silver stains is frequently open to question and the results are sometimes capricious, these

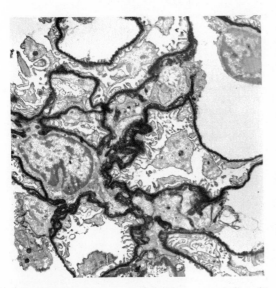

Fig. 1 Normal rat glomerulus, 2590 ×, periodic acid silver, methenamine stain.

stains have a beauty and dramatic quality which rewards the persevering scientist manyfold.

DAVID B. JONES

References

Representative Silver Procedures

Whole neurons—Golgi, C., "Sulla fina struttura dei bulbi olfattorii," *Riv. Sper. di Freniat.,* 1:405–425 (1875).

Axis cylinders—Bodian, D., "A new method for staining nerve fibers and nerve endings in mounted paraffin sections," *Anat. Rec.,* 65:89–97 (1936).

Boutons terminaux—Glees, P., "Terminal degeneration within the central nervous system as studied by a new silver method," *J. Neuropath. Exp. Neurol.,* 5:54–59 (1946).

Microglia and oligodendroglia—McClung, C. E., "Handbook of Microscopic Technique," 3rd ed., R. M. Jones, ed., New York, Paul B. Hoeber, Inc., 1950.

Reticular fibers—Wilder, H. C., "An improved technique for silver impregnation of reticular fiber," *Amer. J. Path.,* 11:817–819 (1935).

Calcium deposits—Von Kóssa, J., "Ueber die in organismus künstlich erzeugbaren verkalkungen," *Beitr. Path. Anat.,* 29:163–202 (1901).

Spirochetes—Warthin, A. S., and A. C. Starry, "A more rapid and improved method of demonstrating spirochetes in tissues," *Amer. J. Syph.,* 4:97–103 (1920).

Vitamin C—Deane, H. W., and A. Morse, "The cytological distribution of ascorbic acid in the adrenal cortex of the rat under normal and experimental conditions," *Anat. Rec.,* 100:127–141 (1918).

Glycogen and Mucous—Gomori, G., "A new histochemical test for glycogen and mucin," *Amer. J. Clin. Pathol., Tech. Sect.,* 10:177–179 (1946).

Electron Microscopy—Movat, H. Z., "Silver impregnation methods for electron microscopy," *Amer. J. Clin. Pathol.,* 35:528–537 (1961).

Basement Membranes—Jones, D. B., "Nephrotic glomerulonephritis," *Amer. J. Pathol.,* 33:313–329 (1957).

See also: GOLGI METHODS, SILVER PROTEINATE STAINS.

SKELETON

An understanding of key features of bone and cartilage is a prime requisite before considering histologic procedures by which these skeletal tissues can be studied.

Cartilage. In common with other connective tissues, cartilage consists of cells (chondrocytes), extracellular fibers, and an amorphous gel-like matrix or ground substance. In cartilage, extracellular fibers and ground substance predominate over the cells, which isolate themselves in small cavities within the matrix. The colloidal character of the matrix becomes vitally important to the nutrition of the cells, since cartilage has no nutrient blood vessels, and contributes significantly to the general firmness and resiliency of the cartilage. Cartilage has the functional capacity for relatively rapid growth and is, indeed, a dominant tissue in most vertebrate skeletons before birth, becoming more limited in distribution with increasing skeletal age. Three kinds of cartilage, hyaline (the most common), elastic, and fibrocartilage, are discernible on the basis of the amount of ground substance and extracellular fibers present. Unlike bone, cartilage usually exists chiefly in an uncalcified condition which is especially the case of small, presumably young, cartilage cells. Calcification of intercellular matrix in older, hypertrophied cartilage can eliminate the diffusion pathway of nutrients to the chondrocytes, leading to cell death.

Bone. Also a specialized connective tissue, bone is similar to cartilage in that it consists of cells, fibers, and extracellular ground substance. Bone cells can be either osteocytes (adult cells) which live in lacunae within the ground substance matrix, or those cells which locate along bone surfaces, called, according to their function, osteoblasts or osteoclasts. Bone differs significantly from cartilage, however, in that its ground substance matrix is the dominant of its three basic connective tissue components. This intercellular matrix is a mixture of collagen fibers and a mucopolysaccharide ground substance which normally calcifies, giving bone a cast-iron tensile strength. Despite its strength and hardness, bone is a dynamic living material constantly being renewed and remodeled throughout the individual's lifetime. Since bone is rigidly calcified, increases in bone size then must come about not by division of cells (cf. cartilage) but by the addition of new layers of bone at bone surfaces, making bones characteristically layered structures.

This complex nature of cartilage and bone indicates that no one technique can adequately demonstrate the many different features and components of skeletal tissue. Consequently, combinations of different preparations are commonly used for the study of either vital or postvital skeletal tissue.

Decalcifying Agents. Staining of decalcified bone sections is a routine histologic method of skeletal study. Decalcifying agents are necessary to dissolve out calcified bone salts so that the specimen can be sectioned on a standard rotary microtome. Chemical decalcification is by far the commonest procedure and can be accomplished by controlled usage of either ordinary fixatives containing weak acids, e.g., Bouin's fluid, or weak solutions of strong acids, as 5% aqueous nitric acid. Very rapid decalcification also can be attained by using phloroglucinol in combination with solutions stronger than 5% HNO_3 and 1–3% HCl. Other decalcifying agents for softening large bone specimens include magnesium citrate solutions as well as 1–5% formic acid in 70% alcohol. In general, decalcifying agents should be used generously and replaced often since bone salts put into solution weaken the agents. Several methods, of which the calcium precipitate technique is most common, can be used to determine if decalcification is complete. In these methods sodium or ammonium oxalate is added to the decalcifying solution containing the specimen. A clear solution with no evidence of a precipitate indicates decalcification is complete. Puncturing of the tissue with a needle is a damaging method and not recommended. X-ray determination of the decalcification endpoint is an excellent procedure yet can be costly in terms of time and equipment. After decalcification the specimen should be washed in running water to remove any residue of the decalcifying agent. Ion-exchange resin solutions and chelating agents, such as ethylene diamine tetracetic acid (EDTA), are also useful in decalcifying small skeletal specimens. Chelating agents are especially good since they maintain good fixation and sharp staining potentials during decalcification. A more time-consuming electrolysis decalcification procedure involves passing a direct current through an electrode to which the bone is attached with the entire unit immersed in a bath of 1% formic acid and 1% HCl.

Decalcified Bone Sections. Sections obtained through decalcification procedures can be cut, after dehydration, from either paraffin, celloidin, or combinations of the two materials. Celloidin is the preferable embedding material for bones of significant size from which undistorted, very thin sections are needed. Paraffin block sectioning can be done on a standard rotary microtome, whereas large, celloidin-embedded specimens must be sectioned on a sliding microtome to yield the best sections. Because of the time required by most celloidin or other nitrocellulose embedding techniques, most laboratories prefer the paraffin or combinations of paraffin and celloidin techniques.

Block Staining of Decalcified Bone. Block staining, using Harris' hematoxylin made with potassium alum, generally has the greatest versatility in staining haversian canals, lacunar margins, cementing lines, canaliculi, and bone cells. Counterstaining with solutions of eosin, basic fuchsin, or picric acids complements the hematoxylin well. Other tissues associated with bone, e.g., periosteum, cartilage, muscles, bone marrow, and blood vessels, in addition to bone, can be vividly demonstrated by using such multichromatic connective tissue stains as Mallory's triple connective tissue stain or a modification of the Masson trichrome stain.

Staining of Collagenic Fibers and Ground Substance. Special note should be given to these substances, present, as mentioned above, in both bone and cartilage. Staining of these two intercellular components of bone and cartilage is especially complicated since both components have approximately the same refractive index and tend to appear homogeneous in most ordinary preparations. Collagen fibers in both cartilage and bone can best be shown after the mucopolysaccharide ground substance is dissolved out of the tissue by appropriate alkaline treatment. The treated tissue can then be stained with any of the common connective tissue stains. Ground substance, on the other hand, can be demonstrated by using any of the metachromatic dyes, e.g., toluidine blue, methylene blue.

Decalcification and Staining of Fixed Dried Bone Sections. This can also be achieved using a relatively rapid technique which omits routine embedding and microtome sectioning. Essential steps of this technique include, in sequence, grinding a thin section of bone, acid decalcification, and staining. Any appropriate combinations of stains can be used for the decalcified ground sections. Hematoxylin and eosin stains will demonstrate an eosinophilic intercellular, collagenous matrix with a basophilia associated with the matrix in areas of non-lamellar bone. This method allows large numbers of stained sections to be produced within a short period of time, and the finished preparations are much superior in clarity and depth of staining to those of undecalcified bone sections. Interestingly, final preparations do not differ markedly from stained sections made by standard methods which require a microtome.

Undecalcified Bone Sections. When it is necessary to demonstrate bony architecture, including haversian systems, canaliculi, and lamellae, it becomes essential that undecalcified bone sections be made so that no portions of the mineralized bony architecture are changed or obliterated as would be in decalcification procedures. Undecalcified bone specimens may range from fresh whole bone to dry bone derived from individual bones, fragments of individual bones, or complete skeletons. Undecalcified bone can be studied by two rather commonly used methods. Cut sections of undecalcified bone specimens can be sliced to desired thicknesses (5 mm is suggested) using commercially available saws, as the Gillings-Hamco thin-sectioning machine. Such thin-sectioning machines usually require that the fixed dehydrated specimen be block-mounted in clear plastic, such as ethyl or methyl methocrylate.

Hand-Ground Undecalcified Sections. This second commonly used method can also be useful in studying cortical bone architecture and is a procedure requiring little elaborate equipment but, instead, a delicate touch. This procedure initially involves well-fixed, thin sections of bone which are individually and carefully ground on optical or metallurgical grinding wheels with finely powdered carborundum, jeweler's rouge, or household cleansing powder. After polishing both sides of the ground section, it is dehydrated, dried, dipped in a plastic solution, dried, affixed to slide with a commercial mounting medium (Permount, Technicon), and covered with a cover glass. Importantly, the plastic solution prevents the escape of air from the section's lacunae and canaliculi, preserves the natural differentiation between tissue components without staining, and allows satisfactory examination with polarized light.

Ground sections of undecalcified bone occasionally may be cleared and stored unmounted in a clearing oil, e.g., oil of wintergreen. Use of oil of wintergreen considerably increases the transparency of the bony matrix. There is some loss of contrast, however, because of the difference in refractive indices. Using appropriately prepared undecalcified bone sections, peculiarities in the density and distribution of mineralized tissues can be accented by such stains as basic fuchsin, silver nitrate, alizarin red S, and metallic sulfides.

Polarization Microscopy. This can be a significant method in analyzing submicroscopic structures of bone material. As a qualitative technique, polarization microscopy has been extremely useful in demonstrating that bone, as a nonhomogeneous biologic material, is anisotropic (doubly refractive) under polarized light because of the coexistence and peculiar arrangement of mineral microcrystals and the basic constituents of the organic matrix. Polarization microscopy of skeletal tissue also has the advantage of being quantitative and flexible enough to permit its application to most bone specimens, e.g., fresh or fixed, decalcified or calcified, stained or unstained. However, that preparative procedures can characteristically alter the polarizing properties of the material must be kept in mind in interpreting results. Cartilage, an isotropic, singly refractive material, does not polarize and this property can be useful in its identification. Theoretically, determination of true polarization requires rotation of the specimen, as false polarization remains optically bright through 360° whereas true polarization occurs only when crystal patterns are parallel to the plane of polarized light.

Bone Formation in Intact or Whole-Mount Skeletons. Bone formation in embryos, fetuses, etc. can be demonstrated by postvital staining with alizarin red S, a derivative of madder. Although the precise action of the dye alizarin red S (sodium alizarin sulfonate) is controversial, alizarin is specific for calcium in bone, showing a definite predilection for that bone, and not cartilage, which is being mineralized during the period of dye administration. Alizarin also can be used effectively as a vital bone marker either in tissue culture (*in vitro*) or *in vivo* experiments with animals.

Use of alizarin red S as a postvital stain to demonstrate the bony skeleton of embryos and small animals has been a classical technique in studies of skeleton development. In general, the specimen is fixed in 95% alcohol, then carefully washed in water, and macerated in a 2% aqueous solution of KOH until the skeleton begins to show through overlying soft tissues which progressively become transparent. As with most steps of this technique, timing is dependent upon the size and fragility of the specimen. (Usually the endpoint for the maceration or clearing procedure is 2–4 hr for smaller animals to 48 hr for larger animals as determined by simple visual inspection.) Flexibility in KOH concentration should be maintained to fit the condition of the specimen. When soft tissues are between the translucent and transparent stages, the specimen should be transferred to a dilute (0.1%) solution of alizarin red S in a 1% aqueous KOH vehicle. Staining may require 30–60 min for smaller animals or 6–12 hr or longer for larger skeletal preparations. Bone should appear deep red to purple at the end of this procedure. Any dye picked up by the soft tissues can be removed by placing the entire specimen in a solution of 1–2% KOH or in a mixture of glycerol, 70% ethanol, and benzyl alcohol. As a final step, the specimen is dehydrated in solutions of 2% KOH, 0.2% formalin, and glycerol for 24 hr followed by another 24 hr period in 2% KOH and glycerol. Finished products showing red-stained skeletal bony parts and transparent soft tissues can either be stored in a mixture of glycerol, 70% ethanol, and benzyl alcohol or embedded in plastic.

Cartilaginous elements in a whole-mount skeletal preparation can be selectively demonstrated using a protocol similar to that of the alizarin red S technique but with the substitution of toluidine blue. Staining with toluidine blue is cartilage-specific and can be combined with alizarin red S in the same specimen to make contrasting whole-mount skeletal preparations showing red-stained bone and blue-stained cartilage.

Vital Staining and Labeling of Growing Bones. This technique, using chemical bone markers, has been successful since the eighteenth century finding that madder fed to young growing animals has an affinity for newly calcified bone. Among the most widely used markers are alizarin red S, alternate injections of alizarin red S and alizarin blue BB, radioactive calcium, lead acetate, and a newer group of reactive dyes including the procion dyes, vinyl sulfones, and remozol dyes. These reactive dyes, unlike the previously mentioned bone markers, appear to form covalent bonds with the protein matrices in bone and allow for greater color combinations over extended periods while permitting decalcification of marked skeletal tissues without loss or displacement of the reactive dye marker. By administering combinations of most vital dyes, singly and in different combinations, it is possible to color-key bone remodeling over a prolonged period of time. Clinical observations that tetracycline antibiotics will deposit *in vivo* at sites of bone formation have introduced still another vital bone marker which can be demonstrated by examination under ultraviolet light, showing a yellow fluorescence of the drug on a background of blue-green autofluorescence of bone. The mechanism of tetracycline fixation to bone has not been fully elucidated. Postvital staining of tissues with tetracycline, as with alizarin red S staining, produces results similar to those seen after administration of the antibiotic to living animals.

Radioactive Materials (Isotopes). These present another vital marker technique devised to determine the amount, rate, and loci of mineral deposition in cartilage and bone tissues. Commonly used isotopes, e.g., ^{32}P, ^{45}Ca, behave chemically and physiologically in the same manner as the element under investigation. Basically, radioautography consists of feeding or injecting growing animals with compounds containing the isotope marker. The animal is subsequently sacrificed, and bone is recovered, sectioned, and laid on appropriate photographic film. After suitable exposure time, the film is developed and those parts of the bone, or cartilage, containing the radioactive material appear as dark areas.

ALPHONSE R. BURDI

References

General references

Baer, M. J., and J. A. Gavan, ed., "Symposium on bone growth as revealed by *in vivo* markers," *Amer. J. Phys. Anthrop.*, **29**:155–310 (1968).

Brain, E. B., "The Preparation of Decalcified Sections," Springfield, Ill., Charles C. Thomas, 1966 (266 pp.).

Emmel, V. M., and E. V. Cowdry, "Laboratory Technique in Biology and Medicine," 4th ed., Baltimore, The Williams & Wilkins Company, 1964 (453 pp.).

Frost, H. M., "Tetracycline-based histological analysis of bone remodeling," *Calc. Tiss. Res.*, **3**:211–237 (1969).

Humason, G. L., "Animal Tissue Techniques," 2nd ed., San Francisco, W. H. Freeman and Company, 1967 (569 pp.).

Jones, R. M., ed., "McClung's Handbook of Microscopical Technique," 3rd ed., New York, Harper Publishing Co., 1950 (790 pp.).

Luna, L. G., ed., "Manual of Histologic Staining Methods of the Armed Forces Institute of Pathology," 3rd ed., New York, The Blakiston Division, McGraw-Hill Book Company, 1968 (258 pp.).

Norton, L. A., W. R. Proffit, and I. C. Bennett, "Effects of tetracycline on bone growth in organ culture," *Growth*, **32**:113–124 (1968).

Seiton, E. C., and M. B. Engel, "Reactive dyes as vital indicators of bone growth," *Amer. J. Anat.*, **126**:373–392 (1969).

Specific references

Burdi, A. R., "Toluidine blue–alizarin red S staining of cartilage and bone in whole-mount skeletons *in vitro*," *Stain Technol.*, **40**:45–48 (1965).

Enlow, D. H., "A plastic-seal method for mounting sections of ground bone," *Stain Technol.*, **29**:21–22 (1954).

Enlow, D. H., "Decalcification and staining of ground thin-sections of bone," *Stain Technol.*, **36**:250–251 (1961).

Frost, H. M., "Staining of fresh, undecalcified, thin bone sections," *Stain Technol.*, **34**:135–146 (1959).

Frost, H. M., "Relation between bone tissue and cell population dynamics, histology and tetracycline labeling," *Clin. Orthop.*, **49**:65–75 (1966).

Frost, H. M., and A. R. Villaneuva, "Tetracycline staining of newly forming bone and mineralizing cartilage *in vivo*," *Stain Technol.*, **35**:135–138 (1960).

Frost, H. M., A. R. Villaneuva, H. Roth, and S. Stanisavljevic, "Tetracycline bone labeling," *J. New Drugs*, **1**:206–216 (1961).

Fullmer, H. M., C. C. Link, Jr., and M. J. Baer, "A stain for bone—illustrating apposition and absorption in two colors," *Stain Technol.*, **39**:71–73 (1964).

Hoyte, D. A. N., "Alizarin as an indicator of bone growth," *J. Anat.*, **94**:432–442 (1960).

Saxen, L., "Effect of tetracycline on osteogenesis *in vitro*," *J. Exp. Zool.*, **162**:269–294 (1966).

Steendijk, R., "Studies on the mechanism of the fixation of the tetracyclines to bone," *Acta Anat.*, **56**:368–382 (1964).

Tapp, E., K. Kovacs, and R. Carroll, "Tetracycline staining of tissues *in vitro*," *Stain Technol.*, **40**:199–203 (1965).

Williams, T. Walley, Jr., "Alizarin red S and toluidine blue for differentiating adult or embryonic bone and cartilage," *Stain Technol.*, **16**:23–25 (1941).

SKIN

The integument or skin covers the entire body and is one of the largest organs (about 18 ft^2 and 9 pounds, in human). Skin *per se* consists of the epidermis and dermis, but the subcutaneous hypodermis should also be included. In addition, there are several kinds of keratinized appendages or derivatives (hairs, feathers, scales, spines, nails, hoofs, horns, etc.) and cutaneous glands (sebaceous, sweat, and mammary). Skin varies considerably in thickness in each individual as well as in different animals. In humans, it is thickest on the back (5 mm) and thinnest on the eyelids (0.5 mm). Direct examination of skin is usually performed with a hand lens or dissecting microscope, using direct or oblique illumination. One should observe its general appearance, color, texture, thickness, moisture, surface oils, number and size of hairs, etc.

Epidermis. Epidermis is usually studied in histological or electron microscopic sections of skin, but direct *in vivo* observations and whole mounts are useful. In so-called "thick skin" the epidermis is prominent (0.15–1.5 mm) and displays several distinct layers. Examples are the weight-bearing surfaces on the hands and feet, skin pad of prehensile tails, chestnut and frog of horses, and snout of many animals. "Thin skin" has a thin epidermis (0.05–0.15 mm) with few layers, and may be glabrous (bare) or covered with hair. Pure epidermal sheets may be prepared for microscopic study as follows: Attach a small piece of skin to a glass slide coated with petroleum jelly or other lubricant, dermal side up. Incubate overnight at 4°C in 1% acetic acid (or 0.5% trypsin dissolved in saline), and gently strip away the dermis with fine forceps. Wash, fix with 10% formalin, stain if desired, and mount the entire specimen. Hair follicles and sebaceous glands often remain intact during this process, but the secretory tubules of sweat glands are usually removed with the dermis. Incubation with 0.1% collagenase in phosphate buffer, pH 7.4, for 3–6 hr at 37°C also provides good results.

Dermis. The dermis (leather when tanned) contains the blood vessels, nerves, and lymphatic capillaries of the skin, and provides the structural support for the epidermis. The dermis varies in thickness in different animals (1–4 mm in humans), but is usually organized into a superficial papillary layer and a deeper reticular layer. Various methods are used to examine the connective tissues, amorphous ground substance, and scattered cells of the dermis, but histological and electron microscopic studies are most common.

Skin is naturally tough, and the connective tissue fibers, cornified layers, and hairs are difficult to section. However, histological preparations remain the fundamental method of study. Select an appropriate site, and remove excess hair with scissors or clippers. Biopsies are often obtained with a rotary punch, but sharp scissors or a scalpel will suffice. Always take small specimens, carefully avoiding damage to the tissues. In animals with thick skin, the biopsy should be trimmed of excess fat and placed directly into a cold fixative. Biopsies from animals with relatively thin skin should first be affixed to coarse blotting paper (dermis down) to prevent curling. The paper should be removed during the dehydration process. Fixation in Zenker-formal is preferred, but Bouin's fluid and 10% neutral formalin are satisfactory. Other fixatives may be required for certain histochemical procedures. Dehydration schedules should be as brief as possible, and clearing in cedarwood oil followed by short immersion in toluene is recommended. After infiltration with soft paraffin for about 1 hr, embed in hard paraffin (56–58°C). Trim the paraffin blocks so that the specimen is oriented somewhat obliquely, and prepare a few trial sections. If any difficulties are encountered, the blocks may be soaked in ice water, but prolonged exposure will cause excessive swelling of the connective tissues. Sectioning should be performed with a cool, freshly sharpened microtome blade, orienting the tissue with the epidermis facing downwards. Recommended stains include: hematoxylin and eosin, toluidine blue buffered to pH 5.0 (glycogen and other carbohydrates), Masson's trichrome (collagen and other elements), Unna's orcein method (elastic fibers), Wilder's silver impregnation method (reticular fibers), and Heidenhain's iron hematoxylin (epidermal fibrils). Histochemical procedures for sulfhydryl groups, amino acids, nucleic acids, etc., are also instructive. Studies of enzymes and lipids invariably require frozen sections. Again, a sharp blade is essential, and the sections should be handled with a brush or glass rod.

Skin is difficult to preserve for electron microscopy. Small pieces of skin must be taken in order to insure adequate penetration (osmium tetroxide, glutaraldehyde, potassium permanganate) through the cornified layers of the epidermis, connective tissues of the dermis, and the hair follicles, etc. However, if the biopsies are too small, most of the structures will show evidence of mechanical injury. If possible, specimens should be obtained from very young animals which have been perfused *in toto* or locally. Place the specimens immediately into a cold fixative for about 30 min, and then carefully dissect out the desired structure (epidermis, dermis, hair follicle, sebaceous gland, sweat gland, nail root, etc.). Routine procedures may then be used. It may be necessary, however, to section several blocks before sections with the desired orientation are obtained.

Hair Roots and Follicles. These are best studied in the furry skin of laboratory animals. In humans, the scalp is often selected (also eyebrow, eyelid, cheek, axilla, mons pubis, etc.). Several types of follicles may be identified: terminal (guard), vellus (body), lanugo (infantile), sinus (vibrissa), tylotrich, etc. In all cases it is essential to understand the cyclic nature of hair growth. Growing hair follicles (anagen) extend deeply into the dermis and hypodermis, and display several prominent features: hair bulb surrounding a hair papilla, thick external root sheath, conspicuous internal root sheath, long hair root with a "keratogenous zone." Resting hair follicles (telogen) are very superficial and show few signs of functional activity.

Hair follicles are most frequently studied in histological and electron microscopic sections. Such sections may also be used to build three-dimensional reconstructions of hair follicles from sheets of balsa wood, etc. When taking biopsies, carefully remove excess hair and wet the skin with dilute detergent or a wetting agent.

Rectangular specimens should be taken, parallel to the slope of the hairs. Thin skin should be mounted first on blotting paper. Paraffin blocks should be mounted and trimmed so as to provide longitudinal sections of follicles; however, cross sections are also useful. Toluidine blue, Masson's trichrome, sulfhydryl, and histoenzymatic methods are particularly instructive. Cleared whole mounts of skin, thick sections (stained or unstained), and teased preparations are useful for visualizing the anatomical relations of the follicles.

Hairs. Hairs may be examined directly in living animals. Most animals display several types of terminal (guard) and vellus (body) hairs, plus tactile hairs (vibrissae, tylotrichs). Indeed, individual hairs may be repeatedly identified by placing a small tattoo in the adjacent skin. For permanent records, the entire hide of an animal may be preserved as follows: After skinning, the pelts are freed of excess fat and muscle, washed, salted out, partially dried with corn meal or sawdust, treated with Boraxo or arsenic, and permanently dried and mounted. Details of this procedure are provided by J. W. Knudsen ("Biological Techniques," New York, Harper and Row, 1966). Individual hairs may be plucked from the skin, washed with ether-alcohol, and observed under a microscope as a dry mount, or as a cleared preparation mounted with balsam. Cuticular scale pattern can be studied in dry mounts of hairs to which a drop of glycerin-alcohol has been applied to the hair tips, or by means of an impression of a hair in nail polish or suitable plastic (Dow thermoplastic; Faxfilm, Brush Development Co., Cleveland). Cross sections of hairs permit studies of the cortex and medulla, and are available in most histological sections of skin. A "Hardy Cross-Sectioning Device" may also be used (Gosnell Mfg. Co., Washington, D.C.). Such methods permit determination of length, diameter, shape, color, groupings, distribution, etc. Comparable methods are available for other keratinized appendages (spines, feathers, scales, horn, etc.).

Nails. Nails are technically found only in primates and humans, but claws and hoofs are comparable structures in other animals. Nails are hard keratinized plates, histologically similar to the stratum lucidum of thick skin. The body of the nail rests on a layer of epidermis, the nail bed, and projects distally as the free edge. Proximally, the nail root is continuous with the epidermal cells of the nail matrix. Growth (about 0.1 mm per day) can be measured by following the migration of a tattoo placed on the nail near the lunula. In such studies, one should also note changes in texture, shape, color, and thickness of the nail, as well as any changes in the associated tissues. Nails may also be dissected free from the digit and processed for structural or chemical analyses. For routine sectioning and staining, fix specimens in cold Bouin's fluid for 48 hr or longer, and remove excess bone and soft tissues. The dehydration should be as brief as possible, and the tissues cleared in cedarwood oil followed by toluene. Embed in hard paraffin, or double embed in celloidin and paraffin. After trimming, the blocks should be soaked in ice water and sectioned with a sharp blade. Masson's trichrome and sulfhydryl methods are particularly recommended. Frozen sections are not difficult, and various histochemical procedures may be employed. Satisfactory electron microscopic sections are difficult to obtain, due to the difficulty in taking small pieces and the poor penetration of most reagents.

Sebaceous Glands. These multilobulated alveolar glands (0.2–2 mm diameter in humans) are usually attached to hair follicles. Modified glands are found in the eyelid (Meibomian and Zeiss glands), external ear canal, lip, nipple of the breast, glans and prepuce of the penis, labia minora, and anus. The preen gland of birds and preputial gland of rodents are also modified sebaceous glands. Whole mounts and thick sections of skin are useful for studying the anatomical relations, size, shape, and distribution of the glands. Specimens are usually fixed in 10% neutral formalin and excess fat and connective tissue gently removed with forceps. The glands may then be stained with Sudan black (Sudan IV, oil blue N, etc.) for sharp visualization. Frozen sections are essential for various histochemical methods for lipids and enzymes. Paraffin sections may be stained with various methods cited for skin. Three-dimensional reconstructions of sebaceous glands are very instructive.

Sweat Glands. These are coiled, simple tubular glands which extend deeply into the dermis and hypodermis. Two types of glands should be recognized, eccrine glands which produce typical perspiration, and apocrine glands which produce a more viscid, turbid secretion. Eccrine glands are distributed over the entire body in humans (except external ear canal, lips, glans and prepuce in males, clitoris and labia minora in females). However, in many animals the eccrine glands are found only on the foot pads. The secretory tubules (0.3–0.4 mm diameter) display two types of cells, dark cells and clear cells, and are invested with slender myoepithelial cells. The excretory duct is relatively straight, but the epidermal sweat duct unit is prominently spiral. Apocrine sweat glands are larger than eccrine glands (3–5 mm diameter), and are usually associated with hair follicles. Apocrine glands are found in the axilla, external ear canal, eyelid (glands of Moll), perianal region, prepuce and scrotum in males, and labia minora and areola of the breast in females. Indeed, the mammary gland itself is a highly modified apocrine gland. Sweat glands are difficult to visualize in whole mounts of skin, but teased preparations and thick sections are recommended for studies of anatomical relations, size, and shape of the glands. For histological studies, Zenker-formal fixation is recommended, but other suitable fixatives may be employed. Useful stains are toluidine blue, periodic acid–Schiff, Masson's trichrome, Wilder's silver impregnation, plus numerous histochemical procedures for enzymes, etc.

MELVIN P. MOHN

References

Butcher, E. O., and R. F. Sognnaes, ed., "Fundamentals of Keratinization," Washington, American Association for the Advancement of Science, 1962.

Carruthers, C., "Biochemistry of Skin in Health and Disease," Springfield, Charles C. Thomas, 1962.

Hurley, H. J., and W. B. Shelley, "The Human Apocrine Sweat Gland in Health and Disease," Springfield, Charles C. Thomas, 1960.

Lyne, A. G., and B. F. Short, eds., "Biology of the Skin and Hair Growth," New York, American Elsevier Publishing, 1965.

Mercer, E. H., "Keratin and Keratinization," Oxford, Pergamon Press, 1961.

Montagna, W., "The Structure and Function of Skin," 2nd ed., New York, Academic Press, 1962.

Montagna, W., and R. A. Ellis, eds., "The Biology of Hair Growth," New York, Academic Press, 1958.

Montagna, W., and W. C. Lobitz, eds., "The Epidermis," New York, Academic Press, 1964.

Montagna, W., R. A. Ellis, and A. F. Silvers, eds.,
"Advances in Biology of Skin. III. Eccrine Sweat
Glands and Eccrine Sweating," Oxford, Pergamon
Press, 1962.
Montagna, W., and R. A. Ellis, eds., "*Ibid.* IV. The
Sebaceous Glands," Oxford, Pergamon Press, 1963.
Montagna, W., and R. L. Dobson, eds., "*Ibid.* IX. Hair
Growth," Oxford, Pergamon Press, 1969.
Montagna, W., J. P. Bentley, and R. L. Dobson, eds.,
"*Ibid.* X. The Dermis," New York, Appleton-Century-
Crofts, 1970.
Pardo-Castello, V., and O. A. Pardo, "Diseases of the
Nails," 3rd ed., Springfield, Charles C. Thomas, 1960.
Rook, A. J., and G. S. Walton, eds., "Comparative
Physiology and Biochemistry of the Skin," Philadelphia,
F. A. Davis, 1965.
Rothman, S., "Physiology and Biochemistry of the Skin,"
Chicago, University of Chicago Press, 1954.
See also: HAIR.

SMOOTH MUSCLE

Both thick and thin filaments of smooth muscle are
routinely visualized in the electron microscope when
freshly excised tissue is fixed with either glutaraldehyde
or formaldehyde. Osmium fixation alone appears to
disrupt thick filaments, and furthermore, the ordered
arrays (120 Å packing) of thin filaments is disturbed.
A fairly wide variety of smooth muscles have been found
to have thick filaments randomly scattered among
individual thin filaments and arrays of thin filaments
(Taenia coli, vas deferens, portal-anterior mesenteric
vein of guinea pig; Taenia coli, vas deferens, small mesenteric artery, main pulmonary artery, and uterus of rabbit;
uterus of humans; colon, bladder, and sphincter pupillae
of rat). A regular packing (660 Å center to center) of
thick filaments in addition to the 120 Å packing of thin
filaments has been found in rabbit portal-anterior mesenteric vein and in guinea pig Taenia coli.

Tissue is cut into 2 × 0.2 mm strips in 2–6% glutaraldehyde or 2% formaldehyde (prepared from paraformaldehyde) in pH 7.2 Krebs-Ringer solution. After 1 hr at
23°C the strips are transferred to 1% OsO₄ in Veronal
buffered at pH 7.4 for 1 hr. The tissue is dehydrated in
graded acetone solutions and embedded in Epon-
Araldite.

ROBERT V. RICE

References

Kelly, R. E. and R. V. Rice, *J. Cell Biol.,* **42**:683–694
(1969).
Rice, R. V., J. A. Moses, G. M. McManus, A. C. Brady,
and L. M. Blasik, *J. Cell Biol.,* **47**:183–196 (1970).
Rice, R. V., G. M. McManus, C. E. Devine, and A. P.
Somlyo, *Nature,* **231**:242–243 (1971).

SPERMATOZOA

Spermatozoa are very small, thin cells a few micra to
some millimeters long, rarely attaining 1 μ in diameter
along most of their length. Their study requires most
refined microscopic and submicroscopic techniques. They
may be examined under either static or dynamic conditions. In the first case this material, prepared in various
ways, may be studied by the light or phase contrast
microscopes (Fig. 1), or by the electron microscope

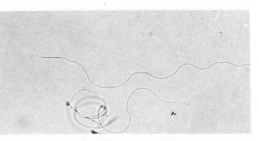

Fig. 1   Spermatozoa of an insect, *Bacillus rossius*, as seen
at the phase contrast microscope. (×160.)

(Fig. 2); in the second case, spermatozoa can best be
studied alive, by dark field microcinematography.

Only the sperm shape can be satisfactorily resolved
by the light microscope. Both the sperm length and the
length ratio between its four segments, namely, the
acrosome, the head (containing the nucleus), the intermediate piece (containing the axial flagellar filament and
the mitochondria), and the tail (containing the axial
flagellar filament alone), can be recognized by phase
contrast or on smears of testicular material stained by
the Pianese method (Malachite green 0.5 g; acid fuchsin
0.1; distilled water 150 ml; 95% ethanol 50 ml).

However, the accurate resolution of the sperm shape
can be achieved only at a submicroscopic level, by means
of the scanning electron microscope (Fig. 4). For this
technique, permitting biopsies for the study of male
sterility, spermatozoa should be placed directly on the
specimen holder, fixed a few minutes in 1–2.5% glutaraldehyde in the most suitable buffer (0.1M, pH 7.2 phosphate buffer for human sperms) then thoroughly rinsed
in distilled water. Spermatozoa adhere to the holder flat
surface and can be shadowed with gold-palladium as
usual.

To get information on the fine morphology of the
various sperm components, the most delicate techniques
of electron microscopy on ultrathin sections should be

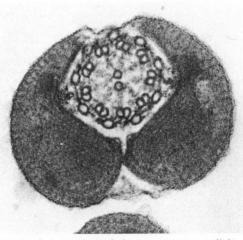

Fig. 2   Cross section of the spermatozoon tail in an
insect, *Notonec)a glauta,* as seen at the electron microscope. (×177,600.)

applied. Light microscopy is of no avail here. A general rule is to work on spermatozoa contained in the seminal vesicles, provided these are fragmented so as to facilitate the fixative penetration. The current double fixation with 2.5% glutaraldehyde buffered at pH 7.2, followed by 1% OsO_4 in the same buffer, is the best. Buffer solutions should be suited to the species under examination: e.g., $0.1M$ phosphate for mammals; Hoyle fluid for insects. As a rule, Epon, or a mixture of equal parts of Araldite and Epon are the embedding media giving the best results. Obviously, special techniques are required for the study of particular organelles, as well as for the level of ultrastructural analysis wished for.

Proteins, lipids, or carbohydrates can be selectively removed by pepsin, chloroform (after dehydration), or amylase, respectively, from unsupported ultrathin sections, or, better, from glutaraldehyde-fixed material; RNAse and DNAse can selectively eliminate either RNA or DNA by the same techniques, thereby enabling the interpretation of the chemical nature of several organelles. Autoradiography at a submicroscopic level is also advantageous and is greatly facilitated by the fact that radioactive isotopes can be incorporated by spermatozoa which are kept alive in natural or artificial culture media.

For the study of the sperm membranes, which are generally highly complex and even multilayered at the level of the acrosome, the freeze-etching technique is most expedient. A fair amount of material being required, one should resort to the study of small clumps of fresh fragmented seminal vesicles or discrete pellets of centrifuged unfixed spermatozoa, merely dipped in 2–10% glycerol in the optimal buffer for the species under study. (See Fig. 3.)

Sonic fragmentation of spermatozoa after light fixation in glutaraldehyde, spread on a common grid and negatively stained with potassium, phosphotungstate, is of great help in the study of the submicroscopic organization of the fibrils composing the axial flagellar filament, or of other highly symmetrical organelles, such as the crystalline proteinaceous mitochondrial derivatives flanking the axial flagellar filament in the insects, and other crystalline components occurring in the flagellum of other much evolved spermatozoa.

An essential prerequisite for the functional interpretation of the structures revealed by electron microscopy is the acquisition of all possible information on the sperm enzyme equipment and on the distribution of the various enzyme activities at the level of individual organelles. A wide range of histochemical tests for the assessment of enzyme activity at a submicroscopical level are available. They are all based on sperm incubation in a suitable medium from which a given enzyme splits an ion which is then precipitated while linked with an electron-opaque chemical vehicle. Generally, lead and silver are used, or else diaminobenzidine and nitro-blue-tetrazolium in the case of the respiratory enzymes. Since spermatozoa are fairly resistant free cells, they can be incubated alive, whereby, without organelle impairment, the best enzyme efficiency is gained. Fixation (always double, in glutaraldehyde followed by osmium) and embedding are performed after incubation.

Sperm motility can be studied only under the light microscope (preferably in a dark field) on accurately preserved living material. Spermatozoa are viable only in a liquid medium, consequently the choice of this medium involves the greatest technical difficulty. In the case of water-dwelling animals commonly discharging their spermatozoa into natural fresh or salt water, this problem is readily solved, since it is enough to prepare samples of sterilized water identical to the natural water in its chemical properties. For terrestrial animals with internal fertilization, the best medium is the spermatic fluid itself in the species (e.g., in mammals) where it is ejaculated in large amounts; otherwise, spermatozoa free of the seminal fluid can be removed only by means of delicate surgical methods. Particular artificial media must be tested for each case. In the insects, for example, satisfactory results are given by the following Goulding mixture:

Sol. A: NaCl: 6.8 g; KCl: 0.20 g; $CaCl_2$: 0.20; $MgCl_2 \cdot 6H_2O$: 0.216 g; $NH_2PO_4 \cdot H_2O$: 9.23 g in 100 ml H_2O
Sol. B: $NaHCO_3$: 0.125 g in 100 ml H_2O
Sol. C: Glucose in 100 ml H_2O to be mixed with H_2O at the time of use at the ratio 1:1:1:7.

Examination of swimming sperms already provides

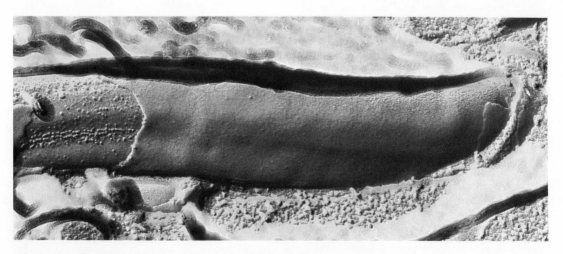

Fig. 3   The other aspect of the spermatozoon of an insect, *Bacillus rossius*, as seen at the electron microscope after the freeze-etching method. ($\times$39,200.)

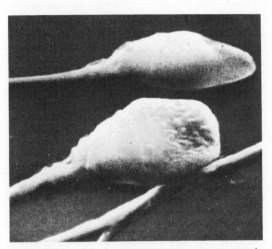

Fig. 4 Heads of the human spermatozoon as seen at the scanning electron microscope. (After Fujita, Miyoshi and Tokunaga, 1970.) (× 5680.)

the first information on these cells (essential for biopsies). However, further study can be based only on micro-cinematography of moving sperms, so as to follow their motion step-by-step. The more rapid the filming (i.e., the greater number of frames per second: about 300 is the optimum as shown by various experiments) the higher its efficiency. Particularly refined devices permit the flow of the data obtained directly to a computer. Otherwise, the coordinates of a sequence of a large number of frames of the spermatozoon position may be plotted.

Other major acquisitions in sperm physiology are made possible by micromethods which are collateral to microscopy; first among them is the measurement of oxygen consumption by means of a microrespirometer.

BACCIO BACCETTI

References

The most important technical directions concerning sperm investigation are available in the volume "Comparative Spermatology," edited by Baccio Baccetti, Proceedings of a Symposium held in Roma-Siena, Accademia dei Lincei–Academic Press, 1970.

STAINING, NOTES ON THE THEORY OF

The question might be asked: What purpose, apart from academic interest, could a valid theory of staining be expected to serve? The answer is simple. The chemical structures of the vast majority of dyes are known. On the other hand very little is known about the chemical structures of histological specimens. If we could fully understand the mechanism of biological staining then dyes could be used much more effectively as research tools for yielding new information regarding both the chemical and physical nature of the components of tissues and cells.

It is, of course, well known that proteins are present everywhere in mammalian tissues. It should be remarked here that this discussion is limited to the theory of mechanism of staining mammalian specimens, owing to

restrictions in the space allotted. However, it is hoped that this article will be of interest in other fields also. It appears that proteins are considered to be the most important biological substances. Like the amino acids (which are the units of which proteins are composed), proteins are amphoteric substances. There are different degrees of amphotericity. For example, some proteins are classed as basic proteins (e.g., histones) since they are predominantly basic in reaction though their molecules also contain groups which are acidic in reaction. The molecules of very many synthetic dyes, used in biological staining, are also amphoteric. Many of them might be regarded as artificial aryl amino acids (6).

Proteins comprise a vast group of substances composed of carbon, hydrogen, oxygen, nitrogen, and usually sulfur. Many of them also contain phosphorus and some contain small proportions of copper, iodine, iron, zinc, etc. Proteins are intimately associated in living cells and tissues with carbohydrates, lipids, hormones, enzymes, vitamins, nucleic acids, etc. The chemistry of proteins is highly complex since there is a greater degree of diversity in their molecular structure and composition than in any other group of biological substances. Proteins exist as cations on the acid side and as anions on the alkaline side of their isoelectric point, which is the pH at which the protein does not migrate in an electric field.

In the field of textile dyeing the fact that some dye molecules, or dye ions, might be too large to penetrate the interstices of the molecules of the textiles to be dyed does not appear to have been recognized until towards the end of the second quarter of this century (1). There are apparently no published records of any attempt to investigate this matter with the thoroughness it appears to deserve.

During the last 35 years or so it has been increasingly recognized that the larger molecules at least are to be considered three-dimensional, not merely two-dimensional as was implied in the earlier structural formulas of compounds. Later this concept led, among other things, to a distinction being drawn between alpha and beta forms of protein chains and the helical structure of nucleic acids. In later work it appears that many molecules can be described as having a "spongy" structure with surface pores which permit access, wholly or partially, of other molecules (and/or ions) to the interior. This interior will contain potentially reactive side chains to which external molecules or ions can attach themselves, but only if the molecule (or ion) as a whole or a suitable side chain of the external molecule (or ion) is able to penetrate a pore. Although the general idea dates back as far as the dialysis of colored solutions, the application of the grading of molecular porosity in relation to dyeing in general and biological staining in particular does not appear to have been taken into account in the manner it deserves.

In the field of biological staining it appears that the writer was the first to suggest that the morphology of a staining molecule or dye ion, which is a physical factor, might be more important than those factors commonly regarded as "chemical" (2, 3, 4). This suggestion was based primarily upon a tabulation of an extensive series of dyes in order of molecular weight in the ionized form (published several years after the tabulation had been made) (3, 4). It should be mentioned, in passing, that the shape and/or size of a molecule is represented here by the term "molecular morphology."

Alcohol (ethanol) is a reagent of considerable interest,

not only in biological microtechnique but in many other spheres. Among other applications, it is used extensively as a solvent for stains. Suggestions as to its possible influence in the mechanism of staining have been made elsewhere (4). A far more important reagent, as far as biological staining is concerned, is water. Some of its lesser known properties, which are highly relevant factors in the theory of staining, will now be described.

It has generally been accepted that the pH of distilled water is 7.0. It also appears to have been generally accepted that spontaneous acidity in distilled water is due to dissolved carbon dioxide (from the atmosphere) which can be expelled by boiling to restore the pH to 7.0. The pH of water used in biological staining is of profound importance in many procedures as variations in its pH are often the cause of unsatisfactory and misleading results.

Some years ago I had reason to carry out investigations which involved a large number of experiments on the pH of water and staining solutions (4, 5). This brought to light some interesting phenomena, highly relevant to the theory of staining. In the first place these experiments appeared to show that freshly distilled (as well as de-ionized) water usually has a pH between 5.0 and 6.0, not 7.0. In the experiments referred to above, whatever the pH of the particular sample of distilled water at 20°C it changed gradually (becoming less and less acid) as its temperature was increased to 100°C at which point the pH had risen by 2.0. That is to say, if at 20°C the sample had a pH of 5.8, then at 100°C its pH became 7.8. Graphs were perfectly linear in every case.

It was also observed that as the temperature of the samples of water fell gradually from 100°C to 20°C the same pH changes took place but in reverse, so that at 20°C the pH readings were exactly the same as those observed before the samples had been heated. Results of further work since carried out by me in this connection appear to indicate that as cooling takes place from 20°C to 4°C the pH of the samples of distilled water becomes more acid. At 4°C the average pH of the samples tested was found to be approximately 4.1. If I may be permitted a little irrelevancy, I would remark that it would be interesting to find out to what degree of acidity does the pH of water fall at subzero temperatures and the influence, if any, pH might have on the mechanism of "frostbite" and its prevention. Irrelevancies apart, the experiments referred to above appear to indicate that spontaneous acidity in distilled water cannot be cured by the old prescription of boiling "to expel the CO_2 absorbed, the 'cause' of acidity found in 'stale' distilled water." Distilled or deionized water as used in the vast majority of biological and chemical laboratories is a slightly acidic substance and buffering is necessary in many staining procedures.

It has also been generally accepted that when acid (anionic) or basic dyes are dissolved in water they do not change the pH of the solvent, except possibly to a negligible degree. It was thought that any slight pH changes which basic dyes might induce in the solvent would be to reduce its pH by an insignificant degree. On the other hand, anionic dyes, if they had any effect at all would raise the pH of the water to an insignificant degree. The results of work carried out several years ago, however, appears to have demonstrated that these beliefs are false (4, 6, 12). Many dyes were found to have a profound effect on the pH of water. For example, the basic dye, Acriflavine, when dissolved in water in the

proportion of 1 part of dye to 100 parts of water, reduced the pH of the latter to 1.4. Similarly picric acid (an anionic dye) reduced the pH to 1.35, whereas other anionic dyes were found to increase the pH very considerably. Amaranth, for example, gave a pH reading of 10.8. Eosin (an anionic dye), on the other hand, had little effect on the pH. Thus the buffering effect of dyes is a factor which has to be taken into account when the mechanism of staining is being considered. This factor has already been put forward in an attempt to explain the mechanism of a number of staining reactions (4, 6, 12). It also appears to indicate that certain granules present in the nucleus of the malarial parasite are not composed of the highly basic protein, protamine, as suggested by Unna (6, 13). Acid (i.e., anionic) and basic (i.e., cationic) dyes combine with proteins to form salts and proteinates. For example, eosin (an anionic dye) combines with proteins, in the presence of water which is on the acid side of their isoelectric points, to form protein eosinates (protein+, eosin−). On the other hand, cationic dyes, e.g., methylene blue, will combine with the same (or other) proteins, in the presence of water which is on the alkaline side of their isoelectric point to form dye proteinates (e.g., protein− methylene blue+). An example of this can be seen when we use eosin–methylene blue (Jenner stain) for blood. The mechanism of this stain has been suggested elsewhere (4). If, after staining, we use water of slightly acid or of neutral reaction for rinsing the preparation, the red blood corpuscles will appear pink or pinkish orange under the microscope. (Incidentally the envelope of the red blood corpuscles is said to be one of the most basic [i.e., alkaline] and one of the least permeable [i.e., of a low degree of porosity] cell elements known [14].) Here we have protein eosinates (protein+ eosin−). If, however, we use rinsing water which, instead of being neutral or slightly acid, is slightly alkaline (pH 7.2–8.2) the red blood corpuscles will appear gray or grayish blue or blue (instead of pink or pinkish orange). Here we have methylene blue proteinates (protein− methylene blue+). Thus it would appear that pH is a very important factor to be taken into account. The fact that a given acid (anionic) dye stains certain elements but not others does not, in all circumstances, necessarily mean that the elements which accept that dye are basic (alkaline) in reaction. Neither can it be taken for granted that, in all circumstances, that all tissue elements which are stained with basic dyes must necessarily be acidic in reaction. It should be mentioned that eosin is a weak acid compared with very many other anionic dyes. This is because it is a carboxylic acid (used in the form of a sodium salt generally), whereas the majority of wholly acid anionic dyes are sulfonic acids (used in the form of sodium salts generally). It should also be noted that unlike tissue elements, which are immobile, dye molecules when applied in solution are mobile and are therefore able to seek out, as it were, and form unions with those accessible tissue elements which are of opposite reaction to the dyes used. That is, an acid dye is most likely to have affinity for accessible tissue elements that are basic in reaction. On the other hand, acidic tissue elements could be expected to show special affinities for basic dyes.

The role of tissue element in this respect is, therefore, a passive one: until acceptable suitors in the form of dye molecules and/or dyeing ions come within their grasp. Marriage, as it were, then takes place in accordance with the Law of Mass Action as suggested elsewhere (10).

The influence of fixatives and mordants has been discussed elsewhere (4, 6, 14) and the arguments need not be repeated here. The same applies to the possible effects of the addition of such substances as acetic acid (6, 14) and/or other substances such as sodium sulfate (14), to certain staining solutions.

The anatomy of dye molecules has also been dealt with at length elsewhere (6, 9). A book of several volumes, rather than a mere article, would be necessary if we were to make a detailed study of all the factors which must be taken into account before a full exposition of the mechanism of staining could be presented, if that were possible in our present state of knowledge. We must content ourselves here with a very brief discussion of some of the more important factors which have to be taken into consideration if we are to acquire a better understanding of the underlying principles of the mechanism of staining. Obviously biological stains themselves must play a not unimportant role in any theory of biological staining.

Biological stains may be described as synthetic organic dyes. There are a few exceptions, however, the most important of these being hematoxylin, whose oxidation product, hematein, is a very important stain of natural origin. At the present time this is a stain which very many histologists would find the greatest difficulty in doing without. There are a number of synthetic dyes which could be used in its place and future generations may have to resort to these if and when the production of hematoxylin ceases. It appears not unlikely that it will eventually share the fate of other natural dyes which were much used in the past but which have since gone out of production. Carmine is the only remaining natural coloring matter used to any extent in biological microtechnique. This stain is of animal origin. Its active principle is carminic acid. There is little doubt that among the multitude of synthetic dyes available today some could be found as satisfactory substitutes for carmine in biological staining. It is more difficult to interpret the results of staining on a rational basis when one uses either of these two natural dyes since they are applied in solutions which, more often than not, contain other chemicals which complicate the issue. If in our investigations we use, wherever possible, simple aqueous solutions of synthetic organic dyes of known structure (and known pH) our task will be lightened, to a considerable extent, in interpreting the results of staining.

Synthetic organic dyes, like the components of biological tissues, can be divided into four main groups: acid, basic, amphoteric, and neutral. Compared with that of tissue-complexes the chemistry of dyes is relatively simple, and is well understood. If nothing were known about the chemical properties of dyes then they would be useless for investigating the chemical nature of histological specimens.

The orthodox methods of classifying dyes according to their chromophores and/or textile usage groups are of doubtless use to the textile dyer. These methods were invented for his benefit in the early days of the dye-making industry. The latter came into being to supply industrial dyers' needs for low-cost dyes. These methods of classification are of very little help, by themselves, to scientific and experimental histologists. Many years ago it occurred to me that there existed a need, if progress were to be made in the rationalization of biological staining, for a completely new method of classifying synthetic dyes for use in biological research. After calculating the molecular weight of a very extensive range of biological dyes and tabulating the latter in order of molecular weight, as mentioned above it appeared to me that in abstract these tables showed promise as research tools in solving a number of problems relating to the mechanism of a variety of differential staining reactions. This compilation showed, for example, that all the dyes that had so far been used, in pairs or trios, by histologists for differentiating keratin, collagen, ordinary cytoplasm, as well as other basic (and acidic) elements of cells and tissues could be divided fairly sharply, by molecular and/or ionic weight, into five classes. Only the second and third classes need be mentioned here. The second class (anionic dyes having molecular weights of around 350–590) stained keratin intensely, while the third class (anionic dyes having molecular weights above 700) stained collagen preferentially. It was also observed that the best stains for ordinary cytoplasm are, with few exceptions, anionic dyes having molecular weights between 350 and 590 (3, 4). All stains that had so far been found acceptable for both keratin and collagen are either wholly acid or amphoteric anionic dyes. However, later work carried out in collaboration with MacConaill (6) appeared to indicate that although the grading of dyes according to molecular weight (in the ionized form) is of help in a number of ways, the application of this supplementary system of classifying biological stains was not sufficient in itself to explain the mechanism of all differential staining techniques. For example, several dyes of very high anionic weight were found to stain ordinary cytoplasm and collagen with equal intensity (4).

In the work just cited it was suggested that the differential staining of RNA and DNA with pyronin G(Y)–methyl green might be due to the difference between the cationic weights of these two dyes. A similar suggestion was put forward independently by Goldstein (7) on the basis of his experimental work; he referred to the tabulation (3) mentioned above as his source of numerical data. Baker and Williams (8, 9) found that a mixture of rhodamine 6G and malachite green (used in a technique devised by these two workers for the differential staining of the two types of nucleic acids) gave essentially the same results as pyronin G(Y) and methyl green. This discovery, they rightly state, does not support the thesis advanced by Gurr (4) that the differential staining of pyronin and methyl green may be due to differences between the cationic weights of these two dyes. While agreeing with these two workers that their experimental evidence, referred to above, does not support my original hypothesis, I would venture to raise one point, as follows. MacConaill (4, 6) had already found that a number of stains did not behave in the manner predicted by the hypothesis four or five years before the publication of the Baker and Williams paper, cited above. MacConaill's experimental evidence was more than sufficient to convince me that my earlier hypothesis could not be considered as all-embracing. By way of explanation of these anomalies the idea then occurred to me that possibly shape or dimensions (rather than the mass of the dye ions) and possibly their pH, were factors to be taken into account. Another idea which occurred to me was that those dyes (which were in the majority of those tested) which behaved as predicted by the hypothesis attached themselves to tissue groups situated in the interior of the tissue pores (interstices of tissue molecules). On the other hand, those dyes which do not behave in accordance with the hypothesis might, it was thought, attach themselves to tissue groups situated at the peripheries of the tissue molecules. In the latter case it would appear that the tissue groups and the dyes would carry opposite charges (i.e., dye⁻, tissue⁺, or dye⁺, tissue⁻), in which case the staining observed would be due to chemical interaction between the dyes and the external tissue groups. The original hypothesis, as explained earlier, was based upon molecular morphology and tissue porosity, both being physical factors.

The discovery of the formation of a new dye, later known as Trifalgic acid, first seen to have been formed within tissues by MacConaill (3, 4, 6, 10), inspired a further reclassification of synthetic dyes by the writer. This reclassification was first presented in outline some years ago (6), and in full quite recently (9). This new system of classification incorporates the earlier ones by the same author (3, 4, 6) as well as the old systems (based on chromophores and usage groups) mentioned here earlier. In this new system dyes are classified according to their salt-forming (or potential salt-forming) side chains, referred to as "colligators" (4, 6, 9). Accordingly dyes can be divided into three main groups: (I) nonionic, (II) cationic, (III) anionic.

These three groups of dyes are subdivided into classes as follows:

Group I (Nonionic)
Class 1: Dyes having acidic colligators
Class 2: Dyes having basic colligators
Class 3: Dyes having neither acid nor basic colligators
Class 4: Amphoteric dyes

Group II (Cationic)
Class 1: Wholly basic
Class 2: Amphoteric

Group III (Anionic)
Subgroup 1 (wholly acid anionic dyes)
Subgroup 1 is divided into six different classes, which need not be described here.
Subgroup 2 (weakly amphoteric anionic dyes)
This is divided into two different classes.
Subgroup 3 (moderately or strongly amphoteric dyes)
This is divided into three different classes.

The members of each class are arranged and described (with their empirical and structural formulas, relative solubilities in 15 different solvents, spectral curves, etc.) in order of molecular (in the case of nonionic dyes), or ionic weight (in the case of ionic dyes). Where known, the pH of the water-soluble dyes are given for 1% solutions. The particular uses of the individual dyes is described (where known) in biological microtechnique, together with references to literature. Although this work in itself will by no means offer the solution to the entire mystery of the mechanism of staining, it might offer a number of ideas which could be of help to investigators in this field. For example, dyes which provide a one-dye differential staining solution are generally anionic dyes belonging to one or the other of three different classes. Members of each class share similar structural and chemical characteristics. The work also shows, among other things, that whereas the differential staining effect of the members of one of these three classes is due to physical factors, that of the others is due mainly to chemical factors, and in some cases, a combination of chemical and physical (e.g., molecular morphology and porosity) factors.

Among the other points brought to light by the new system of dye classification referred to above are:

a. The selective reactions of certain lipid stains (all but a few, notably Nile blue, methylene violet [Bernthsen], p-methoxychrysoidin, p-methoxychrysoidin, etc.) appear to be due to their colligators. For example, Sudan black B has two basic colligators (imino groups) which might well account for its staining of acidic lipids. This could be tested in practice by the substitution of Sudan blue, which, like many other nonionic dyes, possesses basic colligators. Incidentally it should be mentioned that Sudan black B has been found by the writer to possess the property of a nonaqueous acid-base indicator. This fact might be used to explain some of its staining reactions.

The staining of neutral lipids with neutral nonionic dyes (i.e., dyes possessing no colligators of either kind) appears to depend upon purely physical factors.

b. The majority of anionic dyes are really amphoteric and many may be regarded as artificial aryl amino acids. This fact might constitute an important factor in the mechanism of staining. For example, the amphoteric nature of acid fuchsin, which appears to have been first observed by MacConaill (10) and Baker (14) independently several years ago, was employed by the former in the "Falg" technique, a differential stain for tissue elements that are basic in reaction. The mechanism of the Falg technique will be referred to later in this article.

c. There are a number of basic dyes approximately the same cationic weight as pyronin G(Y) which might be worthy of trial in place of the latter in the methyl green–pyronin technique, or a modification thereof, for the differential staining of the two types of nucleic acids. There are also a number of basic dyes having cationic weight approximately the same as methyl green, which might be considered worthy of trial in place of the latter in the technique just mentioned.

d. The majority of cationic dyes are wholly basic in character. Of the few that might be regarded as amphoteric, Rhodamine B is the most outstanding and the most interesting (6, 9, 11). The fact that Rhodamine B is the only cationic dye which possesses a carboxyl group (which is an acidic colligator; all the other amphoteric cationic dyes have only hydroxyl groups as their acidic colligators) might account for some of its unique staining reactions.

It appears that very many important discoveries have come about by accident. Examples of these include William Henry Perkins' discovery, in 1856, of the basic dye Mauveine, which led to his founding of the synthetic dyemaking industry. His first sample of this dye is still in existence today. It appears to have undergone little if any change since it was synthesized by this 18-year-old genius over eleven decades ago (9, 16, 17). A more recent discovery, certainly no less important, was that of penicillin by Alexander Fleming, later developed by Wilfred H. Florey and Ernst Boris Chain and their associates. There are very many other examples of epoch-making discoveries which were evolved by accident. Paul Ehrlich (6) appears to have been the pioneer in laying the foundations on which our present knowledge of the mechanism of staining is based. But this was due to no accident.

This was observed by him to have been formed *in tela* by the interaction of two anionic dyes in his "Falg" staining procedure. His discovery led to others (4, 6, 9) which are highly relevant to the theory of staining. The first of these is that there are two types of counterstaining, namely:

1. complementary secondary staining, and
2. supplementary secondary staining.

The first is exemplified by the well-known hematoxylin-eosin staining procedure. In this case hematoxylin is the primary and eosin the secondary stain. In the Falg technique, acid fuchsin is the primary and light green SF the secondary stain. In the hematoxylin-eosin method the secondary stain (eosin) colors those tissue elements that have not already been colored by the primary stain (hematoxylin). In the Falg technique the secondary stain (light green SF) "stains" the stain (acid fuchsin) which has already stained the tissue elements. This discovery by MacConaill is highly relevant to the theory of staining.

The Falg method was later adopted, with slight modifications, by the late Bernard Squires (a physician with special interests in pediatrics and nutrition) for the detection of protein-calorie malnutrition and state of health in man and animals (15, 18). Following MacConaill's discovery of the blue dye, later called "Trifalgic acid," formed in tissues which had first been treated with acid fuchsin, then with light green,[a] the writer synthesized the dye *in vitro*. Sometime later MacConaill discovered that Trifalgic acid possessed the remarkable property (not shared by either of its two component dyes) of intensifying the effects of elliptically polarized light in birefringent elements of tissues. Other compound dyes synthesized by the writer, to test hypotheses regarding the mechanism of certain staining reactions, were similarly tested by MacConaill and found to give similar but even better results in elliptically polarized (ELP) light microscopy. One of the new compound dyes was produced *in vitro* by the interaction of acid fuchsin with violamine 3B and Sun Yellow G. This was given the name of "Haplo" (synonyms: Chromosome Red; Polarizing Red). Like Trifalgic acid it is a polyanionic polychrome dye (4, 6, 9).

It should be noted that the more common staining techniques are inimical to polarization microscopy. This poses a problem which might be worthy of investigation by some of my readers.

Some years ago I conceived the idea of combining the amphoteric cationic dye, rhodamine B, with the wholly basic cationic dye, Nile Blue. This eventually resulted in my synthesizing *in vitro* a polycationic, polychrome dye by the interaction of the two basic dyes just mentioned. This new dye was given the name of "Rhodanile blue." It was synthesized with the object of providing a new research tool, not to test any hypotheses regarding staining mechanism of any staining reactions already observed. MacConaill found, among other things, that this dye gave even better results as a booster of birefringence in ELP than did the polyanionic dyes referred to above. It was also found that Rhodanile blue had a number of noteworthy properties in ordinary light microscopy, and these have been described elsewhere (6, 9, 11, 19). Leopold and Bryant (9) have devised an elegant technique in which they have employed Rhodanile blue on a wide range of pathological material. An extraordinary range

of structures were, they found, clearly differentiated by this dye, and the possibility of using it for routine staining in lieu of hematoxylin-eosin now seems possible.

Finally I would observe that much practical work remains to be done before we can hope to make further progress in shaping a really acceptable theory of staining.

Obviously any hypotheses put forward as regards the theory of staining, no matter how much thought and time has been devoted to their formulation, will need to be questioned, and tested and retested many times before acceptance. It is to be hoped, however, that some of the ideas and suggestions expressed in this chapter and in the literature cited will provide food for thought and perhaps stimulate further experiment.

EDWARD GURR

References

1. Vickerstaff, T. "The Physical Chemistry of Dyeing," London and Edinburgh, Oliver & Boyd, 1954.
2. Gurr, E., "Methods of Analytical Histology and Histochemistry," London, Leonard Hill, 1958 (p. 21).
3. Gurr, E., "Encyclopedia of Microscopic Stains," London, Leonard Hill; Baltimore, Williams & Wilkins. 1960.
4. Gurr, E., "Staining, Practical and Theoretical," London, Leonard Hill; Baltimore, Williams & Wilkins, 1962a.
5. Gurr, E., "Effect of heat on the pH of water and aqueous solutions of dyes, "*Nature* (London), **195**:4847; 1199–1200 (1962b).
6. Gurr, E., "Rational Use of Dyes in Biology," London, Leonard Hill; Baltimore, Williams & Wilkins, 1965.
7. Goldstein, D. J., "Mechanism of differential staining of nucleic acids," *Nature* (London), **191**:407–408 (1961).
8. Baker, J. R., and G. M. Williams, "The use of methyl green as a histochemical reagent," *Quart. J. Microsc. Sci.,* **106**:3–13 (1965).
9. Gurr, E., "Synthetic Dyes in Biology, Medicine and Chemistry," London and New York, Academic Press, 1971.
10. MacConaill, M. A., and E. Gurr, "The Falg reactions in histology," *Irish J. Med. Sci.,* Ser. 7(412):182–196 (1960).
11. MacConaill, M. A., and E. Gurr, "The histological properties of Rhodanile blue," *Irish J. Med. Sci.,* June, 243–250 (1964).
12. Gurr, E., "pH of anionic dye solutions," *Nature* (London), **202**:4935; 920–921 (1964a).
13. Gurr, E., "The role of eosin in Romanowsky staining of the malaria nucleus," *Nature* (London), **202**: 4936; 1022–1023 (1964b).
14. Baker, J. R., "Principles of Biological Microtechnique," London, Methuen; New York, John Wiley, 1958.
15. Gurr, E., "A fresh look at dyes," *Chem. Brit.,* **4**:7; 301–304 (1967).
16. Gurr, E., "Perkin's original alizarin and after 100 years and more," *J. Soc. Dyers Col.,* **85**:473–474 (1969).
17. "Old stains examined," *Nature* (London), **224**:5224; 1061–1062.
18. MacConaill, M. A., B. T. Squires, and E. Gurr, "Detecting protein malnutrition," *New Scientist,* November, 290 (1966).
19. Gurr, E., "Dyes as research tools in biology and medicine," *Lab. Pract.,* **15**:10; 1128–1131 (1966).
20. Gurr, E., "Synthetic Dyes in Biology, Medicine and Chemistry," New York, Academic Press, 1971.

---

[a]Note: When the two dyes are applied in the reverse order, the blue staining of certain tissue elements does not occur.

STEREOLOGY

Stereology is the three-dimensional interpretation of flat images (sections or projections) using general principles of geometrical probability. The first known stereologists were astronomers of antiquity who interpreted the apparent motions and the changing luminosities of the images of the planets, projected on the vault of the sky, in terms of motions in space. The rather accurate estimates of the sizes and distances from the earth of the moon and the sun by Aristarchus (310–250 B.C.) represent the greatest stereological accomplishments in history.

As a formal, interdisciplinary science, stereology has existed since 1961. Its users are chiefly geologists, metallurgists, and biologists. In these fields we depend, for the identification of structure, on true sections or on translucent slices through three-dimensional masses.

Figures in a plane of section depend in shape and dimensions on the shapes and sizes of the three-dimensional structures cut and on their orientation relative to the cutting plane.

Principle of Dimensional Reduction. This is the basis of thinking for sectional stereology. It is illustrated in Fig. 1 and can be verbalized as follows:

A section through an n-dimensional object is, in general, an $(n-1)$-dimensional figure. Conversely an n-dimensional figure on a sectional plane is, in general, a profile of an $(n+1)$-dimensional object.

In practice, this means that, in most cases, in a sectional plane, a dot is a cut fiber, a line is a trace of a membrane, an area is a profile of a three-dimensional object. These rules apply to slices whose thickness is negligible in comparison with the thickness of the object cut. They do not apply for the extremely rare case that a thin structure lies exactly in the cutting plane.

Stereology as it existed about 1967 dealt chiefly with situations in which the slices are infinitesimally thin. This is section stereology. Projection stereology applies to "thick" slices. Astonishing as it may appear at first sight, an "ultrathin" slice as used in electron microscopy becomes very thick with increasing magnification. Small particles such as synaptic vesicles, ribosomes, glygogene granules, and microtubules have diameters only a fraction of the thickness of an "ultrathin" slice. They produce projected images on the screen or film to which the rules of section stereology do not apply. Since projection stereology is still in its infancy, it will not be discussed

here. The best account on its state in 1970 is found in the book by Underwood (1970).

Stereology has three subdivisions: (1) stereology of shape, (2) quantitative stereology, and (3) stereology of orientation. In quantitative stereology we ask "how big?" and "how many?"

Terms and Symbols. The following terms and symbols are most widely used in the United States. Our European colleagues are still vascillating. The majority of the officers of the International Society for Stereology have, in 1970, voted for their universal acceptance. An official vote by the membership has not yet been taken.

Terms

section	Cut or polished plane surface of an opaque object. A section has no thickness at all.
slice	Histological or mineralogical "section" of finite thickness, but cut from a translucent object.
slab	Portion of an object between two specified cutting planes. A slab may be opaque or it may consist mainly of discarded, thin slices between two relatively distant observed cutting planes.
profile	Section or slice through an individual component of a material, a cell, or an organ.
trace	Section of a two-dimensional structure, such as a membrane, a surface, or an interface. A trace appears as a line.
intercept	Length of that segment of a test line that falls on the area of a profile.
intersection	Point of crossing of a test line with a trace.
axial ratio	Quotient of length over width.

Measured and Counted Quantities in Sections or Slices

P_P	Number of test points falling on profiles of a component divided by total number of test points.
P_L	Number of intersections per length of test line.
n_A	Number of profiles per total test area or average number of profiles per standard test area.
L_L	Sum of intercepts per total length of test lines or average length of intercepts per standard test line.
q or Q	Axial ratio of profile.
d	Diameter of profile (average distance between parallel tangent lines).
T	Number of points of a trace tangent with an array of parallel straight lines.
h	Distance between parallel test lines.

Quantities in Space

V_V	Volume fraction of a component.
S_V	Surface area per unit volume.
L_V	Length of linear structure per unit volume.
T	Thickness of a sheet in space.
t	Thickness of a slice or average distance between cutting planes.
D	Diameter of a particle (average distance between parallel tangent planes).

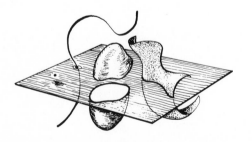

Fig. 1 Dimensional reduction: A section through a solid is an area. A section through a surface or a membrane is or appears as a line. A section through a line or a fiber is or appears as a point. A section through a point does not exist at all.

Stereology of Shape. The shape of a tissue component is the basis for its physiological functioning. For example: a nerve fiber must be cylindrical if it is to be used for impulse conduction; a fat cell approaches most conveniently the shape of a sphere for purposes of storage and shock absorption; the lining of a cavity must have the shape of a sheet; a glandular duct should be cylindrical. In other words, shape is important to function. Therefore its identification must precede quantitative analysis. When applying the rules of dimensional reduction, most tissue elements can be roughly identified, as far as their shapes are concerned, from typical shapes of their profiles. But for precise shape identification profiles must be measured and classified. Many animal organs are composed of structural units (glandular end pieces, lobules, follicles, etc.); and specific cell types contain organelles (nuclei, cisterns of the endoplasmic reticulum, lysosomes, etc.) of similar shapes. We assume that structural units having the same function have a limited range of shapes. It is further assumed that they are distributed in space at random. Most structural units have simple shapes resembling circular cylinders, prolate rotary ellipsoids, spheres, oblate rotatory ellipsoids, tri-

axial ellipsoids, elliptical cylinders and sheets. The shapes of objects of complicated shapes such as kidney podocytes are less easily recognized.

A section through any of them (except the last two mentioned) is an ellipse. The shape of an ellipse can be defined by its axial ratio Q, i.e., the quotient length/width. The axial ratio of a specific sectional ellipse depends on the angle which the cutting plane forms with the axes of the cylinder or ellipsoid. The simplest case is that of spheres. Any section through a sphere is a circle, i.e., an ellipse of $Q = 1$. Thus if all profiles in a sectional plane are circles we know immediately that in space there were spheres (except in the case of an exact transverse section through a bundle of parallel cylinders, such as a nerve).

When bodies of shapes other than spheres are randomly arranged in space a characteristic distribution of axial ratios of the profiles is found for each specific shape. Distribution curves of axial ratios of profiles for each

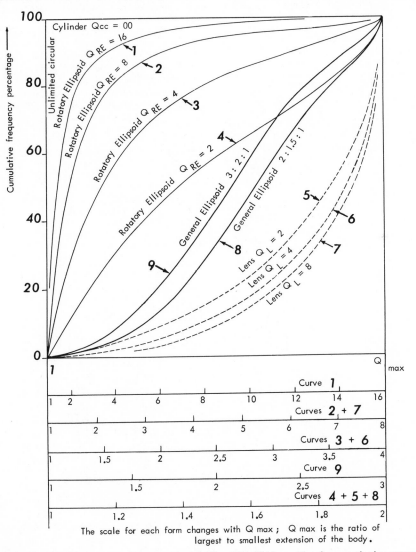

Fig. 2 *Cumulative* distribution curves of axial ratios of ellipses (profiles) resulting from sectioning masses of ellipsoids of equal shape for each curve. Q_{ER} = axial ratios for prolate rotatory ellipsoids; Q_L = axial ratio of oblate rotatory ellipsoids (lens shaped objects).

of the above-mentioned solids have been elaborated. They are shown in Figs. 2 and 3. In order to determine the range of shapes for a mass of solids, one must classify all the profiles in a section or in several sections, construct a cumulative curve, and try to match it with one of the curves shown in Figs. 2 and 3. Due to irregularities of shape in living systems the points of that observational curve will fall between two of our curves. Thus a range of shape, rather than one exact, identical shape, is established.

It is obvious, then, that reconstructions from serial sections are not needed to determine a characteristic, average shape for tissue elements of a specific kind, if the shapes are relatively simple.

The case of circular cylinders (which are very common among tissue components such as tubules, cords, trabeculae, and arteries) does not fit well into the curves of Figs. 2 and 3. Therefore we shall treat it separately. The axial ratio of a section through a circular cylinder Q equals the cosecant of the angle between the axis of the cylinder and the cutting plane (Fig. 4). If circular cylinders, randomly distributed in space, are cut the axial ratios of their profile are distributed so that 75% of them are short ellipses where $1 < Q < 2$ and 25% of them are more oblong. Very long profiles are extremely rare, and if the cylinders are curved, they are completely absent.

Great errors have been made in the past in shape identification because these basic rules have been ignored. For example, the liver was said to consist of circular cylinders. Histologists believed themselves to be always extremely lucky obtaining perfect longitudinal sections only and never a cross section through liver "cords." Equally fortunate seemed the embryologist who thought to have found longitudinal sections of "cords" in every section through embryonic human testicles. Similarly, the flat cisterns of the endoplasmic reticulum were at first thought to be filaments, luckily always located in the cutting plane. In fact it was the clarification of such errors which led to the establishment of stereology as a formal science.

Profiles of sheets, plates, or membranes are always very long stripes sometimes of infinite length, while their width varies from place to place according to the angle of inclination toward the cutting plane. Objects of more complicated shapes are less easily identified from single random sections. After having determined the shapes of organ components one can ask quantitative questions about them.

Quantitative Stereology. In the following the assumption is made that all organ components are randomly arranged in space.

Volume Ratios. These are easily determined by placing a grid of points over a section and counting "hits." (Fig. 5). The number of points (black dots in Fig. 5) falling on profiles of a specific type of organ component is proportional to the volume fraction which this type of component occupies in the entire organ. We express this mathematically proven fact by the equation

$$P_P = V_V \qquad (1)$$

which means: The number of points which *hit* a specific

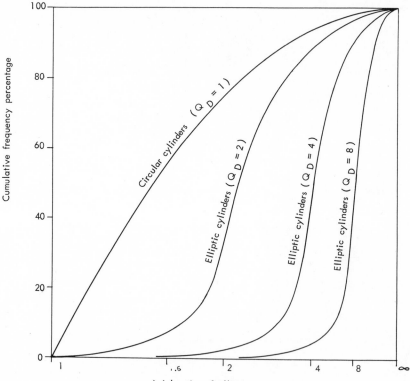

Fig. 3 *Cumulative* distribution curves for axial ratios of sections through elliptical cylinders of specified shapes. Q_D = axial ratio of directrix of cylinder.

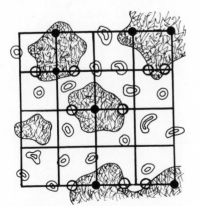

Fig. 4 Dependence of the shape of a sectional ellipse through a circular cylinder on the angle of cutting.

type of tissue in a section divided by the total number of points in the grid equals the volume of that component compared with the volume of the entire organ. As a matter of course, the precision of measurement increases with the frequency of positions of the grid on the section.

Surface in Volume. The surface area of a feature or the interface between features is determined by counting intersection points of straight lines with the traces of that interface (outlines of profiles). In Fig. 5 only the horizontal lines are used for intersection counts, but the vertical lines could be used as well. If the number of intersection points (open circles in Fig. 5) with the outlines of the shaded features is P, then

$$S_V = 2P_L \qquad (2)$$

This means that the surface area of the shaded tissue component per unit volume equals twice the number of intersection points divided by the total length of test lines used. Care must be taken that the unit of measurement is the same on both sides. If V stands for cubic millimeters, L stands for millimeters. If L is measured in microns, V must be given in cubic microns.

Formula (2) applies for structures which extend beyond the field of vision. If a structure is of limited extension, i.e., if it is entirely contained in a single field of vision, then one can obtain its absolute surface area

$$S = 2Pth \qquad (3)$$

where t is the distance in space between cutting planes and h the distance between parallel, equidistant test lines.

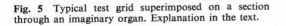

Fig. 5 Typical test grid superimposed on a section through an imaginary organ. Explanation in the text.

long enough to extend beyond the boundary of the object. This formula can be used only when equidistant sections through an entire structure are used. The cutting planes may be few and far apart, but the set of sections must be complete. An example may be a sliced loaf of bread, where a grid of parallel lines may be placed over *every* fifth slice from end to end.

Length in Volume. The total length of linear elements in a unit of space, such as capillaries of a plexus or seminiferous tubules in the testis, can be determined stereologically. Let the small irregularly shaped figures with double outlines in Fig. 5 represent sections through tubules or capillaries or fibers. The total length of all these things in a unit of volume is

$$L_V = 2P_A \qquad (4)$$

which means that the total length of linear elements per unit volume equals twice the number of their profiles counted within the test area A. In the case of Fig. 5 the test area is the area of the measuring square. The absolute total length of linear structures within a volume totally confined to the field of vision is

$$L = 2Pt \qquad (5)$$

where t is, again, the constant distance between successive cutting planes. Here again the object which contains the length must be cut into parallel slices from one end to the other. And a small number of equidistant slices must be subject to a profile count. An example would be a ball of twine embedded in a solid matrix so that it could be sliced.

Number of Particles in Unit Volume. To determine the number of particles in a unit of volume, as, e.g., glomeruli in kidney cortex, one must count their profiles within a test area. That number is, roughly,

$$N_V = \frac{n}{A(\overline{D} + t)} \qquad (6)$$

where N is the number of particles in space, n is the number of their profiles counted within a test area A, $\overline{D}$ is the average diameter of the particles in space, and t is the thickness of the slices (histological "sections") thought to be translucent. For opaque materials such as metals, t does not need to be considered since it equals zero.

For an exact determination of N one should consider that slices cut off a spherical particle near its pole are invisibly thin or unidentifiable, so that from the denominator in formula (6) $2Ah'$ should be subtracted, where h' is the thickness of the largest invisible slice.

Formula (6) implies that the particles to be counted are approximately spherical. For objects of different shapes, other formulas apply.

Size of Spherical Particles. In formula (6) we found the letter $\overline{D}$, the average diameter of the particles in space. This implies that, before their number per unit volume can be determined, their size distribution must be known.

The determination of size distribution, although very important, is the most difficult problem of stereology. Several methods have been employed. They all depend on the following considerations:

Any section through a sphere is a circle whose radius is

$$r = \sqrt{R^2 - h^2}$$

where R is the radius of the sphere and h the distance of the section from the center of the sphere (Fig. 6). As Fig. 7 shows, the percentage of relatively large sectional circles is large, that of small circles very small, *if* all spheres in a mixture are of equal size. Sectional circles through spheres of equal size are distributed as follows:

$$0 < r \leqq \frac{R}{4} \qquad 3.2\%$$

$$\frac{R}{4} < r \leqq \frac{R}{2} \qquad 10.2\%$$

$$\frac{R}{2} < r \leqq \frac{3R}{4} \qquad 20.5\%$$

$$\frac{3R}{4} < r \leqq R \qquad 66.1\%$$

In observational cases, the small circles are less numerous than in this table because near the pole, the profiles fall out of the slices, are too thin to be visible or too small to be identifiable. Most spheroid particles which occur in nature, such as glomeruli, fat cells, and nuclei, are not all alike in size; but they have a size distribution.

All algebraic and analytical methods which have been proposed are not only too coarse for accuracy, but they are also extremely cumbersome. However Elias et al. (1971) have developed standard curves for the determination of the range and distribution of spheres from measurements of their sections. To describe the method (which is both the simplest and the most accurate so far proposed) the reader is referred to their paper.

Perferred Orientation of Tissue Components and Gradients of Density So far we have assumed that the particles to be subjected to stereological analysis are randomly distributed in space. However, in nature this is not always the case. Oblong and flat particles, such as muscle fibers, nerve fibers, kidney tubules, squamous cells, and erythrocytes in fast-streaming blood, tend to exhibit a preferred orientation.

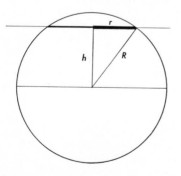

Fig. 6 Dependence of the radius r of a circle which results from cutting a sphere of radius R on its distance h from the center of the sphere.

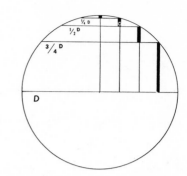

Fig. 7 Frequency distribution (dark lines) of sectional circles of four size classes resulting from cutting spheres of equal size.

Counts on sections through organs with preferentially oriented components (anisotropic organs) are different in different directions of cutting. One can eliminate anisotropy by various sampling methods such as cutting an organ into small cubes and mixing these cubes and then cutting them in directions which have become randomized. Similar rules apply when there are gradients of density as, for example, in the cerebral cortex. The mincing method of fixation for electron microscopy eliminates preferred orientation because the blocks are so small that the worker cannot control the direction of cutting.

The grid shown in Fig. 5 is but one of many patterns which have been proposed. The present writer prefers this pattern because of its simplicity and versatility. A survey of several patterns is given by Sitte (1967).

As in any kind of quantitative work the accuracy of the results increases with the number of observations.

In spite of their astonishing simplicity, the basic formulas of stereology are mathematically correct. Accuracy of results depends only on the thoroughness of the observer and on the nature of the material.

HANS ELIAS

References

Elias, H., "Stereology," New York, Springer-Verlag, 1967.
Elias, H., and D. Schwartz, "Surface areas of the cerebral cortex of mammals determined by stereological methods," *Science,* **166**:111–113 (1969).
Elias, H., A. Hennig, and D. Schwartz, "Stereology: applications to biomedical research," *Physiol. Rev.,* **51**: 158–200 (1971).
Sitte, H., "Morphometrische Untersuchungen an Zellen," pp. 167–198 in "Quantitative Methods in Morphology," E. Weibel and H. Elias, ed., Berlin, Springer, 1967.
Underwood, E. E., "Quantitative Stereology," Reading, Mass., Addison-Wesley, 1970.
Weibel, E. R., and H. Elias, "Quantitative Methods in Morphology," Berlin, Springer-Verlag, 1967.

TARDIGRADA

Introduction. Many chemicals, materials, and methods may be employed in a microscopical study of the tardigrades, but no single procedure has gained eminence. Therefore the reader is encouraged both to try a number of the older techniques and to devise ones of his own. The two monographs, Marcus (1929) and Ramazzotti (1962), are particularly helpful sources of general information on tardigrades.

Killing and Fixing Agents. A number of fixing agents have been employed with similar degrees of success. Marcus (1929) suggested the use of a 4% formalin solution as a killing agent and fixative. Although he too used 4% formalin as a killing agent, Cuenot (1932) also employed a solution containing one part acetic acid, one part water, and one part saturated solution of mercuric chloride for killing and fixing. Riggin (1962) found hot 85% ethanol advantageous because it causes the extension of the animals, rendering them quite suitable for making whole mounts. By diluting a stock preparation of Bouin's solution, Puglia (1964) was able to prepare specimens in a like state of extension. Schuster and Grigarick (1966) used hot water as killing agent and 5% formalin as fixative. Hot water (40–80°C) has also been employed as killing agent by Mehlen (1969) followed by either 4% formalin or 70% ethanol. The use of hot water or hot ethanol avoids the shrinkage and contorting effect that often occurs with killing solutions. Part of that problem, however, may not be that the solution causes much distortion but rather suddenly fixes the animal in whatever position it was in at the time it died. This is an important consideration, as one must be skeptical of measurements made on animals that were not fully extended or that were stretched because of coverslip pressure.

Staining. For vital staining Ramazzotti (1962) recommended the use of neutral red to mark tardigrades being extracted from moss samples and suggested that methylene blue stains nonselectively for muscles and nerve fibers, but must be carefully used in dilute solution because it is often poisonous to tardigrades. To facilitate picking out tardigrades from samples, the animals may be killed and then stained with rose bengal at 100 mg/liter in ethanol or formalin, as suggested by Mason and Yevich (1967) and McGinty and Higgins (1968). In aqueous, rather than alcoholic solution, rose bengal stains living tardigrades very weakly or not at all but readily stains dead ones. This feature may be used for such mortality studies as revival from cryptobiosis. Acetic carmine must be followed microscopically to avoid overstaining. This stain has the inconvenience of destroying a large part of the stylets but serves well to demonstrate spermatozoa, chromosomes, etc. and can be employed without previous fixation of the animal or the egg. May (1948) presented a method of selective staining with tints of yellow, to brown, and to black showing the organization of various cells, in particular the central nervous system. First, a solution of ammonical silver carbonate is prepared: 150 ml of 5% sodium carbonate is poured a little at a time with continual agitation into 50 ml of 10% silver nitrate solution. The precipitate is redissolved by dropwise addition of ammonia, adding a minimal amount for redissolution. The solution should be kept well stoppered in a dark-colored bottle away from light. When the tardigrades are treated with 2% acetic acid complete extension is attained. After washing with water they are put for 24 hr in 10% neutralized formalin. After a quick rinse in water, they should be immersed in a solution composed of 5 parts distilled water and 1 part ammonical silver carbonate solution, then promptly be placed in an oven at 60–65°C for 1–2 min; varying the time in the oven permits selectivity of staining organs and cells. The animals are next washed and put into 10% neutralized formalin; definitive mounting may follow immediately or after storage.

Mountants. Ramazzotti (1962) presented two methods of mounting: Faure's liquid and polyvinyl lactophenol. Faure's liquid has the advantage of preserving the color of acetic carmine, and the polyvinyl lactophenol clears the tissues so that the placoids, claws, and cuticle are readily visible. In Faure's solution the tardigrades often wrinkle but usually return to a normal condition within an hour or so. Both of these mountants tend to overclear the animal with the passage of time. Also, sometimes there is a failure of the mountant to remain adhered to the coverslip during drying and plasticizing of the mountant so that air streaks in, causing partial or complete ruin of the mount.

Gray and Wess's medium (Gray, 1958) is an easy-to-use aqueous mountant, particularly for temporary mounts of the Eutardigrada to observe internal anatomy. As in Faure's solution the tardigrades often wrinkle in Gray and Wess's medium, but usually they return to a normally extended state after an hour or so. For permanent mounts it is recommended that a full-strength solution be used, the cover glass be supported so that the specimens are not crushed, and the preparation be well sealed by a ringing compound or clear nail polish. When ringing was not done, coverslip adherence problems have been encountered after several months.

Hoyer's medium has been employed by some authors: Riggin (1962), Schuster and Grigarick (1965), McGinty and Higgins (1968) who found that it overclears the internal anatomy, specimens tending to disappear after a few months. But this mountant may be used to an advantage when studying the external anatomy, particularly with the Heterotardigardes as in McGinty and Higgins (1968). For internal anatomy, mounting in glycerol is useful. Marcus (1929) suggested clearing in glycerol and mounting in glycerol jelly. Glycerol impregnation may be employed here advantageously: After fixing, the tardigrades are placed in an alcoholic solution of 10% glycerol and left in a dustfree place for a week or so until the ensuing evaporation has left them in a solution of glycerol; this may be hastened by evaporation in an oven or on a hot plate.

Cuenot (1932) recommended clearing in clove oil and mounting in balsam, but this method created considerable distortion of the specimens. After clearing in xylol, Puglia (1964) mounted in Harleco synthetic resin under a 5 mm² coverslip.

Mounting. Depression slides are useful for temporary and permanent whole mount preparations. Higgins (1959) employed them so that the specimens would not be squashed by the coverslip and would remain suitable for length-frequency analysis. To avoid squashing and distortions in studies of internal anatomy or meristic analysis for taxonomy, the coverslip should be supported by a spun ring, chips of broken coverslip, a cell cemented to the slide, or in holes made in a plastic coverslip with a paper punch and cemented to the slide. Double coverslip mounting may be used to advantage in two ways. One way is for sealing mounts made with an aqueous medium such as glycerol jelly: The specimen is mounted on a large cover glass and covered with a smaller one; after the mountant has set, the double cover glass is mounted small cover glass down on a slide and cemented by a resinous mountant. A second way is merely to tape the double cover glass mount to a depression slide. Then, when it is desired to see the animal from the other slide, one may untape the over glass and invert it over the depression (suggested by R. P. Higgins, personal communication). If Cobb aluminum slides are available, the utility of the double cover glass mount is much enhanced.

RONAL H. MEHLEN

References

Cuenot, L., "Tardigrades," *Fauna de France,* **24**:1–96 (1932).

Gray, P., "Handbook of Basic Microtechnique," New York, McGraw-Hill Book Co., 1958 (p. 105).

Higgins, R. P., "Life history of *Macrobiotus islandicus* Richters with notes on the other tardigrades from Colorado," *Trans. Amer. Microsc. Soc.,* **78**(2):137–154 (1959).

McGinty, M. M., and R. P. Higgins, "Ontogenic variation of taxonomic characters of two marine tardigrades with description of *Batillipes bullacaudatus* n. sp.," *Trans. Amer. Microsc. Soc.,* **87**(2):252–262 (1968).

Marcus, E., "Tardigrada," pp. 1–608 in "Klassen und Ordungen des Tierreichs," Vol. V, 1929. Akademische Verlagsgesellschaft, Leipzig.

Mason, W. T., and P. P. Yevich, "The use of phloxine B and rose bengal stains to facilitate sorting benthic samples," *Trans, Amer. Microsc. Soc.,* **86**(2):221–222 (1967).

May, R. M., "Le Vie des Tardigrades," Paris, Gallimard, 1948 (p. 118).

Mehlen, R. H., "New Tardigrada from Texas," *Amer. Midl. Nat.,* **81**(2):395–404 (1969).

Puglia, C. R., "Some tardigrades from Illinois," *Trans. Amer. Microsc. Soc.,* **83**(3):300–311 (1964).

Ramazzotti, G., "Il phylum Tardigrada," *Mem. Dell' Ist. Ital. Di Idrobiol.,* **XIV**:1–595 (1962).

Riggin, G. T., "Tardigrada of S.W. Virginia: with addition of a description of a new species from Florida," Tech. Bull. 152, Va. Agr. Exp. Sta. at Blacksburg, Va., 1962 (145 pp.).

Schuster, R. O., and A. A. Grigarick, "Tardigrada from Western North America," *Univ. Cal. Publ. Zool.,* **76**:1–67 (1965).

Schuster, R. O., and A. A. Grigarick, "Tardigrada from the Galapagos and Cocos Islands," *Proc. Cal. Acad. Sci.,* **34**(5):315–328 (1966).

TEXTILES

The light microscope has been used extensively for many years as a research tool for studies of the appearance and structure of textile fibers, to determine the distribution of fibers within a yarn and to observe the position of yarns in fabrics.

The magnification for this type of work is low when one compares it with that obtained by the electron microscope. Nevertheless much has been revealed by the light microscope and it will continue to play an important part in research work and for routine examinations of textile fibers and products in both qualitative and quantitative work, particularly in the realm of animal fibers where microscopical techniques are essential for fiber identification. Some of these techniques are also applicable to work on vegetable and man-made fibers, although chemical and other tests are of more importance for identification purposes.

Very useful work on the gross structure of animal fibers was published by Solaro (1) in 1914, who gave many illustrations of a wide range of fibers. Improved techniques have been developed which have made more detail visible than was possible for earlier workers to see. Much of the early work was on whole mounts of fibers and therefore some of the detail was missed, particularly the detailed structure of the cuticle.

Patterns formed by the margins of the cuticular scales are a useful indication of the origin of an animal fiber. These patterns may not always be visible on fibers in whole mount because of the presence of pigment, medulla, or dye, and it is then necessary to make casts of fibers in some suitable medium. Various methods have been described by Manby (2), Hardy and Plitt (3), Auber and Appleyard (4), who made casts of fibers in celluloid, gelatin, or plastic films. From these casts much was learned about the arrangement of the scales and the types of patterns their margins form. Hausman (5) published a description of some of these patterns in which he instituted a systematic nomenclature for the types of pattern he had found. These descriptions were much later superseded by those of Wildman (6), who produced a very comprehensive system which corrected some of the impressions given by earlier workers, and extended the list of types by including patterns not formally described elsewhere.

Variations in the scale pattern of animal fibers frequently varies along the length of the fiber; such changes are quickly noticed if casts are made according to the details given in the above references. Where variations occur around the circumference of the fiber they are not shown by these methods, and Wildman (6) devised a method of making rolled impressions so that the whole of the scale pattern could be examined. The technique was modified by Molgaard (7) for use in microscopical studies of the surface of nylon fibers.

Cross-sectioning Fibers. In many instances the traditional method of cutting sections of fibers by embedding in wax or other medium and cutting on the type of microtomes used in histology takes up too much time. Frequently thicker sections than are normal on the traditional microtomes are satisfactory for showing detail of fiber contour, medulla, pigment distribution, and dye penetration in animal fibers, or contour and cell size in vegetable fibers, or contour, skin thickness, pigment distribution, or dye penetration in man-made fibers.

Preston (8) developed a simple plate method which has proved most useful for cutting sections of man-made fibers; it was also described more recently by Ford and Simmens (9) on the introduction of a sectioning kit by the Shirley Institute. Section thickness is limited in this method by the thickness of the plate; therefore the method is useful only for fibers having a good light penetration.

Hardy designed a small hand microtome that has proved to be invaluable for cutting sections of animal and vegetable fibers at a thickness of about 25 μm. In this method fibers are embedded in the microtome with a 3% solution of celloidin (cellulose nitrate) in 50/50 absolute alcohol and diethyl ether. For the stiffer and more rigid fibers such as horse hair or hog bristle a solution of polymerized isobutyl methacrylate in toluene is a better embedding medium. This method is fully described by Wildman (6) and by The Textile Institute (10).

Sections cut in this type of microtome provide adequate detail for the purpose of fiber identification and for some studies of fiber structure. By careful packing of the fibers in the microtome it is possible to obtain sections of animal fibers with a minimum of artifacts. Some difficulty may, however, be experienced in cutting sections of fibers such as wool kemps, deer hair, and other fibers which have only a thin wall of cortex and a very large medulla of the wide lattice type, as such fibers easily crush either in the packing stage or during cutting. Better penetration of the medulla by the celloidin can sometimes be obtained if the bundle of fibers is embedded in the celloidin in a vacuum desiccator before being put into the microtome.

The arrangement of cortical cells in fibers was studied by Appleyard (11), who mounted cross sections, cut on the Hardy microtome, in ortho-chloro-phenol, which has a swelling action on wool; it also has a peculiar phenomenon that before the swelling action begins the cortical cell boundaries become visible. From sections so mounted the cross-sectional shape and size and the arrangement of cortical cells can be studied. Work by Satlow (12) showed that camel hair and wool fibers could be distinguished on the basis of cortical cell size. Later work, using the method just described, showed that an increase in fiber diameter is due to the combination of an increase in the number of cells coupled with an increase in the mean diameter of cells.

Cortical cell diameters were calculated from sections of fibers mounted in ortho-chloro-phenol examined on a projection microscope, and it was found that if the mean cortical cell diameter was plotted against mean fiber diameter the results were grouped around two almost parallel lines, with the mean diameter of cortical cells of camel hair grouped around the lower line, and the mean diameter of cortical cells of all other fibers grouped around the upper line. Thus identification of the fibers on

the basis of cortical cell size is limited to camel hair as being distinct from all other fibers examined.

Similar trials were made on some fur fibers to see if it was possible to distinguish fur fibers from one another in blends. The samples of fur fibers were divided into fur and guard hairs. Four types of fur fiber were examined, namely, Angora rabbit, mink, muskrat, and kolinsky. The results of the first three when plotted were grouped on a line about midway between the lines for camel hair and "other fibers." Only kolinsky was different, this fell much higher than the line for "other fibers" and no differences were found between the beard hairs.

The arrangement of cells within the cortex can be seen to vary in different fibers. Cortical cells have an irregular contour but in some instances the cells may be flattened. Examples of this can be seen in sections of coarse goat hair in which the cells adjacent to the cuticle are frequently flattened (Fig. 1). An extreme example of flattened cortical cells is seen in some coarse wool fibers having a wide lattice-type medulla (Fig. 2). In these fibers the cortex is

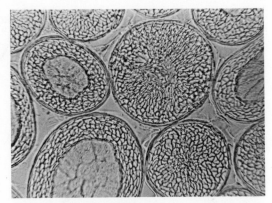

Fig. 1 Cross sections of goat hair showing flattened cortical cells in the peripheral regions of both medullated and nonmedullated fibres. (×280.)

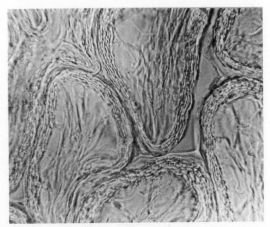

Fig. 2 Cross section of coarse medullated wool fibres showing cortical cells compressed between the medulla and the cuticle. (×330.)

thin and the cortical cells are all flattened. It would appear that this flattening is a result of pressures exerted by the Huxley layer of the fiber follicle and the medulla which, when combined, cause the cells to be compressed. The rigidity of hog bristles may also be associated with the arrangement of cortical cells. A characteristic feature of many hog bristles is that the cells in the peripheral region of the cortex are flattened and arranged with the major axis in a radial direction.

Several methods have been used for preparing sections of yarns and fabrics but most of these have caused some distortion of the fiber arrangement. More recently a technique has been developed by Lomas and Simmens (13) of making sections of cloth to show the position of the yarns in a fabric by a grinding method. Pieces of cloth are embedded on a microscope slide in Araldite and cured in an oven. The hardened specimen is then ground in water on emery papers until it is a suitable thickness to be examined microscopically (Fig. 3).

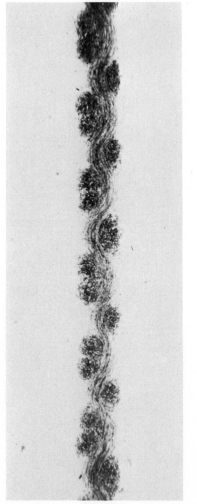

Fig. 3 Ground cross section of 2/1 twill cotton fabric. (×35.) (By courtesy of Lomas & Simmens of the Cotton, Silk and Man-made Fibres Research Association.)

For fiber sections bundles of fibers are pulled through a small hole in a fibrous board material, or plastic sheet about 0.018 in. thick, in the same manner as for sections using the plate method. The bundle on one side is cut flush with the surface of the board, and on the other side the fibers are cut into a short tuft. The flat side is then fixed to the slide with adhesive and the Araldite is run on to the upper surface and cured preparatory to grinding.

There are more elaborate methods of cutting sections of fibers. These include the double embedding techniques developed for histological work where classical rocking or rotary microtomes are used, and the more recent methods of embedding in resins and cutting ultrathin sections with glass knives on the more sophisticated microtomes specially designed for cutting sections thin enough to be examined by the transmission electron microscope.

Research workers using the light microscope are much more concerned with sections of the order of, say, 1–8 μm thick when the following double embedding method can be used to produce excellent sections down to 1 μm thick or even less: fiber bundles are made from five or six fibers which are bound together with thin sewing thread at approximately 5 mm intervals. Each bundle is then suspended vertically and weighted to keep the fibers as straight as possible. The fibers are then coated with successive layers of a 20% solution of poly-isobutyl methacrylate in toluene until the total thickness is sufficient to hold the fibers firmly. After hardening, the "candle" of fibers is cut into small pieces for embedding in a suitable wax, e.g., Gurr's Paramat wax. This embedding technique can be used for both transverse and longitudinal sections. Sections varying in thickness from 1–5 μm thick cut by using the above technique have been used for work on the structure of animal fibers with the light microscope using visible light, ultraviolet light, phase-contrast microscopy, and fluorescence microscopy. Early measurements of the dimensions of the cuticular cells were made on photomicrographs taken with light of wavelength 3650 nm, the shortest wavelength that can be used with glass objectives and therefore at the greatest resolving power of the objective (Appleyard and Greville [14]). This method of taking photomicrographs was described by Johnson (15). Little advantage is to be gained by ultraviolet microscopy because of the high absorption of ultraviolet light, particularly of shorter wavelengths, by keratin.

Following from the work of Horio and Kondo (16) much work was done on the ortho- and para-cortical structure of animal fibers which relates these two segments of the cortex to crimp in fibers. Although these two workers showed up the cortical segments by staining with Janus green, other stains can be used to show the same effect (Fig. 4). Horio and Kondo (17) demonstrated earlier a similar phenomenon in crimped viscose fibers.

From studies of the fiber follicles Auber (18) noted segmentation in the cortex of wool fibers at the pre-keratinization and keratinization levels indicating a bilateral structure in crimped fibers and what he termed periphero-axial structure in coarse, straight fibers such as cowtail hair. Later work has shown that this process of keratinization is associated with the ortho- and para-cortex of the fully keratinized fibers, which has been shown to exist in two distinct segments in the bilateral form, in a periphero-axial form, or with the ortho- and para-cortical cells interspersed.

Fluorescence Microscopy. Although there are references to fluorescence microscopy dating back to the early

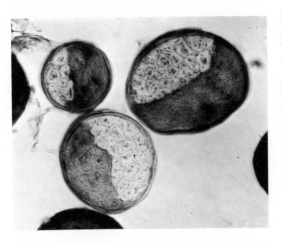

Fig. 4 Cross sections of wool fibres stained with Janus green to show ortho- and para-cortex. ($\times 623$.)

1900s it was not until comparatively recent times that this form of microscopy came to the fore. An excellent paper on fluorescence microscopy techniques was published by Young (19). Essentially it requires a good source of ultra-violet light that will excite an autofluorescent, or stained specimen causing it to emit light of a longer wavelength within the visible range. For work on textile fibers it is necessary to use buffered fluorescent stain solutions to make the fibers fluoresce (Perkin and Appleyard [20]). Both bright and dark field condenser systems can be used, and more recently, a double lighting system has been incorporated so that phase contrast with visible light can be combined with fluorescence. Its use in textiles has so far been limited but it has been used much more extensively in histological and cytological work and in particular fields where the labeled antibody techniques can be used. The detail that is shown in histological specimens by fluorescent stains suggested that they might also be useful for studies on the structure of animal fibers, and for the possible detection of damage or modification to textile fibers during processing. Appleyard and Lees (21) used, for example, acridine orange and rhodamine 3G0 to detect modification to wool fibers after various chemical treatments. Their method consisted of staining samples of chemically treated fibers by boiling them in 0.1 % aqueous solutions of the stains for 5 min, followed by rinsing, drying, and mounting in a suitable medium, usually liquid paraffin or an iso-butyl methacrylate mountant. It was noticed that not only were ortho- and para-cortex and individual cells differentiated with buffered stains, e.g., acridine orange at pH 6.5, but that fibers treated with alkali or acid solutions or hydrogen peroxide were stained completely differently. The treatment given to fibers simulated some of the treatments given in some standard processing procedures, and thus it was shown how this form of microscopy can be of help in investigations into the possible causes of modification to fibers during processing.

Alterations to the scale structure takes place when wool fibers are subject to some of the shrink-resist treatments; the extent of such alteration is made clear when the fibers are subsequently stained with rhodamine 3G0. Normally scale margins are stained deep orange and the

remainder of the fiber is stained lime green. The extent of the deep orange stain indicates the extent to which the scale surface has been damaged by the shrink-resist agent. The greatest modification is seen at the distal edges of the scales, with little or no modification at the proximal end.

Fine structure of fibers is shown to advantage when phase contrast is combined with fluorescence. In this method a stop is inserted beneath the condenser which carries the appropriate annulus needed for the phase-contrast objective, and in addition there is a peripheral annulus which transmits the ultraviolet light; thus phase contrast is obtained with dark ground fluorescence. Figure 5 shows a section of horse tail hair stained with rhodamine 3G0, photographed using a $40\times$ negative phase contrast objective, in which cortical cell boundaries are clearly delineated, and in addition, nuclear spaces are evident in many of the cortical cells. Detailed medullary structure is clearly seen in longitudinal sections stained with these stains. The staining procedure for these is the normal histological technique of staining the sections on the slide in 0.1 % aqueous solution for 2–4 hr, rinsing in distilled water, then taking up quickly through 98 % alcohol, absolute alcohol, and xylol, and then mounted.

Microscopy of Fiber Damage. The causes of fiber damage are frequently a problem to industry and sometimes the microscope can be useful in identifying the type of damage, which may then be helpful in ascertaining the cause. Unfortunately fiber damage may be cumulative; e.g., after some processes the fiber may appear to be undamaged, i.e., no damage to the surface is evident when examined microscopically, yet subsequent processing may cause the fibers to break up. However, some chemical treatments may cause damage which gives the fiber an appearance characteristic of that treatment; e.g., agents used for imparting permanent pleating or crease-resist finishes to fabrics may have a great effect upon the surface structure of the fibers (see Fig. 6); in some instances the scale margins may be etched away leaving an apparently smooth surface or the surface may be severely cracked, and consequently resistance to abrasion will be considerably reduced. Several papers have been published illustrating the microscopical appearance of fiber damage, such as abrasion effects, scorching, microbiological damage including bacterial degradation which is frequently the result of poor storage conditions, fungal growth in the

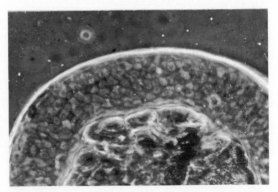

Fig. 5 Cross section of horse tail hair stained with rhodamine 3 GO (combined fluorescence and negative phase contrast). ($\times 456$.)

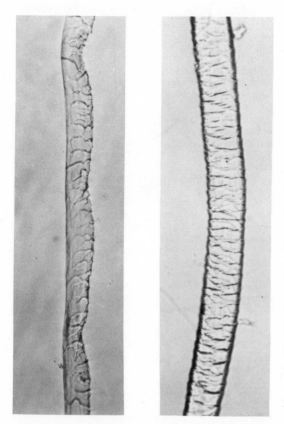

Fig. 6 Left: Wool fiber showing damage caused by 5% monoethanolamine sulfite, followed by steaming for 5 min. (×266.) Right: Wool fiber showing damage of 2% thioglycollic acid followed by steaming for 5 min. (×50.)

fleeces, or the effect of infestation by clothes moth or carpet beetle larvae. It is frequently impossible to asign damage to a specific bacteria or fungus unless the infection has taken place on finished articles. Bacterial and fungal spores are easily removed by washing, so damage to the fibers can be ascribed to such a cause only by the appearance of the fibers rather than by the presence of other evidence.

The identification of bacteria and fungi is, of course, a specialist subject which is generally not within the scope of the textile microscopist.

Several books have been published which illustrate many preparations of animal fibers. Additionally there are other publications dealing with various aspects of damaged fibers. These are useful works of reference but those who work on the subjects will find it advisable to add their own library of photomicrographs to the existing reference books, and in addition build up a collection of samples of known origin.

H. M. APPLEYARD F.T.I.

References

1. A. Solaro, "Studio microscopico e chemico pel reconoscimento della fibre vegetali—lave—peli—pelliccie—seti naturali—set artificiale," Milan, 1914.

2. J. Manby, "Celloid impression of surface structure of animal fibres," *J. Roy. Microsc. Soc.*, **53**:9 (1933).
3. J. I. Hardy and T. M. Plitt, "An improved method for revealing the surface structure of fur fibers," U.S. Dept. Int. Wildlife Circ. No. 7, 1940.
4. L. Auber and H. M. Appleyard, "Surface cells of feather barbs," *Nature*, **168**:763 (1951).
5. L. A. Hausman, "Structural characteristics of the hair of mammals," *Amer. Naturalist*, **54**:496 (1920).
6. A. B. Wildman, "Microscopy of Animal Textile Fibres," Leeds, Wool Ind. Res. Assoc., 1954.
7. J. Molgaard, "Fibre surface replication by rolling," *Nature*, **184**:264 (1959).
8. J. M. Preston, "Modern Textile Microscopy," London, Emmott & Co. Ltd., 1933.
9. J. E. Ford and S. C. Simmens, "Fibre section cutting by the plate method," *J. Text. Inst.*, **1959**:148.
10. Textile Institute, "Identification of Textile Materials," 6th ed., Manchester, Textile Inst., 1970.
11. H. M. Appleyard, "Observations on cortical cell size and arrangement in some animal fibres," *J. Roy. Microsc. Soc.*, **87**:1 (1967).
12. G. Satlow, "Die mikroscopische unterscheidung von. Schaf-und Kamelhaaren," *Z. Ver. Deut. Ing.*, **88**:328 (1944).
13. B. Lomas and S. C. Simmens, "Preparation of cross-sections of textile materials by grinding," *J. Microsc.* **92**:37 (1970).
14. H. M. Appleyard and C. M. Greville, "Cuticle of mammalian hair," *Nature*, **166**:1031 (1950).
15. B. K. Johnson, "Practical Optics," London, Hatton Press Ltd., 1947.
16. M. Horio and T. Kondo, "Crimping of wool fibres," *Text. Res. J.*, **23**:373 (1953).
17. M. Horio and T. Kondo, "Theory and morphology of crimped rayon staple," *Text. Res. J.*, **23**:137 (1953).
18. L. Auber, "The anatomy of follicles producing wool fibres, with special reference to keratinization," *Trans. Roy. Soc. Edinburgh*, LXII:191 (1950–1951).
19. R. Young, "Principles and techniques of fluorescence microscopy," *Quart. J. Microsc. Sci.*, **102**:419 (1961).
20. M. E. A. Perkin and H. M. Appleyard, "The use of fluorescent stains for the study of the structure of animal hairs," *J. Text. Inst.*, **59**(3):117 (1968).
21. H. M. Appleyard and K. Lees, "Observations on untreated and chemically treated wool fibres after staining with fluorescent stains," *J. Text. Inst.*, **56**(1): 38 (1965).

THIAZINE DYES

In the thiazine dyes, as the name indicates, there is a sulfur linkage between two benzene rings. They are therefore analogous to the OXALINE DYES (which see), though very differently used in histology. The best known are thionin, methylene blue, and the various azures. These dyes are excellent nuclear stains but are rarely used for that purpose except in complex staining formulas (see POLYCHROME STAINING FORMULAS); in this case their principal use is in the form of "thiazine eosinates." When methylene blue solutions are boiled with alkalies or acids, there are produced complex mixtures of thiazines which were at one time given the general name of "polychrome methylene blue." Most of the formulas given below are for such preparations.

Anonymous (*Bull. Int. Ass. Med. Mus.*, **26**:13)
FORMULA: water 75, glycerol 20, abs. alc. 5, toluidine blue 1, lithium carbonate 0.5

PREPARATION: Dissolve dye and alkali in water, incubate 24 hr at 37°C and return volume to 75. Add other ingredients.

Borrell see Langeron 1908 (below)

Cobin 1946 (*Tech. Bull.*, **7** :92)
FORMULA: water 100, sodium phosphate, dibasic 2.94, potassium dihydrogen phosphate 3.68, methylene blue 0.76
PREPARATION: Heat the dye and sodium phosphate, dibasic in 30 water on a water bath for 30 min. Cool and add the acid phosphate dissolved in 70 water.

Gatenby and Cowdry 1928 see (below) Nicolle 1871 (note)

Goodpasture cited from Langeron 1934 (Langeron 1934, 498)
FORMULA: water 100, methylene blue 0.25, potassium carbonate 0.25, acetic acid 0.75
PREPARATION: Boil the dye with the carbonate 30 min under reflux. Cool. Add acid, shake to dissolve ppt. Boil 15 min. Cool. Filter.

Jadassohn cited from 1928 Schmorl (Schmorl 1928, 156)
FORMULA: water 100, methylene blue 1, sodium borate 1

Kingsbury and Johannsen 1927 (Kingsbury and Johannsen 1927, 45)
FORMULA: water 90, abs. alc. 10, methylene blue 0.2, potassium hydroxide 0.005

Kuhne cited from 1904a Besson (Besson 1904, 155)
FORMULA: water 100, 95% alc. 30, sodium carbonate 1, methylene blue 0.5

Kuhne cited from 1904b Besson (Besson 1904, 154)
FORMULA: water 100, 95% alc. 10, methylene blue 2, phenol 2

Langeron 1908 (*Arch. Parasitol. Paris*, **12** :135)
PREPARATION: To 100 0.5% silver nitrate add 3% sodium hydroxide until no further ppt. is produced. Wash ppt. thoroughly by decantation and add to it 100 1% methylene blue. Boil 5 min. Cool. Filter.
NOTE: This is "Borrel's Blue." The method of preparation as stated by Langeron 1942, 545, differs little from that of Laveran 1900 (*C.R. Soc. Biol. Paris*, **52** :549).

Langeron 1934 (Langeron 1934, 496)
STOCK SOLUTION: water 100, phenol 0.5, azur II 1
WORKING SOLUTION: water 90, stock 10, 1% potassium carbonate 2
METHOD: [fixed smear]→ stain, 1 min→ water, rinse → dry→ neutral mountant
NOTE: This is also used as a mordant for some polychrome methylene blue eosinate techniques.

Langeron 1942 (Langeron 1942, 610)
REAGENTS REQUIRED: A. water 100, polychrome methylene blue 1; B. glyceric ether 25, water 75
METHOD: [water]→ A, 5 min to 12 hr→ B, until differentiated→ water, thorough wash→ abs. alc. least possible time→ balsam, via cedar oil

Laveran 1900, see Langeron 1908

Löffler 1890 (*Zbl. Bakt.*, **7** :625)
FORMULA: water 80, 95% alc. 20, methylene blue 0.3, potassium carbonate 0.8

PREPARATION: Dissolve dye in alc. Add carbonate solution.
METHOD: see Sahli 1885
Manson cited from 1929 Wenrich (McClung 1929, 408)
FORMULA OF STOCK SOLUTION: water 100, sodium borate 5, methylene blue 2
PREPARATION OF STOCK SOLUTION: Stir the dye into the boiling borax solution. Cool. Filter.
WORKING SOLUTION: water 100, stock I
METHOD: as Langeron 1934

Manwell 1945 (*J. Lab. Clin. Med.*, **30** :1078)
FORMULA: water 100, methylene blue 0.1, 1% sulfuric acid 0.6, potassium dichromate 0.1, 1% potassium hydroxide 2
PREPARATION: Dissolve dye in water. Add acid and dichromate. Autoclave 2 hr at 3 lb or until solution is blue. Add alkali drop by drop shaking till ppt. dissolved. Leave 48 hr. Filter.
NOTE: Manwell (*loc. cit.*) states this method to be a modification of that of Singh, Jaswant, and Bhattacharji 1944 (*Ind. Med. Gaz.*, **79** :102), and refers to it as the "JBS" (*sic*) method.

Martinotti 1910 (*Z. wiss. Mikr.*, **27** :24)
FORMULA: water 75, glycerol 20, 95% alc. 5, toluidine blue 1, lithium carbonate 0.5

Michaelis 1901 (*Zbl. Bakt.*, **29** :763)
FORMULA: water 90, methylene blue 1, 0.4% sodium hydroxide 5, 0.5% sulfuric acid 5
PREPARATION: Add the alkali to the solution of the dye, boil 15 min. Cool. Add acid, filter.

Moschkowsky cited from 1946 Roskin (Roskin 1946, 287)
FORMULA: water 100, sodium borate 2, methylene blue 1

Muller and Chermock 1945 (*J. Lab. Clin. Med.*, **30** :169)
FORMULA: water 70, 95% alc. 30, potassium hydroxide 0.0007, methylene blue 0.44

Nicolle 1871 (*Zbl. med. Wiss.*, 9)
FORMULA: sat. sol. thionine in 50% alc. 10, 1% phenol 90
NOTE: Gatenby and Cowdry 1928 use 50% alc. and increase the phenol solution to a 6 : 1 ratio. Langeron 1942, 539 takes 20 thionin solution to 80 2% phenol. Conn 1936 cites "Thionin, carbol-, Nicollé's" (*sic*) in the index but not in the text.

Proescher and Drueger (*J. Lab. Clin. Med.*, **10** :153)
FORMULA: A, water 100, methylene 1, sodium peroxide 0.025, hydrochloric acid *q.s.*
PREPARATION: Boil dye and peroxide for 15 min. Cool. Adjust to pH 7 with hydrochloric acid.

Roques and Jude 1940 cited from 1949 Langeron (Langeron 1949, 621)
PREPARATION: Grind 1 methylene blue with 10 anhydrous potassium sulfate. Add 100 95% alc. and shake at intervals for some hours. Filter and evaporate filtrate to dryness.

Stoughton 1930 (*Ann. App. Biol.*, **17** :162)
FORMULA: water 95, phenol 5, thionin 0.1

Terry 1922 (*J. Lab. Clin. Med.*, **8** :157)
FORMULA: water 100, methylene blue 0.2, potassium carbonate 0.2

PREPARATION: Boil 2 1/2 min. Cool.

Sahli 1885 (*Z. Wiss. Mikr.*, **2**:14)

REAGENTS REQUIRED: A. water 70, sodium borate 30, sat. sol (*circ.* 4%) methylene blue 1

METHOD: [sections of chrome-fixed or mordanted material]→ water→ A, 1–3 hr→ abs. alc. until differentiated→ balsam, via usual reagents

Singh, Jaswant, and Bhattacharji 1944, see Manwell 1945 (note)

Stévenel 1918 (*Bull. Soc. Path. exotique.*, **11**:870)

FORMULA: A. water 100, methylene blue 1.3, potassium permanganate 2.0

PREPARATION: Dissolve the dye and permanganate each in 50 water. Mix and heat on water bath until ppt. redissolved. Cool. Filter.

Unna 1892 (*Z. wiss. Mikr.*, **7**:483)

PREPARATION OF DRY STOCK: water 100, 95% alc. 20, methylene blue 1, potassium carbonate 1.

PREPARATION: Simmer ingredients until volume reduced to 100. Leave 24 hr. Filter. Evaporate to dryness.

Volkonsky 1933 cited from 1942 Langeron (Langeron 1942, 1105)

FORMULA: water 50, glycerol 50, methylene violet 0.4, azur II 0.1, potassium carbonate 0.1

PETER GRAY

Books Cited

Journal references in the above formulas are given in the standard form. The following books are cited by author and date only.

Besson, A., "Technique Microbiologique et Sérothérapique," Paris, Baillière, 1904.

Gray, P., "The Microtomist's Formulary and Guide," New York and Toronto, The Blakiston Company, 1954.

Kingsbury, B. F., and O. A. Johannsen, "Histological Technique," New York, Wiley, 1927.

Langeron, M., "Précis de Microscopie," 5th ed., Paris, Masson, 1934.

Idem., 6th ed., 1942.

Idem., 7th ed., 1949.

McClung, C. E., "Handbook of Microscopical Technique," New York, Hoeber, 1929.

Roskin, G. E., "Mikroskopecheskaya Technika," Moscow, Sovetskaya Nauka, 1946.

Schmorl, G., "Die Pathologisch-histologische Untersuchungs-methoden," 15th ed., Leipzig, Vogel, 1928.

See also: OXAZINE DYES.

THYMUS

The choice of techniques to be applied to the thymus depends on the purpose of the study. For paraffin sections, Helly's and Bouin's (P.A.F.) fluids are useful; also ice cold Rossman's (1940) fluid for general staining, for the demonstration of metachromasia and glycogen; for frozen sections, Baker's (1944) formol-calcium for lipids. May-Grünwald-Giemsa's solution (Jacobson and Webb, 1952) is an excellent stain for thymic cells. Standard methods for the demonstration of enzyme activities can be employed. For this work, fresh-frozen sections cut in a cryostat, either unfixed or fixed in cold 10% buffered formalin for 30 min at 4°C (Lillie, 1965), give good results. Methods for the demonstration of enzymatic activity have to be modified according to the amount of activity present and this varies with age, strain, and

animal. If a study of the vascular pattern is considered by injection, a negative pressure method is most rewarding (Chillingworth et al., 1936). In sections treated for the demonstration ATPase activity, the endothelial lining of the blood vessels is sharply outlined. Thymuses fixed in 2% osmic acid and buffered with veronal acetate (or other such fixatives) can be prepared for embedding in Epon (Luft, 1961), cut at one micron and stained with toluidine blue. These sections are mounted by heat (hot plate), and stained between one or two minutes (also on a hot plate) with a solution containing 0.25% toluidine blue and 0.25% borax at pH 9, rinsed quickly in distilled water, dehydrated rapidly in alcohols, cleared in xylene, and mounted in permount (Padykula in Smith, 1969). The same material is used for ultrathin sections for study with the electron microscope. Standard methods for staining these sections can be used (Reynolds, 1963).

CHRISTIANNA SMITH

References

Baker, J. R., *Quart. J. Microsc. Sci.*, **85**:1–71 (1944).

Chillingworth, F. P., M. H. Sweet, and J. C. Healy, *Anat. Rec.*, **66**:113–117 (1936).

Jacobson, W., and M. Webb, *Exp. Cell Res.*, **3**:163–183 (1952).

Lillie, R. D., "Histopathologic Technic and Practical Histochemistry," New York, Blakiston, McGraw-Hill, 1965.

Luft, J. H., *J. Biophys. Biochem. Cytol.*, **9**:409–414 (1961).

Reynolds, E. S., *J. Cell Biol.*, **17**:208–212 (1963).

Rossman, D. P., *Amer. J. Anat.*, **66**:277–365 (1940).

Smith, C., *Amer. J. Anat.*, **124**:389–409 (1969).

THYROID

The recent progress on the cytophysiology of the endocrine glands, especially with the thyroid gland, has been made possible with the introduction of several modern

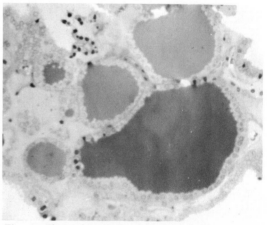

Fig. 1 Photomicrograph of control human thyroid, showing follicles lined by a cubic epithelium and containing a homogenous colloid. Plastic (Epon) thick section, cacodylate-buffered glutaraldehyde and OSO_4 fixation, toluidine blue stain. (× 196.)

techniques. These methods, such as: electron microscopy, autoradiography, histochemistry, immunofluorescence, and the scanning electron microscope, opened a new and rich field of research, marking the "golden age" of thyroid cytology. Most of the findings provided by the light microscope offer us only static information, and we cannot have an accurate image with regard to the thyroid structure without using the modern techniques. This article discusses the best methods to use for the study of thyroid cytology and also their advantages.

Light microscopy is a useful method for the general organization of the thyroid structure and its variation in physiologic or pathologic conditions. Using the light microscope, the general pattern appears in most vertebrates as an encapsulated follicular structure. Follicles are lined with a cubic epithelium and contain in their lumen the colloid with thyroglobulin (Fig. 1). The follicles are separated by connective tissue and sinusoid capillaries. In some species, it is possible to see parafollicular (or C) cells, which never reach the follicular lumen. The best

results are obtained by using formalin, Bouin fluid, or buffered (cacodylate or phosphate) 3% glutaraldehydes, dehydrated and then embedded in epoxy resins (Epon-812) or paraffin. Fresh material (freeze-dried specimens) for cryostat sections can also be used for better evidence of enzymes. Using different staining methods, such as HE, PAS-HE, or trichrome methods, the intracellular colloid droplets, cilia, or microvilli can be seen. Of course, light microscopy is a limited method, due first to the resolution of the light microscope, which is about 0.2 μ, and second to the artifacts induced by different fixatives or embedding media.

Electron Microscopy. Instruments with a higher resolution (2–3 Å) can overcome the inherent difficulties and provide a more detailed ultrastructural pattern, or reveal new organelles. Thus, an irregular apical border with microvilli and large pseudopodia, engulfing colloid from the lumen by an endocytosis process; well-developed rough endoplasmic reticulum (RER), heterogeneous population of secretory granules (colloid droplets and dense

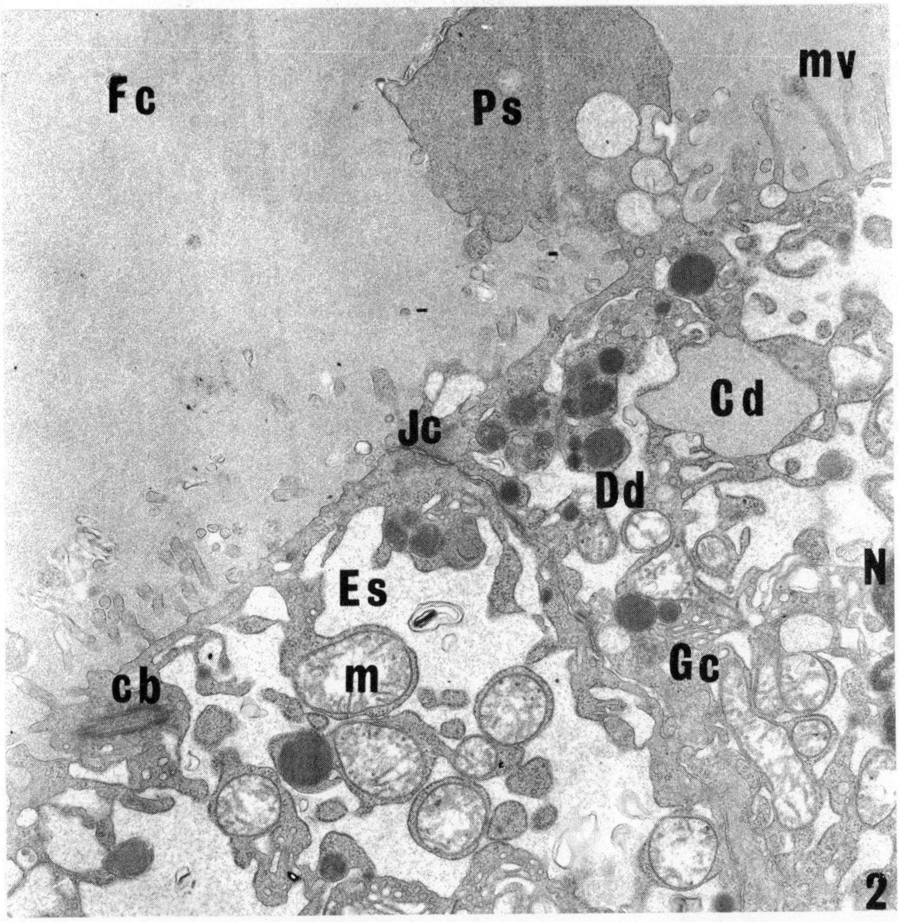

Fig. 2 Electron micrograph of human thyroid (microfollicular adenoma): FC, follicular colloid; Ps, pseudopodia engulfing colloid material; mv, microvilli; Jc, Junctional complex; cb, distal centriole or basal body of cilia; Cd, colloid droplet; m, mitochondrion; Dd, dense droplet or lysosome; Es, ergastoplasmic sac; Gc, Golgi complex; N, nucleus. Cacodylate-buffered glutaraldehyde and OsO$_4$ fixation, Epon embedding, uranyl acetate and lead citrate stain. (× 18,000.)

granules) and mitochondria with evident cristae can be seen (Fig. 2). Fine results, and the possible avoidance of artifacts or tissue damage, can be obtained after a pre-fixation (2–3 hr), by perfusing *in vivo* or *in vitro* with a cacodylate-buffered 3 % glutaraldehyde, followed by fixa-tion (1–2 hr) in buffered (phosphate or veronal acetate) osmium tetroxide (OsO_4), dehydration in alcohol or acetone increasing solutions, and embedment in Epon-812 or polyesters (Vestopal-W); sometimes, it is possible to use Araldite, Maraglas, or even rare Methacrylate. Thin sections (silver-gray) can be obtained by using an Ultrotome equipped with a glass or diamond knife, mounted on coated or uncoated grids, then stained with uranyl acetate and lead citrate.

Electron microscopic histochemistry can demonstrate more evidently the localization of several enzymes (per-oxidase, acid phosphatase, aryl-sulfatase) and their pos-sible role in hormone synthesis. Some enzymes are involved in iodination (peroxidase, catalase), others in transport and release of hormones (acid phosphatase, esterase). Different substrates, such as DAB (diaminobenzidine) for peroxidase or cytidine monophosphate for acid phosphatase, have been used. Acid phosphatase activity is located over the dense granulae or lysosomes, whereas peroxidase is located mostly within apical vesicles and microvillous membranes.

Autoradiographic methods (light and electron micro-scopy) provide a considerable amount of data with regard to the site of hormone synthesis, thyroglobulin iodination, and their intracellular kinetics. Different isotopes, like ³H-leucine (for protein synthesis), ³H-glucose or galactose (for glycoprotein synthesis), radio-iodine (¹²⁵I or ¹³¹I) (for thyroglobulin iodination), and ³H-thymidine (for DNA synthesis) have been used. The incorporation and localization of radioiodine is different in follicles, suggest-ing a heterogeneity of thyroid cells. For light microscopic autoradiography, we can use two procedures: (1) paraffin sections, or (2) Epon-embedding material. In procedure (1), tissue, after 24 hr fixation in formalin, a mixture of formalin and alcohol, or Bouin fluid, is dehydrated and embedded in paraffin; sections of 4–5 μ thick are made and mounted on slides; then dipped in a melted Kodak NTB_2 emulsion and exposed for 1 week to 1 month, or a shorter time for radioiodine (24–48 hr); developed in Kodak D_{19}; and fixed and stained by HE or PAS-HEM. In procedure (2), plastic sections of Epon-embedded material were used, after fixation in glutaraldehyde or osmium tetroxide. Thick sections of 2 μ, mounted on slides, were dipped in a diluted Ilford K5 emulsion, exposed for 2–4 weeks, developed in Kodak D_{19}, and fixed and stained by toluidine or methylene blue. When we used ³H-thymidine, the autoradiographic grains were strictly located over the nuclei (Fig. 3). Grain counts are also possible; they can be reported as per square milli-meter and compared with the background, giving an accurate ratio of isotope distribution.

Electron microscope autoradiography is a more accurate and dynamic method which can provide detailed findings with regard to the isotope localization and also visualize the direction and rate of transport of secretory products. The same isotopes can be used as in light microscope autoradiography. The procedure is the following: After

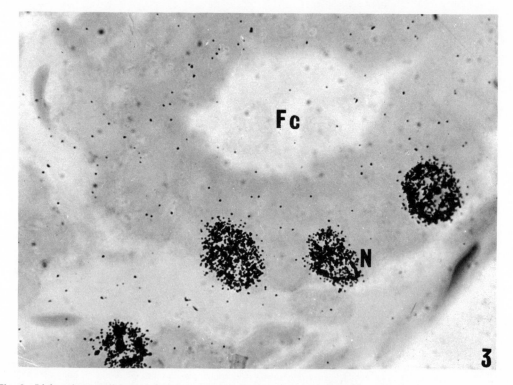

Fig. 3   Light microscopic autoradiography of newt thyroid, using ³H-thymidine, 2 hr, later TSH administration. Heavily labeled nuclei (N) in mitotic cells, and follicular colloid (FC) can be seen. Cacodylate-buffered and glutar-aldehyde OsO_4 fixation, Epon embedding, Ilford K5 emulsion, and toluidine blue stain. (×ob. im.)

isotope administration, small specimens are dropped in 3% glutaraldehyde, then fixed in buffered osmium tetroxide (for 1 hr), dehydrated, and embedded in Epon or methacrylate. Thin sections (silver-gray) are mounted on grids, then a diluted nuclear emulsion is applied. The emulsion, Ilford L$_4$ is handled as suggested by Caro and Tubergen and is applied to the grids by means of a wire loop. Kodak NTE emulsion can also be used. Exposure time varies between 2 and 4 weeks, or more. The exposed grids are developed in Microdol-x, fixed, and stained with uranyl acetate and lead citrate. The developed or radiographic grains are located only on the specific organelles, such as ribosomes in the case of ^{3}H-leucine, Golgi complex in the case of ^{3}H-glucose, or follicular colloid and apical cell border in the case of iodine (Fig. 4). By using radio-iodine (^{125}I or ^{131}I) in electron microscope autoradiography, it is possible to demonstrate that the thyroglobulin iodination takes place almost entirely extra-cellularly, at the active zone located between peripheral colloid and apical cell border. From here, the thyroid hormones (thyroxine and triiodothyronine) are resorbed and released into circulation. The preferred procedure for thyroid electron microscopic autoradiography is best described by Caro and Tubergen, using Ilford L$_4$ emulsion. Autoradiography on the subcellular fractions, using ^{3}H-leucine, showed a higher incorporation of the isotope, namely in the microsomial fraction and much smaller in the nuclei or mitochondria.

Immunofluorescent techniques use labeled fluorescent antiserum (antihuman gamma globulin) on the thyroid cryostat sections and reveal the existence of three different antibodies: antithyroglobulin, microsomial, and nuclear antibodies. These findings have led to the current concept of autoimmune diseases.

Scanning electron microscope (SEM) is useful to show the relief or surface structures of the thyroid gland. SEM can offer interesting and additional results regarding the cytcarchitecture, organization, and growth of thyroid tumors.

According to my knowledge, all these techniques are important and useful for the study of the thyroid gland. For a diagnostic routine, most pathologists use only light microscopy, but for a better understanding of thyroid cytology and pathology, investigations using electron microscope must be linked up with histochemistry, autoradiography, immunofluorescence, and biochemical methods. Although some of these aspects may be considered today only of academic interest, they may have an outstanding importance for the pathology of thyroid in the near future.

A. P. LUPULESCU

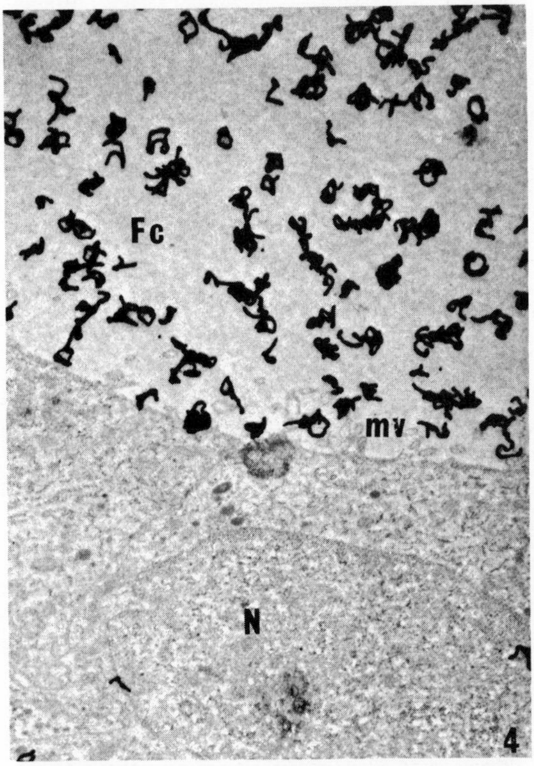

References

Caro, L., and R. Tubergen, "High resolution autoradiography, I. Methods," *J. Cell Biol.*, **15**:173–188 (1962).

Lupulescu, A., D. Andreani, F. Monaco, and M. Andreoli, "Ultrastructure and soluble iodoproteins in human thyroid cancer," *J. Clin. Endocr. Metab.*, **28**:1257–1268 (1968).

Lupulescu, A., and A. Petrovici, "Ultrastructure of the Thyroid Gland," Baltimore, Williams & Wilkins, 1968.

Lupulescu, A., and C. Boyd, "Follicular Adenomas: An ultrastructural and scanning electron misroscopic study," *Arch. Pathol.* **93**:492, 1972.

Pitt-Rivers, R., and W. R. Trotter, "Thyroid Gland," 2 vols., London, Butterworths, 1964.

Sturm, J., and M. J. Karnovsky, "Cytochemical localization of endogenous peroxidase in thyroid follicular cells," *J. Cell Biol.*, **44**:655–666 (1970).

Fig. 4 Electron micrograph autoradiograph of control rat thyroid, using radio-iodine (I^{125}), 6 hr, later administration. The autoradiographic grains are located almost entirely in the follicular colloid (FC), at the apical border and microvillous membrane (mv); N, nucleus. OsO$_4$ fixation, Epon embedding, Ilford L$_4$ emulsion, lead citrate stain ($\times$ 15,400.)

TONGUE[a]

Most of the discussion in this article is based on experience with the preparation of mammalian tongues for microscopic examination. Amphibian and bird tongues are not significantly more difficult to prepare than mammalian tongues provided the same precautions are followed. The most common problem in preparation of this tissue is poor penetration of the tissue by fixing solutions and poor infiltration by the embedding medium. These problems may be more easily understood if we examine the general structure of the tongue.

The tongue is covered with epithelium which varies among the different vertebrate species. Within a given tongue, the epithelium on the dorsal surface is different from that on the ventral surface. The epithelium on the dorsum is usually a stratified epithelium with a fairly tough, durable surface layer. In mammals, the dorsal surface of the tongue is made up primarily of small, pointed filiform papillae lined with stratified squamous

[a]Supported in part by United States Public Health Service Grant No. NS-06181.

keratinizing epithelium. In some mammals, particularly rodents and cats, part of each papilla is covered with epithelium which forms the hard type of keratin, similar to that seen in hair cortex. This tough surface layer is probably the major reason why the tongue is difficult to prepare for microscopic examination. It is not easily penetrated by the routinely used fixing and embedding media.

The lamina propria of tongue varies from the loose connective tissue of mammalian tongues to the rather dense collagenous connective tissue of amphibian tongues. In the latter, the collagen bundles are often arranged in layers with each layer containing collagen fiber bundles at approximately right angles to those in adjacent layers.

Beneath the lamina propria are the closely packed bundles of striated skeletal muscle fibers which make up most of the mass of the tongue. In some animals, cartilage or bone may be found within this muscle, particularly near the root of the tongue. Needless to say, where bone is present, decalcification must be part of the preparation procedure. (This is most easily accomplished by fixing in an acid fixative—Bouin's is recommended.) This region, too, is not easily penetrated by embedding media. The problem is made worse when large pieces of tissue are used.

From a consideration of the histological structure it is clear that when a block of tongue is sectioned, the knife must cut through tissues varying considerably in consistency and hardness, from the tough surface keratin to a softer underlying epithelium and through connective tissue and muscle which are of different consistencies. The two problems most commonly encountered at sectioning are cleavage at the interface between the embedding medium and the dorsal tongue surface, and poor penetration into the center of the specimen. However, when the tissue is well infiltrated with embedding medium, there is little difficulty in sectioning. In the following paragraphs are some simple suggestions that may be helpful to those working with tongue.

Light Microscopy. First, it is best to use small pieces, preferably slices no more than 3 mm thick. If this is not possible because of the nature of the experiment, and the tongue is larger than that of a medium-sized adult rat, then fixation by perfusion rather than by immersion is probably the best way to guarantee adequate preservation. Any of the commonly used fixing mixtures or agents is suitable. Washing and dehydration present no problems provided that adequate time is allowed. For the dehydration of larger pieces, 60–90 min each in 50%, 70%, 95% alcohols, and 2–3 hr in absolute alcohol (with several changes) is not too long. For clearing, toluene or cedarwood oil is preferred over xylene. The tissue may be left overnight in either of these agents without danger of becoming too brittle. Infiltration with paraffin should be gradual with the first stage being a 1 : 1 mixture of cedarwood oil (or toluene) and paraffin. It is recommended that the final paraffin bath be for one hour at reduced atmospheric pressure (in a vacuum oven). Soft paraffins are not recommended. Those used for routine tissue processing are usually adequate.

Electron Microscopy. The routine methods for processing tissue for electron microscope examination sidestep some of the penetration problems mentioned above. The tissue is usually minced into small pieces about 1–2 mm³, and penetration of most agents is fairly complete in pieces of this size. Buffered osmium tetroxide solutions are used as fixing agents by most electron microscopists.

These penetrate tissue very slowly, and this is the primary reason for using such tiny blocks of tissue. However, even with such small specimens the tough keratins in some mammalian tongue epithelia may not be infiltrated properly by the viscous epoxy plastic embedding mixtures. These problems may be circumvented by bringing the tissue from the intermediate agent, usually propylene oxide, through a graded series of propylene oxide–epoxy mixtures in ratios of 2 : 1, 1 : 2, and finally 100% plastic. The first step may last for 1–2 hr, the second from 1 to 2 hr to overnight at room temperature. The specimen may be left in an unsealed vessel to permit gradual evaporation of the propylene oxide from the mixture. The tissue can then be placed in its embedding container and left for another day at room temperature before placing it in a 60°C oven for 24–48 hr to harden.

Summary. Processing the tongue for histologic or ultrastructural examination can present some problems because some of its structures are not readily permeable to the fixing fluids and embedding media commonly used. The tissue is best handled by using common sense methods to guarantee adequate infiltration, and by so doing, to safeguard against sectioning difficulties.

ALBERT I. FARBMAN

TRACERS

Tracer particles of varied sizes have been used in electron microscopy to demonstrate communication between spaces, to study uptake and movement of particles in living cells, and to estimate the size of pores or channels in diffusion barriers. In addition, tracers which can enter narrow channels may reveal fine structural details by a staining effect (Goodenough and Revel, 1970; Tani and Ametani, 1970). *In vivo* tracers are sufficiently nontoxic to be given to living animals, isolated living preparations, or living cells. Some tracers are too toxic to be used in this way and can be used only for *en bloc* treatment during or after fixation.

Colloidal tracer particles are charged and tend to bind to charged components of tissues; so the absence of tracer particles at a particular site may imply a lack of charged groups rather than a lack of access to the tracer. Uncertainties due to regional binding of charged tracers make quantitative deductions about tracer movements impossible. The adsorption of macromolecules onto the surface of colloidal particles makes the effective size of the particles larger than their apparent size in electronmicrographs. Stabilizers added to colloidal solutions similarly increase the size of particles by an unknown amount (Feldherr, 1964). It is worth remembering that the physico-chemical properties of a colloidal particle, which may determine whether the particle will enter a given channel, depend on the nature of the molecules adsorbed to the surface of the particle.

Caution must be exercised when using tracers injected into living mammals to study capillary permeability, as horseradish peroxidase (Clementi and Palade, 1969a; Clementi, 1970) impure ferritin (Bruns and Palade, 1968), colloidal mercuric sulfide (Majno and Palade, 1961), and perhaps some other tracers may release histamine, so producing changes in capillary permeability. When tracers are used to study the permeability of fixed material it must be remembered that fixation may modify the size of pores and channels in diffusion barriers.

Protein tracers obtained in a crystalline state can be

dissolved by sonication but the temperature must be controlled to avoid denaturation (Venkatachalam and Fahimi, 1969). Except where stated, routine methods of fixation, dehydration, and embedding are used for studies with tracers.

Electron opaque tracers

Ferrous gluconate. Brayser et al., (1971) used ferrous gluconate (m.w. 482) as an *in vivo* tracer molecule of very small size to study capillary permeability. Ferrous gluconate forms a true solution. The molecules probably have an effective size of less than 1 nm. Intravenous injection of 250 mg of ferrous gluconate (Hopkin & Williams) per kilogram of body weight, dissolved in bicarbonate-buffered Ringer solution, appears to be well tolerated by a species of lizard; 10 g/kg stops the heart.

The tracer is retained in the tissues by precipitation with orthophosphate ions during fixation in phosphate-buffered glutaraldehyde (pH 7.4) for 2 hr at 4°C. Ferrous gluconate is potentially very useful because no other very small *in vivo* tracer exists.

Ruthenium Red. Ruthenium red is an inorganic compound with a molecular weight of about 860; it carries six positive charges per molecule (Luft, 1966a) and therefore binds to many negatively charged polymeric substances including cell surface coat materials. It forms a true solution. The molecules have an estimated volume of 76.2 nm³ corresponding to spheres 1.13 nm in diameter. The effective diameter is probably smaller as the molecules are thought to have a somewhat elongated rod shape. Ruthenium red is highly toxic and cannot be used as an *in vivo* tracer. Small tissue blocks are usually treated with ruthenium red during fixation, 0.05–0.8% ruthenium red being added to both glutaraldehyde and osmium-containing fixatives which must not be buffered with phosphate ions. Osmium tetroxide appears to be reduced catalytically to osmium black by ruthenium red bound to tissue components. To give this reaction time to build up a large mass of osmium black in regions accessible to the tracer, Luft (1966b) recommends osmicating for 3 hr at room temperature. Although Luft (1966) and Brooks (1969) give methods for purifying ruthenium red, most commercial samples give satisfactory results without further purification (Luft, 1970; personal communication). Ruthenium red does not enter undamaged cells (Fowler, 1970) but can enter cut nerve cell processes (Tani and Ametani, 1970). Under these circumstances the tracer enters the narrow channel in neurotubules (Tani and Ametani, 1970).

Uranyl Acetate at pH 6.0. Aqueous solution of uranyl acetate at pH 6.0 contain small microcrystals less than 2 nm in diameter and large (0.5–1.0 μ) crystals. The microcrystals can be used as an *en bloc* tracer (Matter et al., 1970). Small pieces of tissue are washed thoroughly in 0.05M sodium hydrogen maleate buffer (pH 6.0) after osmication and primary fixation in a suitable aldehyde fixative. They are then immersed for 2 hr at 4°C in 0.5–2% uranyl acetate solution (final pH adjusted to 6.0 with N NaOH) in 0.05M sodium hydrogen maleate buffer and then dehydrated and embedded in epoxy resin. Better results are obtained by treating material with uranyl acetate at pH 6.0 after glutaraldehyde fixation but before postfixation in osmium tetroxide (B.L. Gupta; personal communication). Under the latter conditions the tracer binds to the surface of plasma membranes.

Lanthanum Nitrate. Lanthanum nitrate solutions at slightly alkaline pH contain very small colloidal particles (Goodenough and Revel, 1970). The colloid can be used *en bloc* as a tracer but is too toxic to be given to living material. Although the size of the particles is not known, colloidal lanthanum can enter the very narrow space (approximately 2 nm wide) between the outer leaflets of myelin (Revel and Hamilton, 1969). Colloidal lanthanum does not normally enter undamaged cells after fixation but can enter cut axons (Lane and Treherne, 1970).

In some procedures colloidal lanthanum prepared by adding N NaOH drop by drop to an aqueous 4% lanthanum nitrate solution until the pH reaches 7.7, is added to all fixatives and washing solutions. It is usually only necessary to add lanthanum (final concentration of 1–2%) to the primary fixative and to the osmium tetroxide. Phosphate buffered fixatives must not be used. Very small blocks of tissue or thin slices should not be used as lanthanum can leach out rapidly from these (Revel and Karnovsky, 1967). For this reason rapid dehydration is also advisable.

Colloidal lanthanum penetrates the "gap junctions" found between ependymal cells, astrocytes, and some electrically coupled neurones revealing a hexagonal substructure by a staining effect (Goodenough and Revel, 1970; Brightman and Reese, 1969).

Colloidal Iron Solutions. Saccharated iron oxide with particles measuring 3–5 nm in the electron microscope (Amend and Co. Ltd., New York City) or with particles measuring 2–7 nm (Evans medical Co., Speake, Liverpool 24, Great Britain) have been used as *in vivo* tracers to study permeability and uptake of colloidal particles in vertebrates (Pappas and Tennyson, 1962; Jennings and Florey, 1967). Colloidal iron dextran with rod-shaped particles measuring 34×6 nm (FPL 2,000; Fison's Pharmaceuticals Ltd., 12 Derby Road, Loughborough, Leics. Great Britain) has also been used as an *in vivo* tracer (Trotter, 1970). Though these colloidal solutions seem well tolerated at low doses (about 100–500 mg/kg body weight) by mammals; 5 g/kg of saccharated iron oxide is lethal to mice.

Ferritin. Ferritin is a metalloprotein. The molecule is approximately 11 nm in diameter with an electron-dense core approximately 5 nm in diameter containing six iron atoms (Muir, 1960). Ferritin is a good tracer for injection into living mammals because it is well tolerated when purified, is very electron dense, and remains at high concentration in the circulation for at least 24 hr (Bruns and Palade, 1968). Cadmium present as an impurity in commercial ferritin is highly toxic but can be removed by dialyzing a 5–10% ferritin solution for 36 hr against 0.1M EDTA in 0.07M phosphate buffer at pH 7.2. The dialysis tube is then sealed to prevent dilution of the ferritin and dialysis continued for 24–48 hr at 40°C against several changes of 0.1M phosphate buffer at pH 7.2 (Bruns and Palade, 1968; Clementi and Palade, 1969a). Intravenous injection of 1 ml of purified ferritin (approximately 1 g/kg body weight) is well tolerated by small mammals. Ferritin can be spun down by high-speed centrifugation and resuspended in saline buffer before injection (Clementi and Palade, 1969a).

Colloidal Thorium. Colloidal thorium has been widely used as a tracer for studies on the uptake of colloidal particles by living cells (Padawer, 1969; Kaye and Pappas, 1960). Colloidal thorium stabilized by dextrin ("Thorotrast"; Fellows-Testagar Div., Fellows Medical Mfg. Co. Inc., Detroit, Mich.) contains particles with a diameter of 10–15 nm in electronmicrographs. Thorotrast can be injected directly (Padawer, 1969); purification by dialysis against distilled water before injection (Brandt and Pappas, 1960) is not necessary for most purposes. Thorotrast is well tolerated; doses of 2 ml/kg body weight have been used for intravenous injection.

Colloidal Mercuric Sulfide. Colloidal mercuric sulfide obtained as a dry powder ("Mersulfol," Hille and Co., Chicago, Illinois) has been used as an *in vivo* tracer (Majno and Palade, 1961). The particles have an apparent size of 7–35 nm in electronmicrographs. Mersulfol contains cresol and a stabilizer; intravenous injections of 1.5 g/kg body weight are only fairly well tolerated occasionally producing plumonary edema in rats. Larger doses are badly tolerated.

Colloidal Gold. Colloidal gold solutions have been used as *in vivo* tracers. The size of the colloidal particles depends on the method of preparation or the commercial source. Clementi and Palade (1969b) used colloidal gold particles ("Aurocoloid" TM-198, Abbot laboratory, Teterboro, N.J.) with an apparent size of 3–30 nm to study vascular permeability. Pappas and Tennyson (1962) used Lange's colloidal gold solution (Mager chemicals Inc., Cornwall landing, N.Y.) in a similar study. Feldherr (1962) used colloidal gold prepared by the reduction of chloroauric acid with phosphorus to study the intracytoplasmic transport of colloidal particles injected into the ameba, *Chaos chaos*. The particles prepared by his method had a diameter of 2–5 nm in electronmicrographs. When using Feldherr's procedure for the preparation of colloidal gold it must be remembered that yellow phosphorus in petroleum ether is an extremely dangerous reagent and must be stored in a carefully sealed dark bottle in an explosion-proof refrigerator.

Colloidal carbon. Carbon black particles with an apparent size of 23–48 nm (Pelikan "biological ink"; John Henschel and Co. Inc., Farmingdale L.I., N.Y., or Guenther-Wagner Pelikan-werks, Hanover, Germany) have been used as an *in vivo* tracer (Clementi and Palade, 1969b). Intravenous injection of 1–5 ml of ink (100–500 mg of carbon)/kg body weight is well tolerated by rats although the ink contains fish glue as a stabilizer and phenol as a preservative (Majno et al., 1961).

Enzymes as tracers

Graham and Karnovsky (1966) introduced the use of horseradish peroxidase as an *in vivo* tracer molecule for electron microscopy. The peroxidase is made visible in the electron microscope by treatment with diamino benzidine and hydrogen peroxide before postfixation in osmium tetroxide (DAB method; Graham and Karnovsky, 1966; Clementi and Palade, 1969a; Clementi, 1970). The method is very sensitive by virtue of the cytochemical amplification and enables very small amounts of the enzyme to be detected in the electron microscope.

Horseradish peroxidase (HRP) has a molecular weight of about 40,000. Although the molecule has an effective size (equivalent hydrodynamic diameter—EHD) estimated at 5 nm, it appears to be able to enter the narrow cleft, approximately 3 nm wide in "gap junctions" (Brightman and Reese, 1969). When using HRP to study vascular permeability in living mammals it is important to use low doses (10 mg/kg body weight in mice) of highly purified, high activity enzyme (electrophoretically purified and lyophilized type; HPOFF, RZ 3; Worthington Biochem, Co., Freehold, N.J.) The larger doses (240 mg/kg) required to produce enough reaction product when less active enzyme preparations are used cause increased vascular permeability (Clementi, 1970).

The literature on the use of horseradish peroxidase as a tracer has been reviewed by Straus (1969).

Other peroxidases with different molecular weights have been specially prepared for use in the same way: Graham and Karnovsky (1966) used myeloperoxidase (m.w. 160,000) from human leucocytes, Graham and Kellermeyer (1968) bovine lactoperoxidase (m.w. 82,000) from milk, and Feder (1970) a micro peroxidase (m.w. 1,900) which they prepared from hemoglobin.

Many heme-containing proteins possess peroxidase-like activity which can be detected histochemically by modification of the DAB method. Among those which have been used as tracers are cytochrome c (EHD = 3 nm; Karnovsky and Rice, 1969), beef liver catalase (m.w. 240,000; Venkatachalam and Fahimi, 1969) and bovine hemoglobin (EHD = 5.8 nm; Goldfischer et al., 1970). These proteins are all commercially available. With the improvements in cytochemical techniques that have occurred in recent years the range of enzymes which could be used as tracers is now very wide.

DAVID P. KNIGHT

References

Brandt, P. W., and G. D. Pappas, "An electron microscopic study of pinocytosis in ameba," *J. Biophys. Biochem. Cytol.,* **8**:675–676 (1960).

Brayser, M., J. R. Casley-Smith, and B. Green, "A new small molecular weight tracer for permeability studies with the electron microscope," *Experientia,* **27**:115–116 (1971).

Brightman, M. W., and T. S. Reese, "Junctions between intimately apposed cell membranes in the vertebrate brain," *J. Cell Biol.,* **40**:648–677 (1969).

Brooks, R. E., "Ruthenium red stainable surface layer on lung alveolar cells; electron microscopic interpretation," *Stain Technol.,* **44**:173–177 (1969).

Bruns, R. R., and G. E. Palade, "Studies on blood capillaries 2. Transport of ferritin molecules across the wall of muscle capillaries," *J. Cell Biol.,* **37**:277–299 (1968).

Clementi, F., "Effect of horseradish peroxidase on mice lung capillaries' permeability," *J. Histochem. Cytochem.,* **18**:887–892 (1970).

Clementi, F., and G. E. Palade, "Intestinal capillaries. 1. Permeability to peroxidase and ferritin," *J. Cell Biol.,* **41**:33–58 (1969a).

Clementi, F., and G. E. Palade, "Intestinal capillaries. 2. Structural effects of EDTA and histamine," *J. Cell Biol.,* **42**:706–714 (1969b).

Feder, N., "A heme-peptide as an ultrastructural tracer," *J. Histochem. Cytochem.,* **18**:911–913 (1970).

Feldherr, C. M., "The nuclear annuli as pathways for nucleocytoplasmic exchanges," *J. Cell Biol.,* **14**:65–72 (1962).

Feldherr, C. M., "Binding within the nuclear annuli and its possible effect on nucleocytoplasmic exchanges," *J. Cell Biol.,* **20**:188–192 (1964).

Fowler, B. A., "Ruthenium red staining of rat glomerulus," *Histochemie,* **22**:155–162 (1970).

Goldfischer, S., A. B. Novikoff, A. Albala, and L. Biempica, "Hemoglobin uptake by rat hepatocytes and its breakdown within lysosomes," *J. Cell Biol.,* **44**:513–529 (1970).

Goodenough, D. A., and J.-P. Revel, "A fine structural analysis of intercellular junctions in the mouse liver," *J. Cell Biol.,* **45**:272 (1970).

Graham, R. C., and M. J. Karnovsky, "The early stages of absorption of injected horseradish peroxidase in the proximal tubules of mouse kidney: Ultrastructural cytochemistry by a new technique," *J. Histochem. Cytochem.,* **14**:291–302 (1966).

Graham, R. C., and R. W. Kellermeyer, "Bovine lacto-peroxidase as a cytochemical protein tracer for electron microscopy," *J. Histochem. Cytochem.,* **16**:275–278 (1968).

Jennings, M. A., and H. W. Florey, "An investigation of some properties of endothelium related to capillary permeability," *Proc. Roy. Soc. Brit.,* **167**:39–63 (1967).

Karnovsky, M. J., and D. F. Rice, "Exogenous cytochrome c as an ultrastructural tracer," *J. Histochem. Cytochem.,* **17**:751–752 (1969).

Kaye, G. I., and G. D. Pappas, "Studies on the cornea. 1. The fine structure of the rabbit cornea and uptake and transport of colloidal particles by the cornea 'in vivo'," *J. Cell Biol.,* **12**:457–479 (1960).

Lane, N. J., and J. E. Treherne, "Lanthanum staining of neurotubules in axons from cockroach ganglia," *J. Cell Sci.,* (G.B.), **7**:217–239 (1970).

Luft, J. H., "Fine structure of capillary and endocapillary layer as revealed by ruthenium red," *Fed. Proc. Fedn. Amer. Socs. Exp. Biol.,* **25**:1773–1783 (1966a).

Luft, J. H., "Ruthenium red and violet: Chemistry, purification, methods of use and mechanism of action," Private publication, J. H. Luft, Dept. of Biol. Structure, University of Washington (1966b).

Majno, G., and G. E. Palade, "Studies on inflammation. 1. The effect of histamine and serotonin on vascular permeability: An electron microscope study," *J. Biophys. Biochem. Cytol.,* **11**:571–605 (1961).

Majno, G., G. E. Palade, and G. I. Schoefl, "Studies on inflammation. 2. The site of action of histamine and serotonin along the vascular tree: A topographic study," *J. Biophys. Biochem. Cytol.,* **11**:607–626 (1961).

Matter, A., L. Orci, and C. Rouiller, "A study on the permeability barriers between Disse's space and the bile canaliculus," *J. Ultrastruct. Res.,* **29**(suppl. 11): 1–71 (1970).

Muir, A. R., "The molecular structure of isolated and intracellular ferritin," *Quart. J. Exp. Physiol.,* **45**:192–201 (1960).

Padawer, J., "Uptake of colloidal thorium dioxide by mast cells," *J. Cell Biol.,* **40**:747–760 (1969).

Pappas, G. D., and V. M. Tennyson, "An electron microscopic study of the passage of colloidal particles from blood vessels of the ciliary processes and choroid plexus of the rabbit," *J. Cell Biol.,* **15**:227–239 (1962).

Revel, J.-P., and D. W. Hamilton, "The double nature of the intermediate dense line in peripheral nerve myelin," *Anat. Rec.,* **163**:7–16 (1969).

Revel, J.-P., and M. J. Karnovsky, "Hexagonal array of subunits in intercellular junctions of the mouse heart and liver," *J. Cell Biol.,* **33**:C7–C12 (1967).

Straus, W., "The use of horseradish peroxidase as a marker protein for studies of phagolysosomes, permeability and immunology," *Meth. Achieve. Exp. Pathol.,* **4**:54–91 (1969).

Tani, E., and T. Ametani, "Substructure of microtubules in brain nerve cells as revealed by ruthenium red," *J. Cell Biol.,* **46**:159–165 (1970).

Trotter, C. M., "Phagocytosis of colloidal iron dextran," pp. 65–66 in "Microscopie Electronique," Vol. 3, P. Favard, ed., 1970.

Venkatachalam, M. A., and H. D. Fahimi, "The use of beef liver catalase as a protein tracer for electron microscopy," *J. Cell Biol.,* **42**:480–489 (1969).

TRACHEA

The trachea occurs in most animals as a median air passage and in tetrapods leads from the larynx to the point where the two lung structures diverge. In all classes above the amphibians, the trachea is characteristically supported by bars or incomplete rings of hyaline cartilage spaced at regular intervals along its length. The human trachea is a thin-walled tube about 11 cm long and 2–2.5 cm in diameter extending from the inferior part of the larynx at the level of the sixth cervical vertebra to the level of the upper surface of the fifth thoracic vertebra where it divides into the two bronchi. The wall of the trachea consists of mucosal, submucosal, cartilagenous, and adventitial layers. The mucosa is composed of a ciliated pseudostratified columnar epithelium, a distinct basal lamina, and a highly elastic lamina propria. Three cell types have been described in the epithelium: ciliated columnar cells which extend from the basement lamina to the surface and have oval, rather pale-staining nuclei located in their mid-region; basal cells which lie close to the basement lamina and have spherical, densely staining nuclei; and goblet cells with flat nuclei located at the base of the cell. Cells of the lymphocyte series are commonly found scattered throughout the mucosa with occasional formations of true nodules.

While not obvious in the usual H and E stained section, the boundary between the mucosa and submucosa is marked by the presence of an elastic layer which can be revealed by elastic tissue staining methods such as Taenzer-Unna acid orcein, Verhoeff's iodine iron hematoxylin, Fullmer's orcinol–new fushsin, and Gomori's aldehyde-fuchsin (Lillie). With the last staining method, formalin and Bouin's fluid gives colorless backgrounds, while mercury-containing fixatives result in a pale lilac background. Chromate-containing fixatives should be avoided. Most fixatives are suitable for the other staining methods.

Seromucous glands which are of the tubuloacinous type are found scattered throughout the submucosa and in the dorsal part of the trachea where there is no cartilage, they are also found in the mucosa and in the transverse tracheal muscle, often extending dorsal to the muscle into the adventitia. The excretory opening into the lumen of the trachea is in the form of an ampulla lined with ciliated pseudostratified columnar epithelium which is continuous from the tracheal surface to the adjacent part of the excretory duct. The remainder of the duct is lined with a nonciliated cuboidal epithelium. The highly tortuous tubules are lined with pyramid-shaped mucous cells with basal nuclei. The acini contain both mucous and serous cells.

Topography of tracheal mucous glands is most effectively studied by means of whole mount methods in which whole uncut tissue pieces are stained, followed by clearing, which makes the specimen transparent; or, to facilitate staining, a microdissection can be carried out first resulting in separate mucosal and submucosal layers (Moe, 1952; Tos, 1966). In the mucosa, the goblet cells of the epithelium, the glandular orifices, and parts of the main ducts are stained, while in the submucosa, the remainder of the excretory ducts and the gland mass are stained.

Glandular structures are stained bright red on a pale red background with the periodic acid–Schiff (PAS) method. With PAS–alcian blue, the result is a mixture of colors, ranging from pure red to pure blue (Lamb and Reid, 1969). Since some but not all cells lose their affinity for alcian blue after treatment with sialidase, it has been suggested that some cells produce pure sialomucin, others produce mucin part of which is sialomucin, while still others produce a mucin which is not affected by sialidase (Lamb and Reid, 1969). It has been proposed that sialidase-sensitive and sialidase-resistant sialomucin is produced since acid hydrolysis, which removes all sialic acid, results in a greater reduction of alcian blue staining than treatment with sialadase (Lamb and Reid, 1969). Stains for sulfated mucin, including aldehyde fuchsin–alcian blue (Spicer and Meyer, 1960), alcian blue in aluminum sulfate (Heath, 1961), and a high iron diamine technique (Spicer and Duvenci, 1964) are positive in only a proportion of the acinar cells. However, a variety of sulfate recognized by uptake of radioactive sulfate that stains with alcian blue but not with stains for sulfated mucin has been described in these cells (Lamb and Reid, 1969). In addition, the above-mentioned stains for sulfated mucin are positive in epithelial goblet cells. These cells also show toluidine blue metachromasia resistant to hyaluronidase digestion and azure A metachromasia beginning at pH 1.5 (Korhonen et al., 1969).

The tracheal cartilages function in maintaining a patent airway. The gaps between the ends of the cartilages allow limited constriction and dilatation and are connected by the transverse tracheal muscle. This is a rather thick layer of smooth muscle which is inserted into the elastic fiber bundles which suround the cartilages. The elastic fiber bundles constitute a fibroelastic layer which blends with the perichondrium and is continuous with the fibroelastic tissue in the interspaces between the successive tracheal cartilages.

The adventitia consists of a loose, fibrous connective tissue and contains lymph nodes and the larger branches of blood vessels, nerves, and lymphatics. Lymphatic vessels constitute a delicate network in the mucosa and a much coarser plexus in the submucosa. These lead eventually into the lymph nodes that occur along the length of the trachea. Arteries of the trachea arise mainly from the inferior thyroid arteries. Nerves of the trachea are branches of the vagus, its recurrent branch, and the sympathetic.

WILLIAM A. GIBSON

References

Heath, I. D., *Nature* (London), **191**:1370 (1961).
Korhonen, L. K., E. Holopainen, and M. Paavolainen, *Acta Histochem.*, **32**:57 (1969).
Lamb, D., and L. Reid, *J. Pathol.*, **98**:213 (1969).
Lillie, R. D., "Histopathologic Technic and Practical Histochemistry," 3rd ed., New York, McGraw-Hill Book Co., 1965.
Moe, H., *Stain Technol.*, **27**:141 (1952).
Spicer, S. S., *J. Histochem. Cytochem.*, **13**:211 (1965).
Spicer, S. S., and J. Duvenci, *Anat. Rec.*, **149**:333 (1964).
Spicer, S. S., and D. B. Meyer, *Amer. J. Clin. Pathol.*, **33**:453 (1960).
Tos, M., *Acta Pathol. Microbiol. Scand.*, Supp. 185 (1966).

TRANSMISSION ELECTRON MICROSCOPY

1. Definition. Any form of microscopy involving the passage of a high-velocity homogeneous electron beam through a specimen is, strictly speaking, transmission electron microscopy (TEM). However, this definition includes the formation of an image by scanning a specimen line by line with a very fine, intense electron spot, thus building up an image in the form of a "raster," no imaging lenses being necessary. This is "scanning transmission electron microscopy" (STEM) and is dealt with separately.

TEM is generally understood to refer to the technique whereby a relatively large area (10–10^6 μm^2) of a specimen thin enough to transmit at least 50% of the incident electrons is illuminated by a homogeneous beam, the emergent beam being refracted by a system of imaging lenses to form a real, magnified image of the specimen on a fluorescent screen or photographic emulsion. It is the exact electron counterpart of the transmission light microscope (TLM), and analogously, the term "electron microscope" is generally applied to the TEM.

2. Resolving Power. The sole advantage possessed by the TEM over the TLM is higher resolving power (RP). When a perfect TLM views a perfect specimen, RP is approximately one half-wavelength of the illuminating light used, i.e., 0.25 μm. RP can be improved only by reducing wavelength. Since X-rays cannot be refracted to form images, a different form of radiation is necessary. The only available radiation is the so-called "matter radiation" or "probability waves" associated with high-velocity particles and predicted on theoretical grounds by de Broglie in 1924. The wavelength of this particle-associated radiation is given by the relationship:

$$\lambda = \frac{h}{mv}$$

where m is the mass and v the velocity of the particle, h being Planck's universal quantum number. For an electron accelerated by falling through a potential of 60,000 V (60 kV) and thus traveling at about one-third the velocity of light, $\lambda = 0.05$ Å, which is 100,000 times shorter than λ for green light. A TEM using perfect lenses would therefore in theory be able to resolve 0.025 Å. In practice, however, the RP is about 2.5 Å, a factor of 100 less than that theoretically possible, but nonetheless 1000 times better than th best TLM. The reason for this discrepancy is that present-day electron lenses are still very crude and simple compared to light lenses, because it is not yet possible to correct the aberrations common to both to the same degree of refinement in electron lenses.

3. Magnification. The maximum usable magnification of any microscope is simply the ratio of the RP of the microscope to the RP of the unaided human eye. The comfortable RP of the eye viewing an object at 25 cm is generally taken to be 0.25 mm. Since the RP of the TLM is about 0.25 μm, maximum useful light magnification is $1000\times$, obtained from an objective lens of $100\times$ followed by an eyepiece of $10\times$. The comparable TEM ratio is 0.25 mm/0.25 Å, or 10^6, corresponding to the thousand-fold increase in RP. It is necessary to use a train of at least three imaging lenses to obtain such high magnification, which in practice is attained as follows. An objective lens of $100\times$ is followed by an "intermediate" lens of $25\times$, the final image being projected by a "projector" lens of $100\times$. Further magnification for critical focusing is obtained by viewing the image on the fluorescent screen with a long working distance binocular microscope of $10\times$. The final image is photographed at $250,000\times$, the processed negative being then enlarged a further $4\times$ in a photographic enlarger. This results in a final print (the "electron micrograph") at the desired magnification of $10^6\times$. Higher final magnifications are obviously possible, as with the TLM, but these conceal "empty magnification."

4. Electron Lenses. The ultra-short "matter waves" accompanying high-speed electrons follow the trajectories of the electrons exactly. Since the paths of electrons can be deflected by magnetic or electric fields, it follows that the ultra-short waves are also deflected, i.e., refraction and therefore the formation of magnified images is possible. The trajectories of electrons in refracting spaces were studied by Busch in 1926, who showed that they were exactly analogous to ray paths in geometrical light optics.

Although suitably shaped electrostatic fields can be used as electron lenses, they cannot in practice be made powerful enough for modern RP requirements without breakdown or "arc-over" in the TEM. All modern instruments use exclusively axial magnetic fields formed by the passage of an electric current through a coil of wire or "solenoid," with the exception of the weak electric field used to shape the initial electron beam emerging from the "gun" (*vide infra*).

The principle of an electromagnetic (EM) lens is shown in Fig. 1. The current passing through the coil forms an axially and radially symmetrical magnetic field, which converges to a point, a divergent cone of electrons entering it from a point source, and thus forms a real image on the lens axis. The basic light lens formulas apply; magnification is the ratio of image-lens distance to object-lens distance. Electrons are also deflected sideways, resulting in a spiraling of the trajectories and thus a rotation of the image, but the envelope of trajectories corresponds to "rays of light" in geometrical optics. One difference between light and electron optics is that it is only possible to construct positive lenses; divergent lenses do not exist. By concentrating the magnetic field axially by means of iron shrouds surrounding the wire coil, very short focal length lenses can be constructed; about 1 mm, corresponding to a magnification of about 100×, is the practical limit. An advantage of electromagnetic lenses is that the focal length can be made infinitely variable by varying the coil current. Therefore the lens current control can be used to adjust both magnification and image focus. This convenience, however, brings the disadvantage that image stability depends on lens current stability. It is necessary to stabilize objective lens current to better than 1 part in 50,000 to achieve an RP of 2.5 Å.

5. Lens Aberrations. Since electron lenses obey the laws of geometrical optics, all the aberrations of light lenses (spherical, chromatic, astigmatism, distortion etc.) are suffered by electron lenses, together with further aberrations in magnetic lenses due to the spiral trajectories. In light optics, aberrations can be compensated and in some cases eliminated by combining positive and negative lenses of different glasses having different refractive indices and dispersions. Since there are no negative electron lenses and the only practical refracting medium is a magnetic field, electron lenses cannot be so compensated. Aberrations can be minimized only by roundabout methods.

Spherical aberration, since it increases with the cube of the lens aperture, is the worst enemy. It can be minimized only by reducing lens aperture. The NA of a high-resolution TEM objective lens is between 0.01 and 0.001, compared with the maximum NA of an oil-immersion TLM objective of 1.5. This is the chief reason why the RP of the TEM is a factor of 100 less than the theoretical. Such minute apertures have a negligible diffraction effect since the lens is working so far from its diffraction limit.

Chromatic aberration can be reduced by using a "monochromatic," i.e., constant-velocity, illuminating beam. This entails stabilizing the electron beam to 1 part in 100,000 for a RP of 2.5 Å. Unfortunately, however, many electrons suffer a loss of velocity during their passage through the transmission specimen, and therefore the imaging beam passing into the objective lens is heterochromatic. This effect can also be reduced by reducing objective lens aperture.

Astigmatism is caused by radial asymmetry in a lens, giving rise to a focal length in one plane different from that in another. The effect can be reduced by ultrahigh-precision workmanship in lens manufacture, but sufficient residual astigmatism remains to necessitate a special radial symmetry compensator called the "stigmator." This consists of an adjustable electric or magnetic field which can be applied across the lens in any chosen direction, thus compensating for first order astigmatism. Higher orders remain uncorrectable. To obtain the ultimate RP from the TEM, the stigmator must be adjusted by the operator to a high degree of accuracy.

Image distortion, due to magnification changing across the field from the value at the center, may be positive ("barrel") or negative ("pincushion"). The effect can be compensated by operating two lenses in series, arranging for barrel in one to be compensated by pincushion in the other. This correction is performed on the intermediate and projector lenses, the currents in each being preset for each magnification step by the manufacturer. The effect becomes more pronounced as magnification is decreased, and a three-lens imaging train generally gives an unacceptably distorted image below about 1000 ×. Lower ("scan") magnifications are obtained by using the intermediate lens as a long-focus objective, dispensing with the normal objective which is greatly weakened or switched off.

6. Image Formation. The mechanism of image formation in the TEM (scattering) is quite different from that in the TLM (absorption), although the end result is analogous. When a high-speed electron approaches an atom in the specimen, it becomes deflected through a very small angle by interaction with the electron cloud surrounding the atom. Its velocity is also reduced (chromatic aberration). This process of "inelastic scatter" is the main source of contrast in the TEM. A small proportion of electrons penetrate the electron cloud and approach the

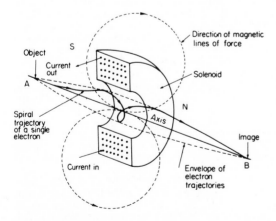

Fig. 1 Principle of an electronmagnetic lens. (From G. A. Meek, "Practical Electron Microscopy for Biologists," Wiley-Interscience, 1970. Reproduced by courtesy of John Wiley and Sons Ltd.)

atomic nucleus. These electrons are scattered through large angles, but tend to lose less energy ("elastic scatter") The total scatter is a direct function of the product of specimen thickness and atomic number ("mass thickness"), and an inverse function of the velocity (voltage) of the electron beam. Thus, increasing specimen thickness or density (atomic number), or reducing accelerating voltage will both increase image contrast.

Scattered electrons should not be allowed to pass through the objective lens and reach the final image, since they will be out of focus due to the uncorrectable chromatic aberration of the lenses. This is achieved by collecting all electrons which are scattered through an angle greater than the aperture angle α of the objective onto a metal disc pierced with a hole (20–50μm diameter) which is placed as close to the optical center of the objective as possible. This "physical objective aperture" also serves to define the NA of the lens. Figure 2 shows how an illuminating cone of electrons focused onto a small area of the specimen is scattered by the specimen. Large-angle ($>\alpha$) scattered electrons are collected by the physical aperture, but small-angle ($<\alpha$) ones give rise to glare, which reduces contrast and resolution of adjacent image points. Both can be increased by reducing the diameter of the objective physical aperture and by reducing the angular aperture of illumination (defocusing condenser; reducing condenser physical aperture.).

7. Depth of Focus and Field. These parameters define the distance along the microscope axis at the image plane and object plane, respectively, along which the final image remains maximally sharp to within the instrumental resolving power. Both are inverse functions of lens aperture. In practice, the "pinhole" apertures of electron lenses give rise to an almost infinite depth of focus below the projector lens, and a depth of field at the specimen of about 1 μm. This is ten times the maximum usable specimen thickness, and therefore the whole of the specimen in the TEM is in sharp focus. Depth of field is a great advantage; in practice, the final image screen or photographic film may be placed anywhere beneath the projector lens. A 35 mm film camera placed close to the projector will record as sharp an image as a large-format plate camera lying beneath the viewing screen. Both cameras can be made to record the same image area. Depth of focus is in general a disadvantage, since the whole thickness of the transmission specimen is simultaneously in focus. All details are imaged sharply one above the other, the resulting confusion leading to loss of resolution and difficulty in image interpretation. The effect, however, makes focusing the image much less critical.

8. The Specimen. The mass thickness of the TEM specimen should lie between 2–10 μg/cm², the optimum for 50 kV being about 5 μg/cm². A specimen of $d = 1$ should therefore be 50 nm (500 Å) thick, but metals must be up to ten times thinner. The specimen may be particulate, in which case a suspension may be dried down on a support film (*vide infra*) for direct examination. Massive specimens must be thinned: biological tissues by embedding in a hard plastic (epoxy resin) followed by slicing on an ultramicrotome using a glass or diamond knife; metals by chemical etching, electropolishing, ion bombardment, vacuum evaporation etc. Since biological tissues are on the whole composed of the same atomic species as organic synthetic resins, "electron stains" conferring differential scattering power to specific tissue components must be added at some stage, either before or after sectioning, or both as is usual. Heavy-metal salts which combine with specific binding sites, e.g., membranes, glycogen, and nucleic acids, are used, the commonest being compounds of osmium, lead, uranium, vandium etc.

9. Specimen Supports. All transmission specimens must be supported in the TEM on a very thin ($<$ 20 nm) electron-translucent film of low atomic number, which in turn is supported on a thin (100 μm thick) metal (copper, gold, platinum) disc or "grid" 3 mm in diameter perforated with round or square holes of about 100 μm side. About 30–80% of the specimen area is visible, the remainder being blocked by the support bars. The thin film may be of a tough plastic (polyvinyl formal or Formvar) or evaporated carbon. Plastic films are formed by evaporating a dilute solution of the plastic on a highly polished surface, e.g., a glass slide, and subsequently floating the film off on a water surface. Carbon films are made by evaporating carbon from an arc *in vacuo* onto a cleaved mica or similar structureless surface and subsequently floating the film on water. The grids are then coated by raising them up beneath the floating film. Epoxy resin sections can be made self-supporting on the grid. Plastic sections or films, being electrical nonconductors, tend to charge up in the electron beam. The resulting electrostatic repulsion of adjacent specimen areas may cause specimen movement or "drift." Carbon support films do not have this disadvantage, but are more difficult to prepare and use.

10. Construction of the TEM. The optical design follows that of the conventional TLM, but for convenience the optical train is turned upside-down so that the electron source is uppermost. A modern tendency is to reduce height by placing the column between the operator's knees; the optical train is then precisely analogous with the TLM. The iron-shrouded coils of insulated wire which energize the magnetic gaps forming the actual

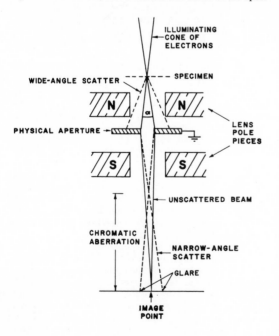

Fig. 2   The scattering of electrons by a specimen.

lenses are manufactured in units which stack one upon the other to form a "column" about 1 m in length (see Figs. 3 and 4). Airtight joints formed by synthetic rubber 0-rings placed between the lens units enable the air, which would otherwise prevent the electrons from passing down the column, to be removed from the beam path. The necessary high internal vacuum (10^{-4} torr) is maintained by vacuum pumps, both diffusion and mechanical, situated below and behind the column.

The components of the column can be conveniently divided as follows:

1. illuminating system
2. specimen manipulation system
3. imaging system
4. image translation system.

Illuminating System. The source of electrons or "cathode" is a hairpin of fine tungsten wire about 2 mm long, maintained at about 2500°K by some 2 W of AC or

Fig. 4   The British AEI-EM-801-A transmission instrument designed especially for biological applications.

DC power. Electrons boil off the white-hot tungsten surface, and are shaped into a conical beam by an electrode system called the "gun." Two further electrodes, the "shield" and the "anode," combine to form an electrostatic collimating lens and accelerator. A suitable accelerating voltage (20–100 kV) is chosen for the specimen under examination, and is applied to the cathode as a negative potential so that the anode may remain at earth potential. The cathode and shield are therefore carried on an insulator. The filament-shield voltage ("cathode bias") is made variable to adjust the total current drawn from the filament, which in turn varies the brightness of the final image.

The divergent electron beam emerging from the anode aperture can in simple instruments be used to illuminate the specimen directly, but sufficient image brightness for high-magnification working is difficult to obtain. As in the TLM, a condenser system is almost invariably interposed between gun and specimen to concentrate the beam on the area of specimen under examination. A single condenser lens suffices for electron-optical work up to 50,000 ×, but for high-resolution work a double condenser system is always used, which will concentrate the beam into an area as small as 1 μm dia.

Specimen Manipulation System. The pierced metal grid

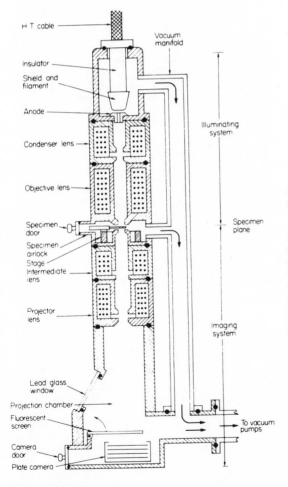

Fig. 3   Schematic diagram of transmission electron microscope. (From G. A. Meek, "Practical Electron Microscopy for Biologists," Wiley-Interscience, 1970. Reproduced by courtesy of John Wiley and Sons Ltd.)

carrying the specimen proper is clamped by the periphery in a suitable holder designed to conduct heat rapidly away; specimen temperature may rise to 200°C. The holder plus grid is introduced into the evacuated specimen chamber through an airlock by means of an insertion tool which is then generally withdrawn after the holder has been placed on the translation stage. A viewing window in the side of the specimen chamber may be necessary for correct placement to be observed. The holder is then free to move with the stage, which is driven from outside the column through airlocks by means of levers and micrometer screws. Two mutually perpendicular stage movements, each of about $\pm$ 1 mm, allow any part of the grid area to be brought to the microscope axis and viewed at the highest magnification. Most instruments provide a "scan" magnification so that the whole grid area may be viewed at about 100 $\times$; suitable specimen areas are then chosen and centered for study at higher magnifications.

Imaging System. Electrons transmitted by the specimen enter the objective lens. Those passing through the physical aperture are imaged at about 100 $\times$ in the intermediate lens object plane 10–20 cm below the specimen. The position of this primary image plane is controlled by the objective lens current (focus control). A second image, which may be magnified or diminished, is formed by the intermediate lens, the current through which controls overall magnification (magnification control). This secondary image, formed in the object plane of the projector lens, is then further magnified, the overall magnification depending on the position of the fluorescent screen or film.

Image Translation System. To render the final image visible to the eye, the energy of the electrons must be translated into visible light. A flat earthed metal plate coated with a thin layer of fluorescent powder plus binder is placed about 30 cm below the projector lens, and is viewed through a thick lead glass window which prevents the irradiation of the operator by X-rays produced by the electrons as they strike the screen. A powder fluorescing in the blue-green is chosen, this being the color to which the eye is most sensitive. Such screens are of relatively low sensitivity (10^{-12} Å/cm²) and cause loss of instrumental resolution due to particle size. A permanent record of the image is therefore made on a fine-grained blue-sensitive photographic emulsion placed directly below the screen. Exposure is controlled by an electrically operated shutter placed below the projector lens; exposure time is of the order of 1 sec. The fluorescent screen must be lifted out of the way before a film is exposed; this action also operates the shutter on modern instruments. A 35 mm camera placed above the screen close to the projector lens is often fitted; care must be taken that resolution is not lost in the photographic grain.

When studying beam-sensitive specimens at high magnifications and low beam currents, an image intensifier is necessary. This is a form of electron amplifier which generally feeds the image signal to a TV display system. Sensitivity is limited to about 10^{-16} Å/cm² by the random particulate nature of the electron beam which causes image "noise" due to the random arrival of electrons. Sensitivity can be further increased by giving long photographic exposures and thus integrating the individual electrons. A TV display system is of particular value if transient phenomena, e.g., the formation of dislocations in metal specimens, are being studied, since such events can be recorded on videotape and replayed as often as necessary for interpretation.

11. Electrical Supplies. The five lenses of the TEM require energizing with 0.5–1 kW, which must be removed from the column by a circulating cold water system to prevent thermal instability. Lens current stability requirements vary from 1 to 50,000 for the objective to 1 in 10,000 for the condenser; the beam voltage must be stabilized to 1 in 100,000. Power is drawn from the domestic mains, and is rectified and stabilized as required by electronic generators and stabilizers, which are solid-state on modern instruments. Power for auxiliary supplies, e.g., vacuum pumps and switch circuits, is generally unstabilized. The total consumption of the instrument is 2–3 kVA.

12. Operation. The operator sits before a console, which in modern instruments contains all the vacuum gear and the transistorized electronics. Most older designs had the electron tube-operated electronics in a separate cabinet due to cooling problems. All the controls fall readily to hand, the beam controls being generally grouped to the left and the lens controls to the right. Meters monitor the most important parameters, e.g., beam current, vacuum, lens currents, and magnification. Adjustments on the column enable the lenses to be adjusted accurately about the microscope axis to minimize aberrations and achieve maximum resolving power. The final image is viewed and critically focused through a 10 $\times$ binocular telescope (long working distance microscope) focused on the tilted screen. If a high-aperture telescope is used, image brightness is not lost in spite of the increased magnification. The image is recorded on a camera placed below the screen. When all the exposures have been used, the camera is removed through an airlock, the films are developed, and the negatives are enlarged in a high-quality photographic enlarger to produce the finished electron micrographs for study and interpretation.

G. A. MEEK

References

Cosslett, V. E., "Modern Microscopy," London, Bell, 1966.
Grivet, P., "Electron Optics," Oxford, Pergamon, 1965.
Haine, M., "The Electron Microscope: The Present State of the Art," London, Spon, 1961.
Kay, D. H., ed., "Techniques for Electron Microscopy," 2nd ed., Oxford, Blackwell, 1965.
Meek, G. A., "Practical Electron Microscopy for Biologists," London and New York, Wiley-Interscience, 1970.
Siegel, B. M., ed., "Modern Developments in Electron Microscopy," New York, Academic Press, 1964.

u

ULTRAMICROBIOPHYSICS

Ultramicrobiophysics combines the techniques of physics and collodial physical chemistry with those of electron microscopy: By this means it is possible to develop all the physical data of one minute living entity (16). It is possible to measure not only the surface charge, but also the local density of the coulombic forces. The surface membrane potentials (electronkinetic Z potentials) are of increasing importance in research (4, 5). It is, however, technically very difficult to measure exactly the quality, density, and intensity of charge on minute parts of an organism in the living condition. This article records electron microscopical methods for doing this.

Previously, Freundlich had used oppositely charged sols to demonstrate mutual coagulation. Lukjanovich et al. (1) used labeling sols to investigate the surface charge of inorganic material (V_2O_5 sol) on an electronoptical scale. The writer and his colleagues have developed these methods to a semiquantative state. These methods were first applied to bull sperm (3,6).

Measuring the Surface Charge of Biocolloids on an Electronoptical Scale. Briefly, this method involves the preparation of colloidal labeling sols (e.g., AgFj that it charges. Living cell suspensions were incubated with these labeling sols and a number of absorbed particles were then topographically plotted on electronmicrographs. Details of this method are given in (7). For use with sperm, it is desirable to have a homodispersed sol of electrodense particles with a diameter of 0.1–0.3 μm and collodial suspensions of AgF give good results and are relatively easy to prepare (8). The concentration and size of the labeling colloids can be determined in an electron microscope (Figs. 1, 2).

For objects smaller than sperm (e.g., virus) it is possible to prepare a labeling sol of smaller particle size or from other materials.

For the work discussed, fresh ejaculates of sperm were obtained from an artifical insemination center. It is impossible to secure good homogenization of sperm and sols without dilution. Although it is possible to secure correct visualization of membrane charges in the presence of the Na^+ and Cl^- ions of saline, experiments were also conducted with isotonic sucrose. However, the residues of both these are too electron dense, and the best results were obtained with an isotonic solution of urea. The solution does not interfere with the membrane charge on the living form and, after the absorption of labeling colloids, the object is readily observable in the electron microscope, since the residue of the diluting solution is not electron dense. The sperm is diluted 30–100 times at body temperature and manually shaken for 5 min. A drop of this suspension is then mounted on a Formvar film on a copper grid and washed in water, though the results vary little from unwashed specimens. The fine structure is not destroyed by the urea as is seen in Figs. 3–6, and in goose sperm investigated with the same technique (Figs. 7,8).

Methods are being developed for the application of these techniques to other objects such as cells, bacteria, and viruses. In the case of relatively larger objects (5–10 μm) that are not strongly anisotropic, it is possible to observe the microcrystalline salt particles in a polarization microscope. This permits photometric measurements of density of the absorbed particles.

The charge density may be determined on an electron micrograph by counting the particles over an area or a section. The Opton-Zeiss apparatus is particularly good for counting the absorbed particles.

Electrophoretic methods may be used to determine natural relationships. The Nernst apparatus (8) for measuring the movement velocities of sols has the many potential errors of microelectrophoretic methods. An alternative to this technique has been developed in our laboratory.

Measuring Dispersion of Internal Material. Imperfections of the human eye, particularly in the perception of gradual increases in blackening, make it difficult to utilize existing equipment, no matter how good, for internal measurements. These measurements are, however, of the upmost importance particularly in electron microscopy. The so-called "densitogram method" (10) involves the integration of photometric curves made by marking parts of equal density. In this way it is possible to show the exact position of the nucleus in a bull sperm head, which is not possible in ultrathin sections.

The photodensitogram method also discloses the double asymmetry of the inner material dispersed in the bull sperm head (Fig. 9). This has made possible (11, 12) the explanation of the swimming stability and stable position of the center of gravity. This was confirmed (13) by semi-incineration on an electron microscopical scale (Fig. 10).

It was also shown that in some pathological conditions, the dispersal of internal material altered the position of the nucleus and this so altered the stability of the sperm that swimming was impossible (14, 15) (Fig. 11).

Thus the correlation of ultramicrobiophysical conditions of material dispersion, cell charged and the like, in addition to morphological appearance, are necessary to the full interpretation of cell methodology.

IMRE VERES

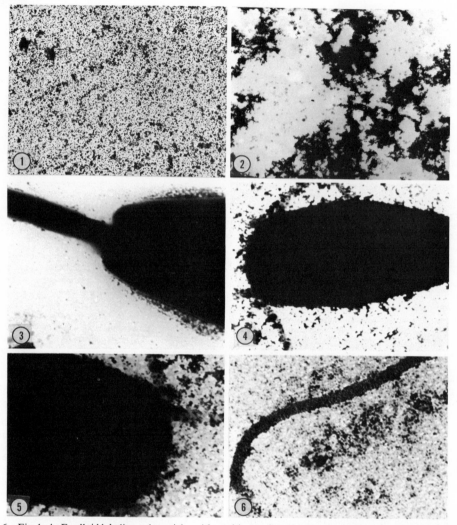

Figs. 1–6 Fig. 1: AgF colloid labeling sol particles with positive surface charge; those prepared with a negative charge are very similar. ($\times$ 3750.) Fig. 2: AgF sol particles with different surface charges after coagulation; this is thus a control of their electrokinetic potentials. ($\times$ 3750.) Fig. 3: Bull sperm head (postnuclear and implantation region) with absorbed positive AgF particles. ($\times$ 5350.) Fig. 4: Frontal region of bull sperm head with many absorbed positive AgF particles, which are in contact with the whole sperm surface. ($\times$ 3750.) Fig. 5: As Fig. 4 but at greater magnification. The concentration of the coulombic surface charges on the frontal part of head is significant, and it may have an important role in guiding the sperm to the ovum. ($\times$ 5000.) Fig. 6: Showing the high concentration of negative charges on the tail end of the bull sperm head. ($\times$ 5350.)

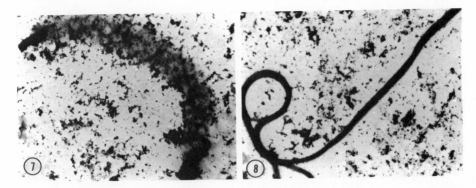

Figs. 7 and 8   Fig. 7: Goose sperm head with absorbed positive AgF particles. The cell surface has a strong negative charge. (×3750.) Fig. 8: After incubation with negative-charged AgF sol, the particles are not absorbed on the goose sperm (tail part). (×3000.)

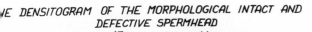

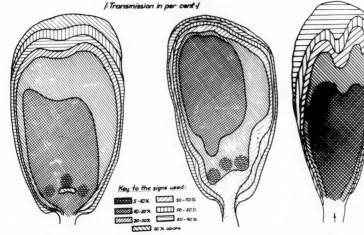

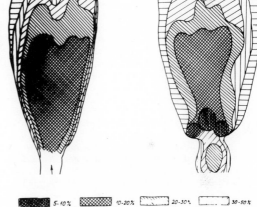

Fig. 9   Densitograms made on the electronmicrographs of a normal and of a pathologically changed (on the right) bull sperm. The double asymmetry of the inner material dispersion, the cause of the swimming stability (i.e., stability of the center of gravity) is not present in some aberrants. (Transmission in percent.)

Fig. 11   Densitograms of two aberrant spermcells show the different changes in the inner material of nucleus.

Fig. 10   The double asymmetry was confirmed after a semi-incineration of the bull sperm head on SiO film. 4,800×

References

1. A. V. Bromberg, et al., *Zsurnal Fizicseszkoj Himii,* **27**:379–388 (1953).
2. I. Veres and A. Öcsényi, *5th Hungarian Conf. Electron microscopy,* Bâlatonszéplak, 1967.
3. I. Veres, *Immun. Sperm. Fertil., 1st Int. Symp.,* Várna, 1967.
4. M. H. Greer and R. F. Baker, *7th Cong. Int. Microsc. Electron.,* Grenoble, 1970.
5. D. Danon et al., *7th Cong. Int. Microsc. Electron.,* Grenoble, 1970.
6. I. Veres, *6th Cong. Int. Reprod. Anim. Insem. Artif.,* Paris, Vol. **I**, 1968.
7. I. Veres, *Mikroskopie,* **23**:166 (1968).
8. A. Buzágh, "Textbook of Colloidchemistry," [Hungarian] Budapest, 1958.
9. I. Veres and W. Krug. *4th European Regional Conf. Electron Microscopy,* Rome, 1968.

10. I. Veres, *Jenaer Rundschau,* **5**:275–278 (1968).
11. I. Veres, *5th Int. Congr. per la Riproducione Animale e la Fecondacione Artificiale Trento,* **4**:582 1964).
12. I. Veres, *Mikroskopie,* **22**:269–273 (1967).
13. I. Veres and A. Öcsényi, *Mikroskopie,* **23**:305–307 (1968).
14. I. Veres and A. Öcsényi, *6th Cong. Int. Reprod Anim. Insem. Artif.,* 209–212, Paris, 1968.
15. I. Veres and A. Öcsényi, *Immun. Sperm. Fertil. 1st Int. Symp.,* Várna, 1967.
16. I. Veres, *7th Cong. Int. Microscopie Electronique* Grenoble, 1970.

ULTRATHIN SECTIONING

Ordinary transmission electron microscopes produce image-forming electrons which, having relatively low energies, are scattered or absorbed by small masses. Since electron microscope images are comparisons among different parts of the specimen with respect to their scattering properties, it follows that useful images can be formed only if parts of the specimen (usually parts containing unwanted information) have little enough mass to transmit a substantial fraction of the electrons that impinge upon them. This consideration requires that most electron microscope specimens be less than 0.1 μ thick. Ultrathin sectioning is a commonly used procedure for obtaining such specimens, especially of biological material.

The requisites for successful ultrathin sectioning are a suitable specimen, a sufficiently sharp knife formed of broken glass or polished diamond, and a microtome capable of reproducibly advancing the specimen or knife the distance of the section thickness desired (ultramicrotome). These necessities were not available until the early 1950s; indeed, it was not until ten years later that ultrathin sectioning became a skill that could be taught as routine.

The sectioning technique involves juxtaposing and aligning the knife edge and the face of the specimen, which has been trimmed accurately to a trapezoidal shape with base about 0.5 mm.

This positioning must be precise and is facilitated by observing through a stereo microscope the relative positions of the knife edge and its reflection on the specimen face. If the specimen and knife can be brought to within a few microns of each other with this rapid coarse adjustment, then a ribbon of sufficiently thin sections can be obtained after a small number of ultrafine advances of the microtome without taking a thick section (0.5 μ or greater), the cutting of which is likely to ruin the knife. Affixed to the knife is a boat containing water to the height of the knife edge. The sections float on the water surface as they are cut and can be lifted from the boat on a specimen screen.

The thickness of the sections is estimated by observing the interference colors produced by light reflected from their surfaces. Silver colored sections, thought to be from 0.06 to 0.09 μ thick, are typically useful specimens.

That ultrathin sectioning remains to a large degree an art is due to the many poorly understood variables involved. The cutting speed (usually 0.5–2.0 mm/sec), included knife angle (40°–70°), knife clearance angle (2°–10°), type and hardness of embedding medium, specimen inhomogeneity, humidity, and temperature are some of the factors that affect sectioning in ways that are not sufficiently clear.

An additional difficulty is that the only decisive assay of the quality of a section is its observation in the electron microscope. Even sections that appear thin and smooth as they are cut may have submicroscopic defects, the two most common of which are chatter and knife damage. Chatter is a fine alternation of thick and thin bands through the section, produced by vibration during sectioning; it is insidious in that it tends to be reproduced in subsequent sections, even though the initial cause is not present. Knife damage is produced by a dull knife which may be sharp enough to cut thin sections, but which smears or scratches the surface of the section.

Ultrathin sectioning is as prevalent in biological electron microscopy as ordinary thin sectioning is in biological light microscopy. It is used less frequently in nonbiological disciplines because of the difficulty in cutting very hard materials such as metals and because the sectioning process itself may deform crystalline materials.

M. D. MASER

References

An informative history of ultrathin sectioning is in:
Porter, K. R., "Ultramicrotomy," in B. M. Siegel, ed., "Modern Developments in Electron Microscopy," New York, Academic Press, 1964.
Descriptions of procedures are in:
Pease, D. C., "Histological Techniques for Electron Microscopy," 2nd ed., New York, Academic Press, 1964.
Sjostrand, F. S., "Electron Microscopy of Cells and Tissues," Vol. 1, Instrumentation and Techniques, New York, Academic Press, 1967.
"Thin Sectioning and Associated Techniques for Electron Microscopy," 2nd ed. with supplement, Norwalk, Conn., Ivan Sorvall Inc., 1967.
The physical theory, among other subjects, of ultramicrotomy, is described in:
Wachtel, A. W., M. E. Gettner, and L. Ornstein, "Microtomy," Chapter 4 in E. Pollister, ed., "Physical Techniques in Biological Research," Vol. 3, Cells and Tissues, 2nd ed., New York, Academic Press, 1966.
See also   PLASTIC EMBEDDING, PARAFFIN SECTIONS.

ULTRAVIOLET MICROSCOPY

Ultraviolet microscopy extends optical microscopy from the visible portion of the electromagnetic spectrum (wavelengths, 700–400 nm) into the invisible ultraviolet region (practically, down to about 240 nm). Special lenses, mirrors, slides, covers, immersion media, sources of ultraviolet energy, and a means of obtaining a visible image are necessary.

While the ultraviolet microscope was originally developed to take advantage of the increased resolution obtainable by using the shorter wavelengths, its main asset is the capability of revealing the distribution and concentration of ultraviolet-absorbing materials within certain microscopic specimens, especially favorable living specimens.

The usual approach to ultraviolet microscopy begins with a desire, or need, to determine the location and concentration of a substance known to absorb UV (at least under certain conditions) within a microscopic specimen. To avoid disappointment and the costly and time-consuming acquisition of equipment, some preliminary consideration may be helpful.

Disappointment may be a result of expecting too much of the method, probably because of unfamiliarity with some of the problems posed by the specimen. Most important of these is the quantity of ultraviolet-absorbing material. Ultraviolet-absorbing compounds must be present in sufficient amount to absorb a measurable portion of the incident energy. How much is that?

This, and several other problems, may be illustrated by an example in which four types of the yeast *Candida utilis* were mixed and photographed at three wavelengths, 333, 297, and 265 nm (Fig. 1). Some of the cells were control cells, others contained an accumulation of uric acid, an accumulation of S-adenosylmethionine, or an accumulation of both uric acid and S-adenosylmethionine (1). Mixing the cells provides a means of obtaining identical conditions for ultraviolet micrography. Since the longer wavelengths are less damaging to living cells than the shorter UV wavelengths, the micrographs were made in the order of decreasing wavelength. At 333 nm the picture resembles one that might be made with a light microscope. Its value lies in revealing solid or refractile structures rather than solutes with a specific absorbance. Note that the cell walls are similar in density (photographic) at all three wavelengths, and that three types of cells can be distinguished at 333 nm (some with clear vacuoles, one with bundles of crystals, and others with an irregular mass in the vacuole), but there is no clue to the identity of the compounds within the cells.

At 297 nm, UV-absorbing material is revealed in some vacuoles. Since uric acid absorbs strongly at 290 nm (Fig. 2) it is suspected that the material in the vacuoles is related to uric acid.

At 265 nm the four types of cells are clearly distinguished because of differences in absorption: (a) control cells have clear vacuoles, (b) cells with uric acid concentrated in the vacuole show crystals in a clear vacuole, (c) cells containing S-adenosylmethionine have vacuoles that are uniformly dark and of circular outline, and (d) cells that contain both uric acid and S-adenosylmethionine have irregular vacuoles.

Cells that contain uric acid crystals in a clear vacuole provide a good example of the problem of the concentration of UV-absorbing materials likely to be found in living specimens. Note that the *crystals* absorb strongly, but that the saturated (supersaturated?) solution surrounding the crystals absorbs only about as much as the distilled water surrounding the yeast cells. From the molar extinction of uric acid, $E_{m (290 nm)} = 12,400$, it can be calculated that the absorbance of a saturated solution of uric acid, measured over a path of 3 μm, the diameter of the vacuole, would be only 0.006. In contrast, S-adenosylmethionine may occur in yeast vacuoles in a concentration of 100–300 μmoles/ml of vacuolar fluid, if it is assumed that the vacuoles comprise about 10% of the cellular space. With $E_{m (260) nm} = 15,400$ the absorbances under these conditions range between 0.47 and 1.4 for a distance of 3 μm. The concentration of S-adenosylmethionine in the vacuole is more than sufficient to reveal its presence, but the concentration of uric acid in solution is not. The 0.1–1, or even 10 μm path of a cell organelle is no match for the 1 cm "light" path of a spectrophotometer cuvette.

It is understandable, therefore, that cellular substances like DPN or ATP, whose concentration usually ranges from 1 to 3 mM, cannot be observed by ultraviolet micrograpny. Their distribution in cells is not uniform;

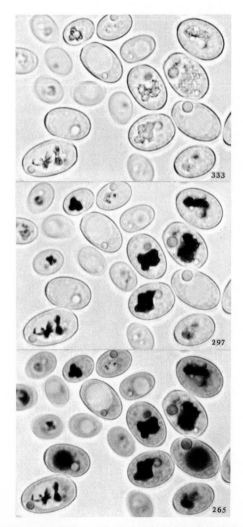

Fig. 1 Four types of the yeast *Candida utilis* mixed and photographed at 333, 297 ,and 265 nm illustrate that ultraviolet microscopy permits observation of the location and concentration of certain naturally occurring chemical substances in the living cell. Details in text. (1500×)

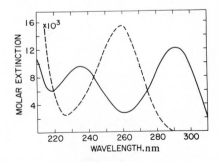

Fig. 2 Ultraviolet absorption of S-adenosylmethionine (broken line) and uric acid (solid line) at pH 6.5.

mitochondria are the site of special concentration. However, the small size of mitochondria causes other optical difficulties which are discussed later.

Two substances with the almost complementary absorption characteristics shown in Fig. 2 would appear to be ideal for UV microscopy, but in actual practice the absorption by the uric acid *crystals* at 265, as well as at 297 nm, makes it less than ideal. If two substances, such as RNA and DNA, have essentially the same UV absorption and are located in the same part of the cell, there is little hope of distinguishing them by UV microscopy. A differentiation is possible, however, if they are in different locations.

Suppose that we wish to demonstrate the yeast nucleus by UV microscopy (2, 3). The problem is complicated by the very low concentration of DNA in yeast (less than 0.1 %). A 265 nm micrograph might be expected to reveal the location, but the prominent spherical lipid body in each cell should not be mistaken for the nucleus. The nucleus appears most often in the angle between vacuole and lipid body, a light ring encircling a darker nucleolus. Contrary to expectation, the nuclei in these cells are visible because they absorb less than their surroundings. Yeast cells contain 20–30 times more RNA than DNA. Most yeast nucleic acid is in the cytoplasm, not in the nucleus. In some instances, the nuclear material of the *living* cell may be in a different chemical state than when extracted, perhaps as insoluble nucleoprotein. For example, the macronuclei of certain protozoa (Euplotes, Paramecium) even though they are comparatively massive, absorb little UV while the organisms are alive and intact, but after serious injury the macronucleus absorbs strongly. Strongly absorbing nuclei have been observed in *C. utilis* cells that we have cultured under magnesium-deficient conditions or in a medium containing an abnormally high concentration of phosphate (3).

The micrographs (Fig. 1) were made at 297 and 265 nm, instead of 290 and 260 nm which are the absorbance maxima of uric acid and DNA, respectively. One reason for this is that a mercury lamp with its characteristic line emission was used as a source of ultraviolet energy, and the lines at 297 and 265 were used because of their spectral purity, intensity, and ease with which they could be identified. It is not always necessary to work in the region of maximum absorption. In fact, it has been found advantageous to make micrographs at 275 nm in order to observe the location of S-adenosylmethionine and, at the same time, observe early stages in the formation of yeast spores that would have been hidden by the stronger absorption at 265 nm. At 275 there was sufficient absorption to reveal the location of the S-AM, but not enough to interfere with the spore images (4).

The use of more than one wavelength permits the microscopist to distinguish between optical densities due to specific absorption by the material of interest and densities due to other factors. Note that the cell walls are dark in all three micrographs (Fig. 1) and that on both sides of the dark wall are parallel light lines. Refraction, diffraction, and interference are occurring, as well as absorption.

Effects due to focus must be recognized. Small refractile granules are especially troublesome. They may appear light or dark in a micrograph depending on how they are focused. The uninitiated may assume that the difference is due to absorption. Note the yeast lipid bodies. One of the large cells contains two lipid bodies at different levels.

One appears dark and the other light. If they were in the same focal plane they would have approximately the same density. Study of a through focus series of images in such cases can be informative. The use of small polystyrene latex spheres has been helpful as a model system to aid in distinguishing fact from artifact (5).

The size and shape of the specimen, and the distribution of structures within the specimen may be important considerations. Yeast mitochondria are about 0.1 μm in diameter, the limit of resolution of the UV microscope. They appear as light circular areas in the cytoplasm near the periphery of the yeasts shown in the UV micrographs and not much more than their location is revealed. Ultraviolet microscopy of objects in the 0.1–1 μm size range is of doubtful value because of the problems of diffraction, refraction, and interference. The use of mounting media of suitable refractive index may help. It might be difficult, however, to find substances compatible with the normal metabolism of the live specimen, and there is probably little that can be done to alter the refractive index within a living cell, although certain pathological conditions might be used advantageously.

Yeast vacuoles about 3 μm in diameter, especially in cells that have been flattened slightly, are better suited to UV microscopy. Overlying and underlying cytoplasm has been displaced in the process of flattening, and the resulting shape, with almost plane-parallel surfaces, approaches that of the ideal spectrophotometer cuvette. The almost homogeneous distribution of the material within the vacuole is desirable.

Cysts of *Entamoeba invadens*, 20 μm in diameter, when studied by UV microscopy, revealed information (6) that had been missed by conventional killing, fixing, and staining techniques. Micrographs of nematode eggs about 75 μm in diameter have been obtained at 265 nm, but little more was revealed than that they contained an ultraviolet-absorbing material. The upper limit of the size of the specimens depends on factors that are too obvious to justify discussion. Interference by overlying or underlying structures is likely to be most important. Optimum size appears to lie in the 5–25 μm range.

Mobility of the specimen is still another problem. While the larger yeasts in the accompanying micrographs were held immobile between the slide and cover during the comparatively long (2–17 sec) photographic exposure necessary to record their images, some of the smaller cells moved, resulting in blurred images. In many instances of light micrography (photomicrography), this problem has been solved by the use of "electronic flash." Comparatively little has been done with pulsed UV in UV micrography.

There have been many attempts to use the ultraviolet microscope as the basis for microspectrophotometry. When it is realized that the absorbance of living specimens is constantly changing and that the specimen is being affected by the UV radiation (7), conventional killing, fixing, and staining techniques are usually resorted to. A light microscope will suffice under these conditions, in most cases. Quantitative measurements of UV absorbance by the material of interest can usually be accomplished by biochemical extraction of large samples (of synchronized cultures if necessary), followed by chromatography and conventional spectrophotometry. The results can then be correlated with the information obtained by ultraviolet microscopy (1, 4, 8). If it is necessary to carry on microspectrophotometry in the ultraviolet, the rules established by Caspersson (9) are valuable:

"The regular series of absorption studies is as a rule along the following lines:

1. Study of the living cell in the ultra-violet microscope.
2. Comparison of that picture with the picture of the frozen dried preparation.
3. Measurements on the frozen dried preparation.
4. Observation and measurements of the same preparation after extractions in different ways to correct for influences of other absorbing substances than the ones in the centre of interest, for instance, for nucleic acid and proteins, lipids of different kinds, mononucleotides, etc."

The microscopist wishing to perform UV microscopy will, most likely, have to assemble his instrument from available components. A modular approach is suggested. The modules should be arranged for the convenience of the operator, rather than for the convenience of assembly. It is very discouraging and painful to have to operate an instrument such as the Color Translating Ultraviolet Microscope, UV-91 (Polaroid) (10) with arms held overhead in order to focus the microscope and manipulate the mechanical stage. To sit at a table with a piece of equipment (monochromator or table support) between the operator's legs is not particularly pleasant, and even the inconvenience of an energy source located *beside* a microscope in order that an accessory first surface aluminized mirror can be used to direct the UV energy into the lens system should not be tolerated. A comparatively simple and reasonably convenient arrangement for an ultraviolet microscope is shown in Fig. 3. The stand was selected because the body tube, with its lenses, remains fixed in relation to the camera, and UV energy can be conveniently directed into the instrument from the rear where the source is out of the way. Focusing is accomplished by the up and down movement of the stage.

Available mechanical stages often are unsatisfactory. It should be possible to introduce a standard-size microscope slide into the optical path of a microscope with the ease and accuracy of loading an "Instamatic" camera or a cassette tape recorder. The author's modification of a Leitz stage is shown in Fig. 4. It permits the rapid exchange of slides with a minimum of focus adjustment. Rapid handling is important with living specimens, and the convenience of having an immersion objective almost in focus each time a new slide is introduced can be appreciated by all microscopists. The modification also permits removal and replacement of the condenser without disturbing its focus adjustment.

Reflecting, combined reflecting-refracting (catadioptric), and refracting UV objectives and condensers have been available. The reflecting and catadioptric objectives are difficult to focus and form poor images of biological materials (11). The microscope illustrated (Fig. 3) is equipped with Zeiss Ultrafluar (refracting) lenses. Since the visible focus and invisible UV focus of these lenses may or may not coincide, it is necessary to determine the UV focus by trial and error with a through focus series of UV micrographs, if the photographic method of obtaining a visible image is to be used. The graduations on the drum of the fine focusing knob are used to make the necessary corrections from the visible focus in green light (546 nm). For example, one of our lenses requires a correction

Fig. 4   Mechanical stage modified by cutting away half the stage and adding a new slide carrier to make slide changing fast and easy. Note that if the slide carrier is inverted, with flat, slide-retaining springs underneath, the specimen plane will be nearly constant regardless of slide thickness and the slide will be cushioned against contact with the objective.

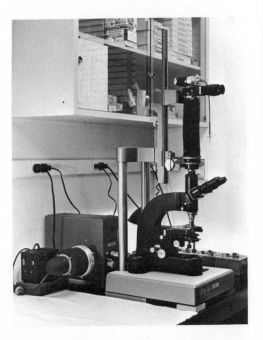

Fig. 3   A convenient, comparatively simple arrangement of components for ultraviolet micrography.

of one division of the drum at 405 nm, three divisions at 297 nm, and five divisions at 265 nm. Another lens requires no correction. The reference point from which the corrections are made is determined by the point at which the visible image on the camera's ground glass appears to disappear. On one side of focus the image is dark in contrast to the background, and on the other side of focus the image is brighter than the background. The transition between the dark and bright image (when the image "disappears," or almost matches the background) is rather sharp, especially with thin preparations in one plane.

The photographic method of converting an ultraviolet image into a visible image is comparatively inexpensive, is readily available to most microscopists, and yields the highest quality images of all the methods available so far. Comparatively rapid access to visible images may be obtained by the use of Polaroid Land film packets, but better images are obtained by the use of Kodak Spectrum Analysis Film No. 1 and conventional photographic processing. This film is quite sensitive to abrasion. The velvet pile light trap of the usual 35 mm film cassette causes scratches. For this reason, the Leica camera, for which special nonscratching cassettes are available, is used. The cassettes are loaded from a daylight-type bulk film loader, also of a nonscratching type.

There are other methods of obtaining a visible image from the ultraviolet image produced by the UV microscope. A photoemissive image-converter tube in which a visible image is produced on a small fluorescent screen has been available. The small image size, which requires magnification, and the graininess of the image limit its usefulness for other than rapid survey. Attempts have been made to use the tube as a focusing device, but since the specimen is constantly exposed to damaging UV radiation while the tube is being used, it may be preferable to focus with visible energy (light) and expose the specimen to UV only at the time of making a micrograph.

Television methods, utilizing ultraviolet-sensitive Vidicon, Image Orthicon, or related tubes, or flying spot systems, using ultraviolet-sensitive photo tubes of high sensitivity, have also been used for image conversion (12).

Both the simple image-converter tube and a closed circuit TV system employing a UV-sensitive Vidicon have been used with the microscope of Fig. 3. The amount of energy available at 265 nm, sufficient to yield good micrographs, was insufficient to yield good TV pictures. The source was a Hanovia-type SH mercury vapor lamp. Replacing this lamp with a more powerful lamp would probably produce better TV pictures, but would also produce greater UV damage to the specimen.

Desired wavelengths are selected by means of a Bausch & Lomb 250 mm grating monochromator equipped with an achromatic UV-transmitting condensing lens at its exit. A servosystem was added to the monochromator to permit rapid selection of a few often-used, wavelengths; e.g., 546 nm for focusing, 297 nm and 265 nm for UV micrography. The alignment of source, monochromator, condensing lens, and microscope conforms with the rules for Koehler illumination familiar to microscopists.

Slides and covers for UV microscopy are of quartz because glass absorbs UV of wavelengths shorter than about 320 nm. In order to distinguish the quartz slides from standard glass microscope slides, it is convenient to have them a different size. Perhaps a standard for quartz slides should be $25 \times 37.5 \times 0.5$ mm with rounded corners (2 mm radius). The brittleness of quartz makes

rounded corners desirable, and the brittleness should also be taken into account for quartz covers. Circles are to be preferred over squares and UV objectives should be designed to be used with quartz covers 0.025 mm thick. The 0.017 mm thickness recommended for use with some UV objectives is too thin for the characteristics of the material. Breakage is high during handling and cleaning (their high cost precludes their being considered disposable). While 0.035 mm thick covers, which are more durable, have been recommended for use with the Ultrafluar lenses, they are too thick to permit focusing anything except the portion of a specimen that may be in direct contact with the surface of the cover. Covers 0.025–0.027 mm thick have been found practical.

Glycerol adjusted to the proper index of refraction for the lens in use is used as an immersion medium, rather than immersion oil. The use of certain salts to adjust and control the refractive index of glycerol has been found to have a harmful effect on the metal mounts of some objectives.

Obviously, the UV absorption of the medium in which specimens are to be mounted should be considered.

A review of ultraviolet microscopy to 1950 can be found in "Medical Physics," Vol. 2, Otto Glasser, ed., Chicago, Year Book Publishers, 1950. A more recent collection of articles of interest to an ultraviolet microscopist is reference 12.

GEORGE SVIHLA

References

1. Svihla, G., J. L. Dainko, and F. Schlenk, "Ultraviolet microscopy of purine compounds in the yeast vacuole," *J. Bacteriol.,* **85**:399–409 (1963).
2. McClary, D. O., W. D. Bowers, Jr., and G. R. Miller, "Ultraviolet microscopy of budding *Saccharomyces*," *J. Bacteriol.,* **83**:276–283 (1962).
3. Janicek, L. E., and G. Svihla, "Ultraviolet micrography in biological research," *J. Biol. Photog. Ass.,* **36**:59–66 (1968).
4. Svihla, G., J. L. Dainko, and F. Schlenk, "Ultraviolet microscopy of the vacuole of *Saccharomyces cerevisiae* during sporulation," *J. Bacteriol.,* **88**:449–456 (1964).
5. Cosslett, A., "Some applications of the ultraviolet and interference microscopes in electron microscopy," *J. Roy. Microscop. Soc.,* **79**:263–271 (1960).
6. Barker, D. C., and G. Svihla, "Localization of cytoplasmic nucleic acid during growth and encystment of *Entamoeba invadens*," *J. Cell Biol.,* **20**:389–398 (1964).
7. Svihla, G., F. Schlenk, and H. L. Dainko, "Some effects of ultraviolet irradiation on yeast cells (*Candida utilis*)," *Rad. Res.,* **13**:879–891 (1960).
8. Svihla, G., and F. Schlenk, "Localization of S-adenosylmethionine in *Candida utilis* by ultraviolet microscopy," *J. Bacteriol.,* **78**:500–505 (1959).
9. Caspersson, T., "The relations between nucleic acid and protein synthesis," "Nucleic Acids. Symposia of the Society for Experimental Biology, Number I," Cambridge, Cambridge University Press, 1947 (pp. 127–151).
10. Shurcliff, W. A., "The Polaroid color-translating ultraviolet microscope," *Lab. Invest.,* **1**:123–128 (1952).
11. Zworykin, V. K., and C. Berkley, "Ultraviolet color-translating television microscopy," *Ann. N.Y. Acad. Sci.,* **97**(2):364–379 (1962).
12. Montgomery, P. O'B., ed., "Scanning techniques in biology and medicine," *Ann. N.Y. Acad. Sci.,* **97**(2):329–526 (1962).

See also OPTICAL MICROSCOPE, RESOLUTION.

URANIUM SALTS

Uranium (as uranyl) salts have been employed both in light microscopy (as in the formol-uranyl technique of Cajal) and in the following applications for electron microscopy.

A. As Positive Stains. *1. Staining of Ultrathin Sections.* (a) Using uranyl acetate: saturated aqueous solutions, at room temperature for 1–2 hr (1) or at 55–60° for 20 min (2,3); saturated solutions in ethanol, at room temperature (4) or at 37° (5;, for 30–90 min; 4–60% (w/v) solutions in methanol for 5–30 min (6). Alcohols are better solvents than water for uranyl acetate and enhance its penetration into the sections. (b) Weinstein et al. (7) have reported on the advantage of sodium zinc uranyl acetate and zinc uranyl acetate over uranyl acetate to stain nuclei of mouse pancreas prepared by freeze-substitution.

2. "En Bloc" Staining. The contrasting of specimens with uranyl salts before embedding can be carried out: (a) Before dehydration: This was first used by Strugger (8) and it is part of the Ryter-Kellenberger procedure (9) and of the rapid method described by Hayat (10). A treatment with a 0.5–2% solution of uranyl acetate in buffer or water, for 30–120 min, is usually employed. Here the staining effect of uranyl acetate is not important since much uranyl is leached out during the subsequent steps of preparation for electron microscopy. However, this treatment is useful because of its fixative action which has been described for nucleic acid-containing structures (9,10) and for procaryotic and eucaryotic cell membranes (11–13). (b) During dehydration: uranyl acetate (1–2%) is added to the usual 70% ethanol washing. A procedure for preferential staining of viruses in plant tissues, using a treatment with uranyl acetate during dehydration after glutaraldehyde fixation, has been described (14).

Staining with uranyl salts is frequently used in combination with lead staining. For the staining of sections, a treatment with uranyl acetate followed by lead (with an intermediary washing with water) is used (15). A procedure was described (16) for simultaneous "en bloc" staining with a mixture, in ethanol-acetone, of uranyl nitrate and lead acetate.

Uranyl ions are known to have high affinity to several chemical groups, as carboxyl and phosphate (17–21), and so may react with, and confer density to, many different cell components. When used to stain specimens fixed by the usual OsO_4 procedures, they produce an overall increase in contrast, with little specificity, except for collagen, which stains strongly (1,22). However, uranyl ions can stain preferentially nucleic acid-containing structures under certain conditions (22–25). The affinity of uranyl ions to nucleic acids seems to be due to the reaction with the phosphate groups (23,26). The high affinity for the phosphate groups of phospholipids (17,21) might explain the fixative action of uranyl ions on cell membranes. The strong staining of collagen seems to be due to a high concentration of polar groups. Glycogen is not significantly contrasted after uranyl staining of sections; moreover it is extracted from tissues during "en bloc" staining with aqueous (unbuffered) solutions of uranyl acetate. Since in these conditions ribosomes are strongly contrasted, they are easily distinguished from glycogen granules (27). The contrasting capacity of uranyl ions depends on the pH of the solutions, increasing when the low pH of aqueous unbuffered solutions (around 4 for 0.5% solutions, depending on the batches) is raised with NaOH or KOH or buffers (some buffers, as cacodilate and phosphate, are incompatible with uranyl salts). Different uranyl ions can be present, depending on the pH of the solutions (26). With uranyl acetate, formate, and nitrate, precipitation occurs at pH values above 5, due to the formation of uranyl hydroxide. This precipitation can be inhibited by EDTA and other complex-forming agents. Uranyl nitrate has a lower contrasting action as compared with uranyl acetate and so is not generally used in spite of its·higher stability.

B. As Negative Stains. Uranyl salts are commonly used as negative stains (see 28 for references). Uranyl acetate can be employed as a 0.5% aqueous solution (pH under 5); or as a uranyl acetate–EDTA complex (pH can be raised up to 7 with NaOH or KOH). Uranyl formate is used in the same way. Uranyl oxalate can be used at pH 5–7 without EDTA and is superior to uranyl acetate and uranyl formate in the study of proteins and viruses.

C. Added to Methacrylate Embedding Mixtures. Uranyl nitrate can be added to methacrylate embedding mixtures in order to reduce polymerization damage (29) or as a substitute for the usual benzoyl peroxide catalyst (30).

As the stability of aqueous or alcoholic solutions of uranyl salts, mainly acetate, is low, it is advisable to use fresh solutions and to centrifuge them before use. The solutions are sensitive to light and must be kept in the dark. Uranyl salts are poisonous and radioactive.

M. T. SILVA

References

1. Watson, M. L., *J. Biophys. Biochem. Cytol.*, **4**:475 (1958).
2. Brody, I., *J. Ultrastr. Res.*, **2**:482 (1959).
3. Sjöstrand, F. S., et al., *J. Ultrastr. Res.*, **7**:504 (1962).
4. Gibbons, I. R., et al., *J. Biophys. Biochem. Cytol.*, **7**:697 (1960).
5. Epstein, M. A., et al., *J. Cell Biol.*, **19**:325 (1963).
6. Stempak, J. G., et al., *J. Cell Biol.*, **22**:697 (1964).
7. Weinstein, R. T., et al., *J. Cell Biol.*, **19**:74A (1963).
8. Strugger, S., *Naturwiss.*, **43**:357 (1956).
9. Ryter, A., et al., *Z. Naturforsch.*, **13b**:597 (1958).
10. Hayat, M. A., *Tissue and Cell*, **2**:191 (1970).
11. Silva, M. T., et al., *Experientia*, **24**:1074 (1968).
12. Silva, M. T., et al., *Biochim. Biophys. Acta* [in press].
13. Terzakis, J. A., *J. Ultrastr. Res.*, **22**:168 (1968).
14. Hill, G. J., et al., *J. Ultrastr. Res.*, **25**:323 (1968).
15. Glauert, A. M., p. 254 in D. H. Kay, ed., "Techniques for Electron Microscopy," London, Blackwell, 1965.
16. Kushida, H., *6th Int Cong. Electron Microsc.*, Kyoto, **II**:39 (1966).
17. Bungenberg de Jong, H. G., p. 259 in H. R. Kruyt, ed., "Colloid Science," Vol. II, Amsterdam, Elsevier, 1949.
18. Passow, H., et al., *Pharmacol. Rev.*, **13**:185 (1961).
19. Parreira, H. C., *J. Colloid Sci.*, **20**:742 (1965).
20. Barton, P. G., *J. Biol. Chem.*, **243**:3884 (1968).
21. Shah, D., *J. Colloid and Interface Sci.*, **29**:210 (1969).
22. Marinozzi, V., *J. Roy. Microsc. Soc.*, **81**:141 (1963).
23. Huxley, H. E., et al., *J. Biophys. Biochem. Cytol.*, **11**:273 (1961).
24. Marinozzi, V., et al., *J. Ultrastr. Res.*, **7**:436 (1962).
25. Bernhard, W., *J. Ultrastr. Res.*, **27**:250 (1969).
26. Zobel, C. R., et al., *J. Biophys. Biochem. Cytol.*, **10**:335 (1961).
27. Vye, M. V., et al., *J. Ultrastr. Res.*, **33**:278 (1970).
28. Mellema, J. E., et al., *Biochim. Biophys. Acta* **140**:180 (1967).
29. Ward, R. T., *J. Histochem. Cytol.*, **6**:398 (1958).
30. Kushida, H., *J. Electronmicrosc.*, **11**:253 (1962).

UROCHORDATA

Fresh Mounts. Many details of urochordate or tunicate morphology and morphogenesis can be discerned by observing living specimens. The development of oviparous solitary ascidians and larvaceans can be followed continuously from fertilization through hatching in fresh mounts, e.g., *Ciona, Ascidia, Styela, Boltenia, Oikopleura*. A complete series of developmental stages of ovoviviparous compound ascidians may be obtained by teasing apart adult colonies, e.g., *Amaroucium, Distaplia, Diplosoma*. Ascidian larvae and stages of metamorphosis are favorable for extended observations or cinemicrographic analyses of morphogenetic movements (tail resorption, rotation of trunk, extension of ampullae, extension and retraction of adhesive papillae, emigration of blood cells into tunic). Bright field or oblique illumination is adequate for many observations but Nomarski differential interference microscopy is superior especially for high-resolution work. Phase contrast microscopy is excellent for sperm, blood cells, (Freeman, 1964) and isolated parts of organs, e.g., cardiac tissue. Rings of filter paper, squares of bolting silk, or plasticine feet placed under a coverslip can be used to obviate or control the compression of specimens. Cold water species (8–16°C) can be maintained in good condition for days, while under continuous observation, with the aid of a thermoelectric cooling stage and perfusion techniques (Cloney et al., 1970).

Paraffin and Celloidin Methods. Most adult and embryonic tissues of ascidians, thaliaceans, and larvaceans can be fixed satisfactorily for histological study in Bouin's fluid made up with sea water or 3 parts Bouin's fluid diluted with 1 part sea water (Scott, 1946; Cloney, 1961). Susa's, Flemming's, Gilson's, Zenker's, Schaudinn's, and other acidic fixatives have been employed by various investigators but none of these are significantly better than Bouin's and they are sometimes more trouble to use. Magnesium sulfate and menthol crystals are widely used as anesthetics to relax ascidians prior to fixation.

When many small specimens are to be embedded in paraffin and precise orientation is important, the following technique is useful. Stain the specimens lightly in an acid dye, e.g., light green, before complete dehydration. After infiltration, transfer them in a warmed pipette to a large drop of molten wax on a glass slide coated with a thin layer of glycerin. Transilluminate the slide under a dissecting microscope and orient the specimens with a hot needle. For transverse sections of ascidian tadpoles for example, the anterior-posterior axis of the animal can be oriented perpendicular to the plane of the glass surface. As many as a dozen $100\mu \times 1000\mu$ larvae can be arranged in a row within one drop of wax and later sectioned, mounted, and stained simultaneously.

As long as the specimens have recognizable axes of symmetry, a principal plane can be oriented parallel to the surface of the slide. After cooling and removal from the slide the flat surface of the planoconvex block serves as a permanent reference plane for later orientation of specimens on the microtome.

To prevent compression when very thin sections are required the following mixture is recommended:

Fisher's Tissuemat, m.p. 60–62°C, 300 g
bleached beeswax,   35 g
dry piccolyte,   45 g

Section in a cryostat or a cold room at 5–10°C (Cloney, 1961).

Epon Embedding for Light Microscopy. The quality of preservation of tunicate tissues obtained by using buffered OsO_4 solutions alone or aldehyde and OsO_4 solutions in double fixations, combined with epoxy embedding methods, is superior to any of the classical paraffin or celloidin techniques and should be used whenever possible. Shrinkage is significantly reduced and the organization of cellular structures is less severely disrupted. The methylene blue and azure II stain of Richardson, Jarrett, and Finke (1960) works well with most tissues. Fading can be retarded by mounting the sections in Cargille's HV immersion oil instead of resinous mounting media. Excellent whole mounts for Nomarski interference microscopy can be made by mounting OsO_4 fixed specimens in Epon directly on slides with coverslips and allowing them to polymerize in the oven.

Electron Microscopy. Finding a suitable fixative for electron microscopy (EM) is a trial and error process that may have to be repeated for different tissues and species. For morphological studies it is desirable to find a fixative that does not effect ruptured membranes, swollen organelles, large empty areas in the cytoplasmic matrix, clumping, excessive extraction, extensive conversion of tubular membrane systems into vacuoles, or the destruction of such things as microtubules, filaments, ribosomes, and inclusions. A technique that will work equally well with all urochordates or even different tissues or stages of development of a single species has not been found.

Osmium tetroxide buffered with sodium bicarbonate originally suggested by J. H. Luft is a satisfactory formula

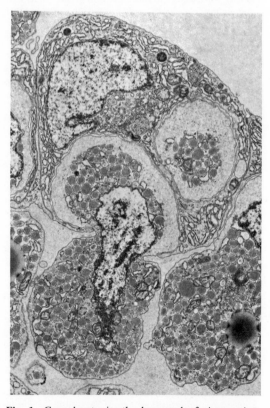

Fig. 1 Granulocyte in the hemocoel of *Amaroucium constellatum* penetrating an epidermal cell during metamorphosis. OsO_4-bicarbonate fixation (From Cloney and Grimm, 1970).

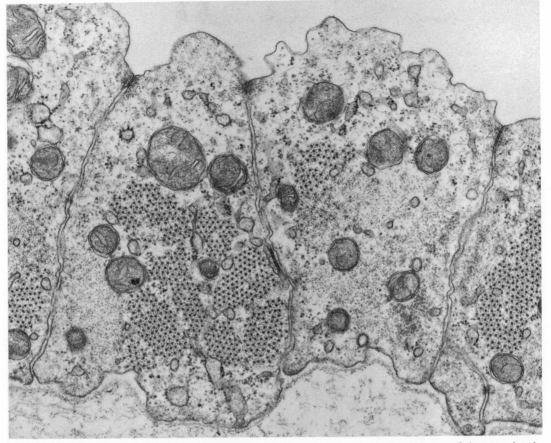

Fig. 2 Transverse section of the myoepithelial part of the ascidian heart showing elements of the sarcoplasmic reticulum in association with arrays of myofilaments. *Corella willmeriana.* Phosphate-buffered glutaraldehyde, post-fixed in bicarbonate-buffered OsO₄. (*Courtesy of L. W. Oliphant, University of Washington.*)

for many tissues. It is recommended for initial investigations of ascidian larvae, stages of metamorphosis (Cloney, 1966, 1969; Cloney and Grimm, 1970), and for adult tissues (Fig. 1). It is useless for early embryonic stages of the aplousobranchs (*Amaroucium, Distaplia* and *Diplosoma*) but satisfactory for the embryos of phlebobranchs (*Ascidia, Chelyosoma, Ascidia, Corella*) and stolidobranchs (*Boltenia, Styela*).

Bicarbonate-Buffered Osmium
2 1/2% sodium bicarbonate,   1 part
buffer adjusted to pH 7.2 with 1*N* HCl
4% osmium tetroxide,   1 part

Sodium bicarbonate buffer is unstable at atmospheric pressures and should be adjusted to the proper pH just before use. After mixing, the pH will rise to about 7.4. Fixations should be carried out in closed containers. Fix in an ice bath for 15 min to 1 hr. Briefly rinse tissues in

water, dehydrate in ethanol, transfer through several changes of propylene oxide, and infiltrate for 5–18 hr sequentially in 1 : 1 and 1 : 3 mixtures of propylene oxide and Epon (Luft, 1961). Embed in Epon. This fixative produces fewer ruptured membranes and does not extract the tissues as much as the more stable S-collidine buffer. Phosphate-buffered OsO₄ (Eakin and Kuda, 1971; Kessel, 1966; Pucci-Minafra, 1965) has also been successfully employed, but Dalton's fixative (Kalk, 1970), OsO₄ buffered with veronal-acetate (Dilly, 1969), and OsO₄ in sea water (Berrill and Sheldon, 1964; Ursprung and Schabtach, 1965) produce extensive artifacts.

Since the introduction of glutaraldehyde as a fixative for electron microscopy it has been used with many combinations of buffers and additives. One of the most satisfactory fixatives for tunicates employs a phosphate buffer and is made slightly hyperosmotic to sea water with sodium chloride (modified from a procedure developed by Dr. Helen Pianka for ctenophores, 1966).

Phosphate-Buffered Glutaraldehyde

Stock Solutions		Final Concentrations	Approximate Osmolality
25% purified glutaraldehyde,	1 part	2.5%	250 mOsm
0.34*M* sodium chloride,	4 parts	0.14*M*	275 mOsm
0.4*M* phosphate buffer,	5 parts	0.2*M*	425 mOsm
			950 mOsm

Fix about 1 hr at room temperature. Rinse briefly in a small amount of the postfixing solution. Postfix in bicarbonate-buffered OsO₄ solution for about 1 hr. Rinse in water and dehydrate in ethanol. Transfer through several changes of propylene oxide and infiltrate for 2–10 hr sequentially in a 1 : 1 and 1 : 2 mixtures of Epon and propylene oxide. Embed in Epon.

Millonig's 0.4M Phosphate Buffer (pH 7.42)
monosodium phosphate (M.W. 138), 11.04 g
sodium hydroxide,   2.68 g
distilled water,   200 ml
Approximately 6.7 ml of 10N sodium hydroxide can be substituted for the dry reagent.

This fixative has provided excellent preservation of the tissues of the larvacean *Oikopleura dioica*, larvae of aplousobranchs, and many post-larval and adult ascidian tissues. When the same fixative was used on the larvae of the stolidobranch *Boltenia villosa*, however, the results were unacceptable and greatly inferior to bicarbonate-buffered OsO₄ alone.

Modifications of the basic formula have been employed by Oliphant and Cloney, 1972, in their study of ascidian cardiac muscle. They found that reducing the concentration of glutaraldehyde to 1 % and lowering the osmolality to 600 mOsm gave superior preservation of the ground substance, organelles, and sarcolemma (Fig. 2).

We have recently obtained excellent preservation of the embryos and larvae of *Distaplia* with the following fixative; 2% glutaraldehyde, 0.2M cacodylate buffer, 0'27M sucrose and 0.05 % ruthenium red (960 mOsm; pH 7.4). Tissues should be post-fixed in bicarbonate-buffered OsO₄. (Cavey and Cloney, 1972).

RICHARD A. CLONEY

References

Berrill, N. J., and H. Sheldon, *J. Cell Biol.,* **23**:644–669 (1964).
Cavey, M. J., and R. A. Cloney, J. Morph., *in press* (1972).
Cloney, R. A., *Amer. Zool.,* **1**:67–87 (1961).
Cloney, R. A., *J. Ultrastruc. Res.,* **14**:300–328 (1966).
Cloney, R. A., *Z. Zellforsch.,* **100**:31–53 (1969).
Cloney, R. A., and L. Grimm, *Z. Zellforsch.,* **107**:157–173 (1970).
Cloney, R. A., J. Schaadt, and J. V. Durden, *Acta Zool.,* **51**:95–98 (1970).
Dilly, P. N., *Z. Zellforsch.,* **95**:331–346 (1969).
Eakin, R. M., and A. Kuda, *Z. Zellforsch.,* **112**:287–312 (1971).
Freeman, G., *J. Exp. Zool.,* **156**:157–184 (1964).
Kalk, M., *Tissue and Cell,* **2**:99–118 (1970).
Kessel, R. G., *Z. Zellforsch.,* **71**:525–544 (1966).
Luft, J., *J. Cell. Biol.,* **9**:409–414 (1961).
Oliphant, L. W. and R. A. Cloney, *Z. Zellforsch.* **129**: 395–412 (1972).
Pucci-Minafra, I., *Acta Embryo et Morph. Exp.,* **8**:289–305 (1965).
Richardson, K. C., I. Jarett, and E. H. Finke, *Stain Technol.* **35**:313–323 (1960).
Scott, F. M., *Biol. Bull.,* **91**:61–80 (1946).
Ursprung, H., and E. Schabtach, *J. Exp. Zool.,* **159**:379–384 (1965).

V

VIRUS

Virus Particles

Available Techniques. For many years the ultramicroscopic size of virus particles was one of the criteria used in their definition. Only the very largest viruses can be seen in the light microscope and then only when the best optical techniques are employed. Consequently most of the information on the structure of viruses and their interaction with living cells has been obtained from the electron microscope.

Viruses contain few, if any, elements of significant electron density, and a straightforward examination of a virus preparation gives little information on size, shape, or structure of the particles. Shadow-casting the specimens from one direction with a suitable heavy metal, alloy or Pt/C enables the particle size to be measured and the shape to be defined (e.g., sphere, cube, brick) Shadow-casting successively from two different directions can be used to distinguish the structure of some icosahedral viruses from other possible geometrical forms (1). Although shadow-casting, by coating the particle surface with metal, may limit the amount of detail seen on the surface, subunit arrays are visible on the surface of shadowed preparations of some viruses (2, 3).

To be effective, shadow-casting requires highly purified preparations containing little or no cell debris. This limitation has resulted in the technique being largely superseded by the negative staining technique (4) which, although optimal in pure preparations, is amazingly effective in partially purified and even crude material. Basically, the technique does not aim to stain the particle but to surround and thus delineate the particle with an electron-dense substance. The feature of negative staining which has led to its universal application in virology is that the stains used, as well as defining the size and shape of the particles, also reveal a wealth of surface detail on the particles. This surface morphology has been shown to be characteristic for each virus group. Thus, negative staining of a virus preparation can reveal the surface architecture of the particle, define its virus group, and distinguish it from other virus groups, an ability which has led to negative staining being used in clinical diagnostic work (5).

Methodology. A particular feature of negative staining is the speed and ease with which it can be carried out. Early techniques involved mixing the virus and stain and spraying drops onto a coated electron microscope mount. Most microscopists now simply transfer a drop of virus suspension to the mount with either a bacteriological loop or a fine pasteur pipette, invert the mount onto a drop of the negative stain of choice, remove after 10 sec, and dry off excess stain with filter paper. Only very small quantities of virus suspension are thus required, but in consequence there must be a high concentration of virus in the original sample. When using the bacteriological loop method, for example, at least 10^9 particles/ml are required for ready visualization of the virus. For the best results an even spreading of the virus suspension over the mount is essential before negative staining. When the virus suspension remains as a small drop on one area of the mount it may be necessary to add a small quantity of protein (usually bovine serum albumen) to facilitate spreading.

Numerous heavy metal salts have been tested as negative stains, but the compound originally used, the potassium salt of phosphotungstic acid (P.T.A.) still remains the best choice for general use. P.T.A. is normally used as a 2% solution neutralized with potassium hydroxide to a pH of 7.2–7.4. More contrast in the specimen is provided by uranyl acetate (1%), but the pH cannot be raised above 5.5. Uranyl formate, although little different from uranyl acetate, can often reveal detail unresolved by either P.T.A. or uranyl acetate (6). However, this compound is unstable in solution and a precipitate forms at above pH 5.5. Ammonium molybate affords the poorest contrast of the four stains discussed here but is becoming increasingly used because of its high tonicity compared to other negative stains (7). In 3% solutions the tonicity is comparable to that of $0.44 M$ sucrose, a level often used in the biochemical isolation of biological components. Hypotonic damage to labile components e.g., viral membranes, isolated under high-tonicity conditions is thus much reduced by this negative stain. The degree of contrast and detail provided by each of these negative stains is illustrated in Fig. 1. The structure of the carbon support films normally used for negative staining work can contribute significantly to the final observed image. Where the finest detail is sought, this contribution can be removed by using holey carbon films (8) and observing the virus supported in only a film of stain over a hole (Fig. 2) (9).

Image Interpretation. It is obvious from the published literature that negative staining can reveal extremely fine detail on the surface of virus particles. How real is this detail? It is always assumed that regularity and symmetry are associated with meaningful structures and that randomness and disorder are equated with artifact. The regularity and symmetry of virus particles seen in negative stains would thus imply that the particles are little distorted from their *in vivo* state. Some credence is given to this view by the observtion that some viruses even after

negative staining and desiccation in the electron microscope retain a degree of biological activity. In addition
X-ray studies of tobacco mosaic virus (10, 11) and turnip
yellow mosaic virus (12, 13) have confirmed that the

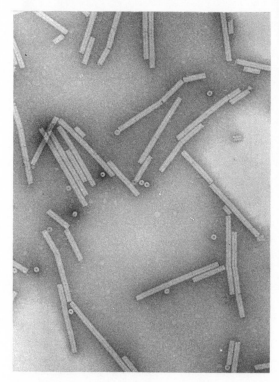

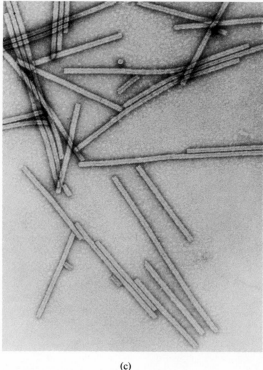

(c)

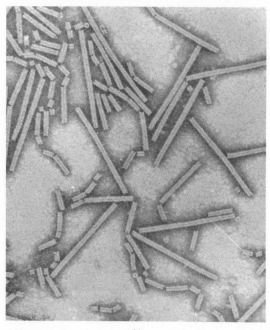

(a)

(b)

Fig. 1 Particles of tobacco mosaic virus in three negative
stains; (a) ammonium molybdate 2%, (b) potassium
phosphotungstate 2%, (c) uranyl acetate 1%. Areas were
photographed at 40,000× and printed under identical
conditions. (×88,000.)

symmetries and substructure revealed by negative staining
are also present outwith the negatively-stained, desiccated
preparations used for electron microscopy.

This does not mean that every image seen is an absolutely faithful representation of the *in vivo* particle.
Desiccation and pH effects can significantly alter the
images of some viruses. For example, T2 tail fibers when
examined in P.T.A. (pH 7.2) are attached to the collar,
when examined in uranyl acetate (pH 5.2) they are
detached and loose (Fig. 3). At first sight, vaccinia virus
apparently exists in two forms, an M and a C form (Fig.
4). However, it has been shown that M can be converted
to C merely by leaving the negatively stained mount
sitting on the bench (14) and that C can be converted to
M by rehydration (15). Most icosahedral viruses retain
their structural integrity on negative staining and no
prefixation is required, but it has been noted that in some
viruses with an outer envelope, this envelope may be
considerably deformed unless prefixation is carried out
(16). Thus, whenever a completely new object is to be
examined by negative staining it is advisable to try all
available stains and to test the effect of prefixation on the
resulting image.

It was generally asssumed for many years that the surface
morphology revealed by negative staining was that of one
surface of the particle, usually the upper surface. This has
now been shown to be true of only a small proportion of

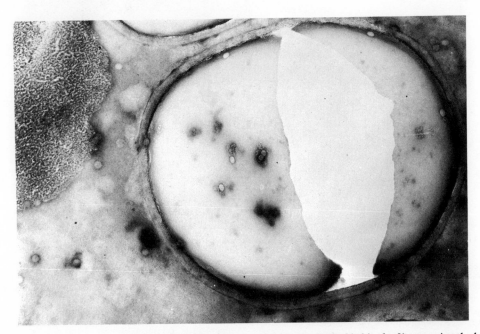

Fig. 2 A hole in a carbon film partly covered by a film of uranyl acetate. Embedded in the film covering the hole are particles of the minute virus of mouse (MVM). Particles can also be seen on the carbon support film adjacent to the hole. (×65,000).

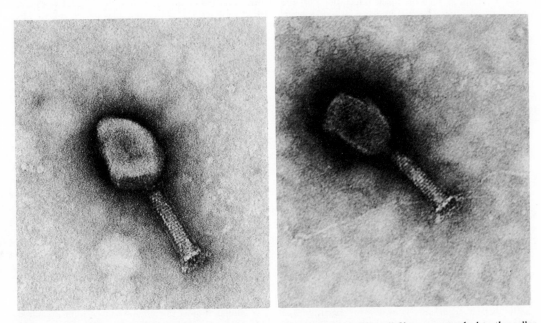

Fig. 3 Left: T2 bacteriophage particle negatively stained in uranyl acetate. The tail fibers are attached to the collar. Right: T2 bacteriophage particle negatively stained in potassium phosphotungstate. The tail fibers are now detached from the collar. (×210,000.)

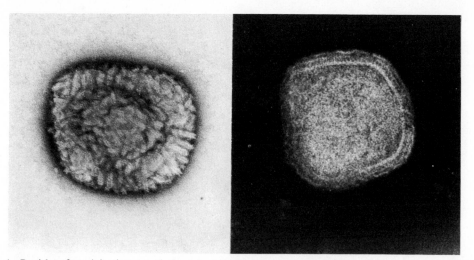

Fig. 4   Particles of vaccinia virus negatively stained in uranyl acetate; left: M-form, right: C-form. ($\times$131,000.)

particles in any preparation. In most particles, the final image is formed by contributions from both upper and lower surfaces, not usually in equal amount (17, 18). This leads to considerable complexities in the interpretation of the negatively stained image and can lead to completely erroneous conclusions. Where the structure observed has a regular repeating feature, e.g., in a helical virus or a bacteriophage tail, the contributions of upper and lower surface can be separated by optical diffraction techniques (19) and reconstructed images formed using either upper or lower contribution (19). If it is suspected that either a linear periodicity or a rotational symmetry element is present in an image, then the linear integration (20) and rotational techniques (21) can be a useful aid in interpretation. These techniques tend to emphasize any regularity present in the image, but great care must be exercised in interpreting the results obtained as spurious structure can be generated especially with the rotational technique. In recent years it has been shown that three-dimensional images of virus particles can be constructed from electron micrographs by applying the analytical techniques of X-ray crystallography to the negatively stained image (22, 23). Although hardly a routine technique, this method can extract, compute, and present all available information from the image of the virus particle. Even with such sophisticated techniques the best resolution attainable with negative staining is 15–20 Å, a long way short of the 2–3 Å resolution now available on commercial electron microscopes.

Particle Counting. The unique surface structure of many virus particles makes them readily distinguishable from other cell organelles and debris and thus permits identification of the virus even in crude, unpurified preparations. This feature enables the particles not only to be seen but also to be counted. Several techniques have been used to count particles in the electron microscope. Most employ latex spheres (24) as a standard whose concentration per milliliter is accurately known and whose image is sufficiently distinct from any known virus particle. The relation of the number of virus particles to the number of latex particles seen in a sample from a mixture of equal volumes of virus, latex, and stain is evaluated. The

accuracy of the figure obtained has been estimated at $\pm 30\%$ (25). One useful advantage of this technique is that damaged and/or defective particles which are not assayable in an infectivity assay can be quantified.

Virus Nucleic Acids

Methodology. Electron microscopy can also be used to examine viral nucleic acids. If a suspension of nucleic acid is examined by the normal procedures of shadowing or negative staining, what little nucleic acid is seen is usually stretched, tangled, and broken. If the nucleic acid is first allowed to adsorb to a film of denatured protein formed on the surface of an ammonium acetate solution (the hypophase), areas of this film picked up, dried in alcohol, and shadowed, then the nucleic acid molecules remain discrete and untangled and have a consistent length from preparation to preparation (Fig. 5). This method of visualizing nucleic acids, the protein monolayer method (26, 27), can be applied to both double-stranded (28, 29) and single-stranded forms (30, 31), although less consistent results are obtained with the single-stranded forms (Fig. 6). To obtain the best results it is necessary to shadow the preparation from at least two directions or while rotating so that all the devious contours of the nucleic acid molecules are equally visible. It is also possible to positively stain the molecules with, e.g., uranyl acetate (32) (Fig. 7).

Molecular Weight Determinations. Examination of numerous double-stranded DNAs has shown that under controlled conditions the molecular weight per unit length of these molecules is independent of their origin (33). Thus the molecular weights of unknown DNAs can be determined from a measurement of their length. Care must be exercised in interpreting the figures obtained, as many factors including the type of protein and the ionic strength of the hypophase can affect the length of the DNA (34). To minimize these effects a standard DNA of known length should be included, if possible, in the same preparation as the unknown DNA to be measured.

Secondary and Tertiary Structure. The protein monolayer technique, by permitting ready visualization of a DNA molecule, provides information on its tertiary

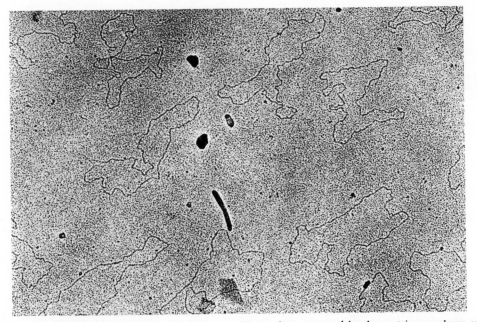

Fig. 5 Circular double-stranded DNA from human papilloma virus, prepared by the protein monolayer method, shadowed with Pt. (×48,000.)

structure, i.e., whether linear, circular, or supercoiled, and on the interconvertability between these various forms (35). Suitable pretreatment of the molecules can also provide some further insight into the secondary structure of double-stranded molecules. Partial denaturation, for example, can cause specific strand separation at regions of high A+T content along a molecule. These regions are readily seen and their positions mapped (36, 37). Where two species of DNA are present with a G+C content difference of at least 6%, the molecules can be distinguished by partial denaturation, the molecule with the lower G+C content showing denatured regions

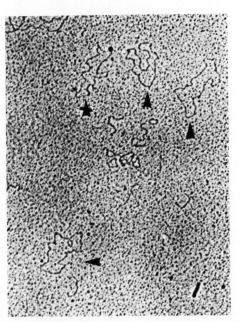

Fig. 6 Circular single-strained DNA from bacteriophage φ X174, shadowed with Pt. (×48,000.)

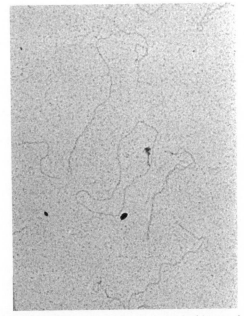

Fig. 7 Circular, supercoiled, and linear double-stranded DNA from polyoma virus prepared by the protein monolayer method and positively stained with uranyl acetate.

before the molecule of higher $G+C$ content (38). Electron microscopic examination of DNA can also be used to map deletion mutants (39). When a mixture of mutant and wild-type DNA is denatured and reannealed, a proportion of the molecules result from a combination of wild-type and mutant strands. Single-stranded loops occur in such molecules, their position and length corresponding to the original deletion. In several double-stranded viral DNA molecules the nucleotide sequences at the beginning of the molecule are repeated at the end (terminal repetition). Specific enzyme digestion and subsequent annealing of such molecules can produce circular molecules which are readily visualized by the protein monolayer method (40). Other viral DNAs are known in which the nucleotide sequences of the molecules are circular permutations of a common sequence. Such molecules can be caused to assume a circular form by alkali denaturation followed by renaturation at neutral pH (41). Electron microscope examination of molecules so treated can simply determine whether or not that particular DNA is circularly permuted.

E. A. C. FOLLETT

References

1. Williams, R. C., and K. M. Smith, "The polyhedral form of the tipula iridescent virus," *Biochim. Biophys. Acta,* 28:464–469 (1958).
2. Bils, R. F., and C. E. Hall, "Electron microscopy of wound-tumor virus," *Virology,* 17:123–130 (1962).
3. Hall, C. E., E. C. Maclean, and I. Tessman, "Structure and dimensions of bacteriophage φX174 from electron microscopy," *J. Mol. Biol.,* 1:192–194 (1959).
4. Brenner, S., and R. W. Horne, "A negative staining method for high resolution electron microscopy of viruses," *Biochim. Biophys. Acta,* 34:103–110 (1959).
5. Cruickshank, J. G., H. S. Bedson, and D. H. Watson, "Electron microscopy in the rapid diagnosis of smallpox," *Lancet,* September 3, 527–530 (1966).
6. Finch, J. T., "The resolution of the substructure of tobacco mosaic virus in the electron microscope," *J. Mol. Biol.,* 8:872 (1964).
7. Muscatello, U., and R. W. Horne, "Effect of tonicity of some negative-staining solutions on the elementary structure of membrane-bounded systems," *J. Ultrastruct. Res.,* 25:73–83 (1968).
8. Harris, W. J., "Holey films for electron microscopy," *Nature,* 196:499–500 (1962).
9. Huxley, H. E., and G. Zubay, "Electron microscope observations on the structure of microsomal particles from *E. coli,*" *J. Mol. Biol.,* 2:10–18 (1960).
10. Klug, A., and D. L. D. Caspar, "The structure of small viruses," *Adv. Virus Res.,* 7:225–325 (1960).
11. Klug, A., and J. E. Berger, "An optical method for the analysis of periodicities in electron micrographs," *J. Mol. Biol.,* 10:565–569 (1964).
12. Klug, A., W. Longley, and R. Leberman, "Arrangement of protein subunits and the distribution of nucleic acid in turnip yellow mosaic virus. I." *J. Mol. Biol.,* 15:315–343 (1966).
13. Finch, J. T., and A. Klug, "Arrangement of protein subunits and the distribution of nucleic acid in turnip yellow mosaic virus. II.," *J. Mol. Biol.,* 15:344–364 (1966).
14. Westwood, J. C. N., W. J. Harris, H. T. Zwartouw, D. H. J. Titmuss, and G. Appleyard, "Studies on the structure of vaccinia virus," *J. Gen. Microbiol.,* 34:67–78 (1964).
15. Harris, W. J., and J. C. N. Westwood, "Phosphotungstate staining of vaccinia virus," *J. Gen. Microbiol..* 34:491–495 (1964).
16. Levy, J. P., M. Boiron, S. Silvestre, and J. Bernard, "The ultrastructure of Rauscher virus," *Virology* 26:146–150 (1965).
17. Klug, A., and J. T. Finch, "Structure of viruses of the papilloma-polyoma type. I. Human wart virus," *J. Mol. Biol.,* 11:403–423 (1965).
18. Finch, J. T., and A. Klug, "Structure of viruses of the papilloma-polyoma type. III. Structure of rabbit papilloma virus," *J. Mol. Biol.,* 13:1–12 (1965).
19. Klug, A., and D. J. De Rosier, "Optical filtering of electron micrographs—reconstruction of one-sided images," *Nature,* 212:29–32 (1966).
20. Markham, R., J. H. Hitchborn, G. J. Hills, and S. Frey, "The anatomy of the tobacco mosaic virus," *Virology,* 22:342–359 (1964).
21. Markham, R., S. Frey, and G. J. Hills, "Methods for the enhancement of image detail and accentuation of structure in electron microscopy," *Virology* 20:88–102 (1963).
22. De Rosier, D. J., and A. Klug, "Reconstruction of 3D structures from electron micrographs," *Nature,* 217:130–134 (1968).
23. Crowther, R. A., L. A. Amos, J. T. Finch, D. J. De Rosier, and A. Klug, "Three dimensional reconstruction of spherical viruses by Fourier synthesis from electron micrographs," *Nature,* 226:421–425 (1970).
24. Diagnostic Products, The Dow Chemical Co., 2030 Abbott Road Center, Midland, Michigan 48640. U.S.A.
25. Watson, D. H., "Electron-micrographic particle counts of phosphotungstate-sprayed virus," *Biochim. Biophys. Acta,* 61:321–332 (1962).
26. Kleinschmidt, A. K., and R. K. Zahn, "Ober deoxyribonucleinsaure molekeln in protein-mischfilmen," *Z. Naturforsch.,* 146:730 (1959).
27. Kleinschmidt, A. K., "Monolayer techniques in electron microscopy of nucleic acid molecules," pp. 361–376 in "Methods in Enzymology," Vol. XIIB, L. Grossman and K. Moldave, ed., New York, Academic Press, 1968.
28. Kleinschmidt, A. K., S. J. Kass, R. C. Williams, and C. A. Knight, "Cyclic DNA of slope papilloma virus," *J. Mol. Biol.,* 13:749–756 (1965).
29. Dunnebacke, T. H., and A. K. Kleinschmidt, "Ribonucleic acid from reovirus as seen in protein monolayers by electron microscopy," *Z. Naturforsch.,* 22:159–164 (1967).
30. Freifelder, D., A. K. Kleinschmidt, and R. L. Sinsheimer, "Electron microscopy of single stranded DNA: Circularity of DNA of bacteriophage φX 174," *Science,* 146:254–255 (1964).
31. Granboulan, N., and M. Girard, "Molecular weight of poliovirus ribonucleic acid," *J. Virol.,* 4:475–479 (1969).
32. Gordon, C. N., and A. K. Kleinschmidt, "High contrast staining of individual nucleic acid molecules," *Biochim. Biophys. Acta,* 155:305–309 (1968).
33. Thomas, C. A., and L. A. Machattie, "The anatomy of viral DNA molecules," *Ann. Rev. Biochem.,* 36:485–518 (1967).
34. Lang, D., H. Bujard, B. Wolff, and D. Russel, "Electron microscopy of size and shape of viral DNA molecules in solutions of different ionic strengths," *J. Mol. Biol.,* 23:163–181 (1967).
35. Follett, E. A. C., and L. V. Crawford, "Electron microscope denaturation of human papillomavirus DNA," *J. Mol. Biol.,* 28:455–459 (1967).
36. Inman, R. B., "A denaturation map of the λ phage DNA molecule determined by electron microscopy," *J. Mol. Biol.,* 18:464–476 (1966).

37. Follett, E. A. C., and L. V. Crawford, "Electron microscope study of the denaturation of human papilloma virus DNA," *J. Mol. Biol.,* **28**:461–467 (1967).

38. Follett E. A. C., and L. V. Crawford, "Electron microscope study of the denaturation of polyoma virus DNA," *J. Mol. Biol.,* **34**:565–573 (1968).

39. Westmoreland, B. C., W. Szybalski, and H. Ris, "Mapping of deletions and substitutions in hetero-duplex DNA molecules of bacteriophage λ by electron microscopy," *Science,* **163**:1343–1348 (1969).

40. MacHattie, L. A., D. A. Ritchie, C. A. Thomas, and C. C. Richardson, "Terminal repetition in permuted T2 bacteriophage DNA molecules," *J. Mol. Biol.,* **23**:355–363 (1967).

41. Thomas, C. A., and L. A. MacHattie, "Circular T2 DNA molecules," *Proc. Nat. Acad. Sci.,* **52**:1297–1301 (1964).

W

WHOLE MOUNTS (BOTANICAL)

The anatomy of higher plants is commonly studied by examining thin slices of organs; such sections are essentially two-dimensional. A three-dimensional approach is to render entire organs (or thick slabs cut from organs) transparent; these may then be immediately studied, or selectively stained to emphasize certain tissues or cell types. Such cleared preparations, aside from permitting the internal architecture to be viewed *in situ*, can be processed easily in large numbers without special equipment or the great expenditure of time needed to make paraffin sections. Leaves, flower parts, many herbaceous stems, and small roots can be routinely cleared. Fruits and thick or heavily lignified organs may clear adequately after slicing into thinner slabs, but clearing techniques also have limitations.

There are numerous published recipes for making cleared preparations; the major variations have been referred to by Lersten (1967). Virtually all processes involve the flushing out of the cell contents followed by dehydration and selective staining by one or two dyes. An exception to this is the method of Galavazi (1965), in which fixed material is cleared without loss of cell contents and examined. It may then be re-embedded in wax and sectioned conventionally. Organs cleared by other methods can also be re-embedded and sectioned, if the cytoplasm is not important.

The simplest method of clearing, which yields rather crude preparations good for gross observations, is to put fresh or preserved material into 75% lactic acid and keep in a 54°C oven until clear (Simpson, 1929). Permanent mounts are possible by ringing the coverslip.

The following, more involved method of clearing adapted from Shobe and Lersten (1967), is effective for a wide range of material. Good results can be obtained from living, preserved, or herbarium material. Use Petri dishes, syracuse watch glasses, or similar containers. Transfer from dish to dish may be made by a spatula, or only one dish can be used and the various reagents merely siphoned off carefully and replaced.

1. Place fresh material in 95% ethanol for about one hour; leave green organs until chlorophyll is mostly gone. Omit this step for dried or liquid-preserved material and start with step 2.
2. Aqueous NaOH (5–10 g/100 ml distilled water). Time varies from a day to a few weeks. Leaching of cell contents is hastened by oven-heating at, e.g., 50°C. If NaOH becomes discolored, pour off and replace.

If dark areas persist, transfer to full-strength chlorine bleach for 2–5 min (watch carefully because some tissues tend to disintegrate).
3. Rinse with three changes of distilled water (about 5 min each).
4. Aqueous chloral hydrate (250 g/100 ml distilled water) for a minimum of several hours. Tissues should become transparent except for lignified areas. Materials may be examined and stored indefinitely in this solution.
5. Repeat step 3. Do not place fragile material directly into pure water, but pass through a dilution series to avoid possible damage from mixing currents.
6. Dehydrate to 95% ethanol. Pass fragile material through a graded series, but most mature material can be placed directly into 95% ethanol (three changes, 5 min each) without harm.
7. Stain for a few seconds to a few minutes in fast green (0.5 g/100 ml 95% ethanol).
8. Rinse briefly in 95% ethanol to remove excess dye.
9. Two changes of 100% ethanol (5 min each).
10. Equal parts of xylene: 100% ethanol (2–5 min).
11. Stain for 2–10 min with safranin O (1 g/100 ml in equal parts of a xylene: 100% ethanol solution).
12. Destain to desired intensity in equal parts of xylene: 100% ethanol.
13. Place in xylene to stop destaining, but change xylene immediately to avoid safranin precipitation.
14. Examine, and if stain is too intense, repeat steps 12 and 13. If stain is not intense enough, pass material backwards in the sequence and restain.
15. Cleared and stained material may be stored unmounted in xylene, but more commonly is permanently mounted on slides using any xylene-soluble mounting medium (e.g., Piccolyte). Plastic embedding is also possible (Vágás, 1969).

Modification of this procedure, particularly regarding times, is frequently necessary for specific items. Only one of the dyes may be used, or other dyes may be substituted. For example, chlorazol black E (0.5 g/100 ml 100% ethanol) is excellent for epidermis and thin leaves. It frequently works well in combination with safranin. Morley (1949) reported the results of 20 different stain combinations used on cleared leaves.

Since the compilation by Lersten (1967), a few papers treating new aspects of clearing technique have appeared (Rodin and Davis, 1967; Morley, 1968; Boke, 1970; Fisher, 1970).

NELS R. LERSTEN

References

Boke, N. H., "Clearing and staining plant materials with lactic acid and pararosaniline hydrochloride," *Proc. Okla. Acad. Sci.*, **49**:1–2 (1970).

Fisher, J. E., "Staining and clearing vascular tissue in Gramineae; a procedure suited for classroom use," *Stain Technol.*, **45**:93–95 (1970).

Galavazi, G., "Clearing and staining plant material *in toto* with phloroglucinol-HCl in methyl benzoate for projection photography and subsequent serial sectioning," *Stain Technol.*, **40**:1–5 (1965).

Lersten, N. R., "An annotated bibliography of botanical clearing methods," *Iowa State J. Sci.*, **41**:481–486 (1967).

Morley, T., "Staining of plant materials cleared in NaOH," *Stain Technol.*, **24**:231–235 (1949).

Morley, T., "Accelerated clearing of plant leaves by NaOH in association with oxygen," *Stain Technol.*, **43**:315–319 (1968).

Rodin, R. J., and R. E. Davis, "The use of papain in clearing plant tissues for wholemounts," *Stain Technol.*, **42**:203–206 (1967).

Shobe, W. R., and N. R. Lersten, "A technique for clearing and staining gymnosperm leaves," *Bot. Gaz.*, **128**:150–152 (1967).

Simpson, J. L. S., "A short method of clearing plant tissues for anatomical studies," *Stain Technol.*, **4**:131–132 (1929).

Vágás, E., "Aufhellung und Einschluss von Pflanzenteilen mit Hilfe von Kunstharzen," *Mikroskopie*, **24**:274–277 (1969).

WHOLEMOUNTS, ZOOLOGICAL MATERIALS

The term wholemount indicates the preservation of an entire, or at least unsectioned object in permanent form. Wholemounts may be prepared as dry wholemounts, in aqueous media, or in resinous media.

Dry Wholemounts. Dry wholemounts are very rarely used today. Their simplest form is that employed by students of micropaleontology, who attach small objects to a 2 × 1 rectangle of black card with any suitable adhesive and bind this under a slide or coverslip. Individual objects may similarly be mounted by attaching a circle of black paper to an ordinary slide and laying on top of this a "cell," which is in effect a washer of the diameter of the coverslip to be used and with a wall thickness suitable to the specimen. These were at one time obtainable in many forms but must now usually be cut by the individual mounter either by sawing sections from suitable tubing or by stamping washer-like forms from cards with ordinary punches.

It is a pity that this form of mounting has fallen into desuetude, since for microfossils, minute insects, many botanical specimens, and indeed, anything that can be dried, is better than any other form of mount. Considerable details of the method are given and several typical examples described in great detail, in Gray, 1954.

Wholemounts in Aqueous Media. There are three types of these; those in liquid preservatives, those in gum-like media, and those in gels. The practice of mounting in fluid preservatives can now be considered a lost art. It is, in fact, the only method by which such invertebrates as rotifers can be preserved in anything like their natural state and there are many other objects that can best be prepared in this manner. Those who would like to revive the art are again referred to Gray, 1954.

Wholemounts in Gum Media. Formulas for many mountants of this type will be found in the article MOUNTANT elsewhere in this volume. They should be more widely known and are even now used extensively by entomologists. They consist for the most part of dispersions of gum arabic in various media designed to preserve the material and to vary the refractive index. Their use is so simple that it can be described in a few words. Any object that is either dry, or in an aqueous fluid, is merely picked up, placed on a slide, a drop of the mountant placed on top, and a coverslip added. The most frequently used is the medium of Berlese (see article referred to), which permits the preparation of excellent whole mounts of insects with no difficulty.

Wholemounts in Gelatin Media. These are also becoming old-fashioned but were at one time in wide use by botanists. They differ from the media described in the last paragraph in that their gelatin content causes them to solidify at room temperature. Almost all have a high glycerol content, and objects are mounted in them after having been exposed to gradually increasing concentrations of this reagent in water. They are primarily of interest for the mounting of algae, and detailed descriptions of this technique will be found in Gray, 1954.

Resinous Media. Resinous media are used for whole mounts not only because they permit mounting stained objects but more particularly because they impart to the specimen a great degree of transparency. This transparency comes from the increase in the index of refraction when the specimen is completely impregnated with the resin. These resins are not, however, miscible with water, hence the water must first be removed (dehydration) and then the dehydrant replaced with some material (clearing agent) with which the resin itself is miscible. Before these operations, invertebrates, the usual subject of whole mounts, must be killed and hardened ("fixed"), and it is customary to stain the specimen in order to bring out those internal structures which would become invisible, were they not colored, through the increase in transparency. All of the following operations must, therefore, be conducted and will be discussed in turn:

1. Narcotizing and fixing
2. Staining
3. Dehydrating
4. Clearing
5. Mounting

Narcotizing and Fixing Specimens. Hard objects such as small arthropods, hairs, and the like may be dehydrated and mounted directly into resinous media. Most objects which are mounted in resinous media are, however, too soft to withstand the process of dehydration and clearing without special treatment. Though hardening and fixing agents were once considered separate, they are now usually combined into a solution known as a fixative. Before dealing with fixatives, however, it is necessary to point out that few small animals, on being plunged into a fixative, will retain their shape, so it is necessary first to narcotize them is some solution which will render them incapable of muscular contraction.

Narcotization may be caused either through the blocking of nerve impulses which cause contraction, or by some treatment which will inhibit the actual contraction of the muscle. For blocking nerve impulses there are a wide range of narcotics available, and making a choice between them must be a matter of experience. It is to be recommended, in the absence of experience, that one of the solutions containing cocaine be first tried, since cocaine is the nearest approach to a universal narcotic known to the author. Should cocaine not be available, crystals of menthol may be sprinkled on the surface of the water containing the specimen. It is very important to distinguish between narcotization and killing, for a good whole mount cannot be made from a specimen which has been permitted to die in the narcotic.

Narcotization should always proceed slowly; i.e., one should add a small quantity of narcotic at the beginning and increase the quantity later, adding the fixative only after the cessation of movement. This is easy to judge in the case of motile forms, which may be presumed to be narcotized shortly after they have fallen to the bottom, but in the case of sessile forms it is necessary to use a fine probe, preferably a hair, to determine the end point of narcotization.

Recommended Narcotics and Fixatives for Specific Objects. It must be pointed out that the primary purpose of fixing an object before making a wholemount is to retain as nearly as possible the natural shape. The fixative selected should, therefore, contain an immobilizing agent as well as a hardening agent. Gray, 1933 (*J. Roy. Microsc. Soc.*, **53**:14), in a discussion of the principles governing the selection of fixatives, came to the conclusion that there were only two good immobilizing agents: heat and osmic acid. It is therefore necessary, when dealing with highly contractile or imperfectly narcotized animals either to select a fixative containing osmic acid or to heat the fixative. Neither osmic acid nor heat is a good hardening agent and neither should, therefore, be used alone. The best hardening agents for objects which are subsequently turned into resinous whole mounts appear to be chromic acid and formaldehyde, used either singly or in combination, and these solutions are usually acidified with acetic acid to assist in the preservation of internal structures, particularly nuclei. The following recommendations, drawn largely from Gray, 1935 (*Microsc. Rec.*, **35**:4), and Gray, 1936 (*Microsc. Rec.*, **37**:10), are to be taken only as suggestions representing the author's opinion and should be used as a basis for further experiment.

Noncontractile Protozoa. These do not require narcotization and may be fixed directly in a weak solution of osmic acid.

Individual Contractile Protozoa. These are very difficult to handle. Ten percent methanol is quite a good narcotic for *Dileptus*, but 1% hydroxylamine seems better for *Spirostomum*. It is the writer's practice to try new forms with the following narcotics in the order given: 10% methanol, 1% hydroxylamine, 1% urethane. There are many forms, however, which do not respond to these narcotics and of which it appears almost impossible to make a good whole mount.

Individual rhizopods, as *Amoeba* and *Difflugia*, are best fixed to a coverslip in the following manner. Take a clean coverslip and smear on it a very slight quantity of fresh egg albumen. Each individual protozoan is placed in the center of a coverslip and left to expand. While this is going on a flask (or kettle) is fitted with a cork. Through this cork is inserted a glass tube the outer end of which has been drawn to a fairly fine point. The water in the flask is boiled to produce a jet of steam. As soon as the protozoan is satisfactorily expanded, the coverslip is picked up very gently and the underside passed momentarily through the jet of steam. This instantly hardens the protozoans in

position and at the same time cements them to the coverslip through the coagulation of the egg white. The coverslip should then be transferred to any standard fixative solution for a few minutes before being washed and stored in alcohol.

Among the Suctoria, *Acineta* and *Dendrocometes* may be prepared by placing them in a good volume of water, sprinkling menthol crystals on the surface, leaving them overnight, and then adding sufficient 40% formaldehyde to bring the total strength to 4%.

Stalked Ciliate Protozoa. These forms are quite easy to fix provided one realizes that double narcotization is necessary: once for the stalk and once for the head. The author's technique is to narcotize with Rousselet's solution until the snapping movements have slowed and then very gently to add weak hydrogen peroxide.

The specimens are then watched under a microscope and the selected fixative—which must contain osmic acid—is flooded onto them at the exact moment when the cilia straighten out and become stationary. This is satisfactory with *Carchesium*, *Zoothamnium*, and *Vorticella*. *Opercularia* and *Epistylis* have noncontractile stalks and one need, therefore, only use hydrogen peroxide. The writer has never made a satisfactory mount of *Scathidium* or *Pyxicola*.

Coelenterata. Hydroids are usually narcotized with menthol. Anthozoa, particularly the small ones likely to be prepared as whole mounts, can be narcotized with menthol though magnesium sulfate is better. Gray's narcotic, prepared by grinding together until fluid 48 gms of menthol and 52 gms of choral hydrate, is also widely used.

Platyhelminthes. Some of the smaller freshwater Turbellaria (e.g., *Vortex*, *Microstomum*) may be narcotized satisfactorily by adding small quantities of 2% chloral hydrate to the water in which they are swimming. Another good technique is to isolate the forms in a watch glass of water and place the watch glass under a bell jar together with a small beaker of ether. The ether vapor dissolves in the water and narcotizes these forms excellently.

Annelida. Small, marine, free-living Polychaetae make excellent whole mounts and do not usually need to be narcotized before killing. They should, however, be stranded on a slide and a very small quantity of the fixative dropped on them, so that they die in a flat condition which makes subsequent mounting possible. Much more realistic mounts are obtained by this means than if they are laboriously straightened before fixing, for they usually contract into the sinuous wave which they show when swimming. There seems to be no certain method of fixing the Nereids with their jaws protruding and one has to rely on chance to obtain one in this condition. The free-swimming larvae of marine polychaetes are very difficult to fix satisfactorily, because the large flotation chaetae usually fall out. The writer prefers for these, as for other marine invertebrate larvae, to concentrate a relatively large quantity of the plankton and then to flood over it three or four times its volume of Bouin's 1900 fixative (see FIXATIVE FORMULAS). The specimens are then allowed to settle, and the fixative is poured off and replaced with 70% alcohol which is replaced daily until it ceases to extract yellow from the specimens. By hunting through a large mass of plankton so fixed, one can usually obtain a considerable number of specimens in a perfectly expanded condition.

Freshwater oligochaetes are best narcotized with chloroform, either by adding small quantities of a saturated solution of chloroform in water, or by placing them in a small quantity of water under a bell jar in which an atmosphere of chloroform vapor is maintained. Leeches are difficult to handle and the author has had most success by placing them in a fairly large quantity of water to which is added, from time to time, small quantities of a saturated solution of magnesium sulfate. As soon as the leeches have fallen to the bottom considerably larger quantities of magnesium sulfate can be added, which will leave the leeches, in a short time, in a perfectly relaxed, but not expanded, condition. They should then be flattened between two slides and fixed in Zenker's 1894 fluid (see FIXATIVE FORMULAS). After the specimens have been fixed sufficiently long to hold their shape when the glass plates are removed, they are transferred for a couple of days to fresh fixative and then washed in running water overnight.

Bryozoa. Marine bryozoans may be narcotized without difficulty by sprinkling menthol on the surface of the water containing them. Subsequent fixation is best in some chromic-acetic mixture, for osmic acid tends to precipitate on the test and blacken the specimen. It may be pointed out that, for taxonomic purposes, dried whole mounts of the test are of more value than are whole mounts with expanded animal. It is usually recommended that freshwater bryozoans be narcotized in some cocaine solution, but the writer has found menthol just as good and much easier to use. Freshwater bryozoans should be fixed directly in 4% formaldehyde since they shrink badly in any other fixative.

Small Crustaceans. These are sometimes prepared as resinous mounts, though the writer prefers to mount them in glycerol jelly. They may be narcotized in weak alcohol and fixed in almost any fixative.

Other Arthropods. Whole mounts of most small arthropods are better made in gum media in the manner described above.

Choice of a Stain. It is now to be presumed that, whatever method of narcotization and fixation has been employed, the specimens to be mounted have been washed free from fixative and accumulated either in water or 70% alcohol. The suggestions which follow are likely to be modified by every individual reading the book; but they are included for the sake of those inexperienced in making whole mounts.

Small Invertebrates and Invertebrate Larvae (see CAR-

MINE). These are best stained in carmine by the indirect process; i.e., by overstaining and subsequent differentiation in acid alcohol. For most specimens the writer prefers Grenacher's 1879 alcoholic-borax-carmine. As an alternative, particularly for marine invertebrates, he has frequently used the two formulas for Mayer's *paracarmine* 1892 (see CARMINE). With these stains available there are very few small invertebrates or invertebrate larvae which cannot be prepared.

Larger Invertebrate Specimens. Larger specimens are better stained by the direct process; i.e., exposed for a considerable length of time to a very weak solution of stain and subsequently not differentiated.

Vertebrate Embryos. These seem to stain more satisfactorily in hematoxylin than in carmine solutions, the author's preference being for the formula of Carazzi 1911 (see HEMATOXYLIN STAINING FORMULAS). This formula is not very well known but may be used whenever the solution of Delafield is recommended.

Dehydration. It is to be presumed that the specimens, plant or animal, stained or unstained, are now accumulated either in distilled water or in 70% alcohol according to the treatment which they have had. It is now necessary to remove the water from them before they can be transferred into a resinous mounting medium. Ethanol is widely used as a dehydrant, and at least in the preparation of whole mounts, only its nonavailability should make any substitute necessary. If substitution is necessary, acetone or methanol, in that order of preference, may be used, but they have the disadvantage of being more volatile than ethanol, and therefore require more care in handling.

Dehydration is carried out by soaking the specimen in gradually increasing strengths of alcohol, it being conventional to employ 30%, 50%, 70%, 90%, 95%, and absolute alcohol. The writer prefers to omit from this series, unless the object is very delicate, both the 30% and the 50% steps in the process, thus starting with direct transfer from water to 70% alcohol. The only difficulty likely to be met in dehydration is in the handling of small specimens, for if they are in specimen tubes it is almost impossible to transfer them from one to the other without carrying over too much weak alcohol. The writer has long since abandoned the use of tubes in favor of the device seen in Fig. 1. This is a short length of glass tube, open at both ends, with a small piece of bolting silk or other fine

Fig. 1   Transferring objects between reagents with cloth bottom tubes. (Figs. 1, 2, 3, and 4) from P. Gray, "Introduction to Basic Microtechnique." Reproduced by permission of the McGraw-Hill Book Company.)

cloth tied across the lower end. The specimens are placed in these little tubes which (see illustration) are transferred from one stender dish to another with a minimum chance of contamination. These tubes are commercially available in England but in America must be either imported or homemade.

There is no means of judging when dehydration is complete save by attempting to clear the object. It is unwise to believe the label on an open bottle or jar if it says *absolute alcohol* because this reagent is hygroscopic and rapidly absorbs water from the air. One should, therefore, keep a quantity of anhydrous copper sulfate at the bottom of the absolute alcohol bottle and cease to regard the alcohol as absolute when the salt starts turning from white to blue. More whole mounts are ruined by being imperfectly dehydrated than by any other method, and even the smallest specimen should have at least 24 hr in absolute alcohol before being cleared.

If the specimen is to be mounted in Canada balsam, or one of the substitutes for it, it must next be cleared.

The Choice of a Clearing Agent. A clearing agent must be some substance which is miscible both with absolute alcohol and with the resinous medium which has been selected for mounting. The ideal substances for this purpose are essential oils, for they impart just as much transparency to the specimen as does the resin used for mounting, so that one has, as it were, a preview of the finished specimen. The use of xylene or benzene, which is so widespread in the preparation of paraffin sections, has tended to spread into the preparation of whole mounts, for which purpose, in the author's opinion, they are worthless. They have a relatively low index of refraction; hence one cannot tell whether or not the slight cloudiness of the specimen is due to imperfect dehydration until after they have been mounted in balsam.

The writer's first choice is terpineol (synthetic oil of lilac), which has advantages possessed by no other oil. It is readily miscible with 90% alcohol, so it will remove from the specimen any traces of water which may remain in it through faulty dehydration, and it has also the property of not making specimens brittle. The odor is very slight and rather pleasant. Oil of cloves is the most widely recommended essential oil for the preparation of whole mounts and it has only two disadvantages: its violent odor and the fact that objects placed in it are rendered brittle. If a small arthropod is cleared in oil of cloves, it is almost impossible to get it into a whole mount without breaking off some appendages. Oil of cloves is, however, miscible with 90% alcohol. Oil of cedar (more correctly oil of cedarwood) has been recommended in the literature and has the advantage of having a pleasant odor and of not rendering objects brittle. Unfortunately it is very sensitive to water, so perfect dehydration in absolute alcohol is necessary before endeavoring to clear with it.

The most satisfactory method of clearing objects subject to shrinkage through violent osmotic change is the "gravitational" method. By this method a small quantity of the essential oil in question is placed at the bottom of a small vial which is then tilted and very carefully filled with absolute alcohol which, naturally, remains on the surface. The object is then removed from absolute alcohol and dropped into the vial where it stops at the interphase between the two fluids. It becomes impregnated with the essential oil, into which it ultimately sinks, by a process of very slow diffusion, and when it has reached the bottom

of the tube, the contents of the tube should carefully be drawn off with an eye dropper type pipette and then the object transferred into fresh essential oil before mounting.

Two clearing agents, which are excellent for unstained specimens, are very little known. These are turpentine and acetic acid. The acid cannot be used with stains for obvious reasons, while the turpentine is a strong oxidizing agent and cannot, therefore, be used after hematoxylin, though it is perfectly safe with carmine. Absolute (glacial) acetic acid is miscible at all proportions both with water and with Canada balsam. Small arthropods may be dropped into acetic acid, left there until they are completely dehydrated, and then transferred directly to balsam. This little-known technique is strongly to be recommended.

Mounting Specimens in Balsam. Nothing is easier than to mount a specimen in a resinous medium, provided that it has been perfectly dehydrated and cleared. A properly made wholemount should be glass-clear, but it will not be clear in balsam unless it is clear in terpineol or clove oil. Not more than one in a thousand wholemounts has this vitreous appearance, and the worker who is accustomed to looking at rather cloudy wholemounts should take the trouble to dehydrate a specimen thoroughly, then to remove the whole of the dehydrating agent with a clearing agent, and then to mount properly in balsam.

The first step, therefore, in making a mount in, say, Canada balsam is to make quite certain that the specimen in its essential oil is glass-clear; the second step is to make certain that one has "natural" Canada balsam and not "dried" balsam which has been dissolved in xylene. Solutions of dried balsam in hydrocarbons are meant for mounting sections and are, for this purpose, superior to the natural balsam. Natural balsam is, however, just as preferable for whole mounts and is just as easy to obtain. If it is found to be too thick for ready use, it may be warmed gently until it reaches the desired consistency. A single small specimen is mounted by placing it in a drop of balsam on a slide and then lowering a coverslip horizontally (Fig. 2) until the central portion touches the drop. The coverslip is then released and pressed very gently until it just touches the top of the object. By this means it is possible to retain the object in the center of the coverslip and also, if one is using natural balsam which does not shrink much in drying, to avoid using cells for any but the largest object. Unfortunately most people are accustomed to mounting sections in thin balsam by the technique shown in Fig. 3; i.e., by touching one edge of the coverslip to the drop and then lowering it from one side. The objection to this is that the balsam, as is seen in the figure, immediately runs into the angle of the coverslip, taking the object with it, and it is difficult to lower the coverslip in such a way that the object is left in the center. If one is mounting thin objects, or deep objects in a cell in which a cavity has been ground, it is desirable to hold the coverslip in place with a clip while the balsam is hardening. This process is seen in Fig. 4, the type of clip there shown being made of phosphor bronze wire, and is far superior, in the writer's opinion, to any other type.

This description presumes that one is using natural Canada balsam, which is unquestionably the best resinous medium in which to prepare whole mounts. Other media are discussed in the article MOUNTANTS.

The best use for solutions of balsam in making whole mounts is in the preparation of delicate specimens or a

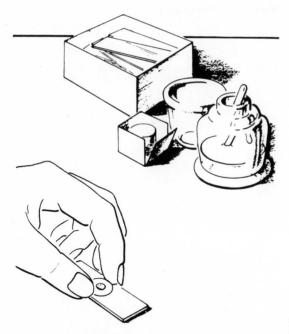

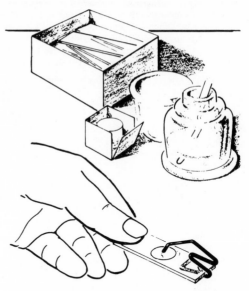

Fig. 4   Balsam whole mount ready for drying. The clip prevents movement of the coverslip or object.

Fig. 2   Applying coverslip to balsam whole mount. Note that the coverslip is held horizontally centered over the drop of balsam.

large number of objects. In this case the objects are transferred from the clearing medium to a tube or dish of the solution of balsam in whatever hydrocarbon has been selected, and the solvent then evaporated. When the balsam which remains has reached a good consistency for mounting, each specimen is taken, together with a drop of balsam, and placed on a slide. A coverslip is then added. By this method a large number of slides may be made in a short time. It is not necessary to use solutions of dried balsam, and the writer prefers, for this purpose, to dilute natural balsam with benzene.

Mounting large objects in a deep cell in Canada balsam is not to be recommended for the reason that the balsam becomes yellow with age and, in thick layers, tends to obscure the specimen. A whole mount of a 96-hr chicken embryo, for example, is of very doubtful value; but if it has to be made it is best first to impregnate it thoroughly with a fairly thin dilution of natural balsam. It is then

placed in the cell, piling the solution up on top, and left in a desiccator. The cell is refilled as the evaporation of the solvent lowers the level. When the cell is finally completely filled with solvent-free balsam it is warmed on a hot table and the coverslip applied directly.

Finishing Balsam Mounts. If a mount has been properly made with natural balsam, and if the size of the drop has been estimated correctly, no finishing is required since no balsam will overflow the edges of the coverslip. Natural balsam takes a long time to harden, and if one has a fairly thick mount the coverslip of which is not supported, drying cannot be hastened by heating as this will liquefy the balsam, causing the coverslip to tip to one side. Mounts in natural balsam are better put away for a month or two before any attempt is made to clean them. The slide should be cleaned, when it is sufficiently hard, first by chipping off any excess balsam with a knife and secondly by wiping away the chips with a rag moistened in 90% alcohol. This will leave over the surface of the slide a whitish film which may then be removed with a warm soap solution, and the slide may be polished before being labeled.

The writer prefers to apply a ring of some cement or varnish around the edge of his balsam mounts for two reasons. In the first place such a ring diminishes the rate of oxidation so that the mounts do not start going brown around the edges. In the second place such slides are less likely to be damaged in students' hands because of the psychological effect produced by a well-finished slide. One must carefully avoid using any cement or varnish which is soluble in, or miscible with, balsam; the author prefers to use one of the numerous cellulose acetate lacquers which are available on the market. With regard to labeling, it may be pointed out that no power on earth will persuade

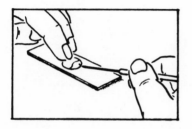

Fig. 3   Wrong way to apply coverslip to balsam whole mount. This method draws the object to one side.

gum arabic, customarily used for attaching labels, to adhere to a greasy or oily slide: the portion of the slide to which the label is to be attached should, therefore, be cleaned more carefully than any other. The writer prefers to moisten both sides of the label, press it firmly to the glass, and write on it only after it is dry.

PETER GRAY

Bibliography

Gray, P., "The Microtomist's Formulary and Guide," Philadelphia, Blakiston, 1954.
Gray, P., "Handbook of Basic Microtechnique," 3rd ed. New York, McGraw-Hill, 1964.
Pantin, C. F., "Notes on Microscopical Technique for Zoologists," Cambridge, The University Press, 1960.
See also   INSECT WHOLEMOUNTS.

X-RAY MICROSCOPY

Introduction

X-ray microscopy may be defined as the use of X-rays to produce a magnified image, or to provide data about individual microscopically identifiable parts of an object. The special properties of X-rays, namely a short wavelength, small but specific absorption, and a simple emission spectrum, define the usefulness of X-ray microscopy. All of these properties have been used to a greater or lesser degree for microscopy and microanalysis of inorganic and biological material.

Five techniques are currently employed, namely, contact microradiography, point projection microscopy, X-ray absorption microanalysis, X-ray fluorescence microanalysis, and electron probe microanalysis. Of these the first two are commonly used to produce images, while all five may be used microanalytically to measure mass of microstructures or to measure local amounts or concentrations of individual chemical elements.

Thus there are the two distinct yet related developments of qualitative and quantitative X-ray microscopy. The former, which has been widely used in biology, dentistry, and medicine, provides information supplementary to that obtained by optical and electron microscopy, and is the subject of this article. The latter, owing to the complexity of its procedures, cannot be considered here, other than to suggest its scope. The simple equation describing X-ray absorption has resulted in reliable measurements of density in single cells. Impressive developments have taken place in X-ray emission analysis. Utilizing the fact that characteristic X-ray wavelengths are associated with each element, elementary microanalysis can be performed by detecting the specific X-ray wavelengths excited when an electron beam (probe analysis) or X-ray beam (fluorescence analysis) strike a specimen. For these aspects, the reader is referred to the proceedings of the international conferences on X-ray microscopy and microanalysis (1–5), "X-ray Microscopy," by Cosslett and Nixon (6), and "X-ray Microscopy in Clinical and Experimental Medicine," by Hall, Röckert, and Saunders (7).

Image Formation

X-rays have a shorter wavelength than light and it follows a greater penetration and theoretically a higher resolving power. At present the resolving power of X-ray microscopy only approaches the limit of optical microscopy (1000 Å). Technical difficulties, largely due to the physical nature of X-rays, hinder further advance toward the ultimate limit set by the wave nature of X-rays. X-rays bear no charge and hence cannot be focused by electromagnetic lenses. Nor can they be usefully refracted optically because the index of refraction for all media is only slightly less than unity; this means, however, that they can be totally reflected if they have a grazing incidence. These facts dictate the procedures which may be used to obtain magnified images with X-rays.

The mirror or reflection microscope (Fig. 1) utilizes the principle of total reflection. A system of adjustable curved mirrors brings reflected X-rays to a focus. X-rays incident at a glancing angle ($< 1°$), are imaged by a vertical mirror to a line, and then by a horizontal mirror back to a point, so giving point-to-point imaging with a resolution of about 1 μ over a limited field (8). Despite attempts to correct the optical aberrations (9) the method is presently too complicated for biological work.

Contact microradiography, which dates from Goby (10), is the simplest method of obtaining an enlarged X-ray image of an object. The object, on being placed in contact with a fine-grain photographic emulsion, is exposed to X-rays from a fine-focus X-ray tube to secure a one-to-one image, which is then examined under a light microscope or enlarged photographically (Fig. 1). The resolution and magnification obtainable are determined by the emulsion grain and recording geometry, i.e., size of the X-ray source, object thickness, tube-object and tube-emulsion distances, and resolving power of the optical microscope. Close contact between object and emulsion is essential to reduce image unsharpness produced by divergent X-rays. Special fine-grain emulsions are used but do not yet wholly meet requirements. Useful magnifications up to 1200× have been demonstrated, but a resolving power better than 0.5 μ has rarely been claimed.

Point projection microscopy is the most direct way of magnifying an object by X-rays. Here an object is placed close to an X-ray point source and its image is projected onto a distant photographic plate. A point source may be obtained by placing a pinhole in front of a relatively large focal spot, or by focusing an electron beam into a fine point on a target. The former "camera obscura" technique has intensity limitations and small apertures (0.5–1 μ) are difficult to make (11). The latter technique has been successfully developed by Cosslett and Nixon of Cambridge University in the projection X-ray microscope (12).

The projection X-ray microscope consists essentially of an electron source and an electron optical system, which are used to produce a small focal spot (1–0.1 μ) on a metal foil or transmission type of target (Fig. 1). The

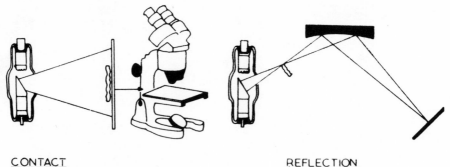

CONTACT
MICRORADIOGRAPHY

REFLECTION
MICROSCOPE

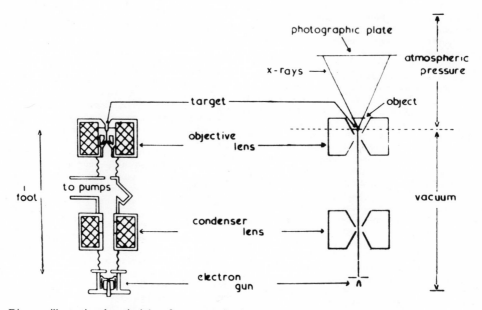

Fig. 1 Diagram illustrating the principles of contact, reflection, and projection X-ray microscopy,

electron optical system consists of a very strong magnetic objective lens, and a weaker condenser lens which controls the demagnification of the electron source. The target lies at the end of the microscope, hence an object can be placed close to the point source of X-rays, with consequent gain in intensity. An enlarged image of an object is projected onto a photographic plate with a resolution about equal to the target spot (1–0.1 μ). X-ray magnification is varied by simply changing the target-object and target-plate distances. Magnifications of $100\times$ or $1000\times$ are obtainable in a short camera length, but much depends on the nature of the specimen. Targets punched from various metal foils (Al, Ag, Cu, Au, W, etc.) operated at suitable voltages (5–30 kV) provide the type of X-radiation required.

Microangiography

Microangiography is a specialized form of X-ray microscopy concerned with the study of fine blood vessels, including the capillaries, filled with a radiopaque solution to provide vessel contrast. The contact and projection methods have been used for microcirculatory studies of skin (13), muscle (14), heart (15), lung (16), liver (17), stomach (18), intestine (19), spleen (20), kidney (21), eye (22), brain (23), bone (24), and teeth (25), etc. Owing to the large literature only these few examples can be cited.

Microangiographic experiments fall into two categories: studies of the vascular bed in injected anatomical specimens, and studies of the microcirculation in the living animal. For injection studies, surgical or autopsy specimens are usually injected with a colloidal suspension of barium (e.g., 7–25% Micropaque* in saline or 5–10% formalin) that replaces the blood in the vessels. The specimens are sectioned grossly (1–10 mm) or by microtome (20–200 μ), to reduce vascular superimposition in the image and also exposure time. In living specimens, vessels are visualized by the injection of a small amount of a solution of high atomic number (e.g., Hypaque* containing iodine; Thorotrast* containing thorium) that mixes, circulates, and is compatible with the animal's blood. In both instances, stereoscopic methods can be used to differentiate overlapping structures. Soft X-rays with a 0.4–4 Å wavelength range are the most suitable for microangiography.

Technique

Injected Specimens. In contact microradiography, the injected specimen is placed on a suitable photographic emulsion (film or plate) and exposed to soft radiation from a microfocus X-ray tube. The emulsion is protected by plastic foil (e.g., Bexphane*) from tissue fluids. An X-ray diffraction tube possessing a beryllium window, copper target, and small focal spot (0.3–1 mm) gives good results. The one-to-one image of vascular detail recorded is enlarged photographically or by microphotography. Contact microangiograms provide good overall vascular delineation in a fairly large and thick block of tissue. Such information is not obtainable by other methods. Stereoscopic views are secured by tilting the specimen slightly (8°) between successive exposures. It is a nondestructive technique, hence vascular areas can be selected for histology and comparison of X-ray and optical appearances.

In projection microscopy, the X-ray microscope is first fitted with a suitable target (Cu, Al) and focused at a selected voltage (10–30 kV) to give the required soft X-ray beam. Focusing is performed by placing a metal grid (1500 silver mesh; 3 μ bars) on the microscope target. The lens controls are then adjusted until a sharp X-ray image of the grid is seen upon a fluorescent viewing screen. The injected specimen, supported on a plastic foil, is positioned over the target, while a photographic plate is placed above the specimen. Exposures at various target-specimen and target-plate distances provide the desired magnification, and vary the size of the vascular field. Contact microangiograms can also be recorded.

Owing to the point source and great depth of field of the projection microscope, there is sharp vessel definition and all blood vessels remain in focus. The volume pattern of a vascular bed is recorded stereoscopically by moving the specimen laterally between successive exposures. The lateral shift (x) is related to the total magnification (M) and interocular distance (65 mm), whence $x = 65/M$. Detailed contact or projection microangiograms can be recorded on standard lantern slide plates. Better resolution is obtained if maximum resolution (Ilford) or spectroscopic plates (Kodak 649) are used. The vascular image is usually enlarged photographically.

Microangiography in Vivo. Microangiography *in vivo* is performed by contact or projection microscopy in a similar manner. The living rabbit ear has been extensively used to study vascular reaction following cold injury, vessel occlusion, and skin grafting (26). The rabbit ear,

surgically prepared tissue flap, or exteriorized organ (ovary, bowel loop), is placed under a microfocus tube or over the projection microscope in suitable relation to a plate holder. A vessel is intubated for the injection of radiopaque, vasomotor drugs, etc. Radiopaque injection is synchronized with rapid, successive X-ray exposures. The radiopaque is injected either at high pressure into a small regional artery (blood displacement technique); or into an artery closer to the heart (circulating slug technique) to avoid vascular spasm and shock in the field under study. The former technique records only the momentary state of the vessels, but as shown by Bellman (27) can yield valuable information. The latter technique probably captures a more physiological picture of the microcirculation.

Vessel movement dictates exposure time, requiring compromise between speed and the grain size of the emulsion. The high-voltage projection microscope (28) can be used to operate an X-ray sensitive vidicon (Matchlett Dynamicon ML 589) in combination with a closed circuit television display. The vessels can then be viewed on a monitor screen and photographed at short exposure times.

In microlymphangiography a radiopaque, usually Thorotrast, is injected directly into the tissues to form a depot area. Successive X-ray exposures then reveal the passage of radiopaque through the lymphatic plexuses and collecting vessels. Microlymphangiography has important medical implications (29).

Examples. Turning to examples, Fig. 2 shows part of a contact microangiogram of a coronal section of an injected human brain. The section (5 mm thick; 5 × 11 cm in area) was first recorded to demonstrate the overall blood supply of the corpus striatum and internal capsule. The area shown was then enlarged photographically to reveal vascular detail. It shows the coarse capillary net which supplies the nerve fibers of the internal capsule (that here forms a diagonal between the arrows) and the slender contributing striate arteries. These vessels are of interest because brain hemorrhage is common in this area. The caudate (C) and lentiform (L) nuclei that flank the internal capsule show a denser capillary net that meets the demands of the numerous neurones therein.

Figures 3 and 4 are projection microangiograms taken of the living rabbit ear after injection of the central artery with radiopaque. Figure 3 reveals an open S-shaped aterio-venous anastomosis connecting a small artery to a larger subcutaneous vein. Such shunts have a caliber of about 0.05–1 mm, and measure 0.5–1 mm in length; the venous end is usually wider and funnel-shaped. Laminar flow is evident in the large veins. The complex vascular bed is evident, and the lower part of the plate faintly shows the passage of radiopaque through the capillary net. This feature is distinctly seen in Fig. 4, which shows both the thoroughfare channels and capillaries proper visualized by the flowing radiopaque. The remarkable density of the capillary net was revealed by prior injection of warm saline and sodium nitrite to induce vasodilatation.

Historadiography

Historadiography is a form of X-ray microscopy concerned with the study of biological tissues and cells. The discovery that X-ray absorption rapidly increases with wavelength led to the construction of special low-voltage X-ray tubes (0.5–5 kV) to produce soft or long wavelength radiation (2–12 Å). Vacuum chambers were used to

---

*Commercial sources of products indicated by an asterisk are given at the end of this article.

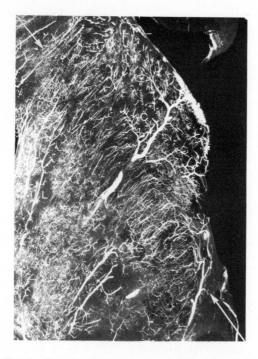

Fig. 2 Contact microangiogram showing the blood supply of the internal capsule (bteween arrows), caudate nucleus (C), and lentiform nucleus (L) of the human brain. Micropaque injection. ($\times$2.8.)

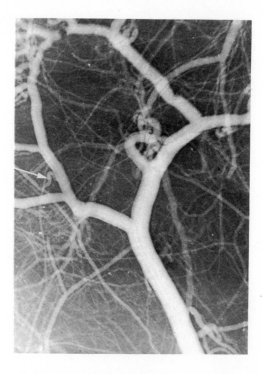

Fig. 3 Projection microangiogram of living rabbit ear showing an open S-shaped arterio-venous anastomosis (see arrow). Note the laminar flow in the large vein, and capillary flow in the lower part of the plate. Thorotrast injection. Compare with Fig. 4. ($\times$27.5.)

reduce air absorption of such rays. Tissue sections mounted on fine-grain emulsion (Lippmann type) were examined *in vacuo* by contact microradiography. Lamarque and Turchini (30) thus established that the differential absorption visible in the image was related to the atomic number of the cell constituents.

Following the work of Lamarque and Turchini, contact historadiography gained impetus due to the ingenious work of the Swedish school, notably by Engström (31), Lindstrom (32), Hyden (33), and their co-workers, such as Greulich (34) and Henke (35). Others, such as Dietrich (36), Röckert (37), Salmon (38), Scott (39), and Sissons (40) furthered the investigation of botanical and mineralized tissue. The advent of the projection X-ray microscope (41) was quickly followed by its application to historadiography, principally by Cunningham (42), Jongebloed (43), Le Poole (44), Oderr (45), Ong Sing Poen (46), and Saunders and van der Zwan (47).

Historadiographic studies have ranged from cell division, features of normal and abnormal cells, investigations of the primary and mineralized tissues, and structures of various parenchymatous organs, to the determination of mass and water content, and elementary analysis.

Equipment. Special equipment is required for historadiography, because the absorption of a histological section is determined by its thickness, density, elementary composition, and the wavelength of the X-ray beam used to examine it. Very thin sections must be used to secure images of high resolution, hence soft X-rays of wavelengths longer than about 8 Å are required to obtain sufficient X-ray absorption and image contrast. The principles of X-ray absorption in microsamples are described in the literature (48), but Engström's classifica-

tion of X-rays may be noted, namely that hard rays range between 0.1–1.0 Å, soft rays from 1 to 10 Å, while ultrasoft are greater than 10 Å.

Histological sections of soft biological tissues consist of organic material, consequently sufficient structural detail can be obtained only by using low-voltage radiation, i.e., soft or ultrasoft X-rays generated in the 0.5–3 kV range. Only the small tubes built specially for contact microradiography operate satisfactorily at these voltages. The X-ray microscope cannot be focussed below 5 kV, owing to poor intensity, but soft radiation can be obtained with an aluminum target operated at 5–15 kV. Mineralized tissues, such as bone and tooth sections, are necessarily examined with hard rays generated in the 15–50 kV voltage range.

Few commercial sealed-off X-ray tubes are suitable for historadiography, hence experimenters have built or adapted demountable tubes for soft and ultrasoft microradiography (49). A continuously evacuated midget tube of this type can be built from two automobile spark plugs (50). A commercial microradiographic unit (Philips CMR 5) which operates at 1–5 kV produces suitably soft X-rays (6–12 Å) (51). Such tubes must be fitted with a vacuum camera to prevent air absorption of the soft rays. Laboratory-built cameras are constructed from clear plastic to permit specimen inspection, and are large enough to hold a high-resolution or spectroscopic plate. The evacuation pump is usually switched off during exposures to eliminate vibration. Another commercial unit (Philips PW) which operates at 2–50 kV (on adding a variable transformer), produces X-rays of 1.8–4 Å,

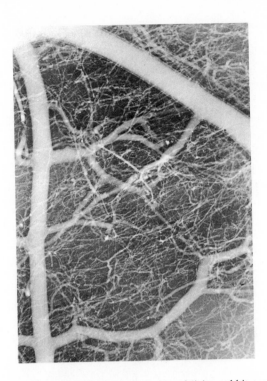

Fig. 4 Projection microangiogram of living rabbit ear showing the thoroughfare channels and capillaries proper. Thorotrast injection. (×25.)

better suited to microradiography of mineralized tissues.

The Cosslett-Nixon X-ray microscope can be used to produce soft or hard radiation, and examine soft or mineralized tissue by contact or projection techniques. The point source (0.1–1 μ) gives a better resolution than any other X-ray tube. The required radiation is obtained by changing the metal foil target (Al, Cu, Ag, Au, W, etc.) and selecting a suitable operating voltage (5–30 kV). Historadiography is usually performed with continuous (polychromatic) radiation, but characteristic radiation such as the soft kα (8.3 Å) of aluminum can be used with advantage (52). Many specimens can be examined under atmospheric conditions, but for those requiring soft radiation a vacuum camera is threaded to the microscope target. The camera interior is stepped to accept specimen rings, and permit variation of the target-specimen distance and magnification. Low magnifications (2–10×) provide sharper definition, and some favor the 2× method of Le Poole and Ong (53).

A standard vacuum spectrograph has been recently adapted for ultrasoft X-ray spectroscopy, in the 10–150 Å region, and applied to ultrasoft qualitative contact microradiography, giving a resolution of about 0.5 μ and useful magnifications up to 1000× (54).

Specimen Preparation. Unstained sections of tissue can be examined by historadiography. The image detail results from differential absorption of X-rays by varying concentrations of organic and inorganic substance present in the tissue. The basic problem is usually one of image contrast. Tissues which undergo keratinization, or mineralization, or which accumulate chemical substances of a "radiopaque" or "radiolucent" nature such as iodine or lipids, naturally provide marked contrast.

Other tissues, presumably composed largely of varying densities or thicknesses of similar material, exhibit little contrast. Histological sections usually consist of organic material of low atomic number, and so have a low mass absorption coefficient. This coefficient increases with atomic number and longer wavelength, hence soft X-rays and chemical stains high on the atomic scale are used to improve contrast.

Freeze-drying and freeze-substitution techniques give sharp image contrast, because they remove water from tissue with little chemical loss. Water must be removed to gain optimal contrast, because tissues contain about 75% water and 25% dry substance (largely protein), and the mass absorption coefficient of tissue and water are similar in this wavelength range. Formalin fixation is often used, but the mercuric chloride content of Zenker increases contrast.

Thick cryostat sections (20–200 μ) mounted on plastic film can be examined in the unstained or stained state, used for stereoscopy, or checking chemical uptake by fresh tissue after *in vivo* injection. Thin sections (1–5 μ) are necessary to obtain tissue images of high resolution-Engström (55) has prepared a useful graph (Fig. 5) showing the optimal voltages, wavelengths, and section thicknesses for high-resolution microradiography of soft tissues.

High-resolution contact microradiography requires close contact between the tissue section and emulsion. The section is best mounted directly on an emulsion protected by a thin film of collodion, obtained by dipping the plate in a dilute (1%) alcohol-ether solution of collodion and drying it upright overnight. The paraffin-embedded section is floated on, allowed to dry, and de-paraffinized. The plate is carried from benzol through descending alcohols to water, and back through ascending alcohols to benzol, to remove paraffin and benzol-insoluble fractions which are X-ray absorbent (56). When dry, the plate is exposed, and the collodion film removed with methyl cellosolve or acetone prior to development. The section can usually be recovered for further histology. More simply, the section can be mounted on a specimen ring bearing a collodion membrane. The ring mount is

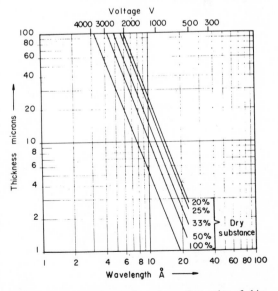

Fig. 5 Suitable voltages for microradiography of thin sections of soft tissues. (Courtesy Prof. A. Engström.)

kept firmly pressed against the emulsion during the exposure.

Stained tissue sections mounted for optical microscopy can be used for historadiography, provided DPX mounting medium has been used (57). The coverslip is removed by soaking overnight in xylene. The slide is then immersed for an hour in equal parts of thick DPX and xylene, drained, and dried upright for an hour. A thin film of DPX containing the section can then be peeled off with a razor blade. Such mounts can be placed directly on the plate emulsion for contact work, or inserted into the beam of the projection microscope. They have a low X-ray absorption, can be repeatedly X-rayed, and remounted for optical microscopy and comparison studies.

Few stains have been tested historadiographically, or comparison made between their optical and X-ray appearance. For example, strong PAS reactivity may or may not be reflected microradiographically, and cell nuclei show striking differences in X-ray opacity not indicated by staining (58). Historadiographic stains known to increase the radiopacity of certain tissue elements are unfortunately still limited in number. The reputed selective affinity of certain tissues for silver is not confirmed by microradiography (59), but the classic heavy-metal stains of Golgi and Cajal give excellent results in historadiography.

Histochemistry with its better understood and more controlled reactions is the most promising approach to the production of artificial contrast for historadiography. Recently it has been realized that the high atomic weight of metals presently used solely for their histochemical color reactions increases beam absorption and image contrast. This has given rise to the technique of X-ray histochemistry, which has a wide range of biological applications (60, 61). Many of the classical histochemical techniques developed for optical microscopy deposit metal in the tissues. The added metal may increase a small amount naturally present, or introduce an entirely new element to mark the site of a specific activity, such as the use of cobalt and lead to indicate the site of acid or alkaline phosphatase enzyme activity. Modifications of these, and new techniques, can be developed to produce radiopacity in the desired location, so extending the uses of historadiography. For example, X-ray histochemistry has been used for simultaneous demonstration of neurones and capillaries in the human brain (62).

Historadiography of mineralized tissues, such as developing teeth and bone, requires specific techniques. Special methods are used to cut thin sections which are examined with harder radiation, for bone salts particularly absorb X-rays of the 1–3 Å range (63, 64). Copper, titanium, and aluminum radiation have proved useful. Formalin-fixed samples of bone or tooth are embedded in methyl methacrylate, gross sectioned with a milling machine, and ground down between glass plates to the desired thickness (10–200 μ). Samples can be decalcified, and the end point checked by microradiography, prior to paraffin embedding and cutting into thin sections. Dental caries (65, 66), also bone changes following intermittent pressure (67) and torsion (68), have been so studied.

Examples. Turning to some examples, Fig. 6 illustrates the historadiographic detail obtainable from a soft tissue, such as human brain, treated with a heavy-metal stain. A section (45 μ thick) of visual cortex (area 17) was prepared by using a longer than usual impregnation in the primary silver bath of a modified Bielschowsky-Gros stain. The plate shows the arrangement of the nerve cells

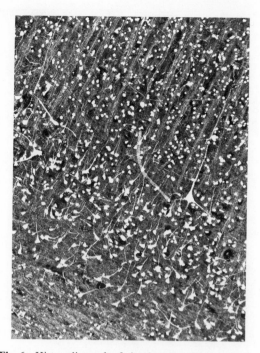

Fig. 6 Historadiograph of visual cortex of human brain showing the deep nerve cell layers and radial axone bundles. Bielschowsky-Gros stain. Recorded by contact method. ($\times$114.)

and fibers in the deepest part of the six-layered visual cortex. From above down are seen small granule cells of layer 4, an isolated pyramidal cell of layer 5, the cell variants of layers 5 and 6, and some horizontal fibers of the optic radiation within the subjacent white matter.

The radial bundles of axons, which pass through the layers of nerve cells, just as depicted in Cajal's drawings (69), are strikingly recorded. Some of the slanting or oblique fibers described by him are also visible. Such features are difficult to record by microphotography owing to the limited focal depth of the optical microscope, so explaining the reliance hitherto on drawings.

Figure 7 depicts an area of a historadiograph, made from a tangential section of human motor cortex (paracentral lobule). The section was stained by Gomori's lead method at pH 5 to show acid phosphatase activity, and recorded on spectroscopic emulsion (Kodak 649) with the projection microscope using minimal magnification. The nerve cell bodies contain appreciable amounts of acid phosphatase for which Gomori developed the classic series of metallic reactions resulting in a deposit of insoluble lead sulfide at the enzyme site. This deposit is markedly X-ray absorbent, so the historadiograph clearly shows the cell bodies of a cluster of Betz cells, and also their basal dendrites.

Figure 8 is another illustration of the technique of X-ray histochemistry. The figure shows part of a historadiograph made from a frozen section (approximately 35 μ thick) of the supra-optic area of the human hypothalamus, prepared by Gomori's acid phosphatase technique, adapted to pH 7 to stain for both the acid phosphatase activity of the nerve cells and the alkaline phosphatase

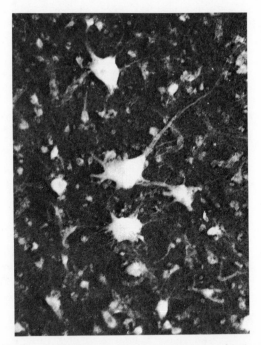

Fig. 7 Historadiograph of human motor cortex showing a cluster of Betz cells and their dendrites. Stained by Gomori's lead method at pH 5 to show acid phosphatase activity. Recorded by projection method. (× 193.) (Courtesy Mary A. Bell.)

Fig. 8 Historadiograph of supra-optic area of human hypothalamus showing the neurosecretory neurones and related capillary vessels. Direct lead incubation at pH 7, after Gomori. Recorded by contact method. (×41.) (Courtesy Mary A. Bell.)

activity of the blood vessels. The endothelial cells of the vessel walls contain alkaline phosphatase, and so by using methods based on Gomori's techniques it has been possible to outline the smaller cerebral vessels with lead (70). The deposits of black lead sulfide is optically visible and markedly X-ray absorbent. The historadiograph, which was recorded on spectroscopic emulsion, therefore shows both the neurosecretory neurones of the supra-optic area and the related vessels of supply, and in particular the capillary net and neurone-vascular pattern.

Conclusion

It will be evident that X-ray microscopy is a profitable adjunct to the methods which a histologist or microscopist might normally be expected to employ, and that it can provide information which is not evident or easily discernible by conventional histological techniques. Moreover, it can record a large field and deal with a specimen far too thick for electron microscopy, and give an image totally different from the light microscope.

R. L. DE C. H. SAUNDERS

*Commercial Sources

Contrast media suitable for microangiography can be obtained from the following firms: Thorotrast (Heyden Chemical Co., New York), Micropaque (Damancy & Co., Ware, Herts, England, and Picker X-ray Co., U.S.A. and Canada), Hypaque (Winthrop Laboratories, Aurora, Ontario, Canada).

Plastic foils suitable for microangiography and histo-radiography mounts, such as Bexphane (25 μ thick), are obtainable from BX Plastics Ltd., Manningtree, England.

References

1. "X-ray Microscopy and Microradiography," Proceedings of a Symposium held at the Cavendish Laboratory, Cambridge, 1956, V. E. Cosslett, Arne Engström, and H. H. Pattee, Jr., ed., New York, Academic Press, 1957 (645 pp.).
2. "X-ray Microscopy and X-ray Microanalysis," Proceedings of the Second International Symposium held at Stockholm, 1959, A. Engström, V. E. Cosslett, and H. H. Pattee, Jr., ed., Amsterdam, Elsevier Publishing Co., 1960 (540 pp.).
3. "X-ray Optics and X-ray Microanalysis," Proceedings of the Third International Symposium held at Stanford University, Stanford, California, 1962, H. H. Pattee, Jr., V. E. Cosslett, and Arne Engström, ed., New York, Academic Press, 1963 (622 pp.).
4. "X-ray Optics and Microanalysis: Optique des Rayons X et Microanalyse," Proceedings of the Fourth International Symposium held at Orsay, France, 1965, R. Castaing, P. Deschamps, and J. Philibert, ed., Paris, Hermann, 1966 (707 pp.).
5. "X-ray Optics and Microanalysis," Proceedings of the Fifth International Congress on X-ray Optics and Microanalysis held at Tubingen, Germany, 1968, G. Möllenstedt, and K. H. Gaukler, ed., Heidelberg, Springer-Verlag, 1968 (612 pp.).
6. Cosslett, V. E., and W. C. Nixon, "Cambridge Monographs on Physics: X-ray Microscopy," Cambridge, The University Press, 1960 (406 pp.).
7. Hall, T., H. O. E. Röckert, and R. L. de C. H. Saunders, "X-ray Microscopy in Clinical and Experimental Medicine," Fort Lauderdale, Florida, C. C. Thomas, 1972.
8. Kirkpatrick, E., and A. V. Baez, "Formation of optical images by X-rays," *J. Opt. Soc. Amer.*, **38**:766–774 (1948).

9. McGee, J. F., "A long-wavelength X-ray reflection microscope," pp. 164–176 in "X-ray Microscopy and Microradiography," New York, Academic Press, 1957.

10. Goby, P., "A new application of X-rays—microradiography," *C. R. Acad. Sci., Paris,* **156**:686–688 (1913).

11. Rovinsky, B. M., V. G. Lutsau, and A. I. Avdeyenko, "X-ray microprojector," pp. 269–277 in "X-ray Microscopy and Microradiography," New York, Academic Press, 1957.

12. Cosslett, V. E., and W. C. Nixon, "An experimental X-ray shadow microscope," *Proc. Roy. Soc. Brit.,* **140**:422–431 (1952).

13. Saunders, R. L. de C. H., and W. Montagna, "X-ray microscopy and microangiography: their potential value," *Arch. Dermatol.,* **89**:451–454 (1964).

14. Saunders, R. L. de C. H., T. Lawrence, and D. Maciver, "Microradiographic studies of the vascular patterns in muscle and skin," pp. 539–550 in "X-ray Microscopy and Microradiography," New York, Academic Press, 1957.

15. Kinley, C. E., and R. L. de C. H. Saunders, "Microangiography in experimental myocardial infarction," *Can. J. Surg.,* **14**:56–65 (1971).

16. Saunders, R. L. de C. H., and V. R. Carvalho, "X-ray microscopy of the microvascular system of the human lung," pp. 109–122 in "X-ray Optics and X-ray Microanalysis," New York, Academic Press, 1963.

17. Bettencourt, J. M., and J. Mirabeau-Cruz, "La Circulation Hepatica Normal y Patologica," Madrid, Libraria Cientifico Medica Espanola, 1963 (329 pp.).

18. Barclay, A. E., "Microarteriography," Oxford, Blackwell Scientific Publications, 1951 (102 pp.).

19. Saunders, R. L., de C. H., R. James, and D. C. d'Arcy, "X-ray microscopy of intestinal villi," *Experentia,* **17**:361–368 (1961).

20. Singh, S., "A microangiographic study of the splenic microcirculation," Abstracts p. 119, Proceedings 9th International Congress Anatomists, Leningrad, 1970.

21. Ljungquist, A., and C. Lagergren, "Normal intrarenal arterial pattern in adult and ageing human kidney: a microangiographical and histological study," *J. Anat.,* **96**:285–300 (1962).

22. Pattee, Jr., H. H., L. K. Garron, W. K. McEwen, and M. L. Feeney, "Stereomicroradiography of the limbal region of human eye," pp. 534–538 in "X-ray Microscopy and Microradiography," New York, Academic Press, 1957.

23. Saunders, R. L. de C. H., W. H. Feindel, and V. R. Carvalho, "X-ray microscopy of the blood vessels of the human brain. Parts I and II," *Med. Biol. Illus.,* **15**:108–122, 234–246 (1965).

24. Göthman, L., "The arterial pattern of the rabbit's tibia after the application of an intramedullary nail," *Acta Chir. Scand.,* **120**:211–219 (1960).

25. Saunders, R. L. de C. H., and H. O. E. Röckert, "Vascular supply of dental tissues, including lymphatics," pp. 199–245 in "Structural and Chemical Organization of Teeth," Vol. 1, New York, Academic Press, 1967.

26. Bellman, S., and E. Velander, "Vascular transformation in experimental tubed pedicles," *Brit. J. Plastic Surg.,* **12**:1–21 (1959).

27. Bellman, S., H. A. Frank, P. B. Lambert, and A. J. Roy, "Studies of collateral vascular responses," *Angiology,* **10**:214–232 (1959).

28. Saunders, R. L. de C. H., and R. V. E. Ely, "Cerebral microangiography and 'in vivo' studies with an x-ray microscope cum x-ray sensitive vidicon," pp. 642–649 in "X-ray Optics and Microanalysis," Paris, Hermann, 1965.

29. Bellman, S., and B. Odén, "Regeneration of surgically divided lymph vessels. An experimental study on the rabbit's ear," *Acta Chir. Scand.,* **116**:99–117 (1959).

30. Lamarque, P., and J. Turchini, "Une nouvelle méthode d'utilisation des rayons x, l'Historadiographie. De quelques applications de l'Historadiographie intéressant plus specialement les Sciences Médicales," Extrait de Montpellier Médical (No. 2, Fev.), 1937 (35 pp.).

31. Engström, A., "X-ray Microanalysis in Biology and Medicine," Amsterdam, Elsevier, 1962 (92 pp.).

32. Lindstrom, B., "Roentgen absorption spectrophotometry in quantitative cytochemistry," *Acta Radiol.,* Supp. 125, 206 pp. (1955).

33. Hyden, H., and S. Larsson, "A new scanning microanalyser for data collection and evaluation from X-ray microradiograms," pp. 51–55 in "X-ray Microscopy and Microanalysis," Amsterdam, Elsevier, 1960.

34. Greulich, R. C., "Application of high resolution microradiography to qualitative experimental morphology," pp. 273–287 in "X-ray Microscopy and Microanalysis," Amsterdam, Elsevier, 1960.

35. Henke, B. L., B. Lundberg, and A. Engström, "Conditions for optimum visual and photometric 'Contrast' in microradiograms," pp. 240–248 in "X-ray Microscopy and Microradiography," New York, Academic Press, 1957.

36. Dietrich, J., "La Microradiographie par contact, méthode d'analyse cytologique," pp. 658–663 in "X-ray Optics and Microanalysis," Paris, Hermann, 1965.

37. Röckert, H. O. E., "A quantitative X-ray microscopical study of calcium in the cementum of teeth," *Acta Odont. Scand.,* **16**, Supp. 25 (1958) (68 pp., 20 figs.).

38. Salmon, J., "Perspectives et bilan des techniques microscopiques par rayons X, derivées de la microradiographie, en particular dans le domaine végétal," pp. 17–67 in "Bulletin de Microscopie Appliquée," Editions Revue d'Optique, 3 Boulevard Pasteur, Paris, 15ᵉ, 1961.

39. Scott, D. B., M. U. Nylen, and M. H. Pugh, "Contact microradiography as an adjunct to electron microscopy," *Norelco Reporter,* **7**:32–35, 43 (1960).

40. Sissons, H. A., J. Jowsey, and L. Stewart, "The microradiographic appearance of normal bone tissue at various ages," pp. 206–215 in "X-ray Microscopy and Microanalysis," Amsterdam, Elsevier, 1960.

41. Nixon, W. C., "X-ray microscopy," pp. 92–102 in "Modern Methods of Microscopy," London, Butterworth Scientific Publications, 1956.

42. Cunningham, G. J., "Microradiography," pp. 155–175 in "Tools of Biological Research" (2nd Series), London, Blackwell, 1960.

43. Jongebloed, W. L., "Application possibilities of projection microradiography," pp. 636–641 in "X-ray Optics and Microanalysis," Paris, Hermann, 1965.

44. Le Poole, J. B., and Ong Sing Poen, "Description of the Delft X-ray microscope," pp. 91–95 in "X-ray Microscopy and Microradiography," New York, Academic Press, 1957.

45. Oderr, C., "Architecture of the lung parenchyma," Studies with a Specially Designed X-ray Microscope, *Amer. Rev. Resp. Dis.,* **90**:401–410 (1964).

46. Ong Sing Poen, "Microprojection with X-rays," Delft, Holland, Hoogland en Waltman, 1959 (132 pp.).

47. Saunders, R. L. de C. H., and L. van der Zwan, "Exploratory studies of tissue by X-ray projection microscopy," pp. 293–305 in "X-ray Microscopy and Microanalysis," Amsterdam, Elsevier, 1960.

48. Lindstrom, B., *See* Reference 32.
49. Ely, R. V., "A high-output rotating anode shock-proof tube (demountable) for X-ray microscopy," pp. 47–50 in "X-ray Microscopy and Microanalysis," Amsterdam, Elsevier, 1960.
50. Lundberg, B., "A simple midget x-ray tube for high resolution microradiography," *Exp. Cell Res.,* **12**:198–200 (1957).
51. Van den Broek, S. L., "A simple contact microradiography unit with sealed-off X-ray tube," pp. 64–71 in "X-ray Microscopy and Microradiography," New York, Academic Press, 1957.
52. Bessen, I. I., "A high-resolution PMR X-ray microscope with electron microscope conversion," *Norelco Reporter,* **4**:119–123. (1957).
53. Nixon, W. C., "Point projection X-ray microscopy," pp. 34–45 in "X-ray Microscopy and Microradiography," New York, Academic Press, 1957.
54. Henke, B. L., "Techniques of low energy X-ray and electron physics," *Norelco Reporter,* **14**:4, 75–83, 98 (1967).
55. Engström, A., "Contact microradiography: a general survey," pp. 24–33 in "X-ray Microscopy and Microradiography," New York, Academic Press, 1957.
56. Greulich, R. C. *See* Reference 34.
57. Bell, M. A., "Histochemical Methods Combined with X-ray Microscopy for the Demonstration of Cellular and Vascular Patterns in the Human Cerebral Cortex," Vols. 1 and 2, M.Sc. Thesis, Dalhousie University, Halifax, N.S., 1969 (127 pp. 62 Figs.).
58. Shackleford, J. M., "Microradiographic interpretations of histologic structure," *Alabama J. Med. Sci.,* **2**:127–136 (1965).
59. Mitchell, G. A. G., "Microradiographic demonstration of tissues treated by metallic impregnation," *Nature,* **165**:429 (1950).
60. Bell, M. A. *See* Reference 57.
61. Bell, M. A., "The technic of X-ray histochemistry applied to neurohistology," *Experientia,* **25**:837–841 (1969).
62. Saunders, R. L. de C. H., M. A. Bell, and V. R. Carvalho, "X-ray histochemistry used for simultaneous demonstration of neurones and capillaries in the human brain," pp. 569–578 in "X-ray Optics and Microanalysis," Berlin, Springer-Verlag, 1969.
63. Engfeldt, B., "Recent observations on bone structure," *J. Bone Jt. Surg.,* **40A**:698–706 (1958).
64. Hallen, O., and H. O. E. Röckert, "The preparation of plane parallel sections of desired thickness of mineralised sections," pp. 169–176 in "X-ray Microscopy and Microanalysis," Amsterdam, Elsevier, 1960.
65. Dreyfus, F., R. M. Frank, and B. Grutmann, "La sclerose dentinaire," *Bull. Group. Int. Rech. Sc. Stomat.,* **7**:207–229 (1964).
66. Dreyfus, F., and R. M. Frank, "Microradiographie et microscopie electronique du cement humain," *Bull. Group. Int. Rech. Sc. Stomat.,* **7**:167–181 (1964).
67. Fyfe, F. W., "Histological effects of intermittent pressure on the rabbit's upper tibial epiphysis," *Anat. Rec.,* **136**:336 (1960).
68. Fyfe, F. W., "Artificial torsion in growing rabbit tibiae," *Proc. Can. Fed. Biol. Soc.,* **3**:24 (1960).
69. Cajal, S. Ramon y, "Histologie du Systéme Nerveux de l'Homme & des Vertebres," Vol. 2, Madrid, Consejo Superior De Investigaciones Cientificas, 1955:
70. Saunders, R. L. de C. H., M. A. Bell, and V. R. Carvalho. *See* Reference 62.

y

YEASTS

The generally accepted definition of yeasts is that they are fungi whose usual and dominant form is unicellular. Some yeasts, such as those used in brewing and baking (*Saccharomyces cerevisiae* and *Saccharomyces carlsbergensis*), rarely if ever exist in any form other than the single cell. Strains of *Endomyopsis*, on the other hand, only occasionally grow in the single-cell form. In terms of numbers of species, yeasts are less numerous than bacteria, filamentous fungi, algae, and protozoa. Authorities recognize some 350 species of yeast which are classified into about 40 genera (Lodder, 1970). Depending on the genus, yeast cells may be spherical, or shaped like oblate spheroids (egg-shaped) or short cylinders. The cells vary in diameter from 3 to 10 μm; to some extent, cell dimensions depend on the conditions under which the cells are grown. Members of the genus *Trigonopsis* are of more than passing interest since they are shaped like triangular lozenges.

Characteristically, yeasts multiply vegetatively by budding, the location of the buds on the mother cell varying to some extent with the genus. The one exception are members of the genus *Schizosaccharomyces*, which multiply by binary fission. Species of *Schizosaccharomyces* are widely used as model organisms in studies on the biochemistry of cell division.

The Dutchman Antonie van Leeuwenhoek (1632–1723) is credited with being the first person to observe yeasts microscopically, and as the quality of light microscopes improved so biologists acquired a greater insight into the cytology of yeast cells. However, it was not until 1897 that the yeast nucleus was first sighted by Schmitz. Thereafter, observations with the light microscope indicated the main cytological features of the yeast cell. However, the resolving power of the light microscope is too low to permit any detailed description of the structure of yeast cell organelles. Microscopists who attempted detailed cytological descriptions of yeasts based on light microscopy invariably entered the lists of controversy.

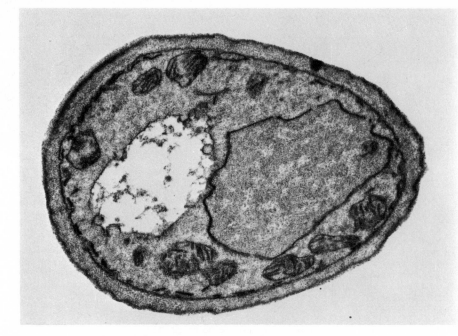

Fig. 1   Electron micrograph of a thin section through a cell of *Saccharomyces cerevisiae*. ($\times$0000.)

Electron microscopy of yeasts has very considerably extended our knowledge of the cytology of yeast cells, and has resolved many of the problems raised by light microscopy. Electron micrographs of shadowed preparations, in the ordinary or the scanning electron microscope, reveal the topography of the birth scar and bud scars on the yeast cell surface. But apart from these features, the surface of the yeast cell is devoid of detail.

Most of the electron microscope studies on the fine structure of yeasts have concentrated on strains of *Sacch. cerevisiae* (Matile et al., 1969). Excellent results have been obtained with thin sections of cells using a primary fixative of potassium permanganate (Luft, 1956; Vitols et al., 1961) followed by uranyl acetate postfixation and embedding in methacrylate. Figure 1 shows an electron micrograph of a thin section through *Sacch. cerevisiae* obtained using this technique. It shows the main cytological features that have been revealed by electron microscopy of yeasts (Fig. 2), including the rigid wall, the plasma or cytoplasmic membrane, mitochondria, vacuoles, nuclei, and storage granules. Permanganate fixation is generally acknowledged to give good preservation of cell membranes. It has been claimed, however, that use of permanganate results in poor preservation of some intracellular organelles. To overcome this problem, Schwab and his colleagues (1970) developed a method using a primary fixative containing acrolein and glutaraldehyde in 50% aqueous dimethylsulfoxide buffered at pH 7.4. The cells were postfixed in buffered osmium

tetroxide. This technique is reported to preserve intracellular structures such as ribosomes and lipid deposits, and was claimed by Schwab and his colleagues to be especially useful in studies on the effect of vitamin deficiencies in yeast cell structure. The value of the method is thought to depend on the permeation effect of dimethylsulfoxide.

In recent years, the freeze-etching technique has been used in studies on electron microscopy of yeast cells, and with on the whole excellent results particularly in the hands of Moor and his colleagues at the Swiss Federal Institute in Zurich. The freeze-etch method, when applied to yeasts, reveals details of membrane structure, especially of endoplasmic reticulum, which cannot be seen in thin sections of stained preparations. Figure 3 shows a micrograph obtained by Moor and his colleagues using the freeze-etch technique.

Several techniques have been devised for disrupting yeast cells and isolating subcellular organelles for biochemical and cytological studies. Basically, there are two types of method for liberating intracellular organelles from yeasts. In the first, yeast cells are disrupted by physical means which, when done in a medium of suitable chemical composition and tonicity, leads to the release of intact organelles including cell walls, nuclei, mitochondria, and vacuoles. Yeast cell walls are mechanically tough, but rapid shaking in the presence of small glass beads (e.g., Ballotini beads, grade 12; 1.2 mm diam.) in a Braun homogenizer will disrupt almost all of the cells in the suspension in about one minute. In general, shaking with glass beads is preferred to using pressure cells or ultrasonics for isolating organelles from yeasts.

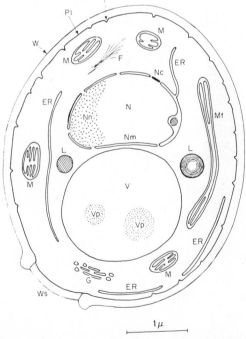

Fig. 2 Diagrammatic representation of the cytology of a resting cell of *Saccharomyces cerevisiae*. ER, endoplasmic reticulum; F, filament; G, Golgi apparatus; L, lipid granule; M, mitochondrion; Mt, thread-like mitochondrion; N, nucleus; Nc, centriolar plaque; Nm, nuclear membrane; Nn, nucleolus; Pi, invagination; Pl, plasma membrane; V, vacuole; Vp, polymetaphosphate granule; W, cell wall; Ws, bud scar.

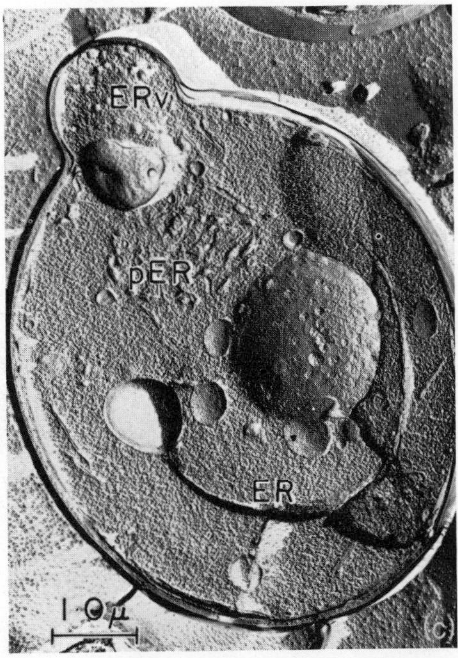

Fig. 3 Electron micrograph of a freeze-etched cell of *Saccharomyces cerevisiae* showing a bud initial, endoplasmic reticulum (ER), proliferating endoplasmic reticulum (pER), and vesicles derived from endoplasmic reticulum (ERv). (From Moor and Robinow, 1969.)

A somewhat more elegant method for releasing intracellular organelles from yeasts can be used with some strains which have cell walls that are susceptible to enzymic lysis. The main structural component in the yeast wall is a branched β-linked glucan. β-Glucanase preparations, including commercial snail-gut juice or those elaborated extracellularly by certain bacteria, streptomycetes and fungi, will lyse the β-glucan in the wall, and possibly other wall components, and allow the release of osmotically sensitive spheroplasts or protoplasts (Phaff, 1971). When yeast spheroplasts or protoplasts are suspended in a solution of much lower tonicity (0.2–0.1M) than that used to protect them (0.8M), they lyse and intracellular organelles are released. These organelles can then be separated by differential centrifugation (Chapell and Hansford, 1969) or better still by centrifuging through a density gradient of sucrose or Ficoll (Matile, 1970). Density-gradient separation is preferred for it leads to the isolation of cleaner organelles. Lysis of yeast spheroplasts is regularly used to isolate mitochondria, and also nuclei (Smitt et al., 1970) and vacuoles (Matile and Wiemken, 1967).

ANTHONY H. ROSE

References

Chappell, J. P., and R. Hansford, pp. 48–49 in "Subcellular Components: Preparation and Fractionation," G. D. Birnie and S. M. Fox ed., London, Butterworths, 1969.

Lodder, J., ed., "The Yeasts; a Taxonomic Study," 2nd ed., London, North-Holland Publ. Co., 1970.

Luft, J. H., *J. Biophys. Biochem. Cytol.,* **2**:799 (1956).

Matile, Ph., pp. 39–49 in "Membranes; Structure and Function," Vol. 20, J. R. Villanueva and F. Ponz, ed., FEBS Meeting, London, Academic Press, 1970.

Matile, Ph., H. Moor, and C. F. Robinow, pp. 219–302 in "The Yeasts," Vol. 1, A. H. Rose and J. S. Harrison, ed., London, Academic Press, 1969.

Matile, Ph., and A. Wiemken, *Arch. Mikrobiol.,* **56**:148 (1967).

Phaff, H. J., pp. 135–209 in "The Yeasts," Vol. 2, A. H. Rose and J. S. Harrison, ed. London, Academic Press, 1971.

Schwab, D. W., A. H. Janney, J. Scala, and L. M. Lewin, *Stain Technol.,* **45**:143 (1970).

Smitt, W. W. S., G. Nannii, Th. H. Rozijn, and G. J. M. Tonino, *Exp. Cell Res.,* **59**:440 (1970).

Vitols, E., R. J. North, and A. W. Linnane, *J. Biophys. Biochem. Cytol.,* **9**:689 (1961).

Z

ZOOM MICROSCOPES

Prior Art. As early as the seventeenth century, microscopists recognized that the magnification at which a specimen should be viewed depended upon both the nature of the specimen and the specific intent of the examination, and many of the early workers went to extremes to provide themselves with a choice of magnifications. Leeuwenhoek, for example, constructed a new microscope of the desired magnification as the need arose. Robert Hooke, who favored the compound microscope, used a four-section draw tube and a removable "middle" lens. By adding or removing this "middle" lens and by changing tube length, he could adjust the magnification within limits to suit his needs.

In a sense, Hooke's microscope might be considered an early version of the modern zoom microscope. Just as is done today, the change in magnification was accomplished by the adjustment of lenses and spacings between the specimen plane and the observer's eye. The similarities end here, however, for whereas the design of the modern zoom microscope is the result of computer optimization throughout the zoom range using multi-element lenses made from a wide variety of modern optical materials, the microscopes of Hooke's day and for many years thereafter were produced by trial and error combinations of uncorrected single elements made from the few materials then available to the lens maker.

During the period from the seventeenth century until about the first quarter of the nineteenth century, lack of knowledge in the theories of lens design and image formation limited the options for changing magnification to draw tubes, interchangeable eyepieces, insertable "middle" lenses, and uncorrected low power low numerical aperture objectives. While these methods proved useful in the lower magnification ranges, the desire to achieve higher magnifications was frustrated by the unsolved problems of spherical and chromatic aberrations and by the little understood problems of diffraction.

The evolution of the theories of achromatization which began late in the eighteenth century brought about profound changes in microscope optics, for it ultimately led to the development of the achromatic doublet. The significance of this development, as it turned out, went far beyond the correction of chromatic aberration because the combinations of crown and flint glasses required for achromatization also contributed to the correction of spherical aberration and coma.

The introduction of the achromatic doublet was followed by great progress during the middle years of the nineteenth century in the development and application of lens design theories and in the development and discovery of new optical materials, both of which helped to revolutionize the design of microscope optics. The gradual understanding of the relationship between diffraction and objective numerical aperture and the realization that numerical aperture imposed an ultimate limit on useful microscope magnification led to the development of objectives having higher and higher numerical aperture, culminating in the immersion objectives with numerical apertures of up to 1.6 and useful microscope magnifications greater than $1000\times$.

The improved performance achieved through the elimination of lens aberrations and increased numerical apertures did impose certain restrictions on the use of these new objectives. Robert Hooke's method for changing magnification by changing tube lengths no longer could be employed, for it was found that significant changes in the state of correction of the higher numerical aperture objectives resulted from even small changes in tube length.

More and more, the experiences of the microscope lens designers of the middle nineteenth century led them to conclude that microscope optics could be designed for one and only one set of specific conditions and that any changes in those conditions led to unacceptable results. Small wonder, then, that the concept of continuously variable magnification which had been useful for many years fell into such disfavor that it was put out of sight for a hundred years.

The reappearance of the continuously variable magnification changer or zoom system in the mid-twentieth century does not mean that the earlier designers erred in their observations that lens aberrations in dynamic optical systems are unstable. The success of the modern zoom microscope merely reflects advance in optical technology which allow the twentieth century lens designer to overcome performance problems which a hundred years previously had been thought to be insurmountable. The continued advancements made in lens design theory between the mid-nineteenth and twentieth centuries, the vast array of new and improved optical materials available to the twentieth century lens designer, and the advent of the high-speed computer all played a part in the development of zooming optical systems. The development, however, is by no means complete. A microscope zoom system capable of covering the entire range of magnifications which can be covered by a set of conventional microscope objectives is probably another hundred years away from a solution, if indeed a solution ever can be found.

Zoom System Classification. While twentieth century

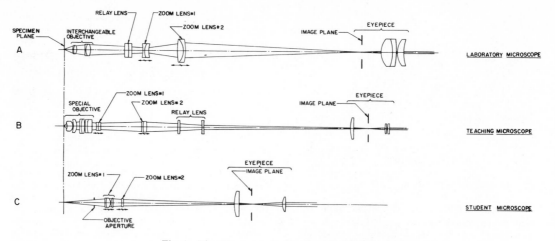

Fig. 1   Three systems of zoom described in text.

microscope zoom systems have not reached the state of development where they can replace the turret of objectives on a standard laboratory microscope, zoom systems of more modest magnification ranges are quite practical when properly integrated into the compound microscope and can be made to serve very useful purposes. Furthermore, zoom systems used in stereomicroscopes do indeed cover most of the useful magnifications range over which stereomicroscopes are used.

Although microscope zoom systems may appear to function as separate and distinct optical subassemblies, in reality either the zoom system and eyepiece work together as a variable-power eyepiece or the zoom system and objective work together as a variable-power objective. The distinction betwen the two concepts will be illustrated by examples.

Variable-Power Eyepieces. Figure 1A illustrates the optical components of a zoom microscope of the *variable-power eyepiece* classification. The objectives for this microscope may be any of the standard achromats, fluorites, apochromats, or flat field objectives which conform to the tube length and objective shoulder position for which the zoom system has been designed. The eyepieces also are of standard design. The relay lens included in Fig. 1A, while not essential to the zoom concept, helps achieve an acceptable overall length for the complete optical system.

The zoom system in Fig. 1A consists of two zoom lens components interposed between the relay lens and eyepiece. The positions of the objectives, relay lens, and eyepiece are permanently fixed on the microscope axis. The two zoom lenses are mounted on mechanical slides along which they can be moved by means of individual cam drives. The motions for the two zoom lenses are calculated to accomplish a continuous change in magnification while maintaining a constant image position.

The free apertures of the zoom lenses are made large enough to permit the objectives to work at their full numerical apertures throughout the zoom range. The upper limit of useful magnification introduced by the zoom system is reached when a further increase results in empty magnification. The usual upper limit is about 2×, although where photomicrography is an important consideration, magnifications as high as 3× are used. Since

the real field coverage varies inversely with magnification, the usual lower limit of magnification is about 1×. The reason for this lower limit is that if lower magnifications were permitted, both the objectives and condenser-illumination system would be forced to cover larger field sizes than those for which they are usually designed. The combination of zoom system and 10× eyepiece in Fig. 1A is equivalent to a continuously variable 10× to 20× eyepiece. Figure 2 shows the manner in which this zoom system is integrated into a typical laboratory microscope.

Zoom systems of the type illustrated in Fig. 1A have virtually no degrading effect on the state of aberration correction of the microscope. The objectives continue to work at the tube length and numerical apertures for which they were designed and their field coverage does not exceed that for which they were designed. Similarly, the demands made on the eyepieces do not exceed those encountered in nonzooming systems. The dynamic conditions under which the zoom lenses work do cause aberration changes in the individual zoom lens components, but it is here that the modern design tools available to the lens designer are utilized. By means of these tools it is a relatively simple task to compensate the aberration changes in one component by introducing equal but opposite changes in the other.

Variable-Power Objectives. The optical system illustrated in Fig. 1B is representative of the *variable-power objective* classification. The distinguishing feature of this optical system which sets it apart from the previously described system is the numerical aperture control exercised by the first zoom lens. When the zoom system is set for lowest microscope magnification, the first zoom lens takes a position close to the objective as shown in Fig. 1B. The free aperture of this lens is smaller than the back aperture of the objective, hence the effective numerical aperture of the complete optical system is less than the rated numerical aperture of the objective. This stopping down of the objective at low power permits the objective to cover the large low-power field and still maintain good aberration control.

As the microscope magnification is increased by means of the cam drives on the two zoom lenses, the first zoom

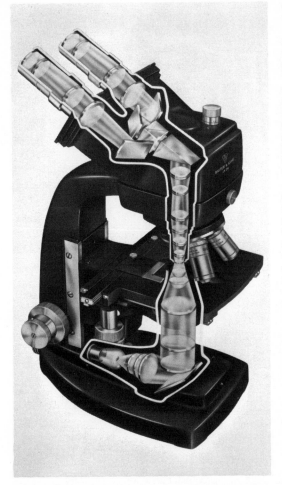

Fig. 2 Optical pathway of typical laboratory zoom microscope.

lens moves away from the objective and therefore is able to transmit a progressively larger percentage of the objective numerical aperture, until finally at the upper limit of the magnification range the full objective numerical aperture is transmitted. It should be noted that this scheme for changing numerical aperture as the microscope magnification is changed is equivalent to the conventional practice of matching particular fixed-power

objective magnifications with particular numerical apertures.

The variable-power objective concept embodied in the microscope illustrated in Fig. 1B is especially well suited to the teaching microscope because of the continuity it provides in the image as the magnification is changed. When the student changes the power of a nonzooming microscope, the large discrete steps between objective magnifications and the temporary loss of the image which occurs when one objective is exchanged for another make it difficult for him to recognize that the small specimen area he sees at high magnification is really a part of the larger specimen area he had seen at the lower magnification. The continuous change in magnification achieved with the zooming microscope eliminates this problem.

The microscope illustrated in Fig. 1B has an objective zoom range from $10\times$ to $50\times$ and a maximum numerical aperture of 0.55. Combined with a $10\times$ eyepiece, the microscope magnification covers the range from $100\times$ to $500\times$. While it would be most desirable to extend this range to cover both lower and higher magnifications, to date (1970) Fig. 1B represents the state of the art for a compound microscope. The objective, which in Fig. 1B is specially designed to match the zoom system, imposes this limit on zoom range. The changes in objective aberrations which occur between low magnification where the field size is maximum and high magnification where numerical aperture is maximum are compensated by the change in numerical aperture and by corresponding changes in aberrations of opposite sign in the zoom system. If the zoom range were increased, changes in objective aberrations would be greater than can be compensated by the zoom system.

The future development of new optical glasses having very high indices of refraction and low dispersions will play a part in extending this zoom range to some degree; however, the development of new design concepts for zoom system is likely to play the more significant role.

From the standpoint of simplicity, the zooming microscope illustrated in Fig. 1C deserves special mention. In this system, the front lens assembly (three elements) serves a dual role in that it is both the objective and first zoom lens. Both zoom lenses are cam driven in the usual fashion. An aperture stop, which is part of and therefore moves with the first zoom lens, controls the numerical aperture as the magnification is changed.

Zooming microscopes of the basic concept represented by Fig. 1C are available in magnification ranges such as $25\times$ to $100\times$ and $50\times$ to $200\times$. These microscopes are especially well suited for use at the elementary school level.

Stereozoom Microscopes. Mention was made earlier that the zoom concept is particularly well suited to

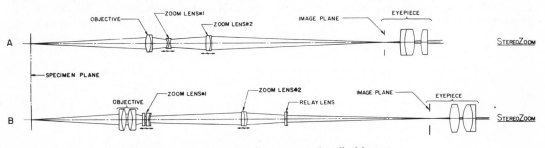

Fig. 3 Two systems of stereozoom described in text.

stereomicroscope applications. This is due to the fact that the upper useful magnification range of stereomicroscopes is much lower than that required for compound microscopes and this obviates the need for high numerical aperture objectives. The modest objective numerical aperture requirements not only simplify the optical design problems, but also lead to less complex optical components and permit greater useful zoom ranges.

The optical system illustrated in Fig. 3A is historically significant in that it is the first stereozoom microscope ever to be manufactured commercially. Since its introduction in the early 1960s by the firm of Bausch & Lomb, virtually every major microscope manufacturer in the world has introduced a competitive model patterned after this original.

The two zoom lenses in Fig. 3A are cam driven in the conventional fashion and the first zoom lens is designed to vary the effective numerical aperture in much the same way as described for the compound microscope in Fig. 1B. With $10\times$ eyepieces, this stereozoom has a magnification range of $7\times$ to $30\times$ and a numerical aperture of about 0.06 at $30\times$. By the addition of a $1/2\times$ or $2\times$ lens attachment in front of the objectives, the magnification ranges become $3.5\times$ to $15\times$ or $14\times$ to $60\times$, respectively, with corresponding changes in numerical aperture.

It should be understood that the optics illustrated in Fig. 3A constitute just one half of an identical pair of optical systems required in the stereozoom. Furthermore, the porro erecting system, which should be inserted in the space between the second zoom lens and the eyepiece has been deleted from the illustration for the sake of simplicity.

The optical system illustrated in Fig. 3B is a second-generation stereozoom. This system covers the microscope magnification range from $10\times$ to $70\times$ with a numerical aperture of 0.10 at $70\times$. As with the older design, the addition of a $1/2\times$ or $2\times$ lens attachment in front of the objective will half or double the magnification values and correspondingly change the numerical aperture throughout the altered zoom ranges.

The fixed-position relay lens, which is not essential to the zoom concept, has been added to lengthen the system. This provides space along the optical axis for the insertion of accessories such as camera optics and vertical illuminators. Figure 4 shows this stereozoom microscope with an attached camera accessory.

Summary. During the period from 1960 to 1970, zoom systems have been instrumental in revolutionizing the concepts of some of our classical microscope designs. Most notable among them have been the stereomicroscope and teaching microscope.

Zoom systems capable of covering most of the useful magnification range of the stereomicroscope are a reality, and this has led to an almost universal acceptance of the stereozoom to the near exclusion of all other type stereomicroscopes.

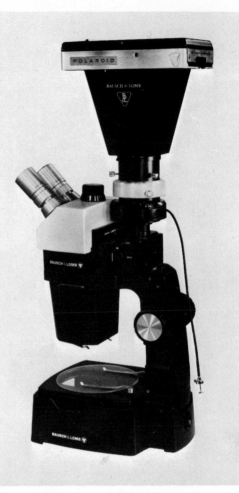

Fig. 4 Typical stereozoom microscope with a photographic attachment.

Zoom systems also have proven their superiority for the compound microscope where the required magnification range is relatively small as in many elementary and high school teaching microscopes.

Extension of the compound microscope zoom range from the present $5\times$ zoom range (1970) to the approximately $40\times$ magnification range of the multi-objective laboratory and research microscopes remains a challenge for perhaps several generations of future lens designers.

HAROLD E. ROSENBERGER

INDEX

(**Boldface** numbers refer to main articles)